Ternary Alloys Based on IV-VI and IV-VI$_2$ Semiconductors

Ternary Alloys Based on IV-VI and IV-VI_2 Semiconductors

Vasyl Tomashyk

CRC Press
Taylor & Francis Group
Boca Raton London New York

CRC Press is an imprint of the
Taylor & Francis Group, an informa business

First Edition published 2022
by CRC Press
6000 Broken Sound Parkway NW, Suite 300, Boca Raton, FL 33487-2742

and by CRC Press
4 Park Square, Milton Park, Abingdon, Oxon, OX14 4RN

CRC Press is an imprint of Taylor & Francis Group, LLC

ISBN: 978-0-367-63923-5 (hbk)
ISBN: 978-0-367-64307-2 (pbk)
ISBN: 978-1-003-12350-7 (ebk)

DOI: 10.1201/9781003123507

Typeset in Times LT Std
by KnowledgeWorks Global Ltd.

Access the Support Material: www.routledge.com/9780367639235

This book is dedicated to the memory of my teacher, my colleague and my friend, Professor Oleg Panchuk (1932–2022)

Contents

All References are available as a downloadable eResource at www.routledge.com/9780367639235

List of Abbreviations

Ac	Acetate-ion, CH_3COO^-
DMF	Dimethylformamide
DSC	Differential scanning calorimetry
DTA	Differential thermal analysis
EMF	Electromotive force
EPMA	Electron probe microanalysis
Et	Ethyl, C_2H_5
M	mol
MeOH	Methanol, CH_3OH
mM	mmol
ppm	Parts per million
SEM	Scanning electron microscopy
STEM	Scanning transmission electron microscopy
(X)	Solid solution based on X
XRD	X-ray diffraction

Preface

A significant volume of semiconductor devices and circuits employs IV-VI and IV-VI_2 semiconductors, the most commonly used crystal material for integrated circuits. The IV-VI semiconductors are among the most interesting materials in solid-state physics. Many of them crystallize in the rock-salt structure, and structural transitions are common for them. The most widely studied compounds in this group are PbTe, PbSe, PbS, SnTe, and GeTe. These materials have small gaps which are usually less than 0.5 eV, hence they are good candidates for devices like infrared lasers and detectors. Despite their simple crystal structure, some of these compounds exhibit ferroelectric, paraelectric, and superconducting behavior. In addition, the temperature dependence of the energy gaps, the energy positions of impurity levels, the high doping levels found, the static dielectric constants, and the electronic structure of some of the alloys of these compounds appear to be anomalous compared to the "conventional behavior" of the diamond and zinc-blende semiconductors. These semiconductors either have already found use or are promising materials for infrared sensors and sources, thermoelectric elements, solar cells, memory elements, etc. The basic characteristics of these compounds, namely, narrow bandgap, high permittivity, relatively high radiation resistance, high mobility of charge carriers, and high bond ionicity, are unique among semiconductor substances.

In most cases, the properties of IV-VI and IV-VI_2 semiconductors could be modified by isovalent or heterovalent foreign impurities (*E*) doping that resulted in forming solid solution or creation of a separate electronic level in the semiconductor lattice. Such ternary IV-*E*-VI materials enlarged and ameliorate the properties of these semiconductors and other materials based on these semiconductors.

The problem of reproducible production of the doped materials with predicted and desired properties cannot be successfully resolved without knowledge of appropriate ternary system phase diagram which is "a map" for technologists. Unfortunately, the main information is scattered in separate papers. The present book is aimed to collect and systematize all available data on the IV-*E*-VI ternary systems. It includes 581 critically compiled ternary systems based on IV-VI and IV-VI_2 semiconductors including the literature data from 2267 papers, the data are illustrated by almost 570 figures. The information is divided into 12 chapters according to the number of possible combinations of Si, Ge, Sn, and Pb with S, Se, and Te. The chapters are structured in the order at first of IV Group element number in the Periodic system increasing, i.e. from Si to Pb compounds, and then in order of the chalcogen number increasing, i.e. from sulfides to tellurides. The same principle is used for further description of the systems in every chapter, i.e. in order of the third component number in the Periodic system increasing.

Every ternary system database description contains brief information in the following order: the diagram type, possible phase transformation and physical chemical interaction of the components, methods of the equilibrium investigation, thermodynamic characteristics, and the method of the samples preparation. Solid and liquid phase equilibriums with vapor are illustrated in some cases also because of their importance for crystal growth from the vapor, from the melt and by the vapor–liquid–solid technique.

The homogeneity range is of a great importance for the crystal defect structure governing. Therefore the reference book collects all such data accessible by now. Besides, this book presents the data on the baric and temperature dependences of the impurities solubility both in the semiconductors lattice and the liquid phase as well as the pressure-composition relationship. As semiconductors and metal mutual solubility are usually small values the illustrating figure presents a restricted concentrations range (in mol. %).

Most of the figures are presented in their original form, although some are a little corrected. If the published data varied essentially, several versions were presented in comparison. The content of system components is presented in mol. % (this is not indicated on the figures). If the original phase diagram is given with mass. %, this is indicated on the figure.

This book is meant for researchers at industrial and national laboratories and for university and graduate students majoring in materials science, solid state chemistry and engineering. It is also suitable for phase relation researchers, inorganic chemists, and solid state physicists.

About the Author

Vasyl Tomashyk is the head of the department of V.Ye. Lashkaryov Institute for Semiconductor Physics of National Academy of Sciences of Ukraine. He graduated from Chernivtsi State University in 1972 (master of chemistry). He is a doctor of chemical sciences (1992), a professor (1999), and author of about 670 publications in scientific journals and conference proceedings and 11 books (6 of them were published by CRC Press), which are devoted to the physical-chemical analysis, the chemistry of semiconductors, and chemical treatment of semiconductor surfaces.

Tomashyk is a specialist of high international level in the field of solid state and semiconductor chemistry, including physical-chemical analysis and technology of semiconductor materials. He was head of research topics within the International program "Copernicus". He is a member of the Materials Science International Team (Stuttgart, Germany, since 1999), which prepares a series of prestigious reference-books under the title *Ternary alloys* and published nine chapters in this series and 35 chapters in Landolt–Börnstein New Series. Tomashyk is actively working with young researchers and graduate students and under his supervision 22 PhD theses were prepared. For many years, he is a professor of Ivan Franko Zhytomyr State University in Ukraine.

1

Systems Based on Silicon Sulfides

1.1 Silicon–Lithium–Sulfur

SiS_2–Li_2S: The phase diagram of this system is given in Figure 1.1 (Ahn and Huggins 1990, 1991). The system has two eutectics which crystallize at 610°C ± 10°C and 680°C ± 10°C and one peritectic at 710°C ± 10°C. Two compounds, **Li_2SiS_3** and **Li_4SiS_4**, are formed in the SiS_2–Li_2S system. Li_2SiS_3 melts congruently at 745°C ± 10°C and depending on cooling conditions from the melt, three different phases of this compound (equilibrium, metastable crystalline and glass) were formed. The equilibrium and metastable crystalline phases crystallize in the orthorhombic structure with the lattice parameters a = 1166.4, b = 673.5, c = 592.6 pm, and a calculated density of 1.97 g·cm^{-3} and a = 1143.6, b = 660.5, c = 648.7 pm, respectively (Ahn and Huggins 1989, 1990, 1991). Metastable crystalline Li_2SiS_3 has a stretched one-dimensional chain structure compared to the equilibrium phase. When the melt cooled down to room temperature continuously in a 5–6 h period, a metastable phase was formed. It transforms into the equilibrium phase after 6 months at room temperature.

When the Li_2SiS_3 melt was cooled down to 780°C just above the melting point and was quenched in water, it became a glassy phase. The glass transition temperature of this phase is about 320°C, the crystallization temperature is about 450°C, and the heat of crystallization is 12.6 kJ·mol^{-1} (Ahn and Huggins 1990, 1991). The recrystallization of the glass phase formed the metastable crystalline phase.

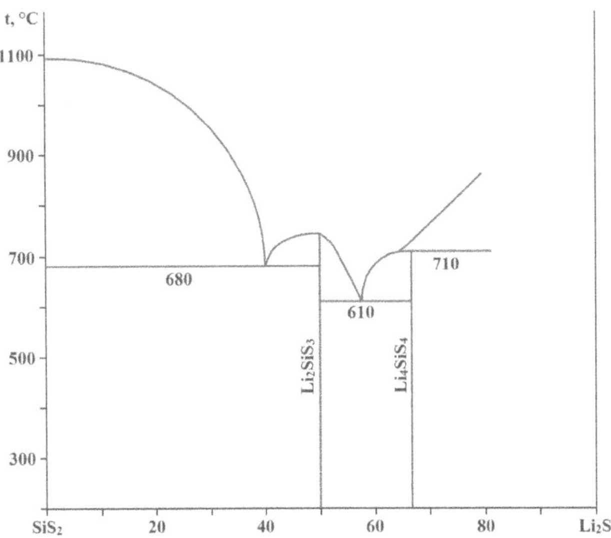

FIGURE 1.1 Phase diagram of the SiS_2–Li_2S system. (From Ahn, B.T., and Huggins, R.A., *Mater. Res. Bull.*, **25**(3), 381, 1990.)

Li_4SiS_4 melts incongruently at 710°C ± 20°C and apparently has two polymorphic modifications. First of them crystallizes in the orthorhombic structure with the lattice parameters a = 1373.5, b = 776.9, c = 614.7 pm, and a calculated density of 1.864 g·cm^{-3} (Ahn and Huggins 1989, 1990, 1991). The second modification crystallizes in the monoclinic structure with the lattice parameters a = 689.34 ± 0.03, b = 776.75 ± 0.03, c = 612.41 ± 0.02 pm, and β = 91.225 ± 0.005° (Murayama et al. 2002a).

In order to prepare various compositions in the SiS_2–Li_2S system, mixtures of SiS_2 and Li_2S were melted in pyrolitic graphite-coated ampoules (Ahn and Huggins 1990, 1991). All the materials were handled in an He-filled glove box to avoid oxidation. Li_2SiS_3 and Li_4SiS_4 were also prepared by the interaction of Li_2Si and $Li_{15}Si_4$, respectively, with sulfur (Weiss and Rocktäschel 1960).

The isothermal section of the Si–Li–S ternary system at room temperature is shown in Figure 1.2 (Ahn and Huggins 1989). To prepare the sample for the investigation of this system, a powder mixture of SiS_2, Si, and Li_2S was reacted at 1000°C for 1 h with the next annealing at 720°C for half a day and cooling down to room temperature for a day.

The glasses with the composition $(Li_2S)_x(SiS_2)_{1-x}$ ($x \le 0.6$) have been prepared by twin roller quenching (Pradel and Ribes 1986). The glass transition temperature for such glasses is within the range of 331°C–341°C. Starting materials were obtained from

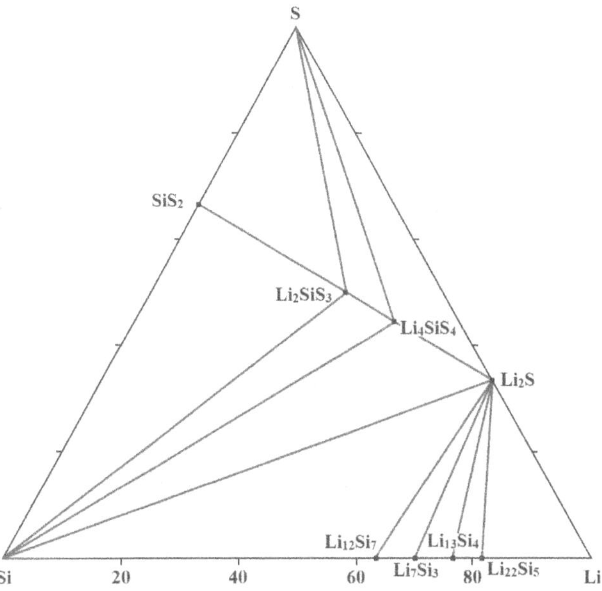

FIGURE 1.2 Isothermal section of the Si–Li–S ternary system at room temperature. (From Ahn, B.T., and Huggins, R.A., *Mater. Res. Bull.*, **24**(8), 889, 1989.)

DOI: 10.1201/9781003123507-1

a mixture of sulfide powders (SiS_2 and Li_2S), placed in vitreous carbon crucible inside a silica tube and sealed under vacuum (10^{-2}–10^{-3} Pa). The powders were melted for 2 h at 1050°C before being quenched at room temperature. The materials obtained were then crushed into small pieces and remelted using a twin roller apparatus. The glasses were obtained as small flakes, orange to brown in color. All these materials are very hygroscopic and the experiments were carried out in an Ar-filled glove box.

1.2 Silicon–Sodium–Sulfur

Four ternary compounds, **Na_2SiS_2**, **Na_4SiS_4**, **$Na_4Si_4S_{10}$**, and **$Na_6Si_2S_6$**, are formed in the Si–Na–S system (Cade et al. 1972; Ribes et al. 1973; Feltz and Pfaff 1983). $Na_4Si_4S_{10}$ crystallizes in the orthorhombic structure with the lattice parameters $a = 1268.1 \pm 0.8$, $b = 1272.0 \pm 0.4$, and $c = 1036.4 \pm 0.5$ pm and the calculated and experimental densities of 2.12 and 2.09 g·cm^{-3}, respectively (Cade et al. 1970; Ribes et al. 1973). To obtain the single crystals of this compound, a mixture of Na_2S and SiS_2 (molar ratio 1:2) was heated in quartz ampoule under vacuum at 800°C for 48 h with next slowly cooling to room temperature.

$Na_6Si_2S_6$ was obtained by a reaction of stoichiometric amounts of the elements in closed silica ampoules under vacuum at 650°C (Feltz and Pfaff 1983). After a 5–h reaction time, the reaction products, which were not yet uniform, were finely ground under inert conditions and subjected again to the described temperature treatment. The repetition gave a uniform reaction product which completely dissolved in MeOH. After concentration of the solution, $Na_6Si_2S_6$ segregated out as finely crystalline colorless precipitates. This compound decomposes due to hydrolysis.

To prepare the glasses in the **SiS_2–Na_2S** system, 2–3 g of sulfides were mixed in the correct proportions and put in a vitreous carbon crucible (Ribes et al. 1980). This crucible was placed inside the silica tube and sealed under a vacuum. After melting at a temperature of 700°C–1000°C, the orange–brown glasses were obtained by cooling down to room temperature or by immersing the silica tube in cold water. The obtained glasses contained up to 60 mol% Na_2S.

1.3 Silicon–Rubidium–Sulfur

The **$Rb_4Si_2S_6$** ternary compound, which crystallizes in the monoclinic structure with the lattice parameters $a = 1323 \pm 2$, $b = 686.4 \pm 0.2$, $c = 953 \pm 1$ pm, and $\beta = 125.15 \pm 0.05°$, is formed in the Si–Rb–S system (Kolb et al. 2004). It was obtained by reacting a mixture of Rb_2S (4.4 mM), Si (4.4 mM), and S powder (13 mM) in alumina crucibles which were sealed into evacuated silica ampoules at 700°C, followed by cooling at a constant rate of 3°C·h^{-1}. The reaction product consisted of colorless prismatic crystals which are sensitive to air and humidity.

1.4 Silicon–Cesium–Sulfur

The **$Cs_4Si_2S_6$** ternary compound, which crystallizes in the monoclinic structure with the lattice parameters $a = 986.9 \pm 0.3$, $b = 713.8 \pm 0.3$, $c = 1132.3 \pm 0.6$ pm, and $\beta = 100.12 \pm 0.04°$, is formed in the Si–Cs–S system (Feldmann et al. 1998). The colorless prisms of this compound were obtained as a result of a solid-state reaction of equimolar amounts of Cs_2S and SiS_2 in a sealed quartz tube at 400°C for 4 days.

1.5 Silicon–Copper–Sulfur

SiS_2–Cu_2S: The phase diagram of this system was constructed by Venkatraman et al. (1995) and Olekseyuk et al. (2005). The more reliable phase diagram in the 0–60 mol% SiS_2 concentration range is presented in Figure 1.3 (Olekseyuk et al. 2005). The eutectic near Cu_2S side contains ≈ 8 mol% SiS_2 and crystallizes at 1062°C. According to the data of Venkatraman et al. (1995), two eutectics in this system contain ≈ 40 and ≈ 90 mol% Cu_2S and crystallize at 567°C ± 10°C and 792°C ± 5°C, respectively.

Three compounds are formed in the SiS_2–Cu_2S system. **Cu_8SiS_6** is a semiconductor (Dogguy 1983), melts congruently at 1186°C [at 1185°C (Cambi and Elli 1961); at 1030°C (Boivin et al. 1967); at 1200° ± 5°C (Venkatraman et al. 1995)] and has very small homogeneity region (Olekseyuk et al. 2005) [has a homogeneity region of 68–80 mol% Cu_2S between 500°C and 900°C (Venkatraman et al. 1995)]. It crystallizes in the orthorhombic structure with the lattice parameters $a = 699.8$, $b = 690.35$, and $c = 976.82$ pm (Venkatraman et al. 1995) [$a = 699.2 \pm 0.2$, $b = 690.1 \pm 0.2$, $c = 976.8 \pm 0.2$ pm (Kuhs et al. 1979); $a = 699.28 \pm 0.06$, $b = 690.00 \pm 0.13$, $c = 977.23 \pm 0.14$ pm and the calculated and experimental densities of 5.15 and 5.05 ± 0.06 g·cm^{-3}, respectively (Levalois and Allais 1981); $a = 699.3 \pm 0.6$, $b = 690.0 \pm 0.1$, $c = 977.2 \pm 0.1$ pm, and an experimental density of 5.0 ± 0.1 g·cm^{-1} (Dogguy 1983); in the cubic structure with the lattice parameter $a = 976$ pm and the calculated and experimental densities of 5.20 and 5.01 g·cm^{-3}, respectively (Hahn et al. 1965); $a = 981$ pm (Thomas 1967)]. According to the data of Kuhs et al. (1979), Cu_8SiS_6 has a phase transition at 63°C and Dogguy (1983) indicated that this compound exists in

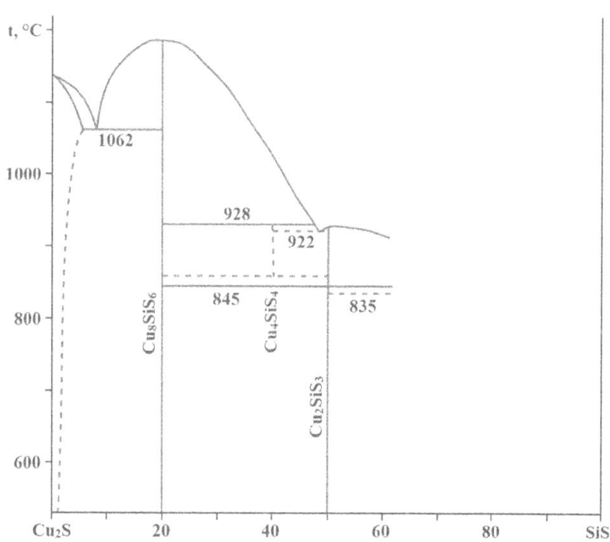

FIGURE 1.3 Phase diagram of the SiS_2–Cu_2S system. (From Olekseyuk, I.D., et al., *J. Alloys Compd.*, **399**(1–2), 149, 2005.)

three polymorphic modifications. The title compound could be prepared by reacting stoichiometric amounts of Cu_2S, Si, and S in an evacuated sealed quartz ampoule for 4 days at 500°C (Ishii et al. 1999) or by reacting stoichiometric amounts of high-purity elements in an evacuated sealed quartz ampoule for about 6 days at temperatures of 600°C–700°C (Kuhs et al. 1979). The synthesis of Cu_8SiS_6 was carried out also from the elements, placed in a silica ampoule, sealed under vacuum (Dogguy 1983). Two methods were used to obtain the crystals: either by simple annealing at fixed temperatures or by gas-phase transport (the agent used was chlorine and the temperature was 1000°C). It could be also obtained by the interaction of SiS_2 and Cu_2S in the temperature region 400°C–800°C (Hahn et al. 1965).

$\mathbf{Cu_2SiS_3}$ is a semiconductor (Chen et al. 1999b), melts congruently at 928°C [at 925°C (Rivet et al. 1963; Rivet 1965); incongruently at 900°C (Venkatraman et al. 1995)] and has a polymorphous transition at 845°C (Olekseyuk et al. 2005) [at 840°C (Rivet et al. 1963; Rivet 1965]. According to the data of Boivin et al. (1967), this compound decomposes at 830°C. Low-temperature modification of this compound crystallizes in the monoclinic structure with the lattice parameters $a = 1151$, $b = 534$, $c = 816$ pm, $\beta = 98.95°$, and the calculated and experimental densities of 3.92 and 3.90 g·cm^{-3}, respectively (Hahn et al. 1966), and high temperature modification crystallizes in the orthorhombic structure with the lattice parameters $a = 1121$, $b = 1204$, and $c = 603$ pm, and a calculated density of 4.10 g·cm^{-3} (Hahn et al. 1966) [$a = 1098.1 \pm 0.3$, $b = 641.6 \pm 0.2$, $c = 604.6 \pm 0.2$ pm (Parthé and Garin 1971)]. According to the date of Rivet et al. (1963) and Rivet (1965), low-temperature modification of Cu_2SiS_3 crystallizes in the tetragonal structure with the lattice parameters $a = 529.0$ and $c = 507.8$ pm and the calculated and experimental densities of 3.89 and 3.63 ± 0.07 g·cm^{-3}, respectively, and high-temperature modification crystallizes in the hexagonal structure with the lattice parameters $a = 368.4$ and $c = 604.4$ pm and the calculated and experimental densities of 3.89 and 3.81 ± 0.05 g·cm^{-3}, respectively. Venkatraman et al. (1995) noted that this compound does not undergo any structural transformation and crystallizes in the monoclinic structure with the lattice parameters $a = 633.3$, $b = 1122.5$, $c = 627.5$ pm, and $\beta = 107.32°$ [$a = 633.2 \pm 0.1$, $b = 1123.0 \pm 0.1$, $c = 627.3 \pm 0.1$ pm, $\beta = 107.49 \pm 0.01°$, and a calculated density of 3.924 g·cm^{-3} (Chen et al. 1999b)]. The samples of Cu_2SiS_3 was prepared from stoichiometric mixtures of the elements by heating (35 days) in thick evacuated quartz tubes (Parthé and Garin 1971). To avoid explosions, the temperature was initially increased very slowly (100°C per day). The chemical reaction was complete after 2 h at 900°C. However, to ensure order of the Cu and Ge atoms, the samples were subsequently post-annealed at 500°C for a week. By means of this method, it was possible to obtain not only the powders, but also single crystals platelets (0.02 mm thick).

Cu_2SiS_3 was also synthesized from a stoichiometric mixture of Cu_2S, Si, and S, which was sealed in a silica tube at a pressure of less than 0.01 Pa (Chen et al. 1999b). The tube was heated gradually to 800°C, where it was kept for a week, then cooled at a rate of 5°C·h^{-1} to 600°C. It was annealed at this temperature for three weeks and quenched in water. The brown crystals with metallic luster were observed in the tube.

The third compound, which contains 40 mol% SiS_2, melts incongruently at 931°C and exists in a limited temperature range, was confirmed by Olekseyuk et al. (2005). It is possible that this compound is $\mathbf{Cu_4SiS_4}$, which might exist over a very narrow temperature range of about 20°C around 850°C (Venkatraman et al. 1995) [according to the data of Boivin et al. (1967) this compound melts at 930°C and decomposes at 830°C]. Cu_4SiS_4 crystallizes in the orthorhombic structure with the lattice parameters $a = 1242$, $b = 1521$, and $c = 1320$ pm (Thomas and Tridot 1967).

In the literature, there is data on the existing two more compounds in the Si–Cu–S ternary system. $\mathbf{Cu_5Si_2S_7}$ is a semiconductor and crystallizes in the monoclinic structure with the lattice parameters $a = 1621.6 \pm 0.9$, $b = 959.7 \pm 0.6$, $c = 631.7 \pm 0.4$ pm, $\beta = 92.38 \pm 0.05°$, and the calculated and experimental densities of 4.3 and 4.1 ± 0.2 g·cm^{-3}, respectively (Dogguy et al. 1982; Dogguy 1983). The synthesis of this compound was carried out from the elements, placed in a silica ampoule, sealed under vacuum. Two methods were used to obtain the crystals: either by simple annealing at fixed temperatures or by gas-phase transport (the agent used was chlorine and the temperature was 1000°C).

Boivin et al. (1967) noted that another compound, $\mathbf{Cu_6Si_2S_7}$, is formed in the SiS_2–Cu_2S system and exists in the temperature interval from 830°C to 910°C, but Venkatraman et al. (1995) indicated that this compound does not exist. According to the data of Thomas and Tridot (1967), this compound crystallizes in the monoclinic structure with the lattice parameters $a = 1623$, $b = 632$, $c = 961$ pm, and $\beta = 92°05'$.

1.6 Silicon–Silver–Sulfur

$\mathbf{SiS_2–Ag_2S}$: The phase diagram of this system, constructed through Differential Thermal Analysis (DTA) and X-Ray Diffraction (XRD), is shown in Figure 1.4 (Venkatraman et al. 1995).

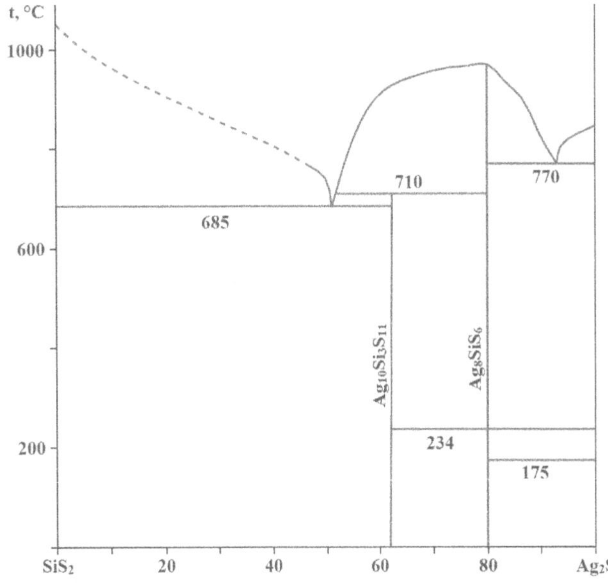

FIGURE 1.4 Phase diagram of the SiS_2–Ag_2S system. (From Venkatraman, M., et al., *Thermochim. Acta*, **249**, 13, 1995.)

The eutectic from the Ag_2S side crystallizes at 770°C and contains ≈ 93 mol% Ag_2S [at 735°C and contains 90 mol% Ag_2S (Cambi and Elli 1961; Gorochov 1968b)]. Two compounds, $\mathbf{Ag_8SiS_6}$ and $\mathbf{Ag_{10}Si_3S_{11}}$, are formed in this system. First of them melts congruently at 970°C (Venkatraman et al. 1995) [at 959°C (Cambi and Elli 1961); at 950°C (Boivin et al. 1967); at 940°C (Gorochov and Flahaut 1967; Gorochov 1968b)], has a polymorphic transformation at 234°C (Gorochov and Flahaut 1967; Gorochov 1968b)], and crystallizes in the orthorhombic structure with the lattice parameters a = 1504.3 ± 0.6, b = 745.2 ± 0.4, c = 1056.5 ± 0.5 pm (Kuhs et al. 1979) [a = 1502.4 ± 0.6, b = 742.8 ± 0.3, c = 1053.3 ± 0.4 pm, and the calculated and experimental densities of 6.124 and 6.09 ± 0.06 g·cm^{-3}, respectively (Krebs and Mandt 1977); in the cubic structure with the lattice parameter a = 1063 pm and an experimental densities of 5.99 g·cm^{-3} (Gorochov and Flahaut 1967; Gorochov 1968b); a = 2100 pm and the calculated and experimental densities of 6.21 and 5.90 g·cm^{-3}, respectively (Hahn et al. 1965)]. Ag_8SiS_6 was synthesized from a mixture (about 8 mM) of Ag_2S, Si, and S (Krebs and Mandt 1977). The mixture was heated for the first 48 h to 860°C–920°C, whereby a temperature gradient over the length of the tube was avoided as far as possible. The product was cooled to room temperature at a rate of 3°C·h^{-1}. The compound is stable to moisture and hydrolyzed only very slowly against nucleophilic attack. This compound was also obtained by the interaction of Ag_2S and SiS_2 at 400°C–800°C (Hahn et al. 1965) or by reacting stoichiometric amounts of high-purity elements in an evacuated sealed quartz ampoule for about 6 days at temperatures of 600°C–700°C (Kuhs et al. 1979).

$Ag_{10}Si_3S_{11}$ melts incongruently at 710°C (Venkatraman et al. 1995) and crystallizes in the triclinic structure with the lattice parameters a = 1241.4 ± 0.6, b = 1347.6 ± 0.6, c = 645.9 ± 0.3 pm, and α = 78.92 ± 0.04°, β = 77.61 ± 0.04°, γ = 68.71 ± 0.04°, and the calculated and experimental densities of 5.158 and 5.16 ± 0.04 g·cm^{-3}, respectively (Mandt and Krebs 1976). It was synthesized from a mixture of Ag_2S, Si, and S at 860°C–920°C (Mandt and Krebs 1976).

In the literature, there is data on the existing two more compounds in the Si–Ag–S ternary system. $\mathbf{Ag_2SiS_3}$ melts at 710°C (Boivin et al. 1967) and crystallizes in the monoclinic structure with the lattice parameters a = 667.09 ± 0.01, b = 665.67 ± 0.02, c = 1317.48 ± 0.03 pm, β = 118.658 ± 0.001°, and a calculated density of 4.3988 ± 0.0004 g·cm^{-3} (Zhbankov et al. 2011) [a = 640.6 and 667.8, b = 642.8 and 666.8, c = 1283.9 and 1346.2 pm, β = 117.243° and 118.025°, a calculated density of 4.8035 and 4.2671 g·cm^{-3} according to the optimization using local density approximation and generalized gradient approximation functional, respectively, and an energy gap of 1.42 eV (Rudysh et al. 2020)]. The synthesis of this compound was performed by co-melting the stoichiometric amounts of pure elements in a vacuum-sealed quartz ampoule. The preliminary synthesis in the flame of oxygas burner flame to complete bonding of the elementary sulfur was performed. The ampoule was heated slowly (20°C·h^{-1}) to 1000°C, kept at this temperature for 6 h with the vibration mixing and cooled at the rate of 10°C·h^{-1} to 400°C. The sample was annealed at this temperature for 300 h, and after that was quenched into water. Venkatraman et al. (1995) noted that Ag_2SiS_3 or $\mathbf{Ag_2SiS_7}$ was not detected in the Si–Ag–S ternary system.

$\mathbf{Ag_6Si_2S_7}$ melts congruently at 735°C (Boivin et al. 1967) [at 750°C (Cambi and Elli 1961)]. This compound was also not detected in the SiS_2–Ag_2S system by Venkatraman et al. (1995).

1.7 Silicon–Magnesium–Sulfur

The $\mathbf{Mg_2SiS_4}$ ternary compound, which crystallizes in the orthorhombic structure with the lattice parameters a = 1264, b = 747, and c = 592 pm (Rocktäschel et al. 1964), is formed in the Si–Mg–S system. It was prepared by the interaction of Mg_2Si with sulfur within the temperature interval from 450°C to 650°C (Weiss and Rocktäschel 1960; Rocktäschel et al. 1964).

1.8 Silicon–Calcium–Sulfur

Three ternary compounds, $\mathbf{CaSiS_3}$, $\mathbf{CaSi_2S_5}$ and $\mathbf{Ca_2SiS_4}$, are formed in the Si–Ca–S system (Weiss and Rocktäschel 1960). Ca_2SiS_4 crystallizes in the orthorhombic structure with the lattice parameters a = 1346, b = 814, and c = 617 pm (Susa and Steinfink 1971b) [a = 1349, b = 818, and c = 621 pm (Rocktäschel et al. 1964)]. $CaSiS_3$, $CaSi_2S_5$ and Ca_2SiS_4 were synthesized by the interaction of CaSi, $CaSi_2$, and Ca_2Si, respectively, with sulfur (Weiss and Rocktäschel 1960). The most successful result for the synthesis of Ca_2SiS_4 was achieved when starting with elemental Si, powdered S, and powdered CaS (Susa and Steinfink 1971b). Stoichiometric ratio of Si and Ca with sulfur or of the CaS were mixed, sealed under vacuum in Vycor tubing, and pre-reacted at temperatures ranging from 200°C to 400°C for 2–4 h. The temperature was then raised and kept in the range 600°C–800°C for about 12 h. After cooling, the vial was examined visually to see whether some of the starting materials had remained unreacted. If the sample looked homogeneous, the vial was opened. If Ca_2SiS_4 was detected in the sample, it was resealed in a Vycor tube and fired at the same or at higher temperatures until it appeared to have completely reacted. In some cases, the temperature was raised to 1330°C after first placing the material in a graphite crucible which was then sealed in a Vycor tube.

1.9 Silicon–Strontium–Sulfur

The $\mathbf{Sr_2SiS_4}$ ternary compound, which crystallizes in the monoclinic structure with the lattice parameters a = 824.8, b = 664.0, c = 655.6 pm, β = 108.36°, and the calculated and experimental densities of 3.23 and 3.18 g·cm^{-3}, respectively, is formed in the Si–Sr–S (Dumail et al. 1970).

1.10 Silicon–Barium–Sulfur

Three ternary compounds, $\mathbf{Ba_2SiS_4}$, $\mathbf{Ba_2Si_2S_5}$ and $\mathbf{Ba_3SiS_5}$, are formed in the Si–Ba–S system. Ba_2SiS_4 decomposes in a vacuum at 900°C (Dumail et al. 1970) and crystallizes in the orthorhombic structure with the lattice parameters a = 893.04 ± 0.09, b = 678.21 ± 0.04, and c = 1201.06 ± 0.08 pm and a calculated density of 3.936 g·cm^{-3} (Lemley 1974) [a = 892.3,

$b = 678.7$, and $c = 1202.6$ pm and the calculated and experimental densities of 3.93 and 3.88 g·cm^{-3}, respectively (Ribes et al. 1969; Dumail et al. 1970); $a = 884$, $b = 676$, and $c = 1193$ pm (Susa and Steinfink 1971b)]. The most successful result for the synthesis of Ba_2SiS_4 was achieved when starting with elemental Si, powdered S, and powdered BaS (Susa and Steinfink 1971b). Stoichiometric ratio of Si and Ba with sulfur or of the BaS were mixed, sealed under vacuum in Vycor tubing, and pre-reacted at temperatures ranging from 200°C to 400°C for 2–4 h. The temperature was then raised and kept in the range 600°C–800°C for about 12 h. After cooling, the vial was examined visually to see whether some of the starting materials had remained unreacted. If the sample looked homogeneous, the vial was opened. If Ba_2SiS_4 was detected in the sample, it was resealed in a Vycor tube and fired at the same or at higher temperatures until it appeared to have completely reacted. In some cases, the temperature was raised to 1330°C after first placing the material in a graphite crucible which was then sealed in a Vycor tube. The amber crystals of the title compound were prepared by heating a stoichiometric mixture of BaS, Si, and S at 1100°C for 26 h (Lemley 1974). Cooling was accomplished stepwise over approximately 3 days. The sample charges were placed in closed graphite crucibles and sealed in evacuated Vycor tubes for firing.

$Ba_2Si_2S_5$ decomposes in a vacuum at 750°C (Dumail et al. 1970).

Ba_3SiS_5 also crystallizes in the orthorhombic structure with the lattice parameters $a = 1212.1 \pm 1.0$, $b = 952.7 \pm 0.8$, and $c = 855.3 \pm 0.8$ pm (Schmitz von 1981). The title compound could be prepared by passing a dried H_2S stream at 1180°C over a mixture of $BaCO_3$ and elemental Si (molar ratio 3:1) in a corundum tube. A pale yellow crystalline substance was obtained.

1.11 Silicon–Zinc–Sulfur

Ternary compounds, including $\mathbf{Zn_4SiS_6}$, were not found in the Si–Zn–S system (Kaldis et al. 1967; Dubrovin et al. 1989a).

1.12 Silicon–Cadmium–Sulfur

SiS_2–CdS: The phase diagram of this system is shown in Figure 1.5 (Odin et al. 1985b). The eutectic composition and temperature are 40 mol% CdS and 986 ± 6°C, respectively, and the peritectic composition is 78 mol% CdS. $\mathbf{Cd_4SiS_6}$ is formed in this system. It melts incongruently at 1235°C ± 8°C (Odin et al. 1985b) and crystallizes in the monoclinic structure with the lattice parameters $a = 1231.0 \pm 0.6$, $b = 704.1 \pm 0.4$, $c = 1233.6 \pm 0.6$ pm, $\beta = 110.38 \pm 0.05°$, and the calculated and experimental density of 4.446 and 4.41 ± 0.01 g·cm^{-3}, respectively (Krebs von and Mandt 1972) [$a = 1234$, $b = 708.9$, $c = 1235$ pm, and $\beta = 110°44'$ (Serment et al. 1968); $a = 1199$, $b = 702$, $c = 1212$ pm, and $\beta = 110.20°$ (Susa and Steinfink 1971b); $\beta = 110.5°$ (Kaldis et al. 1967)]. At 1085°C ± 7°C, polymorphous transition of Cd_4SiS_6 takes place (Odin et al. 1985b). Solid-solution regions based on SiS_2 and CdS do not exceed 1 mol% CdS and 0.7 mol% SiS_2, respectively. Homogeneity region of Cd_4SiS_6 is within the interval 79.4–80.0 mol% CdS.

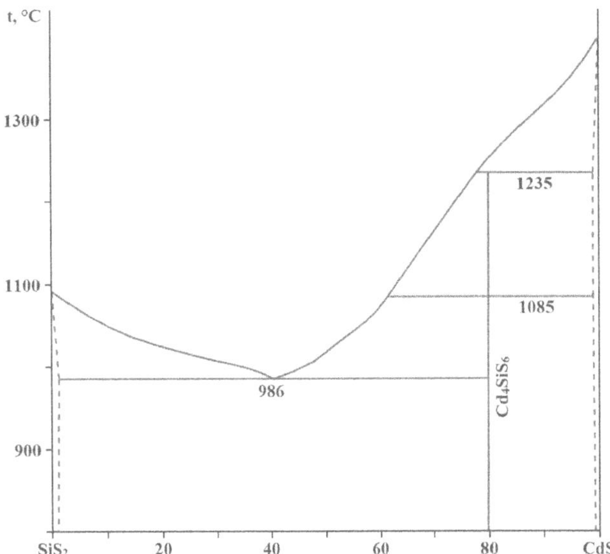

FIGURE 1.5 Phase diagram of the SiS_2–CdS system. (From Odin, I.N., et al., *Zhurn. neorgan. khimii*, **30**(1), 207, 1985.)

This system was investigated using DTA and XRD. The ingots were annealed at 700°C for 720 h (Odin et al. 1985b). Cd_4SiS_6 was obtained by annealing of mixtures of Cd, Si, and S in stoichiometric ratio at 200°C–400°C for 2–4 h and then at 600°C–800°C for 12 h (Susa and Steinfink 1971b). It could be also synthesized by reacting stoichiometric quantities of CdS, Si, and S in evacuated quartz tubes (Kaldis et al. 1967). The tube with the well-mixed starting materials was put in a furnace with a flat temperature profile and was slowly heated up to 850°C for 2 days. This slow heating rate was necessary to avoid explosions. For the preparation of single crystals of the title compound, the chemical transport reactions could be used (Krebs von and Mandt 1972). For this, a stoichiometric mixture of CdS, Si, and S (about 10 mM) with about 60 mg I_2 in an evacuated quartz tube was heated at 1000°C for 12 h, and then exposed to 900°C–700°C for 7 days. Well-formed lemon–yellow crystals were formed in the colder part of the tube.

Si–CdS: The phase diagram of this system is not constructed. According to the data of Odin (1996), it is a eutectic type with eutectic temperature of 1337°C ± 10°C. The mutual solubility of CdS and Si is insignificant. This system was investigated using DTA, metallography, and XRD. The ingots were annealed at temperatures 20°C lower than the temperature of invariant equilibrium, including liquid, for 1000 h.

The field of Si primary crystallization occupies almost all liquidus surfaces of the Si–Cd–S ternary system (Figure 1.6) (Odin 1996). There are also the fields of SiS, α- and β-Cd_4SiS_6, CdS, and SiS_2 primary crystallization and degenerated fields of Cd and β-S primary crystallization. There are three ternary

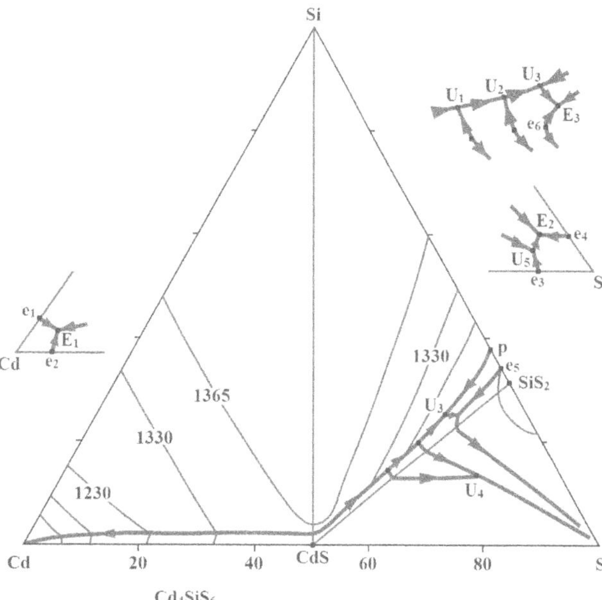

FIGURE 1.6 Liquidus surface of the Si–Cd–S ternary system. (From Odin, I.N., *Zhurn. neorgan. khimii*, **41**(6), 941, 1996.)

eutectics and five transition points in this system: E_1 (317°C) – L ⇔ Cd + Si + CdS; E_2 (113°C) – L ⇔ α-Cd_4SiS_6 + SiS_2 + β-S; E_3 (1004°C) – L ⇔ α-Cd_4SiS_6 + SiS + SiS_2; U1 (1197°C) – L + CdS ⇔ β-Cd_4SiS_6 + Si; U2 (1056°C) – L + β-Cd_4SiS_6 ⇔ Si + α-Cd_4SiS_6; U3 (1043°C)–L+Si⇔α-Cd_4SiS_6+SiS; U4 (1036°C)– L + β-Cd_4SiS_6 ⇔ α-Cd_4SiS_6 + CdS; U5 (116°C) – L + CdS ⇔ α-Cd_4SiS_6 + β-S. The part of the isothermal section at 730°C of the Si–Cd–S ternary system is shown in Figure 1.7.

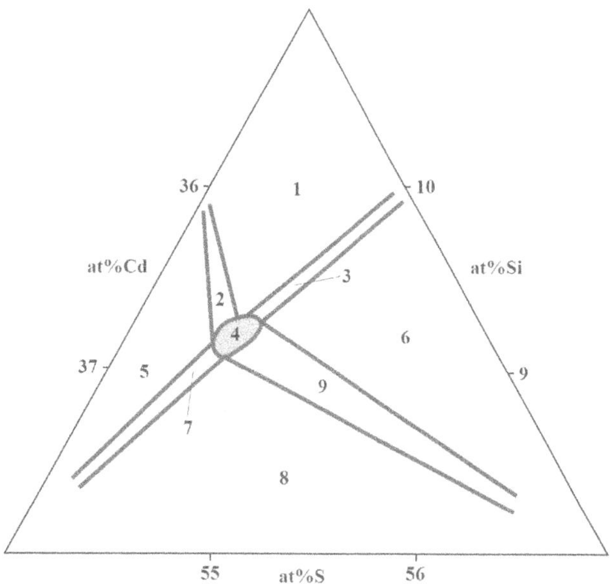

FIGURE 1.7 Part of the Si–Cd–S isothermal section at 730°C: 1, α-Cd_4SiS_6 + Si + SiS_2; 2, α-Cd_4SiS_6 + Si; 3, α-Cd_4SiS_6 + SiS_2; 4, α-Cd_4SiS_6; 5, α-Cd_4SiS_6 + Cd + Si; 6, α-Cd_4SiS_6 + SiS_2 + L; 7, α-Cd_4SiS_6 + CdS; 8, α-Cd_4SiS_6 + CdS + L; and 9, α-Cd_4SiS_6 + L. (From Odin, I.N., *Zhurn. neorgan. khimii*, **41**(6), 941, 1996.)

1.13 Silicon–Mercury–Sulfur

SiS_2–HgS: The phase diagram of this system is not constructed. At 800°C–1000°C, **Hg_4SiS_6** compound is formed in this system (Serment et al. 1968; Gulay et al. 2002b). It crystallizes in the monoclinic structure with the lattice parameters a = 1230.20 ± 0.05, b = 710.31 ± 0.04, c = 1227.91 ± 0.04 pm, and β = 109.721 ± 0.003° (Gulay et al. 2002b) [a = 1229 ± 1, b = 709.6 ± 0.6, c = 1230 ± 1 pm, β = 109°28′ and experimental density of 6.67 ± 0.01 g·cm^{-3} (Serment et al. 1968)]. The parameters of the monoclinic structure were recalculated in the parameters of the rhombohedral and hexagonal structures (rhombohedral structure, a = 1230 ± 1 pm and α = 33°32′ ± 0°10′; hexagonal structure, a = 709.6 ± 0.6 and c = 3480 ± 1 pm). According to the data of Serment et al. (1968), the lattice is practically rhombohedral, but the X-ray spectrum of Hg_4SiS_6 compound well indexes by the parameters both of rhombohedral and hexagonal structures.

The synthesis of the title compound was carried out using high-purity elements (Si was crushed in an agate mortar before weighing) and previously prepared HgS (Gulay et al. 2002b). The mixture was sealed in quartz ampoule and heated to 400°C at a rate of 30°C·h^{-1} and to 830°C at a rate of 10°C·h^{-1}. After 4 h of the synthesis at this temperature, the furnace was cooled to 400°C (10°C·h^{-1}) and then was turned off.

1.14 Silicon–Thallium–Sulfur

SiS_2–Tl_2S: The phase diagram of this system, constructed through DTA, XRD, metallography, and measuring of microhardness, is presented in Figure 1.8 (Lazarev et al. 1983). Three compounds, **Tl_2SiS_3**, **$Tl_2Si_2S_5$**, and **Tl_4SiS_4**, are formed in this system. There are four eutectic points on the phase diagram, which contain 25, 40, 53, and 75 mol% SiS_2 and crystallize at 375°C, 407°C, 550°C, and 607°C, respectively.

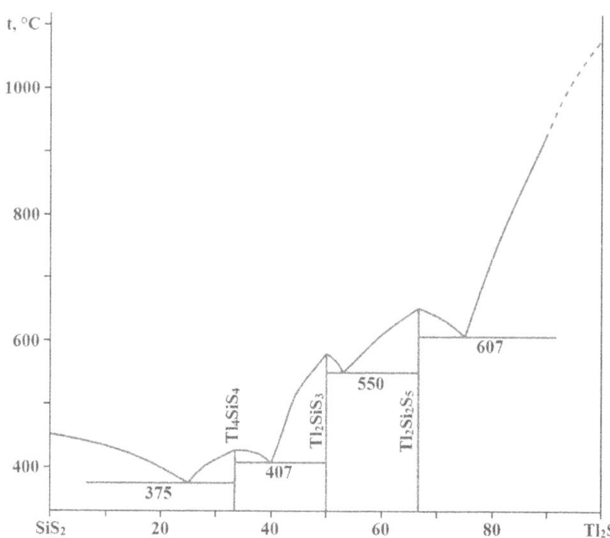

FIGURE 1.8 Phase diagram of the SiS_2–Tl_2S system. (From Lazarev, V.B., et al., *Zhurn. neorgan. khimii*, **28**(8), 2097, 1983.)

Tl_2SiS_3 melts congruently at 580°C (Lazarev et al. 1983, 1984; Peresh et al. 1986) and crystallizes in the triclinic structure with the lattice parameters $a = 669.1 \pm 0.2$, $b = 671.1 \pm 0.2$, $c = 836.7 \pm 0.2$ pm, and $\alpha = 90.69 \pm 0.02°$, $\beta = 111.62 \pm 0.02°$, $\gamma = 111.42 \pm 0.02°$ and the calculated and experimental densities of 5.57 and 5.62 g·cm^{-3}, respectively (Nakamura et al. 1984) [$a = 669.9 \pm 0.5$, $b = 664.5 \pm 0.4$, $c = 838.0 \pm 0.5$ pm, and $\alpha = 90.32 \pm 0.05°$, $\beta = 112.00 \pm 0.05°$, $\gamma = 112.32 \pm 0.05°$, and a calculated density of 5.62 g·cm^{-3} (Eulenberger 1982)]. The enthalpy and entropy of melting of this compound are equal to 30.3 kJ·M^{-1} and 35.5 J·(M·K)$^{-1}$, respectively, and an energy gap is 2.6 eV (Peresh et al. 1986). The homogeneity region of the title compound is given in Figure 1.9a (Lazarev et al. 1984). It can be seen that the homogeneity region does not exceed 3.6 mol%, and the maximum on the liquidus line shifts by 0.1 mol% SiS_2 towards Tl_2S. Tl_2SiS_3 was prepared from the stoichiometric mixture of TlS, Si, and S by reacting at 700°C–800°C for several hours and subsequent annealing at 430°C–460°C for several days in the evacuated silica tube (Nakamura et al. 1984) or by reacting the stoichiometric mixture of Tl, Si, and S in the evacuated silica tube (Eulenberger 1982).

$Tl_2Si_2S_5$ melts congruently at 652°C (Lazarev et al. 1983, 1984; Peresh et al. 1986). The enthalpy and entropy of melting of this compound is equal to 51.6 kJ·M^{-1} and 55.6 J·(M·K)$^{-1}$, respectively (Peresh et al. 1986). The homogeneity region of the title compound is given in Figure 1.9b (Lazarev et al. 1984). It can be seen that the homogeneity region does not exceed 4.1 mol%, and the maximum on the liquidus line shifts by 0.4 mol% SiS_2 towards SiS_2.

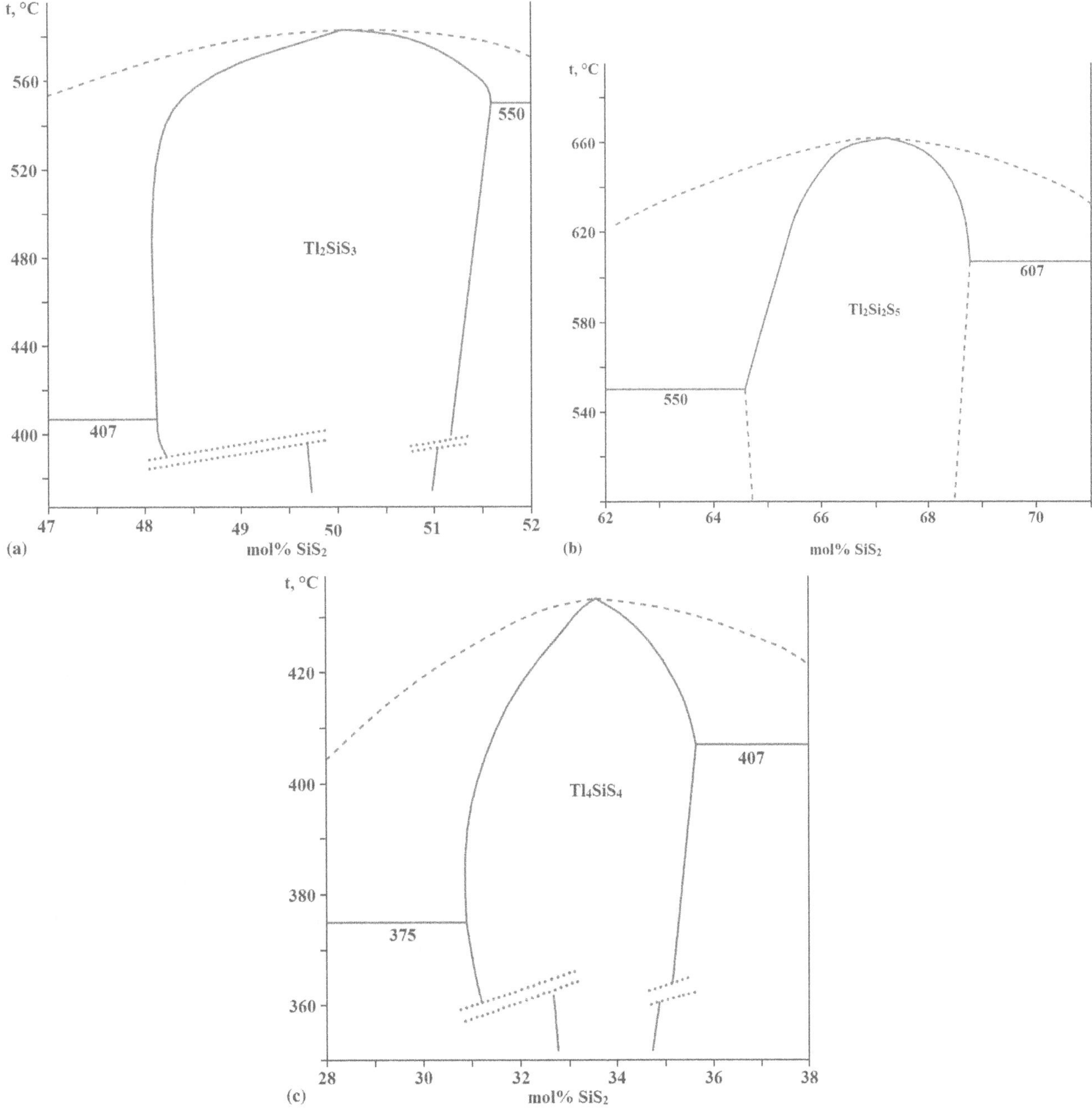

FIGURE 1.9 Homogeneity region of (a) Tl_2SiS_3, (b) $Tl_2Si_2S_5$, and (c) Tl_4SiS_4. (From Lazarev, V.B., et al., *Zhurn. neorgan. khimii*, **29**(6), 1571, 1984.)

Tl_4SiS_4 melts at 427°C (Lazarev et al. 1983, 1984; Peresh et al. 1986) and crystallizes in the monoclinic structure with the lattice parameters a = 1251.8 ± 0.3, b = 1124.1 ± 0.2, c = 756.7 ± 0.2 pm, β = 112.80 ± 0.02°, and a calculated density of 6.59 g·cm^{-3} (Eulenberger 1984a, 1986). The enthalpy and entropy of melting of this compound is equal to 26.0 kJ·M^{-1} and 36.8 J·(M·K)$^{-1}$, respectively, and an energy gap is 2.2 eV (Peresh et al. 1986). The homogeneity region of the title compound is given in Figure 1.9c (Lazarev et al. 1984). It can be seen that the homogeneity region does not exceed 4.7 mol%, and the maximum on the liquidus line shifts by 0.2 mol% SiS_2 towards SiS_2. This compound was prepared from the elements which were sealed in an evacuated quartz ampoule in quantities according to the stoichiometry of the compound (Eulenberger 1984a, 1986). The temperature of the ampoule was gradually raised in an electric tube furnace to 500°C, and then held at 410°C for 83 days. The major part of the reaction product formed a yellow crystalline regulus containing a small amount of a dark glassy material. This compound is stable on exposure to the atmosphere.

1.15 Silicon–Yttrium–Sulfur

The $\mathbf{Y_6Si_{2.5}S_{14}}$ ternary compound, which crystallizes in the hexagonal structure with the lattice parameters a = 975, c = 570 pm, and a calculated density of 3.73 g·cm^{-3}, is formed in the Si–Y–S system (Michelet and Flahaut 1969; Michelet et al. 1975). It was prepared in a vacuum-sealed silica ampoule by the combination of Y_2S_3, Ge, and S. The heating of the ampoule was very progressive to avoid explosions due to too much vapor pressure from sulfur not yet combined: 3 days at 550°C, 1 day at 650°C, 2 days at 750°C, 12 h at 900°C, and 2 h at 1050°C. The cooling was done slowly in the furnace. Annealing to 600°C usually allows obtaining well-crystallized product.

1.16 Silicon–Lanthanum–Sulfur

The $\mathbf{La_2SiS_5}$ ternary compound, which crystallizes in the monoclinic structure with the lattice parameters a = 762.08 ± 0.12, b = 1264.07 ± 0.15, c = 789.98 ± 0.12 pm, β = 101.559 ± 0.013°, and a calculated density of 4.153 g·cm^{-3}, is formed in the Si–La–S (Daszkiewicz et al. 2007) [a = 785.7 ± 0.4, b = 760.6 ± 0.4, c = 1262.7 ± 0.8 pm, and β = 101.55 ± 0.02° (Michelet et al. 1970, 1975)]. It was prepared in the same way as $Y_6Si_{2.5}S_{14}$ was obtained using La_2S_3 instead of Y_2S_3 (Michelet et al. 1975). Single crystals of the title compound were grown by fusion of the elemental constituents in the stoichiometric ratio of La/Si/S= 2:1:5 in an evacuated silica ampoule (Daszkiewicz et al. 2007). The ampoule was heated in a tube furnace with a heating rate of 30°C·h^{-1} to 1150°C and was kept at this temperature for 4 h. It was then cooled down slowly (10°C·h^{-1}h) to 500°C, annealed at this temperature for further 240 h and finally quenched in cold water. The obtained yellow crystals were selected from the brown-colored compact product.

1.17 Silicon–Cerium–Sulfur

Three ternary compounds are formed in the Si–Ce–S system (Gauthier et al. 2003). $\mathbf{Ce_2SiS_5}$ crystallizes in the monoclinic structure with the lattice parameters a = 754.75 ± 0.08, b = 1255.81 ± 0.14, c = 782.86 ± 0.08 pm, β = 101.550 ± 0.013°, a calculated density of 4.280 g·cm^{-3} and an energy gap of 2.34 ± 0.01 eV (Gauthier et al. 2003) [a = 779.8 ± 0.4, b = 754.0 ± 0.4, c = 1252.4 ± 0.4 pm, and β = 101.60 ± 0.02° (Michelet et al. 1970, 1975)]. It was prepared in the same way as $Y_6Si_{2.5}S_{14}$ was obtained using Ce_2S_3 instead of Y_2S_3 (Michelet et al. 1975).

$\mathbf{Ce_4Si_3S_{12}}$ crystallizes in the trigonal structure with the lattice parameters a = 1917.45 ± 0.05 and c = 799.43 ± 0.02 pm, a calculated density of 4.028 g·cm^{-3} and an energy gap of 2.38 ± 0.01 eV (Gauthier et al. 2003) [in the rhombohedral structure with the lattice parameters a = 1136 pm, α = 114°44′ and a calculated density of 4.18 g·cm^{-3} or a = 1914 and c = 795 pm in hexagonal setting (Perez and Duale 1969)]. This compound was obtained by the action of Ce_2S_3 on a mixture of the elementary Si and S in desired proportion in a silica tube sealed under a vacuum (Perez and Duale 1969). At first, the reaction mixture was heated to 800°C with a rate of rise in temperature sufficiently slow to avoid a high pressure of S inside the tube. The temperature was maintained at 800°C for 24 h to allow the formation of SiS_2, and then the mixture was heated to 1050°C for 3 days. The reaction product was quenched with water.

$\mathbf{Ce_6Si_4S_{17}}$, which exists at 1050°C, crystallizes in the triclinic structure with the lattice parameters a = 895.76 ± 0.06, b = 1000.22 ± 0.07, c = 1426.51 ± 0.10 pm, and α = 82.188 ± 0.003°, β = 86.889 ± 0.007°, γ = 89.515 ± 0.004°, a calculated density of 3.934 g·cm^{-3} and an energy gap of 2.56 ± 0.01 eV (Gauthier et al. 2003).

For obtaining Ce_2SiS_5 and $Ce_4Si_3S_{12}$, a mixture of γ-Ce_2S_3, Si, and S was weighed in the stoichiometric proportions and placed in a silica tube sealed under a primary vacuum (0.13 Pa) (Gauthier et al. 2003). The mixture was heated slowly (4°C·h^{-1}) up to 300°C, and was then kept at this temperature for 2 days to allow for the sublimation of S and its subsequent reaction with Si, thus leading to the intermediate SiS_2. Then the temperature was raised slowly (again 4°C·h^{-1}) up to 1050°C and was maintained at this temperature for 4 days before quenching the silica tubes in air. Samples were grounded and then annealed at 700°C for 7 days with 5% sulfur mass excess to avoid decomposition of the phase and hence ensure the homogeneity of the final product.

Several attempts to prepare pure samples showed the need to use both S and Si in excess. Actually, a 10 mol% excess of SiS_2 was used with respect to the targeted formula. For example, Ce_2SiS_5 and $Ce_4Si_3S_{12}$ were obtained from the Ce_2S_3/Si/S with molar ratio 1.0:1.1:2.2 and 2.0:3.3:6.6, respectively. The annealing described above removes the excess SiS_2, which condenses in the cold end of the reaction tube. To eliminate the last traces of SiS_2 and S, samples were washed with ethanol and dried in air. Pure $Ce_6Si_4S_{17}$ samples can be prepared from the mixture of Ce_2S_3/Si/S with a molar ratio 3.0:4.4:8.8 under the synthesis conditions used for Ce_2SiS_5 and $Ce_4Si_3S_{12}$. No other phases were detected at the chosen preparative temperature, but it is worth mentioning that a phase with the composition $\mathbf{Ce_6Si_{2+x}S_{14}}$ exists and is found to occur at a lower temperature (Gauthier et al. 2003).

1.18 Silicon–Praseodymium–Sulfur

Three ternary compounds, $\mathbf{Pr_2SiS_5}$, $\mathbf{Pr_4Si_3S_{12}}$, and $\mathbf{Pr_6Si_4S_{17}}$, are formed in the Si–Pr–S system. First of them crystallizes in the monoclinic structure with the lattice parameters $a = 777.5 \pm 0.4$, $b = 751.4 \pm 0.4$, $c = 1248.9 \pm 0.8$ pm, and $\beta = 101.62 \pm 0.02°$ (Michelet et al. 1970, 1975)]. It was prepared in the same way as $Y_6Si_{2.5}S_{14}$ was obtained using Pr_2S_3 instead of Y_2S_3.

$Pr_4Si_3S_{12}$ crystallizes in the rhombohedral structure with the lattice parameters $a = 1134$ pm, $\alpha = 114°46'$ and a calculated density of 4.21 g·cm^{-3} or $a = 1911$, and $c = 793$ pm in hexagonal setting (Perez and Duale 1969). It was obtained in the same way as $Ce_4Si_3S_{12}$ was synthesized using Pr_2S_3 instead of Ce_2S_3.

$Pr_6Si_4S_{17}$ crystallizes in the triclinic structure with the lattice parameters $a = 890.2 \pm 0.1$, $b = 993.4 \pm 0.1$, $c = 1420.6 \pm 0.2$ pm, and $\alpha = 82.19 \pm 0.01°$, $\beta = 86.94 \pm 0.01°$, $\gamma = 89.40 \pm 0.01°$, and a calculated density of 4.016 g·cm^{-3} (Gulay et al. 2008a). To prepare the title compound, the evacuated silica ampoule with the accounted quantity of the components was heated (30°C·h^{-1}) to 1150°C in a tube furnace. The sample was kept at the maximal temperature during 3 h. Then, they were cooled (10°C·h^{-1}) to 600°C and annealed at this temperature during 240 h. After annealing, the ampoule with the sample was quenched in cold water.

1.19 Silicon–Neodymium–Sulfur

Three ternary compounds, $\mathbf{Nd_2SiS_5}$, $\mathbf{Nd_4Si_3S_{12}}$, and $\mathbf{Nd_6Si_4S_{17}}$, are formed in the Si–Nd–S system. First of them crystallizes in the monoclinic structure with the lattice parameters $a = 774.0 \pm 0.4$, $b = 748.0 \pm 0.4$, $c = 1243.4 \pm 0.4$ pm, and $\beta = 101.66 \pm 0.02°$ (Michelet et al. 1970, 1975). It was prepared in the same way as $Y_6Si_{2.5}S_{14}$ was obtained using Nd_2S_3 instead of Y_2S_3.

$Nd_4Si_3S_{12}$ crystallizes in the rhombohedral structure with the lattice parameters $a = 1131$ pm, $\alpha = 114°46'$ and the calculated and experimental densities of 4.30 and 4.25 g·cm^{-3}, respectively, or $a = 1906$ and $c = 790$ pm in hexagonal setting (Perez and Duale 1969). It was obtained in the same way as $Ce_4Si_3S_{12}$ was synthesized using Nd_2S_3 instead of Ce_2S_3.

$Nd_6Si_4S_{17}$ crystallizes in the triclinic structure with the lattice parameters $a = 888.0 \pm 0.1$, $b = 990.3 \pm 0.1$, $c = 1416.8 \pm 0.2$ pm, and $\alpha = 82.11 \pm 0.01°$, $\beta = 87.04 \pm 0.01°$, $\gamma = 89.31 \pm 0.01°$, and a calculated density of 4.104 g·cm^{-3} (Gulay et al. 2008a). This compound was synthesized in the same way as $Pr_6Si_4S_{17}$ was prepared.

1.20 Silicon–Samarium–Sulfur

Three ternary compounds, $\mathbf{Sm_3Si_{1.25}S_7}$, $\mathbf{Sm_4Si_3S_{12}}$, and $\mathbf{Sm_6Si_4S_{17}}$, are formed in the Si–Sm–S system. $Sm_3Si_{1.25}S_7$ crystallizes in the hexagonal structure with the lattice parameters $a = 995.58 \pm 0.04$, $c = 570.69 \pm 0.04$ pm, and a calculated density of 4.816 g·cm^{-3} (Gulay and Lychmanyuk 2009).

$Sm_4Si_3S_{12}$ crystallizes in the rhombohedral structure with the lattice parameters $a = 1126$ pm, $\alpha = 114°48'$ and a calculated density of 4.48 g·cm^{-3} or $a = 1897$, and $c = 783$ pm in hexagonal setting (Perez and Duale 1969). It was obtained in the same way as $Ce_4Si_3S_{12}$ was synthesized using Sm_2S_3 instead of Ce_2S_3.

$Sm_6S_4iS_{17}$ crystallizes in the triclinic structure with the lattice parameters $a = 883.00 \pm 0.08$, $b = 977.9 \pm 0.1$, $c = 1404.7 \pm 0.1$ pm, and $\alpha = 82.126 \pm 0.008°$, $\beta = 87.338 \pm 0.007°$, $\gamma = 89.018 \pm 0.008°$, and a calculated density of 4.315 g·cm^{-3} (Gulay et al. 2008a). This compound was synthesized in the same way as $Pr_6Si_4S_{17}$ was prepared.

1.21 Silicon–Europium–Sulfur

The $\mathbf{Eu_2SiS_4}$ ternary compound, which crystallizes in the monoclinic structure with the lattice parameters $a = 651.71 \pm 0.06$, $b = 659.54 \pm 0.06$, $c = 821.93 \pm 0.08$ pm, $\beta = 108.437 \pm 0.009°$, and a calculated density of 4.561 g·cm^{-3} (Hartenbach and Schleid 2002) [$a = 652.4 \pm 0.1$, $b = 659.1 \pm 0.1$, $c = 820.5 \pm 0.2$ pm, and $\beta = 108.29 \pm 0.03°$ and a calculated density of 4.563 g·cm^{-3} at 100 K (Johrendt and Pocha 2001)], is formed in the Si–Eu–S system. It was prepared from the elements in two steps (Johrendt and Pocha 2001). First, Eu metal and Si powder were heated to 900°C for 15 h in a quartz tube under an Ar atmosphere. The resulting alloy was grounded in an Ar-filled glove box and then oxidized by the stoichiometric amount of S at 850°C for 50 h. This results in a dark yellow powder containing small single crystals of the title compound.

The conversion of Eu, S, and SiS_2 (molar ratio 2:2:1) with an excess of CsCl as a flux in evacuated silica tube at 850°C for 7 days leads to deep red, plate-shaped single crystals of Eu_2SiS_4, which remain air and water stable for a few days (Hartenbach and Schleid 2002).

1.22 Silicon–Gadolinium–Sulfur

Two ternary compounds, $\mathbf{Gd_4Si_3S_{12}}$ and $\mathbf{Gd_6Si_{2.5}S_{14}}$, are formed in the Si–Gd–S system. First of them crystallizes in the monoclinic structure with the lattice parameters $a = 986.7 \pm 0.2$, $b = 1099.69 \pm 0.19$, $c = 1646.2 \pm 0.4$ pm, $\beta = 102.67 \pm 0.03°$, and a calculated density of 4.185 g·cm^{-3} (Hatscher and Urland 2003) [in the rhombohedral structure with the lattice parameters $a = 1121$ pm, $\alpha = 114°48'$ and a calculated density of 4.66 g·cm^{-3} or $a = 1889$ and $c = 778$ pm in hexagonal setting (Perez and Duale 1969)]. It was obtained in the same way as $Ce_4Si_3S_{12}$ was synthesized using Gd_2S_3 instead of Ce_2S_3 (Perez and Duale 1969). Single crystals of $Gd_4Si_3S_{12}$ were obtained by reaction of the elements in an iodine atmosphere (Hatscher and Urland 2003). Gd metal chips, S and Si powders, and iodine crystals were loaded into fused-silica tubes in a glove box under Ar atmosphere (molar ratio 1:2.26:0.53:0.8). The tube was sealed afterwards, placed into a furnace and exposed to a temperature gradient of 1000°C–850°C for 10 days. Then the ampoule was quenched in water. In the middle of the tube, air-stable white crystals were obtained.

$Gd_6Si_{2.5}S_{14}$ crystallizes in the hexagonal structure with the lattice parameters a = 987, c = 571 pm and a calculated density of 5.04 g·cm^{-3} (Michelet and Flahaut 1969; Michelet et al. 1975). It was prepared in the same way as $Y_6Si_{2.5}S_{14}$ was obtained using Gd_2S_3 instead of Y_2S_3.

1.23 Silicon–Terbium–Sulfur

Two ternary compounds, $\mathbf{Tb_4(SiS_4)_3}$ and $\mathbf{Tb_6Si_{2.5}S_{14}}$, are formed in the Si–Tb–S system. $Tb_4(SiS_4)_3$ crystallizes in the monoclinic structure with the lattice parameters a = 983.6 ± 0.2, b = 1096.4 ± 0.2, c = 1639.1 ± 0.3 pm, β = 102.76 ± 0.02°, and a calculated density of 4.256 g·cm^{-3} (Hatscher and Urland 2002b). Single crystals of this compound were obtained from the elements. Tb metal chips, S and Si powders (molar ratio 1:3.28:1.06) were added into a quartz ampoule in a glove box under Ar atmosphere. This was evacuated and about 50 mg of Br_2 were added. The ampoule was placed in a two-zone furnace for 10 days, and temperature gradient from 1000°C to 800°C was exposed. After cooling, air-stable, colorless crystals were found.

$Tb_6Si_{2.5}S_{14}$ crystallizes in the hexagonal structure with the lattice parameters a = 982, c = 570 pm, and a calculated density of 5.14 g·cm^{-3} (Michelet and Flahaut 1969; Michelet et al. 1975). It was prepared in the same way as $Y_6Si_{2.5}S_{14}$ was obtained using Tb_2S_3 instead of Y_2S_3.

1.24 Silicon–Dysprosium–Sulfur

Two ternary compounds, $\mathbf{Dy_4(SiS_4)_3}$ and $\mathbf{Dy_6Si_{2.5}S_{14}}$, are formed in the Si–Dy–S system. $Dy_4(SiS_4)_3$ crystallizes in the monoclinic structure with the lattice parameters a = 981.3 ± 0.2, b = 1093.87 ± 0.18, c = 1636.0 ± 0.4 pm, β = 102.86 ± 0.03°, and a calculated density of 4.341 g·cm^{-3} (Hatscher and Urland 2002a). To prepare the title compound, Dy metal chips, S and Si powders, and bromine (molar ratio 1:3.28:1.06:~0.3) were added into a quartz ampoule in a glove box under Ar atmosphere. The ampoule was evacuated and heated for 10 days in a temperature gradient from 1000°C to 800°C. After the tube was cooled, a few air-stable, green crystals were obtained.

$Dy_6Si_{2.5}S_{14}$ crystallizes in the hexagonal structure with the lattice parameters a = 975.5, c = 570 pm, and a calculated density of 5.28 g·cm^{-3} (Michelet and Flahaut 1969; Michelet et al. 1975). It was prepared in the same way as $Y_6Si_{2.5}S_{14}$ was obtained using Dy_2S_3 instead of Y_2S_3.

1.25 Silicon–Holmium–Sulfur

The $\mathbf{Ho_3Si_{1.25}S_7}$ ternary compound, which crystallizes in the hexagonal structure with the lattice parameters a = 973.06 ± 0.02 and c = 570.01 ± 0.02, is formed in the Si–Ho–S system (Lychmanyuk et al. 2007). It was prepared by melting high-purity elements in an evacuated silica ampoule. The synthesis was carried out in a tube furnace where the ampoule was heated (30°C·h^{-1}) to 1150°C. The sample was kept at this temperature for 4 h. Then it was slowly cooled (10°C·h^{-1}) to 600°C and annealed at this temperature for 240°C. After annealing, the ampoule with the sample was quenched in cold water.

1.26 Silicon–Thorium–Sulfur

The **ThSiS** ternary compound, which crystallizes in the tetragonal structure with the lattice parameters a = 391.4 and c = 1706 pm and the calculated and experimental densities of 7.42 and 7.37 g·cm^{-3}, respectively, is formed in the Si–Th–S system (Hahn and Stocks 1968; Stocks et al. 1981). To prepare this compound, the elements were weighed under N_2 atmosphere in the stoichiometric ratio and initially heated to about 400°C in an evacuated quartz ampoule. Subsequently, the inhomogeneous product was finely powdered under N_2 and heated to 1050°C in a sealed tungsten crucible, which was enclosed in an evacuated quartz ampoule. Since the preparation is extremely sensitive to traces of oxygen, it proved necessary to add Th as getter metal. Nevertheless, pure preparation was obtained only if no longer than 48 h the mixture was heated at 1050°C. The mixture was annealed at 600°C–700°C for 8–10 days. Then, the sample pressed into pastilles was heated in the same way over different time intervals and at different temperature up to 1400°C in tungsten crucible, wherein above 1050°C additionally, Ar was used as a protective gas. The title compound was obtained as black microcrystalline powder.

1.27 Silicon–Uranium–Sulfur

The **USiS** ternary compound, which crystallizes in the tetragonal structure with the lattice parameters a = 376.7 ± 0.1 and c = 828.5 ± 0.2 pm, and a calculated density of 8.42 g·cm^{-3}, is formed in the Si–U–S system (Haneveld and Jellinek 1969). This compound was prepared in the following way. A small rod of U was placed in a quartz tube and heated in a current of hydrogen at 230°C. The voluminous uranium hydride powder thus produced was dehydrided by heating at 400°C under vacuum. An equivalent proportion of intimate mixture of Si and S was added immediately and the quartz tube was evacuated and sealed. The mixture was heated at 800°C–900°C for 2 or 3 days. Then, the product was slowly cooled to room temperature.

1.28 Silicon–Lead–Sulfur

SiS_2–PbS: According to the data of Elli and Mugnoli (1963), the phase diagram of this system is a eutectic type and the brick-red color ternary compound $\mathbf{Pb_3Si_2S_7}$, which melts congruently, is formed in this system. Another ternary compound, $\mathbf{Pb_2SiS_4}$, melts up to 810°C (Hagenmuller and Pérez 1965) and crystallizes in the monoclinic structure with the lattice parameters a = 647.21 ± 0.05, b = 663.44 ± 0.09, c = 1683.2 ± 0.1 pm, and β = 108.805 ± 0.007° (Iglesias and Steinfink 1973b) [a = 650 ± 2, b = 665 ± 2, c = 1768 ± 5 pm, β = 115.5 ± 0.5°, and the calculated and experimental densities of 5.51 and 5.44 g·cm^{-3}, respectively (Hagenmuller and Pérez 1965)]. It was prepared by the direct reaction of corresponding stoichiometric quantities of the elements in an evacuated, sealed Vycor tube at 800°C for about 2 days (Iglesias and Steinfink 1973b). The obtained crystals were transparent, yellow prisms. Pb_2SiS_4 is very slightly hygroscopic (Hagenmuller and Pérez 1965).

1.29 Silicon–Zirconium–Sulfur

The quite stable **ZrSiS** ternary compound, which crystallizes in the tetragonal structure with the lattice parameters $a = 354.4$ and $c = 805.5$ pm (Haneveld and Jellinek 1964) [$a = 355$ and $c = 807$ pm and the calculated and experimental densities of 4.96 and 4.91 g·cm^{-3}, respectively (Jellinek and Hahn 1962; Onken et al. 1964)], is formed in the Si–Zr–S system. To obtain this compound, stoichiometric amounts of the elements were initially pre-tempered in an evacuated tube at 600°C for 10 days (Onken et al. 1964). Then, the reaction mixture was carefully rubbed in an agate mortar in an N_2-filled box, pressed into a pastille and annealed at 1050°C for several weeks. The ampoule with a tablet was stored in the hottest part of the furnace to obtain the single crystals by sublimation. ZrSiS could be also prepared by heating "SiS" with an equivalent proportion of Zr in an evacuated quartz tube at 1100°C for 2 days (Haneveld and Jellinek 1964).

1.30 Silicon–Hafnium–Sulfur

The **HfSiS** ternary compound, which crystallizes in the tetragonal structure with the lattice parameters $a = 352$ and $c = 800$ pm and the calculated and experimental densities of 8.20 and 7.66 g·cm^{-3}, respectively, is formed in the Si–Hf–S system (Onken et al. 1964). To prepare this compound, stoichiometric amounts of Si, Hf, and S were initially pre-tempered in an evacuated tube at 600°C for 10 days. Then, the reaction mixture was carefully rubbed in an agate mortar in an N_2-filled box, pressed into a pastille and annealed at 1000°C for several weeks. The ampoule with a tablet was stored in the hottest part of the furnace to obtain the single crystals by sublimation.

1.31 Silicon–Antimony–Sulfur

The glass-forming region in the Si–Sb–S ternary system is situated in the S-corner (Hilton et al. 1964). The maximum softening temperature of these glasses is 280°C. All of the glasses reacted rapidly with the atmosphere evolving H_2S. To prepare such glasses, the reactants were weighed out and sealed in a quartz vial at a reduced pressure. The vial was placed in a rocking furnace, slowly heating to 1000°C, and allowed to react and mix while molten for at least 16 h. The glasses were formed by quenching from 1000°C to room temperature in air.

1.32 Silicon–Tantalum–Sulfur

TaS_2 was intercalated by silicon but homogeneous products could not be obtained (Di Salvo et al. 1973).

1.33 Silicon–Selenium–Sulfur

The **SiSSe** ternary compound, which crystallizes in the orthorhombic structure with the lattice parameters $a = 527.1$, $b = 988.0$, and $c = 665.0$ pm, and the calculated and experimental densities of 2.67 and 2.63 g·cm^{-3}, respectively, is formed in the Si–Se–S system (Shevtsova et al. 1987a). This compound was obtained by the method of shock-wave compression of a mixture of elementary components. It was not possible to obtain the title compound by the thermal method.

1.34 Silicon–Tellurium–Sulfur

Ternary compounds were not found in the Si–Te–S system (Shevtsova et al. 1987a).

1.35 Silicon–Chlorine–Sulfur

The $\mathbf{Si_2S_2Cl_4}$ ternary compound, which crystallizes in the monoclinic structure with the lattice parameters $a = 699.8 \pm 0.1$, $b = 631.7 \pm 0.1$, $c = 973.2 \pm 0.1$ pm, $\beta = 97.46 \pm 0.02°$ at 138 K, and a calculated density of 2.040 g·cm^{-3}, is formed in the Si–Cl–S system (Peters et al. 1981). It could be prepared by reacting $SiCl_4$ with the solution of H_2S in pyridine or by interaction of $SiCl_4$ with H_2S at 650°C–950°C.

1.36 Silicon–Bromine–Sulfur

Two compounds, $\mathbf{Si_2S_2Br_4}$ and $\mathbf{Si_4S_6Br_4}$, are formed in the Si–Br–S system (Peters et al. 1981). First of them crystallizes in the orthorhombic structure with the lattice parameters $a = 1279.4 \pm 0.5$, $b = 1160.2 \pm 0.4$, and $c = 668.3 \pm 0.3$ pm, and a calculated density of 2.945 g·cm^{-3}. The second compound crystallizes in the triclinic structure with the lattice parameters $a = 958.2 \pm 0.3$, $b = 960.6 \pm 0.3$, $c = 942.6 \pm 0.3$ pm, and $\alpha = 99.50 \pm 0.02°$, $\beta = 110.88 \pm 0.02°$, $\gamma = 92.59 \pm 0.02°$, and a calculated density of 2.610 g·cm^{-3}. These compounds could be obtained by interaction of $SiBr_4$ with H_2S in the solution of Al_2Br_6, or by reacting $SiBr_4$ with P_4S_{10} or SiS_2.

1.37 Silicon–Iodine–Sulfur

The glass with $\mathbf{SiSI_2}$ composition was obtained in the Si–I–S ternary system (Dembovskiy et al. 1971a). This glass is a dark, viscous mass at room temperature. Glasses of other compositions could not be obtained due to the high vapor pressure of sulfur and iodine.

1.38 Silicon–Manganese–Sulfur

The $\mathbf{Mn_2SiS_4}$ ternary compound, which crystallizes in the orthorhombic structure with the lattice parameters $a = 1268.8 \pm 0.2$, $b = 742.9 \pm 0.2$, and $c = 594.2 \pm 0.1$ pm and a calculated density of 3.157 g·cm^{-3} (Fuhrmann and Pickardt 1989) [a = 1265 ± 1, b = 742.4 ± 0.8, and c = 592.8 ± 0.8 pm and the calculated and experimental densities of 3.17 and 3.11 g·cm^{-3}, respectively (Hagenmuller et al. 1964; Hardy et al. 1965); a = 1261, b = 745, and c = 595 pm (Rocktäschel et al. 1964)], is formed in the Si–Mn–S system. The title compound was obtained by the interaction of Mn_2Si and S in the temperature interval 450°C–650°C (Rocktäschel et al. 1964). Its crystals were grown by chemical transport reaction using I_2 as a transporting agent and MnS, Si, and S_8 as starting materials (Fuhrmann and Pickardt 1989). The reaction was carried out in a sealed quartz ampoule at 650°C/600°C for two weeks. Dark-brown, transparent crystals were obtained. Mn_2SiS_4 slowly decomposes in air with the release of H_2S (Hagenmuller et al. 1964; Hardy et al. 1965).

1.39 Silicon–Iron–Sulfur

The $\mathbf{Fe_2SiS_4}$ ternary compound, which crystallizes in the orthorhombic structure with the lattice parameters a = 1240.7 ± 0.2, b = 719.8 ± 0.1, and c = 581.2 ± 0.1 pm, is formed in the Si–Fe–S system (Vincent et al. 1976). This compound was prepared in vacuum-sealed quartz tubes from pure grounded elements mixed in stoichiometric proportions. The sample was slowly heated to 750°C, maintained at this temperature for a week, and then cooled slowly. The product was often inhomogeneous; a certain amount of SiS_2 and FeS or Fe_7S_8 was found at the end of the tube. In order to obtain a pure compound, the processes of grinding, placing the sample in a tube and heating must be carried out rapidly or in a dry box because SiS_2 hydrolyzes in air. Single crystals of the title compound were obtained by raising the temperature of the sealed tube (5°C·h^{-1}) to 800°C; then cooling it slowly back to 400°C, then raising it slowly again to 900°C, *etc.*, until in the final cycle the maximum temperature was 1100°C and the final temperature was 400°C. In this way, brilliant black irregularly shaped crystal were obtained.

REFERENCES

All References are available as a downloadable eResource at www.routledge.com/9780367639235

2

Systems Based on Silicon Selenides

2.1 Silicon–Sodium–Selenium

Several ternary compounds are formed in the Si–Na–Se system. $\mathbf{Na_4SiSe_4}$ crystallizes in the orthorhombic structure with the lattice parameters a = 1418.2 ± 0.5, b = 920.8 ± 0.3 and c = 712.2 ± 0.3 pm (Preishuber-Pflügl and Klepp 2003). Colorless single crystals of the title compound were obtained by reacting an intimate mixture of 829.5 mg Na_2Se, 124.4 mg Si, and 525.0 mg Se (approximate molar ratio 3:2:3) in a sealed and evacuated silica tube. The sample was gradually heated to 700°C, annealed for 12 h at this temperature and finally allowed to cool to ambient temperature at a controlled rate of 3°C·h^{-1}.

$\mathbf{Na_4Si_4Se_{10}}$ also crystallizes in the orthorhombic structure with the lattice parameters a = 1340.9 ± 0.5, b = 1318.4 ± 0.5, and c = 1086.5 ± 0.4 pm, and the calculated and experimental densities of 3.44 and 3.47 g·cm^{-3}, respectively (Eisenmann et al. 1985c). To prepare this compound, the elements were weighed in stoichiometric proportions under dry, oxygen-free Ar in quartz ampoule with quartz crucible pre-dried at 150°C. In order to compensate for the loss due to the unavoidable reaction with the crucible wall, Na was used empirically in an excess of about 10% by mass. The ampoule was evacuated and sealed. The highly exothermic formation reaction remained under control when heated up to 800°C at a rate of 1°C·min^{-1}. Homogeneous sintered ingots are formed; colorless single-crystalline particles can be split from their cavity-like inclusions. In the open air, $Na_4Si_4Se_{10}$ hydrolyzes extremely rapidly, with the intense smell of H_2Se occurring. The compound could therefore be handled under dry, heavy paraffin oil.

$\mathbf{Na_6Si_2Se_6}$ was obtained by a reaction of stoichiometric amounts of the elements in closed silica ampoules under vacuum at 650°C (Feltz and Pfaff 1983). After a 5-h reaction time, the reaction products, which were not yet uniform, were finely ground under inert conditions and subjected again to the described temperature treatment. The repetition gave a uniform reaction product which completely dissolved in MeOH. After the concentration of solution, $Na_6Si_2Se_6$ segregated out as finely crystalline colorless precipitates. This compound decomposes due to hydrolysis.

$\mathbf{Na_6Si_2Se_7}$ crystallizes in the monoclinic structure with the lattice parameters a = 939.0 ± 0.4, b = 1080.5 ± 0.4, c = 1579.4 ± 0.5 pm, β = 105.0 ± 0.1°, and an experimental density of 3.240 g·cm^{-1} (Eisenmann and Hansa 1993d). It was synthesized from stoichiometric amounts of the elements inevacuated graphitized silica ampoules at 900°C.

$\mathbf{Na_6Si_2Se_8}$ also crystallizes in the monoclinic structure with the lattice parameters a = 727.1 ± 0.3, b = 1700.7 ± 0.5, c = 1378.3 ± 0.4 pm, β = 99.44 ± 0.08°, and the calculated and experimental densities of 3.26 and 3.21 g·cm^{-3}, respectively (Eisenmann et al. 1985a). To prepare this compound, the elements were weighed in stoichiometric proportions under dry, oxygen-free Ar in quartz ampoule with quartz crucible pre-dried at 150°C. The highly exothermic formation reaction remained under control when the batches were slowly heated to 1000°C with a rate of ~100°C·h^{-1}. The good crystalline material was obtained when the samples were annealed for 2 h at this temperature and then cooled at the same rate. It developed brownish-red crystals that decomposed in open air very quickly, with an intense smell of H_2Se. The compound was therefore handled under dry, heavy paraffin oil.

2.2 Silicon–Cesium–Selenium

Three compounds, $\mathbf{Cs_4SiSe_4}$, $\mathbf{Cs_4Si_2Se_8}$, and $\mathbf{Cs_4Si_4Se_{10}}$ are formed in the Si–Cs–Se ternary system. First of them crystallizes in the cubic structure with the lattice parameter a = 1414.7 ± 0.3 pm (Schlirf et al. 2000). The title compound was prepared starting from an inhomogeneous sample obtained by thermal decomposition of CsN_3 mixed with $SiSe_2$ (molar ratio 6:2) at 410°C in a quartz ampoule. Together with additional Se, corresponding to a molar composition Cs/Si/Se = 6:2:6, the raw product was heated to 900°C. After quenching in ice-water, homogenization and subsequent annealing at 580°C for 17 days, the reaction yielded air sensitive, pale yellow single crystals of Cs_4SiSe_4.

$Cs_4Si_2Se_8$ crystallizes in the monoclinic structure with the lattice parameters a = 1514.8 ± 0.3, b = 759.9 ± 0.1, c = 1005.7 ± 0.2 pm, and β = 121.833 ± 0.002° (Chan et al. 2005). The crystals of the title compound were serendipitously formed in a molten chalcogenide reaction of 51.5 mg U, 21.8 mg Si, 94.0 mg Se, and 40.6 mg Cs_2Se_2. The reactants were combined in a fused silica ampoule in an inert atmosphere glove box, sealed under vacuum, and heated to 725°C at a rate of 35°C·h^{-1}. After 200 h of heating, the ampoule was cooled at 3°C·h^{-1} to room temperature. Dimethylformamide was added to dissolve the remaining cesium selenide flux, resulting in well-formed clear block-like crystals of $Cs_4Si_2Se_8$ together with CsU_2Se_6. The same reaction without U leads to $Cs_4Si_2Se_8$.

$Cs_4Si_4Se_{10}$ also crystallizes in the monoclinic structure with the lattice parameters a = 1616.4 ± 0.2, b = 1631.4 ± 0.2, c = 972.3 ± 0.1 pm, and β = 106.683 ± 0.002° (Feng et al. 2005). Its crystals were formed from a molten chalcogenide flux reaction of 127.12 mg Cs_2Se_2, 4.21 mg Si, and 71.06 mg Se. The reactants were combined in a fused silica ampoule in an inert atmosphere glove box, sealed under vacuum, and

DOI: 10.1201/9781003123507-2

heated to 750°C at a rate of 30°C·h^{-1}. After 200 h of heating, the ampoule was cooled at 3°C·h^{-1} to room temperature. Dimethylformamide was added to dissolve remaining cesium selenide flux, resulting in well-formed colorless crystals of the title compound.

2.3 Silicon–Copper–Selenium

$SiSe_2$–Cu_2Se: The phase diagram of this system, constructed through DTA, XRD, and measuring of microhardness and density, is shown in Figure 2.1 (Olekseyuk et al. 1998). The eutectics contain 90 and 40 mol% Cu_2Se and crystallize at 1002°C and 882°C, respectively, and the peritectic transformation takes place at 47 mol% Cu_2Se and 917°C. The eutectoid transformation of the solid solution based on β-Cu_2Se exists at 140°C. The solubility of $SiSe_2$ in Cu_2Se at 580°C reaches up to 5 mol%. The ingots for the investigations were annealing at 580°C for 1000 h.

Two compounds, **Cu_2SiSe_3** and **Cu_8SiSe_6**, are formed in this system. First of them is a semiconductor which melts incongruently at 912°C [at 632.2°C (Chen et al. 1999b)] and has a polymorphic transformation at 617°C (Olekseyuk et al. 1998). α-Cu_2SiSe_3 crystallizes in the tetragonal structure with the lattice parameters $a = 555.0 \pm 0.2$ and $c = 1071.8 \pm 0.4$ pm and an experimental density of 4.61 g·cm^{-3} (Olekseyuk et al. 1998). β-Cu_2SiSe_3 crystallizes in the monoclinic structure with the lattice parameters $a = 666.9 \pm 0.1$, $b = 1179.7 \pm 0.1$, $c = 663.3 \pm 0.1$ pm, $\beta = 107.67 \pm 0.01°$, and a calculated density of 5.237 g·cm^{-3} (Chen et al. 1999b) [$a = 1210$, $b = 562$, $c = 861$ pm, $\beta = 99°$, and the calculated and experimental densities of 5.24 and 5.15 g·cm^{-3}, respectively (Hahn et al. 1966)]. For the preparation of this compound, stoichiometric combinations of Cu_2Se, Si, and Se were mixed in an evacuated sealed silica tube with the addition of eutectic mixtures of KCl/LiCl (Chen et al. 1999b). A flux to reactant ratio of 3:1 was used for the crystal growth experiment. The sample was maintained at 550°C for 6 weeks, followed by quenching. The black prismatic crystals of Cu_2SiSe_3 were isolated by washing the reaction product with deionized water.

Cu_8SiSe_6 has a phase transition at 62°C (Olekseyuk et al. 1998) [at 50°C (Kuhs et al. 1979)]. One of the modification crystallizes in the orthorhombic structure with the lattice parameters $a = 728.35 \pm 0.02$, $b = 721.85 \pm 0.02$, and $c = 1022.81 \pm 0.03$ pm (Ishii et al. 1999) [$a = 727.8 \pm 0.3$, $b = 721.2 \pm 0.3$, $c = 1022.3 \pm 0.4$ pm (Kuhs et al. 1979)] and another crystallizes in the cubic structure with the lattice parameter $a = 1022.96 \pm 0.01$ pm (Olekseyuk et al. 1998) [$a = 1017$ pm and the calculated and experimental densities of 6.37 and 5.97 g·cm^{-3}, respectively (Hahn et al. 1965)]. The title compound could be prepared by reacting stoichiometric amounts of Cu_2Se, Si, and Se in an evacuated sealed quartz ampoule for 4 days at 500°C (Ishii et al. 1999) or by reacting stoichiometric amounts of high-purity elements in an evacuated sealed quartz ampoule for about 6 days at temperatures of 600°C–700°C (Kuhs et al. 1979). It could be also obtained by the interaction of $SiSe_2$ and Cu_2Se in the temperature region 400°C–800°C (Hahn et al. 1965).

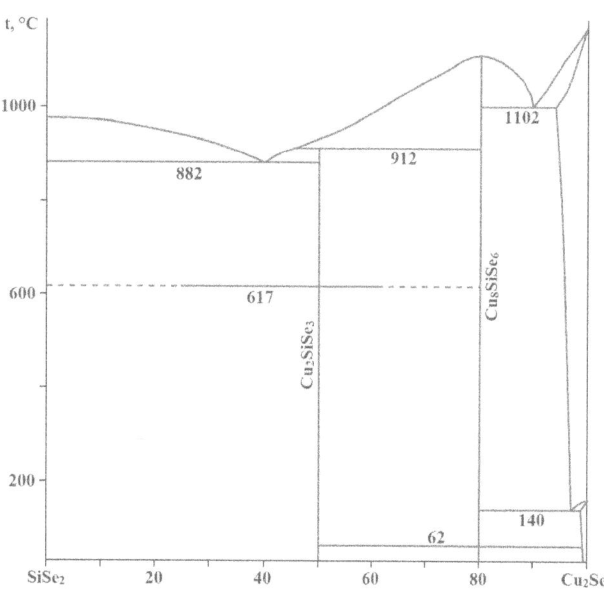

FIGURE 2.1 Phase diagram of the $SiSe_2$–Cu_2Se system. (From Olekseyuk, I.D., et al., *Zhurn. neorgan. khimii*, **43**(3), 516, 1998.)

2.4 Silicon–Silver–Selenium

$SiSe_2$–Ag_2Se: The phase diagram of this system, constructed through DTA and XRD, is shown in Figure 2.2 (Venkatraman et al. 1995). The **Ag_8SiSe_6** ternary compound, which melts congruently at 985°C ± 5°C (Venkatraman et al. 1995) [at 930°C (Gorochov 1968b); at 995°C (Piskach et al. 2006)], is formed in this system. This compound undergoes a polymorphic transformation at 132°C [has two polymorphic transformations at 10°C and 40°C and the thermal effects at 132°C correspond to the phase transition of Ag_2Se (Gorochov 1968b)], and forms a

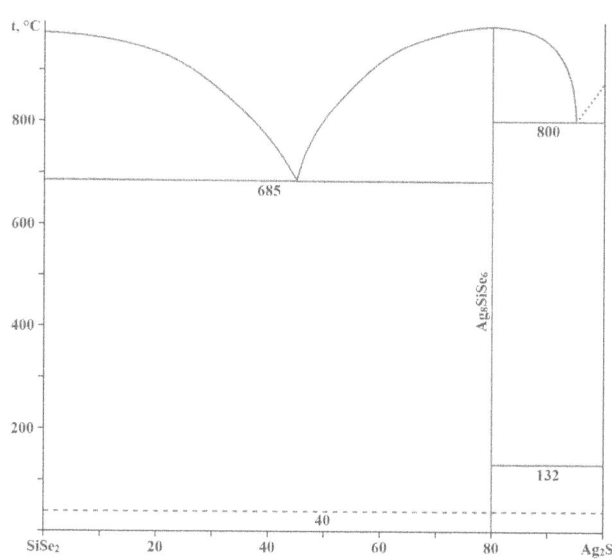

FIGURE 2.2 Phase diagram of the $SiSe_2$–Ag_2Se system. (From Venkatraman, M., et al., *Thermochim. Acta*, **249**, 13, 1995.)

eutectic on either side with $SiSe_2$ and Ag_2Se at 685°C ± 10°C and ≈45 mol% Ag_2Se, and 800°C and ≈95 mol% Ag_2Se, respectively (Venkatraman et al. 1995). Gorochov (1968b) noted that the eutectic from the Ag_2Se side contains 94 mol% Ag_2Se and crystallizes at 850°C, and according to the data of Piskach et al. (2006), the melting point of the eutectic between Ag_8SiSe_6 and Ag_2Se is 848°C and between $SiSe_2$ and Ag_8SiSe_6 – 713°C.

Ag_8SiSe_6 crystallizes in the cubic structure with the lattice parameter a = 1087 pm and the calculated and experimental densities of 7.06 and 6.95 g·cm^{-3}, respectively (Gorochov and Flahaut 1967; Gorochov 1968b) [a = 1086 pm and the calculated and experimental densities of 7.07 and 6.62 g·cm^{-3}, respectively (Hahn et al. 1965)] at room temperature. High-temperature modification of this compound also crystallizes in the cubic structure with the lattice parameter a = 1097 pm at 150°C (Gorochov and Flahaut 1967; Gorochov 1968b). The title compound was prepared by interaction of $SiSe_2$ and Ag_2Se within the temperature interval from 400°C to 800°C (Hahn et al. 1965) or by reacting stoichiometric amounts of high-purity elements in an evacuated sealed quartz ampoule for about 6 days at temperatures of 600°C–700°C (Kuhs et al. 1979).

2.5 Silicon–Magnesium–Selenium

The $\mathbf{Mg_2SiSe_4}$ ternary compound, which crystallizes in the orthorhombic structure with the lattice parameters a = 1327, b = 784, and c = 622 pm, is formed in the Si–Mg–Se system (Rocktäschel et al. 1964). It was prepared by the interaction of Mg_2Si with selenium within the temperature interval from 450°C to 650°C.

2.6 Silicon–Calcium–Selenium

The $\mathbf{Ca_2SiSe_4}$ ternary compound, which crystallizes in the orthorhombic structure with the lattice parameters a = 1408, b = 856, and c = 652 pm, is formed in the Si–Ca–Se system (Rocktäschel et al. 1964). It was prepared by the interaction of Ca_2Si with selenium within the temperature interval from 450°C to 650°C.

2.7 Silicon–Barium–Selenium

The $\mathbf{Ba_2SiSe_4}$ ternary compound, which crystallizes in the monoclinic structure with the lattice parameters a = 918.4 ± 0.5, b = 703.3 ± 0.3, c = 687.2 ± 0.3 pm, β = 109.2 ± 0.1°, and a calculated density of 4.90 g·cm^{-3}, is formed in the Si–Ba–Se system (Brinkmann et al. 1985c). For the preparation of this compound, the elements were weighed in a stoichiometric ratio and pre-dried at 200°C in a corundum crucible inserted in an evacuated quartz ampoule. Barium was added in an excess of about 5%. The mixture was heated to 700°C at a rate of 50°C·h^{-1}. To obtain well-crystalline precipitates, the mixture was cooled at the same rate.

2.8 Silicon–Zinc–Selenium

Ternary compounds in the Si–Zn–Se system were not found (Kaldis et al. 1967).

2.9 Silicon–Cadmium–Selenium

$SiSe_2$–CdSe: The phase diagram of this system, constructed through DTA, metallography, XRD, and microhardness and density measurements, is shown in Figure 2.3 (Odin and Ivanov 1991b). $\mathbf{Cd_4SiSe_6}$ compound is formed in this system, which melts incongruently at 1143°C and has polymorphic transformation at 958°C–964°C (within the homogeneity region) (Odin et al. 1985b; Odin and Ivanov 1991b). The eutectic composition and temperature are 41 mol% CdSe and 912°C, respectively. Homogeneity region of α-Cd_4SiSe_6 along the $SiSe_2$–CdSe system is within the interval of (79.5 ± 0.25)-(80.1 ± 0.15) mol% CdSe. The solubility of $SiSe_2$ in CdSe is equal to 0.4 mol% and the solubility of CdSe in $SiSe_2$ is not higher than 0.6 mol% (Odin and Ivanov 1991b). The ingots were annealed at 730°C for 720 h (Odin et al. 1985b; Odin and Ivanov 1991b).

α-Cd_4SiSe_6 crystallizes in the monoclinic structure with the lattice parameters a = 1282.66 ± 0.05, b = 735.91 ± 0.04, c = 1281.97 ± 0.05 pm, β = 110.052 ± 0.001° and a calculated density of 5.5591 ± 0.0007 g·cm^{-3} (Parasyuk et al. 2003c) [a = 1281, b = 735.2, c = 1280 pm, β = 110.00°, and the calculated and experimental densities of 5.58 and 5.56 g·cm^{-3}, respectively (Odin and Ivanov 1991b); a = 1281.6 ± 0.6, b = 735.5 ± 0.4, c = 1281.1 ± 0.6 pm, β = 110.06 ± 0.05°, and the calculated and experimental density of 5.571 and 5.52 ± 0.01 g·cm^{-3}, respectively (Krebs von and Mandt 1972)].

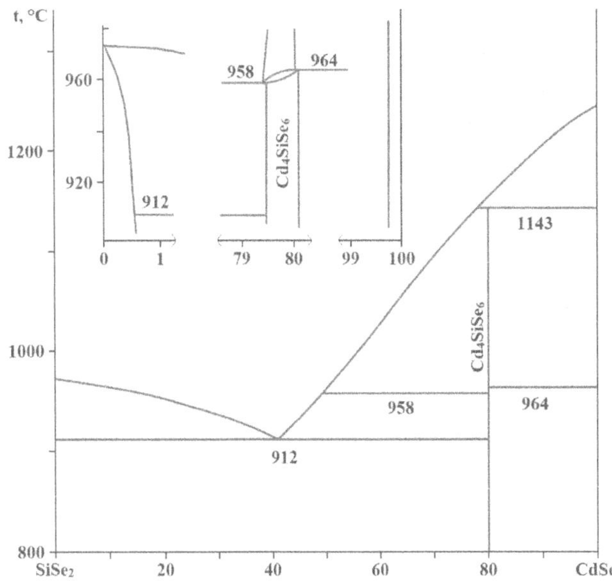

FIGURE 2.3 Phase diagram of the $SiSe_2$–CdSe system. (From Odin, I.N. and Ivanov, V.A., *Zhurn. neorgan. khimii*, **36**(11), 2937, 1991.)

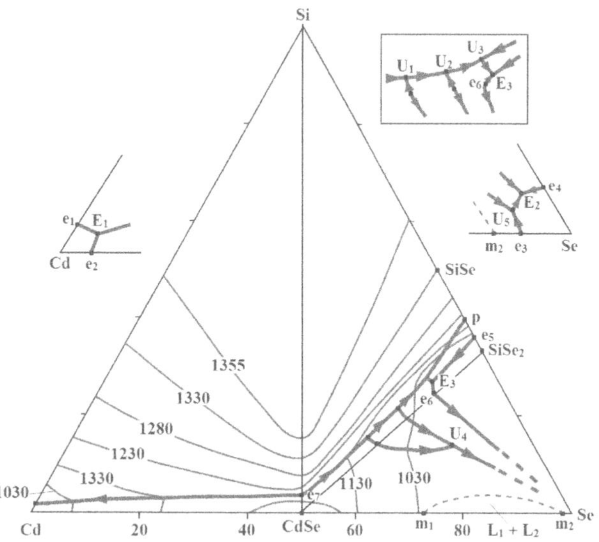

FIGURE 2.4 Liquidus surface of the Si–Cd–Se ternary system. (From Odin, I.N. and Ivanov, V.A., *Zhurn. neorgan. khimii*, **36**(11), 2937, 1991.)

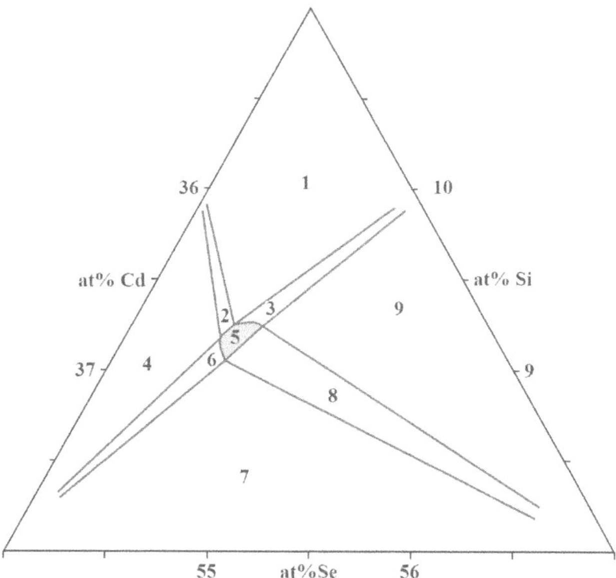

FIGURE 2.5 Part of isothermal section of the Si–Cd–Se ternary system at 730°C: 1, α-Cd_4SiSe_6 + Si + $SiSe_2$; 2, α-Cd_4SiSe_6 + Si; 3, α-Cd_4SiSe_6 + $SiSe_2$; 4, α-Cd_4SiSe_6 + Si + CdSe; 5, α-Cd_4SiSe_6; 6, α-Cd_4SiSe_6 + CdSe; 7, α-Cd_4SiSe_6 + CdSe + L; 8, α-Cd_4SiSe_6 + L; and 9, α-Cd_4SiSe_6 + $SiSe_2$ + L. (From Odin, I.N. and Ivanov, V.A., *Zhurn. neorgan. khimii*, **36**(11), 2937, 1991.)

Cd_4SiSe_6 has been prepared by the chemical transport reactions of CdSe, Si, and Se at 800°C–1000°C using I_2 as a transport agent (Krebs von and Mandt 1972), and its single crystals were grown using chemical transport reactions (Odin and Ivanov 1991b).

Si–CdSe: The phase diagram of this system was constructed using DTA, metallography, XRD, and microhardness and density measurements and is a eutectic type (Odin and Ivanov 1991b). The eutectic composition and temperature are 6 ± 1 at% Si and 1210°C, respectively. The solubility of Si in CdSe is not higher than 0.4 at%. The ingots for investigations were annealed at 730°C for 720 h.

The field of Si primary crystallization occupies the biggest part of the liquidus surface of the Si–Cd–Se ternary system (Figure 2.4) (Odin and Ivanov 1991b). It adjoins with the fields of primary crystallization for SiSe, α- and β-Cd_4SiS_6, CdSe, and Cd phases. The field of β-Cd_4SiS_6 crystallization is tapered with the increase of Se contents. There is also the field of $SiSe_2$ crystallization and degenerated fields of Cd and Se crystallization on the liquidus surface. The immiscibility region exists in the field of CdSe primary crystallization. There are three ternary eutectics and five transition points in this system: E_1 (317°C) – L ⇔ Cd + Si + CdSe; E_2 (213°C) – L_1 ⇔ α-Cd_4SiS_6 + $SiSe_2$ + L_2; E3 (905°C) – L ⇔ α-Cd_4SiS_6 + $SiSe_2$ + SiSe; U1 (1097°C) – L + CdSe ⇔ β-Cd_4SiS_6 + Si; U_2 (959°C) – L + β-Cd_4SiS_6 ⇔ SiSe + Si; U3 (950°C) – L + α-Cd_4SiS_6 ⇔ SiSe + Si; U4 (940°C) – L + β-Cd_4SiS_6 ⇔ α-Cd_4SiS_6 + CdSe; U_5 (217°C) – L_1 + CdSe ⇔ α-Cd_4SiS_6 + L_2. CdSe, $SiSe_2$, α-Cd_4SiS_6, Si, and liquid phases involve in equilibrium at 730°C (Figure 2.5) (Odin and Ivanov 1991b). The homogeneity region of α-Cd_4SiS_6 is elongated along the $SiSe_2$–CdSe quasibinary system.

2.10 Silicon–Mercury–Selenium

$SiSe_2$–HgSe: The phase diagram of this system was constructed through DTA, metallography, and XRD in the range 0–60 mol% $SiSe_2$ (Figure 2.6) (Gulay et al. 2002c; Parasyuk et al. 2003b). It was established that Hg_4SiSe_6 compound is formed in this system. It melts congruently at 771°C, forms eutectics at ~12 (750°C) and ~52 mol% $SiSe_2$ (688°C), has a polymorphous transformation at 435°C (Parasyuk et al. 2003b), and crystallizes in the monoclinic structure with the lattice parameters a = 1281.10 ± 0.04, b = 740.34 ± 0.04, c = 1274.71 ± 0.01 pm, and β = 109.605 ± 0.003° (Gulay et al. 2002).

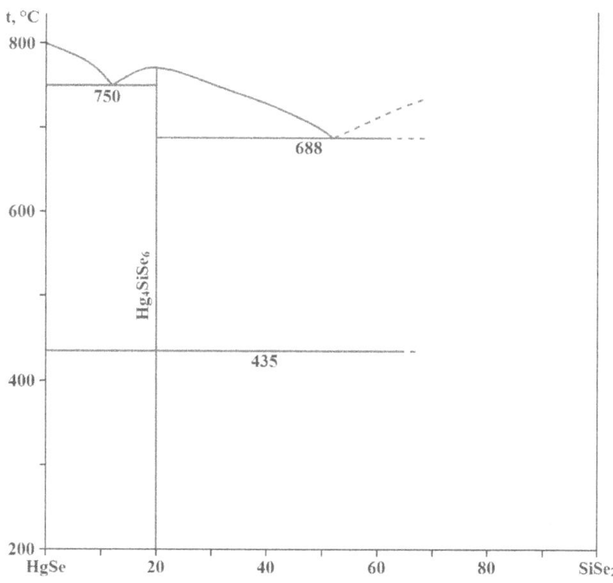

FIGURE 2.6 Phase diagram of the $SiSe_2$–HgSe system. (From Parasyuk, O.V. et al., *J. Alloys Compd.*, **348**(1–2), 157, 2003.)

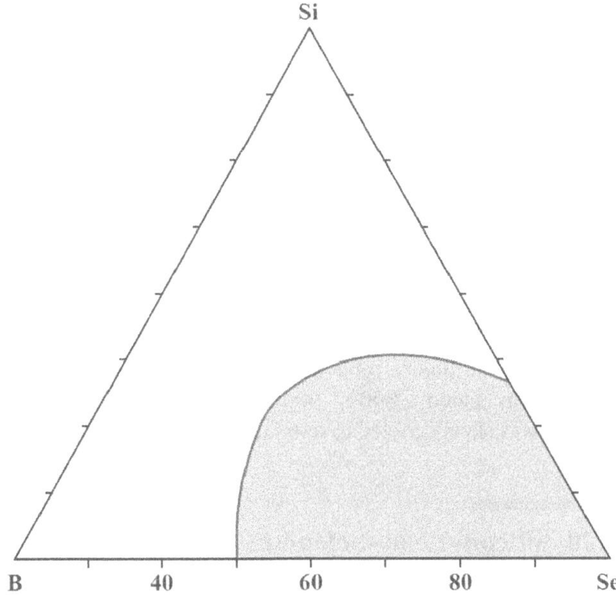

FIGURE 2.7 Glass forming region in the Si–B–Se ternary system. (From Kirilenko, V.V. and Dembovskiy, S.A., *Izv. AN SSSR. Neorgan. mater*, **10**(3), 542, 1974.)

2.11 Silicon–Boron–Selenium

The glass forming region in the Si–B–Se ternary system is given in Figure 2.7 (Kirilenko and Dembovskiy 1974). The obtained black glass-like alloys are unstable in air.

2.12 Silicon–Thallium–Selenium

$SiSe_2$–Tl_2Se: The part of the phase diagram of this system, constructed through DTA, XRD, metallography, and measuring of microhardness, is presented in Figure 2.8 (Lazarev et al. 1983). Two compounds, **Tl_2SiSe_3** and **Tl_4SiSe_4**, are formed in the $SiSe_2$–Tl_2Se system. The eutectics crystallize at 342°C and 374°C, and contain 25 and 40 mol% $SiSe_2$, respectively.

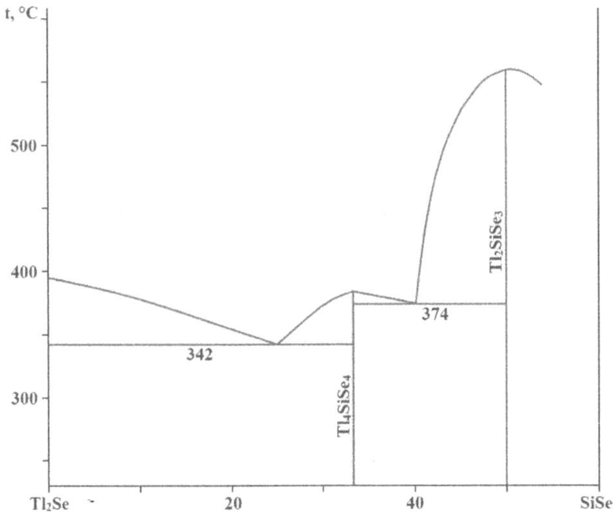

FIGURE 2.8 Part of the phase diagram of the $SiSe_2$–Tl_2Se system (From Lazarev, V.B., et al., *Zhurn. neorgan. khimii*, **28**(8), 2097, 1983.)

Tl_2SiSe_3 melts congruently at 560°C (Lazarev et al. 1983; Peresh et al. 1986) and crystallizes in the triclinic structure with the lattice parameters $a = 687.5 \pm 0.2$, $b = 686.6 \pm 0.2$, $c = 873.1 \pm 0.2$ pm, and $\alpha = 90.50 \pm 0.02°$, $\beta = 111.69 \pm 0.02°$, $\gamma = 113.70 \pm 0.02°$, and a calculated density of 6.49 g·cm^{-3} (Eulenberger 1982). The enthalpy and entropy of melting of this compound are equal to 28.0 kJ·M^{-1} and 33.6 J·(M·K)$^{-1}$, respectively, and an energy gap is 2.1 eV (Peresh et al. 1986). Tl_2SiSe_3 was prepared from the stoichiometric mixture of Tl, Si, and Se by reacting in the evacuated silica tube (Eulenberger 1982).

Tl_4SiSe_4 melts at 384°C (Lazarev et al. 1983; Peresh et al. 1986) and crystallizes in the monoclinic structure with the lattice parameters $a = 1166.4 \pm 0.9$, $b = 727.7 \pm 0.4$, $c = 2490.3 \pm 1.2$ pm, $\beta = 99.93 \pm 0.05°$, and a calculated density of 7.41 g·cm^{-3} (Eulenberger 1984a, 1986). The enthalpy and entropy of melting of this compound is equal to 17.6 kJ·M^{-1} and 26.7 J·(M·K)$^{-1}$, respectively, and an energy gap is 1.7 eV (Peresh et al. 1986). This compound was prepared from the elements which were sealed in an evacuated quartz ampoule in quantities according to the stoichiometry of the compound (Eulenberger 1984a, 1986). The temperature of the ampoule was gradually raised in an electric tube furnace to 700°C within 8 days to achieve complete reaction of silicon. The ampoule was than cooled to 300°C, and solidified reaction product was annealed at this temperature for 1 week. A crystalline gray regulus of metallic appearance was obtained. This compound is stable on exposure to the atmosphere.

2.13 Silicon–Lanthanum–Selenium

The **$La_6Si_4Se_{17}$** ternary compound, which crystallizes in the triclinic structure with the lattice parameters $a = 943.33 \pm 0.04$, $b = 1044.82 \pm 0.04$, $c = 1498.66 \pm 0.06$ pm, and $\alpha = 81.906 \pm 0.002°$, $\beta = 87.475 \pm 0.003°$, $\gamma = 89.499 \pm 0.003°$, is formed in the Si–La–Se system (Marchuk et al. 2012). The samples of the title compound were prepared by sintering the elemental constituents in evacuated quartz tube. The ampoule was first heated in a furnace with a rate of 30°C·h^{-1} up to 1150°C, and then kept at this temperature for 3 h. Afterwards, the sample was cooled slowly (10°C·h^{-1}) down to 500°C, and annealed at this temperature for 720 h. Finally, the ampoule was quenched in air. The so-obtained material was the black powder of $La_6Si_4Se_{17}$.

2.14 Silicon–Cerium–Selenium

The **$Ce_6Si_4Se_{17}$** ternary compound, which crystallizes in the triclinic structure with the lattice parameters $a = 938.3 \pm 0.1$, $b = 1035.6 \pm 0.2$, $c = 1488.4 \pm 0.3$ pm, and $\alpha = 81.94 \pm 0.03°$, $\beta = 87.66 \pm 0.03°$, $\gamma = 89.25 \pm 0.03°$, is formed in the Si–Ce–Se system (Marchuk et al. 2012). It was prepared in the same way as $La_6Si_4Se_{17}$ was synthesized. Dark red powder of the title compound was obtained.

2.15 Silicon–Praseodymium–Selenium

The $Pr_3Si_{1.25}Se_7$ ternary compound, which crystallizes in the hexagonal structure with the lattice parameters a = 1052.68 ± 0.03 and c = 603.96 ± 0.03 pm, and a calculated density of 5.7899 g·cm^{-3}, is formed in the Si–Pr–Se system (Gulay and Lychmanyuk 2008). The samples of this compound were prepared by melting the mixture of high purity elements in evacuated silica ampoules. The synthesis was realized in a tube furnace. The ampoule was heated with a heating rate of 30°C·h^{-1} to 1150°C and kept at this temperature for 3 h. After that, it was cooled slowly (10°C·h^{-1}) to 600°C and annealed at this temperature for 240 h. After annealing the ampoule with the sample was quenched in cold water.

2.16 Silicon–Neodymium–Selenium

The $Nd_3Si_{1.25}Se_7$ ternary compound, which crystallizes in the hexagonal structure with the lattice parameters a = 1047.60 ± 0.03 and c = 602.68 ± 0.03 pm, and a calculated density of 5.9166 g·cm^{-3}, is formed in the Si–Nd–Se system (Gulay and Lychmanyuk 2008). The samples of this compound were prepared in the same way as the samples of $Pr_3Si_{1.25}Se_7$ were obtained.

2.17 Silicon–Samarium–Selenium

The $Sm_3Si_{1.25}Se_7$ ternary compound, which crystallizes in the hexagonal structure with the lattice parameters a = 1041.66 ± 0.06 and c = 598.28 ± 0.06 pm, is formed in the Si–Sm–Se system (Gulay and Lychmanyuk 2008). The samples of this compound were prepared in the same way as the samples of $Pr_3Si_{1.25}Se_7$ were obtained.

2.18 Silicon–Thorium–Selenium

The **ThSiSe** ternary compound, which crystallizes in the tetragonal structure with the lattice parameters a = 401.2 and c = 1750 pm, and the calculated and experimental densities of 7.99 and 7.96 g·cm^{-3}, respectively, is formed in the Si–Th–Se system (Hahn and Stocks 1968; Stocks et al. 1981). To prepare this compound, the elements were weighed under N_2 atmosphere in the stoichiometric ratio and initially heated to about 400°C in an evacuated quartz ampoule. Subsequently, the inhomogeneous product was finely powdered under N_2 and heated to 950°C in a sealed tungsten crucible, which was enclosed in an evacuated quartz ampoule. Since the preparation is extremely sensitive to traces of oxygen, it proved necessary to add Th as getter metal. Nevertheless, pure preparation was obtained only if no longer than 48 h the mixture was heated at 950°C. The mixture was annealed at 600°C–700°C for 8–10 days. Then, the sample pressed into pastilles was heated in the same way over different time intervals and at different temperatures up to 1400°C in a tungsten crucible, wherein above 1050°C additionally, Ar was used as a protective gas. The title compound was obtained as a black microcrystalline powder.

2.19 Silicon–Uranium–Selenium

The **USiSe** ternary compound, which crystallizes in the tetragonal structure with the lattice parameters a = 390 ± 1 and c = 1677 ± 3 pm, and a calculated density of 8.98 g·cm^{-3}, is formed in the Si–U–Se system (Haneveld and Jellinek 1969). This compound was prepared as follows: A small rod of U was placed in a quartz tube and heated in a current of hydrogen at 230°C. The voluminous uranium hydride powder thus produced was dehydrated by heating at 400°C under vacuum. An equivalent proportion of intimate mixture of Si and Se was added immediately and the quartz tube was evacuated and sealed. The mixture was heated at 800°C–900°C for 2 or 3 days. Then, the product was slowly cooled to room temperature.

2.20 Silicon–Lead–Selenium

The Pb_2SiSe_4 ternary compound, which crystallizes in the monoclinic structure with the lattice parameters a = 856.70 ± 0.02, b = 707.45 ± 0.03, c = 1361.60 ± 0.03 pm, and β = 108.355 ± 0.003°, is formed in the Si–Pb–Se system (Iglesias and Steinfink 1973b). It was prepared by heating the elements in the stoichiometric proportions. The mixture was fired in a vacuum-sealed Vycor ampoule until the product looked homogeneous. A single-phase material was obtained after heating at 700°C for 4 days. The crystals were deep ruby red in transmitted light.

2.21 Silicon–Zirconium–Selenium

The quite stable **ZrSiSe** ternary compound, which crystallizes in the tetragonal structure with the lattice parameters a = 362.3 and c = 836.5 pm (Haneveld and Jellinek 1964) [a = 363 and c = 836 pm, and the calculated and experimental densities of 5.98 and 5.33 g·cm^{-3}, respectively (Jellinek and Hahn 1962; Onken et al. 1964); a = 362.4 ± 0.5, c = 836.0 ± 0.8 pm, and the calculated and experimental densities of 5.84 and 5.81 ± 0.01 g·cm^{-3}, respectively, for $Zr_{2.00}Si_{1.80}Se_{1.99}$ (Jeannin and Mosset 1972)], is formed in the Si–Zr–Se system. To obtain this compound, stoichiometric amounts of the elements were initially pre-tempered in an evacuated tube at 600°C for 10 days (Onken et al. 1964). Then, the reaction mixture was carefully rubbed in an agate mortar in a N_2-filled box, pressed into a pastille and annealed at 1000°C for several weeks. The ampoule with tablet was stored in the hottest part of the furnace to obtain the single crystals by sublimation. ZrSiSe could be also prepared by heating "SiSe" with an equivalent proportion of Zr in an evacuated quartz tube at 800°C for 1 or 2 days (Haneveld and Jellinek 1964).

2.22 Silicon–Hafnium–Selenium

The **HfSiSe** ternary compound, which crystallizes in the tetragonal structure with the lattice parameters a = 363 and c = 832 pm, and a calculated density of 8.65 g·cm^{-3}, is formed in the Si–Hf–Se system (Onken et al. 1964). To prepare this compound, stoichiometric amounts of Si, Hf, and Se were initially pre-tempered in an evacuated tube at 600°C for 10 days. Then, the reaction mixture was carefully rubbed in an agate mortar in a N_2-filled box, pressed into a pastille and annealed at 850°C for several weeks. The ampoule with tablet was stored in the hottest part of the furnace to obtain the single crystals by sublimation.

2.23 Silicon–Antimony–Selenium

The glass-forming region in the Si–Sb–Se ternary system is situated in the S-corner (Hilton et al. 1964). The glasses do not show good chemical stability and react with water in the atmosphere, giving off H_2Se. Some of the glasses had high softening points (450°C). To prepare such glasses, the reactants were weighed out and sealed in a quartz vial at a reduced pressure. The vial was placed in a rocking furnace, slowly heating to 1000°C, and allowed to react and mix while molten for at least 16 h. The glasses were formed by quenching from 1000°C to room temperature in the air.

2.24 Silicon–Vanadium–Selenium

The **$VSiSe_3$** ternary compound, which crystallizes in the hexagonal structure with the lattice parameters a = 652.0 ± 0.8 and c = 1955.0 ± 0.9 pm, is formed in the Si–V–Se system (Gopalakrishnan and Nanjundaswamy 1988). It was synthesized if a mixture of V, Si, and Se corresponding to the stoichiometric composition was reacted in an evacuated sealed silica ampoule at 700°C–800°C for varying duration ranging up to 2 weeks. The inner walls of the silica tube were coated with carbon to avoid reaction with the container.

2.25 Silicon–Tellurium–Selenium

Ternary compounds were not found in the Si–Te–S system (Shevtsova et al. 1987a).

2.26 Silicon–Chromium–Selenium

Three quasibinary sections, **$SiSe_2$–CrSe**, **$SiSe_2$–Cr_2Se_3**, and **$SiSe_2$–Cr_3Se_4**, exist in the Si–Cr–Se ternary system (Shabunina and Aminov 2000). The phase diagram of the $SiSe_2$–Cr_2Se_3, constructed through DTA and XRD, is presented in Figure 2.9. The eutectic contains 18 mol% Cr_2Se_3 and crystallizes at 860°C. The solubility of $SiSe_2$ in Cr_2Se_3 at the eutectic temperature reaches up to 10 mol%.

The ternary compounds were not found in the Si–Cr–Se system. However, according to the data of (Gopalakrishnan and Nanjundaswamy 1988), the **$Cr_{1+x}SiSe_3$** ternary compound, which crystallizes in the hexagonal structure with the lattice parameters a = 664.4 ± 0.8 and c = 1964.3 ± 0.7 pm, is formed in this system. It was synthesized if a mixture of Cr, Si, and Se corresponding to the stoichiometric composition was reacted in an evacuated sealed silica ampoule at 700°C–800°C for varying duration ranging up to 2 weeks. The inner walls of the silica tube were coated with carbon to avoid reaction with the container.

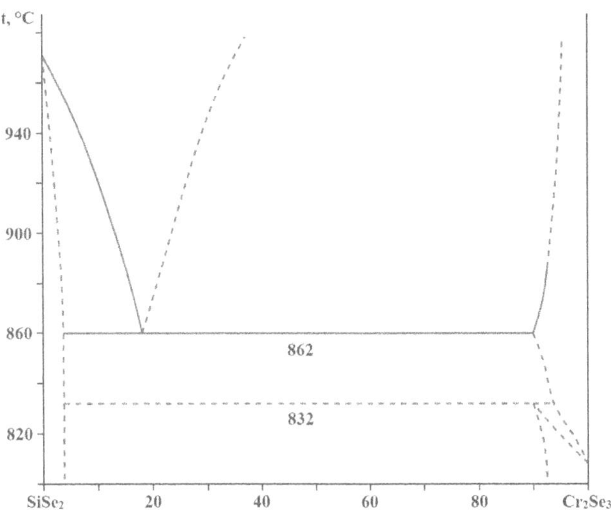

FIGURE 2.9 Phase diagram of the $SiSe_2$–Cr_2Se_3 system. (From Shabunina, and Aminov, T.G., *Zhurn. neorgan. khimii*, **45**(1), 99, 2000.)

The ingots for the investigations were annealed at 630°C for 2 weeks (Shabunina and Aminov 2000).

2.27 Silicon–Iodine–Selenium

The glass-forming region in the Si–I–Se ternary system is shown in Figure 2.10 (Dembovskiy and Popova 1970). The glasses adjacent to the glass formation region crystallize easily and are chemically resistant. The composition corresponding to the supposed compound of $SiSeI_2$ lies almost on

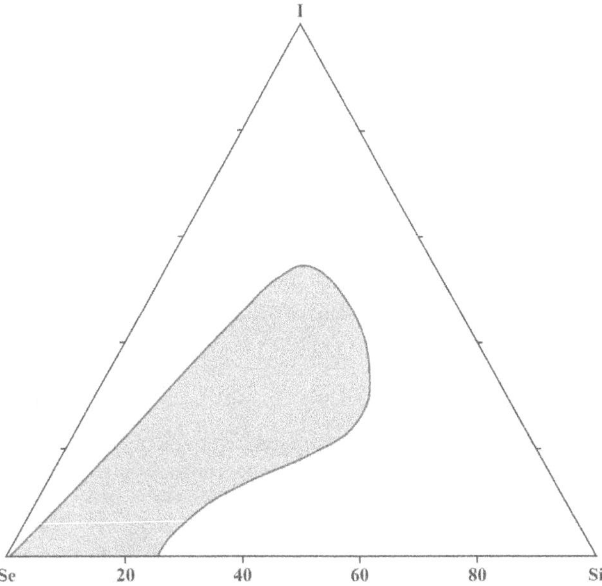

FIGURE 2.10 Glass forming region in the Si–I–Se ternary system. (From Dembovskiy, S.A. and Popova, N.P, *Izv. AN SSSR. Neorgan. mater*, **6**(1), 138, 1970.)

the border of the glass formation region. Such a vast region of glass formation is possible, apparently, only because the $SiSeI_2$ compound is formed that has a chain-like structure and its glass transition temperature is below 0°C. The region of viscous glasses softened at room temperature exists in this system.

2.28 Silicon–Manganese–Selenium

Two compounds, $\mathbf{Mn_{1.5}SiSe_3}$ and $\mathbf{Mn_2SiSe_4}$, are formed in the Si–Mn–Se ternary system. First of them crystallizes in the hexagonal structure with the lattice parameters a = 651.0 ± 0.3 and c = 1371.5 ± 0.4 pm (Gopalakrishnan and Nanjundaswamy 1988). It was synthesized if a mixture of Mn, Si, and Se corresponding to the stoichiometric composition was reacted in an evacuated sealed silica ampoule at 700°C–800°C for varying duration ranging up to 2 weeks. The inner walls of the silica tube were coated with carbon to avoid reaction with the container.

Mn_2SiS_4 crystallizes in the orthorhombic structure with the lattice parameters a = 1330.66 ± 0.08, b = 777.80 ± 0.05, and c = 624.51 ± 0.03 pm, and a calculated density of 4.66 g·cm^{-3} (Jobic et al. 1995) [a = 1305, b = 775, and c = 620 pm (Rocktäschel et al. 1964); a = 1330.5 ± 0.2, b = 777.71 ± 0.08, and c = 624.35 ± 0.08 pm (Bodénan et al. 1996)]. This compound crystallizes at 10.6 GPa in the cubic structure with the lattice parameter a = 803.8 ± 0.9 pm (Grzechnik et al. 1997a, b). Mn_2SiS_4 appeared first as a subproduct of the reaction of Mn, Si, and Se (molar ratio 1:2:4) at 800°C for 2 weeks (Jobic et al. 1995). The pure phase was obtained by heating the elements in stoichiometric proportion at 800°C in an evacuated silica sealed tube (Jobic et al. 1995; Bodénan et al. 1996; Grzechnik et al. 1997a, b). The heating was performed at a rate of 10°C·h^{-1} up to 800°C for 15 days, before cooling to room temperature at 20°C·h^{-1}. The bulk is dark-brown, and tiny yellow crystals were observed in small amount. The presence of a small amount of α-MnSe impurity has been detected by XRD.

High pressure synthesis was carried out using a Walker-type multiple anvil device (Grzechnik et al. 1997a, b). The Raman spectra collected at high pressures indicate the occurrence of a phase transition above approximately 1.8 GPa. The major changes in the structure of Mn_2SiSe_4 at high pressures occur at about 4.0 GPa. An olivine–spinel phase transition in the 2.0–4.0 GPa range at room temperature is most probably accompanied by formation of spinelloid compounds at intermediate pressures.

2.29 Silicon–Iron–Selenium

The $\mathbf{FeSi_2Se_4}$ ternary compound, which crystallizes in the hexagonal structure with the lattice parameters a = 361.5 ± 0.3 and c = 604.8 ± 0.2 pm, is formed in the Si–Fe–Se system (Gopalakrishnan and Nanjundaswamy 1988). It was synthesized if a mixture of Fe, Si, and Se corresponding to the stoichiometric composition was reacted in an evacuated sealed silica ampoule at 700°C–800°C for varying duration ranging up to 2 weeks. The inner walls of the silica tube were coated with carbon to avoid reaction with the container.

2.30 Silicon–Cobalt–Selenium

The $\mathbf{CoSi_2Se_4}$ ternary compound, which crystallizes in the monoclinic structure with the lattice parameters a = 594.5 ± 0.5, b = 358.6 ± 0.8, c = 1110.8 ± 0.8 pm, and β = 91.82 ± 0.04°, is formed in the Si–Co–Se system (Gopalakrishnan and Nanjundaswamy 1988). It was synthesized if a mixture of Co, Si, and Se corresponding to the stoichiometric composition was reacted in an evacuated sealed silica ampoule at 700°C–800°C for varying duration ranging up to 2 weeks. The inner walls of the silica tube were coated with carbon to avoid reaction with the container.

2.31 Silicon–Nickel–Selenium

Two compounds, $\mathbf{NiSi_2Se_4}$ and $\mathbf{Ni_{5.68}SiSe_2}$, are formed in the Si–Ni–Se system. $NiSi_2Se_4$ crystallizes in the monoclinic structure with the lattice parameters a = 603.8 ± 0.9, b = 341.1 ± 0.8, c = 1101.4 ± 0.8 pm, and β = 91.84 ± 0.04° (Gopalakrishnan and Nanjundaswamy 1988). It was synthesized if a mixture of Ni, Si, and Se corresponding to the stoichiometric composition was reacted in an evacuated sealed silica ampoule at 700°C–800°C for varying duration ranging up to 2 weeks. The inner walls of the silica tube were coated with carbon to avoid reaction with the container.

$Ni_{5.68}SiSe_2$ crystallizes in the tetragonal structure with the lattice parameters a = 357.6 ± 0.1, c = 1833.9 ± 0.2 pm, and a calculated density of 7.37 g·cm^{-3} (Isaeva et al. 2007b). The sample of this compound was synthesized starting from the elements by high-temperature ceramic techniques. The thoroughly ground mixture of the elements (the total weight of the sample was 1 g), corresponding to the stoichiometry Ni_7SiSe_2, was placed in quartz ampoule. Then the ampoule was sealed under vacuum (~7 Pa), annealed at 540°C–650°C for 14 days in a vertical furnace and water-quenched. According to the XRD, the sample consisted of four phases, and hence, the phase equilibrium was not achieved. The equilibrium sample containing only $Ni_{5.68}SiSe_2$ and Ni was obtained only after the repeated annealing of pressed pellets (pressure load of about 0.2 GPa) under the same conditions. The total time of annealing was 60 days.

Crystals of the title compound were grown from molten flux starting from $PbCl_2$ (Isaeva et al. 2007b). The stoichiometric mixture of Ni, Si, and Se (molar ratio 7:1:2) (total weight was ~1–2 g, the weight ratio of the charge to the flux was 1:1) was placed in cylindrical quartz ampoules, which were then sealed under vacuum. The ampoule was heated to 850°C for 12 h in a furnace. Then the furnace was slowly cooled in an automatic mode to 500°C at a rate of 1.5°C·h^{-1}. The cooling to room temperature was performed in the switched-off furnace. The sample was ground in a mortar and washed off from the flux by refluxing in hot water for 0.5–1 h. The washed sample contained small golden-colored crystals as thinelongated plates.

2.32 Silicon–Platinum–Selenium

The **PtSiSe** ternary compound, which crystallizes in the orthorhombic structure with the lattice parameters $a = 594.4 \pm 0.2$, $b = 598.8 \pm 0.2$, and $c = 590.0 \pm 0.2$ pm, is formed in the Si–Pt–Se system (Entner and Parthé 1973). It was prepared by heating the elements in evacuated quartz tubes at 600°C and 900°C for about 4 to 6 weeks. Some samples were ground after 1 week and reheated to enhance the attainment of equilibrium.

REFERENCES

All References are available as a downloadable eResource at www.routledge.com/9780367639235

3

Systems Based on Silicon Telluride

3.1 Silicon–Sodium–Tellurium

Two ternary compounds, $\mathbf{Na_6Si_2Te_6}$ and $\mathbf{Na_8Si_4Te_{10}}$ are formed in the Si–Na–Te system. First of them crystallizes in the monoclinic structure with the lattice parameters a = 878.6 ± 0.3, b = 1278.0 ± 0.4, c = 886.4 ± 0.3 pm, β = 119.71 ± 0.05°, and the calculated and experimental densities of 3.686 and 3.70 g·cm^{-3}, respectively (Eisenmann et al. 1981). For its obtaining, stoichiometric amounts of the elements were weighed under Ar into a quartz ampoule pre-dried at 150°C. In order to compensate for the loss due to the reaction with the ampoule wall, Na had to be added in an excess of about 10%. The ampoule was sealed under vacuum (about 10 Pa), heated to 850°C within 5 h and the furnace was switched off after a 1-h annealing. The compound is clearly translucent in the freshly broken state with a reddish color. Upon entry of moist air, it decomposes momentarily, whereby dense tellurium layers form on the vessel walls and crystal surfaces as a result of the air oxidation of the initially formed H_2Te, so that the substance appears shiny metallic gray after standing for a short time.

$Na_8Si_4Te_{10}$ also crystallizes in the monoclinic structure with the lattice parameters a = 1407.3, b = 1284.2, c = 1488.2 pm, β = 99.22, and the calculated and experimental densities of 3.885 and 3.90 g·cm^{-3}, respectively (Eisenmann et al. 1983d). It was prepared by the interaction of the elements in the Ar atmosphere at 550°C.

3.2 Silicon–Potassium–Tellurium

Two compounds, $\mathbf{K_4Si_4Te_{10}}$ and $\mathbf{K_6Si_2Te_6}$, are formed in the Si–K–Te ternary system. First of them crystallizes in the orthorhombic structure with the lattice parameters a = 2125.8 ± 0.8, b = 1200.5 ± 0.7, and c = 1060.8 ± 0.7 pm, and the calculated and experimental densities of 3.85 and 3.79 g·cm^{-3}, respectively (Eisenmann ans Schäfer 1982). To obtain it, stoichiometric amounts of the elements were weighed into a quartz ampoule heated at 200°C under dry, oxygen-free argon, and the ampoule was evacuated and sealed. Because of the unavoidable reaction of the potassium with the inner wall, this element was used with 10% weight excess. The highly exothermic reaction remained under control, when the ampoule was slowly and uniformly heated to 400°C within 6 h and annealed there for 2 h. The temperature was then raised to 650°C over 2 h, held there for 30 min, and then lowered to 350°C. A particularly good crystalline product was obtained by a 14-h annealing at this temperature. The compound forms transparent, pale yellowish crystals. Upon access of traces of moisture, it immediately decomposes with the formation of H_2Te, which oxidizes rapidly in air to Te. The initially shiny crystal surfaces are therefore covered in a short time with a Te layer, so that the compound appears metallic black. For analysis, large crystals were extracted under heavy, dry paraffin oil using a stereomicroscope. These were washed with petroleum ether in an atmosphere of dry, oxygen-free argon, and dried.

$K_6Si_2Te_6$ crystallizes in the monoclinic structure with the lattice parameters a = 965.2 ± 0.5, b = 1362.1 ± 0.8, c = 890.2 ± 0.5 pm, β = 117.34 ± 0.05°, and the calculated and experimental densities of 3.37 and 3.36 g·cm^{-3}, respectively (Dittmar 1977a,b, 1978). It was prepared by melting a stoichiometric mixture of the elements in an evacuated quartz tube which was evacuated and sealed. The ampoule was heated in a corundum crucible to 630°C. After 1 h, the temperature was slowly lowered to 280°C over a period of 6 h, and cooling allowed proceeding further. From the preserved gray-black regulus, the crystals with metallic luster were obtained. They decompose in humid air but can be stored under dry paraffin oil.

3.3 Silicon–Cesium–Tellurium

Two compounds, $\mathbf{Cs_2Si_2Te_6}$ and $\mathbf{Cs_4SiTe_4}$, are formed in the Si–Cs–Te ternary system. First of them crystallizes in the monoclinic structure with the lattice parameters a = 828.5 ± 0.4, b = 1393.5 ± 0.6, c = 1340.4 ± 0.6 pm, β = 100.35 ± 0.12°, and a calculated density of 4.74 g·cm^{-3} (Brinkmann et al. 1985b). For the preparation of $Cs_2Si_2Te_6$, a mixture of the elements (molar ratio of Cs/Si/Te = 2:2:6) was weighed into a pre-dried quartz ampoule with corundum crucible insert, evacuated and sealed, excluding moisture and oxygen. The sample was heated at a rate of 50°C·h^{-1} to 400°C, and kept there for 2 h. Then the temperature was raised to 650°C and then at the same rate cooled down again. The reaction product was a glossy black compact regulus, which was kept under thick-flowed dried paraffin oil to protect against moist air. Under the stereomicroscope, blackish-colored, irregularly broken single crystal particles could be extracted from the crushed regulus.

Cs_4SiTe_4 crystallizes in the cubic structure with the lattice parameter a = 1508.1 ± 0.2 (Schlirf and Deiseroth 2001a). The title compound was prepared starting from an inhomogeneous sample obtained by thermal decomposition of CsN_3 mixed with a pre-reacted mixture of $SiTe_2$ and Te (molar ratio of Cs/Si/Te = 3:1:3) at 370°C in a quartz ampoule. The raw product was heated to 650°C. After quenching in ice water, homogenization and subsequent annealing at 520°C for 16 days, the reaction yielded air sensitive, transparent, red-orange single crystals of Cs_4SiTe_4. An additional phase must be present, but could not be identified.

DOI: 10.1201/9781003123507-3

3.4 Silicon–Copper–Tellurium

The projection of the liquidus surface of the Si–Cu–Te ternary system, constructed through DTA, XRD, and metallography, is given in Figure 3.1 (Dogguy et al. 1979). Five ternary eutectics, E_1, E_2, E_3, E_4, and E_5, which crystallize at 375°C, 652°C, 352°C, 798°C, and 820°C, respectively, and two liquid-liquid immiscibility gaps exist in this system. One of the immiscibility gaps is situated entirely in the ternary system. Some vertical sections were constructed and the reaction scheme was determined. The isothermal section of the Si–Cu–Te ternary system at room temperature is shown in Figure 3.2. The $\mathbf{Cu_2SiTe_3}$ ternary compound is formed in this system. It melts incongruently at 578°C (Dogguy et al. 1979) and apparently has two polymorphic modifications. One of them crystallizes in the monoclinic structure with the lattice parameters $a = 1286$, $b = 607$, $c = 905$ pm, $\beta = 99°$ and the calculated and experimental densities of 5.96 and 5.90 g·cm^{-3}, respectively (Hahn et al. 1966), and another crystallizes in the cubic structure with the lattice parameter $a = 593.4 \pm 0.2$ pm (Olekseyuk et al. 1998) [$a = 593$ pm and the calculated and experimental densities of 5.69 and 5.47 ± 0.03 g·cm^{-3}, respectively (Rivet et al. 1963; Rivet 1965)].

The glass-forming region in the Si–Cu–Te ternary system is presented in Figure 3.3 (Minaev et al. 1980; Minaev 1983) [Dogguy et al. (1979) noted that there is no evidence for a vitreous domain in this system]. Up to 18 at%, Cu can be incorporated into glasses, but alloys with higher tellurium content are more ready to glass formation. There are two endothermic picks on the DTA curves of the glasses, lying within the range of 110°C–160°C, which correspond to two softening temperatures of the glasses. This could indicate on a micro-inhomogeneous glass structure and the presence of at least two glassy phases.

Synthesis of the Si–Cu–Te glassy alloys were carried out from the elements in quartz ampoules under vacuum at 1000°C for 12 h with vigorous stirring with the next quenching in water (Minaev et al. 1980; Minaev 1983). The vitreous state was identified by a characteristic glass wedge, a shell fracture, as well as through XRD, DTA, and metallography. The boundary of the glass formation region was fixed with an accuracy of 2 at%.

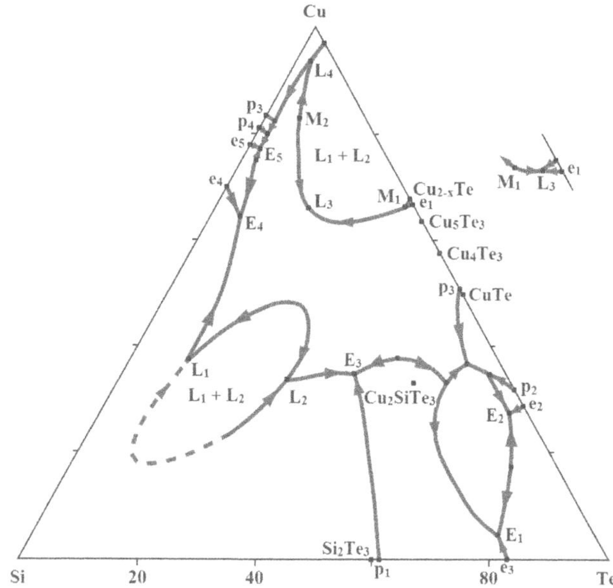

FIGURE 3.1 Projection of the liquidus surface of the Si–Cu–Te ternary system. (From Dogguy, M., et al., *J. Less-Common Metals*, **63**(2), 129, 1979.)

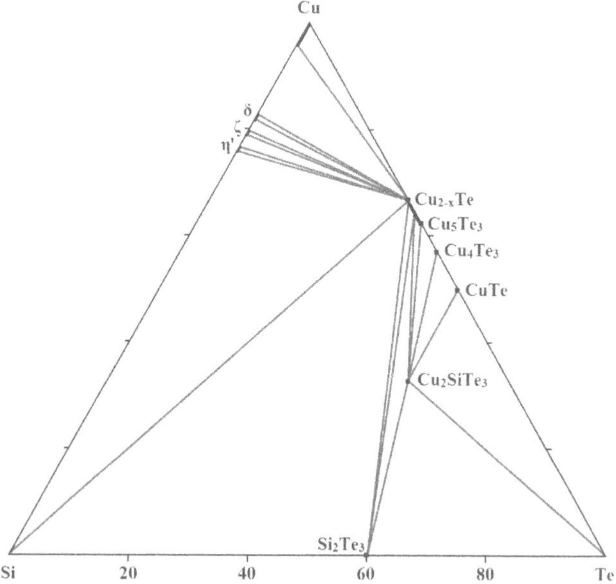

FIGURE 3.2 Isothermal section of the Si–Cu–Te ternary system at room temperature. (From Dogguy, M., et al., *J. Less-Common Metals*, **63**(2), 129, 1979.)

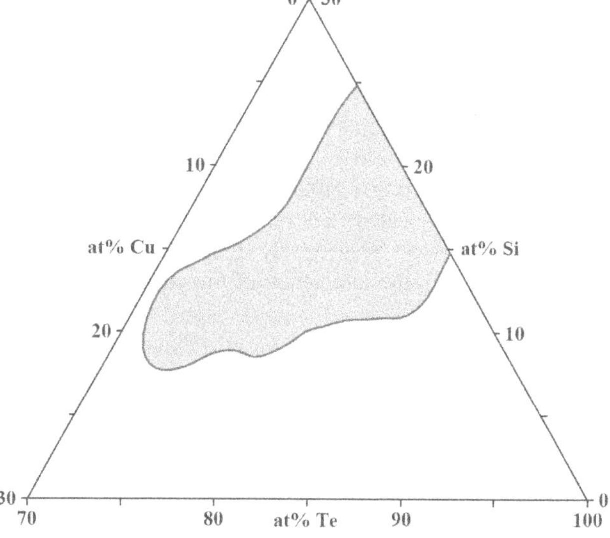

FIGURE 3.3 Glass forming region in the Si–Cu–Te ternary system. (From Minaev, V.S., et al., *Izv. AN SSSR. Neorgan. mater.*, **16**(8), 1481, 1980.)

3.5 Silicon–Silver–Tellurium

"$SiTe_2$"–Ag_2Te: The part of the vertical section was constructed using DTA and XRD by Gorochov and Flahaut (1967), and Gorochov (1968b). The $\mathbf{Ag_8SiTe_6}$ ternary compound, which melts congruently at 870°C and has polymorphic transformations at −78°C and −10°C, is formed in this section (Gorochov and Flahaut 1967; Gorochov 1968b; Boucher et al. 1992). All three modifications crystallize in the cubic structure with the lattice parameter a = 2301.7 ± 0.1 pm for α-Ag_8SiTe_6 at −110°C; a = 2298.4 ± 0.1 pm for β-Ag_8SiTe_6 at −60°C; a = 1152.25 ± 0.07 pm, and a calculated density of 7.192 g·cm^{-3} [a = 1151.5 pm and the calculated and experimental densities of 7.21 and 7.23 g·cm^{-3}, respectively (Gorochov and Flahaut 1967; Gorochov 1968b)] for γ-Ag_8SiTe_6 at room temperature (Boucher et al. 1992). γ-Ag_8SiTe_6 is stable from room temperature to its melting point and was synthesized by direct combination of the elements in an evacuated quartz ampoule (Boucher et al. 1992). Large crystals were easily obtained for a nonstoichiometric 5:1:4 ratio, the samples being heated at 800°C for 7 days and subsequently cooled to 500°C over 6-day period and then to room temperature within a day.

3.6 Silicon–Barium–Tellurium

The $\mathbf{Ba_2SiTe_4}$ ternary compound, which crystallizes in the monoclinic structure with the lattice parameters a = 965.0 ± 0.5, b = 762.6 ± 0.3, c = 746.6 ± 0.3 pm, β = 108.9 ± 0.1°, and the calculated and experimental densities of 5.20 and 5.22 g·cm^{-3}, respectively, is formed in the Si–Ba–Te system (Brinkmann et al. 1985c). For the preparation of this compound, the elements were weighed in a stoichiometric ratio and pre-dried at 200°C in corundum crucible inserted in an evacuated quartz ampoule. Barium was added in an excess of about 5%. The mixture was heated to 1000°C at a rate of 50°C·h^{-1}. To obtain well-crystalline precipitates, the mixture was cooled at the same rate.

3.7 Silicon–Zinc–Tellurium

Si_2Te_3–ZnTe: This section is a non-quasibinary section of Si–Zn–Te ternary system (Figure 3.4) (Odin 1996). Silicon primarily crystallizes from the Si_2Te_3-rich side. Crystallization ends at 818°C by the next transition reaction: L + Si ⇔ ZnTe + β-Si_2Te_3. Si_2Te_3 has polymorphous transformation at 406°C–409°C. The solubility of Si_2Te_3 in ZnTe is not higher than 0.1 mol%, and the solubility of ZnTe in Si_2Te_3 is equal to 0.1–0.2 mol%. This system was investigated using DTA, metallography, and XRD. The ingots were annealed at the temperature 20°C lower than temperatures of invariant equilibria with liquid participation.

Si–ZnTe: The phase diagram is not constructed. According to the data of Odin (1996), this diagram is a eutectic type. The eutectic composition and temperature are 5 ± 1 at% Si and 1267°C ± 10°C, respectively. Mutual solubility of ZnTe and Si is insignificant. This system was also investigated using DTA, metallography, and XRD. The ingots were annealed at temperatures 20°C lower than temperatures of invariant equilibria with liquid participation.

The field of Si primary crystallization occupies almost all Si–Zn–Te ternary system (Figure 3.5) (Odin 1996). There are also the fields of ZnTe, Si_2Te_3, and Zn (degenerated field) primary crystallization in this system. There are two ternary eutectics and one transition point in this system: E_1 (418°C) – L ⇔ ZnTe + Zn + Si; E_2 (407°C) – L ⇔ β-Si_2Te_3 + ZnTe + Te; U (818°C) – L + Si ⇔ ZnTe + β-Si_2Te_3. Ternary compounds are not found in this system (Kaldis et al. 1967; Odin 1996).

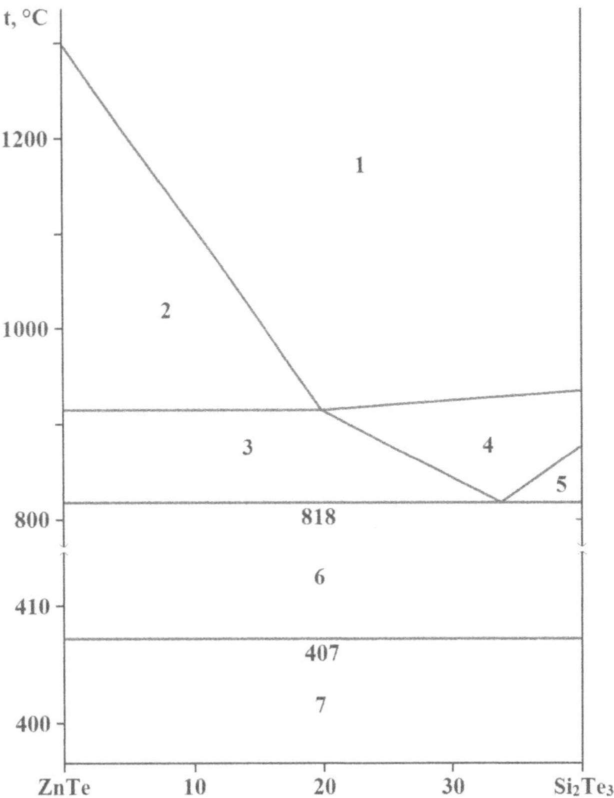

FIGURE 3.4 Phase relations in the Si_2Te_3–ZnTe system: 1, L; 2, L + ZnTe; 3, L + ZnTe + Si; 4, L + Si; 5, L + Si + β-Si_2Te_3; 6, ZnTe + β-Si_2Te_3; and 7, ZnTe + α-Si_2Te_3. (From Odin, I.N., *Zhurn. neorgan. khimii*, **41**(6), 941, 1996.)

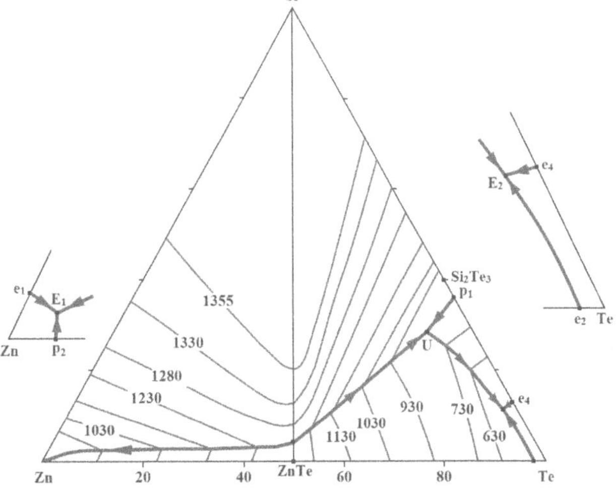

FIGURE 3.5 Liquidus surface of the Si–Zn–Te ternary system. (From Odin, I.N., *Zhurn. neorgan. khimii*, **41**(6), 941, 1996.)

3.8 Silicon–Cadmium–Tellurium

Si_2Te_3–CdTe: This section is a non-quasibinary section of the Si–Cd–Te ternary system (Figure 3.6) (Odin and Ivanov 1991a). Silicon primary crystallizes from the Si_2Te_3-rich side. Crystallization of all melts ends at 810°C by the transition reaction L + Si ⇔ CdTe + β-Si_2Te_3. Polymorphous transformation of Si_2Te_3 takes place at 407°C–409°C. The solubility of Si_2Te_3 in CdTe is not higher than 0.3 mol% and the solubility of CdTe in Si_2Te_3 is equal to 0.2 mol%. This system was investigated using DTA, metallography, XRD, and vapor pressure measurement. The ingots were annealed at 750°C for 1040 h.

Si–CdTe: The phase diagram, constructed using DTA, metallography, XRD, and microhardness measurement, is a eutectic type (Figure 3.7) (Odin and Ivanov 1991a). The eutectic composition and temperature are 5 at% Si and 1083°C, respectively. The solubility of Si in CdTe is not higher than 0.5 at% and the solubility of CdTe in Si is equal to 0.1 at%.

The field of Si primary crystallization occupies the most significant part of the Si–Cd–Te liquidus surface (Figure 3.8) (Odin and Ivanov 1991a). The second-largest area of the liquidus surface represents the field of the CdTe primary crystallization. There are also the fields of Si_2Te_3 and Te crystallization, and the degenerated field of Cd crystallization. The E_1 and E_2 ternary eutectics crystallize at 319°C and 405°C, respectively.

Ternary compounds were not found in the S–Cd–Te ternary system (Kaldis et al. 1967, Odin and Ivanov 1991a).

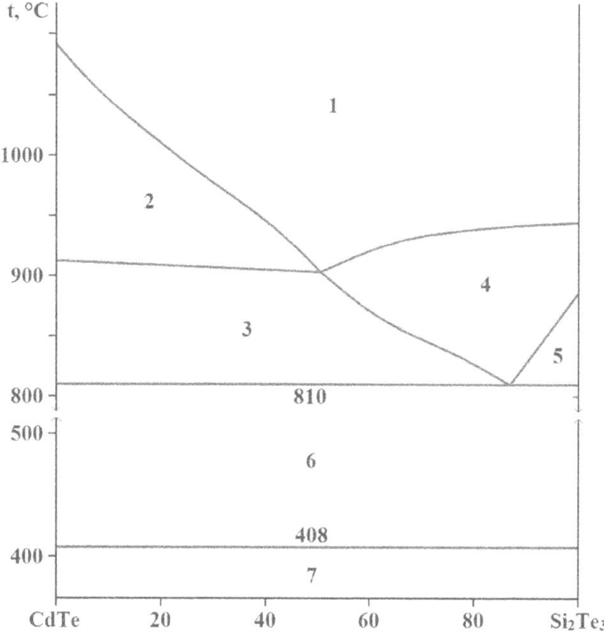

FIGURE 3.6 Phase relations in the Si_2Te_3–CdTe system: 1, L; 2, L + CdTe; 3, L + CdTe + Si; 4, L + Si; 5, L + Si + β-Si_2Te_3; 6, CdTe + β-$S_{i2}Te_3$; and 7, CdTe + α-Si_2Te_3. (From Odin, I.N. and Ivanov, V.A., *Zhurn. neorgan. khimii*, **36**(7), 1837, 1991.)

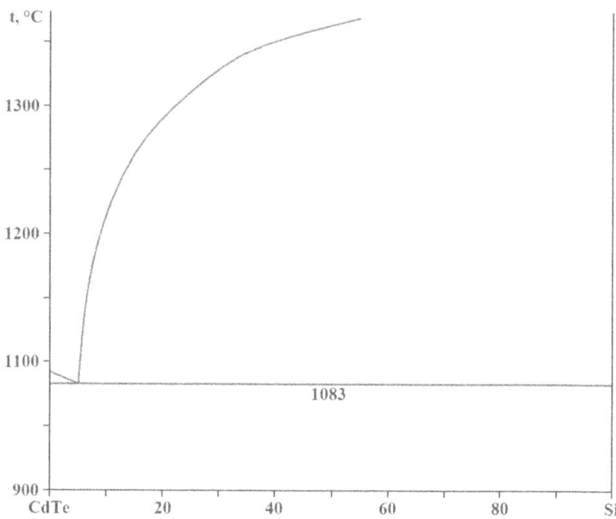

FIGURE 3.7 CdTe–Si phase diagram. (From Odin, I.N. and Ivanov, V.A., *Zhurn. neorgan. khimii*, **36**(7), 1837, 1991.)

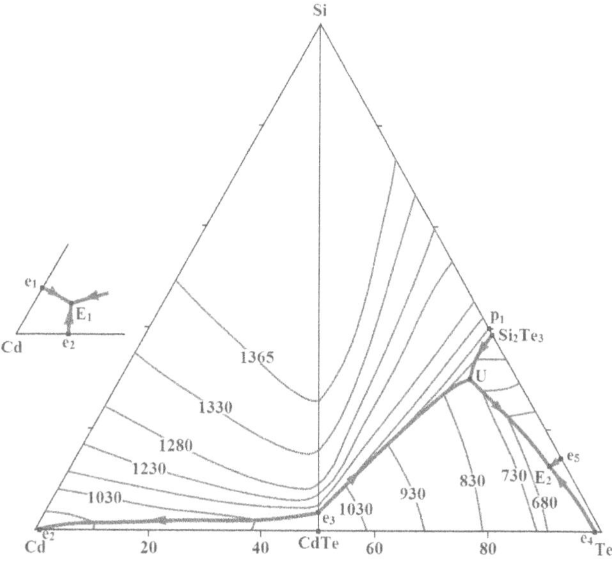

FIGURE 3.8 Liquidus surface of the Si–Cd–Te ternary system. (From Odin, I.N. and Ivanov, V.A., *Zhurn. neorgan. khimii*, **36**(7), 1837, 1991.)

3.9 Silicon–Boron–Tellurium

In the ternary Si–B–Te system, no glass formation region was detected (Kirilenko and Dembovskiy 1974).

3.10 Silicon–Aluminum–Tellurium

The **$AlSiTe_3$** ternary compound, which crystallizes in the hexagonal structure with the lattice parameters a = 683.44 ± 0.06 and c = 699.5 ± 0.1 pm, is formed in the Si–Al–Te system (Sandre et al. 1994). It was prepared by heating the elements in stoichiometric proportions in sealed evacuated silica tube at 650°C for 2 weeks.

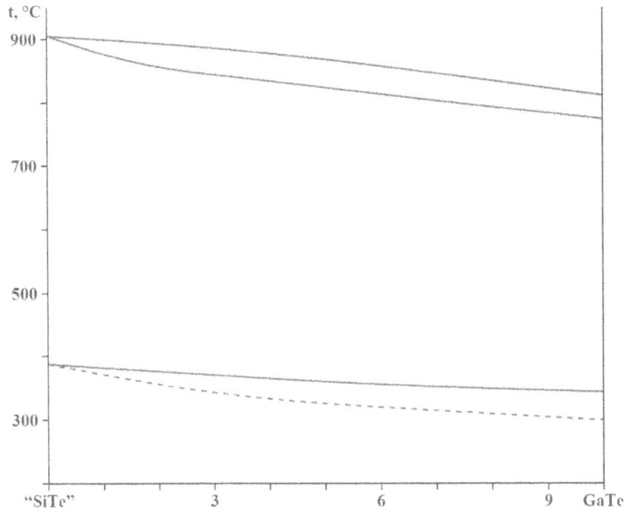

FIGURE 3.9 Part of the "SiTe"–GaTe vertical section. (From Alidzhanov, M.A., et al., *Izv. AN SSSR. Neorgan. mater.*, **11**(4), 762, 1975.)

3.11 Silicon–Gallium–Tellurium

"SiTe"–GaTe: The part of the vertical section, constructed through DTA and microhardness measuring, is presented in Figure 3.9 (Alidzhanov et al. 1975). It can be seen that the solubility of GaTe in "SiTe" is higher than 10 mol%. The solid solutions have a polymorphic transformation in the solid state.

The glass-forming region in the Si–Ga–Te ternary system was presented by Minaev (1983). Synthesis of the glassy alloys was carried out from the elements in quartz ampoules under vacuum at 1000°C for 12 h with vigorous stirring and the next quenching in water. The vitreous state was identified by a characteristic glass wedge, a shell fracture, as well as through XRD and metallography. The boundary of the glass formation region was fixed with an accuracy of 2 at%.

3.12 Silicon–Indium–Tellurium

The glass-forming region in the Si–In–Te ternary system is presented in Figure 3.10 (Minaev 1983). Synthesis of the glassy alloys was carried out from the elements in quartz ampoules under vacuum at 1000°C for 12 h with vigorous stirring and the next quenching in water. The vitreous state was identified by a characteristic glass wedge, a shell fracture, as well as through XRD and metallography. The boundary of the glass formation region was fixed with an accuracy of 2 at%.

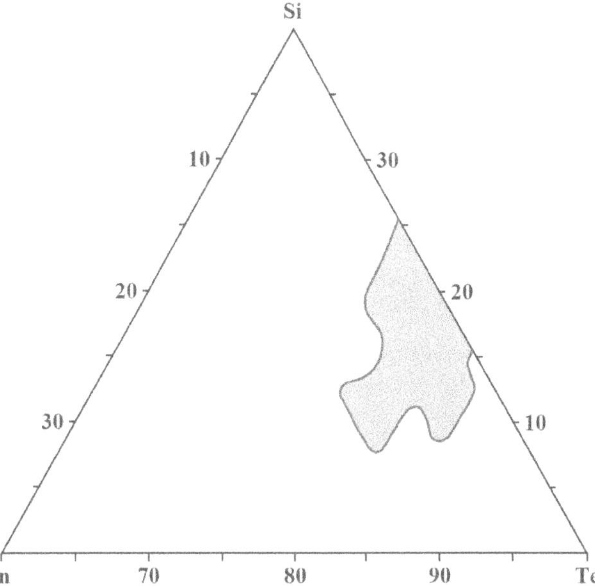

FIGURE 3.10 Glass forming region in the Si–In–Te ternary system. (From Minaev, V.S., *Fiz. i khim. stekla*, **9**(4), 432, 1983.)

3.13 Silicon–Thallium–Tellurium

The $\mathbf{Tl_6Si_2Te_6}$ ternary compound, which crystallizes in the triclinic structure with the lattice parameters $a = 942.35 \pm 0.06$, $b = 966.06 \pm 0.07$, $c = 1038.89 \pm 0.07$ pm, and $\alpha = 89.158 \pm 0.002°$, $\beta = 96.544 \pm 0.002°$, $\gamma = 100.685 \pm 0.002°$, a calculated density of 7.37 g·cm^{-3}, and an energy gap of 0.9 eV is formed in the Si–Tl–Te system (Assoud et al. 2006). The titled compound was prepared starting from the elements taken in the stoichiometric ratio. They were loaded into a silica tube, which was then sealed under vacuum. Subsequently, the tube was heated in a resistance furnace to 800°C within 24 h, and then cooled to 700°C within 15 min, and annealed at this temperature for 200 h. Thereafter the furnace was switched off. The sample consisted mostly of black powder, together with few crystals of metallic luster. The material is not air sensitive at room temperature over a period of a few weeks.

The glass-forming region in the Si–Tl–Te ternary system is presented in Figure 3.11 (Minaev et al. 1981; Minaev 1983). Up

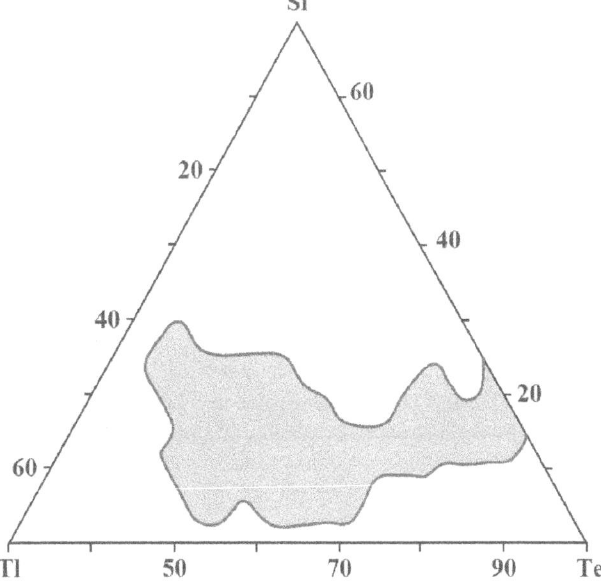

FIGURE 3.11 Glass forming region in the Si–Tl–Te ternary system. (From Minaev, V.S., et al., *Izv. AN SSSR. Neorgan. mater.*, **17**(7), 1305, 1981.)

to 46 at%, Tl can be incorporated into glasses. Synthesis of the Si–Tl–Te alloys was carried out from the elements in quartz ampoules under vacuum at 1100°C for 12 h with vigorous stirring and the next quenching in water. The vitreous state was identified by a characteristic glass wedge, a shell fracture, as well as through XRD and metallography. The boundary of the glass formation region was fixed with an accuracy of 2 at%.

3.14 Silicon–Scandium–Tellurium

The **$ScSiTe_3$** ternary compound, which crystallizes in the hexagonal structure with the lattice parameters $a = 700.65 \pm 0.05$ and $c = 2129.2 \pm 0.1$ pm, is formed in the Si–Sc–Te system (Sandre et al. 1994). It was prepared by heating the elements in stoichiometric proportions in sealed evacuated silica tube at 650°C for 2 weeks.

3.15 Silicon–Thorium–Tellurium

The **ThSiTe** ternary compound, which crystallizes in the tetragonal structure with the lattice parameters $a = 419.0$ and $c = 1818$ pm and the calculated and experimental densities of 8.07 and 7.97 g·cm^{-3}, respectively, is formed in the Si–Th–Te system (Hahn and Stocks 1968; Stocks et al. 1981). To prepare this compound, the elements were weighed under N_2 atmosphere in the stoichiometric ratio and initially heated to about 400°C in an evacuated quartz ampoule. Subsequently, the inhomogeneous product was finely powdered under N_2 and heated to 850°C in a sealed tungsten crucible, which was enclosed in an evacuated quartz ampoule. Since the preparation is extremely sensitive to traces of oxygen, it proved necessary to add Th as getter metal. Nevertheless, pure preparation was obtained only if no longer than 48 h the mixture was heated at 850°C. The mixture was annealed at 600°C–700°C for 8–10 days. Then, the sample pressed into pastilles was heated in the same way over different time intervals and at different temperatures up to 1400°C in tungsten crucible, wherein above 1050°C, additionally Ar was used as a protective gas. The title compound was obtained as black microcrystalline powder.

3.16 Silicon–Germanium–Tellurium

According to the data of Shelimova et al. (1964), the phase diagram of the **"SiTe"–GeTe** system, constructed using DTA and metallography, is a eutectic type with eutectic containing 30 mol% "SiTe" and crystallizing at 685°C. The solubility of "SiTe" in GeTe at the eutectic temperature reaches 3 mol% [less than 1 mol% (Bigvava et al. 1984)] and decreases with temperature decreasing. This solid solution has a polymorphic transformation at 360°C. The ingots were annealing at 600°C for 500 h.

Apparently, the "SiTe"–GeTe is not quasibinary as only two compounds, Si_2Te_3 and $SiTe_2$, are formed in the Si–Te system (Mishra et al. 2016). According to the data of Bigvava et al. (1984), Si_2Te_3 is formed at the content of "SiTe" more than 3 mol%.

The glass-forming region in the Si–Ge–Te ternary system was estimated under standardized conditions, yielding its limits composition as $Si_{1.6-19.6}Ge_{0-19.6}Te_{78-86}$ (Feltz et al. 1972).

3.17 Silicon–Lead–Tellurium

Si–PbTe: The part of the phase diagram of this system, constructed through DTA, XRD, and metallography, is presented in Figure 3.12 (Odin 1994). The eutectic contains 8 mol% Si and crystallizes at 888°C. Noticeable regions of solid solutions based on PbTe and Si were not determined.

Si_2Te_3–PbTe: The phase relations in this system are given in Figure 3.13 (Odin 1994). This vertical section intersects the field of PbTe and Si crystallization. PbTe dissolves up to 1.3 mol% Si_2Te_3 and the noticeable solubility based on Si_2Te_3

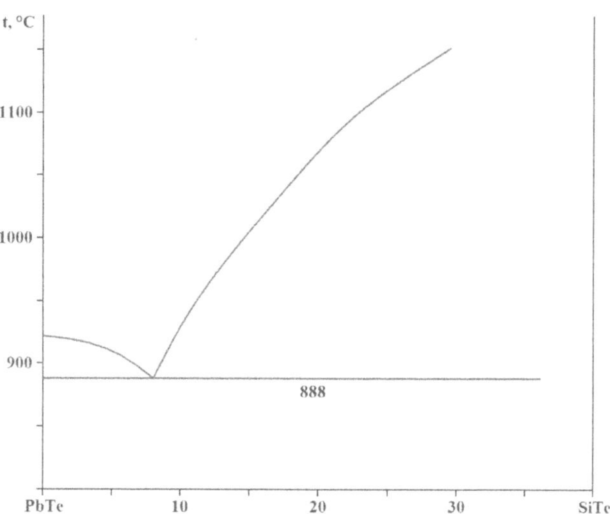

FIGURE 3.12 Part of the phase diagram of the PbTe–Si system. (From Odin, I.N., *Zhurn. neorgan. khimii*, **39**(10), 1730, 1994.)

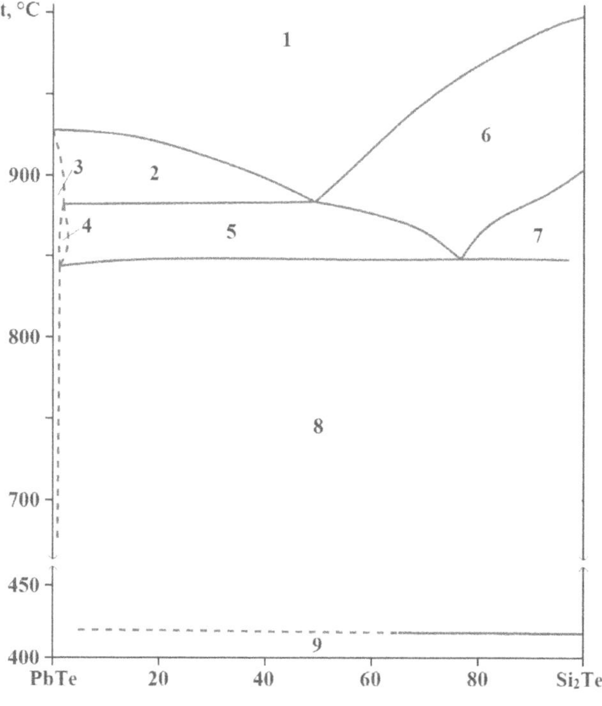

FIGURE 3.13 Phase relations in the Si_2Te_3–PbTe system: 1, L; 2, L + (PbTe); 3, (PbTe); 4, (PbTe) + Si; 5, L + (PbTe) + Si; 6, L + Si; 7, L + Si + β-Si_2Te_3; 8, (PbTe) + β-Si_2Te_3; 9, (PbTe) + α-Si_2Te_3. (From Odin, I.N., *Zhurn. neorgan. khimii*, **39**(10), 1730, 1994.)

was not detected. Si_2Te_3 has a polymorphous transformation at 406°C–409°C.

The liquidus surface of the Si–Pb–Te ternary system is shown in Figure 3.14 (Odin 1994). The field of Si primary crystallization occupies the biggest part of this liquidus surface and the field of Pb primary crystallization is degenerated. The ternary eutectic E_1 crystallizes at 326°C. The isothermal section of this ternary system at 700°C is presented in Figure 3.15 (Odin 1994).

The samples for the investigations from the Si_2Te_3–PbTe–Si region were annealed at 700°C for 1040 h, the samples more reach for Te were annealed at 360°C, and the samples of the Si–PbTe–Pb subsystem were annealed at 310°C (Odin 1994).

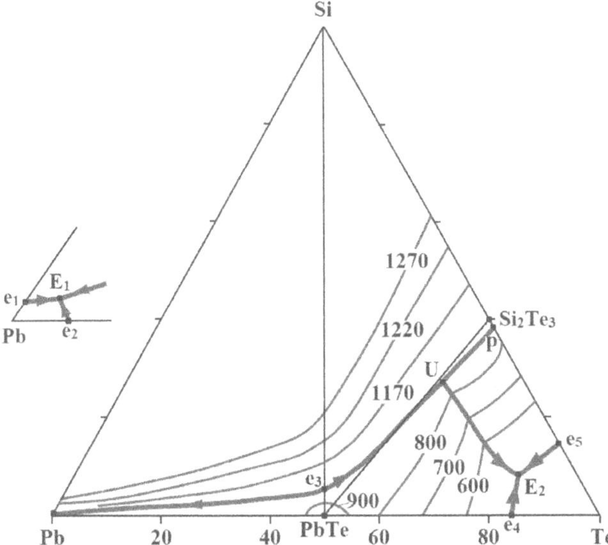

FIGURE 3.14 Liquidus surface of the Si–Pb–Te ternary system. (From Odin, I.N., *Zhurn. neorgan. khimii*, **39**(10), 1730, 1994.)

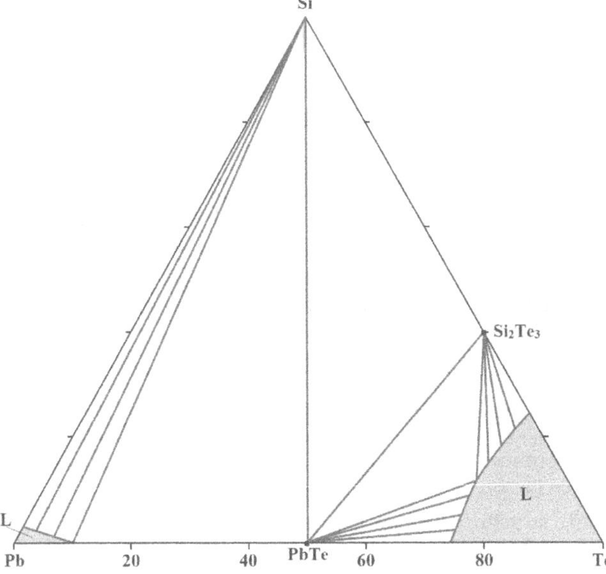

FIGURE 3.15 Isothermal section of the Si–Pb–Te ternary system at 700°C. (From Odin, I.N., *Zhurn. neorgan. khimii*, **39**(10), 1730, 1994.)

3.18 Silicon–Zirconium–Tellurium

The **ZrSiTe** ternary compound, which crystallizes in the tetragonal structure with the lattice parameters $a = 369.03 \pm 0.01$ and $c = 949.6 \pm 0.2$ pm (Wang and Hughbanks 1995) [$a = 369.74 \pm 0.09$, $c = 950.8 \pm 0.2$ pm, and a calculated density of 6.309 g·cm^{-3} (Bensch and Dürichen 1994); $a = 369.2$ and $c = 949.9$ pm (Haneveld and Jellinek 1964); $a = 369$, and $c = 956$ pm, and the calculated and experimental densities of 6.31 and 5.64 g·cm^{-3}, respectively (Jellinek and Hahn 1962; Onken et al. 1964)], is formed in the Si–Zr–Te system. This compound is slowly hydrolyzed in moist air (Haneveld and Jellinek 1964). To obtain ZrSiTe, stoichiometric amounts of the elements were initially pre-tempered in an evacuated tube at 600°C for 10 days (Onken et al. 1964). Then, the reaction mixture was carefully rubbed in an agate mortar in a N_2-filled box, pressed into a pastille and annealed at 900°C for several weeks. The ampoule with tablet was stored in the hottest part of the furnace to obtain the single crystals by sublimation. ZrSiTe could be also prepared by heating "SiTe" with an equivalent proportion of Zr in an evacuated quartz tube at 800°C for 1 or 2 days (Haneveld and Jellinek 1964). The title compound was also accidentally obtained during an attempt to synthesize ZrSnTe (Wang and Hughbanks 1995). The mixture of the Zr, Sn, and Te in a 1:1:1 molar ratio with the transport agent $TeCl_4$ was placed in an evacuated quartz tube. Then the tube was put in a furnace and the temperature was raised to 550°C over 6 h and maintained at 550°C for 4 days. It was then raised to 900°C over 12 h and maintained for 21 days in a 900°C–950°C temperature gradient. The product consisted entirely of black square plate crystals of ZrSiTe.

3.19 Silicon–Hafnium–Tellurium

The **HfSiTe** ternary compound, which crystallizes in the tetragonal structure with the lattice parameters $a = 367$ and $c = 973$ pm, and a calculated density of 8.47 g·cm^{-3}, is formed in the Si–Hf–Te system (Onken et al. 1964). To prepare this compound, stoichiometric amounts of Si, Hf, and Te were initially pre-tempered in an evacuated tube at 600°C for 10 days. Then, the reaction mixture was carefully rubbed in an agate mortar in a N_2-filled box, pressed into a pastille and annealed at 850°C for several weeks. The ampoule with tablet was stored in the hottest part of the furnace to obtain the single crystals by sublimation.

3.20 Silicon–Phosphorus–Tellurium

The $\mathbf{Si_{46-x}P_xTe_y}$ ($y = 7.35$; 6.98; 6.88; $x \le 2y$) phases, which crystallize in the cubic structure with the lattice parameter $a = 997.02 \pm 0.03$ pm for $Si_{31.3}P_{14.7}Te_{7.35}$, $a = 997.94 \pm 0.02$ pm for $Si_{32.4}P_{13.6}Te_{6.98}$, and $a = 998.08 \pm 0.02$ pm for $Si_{33.0}P_{13.0}Te_{6.88}$, is formed in the Si–P–Te system (Zaikina et al. 2007). To prepare these phases, Si, red phosphorus, and Te were mechanically reground, sieved and then mixed in stoichiometric ratio using a ball mill. The obtained powder was pressed into

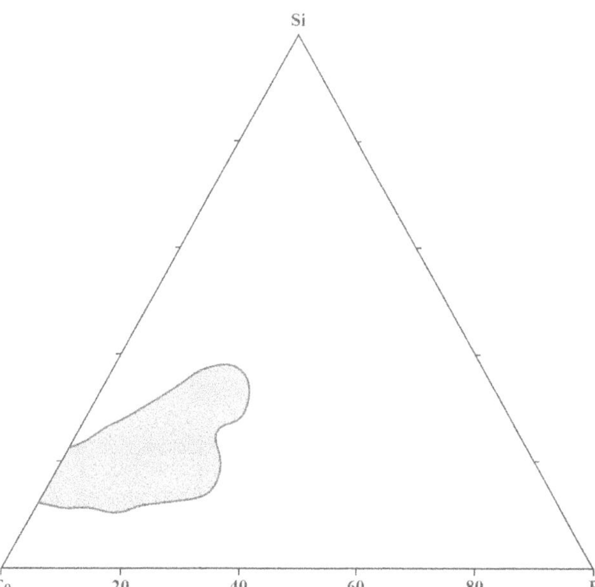

FIGURE 3.16 Glass forming region in the Si–P–Te ternary system. (From Hilton, A.R., et al., *Infrared Phys.*, **4**(4), 213, 1964.)

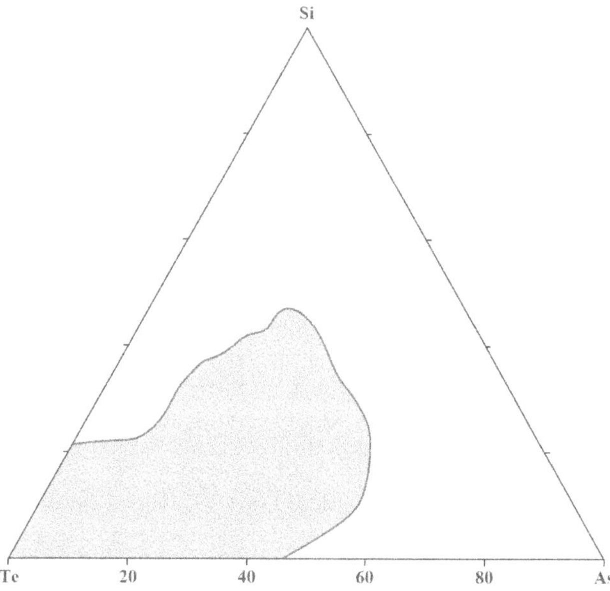

FIGURE 3.17 Glass forming region in the Si–As–Te ternary system. (From Hilton, A.R. and Brau, M.J., *Infrared Phys.*, **3**(2), 69, 1963.)

a pellet in a conventional press, sealed under vacuum in a quartz ampoule and heated to 1100°C with a heating rate of 0.8°C·min^{-1}, annealed for 24 h at this temperature, then cooled in 3 h down to 900°C, and was finally annealed at this temperature for 72 h. After the thermal treatment, the ampoule was slowly cooled to ambient temperature in the furnace. Preparation and storage of the starting materials were carried out in Ar-filled glove box.

The glass-forming region in the Si–P–Te system is shown in Figure 3.16 (Hilton et al. 1964). This region is small and only about one-third to one-half of the glasses are of good quality. The good quality glasses have softening points of only about 200°C. To prepare such glasses, the reactants were weighed out and sealed in a quartz vial at a reduced pressure. The vial was placed in a rocking furnace, slowly heating to 1000°C–1100°C, and allowed to react and mix from 16 to 40 h. The glasses were formed by quenching the samples to room temperature in air. Usually, the sample could be identified as amorphous or crystalline by a simple visual examination.

3.21 Silicon–Arsenic–Tellurium

SiTe–As: No solid solutions were found from the arsenic side (Krebs et al. 1961a).

The glass-forming region in the Si–As–Te system is shown in Figure 3.17 (Hilton and Brau 1963). Several of these glasses, particularly those with high silicon content, showed evidence of a dispersed crystalline phase when examined with the aid of the metallography. To prepare such glasses, the reactants were weighed out and sealed in a quartz vial at a reduced pressure. The vial was placed in a rocking furnace, slowly heating to 1000°C, and allowed to react and mix while molten for at least 16 h. The glasses were formed by quenching from 1000°C to room temperature in air.

3.22 Silicon–Antimony–Tellurium

Three compounds, $\mathbf{SiSb_2Te_4}$, $\mathbf{SiSb_4Te_7}$, and $\mathbf{Si_2Sb_2Te_5}$, which crystallize in the trigonal structure with the lattice parameters a = 425.6, c = 4222.1 pm for the first compound, a = 428.9, c = 2456.2 pm for the second one, and a = 422.1, c = 1748.1 pm for $Si_2Sb_2Te_5$ (Sa et al. 2010), is formed in the Si–Sb–Te ternary system. The lattice parameters were calculated using *ab initio* computational study.

$Si_2Sb_2Te_5$ is a narrow bandgap p-type semiconductor. It is a metastable material. The energy instability of this compound drives its phase separation. The calculations show that $Si_2Sb_2Te_5$ tends to decompose into $SiSb_2Te_4$ or $SiSb_4Te_7$ or Sb_2Te_3 (Sa et al. 2010). Zhang et al. (2007) determined that the bandgap of the amorphous and polycrystalline $Si_2Sb_2Te_5$ is 0.89 and 0.62 eV, respectively.

The glass-forming region in the Si–Sb–Te ternary system is presented in Figure 3.18 (Minaev 1983). Synthesis of the glassy alloys was carried out from the elements in quartz ampoules under vacuum at 1000°C for 12 h with vigorous stirring and the next quenching in water. The vitreous state was identified by a characteristic glass wedge, a shell fracture, as well as through XRD and metallography. The boundary of the glass formation region was fixed with an accuracy of 2 at%.

3.23 Silicon–Niobium–Tellurium

Two ternary compounds, $\mathbf{Nb_2SiTe_4}$ and $\mathbf{Nb_3SiTe_6}$, are formed in the Si–Nb–Te system. Nb_2SiTe_4 crystallizes in the monoclinic structure with the lattice parameters a = 1421.6 ± 0.2, b = 394.40 ± 0.08, c = 633.62 ± 0.13 pm, β = 97.97 ± 0.02°, and a calculated density of 13.674 g·cm^{-3} (Monconduit et al. 1993). It was obtained by heating the elements taken in stoichiometric proportion at 950°C for 12 days. The sealed silica tube was

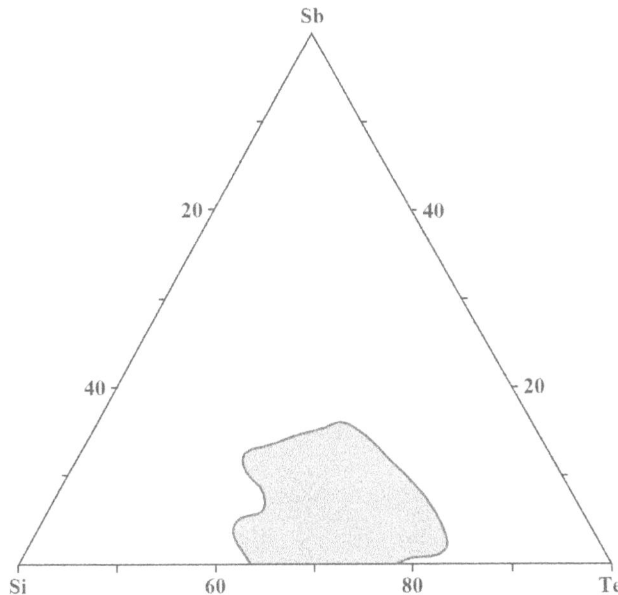

FIGURE 3.18 Glass forming region in the Si–Sb–Te ternary system. (From Minaev, V.S., *Fiz. i khim. stekla*, **9**(4), 432, 1983.)

heated at a rate of 100°C·h^{-1} and subsequently quenched to room temperature.

Nb_3SiTe_6 crystallizes in the orthorhombic structure with the lattice parameters a = 635.3 ± 0.2, b = 1393.8 ± 0.5, and c = 1150.7 ± 0.4 pm, and a calculated density of 6.991 g·cm^{-1} (Li et al. 1992). The plate-like crystals of the title compound were initially obtained in a reaction of 1 g mixture of Nb, Si, and Te (molar ratio 2:1:4). A quartz tube containing the three elements and the transport agent ($TeCl_4$; 10 mg) was sealed under vacuum and placed in a furnace. The tube was slowly heated to 850°C and the reaction was kept at this temperature for 4 days before being cooled to room temperature. Better-quality single crystals used in the structure determination were grown in a reaction at a higher temperature (950°C) for 14 days. The crystals of Nb_3SiTe_6 show metallic luster and are stable in air. The single crystals of the title compound were grown from composite elements by chemical transport reactions in a sealed silica ampoule (Ohno 1999). A small amount of I_2 was added as a transport agent.

3.24 Silicon–Tantalum–Tellurium

Two compounds, $\mathbf{Ta_3SiTe_6}$ and $\mathbf{Ta_4SiTe_4}$, are formed in the Si–Ta–Te ternary system. First of them crystallizes in the orthorhombic structure with the lattice parameters a = 632.9 ± 0.3, b = 1403.1 ± 0.3, and c = 382.58 ± 0.07 pm, and a calculated density of 8.729 g·cm^{-1} (Lee van der et al. 1994c) [a = 632.9, b = 1400, and c = 382.5 pm (Frangis et al. 1998)] for the $TaSi_{0.360}Te$ composition and a = 631.8 ± 0.1, b = 1403.1 ± 0.2, and c = 385.52 ± 0.07 pm, and a calculated density of 8.761 g·cm^{-1} for the $TaSi_{0.414}Te$ composition (Evain et al. 1994). $TaSi_{0.360}Te$ was obtained in an attempt to prepare $TaSi_{0.5}Te_2$ (Lee van der et al. 1994c). Stoichiometric amounts of the elements were placed in an evacuated silica tube. The temperature was raised at 100°C·h^{-1}, then maintained for 10 d at 955°C and finally lowered by exposure of the tube to room-temperature air. $TaSi_{0.414}Te$ was prepared from a mixture of the elemental powders (molar ratio Ta/Si/Te = 7:3:14) (Evain et al. 1994; Frangis et al. 1998). The mixture was ground and loaded into a quartz tube. The tube was evacuated to 1.3 Pa, sealed, and placed in a furnace. The temperature of the furnace was raised from room temperature to 900°C–980°C at 100°C·h^{-1}, kept fixed for 10 days, and then cooled to ambient temperature at 100°C·h^{-1}. The single crystals of Ta_3SiTe_6 were grown from composite elements by chemical transport reactions in a sealed silica ampoule (Ohno 1999). A small amount of I_2 was added as a transport agent.

Ta_4SiTe_4 also crystallizes in the orthorhombic structure with the lattice parameters a = 1053.6 ± 0.3, b = 1827.5 ± 0.5, and c = 479.9 ± 0.1 pm (Badding and DiSalvo 1990) [a = 1054, b = 1828, and c = 480 pm (Li et al. 1990)]. Ta_4SiTe_4 monomer stacks antiprismatically along the c axis to form an infinite chain (Li et al. 1990). The title compound was first synthesized from a mixture of the elements (molar ratio Ta/Si/Te = 2:1:2) in an evacuated quartz tube (Badding and DiSalvo 1990). $TeCl_4$ was included as a transport agent. The sample was heated in a furnace at 600°C for 1 day and removed and placed in another furnace at 1150°C for 4 days, followed by rapid quenching to room temperature. Large needlelike crystals with a metallic luster were formed. The crystal used in the structure determination was taken from a reaction containing a 1-g mixture of the elements in the molar ratio Ta/Si/Te = 4:1:5 with 10 mg of $TeCI_4$; the previously mentioned heating schedule was used. After the structure was determined, a powder sample of Ta_4SiTe_4 was obtained quantitatively by heating a stoichiometric mixture of Ta, Si, and Te in an evacuated quartz ampoule to 1050°C for 2 days. The crystals are very fragile and moderately air sensitive; they decompose on sitting in air after a few days. Finely divided powders often ignite on air exposure.

3.25 Silicon–Chromium–Tellurium

The $\mathbf{Cr_2Si_2Te_6}$ ternary compound, which crystallizes in the hexagonal structure with the lattice parameters a = 674.7 ± 0.8 and c = 2062.2 ± 0.8 pm (Lee et al. 2019) [a = 675.78 ± 0.06 and c = 2066.5 ± 0.3 pm and a calculated density of 5.621 g·cm^{-3} (Marsh 1988; Ouvrard et al. 1988); a = 676.3 ± 0.2 and c = 2058.2 ± 0.3 pm at 50 K and a = 677.3 ± 0.2, and c = 2052.8 ± 0.2 pm at 1.2 K (Carteaux et al. 1991); a = 677 pm according to the first principles calculations within the density-functional theory (Zhuang et al. 2015)], is formed in the Si–Cr–Te system. The powder of the title compound was prepared by heating a mixture of Cr, Si, and Te in the stoichiometric amounts in a sealed evacuated Pyrex tube at 500°C for 10 days, followed by a 10 h slow cooling (Ouvrard et al. 1988; Carteaux et al. 1991) or in a sealed evacuated quartz tube at 900°C for 24 h (Lee et al. 2019). $Cr_2Si_2Te_6$ platelets can be grown in a tellurium melt in the same conditions (Carteaux et al. 1991).

3.26 Silicon–Manganese–Tellurium

The $\mathbf{Mn_3Si_2Te_6}$ ternary compound, which crystallizes in the hexagonal structure with the lattice parameters a = 702.9 ± 0.2 and c = 1425.5 ± 0.3 pm (Vincent et al. 1986)

[a = 702 and c = 1426 pm and a calculated density of 4.9 g·cm^{-3} (Rimet et al. 1981)], is formed in the Si–Mn–Te system. To prepare this compound, a stoichiometric mixture of Mn, and Te was heated in a silica cell sealed under a vacuum of 0,13 Pa for 4 days at 700°C and 1 day at 900°C, then quenched down to room temperature (Rimet et al. 1981; Vincent et al. 1986). The resulting product was a polycrystalline block from which thin platelets of the title compound could be easily cleaved. Single crystals have been grown by the vapor phase transport technique from a mixture of the elements with a slight excess of tellurium and with I_2 as the transport agent; the thermal gradient was from 800°C to 720°C. Complete reaction was not achieved after 8 days; in the cooler zone, thin, roughly hexagonal, shiny dark platelets were found.

3.27 Silicon–Nickel–Tellurium

No ternary compounds were found in the Si–Ni–Te system (Isaeva et al. 2007).

3.28 Silicon–Platinum–Tellurium

The **PtSiTe** ternary compound, which crystallizes in the orthorhombic structure with the lattice parameters a = 611.9 ± 0.2, b = 620.0 ± 0.2, and c = 1237.9 ± 0.4 pm at 113 K, is formed in the Si–Pt–Te system (Mansuetto and Ibers 1994). According to the first-principles calculations carried out within the framework of density functional theory with exchange-correlation functional in the generalized gradient approximation, the lattice parameters are the next: a = 622, b = 1272, and c = 623 pm (Bachhuber et al. 2015). Silver, truncated octahedron crystals of the title compound were synthesized from a combination of Pt, Si, and Te (molar ratio 1:1:5) at 800°C in a sealed quartz tube (Mansuetto and Ibers 1994).

REFERENCES

All References are available as a downloadable eResource at www.routledge.com/9780367639235

4

Systems Based on Germanium Sulfides

4.1 Germanium–Hydrogen–Sulfur

GeS_2–H_2S: Two ternary compounds, **$H_4Ge_4S_{10}$ ($H_2Ge_2S_5$)** and **$H_2Ge_4S_9$**, are formed in this system. First of them decomposes at ca. 250°C (Poling et al. 2003b) and crystallizes in the triclinic structure with the lattice parameters $a = 862.1 \pm 0.4$, $b = 989.9 \pm 0.4$, $c = 1000.9 \pm 0.4$ pm, and $\alpha = 85.963 \pm 0.007°$, $\beta = 64.714 \pm 0.007°$, $\gamma = 89.501 \pm 0.007°$, and a calculated density of 2.652 g·cm^{-3} at 173 K (Poling et al. 2003a). To prepare $H_4Ge_4S_{10}$, GeO_2 was dissolved in a Potassium Hydroxide (KOH) solution and precipitated by neutralization with H_2SO_4 (Willard and Zuehlke 1943). The gelatinous oxide was filtered and washed thoroughly on the filter. The moist oxide was then suspended in absolute EtOH and treated with H_2S. The reaction 4 GeO_2 (hydrated) + 10 H_2S = $H_4Ge_4S_{10}$ + 8 H_2O required 1 h for completion. The title compound is a white amorphous solid, extremely soluble in water. The crystals of this compound were grown from anhydrous liquid H_2S reaction with glassy GeS_2 at room temperature (Poling et al. 2003a). Under anhydrous conditions, GeS_2 (ca. 0.5 g) was reacted with liquid H_2S (ca. 7 g). A reaction time of 2–4 weeks was used to produce single crystals, which appear pale with yellowish tint, nontransparent, prismatic, and non-hydroscopic crystals.

$H_2Ge_4S_9$ decomposes at ca. 360°C (Poling et al. 2003b).

The reaction routes for the formation of these two thiogermanic acids have been investigated for gaseous H_2S reactions with GeS_2 and GeO_2 (Sutherland et al. 2004b). $H_4Ge_4S_{10}$ was shown to form from both GeS_2 and GeO_2. The synthesis route for this compound using GeO_2 as a precursor was found to be H_2S(g) pressure-dependent. While $H_2Ge_4S_9$ did not form from gaseous H_2S reactions with GeS_2, it formed readily with GeO_2 reactions. The role of water in the reaction process was examined. It was suggested that $H_2Ge_4S_9$ cannot be formed from GeS_2 and H_2S(g) regardless of the amount of water added to the reaction.

The reaction of liquid H_2S with GeO_2 produces $Ge_4S_{10}^{-4}$ ions in solution (Poling et al. 2003b). Evaporation of the H_2O–H_2S solution leaves thermally unstable **$H_4Ge_4S_{10}$·$x$$H_2O$** units that decompose into the thermally stable anhydrous $H_4Ge_4S_{10}$ or $H_2Ge_4S_9$ phases. In general, $H_4Ge_4S_{10}$ was obtained from shorter reaction times, whereas $H_2Ge_4S_9$ was suggested for longer reaction times. The Raman, infra-red, and XRD structural evolution as a function of reaction time indicates a strongly kinetically controlled formation rate with presumably the hydrostatic pressure of liquid H_2S as the driving force. Ultimately, $H_2Ge_4S_9$ realized from longer reaction times is approximately 110°C more thermally stable than $H_4Ge_4S_{10}$. For both thiogermanic acids, however, the thermal decomposition product is determined to be glassy GeS_2.

The GeS_2–H_2S system has been systematically investigated to determine the reaction products by Sutherland et al. (2004a). It was suggested that the terminal stoichiometry that results from protonation of GeS_2 with H_2S is consistent with that of the $H_4Ge_4S_{10}$ phase. Two studies were performed, one at high temperatures (~750°C) with moderate pressures (~0.7 MPa) and the other at moderate temperatures (ambient through ~250°C) with higher pressures (2 MPa). Shorter times and lower temperature reactions produce the $H_4Ge_4S_{10}$ protonated thiogermanic acid. Longer times produce an unprotonated low-temperature three-dimensional α-GeS_2. Higher reaction temperatures result in weakly protonated glassy materials.

4.2 Germanium–Lithium–Sulfur

GeS_2–Li_2S: Three ternary compounds, **Li_2GeS_3**, **Li_4GeS_4**, and **Li_8GeS_6** are formed in this system. Li_2GeS_3 crystallizes in the orthorhombic structure with the lattice parameters a = 590, b = 1795, and c = 681 pm (Kanno et al. 2000). To prepare this compound, Li_2S and GeS_2 were weighed, mixed in appropriate molar ratio in an Ar-filled glove box, put into a carbon-coated quartz tube, and heated at a reaction temperature of 600°C–800°C for 8 h. After the reaction, the tube was slowly cooled to room temperature.

Li_4GeS_4 melts at 858°C [at 567°C (Matsushita and Kanatzidis 1998)], and also crystallizes in the orthorhombic structure with the lattice parameters $a = 1400.1 \pm 0.4$, $b = 776.6 \pm 0.2$, and $c = 619.3 \pm 0.2$ pm, a calculated density of 2.255 g·cm^{-3} and an energy gap of 4.13 eV (MacNeil et al. 2014) [$a = 1410.7 \pm 0.6$, $b = 777.0 \pm 0.3$, $c = 616.2 \pm 0.2$ pm, and a calculated density of 2.248 g·cm^{-3} (Matsushita and Kanatzidis 1996, 1998); $a = 1406.58 \pm 0.02$, $b = 775.102 \pm 0.015$, and $c = 614.973 \pm 0.010$ pm (Kanno et al. 2000); $a = 1403.40 \pm 0.03$, $b = 775.48 \pm 0.02$, and $c = 615.023 \pm 0.017$ pm (Murayama et al. 2002b)].

Li_4GeS_4 was synthesized by the next methods. All manipulations were carried out under a dry N_2 or Ar atmosphere in a glove box.

1. The powder of the title compound was prepared via high-temperature solid-state synthesis by weighing and grinding a stoichiometric mixture of Li_2S, Ge, and S in an Ar-filled glove box (MacNeil et al. 2014). The mixture was transferred to a graphite tube which was then flame sealed inside a fused-silica tube under a vacuum of approximately 0.1 Pa. The stoichiometric mixture was heated up to 750°C over the course of 15 h. The temperature was held at 750°C for 96 h, followed by slow cooling to 500°C at 5°C·h^{-1} and

DOI: 10.1201/9781003123507-4

subsequently cooled to room temperature in 3 h. The tube was opened in the Ar-filled glove box. A pale gray, microcrystalline powder was obtained. The compound is moisture sensitive and was stored in a glove box.

2. Single crystals of Li_4GeS_4 were isolated from a lithium polysulfide flux reaction that was intended to produce a quaternary phase (MacNeil et al. 2014). Li_2S (3 mM), Zn (1 mM), Ge (1 mM), and S (8 mM) were weighed out and ground in an Ar-filled glove box for 15 min. The ground mixture was placed into a loosely-capped graphite tube; then the entire assembly was flame-sealed inside a fused-silica tube under a vacuum of approximately 0.1 Pa. In a furnace, the sample was heated to 750°C over 12 h and soaked at that temperature for 96 h. The sample was first slow-cooled to 450°C at 5°C·h^{-1}, and then cooled to room temperature. Pale-orange single crystals of the title compound were obtained.
3. Li_2S, La_2S_3, Ge, and S (molar ratio 2:1:2:6) were well ground and put in a carbon tube which was sealed inside an evacuated carbon-coated fused silica tube (~10^{-3} Pa) (Matsushita and Kanatzidis 1996, 1998). The material was heated to 400°C at 33°C·h^{-1} and kept for 12 h, and then heated to 700°C at 25°C·h^{-1}, kept at this temperature for 200 h and cooled to 500°C at 2°C·h^{-1}. The sample was then quickly cooled to room temperature. The resulting material was isolated with MeOH, acetone and ether to dissolve the polysulfide flux and to give well developed pillar-shaped, clear white to light yellow moisture-sensitive crystals of Li_4GeS_4 along with some unreacted La_2S_3.
4. Li_2S, GeS_2, and S (molar ratio 4:1:4) were mixed, well ground and placed in a carbon tube which was sealed inside an evacuated carbon-coated fused silica tube (~10^{-3} Pa) (Matsushita and Kanatzidis 1998). The mixture was heated to 400°C at 33°C·h^{-1} and kept there for 12 h. It was then heated to 750°C at 25°C·h^{-1} for 200 h, slowly cooled to 500°C at 1°C·h^{-1} followed by quenching to room temperature. The product was isolated with MeOH and ether.
5. Li_2S and GeS_2 were weighed, mixed in appropriate molar ratio in an Ar-filled glove box, put into a carbon-coated quartz tube, and heated at a reaction temperature of 600°C–800°C for 8 h (Kanno et al. 2000). After the reaction, the tube was slowly cooled to room temperature.
6. Li_2S, GeS_2, and P_2S_5 were weighed, mixed in appropriate molar ratio in an Ar-filled glove box, put into a carbon-coated quartz tube and heated at a reaction temperature of 700°C for 8 h (Murayama et al. 2002b). After the reaction, the tube was slowly cooled to room temperature.

To prepare the glasses in the GeS_2–Li_2S system, 2–3 g of sulfides were mixed in the correct proportions and put in a vitreous carbon crucible (Barrau et al. 1980b; Ribes et al. 1980; Souquet et al. 1981). This crucible was placed inside the silica tube and sealed under vacuum. After melting at a temperature 700°C–1000°C, the light yellow glasses were obtained by cooling down to room temperature or by immersing the silica tube in cold water. The obtained glasses contained up to 50 mol% Li_2S, were very hygroscopic and had to be handled in a dry box. The glass transition temperature for these glasses is within the interval 310°C–316°C.

Li_8GeS_6 has phase transformations within the interval from room temperature to 800°C (Putkaradze and Nanobashvili 1966). It was synthesized by the interaction of aqueous solutions of GeS_2 and Li_2S with the next extraction with acetone and drying in an inert gas atmosphere.

4.3 Germanium–Sodium–Sulfur

GeS_2–Na_2S: The part of the phase diagram of this system, constructed through DTA, XRD, and metallography, is presented in Figure 4.1 (Mikhaylo et al. 1989). The samples were annealed at 480°C for 200–250 h. Two eutectics contain 57 and 75 mol% GeS_2 and crystallize at 597°C and 622°C, respectively. The regions of homogeneity for both compounds do not exceed 5 mol%. The field of the solid solution based on GeS_2 reaches up to 6 mol%. Two ternary compounds, **Na_2GeS_3** and **$Na_2Ge_2S_5$**, are formed in this system. Na_2GeS_3 melts congruently at 637°C [at 643°C with melting entropy 36.8 kJ·M^{-1} (Mikaylo et al. 1990); incongruently at 590°C ± 5°C (Olivier-Fourcade et al. 1971, 1972)] and crystallizes in the monoclinic structure with the lattice parameters $a = 697.0 \pm 0.7$, $b = 1528.5 \pm 0.8$, $c = 577.0 \pm 0.1$ pm, and $\beta = 115.7 \pm 0.1°$ (Mikhaylo et al. 1989) [$a = 695.2$, $b = 1523.0$, $c = 572.0$ pm, $\beta = 115.24°$, and the calculated and experimental densities of 2.60 and 2.51 g·cm^{-3}, respectively (Olivier-Fourcade et al. 1971, 1972)]. This compound was synthesized by melting the stoichiometric amounts of Na, GeS_2, and S in alundum crucible encapsulated in the quartz ampoule. Single crystals were grown by the Bridgman

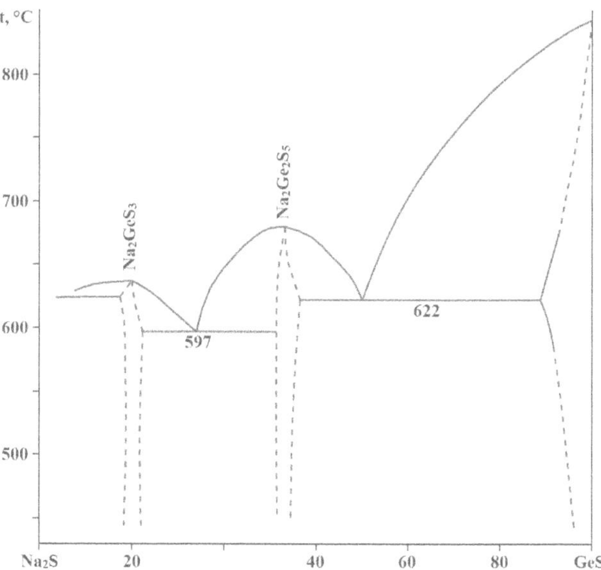

FIGURE 4.1 Part of the phase diagram of the GeS_2–Na_2S system. (From Mikhaylo, I.L. et al., *Zhurn. neorgan. khimii*, **34**(9), 2319, 1989.)

method (Mikaylo et al. 1989, 1990). This compound is moisture sensitive.

$Na_2Ge_2S_5$ melts congruently at 680°C [at 686°C with melting enthalpy 60.9 kJ·M^{-1} (Mikaylo et al. 1990); at 622°C ± 5°C (Olivier-Fourcade et al. 1971, 1972)] and crystallizes in the orthorhombic structure with the lattice parameters a = 1282.0 ± 0.5, b = 1285.2 ± 0.2, and c = 1046.3 ± 0.9 pm (Mikhaylo et al. 1989) [a = 1284.7 ± 0.3, b = 1290.1 ± 0.9, c = 1047.6 ± 0.2 pm, and the calculated and experimental densities of 2.71 and 2.69 g·cm^{-3}, respectively (Philippot and Ribes 1971; Olivier-Fourcade et al. 1971, 1972; Ribes et al. 1973)]. Yellow crystals of Na_2GeS_3 and $Na_2Ge_2S_5$ were obtained by fusing Na, S, and GeS into alundum crucibles placed in evacuated quartz ampoules (Mikhaylo et al. 1989) [white crystals of $Na_2Ge_2S_5$ were prepared by interaction of GeS_2 and Na_2S at 800°C (Philippot and Ribes 1971; Olivier-Fourcade et al. 1971, 1972; Ribes et al. 1973)]. $Na_2Ge_2S_5$ was also synthesized by melting the stoichiometric amounts of Na, GeS_2, and S in alundum crucible encapsulated in the quartz ampoule. Single crystals were grown by the Bridgman method (Mikaylo et al. 1989, 1990). This compound is moisture sensitive.

In the literature, there the data about three other ternary compounds that are formed in this system. $\mathbf{Na_4GeS_4}$ melts incongruently at 666°C ± 5°C and crystallizes in the monoclinic structure with the lattice parameters a = 1981, b = 2925, c = 1096 pm, β = 106°, and the calculated and experimental densities of 2.36 and 2.39 or 2.55 g·cm^{-3}, respectively (Maurin and Ribes 1966; Olivier-Fourcade et al. 1971, 1972).

$\mathbf{Na_6Ge_2S_7}$ also crystallizes in the monoclinic structure with the lattice parameters a = 909.4, b = 1043.7, c = 1546.4 pm, β = 109.49°, and the calculated and experimental densities of 2.69 and 2.70 g·cm^{-3}, respectively (Jumas et al. 1974). It was obtained by interaction of GeS_2 and Na_2S at 650°C with the next slow cooling at a rate of 16°C·h^{-1}. All four compounds are highly hygroscopic.

$\mathbf{Na_8GeS_6}$ has phase transformations within the interval from room temperature to 800°C (Putkaradze and Nanobashvili 1966). It was synthesized by the interaction of aqueous solutions of GeS_2 and Na_2S with the next extraction with acetone and drying in an inert gas atmosphere.

To obtain the glasses in the GeS_2–Na_2S system, 2–3 g of sulfides were mixed in the correct proportions and put in a vitreous carbon crucible (Barrau et al. 1980a; Ribes et al. 1980; Souquet et al. 1981). This crucible were placed inside the silica tube and sealed under vacuum. After melting at a temperature of 700°C–1000°C, the orangish-brown glasses were obtained by cooling down to room temperature or by immersing the silica tube in cold water. The obtained glasses contained up to 60 mol% Na_2S, were very hygroscopic and had to be handled in a dry box. The glass transition temperature for these glasses is within the interval 265°C–278°C.

4.4 Germanium–Potassium–Sulfur

Four ternary compounds, $\mathbf{K_2GeS_3}$, $\mathbf{K_4Ge_4S_{10}}$ ($\mathbf{K_2Ge_2S_5}$), $\mathbf{K_6Ge_2S_6}$, and $\mathbf{K_8GeS_6}$ are formed in the Ge–K–S system. K_2GeS_3 melts congruently at 869°C with melting enthaply 31.9 kJ·M^{-1} (Mikaylo et al. 1990). It was synthesized by melting the stoichiometric amounts of K, GeS_2, and S in alundum crucible encapsulated in the quartz ampoule. Single crystals were grown by the Bridgman method.

$K_4Ge_4S_{10}$ melts at 725°C with melting enthalpy 54.6 kJ·M^{-1} (Mikaylo et al. 1990) and crystallizes in the monoclinic structure with the lattice parameters a = 1516.1 ± 0.3, b = 1519.8 ± 0.2, c = 876.0 ± 0.2 pm, β = 105.36 ± 0.03°, and a calculated density of 2.62 g·cm^{-3} (Klepp and Fabian 1999b). For the preparation of the title compound, stoichiometric amounts of the components were intimately mixed in an Ar-filled glove box and sealed into evacuated silica ampoule. At the beginning of the thermal treatment the sample was kept for 3 days at temperatures between 100°C and 200°C, and then gradually heated to 950°C. After annealing for 5 days, the sample was allowed to reach ambient temperature at a constant cooling rate of 2°C·h^{-1}. The reaction product was of homogeneous appearance. The crushed sample was transparent and virtually colorless. Since it was sensitive to humidity, the sample was handled under inert conditions.

This compound could be also prepared by the interaction of hydrated GeO_2 with H_2S and CH_3COOK (Willard and Zuehlke 1943). The addition of acetone caused the immediate separation of two layers. The lower aqueous layer consisted of a very concentration solution of $K_4Ge_4S_{10}$. The addition of absolute EtOH to this concentrated solution caused the immediate deposition of fine crystalline plates of this compound. The title compound was also synthesized by melting the stoichiometric amounts of K, GeS_2, and S in alundum crucible encapsulated in the quartz ampoule. Its single crystals were grown by the Bridgman method.

$K_6Ge_2S_6$ also crystallizes in the monoclinic structure with the lattice parameters a = 870.5 ± 0.4, b = 1234.6 ± 0.6, c = 822.4 ± 0.4 pm, β = 116.62 ± 0.05°, and the calculated and experimental densities of 2.41 and 2.31 g·cm^{-3}, respectively (Eisenmann et al. 1984b). To obtain the title compound, the elements were weighed in a stoichiometric ratio under moisture and oxygen-free Ar atmosphere in a pre-dried at 150°C quartz ampoule. To compensate the loss of the reaction with the vessel material, potassium was taken in excess of about 10 mass%. The ampoule was melted under vacuum and in Ar-filled corundum tube with a temperature gradient of 50°C·h^{-1} to 650°C and then the furnace was switched off. The transparent colorless crystals were obtained. In the moist air, the decomposition of $K_6Ge_2S_6$ begins immediately, recognizable by the smell of H_2S.

K_8GeS_6 crystallizes in the cubic structure and has phase transformations within the interval from room temperature to 800°C (Putkaradze and Nanobashvili 1966). It was synthesized by the interaction of aqueous solutions of GeS_2 and K_2S with the next extraction with acetone and drying in an inert gas atmosphere.

4.5 Germanium–Rubidium–Sulfur

$\mathbf{GeS_2}$–$\mathbf{Rb_2S}$: Three ternary compounds, $\mathbf{Rb_4Ge_2S_6}$, $\mathbf{Rb_4Ge_4S_{10}}$, and $\mathbf{Rb_8GeS_6}$, are formed in this system. $Rb_4Ge_2S_6$ crystallizes in the monoclinic structure with the lattice parameters a = 1343 ± 5, b = 689.8 ± 0.3, c = 1115 ± 3 pm, and β = 135.06 ± 0.09° (Preishuber-Pflügl und Klepp 2003a). Colorless single crystals of this compound were obtained by reacting an intimate mixture of Rb_2S (875.8 mg), Ge (209.5 mg), and S (139 mg) (approximate molar ratio 3:2:3) in a sealed and evacuated silica tube. The sample was gradually heated to 800°C, held at this temperature for 12 h and allowed to cool to ambient temperature at a controlled rate of 4°C·h^{-1}.

$Rb_4Ge_4S_{10}$ also crystallizes in the monoclinic structure with the lattice parameters a = 1528.2 ± 0.7, b = 1534.1 ± 0.7, c = 906.1 ± 0.4 pm, β = 106.10 ± 0.03°, and a calculated density of 3.11 g·cm^{-3} (Klepp and Fabian 1999b). It was prepared in the same way as $K_4Ge_4S_{10}$ was synthesized.

Rb_8GeS_6 has phase transformations within the interval from room temperature to 800°C (Putkaradze and Nanobashvili 1966). It was synthesized by the interaction of aqueous solutions of GeS_2 and Rb_2S with the next extraction with acetone and drying in an inert gas atmosphere.

4.6 Germanium–Cesium–Sulfur

GeS_2–Cs_2S: Two ternary compounds, **$Cs_4Ge_4S_{10}$** and **Cs_8GeS_6**, are formed in this system. $Cs_4Ge_4S_{10}$ crystallizes in the monoclinic structure with the lattice parameters a = 1571.4 ± 0.3, b = 1585.8 ± 0.2, c = 949.1 ± 0.2 pm, and β = 106.74 ± 0.02° (Klepp and Zeitlinger 2000). The title compound was prepared by reacting Cs_2S (2.104 mM) with GeS_2 (4.19 mM) at 850°C in an evacuated silica ampoule, allowing the melt to attain ambient temperature at a controlled rate of 2°C·min^{-1}. The reaction product was of homogeneous appearance and consisted of polyhedral crystals of globular shape.

Cs_8GeS_6 has phase transformations within the interval from room temperature to 800°C (Putkaradze and Nanobashvili 1966). It was synthesized by the interaction of aqueous solutions of GeS_2 and Rb_2S with the next extraction with acetone and drying in an inert gas atmosphere.

One more ternary compound, **$Cs_4Ge_2S_8$**, is formed in the Ge–Cs–S system (Wu et al. 2003). It crystallizes in the monoclinic structure with the lattice parameters a = 1472.1 ± 0.2, b = 736.4 ± 0.1, c = 982.0 ± 0.1 pm, β = 122.43 ± 0.01°, and a calculated density of 3.450 g·cm^{-3}. It was synthesized from a mixture of Cs_2S_3, Ge, and S (molar ratio 2:2:7). The starting materials were thoroughly mixed and loaded into a silica tube in a N_2-filled glove box. After evacuating to 0.1 Pa the tube was flame-sealed, placed in a furnace and heated to 580°C within 24 h. After 3 d, the tube was cooled at a rate of 2°C·h^{-3} 100°C, followed by rapid cooling to room temperature. The resulting product was washed with dry Dimethylformamide (DMF) and diethyl ether and the residue dried in vacuum. The product consisted of orange platelet-like crystals of the title compound which were slightly contaminated with transparent colorless crystals of Cs_2SO_4. The obtained crystals of $Cs_4Ge_2S_8$ are stable in air for several weeks.

4.7 Germanium–Copper–Sulfur

GeS–Cu_2S: The phase diagram of this system, constructed through DTA, XRD, metallography, and measuring of microhardness, is shown in Figure 4.2 (Dovletov et al. 1977). The eutectics contain 50 and 75 mol% GeS and crystallize at 637°C and 550°C, respectively. The solubility of Cu_2S in GeS and GeS in Cu_2S reaches up 10 and 9 mol%, respectively. The immiscibility region within the interval of 60–80 mol% Cu_2S exists in this system at 725°C. The **$Cu_4Ge_3S_5$** ternary compound, which melts congruently at 675°C and crystallizes in the tetragonal structure with the lattice parameters a = 530 and c = 1048 pm, is formed in the GeS–Cu_2S system.

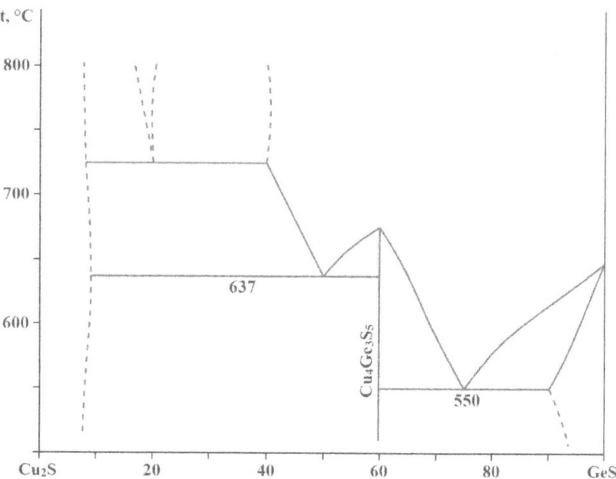

FIGURE 4.2 Phase diagram of the GeS–Cu_2S system. (From Dovletov, K., et al., *Izv. AN SSSR. Neorgan. mater.*, **13**(6), 1092, 1977.)

GeS_2–Cu_2S: The phase diagram of this system, constructed through DTA and XRD, is given in Figure 4.3 (Khanafer et al. 1973). The eutectic contains 5 mol% Cu_2S and crystallizes at 815°C. The thermal effects at 940°C correspond to the decomposition of the phase, stable in the interval 815°C–920°C, and at 556°C to the phase transformation of GeS_2. Mutual solubility of the starting components was not detected.

Four more compounds, **Cu_2GeS_3**, **Cu_4GeS_4**, **$Cu_5Ge_{0.5}S_4$**, and **Cu_8GeS_6**, are formed in the Ge–Cu–S ternary system. Cu_2GeS_3 melts incongruently at 940°C (Khanafer et al. 1973, 1974a) [at 933°C (Rivet et al. 1963; Rivet 1965); at 948°C (Palatnik et al. 1961a); at 955°C (Averkieva et al. 1964); at 956°C (Balanevskaya et al. 1966)] and has a polymorphic transition at 670°C (Rivet et al. 1963; Rivet 1965). α-Cu_2GeS_3 crystallizes in the tetragonal structure with lattice parameters a = 523.6 ± 0.5 and c = 521.9 ± 0.5 pm, an energy gap of 0.3 eV, and the calculated and experimental densities of 4.43 and 4.46 ± 0.4 g·cm^{-3}, respectively (Rivet et al. 1963; Rivet 1965; Khanafer et al. 1973, 1974a) [a = 531.6 and c = 520.5 pm (Averkieva et al. 1964); a = 532.0 and c = 1041 pm and the calculated and experimental densities of 4.43 and 4.34 g·cm^{-3}, respectively (Hahn et al. 1966)]. The lattice of this compound could be also described by the next parameters of the monoclinic structure: a = 746.4 ± 0.5, b = 2238 ± 1, c = 1063 ± 1 pm, and β = 91°52′ (Khanafer et al. 1973, 1974a). β-Cu_2GeS_3 crystallizes in the cubic structure with lattice parameter a = 531.7 pm [a = 530 pm (Palatnik et al. 1961a,b)], and the calculated and experimental densities of 4.36 and 4.45 ± 0.04 g·cm^{-3} [4.39 and 4.36 g cm^{-3} (Palatnik et al. 1961a)], respectively (Rivet et al. 1963; Rivet 1965). The energy gap of this modification is 1.1 eV (Balanevskaya et al. 1966).

According to the data of Parthé and Garin (1971), Cu_2GeS_3 crystallizes in the orthorhombic structure with the lattice

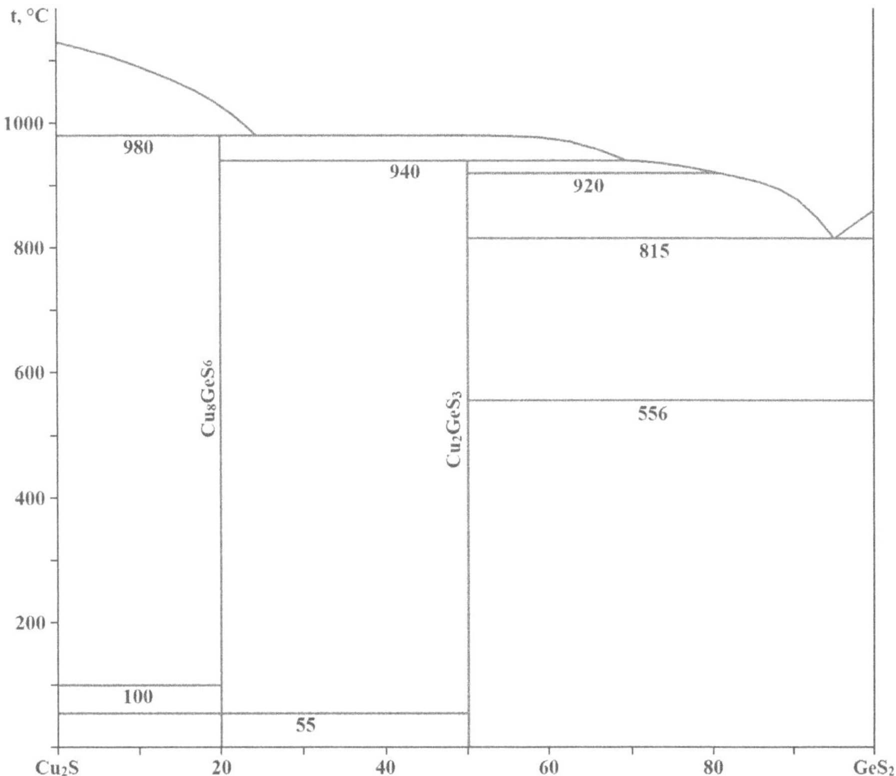

FIGURE 4.3 Phase diagram of the GeS_2–Cu_2S system. (From Khanafer, M., et al., *Bull. Soc. chim. France*, (3), Pt.1, 859, 1973.)

parameters a = 1132.1, b = 376.6, c = 521 pm, and Chalbaud de et al. (1997) noted that this compound crystallizes in the monoclinic structure with the lattice parameters $a = 644.9 \pm 0.2$, $b = 1131.9 \pm 0.3$, $c = 642.8 \pm 0.2$ pm, $\beta = 108.37 \pm 0.02°$, and a calculated density of 4.413 g·cm^{-3}.

The standard thermodynamic functions of formation and standard entropy of the Cu_2GeS_3 compound were calculated from the EMF data by Alverdiev et al. (2018): $\Delta G^0_{298} = -211.3 \pm 2.4$ kJ·mol^{-1}, $\Delta H^0_{298} = -213.7 \pm 2.2$ kJ·mol^{-1}, and $S^0_{298} = 190.3 \pm 5.5$ J·(mol·K)$^{-1}$. According to the thermodynamic simulation, this compound dissociates at the melting temperature forming Cu_2S and GeS_2 binary compounds (Glazov et al. 1978).

Cu_2GeS_3 was prepared from stoichiometric mixtures of the elements by heating (35 days) in thick evacuated quartz tubes (Parthé and Garin 1971). To avoid explosions, the temperature was initially increased very slowly (100°C per day). The chemical reaction was complete after 2 h at 900°C. However, to ensure order of the Cu and Ge atoms, the samples were subsequently post-annealed at 500°C for a week. By means of this method, it was possible to obtain not only the powders, but also small single crystals platelets.

This compound was also obtained by solid state reaction of the elements in stoichiometric proportions in an evacuated and sealed quartz ampoule (Chalbaud de et al. 1997). The ampoule was heated gradually to 600°C where it was kept for 24 h before being brought to 850°C for 2 days. Subsequently, the ampoule was slowly cooled to 600°C. The reacted mixture was annealed at this temperature for 30 days and then allowed to cool to room temperature by turning off the power of the furnace.

The synthesis of Cu_2GeS_3 was also carried out from the elements, placed in a silica ampoule, sealed under vacuum (Khanafer et al. 1974a). An experience has shown that it is necessary, in order to ensure homogeneous preparations, to stand the silica ampoules at a temperature above about 100°C of the melting temperature of the prepared product. The single crystals were obtained either by a multi-week annealing of the products sprayed or not, at a temperature below ca. 50°C of the melting point or by using a temperature gradient of 50°C between the two ends of the ampoule.

To prepare this compound, Cu, Ge, and Se were weighed in stoichiometric mixtures and put in a zirconia grinding jar with zirconia balls (Morihama et al. 2014). A ball-to-powder weight ratio was maintained at 4:1. The milling was conducted in a planetary ball mill for 20 min in a nitrogen atmosphere. The mixed powders were post-heated at 600°C for 30 min in the same atmosphere.

Cu_4GeS_4 is a semiconductor and crystallizes in the monoclinic structure with the lattice parameters $a = 979.0 \pm 0.2$, $b = 1320.5 \pm 0.2$, $c = 994.2 \pm 0.3$ pm, $\beta = 100.90 \pm 0.02°$, and a calculated density of 4.789 g·cm^{-3} (Chen et al. 1999a). It was synthesized using a stoichiometric mixture of Cu_2S (6.67 mM) and GeS_2 (3.33 mM). The sample was ground in an agate mortar and pressed into pellets. The pellets were introduced into a silica tube and sealed at a pressure of less than 0.13 Pa. The tube was held at 700°C for 1 week, then heated gradually to 830°C, where they were kept for 1 month, and then cooled at a rate of 53°C·h^{-1} to 690°C and quenched in cold water. The gray crystals with metallic luster were found in the tube, and they had the shape of prisms. The compound appears to be relatively stable in air and water.

$Cu_5Ge_{0.5}S_4$ (mineral calvertite) is metastable and crystallizes in the cubic structure with the lattice parameter a = 533.7 ± 0.1 pm and a calculated density of 5.239 g·cm^{-3} (Jambor et al. 2007; Piilonen et al. 2008).

Cu_8GeS_6 melts incongruently at 980°C and has a polymorphic transition at 55°C (Khanafer et al. 1973; Aliyeva et al. 2014) [at 92°C (Aliev et al. 1989)]. α-Cu_8GeS_6 crystallizes in the orthorhombic structure with the lattice parameters a = 990.73 ± 0.03, b = 990.73 ± 0.03, and c = 987.03 ± 0.04 pm and a calculated density of 5.358 g·cm^{-3} (Onoda et al. 1999a) [a = 704.45 ± 0.03, b = 696.61 ± 0.03, and c = 986.99 ± 0.05 pm (Ishii et al. 1999; Onoda et al. 2000b; Tomm et al. 2008); in a monoclinic structure representing a slightly distorted cubic structure with the lattice parameters $a = b = c$ = 990 pm, β = 90°, and the calculated and experimental densities of 5.28 and 5.24 ± 0.01 g·cm^{-3}, respectively (Khanafer et al. 1973, 1974a)]. β-Cu_8GeS_6 crystallizes in the cubic structure with the lattice parameter a = 993.5 ± 0.3 pm (Kuhs et al. 1979) [a = 990 pm and the calculated and experimental densities of 5.28 and 5.08 g·cm^{-3} (Hahn et al. 1965); a = 990.9 ± 0.5 pm (Khanafer et al. 1973); a = 995.67 ± 0.02 pm and a calculated density of 7.1935 ± 0.0004 g·cm^{-3} for $Cu_9GeS_{5.421}$ composition (Gulay et al. 2002c)]. The energy gap of this modification is 0.10 eV (Khanafer et al. 1973). The partial and integral thermodynamic functions of α- and β-Cu_8GeS_6 as well as heat and entropy of the polymorphic transformation were calculated based on EMF measurements (Aliyeva et al. 2014): ΔG^0_{298} = −438.9 ± 2.5 kJ·M^{-1}, ΔH^0_{298} = −425.9 ± 4.2 kJ·M^{-1} and S^0_{298} = 536.3 ± 13.1 J·(M·K)$^{-1}$ for α-Cu_8GeS_6, ΔG^0_{400} = −445.3 ± 3.1 kJ·M^{-1}, ΔH^0_{400} = −420.8 ± 5.6 kJ·M^{-1} and S^0_{400} = 552.1 ± 15.8 J·(M·K)$^{-1}$ for β-Cu_8GeS_6, $\Delta H_{phase\ trans.}$ = 5.1 kJ·M^{-1} and $\Delta S_{phase\ trans.}$ = 15.5 J·(M·K)$^{-1}$ [$\Delta H_{phase\ trans.}$ = 8.8 ± 0.4 kJ·M^{-1} (Aliev et al. 1989)].

Cu_8GeS_6 could be prepared by reacting stoichiometric amounts of Cu_2S, Ge, and S in an evacuated sealed quartz ampoule for 4 days at 500°C (Ishii et al. 1999) or by reacting stoichiometric amounts of high-purity elements in an evacuated sealed quartz ampoule for about 6 days at temperatures of 600°C–700°C (Kuhs et al. 1979) [for 24 h at 980°C (Tomm et al. 2008)]. The specimen of this compound was also obtained as a by-product of an experiment attempting to prepare the Cu_4GeS_4. After heating in an evacuated silica tube at 830°C for 1 month and cooling, a small amount of triangular plate-like crystals were observed as the minor product, while the principal product was the gray prismatic crystals of Cu_4GeS_4. The preparation of the $Cu_9GeS_{5.421}$ single crystals was carried out using the recrystallization process at high temperature (Gulay et al. 2002c). High purity elements were used and the synthesis was realized in an evacuated quartz ampoule. A two-zone furnace inclined by 30° was used. Iodine (5 mg·cm^{-3}) sealed in a molybdenic glass capillary was used as a stimulator of the process. After the evacuation and sealing, the capillary was unsealed and I_2 was sublimated into the lower part of the quartz ampoule. After that, the empty reservoir was melted-off from the main part of the ampoule. An independent temperature control of both furnace zones was used. The heating rate was 20°C·h^{-1}. The synthesis temperatures were 700°C for lower zone and 400°C for the upper one. The duration of this stage was 24 h. Then the temperatures were increased simultaneously to 780°C and 1000°C, respectively. After 6 h, the upper zone was cooled (10°C·h^{-1}) to the growing temperature of 730°C. After 170 h of the preparation, both zones were simultaneously cooled (10°C·h^{-1}) to 400°C. The cooling to room temperature was carried out by turning off the furnace. The obtained single crystals exhibited a black color with metallic luster.

The isothermal sections of the Ge–Cu–S ternary system at 700°C and 600°C are shown in Figure 4.4 (Viaene et al. 1968). It is shown that three liquid regions exist at 700°C. Two ternary eutectics, E_1 and E_2, are situated into one of the

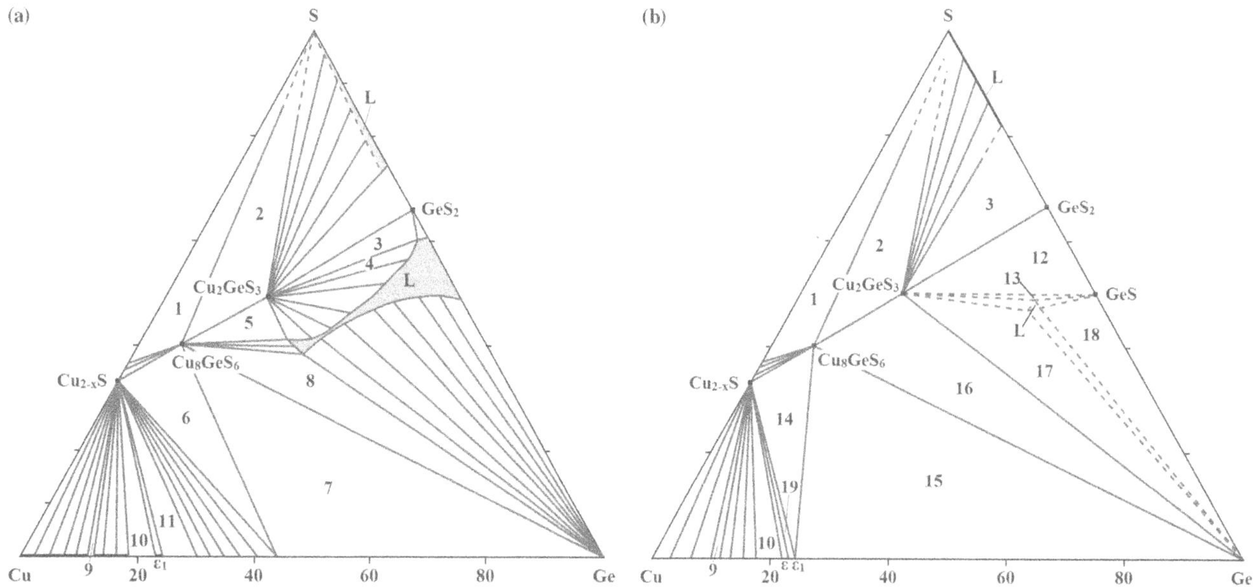

FIGURE 4.4 Isothermal sections of the Ge–Cu–S ternary system at (a) 700°C and (b) 600°C: 1 and 6, $Cu_{2-x}S + \tau_2$ + L; 2 and 5, $\tau_1 + \tau_2$ + L; 3 and 4, τ_1 + L + GeS_2; 7 and 8, τ_2 + L + Ge; 9, $Cu_{2-x}S$ + (Cu_5Ge) + Cu; 10, $Cu_{2-x}S$ + (Cu_5Ge) + ε_1; 11, $Cu_{2-x}S + \varepsilon_1$ + L; 12, τ_1 + GeS_2 + GeS; 13, τ_1 + GeS + L; 14, $Cu_{2-x}S + \tau_2 + \varepsilon$; 15, $\tau_2 + \varepsilon$ + Ge; 16, $\tau_1 + \tau_2$ + Ge; 17, τ_1 + Ge + L; 18, GeS + Ge + L; and 19, $\varepsilon + \varepsilon_1 + Cu_{2-x}S$ (ε and ε_1 are the binary phases in the Cu–Ge system). (From Viaene, W., et al., *C. r. Acad. sci. Sér. D*, **266**(14), 1451, 1968.)

liquid regions. The eutectic E_1 ($L \Leftrightarrow \tau_1 + \tau_2$ + Ge, where τ_1 = Cu_2GeSe_3 and τ_2 = Cu_8GeSe_6) crystallizes within the interval from 600°C to 700°C and the eutectic E_2 ($L \Leftrightarrow \tau_1$ + GeS + Ge) crystallizes below 600°C.

According to the data of Voroshilov et al. (1986) another ternary compound, $\mathbf{Cu_6Ge_2S_6}$, is formed in the Ge–Cu–S system. This compound has a gray metallic color, melts incongruently at 704°C and is characterized by an experimental density of 4.80 g·cm^{-3}.

4.8 Germanium–Silver–Sulfur

$\mathbf{GeS_2–Ag_2S}$: The most reliable phase diagram of this system, constructed through DTA and XRD, is presented in Figure 4.5 (Olekseyuk et al. 2010a). Three compounds, $\mathbf{Ag_2GeS_3}$, $\mathbf{Ag_8GeS_6}$, and $\mathbf{Ag_{10}Ge_3S_{11}}$, are formed in the GeS_2–Ag_2S system [the existence of these compounds was also confirmed at 400°C (Parasyuk et al. 2002)]. Three eutectics contain 7, 47, and 68 mol% GeS_2 and crystallize at 805°C, 639°C, and 632°C, respectively. The phase transformations of Ag_2S take place at 176°C and 571°C. The phase diagram of this system was also constructed by Salaeva et al. (1988) and Chbani et al. (1992) and the part of this phase diagram was determined by Potenza (1961), and Gorochov and Flahaut (1968b).

Ag_2GeS_3 melts congruently at 649°C and has polymorphic transition at 306°C (Olekseyuk et al. 2010a) [at 648°C and 306°C, respectively (Mikolaichuk and Moroz 2010)]. This compound is a phase of variable composition; when it contains by Salaeva stoichiometric Ge, the polymorphic transition becomes below room temperature (Mikolaichuk and Moroz 2010). One of the modification of Ag_2GeS_3 crystallizes in the orthorhombic structure with the lattice parameters a = 1178.95 ± 0.09, b = 707.51 ± 0.05, and c = 634.20 ± 0.05 pm and a calculated density of 4.828 ± 0.001 g·cm^{-3} (Olekseyuk et al. 2010a) [a = 1179.1, b = 707.9,

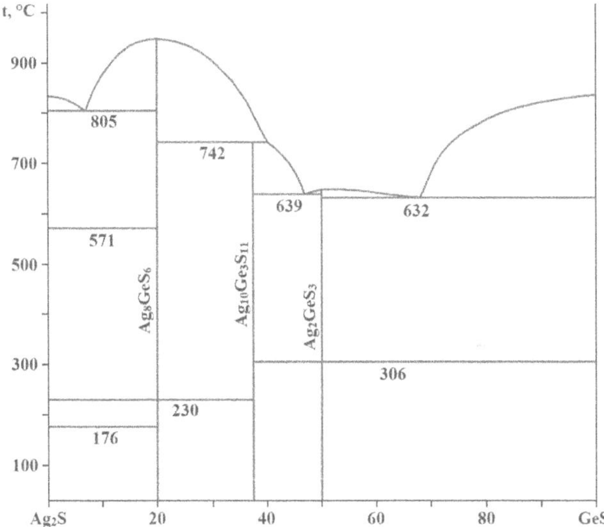

FIGURE 4.5 Phase diagram of the GeS_2–Ag_2S system. (From Olekseyuk, I.D., et al., *Nauk. visnyk Volyns'k. Univ. im. Lesi Ukrainky. Ser. Khim. nauky*, (16), 25, 2010.)

c = 634.4 pm, and the calculated and experimental densities of 4.89 and 4.82 g·cm^{-3}, respectively (Nagel and Range 1978)]. According to the data of Nagel and Range (1978), this compound decomposes rapidly at the temperatures higher than 650°C.

Ag_8GeS_6 melts congruently at 948°C and has a polymorphic transition at 230°C [at 225°C (Kuhs et al. 1979)] (Olekseyuk et al. 2010a) [at 950°C–955°C and 223°C, respectively (Gorochov and Flahaut 1967, 1968b); at 940°C and 200°C, respectively (Salaeva et al. 1988); at 950°C and 215°C, respectively (Chbani et al. 1992); at 950°C and 222°C, respectively (Mikolaichuk and Moroz 2010)]. α-Ag_8GeS_6 crystallizes in the orthorhombic structure with the lattice parameters a = 1516.2 ± 0.4, b = 751.3 ± 0.3, and c = 1058.5 ± 0.4 pm (Bilousov et al. 2006) [a = 1514.9 ± 0.1, b = 747.6 ± 0.2, c = 1058.9 ± 0.1 pm, and the calculated and experimental densities of 6.25 and 6.21 g·cm^{-3}, respectively (Eulenberger 1977a; Chbani et al. 1992)]. β-Ag_8GeS_6 crystallizes in the cubic structure with the lattice parameter a = 1170 pm at 240°C [a = 2119 pm (Hahn et al. 1965)] and the calculated and experimental densities of 6.30 and 6.21 g·cm^{-3}, respectively (Hahn et al. 1965; Gorochov and Flahaut 1967, 1968b). This compound was prepared from a stoichiometric mixture of Ag, Ge, and S under vacuum (Eulenberger 1977). The mixture was slowly heated in a sealed quartz ampoule up to 860°C within 5 days, kept at this temperature for 6 days and cooled to room temperature during 2 days. Single crystals could be isolated from the crystalline dark grey reaction product.

$Ag_{10}Ge_3S_{11}$ melts incongruently at 742°C (Olekseyuk et al. 2010a) and crystallizes in the monoclinic structure with the lattice parameters a = 2624.4 ± 0.4, b = 650.20 ± 0.05, c = 2508.3 ± 0.4 pm, β = 109.910 ± 0.001°, and a calculated density of 5.434 ± 0.002 g·cm^{-3} (Fedorchuk et al. 2013). The synthesis of this compound was performed in two stages (Fedorchuk et al. 2013). At the first stage, the ampoule was heated in the oxygen-gas burner flame to complete bonding of the elementary sulfur. The second stage included rapid heating (50°C·h^{-1}) of the ampoule in a furnace to 900°C and a 6-h homogenization of the melt with periodic vibration mixing. This was followed by the cooling (10°C·h^{-1}) to 400°C, then annealing for 250 h and cooling the alloy with the furnace turned off.

Cambi and Elli (1961) noted that another ternary compound, $\mathbf{Ag_6Ge_2S_7}$, which melts at 675°C, is formed in the GeS_2–Ag_2S.

According to the data of Olekseyuk et al. (2010a), the glasses are formed in the GeS_2–Ag_2S within the interval 0–52 mol% Ag_2S after quenching the melt from 1000°C [0–50 mol% Ag_2S (Souquet et al. 1981; Chbani et al. 1992); 0–55 mol% Ag_2S (Robinel et al. 1983)]. To prepare the glasses in this system, GeS_2 and Ag_2S were mixed in appropriate proportions and put in a vitreous carbon crucible placed inside the silica tube and sealed under vacuum (Souquet et al. 1981; Robinel et al. 1983). After heating at 700°C–900°C for 1–2 h and quenching (in air or in cold water), the tube was then annealed at a temperature 20°C–30°C lower than the glass transition temperature. The reddish-dark glasses were obtained.

Ge–Ag–S: This ternary system includes the next quasibinary sections: GeS_2–Ag_2S, Ag_2S–Ge, Ag_8GeS_6–Ge, and Ag_8GeS_6–S (Salaeva et al. 1988; Chbani et al. 1992). Mikolaichuk and Moroz (2010) noted that the $\mathbf{Ag_2Ge_5S_{10}}$ ternary compound exists in the system. The liquidus projection of the GeS_2–Ag_8GeS_6–Ag–Ge subsystem of the Ge–Ag–S

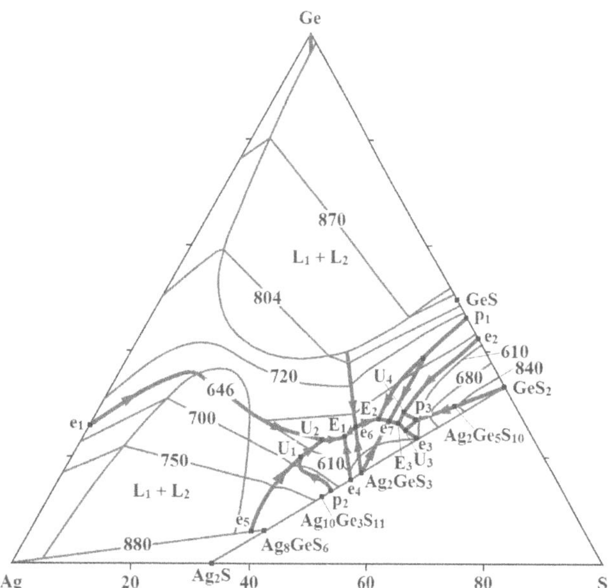

FIGURE 4.6 Liquidus projection of the GeS_2–Ag_8GeS_6–Ag–Ge subsystem of the Ge–Ag–S ternary system. (From Mikolaichuk, A.G., and Moroz, N.V., *Russ. J. Inorg. Chem.*, **55**(1), 87, 2010.)

ternary system (Figure 4.6) consists from eight fields of primary crystallization of Ge, Ag, binary and ternary compounds as well as 17 invariant points (Mikolaichuk and Moroz 2010). The next invariant equilibria exist in the system: U_1 (662°C) – L + α-Ag_8GeS_6 ⇔ Ag + $Ag_{10}Ge_3S_{11}$; U_2 (572°C) – L + Ag ⇔ Ge + $Ag_{10}Ge_3S_{11}$; U_3 (480°C) – L + GeS_2 ⇔ Ag_2GeS_3 + $Ag_2Ge_5S_{10}$; U_4 (492°C) – L + GeS_2 ⇔ GeS + $Ag_2Ge_5S_{10}$; E_1 (464°C) – L + ⇔ $Ag_{10}Ge_3S_{11}$ + Ag_2GeS_3 + Ge; E_2 (407°C) – L + ⇔ Ag_2GeS_3 + GeS + Ge; and E_3 (445°C) – L + ⇔ Ag_2GeS_3 + GeS + $Ag_2Ge_5S_{10}$. The Ge and Ag fields of primary crystallization occupy a considerable part of the liquidus surface and have broad immiscibility regions. The immiscibility region originated from the Ag–Ag_8GeS_6 section is superimposed on the Ag and Ge fields of primary crystallization, crossing the line of their join primary crystallization. In the GeS_2 field of primary crystallization the $Ag_2Ge_5S_{10}$ ternary compound is formed via the reaction of the melt with GeS_2 at 595°C. The liquidus projection of the GeS_2–Ag_2S–S subsystem of the Ge–Ag–S ternary system consists of the field of primary crystallization of Ag_2S, β-Ag_8GeS_6, GeS_2, and S, moreover, the primary crystallization of sulfur is degenerate (Salaeva et al. 1988). The main part of the phase primary crystallizations occurs under immiscibility region. There are two ternary eutectics in this subsystem which are situated near sulfur corner and have the temperature of invariant crystallization ca. 100°C.

The liquidus surface of the Ge–Ag–S ternary system was also constructed by (Movsum-zade et al. 1989). It was noted that two another compounds, $\mathbf{AgGe_2S_4}$ and $\mathbf{Ag_3GeS_2}$, are formed in this system. The existence of these compounds was not confirmed in further studies by other authors.

The glass forming region in the GeS–GeS_2–Ag_2S subsystem is shown in Figure 4.7 (Kawamoto et al. 1973). The glass transition temperature for these glasses is within the interval 228°C–343°C (Feltz and Thieme 1974). When quenched, the ingots of the GeS_2–Ag_8GeS_6–Ag–Ge subsystem yield

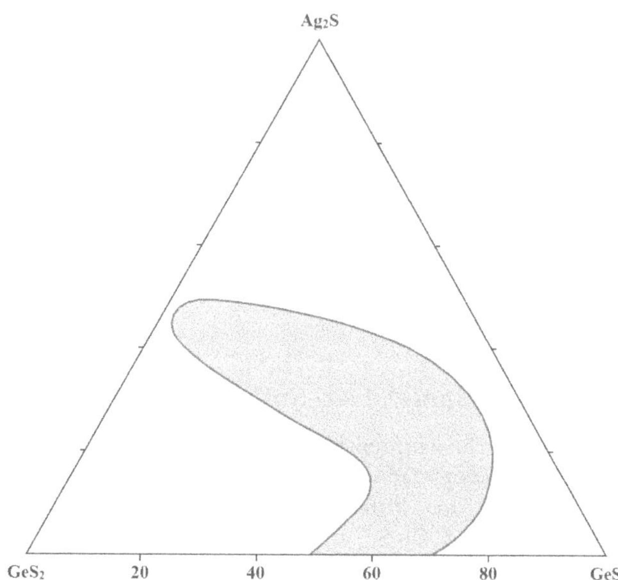

FIGURE 4.7 Glass forming region in the GeS–GeS_2–Ag_2S subsystem. (From Kawamoto, Y., et al., *J. Am. Ceram. Soc.*, **56**(5), 289, 1973.)

materials in a glassy crystalline state (Mikolaichuk and Moroz 2010).

4.9 Germanium–Magnesium–Sulfur

GeS_2–MgS: The $\mathbf{Mg_2GeS_4}$ ternary compound, which crystallizes in the orthorhombic structure with the lattice parameters a = 1277.3 ± 0.5, b = 743.5 ± 0.5, c = 600.5 ± 0.2 pm (Vincent and Perrault 1971) [a = 1284, b = 747, and c = 607 pm (Rocktäschel et al. 1964; Royen et al. 1965)], is formed in this system. This compound is unstable and when heated in air, quantitatively converted into Mg_2GeO_4. It was prepared from pure ground and mixed elements in stoichiometric proportions, then placed in a sealed silica ampoule under vacuum (Vincent and Perrault 1971). After a very slow increasing in temperature, the ampoule was kept at 750°C for 3 weeks and cooled very slowly (5°C·h) to room temperature. It was also prepared by the interaction of Mg_2Ge with sulfur within the temperature interval from 450°C to 650°C (Rocktäschel et al. 1964) or by the interaction of MgS and GeS_2 at 700°C–800°C (Royen et al. 1965).

4.10 Germanium–Calcium–Sulfur

GeS_2–CaS: The $\mathbf{Ca_2GeS_4}$ ternary compound, which crystallizes in the orthorhombic structure with the lattice parameters a = 1360.7 ± 0.3, b = 818.4 ± 0.6, c = 630.0 ± 0.6 pm, and the calculated and experimental densities of 2.66 and 2.58 g·cm^{-3}, respectively (Ribes et al. 1970) [a = 1364, b = 820, and c = 630 pm (Rocktäschel et al. 1964); a = 1361, b = 819, and c = 630 pm (Susa and Steinfink 1971b)], is formed in this system. This compound is unstable and when heated in air, quantitatively converted into Ca_2GeO_4 (Royen et al. 1965; Ribes 1969). Ca_2GeS_4 begins to decompose in a nitrogen atmosphere at 670°C (Ribes 1969). It was prepared by the interaction of

Ca_2Ge with sulfur within the temperature interval from 450°C to 650°C (Rocktäschel et al. 1964)

4.11 Germanium–Strontium–Sulfur

GeS_2–SrS: The **Sr_2GeS_4** ternary compound, which crystallizes in the monoclinic structure with the lattice parameters $a = 823.1 \pm 0.4$, $b = 672.9 \pm 0.3$, $c = 667.2 \pm 0.4$ pm, $\beta = 108.22 \pm 0.03°$, and the calculated and experimental densities of 3.56 and 3.51 g·cm^{-3}, respectively, is formed in this system (Maurin and Ribes 1967; Ribes et al. 1970). This compound is unstable and when heated in air, quantitatively converted into Sr_2GeO_4 (Royen et al. 1965; Ribes 1969). Ca_2GeS_4 begins to decompose in a nitrogen atmosphere at 700°C (Ribes 1969). It was prepared by the interaction of GeS_2 and SrS at 700°C–800°C (Royen et al. 1965; Ribes 1969).

4.12 Germanium–Barium–Sulfur

GeS_2–BaS: Several ternary compounds are formed in this system. **$BaGeS_3$** crystallizes in the orthorhombic structure with the lattice parameters $a = 572.8 \pm 0.5$, $b = 671.1 \pm 0.5$, $c = 1382.0 \pm 0.6$ pm, and the calculated and experimental densities of 3.83 and 3.80 g·cm^{-3}, respectively (Ribes and Maurin 1970). This compound begins to decompose at 605°C (Maurin and Ribes 1967). According to the data of Wu et al. (2015c), **$Ba_2Ge_2S_6$** crystallizes in the monoclinic structure with the lattice parameters $a = 1537.2 \pm 0.6$, $b = 576.11 \pm 0.17$, $c = 1342.0 \pm 0.4$ pm, and $\beta = 115.71 \pm 0.02°$. The composition of this compound correspond to the composition of precedent one. To prepare $Ba_2Ge_2S_6$, a mixture of BaS, Ge, Zn, and S was mixed in the molar ratio 1:1:1:3, and heated to 1050°C in 30 h, kept at 1050°C for 100 h, then cooled at a slow rate of 5°C·h^{-1} to 500°C, and finally cooled to room temperature naturally. Canary yellow crystals were found in the tube, which were stable in air for several weeks.

$BaGe_2S_5$ or **$Ba_2Ge_4S_{10}$** crystallizes in the cubic structure with the lattice parameter $a = 1489.9 \pm 0.2$ and the calculated and experimental densities of 3.55 and 3.48 g·cm^{-3}, respectively (Ribes and Maurin 1970; Ribes et al. 1973). To prepare this compound, a mixture of BaS and GeS_2 (molar ratio 1:2) was heated at 1250°C into a vacuum-sealed silica tube. After slow cooling, orange-colored single crystals were obtained, which slowly decompose in the ambient moisture. $BaGe_2S_5$ begins to decompose at 620°C (Maurin and Ribes 1967).

Ba_2GeS_4 crystallizes in the orthorhombic structure with the lattice parameters $a = 898.3 \pm 1.1$, $b = 687.5 \pm 0.9$, $c = 1222.1 \pm 0.1$ pm, and a calculated density of 4.185 g·cm^{-3} (Wu et al. 2015b) [$a = 895.9 \pm 0.5$, $b = 688.5 \pm 0.5$, $c = 1221.8 \pm 0.4$ pm, and the calculated and experimental densities of 4.19 and 4.15 g·cm^{-3}, respectively (Ribes et al. 1969; Ribes and Maurin 1970); $a = 896$, $b = 689$, and $c = 1223$ pm (Susa and Steinfink 1971b)]. This compound is unstable and when heated in air, quantitatively converted into Ba_2GeO_4 (Royen et al. 1965; Ribes 1969). It begins to decompose in a nitrogen atmosphere at 750°C (Maurin and Ribes 1967; Ribes 1969). To synthesize Ba_2GeS_4, a mixture of BaS and GeS_2 (molar ratio 2:1) was heated to 1050°C during 30 h, left for 72 h, cooled to 600°C (3°C·h^{-1}), and finally cooled to room temperature by switching off the furnace (Wu et al. 2015b). Many yellow block-shaped crystals were found in the ampoule, which are stable in air.

Ba_3GeS_5 also crystallizes in the orthorhombic structure with the lattice parameters $a = 1205.28 \pm 0.09$, $b = 954.97 \pm 0.07$, $c = 859.79 \pm 0.06$ pm, a calculated density of 4.329 g·cm^{-3}, and an energy gap of 3.0 eV (Pan et al. 2014). The synthesis was conducted in Ar-filled glove box or under vacuum. This compound was obtained from a reaction of BaS, $BaCl_2$, and GeS (molar ratio 4:3:1). The reactants were loaded into a graphite crucible and sealed in fused silica tube under vacuum. The mixture was firstly heated to 1000°C at a rate of 30°C·h^{-1} and homogenized at this temperature for 30 h, followed by a slowly cooling down to 400°C at a rate of 30°C·h^{-1}, then the furnace was powered off and the sample was cooled naturally down to ambient temperature. The excessive $BaCl_2$ flux was dissolved into the deionized water. The resulted product was composed of yellow crystals of the targeted compound in excellent crystal quality. Ba_3GeS_5 shows good thermal stability, no significant decomposition processes were identified below 800°C.

4.13 Germanium–Zinc–Sulfur

GeS_2–ZnS: The phase diagram of this system, constructed through DTA and XRD, is a eutectic type (Figure 4.8) (Dubrovin et al. 1989b). The eutectic composition and temperature are 18 mol% ZnS and 810°C ± 5°C. The mutual solubility of ZnS and GeS_2 is not higher than 1 mol%.

According to the data of Hahn and Lorent (1958), **Zn_2GeS_4** was obtained in this system. It crystallizes in the cubic structure of the sphalerite type with lattice parameter $a = 543.6$ pm and the calculated and experimental densities of 3.427 and 3.26 g·cm^{-3}, respectively. This compound was obtained by the heating of 2ZnS + GeS_2 mixture at 500°C–700°C, but later its existence was not confirmed (Kaldis et al. 1967; Dubrovin et al. 1989b). The ingots were annealed at 750°C and 790°C

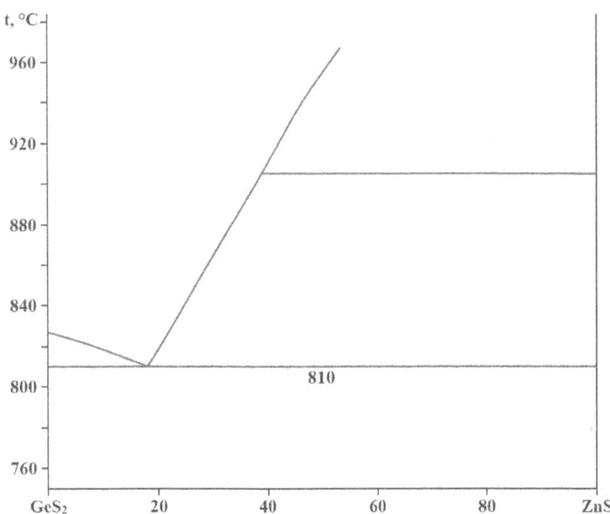

FIGURE 4.8 Phase diagram of the GeS_2–ZnS system. (From Dubrovin, I.V. et al., *Izv. AN SSSR. Neorgan. mater.*, **25**(8), 1386, 1989.)

for 100 and 80 h, respectively (Dubrovin et al. 1989b). Ternary compounds in the Ge–Zn–S system were not found (Kaldis et al. 1967).

4.14 Germanium–Cadmium–Sulfur

GeS–CdS: This section is a non-quasibinary section of Ge–Cd–S ternary system (Figure 4.9) (Odin et al. 1983b). CdS, α-Cd_4GeS_6, and Ge primary crystallize in this system. Thermal effects at 925°C correspond to a joint crystallization of CdS and α-Cd_4GeS_6. Invariant equilibrium L + CdS ⇔ α-Cd_4GeS_6 + Ge takes place at 725°C. Within the interval of 0.5–66.6 mol% CdS, crystallization ends in the ternary transition point at 625°C. The mutual solubility of GeS and CdS is not higher than 0.5 mol% (Odin et al. 1983b). The solubility of GeS in CdS reached 30 mol% when solid solutions were synthesized under 2–3 GPa and 600°C–700°C for 1–4 h (Kobayashi et al. 1979).

The $\mathbf{Cd_4GeS_5}$ compound, which melts incongruently at 605°C (Movsum-zade et al. 1987a) and crystallizes in the monoclinic structure with the lattice parameters a = 1230.3, b = 705.6, c = 1233.5 pm, β = 110°02′ and the calculated and experimental densities of 4.51 and 4.59 g·cm^{-3}, respectively (Nitsche 1964), was not found (Odin et al. 1983b). According to the data of Nitsche (1964), single crystals of this compound in the form of light-yellow polyhedra were grown from the vapor phase by iodine transport.

This system was investigated using DTA, metallography, XRD, and measuring of microhardness (Odin et al. 1983b; Movsum-zade et al. 1987a). The ingots were annealed at 540°C for 1200 h (Odin et al. 1983b) or at the temperatures 50°C–100°C lower than the eutectic temperature (Movsum-zade et al. 1987a).

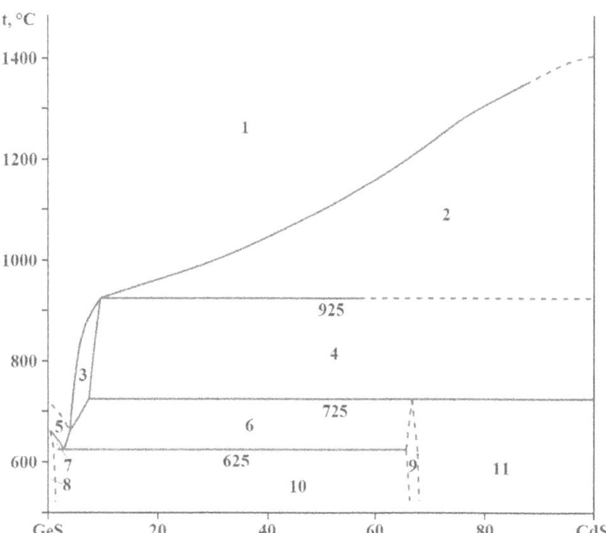

FIGURE 4.9 Phase relations in the GeS–CdS system: 1, L; 2, L + CdS; 3, L + α-Cd_4GeS_6; 4, L + CdS + α-Cd_4GeS_6; 5, L + Ge; 6, L + Ge + α-Cd_4GeS_6; 7, L + Ge + GeS; 8, Ge + GeS; 9, Ge + α-Cd_4GeS_6; 10, GeS + Ge + α-Cd_4GeS_6; and 11, α-Cd_4GeS_6 + Ge + CdS. (From Odin, I.N. et al., *Zhurn. neorgan. khimii*, **28**(9), 2362, 1983.)

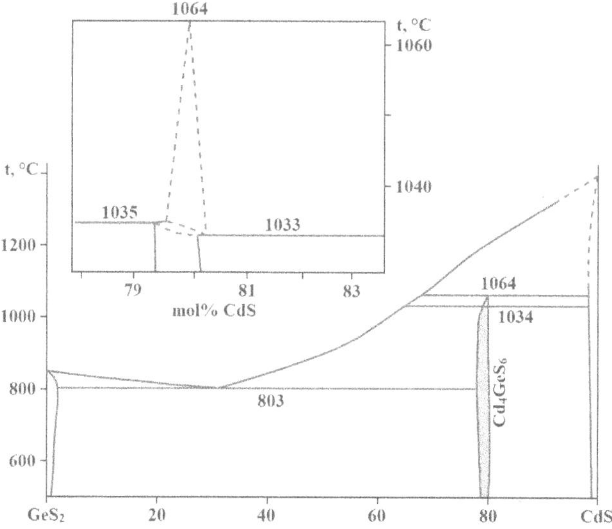

FIGURE 4.10 Phase diagram of the GeS_2–CdS system. (From Odin, I.N. et al., *Zhurn. neorgan. khimii*, **28**(9), 2362, 1983.)

GeS_2–CdS: The phase diagram of this system is shown in Figure 4.10 (Odin et al. 1983b). The peritectic composition is 68 mol% CdS, and eutectic composition and temperature are 31 ± 3 mol% CdS and 803 ± 5°C, respectively (Odin et al. 1983b) [20 mol% CdS, and 800°C (Barnier et al. 1990), 26.47 at% Ge and 700°C (Movsum-zade et al. 1987a)].

The $\mathbf{Cd_4GeS_6}$ ternary compound is formed in this system. It melts incongruently at 1064 ± 7°C (Odin et al. 1983b) [1020°C (Barnier et al. 1990), 775°C (Movsum-zade et al. 1987a)] and crystallizes in the monoclinic structure with the lattice parameters a = 1233.8 ± 0.2, b = 709.3 ± 0.2, c = 1238.1 ± 0.3 pm, and β = 110.097 ± 0.008° (Parasyuk et al. 2005) [a = 1232.6 ± 0.3, b = 708.8 ± 0.2, c = 1236.5 ± 0.3 pm, β = 110.08 ± 0.02°, and the calculated and experimental densities of 4.67 and 4.70 ± 0.05 g·cm^{-3}, respectively (Julien-Pouzol and Jaulmes 1995); a = 1239.5, b = 710.7, c = 1234.9 pm, β = 110.75°, and the calculated and experimental densities of 4.66 and 4.57 g·cm^{-3}, respectively (Motrya et al. 1986a); a = 1234.6 ± 0.2, b = 708.4 ± 0.1, c = 1237.8 ± 0.2 pm, and β = 110.20 ± 0.02° (Susa and Steinfink 1971a); a = 1235, b = 708, c = 1238 pm, and β = 110.2° (Susa and Steinfink 1971b); a = 1236, b = 710.7, c = 1238 pm, β = 110°08′, and an experimental densities of 4.57 g·cm^{-3} (Serment et al. 1968)]. The structure of Cd_4GeS_6 can be described also using a pseudohexagonal structure with the lattice parameters a = 700 and c = 3450 pm (Susa and Steinfink 1971a). This compound has a polymorphous transition at 1034 ± 7°C (Odin et al. 1983b) and decomposes at the heating on the air up to 410°C (Motrya et al. 1987). Its high-temperature modification was not obtained (Odin et al. 1983b). Limits of α-Cd_4GeS_6 homogeneity region at 544°C are shown in Figure 4.11 (Odin and Grin'ko 1991c). This region is strongly elongated along the GeS_2–CdS binary system.

The solubility of GeS_2 in CdS at the peritectic temperature is equal to 1.5 mol% and decreases to 0.5–1.0 mol% at 540°C. Solid solutions based on GeS_2 contain less than 2 mol% CdS at the eutectic temperature [31.27 at% Ge at 700°C (Movsum-zade et al. 1987a)] and approximately 1.5 mol% CdS at 540°C (Odin

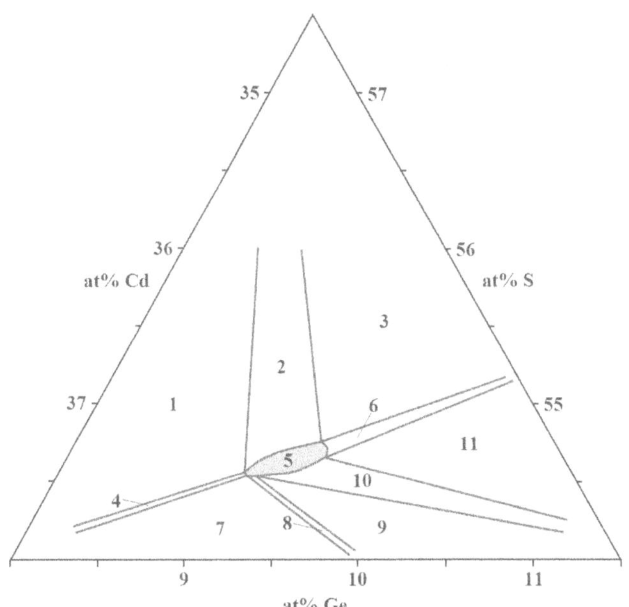

FIGURE 4.11 Limits of α-Cd_4GeS_6 homogeneity region at 544°C in the Ge–Cd–S ternary system: 1, α-Cd_4GeS_6 + CdS + L; 2, α-Cd_4GeS_6 + L; 3, α-Cd_4GeS_6 + GeS_2 + L; 4, α-Cd_4GeS_6 + CdS; 5, α-Cd_4GeS_6; 6, α-Cd_4GeS_6 + GeS_2; 7, α-Cd_4GeS_6 + Ge + CdS; 8, α-Cd_4GeS_6 + Ge; 9, α-Cd_4GeS_6 + GeS + Ge; 10, α-Cd_4GeS_6 + GeS; and 11, α-Cd_4GeS_6 + GeS_2 + GeS. (From Odin, I.N. and Grin'ko, V.V., *Zhurn. neorgan. khimii*, **36**(5), 1332, 1991.)

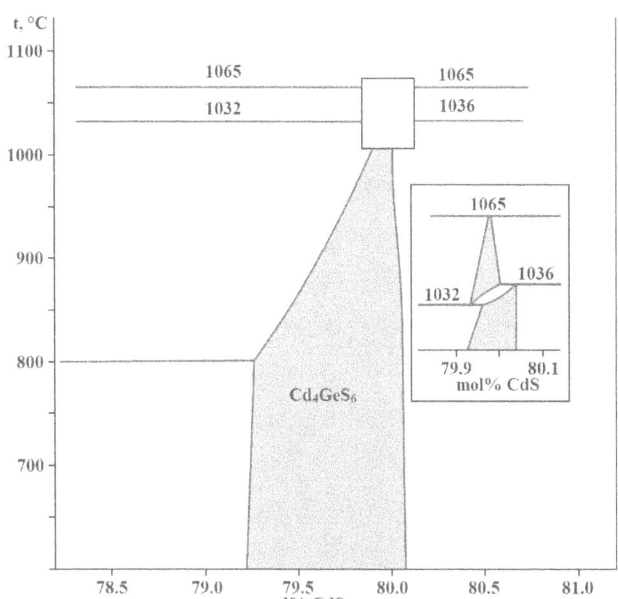

FIGURE 4.12 *T–x* projection of the GeS_2–CdS phase diagram close to Cd_4GeS_6. (From Odin, I.N. and Grin'ko, V.V., *Zhurn. neorgan. khimii*, **36**(5), 1332, 1991.)

et al. 1983b). Homogeneity region of Cd_4GeS_6 at 540°C is equal approximately to 1 mol% (Odin et al. 1983b). Cd_2GeS_4 was not found in this system (Odin et al. 1983b; Movsum-zade et al. 1987a). A glass region is found in the GeS_2–CdS system within the interval from 70 to 100 mol% GeS_2 (Barnier et al. 1990).

Vapor composition in the GeS_2–CdS system corresponds to GeS_2 compound, that is, it lies on this section (Odin and Grin'ko 1991b,c). The vapor above Cd_4GeS_6 consists mainly from GeS and S_2 in ratio 1:0.5, and the Cd concentration in vapor is 0.01 mass% (Motrya et al. 1987; Odin and Grin'ko 1991b,c). The limit of α-Cd_4GeS_6 homogeneity region from the CdS side does not change with temperature and from the GeS_2-side changes at temperatures above eutectic temperature (Figure 4.12) (Odin and Grin'ko 1991c). *(T–x)p* sections of the GeS_2–CdS system at p = 267 and 667 hPa are shown in Figure 4.13, and *(p–x)T* section at 850°C is shown in Figure 4.14 (Odin and Grin'ko 1991b,c).

On *p–T* projection of GeS_2–CdS phase diagram (Figure 4.15), the lines represent two-phase [AB – S(GeS_2)/V, BC – L(GeS_2)/V, BD – L(GeS_2)/S(GeS_2)] and three-phase equilibria [ME – V/S(GeS_2)/S(α-Cd_4GeS_6), EF – S(GeS_2)/L/S(α-Cd_4GeS_6), EB – S(GeS_2)/L/V, EQ – V/L/S(α-Cd_4GeS_6), NG – V/S(α-Cd_4GeS_6)/S(CdS)] (Odin and Grin'ko 1991b). The lines, representing equilibria of condensed phases [L(GeS_2)/S(GeS_2), S(GeS_2)/L/S(α-Cd_4GeS_6)], are practically vertical. The parameters for four-phase equilibria in the GeS_2–CdS system were also determined.

This system was investigated using DTA, metallography, XRD, measuring of microhardness, and using tensimeter (Odin et al. 1983b; Movsum-zade et al. 1987a; Odin and Grin'ko 1991b,c). The ingots were annealed at 540°C for 1200 h (Odin et al. 1983b) [at the temperatures 50°C–100°C lower than eutectic temperature (Movsum-zade et al. 1987a) and at 630°C for 500 h (Odin and Grin'ko 1991b)].

Cd_4GeS_6 was obtained from CdS and GeS_2 at 1000°C (Julien-Pouzol and Jaulmes 1995); by interaction of Ge, S, and CdS at 800°C–855°C using vibration mixing (Motrya et al. 1986a); by heating of Cd, Ge, and S mixtures in stoichiometric ratio at 200°C–400°C for 2–4 h and then at 600°C–800°C for 12 h (Susa and Steinfink 1971a); by heating of mixtures from CdS, Ge, and S at 850°C (Kaldis and Widmer 1965); or by coprecipitation from the water solutions with next long-time heating at 600°C (Kislinskaya 1974). Its single crystals were grown using chemical transport reaction (Kaldis and Widmer 1965; Odin et al. 1983b; Motrya et al. 1986a). Glasses in the GeS_2–CdS system were obtained by quenching the melt from 1000°C (Barnier et al. 1990).

Ge–CdS: The phase diagram of this system, constructed using DTA, metallography, XRD, and measuring of microhardness, is a eutectic type (Figure 4.16) (Movsum-zade et al. 1987a). The eutectic composition and temperature are 53.84 at% Ge and 900°C, respectively. At 925°C, there is an immiscibility region within the interval of 66.66–92.30 at% Ge. The Ge content in the CdS-doped single crystals reaches 0.5 at% (0.25 mass%) (Odin et al. 1999) (GeS was used as a dopant). The ingots were annealed at temperatures 50°C–100°C lower than the eutectic temperature (Movsum-zade et al. 1987a).

Eight fields of primary crystallizations are on the liquidus surface of the Ge–Cd–S ternary system (Figure 4.17) (Movsum-zade et al. 1987a). The field of Cd crystallization is very small, and the field of sulfur crystallization is degenerated. Ternary eutectic E_4 is degenerated from the sulfur-rich side. Considerable part of the liquidus surface is occupied by immiscibility region. The next four-phase equilibria are in the Ge–Cd–S ternary system: E_1 (525°C) – L ⇔ Cd_4GeS_6 + GeS + GeS_2; E_2 (530°C) – L ⇔ GeS + Ge + Cd_4GeS_5; E_3 (300°C) –

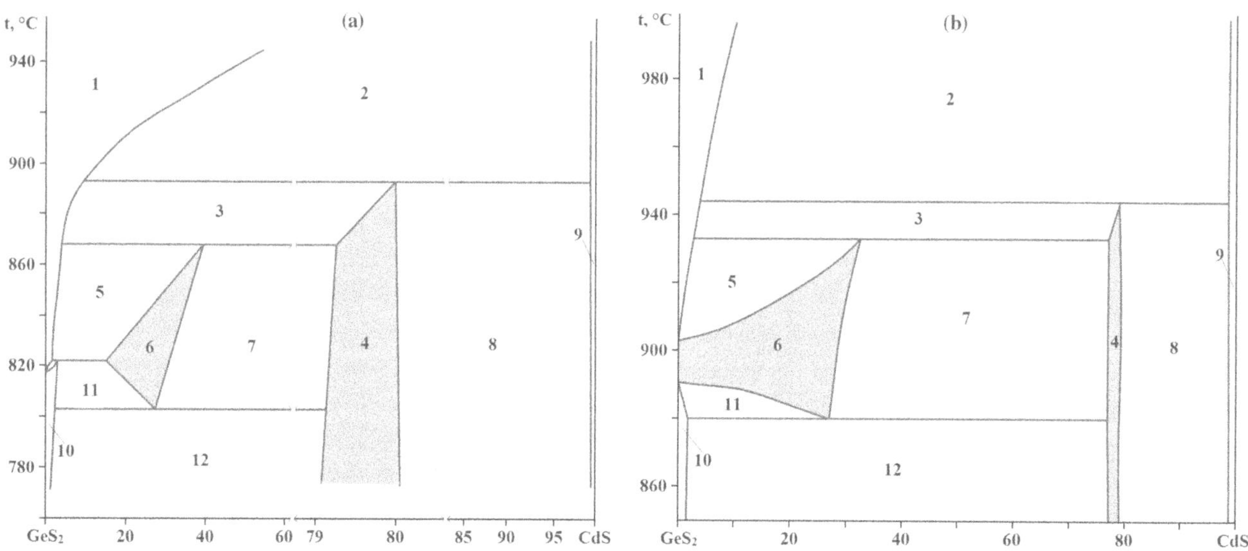

FIGURE 4.13 $(T\text{–}x)_p$ sections of the GeS_2–CdS phase diagram at 267 (a) and 667 hPa (b): 1, V; 2, V + S(CdS); 3, V + S(α-Cd_4GeS_6); 4, S(α-Cd_4GeS_6); 5, V + L; 6, L; 7, S(α-Cd_4GeS_6) + L; 8, S(α-Cd_4GeS_6) + S(CdS); 9, S(CdS); 10, S(GeS_2); 11, S(GeS_2) + L; and 12, S(GeS_2) + S(α-Cd_4GeS_6). (From Odin, I.N. and Grin'ko, V.V., *Zhurn. neorgan. khimii*, **36**(4), 1056, 1991; Odin, I.N. and Grin'ko, V.V., *Zhurn. neorgan. khimii*, **36**(5), 1332, 1991.)

L ⇔ Cd + CdS + Ge; U_1 (–) – L + CdS ⇔ GeS_2 + Cd_4GeS_5; U_2 (770°C) – L + CdS ⇔ Cd_4GeS_6 + Cd_4GeS_5; U_3 (540°C) – L + Cd_4GeS_5 ⇔ Cd_4GeS_6 + GeS; U_4 (575°C) – L + CdS ⇔ Cd_4GeS_5 + Ge.

The glass-forming region in the Ge–Cd–S system is presented in Figure 4.18 (Vassilev and Ivanova 1997). It is seen that up to 10 at% Cd at a molar ratio of S/Ge ≈ 2 can be added in the Ge–S system without crystallization. The attempts to synthesize glassy phases along the GeS_2–CdS line with CdS content above 20 mol% were unsuccessful, even using a smaller weight of the mixtures and ampoule sizes ensuring a higher cooling rate. The glass transition temperature for obtained glasses is within the interval from 239°C to 272°C. The bulk glasses were prepared by heating the mixture of the elements to approximately 1000°C. The synthesis was carried out in evacuated silica ampoule in a rotating furnace and quenched in ice water. The melts were heated first to 200°C and held for 2 h, then to 400°C and 650°C for 7 h, and after that to 1000°C for 2 h.

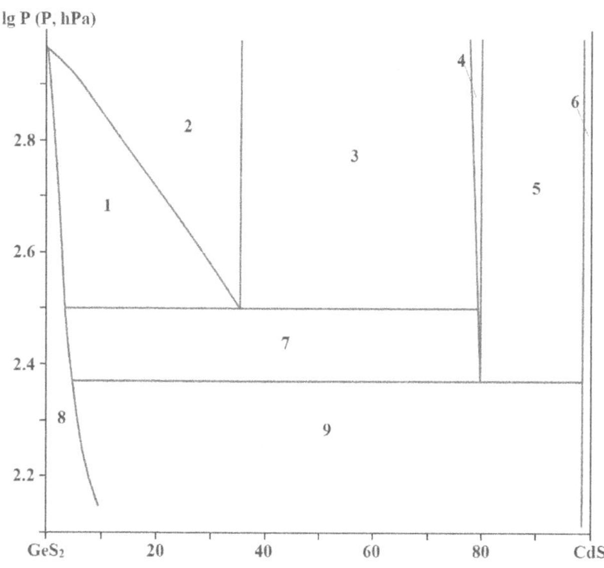

FIGURE 4.14 $(p\text{–}x)_T$ section of the GeS_2–CdS phase diagram at 850°C: 1, L + V; 2, L; 3, L + S(α-Cd_4GeS_6); 4, S(α-Cd_4GeS_6); 5, S(α-Cd_4GeS_6) + (CdS); 6, S(CdS); 7, V S(α-Cd_4GeS_6); 8, V; and 9, V + S(CdS). (From Odin, I.N. and Grin'ko, V.V., *Zhurn. neorgan. khimii*, **36**(4), 1056, 1991.)

4.15 Germanium–Mercury–Sulfur

GeS–HgS: The phase diagram of this system, constructed through DTA, metallography, XRD, and microhardness and density measurements, is shown in Figure 4.19 (Motrya 1991; Voroshilov et al. 1994). The eutectic compositions and temperatures are 24 and 51 mol% HgS and 529 ± 5°C and 510 ± 5°C, respectively. Two ternary compounds are formed in

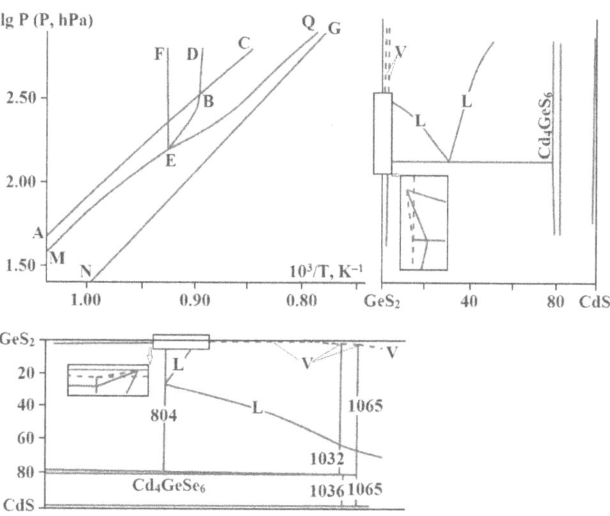

FIGURE 4.15 P_{total}–T–x phase diagram of the GeS_2–CdS system (explanations are in the text). (From Odin, I.N. and Grin'ko, V.V., *Zhurn. neorgan. khimii*, **36**(4), 1056, 1991.)

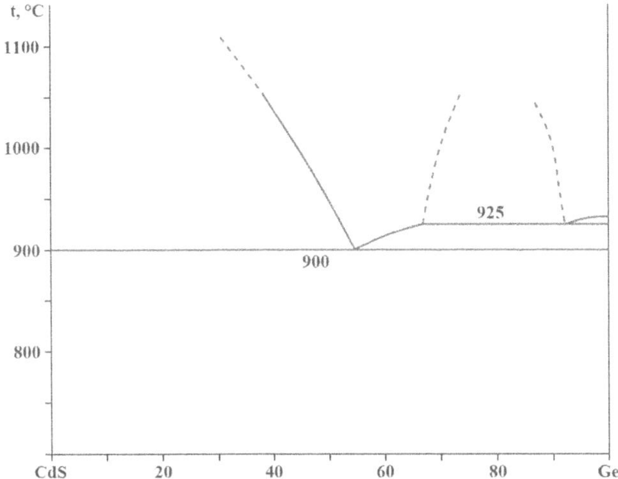

FIGURE 4.16 Phase diagram of the CdS–Ge system. (From Movsum-zade, A.A. et al., *Zhurn. neorgan. khimii*, **32**(4), 1025, 1987.)

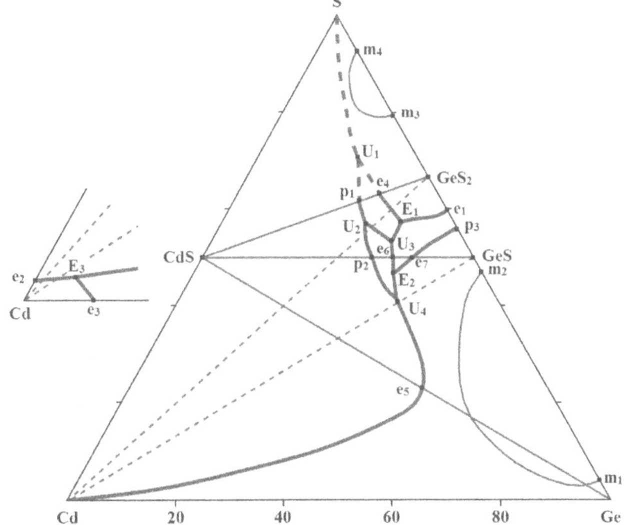

FIGURE 4.17 Projection of the liquidus surface of the Ge–Cd–S ternary system. (From Movsum-zade, A.A. et al., *Zhurn. neorgan. khimii*, **32**(4), 1025, 1987.)

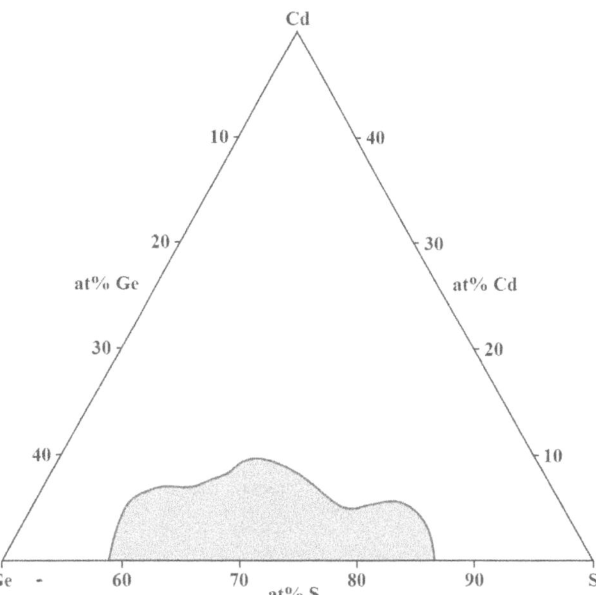

FIGURE 4.18 Glass forming region in the Ge–Cd–S ternary system. (From Vassilev, V.S., and Ivanova, Z.G., *J. Phys. Chem. Solids*, **58**(4), 573, 1997.)

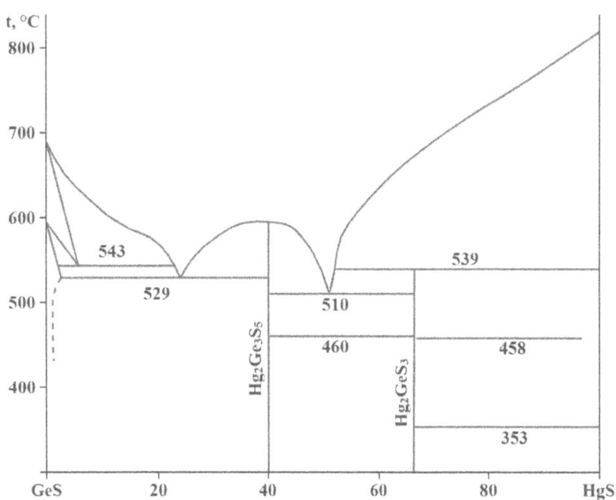

FIGURE 4.19 Phase diagram of the GeS–HgS system. (From Voroshilov, Yu.V. et al., *Ukr. khim. zhurn.*, **60**(1), 27, 1994.)

this system: $\mathbf{Hg_2Ge_3S_5}$ melts congruently at 592 ± 5°C, and $\mathbf{Hg_2GeS_3}$ decomposes peritectically at 539°C and has polymorphous transformation at 425 ± 5°C. The experimental density of $Hg_2Ge_3S_5$ and Hg_2GeS_3 is 4.51 and 6.25 g·cm^{-3}, respectively. At 353 ± 5°C, there is polymorphous transformation of HgS, and polymorphous transformation of solid solutions based on GeS is observed from the GeS-rich side. The ingots for the investigations were annealed at 500°C for 1200 h.

GeS_2–HgS: The most reliable phase diagram of this system is shown in Figure 4.20 (Olekseyuk et al. 2006). Eutectic crystallizes at 582°C [580 ± 5°C (Motrya 1991; Voroshilov et al. 1994)] and contains 44 mol% HgS. $\mathbf{Hg_4GeS_6}$ is formed in this system. It melts incongruently at 723°C [720 ± 5°C (Motrya 1991; Voroshilov et al. 1994)] and has polymorphous transformation at 406°C (395 ± 5°C (Motrya 1991; Voroshilov et al. 1994)]. The peritectic point corresponds to 78 mol% HgS. The polymorphous transitions of HgS and GeS_2 take place at 355°C and 434°C, respectively (Olekseyuk et al. 2006). Glassy ingots were obtained from the GeS_2-rich side (Motrya 1991; Voroshilov et al. 1994).

According to the data of Motrya (1991) and Voroshilov et al. (1994), another ternary compound, $\mathbf{HgGe_2S_5}$, is formed in this system according to the solid-phase reaction at 565 ± 5°C [existence of this compound was not confirmed by Olekseyuk et al. (2001b) and Marchuk et al. (2002)].

Low-temperature modification of Hg_4GeS_6 crystallizes in the monoclinic structure with the lattice parameters a = 1234.51 ± 0.08, b = 716.78 ± 0.06, c = 1234.67 ± 0.07 pm, and β = 109.484 ± 0.005° (Olekseyuk et al. 2006) [a = 1235 ± 1,

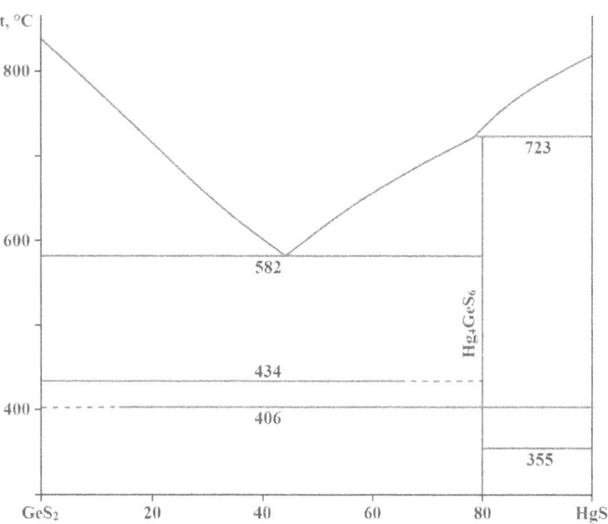

FIGURE 4.20 Phase diagram of the GeS_2–HgS system. (From Olekseyuk, I.D. et al., *J. Alloys Compd.*, **417**(1–2), 131, 2006.)

$b = 717.1 \pm 0.6$, $c = 1238.2 \pm 0.8$ pm, $\beta = 109.2 \pm 0.1°$, and the calculated and experimental densities of 6.92 and 6.90 g·cm^{-3}, respectively (Motrya 1991; Voroshilov et al. 1994); $a = 1234 \pm 1$, $b = 712.7 \pm 0.6$, $c = 1236 \pm 1$ pm, $\beta = 109°27' \pm 0°10'$, and an experimental density of 6.88 g·cm^{-3} (Serment et al. 1968)]. The parameters of the monoclinic structure were recalculated in the parameters of rhombohedral and hexagonal structures (rhombohedral structure, $a = 1236 \pm 1$ pm and $\alpha = 33°30' \pm 0°10'$; hexagonal structure, $a = 712.7 \pm 0.6$ and $c = 3497 \pm 1$ pm) (Serment et al. 1968). The lattice is practically rhombohedral, whereas X-ray spectrum of Hg_4GeS_6 compound well indexes by the parameters both of rhombohedral and hexagonal structure.

According to the data of Hahn and Lorent (1958), $\mathbf{Hg_2GeS_4}$ ternary compound is formed in the GeS_2–HgS system. It crystallizes in the hexagonal structure with the lattice parameters $a = 717$ and $c = 3490$ pm and the calculated and experimental densities of 5.789 and 5.61 g·cm^{-3}, respectively. It can be seen that the lattice parameters of this compound coincide with the lattice parameters of Hg_4GeS_6 in the hexagonal structure (Serment et al. 1968). Therefore, it is possible that composition of Hg_2GeS_4 was determined incorrectly.

This system was investigated using DTA, metallography, XRD, and microhardness and density measurements. The ingots were annealed at 400°C for 250 h (Olekseyuk et al. 2006) [at 500°C for 1200 h (Motrya 1991; Voroshilov et al. 1994)].

At 500°C, four ternary compounds, $Hg_2Ge_3S_5$, Hg_2GeS_3, $HgGe_2S_5$, and Hg_4GeS_6, exist in the Ge–Hg–S ternary system (Figure 4.21) (Motrya 1991; Voroshilov et al. 1994). They are in equilibrium with binary and elementary components and with each other.

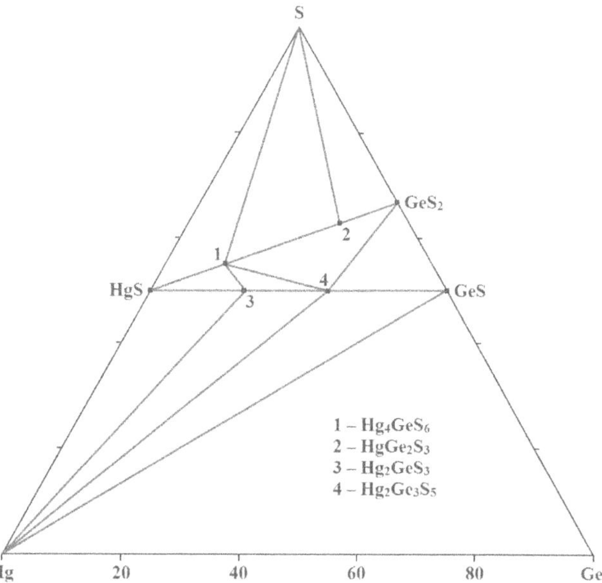

FIGURE 4.21 Isothermal section of the Ge–Hg–S ternary system at 500°C. (From Voroshilov, Yu.V. et al., *Ukr. khim. zhurn.*, **60**(1), 27, 1994.)

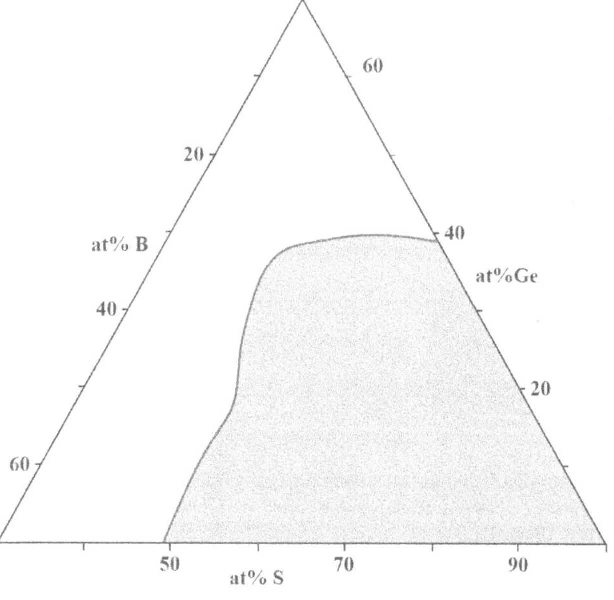

FIGURE 4.22 Glass forming region in the Ge–B–S ternary system. (From Kirilenko, V.V. and Dembovskiy, S.A., *Izv. AN SSSR. Neorgan. mater.*, **10**(3), 542, 1974.)

4.16 Germanium–Boron–Sulfur

The glass forming region in the Ge–B–S ternary system is given in Figure 4.22 (Kirilenko and Dembovskiy 1974). The obtained black glass-like alloys are unstable in air.

4.17 Germanium–Gallium–Sulfur

$\mathbf{GeS_2–Ga_2S_3}$: The phase diagram of this system, constructed through DTA and XRD, is presented in Figure 4.23 (Olekseyuk et al. 2001a). The existence of solid solutions based on the high- and low-temperature modifications of Ga_2S_3 was established. The solubility of GeS_2 in Ga_2S_3 at 450°C is not higher than 3 mol%. The invariant metatectic and eutectic reactions occur in the system with coordinates for the metatectic and eutectic points 930°C, 33 mol% GeS_2, and 744°C, 86 mol% GeS_2 [740°C, 85.7 mol% GeS_2 (Loireau-Lozac'h and Guittard

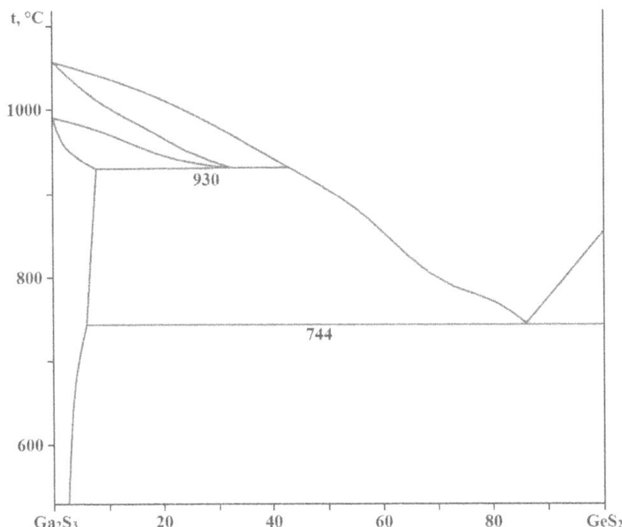

FIGURE 4.23 Phase diagram of the Ge_2S–Ga_2S_3 system. (From Olekseyuk, I.D., *Nauk. visnyk Volyns'k. Univ. im. Lesi Ukrainky, Ser. Khim. nauky*, (6), 56, 2001.)

1975; Nedoshovenko et al. 1986)], respectively. According to the data of Loireau-Lozac'h and Guittard (1975), mutual solubility of the starting components was not detected. The ingots were annealing at 450°C for 1000 h with the next quenching in the ice water (Olekseyuk et al. 2001a).

The homogeneous yellow glasses are formed in this system during quenching of the molten mixtures, containing up to 55 mol% $0.5Ga_2S_3$ from 1100°C in cold water (Loireau-Lozac'h and Guittard 1975). The glasses with the composition of $8GeS_2{\cdot}2Ga_2S_3$ and $9GeS_2{\cdot}Ga_2S_3$ were obtained by quenching of the alloys annealed at 900°C for 10 h in cold water (Olekseyuk et al. 2011a). It was shown that the increasing of germanium content in the glasses leads to the increasing of the energy gap (E_g = 2.46 eV for the first glass and E_g = 2.50 eV for the second one).

The study of coprecipitation in the GeS_2–Ga_2S_3 system does not give an opportunity to judge the nature of the interaction due to the low degree of gallium coprecipitation with germanium disulfide (Chaus et al. 1970). One can only assume that there is an adsorption process or the formation of a number of unstable binary sulfides.

The glass-forming region in the GeS–GeS_2–GaS subsystem is shown in Figure 4.24 (Feltz and Krautwald 1979). The glass transition temperature for these glasses is within the interval 293°C – 354°C. The glasses of the GeS_2–Ga_2S_3 system are high-resistance *p*-type semiconductors (Aleksandrov 1990).

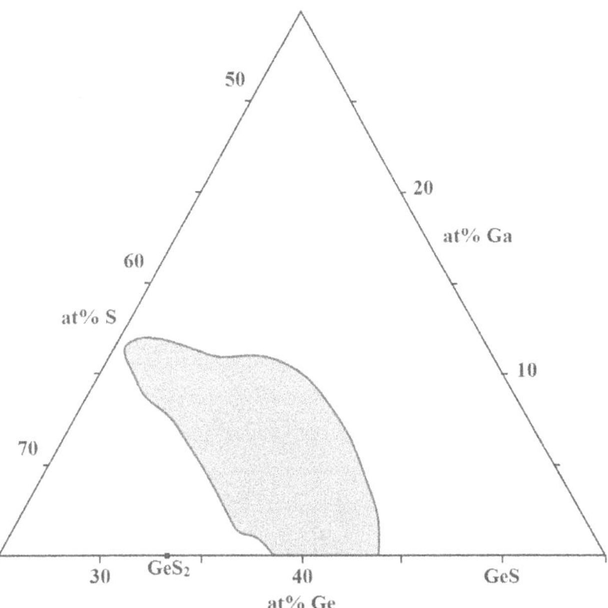

FIGURE 4.24 Glass forming region in the GeS–GeS_2–GaS subsystem. (From Feltz, A., and Krautwald, A., *Z. Chem.*, **19**(2), 78–79 (1979.)

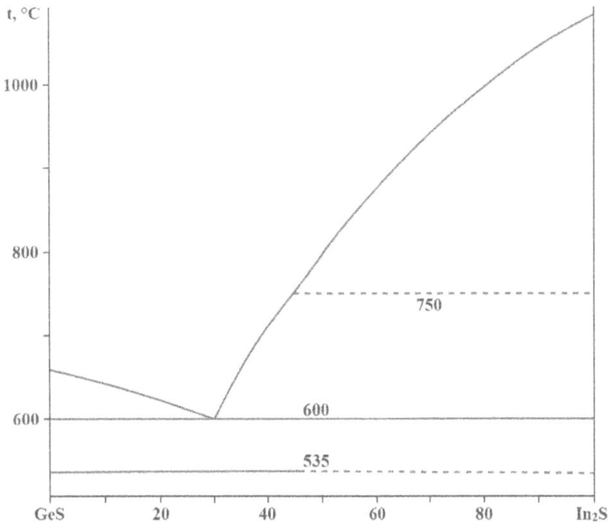

FIGURE 4.25 Phase diagram of the GeS–In_2S_3 system. (From Sarkisov, E.S., et al., *Izv. AN SSSR. Neorgan. mater.*, **6**(1), 184, 1970.)

4.18 Germanium–Indium–Sulfur

GeS–In_2S_3: The phase diagram of this system, constructed through DTA and XRD and chemical analysis, is given in Figure 4.25 (Sarkisov et al. 1970c). The eutectic contains 30 mol% In_2S_3 and crystallizes at 600°C. The effect of polymorphic transformation of GeS at 535°C was recorded only for the samples containing 0–45 mol% In_2S_3. Thermal effects at 750°C are caused by polymorphic transformation of In_2S_3.

GeS_2–In_2S_3: The phase diagram of this system, constructed through DTA and XRD, is a eutectic type (Figure 4.26) (Sarkisov and Lidin 1969; Sarkisov et al. 1971). The eutectic contains 10 ± 2.5 mol% In_2S_3 and crystallizes at 750°C. Small regions of solid solutions based on the binary compounds were found. The next compounds, **$In_2S_3{\cdot}6GeS_2$, $In_2S_3{\cdot}5GeS_2$, $In_2S_3{\cdot}2GeS_2$, $3In_2S_3{\cdot}2GeS_2$,** and **$5In_2S_3{\cdot}2GeS_2$,** are formed in the process of coprecipitation of GeS_2 and In_2S_3 from aqueous solutions (Rudnev and Bil'kevich 1966; Chaus et al. 1970, 1973; Sarkisov et al. 1970a). $In_2S_3{\cdot}6GeS_2$ is X-ray amorphous at room temperature and decomposes to initial sulfides at high temperatures (Rudnev and Bil'kevich 1966; Sarkisov et al. 1970a). Stable ternary compounds, both during coprecipitation and crystallization from the melt, are not formed in the GeS_2–In_2S_3 system.

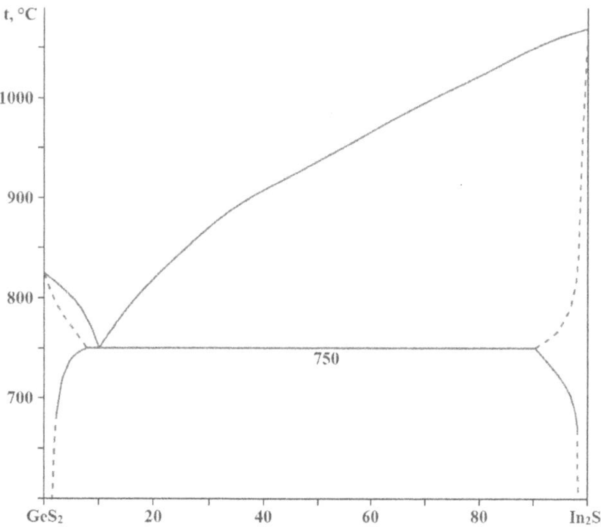

FIGURE 4.26 Phase diagram of the GeS_2–In_2S_3 system. (From Sarkisov, E.S., and Lidin, R.A., *Izv. AN SSSR. Neorgan. mater.*, **5**(5), 985, 1969.)

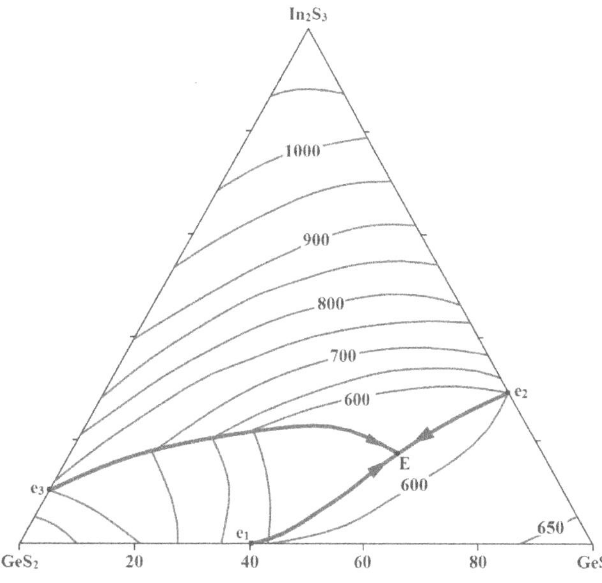

FIGURE 4.27 Liquidus surface of the GeS–GeS_2–In_2S_3 subsystem of the Ge–In–S ternary system. (From Sarkisov, E.S., et al., *Izv. AN SSSR. Neorgan. mater.*, **7**(1), 34, 1971.)

GeS–GeS_2–In_2S_3: The liquidus surface of this subsystem is presented in Figure 4.27 (Sarkisov et al. 1971). Three fields of primary crystallization exist on the liquidus surface, and the biggest part occupies the field of In_2S_3 primary crystallization. The ternary eutectic, *E*, crystallizes at 525°C.

According to the data of Deiseroth and Pfeifer (1993a), the In_4GeS_4 ternary compound, which crystallizes in the cubic structure with the lattice parameter $a = 1212.31$ pm, is formed in the Ge–In–S system. It could be prepared in evacuated quartz ampoule at 650°C from the elements in the stoichiometric amounts.

4.19 Germanium–Thallium–Sulfur

GeS_2–Tl_2S: The most reliable phase diagram of this system, constructed through DTA, XRD, metallography, and measuring of microhardness, is shown in Figure 4.28 (Starosta et al. 1984). The eutectics contain 25, 43, 55, and 73 mol% GeS_2 and crystallize at 375°C, 371°C, 475°C, and 575°C, respectively. Three compounds, **Tl_2GeS_3**, **$Tl_2Ge_2S_5$**, and **Tl_4GeS_4**, are formed in this system. The homogeneity regions of these compounds are given in Figures 4.29 and 4.30 (Peresh et al. 1982; Starosta et al. 1984).

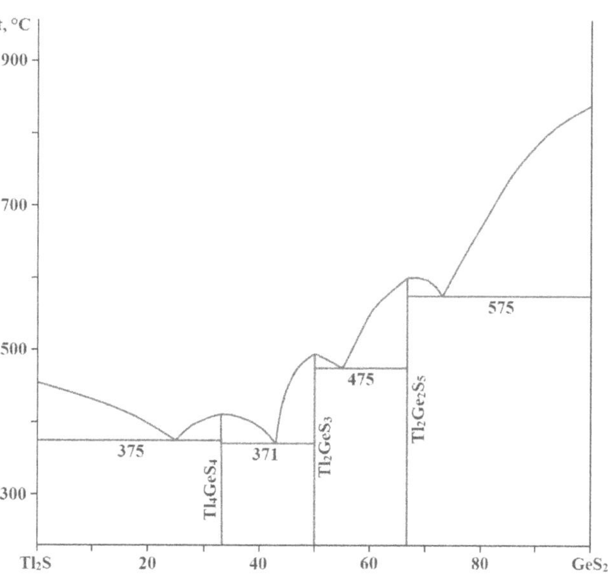

FIGURE 4.28 Phase diagram of the GeS_2–Tl_2S system. (From Starosta, V.I., et al., *Zhurn. neorgan. khimii*, **29**(12), 3131, 1984.)

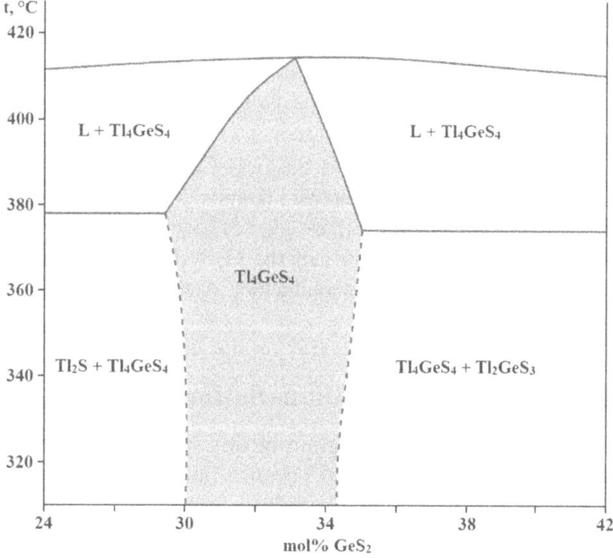

FIGURE 4.29 Homogeneity region of the Tl_4GeS_4 compound. (From Peresh, E.Yu., et al., *Zhurn. neorgan. khimii*, **27**(2), 473, 1982.)

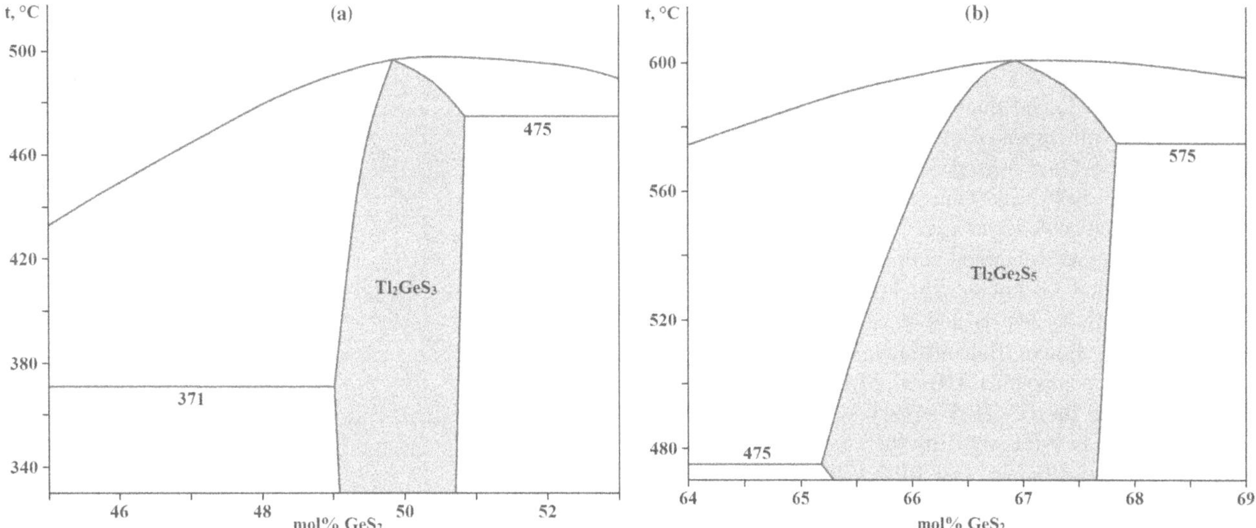

FIGURE 4.30 Homogeneity regions of the (a) Tl_2GeS_3 and (b) $Tl_2Ge_2S_5$ compounds. (From Starosta, V.I., et al., *Zhurn. neorgan. khimii*, **29**(12), 3131, 1984.)

The phase diagram of this system was also constructed by Kulieva and Babanly (1981). According to the data of these authors, three eutectics, which contain 25, 42, and 53 mol% GeS_2 and crystallize at 346°C, 340°C, and 470°C, respectively, and one peritectic at 580°C exist in this system.

Tl_2GeS_3 melts congruently at 490°C (Starosta et al. 1984; Peresh et al. 1986) [at 497°C (Kulieva and Babanly 1981; Babanly et al. 1982; Bokova et al. 2019)] and crystallizes in the triclinic structure with the lattice parameters $a = 671.7 \pm 0.2$, $b = 674.9 \pm 0.2$, $c = 844.8 \pm 0.3$ pm, and $\alpha = 90.28 \pm 0.04°$, $\beta = 111.47 \pm 0.04°$, $\gamma = 113.21 \pm 0.04°$, and the calculated and experimental densities of 5.95 and 5.83 g·cm^{-3}, respectively (Eulenberger 1978) [$a = 806 \pm 4$, $b = 1270 \pm 4$, $c = 676 \pm 6$ pm, and $\alpha = 96°39' \pm 0°20'$, $\beta = 93°10' \pm 0°20'$, $\gamma = 99°31' \pm 0°20'$, and the calculated and experimental densities of 5.675 and 5.66 ± 0.02 g·cm^{-3}, respectively (Isaacs 1975); $a = 682 \pm 1$, $b = 684 \pm 1$, $c = 855 \pm 1$ pm, and $\alpha = 89.1 \pm 0.2°$, $\beta = 112.1 \pm 0.2°$, $\gamma = 113.2 \pm 0.2°$, and the calculated and experimental densities of 5.87 and 5.83 g·cm^{-3}, respectively (Eulenberger and Muller 1974)]. The enthalpy and entropy of melting of this compound are equal to 25.5 kJ·M^{-1} and 33.4 J·(M·K)$^{-1}$, respectively, and an energy gap is 2.3 eV (Peresh et al. 1986). This compound was made from high-purity elements, weighed out in stoichiometric proportions and sealed in an evacuated silica tube, which was then heated slowly past the melting point of the material (Isaacs 1975). The melt was held at approximately 700°C for several hours and was shaken vigorously a number of times to ensure thorough mixing. It was cooled to room temperature by shutting off the furnace. The temperature was then raised to approximately 430°C; the compound was annealed for several days, and then cooled over a period of 2 days to room temperature. Tl_2GeS_3 could be also prepared from the stoichiometric mixture of TlS, Ge, and S by reacting at 700°C–800°C for several hours and subsequent annealing at 430°C–460°C for several days in the evacuated silica tube (Nakamura et al. 1984) or by the interaction of the stoichiometric mixture of GeS_2 and Tl_2S in the evacuated silica tube (Eulenberger and Muller 1974). Its single crystals were grown by the Bridgman method (Starosta et al. 1984).

$Tl_2Ge_2S_5$ melts congruently at 595°C (Starosta et al. 1984; Peresh et al. 1986) [at 596°C (Bokova et al. 2019); melts incongruently at 580°C (Kulieva and Babanly 1981)] and crystallizes in the monoclinic structure with the lattice parameters $a = 1496.7 \pm 0.2$, $b = 1498.0 \pm 0.2$, $c = 881.2 \pm 0.1$ pm, $\beta = 106.98 \pm 0.01°$, and the calculated and experimental densities of 5.03 and 4.96 g·cm^{-3}, respectively (Eulenberger 1976) [$a = 1500 \pm 1$, $b = 1494 \pm 1$, $c = 880 \pm 1$ pm, and $\beta = 107.0 \pm 0.2°$ and the calculated and experimental densities of 5.03 and 4.96 g·cm^{-3}, respectively (Eulenberger and Muller 1974)]. The enthalpy and entropy of melting of this compound are equal to 53.2 kJ·M^{-1} and 61.3 J·(M·K)$^{-1}$, respectively, and an energy gap is 2.6 eV (Peresh et al. 1986). It was prepared by melting together a stoichiometric mixture of GeS_2 and Tl_2S in an evacuated and fused quartz ampoule (10 days, at ca. 500°C) and subsequent annealing at 300°C (Eulenberger and Muller 1974; Eulenberger 1976). The single crystals of $Tl_2Ge_2S_5$ were grown by the Bridgman method (Starosta et al. 1984).

Tl_4GeS_4 melts at 404°C (Starosta et al. 1984; Peresh et al. 1986) [at 415°C (Kulieva and Babanly 1981); at 417°C (Bokova et al. 2019)] and crystallizes in the monoclinic structure with the lattice parameters $a = 1250.1 \pm 0.3$, $b = 1124.8 \pm 0.3$, $c = 760.8 \pm 0.3$ pm, $\beta = 112.21 \pm 0.02°$, and the calculated and experimental densities of 6.83 and 6.74 g·cm^{-3}, respectively (Eulenberger 1977b) [$a = 1244 \pm 1$, $b = 1123 \pm 1$, $c = 759 \pm 1$ pm, $\beta = 111.0 \pm 0.2°$, and the calculated and experimental densities of 6.84 and 6.74 g·cm^{-3}, respectively (Eulenberger and Muller 1974)]. The enthalpy and entropy of melting of this compound are equal to 15.9 kJ·M^{-1} and 23.5 J·(M·K)$^{-1}$, respectively, and an energy gap is 2.1 eV (Peresh et al. 1982, 1986). This compound was prepared by melting together a stoichiometric mixture of GeS_2 and Tl_2S in an evacuated and fused quartz ampoule and subsequent annealing at 300°C (Eulenberger and Muller 1974). The single crystals of Tl_4GeS_4 were grown by the Bridgman method (Starosta et al. 1984).

Tl_2GeS_3, $Tl_2Ge_2S_5$, and Tl_4GeS_4 crystals were also prepared from its constituents by thermal synthesis in the evacuated silica tubes (Bokova et al. 2019). The mixtures were heated slowly up to 700°C and maintained at this temperature for several hours with repeated stirring of the melt to ensure thorough mixing. Once homogeneous, the solids were slowly cooled down to 500°C and annealed at this temperature for 10 days, and then cooled over a period of 2 days to room temperature. The crystals obtained in this way were orange color for Tl_2GeS_3, yellow for $Tl_2Ge_2S_5$, and ruby red for Tl_4GeS_4.

GeS–TlS: This system is non-quasibinary section of the Ge–Tl–S ternary system (Babanly et al. 1985).

Three another compounds, $\mathbf{TlGeS_2}$, $\mathbf{Tl_2GeS_4}$, and $\mathbf{Tl_4Ge_2S_4}$, were obtained in the Ge–Tl–S ternary system. X-ray amorphous $TlGeS_2$ was synthesized by the co-precipitation from aqueous solutions (Rudnev and Bil'kevich 1964). $Tl_4Ge_2S_4$ crystallizes in the orthorhombic structure with the lattice parameters a = 663.3 ± 0.2, b = 882.4 ± 0.2, c = 1397.7 ± 0.4 pm, and a calculated density of 5.54 g·cm^{-3} (Eulenberger 1984a, 1985).

On the liquidus surface of the Ge–Tl–S ternary system, there are two wide immiscibility regions, as well as the surfaces of the primary crystallization of all phases formed in the system (Figure 4.31) (Babanly et al. 1982). Two ternary eutectics, E_1 and E_2, which crystallize at 342°C and 347°C, respectively, and six transition points (U_1 at 292°C, U_2 at 222°C, U_3 at 157°C, U_4 at 237°C, U_5 at 562°C, and U_6 at 387°C) exist in this system. The isothermal section of this ternary system at room temperature (Figure 4.32) was also constructed by Babanly et al. (1982). These authors indicated on the formation of additional compound, $\mathbf{Tl_4GeS_5}$, in this system.

The glass-forming region in the Ge–Tl–S ternary system was investigated by Kirilenko et al. (1978), Dembovskiy et al. (1983), and Babanly et al. (1983), and the part of this region in the GeS–GeS$_2$–Tl$_2$S subsystem is presented in Figure 4.33

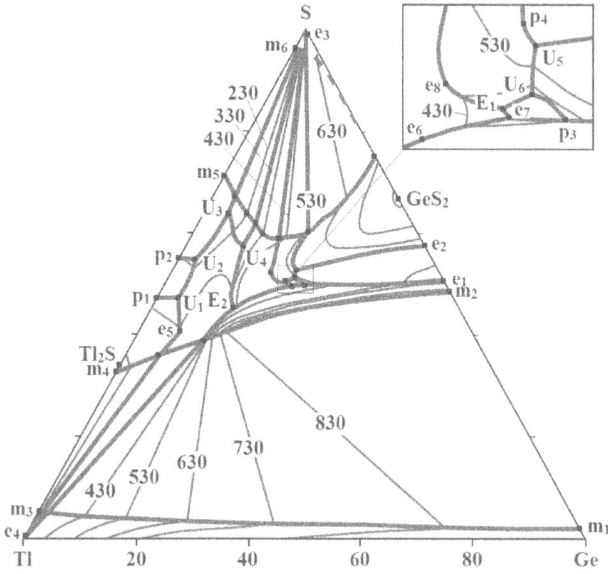

FIGURE 4.31 Liquidus surface of the Ge–Tl–S ternary system. (From Babanly, M.B., et al., *Zhurn. neorgan. khimii*, **27**(9), 2375, 1982.)

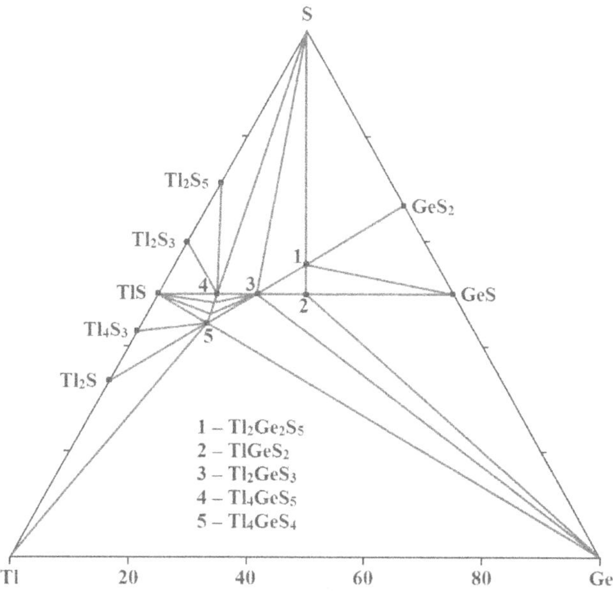

FIGURE 4.32 Isothermal section of the Ge–Tl–S ternary system at room temperature. (From Babanly, M.B., et al., *Zhurn. neorgan. khimii*, **27**(9), 2375, 1982.)

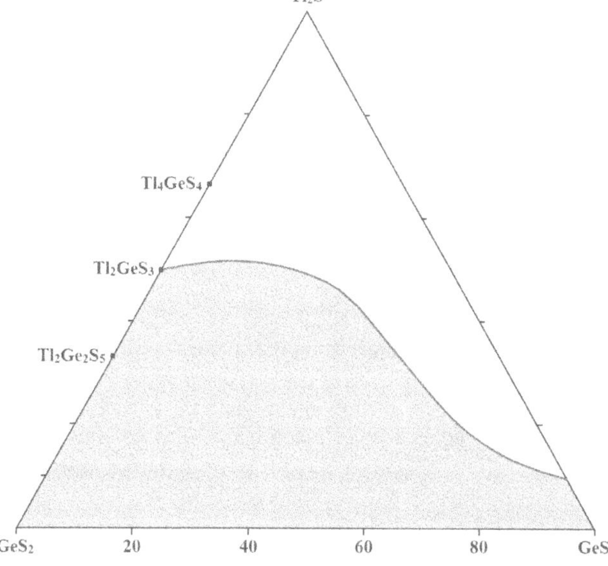

FIGURE 4.33 Glass-forming region in the GeS–GeS$_2$–Tl$_2$S subsystem of the Ge–Tl–S ternary system. (From Bokova, M., et al., *J. Alloys Compd.*, **777**, 902, 2019.)

(Bokova et al. 2019). It is possible to introduce more than 36 mol% Tl into the GeS–GeS$_2$ glassy matrix by the quenching in the ice water. The addition of Tl reduces significantly the glass transition temperature for $(Tl_2S)_x(GeS_2)_{1-x}$ glassy alloys from 481°C for pure GeS$_2$ to 171°C for x = 0.5.

The glassy samples were prepared by the standard melt-quenching technique (Bokova et al. 2019). The appropriate quantities of Tl, Ge, and S were charged and sealed under vacuum (10^{-4} Pa) in a silica tube. The sealed tube was heated slowly to 950°C at 1°C·min^{-1} heating rate and maintained at this temperature for a few days with repeated rotation of the melt. The temperature was kept constant at 500°C and 800°C for about 6 h to avoid the high vapor pressure of sulfur. After homogenization, the glasses were quenched in water at room temperature or in ice water depending on glass-forming ability of the sample. The annealing of the quenched samples was carried out at 20°C–30°C below the glass transition temperature for 24 h.

4.20 Germanium–Yttrium–Sulfur

The $\mathbf{Y_6Ge_{2.5}S_{14}}$ ternary compound, which crystallizes in the hexagonal structure with the lattice parameters $a = 973.0 \pm 0.1$, $c = 582.6 \pm 0.1$ pm, and a calculated density of 4.046 g·cm^{-3}, is formed in the Ge–Y–S system (Michelet and Flahaut 1969; Michelet et al. 1975; Gulay et al. 2006b). It was prepared in a sealed evacuated silica ampoule by reactions of Y_2S_3, Ge, and S. The heating of the ampoule was very progressive to avoid the explosions due to a too strong vapor tension from sulfur not yet combined: 3 days at 550°C, 1 day at 650°C, 2 days at 750°C, 12 h at 900°C, and 2 h at 1050°C. The cooling was done slowly in the furnace. Annealing towards 600°C generally allows obtaining well crystallized product.

4.21 Germanium–Lanthanum–Sulfur

GeS_2–La_2S_3: The phase diagram of this system, constructed through DTA and XRD, is a eutectic type (Figure 4.34) (Sarkisov et al. 1970b; Beskrovnaya et al. 1979). The eutectics contain 7.5, 40, and 57 mol% La_2S_3 and crystallize at 756, 1020, and 1075°C, respectively (Sarkisov et al. 1970b). Two compounds, $\mathbf{La_2GeS_5}$ and $\mathbf{La_2Ge_2S_7}$, are formed in this system (Sarkisov et al. 1968, 1970b; Beskrovnaya et al. 1971b). La_2GeS_5 melts congruently at 1105°C (Sarkisov et al. 1970b) and crystallizes in the monoclinic structure with the lattice parameters $a = 788.7 \pm 0.4$, $b = 767.5 \pm 0.4$, $c = 1272.0 \pm 0.4$ pm, $\beta = 101.40 \pm 0.02°$, and the calculated and experimental densities of 4.55 and 4.45 ± 0.05 g·cm^{-3}, respectively (Michelet et al. 1970, 1975) [$a = 789.3 \pm 0.2$, $b = 764.1 \pm 0.2$, $c = 1270.2 \pm 0.3$ pm, $\beta = 101.39 \pm 0.03°$, and the calculated and experimental densities of 4.55 and 4.45 g·cm^{-3}, respectively (Mazurier and Etienne 1973)]. According to the data of Michelet et al. (1975), this compound is stable only above 600°C, provided however that it maintains a sufficient backpressure of GeS_2 vapor. It was prepared in the same way as $Y_6Ge_{2.5}S_{14}$ was synthesized using La_2S_3 instead of Y_2S_3 (Mazurier and Etienne 1973; Michelet et al. 1975). It is possible to obtain single crystals of La_2GeS_5 by heating the mixture around 1200°C, followed by a slow cooling. This compound was also synthesized by the interaction of the stoichiometric amounts of GeS_2 and La_2S_3 in an evacuated quartz ampoule at 1000°C for 7–10 h (Sarkisov et al. 1968) or synthesized by the interaction of the stoichiometric amounts of Ge, S, and La_2S_3 in an evacuated quartz ampoule at 600°C–700°C for 8–10 h with the next increasing temperature to 1160°C and exposure at this temperature for 2–3 h (Beskrovnaya et al. 1971b).

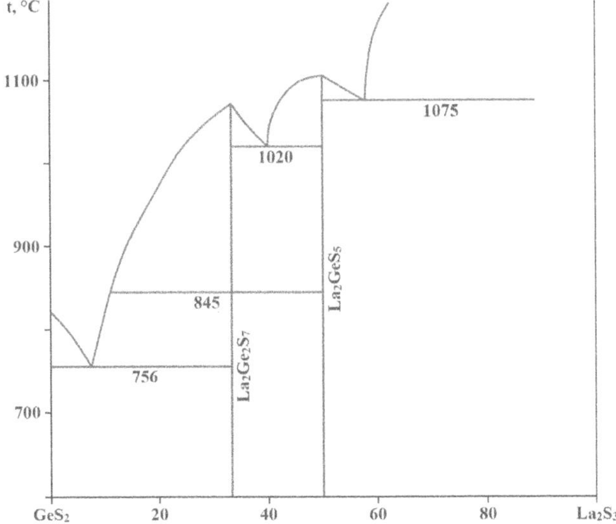

FIGURE 4.34 Phase diagram of the GeS_2–La_2S_3 system. (From Sarkisov, E.S., et al., *Izv. AN SSSR. Neorgan. mater.*, **6**(11), 2054, 1970.)

$La_2Ge_2S_7$ melts congruently at 1071°C and has a polymorphic transition at 845°C (Sarkisov et al. 1970b). This compound could be obtained by the interaction of the stoichiometric amounts of GeS_2 and La_2S_3 at 1000°C for 7–10 h (Sarkisov et al. 1968). It is possible that this compound corresponds to the $\mathbf{La_3Ge_{1.25}S_7}$ compounds, which crystallizes in the hexagonal structure with the lattice parameters $a = 1029.70 \pm 0.15$, $c = 581.20 \pm 0.20$ pm, and a calculated density of 4.555 g·cm^{-3} (Zeng et al. 2008) [$a = 1038$, $c = 581$ pm, and a calculated density of 4.60 g·cm^{-3} (Michelet and Flahaut 1969)]. To prepare this compound, La_2S_3, Al, Ge, and S (molar ratio 0.278:0.556:0.556:1.946) were mixed together and ground thoroughly within a N_2-filled glove box, pressed into pellets, and then sealed in evacuated quartz ampoule, heated in furnace up to 700°C over a period of 37.22 h, annealed at this temperature for 24 h, then heated again up to 1000°C over 15 h, kept at 1000°C for 240 h, cooled to 640°C over 180 h and then to room temperature by switching off the power of the furnace (Zeng et al. 2008). The precursor obtained from the first-step solid-state reaction was mixed with 0.80 g NaBr. Upon regrinding, repelleting, and resealing, the precursor + flux mixture was heated up to 850°C over 54 h, kept at this temperature for 240 h and then slowly cooled to 700°C over 100 h, and finally to room temperature by switching off the furnace powers. Dark red crystals were isolated manually from the residues after the flux was removed by washing with distilled water.

Another ternary compound, $\mathbf{La_4Ge_3S_{12}}$, which crystallizes in the hexagonal structure with the lattice parameters $a = 1940 \pm 2$, $c = 810 \pm 2$ pm (in the rhombohedral structure with the lattice parameters $a = 1153 \pm 2$ and $\alpha = 114.42°$), and the calculated and experimental densities of 4.36 and 4.27–4.29 g·cm^{-3}, respectively, is formed in the GeS_2–La_2S_3 system (Michelet et al. 1966; Mazurier and Etienne 1973; Michelet et al. 1975; Beskrovnaya et al. 1979). Yellow beige crystals of the title compound were prepared in the same way as $Y_6Ge_{2.5}S_{14}$ was synthesized using La_2S_3 instead of Y_2S_3 (Michelet et al. 1975).

4.22 Germanium–Cerium–Sulfur

GeS_2–Ce_2S_3: The phase diagram of this system, constructed through DTA, XRD, and metallography, is given in Figure 4.35 (Gadzhieva et al. 2000). The eutectics contain 22, 42.5, 58.5, and 87 mol% GeS_2 and crystallize at 1100°C, 1050°C, 910°C, and 700°C, respectively. The phase transition of the sold solution based on Ce_2S_3 takes place at 1330°C. Three compounds, **Ce_2GeS_5, $Ce_2Ge_2S_7$,** and **Ce_4GeS_8,** which melt congruently at 1140°C, 995°C, and 1235°C, respectively, are formed in this system. The last compound crystallizes in the hexagonal structure with the lattice parameters a = 1022 and c = 5.82 pm. Its single crystals were grown using chemical transport reactions. Ce_2GeS_5 was synthesized by the interaction of the stoichiometric amounts of Ge, S, and Ce_2S_3 in an evacuated quartz ampoule at 600°C–700°C for 8–10 h with the next increasing temperature to 1160°C and exposure at this temperature for 2–3 h (Beskrovnaya et al. 1971b).

Two another compounds, **$Ce_4(GeS_4)_3$** and **$Ce_6Ge_{2.5}S_{14}$,** are formed in the GeS_2–Ce_2S_3 system. $Ce_4(GeS_4)_3$ crystallizes in the trigonal structure with the lattice parameters a = 1937.5 ± 0.1, c = 802.9 ± 0.2 pm, a calculated density of 4.439 g·cm^{-3}, and an energy gap of 2.3 eV (Choudhury and Dorhout 2008) [in the hexagonal structure with the lattice parameters a = 1936 ± 2, c = 806 ± 2 pm or in the rhombohedral structure with the lattice parameters a = 1149 ± 2 and α = 114°43′ ± 0°12′ and a calculated density of 4.43 g·cm^{-3} (Michelet et al. 1966, 1975)]. Single crystals of this compound were obtained from the reaction of anhydrous $CeCl_3$ and Na_2GeS_3 (Choudhury and Dorhout 2008). Typically 0.36 mM of $CeCl_3$ and 0.28 mM of Na_2GeS_3 were mixed and ground in an agate mortar inside a N_2-filled glove-box. The mixture was then transferred to a graphite crucible, which was then placed inside a fused silica tube; flame sealed in vacuum and placed vertically into a furnace. The furnace was ramped to 825°C at a rate of 20°C·h^{-1}, and held constant at that temperature for 96 h. The furnace was then slowly cooled to ambient temperature at a rate of 5°C·h^{-1}. The as-synthesized product contained orange crystals (major product), a few dark colored hexagonal rods, and a small amount of colorless/white crystalline material. The product mixture was washed with DMF containing 10 vol% H_2O and dispersed in a sonocation bath for 1 min, which removed some white product, presumably NaCl, while leaving some colorless plates, orange, and dark colored crystals unaffected. The colorless plates were GeS_2, while orange and dark colored rod crystals were $Ce_4(GeS_4)_3$ and $NaCe_3GeS_7$, respectively. The powder sample of $Ce_4(GeS_4)_3$ was prepared by heating a stoichiometric combination of pure elements, or GeS_2 as the Ge source, in sealed fused silica ampoules in a furnace. The ampoules were first slowly heated to 550°C (5°Ch^{-1}) and kept at that temperature for 24 h followed by a ramp to 1000°C and kept at that temperature for 8 h, cooled down to 600°C at a rate of 35°Ch^{-1}, and then annealed at that temperature for 5 days before cooling down to room temperature.

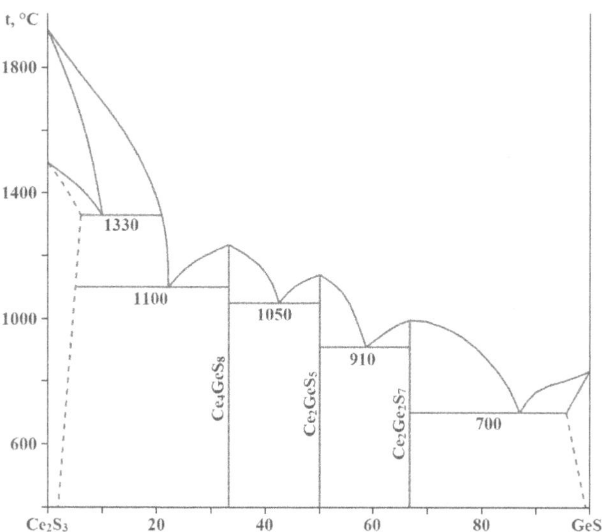

FIGURE 4.35 Phase diagram of the GeS_2–Ce_2S_3 system. (From Gadzhieva, S.R., et al., *Inorg. mater.*, **36**(1), 1, 2000.)

$Ce_6Ge_{2.5}S_{14}$ also crystallizes in the hexagonal structure with the lattice parameters a = 1022, c = 583 pm, and a calculated density of 4.63 g·cm^{-3} (Michelet et al. 1966, 1975). This compound was synthesized in the same way as $Y_6Ge_{2.5}S_{14}$ was prepared using Ce_2S_3 instead of Y_2S_3.

4.23 Germanium–Praseodymium–Sulfur

Several ternary compounds are formed in the Ge–Pr–S system. **$Pr_4Ge_3S_{12}$** crystallizes in the hexagonal structure with the lattice parameters a = 1930 ± 2, c = 801 ± 2 pm (in the rhombohedral structure with the lattice parameters a = 1146 ± 2 and α = 114°45′ ± 0°12′), and a calculated density of 4.50 g·cm^{-3} (Michelet et al. 1966, 1975). **$Pr_6Ge_{2.5}S_{14}$** also crystallizes in the hexagonal structure with the lattice parameters a = 1013, c = 582 pm, and a calculated density of 4.74 g·cm^{-3} (Michelet et al. 1975) [a = 1010, c = 581 pm, and the calculated and experimental densities of 4.83 and 5.0 g·cm^{-3}, respectively (Bakakin et al. 1974). These compounds were synthesized in the same way as $Y_6Ge_{2.5}S_{14}$ was prepared using Pr_2S_3 instead of Y_2S_3 (Michelet et al. 1966, 1975).

According to the data of Beskrovnaya et al. (1977) and Gus'kova and Serebrennikov (1973b,c), three another compounds, **Pr_2GeS_4, Pr_2GeS_5,** and **$Pr_2Ge_2S_7$** were also synthesized in the title ternary system. Pr_2GeS_5 was synthesized by the interaction of the stoichiometric amounts of Ge, S, and Pr_2S_3 in an evacuated quartz ampoule at 600°C–700°C for 8–10 h with the next increasing temperature to 1160°C and exposure at this temperature for 2–3 h (Beskrovnaya et al. 1971b).

4.24 Germanium–Neodymium–Sulfur

Several ternary compounds are formed in the Ge–Nd–S system. **$Nd_4(GeS_4)_3$** crystallizes in the trigonal structure with the lattice parameters a = 1925.0 ± 0.2, c = 794.9 ± 0.1 pm, a calculated density of 4.606 g·cm^{-3}, and an energy gap of 2.62 eV (Choudhury and Dorhout 2008) [in the hexagonal structure with the lattice parameters a = 1924 ± 2, c = 798 ± 2 pm or in the rhombohedral structure with the lattice parameters

$a = 1142 \pm 2$ and $\alpha = 114°44' \pm 0°12'$ and a calculated density of 4.59 g·cm^{-3} (Michelet et al. 1966, 1975)]. Single crystals of this compound were obtained in the same way as the single crystal of $Nd_4(GeS_4)_3$ were grown using anhydrous $NdCl_3$ instead of $CeCl_3$, but is this case the major product was dark orange (Choudhury and Dorhout 2008). The colorless plates were GeS_2, while dark orange and dark colored rod crystals were $Nd_4(GeS_4)_3$ and $NaNd_3GeS_7$, respectively. The powder sample of $Nd_4(GeS_4)_3$ was prepared similarly to the obtaining of the $Ce_4(GeS_4)_3$ powder sample.

$Nd_6Ge_{2.5}S_{14}$ also crystallizes in the hexagonal structure with the lattice parameters $a = 1005$, $c = 582$ pm, and a calculated density of 4.89 g·cm^{-3} (Michelet et al. 1975). This compound was synthesized in the same way as $Y_6Ge_{2.5}S_{14}$ was prepared using Nd_2S_3 instead of Y_2S_3 (Michelet et al. 1975).

According to the data of Gus'kova and Serebrennikov (1973b,c), **Nd_2GeS_4** was also obtained in the Ge–Nd–S ternary system. By the interaction of the stoichiometric amounts of Ge, S, and Nd_2S_3 in an evacuated quartz ampoule at 600°C–700°C for 8–10 h with the next increasing temperature to 1160°C and exposure at this temperature for 2–3 h, Beskrovnaya et al. (1971b) synthesized another ternary compound, **Nd_2GeS_5**, in this system.

4.25 Germanium–Samarium–Sulfur

Several ternary compounds are formed in the Ge–Sm–S system. **$Sm_4Ge_3S_{12}$** crystallizes in the hexagonal structure with the lattice parameters $a = 1919 \pm 2$, $c = 795 \pm 2$ pm (in the rhombohedral structure with the lattice parameters $a = 1139 \pm 2$ and $\alpha = 114°46' \pm 0°12'$), and a calculated density of 4.73 g·cm^{-3} (Michelet et al. 1966, 1975). **$Sm_6Ge_{2.5}S_{14}$** also crystallizes in the hexagonal structure with the lattice parameters $a = 994$, $c = 582$ pm, and a calculated density of 5.12 g·cm^{-3} (Michelet et al. 1975). These compounds were synthesized in the same way as $Y_6Ge_{2.5}S_{14}$ was prepared using SmS_3 instead of Y_2S_3 (Michelet et al. 1975).

According to the data of Beskrovnaya and Serebrennikov (1971) and Gus'kova and Serebrennikov (1973b,c), three another compounds, **Sm_2GeS_4**, **Sm_2GeS_5**, and **$Sm_2Ge_2S_7$** were also synthesized in this ternary system. Sm_2GeS_5 was synthesized by the interaction of the stoichiometric amounts of Ge, S, and Sm_2S_3 in an evacuated quartz ampoule at 600°C–700°C for 8–10 h with the next increasing temperature to 1160°C and exposure at this temperature for 2–3 h (Beskrovnaya et al. 1971b).

4.26 Germanium–Europium–Sulfur

GeS_2–EuS: The phase diagram of this system is a eutectic type with peritectic transformations (Barnier and Guittard 1978). The eutectic contains 25 mol% EuS and crystallizes at 615°C. Two compounds, **$EuGeS_3$** and **Eu_2GeS_4**, are formed in the GeS_2–EuS system. $EuGeS_3$ melts incongruently at 785°C (Barnier and Guittard 1978) and crystallizes in the triclinic structure with the lattice parameters $a = 846.8 \pm 0.3$, $b = 1176 \pm 1$, $c = 838.9 \pm 0.7$ pm and $\alpha = 90.49 \pm 0.11°$, $\beta = 104.56 \pm 0.04°$, $\gamma = 69.52 \pm 0.04°$, and the calculated and experimental densities of 4.25 and 4.22 g·cm^{-3}, respectively (Bugli et al. 1978) [$a = 838.1$, $b = 845.3$, $c = 1183.4$ pm and $\alpha = 111.44°$, $\beta = 99.90°$, $\gamma = 75.39°$, and a calculated density of 4.24 g·cm^{-3} (Barnier and Guittard 1978)]. This compound was prepared by the crystallization of the glass containing 50 mol% EuS at 600°C (Barnier and Guittard 1978).

Eu_2GeS_4 melts incongruently at 1160°C (Barnier and Guittard 1978) and has ferroelectric phase transition at ca. 62°C (Tampier and Johrendt 2001). Both phases crystallize in the monoclinic structure with the lattice parameters $a = 663.9 \pm 0.1$, $b = 667.4 \pm 0.1$, $c = 814.6 \pm 0.1$ pm, $\beta = 108.19 \pm 0.04°$, and a calculated density of 4.889 g·cm^{-3} for α-Eu_2GeS_4 at room temperature [$a = 663.8 \pm 0.1$, $b = 814.6 \pm 0.1$, $c = 667.2 \pm 0.1$ pm, $\beta = 108.20 \pm 0.02°$, and the calculated and experimental densities of 4.89 and 5.2 g·cm^{-3}, respectively (Bugli et al. 1979); $a = 670$, $b = 822$, $c = 667$ pm, and $\beta = 108.1°$ (Barnier and Guittard 1978); $a = 811.2 \pm 1.0$, $b = 671.4 \pm 0.6$, $c = 658.3 \pm 0.9$ pm, $\beta = 107°24' \pm 0°6'$, and the calculated and experimental densities of 4.90 and 4.86 g·cm^{-3}, respectively (Senova and Serebrennikov 1979); $a = 664.3 \pm 0.2$, $b = 667.4 \pm 0.2$, $c = 816.2 \pm 0.2$ pm, $\beta = 108.19 \pm 0.04°$, and a calculated density of 4.878 g·cm^{-3} for β-Eu_2GeS_4 at 107°C (Tampier and Johrendt 2001)]. Powder samples of Eu_2GeS_4 were synthesized from the elements in two steps (Tampier and Johrendt 2001). First, the binary alloy Eu_2Ge was prepared by heating stoichiometric mixtures of Eu metal and Ge powder at 850°C for 10 h in alumina crucibles, sealed in quartz ampoules under Ar atmosphere. Then, after homogenization in an Ar-filled glove box, Eu_2Ge was oxidized with stoichiometric amounts of sulfur at 750°C for 15 h. The samples were homogenized again and reheated to 1050°C for 50 h. This procedure results in bright orange powders of Eu_2GeS_4, which are stable in air. The title compound could be also obtained by union of the EuS and the vitreous variety of GeS_2 (Bugli et al. 1979). The synthesis was carried out under vacuum in a silica ampoule. The heat treatment at 950°C for 24 h was followed by slow cooling. The mixture of Eu_2GeS_4 and Ge was also synthesized at the interaction of GeS and EuS at 520°C (Senova 1979).

The glassy alloys of the GeS_2–EuS system were synthesized by reacting appropriate mixtures of EuS and GeS_2 in evacuated one- or two-compartment silica tubes at 900°C–950°C for at least 30 h, followed by quenching in water (cooling rate 20°C·s^{-1} at 20°C or 30°C·s^{-1} at 0°C) or air (10°C·s^{-1}) or furnace-cooling (from 950°C to 400°C, cooling rate varying from 10 to 2°C/min) (Kompanichenko et al. 2004). The glasses were obtained in the range 70–100 mol % GeS_2. With decreasing GeS_2 content, the color of the glasses changed from yellow-orange to dark red. With increasing EuS content, the glass-transition temperature decreases from 492°C (GeS_2) to 448°C (mixtures). The crystallization temperature also decreases: from 664°C–690°C (GeS_2) to 550°–620°C (mixtures with EuS). The most homogeneous glasses can be prepared at EuS contents below 15 mol%. To obtain the homogeneous glasses in the range 25–30 mol% EuS, an increased cooling rate needed (20–30°C·s^{-1}). According to the data of Barnier and Guittard (1978), the glass samples in this system could be obtained in the region 0–50 mol% EuS. Annealing at 600°C makes it possible to crystallize the obtained glasses. The glasses crystallization takes place within the interval from 515°C to 580°C.

4.27 Germanium–Gadolinium–Sulfur

Several ternary compounds are formed in the Ge–Gd–S system. **$Gd_4Ge_3S_{12}$** is stable up to 1030°C (Stepanets et al. 1973b) and crystallizes in the hexagonal structure with the lattice parameters a = 1909 ± 2, c = 790 ± 2 pm (in the rhombohedral structure with the lattice parameters a = 1133 ± 2 and α = 114°46′ ± 0°12′), and a calculated density of 4.92 g·cm^{-3} (Michelet et al. 1966, 1975). **$Gd_6Ge_{2.5}S_{14}$** also crystallizes in the hexagonal structure with the lattice parameters a = 984, c = 582 pm, and a calculated density of 5.36 g·cm^{-3} (Michelet et al. 1975) [a = 983.8, c = 582 pm, and the calculated and experimental densities of 5.36 and 5.32 g·cm^{-3}, respectively (Stepanets et al. 1973b)]. These compounds were synthesized in the same way as $Y_6Ge_{2.5}S_{14}$ was prepared using Gd_2S_3 instead of Y_2S_3 (Michelet et al. 1975).

According to the data of Gus'kova and Serebrennikov (1973b,c), **Gd_2GeS_4** was also obtained in the Ge–Gd–S ternary system.

4.28 Germanium–Terbium–Sulfur

GeS_2–Dy_2S_3: Two ternary compounds, **Tb_2GeS_5** and **$Tb_6Ge_{2.5}S_{14}$**, are formed in this system. First of them melts at 1010°C–1040°C, has a polymorphic transition at 970°C–1005°C and an energy gap of 2.1 eV (Stepanets et al. 1973a). $Tb_6Ge_{2.5}S_{14}$ crystallizes in the hexagonal structure with the lattice parameters a = 979, c = 582 pm, and a calculated density of 5.45 g·cm^{-3} (Michelet et al. 1975) [a = 978.8, c = 582 pm, and the calculated and experimental densities of 5.45 and 5.41 g·cm^{-3}, respectively (Stepanets et al. 1973b)]. This compound was synthesized in the same way as $Y_6Ge_{2.5}S_{14}$ was prepared using Tb_2S_3 instead of Y_2S_3 (Michelet et al. 1975).

4.29 Germanium–Dysprosium–Sulfur

GeS_2–Dy_2S_3: The **$Dy_6Ge_{2.5}S_{14}$** ternary compound, which crystallizes in the hexagonal structure with the lattice parameters a = 973, c = 582 pm, and a calculated density of 5.59 g·cm^{-3} (Michelet et al. 1975) [a = 973.0, c = 582 pm, and the calculated and experimental densities of 5.54 and 5.52 g·cm^{-3}, respectively (Stepanets et al. 1973b)], is formed in this system. This compound was synthesized in the same way as $Y_6Ge_{2.5}S_{14}$ was prepared using Dy_2S_3 instead of Y_2S_3 (Michelet et al. 1975).

4.30 Germanium–Holmium–Sulfur

GeS_2–Ho_2S_3: Two ternary compounds, **Ho_2GeS_5** and **$Ho_6Ge_{2.5}S_{14}$**, are formed in this system. First of them melts at 1010°C–1040°C and has a polymorphic transition at 970°C–1005°C and an energy gap of 2.5 eV (Stepanets et al. 1973a). $Ho_6Ge_{2.5}S_{14}$ crystallizes in the hexagonal structure with the lattice parameters a = 968.6 ± 0.1, c = 581.9 ± 0.1 pm, and a calculated density of 5.689 g·cm^{-3} (Lychmanyuk et al. 2007) [a = 969, c = 582 pm, and a calculated density of 5.68 g·cm^{-3} (Michelet et al. 1975); a = 971.1, c = 582.5 pm, and the calculated and experimental densities of 5.65 and 5.61 g·cm^{-3}, respectively (Stepanets et al. 1973b); a = 964.80 ± 0.14, c = 579.20 ± 0.12 pm, and a calculated density of 5.774 g·cm^{-3} for the $Ho_6Ge_{2.544}S_{14}$ composition (Zeng et al. 2008)]. To prepare this compound, Ho, Ge, Al, and S (molar ratio 0.556:0.556:0.556:2.780) were mixed together and ground thoroughly within a N_2-filled glove box, pressed into pellets, and then sealed in evacuated quartz ampoule, heated in a furnace up to 360°C over a period of 8 h, then up to 500°C over 9.33 h, annealed at 500°C for 50 h, then heated again up to 850°C over 23.33 h, kept at this temperature for 240 h, cooled to 450°C over 133.33 h and then to room temperature by switching off the power of the furnace. The precursor obtained from the first-step solid-state reaction was mixed with 0.70 g KBr. Upon regrinding, repelleting, and resealing, the precursor + flux mixture was heated up to 900°C over 14.5 h, kept at this temperature for 240 h and then slowly cooled to 650°C over 166.67 h, and finally to room temperature by switching off the furnace power. Dark red crystals were isolated manually from the residues after the flux was removed by washing with distilled water. $Ho_6Ge_{2.5}S_{14}$ was also synthesized in the same way as $Y_6Ge_{2.5}S_{14}$ was prepared using Ho_2S_3 instead of Y_2S_3 (Michelet et al. 1975).

4.31 Germanium–Erbium–Sulfur

GeS_2–Er_2S_3: The **Er_2GeS_5** ternary compound, which crystallizes in the hexagonal structure with the lattice parameters a = 973.1 ± 0.4, c = 582.5 ± 0.4 pm, and the calculated and experimental densities of 6.46 and 6.39 g·cm^{-3}, respectively, is formed in this system (Stepanets et al. 1973c). It was prepared by the interaction of GeS_2 and Er_2S_3 at 1200°C. Gulay et al. (2009) noted that ternary compounds do not exist in this system at 600°C.

According to the data of Zeng et al. (2008, 2009), the **$Er_3Ge_{1.25}S_7$** (this phase lies on the GeS_2–Er_2S_3 section), **$Er_3Ge_{1.333±0.010}S_7$**, **$Er_3Ge_{1.382±0.008}S_7$**, and **$Er_3Ge_{1.5}S_7$** phases exist in the Ge–Er–S system. All these phases crystallize in the hexagonal structure with the lattice parameters a = 961.46 ± 0.14, c = 585.96 ± 0.12 pm, and a calculated density of 5.784 g·cm^{-3} for the first phase, a = 959.30 ± 0.14, c = 584.90 ± 0.12 pm, and a calculated density of 5.869 g·cm^{-3} for the second phase, a = 963.60 ± 0.14, c = 584.60 ± 0.12 pm, and a calculated density of 5.838 g·cm^{-3} for the third one and a = 960.61 ± 0.13, c = 583.46 ± 0.18 pm, and a calculated density of 5.948 g·cm^{-3} for $Er_3Ge_{1.5}S_7$. Single crystals of these phases were obtained *via* a precursor/flux route. Appropriate starting materials (molar ratio Er_2S_3/Al/Ge/S = 1.106:2.223:2.230:7.765, 3.800 g KBr for $Er_3Ge_{1.25}S_7$; molar ratio Er_2S_3/Ag/Ge/S = 0.374:0.249:0.249:0.624, 0.80 g KBr for $Er_3Ge_{1.333±0.010}S_7$; molar ratio Er_2S_3/Ag_2S/Ge/S = 1.164:0.388:0.778:1.559, 2.830 g KBr for $Er_3Ge_{1.382±0.008}S_7$; molar ratio Er_2S_3/GeS/GeS_2 = 2.851:0.950:1.906, 6.383 g KBr for $Er_3Ge_{1.5}S_7$) were mixed together and ground thoroughly within a N_2-filled glove box, pressed into pellets, and then sealed in evacuated quartz ampoules before being heating. The precursors obtained from first-step solid-state reactions ($AlErGeS_5$ for the first and for the second phases, $AgEr_3GeS_7$ for the third phase, and $Er_6Ge_3S_{14}$ for the fourth one) were mixed with KBr flux. Upon regrinding, repelleting, and resealing, the precursor + flux mixtures were heated at 800°C–900°C for 120–240 h, then slowly cooled to 550°C–650°C, and finally to room temperature by switching off the furnace. Dark red crystals were isolated manually from the residues after the fluxes were removed by washing with distilled water.

4.32 Germanium–Ytterbium–Sulfur

Three ternary compounds, $\mathbf{YbGeS_3}$, $\mathbf{Yb_2GeS_4}$, and $\mathbf{Yb_6Ge_{2.5}S_{14}}$, are formed in the Ge–Yb–S system. $YbGeS_3$ crystallizes in the triclinic structure with the lattice parameters a = 816.3, b = 839.5, c = 1159 pm, α = 110.54°, β = 96.70°, and γ = 76.12°, and Yb_2GrS_4 crystallizes in the monoclinic structure with the lattice parameters a = 663.2, b = 812.8, c = 659.3 pm, and β = 107.90° (Rustamov et al. 1981). $Yb_6Ge_{2.5}S_{14}$ crystallizes in the hexagonal structure with the lattice parameters a = 971.2, c = 581.2 pm, and the calculated and experimental densities of 4.07 and 4.04 g·cm^{-3}, respectively (Stepanets et al. 1973b)].

4.33 Germanium–Thorium–Sulfur

The **ThGeS** ternary compound, which crystallizes in the tetragonal structure with the lattice parameters a = 394.11 ± 0.07, c = 1713.95 ± 0.40 pm, and the calculated and experimental densities of 8.40 and 8.32 g·cm^{-3}, respectively (Stocks et al. 1981) [a = 393.0, c = 1712 pm, and the calculated and experimental densities of 8.46 and 8.32 g·cm^{-3}, respectively (Hahn and Stocks 1968)], is formed in the Ge–Th–S system. To prepare this compound, the elements were weighed under N_2 atmosphere in the stoichiometric ratio and initially heated to about 400°C in an evacuated quartz ampoule (Hahn and Stocks 1968; Stocks et al. 1981). Subsequently, the inhomogeneous product was finely powdered under N_2 and heated to 980°C in a sealed tungsten crucible, which was enclosed in an evacuated quartz ampoule. Since the preparation is extremely sensitive to traces of oxygen, it proved necessary to add Th as getter metal. Nevertheless, pure preparation was obtained only if no longer than 48 h the mixture was heated at 980°C. The mixture was annealed at 600°C–700°C for 8–10 days. Then, the sample pressed into pastilles was heated in the same way over different time intervals and at different temperature up to 1400°C in tungsten crucible, wherein above 1050°C additionally Ar was used as a protective gas. The title compound was obtained as black microcrystalline powder.

4.34 Germanium–Uranium–Sulfur

The **UGeS** ternary compound, which crystallizes in the tetragonal structure with the lattice parameters a = 381.12 ± 0.01, c = 830.54 ± 0.05 pm, and a calculated density of 9.43 g·cm^{-3} (Haneveld and Jellinek 1969) [a = 382.0 and c = 832.3 pm (Zygmunt 1977)], is formed in the Ge–U–S system. This compound was prepared in the following way (Haneveld and Jellinek 1969). A small rod of uranium was placed in a quartz tube and heated in a current of hydrogen at 230°C. The voluminous uranium hydride powder thus produced was dehydrided by heating at 400°C under vacuum. An equivalent proportion of GeS was added immediately and the quartz tube was evacuated and sealed. The mixture was heated at 800°C–900°C for 2 or 3 days. Then, the product was slowly cooled to room temperature.

4.35 Germanium–Tin–Sulfur

GeS–SnS: The phase diagram of this system, constructed through DTA and XRD, is given in Figure 4.36 (Magunov and Magunov 1992). The eutectic contains 25 mol% SnS and crystallizes at 630°C. The phase transitions of GeS and SnS were determined within the interval from 525°C to 605°C.

Bletskan et al. (1981) and Elli (1963) indicated that a continuous series of solid solutions is formed in the GeS–SnS, and the phase diagram of this system belongs to the type I according to the Roseboom's classification. The liquidus and solidus are slightly concave toward the compositions axis. Low-temperature solid solutions crystallize in the orthorhombic structure and the high-temperature ones crystallize in the cubic structure. The temperature of polymorphic transition slightly decreases at the increase of the SnS content.

$\mathbf{GeS_2}$**–SnS:** The phase diagram of this system, constructed through DTA, XRD, and metallography, is presented in Figure 4.37 (Feltz et al. 1978). The eutectic contains 66 mol% SnS and crystallizes at 580°C. The composition of peritectic point is 52 mol% SnS. The phase transition of the solid solutions based on SnS takes place at 600°C. The $\mathbf{GeSnS_3}$ ternary compound (mineral stangersite), which melts incongruently at 613°C (Feltz et al. 1978) [at ~610°C (Moh 1975b); at 609°C (Fenner and Mootz 1976)] and crystallizes in the monoclinic structure (Alpen et al. 1975) with the lattice parameters

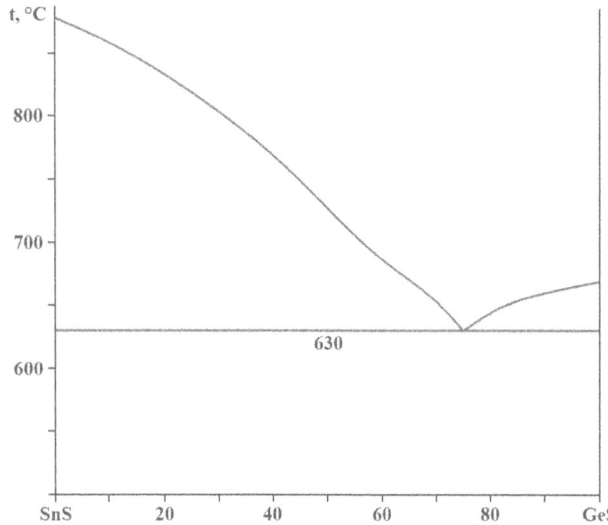

FIGURE 4.36 Phase diagram of the GeS–SnS system. (From Magunov, R.L., and Magunov, I.R., *Zhurn. neorgan. khimii*, **37**(11), 2613, 1992.)

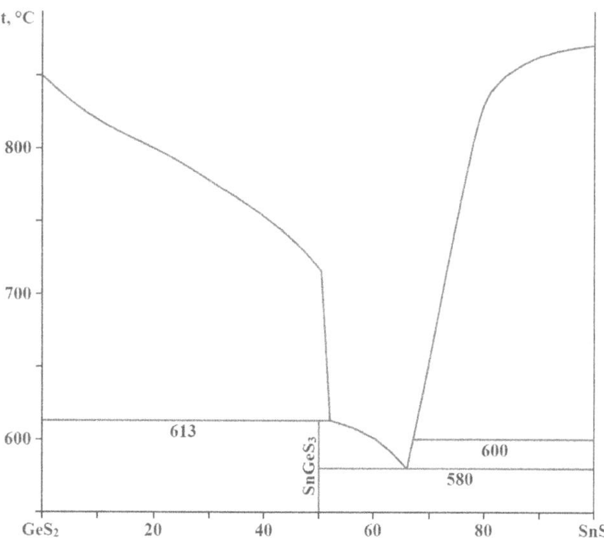

FIGURE 4.37 Phase diagram of the GeS_2–SnS system. (From Feltz, A., et al., *Krist. und Techn.*, **13**(4), 405, 1978.)

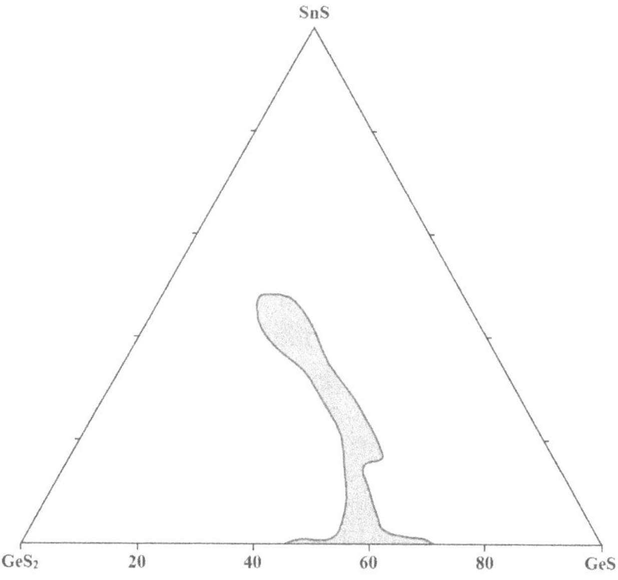

FIGURE 4.38 Glass forming region in the GeS–GeS_2–SnS subsystem. (From Feltz, A., et al., *Z. anorg. und allg. Chem.*, **403**(3), 243, 1974.)

$a = 727.04 \pm 0.15$, $b = 1019.7 \pm 0.2$, $c = 684.63 \pm 0.14$ pm, $\beta = 105.34 \pm 0.03°$, and a calculated density of 3.98 g·cm^{-3} (Sejkora et al. 2001, 2020a,b) [$a = 726.9 \pm 0.1$, $b = 1022.0 \pm 0.2$, $c = 687.3 \pm 0.2$ pm, $\beta = 105.45 \pm 0.01°$, the calculated and experimental densities of 3.879 and 3.71 g·cm^{-3}, respectively, and an energy gap of 2.23 eV (Fenner and Mootz 1974, 1976; Moh 1975b)], is formed in the GeS_2–SnS system. The synthesis of the title compound from the elements or mixture of the binary sulfides was carried out in conventional quartz ampoule (Fenner and Mootz 1976). The starting materials were weighed in stoichiometric amounts. The evacuated ampoule was directly annealed at 500°C, or heated until the sulfide mixture melts up to 1000°C, quenched and annealed at 500°C. $SnGeS_3$ could be also prepared by hydrothermal treatment of binary sulfides mixture in 2 N HCl (the degree of filling was about 60%, the temperature gradient was about 25°C, and the reaction temperature was 450°C).

The synthesis of $SnGeS_3$ was also made by direct alloying the elementary components in evacuated to 0.013 Pa quartz ampoules with vibrational mixing (Bletskan et al. 2003a). The maximum synthesis temperature was 930°C. For polycrystalline alloys to be obtained, the melt was quenched at a rate of 0.2°C·s^{-1}. Single crystals of the title compound were grown by vertically direct crystallization (Bletskan et al. 2003a) from the melt or by sublimation (Feltz et al. 1978). To obtain glassy $SnGeS_3$, the melt was quenched at a rate of 17°C·s^{-1} (Bletskan et al. 2003a). The glass transition temperature for glassy $SnGeS_3$ is 326°C.

According to the data of Moh (1975b), at 500°C a complete solid solution series exists between GeS_2 and SnS. $GeSnS_3$ may coexist with GeS_2, SnS_2, Sn_2S_3 or the Sn-rich portion of the monosulfide solid solution, whereas the Ge-rich solid solution is stable with GeS_2. Although GeS_2 and SnS_2 are connected by a tie line, they show no detectable mutual solubility. Ge is stable with the entire range of monosulfide solid solution.

The glass forming region in the GeS–GeS_2–SnS subsystem is shown in Figure 4.38 (Feltz et al. 1974). The glass transition temperature for these glasses is within the interval 235°C–275°C.

4.36 Germanium–Lead–Sulfur

GeS–PbS: The phase diagram of this system, constructed through DTA and XRD, is presented in Figure 4.39 (Magunov et al. 1992). The eutectics contain 32 and 78 mol% GeS and crystallize at 510°C and 500°C. The **$PbGeS_2$** ternary compound, which melts congruently at 540°C, is formed in this system. GeS dissolves up to 15 mol% PbS and the solubility based on PbS reaches up to 10 mol% GeS. The solid solutions based on GeS are characterized by the polymorphic transition.

GeS_2–PbS: According to the data of Elli and Mugnoli (1963), the phase diagram of this system is a eutectic type with peritectic transformation. In the range of 0–25 mol% GeS_2, the samples are sufficiently stable in humid air, and with a

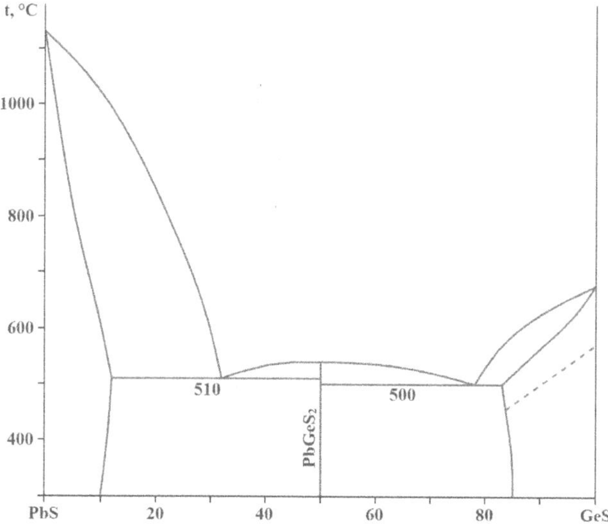

FIGURE 4.39 Phase diagram of the GeS–PbS system. (From Magunov, R.L., et al., *Zhurn. neorgan. khimii*, **37**(11), 2609, 1992.)

higher GeS_2 content, they interact with the moisture in the air with formation of H_2S. Two ternary compounds, **$PbGeS_3$** and **Pb_2GeS_4**, are formed in this system. $PbGeS_3$ melts incongruently at 593°C (Elli and Mugnoli 1963) and crystallizes in the monoclinic structure (Alpen et al. 1975) with the lattice parameters $a = 720.7 \pm 0.4$, $b = 1045 \pm 1$, $c = 683.7 \pm 0.4$ pm, $\beta = 105.62 \pm 0.09°$, and an energy gap of 2.57 ± 0.01 eV (Bletskan et al. 1990) [$a = 722.4 \pm 0.3$, $b = 1044.2 \pm 0.2$, $c = 682.5 \pm 0.2$ pm, $\beta = 105.7 \pm 0.01°$, the calculated and experimental densities of 5.04 and 5.05 g·cm^{-3}, respectively, and an energy gap of 2.4 eV (Ribes et al. 1974; Fenner and Mootz 1976); $a = 727$, $b = 1050$, $c = 688$ pm, and $\beta = 105.0°$ (Fenner and Mootz 1974)]. The synthesis of the title compound from the elements or mixture of the binary sulfides was carried out in conventional quartz ampoule (Fenner and Mootz 1976). The starting materials were weighed in stoichiometric amounts. The evacuated ampoule was directly annealed at 500°C, or heated until the sulfide mixture melts up to 1000°C, quenched and annealed at 500°C. $PbGeS_3$ could be also prepared by hydrothermal treatment of binary sulfides mixture in 2 N HCl (the degree of filling was about 60%, the temperature gradient was about 25°C, and the reaction temperature was 430°C).

The synthesis of $PbGeS_3$ was also made by direct alloying the elementary components in evacuated to 0.013 Pa quartz ampoules with vibrational mixing (Bletskan et al. 1990, 2003a). The maximum synthesis temperature was 930°C. For polycrystalline alloys to be obtained, the melt was quenched at a rate of 0.2°C·s^{-1}. Single crystals of the title compound were grown by vertically direct crystallization from the melt. To obtain glassy $PbGeS_3$, the melt was quenched at a rate of 17°C·s^{-1}. The glass transition temperature for glassy $PbGeS_3$ is 305°C.

Pb_2GeS_4 melts congruently at 621°C (Elli and Mugnoli 1963) and has two polymorphic modifications. α-Pb_2GeS_4 crystallizes in the monoclinic structure with the lattice parameters $a = 801 \pm 3$, $b = 886 \pm 4$, $c = 1091 \pm 3$ pm, $\beta = 114.4 \pm 0.4°$, and an energy gap of 2.29 ± 0.01 eV (Bletskan et al. 1990) [$a = 797.42 \pm 0.06$, $b = 892.55 \pm 0.08$, $c = 1087.61 \pm 0.08$ pm, and $\beta = 114.171 \pm 0.009°$ (Susa and Steinfink 1971a,b); $a = 801 \pm 3$, $b = 886 \pm 4$, $c = 1091 \pm 3$ pm, and $\beta = 114.4 \pm 0.4°$ (Bletskan et al. 2003b)]. β-Pb_2GeS_4 crystallizes the cubic structure with the lattice parameter $a = 1409.6 \pm 0.4$ pm (Poduska et al. 2002).

The most successful result for the synthesis of α-Pb_2GeS_4 was achieved when starting with elemental Ge, powdered S, and powdered PbS (Susa and Steinfink 1971b). Stoichiometric ratios of Ge and Pb with sulfur or of PbS were mixed, sealed under vacuum in Vycor tubing, and pre-reacted at temperatures ranging from 200°C to 400°C for 2–4 h. The temperature was then raised and kept in the range 600°C–800°C for about 12 h. After cooling, the vial was examined visually to see whether some of the starting materials had remained unreacted. If the sample looked homogeneous, the vial was opened. If α-Pb_2GeS_4 was detected in the sample, it was resealed in a Vycor tube and fired at the same or at higher temperatures until it appeared to have completely reacted. In some cases, the temperature was raised to 1330°C after first placing the material in a graphite crucible which was then sealed in a Vycor tube.

It was also synthesized either by melting PbS and GeS_2 in an equimolar ratio or by directly melting the elements in quartz ampoule evacuated to 0.013 Pa under vibrational mixing (Bletskan et al. 2003b). The maximum heating temperature was 830°C. The melts were cooled in a switched-off furnace. The alloys were annealed at 560°C for 200 h. Single crystals were grown by the Bridgman technique.

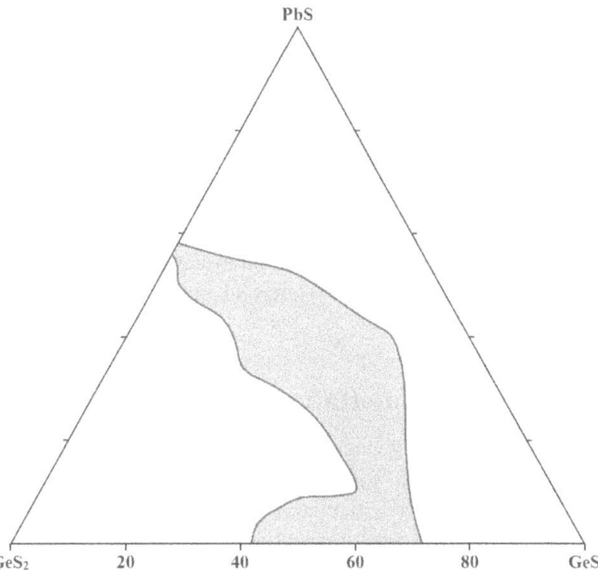

FIGURE 4.40 Glass forming region in the GeS–GeS_2–PbS subsystem. (From Feltz, A., and Voigt, B., *Z. anorg. und allg. Chem.*, **403**(1), 61, 1974.)

β-Pb_2GeS_4 forms consistently when stoichiometric amounts of the elements were sealed in an evacuated ampoule and heated in the appropriate manner (Poduska et al. 2002). After heating at a rate of 50°C·h^{-1} from room temperature to 300°C, the sample remained at this temperature for 4 h. After this soak, the furnace was then ramped at 50°C·h^{-1} to 675°C, after which time the ampoule was removed from the furnace and quenched immediately in cold water. Larger crystals of the desired product tended to form when the sample sat at 675°C for 6–12 h, and then cooled to 625°C over 24 h before the water quenching. The high-temperature phase was present in some samples that were not quenched, but instead were cooled from 675°C to room temperature over 12 h. The α-Pb_2GeS_4 and β-Pb_2GeS_4 are easily distinguished by color.

The **$3PbS{\cdot}2GeS_2$** ternary compound was not determined in the GeS_2–PbS system (Elli and Mugnoli 1963).

The glass forming region in the GeS–GeS_2–PbS subsystem is given in Figure 4.40 (Feltz and Voigt 1974). The glass transition temperature for these glasses is within the interval 263°C–283°C.

4.37 Germanium–Zirconium–Sulfur

The quite stable **ZrGeS** ternary compound, which crystallizes in the tetragonal structure with the lattice parameters $a = 362.6$ and $c = 801.9$ pm (Haneveld and Jellinek 1964) [$a = 361$, $c = 800$ pm, and the calculated and experimental densities of 6.18 and 6.06 g·cm^{-3}, respectively (Onken et al. 1964)], is formed in the Ge–Zr–S system. To obtain this compound, stoichiometric amounts of the elements were initially pre-tempered in an evacuated tube at 600°C for 10 days (Onken et al. 1964). Then, the reaction mixture was carefully rubbed in an agate mortar in a N_2-filled box, pressed into a pastille and annealed at 830°C for several weeks. The ampoule with tablet was stored in the hottest part of the furnace to obtain the single crystals by sublimation. ZrGeS could be also prepared by heating GeS with an equivalent proportion of Zr in an evacuated quartz tube at 700°C–800°C for 2–4 days (Haneveld and Jellinek 1964).

4.38 Germanium–Hafnium–Sulfur

The **HfGeS** ternary compound, which crystallizes in the tetragonal structure with the lattice parameters a = 361, c = 794 pm, and the calculated and experimental densities of 9.09 and 7.76 g·cm^{-3}, respectively, is formed in the Ge–Hf–S system (Onken et al. 1964). To prepare this compound, stoichiometric amounts of Ge, Hf, and S were initially pre-tempered in an evacuated tube at 600°C for 10 days. Then, the reaction mixture was carefully rubbed in an agate mortar in a N_2-filled box, pressed into a pastille and annealed at 850°C for several weeks. The ampoule with tablet was stored in the hottest part of the furnace to obtain the single crystals by sublimation.

4.39 Germanium–Phosphorus–Sulfur

GeS_2–P_4S_3: The phase diagram of this system, constructed through DTA, XRD, metallography, and measuring of microhardness, is presented in Figure 4.41 (Vinogradova and Maysashvili 1979b). The eutectic is degenerated from the P_4S_3-side. The immiscibility region exists in this system at 650°C within the interval 15–92 mol% P_4S_3.

GeS_2–P_4S_7: The phase diagram of this system, constructed through DTA, XRD, metallography, and measuring of microhardness, is given in Figure 4.42 (Vinogradova and Maysashvili 1979b). The eutectic contains 62 mol% P_4S_7 and crystallizes at 260°C. The immiscibility region exists in this system at 550°C within the interval 9–60 mol% P_4S_7.

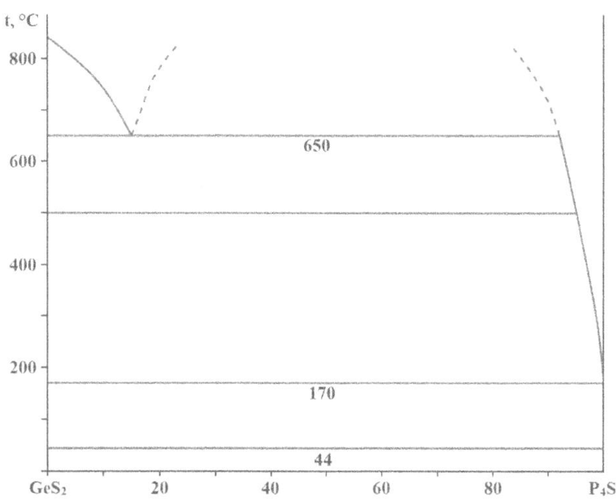

FIGURE 4.41 Phase diagram of the GeS_2–P_4S_3 system. (From Vinogradova, G.Z., and Maysashvili, N.G., *Zhurn. neorgan. khimii*, **24**(4), 1067, 1979.)

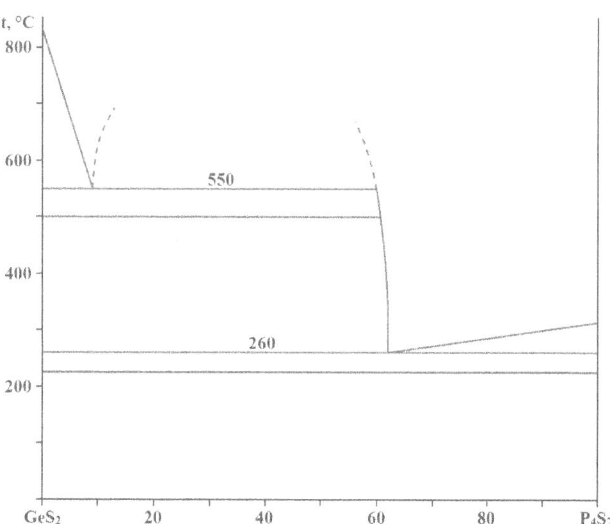

FIGURE 4.42 Phase diagram of the GeS_2–P_4S_7 system. (From Vinogradova, G.Z., and Maysashvili, N.G., *Zhurn. neorgan. khimii*, **24**(4), 1067, 1979.)

GeS_2–P_4S_{10}: The phase diagram of this system, constructed through DTA, XRD, metallography, and measuring of microhardness, is shown in Figure 4.43 (Vinogradova and Maysashvili 1979b). The eutectic contains 34 mol% P_4S_{10} and crystallizes at 250°C. The immiscibility region exists in this system at 550°C within the interval 7–15 mol% P_4S_{10}.

The fields of primary crystallization of GeS_2, sulfur, P_4S_{10}, P_4S_7, P_4S_9, P_4S_5, and P_4S_3 exist on the liquidus surface of the Ge–P–S ternary system Figure 4.44 (Vinogradova and Maysashvili 1979b). The field of the P_4S_3 primary crystallization is degenerated. The P_4S_9 and P_4S_5 compounds melt incongruently and their fields of primary crystallization are small.

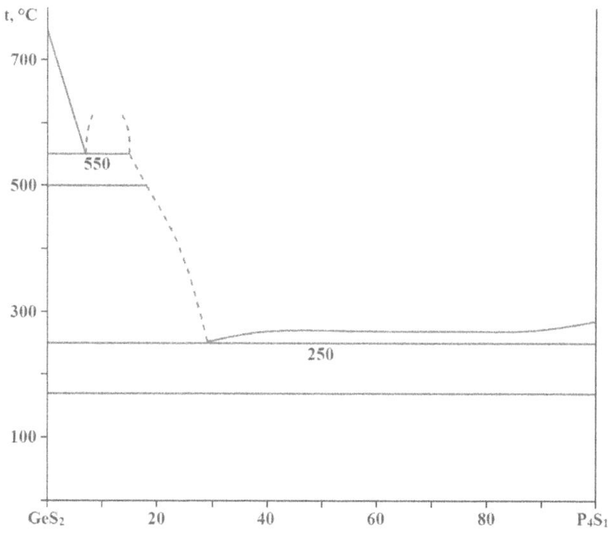

FIGURE 4.43 Phase diagram of the GeS_2–P_4S_{10} system. (From Vinogradova, G.Z., and Maysashvili, N.G., *Zhurn. neorgan. khimii*, **24**(4), 1067, 1979.)

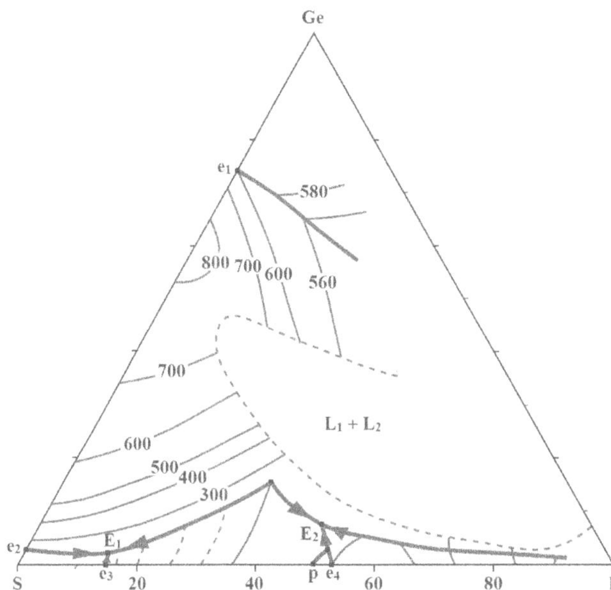

FIGURE 4.44 Liquidus surface of the Ge–P–S ternary system. (From Vinogradova, G.Z., and Maysashvili, N.G., *Zhurn. neorgan. khimii*, **24**(4), 1067, 1979.)

The immiscibility region exists in the field of GeS_2 primary crystallization.

According to the data of Voroshilov et al. (1986) and Potoriy and Milyan (2016), the $\mathbf{Ge_2P_2S_6}$ ternary compound, which melts at 237°C and has an experimental density of 2.20 g·cm^{-3}, is formed in the Ge–P–S system.

The glass-forming region in the Ge–P–S ternary system is given in Figure 4.45 (Hilton et al. 1964; Vinogradova and Maysashvili 1979a). In the area with more than 25 mol% P, considerable sublimation is observed, and the analysis of sublimate shows that its composition varies from pure phosphorus to alloys of the P–S system enriched with phosphorus. The glassy samples were obtained by air quenching from 850°C to 900°C (Vinogradova and Maysashvili 1979a) or from 1000°C (Hilton et al. 1964).

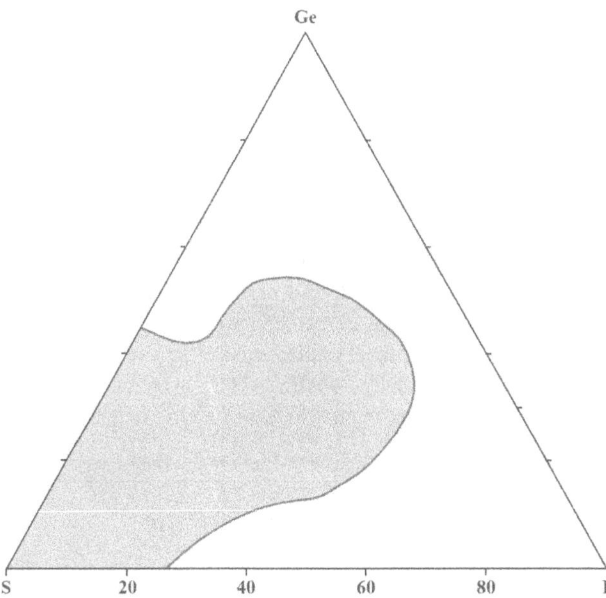

FIGURE 4.45 Glass-forming region in the Ge–P–S system. (From Hilton, A.R., et al., *Infrared Phys.*, **4**(4), 213, 1964.)

4.40 Germanium–Arsenic–Sulfur

$\mathbf{GeS_2–As_2S_3}$**:** The equilibrium phase diagram of this system is shown in Figure 4.46b (Khiminets et al. 1988). It belongs to the peritectic type with limited solubility of the components in the solid state. The fields of solid solutions based on binary compounds are limited by concentrations of 12 mol% As_2S_3 from GeS_2-side and 16 mol% GeS_2 from As_2S_3-side. Peritectic transformation takes place at 449°C. The boundaries of the (α-GeS_2) + (β-GeS_2) + (As_2S_3) three-phase region are approximately determined, which is due to the blurring of the phase transformation and an increase in the glass-forming ability of the alloys when As_2S_3 is introduced into GeS_2. The non-equilibrium phase diagram is presented in Figure 4.46a. In a non-equilibrium state, a continuous series of solid solutions is formed. It was found that if the melt cooling rate exceeds 50°C·s^{-1}, all alloys in the system can be obtained in a glassy state.

The temperature effects of the beginning and end of the samples solidification in the range of 0–62.48 mol% GeS_2 were also determined in this system by Elli et al. (1964). All investigated samples are glassy, sensitive to the air moisture and do not crystallize even after prolonged annealing at 500°C.

The part of the liquidus surface of the Ge–As–S ternary system was constructed by Vinogradova and Dembovskiy (1970, 1971) and is presented in Figure 4.47. The region of the **GeAsS** ternary compound existence was also determined. This compound exist in a glassy form and melts at 670°C. According to the data of Voroshilov et al. (1986), the $\mathbf{Ge_2As_2S_6}$ ternary compound, which also exist in a glassy form and has an experimental density of 3.07 g·cm^{-3}, is formed in the Ge–As–S system. The glass forming region in this system was determined by Myuller et al. (1962).

4.41 Germanium–Antimony–Sulfur

$\mathbf{GeS_2–Sb_2S_3}$**:** The temperature effects of the beginning and end of the samples solidification in the range of 0–68.28 mol% GeS_2 were determined in this system by Elli et al. (1964). The alloys containing up to 42 mol% GeS_2 give cooling curves with distinct and reproducible thermal effects that are in a very narrow range (1°C–22°C). When the content is more than 42 mol% GeS_2 thermal effects are insignificant, and the interval of solidification increases significantly (75°C–118°C). The alloys up to 32 mol% GeS_2 are crystalline, but with a content of 9.87 and 18.63% mol% GeS_2, a glassy phase of GeS_2 was found. In the range of 32–42 mol% GeS_2, the samples are half-glassy. The minimum crystallization temperature of the samples is at 39.97% GeS_2. With a content of more than 42 mol% GeS_2, the samples become glassy. The samples are moisture stable up to 25 mol% GeS_2. The ingots were obtained by slow

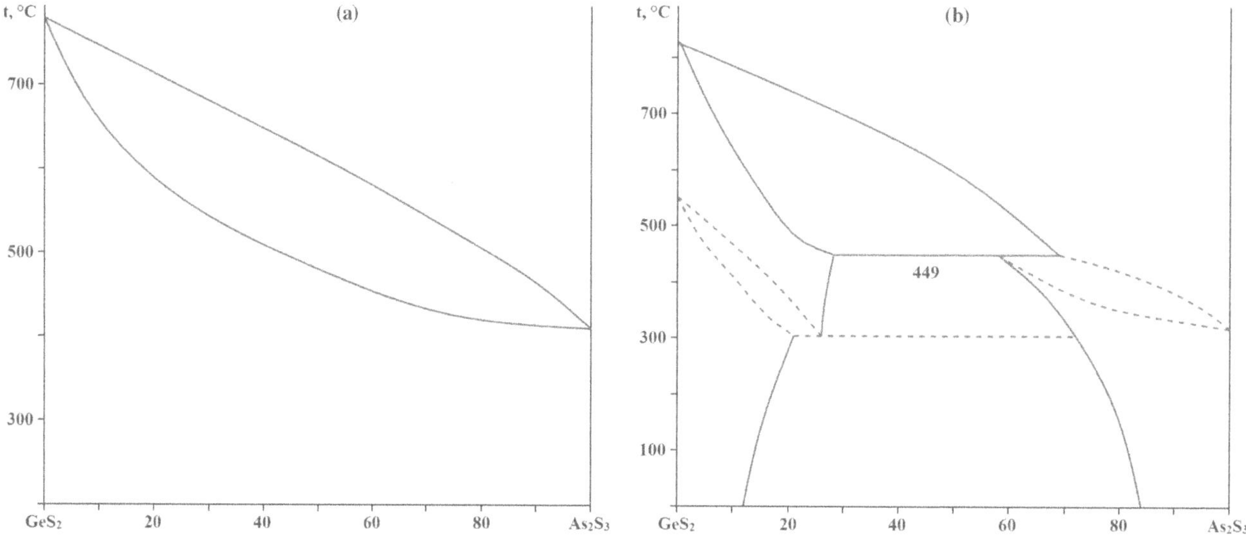

FIGURE 4.46 Nonequilibrium (a) and equilibrium (b) phase diagram of the GeS_2–As_2S_3 system. (From Khiminets, O.V., et al., *Zhurn. neorgan. khimii*, **33**(6), 1541, 1988.)

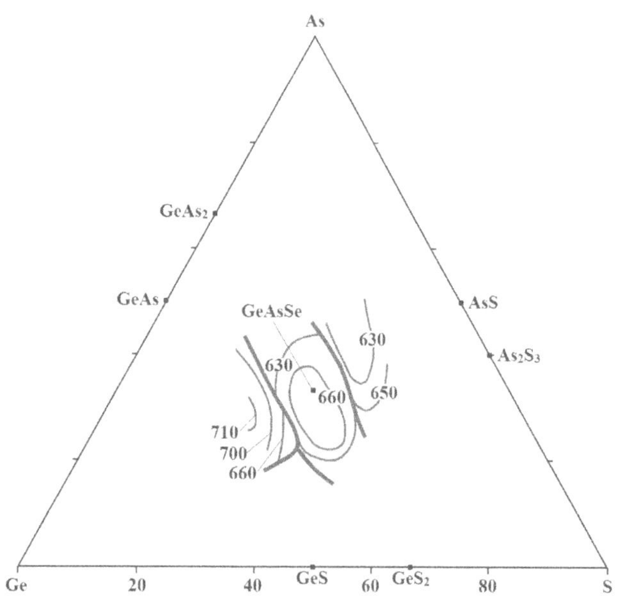

FIGURE 4.47 Part of the liquidus surface of the Ge–As–S system. (From Vinogradova, G.Z., and Dembovskiy, S.A., *Zhurn. neorgan. khimii*, **16**(7), 2036, 1971.)

heating of the components mixture in N_2 atmosphere, saturated by GeS_2 vapor, up to 400°C.

4.42 Germanium–Bismuth–Sulfur

GeS_2–Bi_2S_3: The phase relations in this system were investigated by Odin et al. (1974a). It was shown that this system is non-quasibinary section of the Ge–Bi–S ternary system as GeS_2 melts incongruently. The eutectic contains 48 ± 2 mol% GeS_2 and melts at 492°C ± 3°C. The **$Ge_3Bi_2Te_6$** ternary compound, which melts congruently at 655°C ± 5°C and has polymorphous transformation at 215°C–221°C, is formed in this system. According to the data of Juříček and Frumar (1979), such compound does not exist.

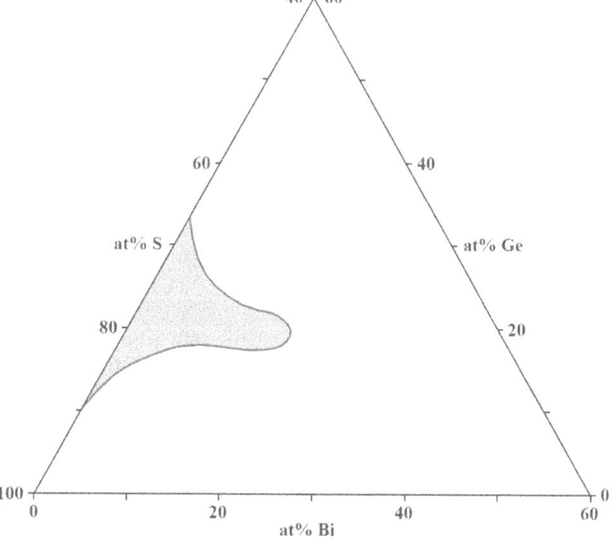

FIGURE 4.48 Glass forming region in the Ge–Bi–S system. (From Frumar, M., et al., *Mater. Res. Bull.*, **11**(11), 1389, 1976.)

The glass forming region in the Ge–Bi–S ternary system is presented in Figure 4.48 (Frumar et al. 1976). The glassy stability depends on the Bi content: with the content of bismuth above 5 mol%, the crystallization of the samples occurs relatively easily with slow cooling. The glassy samples were obtained by quenching from 800°C to 1000°C in water.

4.43 Germanium–Vanadium–Sulfur

The **GeV_4S_8** ternary compound, which crystallizes in the cubic structure with the lattice parameter a = 965.5 ± 0.1 pm and a calculated density of 3.932 g·cm^{-3}, is formed in the Ge–V–S system (Johrendt 1998). It was prepared by heating an appropriate mixture of the elements in a sealed quartz tube under Ar atmosphere. The mixture was heated to 750°C at the rate of 50°C·h^{-1} for 1 day and furnace cooled to room temperature. After homogenization in an Ar-filled glove box, the resulting sample was annealed twice at 800°C for 15 h yielding dark grey powders, which are stable in air. Single crystals of this compound show metallic luster. This compound is a semiconductor within the interval–123°C–27°C.

4.44 Germanium–Niobium–Sulfur

Two intercalation compounds, $\mathbf{Ge_{1/3}NbS_2}$ and $\mathbf{Ge_{1/4}NbS_2}$, are formed in the Ge–Nb–S ternary system (Pocha and Johrend 2002). Both compounds crystallize in the hexagonal structure. First of them is characterized by the lattice parameters $a = 576.7 \pm 0.1$ pm, $c = 1351.8 \pm 0.1$ pm, a calculated density of 4.64 g·cm^{-3}, and crystallizes with a superstructure of *2H*-NbS_2. The lattice parameters of the second compound are the next: $a = 333.9 \pm 0.1$ pm, $c = 3732.6 \pm 0.7$ pm, a calculated density of 4.87 g·cm^{-3}, and the NbS_2-sublattice of this compound represents a *6H*-polymorph. The single crystals of these two compounds have been prepared by heating the mixture of the elements at 800°C or by chemical transport reactions with I_2 as a transport agent at 650°C–800°C.

4.45 Germanium–Tantalum–Sulfur

The $\mathbf{Ge_{1/3}TaS_2}$ intercalation compound is formed in the Ge–Ta–S ternary system (Di Salvo et al. 1973). It crystallizes in the hexagonal structure with the lattice parameters $a = 571.6$ and $c = 1360$ pm and shows the superlattice with $a = \sqrt{3} \times a_0$, were a_0 is the a-parameter of *2H*-TaS_2. This indicates an ordering of Ge between the TaS_2 layers. The superlattice of this compound is quite stable. $Ge_{1/3}TaS_2$ was prepared by the intercalation of TaS_2 with germanium at 550°C.

4.46 Germanium–Selenium–Sulfur

GeS–GeSe: The continuous solid solutions are formed in this system (Koren' and Krasnova 1976; Koren' et al. 1977, 1978, 1983; Sidorov et al. 1976; Marcheva 1979) [according to the data of Bletskan et al. (1976), the solid solutions are formed within the interval from 40 to 100 mol% GeS]. In addition to the thermal effects of liquidus and solidus, small thermal effects were found at 585°C–590°C, due to a phase transition (Koren' et al. 1978; Sidorov et al. 1976). The dependence of the lattice parameters on the composition obeys the Vegard's law, and the energy gap varies smoothly from 1.1 eV (GeSe) to 1.6 eV (GeS) (Koren' and Krasnova 1976; Koren' et al. 1975, 1978).

According to the data of Koren et al. (1983), a spinodal decomposition of the solid solution takes place in this system. The ingots were annealed at 290°C–550°C for 72–96 h after which they were quenched in air. It was shown that the annealing above 470°C did not lead to the solid solution decomposition. At temperatures below 470°C the specimens have been annealed to two crystalline phases, the distance between the boundaries of two-phase region and the one-phase one being larger with decreasing the annealing temperature. The critical decomposition temperature (470°C) corresponds to the $GeS_{0.4}Se_{0.6}$ composition.

The single crystals of GeS_xSe_{1-x} solid solutions were grown by the sublimation method (Bletskan et al. 1976).

$\mathbf{GeS_2}$–$\mathbf{GeSe_2}$: There is a continuous series of solid solutions with a narrow two-phase region in this system (Alekseeva et al. 1978a; Karahanova et al. 1976a). When the composition of this section deviates from stoichiometry, small thermal effects appear at 580°C–590°C. The lattice parameters depending on the composition vary linearly. The ingots of the title system were annealed at 450°C–500°C for 1000 h. The system was investigated through DTA and XRD.

The scheme of the liquidus surface of the Ge–GeS_2–$GeSe_2$ subsystem of the Ge–Se–S ternary system was constructed by Alekseeva et al. (1978a). It was shown that the immiscibility region exists in the field of Ge primary crystallization. The fields of primary crystallization of the GeS_xSe_{1-x} and $GeS_{2x}Se_{2-2x}$ solid solution are also exist on the liquidus surface.

Two glass-forming regions were determined in the Ge–Se–S ternary system (Alekseeva et al. 1978a) [according to the data of Makovskaya and Zhukov (1980), only one glass-forming region exists in this system]. The maximum Ge content in glassy samples reaches up to 30 mol% (Makovskaya and Zhukov 1980). The glass transition temperature of the samples lies in the range of 105°C–200°C, and their crystallization temperature is in the range of 120°C–435°C. All glasses are resistant to air.

Glassy samples were synthesized by step heating with 1-hour exposures at 450°C and 650°C, followed by quenching in air or obtained by slowly cooling the melts (Alekseeva et al. 1978a; Makovskaya and Zhukov 1980).

4.47 Germanium–Tellurium–Sulfur

GeS–GeTe: This system is non-quasibinary section of the Ge–Te–S ternary system (Manéglier-Lacordaire et al. 1973). Mutual solubility of GeS and GeTe was not detected.

$\mathbf{GeS_2}$–GeTe: The phase diagram of this system, constructed through DTA, XRD, and metallography, is presented in Figure 4.49 (Manéglier-Lacordaire et al. 1973, 1975a). The eutectic contains 44 mol% GeS_2 and crystallizes at 560°C ± 4°C. Mutual solubility of the starting components was not detected.

$\mathbf{GeS_2}$–Te: The phase diagram of this system, constructed through DTA, XRD, and metallography, is shown in Figure 4.50 (Manéglier-Lacordaire et al. 1973, 1974, 1975a). The eutectic is degenerated from the Te-side and crystallizes at 450°C ± 4°C. There is an immiscibility region at 785°C in this system. Solid solutions based on the starting components were not found.

Figure 4.51 illustrates the division of the Ge–Te–S ternary system into four triangles of invariance (Manéglier-Lacordaire et al. 1974, 1975a). It also shows the positions of the eutectic values and ternary invariants. No defined ternary compound or solid solution was observed. Two liquid immiscibility gaps were located: a Ge-rich gap inside the GeS–Ge–GeTe triangle resting against a binary immiscibility gap, and a S- or Te-rich

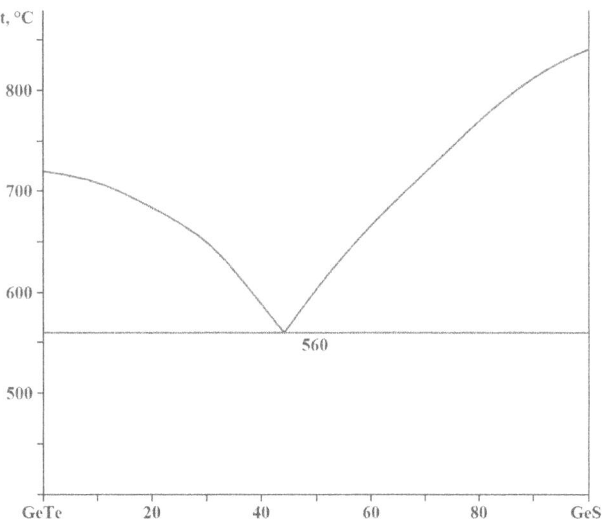

FIGURE 4.49 Phase diagram of the GeS_2–GeTe system. (From Manéglier-Lacordaire, S., et al., *J. Non-Cryst. Solids*, **18**(3), 439, 1975.)

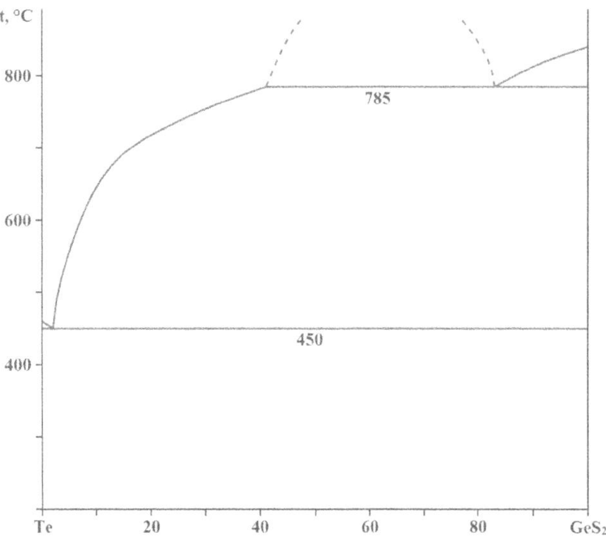

FIGURE 4.50 Phase diagram of the GeS_2–Te system. (From Manéglier-Lacordaire, S., et al., *J. Non-Cryst. Solids*, **18**(3), 439, 1975.)

gap enclosed within the ternary system. The latter miscibility gap is located exactly within the quadrilateral GeS_2–GeTe–Te–S. One transition point (*U*, 584°C) and one ternary eutectic (E_1, 534°C) exist in the GeS_2–Ge–GeTe subsystem. There is another ternary eutectic E_2 in the GeS_2–GeTe–Te subsystem, which crystallizes at 380°C.

The glass-forming region exists in the Ge–Te–S ternary system (Alekseeva et al. 1978b; Makovskaya and Zhukov 1980; Manéglier-Lacordaire et al. 1975a). Depending on the conditions for glassy samples obtaining, the composition of the glasses varies widely. All glasses are resistant to air. The glass transition temperature for these glasses is within the interval 292°C–398°C.

Glassy samples were synthesized by step heating with 1-hour exposures at 450°C and 650°C, followed by quenching in air or obtained by slowly cooling the melts (Alekseeva et al. 1978b; Makovskaya and Zhukov 1980).

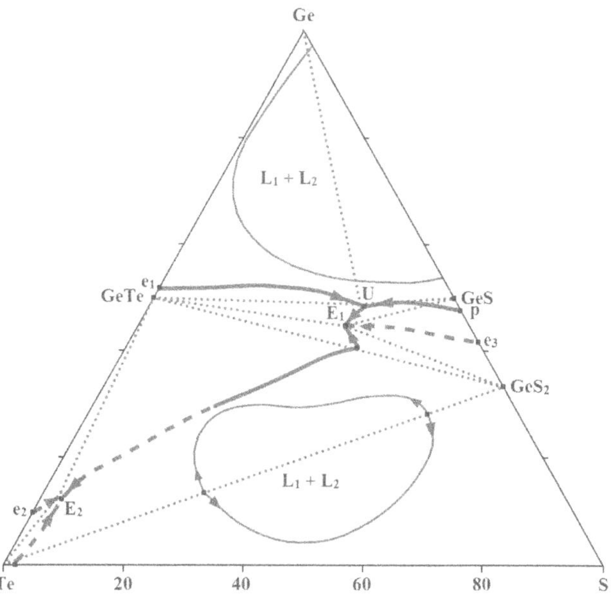

FIGURE 4.51 Projection of the liquidus surface of the Ge–Te–S ternary system. (From Manéglier-Lacordaire, S., et al., *J. Non-Cryst. Solids*, **18**(3), 439, 1975.)

4.48 Germanium–Chromium–Sulfur

GeS_2–Cr_2S_3: The phase diagram of this system, constructed through DTA and XRD (Figure 4.52), is a eutectic type with a degenerate eutectic from GeS_2-side (Shabunina and Aminov 2010). No ternary compound was observed. The solubility of GeS_2 in Cr_2S_3 at 840°C is 4–5 mol% and less than 1 mol% at room temperature.

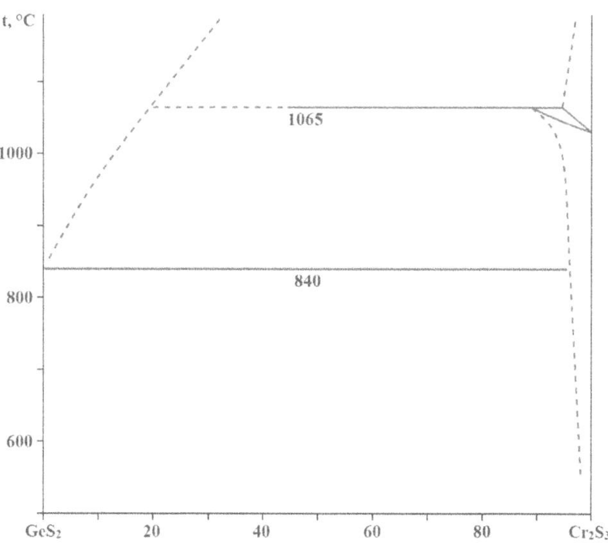

FIGURE 4.52 Phase diagram of the GeS_2–Cr_2S_3 system. (From Shabunina, G.G., and Aminov T.G., *Russ. J. Inorg. Chem.*, **55**(8), 1279, 2010.)

4.49 Germanium–Bromine–Sulfur

GeS_2–$GeBr_4$: The $\mathbf{Ge_4S_6Br_4}$ ternary compound, which melts incongruently at 305°C and crystallizes in the triclinic structure with the lattice parameters a = 880.6 ± 0.2, b = 993.4 ± 0.2, c = 1010.6 ± 0.2 pm, and α = 86.05 ± 0.01°, β = 64.00 ± 0.01°, γ = 89.87 ± 0.01°, and a calculated density of 3.36 g·cm^{-3} (Pohl et al. 1981) [a = 880.6, b = 994.0, c = 1010.4 pm, and α = 86.04°, β = 64.08°, γ = 89.82° (Pohl 1976a,b)], is formed in this system. To prepare the title compound, H_2S was passed for 72 h into a boiling solution of $GeBr_4$ (50 mM) in CS_2 (250 mL; containing catalytic amounts of $AlBr_3$) into a two-necked flask fitted with gas inlet tube and multiple coil condenser, the solvent being repeatedly replaced during the reaction. The solvent was finally removed, unreacted $GeBr_4$ distilled off, and the glassy residue dissolved in CS_2. After 4–5 days at –10°C, the filtered pale yellow solution started to crystallize; crystallization is very slow. Attempts to accelerate the crystallization by a strong concentration of the solution had always resulted in a predominantly amorphous product.

4.50 Germanium–Iodine–Sulfur

GeS–GeI_2: According to the data of Novoselova et al. (1971a), the phase diagram of this system, constructed through DTA, XRD, and metallography, is a eutectic type (Figure 4.53). The eutectic contains 55 mol% GeS and crystallizes at 398°C. The immiscibility region in the liquid state exists within the interval from 25 to 50 mol% GeS at 409°C and the glassy samples are formed at the concentration region 10–80 mol% GeS. Manéglier-Lacordaire et al. (1975b) noted that this system is non-quasibinary section of the Ge–I–S ternary system. The ingots for the investigation were annealed at 370°C for 750 h (Novoselova et al. (1971a).

GeS_2–GeI_2: This system is non-quasibinary section of the Ge–I–S ternary system (Manéglier-Lacordaire et al. 1975b). This section was investigated using DTA, XRD, and metallography.

GeS_2–GeI_4: The $\mathbf{Ge_4S_6I_4}$ ternary compound, which melts incongruently at 310°C and crystallizes in the trigonal structure with the lattice parameters a = 1064.0 ± 0.2 and c = 946.1 ± 0.2 pm, and a calculated density of 3.54 g·cm^{-3} (Pohl et al. 1981), is formed in this system. It was prepared in the same way as $Ge_4S_6Br_4$ was synthesized using GeI_4 instead of $GeBr_4$.

There is an immiscibility region in the GeS–Ge–GeI_2 subsystem, and two transition points at 391°C and 374°C and one ternary eutectic exist in the GeS_2–GeI_2–GeI_4 subsystem of the Ge–I–S system (Manéglier-Lacordaire et al. 1975b). The glass-forming region exists in the Ge–I–S ternary system. As a result of homogenizing annealing, most glasses were unable to crystallize. The resulting glass has a low thermal stability, due to the instability of germanium iodides (Dembovskiy et al. 1971a). The glasses colors depending on the composition vary from light red to light yellow. The glass-forming region elongated to the side of GeI_4 can be explained by the presence of a ternary glass-forming compound $\mathbf{GeSI_2}$. Glass with the composition $GeSI_2$ cannot be crystallized. The glass transition temperature for the glasses obtained is within the interval 165°C–210°C (Manéglier-Lacordaire et al. 1975b). The glassy samples were obtained by the slow cooling (2°C·min^{-1}) (Dembovskiy et al. 1971a).

4.51 Germanium–Manganese–Sulfur

GeS_2–MnS: The phase diagram of this system, constructed through DTA, XRD, and metallography, is presented in Figure 4.54 (Barnier et al. 1984). The eutectic was located in close proximity to GeS_2 and crystallizes at 800°C. The polymorphic transformation of GeS_2 at 550°C takes place in the system. The $\mathbf{Mn_2GeS_4}$ ternary compounds, which melts incongruently at 1145°C (Barnier et al. 1984) and crystallizes in the orthorhombic structure (Duc et al. 1969) with the lattice parameters a = 1277.6, b = 744.1, and c = 603.3 pm

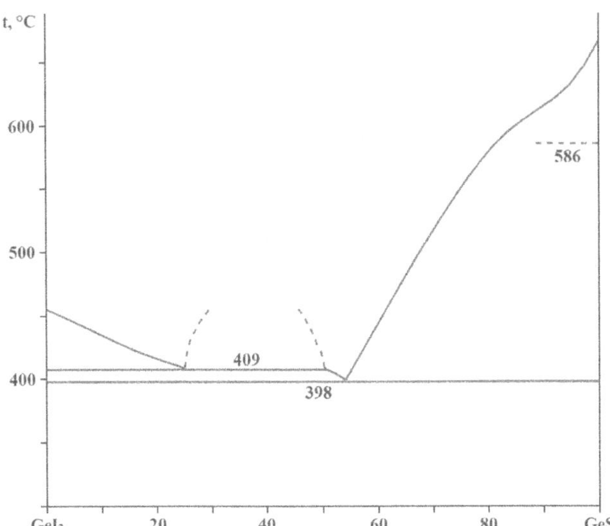

FIGURE 4.53 Phase diagram of the GeS–GeI_2 system. (From Novoselova, A.V., et al., *Izv. AN SSSR. Neorgan. mater.*, **7**(7), 1266, 1971.)

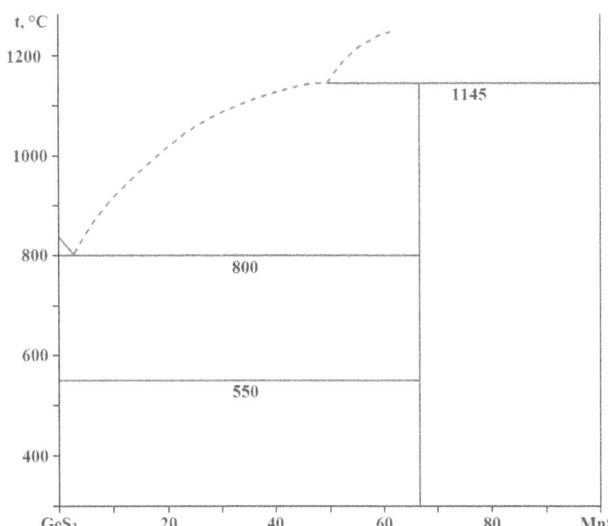

FIGURE 4.54 Phase diagram of the GeS_2–MnS system. (From Barnier, S., et al., *Mater. Res. Bull.*, **19**(7), 837, 1984.)

(Julien-Pouzol et al. 1986) [a = 1279.6 ± 0.2, b = 745.4 ± 0.3, c = 603.4 ± 0.8 pm, and the calculated and experimental densities of 3.59 and 3.56 g·cm^{-3}, respectively (Hagenmuller et al. 1964; Hardy et al. 1965)], is formed in this system (Royen et al. 1965). The title compound was prepared by heating a stoichiometric mixture of GaS_2 and MnS up to 1200°C over 24 h with the next slow cooling during 7 days to room temperature (Barnier et al. 1984; Julien-Pouzol et al. 1986). It was also synthesized from a finely mixed mixture of S, Mn, and Ge in suitable proportions, place in a sealed evacuated silica ampoule (Duc et al. 1969). After a very slow increase of temperature up to 800°C, the thermal treatment was as follows: 800°C for 16 h, then 1050°C for 48 h and finally a slow cooling down to room temperature.

The glassforming region in the GeS_2–MnS system is relatively narrow and ranges from GeS_2 to 20 mol% MnS, for a quenching temperature of 1050°C (Barnier et al. 1984). Between the compositions of GeS_2 and 5 mol% MnS the crystallization of these glasses was not observed.

4.52 Germanium–Iron–Sulfur

GeS–FeS: The phase diagram of this system, constructed through DTA, XRD, and metallography, is presented in Figure 4.55 (Chizhikov et al. 1969; Nikiforov et al. 1976). The eutectic contains 6 mass% FeS and crystallizes at 600°C. The thermal effects on the phase diagram at 322°C and 590°C correspond to the polymorphic transitions of FeS and GeS, respectively. Apparently, the ternary compound **GeS·11FeS**, which melts incongruently at 710°C, is formed in this system.

The isothermal section of the Ge–Fe–S ternary system at 700°C was constructed by Viaene and Moreau (1968). The **Fe_2GeS_4** ternary compound, which decomposes at 800°C and 8 GPa with formation of FeS_2 and crystallizes in the orthorhombic structure with the lattice parameters a = 1246.7 ± 0.2, b = 721.3 ± 0.1, and c = 590.2 ± 0.2 pm at room temperature and a = 1243.8, b = 719.4, and c = 589.5 pm at 4.2 K, is formed in the Ge–Fe–S system (Vincent and Perrault 1971; Vincent and Bertaut 1973; Vincent et al. 1976). Fe_2GeS_4 decomposes at the heating with formation of Fe_2GeO_4 (Royen et al. 1965). At 700°C, $Fe_{1-x}S$, FeS_2, and GeS_2 dissolve less than 2 mol% S, intermetallics binary phases practically do not dissolve the sulfur, and the liquid regions occupy small areas (Viaene and Moreau 1968).

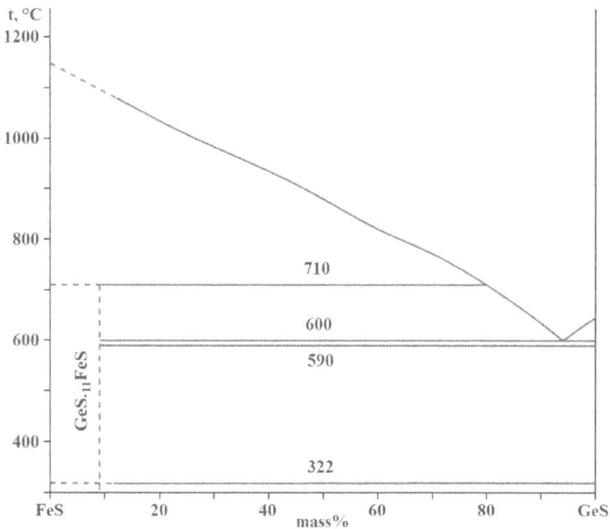

FIGURE 4.55 Phase diagram of the GeS–FeS system. (From Chizhikov, D.M., et al., *Izv. AN SSSR. Neorgan. mater.*, **5**(2), 290, 1969.)

Fe_2GeS_4 was prepared in vacuum-sealed quartz tubes from pure ground elements mixed in stoichiometric proportions (Vincent and Perrault 1971; Vincent and Bertaut 1973; Vincent et al. 1976). The sample was slowly heated to 750°C, maintained at this temperature for 1–3 weeks, and then cooled slowly (5°C·h^{-1}) to room temperature.

4.53 Germanium–Cobalt–Sulfur

The **$CoGe_{1.5}S_{1.5}$** ternary compound, which exists in two polymorphic modifications, is formed in the Ge–Co–S system. First modification of this compound crystallizes in the rhombohedral structure with the lattice parameters a = 1132.287 ± 0.006 and 1133.902 ± 0.007 and c = 1388.82 ± 0.02, and 1390.66 ± 0.02 pm in hexagonal setting at 4.2 K and room temperature, respectively (Vaqueiro et al. 2008) [a = 802.24 ± 0.04 pm and α = 89.91 ± 0.01° or a = 1133.63 ± 0.10 and c = 1391.76 ± 0.09 pm, and a calculated density of 5.55 g·cm^{-3} (Partik et al. 1996)]. The lattice parameters increase smoothly with increasing temperature up to 900°C (Kaltzoglou et al. 2013). Second modification crystallizes in the cubic structure with the lattice parameter a = 801.58 ± 0.01 pm (Vaqueiro and Sobany 2007) [a = 801.7 ± 0.2 pm and the calculated and experimental densities of 5.58 and 5.54 ± 0.03 g·cm^{-3}, respectively (Korenstein et al. 1977)].

To obtain the title compound, mixtures of the elements corresponding to its stoichiometry were ground in an agate mortar prior to sealing into an evacuated (<0.01 Pa) silica tube (Vaqueiro and Sobany 2007; Vaqueiro et al. 2008; Kaltzoglou et al. 2013). The inner wall of the silica tube was coated with a thin layer of carbon by pyrolysis of acetone. The reaction mixture was heated at 500°C–600°C for 2–4 days and then cooled to room temperature at 0.5°C·min^{-1} prior to removal from the furnace. Following re-grinding, the material was sealed into a second silica tube, refired at 700°C for 4–5 days, and cooled to room temperature at 0.1°C·min^{-1}. Polycrystalline samples of $CoGe_{1.5}S_{1.5}$ were also prepared by annealing stoichiometric amounts of the elements in evacuated silica ampoule at 800°C (Partik et al. 1996). Well-formed small single crystals of this compound were grown by the chemical vapor transport method using I_2 as the transporting agent (Korenstein et al. 1977; Partik et al. 1996). $CoGe_{1.5}S_{1.5}$ is p-type semiconductor (Vaqueiro et al. 2008) and decomposes at 1000°C (Korenstein et al. 1977).

4.54 Germanium–Nickel–Sulfur

The **Ni_6GeS_2** ternary compound (mineral nuwaite), which crystallizes in the tetragonal structure with the lattice parameters a = 365.0 and 1814.1 pm, is formed in the Ge–Ni–S system (Ma 2013; Ma and Beckett 2018).

4.55 Germanium–Rhodium–Sulfur

The $\mathbf{RhGe_{1.5}S_{1.5}}$ ternary compound, which exists in two polymorphic modifications, is formed in the Ge–Rh–S system. First modification of this compound crystallizes in the rhombohedral structure with the lattice parameters a = 1167.841 ± 0.009 and 1169.416 ± 0.006 and c = 1436.46 ± 0.02, and 1438.20 ± 0.01 pm in hexagonal setting at 4.2 K and room temperature, respectively (Vaqueiro et al. 2008) [a = 828.2 ± 0.1 pm and α = 89.85 ± 0.01° (Lutz and Kliche 1981)]. Second modification crystallizes in the cubic structure with the lattice parameter a = 828.86 ± 0.01 pm (Vaqueiro and Sobany 2007) [a = 827.46 pm (Lyons et al. 1978)].

To synthesize this compound, mixtures of the elements corresponding to its stoichiometry were ground in an agate mortar prior to sealing into an evacuated (<0.01 Pa) silica tube (Vaqueiro and Sobany 2007; Vaqueiro et al. 2008). The inner wall of the silica tube was coated with a thin layer of carbon by pyrolysis of acetone. The reaction mixtures were heated at 500°C for 1 day, then at 700°C for 1 day and finally at 800°C for 2 days. After cooling at 0.5°C·min^{-1} to room temperature, the materials were ground, sealed into a second silica tube, heated at 800°C for 2 days and then slowly cooled to room temperature, with a cooling rate of 0.5°C·min^{-1}. $RhGe_{1.5}S_{1.5}$ could be also prepared by heating a stoichiometric mixture of the elements in closed silica tubes at 600°C for 14 days and quenching in ice water (Lutz and Kliche 1981), or by direct combination of the elements at 800°C under conditions of ambient pressure, or by heating the partially reacted elements at 600°C and 4 GPa (Lyons et al. 1978). $RhGe_{1.5}S_{1.5}$ is p-type semiconductor (Vaqueiro et al. 2008).

4.56 Germanium–Palladium–Sulfur

The $\mathbf{PdGeS_3}$ ternary compound, which crystallizes in the monoclinic structure with the lattice parameters a = 1421.7 ± 0.6, b = 345.2 ± 0.2, c = 907.9 ± 0.6 pm, β = 106.58 ± 0.04°, and a calculated density of 4.298 g·cm^{-3}, is formed in the Ge–Pd–S system (Johrendt and Tampier 1998). It was synthesized by direct reaction of the elements. A stoichiometric mixture of Ge, Pd, and S was loaded into a silica tube. After the tube had been evacuated and flushed with Ar three times, it was sealed under an Ar atmosphere (ambient pressure). The mixture was heated (50°C·h^{-1}) to 600°C, kept at this temperature for 1 week, and then the furnace was allowed to cool to room temperature. This procedure produced orange transparent fibers of the title compound. Single crystals of the fibrous material were grown in a molten flux of NaSCN.

4.57 Germanium–Iridium–Sulfur

The $\mathbf{IrGe_{1.5}S_{1.5}}$ ternary compound, which exists in two polymorphic modifications, is formed in the Ge–Ir–S system. First modification of this compound crystallizes in the rhombohedral structure with the lattice parameters a = 1172.719 ± 0.008 and c = 1441.45 ± 0.02 pm in hexagonal setting (Vaqueiro et al. 2008). Second modification crystallizes in the cubic structure with the lattice parameter a = 829.37 ± 0.01 pm (Vaqueiro and Sobany 2007) [a = 829.70 pm (Lyons et al. 1978)].

To obtain this compound, mixtures of the elements corresponding to its stoichiometry were ground in an agate mortar prior to sealing into an evacuated (<0.01 Pa) silica tube (Vaqueiro and Sobany 2007; Vaqueiro et al. 2008). The inner wall of the silica tube was coated with a thin layer of carbon by pyrolysis of acetone. The reaction mixtures were heated at 500°C for 1 day, then at 700°C for 1 day and finally at 800°C for 2 days. After cooling at 0.5°C·min^{-1} to room temperature, the materials were ground, sealed into a second silica tube, heated at 800°C for 2 days and then slowly cooled to room temperature, with a cooling rate of 0.5°C·min^{-1}. $IrGe_{1.5}S_{1.5}$ could be also prepared by direct combination of the elements at 800°C under conditions of ambient pressure, or by heating the partially reacted elements at 600°C and 4 GPa (Lyons et al. 1978), or by heating a stoichiometric mixture of the elements in closed silica tubes at 600°C for 14 days with next quenching in ice water (Lutz and Kliche 1981). The title compound is p-type semiconductor (Vaqueiro et al. 2008)

4.58 Germanium–Platinum–Sulfur

The **PtGeS** ternary compound, which crystallizes in the orthorhombic structure with the lattice parameters a = 586.4 ± 0.9, b = 590.2 ± 0.9, and c = 583.5 ± 0.9 pm, is formed in the Ge–Pt–S system (Entner and Parthé 1973). It was prepared by heating the elements in evacuated quartz tubes at 600°C and 900°C for about 4–6 weeks. Some samples were ground after 1 week and reheated to enhance the attainment of equilibrium.

REFERENCES

All References are available as a downloadable eResource at www.routledge.com/9780367639235

5

Systems Based on Germanium Selenides

5.1 Germanium–Lithium–Selenium

The $\mathbf{Li_3Ge_3Se_6}$ ternary compound, which crystallizes in the triclinic structure with the lattice parameters $a = 677.28 \pm 0.03$, $b = 868.04 \pm 0.05$, $c = 1063.00 \pm 0.05$ pm, and $\alpha = 74.519 \pm 0.004°$, $\beta = 76.334 \pm 0.004°$, $\gamma = 69.185 \pm 0.004°$, a calculated density of 4.255 g·cm^{-3} and an energy gap of 2.08 eV, is formed in the Ge–Li–Se system (Li et al. 2017). The crystals and pure phase sample of the title compound were synthesized by solid-state reaction with all elementary substances of Li, Ge and Se (molar ratio 1:1:2). The mixture was weighted and placed into the graphite crucible in the Ar atmosphere, and then flame-sealed into fused silica tube under a high vacuum of 10^{-3} Pa. The tube was moved into a furnace, heated to 850°C in 50 h, dwelled there for 30 h, cooled to 650°C at a rate of 5°Ch^{-1}, then rapidly cooled to room temperature at a rate of 10°Ch^{-1}. All the reactants were stored in a glove box filled with dry Ar without oxygen and moisture.

5.2 Germanium–Sodium–Selenium

$GeSe_2$–Na_2Se: The part of the phase diagram of this system, constructed through DTA, XRD, and metallography, is presented in Figure 5.1 (Mikhaylo et al. 1989). The samples were annealed at 480°C for 200–250 h. Two eutectics contain 56 and 79 mol% GeS_2 and crystallize at 542°C and 560°C, respectively. The regions of homogeneity for both compounds do not exceed 5 mol%. The field of the solid solution based on $GeSe_2$ reaches up to 8 mol%. Two ternary compounds, $\mathbf{Na_2GeSe_3}$ and $\mathbf{Na_2Ge_2Se_5}$, are formed in this system. Na_2GeSe_3 melts congruently at 572°C (Mikhaylo et al. 1989) [at 577°C with a melting entropy 30.3 kJ·M^{-1} (Mikaylo et al. 1990)] and has apparently two polymorphic modifications both of which crystallize in the monoclinic structure with the lattice parameters $a = 836.7 \pm 0.5$, $b = 1192.4 \pm 0.8$, $c = 815.8 \pm 0.5$ pm, $\beta = 118.63 \pm 0.15°$, and a calculated density of 3.52 g·cm^{-3} for the first modification (Eisenmann et al. 1985b) and $a = 709.7 \pm 0.2$, $b = 1606.8 \pm 0.5$, $c = 607.3 \pm 0.2$ pm, $\beta = 113.69 \pm 0.08°$, and the calculated and experimental densities of 3.72 and 3.76 g·cm^{-3}, respectively, for the second one (Eisenmann et al. 1985c). To prepare the title compound, the elements were weighed in stoichiometric proportions under dry, oxygen-free Ar atmosphere in a quartz ampoule with quartz crucible pre-dried at 150°C (Eisenmann et al. 1985c). In order to compensate for the loss due to the unavoidable reaction with the crucible wall, Na was used empirically in an excess of about 10% by mass. The ampoule was evacuated and sealed. The highly exothermic formation reaction remained under control when heated up to 800°C at a rate of 1°C·min^{-1}. Homogeneous sintered ingots are formed; yellow single-crystalline particles can be split from their cavity-like inclusions. Na_2GeSe_3 was also synthesized by melting the stoichiometric amounts of Na, GeS_2, and S in alundum crucible encapsulated in the quartz ampoule. Single crystals were grown by the Bridgman method (Mikaylo et al. 1989, 1990). This compound is moisture sensitive.

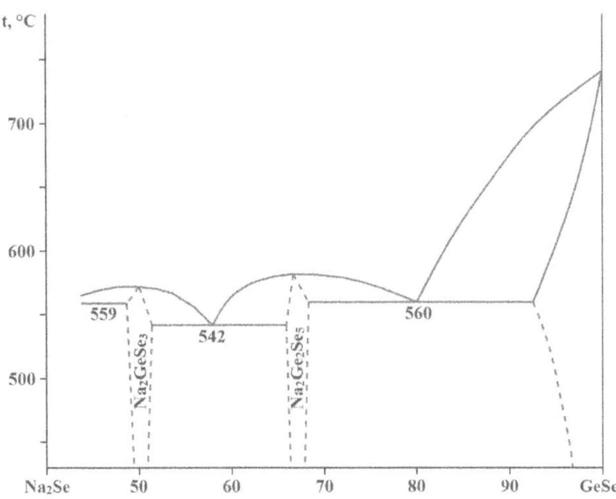

FIGURE 5.1 Part of the phase diagram of the $GeSe_2$–Na_2Se system. (From Mikhaylo, I.L., et al., *Zhurn. neorgan. khimii*, **34**(9), 2319, 1989.)

$Na_2Ge_2Se_5$ melts congruently at 576°C (Chung et al. 2012) [at 582°C (Mikhaylo et al. 1989); at 585°C with a melting enthalpy 71.1 kJ·M^{-1} (Mikaylo et al. 1990)] and has apparently two polymorphic modifications both of which crystallizes in the orthorhombic structure with the lattice parameters $a = 1408.88 \pm 0.07$, $b = 616.51 \pm 0.03$, $c = 1100.72 \pm 0.07$ pm, and a calculated density of 4.071 g·cm^{-3} (Chung et al. 2012) [$a = 1409.5$, $b = 616.7$, $c = 1099.1$ pm, and the calculated and experimental densities of 4.07 and 4.00 g·cm^{-3}, respectively, (Eisenmann et al. 1984a)] for the first modification and $a = 1350.7 \pm 0.5$, $b = 1336.3 \pm 0.5$, $c = 1103.4 \pm 0.5$ pm for the second one (Eisenmann and Hansa 1993i). This compound was synthesized by melting the stoichiometric amounts of Na, GeS_2, and S in alundum crucible encapsulated in the quartz ampoule (Mikaylo et al. 1989, 1990; Eisenmann and Hansa 1993i). Single crystals were grown by the Bridgman method. In the open air, $Na_2Ge_2Se_5$ hydrolyzes extremely rapidly, with the intense smell of H_2Se occurs (Eisenmann et al. 1985c). The compound could therefore be handled under dry, heavy paraffin oil.

Three other ternary compounds are formed in the Ge–Na–Se system. $\mathbf{Na_4GeSe_4}$ crystallizes in the orthorhombic structure with the lattice parameters $a = 2851.8 \pm 0.8$, $b = 944.7 \pm 0.5$,

DOI: 10.1201/9781003123507-5

c = 712.8 ± 0.2 pm (Klepp 1985). It was obtained from a stoichiometric melt of Na_2Se, Ge, and Se at 750°C.

$Na_6Ge_3Se_7$ crystallizes in the monoclinic structure with the lattice parameters a = 945.1, b = 1091.4, c = 1587.4 pm, β = 104.7°, and the calculated and experimental densities of 3.50 and 3.58 g·cm^{-3}, respectively (Eisenmann et al. 1986).

$Na_8Ge_4Se_{10}$ has apparently two polymorphic modifications. First of them crystallizes in the triclinic structure with the lattice parameters a = 707.4 ± 0.5, b = 809.8 ± 0.5, c = 1065.7 ± 0.6 pm, and α = 73.44 ± 0.15°, β = 70.84 ± 0.15°, γ = 81.73 ± 0.15°, and the calculated and experimental densities of 4.07 and 4.00 g·cm^{-3}, respectively (Eisenmann et al. 1985b). The second modification crystallizes in the monoclinic structure with the lattice parameters a = 1310.4 ± 0.5, b = 1205.7 ± 0.5, c = 1410.1 ± 0.5 pm, β = 92.9 ± 0.1°, and the calculated and experimental densities of 3.77 and 3.75 g·cm^{-3}, respectively (Eisenmann et al. 1985c). To prepare the title compound, the elements were weighed in stoichiometric proportions under dry, oxygen-free Ar in quartz ampoule with BN crucible predried at 150°C (Eisenmann et al. 1985c). The ampoule was evacuated and sealed. The highly exothermic formation reaction remained under control when heated up to 750°C at a rate of 2°C·min^{-1}. Homogeneous sintered ingots are formed; orange-yellow single-crystalline particles can be split from their cavity-like inclusions. In the open air, $Na_8Ge_4Se_{10}$ hydrolyzes extremely rapidly, with the intense smell of H_2Se occurs. The compound could therefore be handled under dry, heavy paraffin oil.

5.3 Germanium–Potassium–Selenium

Several ternary compounds are formed in the Ge–K–Se system. **K_2GeSe_4** crystallizes in the monoclinic structure with the lattice parameters a = 1198.9 ± 0.1, b = 827.5 ± 0.1, c = 891.5 ± 0.1 pm, β = 106.38 ± 0.03°, and a calculated density of 3.65 g·cm^{-3} (Sheldrick and Schaaf 1995). To synthesize this compound, K_2CO_3 (2.64 mM), Ge (2.22 mM), and Se (4.42 mM) were slurried in 0.5 mL of MeOH and melted under Ar atmosphere in glass ampoule (degree of filling 5%). The ampoule was heated to 175°C, annealed at this temperature for 16 h, and then cooled down to room temperature at a rate of 1°C·h^{-1}. At the phase boundary, K_2GeSe_4 was obtained in crystalline form.

$K_2Ge_4Se_8$ also crystallizes in the monoclinic structure with the lattice parameters a = 737.52 ± 0.05, b = 1223.9 ± 0.2, c = 1746.8 ± 0.4 pm, and β = 96.02 ± 0.03° (Wang et al. 2004). It was prepared employing the molten-salt reaction in a molar ratio K_2Se_3/Ge/S = 2:1:6 at 500°C for 5 days. Then, the sample was cooled to 200°C (4°C·h^{-1}) with the next quenching in DMF.

$K_4Ge_2Se_6$ melts at 651°C with a melting enthalpy 24.5 kJ·M^{-1} (Mikaylo et al. 1990) and crystallizes in the triclinic structure with the lattice parameters a = 856.4 ± 0.4, b = 744.2 ± 0.4, c = 689.2 ± 0.4 pm, and α = 66.4 ± 0.1, β = 71.8 ± 0.1°, γ = 74.6 ± 0.1°, and an experimental density of 3.480 g·cm^{-3} (Eisenmann and Hansa 1993g). Yellow prisms of the title compound were synthesized from stoichiometric amounts of the elements in evacuated graphitized quartz ampoule at 700°C.

$K_4Ge_4Se_{10}$ melts at 606°C with a melting enthalpy 61.9 kJ·M^{-1} (Mikaylo et al. 1990) and has two polymorphic modifications both of which crystallizes in the monoclinic structure with the lattice parameters a = 1020.2 ± 0.6, b = 1154.4 ± 0.6, c = 980.6 ± 0.6 pm, β = 90.6 ± 0.1°, and an experimental density of 3.590 g·cm^{-3} for the first modification (Eisenmann and Hansa 1993f) and a = 997.96 ± 0.08, b = 970.47 ± 0.08, c = 2318.4 ± 0.2 pm, β = 94.508 ± 0.002°, and a calculated density of 3.669 g·cm^{-3} at 100 ± 2 K for the second one (Choudhury et al. 2007). The first polymorph was prepared from stoichiometric amounts of the elements in evacuated graphitized quartz ampoule at 800°C (Eisenmann and Hansa 1993f). Orange needles of the second polymorph were obtained from a solid-state reaction of K_2Se, Ge, and Se by mixing stoichiometric amounts (molar ratio 1:2:4) (Choudhury et al. 2007). The reactants were loaded into a fused-silica tube under N_2 atmosphere in a glove box. The tube was torch-sealed under vacuum and then placed in a furnace. The sample was heated to 850°C (at 35°C·h^{-1}), held at this temperature for 48 h, and then cooled to room temperature (at 35°C·h^{-1}).

$K_6Ge_2Se_6$ also crystallizes in the monoclinic structure with the lattice parameters a = 907.6 ± 0.4, b = 1285.1 ± 0.6, c = 848.0 ± 0.4 pm, β = 116.76 ± 0.05°, and the calculated and experimental densities of 3.21 and 3.23 g·cm^{-3}, respectively (Eisenmann et al. 1984b). To obtain the title compound, the elements were weighed in a stoichiometric ratio under moisture and oxygen-free Ar atmosphere in a pre-dried at 150°C quartz ampoule. To compensate the loss of the reaction with the vessel material, potassium was taken in excess of about 10 mass%. The ampoule was melted under vacuum and in Ar-filled corundum tube with a heating rate of 50°C·h^{-1} to 650°C and then the furnace was switched off. The transparent colorless crystals were obtained. In the moist air, the decomposition of $K_6Ge_2S_6$ begins immediately, recognizable by the smell of H_2Se.

5.4 Germanium–Rubidium–Selenium

Four ternary compounds are formed in the Ge–Rb–Se system. Three compounds crystallize in the monoclinic structure with the next lattice parameters: a = 739.0 ± 0.2, b = 370.4 ± 0.2, c = 862.3 ± 0.4 pm, β = 104.87 ± 0.07°, and a calculated density of 4.07 g·cm^{-3} for **Rb_2GeSe_4** (Sheldrick and Schaaf 1995); a = 1388.5 ± 0.6, b = 713.3 ± 0.1, c = 998.4 ± 0.5 pm, β = 124.41 ± 0.02° for **$Rb_4Ge_2Se_6$** (Klepp and Fabian 1997); and a = 1609.5 ± 0.9, b = 1609 ± 1, c = 939.0 ± 0.7 pm, β = 105.79 ± 0.02°, and a calculated density of 4.04 g·cm^{-3} for **$Rb_4Ge_4Se_{10}$** (Klepp and Fabian 1999b).

To synthesize Rb_2GeSe_4, Rb_2CO_3 (0.67 mM), Ge (1.35 mM), and Se (2.71 mM) were slurried in 0.5 mL of MeOH and melted under Ar in glass ampoule (degree of filling 5%) (Sheldrick and Schaaf 1995). The ampoule was heated to 190°C, annealed at this temperature for 24 h, and then cooled down to room temperature at a rate of 1°C·h^{-1}. At the phase boundary, the title compound was obtained in crystalline form.

$Rb_4Ge_2Se_6$ was obtained by the reacting a stoichiometric mixture of Rb_2Se, Ge, and Se in a sealed silica tube at 900°C, followed by controlled slow cooling (Klepp and Fabian 1997).

For the preparation of $Rb_4Ge_4Se_{10}$, stoichiometric amounts of the components were intimately mixed in an Ar-filled glove box and sealed into evacuated silica ampoule (Klepp and Fabian 1999b). At the beginning of the thermal treatment the sample was kept for 3 days at temperatures between 100°C and 200°C, and then gradually heated to 950°C. After annealing for 5 days, the sample was allowed to reach ambient temperature at a constant cooling rate of 2°C·h^{-1}. The reaction product was of homogeneous appearance. The crushed sample was transparent and amber color. Since it was sensitive to humidity, the sample was handled under inert conditions.

The $\mathbf{Rb_4Ge_4Se_{12}}$ melts congruently at 345°C and crystallizes in the orthorhombic structure with the lattice parameters a = 1487.0 ± 0.8, b = 1380.0 ± 0.8, c = 1244.5 ± 0.6 pm, and a calculated density of 4.109 g·cm^{-3} (Liu et al. 2017). To prepare this compound, a stoichiometric mixture of Ba, Ge, Se, and RbCl as a reactive flux (molar ratio of 1:4:12:8) was loaded into a sealed silica tube evacuated to 0.013 Pa. The tube was then placed into a furnace, slowly heated to 750°C maintaining that temperature for 96 h, and finally slowly cooled down to 350°C before switching off the furnace. The product was washed by distilled water to remove the excess RbCl, and red single crystals of the title compound were hand-picked under microscope. All starting reactants were handled inside an Ar-filled glove box with controlled oxygen and moisture levels below 0.1 ppm.

5.5 Germanium–Cesium–Selenium

Five ternary compounds are formed in the Ge–Cs–Se system. Four compounds crystallize in the monoclinic structure with the next lattice parameters: a = 1425.2 ± 0.5, b = 740.0 ± 0.3, c = 1033.5 ± 0.4 pm, β = 124.07 ± 0.04°, and a calculated density of 4.232 g·cm^{-3} for $\mathbf{Cs_4Ge_2Se_6}$ (Almsick van and Sheldrick 2005); a = 1527.8 ± 0.4, b = 762.4 ± 0.3, c = 1009.0 ± 0.3 pm, β = 121.71 ± 0.02°, and a calculated density of 4.34 g·cm^{-3} for $\mathbf{Cs_4Ge_2Se_8}$ (Sheldrick and Schaaf 1994a); a = 1634.8 ± 0.9, b = 1649 ± 1, c = 977.1 ± 0.3 pm, β = 107.10 ± 0.03°, and a calculated density of 4.26 g·cm^{-3} for $\mathbf{Cs_4Ge_4Se_{10}}$ (Klepp and Fabian 1999b); and a = 1628.0 ± 0.3, b = 1352.8 ± 0.3, c = 962.1 ± 0.2 pm, and β = 97.70 ± 0.03° for $\mathbf{Cs_6Ge_2Se_6}$ (Schlirf and Deiseroth 2001b).

To synthesize $Cs_4Ge_2Se_6$, Ge (1.0 mM), As (1.0 mM), Se (2.5 mM), and Cs_2CO_3 (1.0 mM) were heated in a sealed glass tube to 190°C in MeOH (0.8 mL) (Almsick van and Sheldrick 2005). After 80 h, the contents were allowed to cool to room temperature at 2°C·h^{-1} to afford orange crystals of $Cs_3AsGeSe_5$, red blocks of Cs_3AsSe_4, and yellow prisms of $Cs_4Ge_2Se_6$.

To prepare $Cs_4Ge_2Se_8$, Cs_2CO_3 (0.78 mM), Ge (1.52 mM), and Se (3.02 mM) were slurried in 0.5 mL of MeOH and melted under Ar into a borosilicate glass ampoule (degree of filling 5%) (Sheldrick and Schaaf 1994a). The ampoule was heated to 190°C and held at this temperature for 1 h. It was then cooled to room temperature at a rate of 1°C·h^{-1}. Orange-yellow prisms of the title compound were obtained. $Cs_4Ge_2Se_8$ decomposes in air with separation of Se.

$Cs_4Ge_4Se_{10}$ was prepared in the same way as $Rb_4Ge_4Se_{10}$ was synthesized (Klepp and Fabian 1999b).

$Cs_6Ge_2Se_6$ was prepared starting from an inhomogeneous sample obtained by thermal decomposition of CsN_3 mixed with $GeSe_2$ (molar ratio 6:2) (Schlirf and Deiseroth 2001b). The reaction was carried out at 360°C in a quartz ampoule. Together with additional Se (molar composition 6Cs/2Ge/6Se), the raw product was heated to 1000°C. After quenching in ice water, homogenization and subsequent annealing at 680°C for 2 weeks the reaction yielded air sensitive, transparent, orange single crystals of $Cs_6Ge_2Se_6$.

The $\mathbf{Cs_4Ge_4Se_{12}}$ melts congruently at 452°C and crystallizes in the orthorhombic structure with the lattice parameters a = 1519.3 ± 0.6, b = 1394.4 ± 0.5, c = 1284.5 ± 0.5 pm, and a calculated density of 4.319 g·cm^{-3} (Liu et al. 2017). To prepare this compound, a stoichiometric mixture of Ba, Ge, Se, and CsCl as a reactive flux (molar ratio of 1:4:12:8) was loaded into a sealed silica tube evacuated to 0.013 Pa. The tube was then placed into a furnace, slowly heated to 750°C maintaining that temperature for 96 h, and finally slowly cooled down to 350°C before switching off the furnace. The product was washed by distilled water to remove the excess CsCl, and red single crystals of the title compound were hand-picked under microscope. All starting reactants were handled inside an Ar-filled glove box with controlled oxygen and moisture levels below 0.1 ppm.

5.6 Germanium–Copper–Selenium

GeSe–Cu$_2$Se: The phase diagram of this system, constructed through DTA, XRD, metallography, and measuring of microhardness, is shown in Figure 5.2 (Dovletov et al. 1977). The $\mathbf{Cu_4Ge_3Se_5}$ ternary compound, which melts congruently at 615°C and crystallizes in the cubic structure with the lattice parameter a = 533 pm, is formed in this system.

GeSe$_2$–Cu$_2$Se: The phase diagram of this system in the region of 15–60 mol% $GeSe_2$, constructed through DTA, XRD, and metallography, is presented in Figure 5.3 (Piskach et al. 2000). The eutectic contains 38 mol% $GeSe_2$ and crystallizes

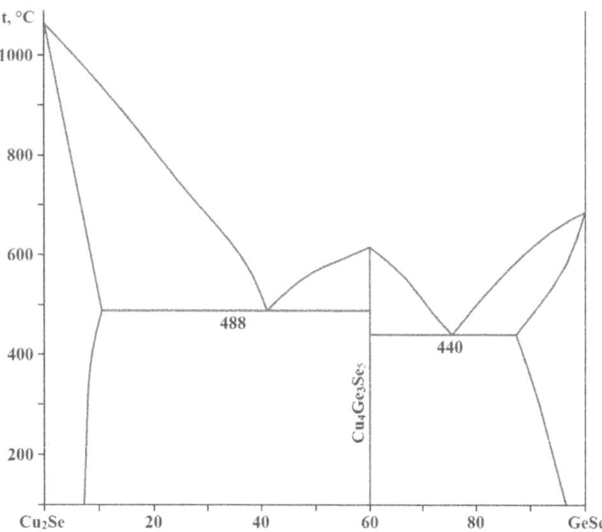

FIGURE 5.2 Phase diagram of the GeSe–Cu$_2$Se system. (From Dovletov, K., et al., *Izv. AN SSSR. Neorgan. mater.*, **13**(6), 1092, 1977.)

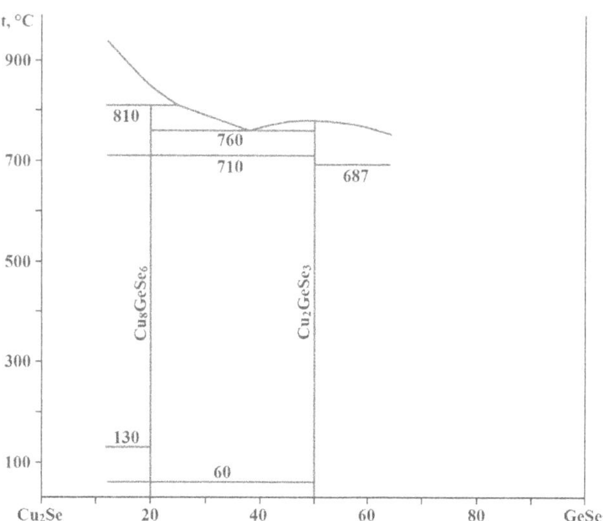

FIGURE 5.3 Phase diagram of the $GeSe_2$–Cu_2Se system in the region of 15–60 mol% $GeSe_2$. (From Piskach, L.V., et al., *J. Alloys Compd.*, **299**(1–2), 227, 2000.)

at 760°C [85 mol% $GeSe_2$ and crystallizes at 700°C (Carcaly et al. 1973)] and a horizontal line at 130°C was interpreted as the temperature of polymorphic transformation of Cu_2Se (Piskach et al. 2000).

The existence of the $\mathbf{Cu_2GeSe_3}$ and $\mathbf{Cu_8GeSe_6}$ ternary compound was confirmed. Cu_2GeSe_3 melts congruently at 780°C [at 788°C (Goryunova et al. 1960a; Averkieva et al. 1964); at 770°C (Palatnik et al. 1961a); at 783°C (Balanevskaya et al. 1966); at 760°C (Rivet et al. 1963; Rivet 1965; Rogacheva et al. 1975b; Zotova and Karagodin 1975b); at 765°C (Villarreal et al. 2003); incongruently at 764°C (Carcaly et al. 1973)] and crystallizes in the orthorhombic structure with the lattice parameters a = 1185.9 ± 0.1, b = 395.1 ± 0.4, c = 548.79 ± 0.02 pm (Ivashchenko et al. 2020; Klymovych et al. 2020; Piskach et al. 2000) [a = 1186.0 ± 0.3, b = 396.0 ± 0.1, c = 548.5 ± 0.2 pm (Parthé and Garin 1971); a = 1187.8 ± 0.8, b = 394.1 ± 0.3, c = 548.5 ± 0.3 pm (Marcano and Nieves 2000); a = 1186.16 ± 0.03, b = 395.25 ± 0.01, c = 548.79 ± 0.01 pm (Villarreal et al. 2003)]; in the monoclinic structure with the lattice parameters a = 673.8 ± 0.6, b = 1182.0 ± 0.0, c = 675.0 ± 0.4 pm, β = 108.41 ± 0.05°, and an energy gap of 0.78 eV (Morihama et al. 2014) [a = 2356 ± 5, b = 782.9 ± 0.7, c = 1119 ± 1 pm, and β = 91°10′ (Carcaly et al. 1973); a = 677.2 ± 0.1, b = 395.60 ± 0.08, c = 395.80 ± 0.08 pm, β = 125.83 ± 0.03°, and a calculated density of 5.622 g·cm^{-3} (Lychmanyuk et al. 2007)]; in the tetragonal structure with lattice parameters a = 559.0 and c = 548.7 pm (Rogacheva et al. 1975b) [a = 558.0 and c = 1096 pm (Goryunova et al. 1960a); a = 557.8 and c = 547.4 pm (Averkieva et al. 1964); a = 559.5, c = 548.2 pm, and the calculated and experimental densities of 5.63 and 5.57 ± 0.08 g·cm^{-3}, respectively (Rivet et al. 1963; Rivet 1965); a = 559.0, c = 1097 pm, and the calculated and experimental densities of 5.63 and 5.46 g·cm^{-3}, respectively (Hahn et al. 1966); a = 559.13 and c = 548.85 pm (Sharma 1970; Sharma and Singh 1974); a = 559.6 and c = 548.5 pm (Rogacheva et al. 1973)]; in the cubic structure with the lattice parameter a = 555 pm and the calculated and experimental densities of 5.65 and 5.65 g·cm^{-3}, respectively (Palatnik et al. 1961a,b)]. The unit cell symmetry of Cu_2GeSe_3 has been found to be very sensitive to Ge concentration (Sharma and Singh 1974). A slight deficiency of Ge lowers the cell symmetry to monoclinic, while a Ge excess raises it to cubic. The composition $Cu_2Ge_{0.85}Se_3$ has a monoclinic unit cell with a = 551.2, b = 559.8, c = 548.6 pm, and β = 89.7°, while the composition $Cu_2Ge_{1.55}Se_3$ is cubic with a = 556.8 [according to the data of Suri et al. (1982), $Cu_2Ge_{1.55}Se_3$ crystallizes in the orthorhombic structure with the lattice parameters a = 393.2, b = 2224.2, c = 1179.6 pm]. The cubic phase of $Cu_2Ge_{1+x}Se_3$ stabilizes for the values of x between 0.5 and 0.6 (Sharma et al. 1975). Homogeneity region of Cu_2GeSe_3 along the Cu_2Se_3–Ge section is within the interval from 49 to 50.75 at% Ge (Rogacheva et al. 1973). The energy gap of Cu_2GeSe_3 is 0.25, 0.77, and 0.94 eV at 0, 77, and 300 K, respectively (Balanevskaya et al. 1966; Endo et al. 1971). According to the thermodynamic simulation, Cu_2GeSe_3 dissociates at the melting temperature forming Cu_2Se and $GeSe_2$ binary compounds (Glazov et al. 1978). The Debye temperature of this compound was estimated to be around 170 K (Marcano and Nieves 2000).

Cu_2GeSe_3 was prepared by direct fusion of the constituent elements in a sealed and evacuated quartz tube (Marcano and Nieves 2000). The ampoules were placed in a vertical furnace. Initially they were heated from room temperature to 300°C at the rate of 30°C·h^{-1}. The mixture was kept at this temperature for 24 h and then heated at 6°C·h^{-1} up to 960°C. Again the mixture was kept at this temperature for 24 h. Then it was heated to 1150°C at the rate of 30°C·h^{-1}. The mixture in the liquid phase was agitated carefully by periodically rocking the furnace for 48 h. It was cooled at a rate of 60°C·h^{-1} up to 900°C and then to 800°C at 10°C·h^{-1} with a dwell time at this temperature of 24 h. The cooling rate from 800 to 650°C was 6°C·h^{-1} with a dwell time of 24 h. The furnace was then cooled to 500°C at 20°C·h^{-1} and the ingot was annealed at this temperature for 200 h. Finally, the furnace was cooled to room temperature at 30°C·h^{-1}.

Cu_8GeSe_6 melts incongruently at 810°C, has two polymorphic transformations at 710°C and 60°C [at 55°C with an enthalpy change of approximation 6 J·g^{-1} and $\Delta H_{phase\ trans.}$ = 8.1 ± 0.4 kJ·M^{-1} (Carcaly et al. 1973; Aliev et al. 1989; Jaulmes et al. 1991; Onoda et al. 1999b)] and is characterized by very small homogeneity region (Piskach et al. 2000). α-Cu_8GeSe_6 crystallizes in the hexagonal structure with the lattice parameters a = 1264.21 ± 0.02 and c = 1175.67 ± 0.03 pm (Klymovych et al. 2020) [a = 1264.38 ± 0.02 and c = 1175.70 ± 0.01 pm at 17°C (Onoda et al. 1999b); a = 1249 and c = 1167 pm (Carcaly et al. 1973); a = 1264.8 ± 0.5, c = 1176 ± 4 pm, and the calculated and experimental densities of 6.60 and 6.60 ± 0.08 g·cm^{-3}, respectively (Jaulmes et al. 1991)]. β-Cu_8GeSe_6 also crystallizes in the hexagonal structure with the lattice parameters a = 729.99 ± 0.04 and c = 1176.79 ± 0.07 pm at 77°C (Onoda et al. 1999b) [a = 728 and c = 1167 pm (Carcaly et al. 1973); a = 730.7 ± 0.3 and c = 1175 ± 1 pm (Jaulmes et al. 1991); a = 732.2 ± 0.1 and c = 1178.5 ± 0.2 pm (Tomm et al. 2008)]. γ-Cu_8GeSe_6 crystallizes in the orthorhombic structure with the lattice parameters a = 704.5, b = 696.6, and c = 987.0 pm (Onoda 2006).

The powder of this compound was prepared by reacting stoichiometric amounts of Cu_2Se, Ge, and Se in an evacuated and sealed silica ampoule at 500°C for 4 days (Onoda et al. 1999b). Cu_8GeSe_6 could be also prepared if the mixture of Ge, Cu, and S in a stoichiometric ratio was placed in a vacuum sealed ampoule which was gradually heated in seven days up to 800°C and then abruptly increased to 1000°C for 3 h (Jaulmes et al. 1991) or by slow heating to 950°C and keeping there for 24 h (Tomm et al. 2008). The preparation was completed by a quenching in the air. To get enough big crystals, it is necessary to make an annealing of about three months at 650°C followed by the next quenching in water.

According to the data of Rogacheva et al. (1975b) and Zotova and Karagodin (1975b), the phase diagram of the GeSe–Cu_2Se system is a eutectic type with formation of Cu_2GeSe_3 and $\mathbf{Cu_6GeSe_5}$ ternary compounds. Two eutectics contain 10 and 60 mol% Cu_2Se [17 and 60 mol% Cu_2Se (Zotova and Karagodin 1975b)] and crystallize at 670°C and 740°C, respectively (Rogacheva et al. 1975b). Cu_2GeSe_3 has a polymorphic transformation at 740°C. The solubility of Cu_2Se in $GeSe_2$ is not higher than 3 mol% and the solubility of $GeSe_2$ in Cu_2Se reaches up to 10 mol%.

Cu_6GeSe_5 melts incongruently at 810°C (Zotova and Karagodin 1975b) [at 800°C (Rogacheva et al. 1975b)] and has a polymorphic transformation at 665°C (Zotova and Karagodin 1975b). It is possible that this compound corresponds to the Cu_8GeSe_6 ternary compound.

According to Voroshilov et al. (1986), the $\mathbf{Cu_6Ge_2Se_6}$ compound, which melts at 534°C and has an experimental density of 5.88 g·cm^{-3}, is also formed in the Ge–Cu–Se ternary system.

5.7 Germanium–Silver–Selenium

$GeSe_2$–Ag_2Se: The phase diagram of this system, constructed trough DTA, XRD, and metallography is presented in Figure 5.4 (Gorochov 1968b). The eutectics contain 13 and 57 mol% $GeSe_2$ [7 and 55 mol% $GeSe_2$ (Krykhovets et al. 2001)] and crystallize at 830°C and 570°C, respectively (Gorochov 1968b; Krykhovets et al. 2001) [5 and 55 mol% $GeSe_2$ and crystallize at 809°C and 559°C, respectively (Borisova et al. 1984); 15 and 56 mol% $GeSe_2$ and crystallize at 810°C and 540°C, respectively (Salaeva et al. 1985c; Movsum-zade et al. 1987b)]. The $\mathbf{Ag_8GeSe_6}$ ternary compound, which melts congruently at 902°C [at 883°C (Borisova et al. 1984)] and has three polymorphic transformations at 70°C (Mikolaychuk and Moroz 1985), at 48°C [at 51°C (Borisova et al. 1984); at 47°C (Bagkheri et al. 2014)] and at −4°C, is formed in this system (Gorochov and Flahaut 1967; Gorochov 1968b; Salaeva et al. 1985c; Krykhovets et al. 2001; Yusibov et al. 2017a)]. The structure of α-Ag_8GeSe_6 was not determined.

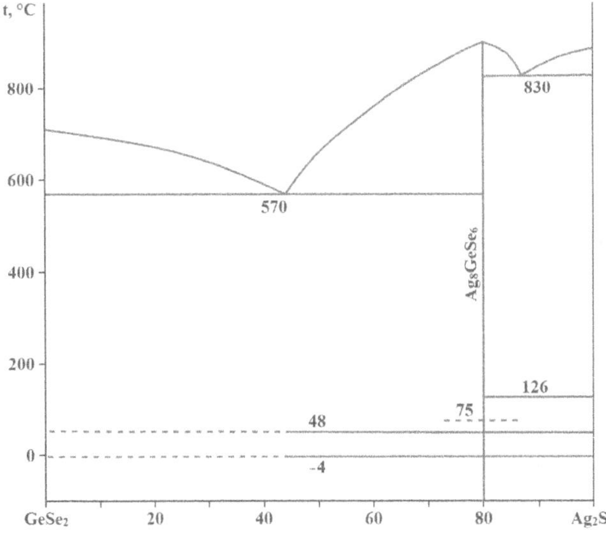

FIGURE 5.4 Phase diagram of the $GeSe_2$–Ag_2Se system. (From Gorochov, O., *Bull. Soc. chim. France*, (6), 2263, 1968.)

β-Ag_8GeSe_6 crystallizes in the orthorhombic structure with the lattice parameters a = 784.43 ± 0.05, b = 773.72 ± 0.05, and c = 1091.41 ± 0.07 pm (Parasyuk et al. 2003a; Bilousov et al. 2006) [a = 782.3, b = 771.2, c = 1088.5 pm, and the calculated and experimental densities of 7.14 and 7.1 g·cm^{-3}, respectively (Carré et al. 1980); a = 784.02, b = 773.22, and c = 1091.18 pm (Alieva et al. 2014; Yusibov et al. 2017a)] and an energy gap of 0.84 eV (Osipishin 1977).

γ-Ag_8GeSe_6 crystallizes in the cubic structure with the lattice parameter a = 1099.3 pm (Yusibov et al. 2017a) [a = 1096 ± 4 pm (Krykhovets et al. 2001); a = 1095 pm and the calculated and experimental densities of 7.13 and 7.07 g·cm^{-3}, respectively (Gorochov and Flahaut 1967; Gorochov 1968b); a = 1091 pm and the calculated and experimental densities of 7.20 and 6.94 g·cm^{-3}, respectively (Hahn et al. 1965)].

The title compound has been examined through EMF concentration chain measurements in the temperature range from 298 to 400 K by Bagkheri et al. (2014). The partial and integral thermodynamic functions of α- and β-Ag_8GeSe_6 (ΔG^0_{298} = 306.0 ± 3.1 kJ·mol^{-1}, ΔH^0_{298} = 285.7 ± 5.7 kJ·mol^{-1}, and ΔS^0_{298} = 694.0 ± 19.2 J·mol^{-1}K^{-1} and ΔG^0_{400} = 316.6 ± 3.4 kJ·mol^{-1}, ΔH^0_{400} = 270.7 ± 4.2 kJ·mol^{-1}, and ΔS^0_{400} = 740.9 ± 13.8 J·mol^{-1}K^{-1}) as well as heat and entropy of its polymorphic transformation (ΔH_{PT} = 15.04 ± 4.72 kJ·mol^{-1} and ΔS_{PT} = 47.0 ± 14.8 J·mol^{-1}K^{-1}) have been calculated.

Ag_8GeSe_6 was synthesized by reacting stoichiometric amounts of high-purity elements in an evacuated (~0.01 Pa) sealed quartz ampoule for about 6 days at temperatures of 600°C–700°C (Kuhs et al. 1979; Yusibov et al. 2017a).

Some authors noted that another ternary compound, $\mathbf{Ag_2GeSe_3}$, which melts incongruently at 574°C (Velásquez-Velásquez et al. 2000) [at 540°C (Balanevskaya et al. 1966); at 550°C (Salaeva et al. 1985c; Movsum-zade et al. 1987b)] and crystallizes in the monoclinic structure with the lattice parameters a = 775.16 ± 0.01, b = 1087.4 ± 0.4, c = 731.8 ± 0.1 pm, and β = 115.82 ± 0.01° (Velásquez-Velásquez et al. 2000), an experimental density of 4.66 g·cm^{-3} and an energy gap of 0.9 eV (Balanevskaya et al. 1966), exists in the $GeSe_2$–Ag_2Se system. No changes in the crystalline structure of this compound were observed in the range of 30°C–330°C using DTA and XRD, but according to the data of EMF technique, a change in the crystalline structure at 262 ± 5°C was observed (Moroz and Prokhorenko 2015a). The next values of the thermodynamic functions in the standard state were determined: $-\Delta G^0_f$ = 104.4 ± 1.5 and 106.7 ± 1.3 kJ·M, and $-\Delta H^0_f$ = −106.7 ± 0.9 and 112.1 ± 0.8 kJ·M for α- and β-Ag_2GeSe_3, respectively.

This compound was synthesized by the direct melting of the constituent elements in the stoichiometric ratio inside an evacuated quartz ampoule (Velásquez-Velásquez et al. 2000). The synthesis profile was conducted first, at a heating rate of 50°C·h^{-1} until a temperature of 900°C was reached, and then the ampoule was kept at this temperature for about 120 h. This was followed by a slow cooling at 6°C·h^{-1} down to room temperature. In this manner homogeneous ingots with a brilliant gray color were obtained. According to the data of Ollitrault-Fichet et al. (1985), Mikolaichuk and Moroz (1987a,b), and Yusibov et al. (2017a), the existence of the Ag_2GeSe_3 and $\mathbf{Ag_8GeSe_5}$ (Salaeva et al. 1984; Movsum-zade et al. 1987b) ternary compounds were not confirmed.

$\mathbf{Ag_8GeSe_6}$**–Ge:** This vertical section is a quasibinary one of the Ge–Ag–Se ternary system (Salaeva et al. 1985c). The eutectic crystallizes at 675°C and contains 40 mol% Ag_8GeSe_6. The separation of liquid alloys occurs on the side of Ge in the range of 3–30 mol% Ag_8GeSe_6 and at a temperature of 750°C [25–35 mol% Ag_8GeSe_6 and 762°C (Mikolaichuk and Moroz 1987b)].

$\mathbf{Ag_8GeSe_6}$**–Se:** This vertical section is also a quasibinary one of the Ge–Ag–Se ternary system (Yusibov et al. 2017a). In the system there are monotectic equilibrium at 700°C and a degenerate eutectic at 220°C.

GeSe–AgSe: The solid solutions with the structure of the NaCl-type were not detected in this system (Gorochov 1968a).

$\mathbf{GeSe_2}$**–Ag:** This system is a non-quasibinary section of the Ge–Ag–Se ternary system (Salaeva et al. 1985b). It crosses the $GeSe_2$–Ag_8GeSe_6–Ge, Ag_8GeSe_6–Ge–Ag_2Se, and Ag_2Se–Ge–Ag subsystems. The vertical sections of **GeSe–Ag**, **GeSe–**$\mathbf{Ag_2Se}$, and $\mathbf{GeSe_2}$**–"**$\mathbf{AgSe_2}$**"** were constructed by Salaeva et al. (1983. 1984, 1985a) and Movsum-zade et al. (1987b).

The liquidus surface of the Ge–Ag–Se system (Figure 5.5) consists of seven fields of primary crystallizations of (Ag), high-temperature modification of Ag_2Se, (Ag_8GeSe_6), (α-GeSe), and ($GeSe_2$), and also elemental Ge and Se (Yusibov et al. 2017a). The latter field is degenerate in the selenium-rich corner of the phase diagram. There are also two wide regions of liquid phase separation in this system. The next invariant equilibria exist in the system: U (472°C) – L + Ge ⇔ (β-Ag_8GeSe_6) + (α-GeSe); E_1 (700°C) – L ⇔ (β-Ag_2Se) + (Ag) + (β-Ag_8GeSe_6); E_2 (612°C) – L ⇔ (Ag) + (β-Ag_8GeSe_6) + Ge; E_3 (412°C) – L ⇔ (β-Ag_8GeSe_6) + ($GeSe_2$) + (α-GeSe); E_4 (217°C) – L ⇔ (β-Ag_2Se) + (β-Ag_8GeSe_6) + Se; and E_5 (212°C) – L ⇔ (β-Ag_8GeSe_6) + ($GeSe_2$) + Se. Yusibov et al. (2017a) also constructed some vertical sections and proposed 3D models of the liquidus surfaces of ($GeSe_2$), (β-Ag_2Se), (β-Ag_8GeSe_6), and immiscibility gaps. The liquidus surface of this ternary system was also constructed by Ollitrault-Fichet et al. (1985) and the liquidus surfaces of the $GeSe_2$–Ag_2Se–Se and $GeSe_2$–Ag_8GeSe_6–Ge subsystem were constructed, respectively, by Salaeva et al. (1983, 1985a) and Mikolaichuk and Moroz (1987b).

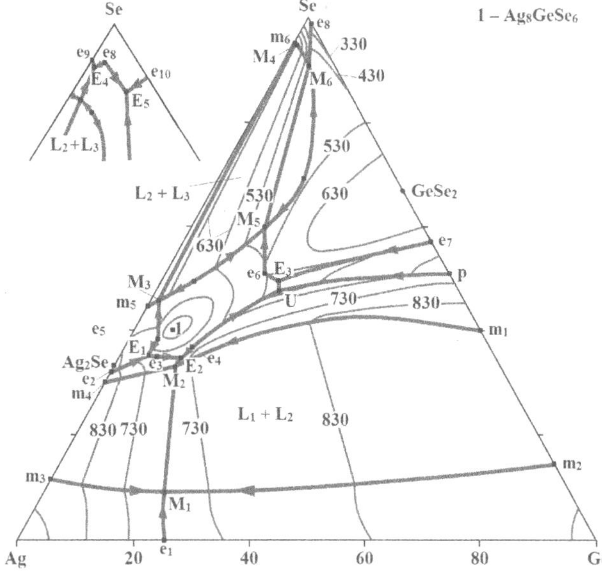

FIGURE 5.5 Liquidus surface of the Ge–Ag–Se ternary system. (From Yusibov, Yu.A., et al., *Russ. J. Inorg. Chem.*, **62**(9), 1223, 2017.)

The isothermal section of the Ge–Ag–Se ternary system at room temperature with and without formation of the Ag_2GeSe_3 compound is given in Figure 5.6 (Moroz and Prokhorenko 2015a; Yusibov et al. 2017a). It is seen, that the composition point of the compound Ag_8GeSe_6 is connected by tie lines to the composition points of all the other compounds and elemental components of the system.

The glass forming region in the $GeSe_2$–Ag_8GeSe_6–GeSe subsystem is presented in Figure 5.7 (Mikolaichuk and Moroz 1987b). The glass transition temperature for these glasses is within the interval 222°C–332°C.

5.8 Germanium–Magnesium–Selenium

The $\mathbf{Mg_2GeSe_4}$ ternary compound, which crystallizes in the orthorhombic structure with the lattice parameters a = 1350 ± 3, b = 785.1 ± 1.5, c = 636.6 ± 1.2 pm, and an energy gap of 2.02 eV, is formed in the Ge–Mg–Se system (Wu et al. 2015c). It was obtained by a direct solid-state reaction as follows: a mixture of MgSe, Ge, and Se (molar ratio 2:1:2) was ground carefully, placed into a silica tube and flame-sealed under 10^{-3} Pa. Subsequently, the tube was heated to 900°C in 30 h, maintained at that temperature for 80 h, and then cooled to 500°C (4°C·h^{-1}), finally dropping to room temperature rapidly. Many red crystals were obtained and they were stable in air for several weeks.

5.9 Germanium–Strontium–Selenium

Two ternary compounds, $\mathbf{Sr_2GeSe_4}$ and $\mathbf{Sr_2Ge_2Se_5}$, are formed in the Ge–Sr–Se system. Sr_2GeSe_4 crystallizes in the orthorhombic structure with the lattice parameters a = 1028.4 ± 0.2, b = 1054.3 ± 0.2, c = 741.1 ± 0.1 pm, and a calculated density of 4.66 g·cm^{-3} (Pocha et al. 2003). It was obtained by the following two-step synthesis. First, binary Sr_2Ge was prepared by heating stoichiometric amounts of Sr and Ge to 800°C in an alumina crucible under Ar atmosphere for 12 h. Then, the binary product was homogenized, mixed with the stoichiometric amount of Se, loaded again in the crucible, and sealed under Ar. The ampoule was rapidly heated to 900°C,

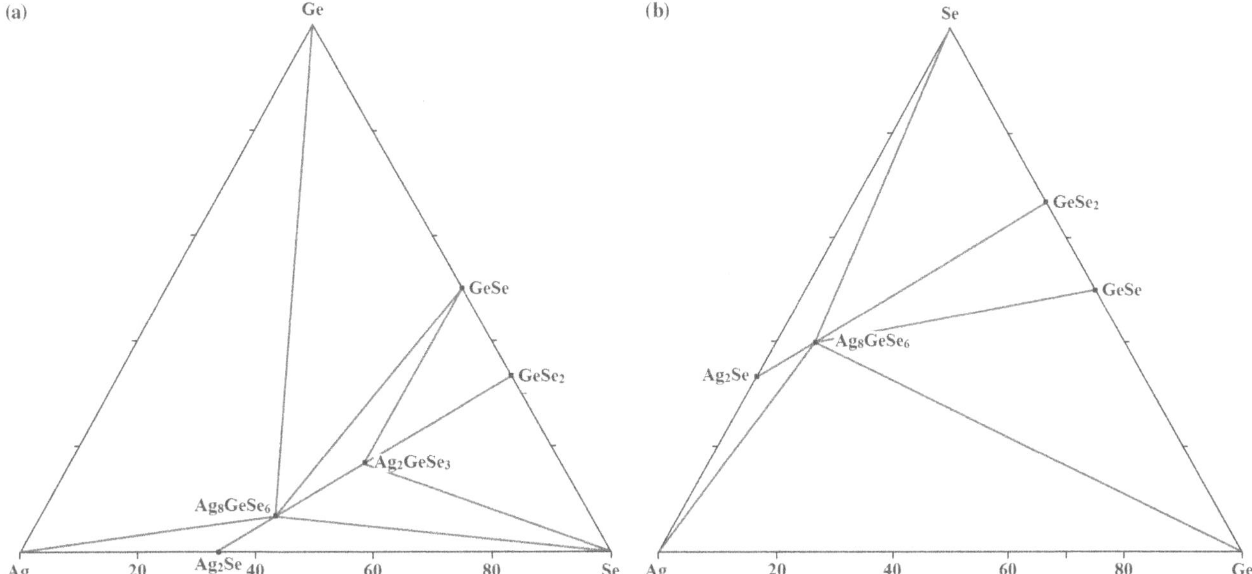

FIGURE 5.6 Isothermal section of the Ge–Ag–Se ternary system at room temperature. (a, from Moroz, M.V., and Prokhorenko, M.V., *Russ. J. Electrochem.*, **51**(7), 697, 2015; b, from Yusibov, Yu.A., et al., *Russ. J. Inorg. Chem.*, **62**(9), 1223, 2017.)

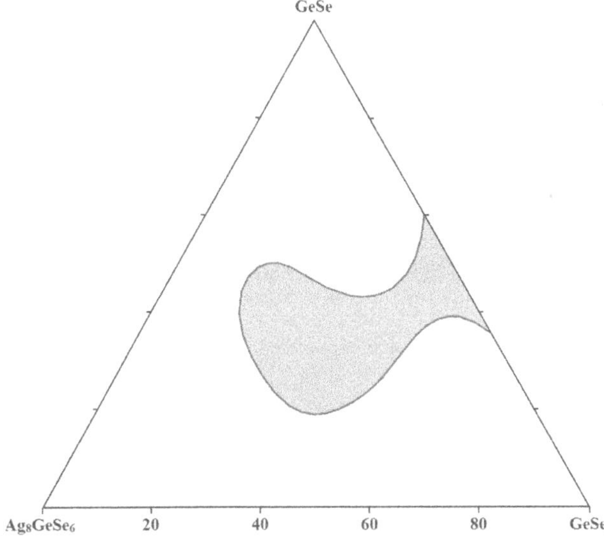

FIGURE 5.7 Glass forming region in the $GeSe_2$–Ag_8GeSe_6–Ge subsystem. (From Mikolaichuk, A.G., et al., *Izv. AN SSSR. Neorgan. mater.*, **23**(8), 1265, 1987.)

kept at this temperature for 2 h and then quenched in water. This results in a yellow-orange powder sample of the title compound, which is stable in air.

$Sr_2Ge_2Se_5$ crystallizes in the monoclinic structure with the lattice parameters $a = 844.5 \pm 0.2$, $b = 1230.2 \pm 0.2$, $c = 917.9 \pm 0.2$ pm, $\beta = 93.75 \pm 0.03°$, and a calculated density of 4.992 g·cm^{-3} (Johrendt and Tampier 2000). To prepare the target compound, a stoichiometric mixture of SrSe (2 mM), Ge (2 mM), and Se (3 mM) was loaded into a corundum crucible. The crucible was transferred into a silica tube and evacuated. It was flushed three times with Ar before being sealed under Ar atmosphere (ambient pressure). The mixture was heated to 750°C at a rate of 50°C·h^{-1} and was kept at this temperature for 50 h. The furnace was then allowed to cool to room temperature. The first product was homogenized in an Ar-filled glove box, reloaded to the crucible and subsequently tempered at 750°C for 50 h. This procedure yielded a brown powder of $Sr_2Ge_2Se_5$, which is stable in air.

5.10 Germanium–Barium–Selenium

Two ternary compounds, $\mathbf{BaGeSe_3}$ and $\mathbf{Ba_2Ge_2Se_5}$, are formed in the Ge–Ba–Se system. $BaGeSe_3$ crystallizes in the orthorhombic structure with the lattice parameters $a = 702$, $b = 1413$, $c = 605$ pm, and the calculated and experimental densities of 4.93 and 4.74 g·cm^{-3}, respectively (Hahn and Schulze 1965). It was prepared by the interaction of BaSe with $GeSe_2$ at the temperatures between 600°C and 1100°C.

$Ba_2Ge_2Se_5$ also crystallizes in the orthorhombic structure with the lattice parameters $a = 1259.4 \pm 0.3$, $b = 917.4 \pm 0.2$, $c = 916.0 \pm 0.2$ pm, and a calculated density of 5.113 g·cm^{-3} (Johrendt and Tampier 2000). This compounds was prepared in the same way as $Sr_2Ge_2Se_5$ was synthesized using BaSe instead of SrSe. The deep red powder of $Ba_2Ge_2Se_5$ showed no reaction when exposed to air or water. Small, red single crystals were transparent only as thin layers.

5.11 Germanium–Zinc–Selenium

$GeSe_2$–ZnSe: The phase diagram of this system is a eutectic type (Figure 5.8) (Olekseyuk et al. 1999b). The eutectic composition and temperature are 16 mol% ZnSe and 720°C. $\mathbf{ZnGeSe_3}$ and $\mathbf{Zn_2GeSe_4}$ ternary compounds (Hahn and Lorent 1958; Radautsan and Ivanova 1961) were not found in this system (Kaldis et al. 1967; Koren' et al. 1984; Olekseyuk et al. 1999b). To obtain the compounds, four methods were used: (1) synthesis at 7 MPa and heating to 2000°C, (2) alloying of

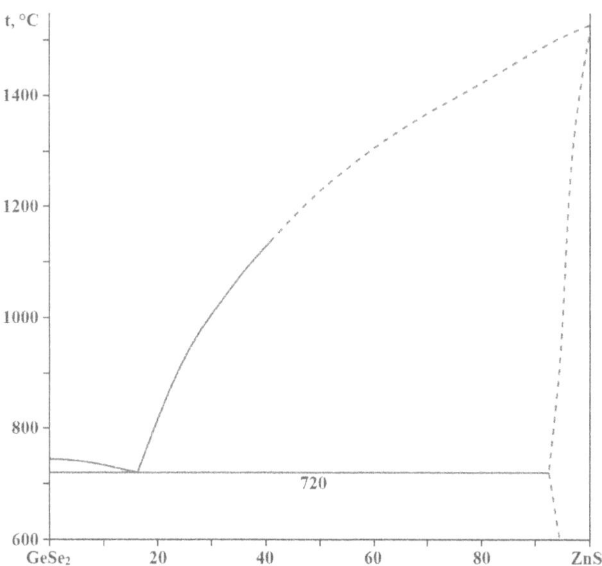

FIGURE 5.8 Phase diagram of the $GeSe_2$–ZnSe system. (From Olekseyuk, I., et al., *Quasiternary chalcogenide systems*, Vol. 1, Luts'k: Vezha Publishing, 168 p., 1999.)

the components with vibration mixing, (3) chemical transport reactions, and (4) sublimation. All ingots were two-phase and contained the mixtures of $GeSe_2$ and ZnSe (Koren' et al. 1984).

The glass-forming region in the Ge–Zn–Se system lies on the Ge–Se side (concentration limits 0–43 at% Ge) and is given in Figure 5.9 (Vassilev et al. 1997). The maximum solubility of Zn in the glasses is 6 at% at Se/Ge molar ratio of 9:1. To obtain the glasses, quartz ampoules with stoichiometric quantities of the starting components, evacuated in advance to a residual pressure of 0.1 Pa and sealed, were heated step-wise to the maximum temperature of the synthesis that does not exceed 1000°C ± 10°C. After vibration agitation of the melt, the latter was quenched in a mixture of water and ice.

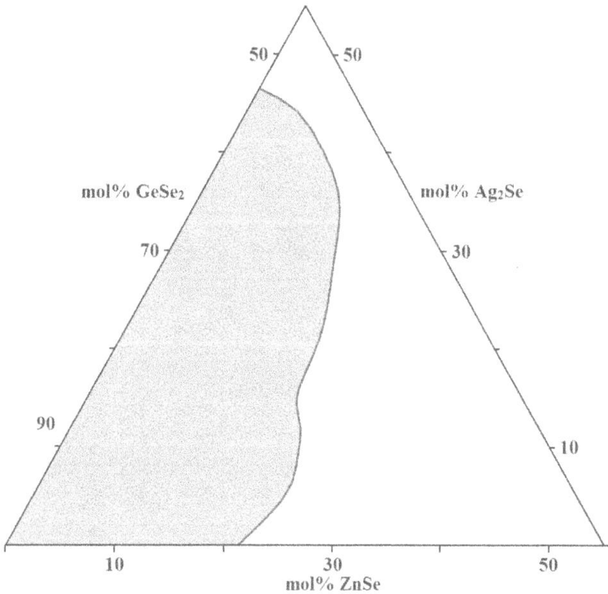

FIGURE 5.9 Glass forming region in the Ge–Zn–Se ternary system. (From Vassilev, V., et al, *J. Mater. Sci.*, **32**(16), 4443, 1997.)

5.12 Germanium–Cadmium–Selenium

GeSe–CdSe: This section is a non-quasibinary section of the Ge–Cd–Se ternary system which intersects the fields of CdSe and Ge primary crystallization (Figure 5.10) (Odin et al. 1985). Secondary crystallization of CdSe and Ge takes place at 683°C and at 668°C there is twice peritectical equilibrium CdSe + Ge + L ⇔ CdSe·9GeSe (γ). Transition point (L + Ge ⇔ γ + β-GeSe) takes place at 660°C. Thermal effects at 625–627°C are determined by the polymorphic transformation of GeSe. Small thermal effects are on the heating curves within the interval of 540–560°C which correspond to the formation of γ from α-GeSe and CdSe. Solubility of CdSe in α-GeSe is equal to 0.5 mol% and solubility of α-GeSe in CdSe is not higher than 1 mol%. This system was investigated by the DTA, metallography and XRD. The ingots were annealed at 500 and 600°C for 1100 and 1000 h respectively.

$GeSe_2$–CdSe: The phase diagram of this system is shown in the Figure 5.11 (Galiulin et al. 1983). The peritectic composition is 49 mol% CdSe and the eutectic composition and temperature are 30 mol% CdSe and 716°C, respectively (Galiulin et al. 1983) [708°C (Quenez and Khodadad 1969a)]. The solubility of CdSe in $GeSe_2$ at 550°C is equal 2.1 mol% (Galiulin et al. 1983) and 1.57 ± 0.05 mol% at 710°C (Odin and Grin'ko 1991d). The solubility of $GeSe_2$ in CdSe at 550°C is not higher than 0.7 mol% (Galiulin et al. 1983) and 0.44 ± 0.06 mol% at 785°C (Odin and Grin'ko 1991d).

Cd_4GeSe_6 compound is formed in this system. It melts incongruently at 863 ± 6°C (Galiulin et al. 1983) [at 840°C (Quenez and Khodadad 1969a)] and crystallizes in a monoclinic structure with lattice parameters a = 1284.7 ± 0.3, b = 740.7 ± 0.2, c = 1285.4 ± 0.2 pm, and β = 109.82 ± 0.01° (Henao et al. 1998a) [a = 1281 ± 2, b = 738 ± 1, c = 1279 ± 2 pm, and

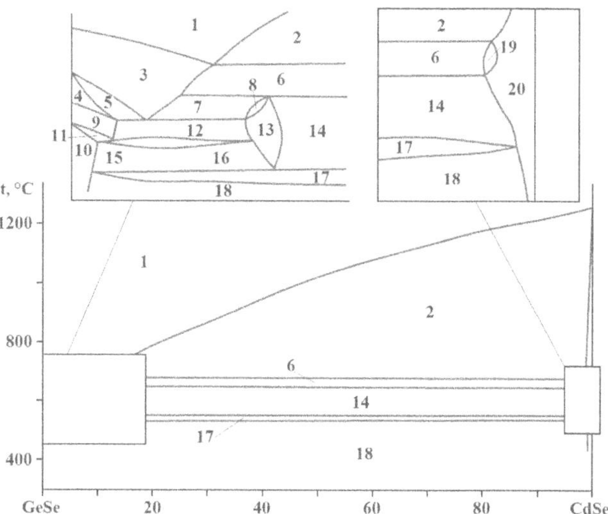

FIGURE 5.10 Phase relations in the GeSe–CdSe system: 1, L; 2, L + CdSe; 3, L + Ge; 4, L + β-GeSe; 5, L + Ge + β-GeSe; 6, L + CdSe + Ge; 7, L + γ + Ge; 8, γ + Ge; 9, β-GeSe; 10, α-GeSe; 11, β-GeSe + α-GeSe; 12, γ + β-GeSe; 13, γ; 14, γ + CdSe; 15, γ + β-GeSe + α-GeSe; 16, α-GeSe + γ; 17, α-GeSe + γ + CdSe; 18, α-GeSe + CdSe; 19, CdSe + Ge; and 20, CdSe. (From Odin, I.N. et al., *Zhurn. neorgan. khimii*, **30**(1), 201, 1985.)

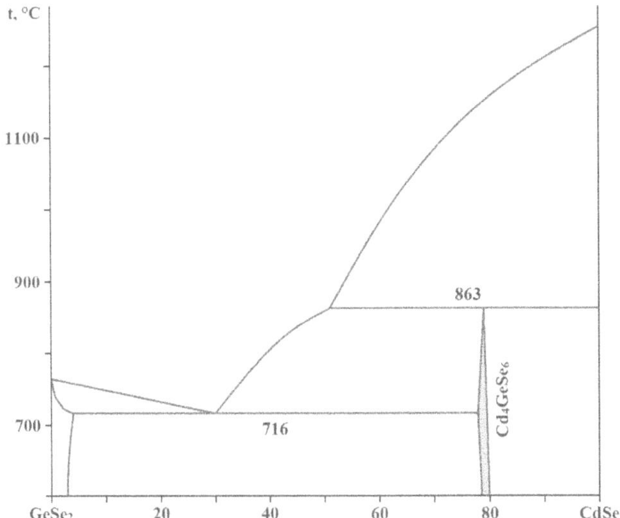

FIGURE 5.11 Phase diagram of the $GeSe_2$–CdSe system. (From Galiulin, E.A. et al., *Zhurn. neorgan. khimii*, **28**(5) 1281, 1983.)

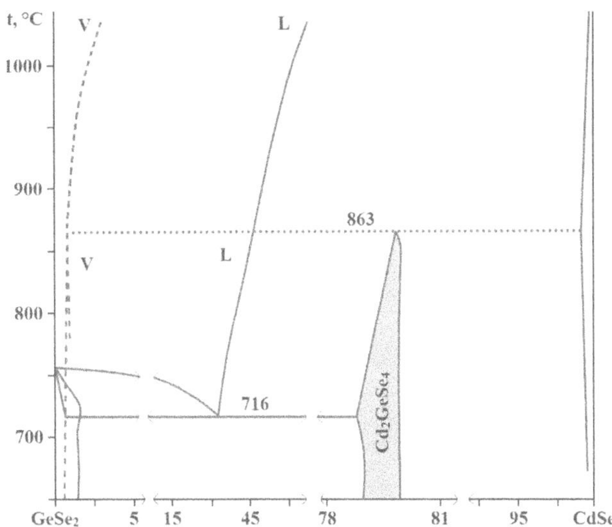

FIGURE 5.12 *T–x* projection of $GeSe_2$–CdSe phase diagram close to Cd_4GeSe_6, CdSe, and $GeSe_2$. (From Odin, I.N. and Grin'ko, V.V., *Zhurn. neorgan. khimii*, **36**(9), 2387, 1991.)

β = 109°34′ ± 0.1′ (Quenez and Khodadad 1969a); a = 1293.7, b = 740.2, c = 1285.2 pm, and β = 110.9° (Motrya et al. 1986a); a = 1282.3, b = 740.9, c = 1280.2 pm, and β = 109.60° (Susa and Steinfink 1971a), calculated and experimental density of 5.75 [5.444 (Hahn and Lorent 1958)] and 5.57 g·cm^{-3} (Motrya et al. 1986a, 1987). Homogeneity region of Cd_4GeSe_6 is within the interval of 78.5–80.3 mol% CdSe (Galiulin et al. 1983) [78.9–80.1 mol% CdSe (Odin and Grin'ko 1991d)].

Cd_4GeSe_6 was synthesized by direct fusion of the constituent elements in a sealed and evacuated quartz ampoule from a mixture of Cd/Ge/Se (molar ratio 2:1:4) during an attempt to prepare **Cd_2GeSe_4** (Henao et al. 1998a). The mixture was heated slowly to 500°C over a period of three days. Subsequently, it was heated slowly to 1100°C and held at this temperature for 2 h. Finally, the reaction mixture was annealed at 500°C for a month. Then, the sample was cooled to room temperature in about a day. Single crystals of Cd_4GeSe_6 were grown by the chemical transport reaction (Motrya et al. 1986a, 1987). This compound decomposes at the heating on the air up to 470°C (Motrya et al. 1986a).

Ternary compounds **$CdGeSe_3$** (Radautsan and Ivanova 1961) and Cd_2GeSe_4 (Hahn and Lorent 1958) were not found in this system (Quenez and Khodadad 1969a; Galiulin et al. 1983). Galiulin et al. (1983) did not confirm the existence of the thermal effects at 450°C near the $GeSe_2$-rich side which characterize devitrification of ingots (Quenez and Khodadad 1969a). According to the data of (Zhao et al. 2005) no glass was obtained in the $GeSe_2$–CdSe system.

Vapor composition in the $GeSe_2$–CdSe system corresponds to $GeSe_2$ compound, i.e. it lies on this section (Odin et al. 1986; Motrya et al. 1986a; Odin and Grin'ko 1991a,d). The vapor above Cd_4GeSe_6 consists mainly from GeSe and Se_2 in ratio of 2:1 and the Cd concentration in vapor is 0.01 mass% (Odin and Grin'ko 1991d). The limit of Cd_4GeSe_6 homogeneity region from the CdSe-side does not change with temperature and from the $GeSe_2$-side changes at the temperatures above a eutectic temperature (Figure 5.12) (Odin and Grin'ko 1991d). $(T–x)_p$-sections of the $GeSe_2$–CdSe system at p = 133 hPa is shown in the Figure 5.13 (Odin and Grin'ko 1991d).

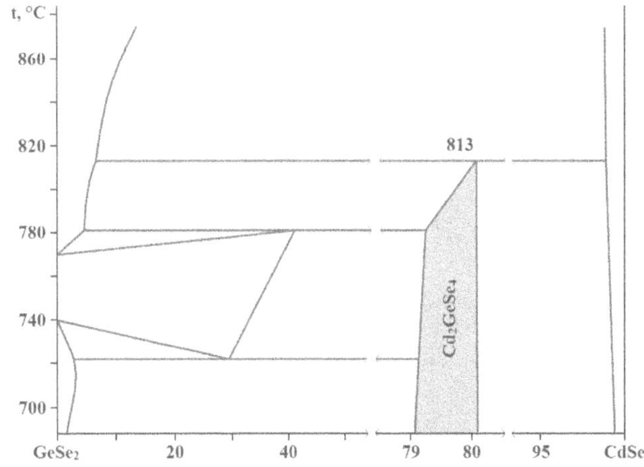

FIGURE 5.13 $(T–x)_p$ section of $GeSe_2$–CdSe phase diagram at 133 hPa. (From Odin, I.N. and Grin'ko, V.V., *Zhurn. neorgan. khimii*, **36**(9), 2387, 1991.)

On *p-T* projection of $GeSe_2$–CdSe phase diagram (Figure 5.14) (Odin et al. 1986) the lines represent two-phase: [*AB* – S($GeSe_2$)/V; *BC* – L($GeSe_2$)/V; *KG* – S(CdSe)/V; *GF* – L(CdSe)/V] and three-phase equilibria [*BD* – S($GeSe_2$)/L/V; *MD* – V/S($GeSe_2$)/S(Cd_4GeSe_6); *DE* – S(Cd_4GeSe_6)/L/V]. The lines of three-phase equilibria intersect in the invariant points: *D* – V/S($GeSe_2$)/L/S(Cd_4GeSe_6); *E* – V/S(Cd_4GeSe_6)/L/S(CdSe). The coordinates of *D* and *E* points are 716°C, 4213 Pa, and 863°C, 71.3 kPa, respectively. The lines, representing equilibria of condensed phases [L($GeSe_2$)/S($GeSe_2$), S($GeSe_2$)/L/S(Cd_4GeSe_6), S(Cd_4GeSe_6)/L/S(CdSe) and S(CdSe)/L(CdSe)], are practically vertical at small pressures. The region of 1000–1250°C is not investigated but it can be confirmed that the line of three-phase equilibrium V/L/S(CdSe) passes through maximum.

This system was investigated using the DTA, metallography, XRD, chemical analysis and tensimeter, as well as measuring of microhardness (Quenez and Khodadad 1969a;

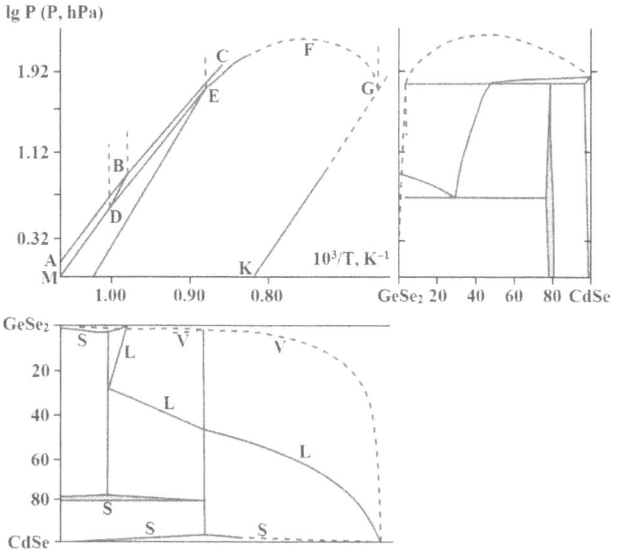

FIGURE 5.14 P_{total}–T–x phase diagram of the $GeSe_2$–CdSe system: 1, $GeSe_2$; 2, 7.0; 3, 60.0; 4, 80.5; and 5, 85.0 mol% CdSe. (From Odin, I.N. et al., *Zhurn. neorgan. khimii*, **31**(5), 1274, 1986.)

Galiulin et al. 1983; Odin et al. 1986; Odin and Grin'ko 1991a,d). The ingots were annealed at 550°C for 750 h (Galiulin et al. 1983) [(at 650°C for 720 h (Odin and Grin'ko 1991d)].

Ge–CdSe: The phase diagram is a eutectic type (Figure 5.15) (Galiulin et al. 1983). The eutectic composition and temperature are 1.5 mol% CdSe and 929°C, respectively. The immiscibility region exists within the interval of 33–90 mol% CdSe with monotectic temperature 1211°C. Mutual solubility of CdSe and Ge is not higher than 0.1 mol%. The Ge content in the CdSe doped single crystals (GeSe was used as a dopant) reaches 0.65 at% (0.25 mass%) (Odin et al. 1999). This system was investigated using the DTA, metallography, and XRD. The ingots were annealed at 550°C for 250 h.

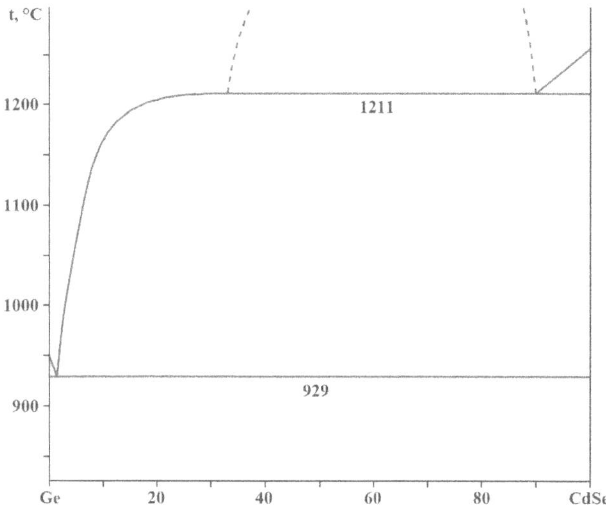

FIGURE 5.15 Phase diagram of the CdSe–Ge system. (From Galiulin, E.A. et al., *Zhurn. neorgan. khimii*, **28**(5), 1281, 1983.)

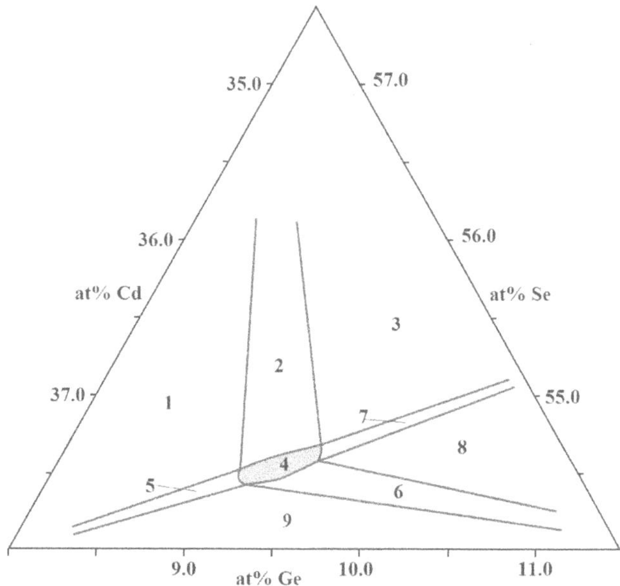

FIGURE 5.16 Homogeneity region of Cd_4GeSe_6 at 544°C in the Ge–Cd–Se ternary system: 1, L + CdSe + Cd_4GeSe_6; 2, L + Cd_4GeSe_6; 3, L + Cd_4GeSe_6 + $GeSe_2$; 4, Cd_4GeSe_6; 5, CdSe + Cd_4GeSe_6; 6, GeSe + Cd_4GeSe_6; 7, $GeSe_2$ + Cd_4GeSe_6; 8, GeSe + $GeSe_2$ + Cd_4GeSe_6; and 9, CdSe + GeSe + Cd_4GeSe_6. (From Odin, I.N. and Grin'ko, V.V., *Zhurn. neorgan. khimii*, **36**(9), 2387, 1991.)

Homogeneity region of Cd_4GeSe_6 at 544°C in the Ge–Cd–Se ternary system is shown in the Figure 5.16 (Odin and Grin'ko 1991d). It is elongated along the CdSe–$GeSe_2$ quasibinary system. **$CdGe_7Se_8$** compound exists at 547°C–552°C and does not appear at 544°C. At 600°C solid phases Cd_4GeSe_6, γ-phase ($CdGe_9Se_{10}$), CdSe, $GeSe_2$, α-GeSe, Ge and liquid are in equilibria (Figure 5.17) (Odin et al. 1985). Liquid regions are situated from Se- and Cd-rich sides and near the eutectics e_3 and E_2.

On the liquidus surface of the Ge–Cd–Se ternary system there are the fields of Cd_4GeSe_6, CdSe·9GeSe (γ-phase), CdSe, $GeSe_2$, α- and β-GeSe, Ge, Se, and Cd primary crystallization (Figure 5.18) (Odin et al. 1985). Phase relations in this system are determined by 19 invariant equilibria with participation of liquid. The next invariant equilibria exist in the system: E_1 (205°C) – L ⇔ Se + CdSe + Cd_4GeSe_6; E_2 (570°C) – L ⇔ α-GeSe + γ + $GeSe_2$; E_3 (311°C) – L ⇔ CdSe + Cd + Ge; U_2 (212°C) – L + CdSe ⇔ Cd_4GeSe_6 + Se; U_3 (590°C) – L + Cd_4GeSe_6 ⇔ γ + $GeSe_2$; U_4 (664°C) – L + CdSe ⇔ Cd_4GeSe_6 + γ; U_5 (660°C) – L + Ge ⇔ β-GeSe + γ; and P (668°C) – L + CdSe + Ge ⇔ γ.

Three immiscibility regions exist on the liquidus surface. One of them is situated in the field of Ge crystallization, the second covers a part of the CdSe crystallization field and the third starts in the CdSe–Se subsystem and occupies small part of the ternary system from the Se-rich side.

The glass-forming region in the Ge–Cd–Se system is situated on the Se-rich of the Ge–Se side of the Gibbs triangle, where up to 12 at% Cd can be introduced into the binary $Ge_{20}Se_{80}$ composition without the formation of the crystallites (Figure 5.19) (Ivanova et al. 1997). The glass transition temperature for obtained glasses varies from 88°C to 277°C,

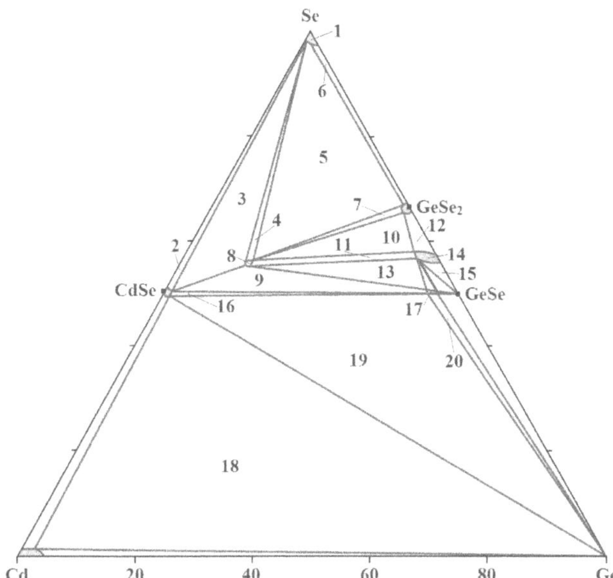

FIGURE 5.17 Isothermal section of the Ge–Cd–Se ternary system at 600°C: 1, L_1; 2, L_1 + CdSe; 3, L_1 + CdSe + Cd_4GeSe_6; 4, L_1 + Cd_4GeSe_6; 5, L_1 + $GeSe_2$ + Cd_4GeSe_6; 6, L_1 + $GeSe_2$; 7, $GeSe_2$ + Cd_4GeSe_6; 8, Cd_4GeSe_6; 9, $CdGe_9Se_{10}$ + Cd_4GeSe_6 + CdSe; 10, L_2 + $GeSe_2$ + Cd_4GeSe_6; 11, L_2 + Cd_4GeSe_6; 12, L_2 + $GeSe_2$; 13, L_2 + Cd_4GeSe_6 + $CdGe_9Se_{10}$; 14, L_2; 15, L_2 + α-GeSe; 16, $CdGe_9Se_{10}$ + CdSe; 17, $CdGe_9Se_{10}$; 18, L + CdSe + Ge; 19, $CdGe_9Se_{10}$ + CdSe + Ge; 20, $CdGe_9Se_{10}$ + Ge; and 21, $CdGe_9Se_{10}$ + α-GeSe + Ge. (From Odin, I.N. et al., *Zhurn. neorgan. khimii*, **30**(1), 201, 1985.)

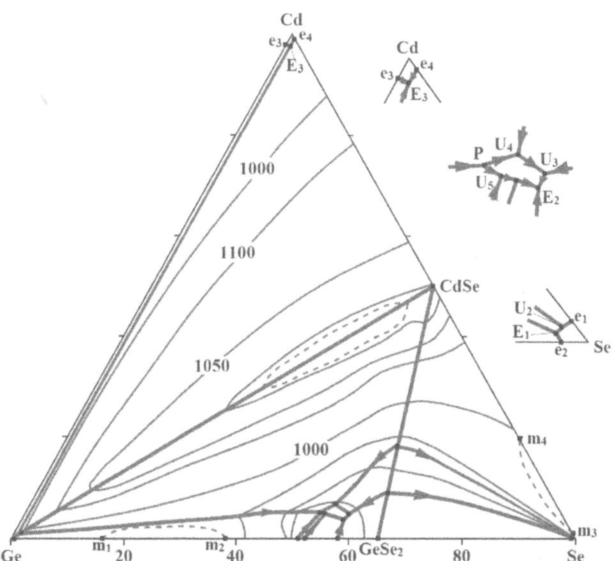

FIGURE 5.18 Liquidus surface of the Ge–Cd–Se ternary system. (From Odin, I.N. et al., *Zhurn. neorgan. khimii*, **30**(1), 201, 1985.)

increasing at larger Cd contents. The bulk glasses were prepared by melting the required amounts of Ge, Cd and Se in evacuated quartz ampoules. A stepwise heating procedure was employed: the melts were held at 300°C for 2 h, at 700°C for 6 h and at about 1000°C for 2 h. Next, the temperature was reduced to 800°C and then the mixtures were quenched in a mixture of ice and salt water.

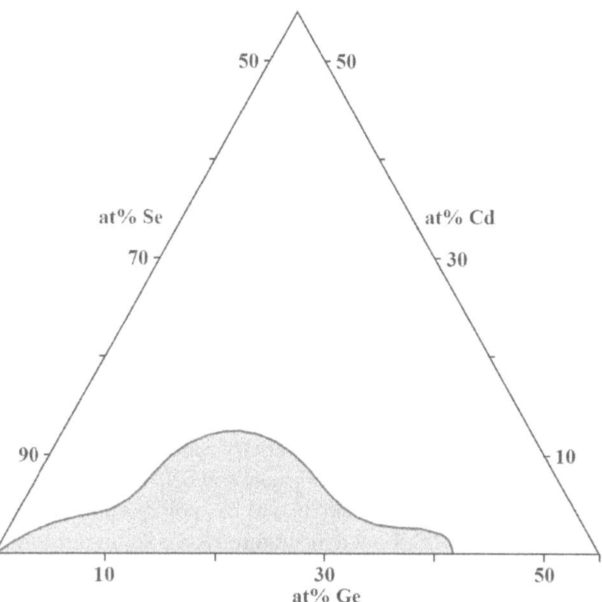

FIGURE 5.19 Glass forming region in the Ge–Cd–Se ternary system. (From Ivanova, Z.G., et al., *J. Non-Cryst. Solids*, **232–234**, 274, 1998.)

5.13 Germanium–Mercury–Selenium

GeSe–HgSe: This system is a non-quasibinary section of the Ge–Hg–Se ternary system (Figure 5.20) (Motrya 1991). The $\mathbf{Hg_2GeSe_3}$ compound is formed, which melts incongruently at 568°C, crystallizes in a cubic structure with lattice parameter a = 1103.4 ± 3 pm and has polymorphous transformation at 537 ± 5°C. Homogeneity region of Hg_2GeSe_3 at the temperature

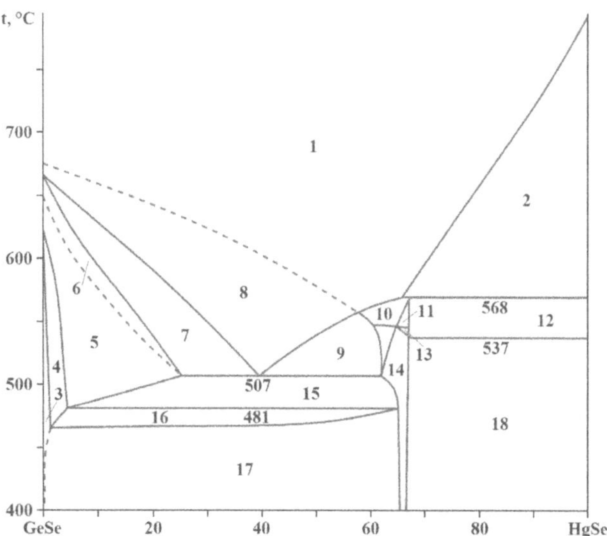

FIGURE 5.20 Phase relations in the GeSe–HgSe system: 1, L; 2, L + HgSe; 3, α-GeSe; 4, α-GeSe + β-GeSe; 5, β-GeSe; 6, L + β-GeSe; 7, L + $GeSe_2$ + β-GeSe; 8, L + $GeSe_2$; 9, L + $GeSe_2$ + α-Hg_2GeSe_3; 10, L + β-Hg_2GeSe_3; 11, L + α-Hg_2GeSe_3; 12, HgSe + β-Hg_2GeSe_3; 13, β-Hg_2GeSe_3; 14, α-Hg_2GeSe_3; 15, β-GeSe + α-Hg_2GeSe_3; 16, α-GeSe + β-GeSe + α-Hg_2GeSe_3; 17, α-GeSe + α-Hg_2GeSe_3; and 18, HgSe + α-Hg_2GeSe_3. (From Motrya, S.F., *Poluch. i svoistva slozhn. poluprovodn.*, Uzhgorod. Gos. Univ. Publish., Kiev, pp. 17–26, 1991.)

of a ternary eutectic (507°C) is within the interval of 63.0–66.7 mol% HgS. The solubility of HgSe in α- and β-GeSe reaches 1 and 28 mol%, respectively, and the solubility of GeSe in HgSe is insignificant. This system was investigated using the DTA, metallography, XRD and measuring of microhardness. The ingots were annealed at 380–430°C for 1200 hours.

The linear dependence of the Electromotive Force (EMF) of galvanic cells in the range of 152°C–182°C were used to calculate the standard thermodynamic values of Hg_2GeSe_3: $-\Delta G^0_{f,\,298} = 30.4 \pm 3.0$ kJ·M and $-\Delta H^0_{f,\,298} = 151.2 \pm 2.9$ kJ·M (Moroz et al. 2017).

$GeSe_2$–HgSe: The phase diagram of this system is shown in the Figure 5.21 (Motrya et al. 1986b; Motrya 1991). The eutectic composition and temperature are 42 mol% HgSe and 600°C, respectively. **Hg_2GeSe_4** ternary compound is formed in this system, which melts incongruently at 620°C (the peritectic composition is 60 mol% HgSe) and has polymorphous transformation at 557°C. Low-temperature Hg_2GeSe_4 modification crystallizes in the tetragonal structure with lattice parameters $a = 567.75 \pm 0.02$, $c = 1125.01 \pm 0.12$ pm, a calculated density of 7.231 g·cm^{-3}, and an energy gap of 1.17 eV (Guo et al. 2018) [$a = 567 \pm 2$ and $c = 1132 \pm 2$ pm, the calculated and experimental densities of 7.20 and 6.97 g·cm^{-3}, respectively (Motrya et al. 1986b); $a = 569.1$ and $c = 1128$ pm, the calculated and experimental densities of 7.179 and 7.09 g·cm^{-3}, respectively (Hahn and Lorent 1958); $a = 567.9$ and $c = 1126.9$ pm (Motrya 1991); $a = 567.86 \pm 0.02$, $c = 1125.79 \pm 0.07$ pm, and a calculated density of 7.2230 ± 0.008 g·cm^{-3} at room temperature (Parasyuk et al. 2003a) and $a = 567.41 \pm 0.17$, $c = 1124.9 \pm 0.5$ pm, and a calculated density of 7.241 g·cm^{-3} at 150 ± 2 K (Dong et al. 2005)].

The linear dependence of the EMF of galvanic cells in the range of 152°C–182°C were used to calculate the standard thermodynamic values of Hg_2GeSe_3: $-\Delta G^0_{f,\,298} = 30.4 \pm 2.8$ kJ·M and $-\Delta H^0_{f,\,298} = 146.3 \pm 2.3$ kJ·M (Moroz et al. 2017).

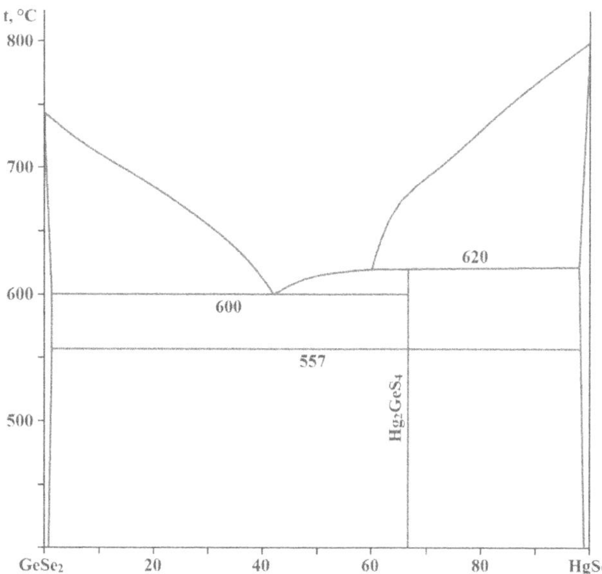

FIGURE 5.21 Phase diagram of the $GeSe_2$–HgSe system. (From Motrya, S.F. et al., *Ukr. khim. zhurn.*, **52**(8), 807, 1986.)

The title compound was prepared by the reaction of HgSe, Ge, and Se with the use of the reactive halide flux technique (Dong et al. 2005). A combination was loaded in a quartz tube (molar ratio 1:1:5) and an $ErCl_3$/NaCl eutectic mixture was then added in a mass ratio of $HgGeSe_6$/($ErCl_3$/NaCl) = 1:2. The tube was evacuated to 1.3 Pa, sealed, and heated gradually (50°C·h^{-1}) to 850°C in a furnace where it was kept for 24 h. The tube was cooled slowly to 100° (4°C·h^{-1}) and quenched. Black shiny needle-shaped crystals were found. The excess halides were removed with distilled water. The crystals are stable in air and water.

Polycrystalline samples of Hg_2GeSe_4 were also synthesized by solid-state reaction technique (Guo et al. 2018). The mixture of HgSe, Ge and Se (molar ratio of 2:1:2) were ground and loaded into fused silica tube, which was then evacuated to a high vacuum (10^{-3} Pa) and sealed. The sample was heated to 500°C for 10 h in a furnace and stayed at that temperature for 120 h. Then it was cooled to room temperature by switching off the furnace. The single crystals of this compound were prepared by spontaneous crystallization method using pure single phase powder. The powder was loaded into fused quartz tube in a glove-box, which were sealed under a high vacuum (10^{-3} Pa) and then placed in a furnace. The samples was heated to 750°C in 15 h and kept at that temperature for 72 h. Then the samples were cooled at a slow rate (3°C·h^{-1}) to 200°C, and finally cooled to room temperature. The air stable block black crystals were found in the tube.

Glass forming region in the $GeSe_2$–HgSe system within the interval from $GeSe_2$ compound to 56 mol% HgSe was determined by Olekseyuk et al. (1999a). The temperatures of softening and crystallization of these glasses decrease at the HgSe addition. Mutual solubility of HgSe and $GeSe_2$ is insignificant (Motrya et al. 1986b).

This system was investigated by the DTA, metallography, XRD and measuring of microhardness and density. The ingots were annealed at 500°C for 1200 h (Motrya et al. 1986b; Motrya 1991; Olekseyuk et al. 1999a).

The Hg_2GeSe_3 and Hg_2GeSe_4 ternary compounds are formed in the Ge–Hg–Se ternary system (Figure 5.22) (Motrya et al. 1986b; Motrya 1991). The regions of solid solutions were not found.

The glass formation was found in the GeSe-HgSe–$GeSe_2$ subsystem in a large range of composition (Figure 5.23) (Feltz et al. 1976; Feltz and Burckhardt 1980).

5.14 Germanium–Boron–Selenium

The glass forming region in the Ge–B–Se ternary system is given in Figure 5.24 (Kirilenko and Dembovskiy 1974). The obtained black glass-like alloys are unstable in air.

5.15 Germanium–Gallium–Selenium

GeSe–GaSe: The results of the investigation of this system are contradictory and it is not possible to select the most reliable ones. According to the data of Sidorov et al. (1977), the phase diagram of the GeSe–GaSe system is of a eutectic type

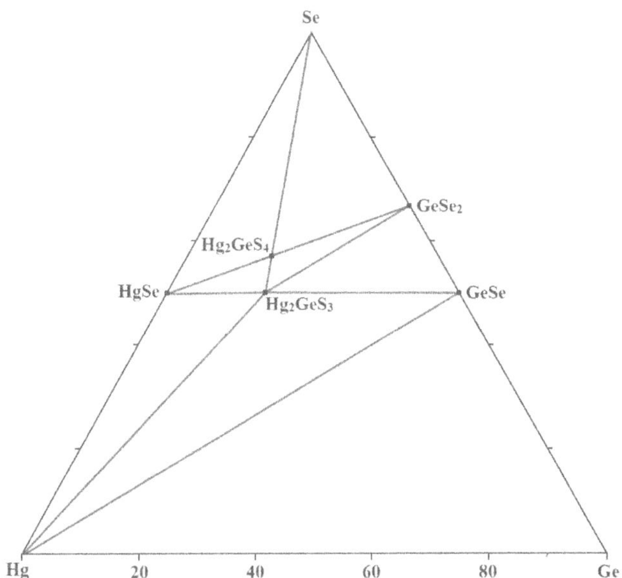

FIGURE 5.22 Isothermal section of the Ge–Hg–Se ternary system at 380°C. (From Motrya, S.F., in *Poluch. i svoistva slozhn. poluprovodn.*, Uzhgorod. Gos. Univ. Publish., Kiev, 17–26, 1991.)

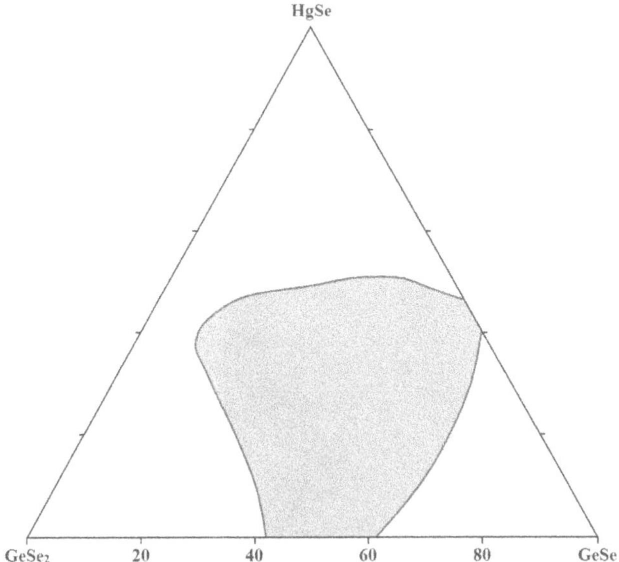

FIGURE 5.23 Glass forming region in the GeSe–HgSe–$GeSe_2$ subsystem. (From Feltz, A., and Burckhardt, W., *Z. anorg. und allg. Chem.*, **461**(1), 35, 1980.)

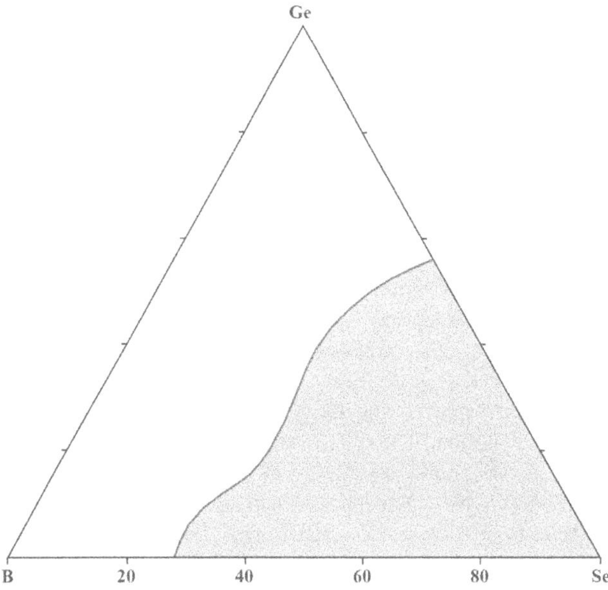

FIGURE 5.24 Glass forming region in the Ge–B–Se ternary system. (From Kirilenko, V.V. and Dembovskiy, S.A., *Izv. AN SSSR. Neorgan. mater.*, **10**(3), 542, 1974.)

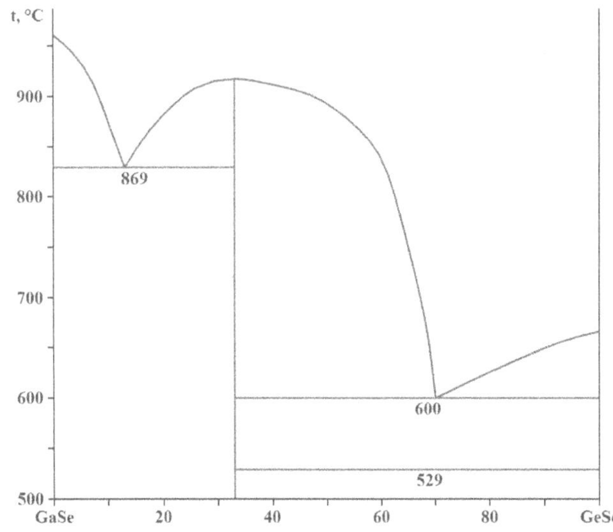

FIGURE 5.25 Phase diagram of the GeSe–GaSe system. (From Sidorov, I.A. et al., *Zhurn. neorgan. khimii*, **22**(5), 1430, 1977.)

(Figure 5.25). The eutectics contain 30 and 87 mol% GaSe. The $\mathbf{Ga_2GeSe_3}$ ternary compound is formed in the system.

According to the data of Rustamov and Movsum-zade (1975) and Movsum-zade and Rustamov (1977), the GeSe–GaSe system is a non-quasibinary section of the Ge–Ga–Se ternary system. This section crosses two subsystems, $GeSe_2$–Ga_2Se_3–Ge and GaSe–Ga_2Se_3–Ge, and is composed of two parts. The liquidus of the part of the section passing through the $GeSe_2$–Ga_2Se_3–Ge subsystem consists of the branches of the Ge and Ga_2Se_3 primary crystallization. The crystallization of this melts end at 560°C as a result of the next invariant reaction: L ⇔ γ-GeSe + Ga_2Se_3 + Ge.

The liquidus of another part of the section passing through the GaSe–Ga_2Se_3–Ge subsystem is composed of two parts: L + Ga_2Se_3 and L + (δ-GaSe). The melts of this part of the section crystallize at the temperature of the ternary eutectic at 855°C, where Ga_2Se_3, Ge and (δ-GaSe) solidify (Movsum-zade and Rustamov 1977). The solubility of GeSe in GaSe at room temperature is 5 mol% and increases up to 6 mol% at 855°C (Rustamov and Movsum-zade 1975; Movsum-zade and Rustamov 1977). The limit solubility of GaSe in GeSe is 5 mol% (Alidzhanov et al. 1982a). According to the data of Thiebault et al. (1975), an immiscibility region exists in the GeSe–GaSe system.

The alloys of the GeSe–GaSe system were annealed at 550°C–600°C (Sidorov et al. 1977) [at 550°C for 300 h (Movsum-zade and Rustamov 1977)]. The system was investigated through DTA, XRD, metallography, and measuring of microhardness (Rustamov and Movsum-zade 1975; Movsum-zade and Rustamov 1977; Sidorov et al. 1977).

GeSe–Ga$_2$Se$_3$: This system is a non-quasibinary section of the Ge–Ga–Se ternary system (Thiebault et al. 1975). The minimum on the liquidus takes place at 77 mol% GeSe and 624°C. The system was investigated through DTA, XRD, and metallography.

GeSe$_2$–GaSe: This system is also a non-quasibinary section of the Ge–Ga–Se ternary system and crosses two subsystems, GeSe$_2$–Ga$_2$Se$_3$–Ge and GeSe–Ga$_2$Se$_3$–Ge (Rustamov and Movsum-zade 1975; Movsum-zade and Rustamov 1977). The solubility of GaSe in GeSe$_2$ and GeSe$_2$ in GeSe is negligible (not higher than 1 mol%). The alloys of this system were annealed at 450°C for 300 h. The system was investigated through DTA, XRD, metallography, and measuring of microhardness.

GeSe$_2$–Ga$_2$Se$_3$: The phase diagram of this system is a eutectic type (Figure 5.26) (Olekseyuk and Parasyuk 1995). The eutectic contains 14 mol% GaSe and crystallizes at 660°C [13.6 mol% GaSe and at 666°C (Thiebault et al. 1975); at 650°C (Rustamov and Movsum-zade 1975)]. The solubility of GeSe$_2$ in Ga$_2$Se$_3$ at the eutectic temperature reaches up to 20 mol% (Olekseyuk and Parasyuk 1995). The **Ga$_2$GeSe$_5$** ternary compound [according to the data of Rustamov and Movsum-zade (1975) and Thiebault et al. (1975) it is stable up to 625°C and crystallizes in the cubic structure with the lattice parameter a = 546.1 pm] was not found in this system (Olekseyuk and Parasyuk 1995). The system was investigated through DTA, XRD, metallography, and measuring of microhardness (Rustamov and Movsum-zade 1975; Thiebault et al. 1975; Olekseyuk and Parasyuk 1995).

Crystallization behavior of $(GeSe_2)_{0.8}(Ga_2Se_3)_{0.2}$ glasses during annealing at 380°C for 10, 25, and 50 h indicates on the possibility of formation of **GeGa$_4$Se$_8$** and Ga$_2$Se$_3$ crystals (Shpotyuk et al. 2014).

GaSe–Ge: The phase diagram of this system is a eutectic type with an immiscibility region (Rustamov and Movsum-zade 1974). The eutectic contains 66 mol% GaSe and crystallizes at 875°C. There is the immiscibility region in this system within the interval of 14–60 mol% GaSe at 920°C. The solubility of Ge in GaSe at room temperature is 2 mol% and increases up to 5 mol% at 800°C. The alloys were annealed at 700°C for 300 h. The system was investigated through DTA, metallography, and measuring of microhardness and density.

The liquidus surface of the Ge–Ga–Se ternary system is given in Figure 5.27 (Rustamov and Movsum-zade 1975). This surface includes the fields of the primary crystallization of Ge, Ga$_2$Se, GaSe, Ga$_2$Se$_3$, GeSe$_2$, and GeSe. The fields of Ga and Se primary crystallization are degenerated. There is also an immiscibility region, four ternary eutectics and two transition points on the liquidus surface. The liquidus surface of this ternary system constructed by Rustamov and Movsum-zade (1975) are somewhat different from those constructed by Thiebault et al. (1975).

A large glass formation region exists in the Ge–Ga–Se ternary system (Ollitrault-Fichet et al. 1977; Giridhar and Mahadevan 1990). It was shown that a narrow region in the neighboring of the ternary eutectic of the GeSe–GeSe$_2$–Ga$_2$Se$_3$ subsystem has a very slow kinetics of crystallization (Ollitrault-Fichet et al. 1977). The crystallization of glasses gives three new phases, which exist in a narrow temperature region and disappear before the solidus. One of these phases has the **GaGeSe$_3$** composition. The temperature of glass transition, T_g, is bigger along the GeSe$_2$–Ga$_2$Se$_3$ section and is rapidly decreasing on all sides of this section. Along this section T_g has a nearly constant value (350°C). Another new phase is **Ga$_4$GeSe$_8$** and the composition of the third phase was not determined. The glass transition temperature for all glasses

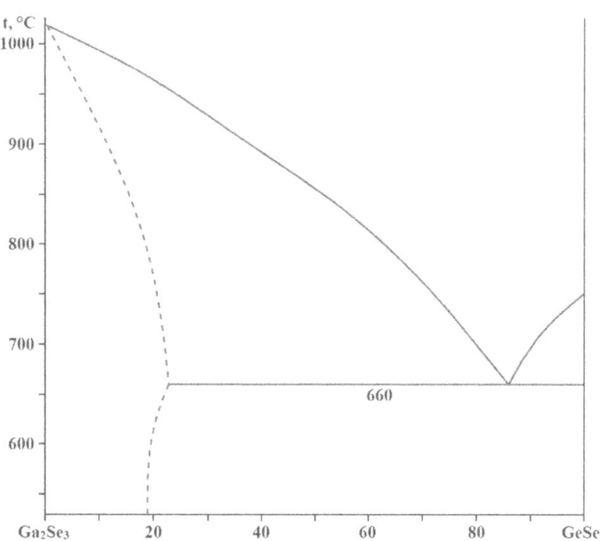

FIGURE 5.26 Phase diagram of the GeSe$_2$–Ga$_2$Se$_3$ system. (From Olekseyuk, I.D., and Parasyuk, O.V. *Zhurn. neorgan. khimii*, **40**(2), 315, 1995.)

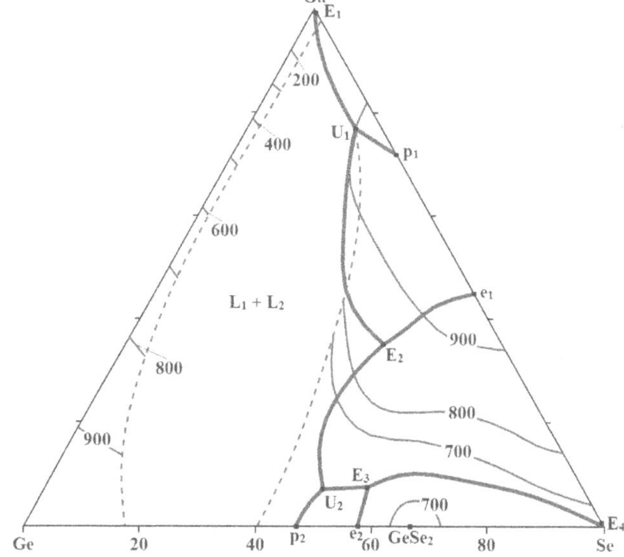

FIGURE 5.27 Liquidus surface of the Ge–Ga–Se ternary system. (From Rustamov, P.G., and Movsum-zade, A.A., *Azerb. khim. zhurn.*, (5), 117, 1975.)

in the Ge–Ga–Se ternary system is within the interval from 246.5°C to 366.5°C (Giridhar and Mahadevan 1990).

Bulk glasses were prepared using pure Ge, Ga, and Se (Giridhar and Mahadevan 1990). Appropriate proportions of the elements were weighed into quartz ampoules which were sealed under vacuum. The ampoules were then heated to 800°C for about 24 h. During this, melting homogeneity was achieved by rotating the ampoules. The ampoules were quenched in ice water to obtain glasses.

5.16 Germanium–Indium–Selenium

GeSe–InSe: The solubility of GeSe in InSe at 450°C is equal to 14 mol% and decreases up to 11 mol% at room temperature (Abdullayev et al. 1984).

GeSe–In_4Se_3: The In_4GeSe_4 ternary compound is formed in this system. It crystallizes in the cubic structure with the lattice parameter $a = 1248.3 \pm 0.1$ pm (Reiner and Deiseroth 1998) [$a = 1247.88 \pm 0.03$ pm (Deiseroth and Pfeifer 1993a)]. The title compound was prepared in evacuated quartz ampoule at 650°C from the elements in the stoichiometric amounts (Deiseroth and Pfeifer 1993a). Single crystals of this compound could be prepared by annealing stoichiometric mixtures of the binary compounds GeSe and In_4Se_3 at 560°C in evacuated quartz ampoules for three weeks (Reiner and Deiseroth 1998).

$GeSe_2$–InSe: This system is a non-quasibinary section of the Ge–In–Se ternary system and crosses the fields of the primary crystallization of $GeSe_2$, γ-In_2Se_3, InSe, and In_5Se_6 (Babaeva 1974; Rustamov and Babaeva 1976). In the range of 0–80% InSe, the alloys crystallize in the form of Ge + $GeSe_2$ + In_2Se_3 three-phase mixtures, and in the range of 80–95 mol% InSe, the alloys contain In_5Se_6, In_2Se_3, and Ge. The ingots of the system were annealed at 490°C for 860 h. The system was investigated through DTA, XRD, metallography, and measuring of microhardness.

$GeSe_2$–In_2Se_3: The phase diagram of this system is a eutectic type with peritectic transformations (Figure 5.28) (Rustamov and Babaeva 1975). The eutectic contains 78 mol% $GeSe_2$ and crystallizes at 610°C. The solubility of $GeSe_2$ in In_2Se_3 at room temperature reaches up to 3 mol%. The system was investigated through DTA, XRD, metallography, and measuring of microhardness.

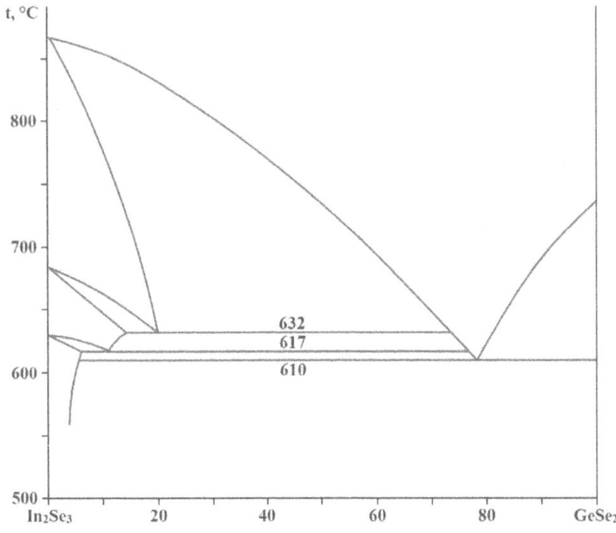

FIGURE 5.28 Phase diagram of the $GeSe_2$–In_2Se_3 system. (From Rustamov, P.G., and Babaeva, B.K., *Zhurn. neorgan. khimii*, **20**(9), 2456, 1975.)

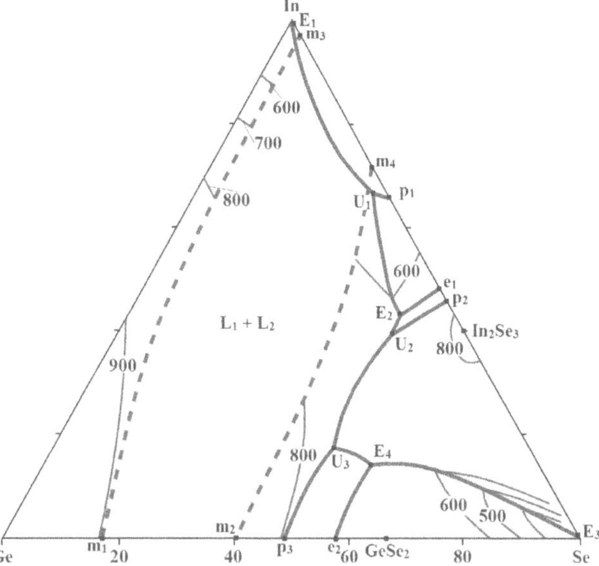

FIGURE 5.29 Liquidus surface of the Ge–In–Se ternary system. (From Rustamov, P.G., and Babaeva, B.K., *Azerb. khim. zhurn.*, (3), 75, 1976.)

The liquidus surface of the Ge–In–Se ternary system is presented in Figure 5.29 (Rustamov and Babaeva 1976). This surface includes the fields of the primary crystallization of In_2Se, InSe, In_5Se_6, In_2Se_3, GeSe, $GeSe_2$, Ge, In, and Se (the fields of In and Se primary crystallization are degenerated). There is also an immiscibility region, four ternary eutectics and three transition points on the liquidus surface. The alloys of the Ge–In–Se ternary system were annealed at 490°C for 860 h.

The $\mathbf{In_xGe_{20}Se_{8-x}}$ (x = 4, 8, 12, 15, 18) bulk glasses were prepared by quenching the melt. The required amounts of Ge, Se, and In were sealed under a vacuum of 7 mPa in quartz ampoules (Rabinal et al. 1995). These ampoules were held at 1000°C for 48 h and homogenized before quenching in an ice-water + NaOH solution. Glasses up to $x = 12$ show single-stage crystallization, whereas the glasses with $x > 15$ show double-stage crystallization. These glasses can be considered as microscopically inhomogeneous, in which microclusters of In_2Se_3 are embedded in the host Ge–Se matrix. The temperature of glass transition is within the interval from 202.9°C to 314.1°C and increases with increases in x, reaches a maximum at $x = 15$ (314.1°C) with the next decreasing to 307.5°C for $x = 18$.

5.17 Germanium–Thallium–Selenium

GeSe–TlSe: This system is a non-quasibinary section of the Ge–Tl–Se ternary system (Babanly and Kulieva 1983). Three ternary compounds, $\mathbf{Tl_2GeSe_3}$, which melts congruently at 437°C [melts incongruently at 441°C (Turkina et al. 1978)], $\mathbf{TlGe_2Se_3}$ and $\mathbf{TlGeSe_2}$, which melt incongruently at 437°C and 337°C, respectively, are formed in this system. Initial and intermediate phases of this section, as well as Ge and solid solutions based on Tl_4GeSe_4, primary crystallize in the system. The samples were annealed at a temperature of 20–30°C below the solidus temperature for 500 h. The system was investigated through DTA, XRD, metallography, and measuring of microhardness and EMF.

GeSe–Tl_2Se: This system is also a non-quasibinary section of the Ge–Tl–Se ternary system (Babanly et al. 1985; Babanly and Kulieva 1986). The **Tl_2GeSe_2** ternary compound, which melts incongruently at 457°C, is formed in this system. The primary crystallization of Ge takes place in the wide region of the concentration. The samples were annealed at a temperature of 20–50°C below the solidus temperature for 400–800 h. The system was investigated through DTA, XRD, metallography, and measuring of microhardness and EMF.

$GeSe_2$–Tl: This system is also a non-quasibinary section of the Ge–Tl–Se ternary system and crosses the Tl_2Se–Ge and **Tl_4GeSe_4**–Ge quasibinary systems (Babanly and Kulieva 1986). The primary crystallization of Ge-based solid solutions occupies the entire composition range and occurs over a wide range according to the monotectic reaction. The system was investigated through DTA, XRD, metallography, and measuring of microhardness.

When the Tl content is more than 22 mol%, glassy samples are formed in this system (Borisova et al. 1976).

$GeSe_2$–TlSe: The phase diagram of this system is a eutectic type with peritectic transformations (Figure 5.30) (Turkina et al. 1979). The eutectic contains 18 mol% $GeSe_2$ and crystallizes at 234°C. Two compounds, **$TlGeSe_3$** and **Tl_2GeSe_4**, which melt incongruently at 440°C and 324°C, respectively, are formed in this system. The thermal effects at 390°C and 377°C are apparently caused by polymorphic transformations of $TlGeSe_3$. In certain alloys containing 60–95 mol% $GeSe_2$, in addition to $GeSe_2$ and $TlGeSe_3$, phase $Tl_2Ge_2Se_5$ was detected by XRD, which may be due to the no equilibrium of these samples. When the TlSe content is more than 22 mol%, glassy samples are formed in this system (Borisova et al. 1976).

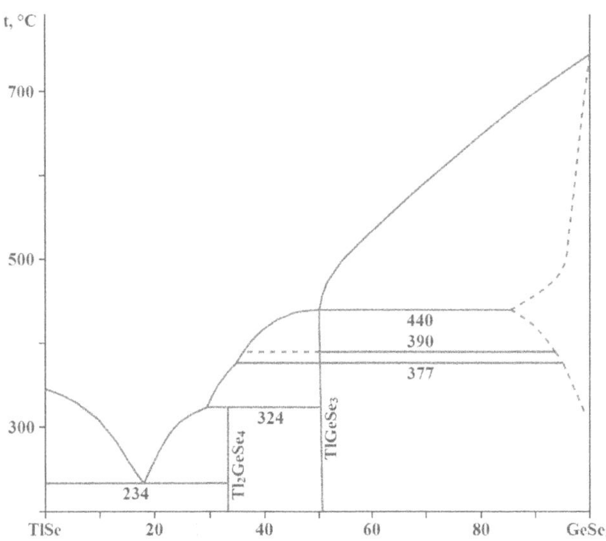

FIGURE 5.30 Phase diagram of the $GeSe_2$–TlSe system. (From Turkina, E.Yu., et al., *Zhurn. neorgan. khimii*, **24**(11), 3134, 1979.)

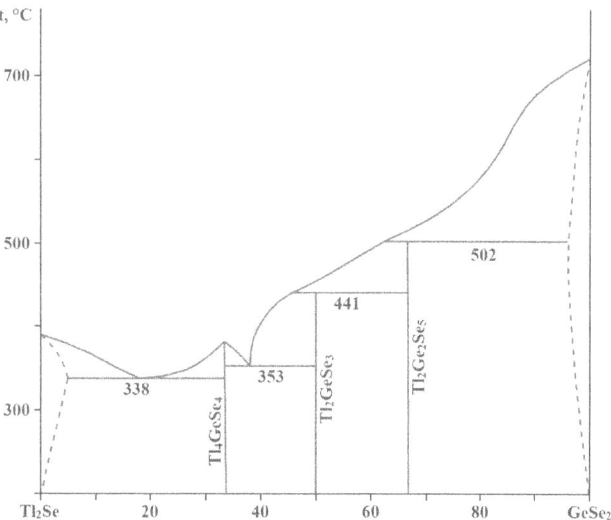

FIGURE 5.31 Phase diagram of the $GeSe_2$–Tl_2Se system. (From Turkina, E.Yu., et al., *Zhurn. neorgan. khimii*, **23**(2), 497, 1978.)

The samples were annealed at 180°C–250°C for 300–600 h. The system was investigated through DTA, XRD, and metallography (Turkina et al. 1979).

According to the data of Babanly and Kulieva (1983), $GeSe_2$–TlSe system is a non-quasibinary section of the Ge–Tl–Se ternary system.

$GeSe_2$–Tl_2Se: The phase diagram of this system is a eutectic type with peritectic transformations (Figure 5.31) (Turkina et al. 1978; Babanly and Kulieva 1983). Two eutectics contain 18 and 38 mol% $GeSe_2$ and crystallize at 338°C and 353°C, respectively (Turkina et al. 1978). Three compounds, Tl_4GeSe_4, which melts congruently at 382°C, Tl_2GeSe_3 and **$Tl_2Ge_2Se_5$**, which melt incongruently at 441°C and 502°C, respectively, are formed in this system. The mutual solubility of the starting components does not exceed 3–5 mol%. At the content of 50–63 mol% $GeSe_2$, glassy samples were obtained in the system.

According to the data of Babanly and Kulieva (1983), Tl_2GeSe_3, $Tl_2Ge_2Se_5$, and Tl_4GeSe_4 melt congruently at 437°C, 497°C, and 377°C, respectively, and four eutectics contain 20, 38, 52, and 67 mol% $GeSe_2$ and crystallize at 327°C, 341°C, 432°C, and 495°C, respectively. The solubility of $GeSe_2$ in Tl_2Se is 2.5 mol% at 280°C and decreases to 2 mol% at room temperature, and the homogeneity region of Tl_4GeSe_4 is within the interval of 31.0–33.5 and 31.5–33.5 mol% $GeSe_2$ at 280°C and room temperature, respectively.

The samples were annealed at 280°C–430°C for 240–650 h. The system was investigated through DTA, XRD, and metallography (Turkina et al. 1978).

Tl_2GeSe_3 crystallizes in the triclinic structure with the lattice parameters $a = 692.5 \pm 0.6$, $b = 693.4 \pm 0.7$, $c = 877.1 \pm 0.6$ pm, and $\alpha = 90.55 \pm 0.07°$, $\beta = 111.42 \pm 0.06°$, $\gamma = 114.45 \pm 0.04°$, a calculated density of 6.81 g·cm^{-3} (Eulenberger 1982) and an energy gap is 1.7 eV (Peresh et al. 1986). This compound was prepared from the stoichiometric mixture of Tl, Ge, and Se by reacting in the evacuated silica tube (Eulenberger 1982).

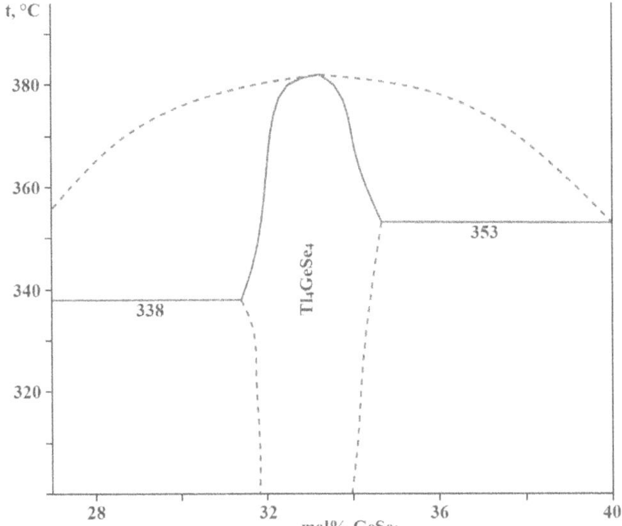

FIGURE 5.32 Homogeneity region of the Tl_4GeSe_4 compound. (From Peresh, E.Yu., et al., *Zhurn. neorgan. khimii*, **27**(2), 473, 1982.)

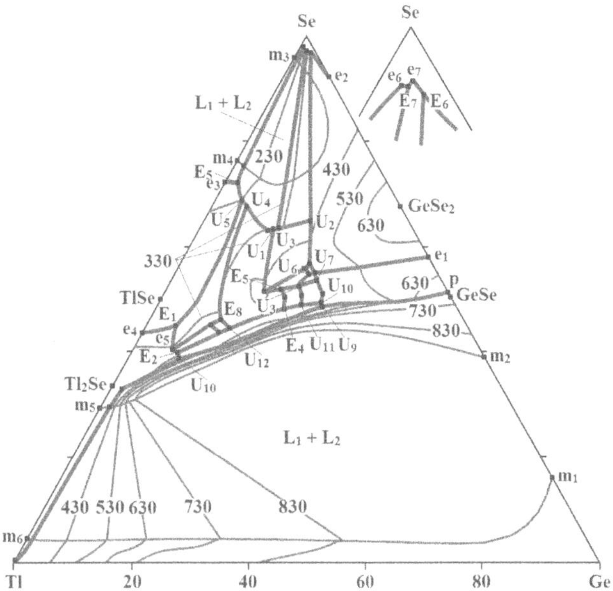

FIGURE 5.33 Liquidus surface of the Ge–Tl–Se ternary system. (From Babanly, M.B., and Kulieva, N.A., *Zhurn. neorgan. khimii*, **31**(9), 2365, 1986.)

$Tl_2Ge_2Se_5$ crystallizes in the monoclinic structure with the lattice parameters $a = 1560.2 \pm 0.6$, $b = 1554.9 \pm 0.5$, $c = 905.2 \pm 0.3$ pm, and $\beta = 107.10 \pm 0.03°$ (Eulenberger 1981a) and an energy gap is 2.0 eV (Peresh et al. 1986).

The homogeneity region of the Tl_4GeSe_4 compound is given in Figure 5.32 (Peresh et al. 1982). This compound also crystallizes in the monoclinic structure with the lattice parameters $a = 1167.00 \pm 0.02$, $b = 731.70 \pm 0.01$, $c = 2560.30 \pm 0.10$ pm, and $\beta = 106.54 \pm 0.01°$, and a calculated density of 7.644 $g \cdot cm^{-3}$ (Glukh et al. 2009). The enthalpy and entropy of melting of this compound is equal to 11.0 $kJ \cdot M^{-1}$ and 16.6 $J \cdot (M \cdot K)^{-1}$, respectively, and an energy gap is 1.6 eV (Peresh et al. 1982, 1986). The title compound was synthesized from $GeSe_2$ and Tl_2Se in evacuated quartz ampoules at 770°C for 96 h and its single crystals were grown using the Bridgman technique (Peresh et al. 1982; Glukh et al. 2009).

The liquidus surface of the Ge–Tl–Se ternary system is given in Figure 5.33 (Babanly and Kulieva 1986). This surface includes 14 fields of the primary crystallization of the phases (the fields of Tl and Se primary crystallization are degenerated). There is also two wide immiscibility region, seven ternary eutectics and 13 transition points on the liquidus surface. The liquidus surfaces of the $GeSe_2$–Tl_4GeSe_4–TlSe and $GeSe_2$–TlSe–Se subsystem were also constructed by Turkina and Orlova (1983) and Turkina and Orlova (1986), respectively. The isothermal section of this ternary system at 130°C is presented in Figure 5.34 (Babanly and Kulieva 1986).

The alloys of the Ge–Tl–Se ternary system were annealed at 160°C and 200°C depending on the composition for 400–800 h (Turkina and Orlova 1986).

The glass-forming region in the Ge–Tl–Se ternary system is given in Figure 5.35 (Borisova et al. 1976). The glass transition temperature for these glasses is within the range from 37°C to 222°C (Linke et al. 1978). According to the data of Dembovskiy (1968), this region is much smaller, which can be explained by the differences in the methodology for preparing

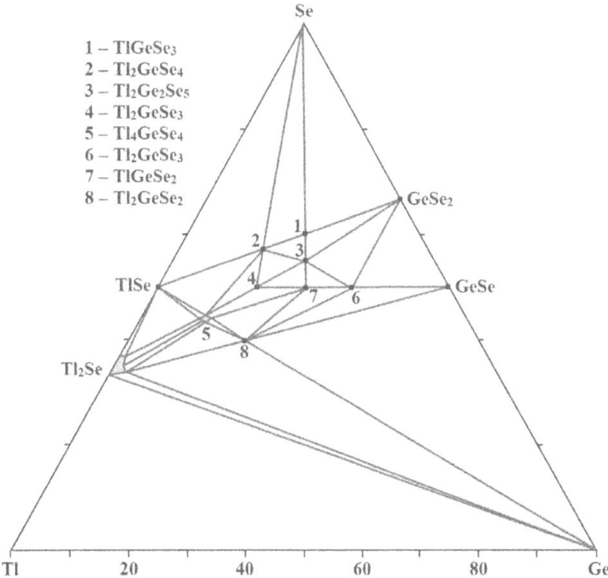

FIGURE 5.34 Isothermal section of the Ge–Tl–Se ternary system at 130°C. (From Babanly, M.B., and Kulieva, N.A., *Zhurn. neorgan. khimii*, **31**(9), 2365, 1986.)

glassy samples. Dembovskiy et al. (1983) noted that only one composition, $Ge_2Tl_{31}Se_{67}$, could be obtained in the glassy state. The glassy samples were obtained by the quenching from 900°C to 950°C in air (Borisova et al. 1976) or from 500°C into water (Dembovskiy 1968).

5.18 Germanium–Yttrium–Selenium

$GeSe_2$–Y_2Se_3: Glassy samples were not obtained in this system (Loireau-Lozac'h and Guittard 1977).

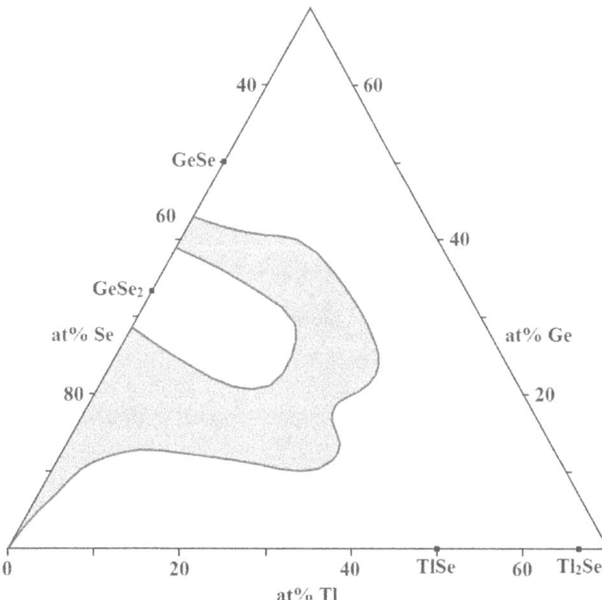

FIGURE 5.35 Glass forming region in the Ge–Tl–Se ternary system. (From Borisova, Z.U., et al., *Izv. AN SSSR. Neorgan. mater.*, **12**(4), 592, 1976.)

La_2GeSe_5, $La_2Ge_2Se_7$ and $La_2Ge_3Se_9$ is equal to 5.82, 5.63 and 5.46 g·cm^{-3}, respectively. La_2GeSe_5 is stable in vacuum up to the melting temperature. All these compounds are semiconductors with an energy gap of 1.68 eV for La_2GeSe_5 and 1.19 eV for $La_2Ge_2Se_7$ (Nasibov et al. 1982). Single crystals of all these compounds could be grown by the chemical transport reactions. Glassy samples were not obtained in this system (Loireau-Lozac'h and Guittard 1977).

The samples were annealed at 500°C–550°C for 400–450 h. The system was investigated through DTA, XRD, metallography, and measuring of the microhardness and density (Rustamov et al. 1973d; Nasibov et al. 1975a).

The $\mathbf{La_3Ge_{1.48}Se_7}$ ternary compounds, which crystallizes in the hexagonal structure with the lattice parameters $a = 1076.56 \pm 0.07$ and $c = 608.01 \pm 0.05$ pm and a calculated density of 5.839 g·cm^{-3}, is formed in the Ge–La–Se ternary system (Daszkiewicz et al. 2010). Powder samples with the nominal compositions $La_4Ge_3Se_7$ was prepared by sintering the elemental constituents in evacuated quartz tubes. The syntheses were carried out in a tube resistance furnace. The ampoules were first heated with a rate of 30°C·h^{-1} up to 1150°C, and then kept at this temperature for 3 h. Afterwards, the sample was cooled slowly (10°C·h^{-1}) down to 600°C, and annealed at this temperature for 720 h. Subsequently, the ampoule was quenched in air.

5.19 Germanium–Lanthanum–Selenium

$GeSe_2$–La_2Se_3: The phase diagram of this system is a eutectic type with peritectic transformations (Figure 5.36) (Rustamov et al. 1973d; Nasibov et al. 1975a). The eutectics crystallize at 875°C and 680°C, with the latter corresponding to a composition of 3 mol% La_2Se_3. Three compounds are formed in this system: $\mathbf{La_2GeSe_5}$ (Beskrovnaya et al. 1971a; Rustamov et al. 1973d; Nasibov et al. 1975a, 1982) which melts congruently at 1030°C, $\mathbf{La_2Ge_2Se_7}$ and $\mathbf{La_2Ge_3Se_9}$, which decomposes according to the peritectic reactions at 925°C and 760°C, respectively (Rustamov et al. 1973d; Nasibov et al. 1975a). La_2GeSe_5 and $La_2Ge_2Se_7$ have polymorphic transformations at 685°C and 605°C, respectively. The experimental density of

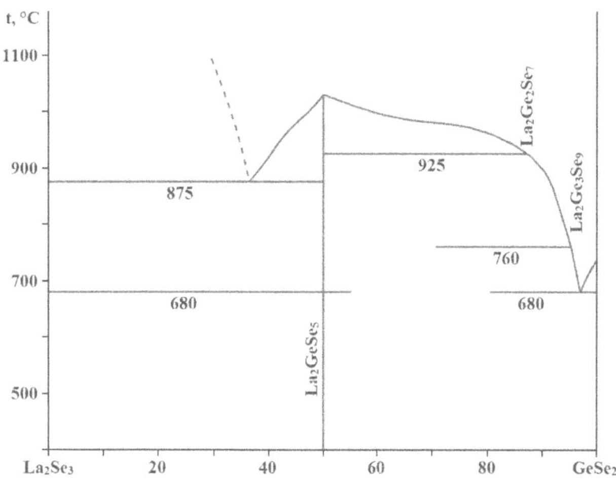

FIGURE 5.36 Phase diagram of the $GeSe_2$–La_2Se_3 system. (From Rustamov, P.G., et al., *Izv. AN SSSR. Neorgan. mater.*, **9**(11), 1900, 1973.)

5.20 Germanium–Cerium–Selenium

$GeSe_2$–Ce_2Se_3: The phase diagram of this system is a eutectic type with peritectic transformation (Figure 5.37) (Rustamov et al. 1972a; Nasibov et al. 1975a). The eutectics crystallize at 1000°C, 975°C and 600°C, with the latter corresponding to a composition of 3 mol% Ce_2Se_3. Three compounds are formed in this system: $\mathbf{Ce_2GeSe_5}$ and $\mathbf{Ce_2Ge_2Se_7}$, which melt congruently at 1075°C and 1020°C, respectively, and $\mathbf{Ce_2Ge_3Se_9}$, which decomposes peritectically at 725°C. Ce_2GeSe_5 and $Ce_2Ge_2Se_7$ have polymorphic transformations at 1025°C and 875°C, respectively. The experimental density of

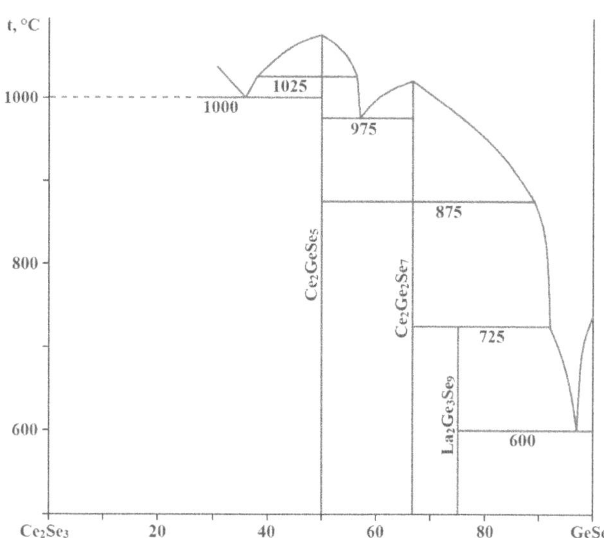

FIGURE 5.37 Phase diagram of the $GeSe_2$–Ce_2Se_3 system. (From Rustamov, P.G., et al., *Izv. AN SSSR. Neorgan. mater.*, **8**(6), 1155, 1972.)

Ce_2GeSe_5, $Ce_2Ge_2Se_7$, and $Ce_2Ge_3Se_9$ is equal to 5.16, 4.77, and 5.33 g·cm^{-3}, respectively. Ce_2GeSe_5 is stable in vacuum up to the melting temperature. All these compounds are semiconductors with an energy gap of 1.61 eV for Ce_2GeSe_5 and 1.21 eV for $Ce_2Ge_2Se_7$ (Nasibov et al. 1982). Single crystals of all these compounds could be grown by the chemical transport reactions. Glassy samples were not obtained in this system (Loireau-Lozac'h and Guittard 1977).

The samples were annealed at 500°C–550°C for 400–450 h. The system was investigated through DTA, XRD, metallography, and measuring of the microhardness and density (Rustamov et al. 1972a; Nasibov et al. 1975a).

The $\mathbf{Ce_3Ge_{1.47}Se_7}$ ternary compounds, which crystallizes in the hexagonal structure with the lattice parameters $a = 1068.4 \pm 0.1$ and $c = 606.11 \pm 0.08$ pm and a calculated density of 5.978 g·cm^{-3}, is also formed in the Ge–Ce–Se ternary system (Daszkiewicz et al. 2010). This compound was obtained in the same way as $La_3Ge_{1.48}Se_7$ was prepared.

5.21 Germanium–Praseodymium–Selenium

$GeSe_2$–Pr_2Se_3: The phase diagram of this system is a eutectic type with peritectic transformation (Figure 5.38) (Alieva et al. 1972; Nasibov et al. 1975a). The eutectics crystallize at 1000°C, 960°C, and 575°C, with the latter corresponding to a composition of 3 mol% Pr_2Se_3. Three compounds are formed in this system: $\mathbf{Pr_2GeSe_5}$ and $\mathbf{Pr_2Ge_2Se_7}$, which melt congruently at 1025°C and 995°C, respectively, and $\mathbf{Pr_2Ge_3Se_9}$, which decomposes peritectically at 700°C. Pr_2GeSe_5 and $Pr_2Ge_2Se_7$ have polymorphic transformations at 880°C and 840°C, respectively. Pr_2GeSe_5 crystallizes in the hexagonal structure with the lattice parameters $a = 1894$ and $c = 802$ pm and the calculated and experimental densities of 4.59 and 4.63 g·cm^{-3}, respectively (Beskrovnaya et al. 1973), and is stable in vacuum up to the melting temperature (Alieva et al. 1972; Nasibov et al. 1975a). All these compounds are semiconductors with an energy gap of 1.86 eV for Pr_2GeSe_5 and 1.42 eV for $Pr_2Ge_2Se_7$ (Nasibov et al. 1982). Single crystals of all these compounds could be grown by the chemical transport reactions. Glassy samples were not obtained in this system (Loireau-Lozac'h and Guittard 1977).

The samples were annealed at 500°C–550°C for 400–450 h. The system was investigated through DTA, XRD, metallography, and measuring of the microhardness and density (Alieva et al. 1972; Nasibov et al. 1975a).

The $\mathbf{Pr_3Ge_{1.49}Se_7}$ ternary compounds, which crystallizes in the hexagonal structure with the lattice parameters $a = 1064.08 \pm 0.09$ and $c = 605.48 \pm 0.07$ pm and a calculated density of 6.059 g·cm^{-3}, is also formed in the Ge–Pr–Se ternary system (Daszkiewicz et al. 2010). This compound was obtained in the same way as $La_3Ge_{1.48}Se_7$ was prepared.

5.22 Germanium–Neodymium–Selenium

$GeSe_2$–Nd_2Se_3: The phase diagram of this system is a eutectic type with peritectic transformation (Figure 5.39) (Nasibov 1978). The eutectics crystallize at 970°C, 940°C, and 600°C. Three compounds are formed in this system: $\mathbf{Nd_2GeSe_5}$ and $\mathbf{Nd_2Ge_2Se_7}$, which melt congruently at 1045°C and 1000°C, respectively, and $\mathbf{Nd_2Ge_3Se_9}$, which decomposes peritectically at 700°C. Nd_2GeSe_5 and $Nd_2Ge_2Se_7$ have polymorphic transformations at 850°C and 775°C, respectively. There is a narrow region of homogeneity based on Nd_2GeSe_5. Low-temperature modification of Nd_2GeSe_5 crystallizes in the tetragonal structure which transforms into hexagonal structure at 850°C and low-temperature modification of $Nd_2Ge_2Se_7$ crystallizes in the hexagonal structure which transforms into tetragonal structure at 775°C. All these compounds are semiconductors. Glassy samples were not obtained in this system (Loireau-Lozac'h and Guittard 1977).

The samples were annealed at 500°C–550°C for 450 h. The system was investigated through DTA, XRD, metallography, and measuring of the microhardness and density (Nasibov 1978).

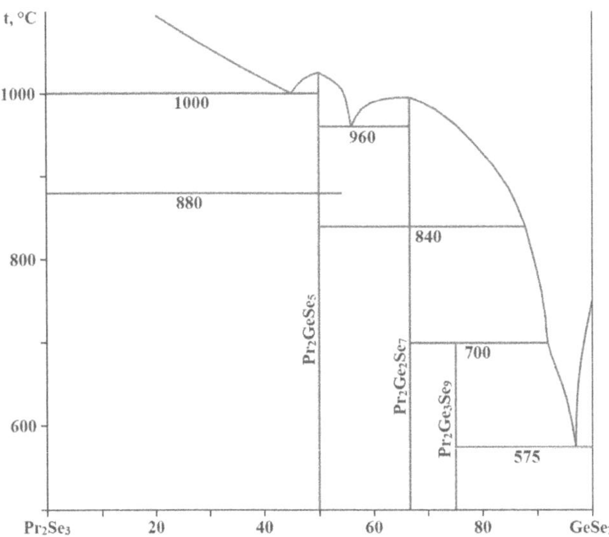

FIGURE 5.38 Phase diagram of the $GeSe_2$–Pr_2Se_3 system. (From Alieva, M.M., et al., *Dokl. AN AzSSR*, **28**(8), 21, 1972.)

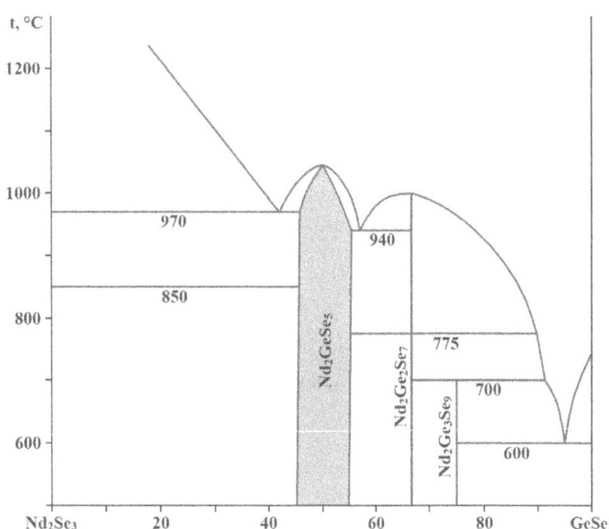

FIGURE 5.39 Phase diagram of the $GeSe_2$–Nd_2Se_3 system. (From Nasibov, I.O., *Uch. zap. Azerb. Univ. Ser. khim. nauk*, (3), 6, 1978.)

5.23 Germanium–Promethium–Selenium

$GeSe_2$–Pm_2Se_3: Glassy samples were not obtained in this system (Loireau-Lozac'h and Guittard 1977).

5.24 Germanium–Samarium–Selenium

$GeSe_2$–Sm_2Se_3: The phase diagram of this system is a eutectic type with peritectic transformation (Figure 5.40) (Nasibov et al. 1981e). The eutectics crystallize at 950°C, 925°C, and 650°C. Three compounds are formed in this system: **Sm_2GeSe_5** and **$Sm_2Ge_2Se_7$**, which melt congruently at 1025°C and 980°C, respectively, and **$Sm_2Ge_3Se_9$**, which decomposes peritectically at 720°C. Sm_2GeSe_5 and $Sm_2Ge_2Se_7$ have polymorphic transformations at 860°C and 800°C, respectively. There is a narrow region of homogeneity based on Sm_2GeSe_5, and the solubility of Sm_2Se_3 in $GeSe_2$ reaches up to 3 mol%. All these compounds are semiconductors with an energy gap of 1.98 eV for Sm_2GeSe_5, 1.68 eV for $Sm_2Ge_2Se_7$, and 1.26 eV for $Sm_2Ge_3Se_9$. According to the data of Olekseyuk et al. (2010c), all these three compounds were not found in the $GeSe_2$–Sm_2Se_3 system. Glassy samples were not obtained in this system (Loireau-Lozac'h and Guittard 1977).

The samples were annealed at 550°C for 650 h. The system was investigated through DTA, XRD, metallography, and measuring of the microhardness and density (Nasibov et al. 1981e).

The **$Sm_3Ge_{1.48}Se_7$** ternary compounds, which crystallizes in the hexagonal structure with the lattice parameters a = 1044.19 ± 0.07 and c = 602.83 ± 0.06 pm and a calculated density of 6.483 g·cm^{-3}, is formed in the Ge–Sm–Se ternary system (Daszkiewicz et al. 2010). This compound was obtained in the same way as $La_3Ge_{1.48}Se_7$ was prepared.

5.25 Germanium–Europium–Selenium

$GeSe$–Eu_2Se_3: Two compounds, **Eu_2GeSe_4** and **$Eu_2Ge_2Se_5$**, are formed in this system. First of them has two polymorphic modifications and both crystallize in the monoclinic structure with the lattice parameters a = 696.4 ± 0.1, b = 705.5 ± 0.2, c = 840.0 ± 0.2 pm, β = 108.12 ± 0.02°, and a calculated density of 5.86 g·cm^{-3} for α-Eu_2GeSe_4 at 23°C and a = 696.9 ± 0.1, b = 705.9 ± 0.2, c = 851.6 ± 0.2 pm, β = 107.99 ± 0.02°, and a calculated density of 5.77 g·cm^{-3} for β-Eu_2GeSe_4 at 400°C (Tampier et al. 2002). Eu_2GeSe_4 has ferroelectric phase transition at 347°C (Tampier and Johrendt 2001).

$Eu_2Ge_2Se_5$ also crystallizes in the monoclinic structure with the lattice parameters a = 842.1 ± 0.4, b = 1223.5 ± 0.4, c = 912.7 ± 0.3 pm, β = 93.67 ± 0.04°, and a calculated density of 5.96 g·cm^{-3} (Tampier et al. 2002).

Powder samples of both compounds were synthesized from the elements in two steps (Tampier et al. 2002). First, binary compounds Eu_2Ge and EuGe were prepared by heating stoichiometric mixtures of Eu and Ge at 850°C for 10 h in alumina crucibles sealed in quartz glass ampoules under an Ar atmosphere. Then, after homogenization in an Ar-filled glove box, both products were oxidized by stoichiometric amounts of Se at 750°C for 15 h. The samples were homogenized again and reheated to 750°C for 50 h. This procedure resulted in deep red powders of Eu_2GeSe_4 and $Eu_2Ge_2Se_5$, respectively. Both compounds are stable in air.

$GeSe_2$–Eu_2Se_3: The phase diagram of this system is a eutectic type with peritectic transformation (Figure 5.41) (Nasibov et al. 1981b). The eutectics crystallize at 920°C, 880°C, and 670°C. Three compounds are formed in this system: **Eu_2GeSe_5** and **$Eu_2Ge_2Se_7$**, which melt congruently at 1000°C and 970°C, respectively, and **$Eu_2Ge_3Se_9$**, which decomposes peritectically. Eu_2GeSe_5 and $Eu_2Ge_2Se_7$ have polymorphic transformations at 870°C and 720°C, respectively. Eu_2GeSe_5 is characterized by a narrow region of homogeneity. The calculated and experimental densities of Eu_2GeSe_5 and $Eu_2Ge_2Se_7$ are equal to

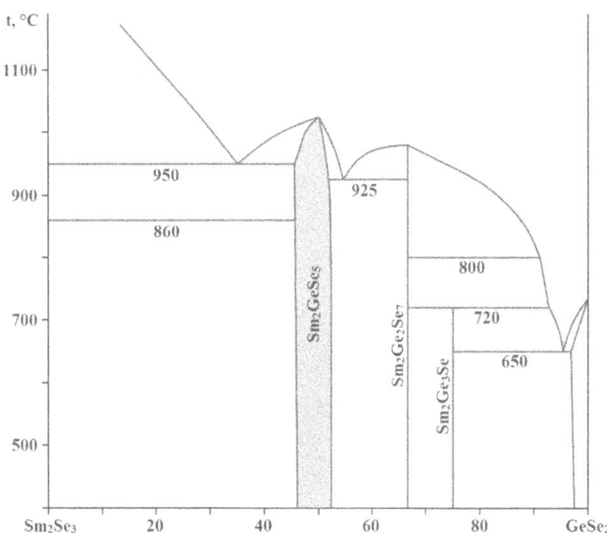

FIGURE 5.40 Phase diagram of the $GeSe_2$–Sm_2Se_3 system. (From Nasibov, I.O., et al., *Zhurn. neorgan. khimii*, **26**(9), 2524, 1981.)

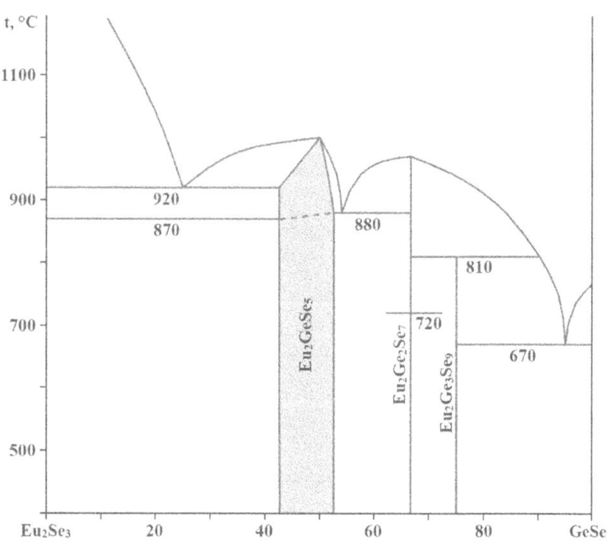

FIGURE 5.41 Phase diagram of the $GeSe_2$–Eu_2Se_3 system. (From Nasibov, I.O., et al., *Zhurn. neorgan. khimii*, **26**(9), 2592, 1981.)

6.40 and 6.34 g·cm^{-3}, and 5.13 and 5.25 g·cm^{-3}, respectively. For $Eu_2Ge_3Se_9$ an experimental density is 5.18 g·cm^{-3}. All these compounds are semiconductors with an energy gap of 2.00 eV for Eu_2GeSe_5, 1.71 eV for $Eu_2Ge_2Se_7$, and 1.38 eV for $Eu_2Ge_3Se_9$. Glassy samples were not obtained in this system (Loireau-Lozac'h and Guittard 1977).

The samples were annealed at 550°C for 450 h. The system was investigated through DTA, XRD, metallography, and measuring of the microhardness and density (Nasibov et al. 1981b).

5.26 Germanium–Gadolinium–Selenium

$GeSe_2$–Gd_2Se_3: The phase diagram of this system is a eutectic type with peritectic transformation (Figure 5.42) (Nasibov et al. 1981d). The eutectics crystallize at 1050°C, 1000°C, 950°C, and 600°C. Four compounds are formed in this system: **Gd_2GeSe_5**, **$Gd_2Ge_2Se_7$**, and **Gd_4GeSe_8**, which melt congruently at 1090°C, 1000°C, and 1125°C, respectively, and **$Gd_2Ge_3Se_9$**, which melts incongruently at 700°C. Gd_2GeSe_5 and Gd_4GeSe_8 each have polymorphic modifications. There are narrow regions of homogeneity based on these two compounds, and the solubility of Gd_2Se_3 in $GeSe_2$ reaches 16.6 mol% at 400°C. All these compounds are semiconductors with an energy gap of 1.83 eV for Gd_2GeSe_5, 1.38 eV for $Gd_2Ge_2Se_7$, and 1.33 eV for $Gd_2Ge_3Se_9$. Glassy samples were not obtained in this system (Loireau-Lozac'h and Guittard 1977).

α-Gd_2GeSe_5 crystallizes in the tetragonal structure with lattice parameters a = 415 and c = 840 pm, $Gd_2Ge_2Se_7$ crystallizes in the hexagonal structure with the lattice parameters a = 679 and c = 1261 pm and $Gd_2Ge_3Se_9$ crystallizes in the orthorhombic structure with the lattice parameters a = 671, b = 1187, c = 2234 pm (Nasibov et al. 1981d).

The samples were annealed at 500–550°C for 400 h. The system was investigated through DTA, XRD, metallography and measuring of the microhardness and density (Nasibov et al. 1981d).

The **$Gd_3Ge_{1.45}Se_7$** ternary compounds, which crystallizes in the hexagonal structure with the lattice parameters a = 1032.5 ± 0.1 and c = 605.06 ± 0.07 pm and a calculated density of 6.716 g·cm^{-3}, is also formed in the Ge–Gd–Se ternary system (Daszkiewicz et al. 2010). This compound was obtained in the same way as $La_3Ge_{1.48}Se_7$ was prepared.

5.27 Germanium–Terbium–Selenium

$GeSe_2$–Tb_2Se_3: The phase diagram of this system is a eutectic type with peritectic transformation (Figure 5.43) (Nasibov et al. 1981a). The eutectics crystallize at 1020°C, 1000°C, 900°C, and 600°C. Four compounds are formed in this system: **Tb_2GeSe_5**, **$Tb_2Ge_2Se_7$**, and **Tb_4GeSe_8**, which melt congruently at 1050°C, 990°C, and 1150°C, respectively, and **$Tb_2Ge_3Se_9$**, which melts incongruently. Tb_2GeSe_5 and Tb_4GeSe_8 have polymorphic transformations at 800°C and 875°C, respectively. There is a narrow region of homogeneity based on Tb_4GeSe_8. All these compounds are semiconductors with an energy gap of 1.98 eV for Tb_2GeSe_5, 1.67 eV for $Tb_2Ge_2Se_7$, and 1.41 eV for $Tb_2Ge_3Se_9$. Glassy samples were not obtained in this system (Loireau-Lozac'h and Guittard 1977).

α-Tb_2GeSe_5 crystallizes in the tetragonal structure with lattice parameters a = 412 and c = 836 pm and the calculated and experimental densities of 6.43 and 6.50 g·cm^{-3}, respectively, $Tb_2Ge_2Se_7$ crystallizes in the hexagonal structure with the lattice parameters a = 675 and c = 1259 pm and the calculated and experimental densities of 5.68 and 5.66 g·cm^{-3}, respectively, and $Tb_2Ge_3Se_9$ crystallizes in the orthorhombic structure with the lattice parameters a = 666, b = 1181, c = 2230 pm and the calculated and experimental densities of 5.31 and 5.36 g·cm^{-3}, respectively (Nasibov et al. 1981a).

The samples were annealed at 550°C for 450 h. The system was investigated through DTA, XRD, metallography and measuring of the microhardness and density (Nasibov et al. 1981a).

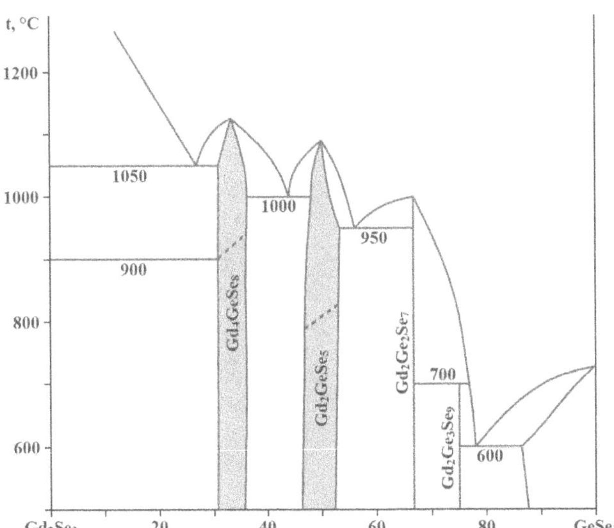

FIGURE 5.42 Phase diagram of the $GeSe_2$–Gd_2Se_3 system. (From Nasibov, I.O., et al., *Izv. AN SSSR. Neorgan. mater.*, **17**(11), 1983, 1981.)

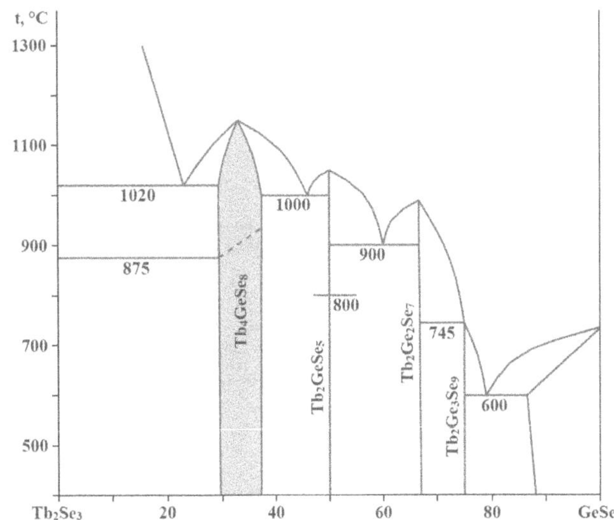

FIGURE 5.43 Phase diagram of the $GeSe_2$–Tb_2Se_3 system. (From Nasibov I.O., et al., *Izv. AN SSSR. Neorgan. mater.*, **17**(8), 1341, 1981.)

The $\mathbf{Tb_3Ge_{1.43}Se_7}$ ternary compounds, which crystallizes in the hexagonal structure with the lattice parameters $a = 1027.45 \pm 0.07$ and $c = 607.43 \pm 0.05$ pm and a calculated density of 6.777 g·cm^{-3}, is also formed in the Ge–Tb–Se ternary system (Daszkiewicz et al. 2010). This compound was obtained in the same way as $La_3Ge_{1.48}Se_7$ was prepared.

5.28 Germanium–Dysprosium–Selenium

$\mathbf{GeSe_2–Dy_2Se_3}$: The phase diagram of this system is a eutectic type with peritectic transformation (Figure 5.44) (Nasibov 1982a). The eutectics crystallize at 1030°C, 1000°C, 900°C, and 650°C. Four compounds are formed in this system: $\mathbf{Dy_2GeSe_5}$, $\mathbf{Dy_2Ge_2Se_7}$, and $\mathbf{Dy_4GeSe_8}$, which melt congruently at 1075°C, 1020°C, and 1170°C, respectively, and $\mathbf{Dy_2Ge_3Se_9}$, which melts incongruently at 800°C. Dy_2GeSe_5 and Dy_4GeSe_8 have polymorphic transformations. There is a narrow region of homogeneity based on Dy_4GeSe_8. All these compounds are semiconductors. Glassy samples were not obtained in this system (Loireau-Lozac'h and Guittard 1977).

The samples were annealed at 550°C for 450 h. The system was investigated through DTA, XRD, metallography, and measuring of the microhardness and density (Nasibov 1982a).

5.29 Germanium–Holmium–Selenium

$\mathbf{GeSe_2–Ho_2Se_3}$: The phase diagram of this system is a eutectic type with peritectic transformation (Figure 5.45) (Nasibov 1982b). The eutectics crystallize at 1000°C, 900°C, 850°C, and 600°C. Four compounds are formed in this system: $\mathbf{Ho_2GeSe_5}$, $\mathbf{Ho_2Ge_2Se_7}$, and $\mathbf{Ho_4GeSe_8}$, which melt congruently at 1060°C, 1000°C, and 1100°C, respectively, and $\mathbf{Ho_2Ge_3Se_9}$, which melts incongruently at 770°C. Ho_2GeSe_5 and Ho_4GeSe_8 have polymorphic transformations 700°C and 650°C, respectively. All these compounds are semiconductors.

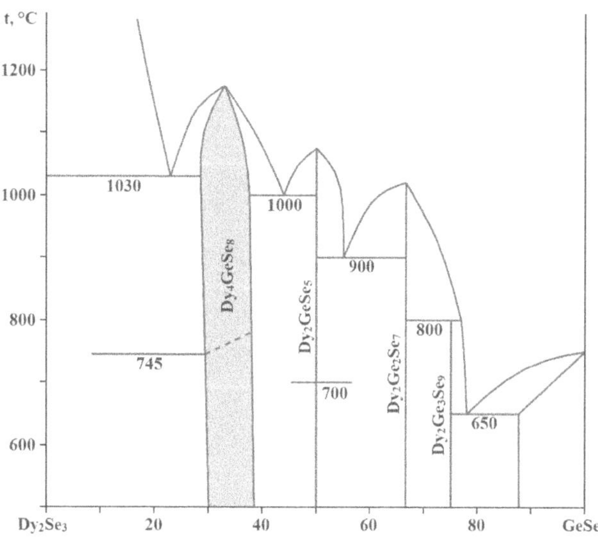

FIGURE 5.44 Phase diagram of the $GeSe_2$–Dy_2Se_3 system. (From Nasibov, I.O., *Zhurn. neorgan. khimii*, **27**(12), 3168, 1982.)

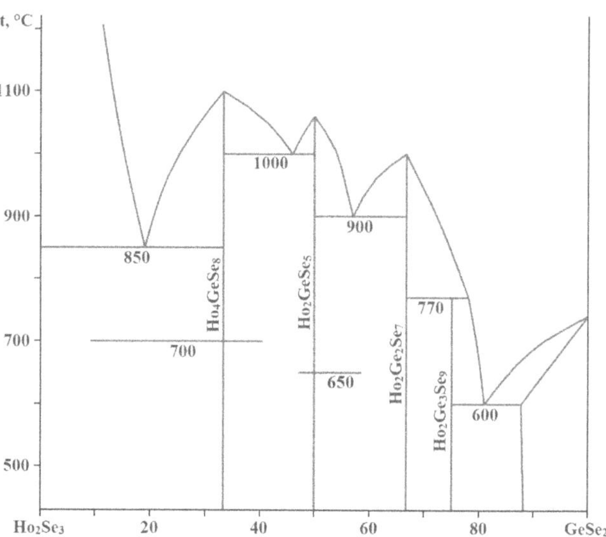

FIGURE 5.45 Phase diagram of the $GeSe_2$–Ho_2Se_3 system. (From Nasibov, I.O., *Izv. AN SSSR. Neorgan. mater.*, **18**(12), 1972, 1982.)

Glassy samples were not obtained in this system (Loireau-Lozac'h and Guittard 1977).

The samples were annealed at 550°C for 450 h. The system was investigated through DTA, XRD, metallography, and measuring of the microhardness and density (Nasibov 1982b).

5.30 Germanium–Erbium–Selenium

$\mathbf{GeSe–Er_2Se_3}$: Glassy samples were not obtained in this system (Loireau-Lozac'h and Guittard 1977).

5.31 Germanium–Ytterbium–Selenium

$\mathbf{GeSe_2–YbSe}$: The phase diagram of this system is a eutectic type with peritectic transformations (Figure 5.46) (Rustamov et al. 1981). The eutectic crystallizes at 530°C and contains

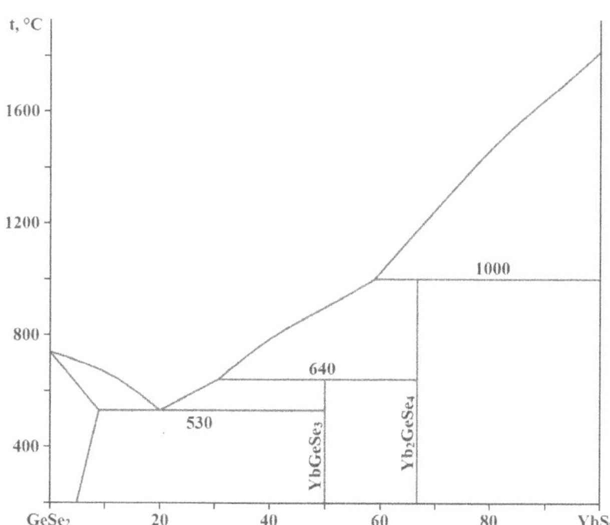

FIGURE 5.46 Phase diagram of the $GeSe_2$–YbSe system. (From Rustamov, P.G., et al., *Izv. AN SSSR. Neorgan. mater.*, **17**(9), 1571, 1981.)

20 mol% YbSe. Two compounds, $\mathbf{YbGeSe_3}$ and $\mathbf{Yb_2GeSe_4}$, which melts incongruently at 640°C and 1000°C, respectively, are formed in this system. The solubility of YbSe in $GeSe_2$ at room temperature reaches up to 3 mol% and the solubility of $GeSe_2$ in YbSe is negligible. $YbGeSe_3$ crystallizes in the triclinic structure with the lattice parameters $a = 834$, $b = 854$, $c = 1168$ pm, and $\alpha = 111.08°$, $\beta = 99.95°$, $\gamma = 75.44°$, and an experimental density of 4.01 $g{\cdot}cm^{-3}$. Yb_2GeSe_4 crystallizes in the monoclinic structure with the lattice parameters $a = 664.4$, $b = 839.0$, $c = 684.0$ pm, $\beta = 108°$, and an experimental density of 4.35 $g{\cdot}cm^{-3}$.

The samples were annealed at 500°C for one month. The system was investigated through DTA, XRD, metallography, and measuring of the microhardness (Rustamov et al. 1981).

GeSe–Yb_2Se_3: The $\mathbf{Yb_2GeSe_5}$ ternary compound, which melts at the temperature higher than 1200°C and has an experimental density of 5.65 $g{\cdot}cm^{-3}$ and an energy gap of 1.5 eV, is formed in this system (Karaev et al. 1969, 1972).

5.32 Germanium–Thorium–Selenium

Two ternary compounds, **ThGeSe** and $\mathbf{Th_2GeSe_5}$, are formed in the Ge–Th–Se system. ThGeSe crystallizes in the tetragonal structure with the lattice parameters $a = 403.3$ and $c = 1757$ pm and the calculated and experimental densities of 8.91 and 8.75 $g{\cdot}cm^{-3}$, respectively (Stocks et al. 1981; Hahn and Stocks 1968)]. To prepare this compound, the elements were weighed under N_2 atmosphere in the stoichiometric ratio and initially heated to about 400°C in an evacuated quartz ampoule. Subsequently, the inhomogeneous product was finely powdered under N_2 and heated to 880°C in a sealed tungsten crucible, which was enclosed in an evacuated quartz ampoule. Since the preparation is extremely sensitive to traces of oxygen, it proved necessary to add Th as getter metal. Nevertheless, pure preparation was obtained only if no longer than 48 h the mixture was heated at 880°C. The mixture was annealed at 600°C–700°C for 8–10 days. Then, the sample pressed into pastilles was heated in the same way over different time intervals and at different temperatures up to 1400°C in tungsten crucible, wherein above 1050°C additionally, Ar was used as a protective gas. The title compound was obtained as a black microcrystalline powder.

Th_2GeSe_5 also crystallizes in the tetragonal structure with the lattice parameters $a = 749.68 \pm 0.04$ and $c = 1363.02 \pm 0.09$ pm and a calculated density of 8.077 $g{\cdot}cm^{-3}$ (Koscielski et al. 2013). To obtain the title compound, a fused-silica tube was loaded with Th (0.129 mM), Ge (0.129 mM), and Se (0.129 mM), evacuated to near 0.01 Pa, flame sealed, and placed in a furnace. It was heated to 1000°C in 24 h, kept at this temperature for 99 h, cooled to 400°C in 198 h, and then rapidly cooled to room temperature in 4 h. The resulting brown powder was loaded with Sb_2Se_3 (0.146 mM) in a fused-silica tube and heated as before. This tube contained brown powder and red-black rods of Th_2GeS_5 were obtained.

5.33 Germanium–Uranium–Selenium

The **UGeSe** ternary compound, which crystallizes in the tetragonal structure with the lattice parameters $a = 393.23 \pm 0.02$ and $c = 1696.9 \pm 0.1$ pm and a calculated density of 9.86 $g{\cdot}cm^{-3}$ (Haneveld and Jellinek 1969) [$a = 393.3$ and $c = 1696.6$ pm (Zygmunt 1977)], is formed in the Ge–U–Se system. This compound was prepared in the following way (Haneveld and Jellinek 1969). A small rod of uranium was placed in a quartz tube and heated in a current of hydrogen at 230°C. The voluminous uranium hydride powder thus produced was dehydrided by heating at 400°C under vacuum. An equivalent proportion of GeSe was added immediately and the quartz tube was evacuated and sealed. The mixture was heated at 800°C–900°C for two or three days. Then, the product was slowly cooled to room temperature.

5.34 Germanium–Tin–Selenium

GeSe–SnSe: A continuous series of solid solutions is formed in this system, and the phase diagram belongs to type I according to the Roseboom's classification Figure 5.47) (Bletskan et al. 1981). The liquidus and solidus are slightly concave toward the compositions axis. Low-temperature solid solutions crystallize in the orthorhombic structure and the high-temperature ones crystallize in the cubic structure. The temperature of polymorphic transition slightly decreases at the increasing of the SnSe content, and the transition itself is diffused. Single crystals of the $Ge_xSn_{1-x}Se$ solid solutions were grown by the slow melt cooling in a horizontal furnace. The system was investigated through DTA, XRD, and metallography.

It should be noted that the constructed phase diagram does not take into account the fact of the GeSe formation from the peritectic reaction, as well as the non-isostructurality of the high temperature modifications of GeSe and SnSe. Therefore the lower part of this phase diagram was omitted.

$GeSe_2$–Sn: This system in a non-quasibinary section of the Ge–Sn–Se ternary system (Baldé et al. 1979). The polythermal section was constructed through DTA, XRD, and metallography.

$GeSe_2$–SnSe: The phase diagram of this system is given in Figure 5.48 (Baldé and Khodadad 1974). The eutectic contains 45.1 mol% $GeSe_2$ and crystallizes at 580°C. The ingots were annealed at 350°C for 1 week and the system was investigated through DTA and XRD. No glass formation was observed in this system (Vassilev et al. 2009).

$GeSe_2$–$SnSe_2$: The phase diagram of this system is a eutectic type (Figure 5.49) (Baldé and Khodadad 1974; Karahanova et al. 1976b). The eutectic contains 50 mol% $SnSe_2$ and crystallizes at 550°C (Karahanova et al. 1976b) [at 569°C (Baldé and Khodadad 1974)]. The maximum mutual solubility of the starting components reaches 5 mol% (Karahanova et al. 1976b).

The ingots were annealed at 450°C for 1000 h (Karahanova et al. 1976b) [at 350°C for one week (Baldé and Khodadad 1974)]. The system was investigated through DTA and XRD (Baldé and Khodadad 1974; Karahanova et al. 1976b).

Ge–SnSe: The phase diagram of this system is a eutectic type (Baldé et al. 1979). The immiscibility region exists in this system at 812°C. The system was investigated through DTA, XRD, and metallography.

The liquidus surface of the Ge–Sn–Se ternary system was constructed by Baldé and Khodadad (1974). The ternary eutectic in the $GeSe_2$–SnSe–$SnSe_2$ subsystem crystallizes at

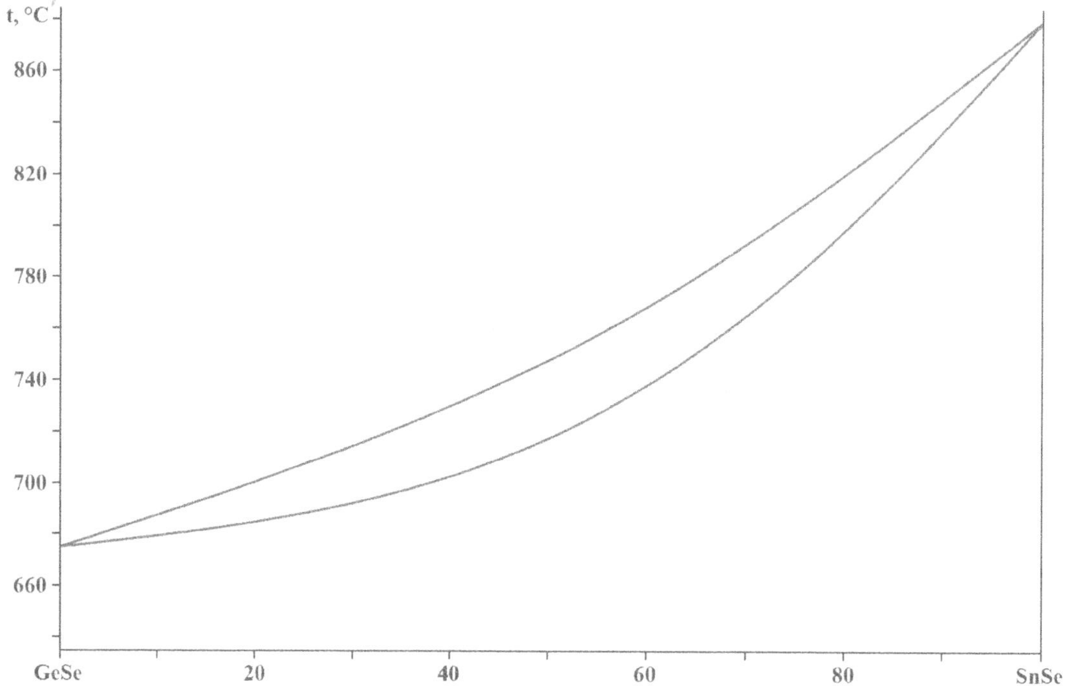

FIGURE 5.47 Phase diagram of the GeSe–SnSe system. (From Bletskan, D.I., et al., *Zhurn. neorgan. khimii*, **26**(3), 761, 1981.)

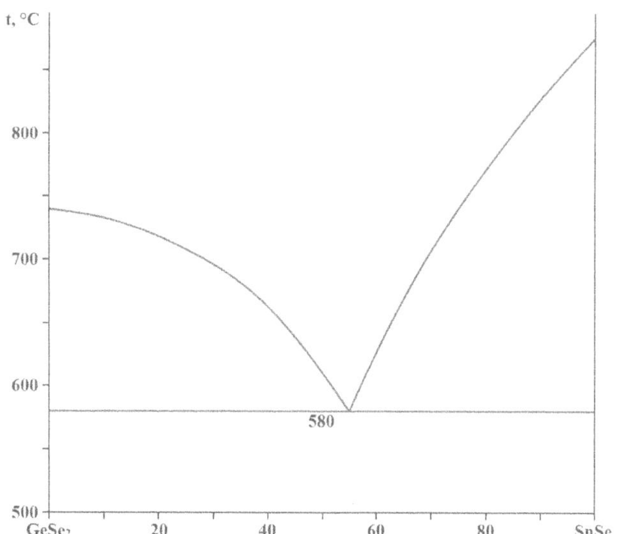

FIGURE 5.48 Phase diagram of the $GeSe_2$–$SnSe_2$ system. (From Baldé, L., and Khodadad, P., *C. r. Acad. sci. Sér. C*, **278**(4), 243, 1974.)

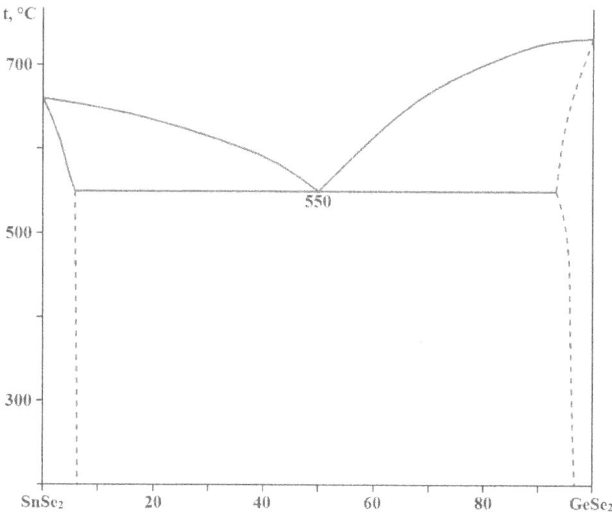

FIGURE 5.49 Phase diagram of the $GeSe_2$–$SnSe_2$ system. (From Karahanova, M.I., et al., *Izv. AN SSSR. Neorgan. mater.*, **12**(8), 1486, 1976.)

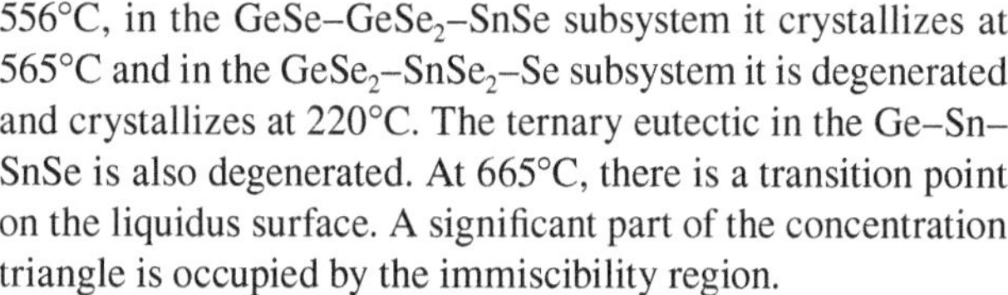

556°C, in the GeSe–$GeSe_2$–SnSe subsystem it crystallizes at 565°C and in the $GeSe_2$–$SnSe_2$–Se subsystem it is degenerated and crystallizes at 220°C. The ternary eutectic in the Ge–Sn–SnSe is also degenerated. At 665°C, there is a transition point on the liquidus surface. A significant part of the concentration triangle is occupied by the immiscibility region.

The glass forming region in the Ge–Sn–Se system is given in Figure 5.50 (Seregin et al. 1972b). Up to 13 at% Sn can be introduced into glassy samples. The most capable of glass formation are alloys containing 75–80 at% Se. The glass forming region was not determined in the GeSe–$GeSe_2$–SnSe subsystem (Feltz et al. 1976).

5.35 Germanium–Lead–Selenium

GeSe–PbSe: This system is a non-quasibinary section of the Ge–Pb–Se ternary system and crosses the field of $Ge_xPb_{1-x}Se$ solid solutions and $GeSe_2$ primary crystallization (Figure 5.51) (Shelimova et al. 1966). The primary crystallization of $Ge_xPb_{1-x}Se$ solid solutions ends at 630°–617°C, and as a result the $Ge_xPb_{1-x}Se + GeSe_2$ eutectic crystallizes. The field of the $GeSe_2$ primary crystallization is adjacent to the three-phase region in which L + $GeSe_2$ ⇔ (γ-GeSe) peritectic reaction takes place. The melt crystallization in the middle part of the section ends at 617°C by the next invariant equilibrium:

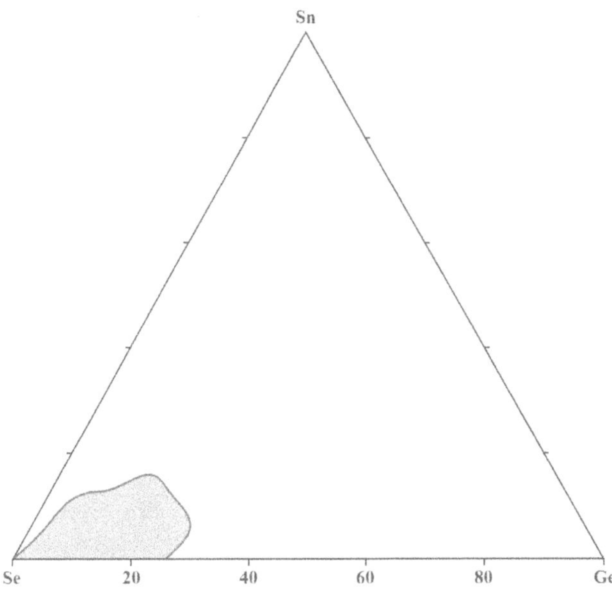

FIGURE 5.50 Glass forming region in the Ge–Sn–Se ternary system. (From Seregin, P.P., et al., *Izv. AN SSSR. Neorgan. mater.*, **8**(3), 567, 1972.)

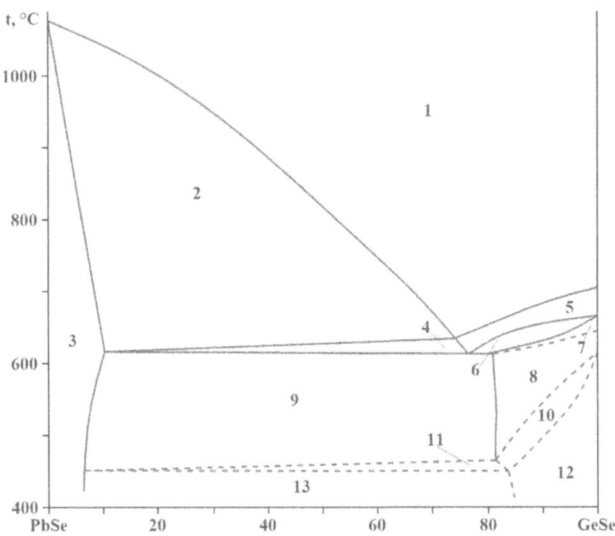

FIGURE 5.51 Phase relations in the GeSe–PbSe system: 1, L; 2, L + (PbSe); 3, PbSe; 4, L + (PbSe) + $GeSe_2$; 5, L + $GeSe_2$; 6, L + $GeSe_2$ + γ-GeSe; 7, L + γ-GeSe; 8, γ-GeSe; 9, (PbSe) + γ-GeSe; 10, γ-GeSe + α-GeSe; 11, (PbSe) + α-GeSe + γ-GeSe; 12, α-GeSe; 13, (PbSe) + α-GeSe. (From Shelimova, L.E., et al., *Izv. AN SSSR. Neorgan. mater.*, **2**(12), 2103, 1966.)

L + $GeSe_2$ ⇔ $Ge_xPb_{1-x}Se$ + (γ-GeSe). The temperature of GeSe polymorphic transformation decreases with increasing the PbSe content. At ca. 460°C, a eutectoid transformation of the (γ-GeSe) with formation of (α-GeSe) and $Ge_xPb_{1-x}Se$ takes place. The eutectoid transformation occurs in a small temperature range due to the presence of homogeneity regions based on GeSe and PbSe. The eutectoid composition is approximately 80 mol% GeSe.

The solubility of GeSe in PbSe at 520°C is not higher than 10 mol% (Shelimova et al. 1966; Kerimov et al. 1986) [9 mol% at 480°C (Krebs and Langner 1964); 40 mol% at 630°C (Nikolić 1969)] and the solubility of PbSe in GeSe at 460°C is 20 mol% (Krebs and Langner 1964; Shelimova et al. 1966).

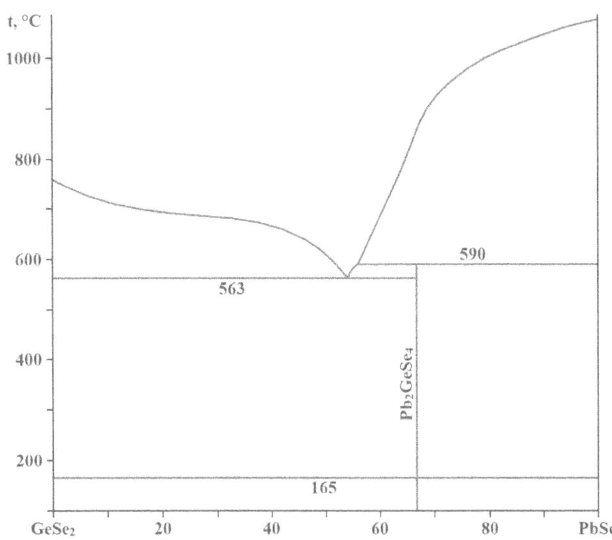

FIGURE 5.52 Phase diagram of the $GeSe_2$–PbSe system. (From Feltz, A., et al., *Krist. und Techn.*, **15**(8), 895, 1980.)

The ingots enriched with PbSe were annealed at 480°C for 80 h, and the ingots enriched with GeSe were annealed at 460°C for 100 h (Krebs and Langner 1964) or at 630°C for three weeks (Nikolić 1969). The alloys with the eutectoid transformation were annealed at 450°C for 3 months (Shelimova et al. 1966). The system was investigated through DTA, XRD, metallography, and measuring of the density.

$GeSe_2$–PbSe: The phase diagram of this system is a eutectic type with peritectic transformation (Figure 5.52) (Feltz et al. 1980b; Ludwig and Feltz 1981). The eutectic contains 54 mol% PbSe and crystallizes at 563°C. The Pb_2GeSe_4 ternary compound, which melts incongruently at 590°C (Feltz et al. 1980b; Ludwig and Feltz 1981) and crystallizes in the cubic structure with the lattice parameter $a = 1457.3 \pm 0.1$ pm (Poduska et al. 2002), is formed in this system. Because of the peritectic decomposition, the best preparation conditions for this compound were given by quenching the melt followed by annealing at 450°C and 465°C for one week. After grinding and preforming, the annealing process has to be repeated at 500°C for one week. The glasses are formed in the region of 49–55 mol% PbSe when the ampoules with the samples were quenched in water.

The glass forming region in the GeSe–$GeSe_2$–PbSe and $GeSe_2$–PbSe–Se subsystem of the Ge–Pb–Se ternary system are given in Figures 5.53 and 5.54 (Feltz and Senf 1975, 1978; Linke et al. 1978; Feltz et al. 1980b; Ludwig and Feltz 1981). The glass transition temperature for these glasses is within the interval 67°C–246°C.

To prepare the glasses, the melting process (heating) of the alloys was followed by annealing at 300°C for 4 weeks (Feltz et al. 1980b). Most of the samples were followed by a second step on annealing at 400°C for 3 weeks. The system was studied by DTA and XRD.

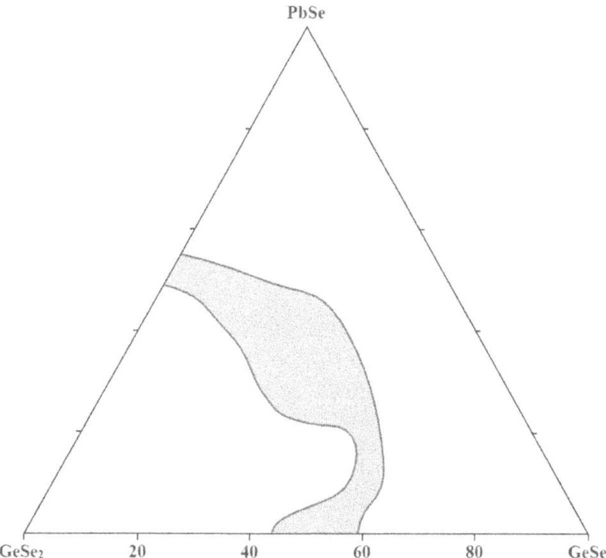

FIGURE 5.53 Glass forming region in the GeSe–GeSe$_2$–PbSe subsystem. (From Feltz, A., and Senf, L., *Z. anorg. und allg. Chem.*, **444**(1), 195, 1978.)

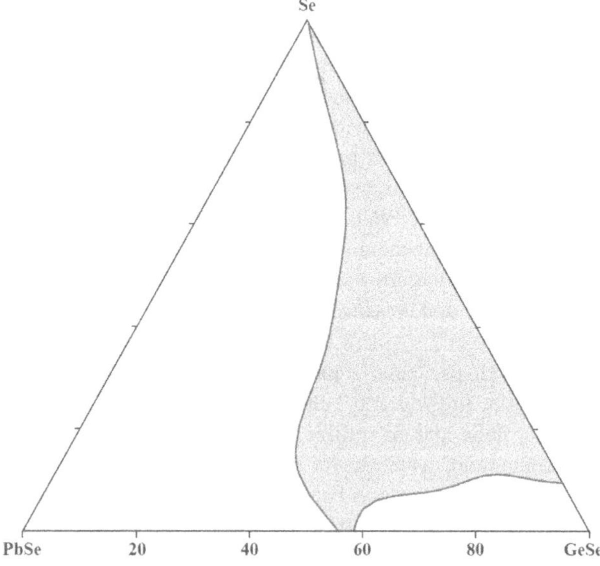

FIGURE 5.54 Glass forming region in the GeSe$_2$–PbSe–Se subsystem. (From Linke, D., et al., *Z. anorg. und allg. Chem.*, **444**(1), 217, 1978.)

5.36 Germanium–Zirconium–Selenium

The quite stable **ZrGeSe** ternary compound, which crystallizes in the tetragonal structure with the lattice parameters $a = 370.6$ and $c = 827.1$ pm (Haneveld and Jellinek 1964) [$a = 369$ and $c = 824$ pm and the calculated and experimental densities of 7.20 and 7.11 g·cm^{-3}, respectively (Onken et al. 1964)], is formed in the Ge–Zr–Se system. To obtain this compound, stoichiometric amounts of the elements were initially pre-tempered in an evacuated tube at 600°C for 10 days (Onken et al. 1964). Then, the reaction mixture was carefully rubbed in an agate mortar in a N$_2$-filled box, pressed into a pastille and annealed at 800°C for several weeks. The ampoule with tablet was stored in the hottest part of the furnace to obtain the single crystals by sublimation. ZrGeSe could be also prepared by heating GeSe with an equivalent proportion of Zr in an evacuated quartz tube at 700°C–800°C for 2–4 days (Haneveld and Jellinek 1964).

5.37 Germanium–Hafnium–Selenium

The **HfGeSe** ternary compound, which crystallizes in the tetragonal structure with the lattice parameters $a = 369$ and $c = 820$ pm and the calculated and experimental densities of 9.80 and 7.89 g·cm^{-3}, respectively, is formed in the Ge–Hf–Se system (Onken et al. 1964). To obtain this compound, stoichiometric amounts of Ge, Hf, and Se were initially pre-tempered in an evacuated tube at 600°C for 10 days. Then, the reaction mixture was carefully rubbed in an agate mortar in a N$_2$-filled box, pressed into a pastille and annealed at 750°C for several weeks. The ampoule with tablet was stored in the hottest part of the furnace to obtain the single crystals by sublimation.

5.38 Germanium–Phosphorus–Selenium

The **Ge$_2$P$_2$Se$_6$** ternary compound, which exists in the glassy state and has an experimental density of 3.90 g·cm^{-3}, is formed in the Ge–P–Se system (Voroshilov et al. 1986). The glass forming region in this system is presented in Figure 5.55 (Hilton et al. 1964; Vinogradova and Maysashvili 1979a).

5.39 Germanium–Arsenic–Selenium

GeSe–As: The phase diagram of this system is a eutectic type with peritectic transformation (Figure 5.56) (Orlova et al. 1984). The eutectic contains 23 mol% As and crystallizes at

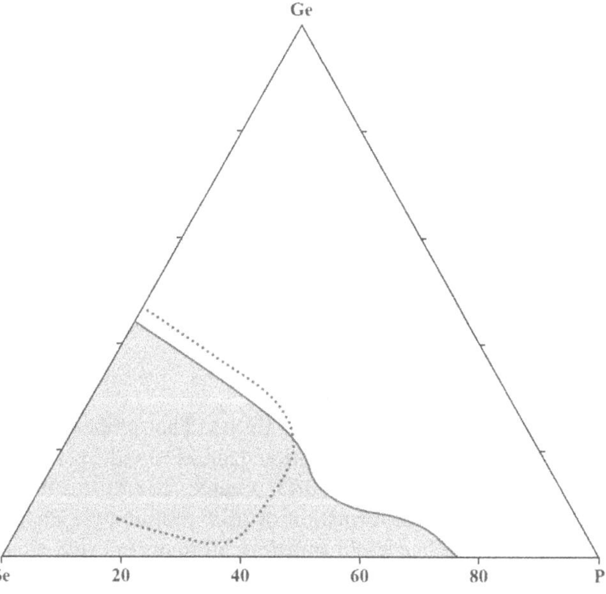

FIGURE 5.55 Glass forming region in the Ge–P–Se system. (Solid line from Vinogradova, and Maysashvili, N.G., *Zhurn. neorgan. khimii*, **24**(4), 1116, 1979; dotted line from Hilton, A.R., et al., *Infrared Phys.*, **4**(4), 213, 1964.)

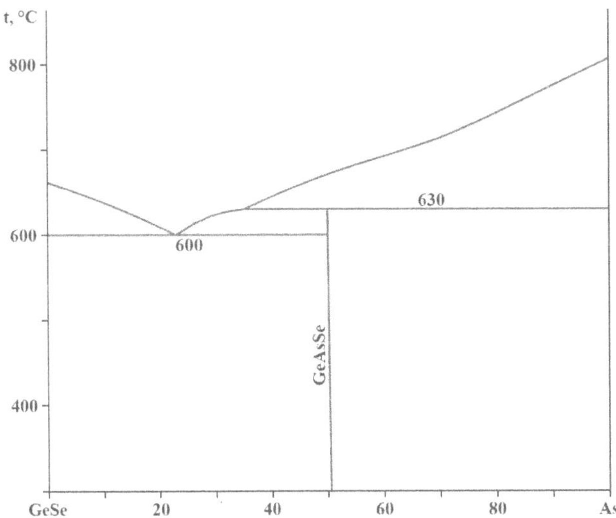

FIGURE 5.56 Phase relations in the GeSe–As system. (From Orlova, T.M., et al, *Vestn. Leningr. Univ.*, (10), 117, 1984.)

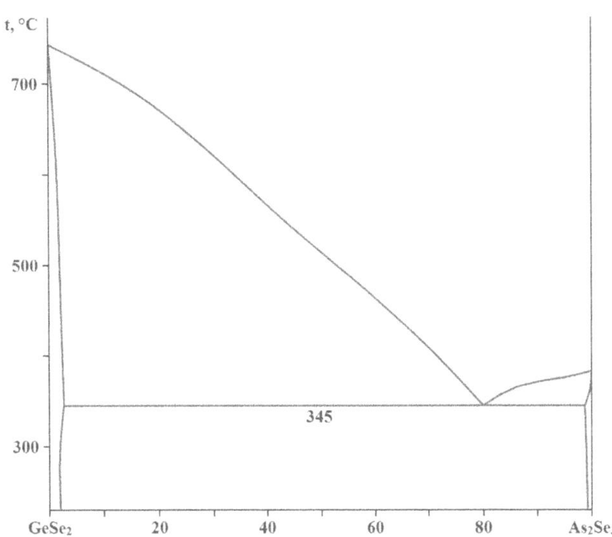

FIGURE 5.57 Phase diagram of the $GeSe_2$–As_2Se_3 system. (From Klymovych, O., et al., *J. Phase Equilib. Dif.*, **41**(2), 157, 2020.)

600°C. The **GeAsSe** ternary compound, which melts incongruently at 630°C (Vinogradova et al. 1968; Orlova et al. 1984) [at 617°C (Pachali et al. 1976)] and crystallizes in the orthorhombic structure with the lattice parameters $a = 506.2 \pm 0.1$, $b = 1011.7 \pm 0.2$, $c = 1168.7 \pm 0.3$ pm, and a calculated density of 5.03 g·cm^{-3} (Hulliger and Siergist 1981) [$a = 507$, $b = 1012$, $c = 1169$ pm (Pachali et al. 1976)], is formed in this system. By reheating the title compound, a glass transformation step appeared at a temperature of about 400°C. On further heating, no recrystallization or melting effects were observed.

The solubility of GeSe in As reaches up to 17.6 mol% (Krebs et al. 1961a). The dependence of the lattice parameters of solid solutions on the composition is nonlinear.

According to the data of Pachali et al. (1976), another ternary compound, **$GeAs_4Se$**, which melts at 747°C and crystallize in the monoclinic structure with the lattice parameters $a = 1237$, $b = 657$, $c = 359$ pm, and $\beta = 101°$, is formed in the GeSe–As system.

The crystals of GeAsSe were grown in a sealed silica tube by an iodine transport reaction (Hulliger and Siergist 1981). In a small temperature gradient around 450°C thin platelets grew within one week. The red transparent crystals showed a very easy cleavage, indicating a layer structure.

Single crystals of GeAsSe and $GeAs_4Se$ were prepared by a solid-vapor phase reaction of GeSe and As, sealed in evacuated silica ampoules, at temperatures between 580°C and 600°C (Pachali et al. 1976). When GeSe and As were heated with a 1:1 stoichiometry at 580°C for 2 h, only the crystals of GeAsSe were obtained as reaction product. Two different compounds, GeAsSe and $GeAs_4Se$, resulted when excess arsenic was present in the reaction ampoule. Finally, a reaction of GeSe and As with a 1:4 stoichiometry at a temperature of 600°C for at least 12 h yielded mainly $GeAs_4Se$.

The samples were annealed at 200°C–500°C for two months, and the system was investigated through DTA, XRD, metallography, and measuring of the microhardness (Krebs et al. 1961a; Vinogradova et al. 1968; Pachali et al. 1976).

$GeSe_2$–As_2Se_3: The phase diagram of this system is a eutectic type (Figure 5.57) (Klymovych et al. 2020). The eutectic contains 20 mol% $GeSe_2$ and crystallizes at 345°C. The mutual solubility of $GeSe_2$ and As_2Se_3 at 240°C is less than 4 mol%. The system was investigated through DTA, XRD, metallography, and measuring of the microhardness.

The part of the liquidus surfaces of the Ge–As–Se system is presented in Figure 5.58 (Vinogradova et al. 1968). Near the Se–As binary system, the formation of three ternary eutectics is possible, which causes the appearance of non-crystallizing glasses in this region. The **$Ge_2As_2Se_6$** ternary compound, which exists in the glassy state and has an experimental density of 4.40 g·cm^{-3}, is formed in this system (Voroshilov et al. 1986).

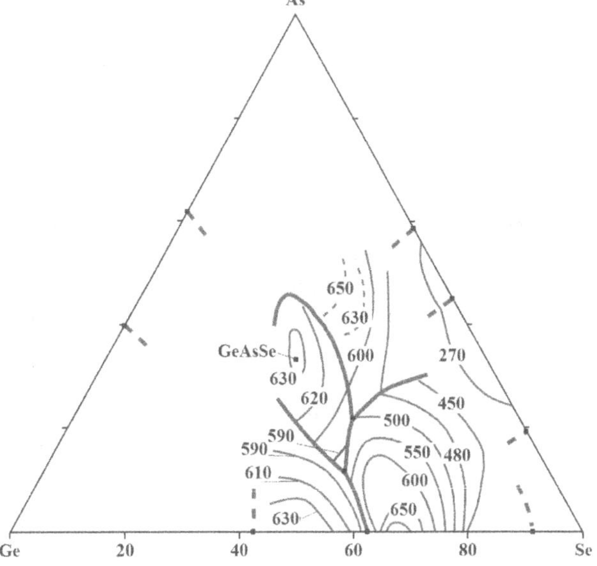

FIGURE 5.58 Part of the liquidus surface of the Ge–As–Se ternary system. (From Vinogradova, G.Z., et al., *Zhurn. neorgan. khimii*, **13**(5), 1444, 1968.)

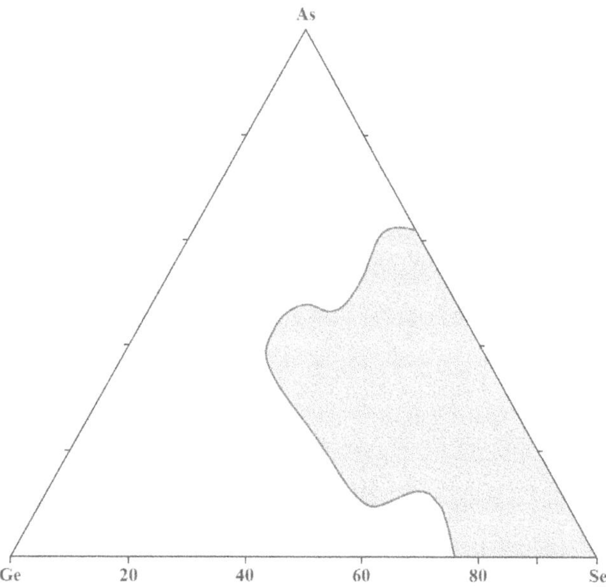

FIGURE 5.59 Glass forming region in the Ge–As–Se system. (From Ayo, L.G., and Kokorina, V.F., *Optiko-mehanich. promyshl.*, (2), 36, 1963.)

The glass forming region in the Ge–As–Se system is shown in Figure 5.59 (Goryunova et al. 1960b; Ayo and Kokorina 1963). The glass transition temperature for these glasses is within the interval 112°C–363°C (Pazin and Borisova 1972).

5.40 Germanium–Antimony–Selenium

GeSe–Sb_2Se_3: The results of the investigation of this system are contradictory: the contradictions mainly concern the existence of a ternary compound. According to the data of Bordas and Clavaguera-Mora (1982), this system is a non-quasibinary section of the Ge–Sb–Se ternary system, and at temperatures below 453°C there is only Sb_2Se_3 and a solid solution based on GeSe in the system.

According to the data of Orlova et al. (1973, 1982) and Odin et al. (1974c), the phase diagram of this system is a eutectic type with peritectic transformation (Figure 5.60). Non-quasibinarity of the section can occur only at temperatures above 670°C in a narrow composition region from GeSe-side, when germanium should be present in the alloys and the branch of its primary crystallization should appear on the diagram. The eutectic contains 56 mol% GeSe and crystallizes at 445°C (Orlova et al. 1973, 1982) [59 mol% GeSe and crystallizes at 427°C (Odin et al. 1974c)]. The **$Ge_4Sb_2Se_7$** ternary compounds, which melts incongruently at 463°C [at 508°C ± 5°C (Odin et al. 1974c)], has polymorphic transition at 400°C and crystallizes in the cubic structure with the lattice parameter $a = 565.5 \pm 0.5$ pm (Orlova et al. 1973, 1982), is formed in the GeSe–Sb_2Se_3 system.

The solubility of GeSe in Sb_2Se_3 reaches 17 mol% at 445°C and the solubility of Sb_2Se_3 in GeSe is 6 mol% at 390°C (Odin et al. 1974c) [8 mol% at 460°C (Orlova et al. 1973)].

The ingots were annealed at 390°C for 1000 h (Odin et al. 1974c) [at 430°C for 200 h at the Sb_2Se_3 content more than

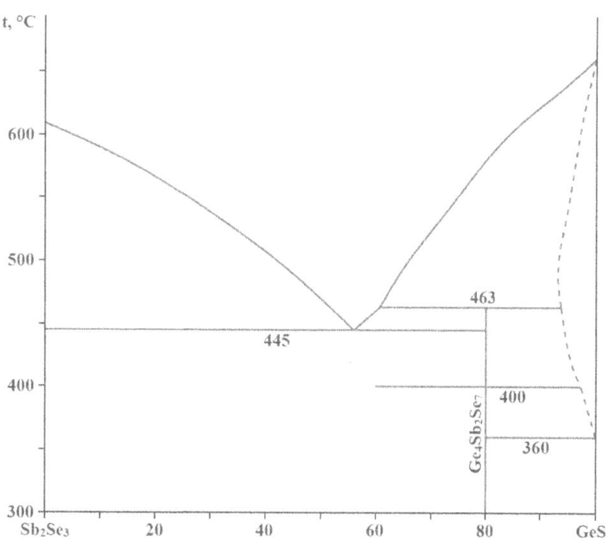

FIGURE 5.60 Phase diagram of the GeSe–Sb_2Se_3 system. (From Orlova, G.M., et al., *Vestn. Leningr. Univ.*, (4), 90, 1973.)

40 mol% and at 440°C and 340°C for 250 and 550 h, respectively, at the Sb_2Se_3 content less than 40 mol% (Orlova et al. 1973, 1982)]. The system was investigated through DTA, XRD, metallography, and measuring of the microhardness and density (Orlova et al. 1973, 1982; Odin et al. 1974c).

$GeSe_2$–Sb_2Se_3: The phase diagram of this system is a eutectic type (Figure 5.61) (Orlova et al. 1973; Bordas and Clavaguera-Mora 1982). The eutectic contains 60 mol% $GeSe_2$ and crystallizes at 485°C (Orlova et al. 1973) [45 mol% $GeSe_2$ and crystallizes at 484°C (Bordas and Clavaguera-Mora 1982)]. The ingots were annealed at 450°C for 250 h (Orlova et al. 1973) [at 400°C–600°C for 500 h (Bordas and Clavaguera-Mora 1982)]. The system was investigated through DTA, XRD, metallography, and measuring of the microhardness and density.

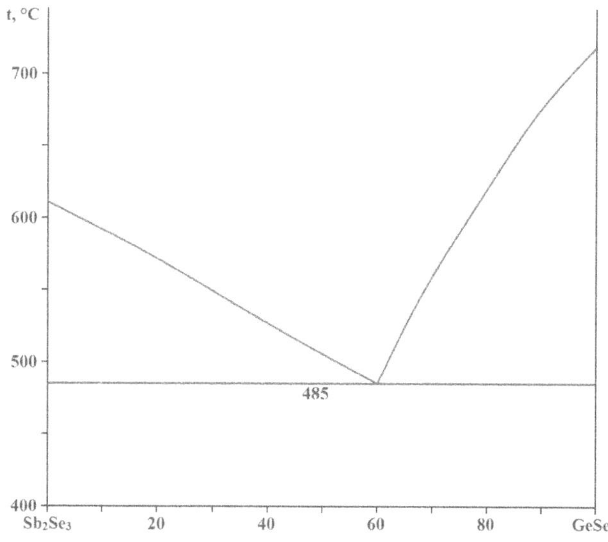

FIGURE 5.61 Phase diagram of the $GeSe_2$–Sb_2Se_3 system. (From Orlova, G.M., et al., *Vestn. Leningr. Univ.*, (4), 90, 1973.)

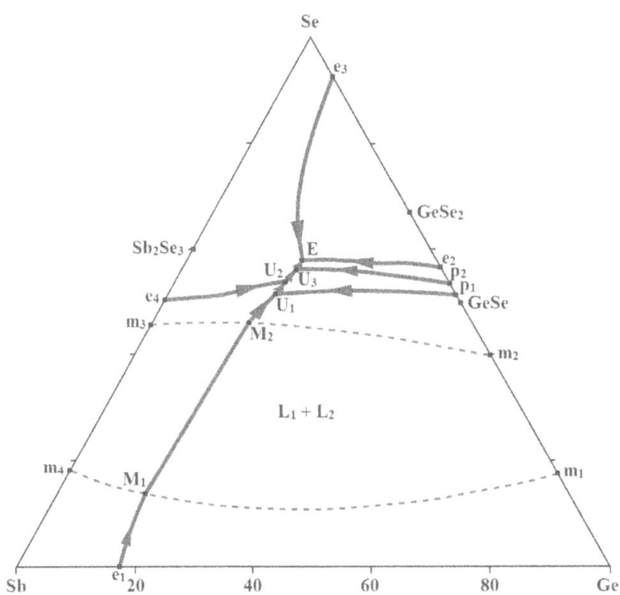

FIGURE 5.62 Projection of the liquidus surface of the Ge–Sb–Se ternary system. (From Bordas, S., and Clavaguera-Mora, M.T., *Thermochim. Acta*, **56**(2), 161, 1982.)

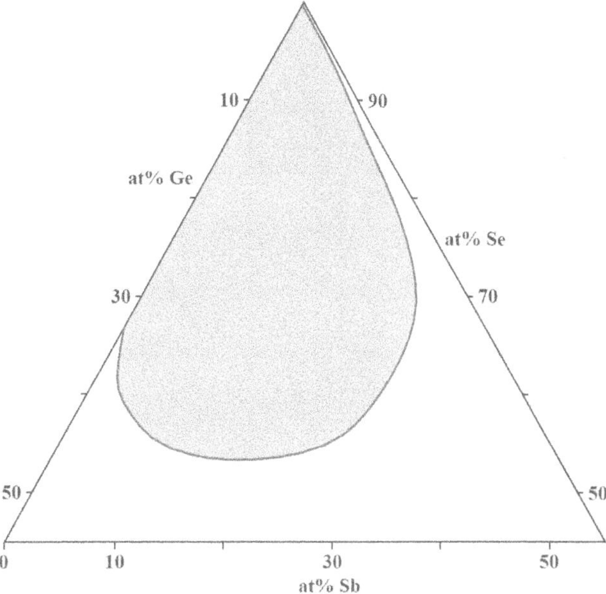

FIGURE 5.63 Glass forming region in the Ge–Sb–Se ternary system. (From Rechtin, M.D., et al., *J. Electron. Mater.*, **4**(2), 347, 1975.)

The large immiscibility region takes place on the liquidus surface of the Ge–Sb–Se ternary system (Figure 5.62) (Bordas and Clavaguera-Mora 1982). At 566°C, a ternary monotectic reaction proceeds (line M_1M_2). There are the transition points at 516°C, 492°C, and 453°C, and the ternary eutectic at 420°C in this system. The high-temperature β-GeSe enters into the ternary system giving a monophasic region with temperature decreasing as Sb content increases. At 313°C and for the composition $Ge_{39}Sb_{13}Se_{48}$, β-GeSe decomposes according to the eutectoid reaction β-GeSe ⇔ α-GeSe + Sb_2Se_3 + (Sb). There is also an invariant in the solid state at 401°C, which corresponds to the peritectoid reaction: β-GeSe + Ge ⇔ α-GeSe + (Sb). The isothermal sections of the Ge–Sb–Se ternary system at the temperatures of all mention above invariant equilibria and some vertical sections were also constructed.

According to the data of Orlova et al. (1982), the ternary eutectic in the GeSe–$GeSe_2$–Sb_2Se_3 subsystem contains 40 mol% GeSe, 31 mol% Sb_2Se_3, and 29 mol% $GeSe_2$ and crystallizes at 420°C and the transition point has a composition 44 mol% GeSe, 26 mol% Sb_2Se_3, and 30 mol% $GeSe_2$ and a temperature 438°C. The ternary eutectic in the $GeSe_2$–Sb_2Se_3–Se subsystem is apparently degenerated.

The GeSe–Sb_2Se_3 system crosses the glass-forming region in the Ge–Sb–Se ternary system (Odin et al. 1974c; Clavaguera-Mora et al. 1975), which is presented in Figure 5.63 (Borisova and Pazin 1965; Hilton et al. 1966b; Haisty and Krebs 1968a,b; Pazin and Borisova 1972; Rechtin et al. 1975). The glass transition temperature for the glasses is within the interval 72°C–343°C (Pazin and Borisova 1972). To prepare the glasses, appropriate mixtures of Ge, Sb, and Se were charged in quartz ampoules, sealed under vacuum of 13 mPa and heated in a rocking furnace at 850°C–950°C for 12–24 h (Rechtin et al. 1975; Abdel-Rahim et al. 1990). During melting of the charge, the ampoules were frequently agitated in order to ensure homogenization of the melt. The next quenching was carried out at temperatures between 550°C and 600°C by use a hand held blower, which was relatively mild quench. This was followed by an annealing at approximately 300°C for 6–8 h and then furnace cooling to room temperature. The investigation of $Ge_{20}Sb_xSe_{80-x}$ (5 ≤ x ≤ 40) glasses showed that the glass transition temperature increases with increasing Sb content (Abdel-Rahim et al. 1990).

5.41 Germanium–Bismuth–Selenium

GeSe–BiSe: According to the data of XRD, metallography, and measuring of the microhardness, the solubility of GeSe in BiSe at 450°C is 5 mol% (Abrikosov et al. 1972). The ingots were annealed at 250°C and 450°C for 10 and 30 days, respectively.

$GeSe_2$–Bi_2Se_3: The phase diagram of this system is a eutectic type (Figure 5.64) (Kozhina et al. 1976). The eutectic contains 60 mol% $GeSe_2$ and crystallizes at 590°C. No mutual solubility of the starting components was found. The ingots were annealed at 450°C for 75 h. The system was investigated through DTA, XRD, metallography, and measuring of the microhardness and density.

The glass forming region in the Ge–Bi–Se ternary system is given in Figure 5.65 (Pazin et al. 1970). The glass transition temperature for the glasses is within the interval 78°C–296°C and increases with Ge and Bi content increasing in the glasses. The glasses exhibit increased crystallization ability with a maximum Bi content.

The glass forming region in the Ge–Bi–Se ternary system was also determined by Pazin and Borisova (1969). It was found that the most favorable for glass formation are the following ratios of the components: 1–10 at% Bi, 20–30 at% Ge, and

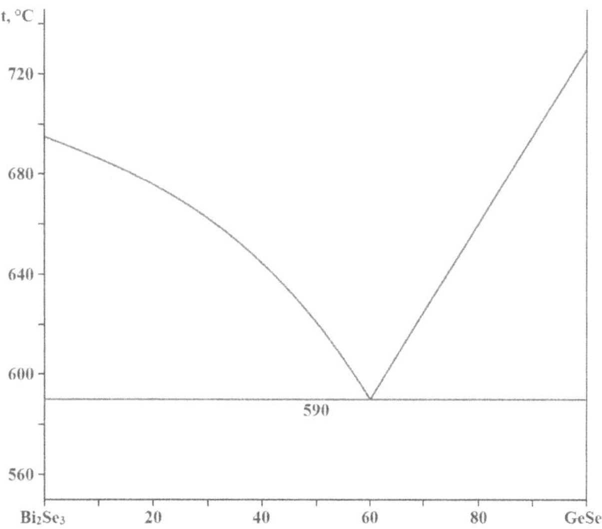

FIGURE 5.64 Phase diagram of the $GeSe_2$–Bi_2Se_3 system. (From Kozhina, I.I., et al., *Vestn. Leningr. Univ.*, (4), 143, 1976.)

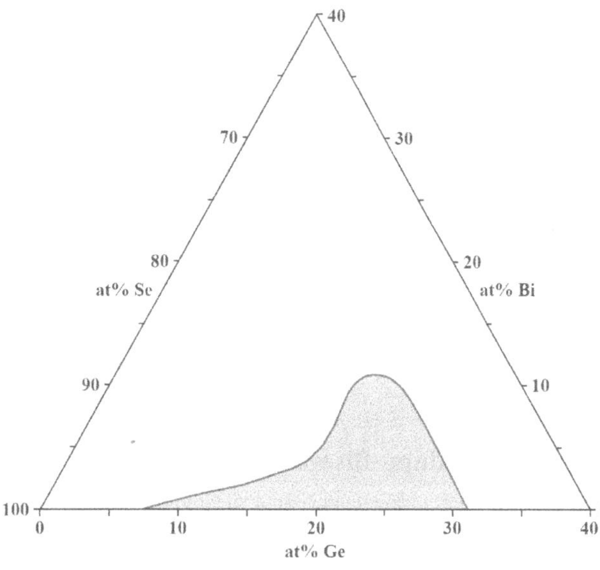

FIGURE 5.65 Glass forming region in the Ge–Bi–Se ternary system. (From Pazin, A.V., et al., *Izv. AN SSSR. Neorgan. mater.*, **6**(5), 884, 1970.)

60–70 at% Se. It was shown that a maximum of 10 at% Bi, 28 at% Ge, and 89 at% Se can be added to the composition of glassy alloys.

5.42 Germanium–Vanadium–Selenium

The $\mathbf{GeV_4Se_8}$ ternary compound, which crystallizes in the cubic structure with the lattice parameter a = 1013.9 ± 0.1 pm and a calculated density of 5.787 g·cm^{-3}, is formed in the Ge–V–Se system (Johrendt 1998). It was prepared by heating appropriate mixture of the elements in sealed quartz tube under Ar atmosphere. An excess amount (≈20%) of GeSe is necessary to suppress the loss of GeSe, which is volatile at temperatures around 800°C. The mixture was heated to 750°C at the rate of 50°C·h^{-1} for one day and furnace cooled to room temperature. After homogenization in an Ar-filled glove box, the resulting sample was annealed twice at 800°C for 15 h yielding dark grey powders, which are stable in air. Single crystals of this compound show metallic luster.

5.43 Germanium–Oxygen–Selenium

The $\mathbf{Ge(SeO_3)_2}$ ternary compounds, which is stable up to 421°C (Gospodinov and Barkov 1982), has two polymorphic modifications (Kong et. al. 2012) and decomposes for GeO_2 and SeO_2 at 569°C (Gospodinov and Bogdanov 1989), is formed in the Ge–O–Se system. α-$Ge(SeO_3)_2$ crystallize in the monoclinic structure with the lattice parameters a = 505.8 ± 0.3, b = 678.1 ± 0.4, c = 729.2 ± 0.4 pm, β = 90.995 ± 0.008°, and a calculated density of 4.336 g·cm^{-3} and β-$Ge(SeO_3)_2$ crystallize in the cubic structure with the lattice parameter a = 776.1 ± 0.3 pm and a calculated density of 4.640 g·cm^{-3} (Kong et. al. 2012). At high temperatures this compound decomposes forming GeO_2 and SeO_2 (Gospodinov and Barkov 1982; Gospodinov and Bogdanov 1989).

To prepare this compound, a mixture of GeO_2 (0.30 mM), SeO_2 (1.0 mM), and H_2O (2 mL) was sealed in an autoclave equipped with a Teflon liner (23 mL) and heated at 200°C for 5 days, followed by slow cooling to room temperature at a rate of 4°C·h^{-1} (Kong et. al. 2012). Colorless prism-shaped crystals of α-$Ge(SeO_3)_2$ and octahedron-shaped crystals of β-$Ge(SeO_3)_2$ were obtained simultaneously. The pure phase of α-$Ge(SeO_3)_2$ in ca. 28% yield (based on Ge) was obtained by heating a mixture of GeO_2 (0.2 mM), SeO_2 (1.0 mM), and H_2O (2 mL) at 200°C for 5 days. Octahedral crystals of β-$Ge(SeO_3)_2$ along with α-$Ge(SeO_3)_2$ and GeO_2 powders were obtained by heating a mixture of GeO_2 (0.32 mM), SeO_2 (1.0 mM), and 0.1 M HCl solution (2 mL) at 200°C for 5 days. The pure phase of β-$Ge(SeO_3)_2$ in ca. 15% yield (based on Ge) was isolated by filtration. It was found that the molar ratio of GeO_2/SeO_2 used in the reactions is very important to the chemical compositions and structures of the products formed. If the molar ratio was less than 1:5, only plained solution formed. Between 1:5 and 2:5, a mixture of α-$Ge(SeO_3)_2$, β-$Ge(SeO_3)_2$ and GeO_2 was obtained. When this ratio was larger than 2:5, just GeO_2 powder appeared. Furthermore, using 0.1 M HCl solution instead of water may help the crystal growth of β-$Ge(SeO_3)_2$, which makes it easier to separate the β-$Ge(SeO_3)_2$ from the mixture. $Ge(SeO_3)_2$ could be also obtained at the interaction of GeO_2 and H_2SeO_3 at the components ratio of 3.0–5.0 M Se for 1 M of Ge (Shokol et al. 1967) or its continued crystallization at 100°C in aqueous solution of H_2SeO_3 (Gospodinov and Bogdanov 1989).

5.44 Germanium–Tellurium–Selenium

GeSe–Te: This system is a non-quasibinary section of the Ge–Te–Se ternary system and is divided by the $GeSe_2$–GeTe system in two parts with the invariant equilibria at 514°C and at 360°C (Bordas et al. 1980). At 660°C, the peritectic transformation Ge ⇔ GeSe + L takes place.

GeSe–Ge$_{0.98}$Te: This system is also a non-quasibinary section of the Ge–Te–Se ternary system (Figure 5.66) (Shelimova

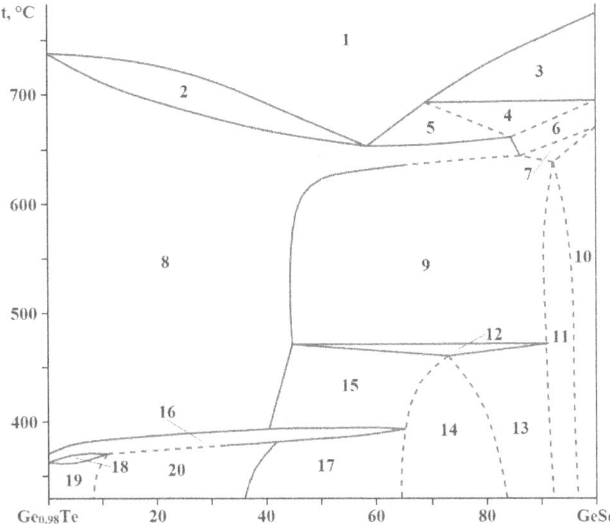

FIGURE 5.66 Phase relations in the GeSe–$Ge_{0.98}Te$ system: 1, L; 2, L + $GeSe_xTe_{1-x}$; 3, L + Ge; 4, L + Ge + $GeSe_xTe_{1-x}$; 5, L + $GeSe_xTe_{1-x}$; 6, $GeSe_xTe_{1-x}$ + Ge; 7, $GeSe_xTe_{1-x}$ + Ge + γ'-GeSe; 8, $GeSe_xTe_{1-x}$; 9, $GeSe_xTe_{1-x}$ + γ'-GeSe; 10, Ge + γ'-GeSe; 11, γ'-GeSe; 12, $GeSe_xTe_{1-x}$ + γ'-GeSe + $GeSe_{0.75}Te_{0.25}$; 13, $GeSe_{0.75}Te_{0.25}$ + γ'-GeSe; 14, $GeSe_{0.75}Te_{0.25}$; 15, $GeSe_xTe_{1-x}$ + $GeSe_{0.75}Te_{0.25}$; 16, α-GeTe + $GeSe_xTe_{1-x}$; 17, α-GeTe + $GeSe_{0.75}Te_{0.25}$; 18, α-GeTe + $GeSe_xTe_{1-x}$ + γ'-GeSe; 19, α-GeTe + γ'-GeSe; 20, α-GeTe. (From Abrikosov, N.H., et al., *Izv. AN SSSR. Neorgan. mater.*, **24**(1), 46, 1988.)

et al. 1966; Bordas et al. 1980; Abrikosov et al. 1988b). In a wide composition range, the crystallization ends with the formation of a continuous series of solid solutions (β) between high-temperature modifications of GeTe and GeSe. In the alloys enriched by GeSe, the β-solid solution is formed as a result of the next peritectic reaction: L + Ge ⇔ β. The boundary between the phase regions β and β + Ge is in the range of compositions that correspond to the intersection of the homogeneity interface of a solid solution saturated with germanium with the GeSe–$Ge_{0.98}Te$ section. The low-temperature orthorhombic modification of GeSe (γ') is formed by the peritectic reaction in the temperature range 651°C–627°C. GeSe is adjacent to the region Ge + γ' associated with its deviation from stoichiometry. The extent of this region decreases with temperature decreasing. The **$GeSe_{0.75}Te_{0.25}$** ternary compound (ε-phase) is formed according to the peritectic reaction β + γ' ⇔ ε within the interval of temperatures 452°C–462°C (Shelimova et al. 1966; Abrikosov et al. 1984a, 1985a, 1988b). This compound is the only such kind of compounds in the $A^{IV}B^{VI}$–$A^{IV}B^{VI}$ systems (Abrikosov et al. 1988b). Low-temperature α-GeTe with the orthorhombic structure is formed at 379°C–387°C as a result of the peritectoid reaction β + ε ⇔ α. Near $Ge_{0.98}Te$, (α + γ) and (α + β) two-phase regions are divided by the (α + β + γ) three-phase region. The temperature of the α → β phase transition decreases with the increasing of GeTe content (Abrikosov et al. 1984b, 1988b).

$GeSe_{0.75}Te_{0.25}$ is stable at the temperatures below 400°C and has two polytypic modifications, which crystallize in the hexagonal structure with the lattice parameters $a_1 = a_2 = 382$, $c_1 = 1562$ and $c_2 = 4687$ pm and an experimental density of 6.65 ± 0.05 g·cm^{-3} (Muir and Cashman 1966, 1967; Muir and Beato 1973) [$a_2 = 381.7 \pm 0.05$ and $c_2 = 4695 \pm 3$ pm (Abrikosov et al. 1988b)]. In the single-phase region, the lattice parameters of the compound change according to Vegard's law. This compound is a degenerate semiconductor with a bandgap much less than 1.0 eV (Muir and Cashman 1967). Single crystals of this compound were grown by the chemical transport reactions (Muir and Cashman 1966, 1967).

According to the data of Shevtsova et al. (1987b), a shock compression of the components between the temperature −180°C and +340°C leads to the formation of the **GeSeTe** ternary compound.

The solubility of GeTe in GeSe reaches 8 mol% (Muir and Cashman 1967) [5 mol% at 500°C (Krebs and Langner 1964); less than 10 mol% at 425°C (Shelimova et al. 1966)] and the solubility of GeSe in GeTe is 50 mol% (Muir and Cashman 1966) [63 mol% at 500°C (Krebs and Langner 1964)].

The ingots of the GeSe–$Ge_{0.98}Te$ were annealed at 500°C for 500 h and then slowly cooled to 300°C and annealed for 1000 h with next cooling to room temperature at a rate of 2°C·h^{-1} (Abrikosov et al. 1988b). The system was investigated through DTA, XRD, metallography, and measuring of the microhardness and density (Muir and Cashman 1966, 1967; Shelimova et al. 1966; Muir and Beato 1973; Abrikosov et al. 1988b).

$GeSe_2$–Te: The phase diagram of this system is shown in Figure 5.67 (Obraztsov 1972; Bordas et al. 1977, 1980). The eutectic contains 7 mol% $GeSe_2$ and crystallizes at 440°C (Bordas et al. 1977, 1980) [10 mol% $GeSe_2$ and crystallizes at 435°C (Obraztsov 1972)]. There is a liquid-liquid immiscibility region which extends from 28 to 86 at% Te, and gives the monotectic reaction at 560°C.

The ingots were annealed at 500°C for 3 days and the system was investigated through DTA, XRD, metallography, and measuring of the microhardness (Obraztsov 1972; Bordas et al. 1977, 1980).

$GeSe_2$–GeTe: The results of the investigation of this system are contradictory. According to the data of Bordas et al. (1980),

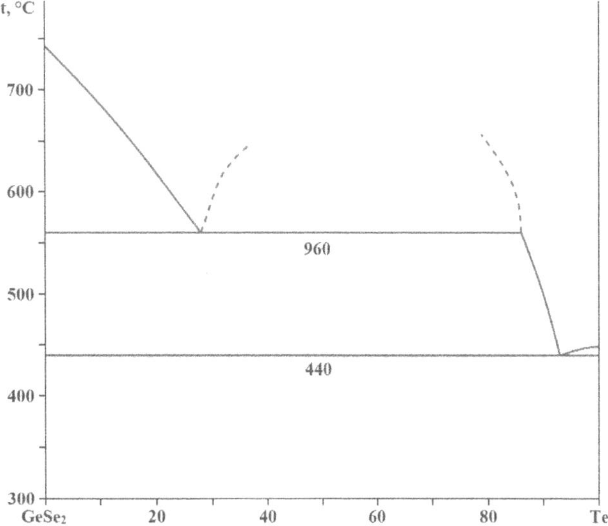

FIGURE 5.67 Phase diagram of the $GeSe_2$–Te system. (From Bordas, S., et al., *Thermochim. Acta*, **37**(2), 197, 1980.)

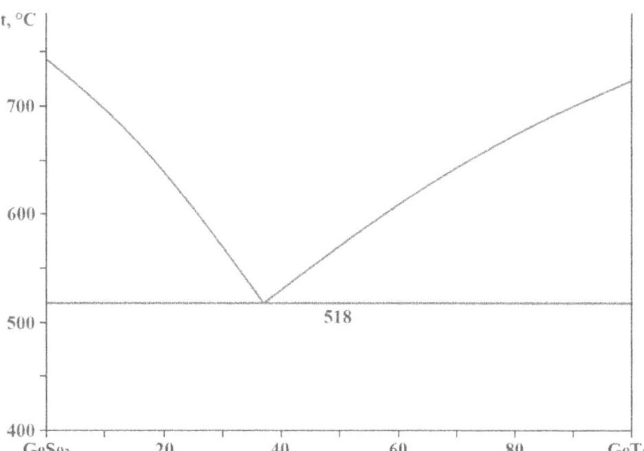

FIGURE 5.68 Phase diagram of the $GeSe_2$–GeTe system; 1, L; 2, L + β-GeSe; 3, L + β-GeSe + Ge; 4, β-GeSe + Ge; 5, β-GeSe; 6, L + β-GeSe; 7, L + $(GeSe_2)$;. 8, L + β-GeSe + $(GeSe_2)$; 9, $(GeSe_2)$ + β-GeSe, 10, α-GeSe + β-GeSe + Ge; 11, α-GeSe + β-GeSe; 12, α-GeSe + Ge; 13, α-GeSe; 14, α-GeSe + β-GeSe + γ-GeSe; 15, + β-GeSe + γ-GeSe; 16, + α-GeSe + β-GeSe; 17, α-GeSe + β-GeSe + γ-GeSe; 18, α-GeSe + γ-GeSe; 19, γ-GeSe; 20, α-GeSe + γ-GeSe; 21, α-GeSe; 22, $(GeSe_2)$. (From Bordas, S., et al., *Thermochim. Acta*, **37**(2), 197, 1980.)

the phase diagram of this system is a eutectic type (Figure 5.68). The eutectic occurs at 37 mol% GeTe and 518°C.

According to the data of Abrikosov et al. (1985b), this system is a non-quasibinary section of the Ge–Se–Te ternary system due to a deviation from stoichiometry of GeTe (Figure 5.69). The next monovariant reactions take place in this system: L ⇔ β-GeSe + $GeSe_2$ at ca. 537°C (1); β-GeSe + Ge ⇔ α-GeSe at 397°–427°C (2); β-GeSe + $GeSe_2$ ⇔ α-GeSe at 353°C–367°C (3); α-GeSe + β-GeSe ⇔ γ–GeSe at 359°C–367°C (4); and β-GeSe ⇔ α-GeSe + γ-GeSe at 339°–347°C (5). The section twice intersects the region of the α-phase existence.

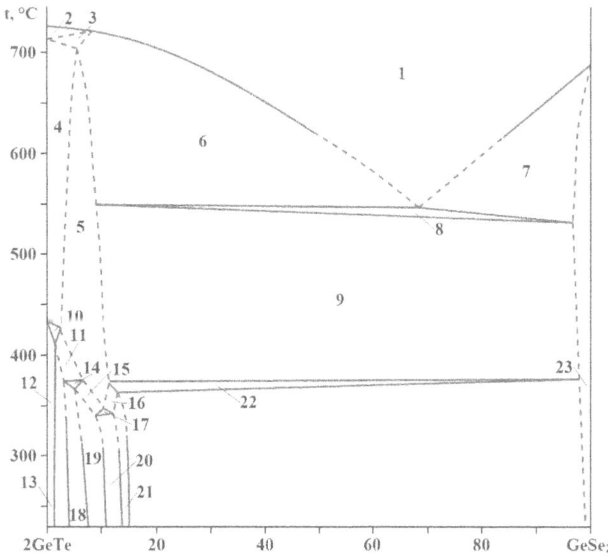

FIGURE 5.69 Phase relations in the $GeSe_2$–2GeTe system. (From Abrikosov, N.H., et al., *Izv. AN SSSR. Neorgan. mater.*, **21**(10), 1659, 1985.)

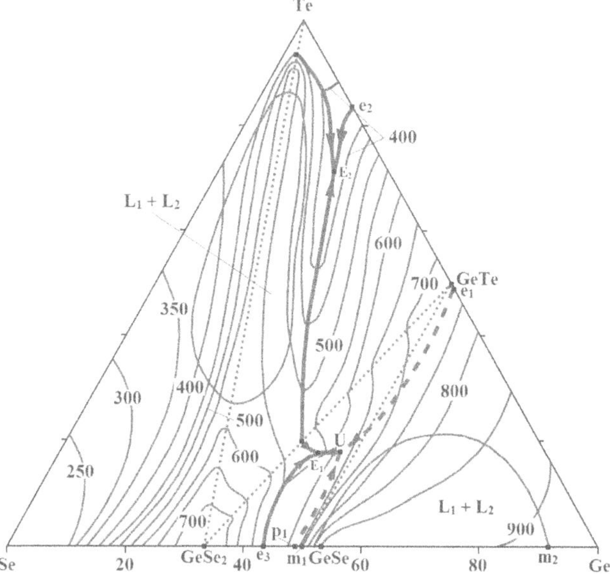

FIGURE 5.70 Liquidus surface of the Ge–Te–Se ternary system. (From Bordas, S., et al., *Thermochim. Acta*, **37**(2), 197, 1980.)

The liquidus surface and some vertical sections of the Ge–Te–Se ternary system were constructed by (Bordas et al. 1980). The liquidus isotherms, main invariant points, monovariant lines and liquid immiscibility regions of the system are shown in Figure 5.70. The $GeSe_2$–Te–Se subsystem shows liquidus profiles which decrease in temperature with increasing Se content. The thermal stability of the alloys increases with increasing the Te content and decreases with increasing the $GeSe_2$ content. Thermal decomposition begins near 300°C for large Se content and near 450°C–500°C for large Te or $GeSe_2$ content. Alloys with composition in the liquid immiscibility gap have lower stability than the surrounding ones.

The GeTe–$GeSe_2$–Te subsystem has an invariant solidus surface at 360°C that corresponds to the eutectic reaction (E_1) L ⇔ GeTe + $GeSe_2$ + Te. The composition of this eutectic point is 19 at% Ge, 72 at% Te, and 9 at% Se. Thermal stability increases generally with Te content and is also high near the ternary eutectic. Temperatures for the beginning of thermal decomposition are in the range 460°C–520°C.

The GeTe–GeSe–$GeSe_2$ subsystem has a ternary eutectic E_2 of composition 44 at% Ge, 18 at% Te, and 38 at% Se, which corresponds to the invariant reaction L ⇔ GeTe + GeSe + $GeSe_2$ at 514°C, and a transition point (U_1) L_1 + Ge ⇔ GeSe + GeTe at 635°C. The approximate composition of this point is 48 at% Ge, 18 at% Te, and 34 at% Se.

In the GeSe–Ge–GeTe subsystem there is a large liquid immiscibility region based on that existing in the Ge–Se system. Thermal stability is very low for these alloys; the decomposition begins at 500°C–600°C and, once initiated, is very rapid. For this reason, the determination of the monovariant equilibrium curves extending from the transition point U to both the Ge–Te binary eutectic e_1 and the Ge–Se peritectic p_1 (dotted lines in Figure 5.70) is approximate.

The isothermal section of the Ge–Te–Se ternary system at 300°C is presented in Figure 5.71 (Abrikosov et al. 1985b).

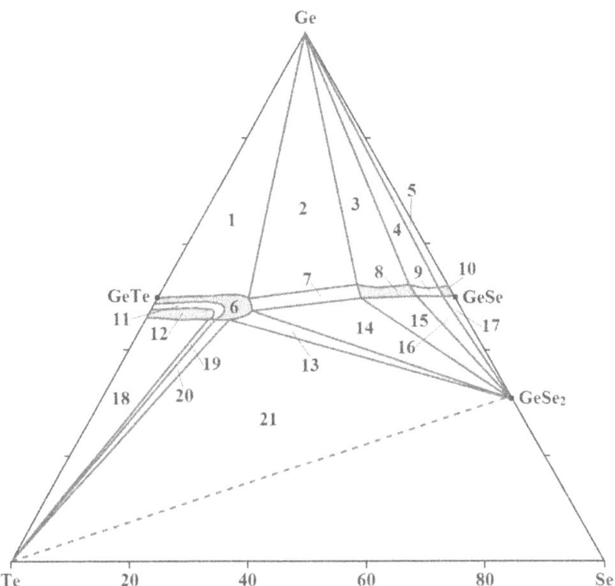

FIGURE 5.71 Isothermal section of the Ge–Te–Se ternary system at 300°C: 1, α-GeTe + Ge; 2, α-GeTe + ε + Ge; 3, ε + Ge; 4, ε + Ge + GeSe; 5, Ge + GeSe; 6, α-GeTe; 7, α-GeTe + ε; 8, ε, 9, ε + GeSe; 10, GeSe, 11, α-GeTe + γ-GeTe; 12, γ-GeTe; 13, α-GeTe + $GeSe_2$; 14, α-GeTe + ε + $GeSe_2$; 15, ε + $GeSe_2$; 16, ε + $GeSe_2$ + GeSe; 17, $GeSe_2$ + GeSe; 18, γ-GeTe + Te; 19, α-GeTe + γ-GeTe + Te; 20, α-GeTe + Te; 21, α-GeTe + Te + $GeSe_2$. (From Abrikosov, N.H., et al., *Izv. AN SSSR. Neorgan. mater.*, **21**(10), 1659, 1985.)

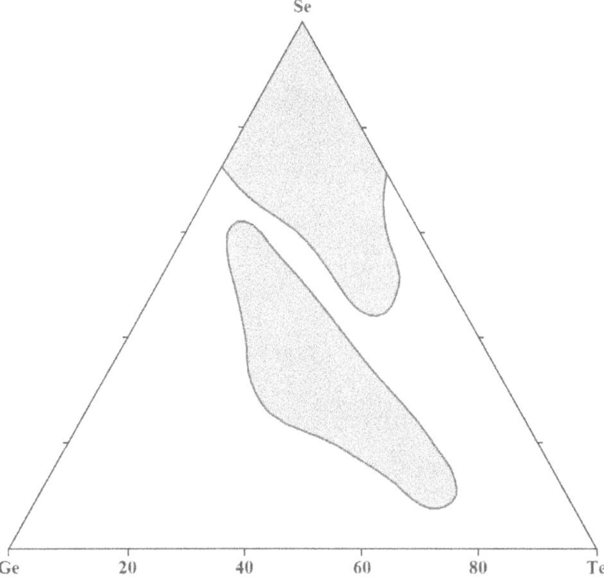

FIGURE 5.72 Glass forming regions in the Ge–Te–Se ternary system. (From Makovskaya, Z.G., and Zhukov, E.G., Izv. *AN SSSR. Neorgan. mater.*, **16**(2), 251, 1980.)

With increasing of the Se content, there is an expansion of the existence region of the α-modification and "pinching out" of the γ-phase (Shelimova and Karpinskiy 1991a). While maintaining a constant deviation from stoichiometry and with variable Se content in $Ge_{0.49}(Se_xTe_{1-x})_{0.51}$ alloys, $\gamma \rightarrow (\gamma + \alpha)$ with increasing x a concentration phase transition is observed at $x \approx 0.12$. Inside the region of solid solutions based on GeTe, the boundaries of the existence regions of α- and γ-phases are outlined (Abrikosov et al. 1985b). It is seen that with an increase in the selenium content, the region of existence of the α-phase expands and decreases for the γ-phase.

There are two glass forming regions in the Ge–Te–Se ternary system (Figure 5.72) (Pazin et al. 1972; Feltz et al. 1980a; Makovskaya and Zhukov 1980; Bordas et al. 1985). Alloys lying between the regions of glass formation are disposed to delamination. The glass transition isotherms for these glasses are presented in Figure 5.73 (Sarrach et al. 1976).

The investigations on $(GeTe)_{1-x}Se_x$ (x = 0.02, 0.10, 0.20 and 0.50) bulk shows that the crystalline structure of GeTe alloys does not affect up to $x = 0.02$ (Vinod et al. 2012). With increasing amount of Se content the alloys gets modified into a homogeneous amorphous structure.

Bulk samples of the glasses have been prepared by melt quenching technique (Pazin et al. 1972; Feltz et al. 1980a; Makovskaya and Zhukov 1980; Bordas et al. 1985; Acharya et al. 1999; Vinod et al. 2012). High purity elements were taken in the stoichiometric ratio into quartz ampoules. These were evacuated and sealed under the vacuum of 1 mPa. The sealed ampoules were put into furnace and heated up to 1000°C with a heating rate of 100°C·h^{-1}, and kept at that temperature for 24 h. To ensure the homogeneity of the molten materials, the ampoules were rotated for 12 h and quenched into ice cooled water.

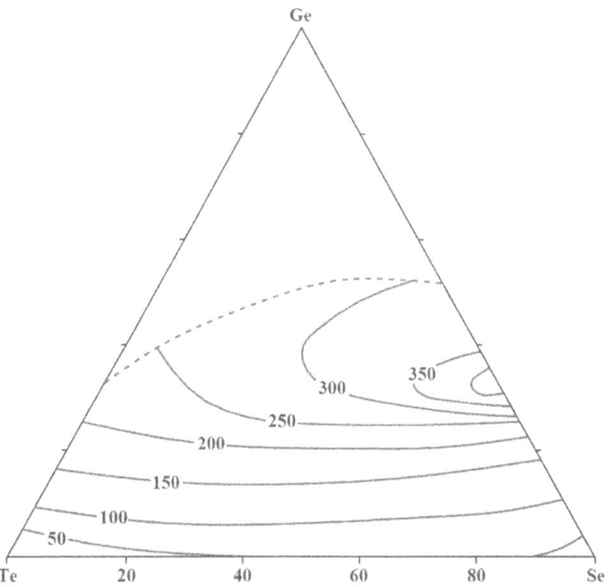

FIGURE 5.73 Glass transition isotherms for the glasses of the Ge–Te–Se ternary system. (From Sarrach, D.J., et al., *J. Non-Cryst. Solids*, **22**(2), 245, 1976.)

5.45 Germanium–Chromium–Selenium

$GeSe–Cr_3Se_4$: The phase relations in this system are given in Figure 5.74 (Shabunina et al. 2003). The section crosses the fields of primary crystallization of Ge and two Cr_3Se_4 modifications. The point of Ge and Cr_3Se_4 joint crystallization is at 8 mol% Cr_3Se_4 and 670°C. The region of Cr_3Se_4 primary crystallization takes place in the concentration range of 8–19 mol% Cr_3Se_4 and the temperature interval from 870°C to 670°C. The solubility of GeSe in Cr_3Se_4 reaches 10 mol% at 870°C and

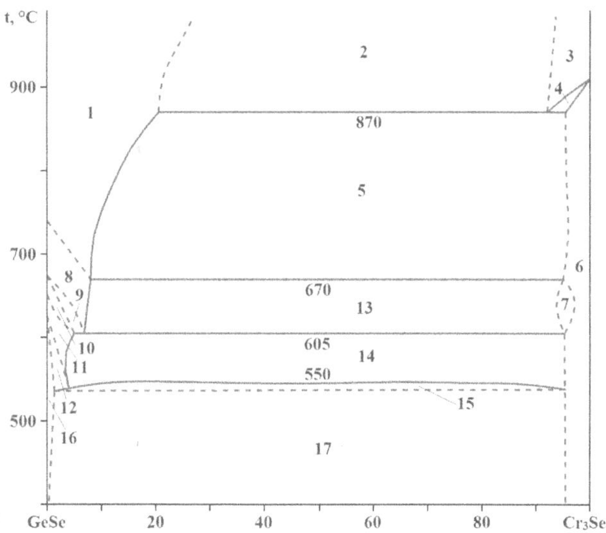

FIGURE 5.74 Phase diagram of the GeSe–Cr_3Se_4 system: 1, L, 2, L + γ, 3, γ, 4, γ + (Cr_3Se_4); 5, L + (Cr_3Se_4); 6, (Cr_3Se_4); 7, Ge + (Cr_3Se_4); 8, L + Ge; 9, L + Ge + β-GeSe; 10, L + β-GeSe; 11, L, β-GeSe; 12, α-GeSe + β-GeSe; 13, L + Ge + (Cr_3Se_4); 14, β-GeSe + (Cr_3Se_4); 15, α-GeSe + β-GeSe + (Cr_3Se_4); 16, α-GeSe; 17, α-GeSe + (Cr_3Se_4). (From Shabunina, G.G., et al., *Zhurn. neorgan. khimii*, **48**(9), 1542, 2003.)

decreases to 5 mol% at room temperature. This system was investigated through DTA and XRD.

$GeSe_2$–Cr_2Se_3: The phase diagram of this system is a eutectic type (Figure 5.75) (Shabunina et al. 2003). The eutectic is degenerated from the $GeSe_2$ side. The **$Cr_4Ge_3Se_{12}$** ternary compound, which melts incongruently at 810°C and crystallizes in the monoclinic structure with the lattice parameters $a = 1538.1 \pm 0.2$, $b = 523.3 \pm 0.1$, $c = 640.50 \pm 0.08$ pm, and $\beta = 98.051 \pm 0.016°$, is formed in this system. The solubility of $GeSe_2$ in Cr_2Se_3 is 13 mol% at 810°C and decreases to 10 mol% at room temperature. This system was investigated through DTA and XRD.

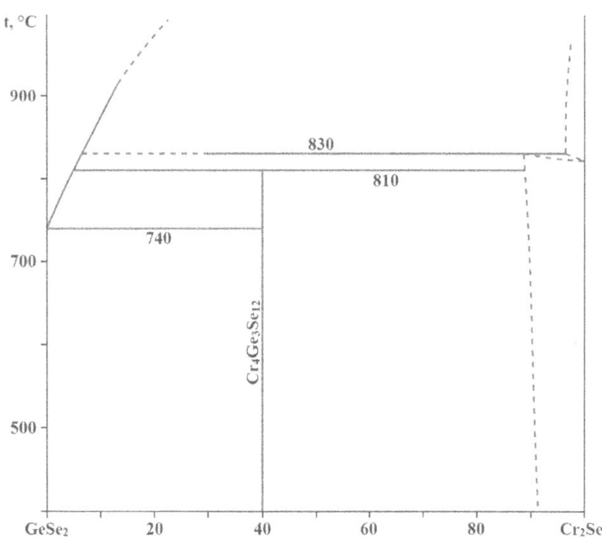

FIGURE 5.75 Phase diagram of the $GeSe_2$–Cr_2Se_3 system. (From Shabunina, G.G., et al., *Zhurn. neorgan. khimii*, **48**(9), 1542, 2003.)

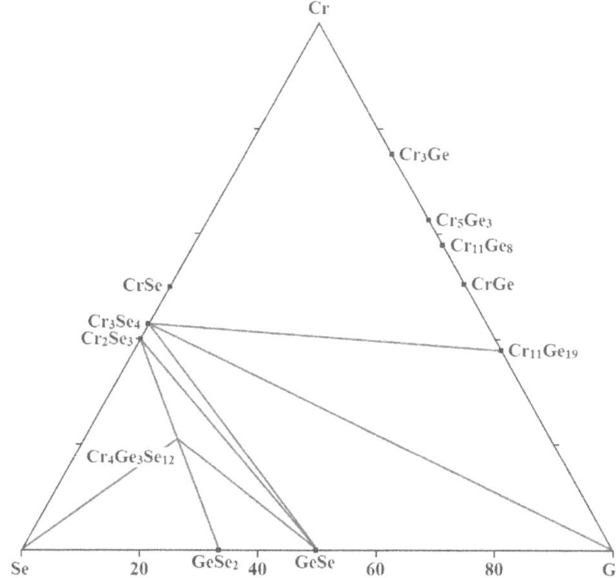

FIGURE 5.76 Isothermal section of the Ge–Cr–Se ternary system at room temperature. (From Shabunina, G.G., et al., *Zhurn. neorgan. khimii*, **48**(9), 1542, 2003.)

Ge–Cr_3Se_4: The phase diagram of this system is a eutectic type (Shabunina et al. 2003). The eutectic contains ca. 9 mol% Cr_3Se_4 and crystallizes at 870°C. The solubility of Ge in Cr_3Se_4 is equal to 12 mol% at 870°C and decreases to 9–10 mol% at room temperature.

The isothermal section of the Ge–Cr–Se ternary system is shown in Figure 5.76 (Shabunina et al. 2003). This system was investigated through DTA and XRD.

5.46 Germanium–Iodine–Selenium

GeSe–GeI_2: The phase diagram of this system is a eutectic type (Figure 5.77) (Danilov et al. 2006a,b). The eutectic contains 15 mol% GeSe and crystallizes at 420°C. The solubility

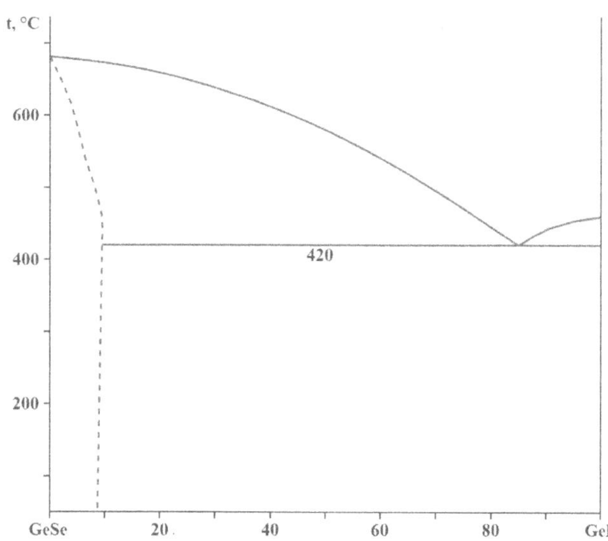

FIGURE 5.77 Phase diagram of the GeSe–GeI_2 system. (From Danilov, D.N., et al., *Inorg. Mater.*, **42**(5), 474, 2006.)

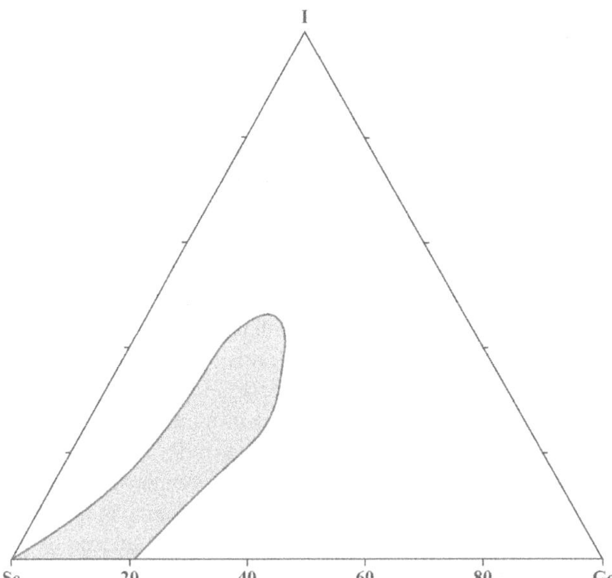

FIGURE 5.78 Glass forming region in the Ge–I–Se ternary system. (From Dembovskiy, S.A., and Popova N.P., *Izv. AN SSSR. Neorgan. mater.*, **6**(1), 138, 1970.)

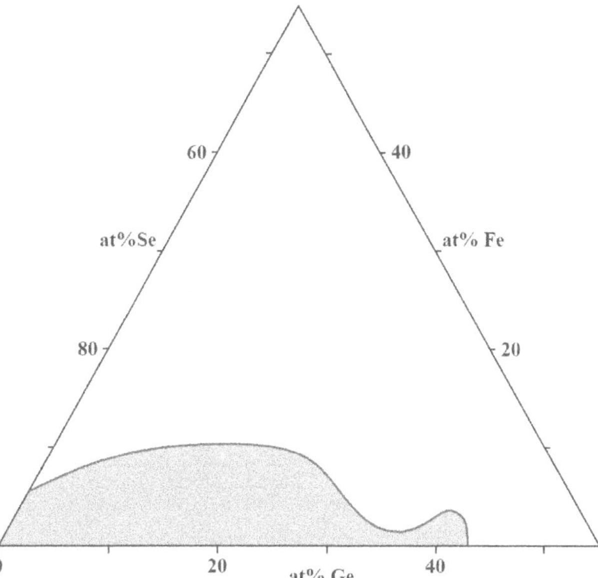

FIGURE 5.79 Glass forming region in the Ge–Fe–Se ternary system. (From Ivanova, Z.G., and Vassilev, V.S., *J. Phys. Chem. Solids*, **58**(9), 1347, 1997.)

of GeI_2 in GeSe reaches 9 mol% at 300°C. The samples were annealed at 300°C for 150–200 h in an Ar atmosphere. The system was investigated through DTA and XRD.

Three ternary compounds were found in the Ge–I–Se system: $\mathbf{GeSeI_2}$, which melts at 170°C, $\mathbf{Ge_2Se_3I_2}$, which melts at 350°C and $\mathbf{Ge_2SeI_6}$ (Dembovskiy and Popova 1970).

The glass forming region in the Ge–I–Se ternary system is presented in Figure 5.78 (Dembovskiy and Popova 1970). Glasses adjacent to the glass formation boundary crystallize easily and are chemically unstable.

5.47 Germanium–Manganese–Selenium

$GeSe_2$–MnSe: The Mn_2GeSe_4 ternary compound, which crystallizes in the orthorhombic structure with the lattice parameters a = 1335.0 ± 0.3, b = 776.5 ± 0.2, c = 630.7 ± 0.1 pm, and a calculated density of 5.062 g·cm^{-3}, is formed in this system (Deiseroth et al. 2005). Homogeneous samples of the title compound can be prepared by direct reaction of intimate stoichiometric mixtures of the elements in evacuated sealed silica ampoules. To ensure a controlled reaction, the starting mixture was slowly heated (~20°C·h^{-1}) up to 600°C and kept at this temperature for 20 days. Subsequent cooling to room temperature (~20°C·h^{-1}) yielded a dark-grey raw product, which was grinded again. A final transport reaction under optimized conditions (temperature gradient 600°C/200°C, evacuated silica ampoule, $l \approx 15$ cm) in the presence of traces of I_2 yielded a well-crystallized solid with grey-metallic luster on the low-temperature side of the ampoule. The samples are not significantly air sensitive.

5.48 Germanium–Iron–Selenium

$GeSe_2$–FeSe: The $\mathbf{Fe_2GeSe_4}$ ternary compound, which crystallizes in the orthorhombic structure with the lattice parameters a = 1306.9 ± 0.1, b = 755.9 ± 0.1, c = 620.37 ± 0.06 pm, and a calculated density of 5.42 g·cm^{-3}, is formed in this system (Henao et al. 1998b). A polycrystalline sample of Fe_2GeSe_4 was synthesized by direct fusion of the elements in a stoichiometric ratio in a sealed and evacuated quartz ampoule. The mixture was heated slowly to 500°C over a period of 3 days. It was kept at this temperature for another 3 days. Subsequently, it was heated to 1100°C and remained at this temperature for 2 h. Finally, the reaction mixture was annealed at 500°C for a month. Then, the sample was cooled to room temperature in about a day.

The glass-forming region in the Ge–Fe–Se ternary system is limited to 10 at% Fe (Figure 5.79) (Ivanova and Vassilev 1997). The glass transition temperature varies from 116°C to 300°C. The glasses were prepared by melting the mixtures of the elements in evacuated to 10^{-2} Pa fused silica ampoules. A stepwise heating process in a rotating furnace was applied: at 250°C for 2 h, at 800°C for 8 h, and at 1000°C for 2 h. Then the temperature was reduced to 900°C at 15°C·min^{-1} and the mixture was quenched in iced water.

5.49 Germanium–Cobalt–Selenium

$GeSe_2$–CoSe: The phase diagram of this system, constructed through DTA, XRD, metallography and measuring of microhardness, is given in Figure 5.80 (Movsum-zade et al. 1984b). The eutectic contains 1 mol% CoSe and crystallizes at 700°C. Two compounds, $\mathbf{CoGeSe_3}$ and $\mathbf{CoGeS_4}$, which forms according to the solid-state reactions at 550°C and 660°C, respectively, are formed in this system. The temperature of CoSe phase transformation decreases from 842°C to 820°C with increasing of $GeSe_2$ content. There is an immiscibility region within the interval of 5–21 mol% $GeSe_2$ at 900°C. The solubility of $GeSe_2$ in CoSe is 2 mol% at room temperature and the solubility of CoSe in $GeSe_2$ is negligible. The alloys were annealed at 500°C for 200 h.

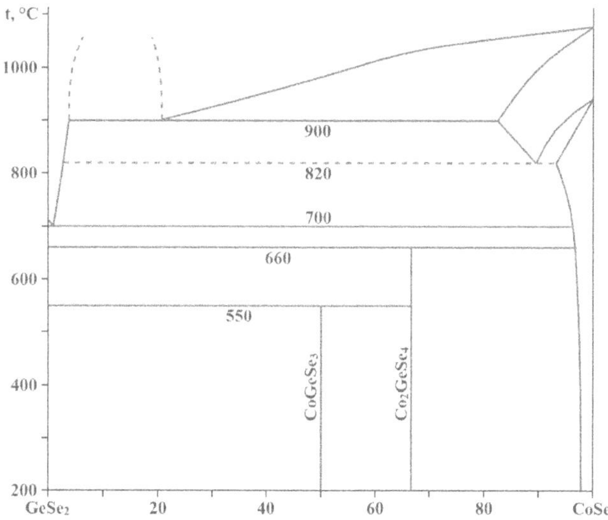

FIGURE 5.80 Phase diagram of the $GeSe_2$–CoSe system. (From Movsum-zade, A.A., et al., *Zhurn. neorgan. khimii*, **29**(11), 2891, 1984.)

The $\mathbf{CoGe_{1.5}Se_{1.5}}$ ternary compound, which decomposes at 800°C (Korenstein et al. 1977) and exists in two polymorphic modifications, is also formed in the Ge–Co–Se system. First modification of this compound crystallizes in the rhombohedral structure with the lattice parameters a = 829.99 ± 0.07 pm and α = 89.94 ± 0.01° or a = 1173.00 ± 0.08 and c = 1441.1 ± 0.3 pm in hexagonal setting and a calculated density of 6.643 g·cm^{-3} (Partik et al. 1996)]. Second modification crystallizes in the cubic structure with the lattice parameter a = 830.6 ± 0.8 pm and a = 830.1 ± 0.8 pm for $CoGe_{1.452}Se_{1.379}$ and $CoGe_{1.431}Se_{1.385}$ composition, respectively, and an energy gap of 0.336 eV (Nolas et al. 2003) [a = 822.9 ± 0.2 pm and the calculated and experimental densities of 6.65 and 6.62 ± 0.03 g·cm^{-3}, respectively (Korenstein et al. 1977)].

Polycrystalline samples of the title compound were prepared by annealing stoichiometric amounts of the elements in evacuated silica ampoule at 800°C (Partik et al. 1996). Well-formed small single crystals of this compound were grown by the chemical vapor transport method using I_2 as the transporting agent (Korenstein et al. 1977; Partik et al. 1996). The cubic modification was prepared by mixing and reacting high purity constituent elements at 700°C inside BN crucibles that were themselves sealed inside fused quartz ampoules under a N_2 atmosphere for several days (Nolas et al. 2003).

5.50 Germanium–Nickel–Selenium

The $\mathbf{Ni_{5.45-5.46}GeSe_2}$ ternary compound, which crystallizes in the tetragonal structure with the lattice parameters a = 359.9 ± 0.1, c = 1827.9 ± 0.5 pm at 140 K, and a = 360.6 ± 0.1, c = 1827.6 ± 0.5 pm, and a calculated density of 7.73 g·cm^{-3} at room temperature (Isaeva et al. 2007b) [a = 361.17 ± 0.04, c = 1828.1 ± 0.3 pm (Deiseroth et al. 2007a)], is formed in the Ge–Ni–Se system.

The sample of this compound was synthesized starting from the elements by high-temperature ceramic techniques (Isaeva et al. 2007b). The thoroughly ground mixture of the elements (the total mass of the sample was 1 g), corresponding to the stoichiometry Ni_7GeSe_2, was placed in quartz ampoule. Then the ampoule was sealed under vacuum (~7 Pa), annealed at 540°C–650°C for 14 days in a vertical furnace and water-quenched. According to the XRD, the sample consisted of four phases, and hence, the phase equilibrium was not achieved. The equilibrium sample containing only $Ni_{5.46}GeSe_2$ and Ni was obtained after the repeated annealing of pressed pellets (pressure load of about 0.2 GPa) under the same conditions. The total time of annealing was 90 days.

Crystals of the title compound were grown from molten flux starting from $PbCl_2$ (Isaeva et al. 2007b). The stoichiometric mixture of Ni, Ge, and Se (molar ratio 7:1:2) (total mass was ~1–2 g, the mass ratio of the charge to the flux was 1:1) was placed in cylindrical quartz ampoules, which was then sealed under vacuum. The ampoule was heated to 850°C for 12 h in a furnace. Then the furnace was slowly cooled in an automatic mode to 500°C at a rate of 1.5°C·h^{-1}. The cooling to room temperature was performed in the switched-off furnace. The sample was ground in a mortar and washed off from the flux by refluxing in hot water for 0.5–1 h. The washed sample contained small golden-colored crystals of the title compound as thin elongated plates.

Single crystals of $Ni_{5.46}GeSe_2$ could be prepared in a solid-state reaction starting from an intimate mixture of pure elements (molar ratio of Ni/Ge/Se was 5:1:2) and iodine as a transport agent, which was filled into a dry, evacuated and sealed quartz ampoule (Deiseroth et al. 2007a). The ampoule was placed with one end in the center of a tube furnace at 800°C for 5 days. The product consists of black square-shaped crystals which form preferably at the cold end of the ampoule at an estimated temperature of 600°C. This compound can be handled in air.

5.51 Germanium–Rhodium–Selenium

The $\mathbf{RhGe_{1.5}Se_{1.5}}$ ternary compound, which apparently exists in two polymorphic modifications, is formed in the Ge–Rh–Se system. First modification of this compound crystallizes in the rhombohedral structure with the lattice parameters a = 854.6 ± 0.1 pm and α = 89.86 ± 0.01° (Lutz and Kliche 1981)]. Second modification crystallizes in the trigonal structure with the lattice parameter a = 1206.01 ± 0.08 and c = 1482.6 ± 0.1 pm and a calculated density of 7.04 g·cm^{-3} (Liang et al. 2011).

The samples of the title compound were synthesized in two steps (Liang et al. 2011). First, stoichiometric amounts of powders of Rh, Ge, and Se were cold pressed to pellets and heated at 600°C for 120 h in carbon-glass crucibles inside the sealed quartz tubes. The quartz tube after evacuation was filled with Ar to a pressure of 60 kPa at room temperature. In the second step, the resulting mixture was powdered using a ball-mill and treated at 550°C for 2 h under spark-plasma-sintering conditions in Ar atmosphere. $RhGe_{1.5}Se_{1.5}$ could be also prepared by heating a stoichiometric mixture of the elements in closed silica tubes at 600°C for 14 days with next quenching in ice water (Lutz and Kliche 1981). The title compound is p-type semiconductor (Liang et al. 2011).

5.52 Germanium–Iridium–Selenium

The $\mathbf{IrGe_{1.5}Se_{1.5}}$ ternary compound, which crystallizes in the cubic structure with the lattice parameter a = 855.91 pm (Lyons et al. 1978), is formed in the Ge–Ir–Se system. This compound could be prepared by direct combination of the elements at 800°C under conditions of ambient pressure, or by heating the partially reacted elements at 600°C and 4 GPa (Lyons et al. 1978), or by heating a stoichiometric mixture of the elements in closed silica tubes at 600°C for 14 days with next quenching in ice water (Lutz and Kliche 1981).

5.53 Germanium–Platinum–Selenium

The **PtGeSe** ternary compound, which crystallizes in the orthorhombic structure with the lattice parameters $a = 601.5 \pm 0.4$, $b = 607.2 \pm 0.1$, and $c = 599.2 \pm 0.1$ pm, and a calculated density of 10.52 g·cm^{-3}, is formed in the Ge–Pt–Se system (Entner and Parthé 1973). It was prepared by heating the elements in evacuated quartz tubes at 600°C and 900°C for about four to six weeks. Some samples were ground after one week and reheated to enhance the attainment of equilibrium.

REFERENCES

All References are available as a downloadable eResource at www.routledge.com/9780367639235

6

Systems Based on Germanium Telluride

6.1 Germanium–Lithium–Tellurium

The $\mathbf{LiGeTe_2}$ ternary compound, which crystallizes in the triclinic structure with the lattice parameters a = 725.5 ± 0.3, b = 913.2 ± 0.4, c = 1134.0 ± 0.4 pm, and α = 75.75 ± 0.05°, β = 77.11 ± 0.05°, γ = 70.77 ± 0.05°, and a calculated density of 4.90 g·cm^{-3}, is formed in the Ge–Li–Te system (Eisenmann et al. 1983c). To obtain this compound, the stoichiometric mixture of Li_9Ge_4, Ge, and Te was heated slowly in a Ta crucible under an Ar atmosphere in a closed iron bomb with a rate of 50°·h^{-1} to 450°C, held there for 1 h, and then switched off the furnace. The result was a metallic dark-appearing regulus, which quickly disintegrated in moist air to form gray decomposition products of previously unknown stoichiometry. The substance was therefore handled under vigorously dried, heavy paraffinoil.

6.2 Germanium–Sodium–Tellurium

Two ternary compounds, $\mathbf{Na_6Ge_2Te_6}$ and $\mathbf{Na_8Ge_4Te_{10}}$, are formed in the Ge–Na–Te system. First of them crystallizes in the monoclinic structure with the lattice parameters a = 889.9 ± 0.4, b = 1294.6 ± 0.6, c = 890.8 ± 0.4 pm, β = 119.87 ± 0.05°, and a calculated density of 3.91 g·cm^{-3} (Eisenmann et al. 1984b). To obtain the title compound, the elements were weighed in a stoichiometric ratio under moisture and oxygen-free Ar atmosphere in a pre-dried at 150°C quartz ampoule. To compensate the loss of the reaction with the vessel material, potassium was taken in excess of about 10 mass%. The ampoule was melted under vacuum and in Ar-filled corundum tube with heating rate of 50°C·h^{-1} to 650°C and then the furnace was switched off. The crystals with a reddish hue were obtained. In the moist air, the decomposition of $K_6Ge_2Te_6$ begins immediately, recognizable by the smell of H_2Te. H_2Te is rapidly oxidized to Te, which precipitates on the crystal surfaces, so that in short time the crystals appear quite dark.

$Na_8Ge_4Te_{10}$ has two polymorphic modifications (Eisenmann et al. 1983b, 1983d). One of them crystallizes in the triclinic structure with the lattice parameters a = 1070.1 ± 0.5, b = 963.9 ± 0.5, c = 793.2 ± 0.4 pm, and α = 67.9 ± 0.1°, β = 70.7 ± 0.1°, γ = 68.3 ± 0.1°, and a calculated density of 4.236 g·cm^{-3}. The second modification crystallizes in the monoclinic structure with the lattice parameters a = 1412.8, b = 1293.8, c = 1502.8 pm, and β = 92.71°. This compound was prepared in the same way as $Na_6Ge_2Te_6$ was obtained.

6.3 Germanium–Potassium–Tellurium

Two ternary compounds, $\mathbf{K_2GeTe_4}$ and $\mathbf{K_2Ge_2Te_6}$, are formed in the Ge–K–Te system. K_2GeTe_4 melts at 405°C ± 10°C and crystallizes in the monoclinic structure with the lattice parameters a = 1270.3 ± 0.6, b = 868.0 ± 0.4, c = 982.9 ± 0.4 pm, β = 104.8 ± 0.1°, and the calculated and experimental densities of 4.19 and 4.23 g·cm^{-3}, respectively (Eisenmann et al. 1984c). This compound was prepared in the same way as $Na_6Ge_2Te_6$ was synthesized.

$K_2Ge_2Te_6$ also crystallizes in the monoclinic structure with the lattice parameters a = 1601.0 ± 0.8, b = 1361.9 ± 0.8, c = 971.3 ± 0.5 pm, β = 95.19 ± 0.05°, and the calculated and experimental densities of 3.61 and 3.60 g·cm^{-3}, respectively (Dittmar 1979). To obtain this compound, stoichiometric amounts of the elements were heated in the evacuated quartz tube to 630°C, annealed for 1 h and then cooled overnight to room temperature. Crystal pieces of metallic luster could be isolated from the gray-black regulus obtained.

6.4 Germanium–Rubidium–Tellurium

The $\mathbf{Rb_2GeTe_4}$ ternary compound with a calculated density of 4.53 g·cm^{-3} is formed in the Ge–Rb–Te system (Sheldrick and Schaaf 1995). To synthesize this compound, Rb_2CO_3 (0.76 mM), Ge (1.51 mM), and Te (3.03 mM) were slurried in 0.5 mL of MeOH and melted under Ar atmosphere in glass ampoule (degree of filling 5%). The ampoule was heated to 200°C, annealed at this temperature for 10 h, and then cooled down to room temperature at a rate of 1°C·h^{-1}. At the phase boundary, Rb_2GeTe_4 was obtained in crystalline form.

6.5 Germanium–Cesium–Tellurium

Three compounds, $\mathbf{Cs_2GeTe_4}$, $\mathbf{Cs_4GeTe_6}$, and $\mathbf{Cs_2Ge_2Te_6}$, are formed in the Ge–Cs–Te system. Cs_2GeTe_4 crystallizes in the orthorhombic structure with the lattice parameters a = 1260.2 ± 0.3, b = 1150.1 ± 0.2, c = 769.4 ± 0.2 pm, and a calculated density of 5.06 g·cm^{-3} (Sheldrick and Schaaf 1995). To prepare this compound, Cs_2CO_3 (0.69 mM), Ge (1.39 mM), and Te (2.77 mM) were slurried in 0.5 mL of MeOH and melted under Ar atmosphere in glass ampoule (degree of filling 5%). The ampoule was heated to 160°C, annealed at this temperature for 10 h, and then cooled down to room temperature at a rate of 1°C·h^{-1}. At the phase boundary, the title compound was obtained in crystalline form.

DOI: 10.1201/9781003123507-6

Cs_4GeTe_6 crystallizes in the monoclinic structure with the lattice parameters $a = 1784.5 \pm 0.8$, $b = 1398.1 \pm 0.6$, $c = 786.7 \pm 0.4$ pm, $\beta = 101.6 \pm 0.1°$, and a calculated density of 4.73 g·cm^{-3} (Brinkmann et al. 1985a). For the preparation of this compound, Ge and Te with $Cs(CH_3COO)$ (molar ratio 1:6:4) were weighed in a corundum crucible and placed in a corundum tube under Ar and heated at a rate of 50°C·h^{-1} to 900°C and cooled again at the same speed. The water resulting from the pyrolytic decomposition of the acetate was transferred to a connected flask with P_2O_5. The reaction product was a dull gray compact regulus with black glossy fracture surfaces. To avoid decomposition in moist air, the preparation was handled under dry, heavy paraffin oil. On the surface of the regulus and in caverns, fine, strip-like crystals were observed, which were often fused into bundles.

$Cs_2Ge_2Te_6$ also crystallizes in the monoclinic structure with the lattice parameters $a = 1702.7 \pm 0.2$, $b = 1423.7 \pm 0.1$, $c = 1010.4 \pm 0.1$ pm, $\beta = 96.701 \pm 0.008°$, and a calculated density of 4.665 g·cm^{-3} (Friede and Jansen 1999b). It has been prepared at the reacting of the elements in a stoichiometric ratio at 650°C for 1 h in an evacuated silica glass tube. To reduce losses of Cs, it was added in excess of 10%.

6.6 Germanium–Copper–Tellurium

GeTe–Cu: According to the data of DTA, XRD, metallography, and measuring of microhardness, this system is a non-quasibinary section of the Ge–Cu–Te ternary system and crosses the fields of Cu, Ge, $Cu_{2-x}Te$, and GeTe primary crystallization (Dogguy et al. 1977). The maximum solubility of Cu in GeTe at 550°C is 4.4 mol% (Gogishvili et al. 1988a). According to the electrical measurements, the solubility of Cu in $Ge_{0.975}Te$ at 400°C is 4.7×10^{20} cm^{-3}, and at 500°C is 5.6×10^{20} cm^{-3} (Gogishvili et al. 1982). In the region of low-temperature GeTe modification, the solubility of Cu is approximately 5.6×10^{20} cm^{-3} at 200°C and 300°C.

GeTe–Cu_2Te: The results of the investigation of this system are contradictory. According to the data of Abrikosov et al. (1970b) and Dovletov et al. (1974b), the phase diagram is a eutectic type (Figure 6.1). The eutectic contains 38 mol% Cu_2Te and crystallizes at 565°C (Abrikosov et al. 1970a) [at 570°C (Dovletov et al. 1974b)]. The solubility of GeTe in Cu_2Te and Cu_2Te in GeTe at the eutectic temperature reach 10–12 mol% and 5–6 mol%, respectively. With temperature decreasing, the solubility of Cu_2Te in GeTe decreases and at 550°C is 4 mol% (Abrikosov et al. 1980d). Thermal effects at 354°C and 304°C correspond to the polymorphic transformation of Cu_2Te (Abrikosov et al. 1970a; Dovletov et al. 1974b).

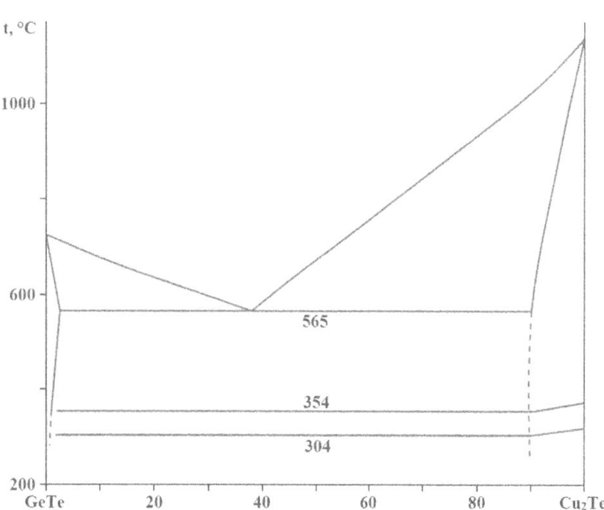

FIGURE 6.1 Phase diagram of the GeTe–Cu_2Te system. (From Abrikosov, N.H., et al., *Izv. AN SSSR. Neorgan. mater.*, **6**(5), 864, 1970.)

According to the Dogguy et al. (1977), GeTe–Cu_2Te is a non-quasibinary section of the Ge–Cu–Te ternary system since composition of $Cu_{2-x}Te$ at temperatures below the melting temperature changes in the direction of increasing tellurium content. Therefore, this section intersects several three-phase regions from the Cu_2Te-side in the Ge–Cu–Te system. On a vertical section, polymorphic transformations of $Cu_{2-x}Te$ are observed at 150°C, 250°C, 325°C, 350°C, and 420°C. The invariant line at 55°C corresponds to the ternary eutectic, and the thermal effects at 400°C correspond to the polymorphic transformation of GeTe.

The melts of the GeTe–Cu_2Te system are liquid semiconductors (Dovletov and Tashliev 1978).

This system was investigated using DTA, XRD, and metallography (Abrikosov et al. 1970a; Dovletov et al. 1974b; Dogguy et al. 1977). The ingots were annealed at 500°C for 1 month (Abrikosov et al. 1970a) or at the temperatures below the eutectic temperature for more than 250 h (Dovletov et al. 1974b).

A part of the isothermal section of the Ge–Cu–Te ternary system in the Cu–Cu_2Te–Cu_3Ge region at 130°C was constructed by Yusibov et al. (1991) and it was shown that Cu_2Te is in equilibrium with solid solutions based on copper, as well as ε_1 and ζ phases of the Cu–Ge system. There is an immiscibility region in the ternary system. The Cu_2Te–$Cu_{0.77}Ge_{0.23}$ (ε_1) and Cu_2Te–$Cu_{0.87}Ge_{0.73}$ (ζ) vertical sections were also constructed by these authors.

The projection of the liquidus surface of the Ge–Cu–Te ternary system is presented in Figure 6.2 (Dogguy et al. 1977).

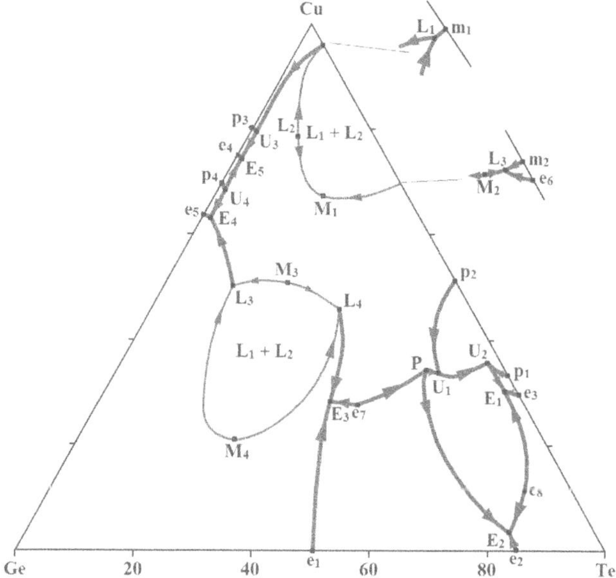

FIGURE 6.2 Projection of the liquidus surface of the Ge–Cu–Te ternary system. (From Dogguy, M., et al., *J. Less-Common Metals*, **51**(2), 181, 1977.)

The next invariant equilibria exist in the system: U_3 (823°C) – L + Cu ⇔ ζ + β-$Cu_{2-x}Te$; E_5 (740°C) – L ⇔ ε + ζ + β-$Cu_{2-x}Te$; U_4 (700°C) – L + ε ⇔ $ε_2$ + β-$Cu_{2-x}Te$; E_4 (638°C) – L ⇔ ε + β-$Cu_{2-x}Te$ + Ge; E_3 (555°C) – L ⇔ β-$Cu_{2-x}Te$ + β-GeTe + Ge; P (500°C) – L + β-$Cu_{2-x}Te$ + β-GeTe ⇔ Cu_2GeTe_3; U_1 (495°C) – L + β-$Cu_{2-x}Te$ ⇔ Cu_2GeTe_3 + Cu_4Te_3; U_2 (418°C) – L + Cu_4Te_3 ⇔ CuTe + Cu_2GeTe_3; E_2 (360°C) – L ⇔ α-$Cu_{2-x}Te$ + Te + Cu_2GeTe_3; and E_1 (338°C) – L ⇔ CuTe + Te + Cu_2GeTe_3 (ε, $ε_2$ and ζ are the binary phases of the Cu–Ge system).

The regions of the solid solutions based on GeTe and Cu_2Te (Figures 6.3 and 6.4) are elongated along the GeTe–Cu_2Te section (Abrikosov et al. 1970a, 1973a).

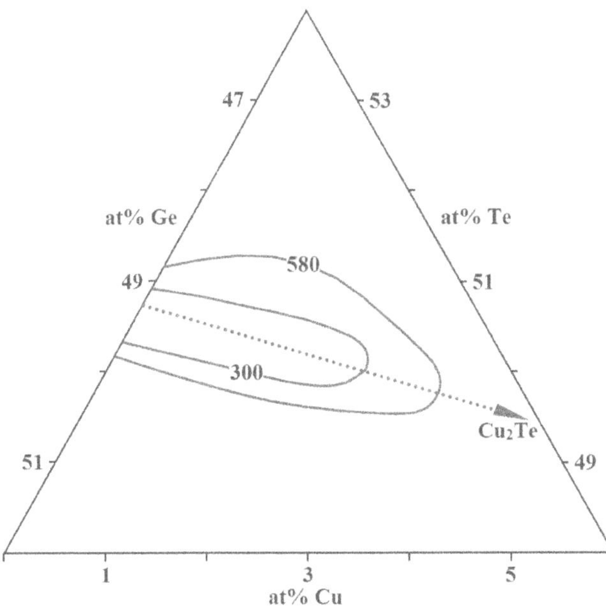

FIGURE 6.3 Solid solution region based on GeTe in the Ge–Cu–Te system at 300°C and 580°C. (From Abrikosov, N.H., et al., *Izv. AN SSSR. Neorgan. mater.*, **6**(5), 864, 1970.)

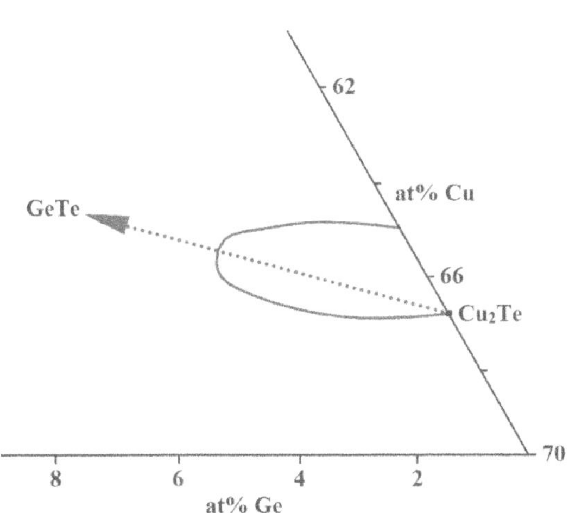

FIGURE 6.4 Solid solution region based on Cu_2Te in the Ge–Cu–Te system at 300°C. (From Abrikosov, N.H., et al., *Izv. AN SSSR. Neorgan. mater.*, **9**(1), 129, 1973.)

The $\mathbf{Cu_2GeTe_3}$ ternary compound, which melts at 504°C (Palatnik et al. 1961b; Villarreal et al. 2003) [at 595°C (Rivet et al. 1963; Rivet 1965); at 492°C (Averkieva et al. 1964)] and has a phase transition at 357°C (Villarreal et al. 2003), is formed in the Ge–Cu–Te system. One of the modification of this compound crystallizes in the orthorhombic structure with the lattice parameters a = 1264.06 ± 0.06, b = 421.15 ± 0.02, c = 592.61 ± 0.02 pm (Villarreal et al. 2003; Delgado et al. 2004). The second modification crystallizes in the tetragonal structure with the lattice parameters a = 595.9 and c = 592.9 pm (Sharma et al. 1977a) [a = 595.6, c = 592.6 pm, and the calculated and experimental densities of 6.13 and 5.95 ± 0.05 g·cm^{-3}, respectively (Rivet et al. 1963; Rivet 1965); a = 591.6, c = 1185 pm, and the calculated and experimental densities of 6.17 and 6.05 g·cm^{-3}, respectively (Hahn et al. 1966)] or in the cubic structure with the lattice parameters a = 595 pm and the calculated and experimental densities of 6.11 and 6.14 g·cm^{-3}, respectively (Palatnik et al. 1961a, b).

Light-gray crystals of Cu_2GeTe_3 were synthesized by direct fusion of its constituent elements, weighed in appropriate stoichiometric ratios inside quartz ampoule sealed in vacuum (Villarreal et al. 2003; Delgado et al. 2004). The ampoule was previously subjected to pyrolisis in order to avoid reaction of the starting materials with the quartz. In the first stage, the reaction process was carried out inside a furnace (in vertical position) heated at a rate of 60°C·h^{-1} up to 900°C. Then the compound was kept at that temperature for 276 h. Finally, the sample was cooled to room temperature at a rate of 30°C·h^{-1} or at a rate of 6°C·h^{-1} during 2 days.

According to the data of Glazov et al. (1976), the Cu_2GeTe_3 ternary compound was not found at the investigation of the phase equilibria along the Cu_2Te–"$GeTe_2$" section.

The glass forming region in the Ge–Cu–Te ternary system is presented in Figure 6.5 (Minaev 1983). Synthesis of the glassy alloys was carried out from the elements in quartz ampoules under vacuum at 1000°C for 12 h with vigorous stirring and

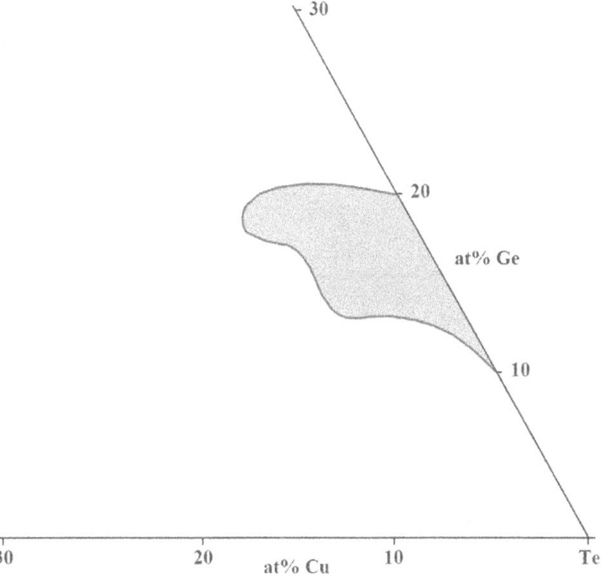

FIGURE 6.5 Glass forming region in the Ge–Cu–Te ternary system. (From Minaev, V.S., *Fiz. i khim. stekla*, **9**(4), 432, 1983.)

the next quenching in water. The vitreous state was identified by a characteristic glass wedge, a shell fracture, as well as through XRD, DTA, and metallography. The boundary of the glass formation region was fixed with an accuracy of 2 at%.

6.7 Germanium–Silver–Tellurium

GeTe–Ag_2Te: This system is a non-quasibinary section of the Ge–Ag–Te ternary system (Blachnik and Gather 1978a; Plachkova et al. 1981, 1983, 1984). The temperature of the β-Ag_2Te → γ-Ag_2Te transition increases from 803°C to 830°C and the temperature of the α-Ag_2Te → β-Ag_2Te transition decreases from 147°C to 139°C at the increase of the GeTe content. The **$8Ag_2$Te·3GeTe** ternary compound, which melts incongruently at 610°C and has an experimental density of 7.34 g·cm^{-3}, is formed in this system. This compound could be the solid solution based on **Ag_8GeTe_6**. The mutual solubility of GeTe and Ag_2Te is not higher than 2–3 mol%.

According to the data of Dovletov et al. (1974b), the phase diagram of the GeTe–Ag_2Te system is a eutectic type, which is less likely. The eutectic contains 40 mol% 3GeTe (the system is considered as 3GeTe–2Ag_2Te) and crystallizes at 600°C. The solubility of 3GeTe in 2Ag_2Te reaches 10 mol% and the solubility of 2Ag_2Te in 3GeTe is equal 15 mol%. The polymorphic transformation of Ag_2Te takes place at 160°C.

The isothermal section of the Ge–Ag–Te ternary system at room temperature is presented in Figure 6.6 (Blachnik and Gather 1978a). The triangulation of the system is determined by the Ag_8GeTe_6 compound. The region of the GeTe-based solid solutions is oriented along the GeTe–Ag_2Te section (Figure 6.7); moreover, with an Ag content of up to 0.25 at%, the properties of the alloys change differently than with a larger amount of Ag (Rogacheva et al. 1985b). According to

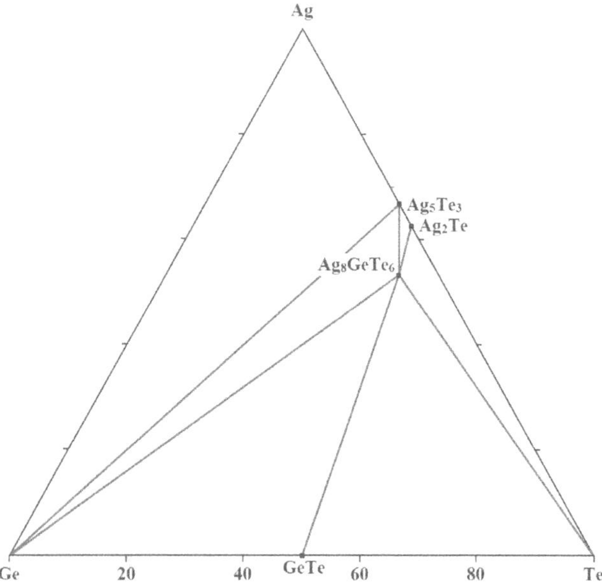

FIGURE 6.6 Isothermal section of the Ge–Ag–Te ternary system at room temperature. (From Blachnik, R., and Gather, B., *J. Less-Common Metals*, **60**(1), 25, 1978.)

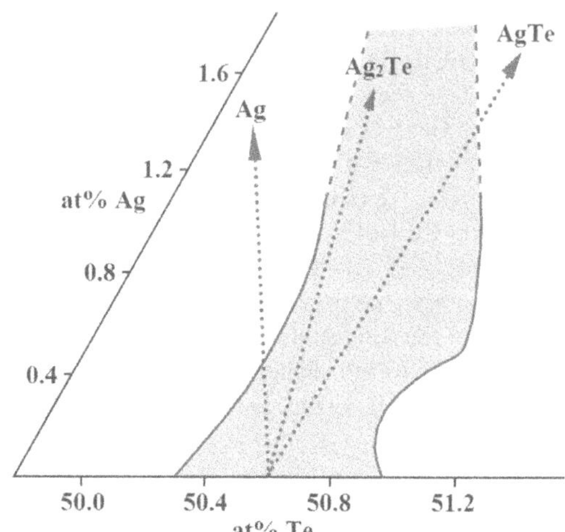

FIGURE 6.7 Homogeneity region based on GeTe in the Ge–Ag–Te ternary system at 550°C. (From Rogacheva, E.I., et al., *Izv. AN SSSR. Neorgan. mater.*, **21**(3), 397, 1985.)

the data of Gorochov (1968a), the solid solutions based on GeTe along the **GeTe–AgTe** section were not found.

The Ag_8GeTe_6 ternary compound, which melts incongruently at 645°C and has three polymorphic modifications (Gorochov and Flahaut 1967; Gorochov 1968b; Rysanek et al. 1976; (Katty et al. 1981), is formed in this system. α- and β-Ag_8GeTe_6 are stable below −52°C and −29°C, respectively, and γ-Ag_8GeTe_6 is stable at ambient conditions. According to the data of Katty et al. (1981), the phase transitions take place at 173.3 K, 223.6 K (ΔH_{tr} = 2730 ± 100 J·M^{-1}) and 245.3 K (ΔH_{tr} = 1740 J·M^{-1}).

γ-Ag_8GeTe_6 crystallizes in the cubic structure with the lattice parameter a = 1156.56 ± 0.04 pm and a calculated density of 7.304 g·cm^{-3} (Boucher et al. 1993) [a = 1157.0 pm and the calculated and experimental densities of 7.29 and 7.22 g·cm^{-3}, respectively (Gorochov and Flahaut 1967; Gorochov 1968b); a = 1156.6 ± 0.2 pm and the calculated and experimental densities of 7.30 and 7.22 g·cm^{-3}, respectively (Rysanek et al. 1976); a = 1158 ± 2 pm and a calculated density of 7.28 g·cm^{-3} (Geller 1979); a = 1156 ± 6 pm and the calculated and experimental densities of 7.31 ± 0.7 and 7.22 g·cm^{-3}, respectively (Katty et al. 1981)]. According to the data of Unterrichter and Range (1978), this compound (may be β-Ag_8GeTe_6) crystallizes in the rhombohedral structure with the lattice parameters a = 817.6 ± 0.2 pm and α = 60.0° and a calculated density of 7.307 g·cm^{-3}.

γ-Ag_8GeTe_6 was prepared from a mixture of the elements taken in stoichiometric proportion and placed in a carbon tube which was enclosed in an evacuated silica tube (Geller 1979; Kuhs et al. 1979; Boucher et al. 1993). After a heating the mixture at 800°C for a week, the sample was quenched in melting ice to prevent any peritectic decomposition. Additionally, a 1 week annealing was made. Single crystals were grown using chemical transport reactions with I_2 as a transport agent.

According to the data of Balanevskaya et al. (1966), the Ag_2GeTe_3 ternary compound, which melts at 330°C, has an experimental density of 5.12 g·cm^{-3} and an energy gap of 0.25 eV, is also formed in the Ge–Ag–Te system, but this compound was not found by Blachnik and Gather (1978a).

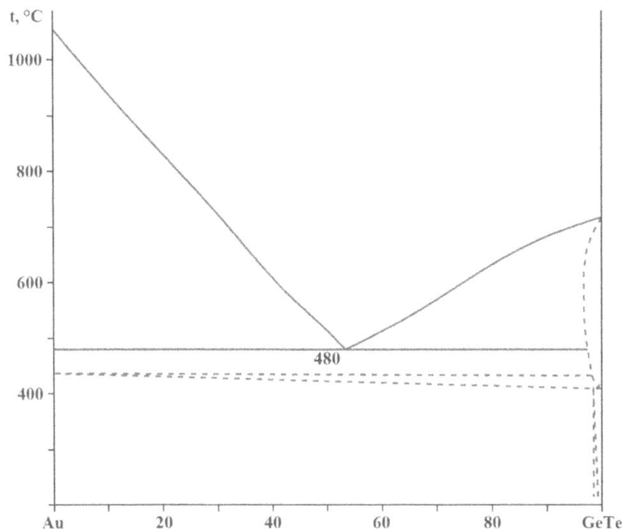

FIGURE 6.8 Phase diagram of the GeTe–Au system. (From Legendre, B., et al., *C. r. Acad. sci. Sér. C*, **284**(12), 451, 1977.)

This system was investigated using DTA, XRD, and metallography (Gorochov 1968b; Dovletov et al. 1974b; Blachnik and Gather 1978a; Plachkova et al. 1983, 1984). The ingots were annealed at the temperatures below the eutectic temperature for more than 250 h (Dovletov et al. 1974b).

6.8 Germanium–Gold–Tellurium

GeTe–Au: The phase diagram of this system is a eutectic type (Figure 6.8) (Legendre et al. 1977). The eutectic contains 46.66 mol% Au and crystallizes at 480°C.

GeTe–AuTe$_2$: The phase diagram of this system is also a eutectic type (Figure 6.9) (Legendre et al. 1977). The eutectic contains 40 mol% GeTe and crystallizes at 400°C.

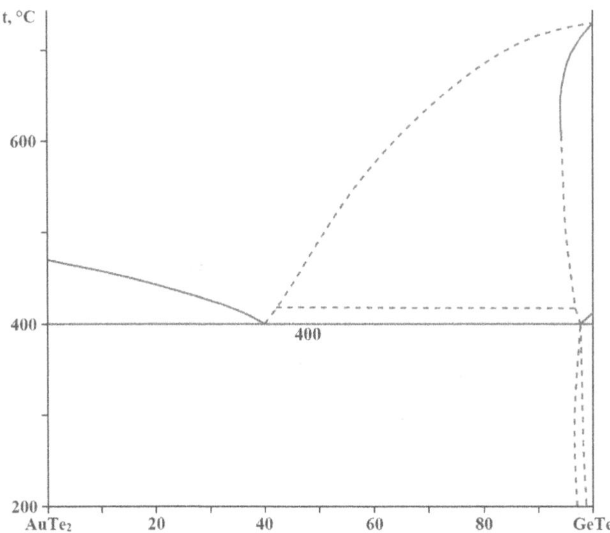

FIGURE 6.9 Phase diagram of the GeTe–AuTe$_2$ system. (From Legendre, B., et al., *C. r. Acad. sci. Sér. C*, **284**(12), 451, 1977.)

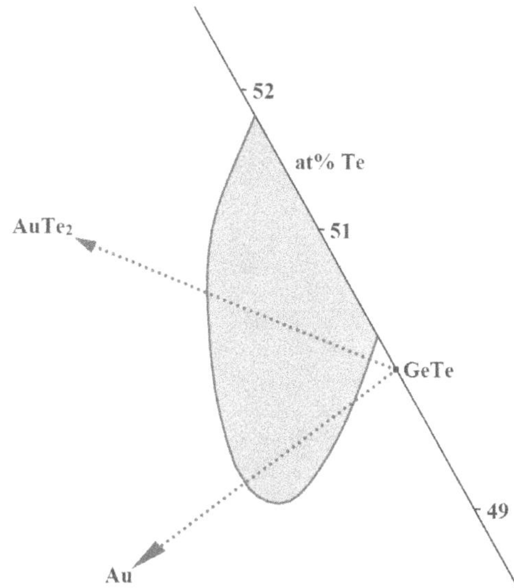

FIGURE 6.10 Region of the solid solutions based on GeTe in the Ge–Au–Te ternary system at 300°C. (From Legendre, B., et al., *C. r. Acad. sci. Sér. C*, **284**(12), 451, 1977.)

The region of the solid solutions based on GeTe in the Ge–Au–Te ternary system at 300°C is shown in Figure 6.10 (Legendre et al. 1977). This system was investigated using DTA, XRD, metallography, and measuring of the microhardness. The ingots were annealed at 300°C for 6000 h.

6.9 Germanium–Barium–Tellurium

The **Ba$_2$Ge$_2$Te$_5$** ternary compound, which crystallizes in the orthorhombic structure with the lattice parameters $a = 1339.7 \pm 0.5$, $b = 917.7 \pm 0.3$, $c = 995.0 \pm 0.3$ pm, and the calculated and experimental densities of 5.74 and 5.76 g·cm^{-3}, respectively, is formed in the Ge–Ba–Te system (Brinkmann et al. 1984).

6.10 Germanium–Zinc–Tellurium

GeTe–ZnTe: The phase diagram of this system is a eutectic type (Figure 6.11) (Glazov et al. 1970, 1972, 1975). The eutectic temperature is 705°C. Interaction of the low-temperature GeTe modification with ZnTe is described by the eutectoid equilibrium at 380°C. Maximum mutual solubility of ZnTe and GeTe is not higher than 5 mol% (Glazov et al. 1972). According to the data of Kutsia and Stavrianidis (1983), the solubility of ZnTe in GeTe at 530°C and 230°C is 2.5 and 2 mol%, respectively Ternary compounds in the Ge–Zn–Te system were not found (Kaldis et al. 1967).

This system was investigated using DTA, metallography, and XRD (Glazov et al. 1970, 1972, 1975; Kutsiya and Stavrianidis 1983). The ingots were annealed at 530°C, 430°C, 330°C, and 230°C for 200, 400, 600, and 800 h, respectively.

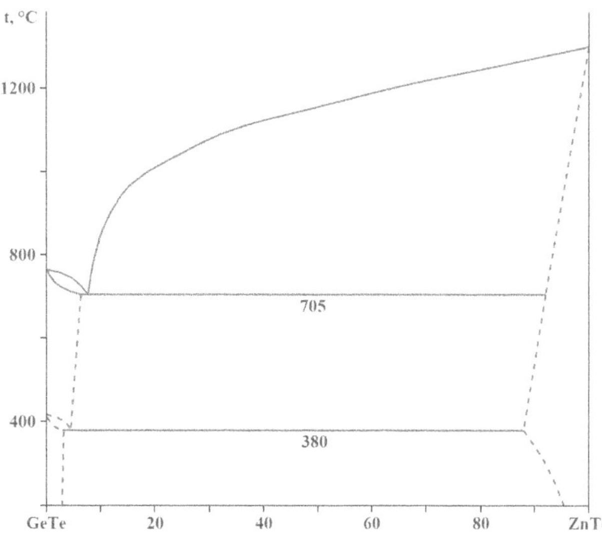

FIGURE 6.11 Phase diagram of the GeTe–ZnTe system. (From Glazov, V.M. et al., *Izv. AN SSSR. Neorgan. mater.*, **6**(3), 569, 1970.)

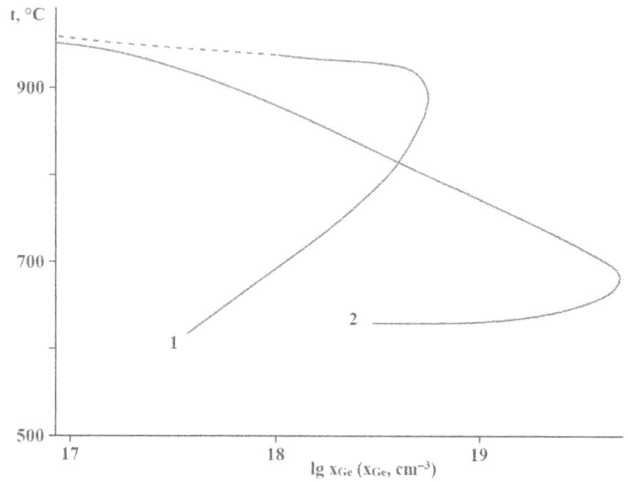

FIGURE 6.13 Temperature dependences of Ge solubility in CdTe: 1, p_{Cd}^{max}; 2, p_{Cd}^{min}. (From Panchuk, O.E. and Shcherbak, L.P., *Izv. AN SSSR. Neorgan. mater.*, **15**(8), 1339, 1979.)

6.11 Germanium–Cadmium–Tellurium

"$CdGeTe_3$"–CdTe: The mixture of two or more phases with different crystal lattices is formed in this system at the alloying of "$CdGeTe_3$" and CdTe (Radautsan and Ivanova 1961). This system was investigated using metallography, XRD, and measuring of the microhardness.

Ge–CdTe: The phase diagram belongs to the eutectic type (Figure 6.12) (Panchuk et al. 1976; Dichi et al. 1995; Shcherbak et al. 1997). The monotectic lies at 1074°C in the 4–83 at% Ge composition range (Shcherbak et al. 1997) [the immiscibility region exists within the interval of 7–93 at. % Ge with monotectic temperature 1050°C (Panchuk et al. 1976)]. Dichi et al. (1995) did not observe the immiscibility region, but they did not use the metallography. The position of the liquidus on CdTe–Ge phase diagram indicates the presence of immiscibility region. Panchuk et al. (1976) found four phases (CdTe, GeTe, Cd, and Ge) in the CdTe–Ge system that can be explained by the nonequilibrium state of the alloys. However, the stability of monotectic temperature is an evidence of this system quasibinarity. The eutectic composition and temperature are 2.5 mol% CdTe and 924°C, respectively (Shcherbak et al. 1997) [3 mol% CdTe and 908°C (Panchuk et al. 1976); according to the data of Dichi et al. (1995), the eutectic is degenerated and crystallizes at 921°C]. Thermal effects at 720°C from Panchuk et al. (1976) correspond to the melting of GeTe that is formed as a result of the nonequilibrium in this system.

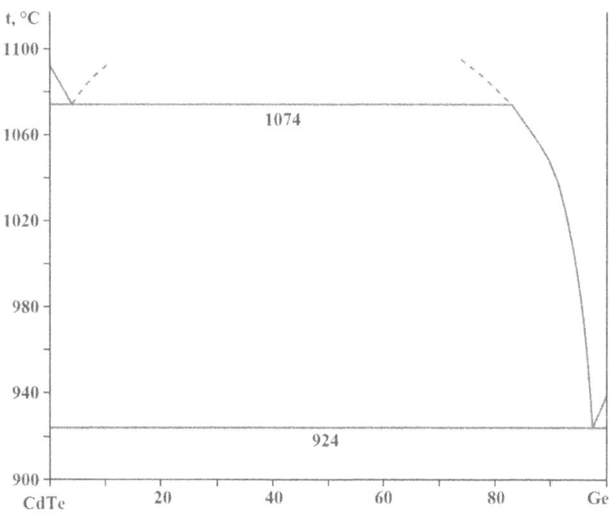

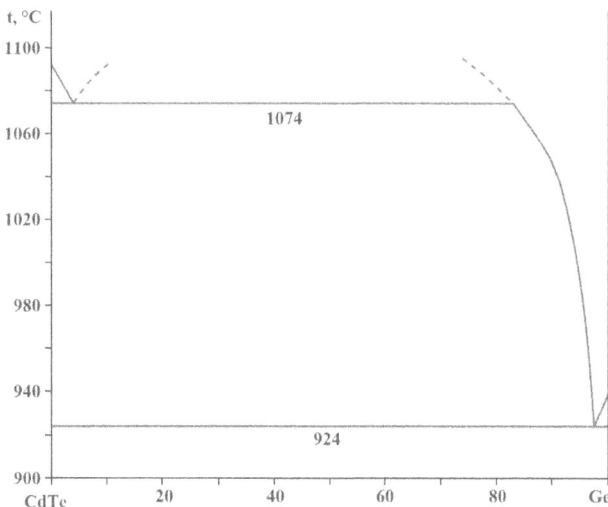

FIGURE 6.12 Phase diagram of the CdTe–Ge system. (From Shcherbak, L. et al., *Calphad*, **21**(4), 463, 1997.)

The maximum mutual solid solubility of CdTe and Ge is less than 1 mol% (Shcherbak et al. 1997). Temperature dependences of Ge solubility in CdTe at maximum and minimum cadmium vapor pressure are retrograde (Figure 6.13) (Panchuk and Shcherbak 1979). The maximum solubility of Ge at minimum and maximum cadmium vapor pressure (4.10^{19} and 4.10^{18} cm^{-3}) corresponds to 680°C and approximately 900°C, respectively. The solubility of Ge (cm^{-3}) in CdTe at p_{Cd}^{max} up to maximum x_{Ge} value can be described by the next equation: $x_{Ge} = 1.48.10^{23} \exp(E/kT)$, where $E = -1.00$ eV.

This system was investigated using DTA, DSC, metallography, XRD, and microhardness measurement supplemented by optical and scanning electron microscopy combined with EPMA (Panchuk et al. 1976; Dichi et al. 1995; Shcherbak et al. 1997). The solubility of Ge in CdTe was determined using radioactive isotopes (Panchuk and Shcherbak 1979).

GeTe–CdTe: The results of this system investigation are contradictory. According to the data of Glazov et al. (1970, 1972, 1975) and Dichi et al. (1995), the phase diagram is a eutectic type (Figure 6.14a). The eutectic composition and temperature are 3.5 mol% CdTe and 723°C, respectively (Dichi et al. 1995) [700°C (Glazov et al. 1970, 1972, 1975)]. The solubility of CdTe in GeTe at 230°C and 530°C is equal to 2 and 3 mol%, respectively (Kutsia and Stavrianidis 1983). This section loses its quasibinary character in the vicinity of the GeTe solid solution (Dichi et al. 1995). The plateau at 400°C

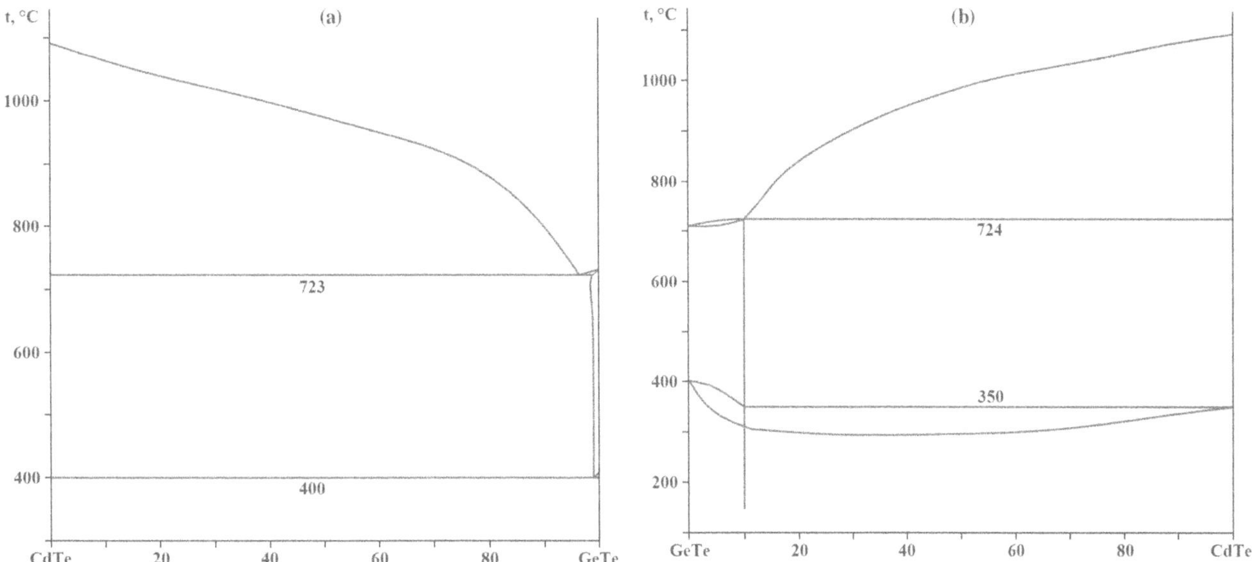

FIGURE 6.14 Phase diagram of the CdTe–GeTe system. (a, from Dichi, E. et al., *J. Alloys Compd.*, **217**(2), 193, 1995; b, from Quenez, P. and Khodadad, P., *Bull. soc. Chim. France*, (1), 3, 1969.)

characterizes the phase transition of GeTe according to the reaction of *U*-type as follows: β-GeTe + Ge ⇔ α-GeTe + CdTe.

According to the data of Quenez and Khodadad (1969b), the phase diagram is a peritectic type with the peritectic temperature 724°C (Figure 6.14b). Two regions of α- and β-solid solutions, based on GeTe and limited by the $CdGe_9Te_{10}$ composition, were found in the GeTe–CdTe system. α-β-Phase transition for the $CdGe_9Te_{10}$ alloy takes place at 350°C.

This system was investigated using DTA, DSC, XRD, and metallography (Quenez and Khodadad 1969b; Glazov et al. 1970, 1972, 1975; Kutsia and Stavrianidis 1983; Dichi et al. 1995). The ingots were annealed at 530°C, 430°C, 330°C, and 230°C for 200, 400, 600, and 800 h, respectively (Kutsia and Stavrianidis 1983).

The scheme of Ge–Cd–Te liquidus surface is shown in Figure 6.15 (Dichi et al. 1995). No ternary compound was observed in this system. So the ternary system is divided into three sub-ternaries (CdTe–GeTe–Te, CdTe–GeTe–Ge, and CdTe–Cd–Ge), each of which is characterized by a ternary invariant. Two ternary eutectics and one transition point have been found. A ternary metatectic reaction has been observed owing to the occurrence of the phase transition in the GeTe. The glass area covers largely the sub-ternary CdTe–GeTe–Te system.

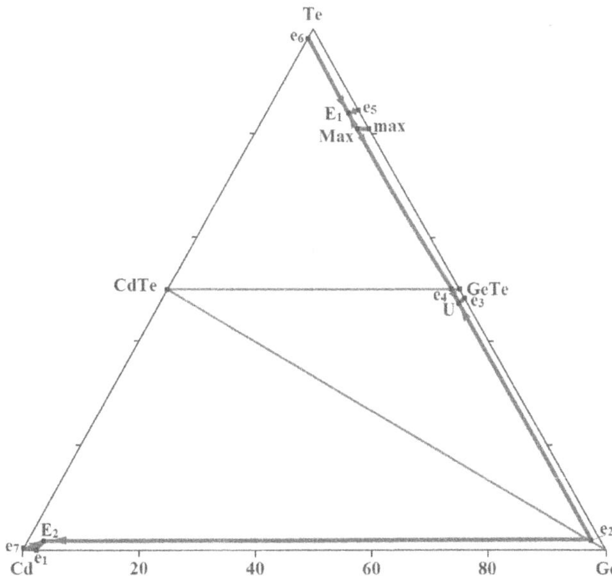

FIGURE 6.15 Scheme of the Ge–Cd–Te liquidus surface. (From Dichi, E. et al., *J. Alloys Compd.*, **217**(2), 193, 1995.)

6.12 Germanium–Mercury–Tellurium

GeTe–HgTe: The phase diagram is a eutectic type (Figure 6.16) (Glazov et al. 1970, 1972, 1975). The eutectic temperature is 560°C. At 340°C, there is a eutectoid transformation in this system. The mutual solubility of GeTe and HgTe is not higher

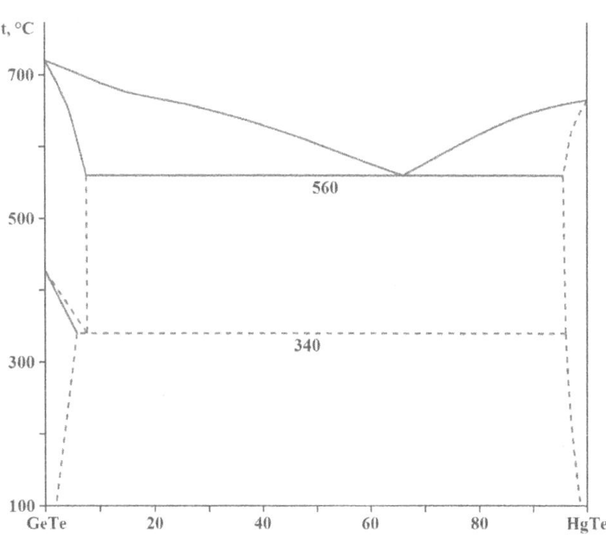

FIGURE 6.16 Phase diagram of the GeTe–HgTe system. (From Glazov, V.M. et al., *Izv. AN SSSR. Neorgan. mater.*, **6**(3), 569, 1970.)

than 4 mol%. This system was investigated using DTA, metallography, and XRD.

6.13 Germanium–Boron–Tellurium

The glass forming region in the Ge–B–Te ternary system was not found (Kirilenko and Dembovskiy 1974).

6.14 Germanium–Gallium–Tellurium

GeTe–Ga: This system is a non-quasibinary section of the Ge–Ga–Te ternary system (Zargarova 1972; Zargarova and Akperov 1972; Kra et al. 1978). Liquidus determines the Ge primary crystallization and has a maximum related to the intersection of the GeTe–GaGeTe$_2$–Ge and Ge–GaGeTe$_2$–Ga subsystems. The alloys of the first subsystem end the crystallization in the ternary eutectic at 630°C. The **GaGeTe** ternary compound, which melts congruently at 800°C and crystallizes in the rhombohedral structure with the lattice parameters a = 405 and c = 3465 pm (in hexagonal setting) and the calculated and experimental densities of 5.46 and 5.3 g·cm^{-3}, respectively (Kra et al. 1978), and an energy gap of 1.12 eV (Fenske and Schnering 1983a, b; Kucek et al. 2013), is formed in the GeTe–Ga system. The synthesis of this compound was carried out in a horizontal furnace at a temperature of 950°C for 2 h (Fenske and Schnering 1983a, b; Kucek et al. 2013). The growth of the crystal was carried out in the same ampoule using a modified Bridgman technique. A conical quartz ampoule containing the synthesized polycrystalline material was placed in the upper hot part of the Bridgman furnace, where it was annealed at 980°C for 24 h. It was then lowered into a temperature gradient 80°C·cm^{-1} at a rate 1.3 mm·h^{-1}. This procedure provided ingots 50 mm in length and 10 mm in diameter.

The **Ga$_2$GeTe** ternary compound was not found in this system (Zargarova 1972; Zargarova and Akperov 1972).

This system was investigated using DTA, XRD, metallography, and measuring of microhardness and density (Zargarova 1972; Zargarova and Akperov 1972; Kra et al. 1978). The ingots were annealed at 640°C for 1–4 weeks (Kra et al. 1978).

GeTe–GaTe: The results of this system investigation are contradictory. The most reliable are the information that the system is a non-quasibinary section of the Ge–Ga–Te ternary system (Kra et al. 1977, 1978).

According to the data of Zargarova and Akperov (1972), the phase diagram of this system is a eutectic type. The eutectics contain 38 and 58 mol% GaTe and crystallize at 610°C and 690°C, respectively. A peritectic point exists at 62 mol% GaTe and 630°C. From the GeTe-side at 340°C, a eutectoid transformation is observed due to the polymorphic transformation of solid solutions. The solubility of GaTe in GeTe reaches 6–7 mol% (Nasirov et al. 1968a; Zargarova and Akperov 1972; Sysoeva et al. 1970). Two compounds, **GaGeTe$_2$** and **Ga$_2$GeTe$_3$**, were found in the GeTe–GaTe system (Nasirov et al. 1968a, 1970; Khalilov et al. 1969; Zargarova and Akperov 1972). Fist of them melts congruently (Nasirov et al. 1968a; Zargarova and Akperov 1972) and crystallizes in the hexagonal structure with the lattice parameters a = 402 and c = 1682 pm (Nasirov et al. 1970). The second compound melts incongruently (Nasirov et al. 1968a; Zargarova and Akperov 1972) and also crystallizes in the hexagonal structure with the lattice parameters a = 405 and c = 1687 pm (Nasirov et al. 1970). Single crystals of both compounds were grown by the Bridgman method (Nasirov et al. 1970). According to the data of Kra et al. (1977), these compounds were not found in the system.

This system was investigated using DTA, XRD, metallography, and measuring of microhardness and density (Zargarova 1972; Zargarova and Akperov 1972; Sysoeva et al. 1970). The ingots were annealed at 640°C for 1–4 weeks (Kra et al. 1978) [at 200°C for 200 h (Khalilov et al. 1969)].

GeTe–Ga$_2$Te$_3$: The phase diagram of this system is a eutectic type (Figure 6.17) (Abrikosov et al. 1976; Babayev 2002). The eutectic contains 30 mol% Ga$_2$Te$_3$ and crystallizes at 636°C [26 mol% Ga$_2$Te$_3$ and crystallizes at 650°C (Kra et al. 1977, 1978); 30 mol% Ga$_2$Te$_3$ and crystallizes at 630°C (Rogacheva et al. 1974b)]. With the increasing of Ga$_2$Te$_3$ content, the temperature of the GeTe polymorphic transformation decreases to 320°C (Abrikosov et al. 1976) [to 325°C (Rogacheva et al. 1974b)]. The solubility of Ga$_2$Te$_3$ in GeTe at 600°C reaches 2 mol% and decreases to 1 mol% at 250°C (Abrikosov et al. 1976) [reaches 5 mol% (Sysoeva et al. 1970); is not higher than 0.5 mol% (Rogacheva et al. 1974b)]. The solubility of GeTe in Ga$_2$Te$_3$ at 600°C reaches 5 mol% and decreases to 2 mol% at 250°C (Abrikosov et al. 1976).

This system was investigated using DTA, XRD, metallography, and measuring of microhardness and density (Rogacheva et al. 1974b; Abrikosov et al. 1976; Kra et al. 1977, 1978). The ingots were annealed at 250°C, 400°C, 500°C, and 600°C (Abrikosov et al. 1976) [at 640°C for 1–4 weeks (Kra et al. 1978); at 300°C for 480 h Rogacheva et al. 1974b].

The region of solid solutions based on GeTe in the Ga–Ge–Te ternary system at 600°C (Figure 6.18) is situated along the GeTe–Ga$_2$Te$_3$ section (Abrikosov et al. 1976; Abrikosov and Danilova-Dobryakova 1977; Melikhova et al. 1980b). In the region of the solid solution, the rhombohedral structure of

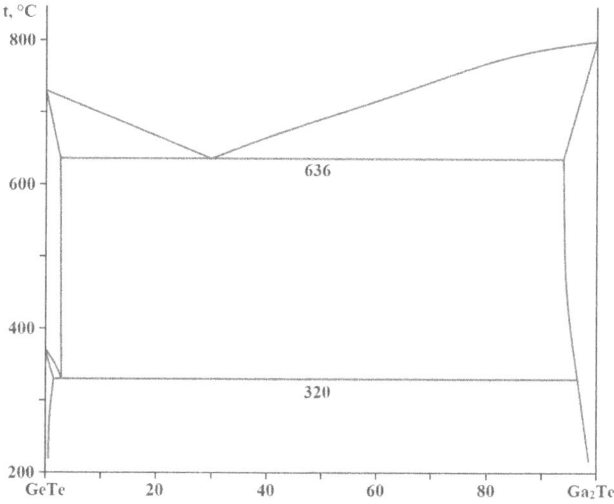

FIGURE 6.17 Phase diagram of the GeTe–Ga$_2$Te$_3$ system. (From Abrikosov, N.H., et al., *Izv. AN SSSR. Neorgan. mater.*, **12**(4), 605, 1976.)

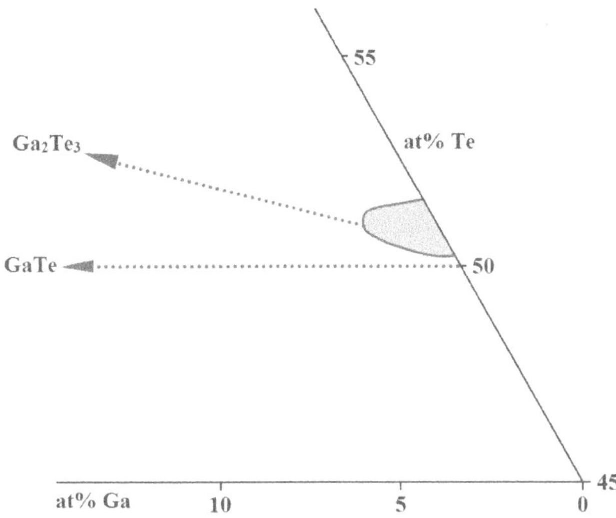

FIGURE 6.18 Region of solid solutions based on GeTe in the Ga–Ge–Te ternary system at 600°C. (From Abrikosov, N.H., et al., *Izv. AN SSSR. Neorgan. mater.*, **12**(4), 605, 1976.)

GeTe is retained with a slight increase in the angle of rhombohedrality α and a slight decrease in the volume of the elementary lattice.

The liquidus surface of the Ge–Ga–Te ternary system (Figure 6.19) includes ten fields of primary crystallization of Ga (this field is degenerated), Ge, Te, GeTe, Ga_3Te_2, GaTe, Ga_2Te_3, $GaTe_3$, $GaGeTe_2$, and Ga_2GeTe_3 (Babayev 2004a; Zargarova and Akperov 1972). There are six ternary eutectics, four transition points and immiscibility region on the liquidus surface. According to the data of Babayev and Bababeyli (2006), the elliptic form of the immiscibility region has been due to stability, prevalence and theoretical advantage of such form.

The excess enthalpies of the liquid Ga–Ge–Te alloys were measured in a heat-flow calorimeter at 930°C (Gather et al. 1987). The enthalpy surface (Figure 6.20) is characterized by

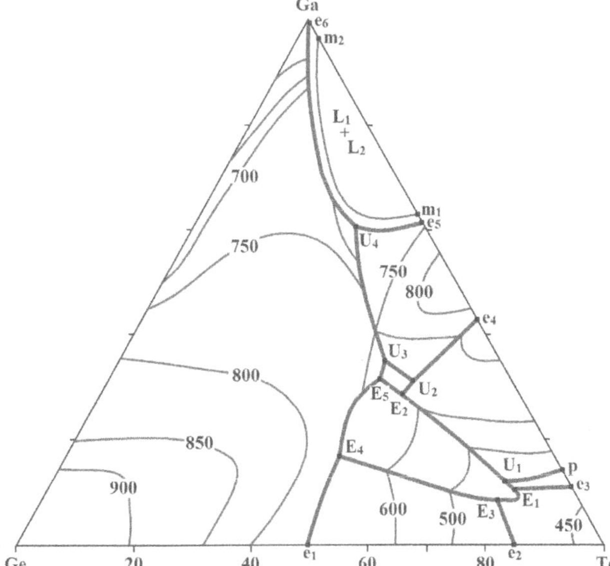

FIGURE 6.19 Liquidus surface of the Ge–Ga–Te ternary system. (From Zargarova, M.I., and Akperov, M.M., *Azerb. khim. zhurn.*, (2), 124, 1972.)

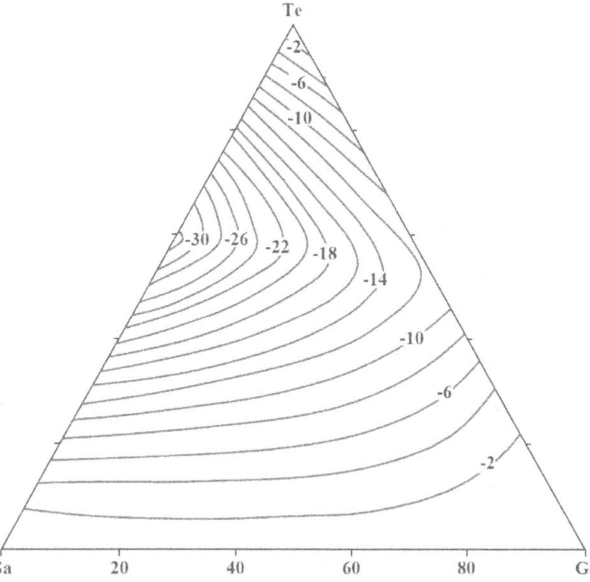

FIGURE 6.20 Isoenthalpic lines (in kJ·mol^{-1}) in the Ge–Ga–Te ternary system at 930°C. (From Gather, B., et al, *J. Less-Common Metals*, **136**(1), 183, 1987.)

a valley stretching from the exothermic minimum in the Ga–Te system to the minimum of the Ge–Te systems. The minimum in the ternary systems was found in this valley, i.e. on the sections GeTe–Ga_2Te_3. A comparison of the experimental data with those calculated from the excess enthalpies of the constituent binaries with the aid of the Bonnier model, reveals only small deviations.

According to the data of Drasar et al. (2015), another ternary compound, $\mathbf{Ga_7Ge_3Te_{11}}$, which crystallizes in the orthorhombic structure with the lattice parameters $a = 3890.3 \pm 0.6$, $b = 408.77 \pm 0.07$, $c = 1537.3 \pm 0.3$ pm, a calculated density of 5.733 g·cm^{-3}, and an energy gap of ≈0.7 eV, is formed in the Ga–Ge–Te ternary system. The synthesis of this compound was carried out in a horizontal furnace at a temperature of 900°C for 2 h. The growth of the crystal was carried out in the same ampoule using a transport in a small temperature gradient at an average temperature of 690°C. Various compositions were examined to find a composition suitable for the growth of single crystals. The largest single crystals were obtained for the composition $Ga_8Ge_2Te_{11}$.

6.15 Germanium–Indium–Tellurium

GeTe–In: This system is a non-quasibinary section of the Ge–In–Te ternary system (Zargarova 1972; Zargarova and Akperov 1973). The $\mathbf{In_2GeTe}$ ternary compound was not found in this system.

GeTe–InTe: The phase diagram of this system is a eutectic type with a peritectic transformation (Figure 6.21) (Nasirov et al. 1968b; Zargarova and Akperov 1973). The solubility of InTe in GeTe reaches 15 mol% (Nasirov et al. 1968b) and decreases to 10 mol% at room temperature (Zargarova and Akperov 1973; Sysoeva et al. 1970) [8 mol% at 400°C (Woolley 1965)]. The solubility of GeTe in InTe at room temperature

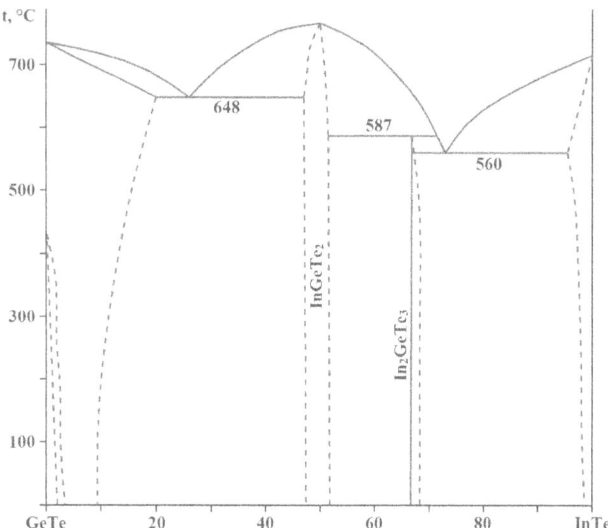

FIGURE 6.21 Phase diagram of the GeTe–InTe system. (From Zargarova, M.I., et al., *Izv. AN SSSR. Neorgan. mater.*, **9**(7), 1138, 1973.)

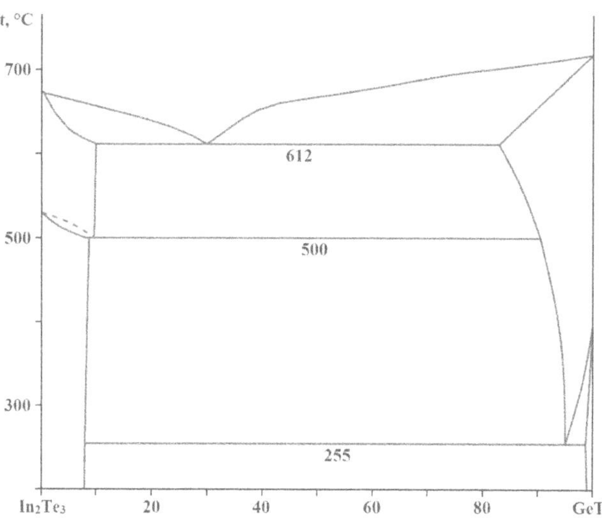

FIGURE 6.22 Phase diagram of the GeTe–In$_2$Te$_3$ system. (From Abrikosov, N.H., et al., *Izv. AN SSSR. Neorgan. mater.*, **10**(8), 1426, 1974.)

is not higher than 3 mol% (Zargarova and Akperov 1973) [10 mol% (Nasirov et al. 1968b); 5 mol% at 400°C (Woolley 1965)]. The solid solutions based on GeTe have a polymorphic transformation. According to the data of Zhigareva et al. (1981) and Bigvava et al. (1983), In$_2$Te$_3$ appears in the system at the increasing of the InTe content.

Two compounds, **InGeTe$_2$** and **In$_2$GeTe$_3$**, are formed in this system. First of them melts congruently (Zargarova and Akperov 1973) and crystallizes in the cubic structure with the lattice parameter a = 602 pm (Nasirov et al. 1970). The second compound melts incongruently (Zargarova and Akperov 1973) and also crystallizes in the cubic structure with the lattice parameter a = 606 pm (Nasirov et al. 1970). Single crystals of these compounds were grown by the Bridgman method (Nasirov et al. 1970).

This system was investigated using DTA, XRD, metallography, and measuring of microhardness and density (Woolley 1965; Zargarova and Akperov 1973; Sysoeva et al. 1970). The ingots were annealed at 200°C for 200 h (Zargarova and Akperov 1973) [at 400°C for 4 days (Woolley 1965)].

GeTe–In$_2$Te$_3$: The phase diagram of this system is a eutectic type (Figure 6.22) (Abrikosov et al. 1974; Rogacheva et al. 1974a; Babayev 2002). The eutectic contains 30 mol% GeTe and crystallizes at 612°C (Abrikosov et al. 1974) [29 mol% GeTe and crystallizes at 615°C (Rogacheva et al. 1974a); 12 mol% GeTe and crystallizes at 617°C (Babayev 2002)]. Addition of GeTe to In$_2$Te$_3$ decreases the phase transformation temperature from 530° for In$_2$Te$_3$ to 500° at 10 mol% GeTe. Introduction of In$_2$Te$_3$ lowers the temperature of phase transition of GeTe to 255°C (Abrikosov et al. 1974) [to 220°C (Rogacheva et al. 1974a)]. The eutectoid point corresponds to a composition of 5 mol% In$_2$Te$_3$. The solubility of In$_2$Te$_3$ in GeTe is 15 mol% at 600°C [56 mol% at the eutectic temperature (Rogacheva et al. 1974a)], 5 mol% at 260°C and 2 mol% [5 mol% (Rogacheva et al. 1974a)] at 200°C, and the solubility of GeTe in In$_2$Te$_3$ reaches 10 mol% at 600°C and 8 mol% at 400°C (Abrikosov et al. 1974).

According to the data of Rosenthal et al. (2013), the **Ge$_{12}$In$_2$Te$_{15}$** ternary compound, which crystallizes in the cubic structure with the lattice parameter 594.723 ± 0.004 pm and a calculated density of 6.347 g·cm^{-3}, is formed in the GeTe–In$_2$Te$_3$ system. This compound was prepared by melting stoichiometric amounts of the elements in silica glass ampoule sealed under Ar atmosphere at 950°C for 2 h. The sample was quenched in water and annealed at 590°C for 3 days with subsequent quenching in water once again.

This system was investigated using DTA, XRD, metallography, and measuring of microhardness (Abrikosov et al. 1974; Rogacheva et al. 1974a). The ingots were annealed at 200°C, 400°C, 500°C, and 600°C.

The liquidus surface of the Ge–In–Te ternary system (Figure 6.23) consists from 12 fields of primary crystallization of Ge, Te, In, GeTe, In$_9$Te$_7$, InTe, In$_3$Te$_4$, In$_2$Te$_3$, In$_3$Te$_5$, In$_2$Te$_5$,

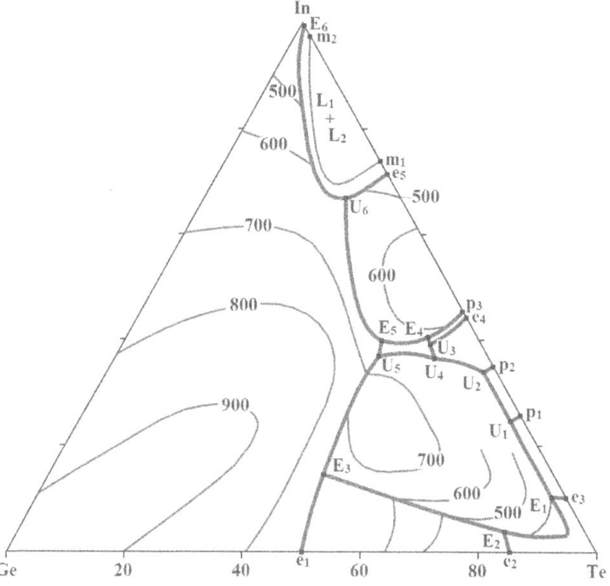

FIGURE 6.23 Liquidus surface of the Ge–In–Te ternary system. (From Zargarova, M.I., and Akperov, M.M., *Izv. AN SSSR. Neorgan. mater.*, **9**(7), 1138, 1973.)

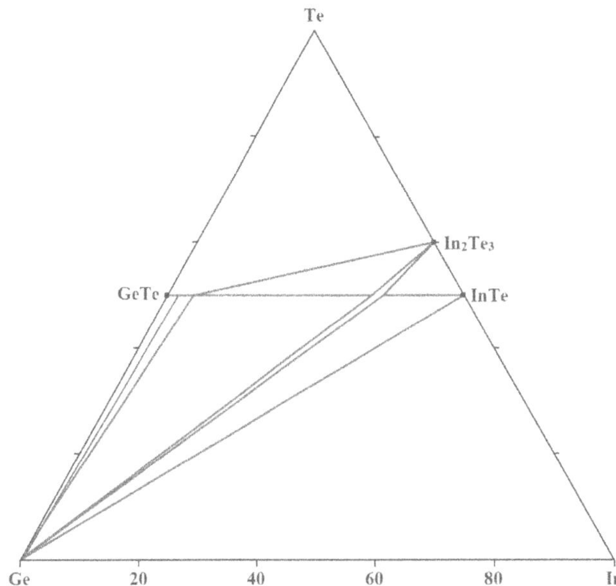

FIGURE 6.24 Isothermal section of the Ge–In–Te ternary system in the GeTe–In_2Te_3–InTe–Ge region at 400°C. (From Woolley, J.C., *J. Electrochem. Soc.*, **112**(9), 906, 1965.)

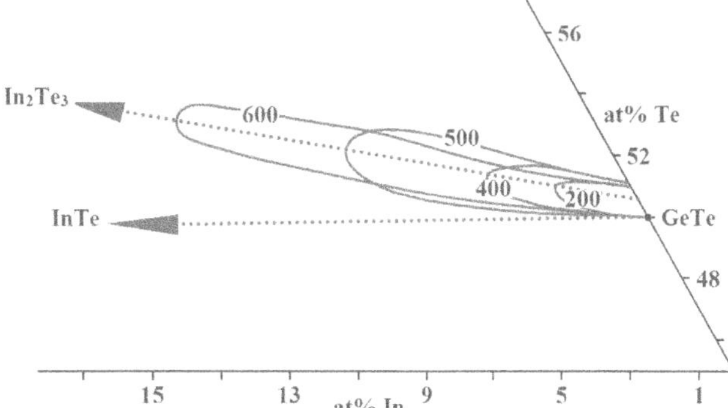

FIGURE 6.25 Region of solid solutions based on GeTe in the Ga–In–Te ternary system at 200°C, 400°C, 500°C, and 600°C. (From Abrikosov, N.H., and Danilova-Dobryakova, G.T. et al., *Izv. AN SSSR. Neorgan. mater.*, **12**(7), 1204, 1976.)

InGeTe$_2$, and In$_2$GeTe$_3$ (Zargarova and Akperov 1973). Near In-corner, there is an immiscibility region adjacent to the In–Te system. The largest area is occupied by the field of Ge primary crystallization. There are six ternary eutectics and six transition points in this ternary system: E_1 (350°C) – L ⇔ InGeTe$_2$ + In$_2$Te$_5$ + Te; E_2 (330°C) – L ⇔ InGeTe$_2$ + GeTe + Te; E_3 (630°C) – L ⇔ InGeTe$_2$ + GeTe + Ge; E_4 (530°C) – L ⇔ In$_2$GeTe$_3$ + In$_2$Te$_5$ + InTe; E_5 (400°C) – L ⇔ In$_2$GeTe$_3$ + InTe + Ge; E_6 (150°C) – L ⇔ In + Ge + In$_9$Te$_7$; U_1 (600°C) – L + In$_2$Te$_3$ ⇔ In$_3$Te$_5$ + InGeTe$_2$; U_2 (420°C) – L + In$_3$Te$_5$ ⇔ In$_2$Te$_5$ + InGeTe$_2$; U_3 (540°C) – L + In$_2$Te$_3$ ⇔ In$_3$Te$_4$ + In$_2$GeTe$_3$; U_4 (550°C) – L + InGeTe$_2$ ⇔ In$_2$GeTe$_3$ + In$_2$Te$_3$; U_5 (430°C) – L + InGeTe$_2$ ⇔ In$_2$GeTe$_3$ + Ge; and U_6 (430°C) – L + InTe ⇔ In$_9$Te$_7$ + Ge.

The isothermal section of the Ge–In–Te ternary system in the GeTe–In$_2$Te$_3$–InTe–Ge region at 400°C is presented in Figure 6.24 (Woolley 1965).

The region of the solid solutions based on GeTe in the Ge–In–Te ternary system (Figure 6.25) is situated along the GeTe–In$_2$Te$_3$ section and its width lies within 2 at% Te (Abrikosov and Danilova-Dobryakova 1976; Palatnik et al. 1977a; Rogacheva et al. 1977).

6.16 Germanium–Thallium–Tellurium

GeTe–Tl: This system is a non-quasibinary section of the Ge–Tl–Te ternary system (Kulieva et al. 1982; Kulieva and Babanly 1982). The biggest part of the section occupies an immiscibility region.

GeTe–TlTe: The results of this system investigation are contradictory. The most reliable are the information that the system is a non-quasibinary section of the Ge–Tl–Te ternary system (Figure 6.26) (Abba Toure et al. 1987, 1991). The **Tl$_2$GeTe$_3$** ternary compound, which melts incongruently at 378°C

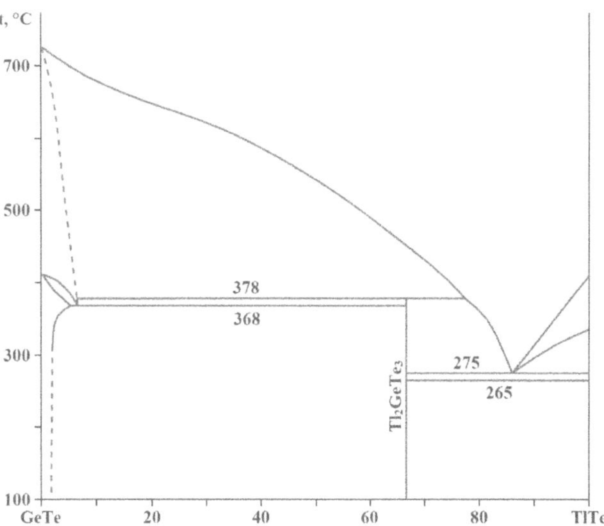

FIGURE 6.26 Phase equilibria in the GeTe–TlTe system. (From Abba Toure, A., et al., *C. r. Acad. sci., Sér. II*, **304**(9), 411, 1987.)

[at 300°C (Nasirov et al. 1969)], is formed in the Ge–Tl–Te system. The **TlGeTe$_2$** ternary compound was not found. At 368°C, the polymorphic transformation of solid solutions based on GeTe takes place. From the TlTe-side, at 275°C a peritectic reaction proceeds with the formation of Tl$_5$Te$_3$, and at 265°C a ternary eutectic crystallized.

Tl$_2$GeTe$_3$ crystallizes in the orthorhombic structure with the lattice parameters $a = 830.3 \pm 0.4$, $b = 2151.4 \pm 0.9$, $c = 845.3 \pm 0.6$ pm, and a calculated density of 7.61 g·cm^{-3} (Klepp and Ecker 1995) [a = 413, b = 1202, c = 848 pm, and the calculated and experimental densities of 6.82 and 6.95 g·cm^{-3}, respectively (Abba-Toure et al. 1987, 1991)]. The black lustrous single crystals of this compound were prepared from the melt (Klepp and Ecker 1995). Stoichiometric mixture of powdered Tl$_5$Te$_3$, Ge, and Te was sealed into evacuated silica ampoule and allowed to react at 750°C for 3 days. The thermal treatment was finished by a controlling cooling (2.5°C·h^{-1}) to room temperature.

According to the data of (Nasirov et al. 1969), the phase diagram of this system is a eutectic type with two peritectic transformations. The eutectic contains 14 mol% GeTe and crystallizes at 240°C. Two compounds, $TlGeTe_2$ (melts incongruently at 370°C) and Tl_2GeTe_3, are formed in the GeTe–TlTe system (Nasirov et al. 1969; Kulieva and Babanly 1983). The solubility of TlTe in GeTe reaches 6 mol% at room temperature (Nasirov et al. 1969) (is not higher than 3 mol% (Sysoeva et al. 1970).

This system was investigated using DTA, XRD, metallography, and measuring of microhardness and EMF of concentration chains (Nasirov et al. 1969; Kulieva and Babanly 1983). The ingots were annealed at 200°C for 8 days.

GeTe–Tl_2Te: The phase diagram this system is a eutectic type (Figure 6.27) (Kulieva et al. 1982; Kulieva and Babanly 1983). The eutectic from the Tl_2Te-side is degenerated and the second eutectic contains 60 mol% GeTe and crystallizes at 410°C. The peritectic point contains 57 mol% GeTe. Two compounds, **Tl_2GeTe_2**, which melts incongruently at 417°C [at 422°C (Abba-Touré et al. 1991)] and has polymorphous transformation at 325°C, and **Tl_8GeTe_5**, which melts congruently at 480°C [at 470°C (Abba-Touré et al. 1991)] and has polymorphous transformation at 340°C, are formed in this system. There are limited regions of the solid solutions based on Tl_8GeTe_5 and GeTe.

Tl_2GeTe_2, or $Tl_4Ge_2Te_4$, crystallizes in the triclinic structure with the lattice parameters $a = 947.1 \pm 0.2$, $b = 971.4 \pm 0.2$, $c = 1038.9 \pm 0.2$ pm, and $\alpha = 89.39 \pm 0.01°$, $\beta = 97.27 \pm 0.01°$, and $\gamma = 100.79 \pm 0.01°$ (Eulenberger 1984a). Tl_8GeTe_5 crystallizes in the tetragonal structure with the lattice parameters $a = 894$ and $c = 1240$ pm (Nasirov et al. 1969; Kulieva and Babanly 1983).

This system was investigated using DTA, XRD, metallography, and measuring of EMF of concentration chains (Nasirov et al. 1969; Kulieva and Babanly 1983). The ingots were annealed at temperatures for 30–50°C below the solidus temperatures for 400 h.

The **GeTe–Tl_2Te_3** and **GeTe–Tl_5Te_3** systems are non-quasibinary sections of the Ge–Tl–Te ternary system (Kulieva and Babanly 1982).

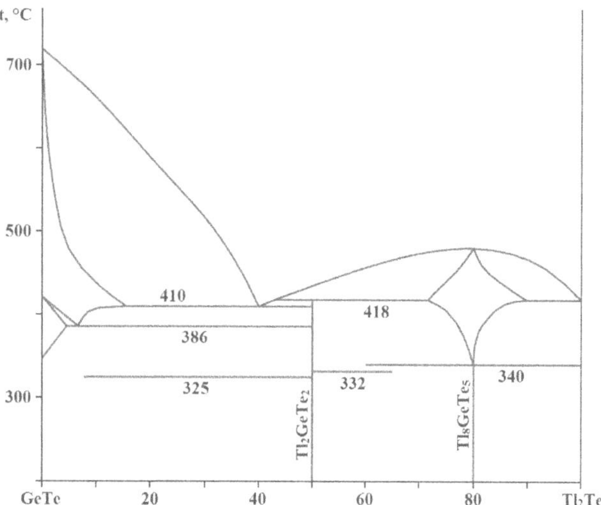

FIGURE 6.27 Phase diagram of the GeTe–Tl_2Te system. (From Kulieva, N.A., and Babanly, M.B. *Azerb. khim. zhurn.*, (1), 121, 1983.)

Ge–Tl_5Te_3: The phase diagram is a eutectic type (Abba-Touré et al. 1991). The eutectic is degenerated from Tl_5Te_3-side and crystallizes at 430°. The wide immissibity region exists at 930°C. The Tl_8GeTe_5 ternary compound, which crystallizes in the monoclinic structure with the lattice parameters $a = 1589$, $b = 898$, $c = 895$ pm, and $\beta = 124.01°$, is formed in this system.

There are 17 invariant points, one ternary peritectic, six ternary eutectics and ten transition points, on the liquidus surface of the Ge–Tl–Te ternary system (Abba-Touré et al. 1991). The part of the liquidus surface and the isothermal section at 230°C in the region GeTe–Tl_2Te–Tl–Ge were also constructed by Kulieva et al. (1982). It was shown that the liquidus surface in this region is completely covered by the region of Ge primary crystallization, which proceeds through a monotectic reaction in a wide range of concentrations. The solubility based on all phases below 230°C is practically absent. The liquidus surface of the GeTe–Tl_2Te–Te and the isothermal section at room temperature were also constructed by Kulieva and Babanly (1983).

Three another compounds, **$TlGeTe_3$**, **Tl_2GeTe_5**, and **$Tl_6Ge_2Te_6$**, were also found in the Ge–Tl–Te ternary system. $TlGeTe_3$ melts incongruently at 330°C (Abba-Touré et al. 1991).

Tl_2GeTe_5 melts congruently at 322°C [at 265°C–270°C (Sharp et al. 1999)] and crystallizes in the orthorhombic structure with the lattice parameters $a = b = 1165.7 \pm 0.4$, $c = 1491.7 \pm 0.5$ pm, and a calculated density of 8.19 g·cm^{-3} (Abba-Touré et al. 1990, 1991) [in the tetragonal structure with the lattice parameters $a = 824.3 \pm 0.4$, $c = 1491.7 \pm 0.5$ pm (Marsh 1990)]. The single crystals of this compound were prepared by slow cooling of the molten mixtures (Abba-Touré et al. 1990). Tl, Ge, and Te, in stoichiometric quantity, were intimately mixed, then introduced into a glass ampoule which was sealed under vacuum (0.1 Pa). This mixture was gradually brought to the temperature of 350°C from which it was cooled at a rate of 1°C·h^{-1} to the temperature of 200°C. The single crystals were in the form of black parallelepiped plates.

$Tl_6Ge_2Te_6$ melts incongruently at 416°C (Abba-Touré et al. 1991) [at 390°C (Voroshilovet al. 1986)] and crystallizes in the triclinic structure with the lattice parameters $a = 947.06 \pm 0.17$, $b = 971.40 \pm 0.15$, $c = 1038.90 \pm 0.16$ pm, and $\alpha = 89.387 \pm 0.012°$, $\beta = 97.270 \pm 0.013°$, $\gamma = 100.789 \pm 0.014°$, and a calculated density of 7.68 g·cm^{-3} (Eulenberger 1984b) [$a = 946$, $b = 970$, $c = 1038$ pm, and $\alpha = 89.40°$, $\beta = 97.32°$, $\gamma = 100.81°$, and a calculated density of 7.68 g·cm^{-3} (Abba-Touré et al. 1990)] and an energy gap of 0.9 eV (Assoud et al. 2006). The experimental density of $Tl_6Ge_2Te_6$ is 7.61 g·cm^{-3} (Voroshilov et al. 1986). The titled compound was prepared starting from the elements taken in the stoichiometric ratio. They were loaded into a silica tube, which was then sealed under vacuum. Subsequently, the tube was heated in a resistance furnace to 800°C within 24 h, and then cooled to 700°C within 15 min, and annealed at this temperature for 200 h. Thereafter the furnace was switched off. The sample consisted mostly of black powder, together with few crystals of metallic luster. The material is not air sensitive at room temperature over a period of a few weeks (Assoud et al. 2006).

According to the data of Kulieva et al. (1982) and Kulieva and Babanly (1982), the $Tl_6Ge_2Te_6$ ternary compound is not formed in the Ge–Tl–Te system, but Eulenberger (1984b) noted that the compound Tl_2GeTe_2 from Kulieva et al. (1982) and Kulieva and Babanly (1982) is identical to $Tl_6Ge_2Te_6$.

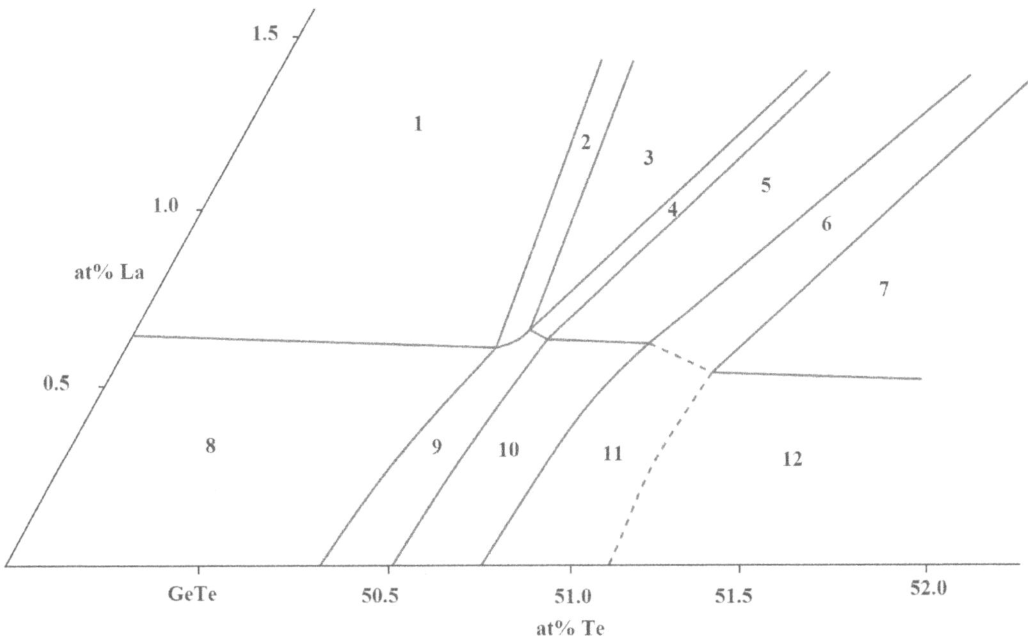

FIGURE 6.28 Part of the Ge–La–Te isothermal section at 300°C near GeTe: α-GeTe + Ge + LaTe; 2, α-GeTe + LaTe; 3, α-GeTe + LaTe + $LaTe_{2-x}$; 4, α-GeTe + $LaTe_{2-x}$; 5, α-GeTe + γ-GeTe + $LaTe_{2-x}$; 6, γ-GeTe + $LaTe_{2-x}$; 7, γ-GeTe + $LaTe_{2-x}$ + Te; 8, α-GeTe + Ge; 9, α-GeTe; 10, α-GeTe + γ-GeTe; 11, γ-GeTe; 12, γ-GeTe + Te. (From Shelimova, L.E., et al., *Izv. AN SSSR. Neorgan. mater.*, **27**(11), 2285, 1991.)

6.17 Germanium–Scandium–Tellurium

GeTe–ScTe: The solubility of ScTe in GeTe reaches 5 mol% at 500°C (Abrikosov et al. 1988a; Shelimova et al. 1993a). Substitution of Ge with Sc leads to a significant decrease in the degree of rhombohedral distortion of α-GeTe (at 5 mol% Sc the phase transition temperature is 280°C). The ingots were annealed at for 500°C for 700 h.

6.18 Germanium–Lanthanum–Tellurium

GeTe–LaTe: The solubility of LaTe in GeTe is negligible (ca. 1 mol%) (Shelimova et al. 1993a).

The part of the Ge–La–Te isothermal section at 300°C near GeTe is shown in Figure 6.28 (Shelimova et al. 1991). It is seen that the boundary of La solubility in GeTe passes between 0.50 and 0.75 at%. With increasing La content, the boundary of the homogeneity region shifts towards an increase in Te content. With an increase in the La content, the boundary between the phase regions [(α-GeTe) + (γ-GeTe)] and (γ-GeTe) shifts toward an increase in the Te content, i.e. the region of (γ-GeTe) decreases with increasing of La content in the solid solution.

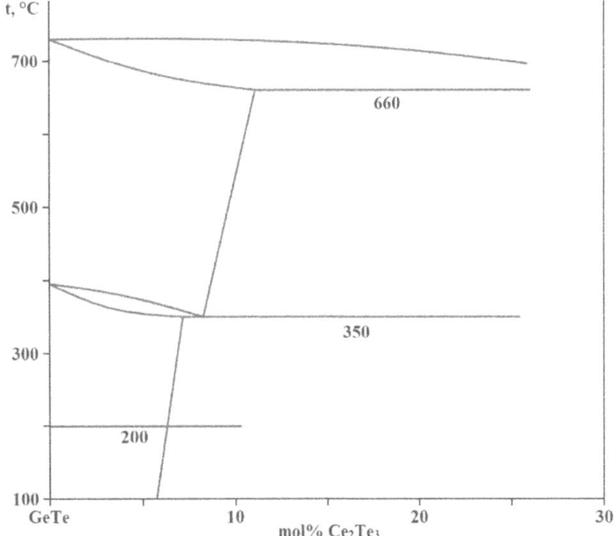

FIGURE 6.29 Part of the phase diagram of the GeTe–Ce_2Te_3 system. (From Kurbanova, R.D., et al., *Azerb. khim. zhurn.*, (2), 43, 2006.)

6.19 Germanium–Cerium–Tellurium

GeTe–Ce_2Te_3: The part of the phase diagram of this system from the GeTe-side is presented in Figure 6.29 (Kurbanova et al. 2006). The solubility of Ce_2Te_3 in GeTe is equal ~10 mol% at 675°C, ~8 mol% at 400°C, ~6 mol% at 200°C, and 5 mol% at room temperature. This system was investigated using DTA, XRD, metallography, and measuring of microhardness. The ingots were annealed at 130°C–400°C for 200 h.

6.20 Germanium–Samarium–Tellurium

GeTe–SmTe: The phase diagram of this system is given in Figure 6.30 (Aliev et al. 1986). The $\mathbf{Sm_5Ge_2Te_7}$ ternary compound, which melts incongruently at 940°C, is formed in the system. The composition and temperature of the peritectic point is 8 mol% GeTe and 750°C, respectively. The solubility

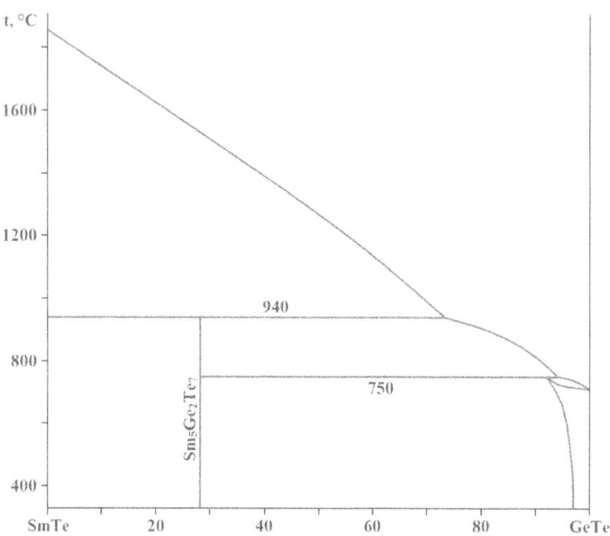

FIGURE 6.30 Phase diagram of the GeTe–SmTe system. (From Aliev, O.M., et al., *Zhurn. neorgan. khimii*, **31**(10), 2628, 1986.)

of SmTe in GeTe is 8 mol% at 750°C and decreases to 3 mol% at room temperature. This system was investigated using DTA, XRD, metallography, and measuring of microhardness. The ingots were annealed at 430°C–530°C for 240–300 h.

GeTe–Sm_2Te_3: The phase diagram of this system is a eutectic type with peritectic transformation Figure 6.31 (Aliev et al. 1986; Mukhtarova 2019). The **Sm_2GeTe_4** ternary compound, which melts incongruently at 730°C, is formed in this system. The eutectic contains 12 mol% Sm_2Te_3 and crystallizes at 650°C. The solubility of Sm_2Te_3 in GeTe is 2 mol% at room temperature. This system was investigated using DTA, XRD, metallography, and measuring of microhardness. The ingots were annealed at 730°C for 500 h.

The $Ge_{0.8}Te_{0.2}$–$Sm_{0.8}Te_{0.2}$ and $Ge_{0.84}Te_{0.16}$–$Sm_5Ge_2Te_7$ polythermal section of the Ge–Sm–Te ternary system were constructed by Mukhtarova (2010) and Mukhtarova et al. (2011).

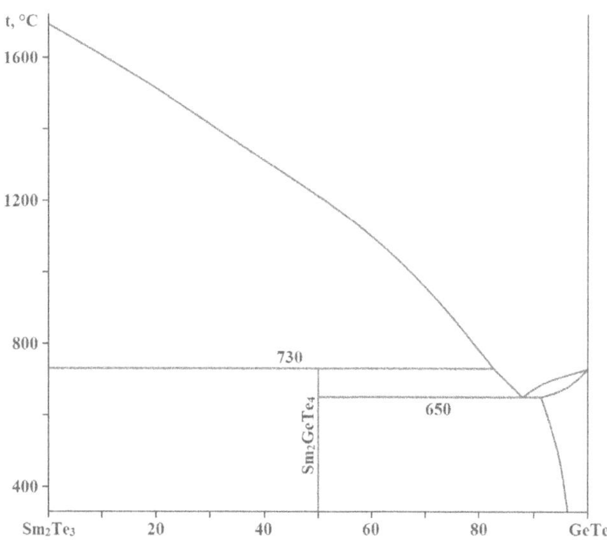

FIGURE 6.31 Phase diagram of the GeTe–Sm_2Te_3 system. (From Aliev, O.M., et al., *Zhurn. neorgan. khimii*, **31**(10), 2628, 1986.)

6.21 Germanium–Gadolinium–Tellurium

GeTe–GdTe: The solubility of GdTe in GeTe is ~4.5 mol% (Shvangiradze and Kutsiya 1999).

6.22 Germanium–Terbium–Tellurium

GeTe–TbTe: The solubility of TbTe in GeTe is ~4.5 mol% (Shvangiradze and Kutsiya 1999).

6.23 Germanium–Dysprosium–Tellurium

GeTe–DyTe: The solubility of DyTe in GeTe is ~4.5 mol% (Shvangiradze and Kutsiya 1999).

6.24 Germanium–Holmium–Tellurium

GeTe–HoTe: The solubility of HoTe in GeTe is ~4.5 mol% (Shvangiradze and Kutsiya 1999).

6.25 Germanium–Thulium–Tellurium

GeTe–TmTe: The solubility of TmTe in GeTe is ~4.5 mol% (Shvangiradze and Kutsiya 1999).

6.26 Germanium–Ytterbium–Tellurium

GeTe–Yb_3Ge_5: The phase diagram of this system is presented in Figure 6.32 (Mukhtarova et al. 2007). The **$Yb_3Ge_7Te_2$** ternary compound, which melts congruently at 960°C and has an experimental density of 8.30 g·cm^{-3}, is formed in this system. The eutectics contain 6 and 46 mol% Yb_3Ge_5 and crystallize

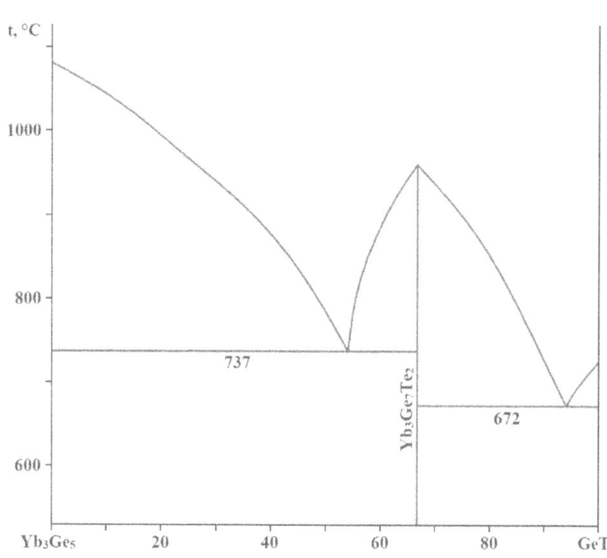

FIGURE 6.32 Phase diagram of the GeTe–Yb_3Ge_5 system. (From Mukhtarova, Z.M., et al., *Azerb. khim. zhurn.*, (4), 155, 2007.)

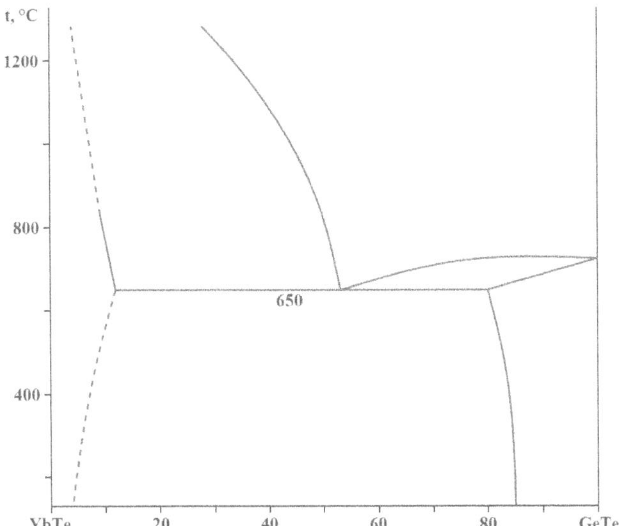

FIGURE 6.33 Phase diagram of the GeTe–YbTe system. (From Mukhtarova, Z.M., et al., *Zhurn. neorgan. khimii*, **30**(5), 1332, 1985.)

at 672°C and 737°C, respectively. This system was investigated using DTA, XRD, metallography, and measuring of microhardness and density. The ingots were annealed at 430°C for 600 h.

GeTe–YbTe: The phase diagram of this system is a eutectic type Figure 6.33 (Mukhtarova et al. 1985). The eutectic contain 47 mol% YbTe and crystallize at 650°C. The solubility of YbTe in GeTe at the eutectic temperature is 20 mol% and 15 mol% at room temperature. This system was investigated using DTA, XRD, and measuring of microhardness. The ingots were annealed at 550°C for 340 h.

6.27 Germanium–Lutetium–Tellurium

GeTe–LuTe: The solubility of LuTe in GeTe is ~4.5 mol% (Shvangiradze and Kutsiya 1999).

6.28 Germanium–Thorium–Tellurium

The **ThGeTe** ternary compound, which crystallizes in the tetragonal structure with the lattice parameters $a = 419.1$ and $c = 1822$ pm and the calculated and experimental densities of 8.97 and 8.65 g·cm^{-3}, respectively (Stocks et al. 1981; Hahn and Stocks 1968)], is formed in the Ge–Th–Te system. To prepare this compound, the elements were weighed under N_2 atmosphere in the stoichiometric ratio and initially heated to about 400°C in an evacuated quartz ampoule. Subsequently, the inhomogeneous product was finely powdered under N_2 and heated to 800°C in a sealed tungsten crucible, which was enclosed in an evacuated quartz ampoule. Since the preparation is extremely sensitive to traces of oxygen, it proved necessary to add Th as getter metal. Nevertheless, pure preparation was obtained only if no longer than 48 h the mixture was heated at 800°C. The mixture was annealed at 600°C–700°C for 8–10 days. Then, the sample pressed into pastilles was heated in the same way over different time intervals and at different temperatures up to 1400°C in tungsten crucible, wherein above 1050°C additionally Ar was used as a protective gas. The title compound was obtained as black microcrystalline powder.

6.29 Germanium–Uranium–Tellurium

The **UGeTe** ternary compound, which crystallizes in the tetragonal structure with the lattice parameters $a = 410.58 \pm 0.01$ and $c = 1760.14 \pm 0.05$ pm and a calculated density of 9.81 g·cm^{-3} (Haneveld and Jellinek 1969) [$a = 411.0$ and $c = 1759.9$ pm (Zygmunt 1977)], is formed in the Ge–U–Te system. This compound was prepared in the following way (Haneveld and Jellinek 1969). A small rod of uranium was placed in a quartz tube and heated in a current of hydrogen at 230°C. The voluminous uranium hydride powder thus produced was dehydrated by heating at 400°C under vacuum. An equivalent proportion of GeTe was added immediately and the quartz tube was evacuated and sealed. The mixture was heated at 800°C–900°C for 2 or 3 days. Then, the product was slowly cooled to room temperature. This compound is stable in air.

6.30 Germanium–Tin–Tellurium

GeTe–SnTe: The first investigations of this system showed (Abrikosov et al. 1958b; Krebs et al. 1961b; Mazelsky et al. 1962; Bierly et al. 1963) that it could be considered as quasibinary in the first approximation. At high temperatures, a continuous series of solid solutions is formed in the system. Later investigations showed (Abrikosov and Shelimova 1986) that the $Ge_{1-x}Sn_xTe$ alloys are not single-phase and, depending on x, contain either Ge precipitates ($0 \le x \le 0.8$) or Sn with a small amount of Ge ($0.92 \le x \le 0.97$), i.e. the system can be considered as non-quasibinary due to deviations from stoichiometry in GeTe and SnTe. According to the data of (Volykhov et al. 2006), the behavior of this system could be described in terms of a four-parameter model for the excess Gibbs energy.

The phase diagram of the GeTe–SnTe system is presented in Figure 6.34 (Yashina and Leute 2000; Yashina et al. 2006). The liquidus and solidus has a minimum at 29 mol% SnTe and 700°C [at 16.7 mol% SnTe and 700°C (Abrikosov et al. 1958b); at 40 mol% SnTe and 700°C (Baldé et al. 1995); at 31.5 mol% SnTe (Volykhov et al. 2006)]. According to the calculation from Gibbs energy functions (Yashina and Leute 2000), below 268°C the $Ge_{1-x}Sn_xTe$ solid solutions has spinodal decomposition (critical point contains 50 mol% SnTe). The calculated spinodal miscibility gap is so small that it is very improbable that it can be separated experimentally from the adjacent broad structural miscibility gap. The monotectic reaction takes place at 241°C.

The $Ge_{1-x}Sn_xTe$ solid solutions exhibit a positive deviation from ideality (Volykhov et al. 2006), and the GeTe–SnTe system must be described by a regular behavior (Yashina et al. 2006).

Very precise X-ray measurements of the spontaneous lattice strain around the paraelectric-ferroelectric transition in the $Ge_{1-x}Sn_xTe$ solid solutions have revealed a crossover from continuous to discontinuous transitions as x decreases below 0.72 (Clarke 1978). The tricritical behavior was analyzed in terms of Landau theory and the free energy expansion coefficients

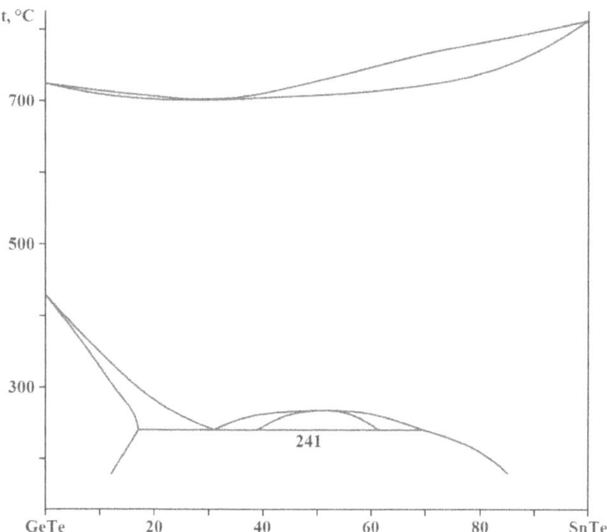

FIGURE 6.34 Phase diagram of the GeTe–SnTe system. (From Yashina, L., and Leute, V., *J. Alloys Compd.*, **313**(1–2), 85, 2000.)

have been calculated up to sixth order in good agreement with calculations on pure SnTe.

When studying the influence of deviations from stoichiometry according to tellurium on the homogeneity region of solid solutions and the nature of phase transformations, a low-temperature anomaly was discovered, which is associated with an irreversible phase transformation, the temperature of which increases in the region of solid solutions (Abrikosov et al. 1980b).

The *p-x*-phase diagram of the GeTe–SnTe system is given in Figure 6.35a (Serebryanaya 1991). It was established that at the increasing of the GeTe content the pressure of $\beta \Leftrightarrow \gamma$ phase transition monotonically increases from 1.2 to 11 GPa (Serebryanaya 1991; Kabalkina et al. 1967). There is also $\alpha \Leftrightarrow \beta$ phase transition in this system. An increase in pressure leads to a decrease in the temperature of this polymorphic transformation (Abrikosov et al. 1969b).

The single crystals of the $Ge_{1-x}Sn_xTe$ solid solutions could be grown by the Bridgman technique (Lefkowitz and Shields 1970; Rehwald and Lang 1975).

By means of an alternating current calorimetry the heat capacity has been measured in several $Ge_{1-x}Sn_xTe$ crystals with x ranging from 0.62 to 0.91 (Hatta and Rehwald 1977). The anomalous specific heat near the transition temperature can be separated into two parts. One shows a step-like behavior; the height of the step becomes larger with decreasing x. The other contribution is a logarithmic critical behavior; its critical region becomes broader with decreasing x as well. These characteristics might be related to a tricritical point. It has been found that in $Ge_{0.21}Sn_{0.79}Te$ crystals the anomalous specific heat is proportional to the anomalous elastic stiffness.

This system was investigated using DTA, XRD, metallography, and measuring of microhardness (Abrikosov et al. 1958b; Abrikosov and Shelimova 1986) and calculated from Gibbs energy functions (Yashina and Leute 2000; Yashina et al. 2006). The ingots were annealed at 450°C for 320 h (Abrikosov et al. 1958b), at 400°C for 90–100 h (Krebs et al. 1961b), and at 500°C for 600 h (Abrikosov and Shelimova 1986).

Ge–SnTe: The phase diagram of this system is a eutectic type (Figure 6.36) (Baldé et al. 1995). The eutectic contain 85 mol% SnTe and crystallize at 752°C. The solubility of SnTe in Ge is negligible and the solubility of Ge in SnTe is more significant. The system was investigated using DTA, XRD, and metallography. This phase diagram was also constructed by Schlieper and Blachnik (1998) using calculated and experimental excess enthalpies.

The liquidus surface of the GeTe–SnTe–Te subsystem of the Ge–Sn–Te ternary system deduced after the assessment of the existing data is shown in Figure 6.37 (Shtanov et al. 2009).

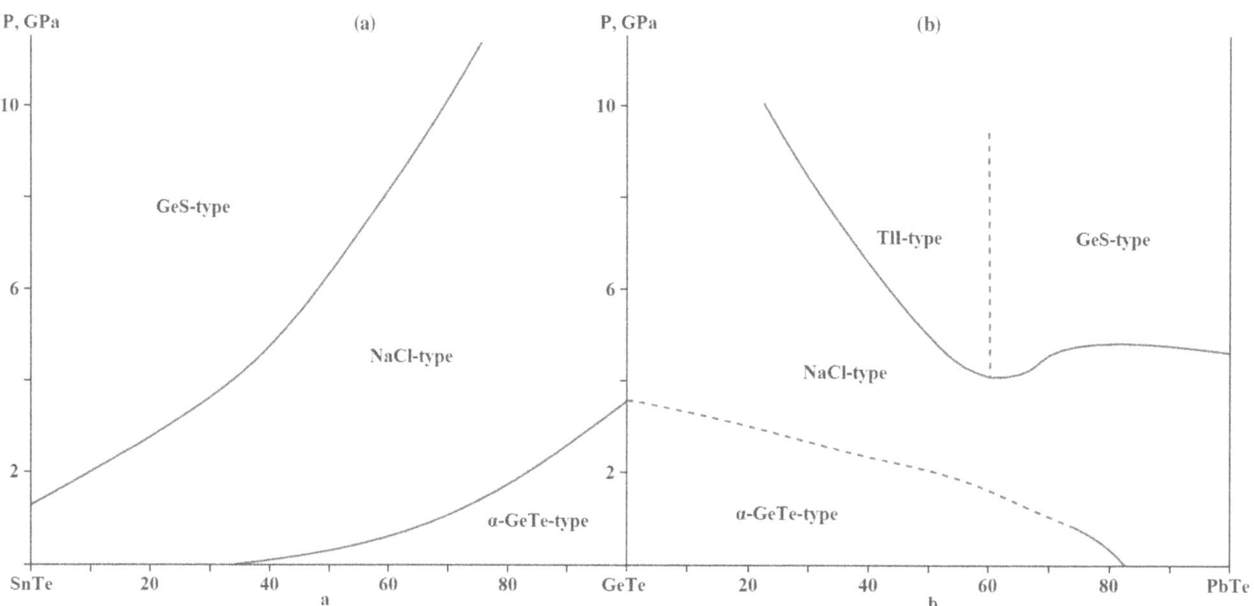

FIGURE 6.35 *p-x*-Phase diagrams of (a) the GeTe–SnTe and (b) GeTe–PbTe systems. (From Serebryanaya, A.R., *Izv. AN SSSR. Neorgan. mater.*, **27**(8), 1611, 1991.)

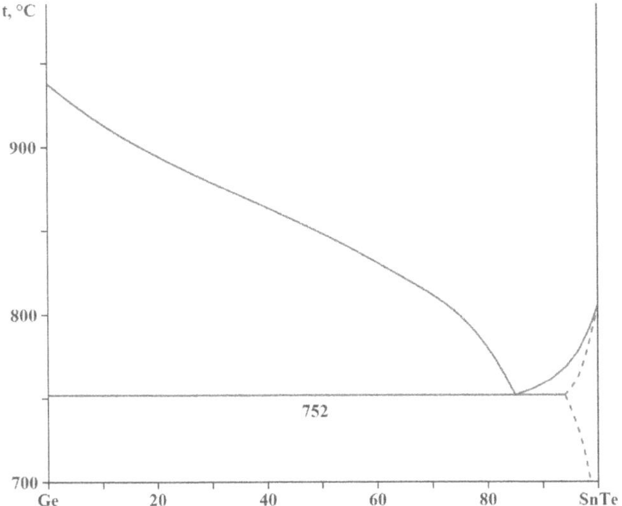

FIGURE 6.36 Phase diagram of the Ge–SnTe system. (From Baldé, L., et al., *J. Alloys Compd.*, **216**(2), 285, 1995.)

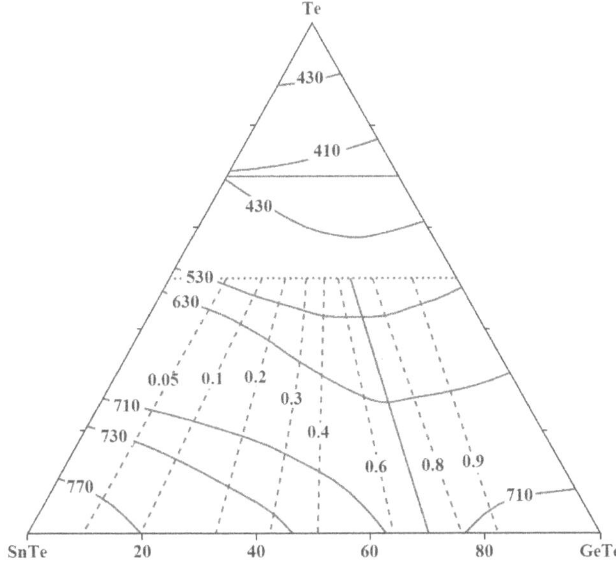

FIGURE 6.37 Liquidus surface of the GeTe–SnTe–Te ternary system with the solidus isoconcentrates for the $Ge_xSn_{1-x}Te_x$ solid solution: 1, $x = 0.05$, 2, $x = 0.1$, 3, $x = 0.2$, 4, $x = 0.3$, 5, $x = 0.4$, 6, $x = 0.6$, 6, $x = 0.8$, 6, $x = 0.9$. (From Shtanov, V.I., et al., *J. Alloys Compd.*, **476**(1–2), 812, 2009.)

The solid lines show the projection of the liquidus isotherms. Figure 6.37 also illustrates the line of the three-phase equilibria of the β-$Ge_xSn_{1-x}Te$+Te crystallization. On the basis of the obtained data, it was found that the minimum for the line of this crystallization lies at 330°C and ~7 mol% GeTe. The dashed lines correspond to the isomole-fraction (or isomolarity) lines of β-$Ge_xSn_{1-x}Te$. They indicate the melt compositions at equilibrium with the β-$Ge_xSn_{1-x}Te$ of a given mole fraction $x = 0.05; 0.1; 0.2; 0.3; 0.4; 0.6; 0.8; 0.9$ at different temperatures.

The excess enthalpies of liquid alloys in the Ge–Sn–Te ternary system were determined in a heat-flow calorimeter by Schlieper and Blachnik (1998). The system was described by means of the association model. The calculated isoenthalpic lines are presented in Figure 6.38. The largest values of the

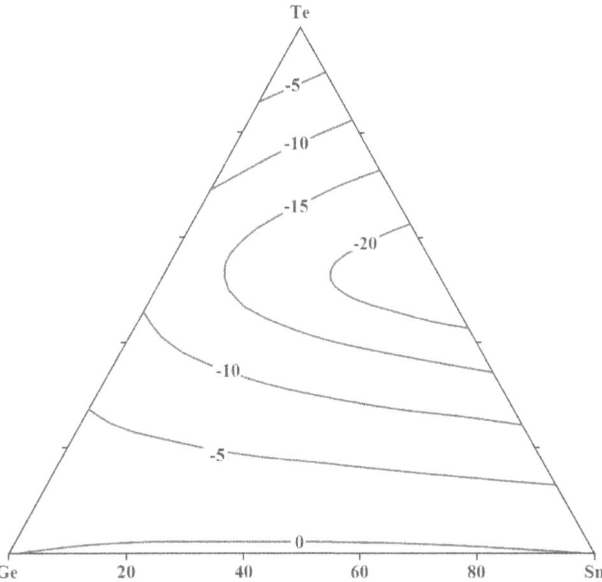

FIGURE 6.38 Calculated isoenthalpic lines (in kJ·mol⁻¹) of the Ge–Sn–Te ternary system. (From Schlieper, A., and Blachnik, R., *Z. Metallkd.*, **89**(1), 3, 1998.)

exothermic enthalpies were found on a line stretching from GeTe to SnTe.

The isothermal sections of the Ge–Sn–Te ternary system at 400°C and at room temperature as well as six vertical sections and the reaction scheme were determined by Baldé et al. (1995). The part of the isothermal section at 200°C near GeTe was constructed through XRD and metallography using the alloys annealed at this temperature for 2000 h by Abrikosov et al. (1990) and Shelimova and Karpinskiy (1991d, 1992) and is shown in Figure 6.39. It is seen that an increase in Sn leads to a decrease in the region of existence of the γ-phase (solid solution based on GeTe with an orthorhombic structure) and to an expansion of the region of the existence of the α-phase (solid solution based on GeTe with a rhombohedral structure). The last region at $x > 0.25$ becomes the only low-temperature modification, and in the composition region with $x = 0.30–0.35$ it transforms into the β-phase (solid solution based on GeTe with a cubic structure of NaCl-type). An increase in the Sn content also leads to a decrease in the temperature of the $\gamma \rightarrow \beta$ phase transition. The temperatures of phase transitions in solid solutions based on GeTe depend not only on the degree of cationic or anionic substitution, but also on the degree of deviation from stoichiometry.

6.31 Germanium–Lead–Tellurium

GeTe–PbTe: The first investigations of this system showed (Woolley and Nikolić 1965; Karbanov et al. 1969; Bigvava et al. 1972; Hohnke et al. 1972; Parker et al. 1974) that it could be considered as quasibinary in the first approximation. At high temperatures, a continuous series of solid solutions are formed in the system, which undergo decomposition in the solid state. The fact that two phases were detected in the $Ge_xPb_{1-x}Te$ samples quenched from 650°C (Shelimova et al. 1964) indicates the

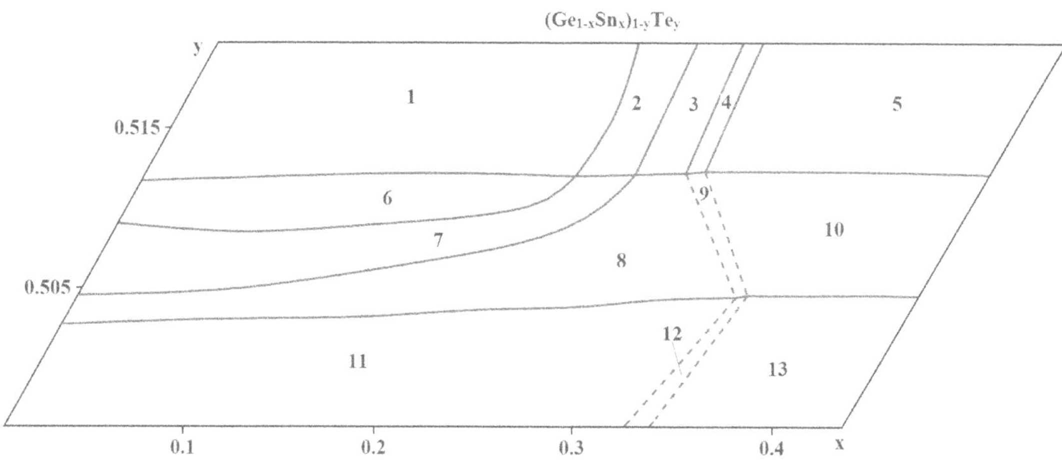

FIGURE 6.39 Part of the isothermal section of the Ge–Sn–Te ternary system at 200°C near GeTe: 1, γ-GeTe + Te; 2, α-GeTe + γ-GeTe + Te; 3, α-GeTe + Te; 4, α-GeTe + β-GeTe + Te; 5, β-GeTe + Te; 6, γ-GeTe; 7, α-GeTe + γ-GeTe; 8, α-GeTe; 9, α-GeTe + β-GeTe; 10, β-GeTe; 11, α-GeTe + Ge; 12, α-GeTe + β-GeTe + Ge; 13, β-GeTe + Ge. (From Abrikosov, N.H., et al., *Izv. AN SSSR. Neorgan. mater.*, **26**(7), 1416, 1990.)

decomposition of solid solutions during quenching (Hohnke et al. 1972). An immiscibility region between $Ge_{0.025}Pb_{0.975}Te$ to almost pure GeTe was found to exist at room temperature (Massimo and Cadoff 1976). Complete solubility does exist at and above 600°C, but it was found that quenched supersaturated alloys were quite unstable upon heating to as low as 200°C, decomposing to GeTe and $Ge_{1-x}Pb_xTe$.

More recent studies have shown (Abrikosov and Shelimova 1986) that $Ge_xPb_{1-x}Te$ stoichiometric alloys are not single-phase; in all alloys, with the exception of the composition with $x = 0.9$, germanium was observed. The amount of the second phase decreases with increasing PbTe content, which indicates that the boundary of the homogeneity region saturated with metal approaches the GeTe–PbTe stoichiometric section. According to the data of (Volykhov et al. 2006), the behavior of this system could be described in terms of a four-parameter model for the excess Gibbs energy. The $Ge_{1-x}Pb_xTe$ solid solutions exhibit a large deviation from ideality.

The phase diagram of the GeTe–PbTe system is shown in Figure 6.40 (Yashina and Leute 2000; Yashina et al. 2006). The liquidus and solidus has a minimum at 26 mol% PbTe and 692°C [at 20 mol% SnTe and 695°C (Hohnke et al. 1972); at about 681°C (Parker et al. 1974)]. According to the calculation from Gibbs energy functions (Yashina and Leute 2000), below 572°C the $Ge_{1-x}Pb_xTe$ solid solutions has spinodal decomposition (critical point contains 40 mol% PbTe). The monotectic reaction takes place at 426°C. Good agreement between experimental data and calculated curves has been obtained.

According to the data of Abrikosov and Shelimova (1986), at 367°C an invariant eutectoid reaction proceeds in alloys associated with the polymorphic transformation of GeTe. The temperature of this phase transition sharply increases with increasing GeTe content (Abrikosov et al. 1980c, 1981a).

The incorporation of 3 mol% PbTe into GeTe shifts the boundary of the homogeneity region of the $Ge_{1-x}Pb_xTe_{1+y}$ solid solutions from $y = 0.044$ to $y = 0.075$ (Gogishvili et al. 1988c). The solubility of PbTe in GeTe depends on the tellurium content and increases with an increase in the deviation of GeTe stoichiometry to tellurium (Gogishvili et al. 1983). The boundaries

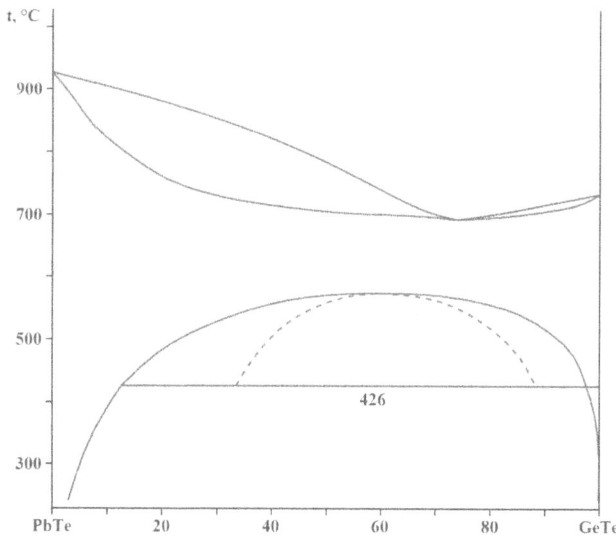

FIGURE 6.40 Phase diagram of the GeTe–PbTe system. (From Yashina, L., and Leute, V., *J. Alloys Compd.*, **313**(1–2), 85, 2000.)

of the homogeneity region of the $Ge_xPb_{1-x}Te$ solid solution at $x = 0.02$ and 0.06 were established by Kalinin et al. (1991).

The *p-x*-phase diagram of the GeTe–PbTe system is given in Figure 6.35b (Serebryanaya 1991). A no monotonic concentration dependence of the pressures of the β ⇔ γ-transition with a minimum at 37.5 mol% GeTe and 4.3 GPa was found in the system. It can be assumed that this minimum is due to a change in the structural types for γ-phase: GeS-type → TlI-type (both structures are orthorhombic).

Ge–PbTe: The phase diagram of this system, constructed through DTA, XRD, metallography, and measuring of the microhardness, is a eutectic type (Figure 6.41) (Burmistrova et al. 1982). The eutectic contains 25 at% Ge and crystallizes at 865°C. The solubility of Ge in PbTe is 0.22 at% at the eutectic temperature and 0.1 at% at room temperature and the solubility of PbTe in Ge at the same temperatures is 0.74 and 0.45 mol%, respectively. The ingots for the investigations were annealed at 850°C for 300 h.

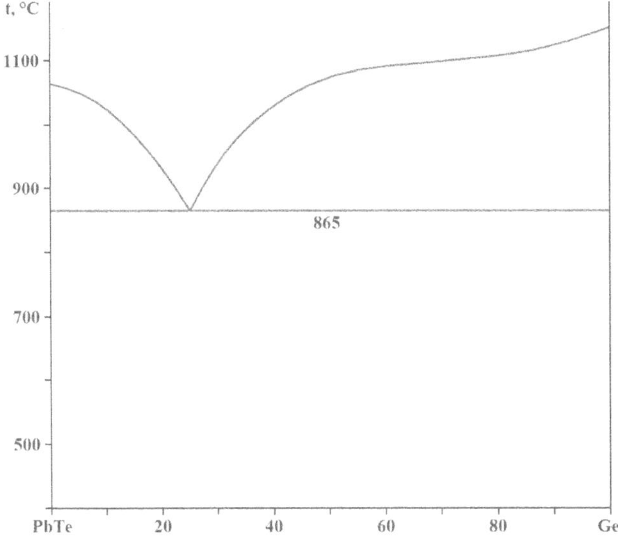

FIGURE 6.41 Phase diagram of the Ge–PbTe system. (From Burmistrova, N.P., et al., *Izv. AN SSSR. Neorgan. mater.*, **18**(5), 876, 1982.)

Ge_2PbTe, Ge_3PbTe_3, Ge_2PbTe_3, Ge_3PbTe_2, Ge_3PbTe_4, and Ge_4PbTe_4 complex molecules were found in vapor above the samples of the GeTe–PbTe system (Karbanov et al. 1968).

This system was investigated using DTA, XRD, and metallography (Shelimova et al. 1964; Bigvava et al. 1972; Hohnke et al. 1972; Abrikosov and Shelimova 1986) and calculated from Gibbs energy functions (Yashina and Leute 2000; Yashina et al. 2006). The ingots were annealed at 450°C for 650 h (Shelimova et al. 1964; Abrikosov and Shelimova 1986) [at 600°C for 1 month (Parker et al. 1974); at 500°C for 1500–2000 h (Abrikosov et al. 1980c, 1981a)].

The liquidus isotherms of the Ge–Pb–Te ternary system was constructed by Kalinin et al. (1991).

6.32 Germanium–Titanium–Tellurium

The **$TiGeTe_6$** ternary compound, which crystallizes in the monoclinic structure with the lattice parameters a = 1397.2 ± 2.6, b = 390.9 ± 0.9, c = 1745.4 ± 3.2 pm, β = 104.95 ± 0.05° at 113 K, and a calculated density of 6.39 g·cm^{-3}, is formed in the Ge–Ti–Te system (Mar and Ibers 1993). This compound appears to undergo a metal-to-semiconductor transition below 165 K. To prepare $TiGeTe_6$, powders of the elements in various ratios (total mass 0.25 g) were ground together and loaded into quartz tubes that were then evacuated and placed in a furnace. In the most successful heating profile for its preparing, the mixtures were heated to 650°C over 1 day, kept at 650°C for one another day, heated to 900°C over 6 h, kept at this temperature for 4 days or longer, and slowly cooled to room temperature over 10 days. Because of the layered nature of the structures, black crystals of the ternary compound generally grow as flat needles that are very thin and easily bent. The product is invariably contaminated with binary compounds. Crystals of the title compound usually grow in large aggregates or within a Te melt.

6.33 Germanium–Zirconium–Tellurium

Two ternary compounds, **ZrGeTe** and **$ZrGeTe_4$**, are formed in the Ge–Zr–Te system. First of them crystallizes in the tetragonal structure with the lattice parameters a = 386.6 and c = 859.9 pm (Haneveld and Jellinek 1964); a = 383 and c = 860 pm and the calculated and experimental densities of 7.66 and 7.51 g·cm^{-3}, respectively (Onken et al. 1964)]. This compound is slowly hydrolyzed in most air (Haneveld and Jellinek 1964). To obtain ZrGeTe, stoichiometric amounts of the elements were initially pre-tempered in an evacuated tube at 600°C for 10 days (Onken et al. 1964). Then, the reaction mixture was carefully rubbed in an agate mortar in a N_2-filled box, pressed into a pastille and annealed at 780°C for several weeks. The ampoule with tablet was stored in the hottest part of the furnace to obtain the single crystals by sublimation. ZrGeTe could be also prepared by heating GeTe with an equivalent proportion of Zr in an evacuated quartz tube at 700°C–800°C for 2–4 days (Haneveld and Jellinek 1964).

$ZrGeTe_4$ is practically prone to form as hair like fibers and crystallizes in the orthorhombic structure with the lattice parameters a = 397.6 ± 0.4, b = 1587.6 ± 1.6, and c = 1094.8 ± 1.2 pm at 113 K (Mar and Ibers 1993). This compound is probably non-stochiometric and is a semiconductor (a calculated value of energy gap is 0.11 ± 0.03 eV). $ZrGeTe_4$ was prepared in the same way as $TiGeTe_6$ was synthesized.

6.34 Germanium–Hafnium–Tellurium

Two ternary compounds, **HfGeTe** and **$HfGeTe_4$**, are formed in the Ge–Hf–Te system. First of them crystallizes in the tetragonal structure with the lattice parameters a = 387 and c = 850 pm and a calculated density of 9.88 g·cm^{-3} (Onken et al. 1964). HfGeTe was prepared in the same way as ZrGeTe was synthesized, but the reaction mixture was annealed at 650°C (instead of 780°C) for several weeks.

$HfGeTe_4$ crystallizes in the orthorhombic structure with the lattice parameters a = 396.3 ± 0.3, b = 1587.5 ± 1.0, c = 1094.1 ± 0.7 pm at 113 K, and a calculated density of 7.35 g·cm^{-3} (Mar and Ibers 1993). This compound is probably non-stochiometric and is a semiconductor (a calculated value of energy gap is 0.18 ± 0.02 eV). $HfGeTe_4$ was prepared in the same way as $TiGeTe_6$ was synthesized.

6.35 Germanium–Phosphorus–Tellurium

The glass forming region in this system is presented in Figure 6.42 (Hilton et al. 1966a; Savage and Nielsen 1966). The glass transition temperature for these glasses is within the of 130°C–190°C. According to the data of Vinogradova and Maysashvili (1979a), no more than 1–2 at% P can be introduced into the glassy alloys.

6.36 Germanium–Arsenic–Tellurium

GeTe–GeAs: The phase diagram of this system, constructed through DTA and XRD, is a eutectic type (Figure 6.43) (Shu et al. 1987b). The eutectic contains 60 mol% GeTe and crystallizes at

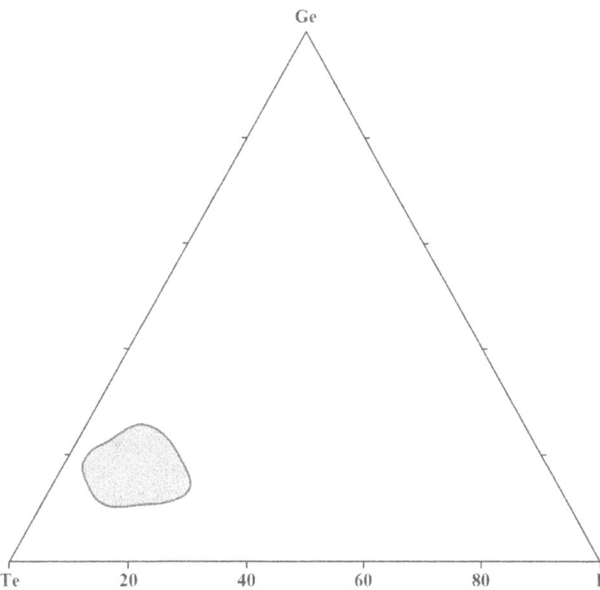

FIGURE 6.42 Glass forming region in the Ge–P–Te ternary system. (From Hilton, A.R., et al., *Infrared Phys.*, **6**(4), 183, 1966.)

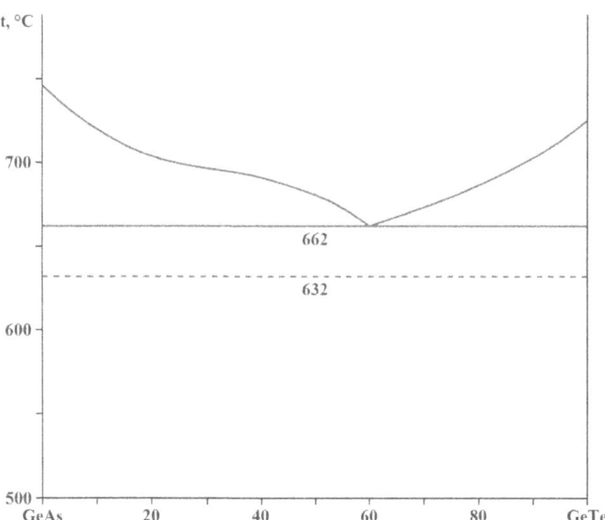

FIGURE 6.43 Phase diagram of the GeTe–GeAs system. (From Shu, H.W., et al., *J. Solid State Chem.*, **69**(1), 55, 1987.)

662°C. At 632°C, some thermal effects due to the presence of the ternary peritectic in the GeTe–GeAs–GeAs$_2$ subsystem are manifested. This ingots were annealed at 350°C for 2 weeks.

GeTe–GeAs$_2$: This system is a non-quasibinary section of the Ge–As–Te ternary system, since GeAs is primarily crystallized in the middle composition range (Shu et al. 1987b).

GeTe–As: This system is also a non-quasibinary section of the Ge–As–Te ternary system (Shu et al. 1987b). Two isotherm lines of monotectic equilibria from the GeTe-side at 619°C and 551°C exist in the system. From the As-side, the third line of monovariant equilibrium begins to appear at 670°C. According to the data of Krebs et al. 1961a, the solubility of GeTe in As reaches 22.7 mol%.

GeTe–As$_2$Te$_3$: The phase diagram of this system is a eutectic type with peritectic transformations (Figure 6.44) (Abrikosov

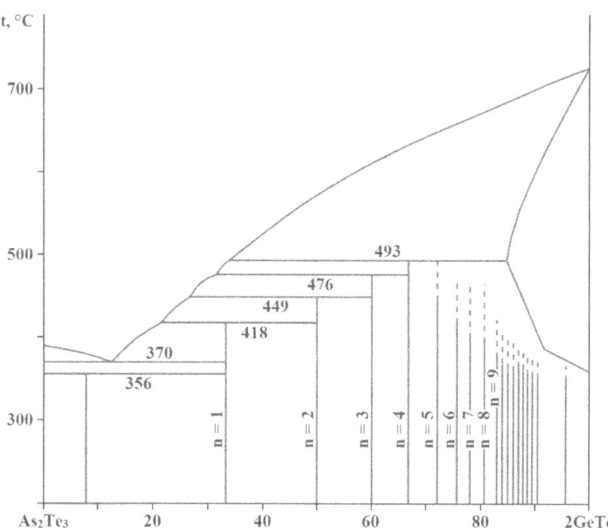

FIGURE 6.44 Phase diagram of the GeTe–As$_2$Te$_3$ system. (From Shu, H.W., et al., *J. Solid State Chem.*, **69**(1), 48, 1987.)

et al. 1977; Shu et al. 1986, 1987a). The eutectic contains 21.4 mol% GeTe and crystallizes at 370°C (Shu et al. 1986, 1987a) [contains 25 mol% GeTe and crystallizes at 355°C (Abrikosov et al. 1977)]. A whole series of the layered compounds with the general formula **Ge$_n$As$_2$Te$_{3+n}$** having a close crystalline structure is formed in the system. These compound were synthesized with $n \le 15$ (Shu et al. 1986, 1987a) [with $n \le 18$ (Kuypers et al. 1987, 1988a, b)]. The first of these compounds (up to $n = 8$) are characterized individually (Table 6.1), and for large n it becomes more difficult to isolate them. On the phase diagram, the monovariant lines of peritectic decomposition of the first four compounds exist: they melt incongruently at 418°C (**GeAs$_2$Te$_4$**), at 449°C [at 410°C (Abrikosov et al. 1977)] (**Ge$_2$As$_2$Te$_5$**), at 476°C (**Ge$_3$As$_2$Te$_6$**), and at 493°C [at 490°C (Abrikosov et al. 1977)] (**Ge$_4$As$_2$Te$_7$**) (Shu et al. 1986). At $5 \le n \le 8$, these compounds decompose in the solid state at temperatures of about 490°C, and at $n > 8$ they decompose in the temperature range 380°C–400°C with the formation of GeTe-based solid solution, which exist only at elevated temperatures. These compounds are layered, where one layer As$_2$Te$_3$ accounts for n layers GeTe.

The Ge$_n$As$_2$Te$_{3+n}$ ternary compounds were obtained by direct union of the three elements in a vacuum-sealed ampoule (Shu et al. 1986, 1987a; Jaulmes et al. 1987). Heating was carried out in three stages: 5–8 h at 800°C, 2 h at 1000°C, then 2 weeks at 350°C. This latter heating is necessary for the formation of the compound, the decomposition of which takes place around 490°C. The mass obtained is powdery but single crystals of sufficient size for the determination of the structure can be removed.

Three more compounds were detected in the GeTe–As$_2$Te$_3$ system. The **GeAs$_4$Te$_7$** metastable compound, not reflected in the state diagram, decomposes at 233°C to form α-As$_2$Te$_3$ and GeAs$_2$Te$_4$ (Shu et al. 1986, 1987a) [this compound melts incongruently at 370°C (Abrikosov et al. 1977)]. It crystallizes in the trigonal structure with the lattice parameters $a = 406.9$ and $c = 2331$ pm (Shu et al. 1988).

The second phase with an unknown structure exists near 15 mol% GeTe (Shu et al. 1986, 1987a). It decomposes in the solid state at 356°C and can only be obtained by prolonged

TABLE 6.1

Properties of the $Ge_nAs_2Te_{3+n}$ Ternary Compounds with Trigonal Structure

n	1	2	3	4	5	8
Compound	$GeAs_2Te_4$	$Ge_2As_2Te_5$	$Ge_3As_2Te_6$	$Ge_4As_2Te_7$	$Ge_5As_2Te_8$	$Ge_8As_2Te_{11}$
a, pm	408.3	408.4	410.2	410.6	411.2 (411.4 ± 0.1)[a]	411.7
c, pm	4038	1701	6159	7217	2754 (2756)[a]	3794
Calculated density, g·cm^{-3}	6.26	6.30	6.29	6.30	6.31 (6.30 and experimental value 6.2 ± 0.2)[a]	6.37
Melting temperature, °C	418 (incongr.)	449 (incongr.)	476 (incongr.)	493 (incongr.)	–	–

Source: Shu et al. (1986, 1988).

[a] Jaulmes et al. (1987).

annealing at 300°C–330°C. The third phase, which is different in structure, resembles an orthorhombic modification of γ-GeTe and the composition of this phase is near 97.5 mol% GeTe. It decomposes at 370°C to form a solid solution with a structure of NaCl-type and can be obtained by annealing at 350°C products obtained at higher temperatures.

The solubility of GeTe in As_2Te_3 is not higher than 5.8 mol% and the solid solutions based on low-temperature α-GeTe were not determined (Shu et al. 1986). According to the data of Abrikosov et al. (1977), the solubility of GeTe in As_2Te_3 reaches 8 mol% at 300°C and the solubility of As_2Te_3 in GeTe is 10 and 4 mol% at 500°C and 350°C, respectively.

This system was investigated through DTA, XRD, and metallography, and the ingots were annealed at 300°C, 330°C, 350°C, and 470°C for 300–1000 h (Abrikosov et al. 1977; Shu et al. 1986, 1987a).

The projection of the liquidus surface of the Ge–As–Te ternary system is given in Figure 6.45 (Shu et al. 1987b). The reaction scheme is also presented by the authors. There are three ternary eutectics and ten transition points on the liquidus surface. The liquidus surface of this system was also constructed by Vinogradova et al. (1975, 1976) and apart from the stable phases two metastable phases, named *X* and *Y*, were found. The phase *X* decomposes at the annealing forming the solid solution based on GeTe and As. The composition of these phases is near As–Te binary system.

The glass forming region in this system is given in Figure 6.46 (Hilton et al. 1966a). The glass transition temperature for these glasses is within the interval of 135°C–270°C. According to the data of Savage and Nielsen (1966) and Panus and Borisova (1966, 1967), only one glass forming region which coincide with one of the region from Hilton et al. (1966a) exists in this system. The obtaining glasses have enhanced crystallization properties (Panus and Borisova 1967). The crushed glassy alloys crystallize at a lower temperature than cast alloys.

6.37 Germanium–Antimony–Tellurium

GeTe–Sb: According to the first investigations of Krebs et al. (1961a), the solubility of GeTe in Sb reaches 23 mol%, but such significant solubility has not been confirmed by further studies (Bordas et al. 1986).

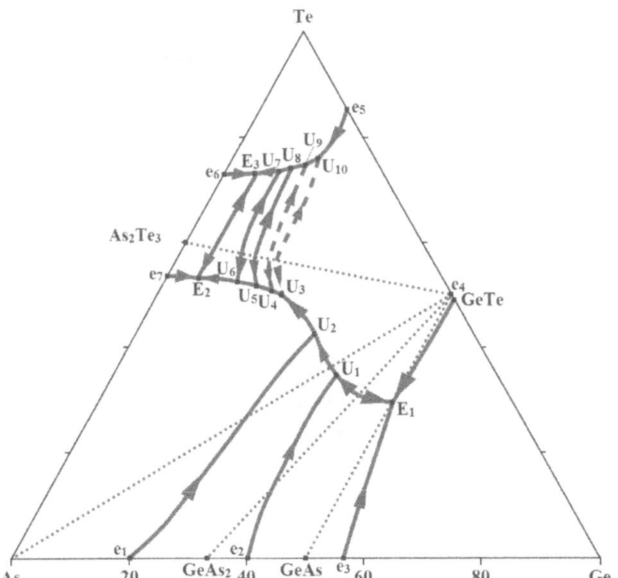

FIGURE 6.45 Projection of the liquidus surface of the Ge–As–Te ternary system. (From Shu, H.W., et al., *J. Solid State Chem.*, **69**(1), 55, 1987.)

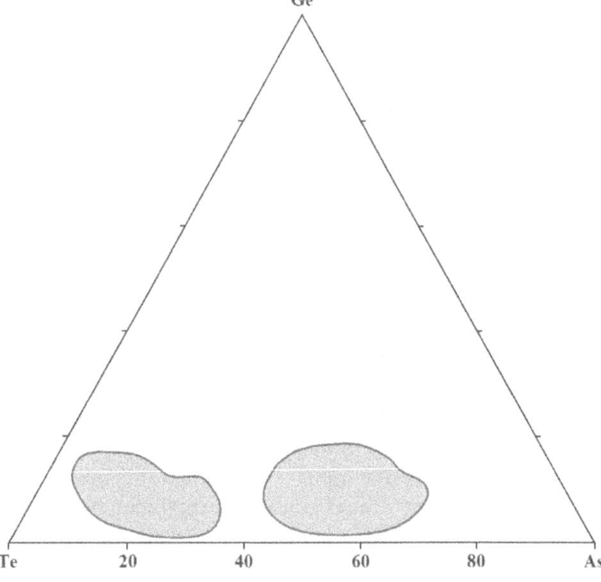

FIGURE 6.46 Glass forming region in the Ge–As–Te ternary system. (From Hilton, A.R., et al., *Infrared Phys.*, **6**(4), 183, 1966.)

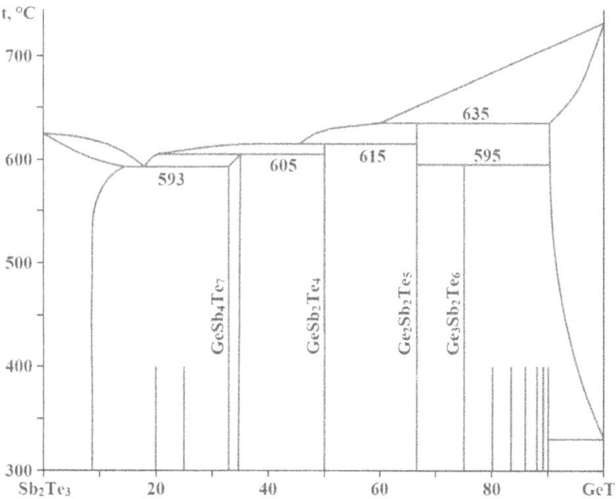

FIGURE 6.47 Phase diagram of the GeTe–Sb_2Te_3 system. (From Shelimova, L.E., et al., *Inorg. mater.*, **36**(3), 235, 2000.)

GeTe–SbTe: This system is a non-quasibinary section of the Ge–Sb–Te ternary system (Bordas et al. 1986) as Sb_2Te_3 crystallizes primarily from the SbTe-side. A single-phase solid solution based on GeTe crystallizes with an excess of Te in the lattice compared with stoichiometry (Bigvava et al. 1985).

GeTe–Sb_2Te_3: The phase diagram of this system is given in (Figure 6.47) (Abrikosov and Danilova-Dobryakova 1965b; Shelimova et al. 2000c) and shows the ***n*GeTe·*m*Sb_2Te_3** homologous series of ternary layered compounds (Frangis et al. 1989; Karpinsky et al. 1998a). The eutectic contains 18 mol% GeTe and crystallizes at 593°C. The melting behavior of these compounds is in most cases unknown. They have many-layered, long-period structures, in which the number of layers per unit cell depends in a systematic way on the relative amounts of the constituents tellurides (Shelimova et al. 2000c). Two kinds of crystalline phase exist: one is a metastable phase with a NaCl-type cubic structure and the other is a spectrum of stable phases with homologous structure (Matsunaga et al. 2008b). The NaCl-type metastable phase becomes unstable with increased Sb_2Te_3 content. The structure of room-temperature modifications exhibits trigonal twin domains, each comprising a stacking-disordered sequence distorted NaCl-type slabs with variable thickness (Urban et al. 2015a).

Several members of the series of *n*GeTe·Sb_2Te_3 compounds were prepared by mixing the constituent elements and heating in evacuated quartz tubes first at 800°C for a few hours and subsequently at 350°C for 2 weeks (Kuypers et al. 1988a). For higher *n* values (n = 11, 12, 15, 18) of the series and for a particular low *n* value (n = 4) the second part of the heat treatment was extended to 3 weeks at 300°C to increase the possibility of forming members of the homologous series corresponding with higher *n* values and of obtaining more regular structures. It is necessary to mention that a compound prepared in the way described above always contains several members of the homologous series, with *n* values varying around that corresponding with the overall composition. By means of electron diffraction and high-resolution electron microscopy it was shown that substitution of As in the homologous series of mixed layer compounds *n*GeTe·As_2Te_3 by Sb gives rise to a new, structurally identical, series of mixed layer compounds *n*GeTe·Sb_2Te_3. The duration of the heat treatment required to obtain regular structures increases with *n* and depends on the substituted element.

The solubility of GeTe in Sb_2Te_3 at 550°C is 10 mol% and decreases to 5 mol% at 300°C, and the solubility of Sb_2Te_3 in GeTe is equal 10 mol% at 550°C and 5 mol% at 400°C (Abrikosov et al. 1973b; Abrikosov and Danilova-Dobryakova 1965b; Shelimova et al. 2000c).

The next compounds were determined in the GeTe–Sb_2Te_3 system. **$GeSb_2Te_4$** melts incongruently at 615°C [according to the data of Skoropanov et al. (1985b), this compound melts congruently], its melting entropy and enthalpy are 160 J·(M.K)$^{-1}$ and 142 kJ·M^{-1}, respectively, and the formation enthalpy is −92 kJ·M^{-1} (Abrikosov and Danilova-Dobryakova 1965b; Skums et al. 1985; Shelimova et al. 2000c). This compound has two polymorphic modifications. The stable phase crystallizes in the tetragonal structure with the lattice parameters $a = 4227.21 \pm 0.01$ and $c = 4168.6 \pm 0.2$ pm at 600°C (Matsunaga and Yamada 2004) [a = 424 ± 1, c = 4112.5 ± 0.3 pm, and a calculated density of 6.467 g·cm^{-3} (Karpinsky et al. 1998a, b; Shelimova et al. 2000b, c); in the trigonal structure with the lattice parameters a = 425 and c = 4100 pm (Kooi and De Hosson 2002)]. The metastable phase crystallizes in the cubic structure with the lattice parameter $a = 604.30 \pm 0.09$ at room temperature (Matsunaga and Yamada 2004) [a = 602.6 pm (experimental value) and a = 592.2 and 607.0 pm according to the calculations based on the density approximation with local density approximation and general approximation, respectively (Sun et al. 2007); in the hexagonal structure with the lattice parameters a = 421 and c = 4060 pm (Petrov et al. 1968; Hulliger 1976)].

The specimen of the metastable phase for diffraction measurement was made using the following method (Matsunaga and Yamada 2004). First, a thin film of $GeSb_2Te_4$ with a thickness of approximately 0.5 μm was formed by sputtering on a glass disk. The film was amorphous just after its formation and crystallized into the metastable crystal phase by means of laser irradiation and was then scraped off with a spatula to create a powder. The metastable phase retains its structure up to around 230°C. Once changed into the stable phase, the structure remains stable over a wide temperature range of 90°C to around the melting point, with no transformation to the metastable phase observed. An intermediate phase could exist in the transition process from the solid to the liquid phase. $GeSb_2Te_4$ could be also synthesized by the sublimation (Agaev and Talybov 1966) and its single crystals could be grown by the Bridgman method (Frumar et al. 1972b).

$GeSb_2Te_4$ was also produced by mixing the pure Ge, Sb, and Te in an evacuated quartz tube at 750°C (Kooi and De Hosson 2002). The melt was solidified using three different cooling rates: (1) furnace cooled, (2) pulled out of the furnace on a metal plate at room temperature, and (3) quenching the liquid material into water (dispersing the melt into small solid pieces). Pieces of the alloy according to (1) and (2) were also annealed 24 h at 400°C. The crystals were grown from molten $Ge_2Sb_2Te_5$.

$GeSb_4Te_7$ melts congruently at 605°C [according to the data of Skoropanov et al. 1985b, this compound melts congruently], its melting entropy and enthalpy are 273 J. (M·K)$^{-1}$ and 240 kJ·M^{-1}, respectively, and the formation enthalpy is –140 kJ·M^{-1} (Skums et al. 1985). It crystallizes in the tetragonal structure with the lattice parameters a = 423.60 ± 0.02 and c = 2376.1 ± 0.2 pm at 90 K (Matsunaga et al. 2008a) [in the trigonal structure with the lattice parameter a = 425.0, c = 2385.2 ± 0.3 pm, and a calculated density of 6.503 g·cm^{-3} (Karpinsky et al. 1998a, b; Shelimova et al. 2000b, c); in the cubic structure with the lattice parameter a = 604.1 pm (experimental value) and a = 599.3 and 608.8 pm according to the calculations based on the density approximation with local density approximation and general approximation, respectively (Sun et al. 2007); in the hexagonal structure with the lattice parameters a = 421 and c = 2365 pm (Petrov et al. 1968; Hulliger 1976)]. The title compound was prepared by melting a stoichiometric mixture of Ge, Sb, and Te in a silica tube filled with Ar at 800°C and then quenching it in ice water (Matsunaga et al. 2008a). The polycrystalline alloys were annealed at 450°C. The resulting alloy ingot was annealed at 450°C for 32 days or at 800°C for 5 h in an Ar atmosphere. If there are no defects in the crystal and structure, this compound can be regarded as a semiconductor with a very narrow band gap.

$GeSb_6Te_{10}$ crystallizes in the trigonal structure with the calculated lattice parameters a = 423.6 and c = 10150.87 pm at 92 K (Matsunaga et al. 2010) [a = 425.4 and c = 10202 ± 1 pm and a calculated density of 6.512 g·cm^{-3} (Karpinsky et al. 1998a, b; Shelimova et al. 2000b, c)]. This compound is a semiconductor with a very narrow band gap (Matsunaga et al. 2010). It was obtained by melting a stoichiometric mixture of Ge, Sb, and Te in an Ar atmosphere and then quenching it in ice water. The resulting alloy ingots were respectively annealed at 500°C for 32 days, at 180°C for 9 days and at 503°C for 5 days in an Ar atmosphere.

$GeSb_8Te_{13}$ crystallizes in the trigonal structure with the calculated lattice parameters a = 425.7, c = 13256 pm, and a calculated density of 6.515 g·cm^{-3} (Karpinsky et al. 1998a, b; Shelimova et al. 2001a).

$Ge_2Sb_2Te_5$ melts congruently at 630°C (Abrikosov and Danilova-Dobryakova 1965b). It has two polymorphic modifications, and a metastable phase transforms into stable phase at around 230°C (Matsunaga et al. 2004). Zhang et al. (2007) found that the band gap of the amorphous $Ge_2Sb_2Te_5$ is 0.73 eV. The stable phase crystallizes in the trigonal structure with the lattice parameters a = 422.57 ± 0.01, c = 1728.09 ± 0.18 pm, and a calculated density of 6.35 g·cm^{-3} (Urban et al. 2013) [a = 422.47 ± 0.01 and c = 1723.91 ± 0.04 pm (Matsunaga et al. 2004); a = 422.4, c = 1729.9 ± 0.1 pm, and a calculated density of 6.441 g·cm^{-3} (Karpinsky et al. 1998a, b; Shelimova et al. 2000b, c); a = 425 and c = 1827 pm (Kooi and De Hosson 2002); in the hexagonal structure with the lattice parameters a = 420 and c = 1696 pm (Petrov et al. 1968; Hulliger 1976)]. The metastable phase crystallizes in the cubic structure with the lattice parameter a = 602.93 ± 0.01 pm (Matsunaga et al. 2004) [a = 601.17 ± 0.05 pm and the calculated and experimental densities of 6.28 g·cm^{-3} (20% vacancy), 6.97 g·cm^{-3} (no vacancy) and 5.86 g·cm^{-3} (as-deposited amorphous phase), 6.13 g·cm^{-3} (for annealed crystalline phase), respectively (Nonaka et al. 2000); a = 602.48 ± 0.02 pm (Yamada and Matsunaga 2000); a = 601.1 pm (experimental value) and a = 591.3 and 605.9 pm according to the calculations based on the density approximation with local density approximation and general approximation, respectively (Sun et al. 2007)]. The cubic modification was prepared through the following process (Yamada and Matsunaga 2000): deposition of the film on a glass disk substrate by sputtering a Ge–Sb–Te alloy target in Ar gas, crystallization of the film using a laser exposure instrument, and manually scratching of the crystallized film from the glass substrate using a spatula. Urban et al. (2013) was not determined the cubic high-temperature phase of this compound.

$Ge_2Sb_2Te_5$ was also produced in the same way as $GeSb_2Te_4$ was synthesized (Kooi and De Hosson 2002). Single crystals were prepared by chemical vapor transport (Urban et al. 2013). A stoichiometric mixture of Ge, Sb, and Te was sealed in a silica glass ampoule under dry argon and heated to 950°C for 2 h. After quenching in water, the product was crushed and 20 mass% SbI_3 were added as a transport agent. The mixture was sealed in a silica glass ampoule and placed in a two-zone furnace. Chemical transport took place from 585°C to 525°C for 17.5 h. The obtained crystals were washed with acetone.

The as-deposited $Ge_2Sb_2Te_5$ fully converts to a stressed metastable cubic phase (at ca. 135°C) at pressures below 3 GPa and remains cubic up to 240°C (Krbal et al. 2021). At higher pressures, the as-deposited phase partially crystallizes directly into the stable hexagonal phase at a significantly lower temperature (110°C), however a significant volume fraction of the amorphous phase remains even for temperatures as high as 240°C. In stark contrast, the pressure-induced amorphous phase crystallizes at ambient conditions at a lower temperature than its as-deposited counterpart. Furthermore, the pressure-induced amorphous phase also fully transforms directly into the hexagonal modification at pressures up to ca. 5 GPa. At higher pressure (8.5 GPa), an orthorhombic phase is formed. Different from the as-deposited phase, the crystallization temperature of pressure-induced amorphous $Ge_2Sb_2Te_5$ increases with pressure.

$Ge_{2.8}Sb_2Te_{5.8}$ crystallizes in the cubic structure with the lattice parameter a = 604.9 ± 0.6 and a calculated density of 6.1214 g·cm^{-3} (Urban et al. 2015a). The crystals of this compound were obtained via gas phase. A mixture of Ge, Sb, and Te (molar ratio 1.44:2:5) were fused at 950°C and subsequently quenched in water. About 200 mg of the powdered product was sealed in a silica ampoule (Ar atmosphere), kept at 628°C for 4 h and then slowly cooled to 618°C without further annealing. In addition to the slow cooling, the natural temperature gradient of the tube furnace was probably beneficial for crystals growth.

$Ge_3Sb_2Te_6$ forms by the peritectoid reaction β-GeTe + $Ge_2Sb_2Te_5$ ⇔ $Ge_3Sb_2Te_6$ (Shelimova et al. 2000c). It crystallizes in the trigonal structure with the lattice parameters a = 425 and c = 6260 pm (Kooi and De Hosson 2002) [a = 421.4, c = 6223.4 ± 0.5 pm, and a calculated density of 6.422 g·cm^{-3} (Karpinsky et al. 1998a, b; Shelimova et al. 2000b, c); in the tetragonal structure with the lattice parameters a = 421.28 ± 0.02 and c = 6230.9 ± 0.3 pm at room temperature and a = 419.78 ± 0.01 and c = 6202.1 ± 0.2 pm at 90 K (Matsunaga et al. 2007a); a = 419.1, c = 6028.1, and a = 428.7, c = 6312.3 pm according to the calculations based on the density approximation with local density approximation and general approximation, respectively (Sa et al. 2010b); in the cubic structure with the lattice parameter $a = 611.40 \pm 0.06$ pm at 600°C (Schneider et al. 2010a); $a = 600.2$ pm (experimental value) and $a = 594.3$ and 607.9 pm according to the calculations based on the density approximation with local density approximation and general approximation, respectively (Sun et al. 2007)]. This compound is a *p*-semiconductor with a very narrow band gap (Matsunaga et al. 2007a; Sa et al. 2010b). It was produced in the same way as $GeSb_2Te_4$ was synthesized (Kooi and De Hosson 2002; Schneider et al. 2010a). The nanostructured samples were obtained by reheating the obtained bulk samples to 500°C and quenching in water again (Schneider et al. 2010a).

$Ge_4Sb_2Te_7$ has two polymorphic modifications. First of them crystallizes in the trigonal structure with the lattice parameters $a = 418.9 \pm 0.1$ and $c = 6217 \pm 2$ pm and a calculated density of 6.373 ± 0.003 g·cm^{-3} (Schneider and Oeckler 2008) [$a = 420.7$, $c = 7279 \pm 1$ pm, and a calculated density of 6.408 g·cm^{-3} (Karpinsky et al. 1998a, b; Shelimova et al. 2000b, c)]. The second modification is high-temperature stable phase and crystallizes in the cubic structure with the lattice parameter $a = 598.8 \pm 0.2$ pm and a calculated density of 6.307 g·cm^{-3} (Rosenthal et al. 2014). Bulk samples of this compound were prepared by melting stoichiometric amounts of the pure elements in sealed silica ampoules under Ar atmosphere at 950°C (Schneider and Oeckler 2008; Rosenthal et al. 2014). After quenching to room temperature in water, the samples were annealed at 550°C for 100 h under Ar.

$Ge_5Sb_2Te_8$ crystallizes in the trigonal structure with the lattice parameters $a = 420.1$ and $c = 2778$ pm (Karpinsky et al. 1998a, b). The octahedral crystals were grown in the stability range the high-temperature phase by a chemical transport reactions (Urban et al. 2015a). Quenched and powdered **$Ge_{12}Sb_2Te_{15}$** (typically 120 mg, obtained by quenching a melt with stoichiometric composition) and SbI_3 (ca. 50 mg) as transport agent and Sb source were sealed in silica ampoule. Chemical transport was carried out from 600°C to 520°C within 1 day. Subsequently, the ampoule was quenched in air.

$Ge_6Sb_2Te_9$ has two polymorphic modifications. Low-temperature modification crystallizes in the trigonal structure with the lattice parameters $a = 417.59 \pm 0.02$, $c = 9339.6 \pm 0.4$ pm at 90 K (Matsunaga et al. 2008b) [$a = 419.7$ and $c = 9375$ pm (Karpinsky et al. 1998a, b)]. High-temperature modification of this compound is formed at 550°C and crystallizes in the cubic structure with the lattice parameter $a = 600$ pm (Schneider et al. 2012). According to the data of Matsunaga et al. (2008b), $Ge_6Sb_2Te_9$ maintains a NaCl-type structure from 92 K to near the melting point.

This compound is *p*-type semiconductor (Schneider et al. 2012).

$Ge_7Sb_2Te_{10}$ has two polymorphic modifications. First of them crystallizes in the trigonal structure with the lattice parameters $a = 423.7 \pm 0.3$, $c = 1029 \pm 1$ pm, and a calculated density of 6.314 g·cm^{-3} (Schneider et al. 2010b) [$a = 419.4$ and $c = 10442$ pm (Karpinsky et al. 1998a, b)]. The second modification is a high-temperature stable phase and crystallizes in the cubic structure with the lattice parameter $a = 599.160 \pm 0.006$ pm and a calculated density of 6.379 g·cm^{-3} (Rosenthal et al. 2014). This compound was synthesized under Ar or N_2 atmosphere in sealed silica glass ampoule (Schneider et al. 2010b; Rosenthal et al. 2014). Stoichiometric mixture of the elements was melted (minimum 1 h at 900°C–950°C) and quenched in water. The compact metallic gray ingot was annealed in the existence range of the cubic phase, i.e. between 560°C and 610°C, for up to 10 days or at 500°C for 20 h and quenched in water again.

$Ge_8Sb_2Te_{11}$ has two polymorphic modifications. The stable low-temperature modification crystallizes in the rhombohedral structure with the lattice parameters $a = 424.75 \pm 0.03$ pm and $\alpha = 59.314 \pm 0.003°$ at 92 K (Matsunaga et al. 2008b) [in the trigonal structure with the lattice parameters $a = 420.3$ and $c = 1045.8$ pm at 92 K (Schneider et al. 2010b); $a = 419.1$ and $c = 3824$ pm (Karpinsky et al. 1998a, b); $a = 412.3$ and $c = 2756.9$ pm (Schürmann et al. 2011)]. The metastable high-temperature modification crystallizes in the cubic structure with the lattice parameters $a = 605.80 \pm 0.01$ pm (Matsunaga et al. 2008b). Bulk sample of this compound was prepared by heating the stoichiometric mixture of the powered Ge, Sb, and Te in evacuated pre-flushed (nitrogen) sealed silica ampoule to 500°C with a rate of 50°C·h^{-1} or to 950°C (Schneider et al. 2010b; Schürmann et al. 2011). After annealing for 3 days at 500°C or at 950°C for ca. 2 h, the ampoule was quenched in water. Single-phase samples could not be obtained (Schneider et al. 2010b). This compound is a semiconductor (Matsunaga et al. 2008b).

$Ge_9Sb_2Te_{12}$ crystallizes in the trigonal structure with the lattice parameters $a = 418.8$ and $c = 12519$ pm (Karpinsky et al. 1998a, b).

$Ge_{10}Sb_2Te_{13}$ has two polymorphic modifications, and the transformation takes place at about 160°C (Matsunaga et al. 2006). The low-temperature modification crystallizes in the rhombohedral structure with the lattice parameters $a = 425.82 \pm 0.02$ pm and $\alpha = 59.223 \pm 0.004°$ at room temperature. The high-temperature modification crystallizes in the cubic structure with the lattice parameter $a = 600.28 \pm 0.05$ pm at 200°C. The specimens for diffraction measurement were made using the following method. First, GeTe–Sb_2Te_3 thin films with a thickness of about 0.3 μm were deposited onto glass disks with a diameter of 120 mm by DC magnetron sputtering from compound targets in an Ar gas atmosphere. The films were amorphous just after their formation and were crystallized by laser irradiation into cubic or rhombohedral NaCl-type structures. Each film was then powdered by being scraped with a spatula.

$Ge_{11}Sb_2Te_{14}$ crystallizes in the cubic structure with the lattice parameter $a = 597.9 \pm 0.6$ and a calculated density of 6.337 g·cm^{-3} (Urban et al. 2015a). To prepare this compound, a mixture of Ge, Sb, and Te (molar ratio 1.44:2:5) were fused at 950°C and subsequently quenched by cooling the ampoule in water. About 200 mg of the powdered product was sealed in a silica ampoule (Ar atmosphere), kept at 628°C for 4 h and then slowly (within 6 h) cooled to 618°C. After holding this temperature for 75 h, the ampoule was quenched in air. The octahedral crystals of target compound were obtained.

$Ge_{12}Sb_2Te_{15}$ has two polymorphic modifications. First of them crystallizes in the trigonal structure with the lattice parameters 421.28 ± 0.01, $c = 1039.71 \pm 0.07$ pm, and a calculated density of 6.295 g·cm^{-3} (Rosenthal et al. 2013) [$a = 423.7$ and $c = 1036$ pm (Schneider et al. 2010b); in the hexagonal structure with the lattice parameters $a = 425$ and $c = 1052$ pm at 400°C (Schneider et al. 2012)]. The second modification crystallizes in the cubic structure with the lattice parameter $a = 608.26 \pm 0.05$ pm at 600°C (Schneider et al. 2010a)]. This compound was prepared by melting stoichiometric amounts of the elements in silica glass ampoule sealed under Ar or N_2 atmosphere at 900°C–950°C for 2 h (Schneider et al. 2010b; Rosenthal et al. 2013). The sample was quenched in water and annealed at 500°C–590°C for 1–3 days with subsequent quenching in water once again. Single crystals of this compound were grown by the chemical transport reactions using iodine as a transport agent (Schneider et al. 2010a). The nanostructured samples were obtained by reheating the obtained bulk samples to 500°C and quenching in water again (Schneider et al. 2010a).

High-resolution TEM on quenched samples of $Ge_{12}Sb_2Te_{15}$ reveal nanoscale twin domains (Schneider et al. 2010b). Cation defects form planar domain boundaries. The metastability of the samples was proved by *in situ* temperature-dependent powder diffraction experiments, which upon heating show a slow phase transition to a trigonal layered structure at ca. 325°C. The stable cubic high-temperature modification is formed at about 475°C. Powder diffraction on samples with defined particle sizes reveal that the formation of the stable superstructure phase is influenced by stress and strain induced by the twinning and volume change due to the cubic → rhombohedral phase transition upon quenching. This compound is metastable and is *p*-type semiconductor (Schneider et al. 2012).

$Ge_{15}Sb_2Te_{18}$ crystallizes in the hexagonal structure with the lattice parameters $a = 422$ and $c = 1057$ pm (Schneider et al. 2012). This compound is metastable and is *p*-type semiconductor.

The GeTe–Sb_2Te_3 system was investigated through DTA, XRD, metallography and measuring of the microhardness and the ingots were annealed at 500°C for 1000 h (Abrikosov et al. 1973b; Abrikosov and Danilova-Dobryakova 1965b; Shelimova et al. 2000c).

Two more ternary compounds, **$GeSb_4Te_4$** and **$Ge_{2-x}Sb_{2+x}Te_5\cdot Sb_8$** ($x = 0.43$) or **≈$Ge_9Sb_{61}Te_{29}$**, were found in the Ge–Sb–Te system. First of them melts incongruently at ca. 540°C and crystallizes in the trigonal structure with the lattice parameters $a = 424.66 \pm 0.02$, $c = 1748.3 \pm 0.3$ pm, and a calculated density of 6.507 ± 0.001 g·cm^{-3} (Schneider and Oeckler 2010). This compound can be obtained by quenching a stoichiometric melt and subsequent annealing it at 500°C for 150 h.

$Ge_{2-x}Sb_{2+x}Te_5\cdot Sb_8$ is kinetically stable at room temperature, also melts incongruently at 529°C and crystallizes in the trigonal structure with the lattice parameters $a = 425.8 \pm 0.1$, $c = 9723 \pm 2$ pm, and a calculated density of 6.598 g·cm^{-3} (Schneider et al. 2009). It is a degenerate semiconductor. To prepare this compound, mixture of the powdered elements was heated in a sealed silica glass ampoule under Ar atmosphere with a rate of 15°C·min^{-1} to 850°C. This temperature was maintained for 4 h, and then the melts were cooled to 350°C with a rate of 0.1°C·min^{-1}. After annealing at 350°C for 15 h, the sample was cooled to room temperature by switching off the furnace. Initially, the crystals of $Ge_{2-x}Sb_{2+x}Te_5\cdot Sb_8$ were isolated from ingots with the nominal compositions $Ge_{12}Sb_{59}Te_{29}$.

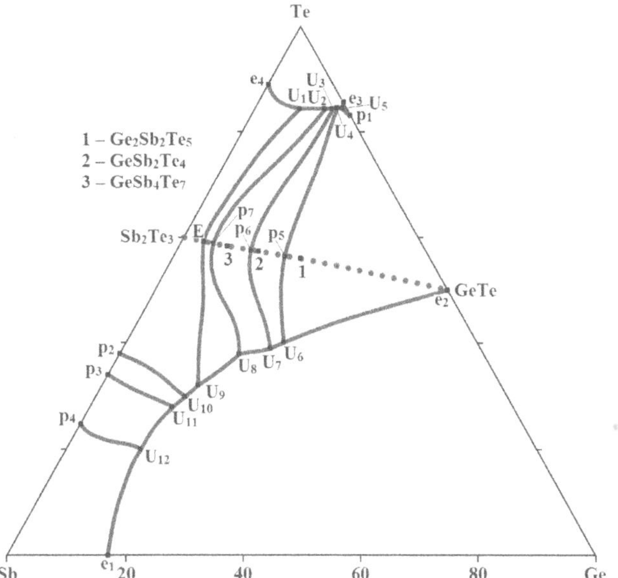

FIGURE 6.48 Projection of the liquidus surface of the Ge–Sb–Te ternary system. (From Shestakov, V.A., et al., *Zhurn. neorgan. khimii*, **46**(5), 836, 2001.)

The projection of the liquidus surface of the Ge–Sb–Te ternary system is given in Figure 6.48 (Abrikosov and Danilova-Dobryakova 1970a; Legendre et al. 1984; Bordas et al. 1986; Shestakov et al. 2001). It consists from 11 fields of primary crystallization, and the next invariant equilibria exist in the system (Shestakov et al. 2001): E (593°C) – L ⇔ (Sb_2Te_3) + ($GeSb_4Te_7$); U_1 (406.2°C) – L + (Sb_2Te_3) ⇔ ($GeSb_4Te_7$) + Te; U_2 (399.4°C) – L + ($GeSb_4Te_7$) ⇔ $GeSb_2Te_4$ + Te; U_3 (396°C) – L + $Ge_2Sb_2Te_5$ ⇔ $Ge_2Sb_2Te_5$ + Te; U_4 (393°C) – L + $GeSb_2Te_4$ ⇔ β-GaTe + Te; U_5 (391°C) – L + β-GeTe ⇔ α-GaTe + Te; U_6 (615.3°C) – L + β-GeTe ⇔ $Ge_2Sb_2Te_5$ + Ge; U_7 (600°C) – L + $Ge_2Sb_2Te_5$ ⇔ $GeSb_2Te_4$ + Ge; U_8 (561.9°C) – L + $GeSb_2Te_4$ ⇔ ($GeSb_4Te_7$) + Ge; U_9 (540.4°C) – L + ($GeSb_4Te_7$) ⇔ (Sb_2Te_3) + Ge; U_{10} (534°C) – L + (Sb_2Te_3) ⇔ γ + Ge; U_{11} (526.7°C) – L + γ ⇔ δ + Ge; and U_{12} (532.1°C) – L + Sb ⇔ δ + Ge, where γ and δ – the phases in the Sb–Te binary system. The authors also presented a mathematical model of the liquidus surface.

The isothermal section the Ge–Sb–Te ternary system at 350°C (Figure 6.49) was constructed through XRD, metallography, and microhardness measurements (Shelimova et al. 2001a). This section contains the phase fields of seven *n*GeTe·*m*Sb_2Te_3 compounds and the regions of the solid solutions based on GeTe and Sb_2Te_3. The region of the GeTe-based solid solutions in this ternary system is elongated along the GeTe–Sb_2Te_3 section (Figure 6.50) (Abrikosov and Danilova-Dobryakova 1974).

The glass forming regions in the Ge–Sb–Te ternary system is presented in Figure 6.51 (Minaev 1983; Lebaudy et al. 1991). Synthesis of the glassy alloys was carried out from the elements in quartz ampoules in a horizontally rotating furnace under vacuum at 1000°C for 12 h with vigorous stirring and the next quenching in water. The glass transition temperature of the amorphous alloys is within the interval from 108°C to 127°C. The vitreous state was identified by a characteristic glass wedge, a shell fracture, as well as through XRD, DTA,

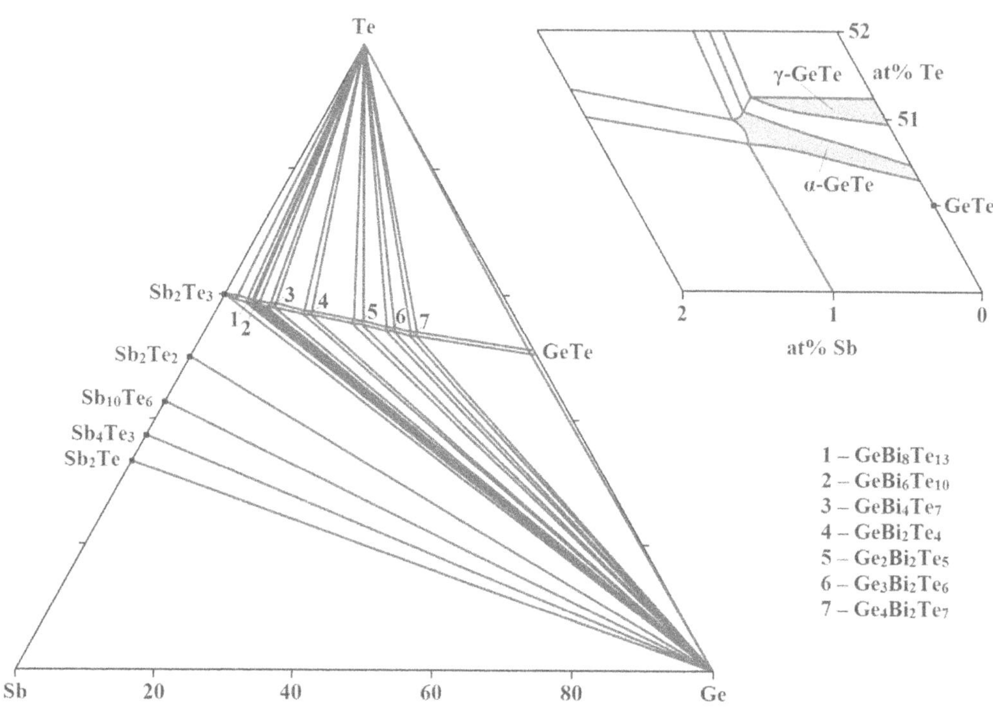

FIGURE 6.49 Isothermal section of the Ge–Sb–Te ternary system at 350°C. (From Shelimova, L.E., et al., *Inorg. Mater.*, **37**(4), 342, 2001.)

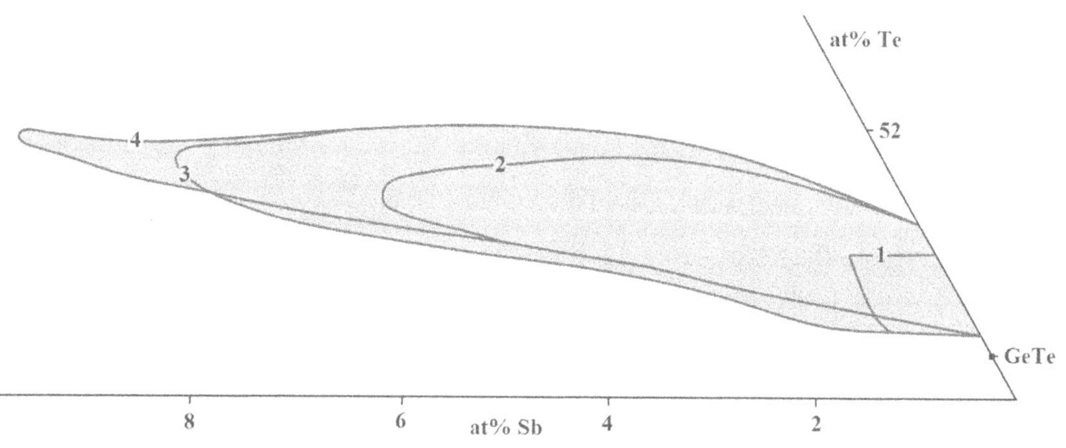

FIGURE 6.50 Regions of the GeTe-based solid solutions in the Ge–Sb–Te ternary system at (1) 250°C, (2) 400°C, (3) 500°C, and (4) 600°C (From Abrikosov, N.H., and Danilova-Dobryakova, G.T., *Izv. AN SSSR. Neorgan. mater.*, **10**(5), 811, 1974.)

and metallography. The boundary of the glass formation region was fixed with an accuracy of 2 at%.

6.38 Germanium–Bismuth–Tellurium

GeTe–BiTe: This system is a non-quasibinary section of the Ge–Bi–Te ternary system and intersects the fields of primary crystallization of (Bi_2Te_3), **$GeBi_4Te_7$**, Ge, and (GeTe) (Abrikosov and Danilova-Dobryakova 1970b).

$GeTe$–Bi_2Te_3: The phase diagram of this system is given in (Figure 6.52) (Abrikosov and Danilova-Dobryakova 1965a; Shelimova et al. 1997, 2000c, 2004a) and shows the ***n*GeTe·*m*Bi$_2$Te$_3$** homologous series of ternary layered compounds (Frangis et al. 1989; Karpinsky et al. 1998c; Shelimova et al. 2000a). The eutectic contains 25 mol% GeTe and crystallizes at 552°C. The melting behavior of these compounds is in most cases unknown. They have many-layered, long-period structures, in which the number of layers per unit cell depends in a systematic way on the relative amounts of the constituents tellurides (Shelimova et al. 2000a,c).

Several members of the series of nGeTe·Bi_2Te_3 compounds were prepared by mixing the constituent elements and heating in evacuated quartz tubes first at 800°C for a few hours and subsequently at 350°C for 2 weeks (Kuypers et al. 1988a). For higher n values (n = 11, 12, 15, 18) of the series and for a particular low n value (n = 4) the second part of the heat treatment was extended to 3 weeks at 300°C to increase the possibility of forming members of the homologous series corresponding with higher n values and of obtaining more regular structures.

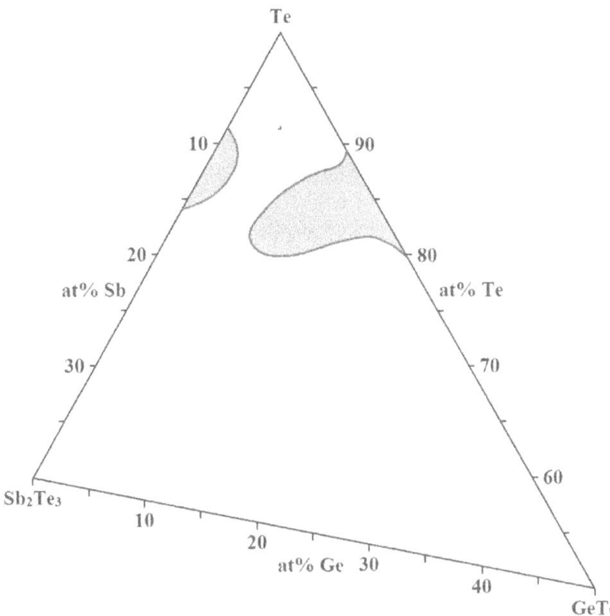

FIGURE 6.51 Glass forming regions in the Ge–Sb–Te ternary system. (From Lebaudy, P., et al., *Mater. Sci. Eng. A*, **132**, 273, 1991.)

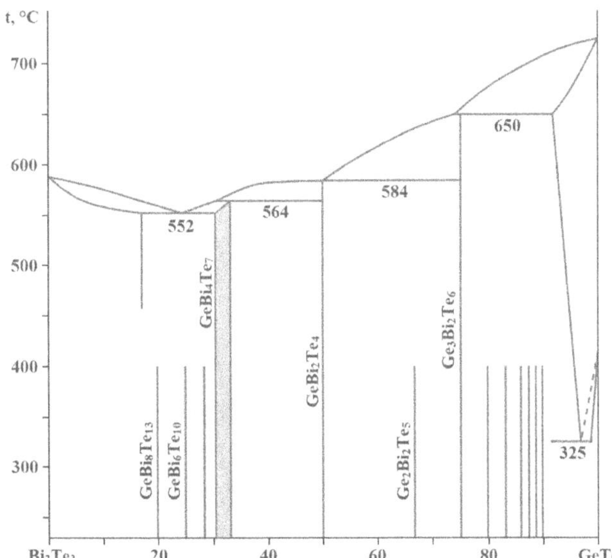

FIGURE 6.52 Phase diagram of the GeTe–Bi_2Te_3 system. (From Shelimova, L.E., et al., *Inorg. mater.*, **40**(5), 451, 2004.)

It is necessary to mention that a compound prepared in the way described above always contains several members of the homologous series, with *n* values varying around that corresponding with the overall composition. The GeTe-rich and Bi_2Te_3-rich members differ significantly in structure components (Shelimova et al. 2000a,c).

The solubility of GeTe in Bi_2Te_3 at 500°C reaches 17.4 mol% (Abrikosov et al. 1965a; Abrikosov and Danilova-Dobryakova 1972b), and the solubility of Bi_2Te_3 in GeTe is 4 mol% at 550°C (Abrikosov et al. 1980d). With increasing of Bi_2Te_3 content, the temperature of GeTe polymorphic transformation decreases and at 325°C the eutectoid decomposition of a solid solution based on GeTe takes place (Abrikosov and Danilova-Dobryakova 1965a; Shelimova et al. 2000c, 2004a). The eutectoid point corresponds to the composition of 97.3 mol% GeTe.

The next compounds were determined in the GeTe–Bi_2Te_3 system. **$GeBi_2Te_4$** melts incongruently at 584°C (Abrikosov and Danilova-Dobryakova 1965a; Shelimova et al. 2000c, 2004a) [melts congruently at 587°C with melting enthalpy of 139 kJ·M^{-1} (Valevskiy et al. 1983; Skoropanow et al. 1987)]. Its formation enthalpy is –114 kJ·M^{-1} (Skoropanov et al. 1983, 1985b; Skoropanow et al. 1987). This compound crystallizes in the trigonal structure with the lattice parameters $a = 432.3 \pm 0.1$, $c = 4129.3 \pm 1.5$ pm, and an energy gap of 0.19 eV (Kuznetsova et al. 2000) [$a = 432.2 \pm 0.5$, $c = 4127 \pm 2$ pm and a calculated density of 7.512 g·cm^{-3} (Karpinsky et al. 1998b, c; Shelimova et al. 2000c, 2001b, 2004a); in the hexagonal structure with the lattice parameters a = 428 and c = 3920 pm (Agaev and Semiletov 1965), an energy gap of 0.23 eV (Abrikosov and Danilova-Dobryakova 1965a) and an experimental density of 7.54 g·cm^{-3} (Petrov and Imamov 1970); a = 432.4 and c = 4138 pm (Shelimova et al. 1993b)]. The energy gap of $GeBi_2Te_4$ is 0.20 eV at 0 K (Konstantinov et al. 1999). Using the dilatometer, the phase transition temperature of this compound was determined depending on the degree of deviation from stoichiometry (Shelimova and Kretova 1993). It was established that the temperature of the phase transition decreases with increasing concentration of Ge, Bi, and Te excess atoms. It was determined that this phase transition belongs to the phase transitions of the type I. Single crystals of $GeBi_2Te_4$ were obtained by sublimation (Agaev and Semiletov 1965).

This compound was prepared by melting a charge in evacuated sealed quartz ampoule coated inside with a pyrolytic graphite layer (Kuznetsova et al. 2000). The charge from required amounts of Bi_2Te_3 and GeTe was melted at 680°C–730°C for 5–6 h with periodical vibration followed by quenching rapidly in water. The polycrystalline ingot was homogenized at 510°C for 750 h in evacuated sealed quartz ampoules followed by rapid quenching.

By means of an XRD study of $Ge_{1\pm\delta}Bi_2Te_4$($-0.06 \le \delta \le 0.20$) single crystals and powders, it was shown that excess with respect to stoichiometry Ge atoms enter in the van der Waals gaps between the packets in a strictly ordered manner (Karpinskiy et al. 1993). As a result, a 36-layer structure is formed, characterized by the alternation of one germanium layer and five seven-layer packets. With a low deviation from stoichiometry ($\delta > 0$), fragments of 21- and 36-layer structures coexist in the crystal, alternating in the direction of the *c* axis.

$GeBi_4Te_7$ melts incongruently at 564°C (Abrikosov and Danilova-Dobryakova 1965a; Shelimova et al. 2000c, 2004a) [melts congruently at 574°C (Valevskiy et al. 1983; Skoropanow et al. 1987)]. Its formation and melting enthalpies are –191 and 263 kJ·M^{-1}, respectively (Valevskiy et al. 1983; Skoropanov et al. 1983, 1985b; Skoropanow et al. 1987). The title compound crystallizes in the trigonal structure with the lattice parameters $a = 435.1 \pm 0.3$, $c = 2391 \pm 2$ pm, and an energy gap of 0.21 eV (Kuznetsova et al. 2000) [$a = 434.8 \pm 0.2$ and $c = 2391 \pm 2$ pm and a calculated density of 7.638 g·cm^{-3} (Karpinsky et al. 1998b, c; Shelimova et al. 2000c, 2001b, 2004a); in the hexagonal structure with the lattice parameters a = 435.2,

c = 2392.5 pm, and an experimental density of 7.61 [8.03 (Petrov and Imamov 1970)] g·cm^{-3} (Zhukova and Zaslavskiy 1970a, 1971); a = 436 and c = 2411 pm (Agaev et al. 1968)]. The energy gap of $GeBi_4Te_7$ is 0.22 eV at 0 K (Konstantinov et al. 1999). Using the dilatometer, the phase transition temperature of this compound was determined depending on the degree of deviation from stoichiometry (Shelimova and Kretova 1993). It was established that the temperature of the phase transition decreases with increasing concentration of Ge, Bi, and Te excess atoms. It was determined that this phase transition belongs to the phase transitions of the type I.

This compound was prepared in the same way as $GeBi_2Te_4$ was obtained (Kuznetsova et al. 2000).

$GeBi_6Te_{10}$ crystallizes in the trigonal structure with the lattice parameters a = 435.9 and c = 10201 ± 2 pm and a calculated density of 7.769 g·cm^{-3} (Karpinskiy et al. 2000; Shelimova et al. 2000c, 2001b, 2004a) [a = 435.15 and c = 10187.39 pm at 92 K (Matsunaga et al. 2010); a = 435.9 and c = 10227 pm (the predicted values) (Karpinsky et al. 1998b, c)]. This compound is a semiconductor with a very narrow band gap (Matsunaga et al. 2010).

$GeBi_8Te_{13}$ crystallizes in the trigonal structure with the lattice parameters a = 436.3 and c = 13268.2 pm and a calculated density of 7.795 g·cm^{-3} (Shelimova et al. 2000c, 2001b) [a = 436.5 and c = 13276 pm (the predicted values) (Karpinsky et al. 1998b, c)].

$Ge_2Bi_2Te_5$ has two polymorphic modifications. First of them crystallizes in the trigonal structure with the lattice parameters a = 428.072 ± 0.012, c = 1726.51 ± 0.06 pm, and a calculated density of 7.28 g·cm^{-3} at 87 K and a = 430.040 ± 0.013, c = 1736.59 ± 0.06 pm, and a calculated density of 7.17 g·cm^{-3} at 293 K and (Matsunaga et al. 2007b) [a = 428.6 ± 0.2, c = 1735 ± 1 pm, and a calculated density of 7.244 g·cm^{-3} (Karpinsky et al. 1998b, c; Shelimova et al. 2000c, 2001b)]. Metastable phase crystallizes in the cubic structure with the lattice parameter a = 609.32 ± 0.05 and 611.09 ± 0.09 pm and a calculated density of 6.98 and 6.96 g·cm^{-3} at 87 and 293 K, respectively (Matsunaga et al. 2007b).

$Ge_2Bi_{10}Te_{17}$ crystallizes in the trigonal structure with the lattice parameters a = 435.5 and c = 17371.1 pm (Karpinskiy et al. 2000; Shelimova et al. 2000c, 2001b).

$Ge_3Bi_2Te_6$ melts incongruently at 650°C and has a homogeneity region within the interval of 30–35 mol% GeTe (Abrikosov and Danilova-Dobryakova 1965a; Shelimova et al. 2000c, 2004a) [at 585°C (Valevskiy et al. 1983; Skoropanow et al. 1987)]. Its formation and melting enthalpy are −234 and 213 kJ·M^{-1}, respectively (Valevskiy et al. 1983; Skoropanov et al. 1983, 1985b; Skoropanow et al. 1987). This compound crystallizes in the trigonal structure with the lattice parameters a = 427.8 ± 0.1, c = 6256.7 ± 0.7 pm (Kuznetsova et al. 2000) [a = 426.8 ± 0.1, c = 6259 ± 2 pm, and a calculated density of 7.111 g·cm^{-3} (Karpinsky et al. 2000; Shelimova et al. 2000c, 2001b); a = 426.8 ± 0.1, c = 6264 ± 2 pm (Karpinsky et al. 1998b, c); in the hexagonal structure with the lattice parameters a = 421 ± 2, c = 6100 ± 50 pm, and an experimental density of 7.15 g·cm^{-3} (Petrov and Imamov 1970)]. The energy gap of $Ge_3Bi_2Te_6$ is 0.24 eV at 0 K (Konstantinov et al. 1999). Using the dilatometer, the phase transition temperature of this compound was determined depending on the degree of deviation from stoichiometry (Shelimova and Kretova 1993). It was established that the temperature of the phase transition decreases with increasing concentration of Ge, Bi, and Te excess atoms. It was determined that this phase transition belongs to the phase transitions of the type I.

This compound was prepared in the same way as $GeBi_2Te_4$ was obtained (Kuznetsova et al. 2000).

$Ge_4Bi_2Te_7$ has two polymorphic modifications. First of them crystallizes in the trigonal structure with the lattice parameters a = 428.18 ± 0.16, c = 1048.8 ± 0.4 pm (Urban et al. 2015b) [a = 425.9 ± 0.1, c = 7313 ± 1 pm, and a calculated density of 6.989 g·cm^{-3} (Shelimova et al. 2000c, 2001b); a = 417.1 and c = 7284.7 pm (Schürmann et al. 2011); a = 425.2 and c = 7339.2 pm (the predicted values) (Karpinsky et al. 1998b, c)]. Metastable modification, which is highly disordered, crystallizes in the cubic structure with the lattice parameter a = 605.5 ± 0.2 and a calculated density of 6.807 g·cm^{-3} (Urban et al. 2015b).

$Ge_5Bi_2Te_8$ crystallizes in the trigonal structure with the lattice parameters a = 423.8 ± 0.2, c = 2789 ± 4 pm, and a calculated density of 6.936 g·cm^{-3} (Shelimova et al. 2001b) [a = 424.1 and c = 2802.4 pm (the predicted values) (Karpinsky et al. 1998b, c)].

$Ge_6Bi_2Te_9$ crystallizes in the trigonal structure with the lattice parameters a = 417.5, c = 9367.7 pm (Schürmann et al. 2011) [a = 423.2, c = 9474.8 pm (the predicted values) (Karpinsky et al. 1998b, c)].

$Ge_7Bi_2Te_{10}$ crystallizes in the trigonal structure with the lattice parameters a = 422.5, c = 10542 pm (the predicted values) (Karpinsky et al. 1998b, c).

$Ge_8Bi_2Te_{11}$ crystallizes in the trigonal structure with the lattice parameters a = 421.9, c = 3869.6 pm (the predicted values) (Karpinsky et al. 1998b, c).

$Ge_9Bi_2Te_{12}$ crystallizes in the trigonal structure with the lattice parameters a = 421.4, c = 12676 pm (the predicted values) (Karpinsky et al. 1998b, c).

Two non-stoichiometric phases, **$Ge_{2-x}Bi_{2+x}Te_5$** and **$Ge_{4-x}Bi_{2+x}Te_7$**, which crystallizes in the trigonal structure with the lattice parameters a = 431.0 ± 0.1, c = 1738.4 ± 0.3 pm (Shelimova et al. 1996) [a = 430.5 ± 0.5, c = 1737.2 ± 1.5 pm for **$Ge_{1.5}Bi_{2.5}Te_5$** (Karpinsky et al. 1998c)], and a = 426.0 ± 0.2, c = 7328 ± 1 pm for **$Ge_{3.66}Bi_{2.34}Te_7$** (Shelimova et al. 1996), respectively, have been found in the Ge–Bi–Te ternary system at 500°C.

To obtain the nGeTe·$m$$Bi_2Te_3$ compounds (n = 1–9, m = 1–4), the starting materials (Ge, Bi, and Te) were sealed in evacuated quartz ampoules and melted at 800°C for 5 h (Karpinsky et al. 1998c; Konstantinov et al. 1999; Shelimova et al. 2001b, 2004a). After synthesis, the ampoules were air-quenched or annealed at 500°C for 1000 h and then at 350°C for 1200 h or at 300°C for 1500 h. X-ray powder diffraction of high members of the homologous series demonstrates that the alloys enriched by GeTe ($3 < n/m \le 9$) contain a mixture of domains of different members of the homologous series.

$GeBi_6Te_{10}$ was obtained by melting a stoichiometric mixture of Ge, Sb, and Te in an Ar atmosphere and then quenching it in ice water (Matsunaga et al. 2010). The resulting alloy ingots were respectively annealed at 500°C for 32 days, at 180°C for 9 days and at 503°C for 5 days in an Ar atmosphere.

To prepare $Ge_4Bi_2Te_7$, a mixture of Ge, Bi, and Te (molar ratio 3:2:6; total amount 280.2 mg) was melted in a sealed silica glass ampoule under dry Ar atmosphere at 950°C for 2 h and quenched in water (Urban et al. 2015b). For crystal growth, 256.1 mg of the product was sealed in a silica glass ampoule with 18 mass% I_2 as transport agent under vacuum. Chemical vapor transport was carried out in a two-zone furnace from 650°C to 570°C for 18 h. A crystal was isolated under an optical microscope and washed with acetone.

Bulk sample of the $Ge_8Bi_2Te_{11}$ compound was prepared by heating the stoichiometric mixture of the powered Ge, Sb, and Te in evacuated pre-flushed (nitrogen) sealed silica ampoule to 500°C with a rate of 50°C·h^{-1} or to 950°C (Schürmann et al. 2011). After annealing for 3 days at 500°C or at 950°C for ca. 2 h, the ampoule was quenched in water.

The GeTe–Bi_2Te_3 system was investigated through DTA, XRD, metallography, and measuring of the microhardness (Abrikosov and Danilova-Dobryakova 1965a; Shelimova et al. 2000c, 2004a). The ingots were annealed at 500°C–550°C for 1000–1500 h (Abrikosov and Danilova-Dobryakova 1965a, 1972b; Shelimova et al. 2000c, 2004a) [at 550°C for 600 h (Abrikosov et al. 1980d)].

The liquidus surface of the GeTe–Bi_2Te_3–Bi–Ge subsystem of the Ge–Bi–Te ternary system is presented in Figure 6.53 (Abrikosov and Danilova-Dobryakova 1970b). It consists from nine fields of the primary crystallization of phases, with the largest region occupied by the field of Ge primary crystallization. There are 7 transition points and one ternary eutectic in this subsystem. The minimum on the liquidus surface is at the point of ternary eutectic with a crystallization temperature of 254°C, which practically coincides with the Bi melting point.

The excess enthalpies of liquid alloys of the Ge–Bi–Te ternary system were determined in a heat-flow calorimeter for five sections Bi_xGe_{1-x}–Te (x = 0.2, 0.4, 0.5, 0.6, and 0.8) by

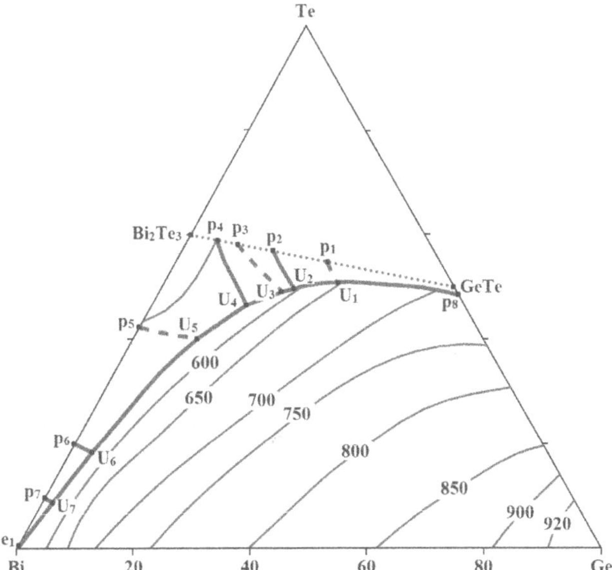

FIGURE 6.53 Liquidus surface of the GeTe–Bi_2Te_3–Bi–Ge subsystem of the Ge–Bi–Te ternary system. (From Abrikosov, N.H., and Danilova-Dobryakova G.T., *Izv. AN SSSR. Neorgan. mater.*, **6**(10), 1798, 1970.)

Blachnik et al. (1997a). The enthalpy surface is characterized by a valley stretching from the exothermic minimum in the Ge–Te system to the exothermic minimum in the Bi–Te system. For the description of the thermodynamic functions in the ternary system the equation of Bannier was taken using additional ternary coefficients. The calculated curves obtained from optimization are in good agreement with the experimental data.

Two isothermal sections of the Ge–Bi–Te ternary system at 300°C (Figure 6.54) and 500°C (Figure 6.55) have been constructed using metallography and XRD (Shelimova et al. 1993b, 1996, 1997). The ingots were annealed at 500°C for 1000 h and at 300°C for 1500 h. Two non-stoichiometric $Ge_{2-x}Bi_{2+x}Te_5$ (D) and $Ge_{4-x}Bi_{2+x}Te_7$ (F) phases have been found at 500°C in addition to the $Ge_3Bi_2Te_6$ (A), $GeBi_2Te_4$ (B), and $GeBi_4Te_7$ (C) compounds. The isothermal section at 300°C has been constructed taking into account the formation of high members of nGeTe·$m$$Bi_2Te_3$ homologous series (n = 1 to 9; m = 1 to 4) and ordered intermediate phases in the Bi–Te binary system. The region of the solid solution based on GeTe is located along the GeTe–Bi_2Te_3 section (Abrikosov and Danilova-Dobryakova 1972a; Abrikosov et al. 1981b; Rogacheva et al. 1986; Shelimova et al. 1993b, 1996, 1997).

6.39 Germanium–Niobium–Tellurium

The $\mathbf{Nb_3Ge_xTe_6}$ ($x \approx 0.9$) ternary compound, which apparently has two polymorphic modifications, is formed in the Ge–Nb–Te system. First of them crystallizes in the orthorhombic structure with the lattice parameters a = 643.18 ± 0.05, b = 1391.98 ± 0.11, c = 1154.07 ± 0.05 pm, and a calculated density of 7.204 g·cm^{-3} (Monconduit et al. 1992) [a = 643.5 ± 0.2, b = 1391.5 ± 0.4, c = 1152.4 ± 0.2 pm, and a calculated density of 7.188 g·cm^{-3} (Li and Carrol 1992); a = 643.5 ± 0.1, b = 1400.6 ± 0.2, c = 390.72 ± 0.04 pm, and a calculated density of 7.152 g·cm^{-3} (Lee van der et al. 1994d)]. The second modification crystallizes in the monoclinic structure with the lattice parameters a = 644.37 ± 0.09, b = 1600.4 ± 0.2, c = 389.36 ± 0.04 pm, β = 119.10 ± 0.01°, and a calculated density of 7.14 g·cm^{-3} (Lee van der et al. 1994a).

Some methods were used to prepare the orthorhombic modification of the title compound.

1. It was obtained as a side product of the $\mathbf{NbGe_{1/3}Te_2}$ synthesis (Lee van der et al. 1994d). Stoichiometric amounts of the elements for the latter compound were sealed in an evacuated silica tube. The temperature of the tube was raised to 430°C and maintained for several hours, then raised to 830°C, maintained for 2 days, and finally raised, step by step, to 1010°C. After 10 days, the tube was cooled by exposure to air. Crystals of composition $NbGe_{1/3}Te_2$, $\mathbf{NbGe_{2/5}Te_2}$, and $\mathbf{NbGe_{3/7}Te_2}$ (or $Nb_3Ge_xTe_6$ with $x \approx 0.9$) were most frequently found, with the c axes of the orthorhombic unit cell very close to the ratio 3:5:7, respectively.
2. Thin plate-like crystals of this modification were initially obtained from a reaction in which Nb, Ge,

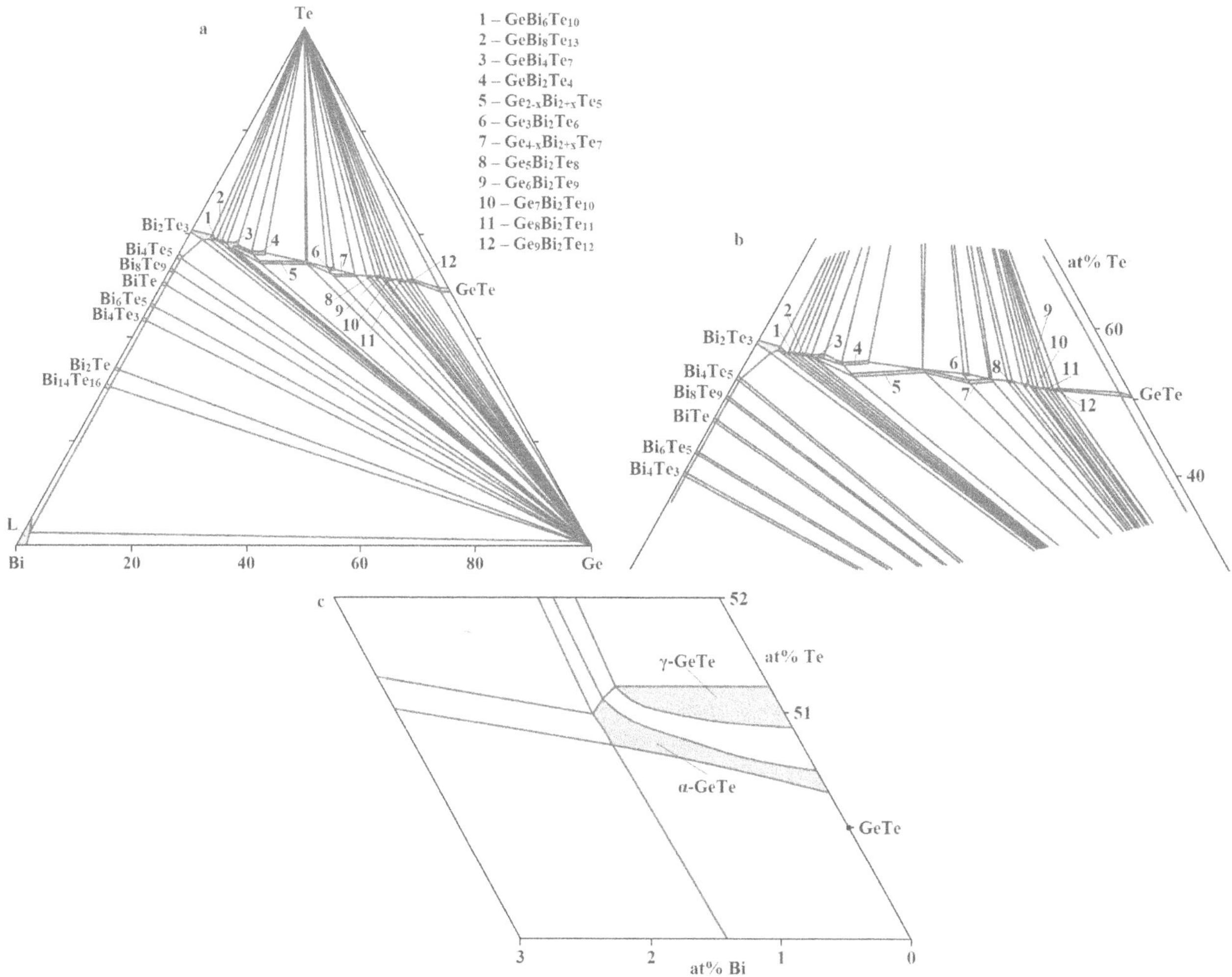

FIGURE 6.54 Isothermal sections of (a) the Ge–Bi–Te ternary system at 300°C, (b) its fragment in the 40–70 at% Te composition range and (c) its schematic fragment near the $Ge_{1-\delta}Te$ compound. (From Shelimova, L E., et al., *J. Alloys Compd.*, **243**(1–2), 194, 1996.)

and Te were directly mixed with a molar ratio of 2:1:4 (Li and Carrol 1992). A quartz tube containing the three elements and the transport agent ($TeCl_4$) was evacuated and sealed. The reaction container was slowly heated to 850°C and maintained at this temperature for 6 days. The tube was then cooled to room temperature by rapid quenching. The better quality crystals could be obtained from a reaction in which Nb, Ge, and Te (molar ratio 3:1:5) were mixed along with a small amount of $TeCl_4$. The tube was evacuated, sealed and placed in a tube furnace. The reaction temperature was brought up to and kept at 850°C for 11 days. The very thin plates of the $Nb_3Ge_xTe_6$ crystals show metallic luster and are stable in air. Polycrystalline samples were prepared at 900°C for 2 days.

3. This modification could be also prepared by direct combination of Nb, Ge, and Te (molar ratio 3:1:6) in sealed under vacuum silica tube (Monconduit et al. 1992). The temperature was raised at a rate of 100°C·h^{-1} to 1000°C. After a 10-day reaction, the tube with the sample was quenched in air.

The monoclinic modification was obtained at the heating the stoichiometric mixture of the elements in evacuated silica ampoule several hours at 430°C, 2 days at 630°C, 2 days at 830°C, and then 10 days at 1010°C, followed by quenching in air (Lee van der et al. 1994a).

6.40 Germanium–Tantalum–Tellurium

The $\mathbf{TaGe_{4/11}Te_2}$ ternary compound, which crystallizes in the orthorhombic structure with the lattice parameters a = 643.94 ± 0.05, b = 1402.5 ± 0.2, c = 384.56 ± 0.05 pm, and a calculated density of 8.83 g·cm^{-3} (Boucher et al. 1996) [a = 643.96 ± 0.08, b = 1404.4 ± 0.2, c = 385.22 ± 0.05 pm, and a calculated density of 8.818 cm^{-3} (Lee van der et al. 1994b)], is formed in the Ge–Ta–Te system. It was obtained in an attempt to prepare $\mathbf{TaGe_{3/7}Te_2}$ by direct combination of the elements in stoichiometric proportion (Boucher et al. 1996). The mixture was placed under vacuum in a sealed silica tube and heated at 950°C K for 10 days. Subsequent cooling to room temperature was achieved at a speed of 100°C·h^{-1}. The title compound was also obtained during an effort to synthesize $\mathbf{TaGe_{1/3}Te_2}$

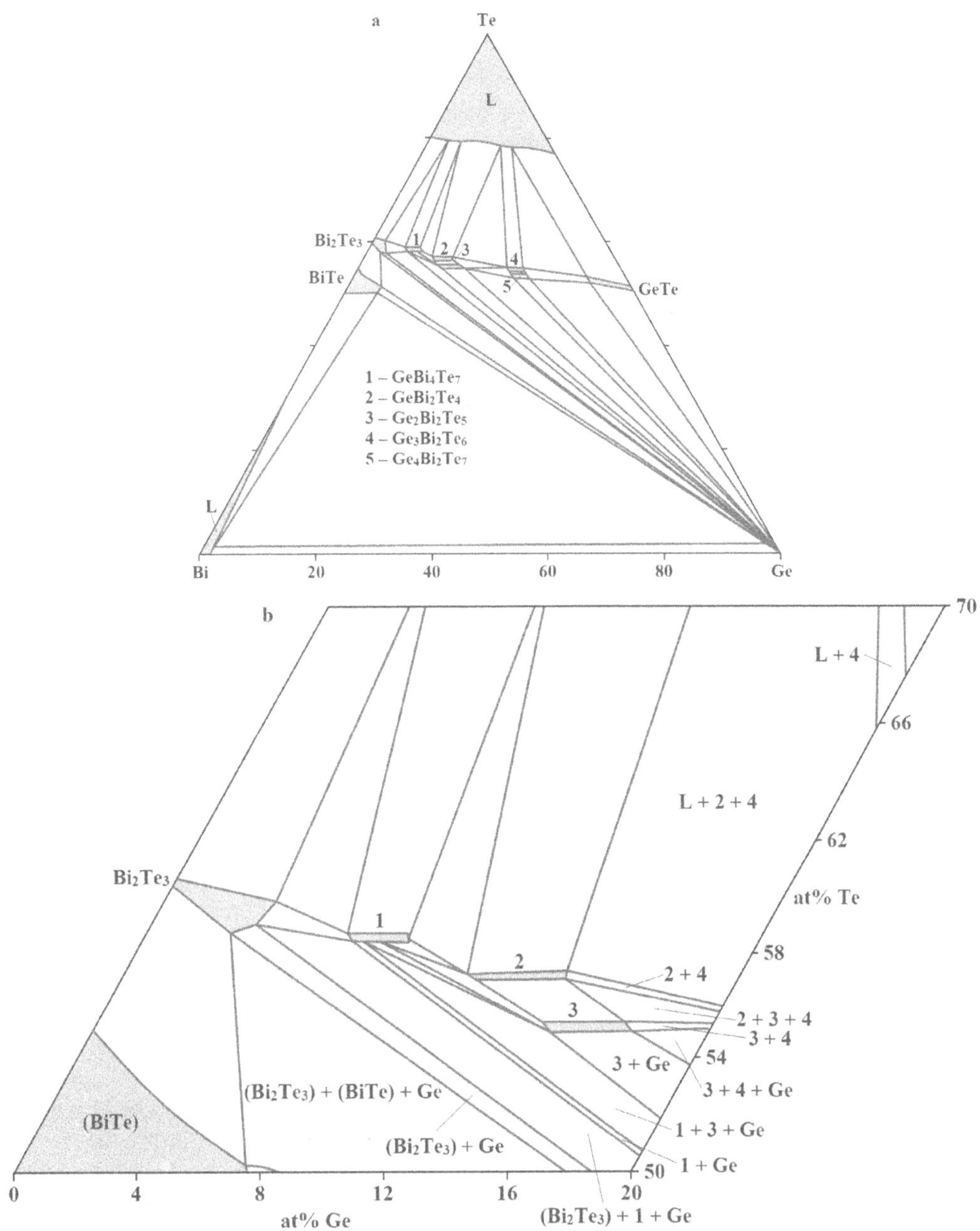

FIGURE 6.55 Isothermal sections of (a) the Ge–Bi–Te ternary system at 500°C and (b) its fragment in the 0–20 at% Ge and 50–70 at% Te composition range. (From Shelimova, L E., et al., *J. Alloys Compd.*, **243**(1–2), 194, 1996.)

(Lee van der et al. 1994b). Stoichiometric amounts of the elements for the latter compound were sealed in an evacuated silica tube. The temperature of the tube was raised to 430°C at a rate of 150°C·h^{-1}. After 22 days the tube was cooled to room temperature at the same rate. Very thin platelets of $TaGe_{4/11}Te_2$ with a metallic luster were found in the batch.

6.41 Germanium–Oxygen–Tellurium

GeO_2–Te: The phase diagram of this system is characterized by the degenerate eutectic near Te at 451°C ± 5°C, and the immiscibility region at 1110°C ± 8°C exists in the entire concentration range (Mikhaylov et al. 1991a).

GeO_2–TeO_2: The phase diagram of this system, constructed using DTA, XRD and metallography, is a eutectic type (Dimitriev et al. 1976). The eutectic contains 33 mol% GeO_2 and crystallizes at 685°C ± 10°C. The composition from 15 to 100 mol% GeO_2 were obtained as stable glasses and these between 10 to 15 mol% GeO_2 as glass plus crystals from the conventional rates of cooling applied. The glass transition temperature of obtained glasses is within the interval from 310°C to 340°C.

The **$Ge(TeO_3)_2$** ternary compound, which crystallizes in the monoclinic structure with the lattice parameters $a = 522.01 \pm 0.08$, $b = 697.30 \pm 0.13$, $c = 732.52 \pm 0.15$ pm, and $\beta = 91.66 \pm 0.02°$, is formed in the Ge–O–Te system (Jennene et al. 2009). Small single crystals of this compound were obtained accidentally in experiments initially intended to synthesize new oxyfluorotellurates (IV). An equimolar mixture of GeO_2 and TeO_2 was dissolved in hydrofluoric acid (40%) in a Teflon beaker and heated at 180°C. Then, after slow evaporation, the product was crushed and heated at 400°C in a Pt crimped tube for 48 h. Small colorless tablets of the title compound, air stable and suitable for structural study, were obtained.

On the liquidus surface of the GeO_2–TeO_2–Te subsystem of the Ge–O–Te ternary system, an extensive immiscibility region is observed, extending from TeO_2–Te system to GeO_2–Te system (Mikhaylov et al. 1991b). The ternary eutectic is degenerated from Te and crystallizes at 451°C ± 5°C.

Based on X-ray studies, it was found that oxygen interacts with dispersed GeTe in three stages, which are determined by the temperature of the "gas-solid" system (Gogishvili et al. 1988b). At room temperature, oxygen forms a chemisorbed layer on the surface of the powder. An increase in temperature to 300°C leads to an increasing in the internal energy of chemisorbed particles and their penetration into the surface layers of the sample, which causes displacement of tellurium into the second phase, a change in the cation-anion ratio and a decreasing in the lattice symmetry. At temperatures above the phase transformation temperature of GeTe, oxygen leaves the matrix, forming GeO_2.

According to the data of Pashinkin et al. (1978), in the Ge–O–Te system at 630°C the solid phases are GeTe, GeO_2, and Ge. The ternary phases in the system do not form. The triple point lies at very low partial pressures of O_2 and TeO_2, therefore, the reaction of the metallic Ge release is practically not realized. At low partial pressures, thermal dissociation of TeO_2 is observed, and at certain ratios of partial pressures of O_2 and TeO_2, the vapor pressure TeO_2 becomes saturated, which leads to tellurium condensation.

6.42 Germanium–Chromium–Tellurium

GeTe–CrTe: The solubility of CrTe in GeTe at 630°C and 230°C is 4.5 and 2.5 mol%, respectively (Stavrianidis et al. 1984) [according to the data of Alidzhanov et al. (1981), the solubility of CrTe in GeTe reaches up to 15 mol%]. The ingots were annealed 630°C, 480°C, 230°C at 192, 192, and 288 h, respectively (Stavrianidis et al. 1984).

GeTe–Cr_3Te_4: The solubility of Cr_3Te_4 in GeTe reaches up to 1 mol% (Keyyan et al. 1987). The second phase released after reaching the solubility limit is Cr_3Te_4, which indicates on the quasibinarity of the system. The ingots were annealed 630°C, 480°C, 230°C at 192, 192, and 288 h, respectively.

The **$Cr_2Ge_2Te_6$** ternary compound, which crystallizes in the trigonal structure with the lattice parameters $a = 683.00 \pm 0.01$ and $c = 2056.91 \pm 0.06$ pm (Yang et al. 2016) [$a = 682.75 \pm 0.04$, $c = 2056.19 \pm 0.09$ pm, and a calculated density of 6.091 g·cm^{-3} at 293 K, $a = 680.9 \pm 0.1$ and $c = 2044.4 \pm 0.3$ pm at 270 K, $a = 681.96 \pm 0.07$ and $c = 2037.1 \pm 0.3$ pm at 5 K, or $a =$ 790.7 and $\alpha = 51.16°$ in rhombohedral setting (Carteaux et al. 1995); $a = 683$ pm according to the first-principles calculations within the density-functional theory (Zhuang et al. 2015)], is formed in the Ge–Cr–Te system. The title compound was prepared by heating a stoichiometric mixture of the elements in a sealed evacuated quartz tube to 700°C for 20 d, followed by cooling for 10 h (Carteaux et al. 1995). A fine dark-grey crystalline powder was obtained, in the bulk of which small hexagonal single-crystal platelets were found. Single crystals up to a centimeter size were yielded by coalescence of powdery $Cr_2Ge_2Te_6$ in a tellurium flux at 700°C within 100 h (molar ratio 1:30), followed by very slow cooling (5°C·h^{-1}).

Polycrystalline $Cr_2Ge_2Te_6$ samples were also synthesized by the conventional solid state reaction in evacuated quartz tubes (Yang et al. 2016). The spark plasma sintering technique was applied to prepare fully dense bulk samples. The starting materials (Cr, Ge, and Te) were weighed in a stoichiometric ratio and sealed in a quartz ampoule under a vacuum of 10 mPa. The quartz ampoule was heated slowly to 700°C and held there for 30 h. The furnace was subsequently switched off and cooled down to room temperature naturally. In order to react completely, the above heating procedure was repeated three times.

6.43 Germanium–Iodine–Tellurium

The glass forming region in this ternary system was estimated under standardized conditions, yielding its limits composition as $Ge_{17-23}I_{1-12}Te_{88-82}$ (Feltz et al. 1972). The glass transition temperature of these alloys was determined to be 125°C ± 10°C.

6.44 Germanium–Manganese–Tellurium

GeTe–MnTe: The phase diagram of this system is given in Figure 6.56 (Johnston and Sestrich 1961). At high temperatures, a continuous series of solid solutions with a NaCl-type structure forms in the system. The lattice parameter in the region of a high-temperature solid solution varies linearly depending on the composition. With decreasing temperature, the solid solutions decompose. The MnTe-based solid solutions crystallize in the NiAs-type structure. The calculated

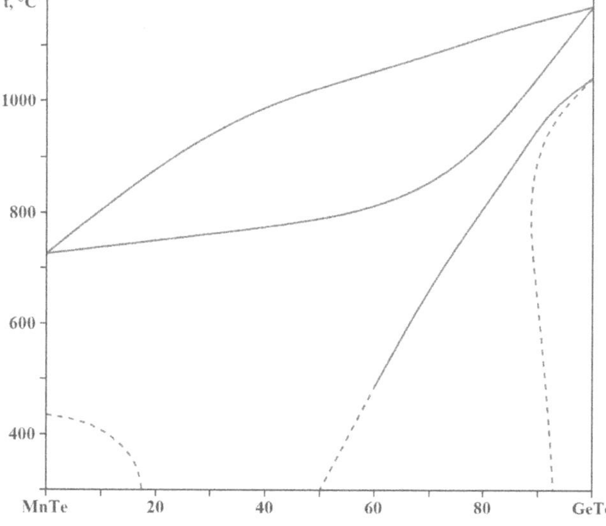

FIGURE 6.56 Phase diagram of the GeTe–MnTe system. (From Johnston, W.D., and Sestrich, D.E., *J. Inorg. Nucl. Chem.*, **19**(3–4), 229, 1961.)

solidus curve is in good agreement with experimental data (Steininger 1970b).

The concentration dependence of the lattice parameter for the solid solutions based on α- and β-GeTe is significantly different (Zhigareva et al. 1981; Abrikosov et al. 1985c).

With an increase in MnTe content, the temperature of polymorphic transformation of solid solutions based on GeTe sharply decreases and at 20 mol% MnTe is 100°C (Johnston and Sestrich 1961; Dudkin et al. 1978, 1979; Abrikosov et al. 1980a,c). At this temperature, the solubility of MnTe in GeTe is 7 mol% (Dudkin et al. 1978, 1979) [according to the data of Chomentowski et al. (1965), Alidzhanov et al. (1981), and Shelimova et al. (1993a), the solubility of MnTe in GeTe reaches 15–20 mol%]. The GeTe-based solid solution region in the Ge–Mn–Te ternary system is oriented along the GeTe–MnTe section (Melikhova et al. 1980a). This system was investigated through DTA, XRD, and metallography and the ingots were annealed at 700°C and 800°C for 24–48 h (Johnston and Sestrich 1961).

The $\mathbf{Mn_2GeTe_4}$ ternary compound, which crystallizes in the orthorhombic structure with the lattice parameters a = 1394.14 ± 0.04, b = 813.82 ± 0.03, and c = 656.40 ± 0.02 pm (Delgado et al. 2009b) [a = 1395.0 ± 0.02, b = 811.5 ± 0.1, and c = 659.2 ± 0.2 pm (Delgado et al. 2009a); a = 1360.0 ± 0.1, b = 1074.5 ± 0.1, and c = 777.5 ± 0.1 pm (Quintero et al. 2007, 2009)], is formed in the Ge–Mn–Te system. The samples of the title compound were synthesized by direct fusion of stoichiometric quantities of Mn, Ge, and Te in sealed, evacuated quartz ampoules (Delgado et al. 2009a, b; Quintero et al. 2007, 2009). The fusion process (14 days) was carried out into a furnace (vertical position) heated up to 1050°C or at 1150°C for 3 h. Then, the temperature was gradually lowered to 500°C and the furnace was turned off and the ingots were cooled to room temperature, or cooled to 550°C, annealed at this temperature for 1 month with the next very slowly cooling to room temperature.

The isotherm of manganese solubility in GeTe at 550°C was constructed by Rogacheva et al. (1998a). It was shown that it is oriented along the GeTe–MnTe section. Within the homogeneity region, two subregions with different dependence of properties on the Mn content can be distinguished. In the range of 0.5–1 at% Mn, anomalies were found in the concentration dependences of properties. The ingots for the investigations were annealed at 500°C for 200 h with the next quenching in water.

6.45 Germanium–Iron–Tellurium

GeTe–FeGe$_2$: The phase diagram of this system, constructed through DTA and metallography, is a eutectic type (Figure 6.57) (Akperov 1981). The eutectic contains 4 mol% $FeGe_2$ and crystallizes at 700°C. The eutectoid transformation α-GeTe → β-GeTe takes place at 370°C. The solubility of $FeGe_2$ in GeTe at room temperature is not higher than 1 mol%.

GeTe–FeTe: The solubility of FeTe in GeTe reaches 15 mol% (Alidzhanov et al. 1981).

At the temperatures lower than the solidus surface, the triangulation of the Ge–Fe–Te ternary system is determined by two compounds, GeTe and $\mathbf{Fe_3GeTe_2}$ (Figure 6.58) (Abrikosov et al. 1985d). At 450°C, Fe_3GeTe_2 is in equilibrium with five binary compounds: Fe_3Ge, Fe_5Ge_3, GeTe, $FeTe_2$, and Fe_2Te. The regions of solid solutions based on binary compounds do not exceed 1 mol%, and on the basis of Fe_3GeTe_2, the homogeneity regions were not formed. Fe_3GeTe_2 has a polymorphic transformation at 207°C.

It should be mention that according to Massalski (1990), Fe_2Te binary compound is not exist in the Fe–Te system and $FeTe_2$ has only very small homogeneity region.

This compound melts incongruently at 837°C and crystallizes in the hexagonal structure with the lattice parameters a = 399.1 ± 0.1, c = 1682 ± 3 pm, and a calculated density of 7.3 g·cm^{-3} (Deiseroth et al. 2006). The homogeneous gray,

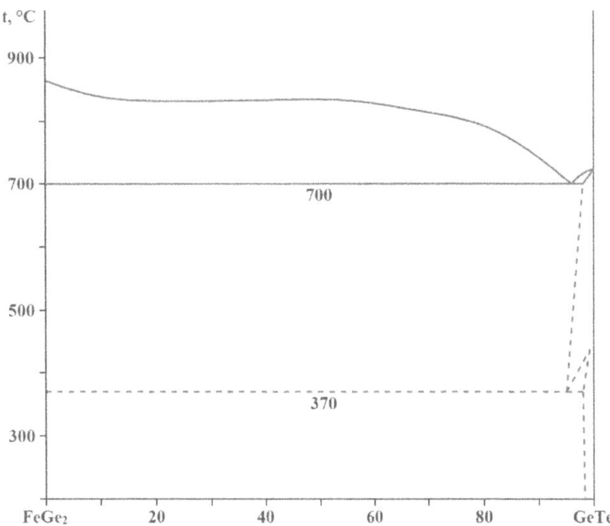

FIGURE 6.57 Phase diagram of the GeTe–FeGe$_2$ system. (From Akperov, M.M., *Izv. AN SSSR. Neorgan. mater.*, **17**(10), 1914, 1981.)

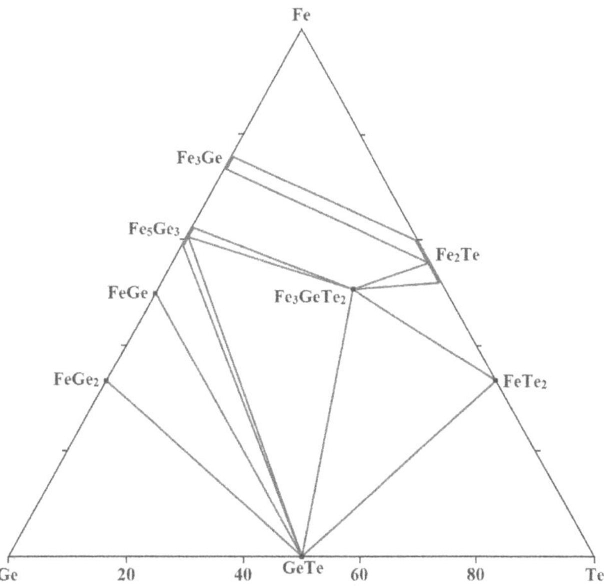

FIGURE 6.58 Isothermal sections of the Ge–Fe–Te ternary system at 450°C. (From Abrikosov, N.H., et al., *Izv. AN SSSR. Neorgan. mater.*, **21**(10), 1680, 1985.)

air-stable Fe_3GeTe_2 was prepared by a direct solid-state reaction starting from an intimate mixture of the pure elements (Fe/Ge/Te molar ratio was 3:1:2). The mixture of the starting materials was filled into a dry quartz ampoule, which was evacuated and sealed, heated up to 625°C, and kept at this temperature for 2 weeks.

Another ternary compound, $\mathbf{Fe_2GeTe_4}$, is also formed in the Ge–Fe–Te ternary system. It crystallizes in the orthorhombic structure with the lattice parameters a = 1365.53 ± 0.06, b = 789.70 ± 0.03, and c = 648.17 ± 0.02 pm (Delgado et al. 2009b) [a = 1352.0 ± 0.2, b = 1069.9 ± 0.1, and c = 775.7 ± 0.1 pm (Quintero et al. 2007, 2009)]. This compound was synthesized in an analogous way as Mn_2GeTe_4 was prepared using Fe instead of Mn.

6.46 Germanium–Cobalt–Tellurium

GeTe–Co: This system is a non-quasibinary section of the Ge–Co–Te ternary system (Kahramanov et al. 1981). Liquidus includes a wide immiscibility region, limited by the interval of 47–71 at% Co, and the fields of GeTe, Co_2Ge, and Co primary crystallization. The $\mathbf{Co_2Ge_3Te_3}$ ternary compound is formed over this section (Abrikosov and Petrova 1978). It is an n-type semiconductor with an energy gap of 0.77 eV (Návrátil et al. 2004a, b; Vaqueiro et al. 2006), undergoes an order–disorder transition at ca. 600°C (Kaltzoglou et al. 2013), and apparently has two polymorphic modifications. First of them crystallizes in the trigonal structure with the lattice parameters a = 1233.239 ± 0.003 and c = 1510.650 ± 0.007 pm (Vaqueiro et al. 2010) [in the rhombohedral structure with the lattice parameters a = 1230.09 ± 0.03 and c = 1506.78 ± 0.08 pm at 4.2 K, and a = 1232.70 ± 0.05 and c = 1510.2 ± 0.1 pm at 293 K in hexagonal setting (Vaqueiro et al. 2006)]. The second modification crystallizes in the cubic structure with the lattice parameter a = 870.98 pm (Vaqueiro and Sobany 2007) [a = 871.74 pm (Návrátil et al. 2004a, b); a = 872.7 pm (Domashevskaya et al. 2015)].

$Co_2Ge_3Te_3$ was prepared from the stoichiometric mixture of Co, Ge, and Te (Vaqueiro et al. 2006, 2010; Kaltzoglou et al. 2013; Domashevskaya et al. 2015). The mixture was ground in an agate mortar prior to sealing in an evacuated (13 mPa) silica tube. The inner wall of the silica tube was coated with a thin layer of carbon by pyrolysis of acetone. The reaction mixture was heated at 430°C for 2 days, followed by another 2 days at 580°C. The sample was cooled to room temperature at 0.1°C·min^{-1}, prior to removal from the furnace. Following regrinding, the sample was sealed into a second carbon-coated silica tube and re-fired at 580°C for 2 days, followed by cooling to room temperature at 0.1°C·min^{-1}.

The title compound could be also prepared as following (Návrátil et al. 2004a, b). Starting GeTe and Co were ground and mixed together in agate mortar under pure acetone. After blowing of acetone away by Ar, the starting powder was loaded in a steel die and cold pressed into cylindrical pellets. The pellets were sealed into evacuated quartz ampoules. The ampoules were then heated at 600°C for 3 days. The resulted material was once more grounded under acetone and the above mentioned process was repeated. The prepared samples were found to be approximately 80% of theoretical density. The composition of the obtained samples was $CoGe_{1.7}Te_{1.47}$. The deviation from stoichiometric composition was due to the specific way of preparation.

GeTe–CoGe$_2$: This system is also a non-quasibinary section of the Ge–Co–Te ternary system that intersects the fields of GeTe, $CoGe_2$, and CoGe primary crystallization (Abrikosov and Petrova 1981). The region of GeTe primary crystallization lies in the range of 0–2 mol% $CoGe_2$, and that of $CoGe_2$ is in the range of 2–10 mol% $CoGe_2$. In the region of CoGe primary crystallization, lying in the range of 10–100 mol% $CoGe_2$, an immiscibility region is observed in the range of 10–75 mol% $CoGe_2$. $CoGe_2$ crystallizes secondary in the system, and the crystallization of the alloys ends at 716°C with the formation of the eutectic GeTe + $CoGe_2$. The eutectic point is near the composition of 2 mol% $CoGe_2$. The mutual solubility of the components is not higher than 1 mol%.

GeTe–Co$_2$Ge: This system is also a non-quasibinary section of the Ge–Co–Te ternary system that intersects the fields of GeTe, $Co_2Ge_3Te_3$, CoGe, and Co_2Ge primary crystallization (Abrikosov et al. 1982). The region of GeTe primary crystallization is situated up to 1 mol% Co_2Ge. The primary crystallization of $Co_2Ge_3Te_3$ is within the interval from 1 to 2 mol% Co_2Ge. The CoGe-based phase crystallizes primarily in the concentration range of 2–30 mol% Co_2Ge. In the rest alloys of the section, Co_2Ge crystallizes primarily. In the range of 30–80 mol% Co_2Ge, there is an immiscibility region in the liquid state, and the temperature of the onset of the monotectic reaction increases with increasing Co_2Ge content in the alloys. Monotectic reaction ends at 996°C. Four invariant reactions take place in this section. Solid solutions based on GeTe and Co_2Ge contain no more than 1 mol% of the second component.

According to the data of Zargarova et al. (1976b), the phase diagram of the GeTe–Go_2Ge system is a eutectic type. The eutectic contains 1.39 at% Co and crystallizes at 700°C. There is an immiscibility region within the interval from 24 to 27 at% Co.

GeTe–CoTe: The solubility of CoTe in GeTe reaches 15 mol% (Alidzhanov et al. 1981).

GeTe–CoTe$_2$: This system is also a non-quasibinary section of the Ge–Co–Te ternary system that intersects four fields of the primary crystallization of next phases: solid solution based on GeTe containing less than 1 mol% CoTe within the interval of 0–3 mol% $CoTe_2$, solid solution based on $Co_2Ge_3Te_3$ within the interval of 3–5 mol% $CoTe_2$, solid solution based on CoGe within the interval of 5–25 mol% $CoTe_2$, and solid solution based on $CoTe_2$ within the interval of 25–100 mol% $CoTe_2$ (Abrikosov and Petrova 1981). There are three invariant reactions at 715°C, 690°C, and 670°C in this section. According to the data of Stavrianidis et al. (1979), the immiscibility region within the interval from 20 to 92.5 mol% GeTe exists in this system but the later investigations (Abrikosov and Petrova 1981) do not confirmed its existence.

The solubility of GeTe in $CoTe_2$ at 380°C, 540°C, 600°C, 650°C, and 665°C is 3.0, 3.5, 5.0, 7.5, and 8.5 mol%, respectively (Stavrianidis et al. 1979).

GeTe–Co$_3$Te$_4$: This system is also a non-quasibinary section of the Ge–Co–Te ternary system that intersects the same fields of primary crystallization as the GeTe–$GoTe_2$ system intersects (Abrikosov and Petrova 1981). The solubility of

GeTe in Go_3Te_4 at 600°C reaches 22 mol% and GeTe dissolves less than 1 mol% Go_3Te_4.

According to the data of Kahramanov et al. (1981), the phase diagram of the GeTe–Go_2Ge system is a eutectic type with a peritectic transformation. The eutectic crystallizes at 800°C. A ternary phase, which melts incongruently at 800°C, was determined within the interval of 37.5–39.5 at% Co.

Another ternary compound, **CoGeTe**, is also formed in the Ge–Co–Te ternary system. It melts incongruently at about 725°C, is *n*-type semiconductor (Laufek et al. 2008b; Vaqueiro et al. 2009) and crystallizes in the orthorhombic structure with the lattice parameters a = 619.30 ± 0.04, b = 623.26 ± 0.04, and c = 1112.89 ± 0.07 pm (Vaqueiro et al. 2009) [a = 618.92 ± 0.04, b = 622.85 ± 0.04, c = 1112.40 ± 0.06 pm, and an energy gap of ≈ 0.6 eV (Laufek et al. 2007a, 2008b); in tetragonal structure with the lattice parameters a = 620.4 ± 1.0 and c = 1127.3 ± 3.0 pm (Domashevskaya et al. 2015)].

The title compound was prepared by conventional solid-state reactions (Vaqueiro et al. 2009). The stoichiometric mixture of the elements was sealed in evacuated silica tubes (<13 mPa). The inner wall of the silica tube was coated with a thin layer of carbon by pyrolysis of acetone. The sealed tube was placed into a furnace at 440°C for 36 h, heated at 0.5 C·min^{-1} to 580°C, and held at this temperature for 10 days. The tube was then cooled to room temperature at ca. 5 C·min^{-1} prior to removal from the furnace. Following regrinding, the material was sealed again into silica tube, annealed at 580°C for 2 days, and then cooled to room temperature at 0.5 C·min^{-1}, prior to removal from the furnace.

CoGeTe could be also prepared as following (Laufek et al. 2007a, 2008b). Stoichiometric amounts of Co, Ge, and Te were sealed in evacuated quartz tubes and heated at 1150°C for 4 h. Following this, the sample was ground using the agate mortar and pestle, and sealed again under vacuum in quartz tube. The mixture was heated at 550°C for 3 days. The resultant material was once again ground and heated at 670°C. After long-term annealing, the sample was quenched in cold water.

Single crystals of CoGeTe were prepared by loading a silica tube with a mixture of Co, Ge, and Te, and ca. 10 mg of I_2 (Vaqueiro et al. 2009). The silica tube was sealed under vacuum (<13 mPa), and heated at 700°C for 6 days in a chamber furnace, under its natural temperature gradient. The reaction product consisted of a black powder and a small number of black crystals, which were located at the hot end of the tube.

The liquidus surface of the Ge–Co–Te ternary system (Figure 6.59) consists of ten fields of primary crystallization of phases and an immiscibility gap (Abrikosov et al. 1978; Kahramanov et al. 1981). At 600°C, the triangulation of this ternary system is defined by the $Co_2Ge_3Te_3$ compound (Figure 6.60): it is in equilibrium with all binary compounds with exception of Co_2Ge and $CoGe_2$, which are in equilibrium with γ-phase of the Co–Te binary system (Abrikosov et al. 1978).

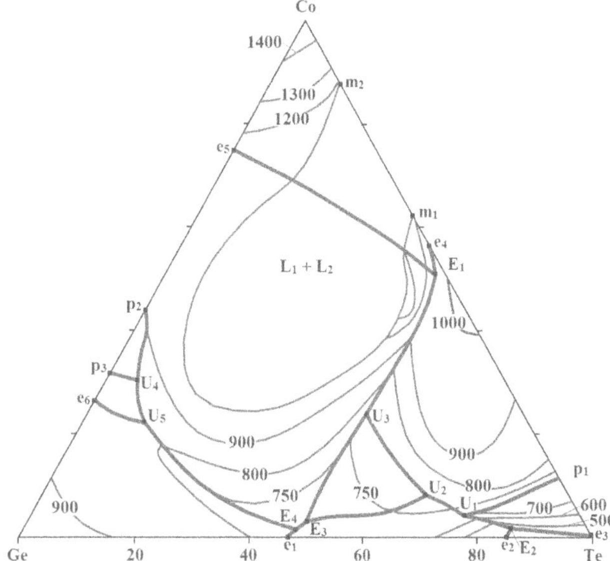

FIGURE 6.59 Liquidus surface of the Ge–Co–Te ternary system. (From Kahramanov, K.Sh., et al., *Izv. AN SSSR. Neorgan. mater.*, **17**(1), 43, 1981.)

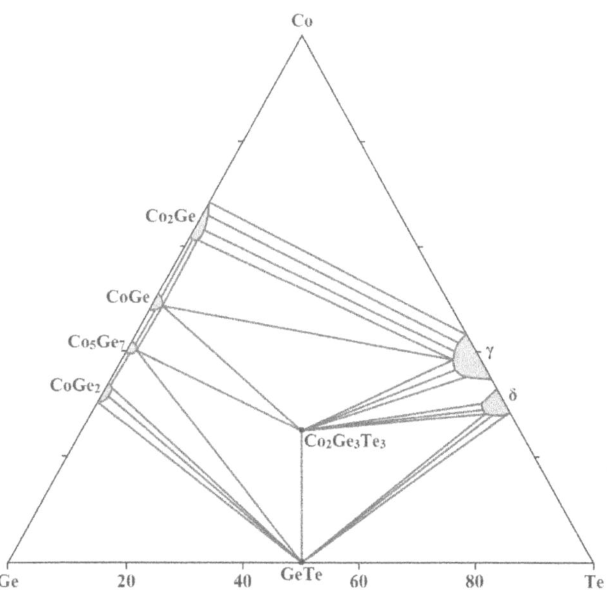

FIGURE 6.60 Isothermal sections of the Ge–Co–Te ternary system at 600°C. (From Abrikosov, N.H., et al., *Izv. AN SSSR. Neorgan. mater.*, **18**(3), 376, 1982.)

6.47 Germanium–Nickel–Tellurium

GeTe–NiTe: The part of the phase diagram of this system, constructed using DTA, metallography, and measuring of microhardness, is presented in Figure 6.61 (Alidzhanov et al. 1982b). The solubility of NiTe in GeTe increases with temperature increasing from 8 mol% at room temperature to 20 mol% at 600°C. The alloys were annealed at 200°C, 275°C, 400°C, and 500°C for 600 h. According to the data of Alidzhanov et al. (1981), the solubility of NiTe in GeTe is not higher than 15 mol%.

GeTe–NiTe₂: The part of the phase diagram of this system from the GeTe-side up to 22 mol% $NiTe_2$ was constructed by Alieva et al. (1997) using DTA, XRD, metallography, and measuring of microhardness. The solid solutions with two structural modifications are formed based on GeTe. The solubility of $NiTe_2$ in α-GeTe reaches 8 mol% and 17 mol% in β-GeTe.

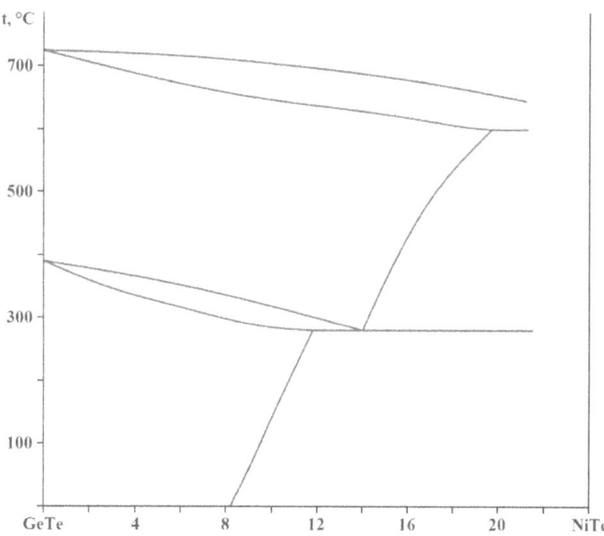

FIGURE 6.61 Part of the phase diagram of the GeTe–NiTe system. (From Alidzhanov, M.A., et al., *Izv. AN SSSR. Neorgan. mater.*, **18**(9), 1587, 1982.)

GeTe–Ni_3Te_4: The phase diagram of this system is given in Figure 6.62 (Kurbanova et al. 2007). The eutectic contains 18 mol% Ni_3Te_4 and crystallizes at 415°C. The **Ni_3GeTe_5** ternary compound, which melts incongruently at 640°C, crystallizes in the hexagonal structure with the lattice parameters a = 1340.8 and c = 1617.5 pm, and has entropy, enthalpy and Gibbs energy of formation and isobaric heat capacity of 36 ± 5 J·M^{-1}K^{-1}, –254 ± 25 kJ·M^{-1}, –223 ± 20 kJ·M^{-1}, and 243 ± 10 J·M^{-1}K^{-1}, respectively, is formed in the GeTe–Ni_3Te_4 system. The (α-GeTe) ⇔ (β-GeTe) + Ni_3GeTe_5 eutectoid transformation takes place at 380°C. The solubility of Ni_3Te_4 in GeTe reaches 6, 4.5 and 4 mol% at 300°C, 150°C and room temperature, respectively. The solubility of GeTe in Ni_3Te_4 was not determined.

Another ternary compounds, **Ni_3GeTe_2** and **$Ni_{5.45}GeTe_2$**, are also formed in the Ge–Ni–Te ternary system. First of them melts incongruently at 879°C and crystallizes in the hexagonal structure with the lattice parameters a = 391.1 ± 0.1 and c = 1602.0 ± 0.3 pm and a calculated density of 7.89 g·cm^{-3} (Deiseroth et al. 2006). The homogeneous gray, air-stable Ni_3GeTe_2 was prepared by a direct solid-state reaction starting from an intimate mixture of the pure elements (Ni/Ge/Te molar ratio was 3:1:2). The mixture of the starting materials was filled into a dry quartz glass ampoule, which was evacuated and sealed, heated up to 625°C, and kept at this temperature for 2 weeks.

$Ni_{5.45}GeTe_2$ crystallizes in the tetragonal structure with the lattice parameters a = 368.58 ± 0.03 and c = 1907.2 ± 0.3 pm (Deiseroth et al. 2007a) [a = 368.8 ± 0.1, c = 1902.7 ± 0.2 pm, and a calculated density of 8.35 g·cm^{-3} (Isaeva et al. 2007b)].

The sample of this compound was synthesized starting from the elements by high-temperature ceramic techniques (Isaeva et al. 2007b). The thoroughly ground mixture of the elements (the total mass of the sample was 1 g), corresponding to the stoichiometry Ni_7GeTe_2, was placed in quartz ampoule. Then the ampoule was sealed under vacuum (~7 Pa), annealed at 540°C–650°C for 14 days in a vertical furnace and water-quenched. According to the XRD, the sample consisted of four phases, and hence, the phase equilibrium was not achieved. The equilibrium sample containing only $Ni_{5.46}GeTe_2$ and Ni was obtained after the repeated annealing of pressed pellets (pressure load of about 0.2 GPa) under the same conditions. The total time of annealing was 90 days.

Crystals of the title compound were grown from molten flux starting from KI (Isaeva et al. 2007b). The stoichiometric mixture of Ni, Si, and Te (molar ratio 7:1:2; total mass was ~1–2 g; the mass ratio of the charge to the flux was 1:1) was placed in cylindrical quartz ampoules, which was then sealed under vacuum. The ampoule was heated to 850°C for 12 h in a furnace. Then the furnace was slowly cooled in an automatic mode to 500°C at a rate of 1.5°C·h^{-1}. The cooling to room temperature was performed in the switched-off furnace. The sample was ground in a mortar and washed off from the flux by refluxing in hot water for 0.5–1 h. The washed sample contained small golden-colored crystals of the title compound as thin elongated plates.

Single crystals of $Ni_{5.45}GeTe_2$ could be prepared in a solid-state reaction starting from an intimate mixture of pure elements (molar ratio of Ni/Ge/Te was 5:1:2) and iodine as a transport agent, which was filled into a dry, evacuated and sealed quartz ampoule (Deiseroth et al. 2007a). The ampoule was placed with one end in the center of a tube furnace at 800°C for 5 days. The product consists of black square-shaped crystals which form preferably at the cold end of the ampoule at an estimated temperature of 600°C. This compound can be handled in air.

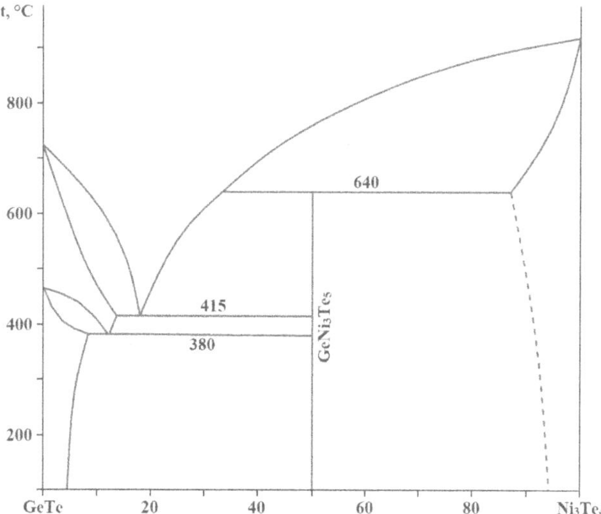

FIGURE 6.62 Phase diagram of the GeTe–Ni_3Te_4 system. (From Kurbanova, R.D., et al., *Azerb. khim. zhurn.*, (2), 102, 2007.)

6.48 Germanium–Rhodium–Tellurium

The **RhGeTe** ternary compound, which is n-type semiconductor and crystallizes in the orthorhombic structure with the lattice parameters a = 635.79 ± 0.01, b = 638.79 ± 0.01, and c = 1149.02 ± 0.03 pm (Vaqueiro et al. 2009), is formed in the Ge–Rh–Te system. The title compound was prepared by conventional solid-state reactions. The stoichiometric mixture of the elements was sealed in an evacuated silica tube (<13 mPa). The inner wall of the silica tube was coated with a thin layer of carbon by pyrolysis of acetone. The sealed tube was placed into a furnace at 440°C for 36 h, heated at 0.5 C·min^{-1} to 580°C, and held at this temperature for 10 days. The tube was then cooled to room temperature at ca. 5°C·min^{-1} prior to removal from the furnace. Following regrinding, the material was sealed again into silica tube, annealed at 580°C for 2 days, and then cooled to room temperature at 0.5 C·min^{-1}, prior to removal from the furnace.

6.49 Germanium–Palladium–Tellurium

The Pd-rich part of the isothermal section of the Ge–Pd–Te ternary system at 480°C was constructed by El-Boragy and Schubert (1971) and is presented in Figure 6.63. Four homogeneous regions were determined in the system: *W*-phase between compositions $Ge_{13}Pd_{72}Te_{15}$ and $Ge_4Pd_{70}Te_{26}$, *X*-phase between compositions $Ge_7Pd_{72}Te_{21}$ and $Ge_4Pd_{72}Te_{24}$, *Y*-phase between compositions $Ge_{16}Pd_{75}Te_9$ and $Ge_5Pd_{73}Te_{22}$, and *Z*-phase between compositions $Ge_6Pd_{75}Te_{19}$ and $Pd_{20}Te_7$.

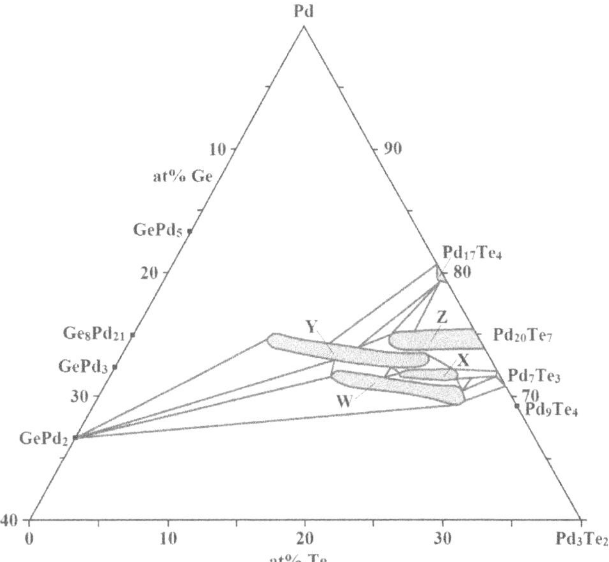

FIGURE 6.63 Pd-rich part of the isothermal sections of the Ge–Pd–Te ternary system at 480°C. (From El-Boragy, and M., Schubert, K, *Z. Metallkd.*, **62**(4), 314, 1971.)

6.50 Germanium–Iridium–Tellurium

The $\mathbf{Ir_2Ge_3Te_3}$ ternary compound, which crystallizes in the cubic structure with the lattice parameter $a = 895.621 \pm 0.004$ pm (Vaqueiro et al. 2010) [$a = 896.32 \pm 0.01$ pm (Vaqueiro et al. 2007)], is formed in the Ge–Ir–Te system. This compound was prepared from appropriate stoichiometric mixture of Co, Ir, and Te (Vaqueiro et al. 2007, 2010). The mixture was ground in an agate mortar prior to sealing in an evacuated (<13 mPa) silica tube. The inner wall of the silica tube was coated with a thin layer of carbon by pyrolysis of acetone. The reaction mixture was heated at 430°C for 2 days, followed by another 2 days at 580°C. The sample was cooled to room temperature at 0.1°C·min^{-1}, prior to removal from the furnace. Following re-grinding, sample was sealed into a second carbon-coated silica tube and re-fired at 580°C for 2 days, followed by cooling to room temperature at 0.1°C·min^{-1}.

REFERENCES

All References are available as a downloadable eResource at www.routledge.com/9780367639235

7

Systems Based on Tin Sulfides

7.1 Tin–Lithium–Sulfur

Three compounds, $\mathbf{LiSnS_2}$, $\mathbf{Li_2SnS_3}$, and $\mathbf{Li_4SnS_4}$, are formed in the Sn–Li–S ternary system. First of them crystallizes in the trigonal structure with the lattice parameters a = 367 and c = 790 pm (Le Blanc et al. 1969; Le Blanc and Rouxel 1972). This compound was prepared by reacting SnS_2 with a solution of Li in liquid ammonia. It is highly hygroscopic and decomposes at temperatures above 200°C.

Li_2SnS_3 is stable in air at ambient conditions, melts at ca. 750°C and crystallizes in the monoclinic structure with the lattice parameters a = 639.64 ± 0.05, b = 1108.64 ± 0.09, c = 1240.5 ± 0.1 pm, β = 99.867 ± 0.005°, a calculated density of 3.507 g·cm^{-3}, and an energy gap of 2.38 eV (Brant et al. 2015). It was prepared by grinding stoichiometric amounts of Li_2S (1 mmol), Sn (1 mmol), and S (2 mmol) using an agate mortar and pestle in an Ar-filled glove box. Inside of a fused silica tube, the mixture was contained in a graphite crucible. The tube was sealed under a vacuum of ~0.1 Pa. The sample was heated at a rate of 50°C/h to 750°C and held at 750°C for 96 h, slowly cooled to 500°C in 125 h and then allowed to cool to room temperature. The product is a dark-green polycrystalline powder.

Li_4SnS_4 melts at 958°C and crystallizes in the orthorhombic structure with the lattice parameters a = 1430.5 ± 0.4, b = 789.9 ± 0.2, and c = 632.8 ± 0.2 pm, a calculated density of 2.552 g·cm^{-3} and an energy gap of 3.54 eV (MacNeil et al. 2014) [a = 1381.2 ± 0.3, b = 796.24 ± 0.16, c = 636.70 ± 0.13 pm, and a calculated density of 2.60 g·cm^{-3} at 100 K (Kaib et al. 2012)].

To obtain a powder of Li_4SnS_4, a stoichiometric mixture of Li_2S, Sn, and S was grinding using an agate mortar and pestle in an Ar-filled glove box (MacNeil et al. 2014). The mixture was transferred to a graphite tube which was then flame sealed inside a fused-silica tube under a vacuum of approximately 0.1 Pa. The stoichiometric mixture was heated up to 750°C over the course of 15 h. The temperature was held at 750°C for 96 h, followed by slow cooling to 500°C in 72 h and subsequently cooled to room temperature in 3 h. The tube was opened in the Ar-filled glove box. A pale gray, microcrystalline powder was obtained. The compound is moisture sensitive.

Single crystals of this compound were prepared as follows (MacNeil et al. 2014). Li_2S (3.5 mM), Fe (1 mM), Sn(1 mM), and S (8 mM) were ground in an Ar-filled glove box for 15 min. The ground mixture was placed into a loosely-capped graphite tube; then the entire assembly was flame-sealed inside a fused-silica tube under a vacuum of approximately 0.1 Pa. In a furnace, the sample was heated to 750°C over 12 h and soaked at that temperature for 96 h. The sample was first slow-cooled to 500°C at 5°C·h^{-1}, and then cooled to room temperature in 3 h. Under a nitrogen atmosphere, the sample was washed numerous times with ethanol in order to remove the excess flux, then washed with heptane and allowed to dry under nitrogen. Pale yellow single crystals were obtained.

Li_4SnS_4 could be also prepared as follows. A stoichiometric mixture of Li_2S (14.56 mM), Sn (7.28 mM), and S (14.56 mM) was slowly heated to 400°C in a silica-ampoule using a heat gun, followed by melting and heating for another 30 min using a gas burner (Kaib et al. 2012). To gain single-crystalline products, the reaction was performed in a furnace with a definite temperature program: heating to 650°C with a rate of 18°C·h^{-1}, keeping for 24 h at 650°C, and cooling to room temperature with a rate of 4°C·h^{-1}. It was obtained as a mixture of amorphous and polycrystalline powder in addition to a small amount of clean single crystals. The compound could be obtained as phase pure, polycrystalline powdery upon complete dehydration of $Li_4SnS_4 \cdot 13H_2O$ at 320°C under reduced pressure of ca. 10^{-6} Pa for 4 h. All reaction steps were performed with strong exclusion of air and external moisture.

7.2 Tin–Sodium–Sulfur

Several ternary compounds are formed in the Sn–Na–S system. $\mathbf{NaSnS_2}$ crystallizes in the trigonal structure with the lattice parameters a = 369 and c = 2554 pm (Le Blanc et al. 1969; Le Blanc and Rouxel 1972). This compound was prepared by reacting SnS_2 with a solution of Na in liquid ammonia. It is highly hygroscopic and decomposes at temperatures above 200°C.

$\mathbf{Na_2SnS_3}$ has a polymorphic transformation at 540°C. Low-temperature modification crystallizes in the monoclinic structure with the lattice parameters a = 664.0 ± 0.3, b = 1149.9 ± 0.5, c = 1343.9 ± 0.6 pm, β = 99.47 ± 0.03°, and the calculated and experimental densities of 3.43 and 3.39 g·cm^{-3}, respectively (Vermot-Gaud-Daniel and Jumas 1972) [in the hexagonal structure with the lattice parameters a = 664.0 ± 0.3 and c = 3975 ± 2 pm (Mark et al. 1974)]. A well-crystallized powder of α-Na_2SnS_3 could be obtained by annealing a stoichiometric mixture of SnS_2 and Na_2S at 500°C for 3 days, and its single crystals were grown by the treatment at 700°C for 2 days with the next slow cooling (16°C·h^{-1}) (Vermot-Gaud-Daniel and Jumas 1972). High-temperature modification of α-Na_2SnS_3 was prepared by the quenching of the sample from 600°C.

$\mathbf{Na_4SnS_4}$ crystallizes in the tetragonal structure with the lattice parameters a = 783.7 ± 0.3 and c = 695.0 ± 0.3 pm and the calculated and experimental densities of 2.637 and 2.64 g·cm^{-3}, respectively (Vermot-Gaud-Daniel and Jumas 1972; Jumas et al. 1973, 1975b). A well-crystallized powder of this compound was obtained by annealing a stoichiometric mixture of

DOI: 10.1201/9781003123507-7

SnS_2 and Na_2S at 500°C for 3 days, and its single crystals were grown by the treatment at 850°C for 1 day with the next slow cooling (16°C·h^{-1}) (Vermot-Gaud-Daniel and Jumas 1972).

$Na_4Sn_3S_8$ crystallizes in the monoclinic structure with the lattice parameters $a = 1142.7 \pm 0.3$, $b = 733.7 \pm 0.2$, $c = 1762.1 \pm 0.4$ pm, $\beta = 95.27 \pm 0.02°$, and the calculated and experimental densities of 3.233 and 3.23 g·cm^{-3}, respectively (Jumas et al. 1975a). A well-crystallized powder of this compound was obtained by annealing a stoichiometric mixture of SnS_2 and Na_2S at 500°C for 3 days (Vermot-Gaud-Daniel and Jumas 1972; Jumas et al. 1975a).

$Na_6Sn_2S_7$ also crystallizes in the monoclinic structure with the lattice parameters $a = 939.4 \pm 0.6$, $b = 1071.6 \pm 0.6$, $c = 1565.8 \pm 0.8$ pm, $\beta = 109.97 \pm 0.04°$, and the calculated and experimental densities of 4.21 and 4.12 g·cm^{-3}, respectively (Krebs and Schiwy 1973; Jumas et al. 1974). A well-crystallized powder of this compound was obtained by annealing a stoichiometric mixture of SnS_2 and Na_2S at 500°C–700°C for 3 days (Vermot-Gaud-Daniel and Jumas 1972; Krebs and Schiwy 1973; Jumas et al. 1974). This compound is highly hygroscopic (Jumas et al. 1974).

7.3 Tin–Potassium–Sulfur

Several ternary compounds are formed in the Sn–K–S system. **$KSnS_2$** crystallizes in the trigonal structure with the lattice parameters $a = 367$ and $c = 2561$ pm (Le Blanc et al. 1969; Le Blanc and Rouxel 1972). This compound was prepared by reacting SnS_2 with a solution of K in liquid ammonia. It is highly hygroscopic and decomposes at temperatures above 200°C.

$K_2Sn_2S_5$ crystallizes in the monoclinic structure with the lattice parameters $a = 1180.4 \pm 0.3$, $b = 780.8 \pm 0.1$, $c = 1153.9 \pm 0.1$ pm, $\beta = 108.35 \pm 0.01°$, a calculated density of 3.33 g·cm^{-3}, and an energy gap of 2.36 eV (Liao et al. 1993) [$a = 1107.2 \pm 0.5$, $b = 780.6 \pm 0.3$, $c = 1151.7 \pm 0.5$ pm, $\beta = 108.43 \pm 0.02°$, and a calculated density of 3.35 g·cm^{-3} (Klepp 1992a)]. To prepare this compound, a mixture of Sn (1 mM), K_2S (2 mM), and S (8 mM) was loaded into a Pyrex tube in a N_2 glove box (Liao et al. 1993). The tube was evacuated and flame sealed at the pressure of 0.13 Pa. The mixture was heated at 320°C for 4–6 days. Upon cooling of the mixture to room temperature at the rate of 2°C·h^{-1}, small yellow crystals were formed. The product was washed with water to remove excess K_2S and then washed with CS_2 to remove unreacted sulfur.

$K_2Sn_2S_8$ also crystallizes in the monoclinic structure with the lattice parameters $a = 958.0 \pm 0.8$, $b = 1000.4 \pm 0.5$, $c = 1413.1 \pm 0.7$ pm, $\beta = 107.82 \pm 0.06°$, a calculated density of 2.95 g·cm^{-3}, and an energy gap of 2.08 eV (Liao et al. 1993). This compound was prepared by the following way. A mixture of Sn (1 mM), K_2S (2 mM), and S (8 mM) was loaded into a Pyrex tube in a N_2 glove box. The tube was evacuated and flame sealed at a pressure of 0.13 Pa. The mixture was heated at 250°C or 275°C for 4–6 days. Upon cooling of the mixture to room temperature at the rate of 2°C·h^{-1}, small orange chunky crystals were formed. The crystals obtained at 275°C are larger in size. The product was washed with water to remove excess K_2S and then washed with CS_2 to remove unreacted sulfur.

$K_2Sn_4S_9$ melts congruently, crystallizes in the orthorhombic structure and has an energy gap of 2.66 eV (Marking et al. 1998). Thin plates of this compound can be prepared through sulfur-rich polychalcogenide flux reactions kept for 4 days at 500°C and then cooled to 200°C at 4°C·h^{-1}. Thin poor quality yellowish-gold "mica-like" crystals of $K_2Sn_4S_9$ were also prepared as the major phase by reaction of K_2S + 3Sn + 12S mixture at 500°C. Direct combination with a slight excess of K_2S_4 flux at 750°C for 3 days resulted in nearly pure microcrystalline $K_2Sn_4S_9$. Product from the flux and direct combination reactions was washed with dimethylformamide (DMF) to remove excess alkali metal polysulfide flux and other soluble byproducts. A brief wash with H_2O removed small amounts of water-soluble phases from the products. Acetone and diethyl ether were used to rinse and dry the products. This compound is stable in water and no decomposition can be detected in samples left under ambient atmospheric conditions for extended periods of time (at least 2 years). All manipulations were carried out under a nitrogen atmosphere.

7.4 Tin–Rubidium–Sulfur

Several ternary compounds are formed in the Sn–Rb–S system. **$RbSnS_2$** crystallizes in the trigonal structure with the lattice parameters $a = 376$ and $c = 2443$ pm (Le Blanc et al. 1969; Le Blanc and Rouxel 1972). This compound was prepared by reacting SnS_2 with a solution of Rb in liquid ammonia. It is highly hygroscopic and decomposes at temperatures above 200°C.

$Rb_2Sn_2S_8$ has two polymorphic modifications (Liao et al. 1993). α-$Rb_2Sn_2S_8$ crystallizes in the monoclinic structure with the lattice parameters $a = 978.8 \pm 0.3$, $b = 997.8 \pm 0.3$, $c = 1436.0 \pm 0.2$ pm, $\beta = 106.70 \pm 0.02°$, a calculated density of 3.29 g·cm^{-3}, and an energy gap of 2.22 eV. β-$Rb_2Sn_2S_8$ crystallizes in the orthorhombic structure with the lattice parameters $a = 998.7 \pm 0.5$, $b = 1963.5 \pm 0.3$, $c = 1374.7 \pm 0.3$ pm, a calculated density of 3.28 g·cm^{-3}, and an energy gap of 2.14 eV.

α-$Rb_2Sn_2S_8$ was prepared as follow. A mixture of Sn (1 mM), Rb_2S (2 mM), and S (8 mM) was loaded into a Pyrex tube in a N_2 glove box. The tube was evacuated and flame sealed at a pressure of 0.13 Pa. The mixture was heated at 270°C for 4 days. Pure α-$Rb_2Sn_2S_8$ was made and the yield was improved when the reaction temperature was increased to 330°C and the Sn/Rb_2S/S molar ratio was 1:2:12. Upon cooling of the mixture to room temperature at the rate of 2°C·h^{-1}, well-formed orange needlelike crystals were obtained. The product was washed with water to remove excess Rb_2S and then washed with CS_2 to remove unreacted sulfur.

To prepare β-$Rb_2Sn_2S_8$, a mixture of Sn (1 mM), Rb_2S (2 mM), and S (8 mM) was loaded into a Pyrex tube in a N_2 glove box. The tube was evacuated and flame sealed at the pressure of 0.13 Pa. The mixture was heated at 450°C for 4 days. Upon cooling the mixture to room temperature at the rate of 3°C·h^{-1}, orange thin plate-like crystals were obtained. The product was washed with water to remove excess Rb_2S and then washed with CS_2 to remove unreacted sulfur.

$Rb_2Sn_4S_9$ melts congruently and crystallizes in the orthorhombic structure with the lattice parameters $a = 1189.10 \pm 0.07$, $b = 1188.67 \pm 0.11$, and $c = 2591.5 \pm 0.2$ pm, a calculated density of 3.387 g·cm^{-3}, and an energy gap of 2.65 eV (Marking et al. 1998). Thin plates of $Rb_2Sn_4S_9$ can be prepared through sulfur-rich polychalcogenide flux reactions kept for 4 days at 500°C and then cooled to 200°C at 4°C·h^{-1}. Yellow-gold flaky plates of $Rb_2Sn_4S_9$ were also prepared by reaction of Rb_2S + 4Sn + 16S mixture at 500°C. Direct combination of Rb_2S + 4Sn + 8S at 750°C for 3 days results in nearly pure microcrystalline $Rb_2Sn_4S_9$ after isolation. Products from the flux and direct combination reactions were washed with DMF to remove excess alkali metal polysulfide flux and other soluble byproducts. A brief wash with H_2O removed small amounts of water-soluble phases from the products. Acetone and diethyl ether were used to rinse and dry the products. This compound is stable in water and no decomposition can be detected in samples left under ambient atmospheric conditions for extended periods of time (at least 2 years). All manipulations were carried out under a nitrogen atmosphere.

Rb_4SnS_4 crystallizes in the cubic structure with the lattice parameter a = 1342.24 ± 0.07 (Klepp and Fabian 2002). Colorless crystals of the title compound were obtained by prolonged annealing of intimate mixture of Rb_2S, Sn, and S (molar ratio 2:1:2) in a sealed silica ampoule at 400°C, followed by controlled cooling to ambient temperature at a rate of 3°C·h^{-1}.

$Rb_6Sn_2S_7$ crystallizes in the orthorhombic structure with the lattice parameters a = 998.2 ± 0.4, b = 1345 ± 1, c = 1520 ± 1 pm, and a calculated density of 3.18 g·cm^{-3} (Klepp and Fabian 1999a). This compound was obtained be reacting a stoichiometric melt of Rb_2S, Sn, and S. Rb_2S (3.044 mM), Sn (1.522 mM), and S (3.041 mM) were weighed in the Ar-filled glove box, mixed well and melted under a vacuum of 0.01 Pa into a quartz ampoule. The sample was heated in a furnace to 700°C, kept at this temperature for 4 days and then cooled to room temperature at a rate of 2°C·h^{-1}. The product consisted from colorless, moisture sensitive crystals.

7.5 Tin–Cesium–Sulfur

Several ternary compounds are formed in the Sn–Cs–S system. **$CsSnS_2$** is amorphous to X-rays and was prepared by reacting SnS_2 with a solution of Cs in liquid ammonia. It is highly hygroscopic and decomposes at temperatures above 200°C (Le Blanc et al. 1969; Le Blanc and Rouxel 1972).

Cs_2SnS_{14} crystallizes in the monoclinic structure with the lattice parameters a = 696.4 ± 0.6, b = 1866 ± 1, c = 1480 ± 1 pm, β = 99.39 ± 0.01°, and a calculated density of 2.92 g·cm^{-3} (Liao et al. 1993). This compound was prepared in the following way. A mixture of Sn (0.5 mM), Cs_2S (1 mM), and S (3.0–5.0 mM) was loaded into a Pyrex tube in a N_2 glove box. The tube was evacuated and flame sealed at a pressure of 0.13 Pa. The mixture was heated at 250°C–275°C for 4–6 days with the next cooling of the mixture to room temperature at the rate of 2°C·h^{-1}. The product was washed by methanol under N_2 to remove excess Cs_2S_x flux. All sulfur ratios gave a mixture of red crystals of the title compound and yellow crystalline **$Cs_2Sn_2S_6$**. The crystals of Cs_2SnS_{14} dissolve slowly in water but are stable in air.

$Cs_2Sn_2S_6$ crystallizes in the triclinic structure with the lattice parameters a = 728.9 ± 0.4, b = 759.7 ± 0.3, c = 679.6 ± 0.3 pm and α = 114.80 ± 0.03°, β = 108.56 ± 0.04°, γ = 97.54 ± 0.04°, a calculated density of 3.78 g·cm^{-3}, and an energy gap of 2.44 eV (Liao et al. 1993). Pure yellow thin plate-like crystals of this compound was prepared in the following way. A mixture of Sn (0.5 mM), Cs_2S (0.5 mM), and S (4 mM) was loaded into a Pyrex tube in a N_2 glove box. The tube was evacuated and flame sealed at a pressure of 0.13 Pa. The mixture was heated at 400°C for 4 days with the next cooling of the mixture to room temperature at the rate of 4°C·h^{-1}. The product was washed with DMF to remove excess Cs_2S_x followed by washing with ether.

$Cs_2Sn_3S_{11}$ or **$Cs_2Sn_3S_7$·½S_8** melts incongruently at 436°C, crystallizes in the monoclinic structure with the lattice parameters a = 2235.0 ± 0.6, b = 1323.8 ± 0.2, c = 1630.1 ± 0.4 pm, β = 124.424 ± 0.002°, a calculated density of 3.254 g·cm^{-3}, and an energy gap of 2.64 eV (Marking and Kanatzidis 1995). High-quality crystals of the title compound were synthesized through the reactive flux techniques. A mixture of Sn, Cs_2S, and S (molar ratio 2:2:18) was loaded into a Pyrex tube in a N_2 glove box. The tube was evacuated to a residual pressure of <0.13 Pa and flame-sealed. The mixture was heated to 400°C over 16 h, isothermal at that temperature for 5 days, and cooled to 180°C at 4°C·h^{-1}. The product was washed with DMF to remove Cs_2S_x flux. A brief wash with CS_2 was used to remove any unreacted sulfur from the product. Methanol, acetone, and diethyl ether were successively used to rinse and dry the product. The crystals can range in color from yellow with a slight green tint to bright yellow-green. No decomposition of product was detected in samples left under ambient atmospheric conditions for periods exceeding 3 months. All manipulations were carried out under a N_2 atmosphere.

$Cs_2Sn_4S_9$ melts congruently and crystallizes in the orthorhombic structure with the lattice parameters a = 1340.1 ± 0.4, b = 1197.3 ± 0.2, and c = 1203.4 ± 0.2 pm, a calculated density of 3.540 g·cm^{-3}, and an energy gap of 2.66 eV (Marking et al. 1998). Thin plates of $Cs_2Sn_4S_9$ was prepared through sulfur-rich polychalcogenide flux reactions kept for 4 days at 500°C and then cooled to 200°C at 4°C·h^{-1}. Gold flaky plates of $Cs_2Sn_4S_9$ can be also prepared as the major phase by reaction of Cs_2S + 2Sn + 12S mixture at 500°C. $Cs_2Sn_4S_9$ was prepared as a pure microcrystalline phase by direct combination of Cs_2S + 4Sn + 8S at 720°C for 1 day. Products from the flux and direct combination reactions were washed with DMF to remove excess alkali metal polysulfide flux and other soluble byproducts. A brief wash with H_2O removed small amounts of water-soluble phases from the products. Acetone and diethyl ether were used to rinse and dry the products. This compound is stable in water and no decomposition can be detected in samples left under ambient atmospheric conditions for extended periods of time (at least 2 years). All manipulations were carried out under a nitrogen atmosphere.

7.6 Tin–Copper–Sulfur

SnS–Cu_2S: The phase diagram of this system constructed through DTA is a eutectic type without a solid solutions based on binary compounds (Figure 7.1) (Khanafer et al. 1974b; Koike 1993). The eutectic contains 45 mol% SnS and crystallizes at 490°C.

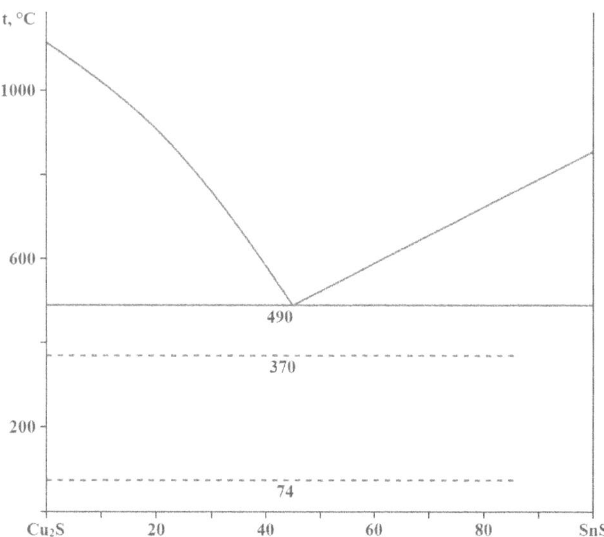

FIGURE 7.1 Phase diagram of the SnS–Cu_2S system. (From Koike, K., *Shigen-to-Sozai*, **109**(1), 23, 1993.)

The vapor pressure of SnS of the SnS–Cu_2S system was measured by means of the transportation method within a temperature range from 900°C to 1100°C, and activities of SnS and Cu_2S was calculated (Koike 1993). It can be seen that both activities exhibited a negative departure from the ideal behavior, but they were little changed by differences in temperature (Figure 7.2). The activities coefficients at the infinite dilution of the components at 1100°C were estimated as γ^0(SnS) = 0.29 and γ^0(Cu_2S) = 0.039.

SnS_2–Cu_2S: The results of the investigation of this system are contradictory. The phase diagram of this system constructed by Fiechter et al. (2003) through DTA and XRD is shown in Figure 7.3a. Four ternary phases were found in this system. With the exception of **$Cu_4Sn_3S_8$**, which is only stable above 685°C, the other three phases are stable at room temperature. It has been observed that they coexist with each other. The system is characterized by four eutectic points, and series of solid-solid phase transitions for the compositions more than 50 mol% SnS_2 as well as a wide homogeneity range of $Cu_2Sn_{3+x}S_{7+2x}$ at elevated temperatures. The first eutectic crystallizes at 800°C and contains 20–25 mol% SnS_2. The position of this eutectic point could not be determined more accurately. The second eutectic point is located at 38 mol% SnS_2. In the following part of the phase diagram, two lines at about 675° and 685°C appear that connect **Cu_2SnS_3** with SnS_2 via **$Cu_2Sn_3S_7$**. A line at 770°C connects $Cu_4Sn_3S_8$, which exists in a temperature range from 685°C to 785°C, via $Cu_2Sn_3S_7$ with SnS_2. The third eutectic is located at 72.5 mol% SnS_2 and crystallizes at 793°C, and the fourth eutectic crystallizes at 790°C and contains 83 mol% SnS_2. $Cu_2Sn_3S_7$ has three solid-solid phase transitions at 675°C, 685°C, and 770°C. SnS_2 possesses one solid-solid transition at 718°C.

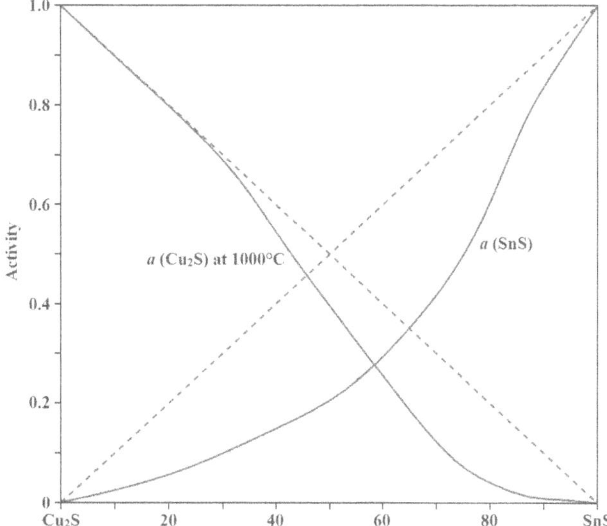

FIGURE 7.2 Activities of SnS and Cu_2S in the SnS–Cu_2S at 1100°C. (From Koike, K., *Shigen-to-Sozai*, **109**(1), 23, 1993.)

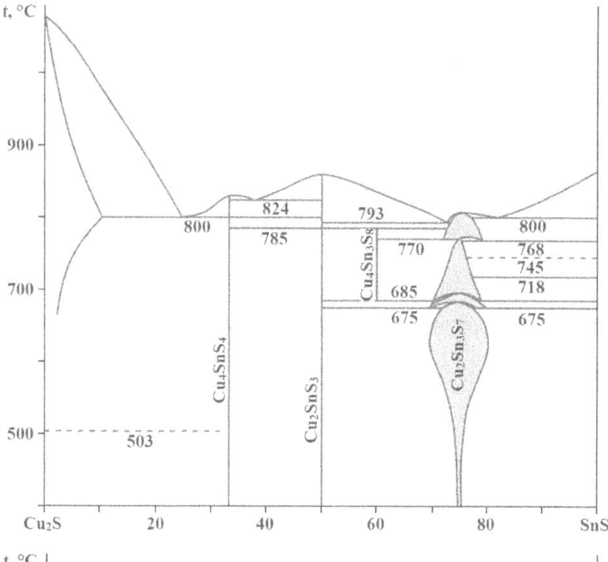

FIGURE 7.3 Phase diagram of the SnS_2–Cu_2S system. (From Fiechter, S., et al., *J. Phys. Chem. Solids*, **64**(9–10), 1859, 2003.)

Piskach et al. (1998) noted that in the SnS_2–Cu_2S system there are only three compounds, Cu_2SnS_3, **$Cu_2Sn_4S_9$**, and **Cu_4SnS_4** (Figure 7.3b). Two eutectics melt at 820°C and

788°C. According to the data of Khanafer et al. (1974b), four compounds, Cu_2SnS_3, $Cu_2Sn_4S_9$, Cu_4SnS_4, and $Cu_4Sn_3S_8$, are formed in this system. The $\mathbf{Cu_8SnS_6}$ phase is a mixture of Cu_4SnS_4 and $Cu_{1.96}S$ and the $\mathbf{Cu_2Sn_2S_5}$ compound is a mixture of Cu_2SnS_3 and $Cu_2Sn_4S_9$. Some characteristics of the compounds, forming in the SnS_2–Cu_2S, system are given below.

Cu_2SnS_3 melts congruently at approximately 856°C [at 845°C (Palatnik et al. 1961a; at 855°C (Averkieva et al. 1964); at 838°C (Rivet 1963, 1965); at 874°C (Balanevskaya et al. 1966); at 850°C (Khanafer et al. 1974a, b); at 860°C (Piskach et al. 1998)] and has an energy gap of 1.11–1.34 eV (Neves et al. 2016a, b) [1.1 eV (Tiwari et al. 2014); 0.93 eV (Fiechter et al. 2003); 0.87 eV (Nomura et al. 2013a, b)]. Apparently, this compound has some polymorphic modifications: Moh (1963, 1975) noted that this compound has a polymorphic transformation at 775°C–780°C. One of them crystallizes in the monoclinic structure with the lattice parameters $a = 665.4 \pm 0.8$, $b = 1153.4 \pm 0.2$, $c = 665.9 \pm 0.4$ pm, $\beta = 109.40 \pm 0.05°$, and an energy gap of 0.87 eV (Nomura et al. 2013a, b) [$a = 664.0$, $b = 1153$, $c = 1991$ pm, and $\beta = 109.45°$ (Wu et al. 1986; Moh 1975); $a = 665.3 \pm 0.1$, $b = 1153.7 \pm 0.2$, $c = 666.5 \pm 0.1$ pm, $\beta = 109.39 \pm 0.03°$, and a calculated density of 4.708 $g{\cdot}cm^{-3}$ (Onoda et al. 2000b)]. The second modification crystallizes in the tetragonal structure with the lattice parameters $a = 541.9$ and $c = 1084.2$ (Tiwari et al. 2014) [$a = 541.3 \pm 0.1$ and $c = 1082.4 \pm 0.1$ and a calculated density of 4.774 $g{\cdot}cm^{-3}$ (Chen et al. 1998); $a = 542.6$ and $c = 1088$ pm and the calculated and experimental densities of 4.71 and 4.62 $g{\cdot}cm^{-3}$, respectively (Hahn et al. 1966)]. The third modification crystallizes in the cubic structure with the lattice parameter $a = 544.5 \pm 0.5$ pm and the calculated and experimental densities of 4.69 and 4.68 ± 0.01 $g{\cdot}cm^{-3}$, respectively (Rivet 1963, 1965; Khanafer et al. 1974a, b; Moh 1975) [$a = 543$ pm (Palatnik et al. 1961a, b); $a = 542.5$ pm (Averkieva et al. 1964)]. The fourth modification, mineral mohite, crystallizes in the triclinic structure with the lattice parameters $a = 664$, $b = 1151$, $c = 1993$ pm, and $\alpha = 90°$, $\beta = 109°45'$, $\gamma = 90°$ (Wang 1974; Kovalenker et al. 1982; Fleischer and Pabst 1983). According to the thermodynamic simulation, Cu_2SnS_3 dissociates at the melting temperature forming Cu_2S and SnS_2 binary compounds (Glazov et al. 1978).

The standard enthalpy of Cu_2SnS_3 formation from CuS and SnS has been determined by Stolyarova et al. (2020): $2CuS + SnS \rightarrow Cu_2SnS_3$, $\Delta H^0_{298} = -51.01 \pm 1.26$ $kJ{\cdot}M^{-1}$. The standard enthalpy of Cu_2SnS_3 formation from elements has been calculated using reference data: $\Delta H^0_{298} = -263.79 \pm 2.28$ $kJ{\cdot}M^{-1}$.

The monoclinic modification of Cu_2SnS_3 was prepared using a stoichiometric mixture of Cu_2S (5 mM) and SnS_2 (5 mM) (Onoda et al. 2000b). The sample was sealed in a silica tube at a pressure of less than 0.13 Pa. The tube was gradually heated to 853°C where they were kept for 2 days. They were then cooled to 600°C at a rate of 5°C·h^{-1}, followed by annealing at that temperature for 2 weeks and quenching in cold water. The gray needlelike crystals were isolated from the solidified mass mechanically. The powders of monoclinic Cu_2SnS_3 were also successfully synthesized by combining mechanical alloying and spark plasma sintering (Neves et al. 2016a, b). XRD revealed the formation of this compound after 2 h of mechanical alloying. Spark plasma sintering of the powder particles was performed at 600°C yielding high densification without much sulfur loss or oxidation. The monoclinic Cu_2SnS_3 is quite stable upon spark plasma sintering under a pressure of 50 MPa.

Samples of the tetragonal Cu_2SnS_3 were prepared by solid state reaction between Cu $(CH_3COO)_2$, $SnCl_2{\cdot}2H_2O$, and a sulfur source (Tiwari et al. 2014). The metal salts are pulverized and blended with a sulfur source in molecular proportion to yield a homogenous paste. Five different sulfur sources, such as, thiourea, thioacetamide, dimethylthiourea, Na_2S, and elemental S were utilized, but only thiourea was found to yield the title compound without any unwanted CuS_x or SnS_x. The paste was heated in a furnace at 200°C in air for 1 h. On heating the pastes are converted to black powders. This modification could be also synthesized by a direct reaction of Cu, Sn, and S taken in stoichiometric quantities, at 1050°C for 2 days (Chen et al. 1998). The sample was slowly cooled at a rate of 5°C·h^{-1} to 700°C and annealed there for 2 days with the next quenching in air.

The synthesis of cubic Cu_2SnS_3 was carried out from the elements, placed in a silica ampoule, sealed under vacuum (Khanafer et al. 1974a). An experience has shown that it is necessary, in order to ensure homogeneous preparations, to stand the silica ampoules at a temperature above about 100°C of the melting temperature of the prepared product. The single crystals were obtained either by a multi-week annealing of the products sprayed or not, at a temperature below ca. 50°C of the melting point or by using a temperature gradient of 50°C between the two ends of the ampoule.

$Cu_2Sn_3S_7$ melts congruently at approximately at 803°C, has an energy gap of 0.93 eV (Fiechter et al. 2003), and crystallizes in the monoclinic structure with the lattice parameters $a = 1268$, $b = 735.1$, $c = 1276$ pm, and $\beta = 109.60°$ (Wu et al. 1986) [$a = 1275$, $b = 734$, $c = 1271$ pm, and $\beta = 109°30'$ (Moh 1975)].

$Cu_2Sn_4S_9$ melts incongruently at 670°C (Piskach et al. 1998) [at 665°C (Khanafer et al. 1974b)] and crystallizes in the cubic structure with the lattice parameter $a = 1040 \pm 1$ pm and the calculated and experimental densities of 5.10 and 5.03 $g{\cdot}cm^{-3}$, respectively (Khanafer et al. 1974b).

Cu_4SnS_4 melts congruently at approximately at 833°C (Fiechter et al. 2003) [melts incongruently at 810°C (Khanafer et al. 1974a, b; Piskach et al. 1998)], has polymorphic transformation at −41°C (Khanafer et al. 1974b; Mahy et al. 1985; Suzumura et al. 2015; Choudhury et al. 2017) [at −37°C (Anzai and Fukazawa 1986); at ~ −43°C (Goto et al. 2013)], and an energy gap of 0.92 eV (Choudhury et al. 2017) [0.11–0.12 eV for α-Cu_4SnS_4 and 0.03–0.04 eV for β-Cu_4SnS_4 (Khanafer et al. 1974a); 0.39 eV, the calculated value (Goto et al. 2013)]. The first order transformation is associated with a space group change: the room-temperature phase is primitive orthorhombic, whilst the low-temperature phase appears as *A*-face centered orthorhombic (Mahy et al. 1985). In the low-temperature phase the *a* and *c* axes gradually increase with increase in temperature, while the *b* axis shows little dependence on temperature (Anzai and Fukazawa 1986). The *b* and *c* axes suddenly contract at −37°C, and the *a* axis suddenly expands at this temperature. The changes in the *a*, *b*, and *c* parameters between −38°C and −35°C are +6.7, −6.8 and −1.1 pm, respectively. Above −37°C, the lattice parameters gradually increase with further increase in temperature. Later investigations showed that the low-temperature modification crystallizes in the monoclinic structure, while the high-temperature modification crystallizes in the orthorhombic structure with the next lattice parameters: α-Cu_4SnS_4 – a = 1005.8 ± 0.6, b = 1348.9 ± 0.8, c = 1008.5 ± 0.6 pm, β = 100.840°, and a calculated density of 4.953 g·cm^{-3} at 170 ± 2 K (Choudhury et al. 2017) [a = 1014.9, b = 1345.8, c = 1027.8 pm, and β = 100.66° (Goto et al. 2013); a = 1001.575 ± 0.015, b = 1346.981 ± 0.002, c = 1010.091 ± 0.015 pm, β = 100.8035 ± 0.0009°, experimental results, and a = 1009.1208, b = 1353.1941, c = 1028.0067 pm, β = 100.750°, calculated results (Suzumura et al. 2015)]; β-Cu_4SnS_4 – a = 1356.6 ± 0.2, b = 768.9 ± 0.3, c = 641.6 ± 0.2 pm, and a calculated density of 4.972 g·cm^{-3} at room temperature (Choudhury et al. 2017) [a = 1370 ± 1, b = 775.0 ± 0.5, c = 645.4 ± 0.5 pm, and the calculated and experimental densities of 4.86 and 5.05 ± 0.01 g·cm^{-3}, respectively (Khanafer et al. 1974a, b); a = 1351, b = 768, c = 641 pm (Wang 1974); a = 1351, b = 768, c = 641 pm (Moh 1975); a = 1355.8 ± 0.3, b = 768.1 ± 0.1, c = 641.2 ± 0.1 pm, and the calculated and experimental densities of 4.86 and 5.05 ± 0.01 g·cm^{-3}, respectively (Jaulmes et al. 1977); a = 1350, b = 766, c = 639.5 pm (Wu et al. 1986); a = 1360.8, b = 769.0, c = 648.2 pm (Goto et al. 2013); a = 1356.460 ± 0.014, b = 768.999 ± 0.008, and c = 641.601 ± 0.006 pm, experimental results, and a = 1360.13, b = 768.862, and c = 649.197 pm, calculated results (Suzumura et al. 2015)].

Cu_4SnS_4 was synthesized from the stoichiometric mixture of Cu, Sn, and S, which was loaded in a fused quartz ampoule inside N_2-filled glove box (Choudhury et al. 2017). The ampoule was then taken out from the glove box with the help of an adapter and hooked up in a Schlenk line, evacuated, flame-sealed, and placed in a furnace. The furnace temperature was increased to 450°C at a slow rate of 3°C·h^{-1} to avoid building up of excess sulfur vapor pressure and held constant at this temperature for 72 h. After that, the temperature of the furnace was further raised to 850°C, held constant at that temperature for another 72 h, slowly cooled to 600°C, held there for 1 h and finally decreased to room temperature at a rate of 200°C·h^{-1}. The as-synthesized product contained an ingot and on the surface of which was embedded diamond shaped crystals. These crystals were appropriate for single-crystal structure determination. Polycrystalline samples of the title compound could be also prepared from a stoichiometric mixture of the elements, which was slowly heated to 900°C, kept for 24 h, and then cooled in the furnace (Suzumura et al. 2015). After being powdered roughly, pressed into pellet, and sealed in a vacuum quartz ampoule again, it was heated at 750° for 48 h to be in a single phase. To prepare this compound, a stoichiometric mixture of the elements could be also heated at 800°C for 96 h in an evacuated silica ampoule with the next grinding, pressing into pellets and heating again at 700°C for 40 h in an evacuated silica ampoule (Goto et al. 2013).

$Cu_4Sn_3S_8$ melts incongruently at 790°C and has a polymorphic transformation at 625°C (Khanafer et al. 1974b).

Some other compounds were found in the SnS_2–Cu_2S system. **$Cu_4Sn_7S_{16}$** crystallizes in the trigonal structure with the lattice parameters a = 737.88 ± 0.02, c = 3603.2 ± 0.2 pm, and an energy gap of 0.8 eV (Jemetio et al. 2006) [a = 737.2 ± 0.1, c = 3601.0 ± 0.7 pm, a calculated density of 4.697 g·cm^{-3}, and an energy gap of 0.42 eV (Chen et al. 1998)]. Single crystals of the title compound were prepared by direct reaction of Cu, Sn, and S (molar ratio 4:7:16) (Jemetio et al. 2006). The mixture was filled in silica tube, sealed under vacuum and placed into a furnace. The sample was slowly heated to 450°C within 120 h, then to 850°C within 48 h, and subsequently annealed for 120 h at 850°C, followed by cooling to room temperature over 24 h. The compact product appeared black and contained black plate-like crystals of $Cu_4Sn_7S_{16}$. All preparative steps were carried out in an argon-filled glove box.

$Cu_4Sn_7S_{16}$ could be also synthesized from a stoichiometric mixture of Cu_2S and SnS_2 (Chen et al. 1998). The sample was ground in an agate mortar and pressed into pellets. The pellets were introduced into a silica tube and sealed at a pressure of less than 0.13 Pa. The tube was held at 450°C for 3 days, then heated gradually to 850°C, where they were kept for 3 days, then cooled at a rate of 5°C·h^{-1} to 600°C and quenched in air. The gray crystals with metallic luster were found in the tube.

$Cu_4Sn_{15}S_{32}$ or **$CuSn_{3.75}S_8$** crystallizes in the cubic structure with the lattice parameter a = 1039.3 ± 0.2 and the calculated and experimental densities of 4.53 and 4.60 g·cm^{-3}, respectively (Jaulmes et al. 1982b).

Cu_8SnS_6 has a polymorphic transformations at 55°C with $\Delta H_{phase\ trans.}$ = 5.1 ± 0.3 kJ·M^{-1} (Aliev et al. 1989).

Besides the compounds which exist in the SnS_2–Cu_2S system, some other compounds were found in the Sn–Cu–S ternary system. **Cu_3SnS_4** (mineral kuramite) crystallizes in the tetragonal structure with the lattice parameters a = 544.5 ± 0.5, c = 1075 ± 2 pm, and an experimental density of 4.56 g·cm^{-3} (Kovalenker et al. 1979; Fleischer et al. 1980).

Cu_4SnS_6 crystallizes in the trigonal structure with the lattice parameters a = 373.9 ± 0.1, c = 3294.1 ± 0.7 pm, and a calculated density of 4.707 g·cm^{-3} (Chen et al. 1999c) [a = 375.6 ± 0.8, c = 3291 ± 4 pm, and a calculated density of 4.53 g·cm^{-3} for mineral erazoite (Schlüter et al. 2015; Belakovskiy et al. 2017)]. Single crystals of this compound were grown by a two-step process (Chen et al. 1999c). First, the precursor, Cu_4SnS_4, was prepared by reacting a stoichiometric mixture of Cu_2S (4.0 mM) and SnS_2 (2.0 mM) in an evacuated silica tube at 600°C for 1 week. The resulting product was reacted with excess sulfur at 400°C to get the homogeneous Cu_4SnS_6 material. A eutectic flux of 54 mass% KCl and 46 mass% LiCl was employed with a flux-to-charge ratio of 3:1 for the crystal growth experiment. The mixture was placed in a silica tube and sealed at a pressure of less than 0.13 Pa. The tube was heated gradually to 495°C, at which temperature they were maintained for 2 months, then cooled at a rate of 3°C·h^{-1} to 300°C and quenched in air. Gray crystals with metallic luster were isolated by washing the reaction product with deionized water. The crystals of the title compound were soft and easily deformed, and they appeared to be thin plates of hexagonal habit. In addition, a few large crystals of Cu_9S_5 were observed in the reaction product.

The **$Cu_5Sn_2S_7$** ternary compound is stable in the temperature range between 400°C and 650°C (Wu et al. 1986).

$Cu_{10}Sn_2S_{13}$ is stable up to 641°C ± 3°C and crystallizes in the tetragonal structure with the lattice parameters a = 954.0, c = 1093 pm (Wu et al. 1986).

The isothermal section of the Sn–Cu–S ternary system at 500°C is shown in Figure 7.4 (Wu et al. 1986). It is seen, that a metallic liquid develops only in the Sn-rich portion of the

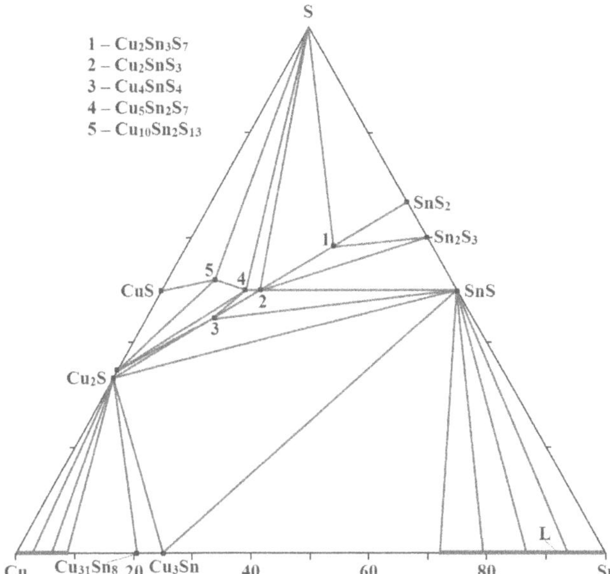

FIGURE 7.4 Isothermal section of the Sn–Cu–S ternary system at 500°C. (From Wu, D., et al., *Mineralog. Mag.*, **50**(356), 323, 1986.)

system, and is isolated from sulfide assemblages, except SnS, by the join SnS–Cu_3Sn. In the S-rich portion, liquid sulfur exists in equilibrium with sulfides.

The isothermal section of the Sn–Cu–S ternary system at 600°C was constructed by (Moh 1975). It was shown that three ternary compounds are stable: Cu_4SnS_4, Cu_2SnS_3, and a phase ranging in composition from $Cu_2Sn_2S_5$ to $Cu_2Sn_3S_7$; all three are situated on the SnS_2–Cu_2S join. In the central portion, ternary liquid occurs with approximately 21 mass% S, but a variable Cu/Sn ratio, and it appears to be immiscible with a Sn melt which contains maximum of 18 mass% Cu. α-Cu and β, γ, ζ and ε phases from the Cu–Sn binary system may coexist with Cu_2S, ε also with the ternary liquid. In addition, this liquid may coexist with Cu_2S, Cu_4SnS_4, Cu_2SnS_3, or with SnS. Cu_2SnS_3 is stable with SnS or Sn_2S_3; Sn_2S_3 can equilibrate with the $Cu_2Sn_2S_5$–$Cu_2Sn_3S_7$ solid solution. With the exception of Cu_4SnS_4, the ternary compounds as well as SnS_2 or the S-rich end of the Cu_2S solid solution can all coexist with liquid S.

7.7 Tin–Silver–Sulfur

SnS–Ag_2S: The phase diagram of this system was constructed using DTA, isothermal annealing, and hydrothermal synthesis by Nekrasov et al. (1976). No ternary compounds were found in the system. The solubility of SnS in β-Ag_2S reaches 6 mol% at 200°C and 21 mol% at 519°C, and the solubility of Ag_2S in SnS is 1.5–2 mol%.

SnS_2–Ag_2S: The phase diagram of this system, constructed using DTA and XRD, is a eutectic type (Gorochov 1968; Elli and Canepari 1963). The eutectics contain 6 and 50 mol% SnS_2 and crystallize at 750°C [768°C (Elli and Canepari 1963)] and 635°C, respectively (Gorochov 1968). The thermal effects at 169°C and 600°C are caused by polymorphic transformations of Ag_2S. The **Ag_8SnS_6** ternary compound is formed in the system which melts congruently at 839°C (Gorochov and Flahaut 1967; Gorochov 1968) [at 822°C (Elli and Canepari 1963)] and has a polymorphic transformation at 172°C (Gorochov and Flahaut 1967; Gorochov 1968) [at 610°C (Elli and Canepari 1963)]. It is possible that this compound has three polymorphic modifications. High-temperature modification crystallizes in the cubic structure with the lattice parameter a = 1085 pm and an experimental density of 6.28 g·cm^{-3} at 200°C (Gorochov and Flahaut 1967; Gorochov 1968).

Low-temperature modification also crystallizes in the cubic structure with the lattice parameter a = 2143 pm and the calculated and experimental densities of 6.31 and 5.95 g·cm^{-3}, respectively (Hahn et al. 1965; Gorochov 1968).

Third modification of Ag_8SnS_6 crystallizes in the orthorhombic structure with the lattice parameters $a = 1533.4 \pm 0.3$, $b = 756.2 \pm 0.1$, and $c = 1072.4 \pm 0.2$ pm (Bagheri et al. 2014) [a = 1529.3, b = 754.6, and c = 1071.5 pm (Li et al. 2000); a = 1582, b = 1242, and c = 689 pm (Nekrasov et al. 1976)]. This modification can be prepared from stoichiometric mixtures of $AgNO_3$ (11.29 mM), $SnCl_2{\cdot}2H_2O$ (1.41 mM), and S (8.44 mM) (Li et al. 2000). The mixture was loaded into a 100-mL Teflon liner autoclave, which was then filled with anhydrous ethylenediamine up to 90% of the total volume. The autoclave was sealed and maintained at 100°C for 10 h and then cooled to room temperature naturally. The product was filtered and washed with distilled water and absolute ethanol several times to remove the by-products. The precipitates were dried in vacuum at 60°C for 4 h. It was also synthesized from the elements (Bagheri et al. 2014). The mixture was placed into quartz ampoule evacuated to ~10^{-2} Pa and sealed. Synthesis was carried out in a two-zone furnace. The hot zone was heated up to temperature ~30–50°C above the melting point of the compound and the cold zone was heated to 380°C.

Besides Ag_8SnS_6, another compound, **Ag_2SnS_3**, is formed in the SnS_2–Ag_2S system (Nekrasov et al. 1976). It melts at 640°C [at 650°C (Wold and Brec 1976)] and has polymorphic transformation at 256°C (Nekrasov et al. 1976). First modification of this compound crystallizes in the monoclinic structure with the lattice parameters a = 803, b = 1081.5, c = 508.5 pm, and β = 108.28° (Belandria et al. 2000) [a = 654, b = 1134, c = 1269 pm, and β = 96°5′ (Nekrasov et al. 1976)] and an energy gap of 0.5 eV (Balanevskaya et al. 1966). Second modification crystallizes in the hexagonal structure with the lattice parameters a = 1136, c = 1369 pm, and the calculated and experimental densities of 6.47 and 6.4 g·cm^{-3}, respectively (Nekrasov et al. 1976).

It was synthesized by direct melting of the elements in the stoichiometric ratios inside evacuated and sealed quartz ampoule (Wold and Brec 1976; Belandria et al. 2000). The top temperature varied between 650°C and 960°C.

Some other ternary compounds were found in the Sn–Ag–S system. **$AgSnS_2$** crystallizes in the cubic structure with the lattice parameter $a = 550.6 \pm 0.2$ pm and the calculated and experimental densities of 5.78 and 5.75 g·cm^{-3}, respectively (Wold and Brec 1976). It was prepared by heating a mixture of SnS_2 and Ag at 1200°C and 6.5 GPa.

$Ag_4Sn_3S_8$ crystallizes in the cubic structure with the lattice parameter $a = 1080.898 \pm 0.007$ and $a = 1089.01 \pm 0.01$ pm at room temperature and 300°C, respectively (Hull et al. 2005) [$a = 1080.13 \pm 0.04$ pm and the calculated and experimental densities of 5.39 and 5.41 g·cm^{-3}, respectively, for $Ag_{3.8}Sn_3S_8$

composition (Amiel et al. 1995)]. This compound was prepared by solid state synthesis. Stoichiometric amounts of Ag_2S, Sn, and S were used as starting materials, with a small excess of sulfur (Amiel et al. 1995; Hull et al. 2005). The mixture was then ground in an agate mortar and pressed into pellets, which were placed in a quartz ampoule which was then evacuated. The sample was annealed for 1 week at 600°C, cooled by switching off the furnace, reground and annealed at 400°C–600°C for 5 days. The single crystals were grown from a starting mixture with a molar ratio of Ag/Sn/S equal to 2:3:7 (Amiel et al. 1995). Such mixture gave, after 2 weeks at 600°C, single crystals of $Ag_4Sn_3S_8$ together with SnS_2.

7.8 Tin–Magnesium–Sulfur

The $\mathbf{Mg_2SnS_4}$ ternary compound, which crystallizes in the orthorhombic structure with the lattice parameters a = 1296.7 ± 0.6, b = 751.1 ± 0.5, and c = 622.6 ± 0.4 pm, a calculated density of 3.237 g·cm^{-3} and an energy gap of ≈2.05 eV (Wu et al. 2015b) [a = 1293, b = 752, and c = 616 pm (Rocktäschel et al. 1964)], is formed in the Sn–Mg–S system. To prepare this compound, a mixture of MgS (2 mM) and SnS_2 (1 mM) were heated to 850°C during 30 h, left for 72 h, cooled to 400°C at a rate of 3°C·h^{-1}, and finally cooled to room temperature by switching off the furnace (Wu et al. 2015b). Many red block-shaped crystals were found in the ampoule, which are stable in air. A polycrystalline sample of this compound was synthesized using solid-state reaction techniques. The mixtures of MgS and SnS_2 in a molar ratio of 2:1 were heated to 700°C during 20 h, kept at that temperature for 72 h, and then the furnace was turned off.

7.9 Tin–Calcium–Sulfur

SnS–CaS: The phase diagram of this system is a eutectic type (Murach 1947). The eutectic contains 10 mol% CaS and crystallizes at 830°C.

SnS_2–CaS: Two ternary compounds, $\mathbf{CaSnS_3}$ and $\mathbf{Ca_2SnS_4}$, are formed in this system. $CaSnS_3$ crystallizes in the orthorhombic structure with the lattice parameters a = 668.7, b = 708.4, c = 1128.6 pm, and an energy gap of 1.72 eV (Shaili et al. 2021). The first step of this compound preparation consists of preparing the $CaSnO_3$ thin film. An aqueous solution was prepared by mixing 0.02 M $Ca(NO_3)_2{\cdot}4H_2O$ and 0.02 M $SnCl_4{\cdot}5H_2O$ in 25 mL of distilled water at room temperature. The mixture was stirred at 60°C for 1 h resulting a clear homogenous solution. The stirred mixture was deposited on the surface of the substrates with a flow rate of 0.9 mL·min^{-1}. After the deposition, the thin films were post-annealed at 300°C for 30 min and then the temperature was slowly decreased to the room temperature. In the second step the thin films were calcined in air at 800°C for 2 h with a heating rate of 15°C·min^{-1} to obtain $CaSnO_3$. The resulted films were white and transparent. In the sulfurization step, the samples were placed in a quartz tube next to a boat contains a 5 mg of sulfur powder, and the quartz tube was placed inside a furnace. After closing the quartz tube, it was purged with a flow of Ar and the temperature was increased with a rate of 15°C·min^{-1} until it reached 400°C or 500°C. The samples were left under the previous conditions for 2 h. Finally, the temperature was dropped under constant Ar flow until it reached the room temperature and the Ar flow was kept constant for another 2 h. The exhaust gas from the tube was bubbled though 2 M of NaOH solution during the reaction.

Ca_2SnS_4 also crystallizes in the orthorhombic structure with the lattice parameters a = 1389.1 ± 0.3, b = 817.57 ± 0.16, c = 653.84 ± 0.13 pm, a calculated density of 2.92564 g·cm^{-3}, and an energy gap of 2.32 ± 0.03 eV (Zhou et al. 2016) [a = 1374, b = 823, and c = 644 pm (Rocktäschel et al. 1964); a = 1388, b = 818, and c = 651 pm (Susa and Steinfink 1971b)]. To prepare this compound, a mixture of CaS and SnS_2 in the molar ratio of 2:1 was ground and loaded into fused-silica tube under an Ar atmosphere in a glove box, which was sealed under 10^{-3} Pa and then placed in a furnace (Zhou et al. 2016). The sample was heated to 950°C within 15 h and kept at that temperature for 80 h, then slowly cooled to 300°C at a rate of 3°C·h^{-1}, and finally cooled to room temperature. The product consisted of block red crystals of Ca_2SnS_4.

The title compound could be also obtained when starting with elemental Sn, and powdered S and CaS (Susa and Steinfink 1971b). Stoichiometric ratio of Sn and Ca with sulfur or CaS were mixed, sealed under vacuum in Vycor tubing, and pre-reacted at temperatures ranging from 200°C to 400°C for 2–4 h. The temperature was then raised and kept in the range 600°C–800°C for about 12 h. After cooling, the vial was examined visually to see whether some of the starting materials had remained unreacted. If the sample looked homogeneous, the vial was opened. If Ca_2SnS_4 was detected in the sample, it was resealed in a Vycor tube and fired at the same or at higher temperatures until it appeared to have completely reacted. In some cases the temperature was raised to 1330°C after first placing the material in a graphite crucible which was then sealed in a Vycor tube.

7.10 Tin–Strontium–Sulfur

SnS_2–SrS: The phase diagram of this system, constructed through DTA and XRD, is a eutectic type with peritectic transformations (Figure 7.5) (Guittard et al. 1978a). The eutectic contains 70 mol% SnS_2 and crystallizes at 745°C. In the range of 70–95 mol% SnS_2, weak thermal effects were observed at 700°C, due to the presence of a eutectic in the Sn–S binary system between SnS and SnS_2 compounds at 705°C. Two compounds, $\mathbf{Sr_2SnS_4}$ and $\mathbf{Sr_3Sn_2S_7}$, which melt incongruently at 990°C and 810°C, respectively, were found in this system. Both compounds crystallize in the orthorhombic structure with the lattice parameters a = 997.7 ± 0.1, b = 1031.1 ± 0.2, c = 724.3 ± 0.1 pm, and a calculated density of 3.76 g·cm^{-3} (Pocha et al. 2003) [a = 1590, b = 830, c = 1162 pm, and a calculated density of 3.66 g·cm^{-3} (Guittard et al. 1978a)] for Sr_2SnS_4 and a = 1144, b = 1242, c = 393 pm, and a calculated density of 4.31 g·cm^{-3} (Guittard et al. 1978a) for $Sr_3Sn_2S_7$.

Sr_2SnS_4 obtained by the following two-step synthesis (Pocha et al. 2003). First, Sr_2Sn was prepared by heating stoichiometric amounts of Sr and Sn to 800°C in an alumina crucible under

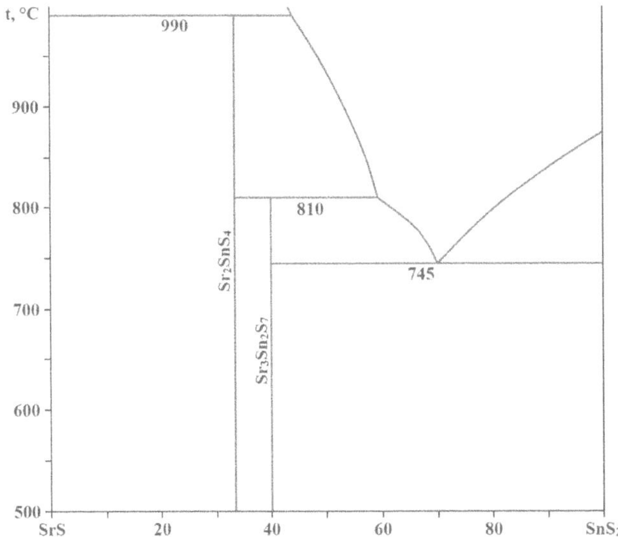

FIGURE 7.5 Phase diagram of the SnS_2–SrS system. (From Guittard, M., et al., *C. r. Acad. sci. Sér. C*, **287**(6), 239, 1978.)

Ar atmosphere for 12 h. Then, the binary product was homogenized, mixed with the stoichiometric amount of sulfur, loaded again in the crucible, and sealed under Ar. The ampoule was rapidly heated to 900°C (furnace preheated to 900°C), kept at this temperature for 2 h and then quenched in water. This results in a yellow-orange powder sample of Sr_2SnS_4, which is stable in air.

Two another ternary compounds, $\mathbf{SrSnS_3}$ and $\mathbf{Sr_2SnS_5}$, are formed in the Sn–Sr–S system. First of them also crystallizes in the orthorhombic structure with the lattice parameters a = 826.4 ± 0.2, b = 1411.6 ± 0.3, c = 386.7 ± 0.1 pm (Yamaoka and Okai 1970). To prepare this compound, an equimolar mixture of SrS and SnS_2 or 1:1:2 mixture of SrS, Sn, and S was sealed in a Pt capsule. A piston-cylinder type of high pressure apparatus was used with the furnace assembly in the pressure cell. Pressure was applied to the sample first, and then the temperature was raised to 500°C–800°C at a rate of 2–3°C·min^{1} and held for 15–20 h. The specimen was quenched to room temperature before the pressure was released. Almost perfect single crystals were obtained at a pressure of 2 GPa.

Sr_2SnS_5 melts incongruently at 908°C and also crystallizes in the orthorhombic structure with the lattice parameters a = 412, b = 1580, c = 1154 pm, and the calculated and experimental densities of 4.00 and 3.88 g·cm^{-3}, respectively (Guittard et al. 1983). Red crystals of this compound were obtained in vacuum-sealed quartz ampoule by reaction of 2SrS + SnS_2 + S mixture. The heating was carried out gradually, so as to reach in 5 days the temperature of 800°C, which is maintained for 3 days.

7.11 Tin–Barium–Sulfur

SnS–BaS: Hervieu et al. (1967) highlighted the existence in this system of a definite compound $\mathbf{BaSnS_2}$ as well as two domains of solid solution represented by the formulas $\mathbf{Ba_xSn_{1-x}S}$ ($0 \le x \le 0.245$) and $\mathbf{Ba_{3x}Sn_{3(1-x)}S}$ ($0.295 \le x \le 0.333$) the latter existing in two allotropic forms α and β marked by an order ⇔ disorder transformation at ~680°C. The powder diagrams made it possible to find for this domain an orthorhombic symmetry, the values of the lattice parameters for the upper limit phase $\mathbf{BaSn_2S_3}$ being for the form α: a = 1158 ± 1, b = 1408 ± 1, and c = 1518 ± 1 pm and the calculated and experimental densities of 5.055 and 4.828 ± 0.005 g·cm^{-3}, respectively, and for the form β: a = 1158.3 ± 0.8, b = 703.8 ± 0.7, and c = 758.9 ± 0.7 pm and the calculated and experimental densities of 5.055 and 4.819 g·cm^{-3}, respectively. According to the data of Del Bucchia et al. (1980), this compound could crystallize in the monoclinic structure with the lattice parameters a = 613.9 ± 0.4, b = 1349.3 ± 0.9, c = 1223.5 ± 0.8 pm, β = 108.49 ± 0.04°, and a calculated density of 4.88 g·cm^{-3}. To prepare the monoclinic $BaSn_2S_3$, a BaS + 2SnS stoichiometric mixture was heated to 800°C for a week and then cooled slowly. The product obtained was in the form of a compact black pellet, highlighting many single crystals which grinds give a brown powder.

Apparently, $BaSnS_2$ has two polymorphic modifications: one of them crystallizes in the orthorhombic structure with the lattice parameters a = 1200.0 ± 0.4, b = 838.1 ± 0.8, c = 460.6 ± 1.0 pm, and the calculated and experimental densities of 4.612 and 4.586 g·cm^{-3}, respectively (Hervieu et al. 1967) and another modification crystallizes in the monoclinic structure with the lattice parameters a = 608.48 ± 0.03, b = 1213.96 ± 0.08, c = 623.56 ± 0.02 pm, β = 97.058 ± 0.004°, and a calculated density of 4.65 g·cm^{-3} (Iglesias and Steinfink 1973a). The monoclinic modification was prepared from mixture of BaS and SnS (molar ratio 1:1) fired at 750°C for 2 weeks with next cooling it by cutting off power to the furnace (Iglesias and Steinfink 1973a). The material appeared as yellow-orange plate like crystals with some metallic luster.

$\mathbf{SnS_2}$–BaS: Six ternary compounds, $\mathbf{BaSnS_3}$, $\mathbf{BaSn_2S_5}$, $\mathbf{Ba_2SnS_4}$, $\mathbf{Ba_3Sn_2S_7}$, $\mathbf{Ba_6Sn_7S_{20}}$, and $\mathbf{Ba_7Sn_3S_{13}}$ are formed in this system. $BaSnS_3$ begin to decompose in vacuum at 400°C and crystallizes in the monoclinic structure with the lattice parameters a = 2449, b = 636, c = 2311 pm, β = 90.15°, and the calculated and experimental densities of 3.90 or 4.55 and 4.14 g·cm^{-3}, respectively (Jumas et al. 1971a). It was prepared by heating a stoichiometric mixture of SnS_2 and BaS at 700°C for 48 h.

This compound obtained at high pressures crystallizes in the orthorhombic structure with the lattice parameters a = 852.7 ± 0.2, b = 1451.5 ± 0.5, c = 393.3 ± 0.1 pm (Yamaoka and Okai 1970). To prepare the title compound, an equimolar mixture of BaS and SnS_2 or 1:1:2 mixture of BaS, Sn, and S was sealed in a Pt capsule. A piston-cylinder type of high pressure apparatus was used with the furnace assembly in the pressure cell. Pressure was applied to the sample first, and then the temperature was raised to 500°C–800°C at a rate of 2–3°C·min^{-1} and held for 15–20 h. The specimen was quenched to room temperature before the pressure was released. Almost perfect single crystals were obtained at a pressure of 2 GPa.

$BaSn_2S_5$ is thermally stable up to 600°C under N_2 atmosphere and crystallizes in the orthorhombic structure with the lattice parameters a = 667.4 ± 0.3, b = 1060.7 ± 0.5, c = 1139.4 ± 0.5 pm, a calculated density of 4.407 g·cm^{-3}, and an energy gap of ≈2.35 eV (Luo et al. 2013). It was synthesized by using a conventional high-temperature solid-state reaction method in an evacuated silica tube. The crystals were prepared from a mixture of BaS (0.93 mM), Sn (1.87 mM), and S (3.74 mM) which were mixed roughly. The sample was transferred into a predried graphite crucible and flame-sealed in an evacuated silica tube under 10^{-2} Pa. The tube was then placed in a resistance furnace and heated to 750°C within 50 h (holding for 30 h), and the sample was slowly cooled to 500°C in 150 h, followed by cooling to room temperature in 50 h. As the temperature decreased, bulk crystals with a yellow color, which were stable in air and moisture conditions, were obtained.

Ba_2SnS_4 has a polymorphic transformation at 740°C and decomposes in vacuum at 550°C. α-Ba_2SnS_4 crystallizes in the monoclinic structure with the lattice parameters $a = 848.1 \pm 0.4$, $b = 852.6 \pm 0.3$, $c = 1228.0 \pm 0.5$ pm, $\beta = 112.97 \pm 0.03°$, and the calculated and experimental densities of 4.24 and 4.09 g·cm^{-3}, respectively (Jumas et al. 1971a, 1975b) [in the orthorhombic structure with the lattice parameters $a = 1782.3 \pm 0.3$, $b = 735.9 \pm 0.1$, $c = 1261.3 \pm 0.2$ pm, and a calculated density of 4.18 g·cm^{-3} (Susa and Steinfink 1971b].

The title compound was obtained when starting with elemental Sn, and powdered S and BaS (Susa and Steinfink 1971b). Stoichiometric ratio of Sn and Ba with sulfur or BaS were mixed, sealed under vacuum in Vycor tubing, and pre-reacted at temperatures ranging from 200°C to 400°C for 2–4 h. The temperature was then raised and kept in the range 600°C–800°C for about 12 h. After cooling, the vial was examined visually to see whether some of the starting materials had remained unreacted. If the sample looked homogeneous, the vial was opened. If Ba_2SnS_4 was detected in the sample, it was resealed in a Vycor tube and fired at the same or at higher temperatures until it appeared to have completely reacted. In some cases the temperature was raised to 1330°C after first placing the material in a graphite crucible which was then sealed in a Vycor tube. It was also prepared by heating a stoichiometric mixture of SnS_2 and BaS at 800°C for 48 h (Jumas et al. 1971a).

$Ba_3Sn_2S_7$ begin to decompose in vacuum at 440°C and crystallizes in the monoclinic structure with the lattice parameters $a = 1107.3 \pm 0.3$, $b = 677.1 \pm 0.2$, $c = 1870.3 \pm 0.9$ pm, $\beta = 100.77 \pm 0.02°$, and the calculated and experimental densities of 4.21 and 4.12 g·cm^{-3}, respectively (Jumas et al. 1971a, b, 1974). It was prepared by heating a stoichiometric mixture of SnS_2 and BaS at 750°C–1000°C for 48 h with the next slowly cooling (16°C·h^{-1}).

$Ba_6Sn_7S_{20}$ crystallizes in the monoclinic structure with the lattice parameters $a = 2451.9 \pm 0.9$, $b = 635.5 \pm 0.2$, $c = 2310.9 \pm 0.8$ pm, $\beta = 90.101 \pm 0.005°$, and a calculated density of 4.236 g·cm^{-3} (Luo et al. 2013). It was prepared by using a conventional high-temperature solid-state reaction method in an evacuated silica tube. The crystals were obtained from a mixture of BaS (1.31 mM), Sn (1.52 mM), and S (3.04 mM) which were mixed roughly. The sample was transferred into a pre-dried graphite crucible and flame-sealed in an evacuated silica tube under 10^{-2} Pa. The tube was then placed in a resistance furnace and heated to 700°C within 50 h (holding for 30 h), and the sample was slowly cooled to 500°C in 150 h, followed by cooling to room temperature in 50 h. After cooling, a lot of bulk crystals with a yellow color were obtained. The crystals are stable in air and moisture conditions.

$Ba_7Sn_3S_{13}$ crystallizes in the orthorhombic structure with the lattice parameters $a = 1238.6 \pm 0.5$, $b = 2417.3 \pm 1.1$, $c = 887.2 \pm 0.4$ pm, a calculated density of 4.336 g·cm^{-3}, and an energy gap of 3.0 eV (Duan et al. 2017). To synthesized this compound, a total 0.3 g of Ba, Sn, and S (molar ratio 7:3:13) were mixed in a silica crucible in the glove box. The silica crucible was embedded within another larger silica tube, and then this silica tube was flame-sealed under vacuum (10^{-3} Pa). The tube was heated to 950°C in 72 h and holding 100 h at this temperature. Subsequently, they were cooled to 300°C with 100 h before the furnace was turned off. Finally, the light yellow crystals were obtained which are stable in the air for at least 10 months at the room temperature. All manipulations were accomplished in the Ar-filled glove box.

According to the data of Jumas et al. (1971a), two another compounds, $\mathbf{Ba_2Sn_3S_8}$ and $\mathbf{Ba_3SnS_5}$, could be formed in the SnS_2–BaS system.

Another four compounds, $\mathbf{BaSn_2S_4}$, $\mathbf{Ba_7Sn_5S_{15}}$, $\mathbf{Ba_8Sn_4S_{15}}$, and $\mathbf{Ba_{12}Sn_4S_{23}}$, are formed in the Sn–Ba–S ternary system. $BaSn_2S_4$ crystallizes in the tetragonal structure with the lattice parameters $a = 550$ and c = 813 pm (Bok and Boeyens 1957).

$Ba_7Sn_5S_{15}$ is thermally stable up to 750°C under N_2 atmosphere and crystallizes in the hexagonal structure with the lattice parameters $a = 2512.34 \pm 0.04$, $c = 841.20 \pm 0.02$ pm, a calculated density of 4.411 g·cm^{-3}, and an energy gap of ≈2.29 eV (Luo et al. 2013). It was obtained by using a conventional high-temperature solid-state reaction method in an evacuated silica tube. The crystals were synthesized from a mixture of BaS (2.75 mM), Sn (1.96 mM), and S (3.15 mM) which were mixed roughly. The sample was transferred into a pre-dried graphite crucible and flame-sealed in an evacuated silica tube under 10^{-2} Pa. The tube was then placed in a resistance furnace and heated to 800°C within 50 h (holding for 30 h), and the sample was slowly cooled to 500°C in 150 h, followed by cooling to room temperature in 50 h. The bulk crystals with a red color were obtained. They are stable in air and moisture conditions.

$Ba_8Sn_4S_{15}$ is thermally stable up to 850°C under a N_2 atmosphere and crystallizes in the orthorhombic structure with the lattice parameters $a = 2871.5 \pm 0.4$, $b = 850.84 \pm 0.13$, $c = 2356.7 \pm 0.4$ pm, a calculated density of 4.403 g·cm^{-3}, and an energy gap of 2.31 eV (Luo et al. 2014b). To prepare the title compound, BaS (2.34 mM), Sn (1.17 mM), and S (2.05 mM) were mixed roughly and loaded into a graphite crucible sealing in the evacuated silica tube, which was heated to 800°C within 30 h (holding for 48 h). Then, the sample was slowly cooled to 450°C in 100 h and finally to room temperature in 15 h. Bulk yellow crystals were obtained, and they were stable in air and moisture conditions for several months. The pure phase of $Ba_8Sn_4S_{15}$ was produced using a stoichiometry mixture of BaS, SnS, and SnS_2 (molar ratio 8:1:3). The mixture was heated to 700°C in 20 h and kept for 48 h and then cooled down to room temperature in 10 h. The product was reground and heated again by the same procedure (except that the highest temperature was 800°C) to improve the homogeneity and purity.

$Ba_{12}Sn_4S_{23}$ crystallizes in the monoclinic structure with the lattice parameters $a = 1340.6 \pm 0.5$, $b = 1242.3 \pm 0.6$, $c = 2685 \pm 2$ pm, $\beta = 95.820 \pm 0.003$, a calculated density of 4.271 g·cm^{-3}, and an energy gap of 2.98 eV (Duan et al. 2017). To synthesized this compound, a total 0.3 g of Ba, Sn, and S (molar ratio 12:4:23) were mixed in a silica crucible in the glove box. The silica crucible was embedded within another larger silica tube, and then this silica tube was flame-sealed under vacuum (10^{-3} Pa). The tube was heated to 950°C in 72 h and holding 100 h at this temperature. Subsequently, they were cooled to 300°C with 100 h before the furnace was turned off. Finally, the light yellow crystals were obtained which are stable in the air for at least 10 months at the room temperature. All manipulations were accomplished in the Ar-filled glovebox.

7.12 Tin–Zinc–Sulfur

SnS–ZnS: The ZnS solubility in liquid tin sulfide is 4 mass% at 940°C and increases to 10 mass% at 1080°C (Chernyshev and Babanski 1977; Babanski et al. 1980). α-ZnS, which was recrystallized from the solution in SnS melt, contains 6×10^{-3} at% Sn. The solubility of ZnS in SnS was determined by loss in ZnS mass at its dissolution in liquid tin sulfide (Chernyshev and Babanski 1977).

SnS_2–ZnS: The phase diagram is a eutectic type (Figure 7.6) (Olekseyuk et al. 2004). The eutectic point coordinates are 2.5 mol% ZnS and 807°C. The solubility does not exceed 1 mol% of ZnS in SnS_2 and 5 mol% of SnS_2 in ZnS.

Sn–ZnS: The ZnS solubility in liquid tin is shown in Figure 7.7 (Rubenstein 1968; Kimura 1971; Babanski et al. 1980). A part of the Sn–Zn–S liquidus surface on the Sn-rich side was presented in Figure 7.8 (Kimura 1971). Liquidus temperatures were determined by high-temperature filtration (Rubenstein 1968) and visual observation of solid phase appearance and disappearance for heating and cooling of ingots (accuracy of measurement was ±5°C) (Kimura 1971).

At 300°C–500°C and 100 MPa, four phases in hydrothermal conditions exist in this system: α- and β-ZnS, SnS, and SnS_2 (Nekrasov et al. 1977). At 500°C, α-ZnS dissolves not higher than 0.086 mol% SnS in equilibrium with SnS and Sn and not higher than 0.018 mol% SnS in equilibrium with SnS and SnS_2. Ternary compounds were not found in the Sn–Zn–S system.

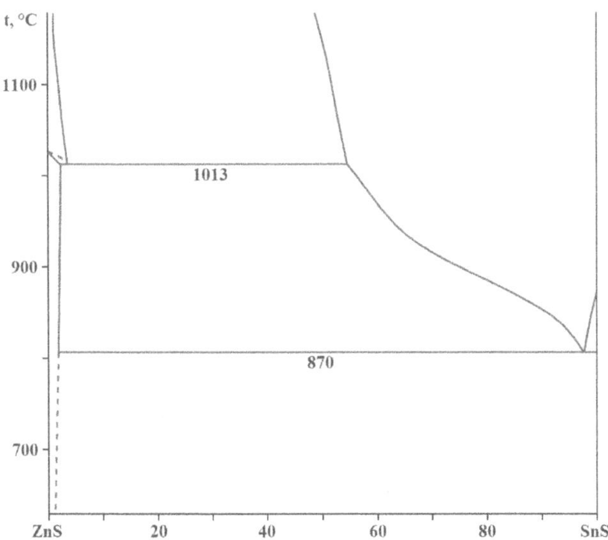

FIGURE 7.6 Phase diagram of the SnS_2–ZnS system. (From Olekseyuk, I.D. et al., *J. Alloys Compd.*, **368**(1–2), 135, 2004.)

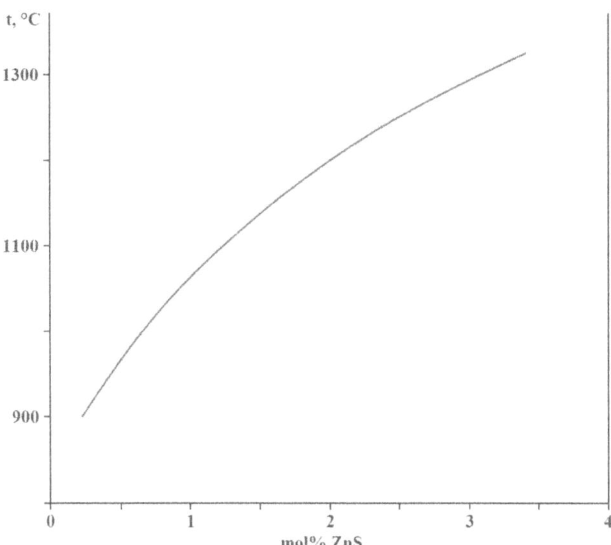

FIGURE 7.7 Temperature dependence of ZnS solubility in liquid Sn. (From Rubenstein, M., *J. Cryst. Growth*, **3–4**, 309, 1968.)

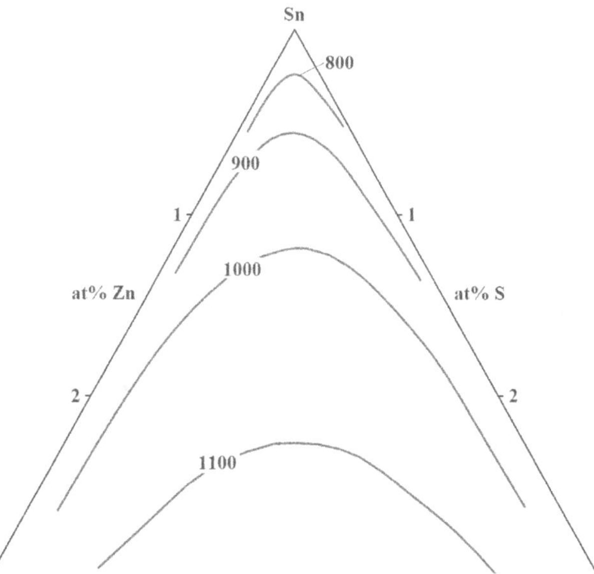

FIGURE 7.8 Part of the Sn–Zn–S liquidus surface on the Sn-rich side. (From Kimura, S., *J. Chem. Thermodyn.*, **3**(1), 7, 1971.)

7.13 Tin–Cadmium–Sulfur

SnS–CdS: The phase diagram is a eutectic type (Figure 7.9) (Galiulin et al. 1981). The eutectic composition and temperature are 13 mol% CdS and 831°C, respectively (Galiulin et al. 1981) [27 mol% CdS and 750°C (Zargarova et al. 1985)]. In the solid state, there is a phase transition of solid solutions based on SnS. At 650°C, the mutual solubility of CdS and SnS is not higher than 1 mol% (Galiulin et al. 1981) (according to the data of Zargarova et al. 1985, the solubility of CdS in SnS at room temperature reaches 4 mol%). The solubility of SnS in CdS reaches 30 mol% when solid solutions were synthesized under 2–3 GPa and 600°C–800°C for 1–4 h (Kobayashi et al. 1979).

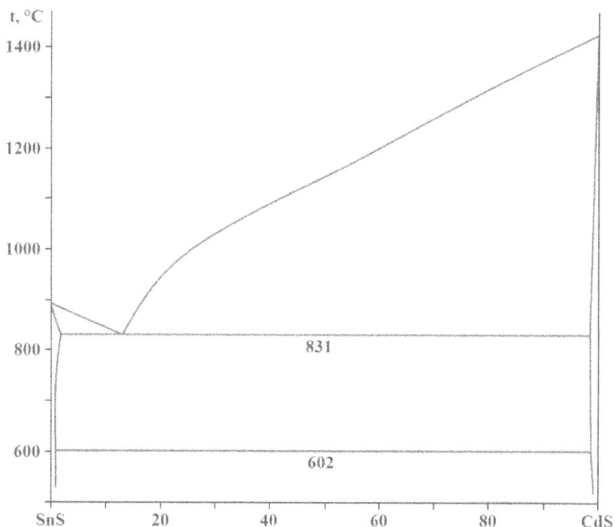

FIGURE 7.9 Phase diagram of the SnS–CdS system. (From Galiulin, E.A., et al., *Zhurn. neorgan. khimii*, **26**(7), 1881, 1981.)

This system was investigated using DTA, metallography, XRD, and measuring of microhardness (Galiulin et al. 1981; Zargarova et al. 1985). The ingots were annealed at 650°C for 250 h (Galiulin et al. 1981) [at 150°C for 200 h (Zargarova et al. 1985)].

SnS_2–CdS: The results of this system investigation are contradictory. According to the data of Piskach et al. (1998), the phase diagram is of a eutectic type (Figure 7.10a). The eutectic point coordinates are 22 mol% CdS and 770°C. Intermediate phases have not been found in this system.

According to the data of Zargarova et al. (1985), **$CdSnS_3$**, **$CdSn_2S_5$**, and **Cd_2SnS_4** ternary compounds are formed in this system (Figure 7.10b). $CdSn_2S_5$ melts congruently at 800°C, and $CdSnS_3$ and Cd_2SnS_4 melt incongruently correspondingly at 685°C and 785°C. The eutectic compositions and temperatures are 55 and 88 mol% SnS_2 and 665°C and 700°C, respectively.

Kislinskaya (1974) noted that the limited solid solutions form at the heating up to 400°C of CdS and SnS_2 mixtures obtained by coprecipitation from water solutions.

This system was investigated using DTA, metallography, XRD, and measuring of microhardness. The ingots were annealed at 500°C for 200 h (Zargarova et al. 1985).

Sn–CdS: The phase diagram is a eutectic type (Figure 7.11) (Alieva et al. 1983; Zargarova et al. 1985). The eutectic composition and temperature are 82 at% Sn [70 at% Sn (Alieva et al. 1983)] and 205°C, respectively (Zargarova et al. 1985). The mutual solubility of CdS and Sn is insignificant.

According to the data of Rubenstein (1968) and Kimura (1971), the eutectic is degenerated from the Sn-rich side (Figure 7.12). The solubility of CdS in liquid tin at 900°C–1250°C can be described as follows (Buehler and Bachmann 1976): $\ln x_{CdS} = E/RT + 3.795$, where E – activation energy ($E = -70$ kJ/mol).

The Sn content in the CdS-doped single crystals reaches 0.11–0.15 at% (Odin et al. 1999) (SnS was used as a dopant).

This system was investigated using DTA, metallography, XRD, and measuring of microhardness (Alieva et al. 1983; Zargarova et al. 1985). The ingots were annealed at 650°C for 250 h (Galiulin et al. 1981) [at 150°C for 200 h (Zargarova et al. 1985)].

Eleven fields of primary crystallizations are on the liquidus surface of the Sn–Cd–S ternary system (Figure 7.13) (Zargarova et al. 1985). The field of CdS primary crystallization occupies considerable part of the liquidus surface, and the field of sulfur crystallization is degenerated. The next invariant equilibria exist in the Sn–Cd–S ternary system: E_1 (142°C) – L ⇔ CdS + Cd + β(Sn); E_2 (182°C) – L ⇔ CdS + SnS + Sn; E_3 (477°C) – L ⇔ $CdSnS_3$ + SnS + $CdSn_2S_5$; E_4 (537°C) – L ⇔ $CdSn_2S_5$ + SnS + Sn_3S_4; U_1 (587°C) – L + CdS ⇔ Cd_2SnS_4 + Sn_3S_4; U_2 (557°C) – L + Cd_2SnS_4 ⇔ $CdSnS_3$ + SnS; U_3 (592°C) – L + Sn_2S_3 ⇔ Sn_3S_4 + $CdSn_2S_5$; U_4 (647°C) – L + SnS_2 ⇔ Sn_2S_3 + $CdSn_2S_5$; and U_5 (175°C) – L + Sn ⇔ CdS + β(Sn).

Kimura (1971) determined liquidus temperatures of Sn–Cd–S ternary system from the Sn-rich side and constructed a part of Sn–Cd–S liquidus surface on the Sn-rich side, which is

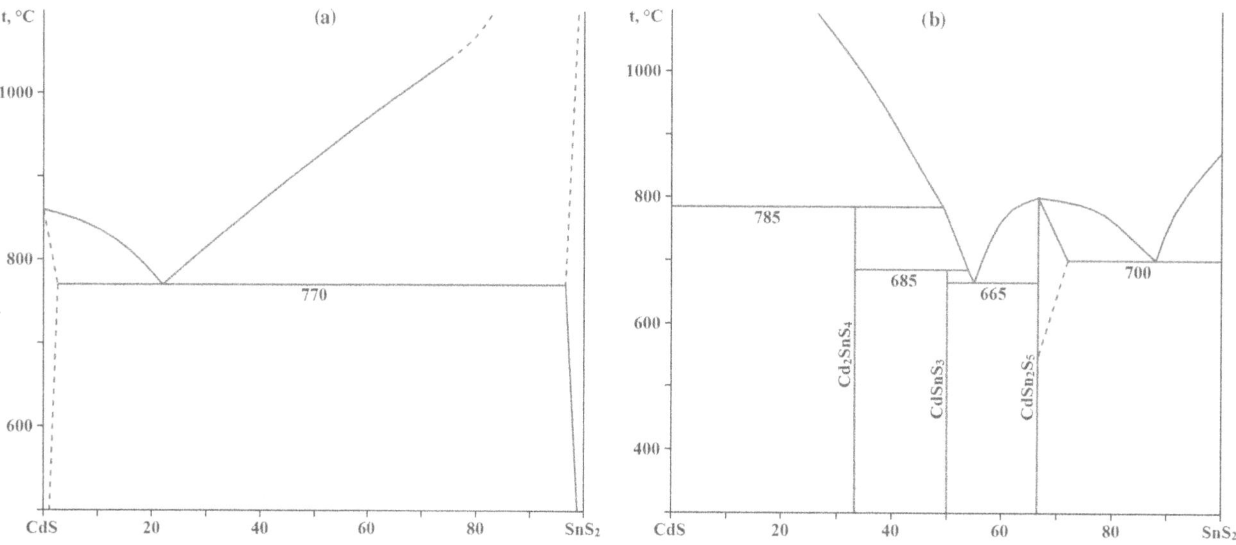

FIGURE 7.10 Phase diagram of the SnS_2–CdS system. (a, from Piskach, L.V. et al., *J. Alloys Compd.*, **279**(2), 142, 1998; b, from Zargarova, M.I. et al., *Zhurn. neorgan. khimii*, **30**(5), 1279, 1985.)

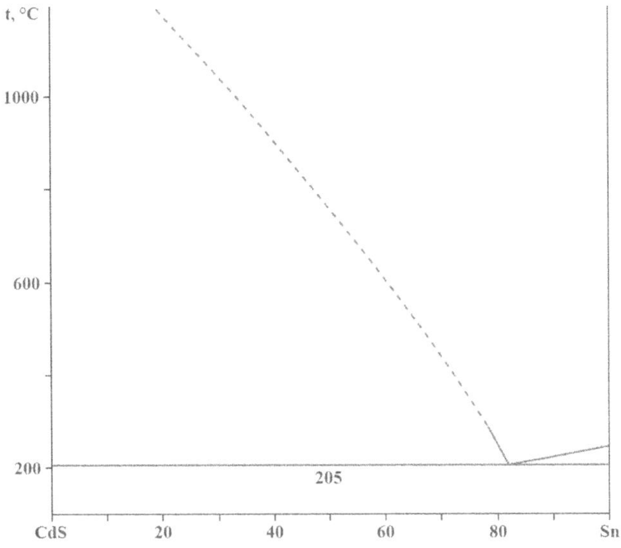

FIGURE 7.11 Phase diagram of the Sn–CdS system. (From Zargarova, M.I. et al., *Zhurn. neorgan. khimii*, **30**(5), 1279, 1985.)

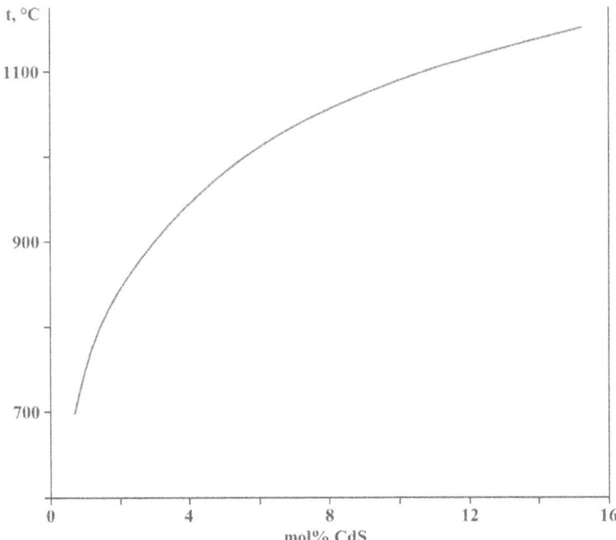

FIGURE 7.12 Temperature dependence of CdS solubility in liquid Sn. (From Rubenstein, M., *J. Cryst. Growth*, **3–4**, 309, 1968).

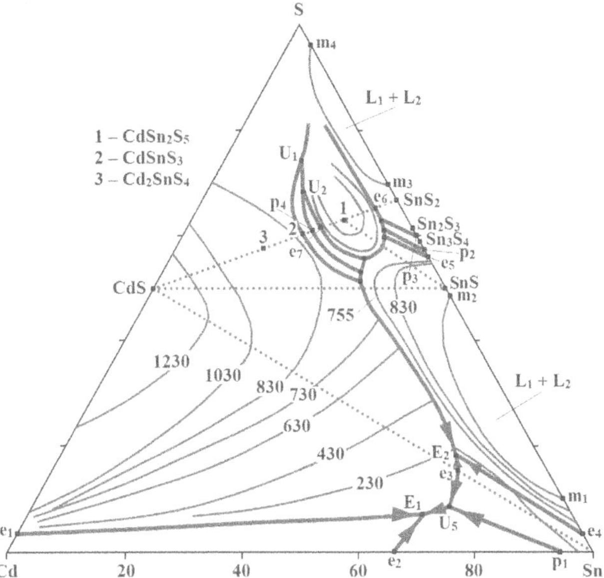

FIGURE 7.13 Liquidus surface of the Sn–Cd–S ternary system. (From Zargarova, M.I. et al., *Zhurn. neorgan. khimii*, **30**(5), 1279, 1985.)

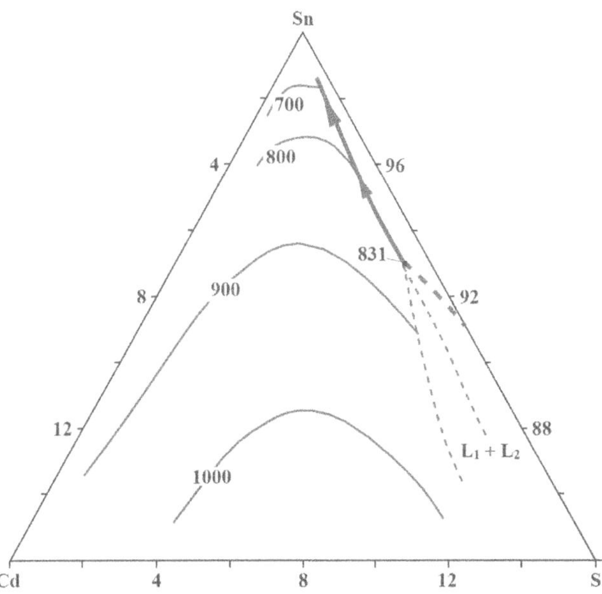

FIGURE 7.14 Part of the Sn–Cd–S liquidus surface on the Sn-rich side. (From Kimura, S., *J. Chem. Thermodyn.*, **3**(1), 7, 1971.)

shown in Figure 7.14. According to the data of Kimura (1971), and Buehler and Bachmann (1976), a border between the fields of CdS and SnS primary crystallization passes across the two-phase region. Liquidus temperatures were determined by visual observation of solid phase appearance and disappearance for heating and cooling of ingots (accuracy of measurement was ±5°C) (Kimura 1971) and by the method of high-temperature filtration (Rubenstein 1968).

7.14 Tin–Mercury–Sulfur

SnS–HgS: The phase diagram is a eutectic type (Figure 7.15) (Motrya et al. 1988; Motrya 1991). The eutectic composition and temperature are 52 mol% HgS and 562°C, respectively. Thermal effects at 585°C correspond to the polymorphous transition of SnS, and at 347°C, there is also a polymorphous transformation of HgS. The solubility of HgS in SnS and SnS in HgS at the eutectic temperature is not higher than 2 and 3 mol%, respectively. At 500°C, SnS dissolves less than 1.5 mol% HgS and HgS dissolves not higher than 2 mol% SnS. This system was investigated using DTA, metallography, XRD, and microhardness measurement. The ingots were annealed at 500°C for 1200 h.

SnS_2–HgS: The phase diagram is a eutectic type (Figure 7.16) (Motrya et al. 1988; Motrya 1991). The eutectic composition and temperature are 48 mol% HgS and 647°C, respectively. At 347°C, there is polymorphous transformation of HgS. The solubility of HgS in SnS_2 at the eutectic temperature is not higher

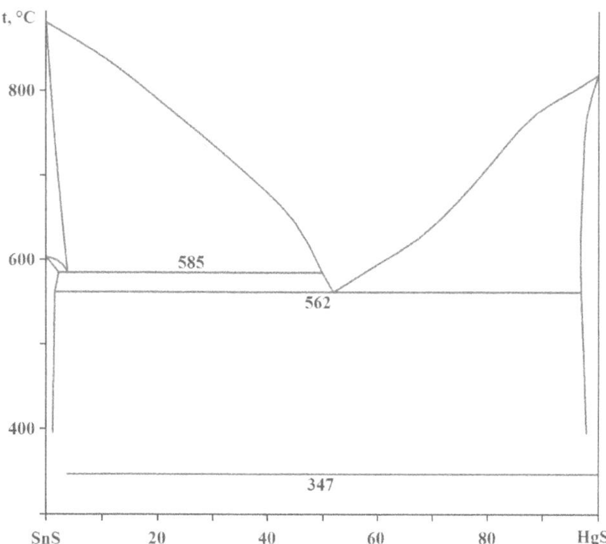

FIGURE 7.15 Phase diagram of the SnS–HgS system. (From Motrya, S.F. et al., *Zhurn. neorgan. khimii*, **33**(8), 2103, 1988.)

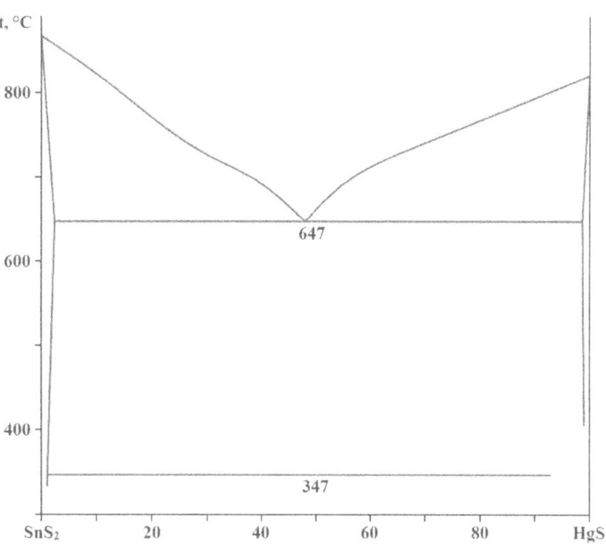

FIGURE 7.16 Phase diagram of the SnS_2–HgS system. (From Motrya, S.F. et al., *Zhurn. neorgan. khimii*, **33**(8), 2103, 1988.)

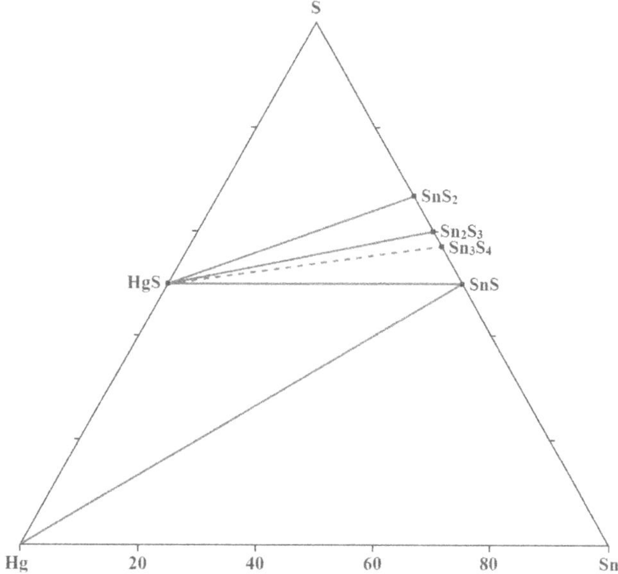

FIGURE 7.17 Isothermal section of the Sn–Hg–S ternary system at 500°C. (From Voroshilov, Yu.V. et al., *Zhurn. neorgan. khimii*, **38**(6), 1061, 1993.)

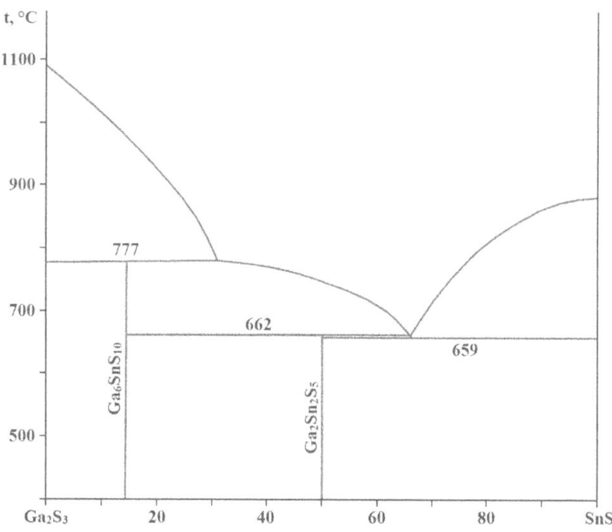

FIGURE 7.18 Phase diagram of the $2SnS$–Ga_2S_3 system. (From Thevet, F., et al., *C. r. Acad. sci., Sér. 2*, **293**(4), 275, 1981.)

than 2 mol% and the solubility of SnS_2 in HgS at 500°C is less than 1 mol%. This system was investigated using DTA, metallography, XRD, and microhardness measurement. The ingots were annealed at 500°C for 1200 h.

Ternary compounds were not found in the Sn–Hg–S ternary system (Voroshilov et al. 1993). At 500°C, mercury sulfide is in equilibrium with binary phases of Sn–S binary system, and Hg is in equilibrium with SnS (Figure 7.17) (Motrya 1991; Voroshilov et al. 1993).

7.15 Tin–Gallium–Sulfur

SnS–Ga_2S_3: The phase diagram of the $2SnS$–Ga_2S_3 system, constructed through DTA and XRD, is a eutectic type with peritectic transformations (Figure 7.18) (Thévet et al. 1981). The eutectic contains 66 mol% 2SnS and crystallizes at 659°C. Two compounds, $\mathbf{Ga_2Sn_2S_5}$ and $\mathbf{Ga_6SnS_{10}}$, are formed in this system. The glass forming region was determined within the interval of 20–60 mol% 2SnS.

$Ga_2Sn_2S_5$ melts incongruently at 662°C and crystallizes in the orthorhombic structure with the lattice parameters $a = 1241 \pm 1$, $b = 622 \pm 1$, and $c = 1088 \pm 1$ pm and a calculated density of 4.25 g·cm^{-3} (Mazurier et al. 1983) [$a = 623.3 \pm 0.7$, $b = 1244 \pm 1$, and $c = 1088 \pm 1$ pm and the calculated and experimental densities of 4.23 ± 0.01 and 4.30 g·cm^{-3}, respectively (Thévet et al. 1981). This compound was prepared by the interaction of SnS and Ga_2S_3 at 630°C and its single crystals were grown by the annealing its powder at 600°C for 1 month (Thévet et al. 1981; Mazurier et al. 1983)].

Ga_6SnS_{10} melts incongruently at 777°C (Thévet et al. 1981).

According to the data of Luo et al. (2014a), another ternary compound, **Ga_4SnS_7**, which is thermally stable up to 700°C under a N_2 atmosphere, is formed in the SnS–Ga_2S_3 system. It crystallizes in the monoclinic structure with the lattice parameters $a = 726.9 \pm 0.5$, $b = 636.1 \pm 0.4$, $c = 1240.8 \pm 0.8$ pm, $\beta = 106.556 \pm 0.011°$, a calculated density of 3.757 g·cm^{-3}, and an energy gap of 3.10 eV. To prepare the title compound, stoichiometric amounts of Sn (0.80 mM), Ga (3.22 mM), and S (5.63 mM) were mixed roughly and loaded into a silica tube, which was sealed under vacuum. The mixture was heated to 850°C within 48 h, kept at this temperature for 96 h, and cooled to 400°C in 100 h, followed by cooling to room temperature in 24 h. Light-yellow crystals were obtained, which were stable in air for months.

SnS_2–Ga_2S_3: When studying the coprecipitation of SnS_2 and Ga_2S_3, it was found that at low concentrations of Ga_2S_3, the formation of a solid solution is observed, and at a molar ratio of Ga and Sn of 0.6–0.7, the **$3SnS_2 \cdot Ga_2S_3$**, **$SnS_2 \cdot Ga_2S_3$**, and **$SnS_2 \cdot 3Ga_2S_3$** double sulfides are formed (Chaus and Sheka 1963; Chaus et al. 1967).

7.16 Tin–Indium–Sulfur

SnS–In_2S_3: The phase diagram of this system, constructed through DTA and XRD, is a eutectic type with peritectic transformations (Figure 7.19) (Adenis and Olivier-Fourcade 1985; Adenis et al. 1986). The eutectic contains 22 mol% In_2S_3 and crystallizes at 750°C. Three compounds, **In_2SnS_4** or **$In_{11}Sn_{5.5}S_{22}$**, **$In_2Sn_2S_5$**, and **$In_{12}Sn_5S_{23}$**, are formed in the SnS–In_2S_3 system. At 803°C, the peritectoid transformation of In_2S_3-based solid solutions occurs in the system, and at 860°C, peritectic decomposition of these solid solutions takes place. At 744°C, a metastable eutectic crystallizes, which contains 25 mol% In_2S_3 and complicates the formation of In_2SnS_4. The region of solid solutions from In_2S_3 side reaches 8 mol% SnS, and from the SnS side there is no homogeneity region.

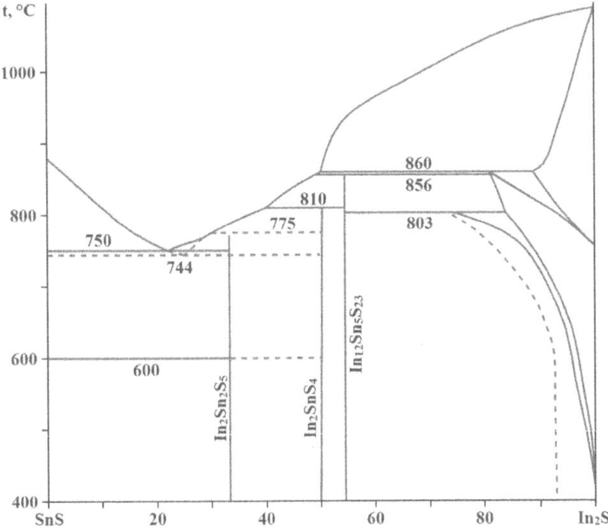

FIGURE 7.19 Phase diagram of the SnS–In_2S_3 system. (From Adenis, C., et al., *J. Solid State Chem.*, **65**(2), 251, 1986.)

In_2SnS_4 or $In_{11}Sn_{5.5}S_{22}$ melts incongruently at 810°C and crystallizes in the monoclinic structure with the lattice parameters $a = 2861$, $b = 385$, $c = 1583$ pm, and $\beta = 109°$ (Adenis and Olivier-Fourcade 1985; Adenis et al. 1986) [$a = 1563.6 \pm 0.6$, $b = 1462.7 \pm 0.7$, $c = 384.7 \pm 0.1$ pm, $\beta = 97.38 \pm 0.04°$, and the calculated and experimental density of 4.99 and 4.98 g·cm^{-3}, respectively, for $In_{11}Sn_{5.5}S_{22}$ composition (Likforman et al. 1988a, b, c, 1989)]. The single crystals of the title compound were extracted from a polycrystalline mass from a sample of In_2SnS_4 composition prepared from a mixture of In_2S_3 and SnS in a vacuum-sealed ampoule (Likforman et al. 1988c). The temperature brought in 24 h to 930°C is maintained 48 h, and then lowered to 780°C. This heat treatment is followed by a quenching. The compound such obtained is pure. The single crystals of this compound were also grown by the chemical transport reactions (Adenis and Olivier-Fourcade 1985; Adenis et al. 1986).

$In_2Sn_2S_5$ melts incongruently at 775°C and the single crystals of this compound were grown by the chemical transport reactions (Adenis and Olivier-Fourcade 1985; Adenis et al. 1986).

$In_{12}Sn_5S_{23}$ melts incongruently at 856°C and crystallizes in the monoclinic structure with the lattice parameters $a = 2800$, $b = 385$, $c = 1558$ pm, and $\beta = 122°$ (Adenis and Olivier-Fourcade 1985; Adenis et al. 1986).

Three another ternary compounds were synthesized in the SnS–In_2S_3 system. **$In_{10}Sn_6S_{21}$** crystallizes in the monoclinic structure with the lattice parameters $a = 1556.6 \pm 0.3$, $b = 384.5 \pm 0.1$, $c = 2757.8 \pm 0.7$ pm, $\beta = 95.44 \pm 0.02°$, and an experimental density of 5.12 ± 0.02 g·cm^{-3} (Likforman et al. 1989, 1990). The single crystals of this compound were extracted from a polycrystalline mass from a sample of In_2SnS_4 composition prepared from a mixture of In_2S_3 at SnS in a vacuum-sealed ampoule (Likforman et al. 1990). The temperature brought in 24 h to 930°C was maintained 24 h, and then lowered to 780°C. This heat treatment is followed by a quenching. After determination of the crystal structure which led to the formula $In_{10}Sn_6S_{21}$, the compound was prepared according to the same protocol from a mixture of composition $5In_2S_3 + 6SnS$ which makes it possible to obtain the pure compound. The single crystals of this compound were also grown by the chemical transport reactions (Adenis and Olivier-Fourcade 1985; Adenis et al. 1986).

$In_{14}Sn_5S_{26}$ crystallizes in the monoclinic structure with the lattice parameters $a = 3791.7 \pm 0.4$, $b = 1375.8 \pm 0.3$, $c = 384.33 \pm 0.06$ pm, $\beta = 91.20 \pm 0.01°$, and the calculated and experimental densities of 5.03 and 4.98 g·cm^{-3}, respectively (Likforman et al. 1988a, b, 1989). The single crystals were extracted from a polycrystalline mass from a sample of overall composition $In_6Sn_2S_{11}$ prepared from a mixture of In_2S_3 and SnS in a vacuum-sealed ampoule (Likforman et al. 1988b). The temperature is brought in 24 h to 900°C and maintained 48 h. This heating is followed by quenching. After determining the crystal structure which leads to the formula $In_{14}Sn_5S_{26}$, the compound was prepared according to the same protocol from a mixture of composition $7In_2S_3 + 5SnS$ which makes it possible to obtain the compound in pure form. It should be noted that this phase is only obtained by quenching; by slow cooling another phase was obtained. The structure described here therefore corresponds to high-temperature form.

$In_{18}Sn_7S_{34}$ crystallizes in the orthorhombic structure with the lattice parameters $a = 1512.2 \pm 0.6$, $b = 2270 \pm 1$, $c = 383.3 \pm 0.2$ pm, and the calculated and experimental densities of 5.03 ± 0.02 and 5.05 g·cm^{-3}, respectively (Likforman et al. 1987, 1989). Single crystals of the title compound were extracted from a polycrystalline mass from a sample of In_2SnS_4 composition, prepared from a mixture of In_2S_3 and SnS in a vacuum-sealed ampoule. The temperature was brought in 24 h to 900°C and maintained 24 h; heating followed by regular cooling in 48 h until room temperature. After the structural determination, the compound was prepared according to the same protocol from a mixture of composition $9In_2S_3 + 7SnS$, which makes it possible to obtain the compound in pure phase. Black needles of $In_{18}Sn_7S_{34}$ were obtained.

SnS_2–In_2S_3: The **$In_{16}Sn_4S_{32}$** ternary compound, which has two polymorphic modifications, is formed in this system. There is no important structural difference between the high-temperature (black) and the low-temperature (orange-red) forms of this compound (Dedryvère et al. 2001). The main difference comes from the presence of sulfur vacancies in the high-temperature form. Both modifications crystallize in the cubic structure with the lattice parameter $a = 1071.9 \pm 0.2$ pm [$a = 1075.5 \pm 0.5$ pm (Elidrissi-Moubtassim et al. 1990)] for high-temperature modification and $a = 1073.1 \pm 0.3$ pm [$a = 1069.4 \pm 0.5$ pm (Elidrissi-Moubtassim et al. 1990)] for low-temperature modification (Dedryvère et al. 2001). This compound was prepared by solid state reaction between the binary compounds SnS_2 and In_2S_3, mixed in the appropriate ratio and sealed under vacuum in quartz tubes (Dedryvère et al. 2001). The mixtures were heated at 750°C and 800°C, for different periods of time, and then slowly cooled at ca. 1°C·min^{-1} to the desired temperature (610°C, 550°C, 400°C, and 300°C). The final sample was then obtained by quenching in ice-water bath from this last temperature.

When studying the coprecipitation of SnS_2 and In_2S_3, it was found (Rudnev and Dzhumaev 1964; Chaus et al. 1967) that in the system there are two regions of solid solutions: at small and high In_2S_3 concentrations.

Three another ternary compounds are formed in the Sn–In–S system. **In_4SnS_4** crystallize in the cubic structure with the lattice parameter $a = 1234.75 \pm 0.08$ pm and a calculated density of 4.981 g·cm^{-3} (Deiseroth and Pfeifer 1991).

$In_5Sn_{0.5}S_7$ crystallizes in the monoclinic structure with the lattice parameters $a = 904.7 \pm 0.2$, $b = 1701.3 \pm 0.2$, $c = 387.3 \pm 0.1$ pm, $\beta = 101.72 \pm 0.05°$, and the calculated and experimental densities of 4.88 and 4.89 g·cm^{-3}, respectively (Likforman et al. 1984, 1989). The single crystals were extracted from a polycrystalline mass of this compound, prepared by an interaction of the three elements in a vacuum-sealed ampoule. The temperature was slowly brought to 900°C and maintained for 24 h. This heating was followed by slow cooling. The crystals were black with a brilliant appearance and appear in the form of the parallelepipeds.

$In_6Sn_8S_{19}$ also crystallizes in the monoclinic structure with the lattice parameters $a = 2925.3$, $b = 381.9$, $c = 1552.8$ pm, and $\beta = 121.85°$ (Adenis et al. 1988). It was prepared by a solid-state reaction of the In_2S_3, SnS, and SnS_2 mixture with the next annealing at 800°C for 1 week.

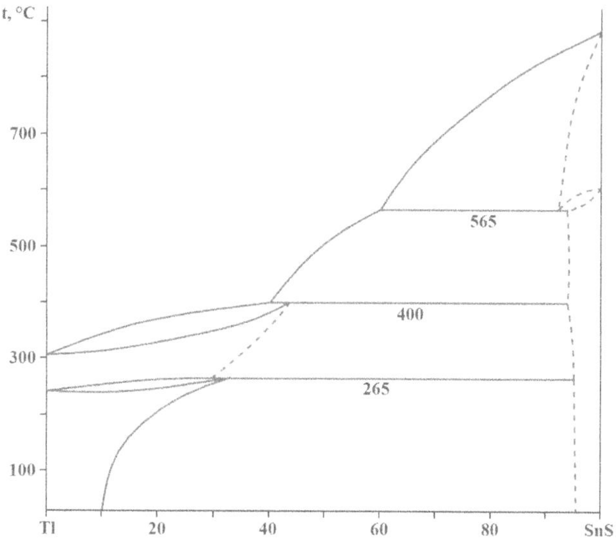

FIGURE 7.20 Phase diagram of the SnS–Tl system. (From Malakhovskaya-Rosokha, T.A., et al., *Russ. J. Inorg. Chem.*, **56**(1), 118, 2011.)

7.17 Tin–Thallium–Sulfur

SnS–Tl: The phase diagram of this system, constructed through DTA and XRD, is given in Figure 7.20 (Malakhovskaya-Rosokha et al. 2011). There are three invariant processes in this system: peritectic process at 400°C, peritectoid process at 265°C, and metatectic process due to the polymorphic transformation of SnS at 565°C.

SnS–Tl_2S: The phase diagram of this system, constructed through DTA and XRD, is a eutectic type with peritectic transformations (Figure 7.21) (Gotuk et al. 1979b; Malakhovskaya-Rosokha et al. 2011). The eutectic contains 48 mol% SnS and crystallizes at 340°C (Malakhovskaya-Rosokha et al. 2011).

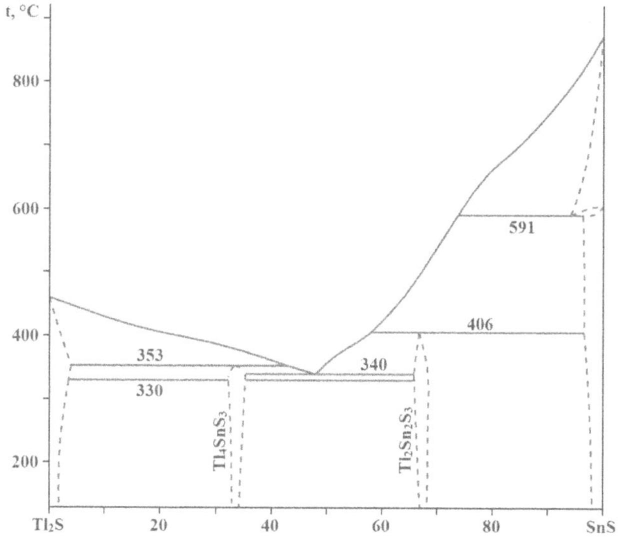

FIGURE 7.21 Phase diagram of the SnS–Tl_2S system. (From Malakhovskaya-Rosokha, T.A., et al., *Russ. J. Inorg. Chem.*, **56**(1), 118, 2011.)

The peritectics contain 43 and 58 mol% SnS, and the metatectic process takes place at 73 mol% SnS and 591°C. The thermal effects at 330°C probably correspond to a polymorphic transformation of $\mathbf{Tl_4SnS_3}$.

According to the data of Gotuk et al. (1979b), two eutectics exist in the SnS–Tl_2S system: one of them contains 40 mol% SnS and crystallizes at 345°C and the second eutectic is degenerated and practically coincides with the melting temperature of Tl_4SnS_3. The peritectic point contains 60 mol% SnS, and the thermal effects at 595°C correspond to the phase transition in SnS. The solubility of SnS in Tl_2S at 350°C is 2 mol%, and the homogeneity regions based on other phases were not found.

Two compounds, $\mathbf{Tl_2Sn_2S_3}$ and Tl_4SnS_3, are formed in the SnS–Tl_2S system. $Tl_2Sn_2S_3$ melts incongruently at 406°C (Babanly et al. 1979a; Gotuk et al. 1979b; Malakhovskaya-Rosokha et al. 2011) and crystallizes in the monoclinic structure with the lattice parameters $a = 1388.7 \pm 0.7$, $b = 774.2 \pm 0.4$, $c = 726.7 \pm 0.4$ pm, $\beta = 105.39 \pm 0.03°$, and the calculated and experimental densities of 6.55 and 6.54 g·cm^{-3}, respectively (Del Bucchia et al. 1982). The absolute value of the formation enthalpy of $Tl_2Sn_2S_3$ from the binary chalcogenides is −55.0 kJ·equiv^{-1} (Babanly and Zinchenko 1992). The single crystals of the title compound were prepared from the stoichiometric mixture of Tl_2S and SnS (Del Bucchia et al. 1982). Initially, the mixture placed in a vacuum-sealed glass ampoule was brought to 350°C for 1 week. After this treatment during which the phase $Tl_2Sn_2S_3$ was formed, the temperature was raised until the fusion (425°C) for 1 day. A slow cooling (2°C·h^{-1}) then allows obtaining single crystals. They are in the form of small, more or less flattened sticks of black color, the elongation axis of which corresponds to the direction [011].

Tl_4SnS_3 melts incongruently at 353°C (Malakhovskaya-Rosokha et al. 2011; Peresh et al. 2015) [at 341°C (Malakhovs'ka et al. 2007); melts congruently at 350°C, has a polymorphic transformation at 327°C (Babanly et al. 1979a; Gotuk et al. 1979b)] and crystallizes in the tetragonal structure with the lattice parameters $a = 831.4$ and $c = 1264.7$ pm (the predicted values) (Piasecki et al. 2017) and the calculated and experimental densities of 7.86 and 7.84 g·cm^{-3}, respectively (Malakhovs'ka et al. 2007). The melting entropy of Tl_4SnS_3 is 105 kJ·M^{-1} (Malakhovs'ka et al. 2007) and the absolute value of the formation enthalpy from the binary chalcogenides is −55.0 kJ·equiv^{-1} (Babanly and Zinchenko 1992). This compound is a narrow band gap semiconductor (Piasecki et al. 2017).

The single crystals of the title compound were obtained using directional crystallization of stoichiometric composition performed by Bridgman-Stockbarger method (Malakhovs'ka et al. 2007; Piasecki et al. 2017). For conversion to the liquid state, the sample was heated up to 400°C, maintained at this temperature for 24 h, and then was slowly cooled to the annealing temperature in the growth zone decreased to 180°C and maintained for 72 h. The crystallization front displacement rate was equal to 0.1–0.3 mm·h^{-1}, a temperature gradient in the zone of crystallization was equal to 2–4°C·mm^{-1}, and the rate of cooling to ambient temperature after annealing was 30°C·h^{-1}.

SnS–$\mathbf{Tl_2SnS_3}$: The phase diagram of this system, constructed through DTA and XRD, is a eutectic type with a metatectic transformation (Figure 7.22) (Malakhovskaya-Rosokha et al. 2011). The eutectic contains 25 mol% SnS and crystallizes at 338°C and the metatectic process takes place at 65 mol% SnS and 589°C.

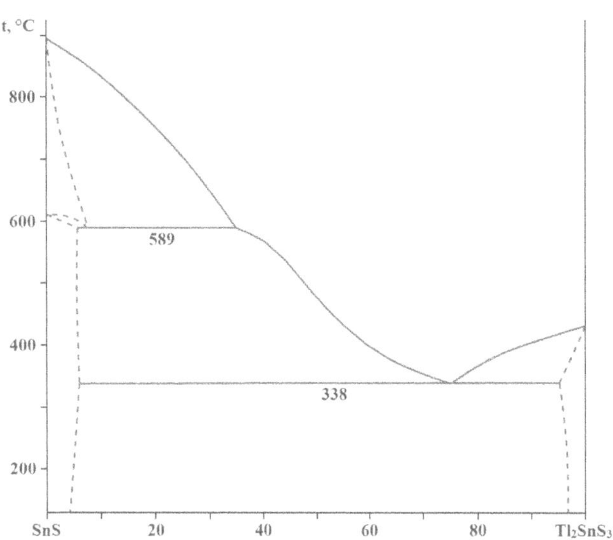

FIGURE 7.22 Phase diagram of the SnS–Tl_2SnS_3 system. (From Malakhovskaya-Rosokha, T.A., et al., *Russ. J. Inorg. Chem.*, **56**(1), 118, 2011.)

SnS–$\mathbf{Tl_4SnS_4}$: The phase diagram of this system, constructed through DTA and XRD, is a eutectic type with a metatectic transformation and is similar to the phase diagram of the SnS–Tl_2SnS_3 system (Malakhovskaya-Rosokha et al. 2011). The eutectic contains 40 mol% SnS and crystallizes at 399°C and the metatectic process takes place at 75 mol% SnS and 567°C.

$\mathbf{SnS_2}$–$\mathbf{Tl_2S}$: The phase diagram of this system, constructed through DTA and XRD, is given in Figure 7.23 (Ajavon et al. 1983). Three eutectics contain 21, 42, and 57 mol% SnS_2 and crystallize respectively at 333°C, 357°C, and 397°C, and the peritectic point is at 61 mol% SnS_2 and 457°C. Three compounds, $\mathbf{Tl_2SnS_3}$, $\mathbf{Tl_4SnS_4}$, and $\mathbf{Tl_4Sn_5S_{12}}$, are formed in this system. Tl_2SnS_3 melts congruently at 420°C [at 426°C (Peresh

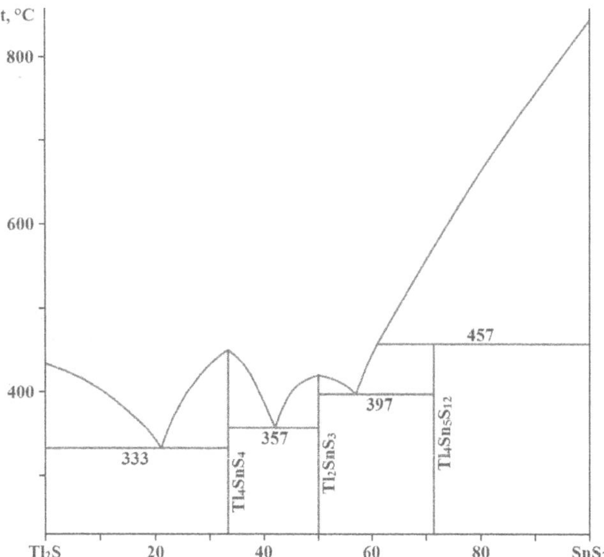

FIGURE 7.23 Phase diagram of the SnS_2–Tl_2S system. (From Ajavon, Auté-Lô, et al. *Rev. chim. minér.*, **20**(3), 421, 1983.)

et al. 1986, 2015)] and crystallizes in the monoclinic structure with the lattice parameters a = 741.9, b = 384.2, c = 2326.8 pm, β = 95.47°, and the calculated and experimental densities of 6.30 and 6.27 [6.34 (Lazarev et al. 1985)] g·cm^{-3}, respectively (Ajavon et al. 1983) [a = 2303 ± 1, b = 383.4 ± 0.1, c = 737.9 ± 0.3 pm, and β = 94.07 ± 0.05° (Klepp 1984)] and an energy gap of 1.7–1.8 eV (Lazarev et al. 1985; Peresh et al. 1986) [1.44 eV (Ibañez et al. 1986)]. The enthalpy and entropy of melting of this compound is equal to 43.2 kJ·M^{-1} and 61.5 J·(M·K)$^{-1}$, respectively (Peresh et al. 1986). The absolute value of the formation enthalpy from the binary chalcogenides is –20.5 kJ·equiv^{-1} (Babanly and Zinchenko 1992). The homogeneity region of Tl_2SnS_3 at 150°C is not higher than 0.5 mol% (49.7–50.2 mol% SnS_2) (Lazarev et al. 1985). Black, fiber-like crystals of this compound were obtained from the melt using high purity elements (Klepp 1984).

Tl_4SnS_4 melts congruently at 450°C [at 468°C (Peresh et al. 1986, 2015)] and also crystallizes in the monoclinic structure with the lattice parameters a = 835.7, b = 824.6, c = 1533.4 pm, β = 103.69°, and the calculated and experimental densities of 6.91 and 6.89 [6.90 (Lazarev et al. 1985)] g·cm^{-3}, respectively (Ajavon et al. 1983; Piffard et al. 1984a) [a = 839.5 ± 0.6, b = 828.0 ± 0.3, c = 1539.8 ± 1.2 pm, β = 103.69 ± 0.06°, and a calculated density of 6.80 g·cm^{-3} (Klepp and Eulenberger 1984)] and an energy gap of 2.0 eV (Lazarev et al. 1985; Peresh et al. 1986) [1.31 eV (Ibañez et al. 1986)]. The enthalpy and entropy of melting of this compound is equal to 46.1 kJ·M^{-1} and 62.2 J. (M·K)$^{-1}$, respectively (Peresh et al. 1986). The absolute value of the formation enthalpy from the binary chalcogenides is –20.5 kJ·equiv^{-1} (Babanly and Zinchenko 1992). The homogeneity region of Tl_4SnS_4 at 150°C is not higher than 1.4 mol% (32.3–33.7 mol% SnS_2) (Lazarev et al. 1985). This compound was prepared by melting the mixture of Sn, Tl, and S at 600°C in the evacuated quartz ampoule (Piffard et al. 1984a). The cooling from 600°C to 200°C at a rate of 2°C·h^{-1} gave the possibility to obtain single crystals.

$Tl_4Sn_5S_{12}$ melts congruently at 457°C and crystallizes in the triclinic structure with the lattice parameters a = 1708.5, b = 735.5, c = 963.9 pm, and α = 67.23°, β = 75.05°, γ = 89.79°, the calculated and experimental densities of 5.57 and 5.54 g·cm^{-3}, respectively (Ajavon et al. 1983; Piffard 1984b), and an energy gap of 1.51 eV (Ibañez et al. 1986). Red single crystals of this compound were obtained by slow heating the mixture of Sn, Tl, and S in the evacuated quartz ampoule to 600°C, annealing at this temperature for 3 days and slow cooling (2°C·h^{-1}) to 200°C (Piffard et al. 1984a). According to the data of Olekseyuk et al. (2011b), $Tl_4Sn_5S_{12}$ was not found in the Sn–Tl–S ternary system at 250°C (annealing for 250 h).

The phase diagrams of the **Tl_2SnS_3–S** and **Tl_4SnS_4–S** are the eutectic type with degenerate eutectics from the S-side and characterized by the liquid immiscibility regions: 19–48 mol% Tl_2SnS_3 at 358°C and 18–45 mol% Tl_4SnS_4 at 251°C (Malakhovskaya-Rosokha et al. 2011). The eutectics crystallize at 110°C and 112°C for the first and the second systems, respectively.

According to the date of Eulenberger (1981b), another compound, **$Tl_2Sn_2S_5$**, which crystallizes in the monoclinic structure with the lattice parameters a = 1111.5 ± 0.2, b = 772.3 ± 0.1, c = 1149.2 ± 0.2 pm, β = 108.60 ± 0.01°, and a calculated density of 5.73 g·cm^{-3}, is formed in the SnS_2–Tl_2S system. This compound is a semiconductor with an energy gap of 1.17 eV (Ibañez et al. 1986).

Gutierrez-Zorilla et al. (1982) synthesized in the Sn–Tl–S ternary system by the interaction of Sn, Tl, and S at 650°C–700°C the **Tl_2SnS_4** compound. It crystallizes in the orthorhombic structure with the lattice parameters a = 1141.3 ± 0.9, b =413.1 ± 0.2, and c = 1475.3 ± 1.2 pm.

7.18 Tin–Yttrium–Sulfur

According to the date of Mel'chenko and Serebrennikov (1973a), the **$Y_2Sn_2S_5$** ternary compound is formed in the Sn–Y–S system. No ternary compounds were found in the **SnS_2–Y_2S_3** system (Marchuk et al. 2007; Shemet et al. 2006d).

7.19 Tin–Lanthanum–Sulfur

SnS_2–La_2S_3: The phase diagram of this system, constructed through DTA and XRD, is shown in Figure 7.24 (Guittard et al. 1976). The system is most likely quasi-binary. However, DTA curves cannot be fully interpreted due to spurious peaks. The partly dissociation of SnS_2 leads to the formation of LaS_2 and the appearance of the thermal effects at 745°C is a characteristic of the **SnS_2–LaS_2** system. The eutectic crystallizes at 830°. The **La_2SnS_5** ternary compound, which melts incongruently at 1085°C and crystallizes in the orthorhombic structure with the lattice parameters a = 1126 ± 5, b = 789 ± 3, and c = 399 ± 1 pm (Guittard et al. 1976) [a = 1122 ± 1, b = 791.5 ± 0.5, and c = 396 ± 1 pm (Jaulmes 1974)], is formed in the SnS_2–La_2S_3 system (Mel'chenko and Serebrennikov 1973a, b).

7.20 Tin–Cerium–Sulfur

SnS_2–Ce_2S_3: The **Ce_2SnS_5** ternary compound, which melts incongruently at 994°C and crystallizes in the orthorhombic structure with the lattice parameters a = 1124 ± 5, b = 786 ± 3, and c = 395 ± 1 pm, is formed in this system (Guittard et al. 1976).

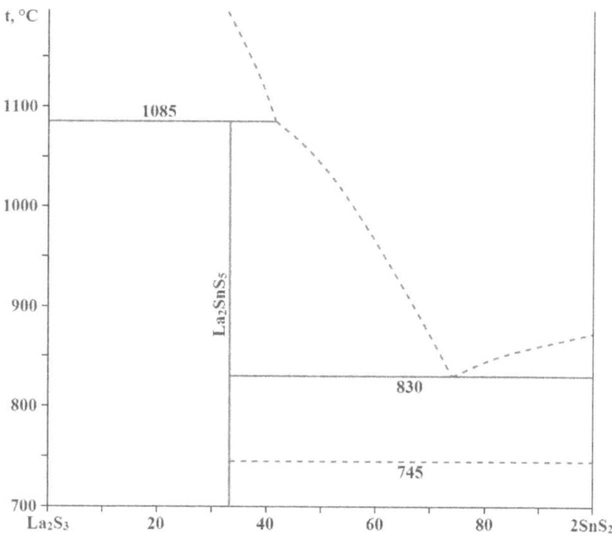

FIGURE 7.24 Phase diagram of the 2SnS_2–La_2S_3 system. (From Guittard, M., et al., *Mater. Res. Bull.*, **11**(9), 1073, 1976.)

7.21 Tin–Praseodymium–Sulfur

SnS_2–Pr_2S_3: The **Pr_2SnS_5** ternary compound, which melts incongruently at 770°C and crystallizes in the orthorhombic structure with the lattice parameters a = 781.95 ± 0.09, b = 1121.45 ± 0.14, c = 396.42 ± 0.05 pm, and a calculated density of 5.382 g·cm^{-3} (Daszkiewicz et al. 2008) [a = 1117 ± 5, b = 783 ± 3, and c = 393 ± 1 pm (Guittard et al. 1976)], is formed in this system (Mel'chenko and Serebrennikov 1973a, b). The title compound was prepared by sintering of the elements in the molar ratio Pr/Sn/S = 2:1:5 in evacuated quartz ampoule in a furnace (Daszkiewicz et al. 2008). The ampoule was heated (30°C·h^{-1}) to a maximum temperature of 1150°C and kept at this temperature for 4 h. Afterwards it was slowly cooled (10°C·h^{-1}) to 600°C and annealed at this temperature for 200 h. After annealing, the sample was quenched in cold water.

7.22 Tin–Neodymium–Sulfur

SnS–Nd_2S_3: This system has been investigated at 1100°C (Nikolaev et al. 2015). Nd_2S_3 solubility in molten SnS at this temperature was determined to be 4.4 ± 0.4 mol %. The diffusion coefficient of Nd was estimated at 1.2 × 10^{-8} m^2·s^{-1}. Nd_2S_3 dissolution in SnS has a negligible effect on the SnS vapor pressure over saturated liquid solution in comparison with pure molten SnS. Using solvent evaporation the crystals of $Nd_{9.5}Sn_{1.8}S_{16}$ high-temperature solid solution have been grown. This solid solution crystallizes in the cubic structure with the lattice parameter a = 856.0 ± 0.1 pm, the calculated and experimental densities of 5.55 ± 0.07 and 5.58 ± 0.01 g·cm^{-3}, respectively, and an energy gap of 2.0 ± 0.1 eV.

SnS_2–Nd_2S_3: The **Nd_2SnS_5** ternary compound, which melts incongruently at 1020°C, has a polymorphic transformation at 820°C (Mel'chenko and Serebrennikov 1973a, b) and crystallizes in the orthorhombic structure with the lattice parameters a = 777.21 ± 0.14, b = 1121.8 ± 0.2, c = 392.72 ± 0.09 pm, and a calculated density of 5.504 g·cm^{-3} (Daszkiewicz et al. 2008) [a = 1115 ± 5, b = 778 ± 3, and c = 392 ± 1 pm (Guittard et al. 1976)], is formed in this system. This compound was prepared in the same way as Pr_2SnS_5 was obtained using Nd instead of Pr (Daszkiewicz et al. 2008).

7.23 Tin–Samarium–Sulfur

SnS_2–Sm_2S_3: The **Sm_2SnS_5** ternary compound, which melts incongruently at 1000°C and crystallizes in the orthorhombic structure with the lattice parameters a = 1127.6 ± 0.8, b = 777.3 ± 0.5, c = 389.5 ± 0.3 pm (Julien-Pouzol and Jaulmes 1979) [a = 1118 ± 5, b = 776 ± 3, and c = 390 ± 1 pm (Guittard et al. 1976)], is formed in this system.

7.24 Tin–Europium–Sulfur

Several ternary compounds are formed in the Sn–Eu–S system. **$EuSnS_3$** was synthesized by Guittard et al. (1976) by heating a mixture of SnS_2 and EuS.

Eu_2SnS_4 melts incongruently at 930°C (Flahaut et al. 1979) and crystallizes in the orthorhombic structure with the lattice parameters a = 1118.7 ± 0.2, b = 876.8 ± 0.2, c = 753.8 ± 0.2 pm, and a calculated density of 4.95 g·cm^{-3} (Pocha et al. 2003) [a = 1524, b = 888, c = 1130 pm, and a calculated density of 4.77 g·cm^{-3} (Flahaut et al. 1979)]. It was synthesized by heating stoichiometric mixture of the elements in an alumina crucible, sealed in a quartz ampoule under Ar atmosphere (Pocha et al. 2003). The temperature was first slowly raised (~50°Ch^{-1}) up to 900°C and kept for 15 h. The sample was subsequently homogenized in an Ar-filled glove box and reheated to 900°C for 20 h. This procedure resulted in red brown crystalline powder of the title compound, which is stable in air.

Eu_2SnS_5 melts incongruently at 978°C and crystallizes in the orthorhombic structure with the lattice parameters a = 410.0 ± 0.1, b = 1562.1 ± 0.3, c = 1150.7 ± 0.2 pm, and the calculated and experimental densities of 5.11 and 5.15 g·cm^{-3}, respectively (Jaulmes et al. 1982a) [a = 410, b = 1562, c = 1151 pm, and the calculated and experimental densities of 5.25 and 5.15 g·cm^{-3}, respectively (Guittard et al. 1983)]. Black crystals of this compound were obtained in vacuum-sealed quartz ampoule by reaction of 2EuS + SnS_2 + S mixture (Guittard et al. 1983). The heating was carried out gradually, so as to reach in 5 days the temperature of 800°C, which is maintained for 3 days. According to the data of Flahaut et al. (1979), this phase is hypothetical and its composition corresponds to **$Eu_4Sn_2S_9$** or **$Eu_2^{II}Eu_2^{III}Sn_2S_9$**, which have the same lattice parameters.

$Eu_3Sn_2S_7$ melts incongruently at 900°C and crystallizes in the orthorhombic structure with the lattice parameters a = 1154.2 ± 0.4, b = 1269.0 ± 0.5, c = 397.4 ± 0.1 pm, and the calculated and experimental densities of 5.24 and 5.18 g·cm^{-3}, respectively (Jaulmes and Julien-Pouzol 1977b; Flahaut et al. 1979). This compound was obtained by heating at 880°C in a vacuum-sealed silica ampoule of a mixture of EuS and SnS_2 in stoichiometric proportions (Jaulmes and Julien-Pouzol 1977b).

$Eu_5Sn_3S_{12}$ or **$Eu_3^{II}Eu_2^{III}Sn_3S_{12}$** crystallizes in the orthorhombic structure with the lattice parameters a = 390.8 ± 0.1, b = 2011.5 ± 0.4, c = 1145.1 ± 0.2 pm, and a calculated density of 5.536 g·cm^{-3}, (Jakubcová et al. 2007) [a = 392.4 ± 0.1, b = 1150.9 ± 0.1, c = 2021.9 ± 0.1 pm, and the calculated and experimental densities of 5.46 and 5.28 g·cm^{-3}, respectively (Jaulmes and Julien-Pouzol 1977a; Flahaut et al. 1979)]. Thermodynamic functions, entropy S^0_{298} = 724.0 J·(M·K)$^{-1}$ and enthalpy H^0_{298} = 97.7 kJ·M^{-1}, have been calculated from experimental specific heat data by Volkonskaya et al. (1980). This compound was synthesized by heating the elements or binary sulfides in alumina crucibles, sealed in silica ampoules under an Ar atmosphere (Jakubcová et al. 2007). At first the temperature was raised slowly (50°C·h^{-1}) to 900°C and kept there for 24 h. After cooling, the sample was homogenized and heated to 900°C for 24 h. This procedure was repeated once more and resulted in homogeneous dark powder, which was not sensitive to air.

According to the data of Senova (1979), in the **SnS–EuS** system during heat treatment of mixtures of SnS and EuS in various ratios, a mixture of the initial sulfides was detected.

7.25 Tin–Gadolinium–Sulfur

SnS_2–Gd_2S_3: The **Gd_2SnS_5** ternary compound, which melts incongruently at 1050°C and crystallizes in the orthorhombic structure with the lattice parameters a = 733.30 ± 0.17, b = 1129.0 ± 0.2, c = 382.17 ± 0.09 pm, and a calculated density of 5.907 g·cm^{-3} (Daszkiewicz et al. 2008) [a = 1116 ± 5, b = 775 ± 3, and c = 388 ± 1 pm (Guittard et al. 1976)], is formed in this system (Mel'chenko and Serebrennikov 1973a). This compound was prepared in the same way as Pr_2SnS_5 was synthesized using Gd instead of Pr (Daszkiewicz et al. 2008).

7.26 Tin–Terbium–Sulfur

SnS_2–Dy2S_3: The **Dy_2SnS_5** ternary compound, which has a polymorphic transformation at 820°C (Gus'kova and Serebrennikov 1973a) and crystallizes in the orthorhombic structure with the lattice parameters a = 771.7 ± 0.2, b = 1124.60 ± 0.15, c = 380.56 ± 0.08 pm, and a calculated density of 6.002 g·cm^{-3} (Daszkiewicz et al. 2008) [a = 1115 ± 5, b = 775 ± 3, and c = 387 ± 1 pm (Guittard et al. 1976)], is formed in this system. This compound was prepared in the same way as Pr_2SnS_5 was synthesized using Tb instead of Pr (Daszkiewicz et al. 2008). Tb_2SnS_5 crystallizes well in the form of brown needle crystals and is stable in air.

7.27 Tin–Dysprosium–Sulfur

SnS_2–Dy_2S_3: The **Dy_2SnS_5** ternary compound, which crystallizes in the orthorhombic structure with the lattice parameters a = 1114 ± 5, b = 775 ± 3, and c = 386 ± 1 pm (Guittard et al. 1976)], is formed in this system (Mel'chenko and Serebrennikov 1973a).

7.28 Tin–Holmium–Sulfur

SnS_2–Ho_2S_3: No ternary compounds were found in this system (Guittard et al. 1976).

7.29 Tin–Erbium–Sulfur

Er_2S_3: The **Er_2SnS_5** ternary compound, which has a polymorphic transformation at 820°C, is formed in this system (Gus'kova and Serebrennikov 1973a). This compound crystallizes in the form of brown needle crystals and is stable in air.

7.30 Tin–Ytterbium–Sulfur

Five ternary compounds, **$YbSnS_3$**, **Yb_2SnS_4**, **$Yb_3Sn_2S_7$**, **$Yb_4Sn_2S_9$**, and **$Yb_5Sn_3S_{12}$**, which is semiconductors, are formed in the Sn–Yb–S system (Rustamov et al. 1981). Four of them crystallize in the orthorhombic structure with the lattice parameters a = 1520, b = 880, c = 1125 pm, and an experimental density of 4.76 g·cm^{-3} for Yb_2SnS_4, a = 1150, b = 1258, c = 395 pm, and an experimental density of 5.25 g·cm^{-3} for $Yb_3Sn_2S_7$, a = 406, b = 1560, c = 1128 pm, and an experimental density of 5.62 g·cm^{-3} for $Yb_4Sn_2S_9$, and a = 390, b = 1146, c = 2018 pm, and an experimental density of 5.36 g·cm^{-3} for $Yb_5Sn_3S_{12}$.

7.31 Tin–Lead–Sulfur

SnS–PbS: The phase diagram of this system, constructed through DTA, XRD, metallography, measuring of microhardness, and hydrothermal synthesis (Nekrasov et al. 1974; Latypov et al. 1976b; Leute et al. 1994) and optimized using a four parameter for the excess Gibbs energy by Volykhov et al. (2008), is given in Figure 7.25. This system proved to be

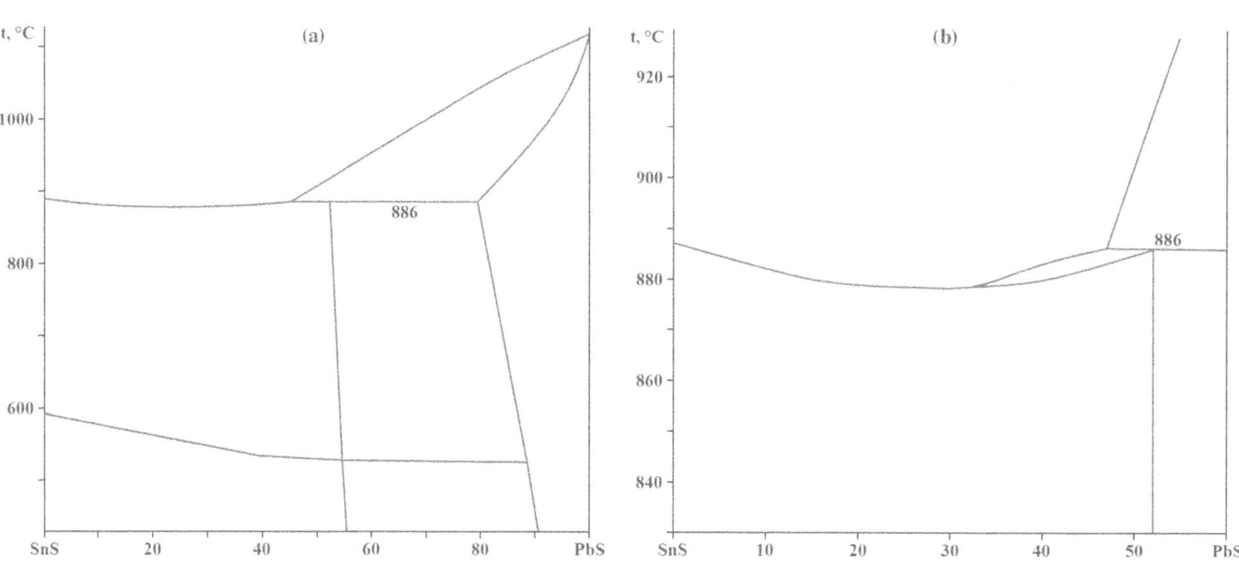

FIGURE 7.25 Phase diagram of the SnS–PbS system (a) and its enlarged part from the SnS-rich side (b). (From Volykhov, A.A., et al., *Inorg. mater.*, **44**(4), 345, 2008.)

azeotropic with the azeotropic point at 67 mol% SnS and ca. 874°C [84 mol% SnS and 824°C (Nekrasov et al. 1974]. At temperatures below 530°C the phase diagram is characterized by a broad miscibility gap between the PbS-rich cubic phase and the SnS-rich orthorhombic phase. The solubility of SnS in PbS is 10 mol% at 20°C and 15 mol% at 700°C (Latypov et al. 1976b) [20 mol% at 400°C (Bigvava et al. 1974); 2.5, 6, and 15 mol% at 300°C, 700°C, and 836°C, respectively (Nekrasov et al. 1974); 10 mol% at 720°C (Krebs and Langner 1964)]. The unit cell parameter of PbS-based solid solutions decreases from 593.18 to 588.7 pm for an alloy saturated with SnS (Kuznetsov and Li 1964). The solubility of PbS in SnS reaches 45 mol% at the peritectic temperature (Latypov et al. 1976b) [25 mol% at 400°C (Bigvava et al. 1974); 50 mol% at room temperature, 51 mol% at 400°C, 56 mol% at 800°C and 56.5 mol% at 836°C (Nekrasov et al. 1974); 55 mol% at 640°C (Krebs and Langner 1964)].

According to the data of Murach (1947), Bigvava et al. (1974), and Kuznetsov and Li (1964), **$PbSnS_2$** ternary compound is formed in this system. It crystallizes in the orthorhombic structure with the lattice parameters a = 408.2, b = 427.0, and c = 1141.7 pm (Chang and Brice 1971). At 400°C, the homogeneity region of this compound is within the interval of 30–52 mol% PbS (Bigvava et al. 1974; Kuznetsov and Li 1964). When dissolving SnS, the parameters of the $PbSnS_2$ unit cell decrease (the saturation limit is not set), and when dissolving PbS they increase to the values a = 404.8, b = 435.4, and c = 1135.2 pm (Kuznetsov and Li 1964).

Hayashi et al. (2001) noted, that the solid solution area is between $Pb_{1.060}Sn_{0.940}S_2$ and SnS at 700°C. $PbSnS_2$ coexisting with PbS was synthesized by hydrothermal crystallization at 300°C, 400°C, and 450°C. The composition of solid solutions was $Pb_{1.140}Sn_{0.860}S_2$ at 300°C, $Pb_{1.114}Sn_{0.886}S_2$ at 400°C, and $Pb_{1.124}Sn_{0.876}S_2$ at 450°C; there compositions shift towards the PbS end member from $PbSnS_2$. There is a linear relationship between cell dimensions and composition of solid solution (Chang and Brice 1971; Hayashi et al. (2001). The melting temperature of the solid solution decreases with increasing amounts of PbS (Chang and Brice 1971).

This compound also exists as mineral teallite, which crystallizes in the orthorhombic structure with the lattice parameters $a = 404 \pm 2$, $b = 428 \pm 2$, $c = 1133 \pm 2$ pm, and an experimental density of 6.36 g·cm^{-3} (Hofmann 1935).

The standard entropy of formation (S^0_{298} = 169.1 J·M^{-1}K^{-1}), entropy (ΔS^0_{298} = 4.32 J·M^{-1}K^{-1}), enthalpy (ΔH^0_{298} = 258.4 kJ·M^{-1}) and Gibbs energy of formation (ΔG^0_{298} = 257.2 kJ·M^{-1}) for $PbSnS_2$ were calculated by Mammadov et al. (2014).

$Pb_{2-x}Sn_xS_2$ nanocrystals ($0.40 < x < 1$) were obtained by Soriano et al. al. (2012b). They are single-phase solid solutions with cubic NaCl-type structure with the lattice parameter a = (593.4 ± 0.1)–(594.3 ± 0.1) pm and an energy gap of 0.52–0.57 eV. All syntheses were carried out under air-free conditions using a Schlenck line. In a typical synthesis, $Pb(CH_3COO)_2{\cdot}3H_2O$ (0.400 mM), and $Sn(CH_3COO)_2$ (8 mM), 1-octadecene (8.00 mL), oleic acid (5.0 mL), and of oleylamine (5.0 mL) were placed in a 100 mL Schlenk flask and dried under vacuum at 100°C for 2–3 h with constant stirring to obtain a homogeneous pale yellow solution. The reaction flask was flushed with N_2, and the temperature was raised to 190°C. At 190°C, the prepared S precursor was swiftly injected. A 0.100 M sulfur precursor was prepared by dissolving of S (0.80 mM) in 1-octadecene (8.00 mL) at 100°C for 2 h under a N_2 atmosphere. Upon injection, a dark brown colored mixture was observed. The temperature of the resulting mixture was maintained at 170°C–180°C for 30, 60, or 120 s. The reaction was then cooled to room temperature using an ice-water bath. The resulting nanocrystals were precipitated by the addition of a solution of ethanol/hexanes (volume ratio 1:1) followed by centrifugation. Discarding the supernatant liquid gave a dark brown pellet of $PbSnS_2$ nanocrystals. The resulting nanocrystals were re-suspended in hexanes and re-precipitated with excess ethanol. The purified nanocrystals can be dispersed in nonpolar organic solvents such as hexanes, chloroform, or carbon tetrachloride to form stable colloidal suspensions.

According to the data of Morozov and Li (1963), the phase diagram of the SnS–PbS system belongs to the type III according to the Roseboom classification with a minimum at 17 mol% PbS and 838°C. In the range of 5–50 mol% PbS, additional thermal effects at 428°C and 540°C appear which, apparently, correspond to the beginning of the formation of a high-temperature phase and the end of the transition to a single-phase state with the NaCl-type structure (Morozov and Li 1963; Bigvava et al. 1974).

The ingots for investigations were annealed at 400°C for 100 h (Latypov et al. 1976b) [at 650°C for 100 h (Bigvava et al. 1974); at 400°C for 42 days, at 700°C for 10 days and at 800°C for 7 days with the next quenching in air (Nekrasov et al. 1974); at 700°C for 24 h (Kuznetsov and Li 1964); at 720°C from the PbS-side and at 640°C from the SnS-side for 70 h (Krebs and Langner 1964)].

SnS_2–PbS: The **$PbSnS_3$** ternary compound, which melts incongruently at 740°C [melts congruently at 730°C (Sachdev and Chang 1975)] and crystallizes in the orthorhombic structure with the lattice parameters a = 873.8, b = 1405.2, c = 379.2 pm, the calculated and experimental densities of 6.01 and 5.96 g·cm^{-3}, respectively, and an energy gap of 1.05 eV (Jumas et al. 1972; Fenner and Mootz 1976) [$a = 874.0 \pm 0.2$, $b = 1407.9 \pm 0.2$, $c = 379.6 \pm 0.1$ pm (Yamaoka and Okai 1970); a = 874, b = 1403, c = 376 pm (Sachdev and Chang 1975)], is formed in this system (Alpen et al. 1975; Marchuk et al. 2007). This compound also exists as mineral suredaite, which crystallizes in the orthorhombic structure with the lattice parameters $a = 882.21 \pm 0.03$, $b = 377.28 \pm 0.03$, $c = 1400.76 \pm 0.03$ pm, and the calculated and experimental densities of 5.615 and 5.54–5.88 g·cm^{-3}, respectively (Paar et al. 2000; Mandarino 2002). According to the data of Sachdev and Chang (1975), this compound can take 2 mol% SnS_2 in excess along the SnS_2–PbS binary system.

The synthesis of $PbSnS_3$ from the elements or mixture of the binary sulfides was carried out in conventional quartz ampoule (Fenner and Mootz 1976). The starting materials were weighed in stoichiometric amounts. The evacuated ampoule was directly annealed at 600°C, or heated until the sulfide mixture melts up to 1000°C, quenched and annealed at 600°C. The title compound could be also prepared by hydrothermal treatment of binary sulfides mixture in 2 N HCl (the degree of filling was about 60%, the temperature gradient was about 25°C, and the reaction temperature was 430°C). This compound could be also prepared from an equimolar mixture of PbS and SnS_2 or 1:1:2 mixture of PbS, Sn, and S which was sealed in a Pt capsule (Yamaoka and Okai 1970). A piston-cylinder type of high pressure apparatus was used with the furnace assembly in the pressure cell. Pressure was applied to the sample first, and then the temperature was raised to 500°C–800°C at a rate of 2–3°C·min^{-1} and held for 15–20 h. The specimen was quenched to room temperature before the pressure was released. Almost perfect single crystals were obtained at a pressure of 2 GPa.

The $\mathbf{Pb_2SnS_4}$ and $\mathbf{PbSn_2S_4}$ were not found in the SnS_2–PbS system (Bok and Boeyens 1957).

When studying the coprecipitation in the SnS_2–PbS system, it was found that with a Pb/Sn ratio of up to 0.2, the formation of a solid solution based on SnS_2 is possible (Toptygina et al. 1979).

$\mathbf{Sn_2S_3}$–PbS: The phase diagram of this system is a eutectic type (Gaudin and Hamlyn 1938).

$\mathbf{Sn_2S_3}$–$\mathbf{PbSnS_3}$: A complete solid solution is formed in this system (Sachdev and Chang 1975).

7.32 Tin–Titanium–Sulfur

Several ternary compounds are formed in the Sn–Ti–S system. $\mathbf{SnTiS_3}$ and $\mathbf{SnTi_2S_5}$ crystallize in the orthorhombic structure with the lattice parameters a = ~2310, b = ~579, c = ~2320 pm, and a = 2279.7 ± 2.5, b = 581.0 ± 0.7, c = 3513 ± 4 pm, respectively (Guemas et al. 1988). To prepare these compounds, Sn, Ti, and S were mixed in the 1:1:3 and 1:2:5 proportions in relation with the ratios SnS/TiS_2 of 1:1 or 1:2, respectively. The mixtures were put into silica tubes, sealed under vacuum and subsequently heated up to 900°C for about 15 days after 2 days at 300°C. Cooling down took place within a few hours. Samples were then ground and reheated up to 900°C with a small amount of iodine (ca. 5 mg·cm^{-3}). The preparations led to crystallized $SnTiS_3$ or $SnTi_2S_5$.

$\mathbf{SnTi_2S_4}$ was obtained at the interaction of a stoichiometric mixture of SnS_2 and TiS at 750°C–800°C for 1 week (Jumas et al. 1977b).

$\mathbf{Sn_{1.2}Ti_{0.8}S_3}$ also crystallizes in the orthorhombic structure with the lattice parameters a = 889.9 ± 0.4, b = 360.5 ± 0.2, c = 1350.6 ± 0.6 pm (Gressier et al. 1987). To synthesize the title compound, Sn, Ti, and S powders were mixed in ratio 1:1:3. The mixture was put into silica tubes together with a small quantity of iodine (<5 mg·cm^{-3}) to be used as a carrier reagent. The tubes were sealed under vacuum and subsequently heated to 750°C–800°C for about 15 days. Cooling took place within a few hours. Single-crystalline parallelepiped shining-black needles of this compound were obtained.

$\mathbf{(SnS)_{1.20}TiS_2}$ is a misfit layer compound built of alternate double layers of SnS and sandwiches of TiS_2, which crystallizes in the triclinic structure with the lattice parameters a = 568.3 ± 0.1, b = 583.2 ± 0.1, c = 1168.0 ± 0.5 pm, α = 95.85 ± 0.03°, β = 94.78 ± 0.03°, γ = 90.03 ± 0.03° for SnS-sublattice, and a = 341.2 ± 0.1, b = 583.5 ± 0.1, c = 2328.9 ± 0.3 pm, α = 95.86 ± 0.01°, β = 90.30 ± 0.01°, γ = 90.01 ± 0.01° for TiS_2-sublattice (Wiegers et al. 1990b, 1991; Wiegers and Meerschaut 1992). A powder sample of this compound was synthesized from the elements in the stoichiometric ratio. The mixture was heated at 800°C in an evacuated quartz tube for 7 days. Single crystals were obtained by vapor transport using chlorine as a carrier reagent, for which about 1 mass% $(NH_4)_2PbCl_6$ was used. Crystals grow as thin platelets at the cold part of the gradient of 720°C–650°C. This compound could be also prepared at the reaction of the ternary oxides with CS_2 or H_2S vapor at 1000°C–1300°C.

7.33 Tin–Zirconium–Sulfur

SnS–$\mathbf{ZrS_2}$: The $\mathbf{ZrSnS_3}$ ternary compound, which crystallizes in the orthorhombic structure with the lattice parameters a = 918.8 ± 0.1, b = 371.7 ± 0.1, c = 1383.9 ± 0.1 pm, and a calculated density of 4.302 g·cm^{-3} (Meetsma et al. 1993), is formed in this system.

$\mathbf{SnS_2}$–$\mathbf{ZrS_2}$: A continuous series of solid solutions is formed in the system (Al-Alamy and Balchin 1973). The lattice parameters, depending on the composition, vary with a slight deviation from Vegard's law.

7.34 Tin–Hafnium–Sulfur

SnS–$\mathbf{HfS_2}$: The $\mathbf{HfSnS_3}$ ternary compound, which crystallizes in the orthorhombic structure with the lattice parameters a = 913.9 ± 0.1, b = 369.4 ± 0.1, c = 1387.5 ± 0.4 pm, and a calculated density of 5.578 g·cm^{-3} (Wiegers et al. 1989a, b), is formed in this system. Powder sample of this compound was obtained by heating the elements in an evacuated quartz ampoule. An intimate mixture of the elements was first slowly heated to about 400°C for 2 days and then heated at 750°C for 10 days. Single crystals were grown by vapor transport in a temperature gradient of 770°C–710°C or of 720°C–650°C over 12 days.

7.35 Tin–Phosphorus–Sulfur

SnS–"$\mathbf{P_2S_4}$": The part of the phase diagram of this system, constructed using DTA and XRD, is given in Figure 7.26 (Voroshilov et al. 1992). The eutectic contains 80 mol% SnS

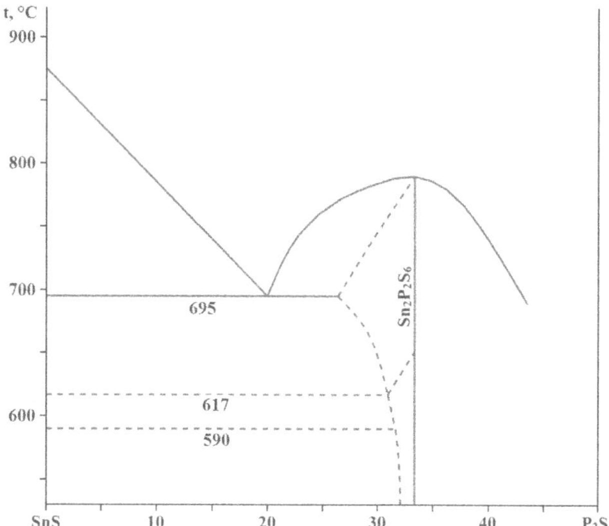

FIGURE 7.26 Part of the phase diagram of the SnS–"P_2S_4" system. (From Voroshilov, Yu.V., et al., *Ukr. khim. zhurn.*, **58**(3), 216, 1992.)

and crystallizes at 695°C. The polymorphic transformation of SnS takes place at 590°C. The $\mathbf{Sn_2P_2S_6}$ ternary compound, which melts congruently at 790°C (Potoriy et al. 1990; Voroshilov et al. 1992; Potoriy and Milyan 2016), is formed in this system. Above 590°C, this compound has a homogeneity region growing up to 6 mol% at 695°C (Voroshilov et al. 1992; Prits et al. 2011b). According to the nuclear gamma resonance data, in the Sn–P–S ternary system, in addition to the $Sn_2P_2S_6$ compound, there are two more compounds of Sn^{2+} (Seregin et al. 1972c).

SnS_2–PS: The part of the phase diagram of this system, constructed using DTA, XRD, and measuring of the microhardness, is given in Figure 7.27 (Potoriy et al. 1990; Voroshilov et al. 1991, 1992). The eutectic contains 65 mol% SnS_2 and crystallizes at 665°C. Two compounds, $Sn_2P_2S_6$ and $\mathbf{Sn_3P_2S_8}$, are formed in this system. $Sn_2P_2S_6$ is at the intersection of the SnS–"P_2S_4" and SnS_2–PS systems, and $Sn_3P_2S_8$ melts incongruently at 707°C. The thermal behavior of $Sn_2P_2S_6$ in the temperature range 20°C–830°C was studied using differential scanning calorimetry (Sharpataya et al. 1992). It was shown (Figure 7.28) that the first polymorphic transition of this compound takes place at 64°C with the transition enthalpy of 566 ± 10 $J{\cdot}M^{-1}$. The second polymorphic transformation was observed at 624°C–633°C with the transition enthalpy of 5.34 ± 0.27 $kJ{\cdot}M^{-1}$. According to the data of Sharpataya et al. (1992), this compound melts at 771°C with the melting enthalpy of 103.9 ± 0.3 $kJ{\cdot}M^{-1}$, and the melting is accompanied by a delamination of the resulting liquid. The energy gap of $Sn_2P_2S_6$ is 2.3 eV (Gurzan et al. 1977). This compound decomposes at 670°C forming tin and phosphorus sulfides (Nitsche and Wild 1970).

One of the $Sn_2P_2S_6$ polymorphic modifications crystallizes in the monoclinic structure with the lattice parameters $a = 936.2 \pm 0.2$, $b = 749.3 \pm 0.1$, $c = 655.0 \pm 0.3$ pm, $\beta = 91.17 \pm 0.03°$, and a calculated density of 3.55 $g{\cdot}cm^{-3}$ (Scott et al. 1992) [$a = 937.8 \pm 0.5$, $b = 748.8 \pm 0.5$, $c = 631.5 \pm 0.5$ pm, $\beta = 91.15 \pm 0.05°$, and the calculated and experimental densities of 3.57 and 3.54 $g{\cdot}cm^{-3}$, respectively (Dittmar and Schäfer 1974); $a = 937 \pm 7$, $b = 745.0 \pm 0.1$, $c = 651 \pm 5$ pm, $\beta = 91 \pm 1°$, and $a = 937 \pm 1$, $b = 748.22 \pm 0.02$, $c = 652.2 \pm 0.9$ pm, $\beta = 91.1 \pm 0.2°$ for two different samples (Bourdon et al. 1997)].

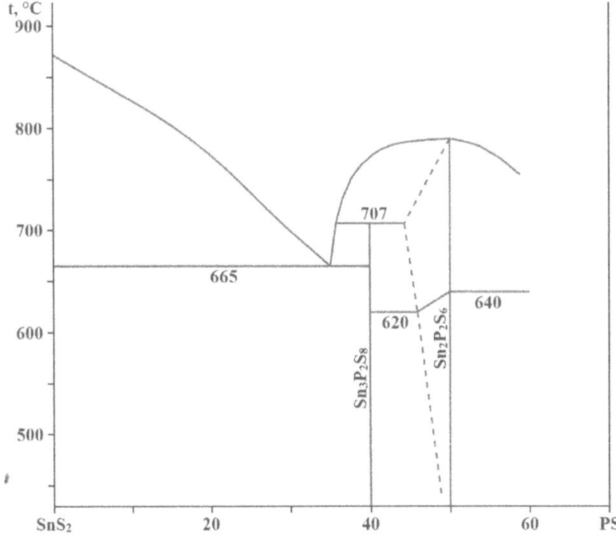

FIGURE 7.27 Part of the phase diagram of the SnS_2–PS system. (From Potoriy, M.V., et al., *Izv. AN SSSR. Neorgan. mater.*, **26**(11), 2363, 1990.)

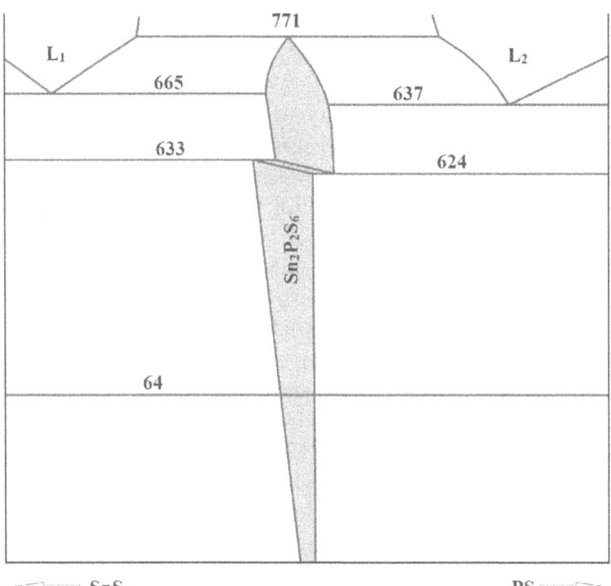

FIGURE 7.28 Fragment of the phase diagram of the SnS_2–PS system in the region of the $Sn_2P_2S_6$ compound. (From Sharpataya, G.A., et al., *Neorgan. mater.*, **28**(1), 25, 1992.)

Second polymorphic modification of this compound also crystallizes in the monoclinic structure with the lattice parameters $a = 651.3$, $b = 748.8$, $c = 1152.5$ pm, $\beta = 125.56°$, and $a = 651.6$, $b = 747.9$, $c = 1124$ pm, $\beta = 123.99°$, and an experimental density of 3.60 $g{\cdot}cm^{-3}$ for two different samples (Potoriy et al. 1990; Voroshilov et al. 2016) [$a = 643$, $b = 747$, $c = 1103$ pm, $\beta = 122.2°$, and the calculated and experimental densities of 3.64 and 3.32 $g{\cdot}cm^{-3}$, respectively (Nitsche and Wild 1970; Klingen et al. 1973); $a = 652.9 \pm 0.2$, $b = 748.5 \pm 0.2$, $c = 1131.3 \pm 0.3$ pm, $\beta = 124.11 \pm 0.03°$, and the calculated and experimental densities of 3.57 and 3.56 $g{\cdot}cm^{-3}$, respectively (Carpentier and Nitsche 1974); $a = 633$, $b = 760$, $c = 1142$ pm, and $\beta = 124.8°$ (Vysochanskiy et al. 1979)]. According to the data of Klingen et al. (1970, 1973), $Sn_2P_2S_6$ crystallizes in the monoclinic structure with the lattice parameters $a = 599$, $b = 1036$, $c = 680$ pm, $\beta = 107.1°$, and the calculated and experimental densities of 4.05 and 4.00 $g{\cdot}cm^{-3}$, respectively.

Third polymorphic modification of $Sn_2P_2S_6$ crystallizes in the trigonal structure with the lattice parameters $a = 599.69 \pm 0.03$, $c = 1941.8 \pm 0.2$ pm, and a calculated density of 3.073 $g{\cdot}cm^{-3}$ (Prits et al. 2006) [$a = 598$, $c = 1932$ pm, and the calculated and experimental densities of 4.09 and 4.01 $g{\cdot}cm^{-3}$, respectively (Klingen et al. 1970, 1973); $a = 599.9 \pm 0.2$, $c = 1942.4 \pm 0.4$ pm (Wang et al. 1995)].

$Sn_2P_2S_6$ may be synthesized at room temperature (Bourdon et al. 1997). The method is based on a metathesis mechanism in aqueous media between the Sn^{2+} and $P_2S_6^{6-}$ ions. The cations were furnished by $SnCl_2$ and the anions by $Li_4P_2S_6$, each salt being used in the form of a 0.1 M water solution. Gradual pouring of the $SnCl_2$ solution into that of $Li_4P_2S_6$ resulted in a brown gel of the title compound. Twice as much $SnCl_2$ than required by stoichiometry (Sn/P = 2) was employed to favor the reaction of all available $P_2S_6^{6-}$ anions. A few syntheses with Sn/P ratio close to unity were also carried out. Water was eliminated from the gel using a rotary evaporator and the product was subsequently washed in acetone or alcohol. Upon filtration, the solvent and LiCl were removed and a dark brown powder was obtained. The washing and filtration were repeated two to three times to pull out residual water and salt. Finally, the $Sn_2P_2S_6$ powder was dried under vacuum at 120°C.

Single crystals of this compound were grown through chemical transport reactions (Klingen et al. 1970; Nitsche and Wild 1970; Gurzan et al. 1977; Carpentier and Nitsche 1974). Large crystals of $Sn_2P_2S_6$ were grown from the elements using iodine vapor transport (Scott et al. 1992). Stoichiometric amounts of Sn, P, and S (4 g total) were lightly ground together and then transferred to a quartz tube. Iodine crystals (0.225 g) were added to the reaction tube. The sample was evacuated, sealed, and placed in a furnace. The growth zone of the tube was set to 700°C for 1 day to allow reverse transport out of the growth zone. Then the growth zone was lowered to 600°C and the sample zone was raised to ~630°C. After 4 days, large transparent, orange or yellow, crystals of $Sn_2P_2S_6$ were obtained.

Large crystals of $Sn_2P_2S_6$ with rhombohedral structure were grown from the elements (Wang et al. 1995). Stoichiometric amounts of Sn, P, and S (10 g total) were lightly ground together and then transferred to a quartz reactor. The sample was evacuated, sealed, and placed in a tube furnace. A gradient was established in the furnace with a maximum temperature of 450°C. After several days, large single crystals of the title compound were produced. The crystals were red, lustrous, and transparent.

$Sn_3P_2S_8$ crystallizes in the cubic structure with the lattice parameter a = 1355.5 pm and an experimental density of 3.93 g·cm^{-3} (Potoriy and Milyan 2016; Voroshilov et al. 2016).

The isothermal section of the Sn–P–S ternary system at 500°C is presented in Figure 7.29 (Voroshilov et al. 1992). Besides $Sn_2P_2S_6$ and $Sn_3P_2S_8$, another ternary compound, $\mathbf{Sn_3P_4S_{13}}$, is formed in the system (Voroshilov et al. 1992, 2016). It melts congruently at 465°Cand crystallizes in the orthorhombic structure with the lattice parameters a = 590.1 ± 0.5, b = 647.6 ± 0.1, and c = 1200.5 ± 0.4 pm and the calculated and experimental densities of 3.24 and 3.04 g·cm^{-3}, respectively (Voroshilov et al. 1992; Potoriy and Milyan 2016). It is possible that the forth compound, $\mathbf{Sn_3P_4S_{10}}$, could exist in the Sn–P–S ternary system (Voroshilov et al. 1992).

7.36 Tin–Arsenic–Sulfur

SnS–As$_2$S$_3$: The $\mathbf{SnAs_2S_4}$ ternary compound, which crystallizes in the tetragonal structure with the lattice parameters a = 437.4 and c = 319.9 pm, is formed in this system (Bok and Boeyens 1957). It was prepared by the heating a mixture of Sn, As, and S in the stoichiometric ratio at 800°C for 18 h.

A glass forming region was not detected in the Sn–As–S ternary system (Seregin et al. 1972c).

7.37 Tin–Antimony–Sulfur

SnS–Sb: The phase diagram of this system, constructed through DTA, XRD, metallography, and measuring of the microhardness, is a eutectic type (Figure 7.30) (Kurbanova et al. 1987). The eutectic contains 4 mol% SnS and crystallizes at 615°C. The polymorphic transition of the solid solution based on SnS takes place at 575°C. The area of solid solutions based on SnS reaches 2 mol% Sb. The ingots for investigations were annealed at 350°C for 200 h.

SnS–Sb$_2$S$_3$: The phase diagram of this system, constructed through DTA, XRD, metallography and measuring of the microhardness, is a eutectic type with peritectic transformation (Figure 7.31) (Novoselova et al. 1972; Nekrasov et al. 1975).

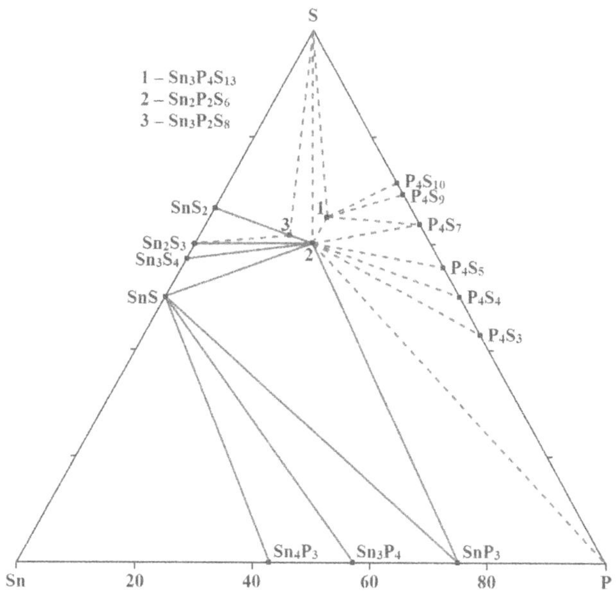

FIGURE 7.29 Isothermal section of the Sn–P–S ternary system at 500°C. (From Voroshilov, Yu.V., et al., *Ukr. khim. zhurn.*, **58**(3), 216, 1992.)

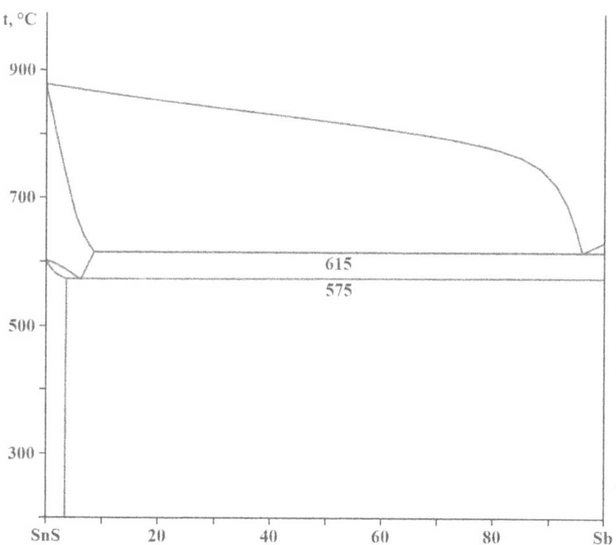

FIGURE 7.30 Phase diagram of the SnS–Sb system. (From Kurbanova, R.D., et al., *Izv. AN SSSR. Neorgan. mater.*, **23**(11), 1796, 1987.)

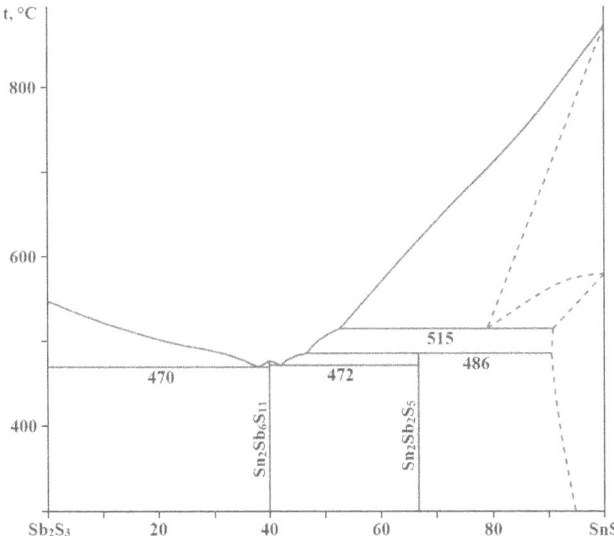

FIGURE 7.31 Phase diagram of the SnS–Sb_2S_3 system. (From Novoselova, A.V., et al., *Izv. AN SSSR. Neorgan. mater.*, **8**(1), 173, 1972.)

The eutectics contain 38 and 42 mol% SnS and crystallize at 470°C and 472°C, respectively. Two compounds, $\mathbf{Sn_2Sb_2S_5}$, which melts incongruently at 486°C, and $\mathbf{Sn_2Sb_6S_{11}}$, which melts congruently at 477°C, are formed in this system [according to the data of Wang and Eppelsheimer (1976), $Sn_2Sb_6S_{11}$ was not found in the SnS–Sb_2S_3 system]. With increasing Sb_2S_3 content, the temperature of the polymorphic transformation of solid solutions based on SnS decreases to 515°C. The solubility of Sb_2S_3 in SnS at 400°C reaches 7 mol%, and the solubility of SnS in Sb_2S_3 is negligible. The ingots for investigations were annealed at 400°C for 500 h (Novoselova et al. 1972).

According to the data of Wang and Eppelsheimer (1976) and Eppelsheimer (1981), solid solutions of the general formula $(Sn,Sb)_3S_4$ are formed in the system with a Sn/Sb ratio of 5: 2 to 3: 2. The melting temperature of these solid solutions is 614°C and they crystallize in the cubic structure with the lattice parameter a = 1144.81 pm and an experimental density of 4.78 g·cm^{-3}.

Beside $Sn_2Sb_2S_5$ and possibly $Sn_2Sb_6S_{11}$, some other compounds are formed in the SnS–Sb_2S_3 system. $\mathbf{SnSb_2S_4}$ melts incongruently at 501°C (Wang and Eppelsheimer 1976) [at 493°C (Sachdev and Chang 1975)] and apparently has two polymorphic modifications. One of them crystallizes in the orthorhombic structure with the lattice parameters a = 2564.1 ± 0.5, b = 2038.1 ± 0.4, c = 389.73 ± 0.15 pm, and a calculated density of 4.80 g·cm^{-3} (Smith and Parise 1985). Second modification crystallizes in the monoclinic structure with the lattice parameters a = 2071.1, b = 390.34, c = 2227.6 pm, and β = 96°11.1′ (Eppelsheimer 1981) [a = 2070, b = 390, c = 2246 pm, β = 96°10′, and an experimental density of 4.57 g·cm^{-3} (Wang and Eppelsheimer 1976)].

The standard entropy of formation (S^0_{298} = 259.1 J·M^{-1}K^{-1}), entropy (ΔS^0_{298} = 4.84 J·M^{-1}K^{-1}), enthalpy (ΔH^0_{298} = 351.8 kJ·M^{-1}) and Gibbs energy of formation (ΔG^0_{298} = 350.5 kJ·M^{-1}) for $SnSb_2S_4$ were calculated by Mammadov et al. (2014).

This compound was synthesized from the elements combined in stoichiometric proportion in an evacuated silica ampoule (Smith and Parise 1985). The charge was held at 800°C for 15 h, then at 470°C for 400 h.

$\mathbf{SnSb_4S_7}$ crystallizes in the monoclinic structure with the lattice parameters a = 1133.1 ± 0.4, b = 386.5 ± 0.2, c = 1394.0 ± 0.5 pm, β = 105.281 ± 0.006°, and a calculated density of 4.72 g·cm^{-3} (Topa et al. 2010c). The crystals of this compound were prepared at 650°C in an evacuated silica ampoule, in a mixture with crystals of Sb_2S_3 and $\mathbf{Sn_4Sb_6S_{13}}$.

$Sn_2Sb_2S_5$ melts incongruently at 516°C (Wang and Eppelsheimer 1976) and crystallizes in the orthorhombic structure with the lattice parameters a = 1959 ± 3, b = 393.8 ± 0.3, and c = 1142.6 ± 0.3 pm (Smith and Hyde 1983) [a = 1966, b = 396, and c = 1142 pm (Wang and Eppelsheimer 1976); a = 1959.5, b = 394.09, and c = 1142.0 pm (Eppelsheimer 1981)].

The standard entropy of formation (S^0_{298} = 336.2 J·M^{-1}K^{-1}), entropy (ΔS^0_{298} = 4.40 J·M^{-1}K^{-1}), enthalpy (ΔH^0_{298} = 473.8 kJ·M^{-1}) and Gibbs energy of formation (ΔG^0_{298} = 472.5 kJ·M^{-1}) for $Sn_2Sb_2S_5$ were calculated by Mammadov et al. (2014).

The title compound was synthesized at the interaction of the stoichiometric amounts of SnS and Sb_2S_3 at 495°C (Smith and Hyde 1983).

The standard entropy of formation (S^0_{298} = 700.1 J·M^{-1}K^{-1}), entropy (ΔS^0_{298} = 14.66 J·M^{-1}K^{-1}), enthalpy (ΔH^0_{298} = 921.4 kJ·M^{-1}) and Gibbs energy of formation (ΔG^0_{298} = 917.2 kJ·M^{-1}) for $Sn_2Sb_6S_{11}$ were calculated by Mammadov et al. (2014).

$\mathbf{Sn_3Sb_2S_6}$ melts incongruently at 528°C (Wang and Eppelsheimer 1976) [at 565°C (Sachdev and Chang 1975)] and crystallizes in the orthorhombic structure with the lattice parameters a = 2315.4 ± 0.2, b = 396.31 ± 0.04, and c = 3491.0 ± 0.4 pm (Parise et al. 1984) [a = 2313, b = 396, and c = 3501 pm and an experimental density of 4.85 g·cm^{-3} (Wang and Eppelsheimer 1976); a = 2315.4, b = 395.87, and c = 3488.2 pm (Eppelsheimer 1981); a = 2318 ± 1, b = 396.5 ± 0.1, and c = 3494 ± 1 pm and a calculated density of 4.91 g·cm^{-3} (Smith 1984)].

To obtain this compound, SnS and Sb_2S_3 (molar ratio 3:1) were melted together in an evacuated silica ampoule at 850°C for 4 h and then annealed at 490°C for 3 d (Smith 1984). It could be also prepared by combining the elements in stoichiometric ratio in evacuated silica glass ampoule and heating at 480°C for 72 and 168 h respectively (Parise et al. 1984).

According to the data of Eppelsheimer (1974), $\mathbf{Sn_4Sb_2S_7}$ ternary compound is also formed in the SnS–Sb_2S_3 system.

$Sn_4Sb_6S_{13}$ crystallizes in the monoclinic structure with the lattice parameters a = 2431 ± 1, b = 391.5 ± 0.5, c = 2349 ± 3 pm, β = 94.05 ± 0.05°, and a calculated density of 4.82 g·cm^{-3} (Jumas et al. 1980). Single crystals of this compound were grown by the chemical transport reactions.

$\mathbf{Sn_6Sb_{10}S_{21}}$ crystallizes in the monoclinic structure with the lattice parameters a = 4499.5 ± 0.5, b = 390.23 ± 0.05, c = 2061.3 ± 0.3 pm, β = 96.21 ± 0.01°, and a calculated density of 4.80 g·cm^{-3} (Parise and Smith 1984a, b). Black, needle-shaped crystals of this compound were grown from a charge of Sn, Sb, and S (to give composition $SnSb_2S_4$) sealed in an evacuated silica glass ampoule and heated at 490°C for 120 h. The title compound was previously reported as $SnSb_2S_4$.

$\mathbf{SnS_2–Sb_2S_3}$**:** The phase diagram of this system, constructed through DTA, XRD, metallography, and measuring of the microhardness, is presented in Figure 7.32 (Rustamov et al. 1987). $\mathbf{SnSb_2S_5}$ ternary compound, which melts incongruently at 460°C, is formed in this system. The eutectic contains 62 mol% Sb_2S_3 and crystallizes at 445°C [493°C (Eppelsheimer

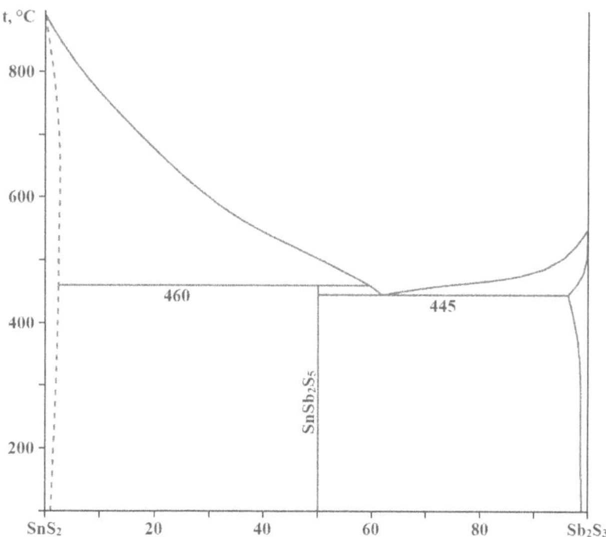

FIGURE 7.32 Phase diagram of the SnS_2–Sb_2S_3 system. (From Rustamov, P.G., et al., *Dokl. AN AzSSR*, **43**(1), 42, 1987.)

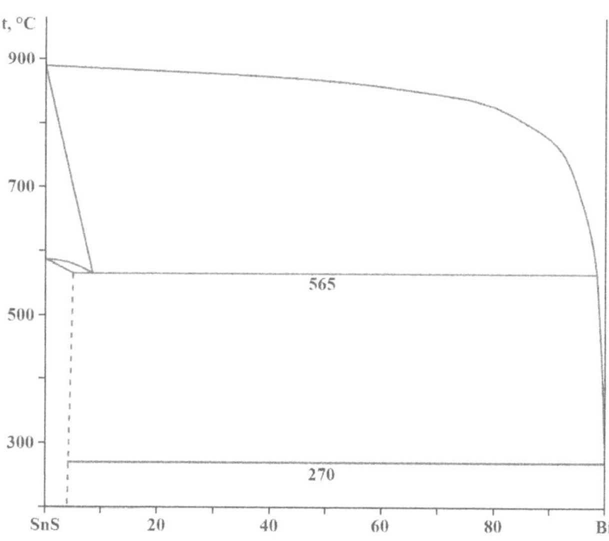

FIGURE 7.33 Phase diagram of the SnS–Bi system. (From Movsum-zade, A.A., et al., *Azerb. khim. zhurn.*, (4), 134, 1983.)

1974)]. The solubility of SnS_2 in Sb_2S_3 at room temperature is not higher than 2 mol%. The ingots for investigations were annealed at 350°C for 150 h.

According to the data of Gaudin and Hamlyn (1938) and Nekrasov et al. (1975), no ternary compounds were found in the SnS_2–Sb_2S_3 system. Nekrasov et al. (1975) noted, that the phase diagram of the SnS_2–Sb_2S_3 system is a eutectic type. The eutectic contains 22 mol% SnS_2 and crystallizes at 516°C. The solubility of SnS_2 in Sb_2S_3 at 400°C is 8–8.5 mol% and increases to 16–16.5 mol% at 500°C.

SnS–SnS_2–Sb_2S_3: The solid solution containing 62.5–74 mol% SnS, 10.0–15.5 mol% SnS_2 and 13.0–25.0 mol% Sb_2S_3 is formed in this subsystem (Sachdev and Chang 1975). This phase forms equilibrium with all phases in this subsystem. The $\mathbf{Sn^{+2}_{0.70}Sn^{+4}_{0.15}Sb_{0.30}S_{1.45}}$ solid solution melts incongruently at 627°C forming Sn_2S_3 and liquid.

At 510°C, the Sn–Sb–S ternary system has 10 three-phase regions formed by SnS_2, Sn_2S_3, Sb_2S_3, $Sn_4Sb_2S_7$, $SnSb_2S_4$, SnS-based solid solutions, liquid S, Sn-based molten metal, Sb-based solid solution and melt inside the concentration triangle (Eppelsheimer 1974). The $\mathbf{3Sn_2S_3{\cdot}Sb_2S_3}$ or $\mathbf{4SnS{\cdot}Sn_2S_3{\cdot}Sb_2S_3}$ ternary compound is also formed in the Sn–Sb–S system (Gaudin and Hamlyn 1938).

7.38 Tin–Bismuth–Sulfur

SnS–Bi: The phase diagram of this system, constructed through DTA, XRD, metallography, and measuring of the microhardness, is a eutectic type (Figure 7.33) (Movsum-zade et al. 1983). The eutectic is degenerated from the Bi-side and crystallizes at 270°C. The solubility of Bi in SnS is 5 mol% at 565°C and 3 mol% at room temperature (Movsum-zade et al. 1983; Rustamov et al. 1985d). Substitution of tin with bismuth in the SnS crystal lattice leads to a decrease of the energy gap. The ingots for the investigations were annealed at 200°C for 300 h (Movsum-zade et al. 1983) or at 500°C for 500 h (Rustamov et al. 1985d).

SnS–Bi_2S_3: The phase diagram of this system, constructed through DTA, XRD, metallography, and measuring of the microhardness, is given in Figure 7.34 (Rustamov et al. 1985b). The eutectics contain 37 and 64 mol% SnS and crystallize at 600°C and 575°C, respectively. The solubility of Bi_2S_3 in SnS is not higher than 5 mol% (Rustamov et al. 1985b,d) [10 mol% at 500°C–600°C and decreases to 5 mol% at 400°C (Gospodinov et al. 1971)], and the solubility of SnS in Bi_2S_3 reaches 9 mol% (Rustamov et al. 1985b) [20 mol% at 400°C (Gospodinov et al. 1971)]. The $\mathbf{SnBi_2S_4}$ ternary compound, which melts congruently at 645°C and crystallizes in the hexagonal structure with the lattice parameters $a = 1350$, $c = 400$ pm, and an energy gap of 0.54 eV, is formed in this system (Rustamov et al. 1985b). Single crystals of this compound were grown by the chemical transport reactions using iodine as a carrier reagent.

According to the data of Gospodinov et al. (1971), the phase diagram of the SnS–Bi_2S_3 system is a peritectic type with the formation of the solid solutions with a minimum at 45 mol%

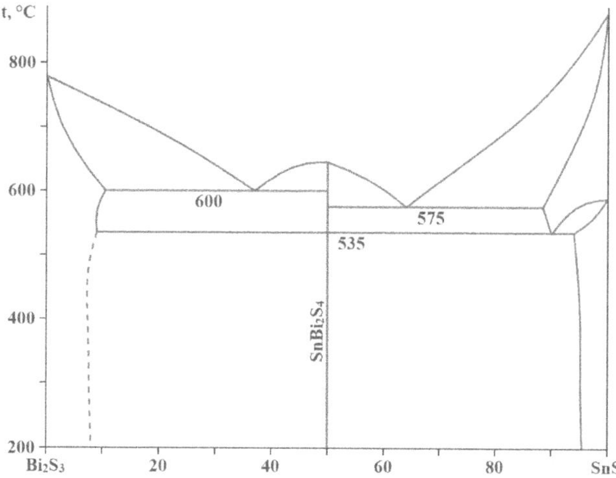

FIGURE 7.34 Phase diagram of the SnS–Bi_2S_3 system. (From Rustamov, P.G., et al., *Izv. AN SSSR. Neorgan. mater.*, **21**(11), 1865, 1985.)

SnS and 529°C. The solid solutions based on Bi_2S_3 decompose peritectically at 600°C with the formation of the solid solutions based on high-temperature SnS and a melt. Increasing of the Bi_2S_3 content leads to the decreasing of the temperature of the polymorphic transition of the solid solutions based on SnS from 590°C to 560°C.

Kurbanova et al. (1997) noted that the phase diagram of this system is a eutectic type with monotectic transformation at 662°C. The eutectic contains 18 mol% Bi_2S_3 and crystallizes at 602°C. The $\mathbf{Sn_4Bi_2S_{11}}$ ternary compound is formed in the subsolidus part, decomposing incongruently at 577°C. The solubility of SnS_2 in Bi_2S_3 at room temperature reaches 5 mol%.

The ingots for the investigations were annealed at 400°C for 300 h (Rustamov et al. 1985b) [for 550 h (Gospodinov et al. 1971)], at 500°C for 500 h (Rustamov et al. 1985d) or at 400°C for 300 h (Kurbanova et al. 1997)].

$\mathbf{SnS_2}$–Bi: This system is a non-quasibinary section of the Sn–Bi–S ternary system (Kurbanova et al. 1986).

The liquidus surface of the Sn–Bi–S ternary system was constructed by Kurbanova et al. (1986). There six ternary eutectics and two transition points on the liquidus surface.

7.39 Tin–Niobium–Sulfur

Several ternary compounds are formed in the Sn–Nb–S system. $\mathbf{SnNbS_3}$ crystallizes in the orthorhombic structure with the lattice parameters a = 2285.1 ± 0.5, b = 575.7 ± 0.1, c = 1178.3 ± 0.3 pm (Guemas et al. 1988) [a = 3978, b = 575, c = 1176 pm (Kuypers et al. 1989); a = 2286 ± 1, b = 571 ± 3, c = 1175 ± 3 pm (Hernán et al. 1991a); in the tetragonal structure with the lattice parameters a = 405 and c = 1177 pm (Maaren van 1972)]. To prepare the title compound, Sn, Nb, and S were mixed in the 1:1:3 proportion in relation with the ratios SnS/NbS_2 of 1:1 (Maaren van 1972; Guemas et al. 1988). The mixtures were put into silica tubes, sealed under vacuum and subsequently heated up to 900°C for about 15 days after two days at 300°C. Cooling down took place within a few hours. Samples were then ground and reheated up to 900°C with a small amount of iodine (ca. 5 mg·cm^{-3}). The preparations led to crystallized $SnNbS_3$. Single crystals of this compound were grown by the chemical transport reactions using Cl_2 as carrier reagent (Hernán et al. 1991a).

$\mathbf{(SnS)_{1.17}NbS_2}$ is a misfit layer compound, which belongs to the class of the intergrowth compounds (Smaalen van 1989), is formed in this system. In the simplest approximation, its structure can be described as that of two interpenetrating sublattices. The true structure consists of two interpenetrating, incommensurately modulated structures. The title compound is consisted of two-atom-thick layers of SnS and three-atom-thick layers of NbS_2, which alternate along the c axis (Meetsma et al. 1989). It crystallizes in the orthorhombic structure with the lattice parameters a = 567.3 ± 0.1, b = 575.0 ± 0.1, c = 1176.0 ± 0.1 pm for SnS-sublattice, and a = 332.1 ± 0.1, b = 575.2 ± 0.1, c = 1176.3 ± 0.1 pm for NbS_2-sublattice (Wiegers et al. 1988, 1989c, 1992; Meetsma et al. 1989; Kuypers et al. 1990; Wiegers and Meerschaut 1992).

Two methods were used for the syntheses of $(SnS)_{1.17}NbS_2$, namely the direct combination of the elements or the binary compounds at temperatures of 800°C–1200°C and the reaction of the ternary oxides with CS_2 or H_2S vapor at 1000°C–1300°C (Wiegers et al. 1989c, 1992). In order to obtain single crystals of this compound the vapor transport was used (Kuypers et al. 1990).

The $\mathbf{Sn_{1/3}NbS_2}$ intercalate compound crystallizes in the hexagonal structure with the lattice parameters a = 577.8 ± 0.1, c = 1439.4 ± 0.1 pm, and a calculated density of 4.707 g·cm^{-3} (Fang et al. 1996) [a = 547.3, c = 1468 pm (Eppinga and Wiegers 1977); a = 577.8, c = 1441 pm (Eppinga and Wiegers 1980)]. The preparation of this compound from the elements and from 2*H*-NbS_2 and Sn lead to the same product. The mixtures were heated in evacuated quarts ampoules at 700°C–900°C for 1–4 weeks (Eppinga and Wiegers 1977, 1980). It could be also obtained by direct exposure of NbS_2 to the Sn vapor.

The $\mathbf{SnNbS_2}$ intercalate compound crystallizes in the hexagonal structure with the lattice parameters a = 332.4, c = 1737 pm (Eppinga and Wiegers 1977). The preparation of this compound from the elements and from 2*H*-NbS_2 and Sn lead to the same product. The mixtures were heated in evacuated quarts ampoules at 900°C for 1–2 weeks.

The $\mathbf{Sn_3Nb_2S_6}$ layered crystals were obtained in an evacuated quartz tube at 900°C, using iodine as a transport agent (Prodan and Boswell 1980). These crystals include intergrowth thin alternating layers of $SnNbS_2$ and epitaxially deformed SnS_2. The SnS_2 has a monoclinic structure with a = 1578, b = 567 pm, and β = 95° (c parameter was not indicated).

7.40 Tin–Tantalum–Sulfur

Several ternary compounds are formed in the Sn–Ta–S system. $\mathbf{SnTaS_3}$ crystallizes in the tetragonal structure with the lattice parameters a = 406 and c = 1187 pm (Maaren van 1972; Wiegers et al. 1989a). To prepare the title compound, Sn, Ta, and S were mixed in the 1:1:3 proportions in evacuated quartz ampoule heated at about 800°C for 1–2 weeks. In order to obtain single crystals the vapor transport was used. It turned out that those crystals could be grown using Cl_2 as a carrier reagent in a temperature gradient of 850°C–780°C; the crystals grow at the low-temperature side of the tube over 10 days. To the sample to be transported (about 300 mg) about 5–10 mg of $(NH_4)_2PbCl_6$ was added. The crystals have a metallic luster.

The $\mathbf{(SnS)_{1.15}TaS_2}$ layered composite compound is consisted of alternate double layers of SnS and sandwiches of TaS_2. It crystallizes in the orthorhombic structure with the lattice parameters a = 574.9 ± 0.4, b = 573.7 ± 0.4, c = 1187.55 ± 0.24 pm for SnS-sublattice, and a = 575.0 ± 0.4, b = 330.8 ± 0.3, c = 2376.0 ± 0.5 pm for TaS_2-sublattice (Gotoh et al. 1993) [a = 573.9, b = 571.5, c = 1186 pm for SnS-sublattice, and a = 331.1, b = 571.5, c = 1186 pm for TaS_2-sublattice (Wiegers et al. 1988, 1989a,c; Kuypers et al. 1990); a = 572.0, b = 574.2, c = 1188.1 pm for SnS-sublattice, and a = 331.6, b = 574.2, c = 2376.2 pm for TaS_2-sublattice in the case of the $(SnS)_{1.16}TaS_2$ composition (Wiegers and Meerschaut 1992)]. Two methods were used for synthesis of this misfit layer compound, namely direct combination of the elements or the binary compounds at temperatures 800°C–1200°C and reaction of the ternary oxide with CS_2 or H_2S vapor at 1000°C–1300°C (Wiegers and Meerschaut

1992). The mixtures of the elements were heated in evacuated quartz ampoule at about 800°C for 1–2 weeks (Wiegers et al. 1989c, 1990b). The single crystals were prepared by vapor transport using Cl_2 as a carrier reagent (Kuypers et al. 1990). It could be also prepared at the reaction of the ternary oxides with CS_2 or H_2S vapor at 1000°C–1300°C.

The $\mathbf{Sn_{1/3}TaS_2}$ intercalation compound crystallizes in the hexagonal structure with the lattice parameters a = 575.6 pm, c = 1451 pm, and shows the superlattice with $a = \sqrt{3} \times a_0$, were a_0 is the a-parameter of *2H*-TaS_2 (Eppinga and Wiegers 1980) [a = 569.8 pm, c = 1435 pm (Di Salvo et al. 1973; Eppinga et al. 1976; Eppinga and Wiegers 1977)]. The superlattice of this compound is quite stable. The powder of $Sn_{1/3}TaS_2$ was prepared by heating the elements in an evacuated quartz tube at 900°C for 1–2 weeks (Eppinga et al. 1976). It could be also obtained by the intercalation of TaS_2 with tin vapor at 900°C (Di Salvo et al. 1973; Eppinga and Wiegers 1980).

The $\mathbf{SnTaS_2}$ intercalation compound also crystallizes in the hexagonal structure with the lattice parameters a = 330.7 pm, c = 1744.2 pm (Eppinga and Wiegers 1977) [a = 328 pm, c = 1740 pm (Di Salvo et al. 1973; Eppinga et al. 1976)]. The powder of this compound was prepared by heating the elements in an evacuated quartz tube at 900°C for 1–2 weeks (Eppinga et al. 1976). It could be also obtained by the intercalation of TaS_2 with tin vapor at 850°C (Di Salvo et al. 1973).

The $\mathbf{Sn_2TaS_4}$ ternary compound was not determined in the Sn–Ta–S system (Bok and Boeyens 1957).

The $\mathbf{Sn_3Ta_2S_6}$ layered crystals were obtained in an evacuated quartz tube at 900°C, using iodine as a transport agent (Prodan and Boswell 1980). These crystals include intergrowth thin alternating layers of $SnTaS_2$ and epitaxially deformed SnS_2. The SnS_2 has a monoclinic structure with a = 1578, b = 567 pm, and β = 95° (c parameter was not indicated).

7.41 Tin–Oxygen–Sulfur

Several ternary compounds are formed in the Sn–O–S system. The **SnOS** was obtained at the interaction of SnO and S at 280°C (Batsanov and Shestakova 1966). It has an experimental density of 5.31 g·cm^{-3}.

$\mathbf{SnSO_4}$ crystallizes in the orthorhombic structure with the lattice parameters a = 879.9 ± 0.1, b = 531.9 ± 0.1, c = 711.5 ± 0.1 pm, and the calculated and experimental densities of 4.185 and 4.21 g·cm^{-3}, respectively (Donaldson and Puxley 1972) [a = 881 ± 2, b = 717 ± 2, c = 535 ± 1 pm, and the calculated and experimental densities of 4.22 and 4.21 ± 0.01 g·cm^{-3}, respectively (Donaldson and Moser 1960); a = 879.90 ± 0.11, b = 531.90 ± 0.06, c = 711.50 ± 0.09 pm, and the calculated and experimental densities of 4.185 and 4.15 g·cm^{-3}, respectively (Rentzeperis 1962)]. Rentzeperis (1962) noted that this compound is highly hygroscopic but according to the data of Donaldson and Puxley (1972), it is not hygroscopic.

To prepare this compound, Sn (37 g) was added to a solution of $CuSO_4$ (37 g) in water (300 mL) and H_2SO_4 (15 mL), the solution being kept under oxygen-free nitrogen (Carson 1926; Donaldson and Moser 1960). The solution was boiled, Cu being deposited. Eventually the solution became colorless and the deposit became grey owing to re-deposition of Sn on the surface of Cu. At this point the solution was filtered rapidly and concentrated to about 50 mL in a vacuum rotary evaporator. The mother-liquor was filtered off, and $SnSO_4$ (~25 g) washed with alcohol and ether and dried at 100°C.

$SnSO_4$ can also be obtained from its solution by precipitation with organic solvents, but the final product is less pure and is often yellowish owing to the presence of Sn (IV) (Donaldson and Moser 1960). It can be kept for several months in contact with air, and much longer in the dark under a vacuum. Pure $SnSO_4$ has been stored out of direct sunlight for 20 months in stoppered colorless Pyrex glass or brown glass. Complete decomposition into SnO_2 and SO_2 occurred at 378°C and there was little decomposition below this temperature.

$\mathbf{Sn_2OSO_4}$ or $\mathbf{SnO \cdot SnSO_4}$ crystallizes in the tetragonal structure with the lattice parameters a = 1093.0 ± 0.2, c = 893.1 ± 0.2 pm, and the calculated and experimental densities of 4.35 and 4.26–4.27 g·cm^{-3}, respectively (Wernfors 1961). Its crystals were obtained in the following way. A stock solution of $SnSO_4$ was prepared by slowly passing a saturated solution of $CuSO_4$ through a column of granular Sn so that complete reduction of Cu (II) to copper metal occurred. The use of this method eliminates the oxidation of Sn (II) by air. Series of solutions with different Sn (II) ion concentration and different values of pH were then prepared. Portions of the stock solution were diluted and were made slightly acidic with H_2SO_4. For some of the solutions the pH was increased by adding small amounts of ammonia or freshly precipitated and washed $Sn(OH)_2$. The solutions were sealed in thick-walled Pyrex tubes and heated in a furnace at temperatures ranging from 100°C to 150°C for periods of from 1 day to 3 or 4 weeks. On completion of the period of heating the tube were quickly cooled and opened, and the crystals were filtered off, washed with H_2O several times and dried in air. In all cases the title compound was formed in spite of the great variations in pH and concentration. According to the data of Carson (1926), $SnO \cdot SnSO_4$ is formed in the solution at the interaction of $SnSO_4$ and NaOH at 100°C.

$\mathbf{Sn_3O_2SO_4}$ or $\mathbf{2SnO \cdot SnSO_4}$· crystallizes in the orthorhombic structure with the lattice parameters a = 502 ± 1, b = 1380 ± 2, c = 1543 ± 2 pm, and the calculated and experimental densities of 4.51 and 4.48 g·cm^{-3}, respectively (Davies and Donaldson 1967). This compound was obtained at the decomposition of $\mathbf{Sn_3(OH)_2OSO_4}$ at 230°C. It decomposes at 405°C with slight explosion to yield predominantly SnO_2 and SO_2.

The results of the examination in the Sn–O–S ternary system were presented in several isothermal sections (Moh 1974). At 250°C, oxygen and SO_2 are miscible gases, S and Sn are liquids, but all other compounds are solids. SnS can remain stable with the three oxides, SnO, Sn_3O_4, and SnO_2, whereas SnO_2 can coexist with all three tin sulfides, SnS, Sn_2S_3, and SnS_2. Above 270°C, the tie line between SnO and SnS disappears, while the tie line Sn_3O_4–SnS is still stable. Above 450°C, the only stable solids are SnO_2 and the tin sulfides. At 738°C, a binary eutectic liquid appears in the Sn–S system which coexists also with SnO_2, while a tie line between SnS and SnO_2 persists. At ~800°C, a ternary liquid appears in the monovariant field Sn_{liq}–SnO_2–SnS which has an approximate composition of 50 at% Sn, 40 at% S, and 10 at% oxygen. With further increase in temperature, the compositional field of this liquid enlarges towards the SnO composition, and above 1080°C, it forms a single liquid field stretching from SnO composition to molten tin sulfides. There is no detectable solubility of any sulfides of Sn, nor do sulfides dissolve measurable amounts of SnO_2. Oxidation experiments of sulfides showed that, in an excess of air, SnS is gradually oxidized in order Sn_2S_3–SnS_2–SnO_2 between 400°C and 600°C. Sulfurization of SnO_2 gave SnS_2, in some cases after metastable Sn_2S_3. By the method used, no $SnSO_4$ or any similar sulfate was not prepared.

7.42 Tin–Selenium–Sulfur

SnS–SnSe: The phase diagram of this system, constructed through DTA and XRD, and optimized using a four parameter for the excess Gibbs energy by Volykhov et al. (2008), is presented in Figure 7.35 (Karahanova et al. 1976c). A complete series of solid solutions exists in the system. Experimental runs made at higher temperatures show a continuous melting of the solid solutions with a minimum at 853°C and 50 mol% SnS (Albers et al. 1962; Karahanova et al. 1976c; Liu and Chang 1992). The separation of liquidus and solidus is very narrow, within a maximum of 5°C. The temperature of the phase transition of the $\mathbf{SnS_xSe_{1-x}}$ solid solutions varies almost linearly in going from SnS to SnSe. The two-phase region was not detected experimentally, presumably because it is very narrow. Lattice parameters *a* and *c* of this solid solutions show negative deviation from Vegard's law, while parameter *b* has a linear relationship. The ingots for investigations were annealed at 500°C for 500 h (Karahanova et al. 1976c). Single crystals of the SnS_xSe_{1-x} solid solutions were grown by directional crystallization (Albers et al. 1962) and using chemical transport reactions (Karahanova et al. 1976c).

SnS_2–$SnSe_2$: The phase diagram of this system, constructed through DTA, XRD, and metallography, is given in Figure 7.36 (Shevchuk and Olekseyuk 2008). In the entire concentration range, solid solutions with a hexagonal structure are formed (Rimmington and Balchin 1971; Katty et al. 1989; Liu and Chang 1992; Harbec et al. 1978). Lattice parameter *a* changes linearly with composition, whereas a positive deviation from Vegard's law shown in *c* parameter. The ingots for investigations were annealed at 450°C for 960 h with the next quenching in water (Shevchuk and Olekseyuk 2008).

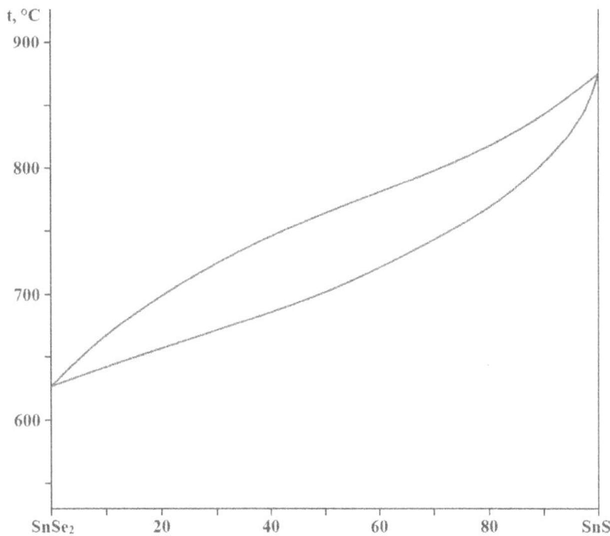

FIGURE 7.36 Phase diagram of the SnS_2–$SnSe_2$ system. (From Shevchuk, M.V., and Olekseyuk, I.D., *Nauk. Visnyk Volyns'k. Nats. Univ. im. Lesi Ukrainky. Ser. Khim. nauky*, (16), 56, 2008.)

The single crystals of the $\mathbf{SnS_{2x}Se_{2(1-x)}}$ solid solutions could be grown by the chemical transport reactions and using the Bridgman techniques (Katty et al. 1989). For this, the best results were obtained with Br_2. For $x = 0.5$, the lattice parameters are the next: $a = 370.8$, $c = 1235.8$ pm, and $a = 366.4$, $c = 1246.9$ pm for two different samples.

The phase relations in the Sn–Se–S ternary system at 500°C are shown in Figure 7.37 (Liu and Chang 1992). It is shown that Sn_2S_3 has a range of solid solutions extending to a

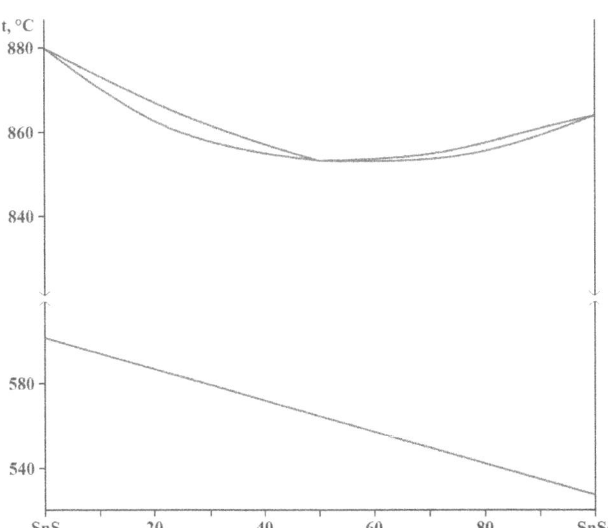

FIGURE 7.35 Phase diagram of the SnS–SnSe system. (From Karahanova, M.I., et al., *Izv. AN SSSR. Neorgan. mater.*, **12**(5), 942, 1976.)

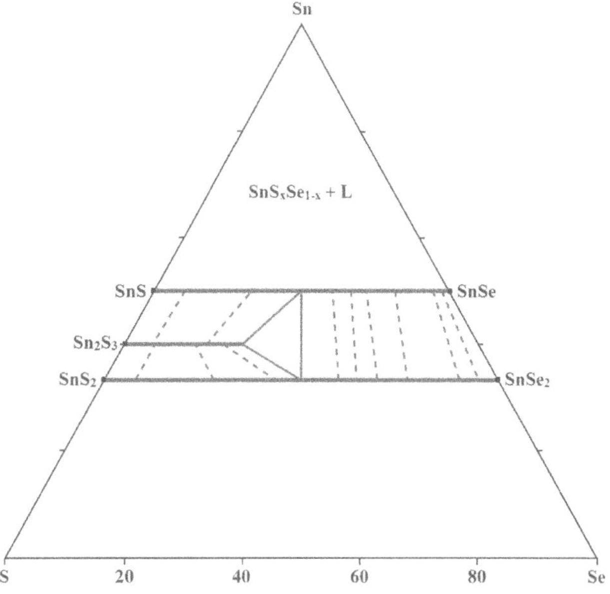

FIGURE 7.37 Phase relations in the Sn–Se–S ternary system at 500°C. (From Liu, H., and Chang, L.L.Y., *J. Alloys Compd.*, **185**(1), 183, 1992.)

composition Sn_2S_2Se, which forms a three-phase assemblage with $SnS_{0.5}Se_{0.5}$ and $Sn(S_{0.5}Se_{0.5})_2$.

7.43 Tin–Sulfur–Tellurium

SnS–SnTe: The phase diagram of this system, constructed through DTA and XRD, and optimized using a four parameter for the excess Gibbs energy by Volykhov et al. (2008), is a eutectic type (Nasirov and Feiziev 1969; Liu and Chang 1992) (Figure 7.38). The eutectic contains ca. 40 mol% SnS and crystallizes at 729°C [the calculated position of the eutectic is 707°C and 51 mol% SnTe (Volykhov et al. 2008). At 500°C, the solubility of SnS in SnTe reaches up to 2.5 mol% and no solid solutions of SnTe was detected in SnS (Liu and Chang 1992). The ranges of the solid solutions are 5.0 mol% SnTe in SnS and 8.5 mol% SnS in SnTe at 700°C. According to the data of Nasirov and Feiziev (1969), the solubility of SnS in SnTe reaches 10 mol% [40 mol% (Vasiliev et al. 1976b)], and the solubility of SnTe in SnS is 40 mol% (Nasirov and Feiziev 1969) [20 mol% (Vasiliev et al. 1976b)]. The ingots for investigations were annealed at 1000°C for 8 h and then at 450°C for 40 h (Vasiliev et al. 1976b).

Phase relations in the Sn–S–Te ternary system at 500°C were determined by Liu and Chang (1992). It was shown that tellurium liquid forms equilibrium assemblages with all tin chalcogenides stable in the system. At 300°C, the region SnS–SnS_2–SnTe–Te consists of no liquid phase.

7.44 Tin–Chromium–Sulfur

The $\mathbf{Cr_2Sn_{1-x}S_{4-x}}$ ($x \approx 0.3$) ternary compound, which crystallizes in the hexagonal structure with the lattice parameters $a = 2132.2 \pm 0.1$ and $c = 346.59 \pm 0.06$ pm (Fukuoka et al. 1995)

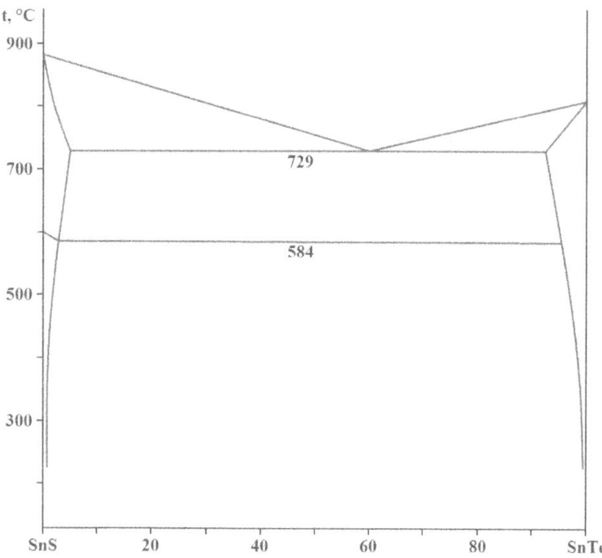

FIGURE 7.38 Phase diagram of the SnS–SnTe system. (From Nasirov, Ya.N., and Feiziev, Ya.S., *Izv. AN SSSR. Neorgan. mater.*, **5**(2), 380, 1969.)

[$a = 2132.5$ and $c = 346.90$ pm for Cr_2SnS_4 composition (Sleight and Frederick 1973)] is formed in the Sn–Cr–S system (Jumas et al. 1977). To prepare this compound, Sn, Cr, and S (molar ratio 1:2:4) were mixed in an agate mortar and pressed into pellets in a glove box (Fukuoka et al. 1995). After being sealed in an evacuated silica ampoule, the sample was heated at 800°C–900°C for several days and quenched in cold water. In the reaction products, two parts could be distinguished. The main product was black, needle-like crystals. It was to be a mixture of $Cr_2Sn_{1-x}S_{4-x}$ and small amounts of Cr_2S_3, which could not be completely separated from the title compound. A second part, a small mass fraction, was found to be a mixture of tin sulfides. The best sample containing the smallest amount of Cr_2S_3 was obtained from starting mixture having a composition with a slight excess of sulfur. A rapid heating was necessary for synthesis of pure $Cr_2Sn_{1-x}S_{4-x}$.

Density-functional theory calculations were reported by Pielnhofer et al. (2014b) on a hypothetical $\mathbf{Cr_3Sn_2S_2}$ compound from scalar relativistic spin-polarized density-functional theory calculations including predictions on crystal and electronic structures. It was shown that this compound could crystallizes in the rhombohedral structure with the lattice parameters a = 565 pm and $\alpha = 56.7°$.

7.45 Tin–Molybdenum–Sulfur

Several ternary compounds are formed in the Sn–Mo–S system. $\mathbf{Sn_{0.854}Mo_6S_8}$ crystallizes in the rhombohedral structure with the lattice parameters $a = 651.00 \pm 0.10$ pm, $\alpha = 89.47 \pm 0.02°$, and a calculated density of 5.619 g·cm^{-3} (Le Lay et al. 1992) [a = 652 pm, α = 89°44′ or a = 919 and c = 1134 pm in hexagonal setting (Chevrel et al. 1971; Gardinier and Chang 1978); these authors identified the composition of this compound as $\mathbf{SnMo_5S_6}$]. To prepare $Sn_{0.854}Mo_6S_8$, about 2 g of a mixture of SnS, MoS_2, and Mo, whose nominal composition was $Sn_{0.66}Mo_6S_6$, were placed in analumina crucible. This ensemble was then put into an out-gassed Mo container which was sealed in a 250 kPa argon atmosphere. The container was then heated in a furnace, in a 13 Pa vacuum. The technique used ensures minimum O_2 contamination. The sample was heated from room temperature to 1300°C in 40 min, then heated to 1600°C in 1 h, held at 1600°C for 5 min, cooled from 1600°C to 600°C in 16.7 h, and finally furnace cooled to room temperature.

$\mathbf{SnMo_6S_7}$ crystallizes in the cubic structure with the lattice parameter a = 653 pm and an experimental density of 5.69 ± 0.3 g·cm^{-3} (Espelund 1967). It was prepared from a mixture of finely powdered Mo, Sn, and S, which was sealed in quartz ampoule under vacuum and heated up to 900°C.

$\mathbf{Sn_xMo_{15}S_{19}}$ crystallizes in the hexagonal structure with the lattice parameters a = 932.8, 933.6 and 933.1, and c = 1867.8, 1869.8 and 1872.8 pm at x = 2.05, 2.20 and 2.53, respectively (Tarascon and Hull 1986). These phases are thermally stable up to 820°C. They were obtained by the diffusion of Sn into $Mo_{15}S_{19}$ at 420°C.

The isothermal sections of the Sn–Mo–S ternary system at 800°C and 500°C are given in Figure 7.39 (Gardinier and Chang 1978). Two liquids, one metallic, stable at the tin apex, and one sulfide, stable between SnS and SnS_2, were identified at 800°C.

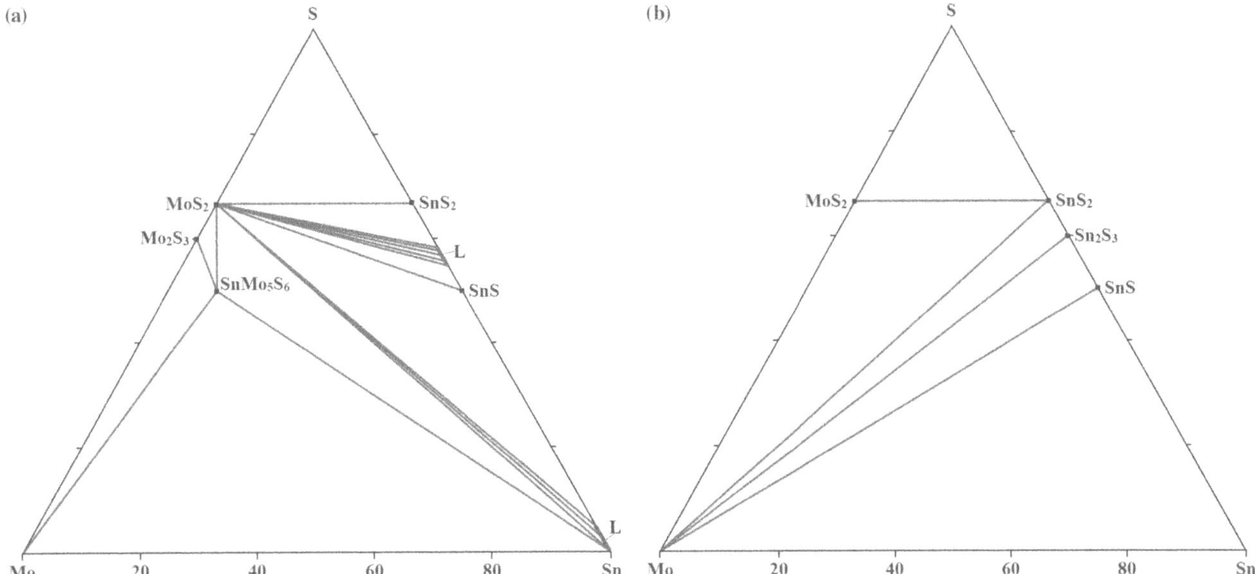

FIGURE 7.39 Isothermal sections of the Sn–Mo–S ternary system at (a) 800°C and (b) 500°C. (From Gardinier, C.F., and Chang, L.L.Y., *J. Less-Common Metals*, **61**(2), 221, 1978.)

The solid phases present were MoS_2, Mo_2S_3, SnS_2, SnS, Mo, and $SnMo_5S_6$. No solid solution was found between SnS_2 and MoS_2. Both Mo_2S_3, and $SnMo_5S_6$ disappear from the equilibrium assemblages at 500°C and the three tin sulfides, SnS, Sn_2S_3, and SnS_2, coexist with Mo, instead of with MoS_2 as at 800°C.

7.46 Tin–Tungsten–Sulfur

The isothermal sections of the Sn–W–S ternary system at 800°C and 500°C are presented in Figure 7.40 (Gardinier and Chang 1978). Both the metallic and the sulfide liquid are stable phases and form equilibrium assemblages with the solid phases WS_2, SnS, SnS_2, and W. No solid solution was found between SnS_2 and WS_2. At 500°C, liquid tin sulfide disappears and WS_2 forms equilibrium assemblages with all tin sulfides.

7.47 Tin–Fluorine–Sulfur

SnS-SnF$_2$: The phase diagram of this system, constructed through DTA and XRD, is a eutectic type (Figure 7.41) (Moskvin et al. 1990). The eutectic contains <1 mol% SnS and crystallizes at 215°C. There are the phase transitions of SnS at 150°C and 596°C on the diagram. An immiscibility region exists at 800°C within the interval from 15 to 90 mol% SnS.

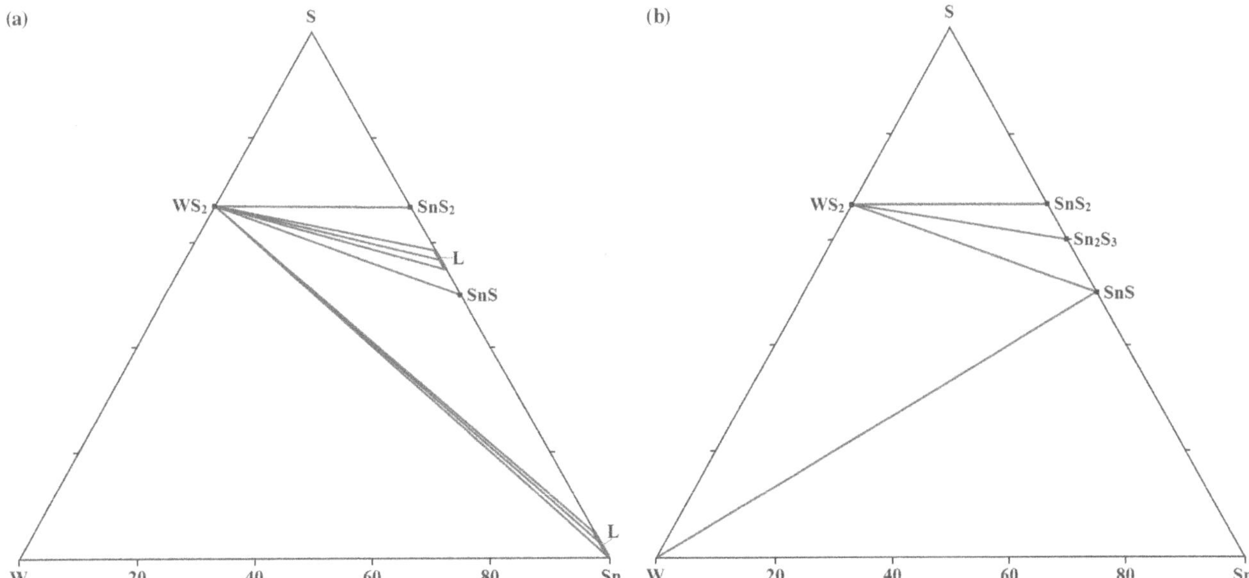

FIGURE 7.40 Isothermal sections of the Sn–W–S ternary system at (a) 800°C and (b) 500°C. (From Gardinier, C.F., and Chang, L.L.Y., *J. Less-Common Metals*, **61**(2), 221, 1978.)

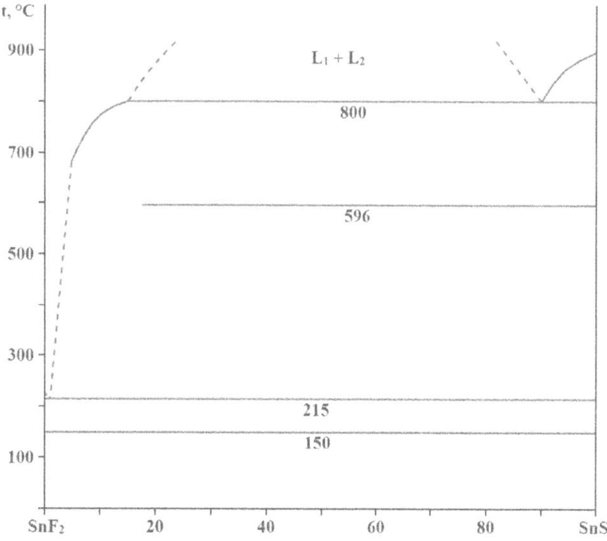

FIGURE 7.41 Phase diagram of the SnS–SnF$_2$ system. (From Moskvin, A.L., et al., *Zhurn. neorgan. khimii*, **35**(6), 1567, 1990.)

7.48 Tin–Chlorine–Sulfur

SnS–SnCl$_2$: The phase diagram of this system, constructed through DTA, is a eutectic type (Figure 7.42) (Morozov and Li 1963; Blachnik and Kasper 1974). The eutectic contains 3.2 mol% SnS and crystallizes at 244°C (Blachnik and Kasper 1974) [2.8 mol% SnS and 240°C (Morozov and Li 1963)].

According to the data of Batsanov and Shestakova (1966), the **Sn$_2$S$_3$Cl$_2$** ternary compound with an experimental density of 3.48 g·cm^{-3}, is formed in the Sn–Cl–S system. It was prepared by melting sulfur together with SnCl$_2$ at 140°C–150°C for 8 h.

7.49 Tin–Bromine–Sulfur

SnS–SnBr$_2$: The phase diagram of this system, constructed through DTA and XRD, is a eutectic type (Figure 7.43) (Thévet et al. 1973; Blachnik and Kasper 1974). The eutectic contains 5 mol% SnS and crystallizes at 195°C (Thévet et al. 1973) [4.3 mol% SnS and 230°C (Blachnik and Kasper 1974)]. Thermal effects at 590°C are due to the SnS polymorphic transformation. The **Sn$_2$SBr$_2$** ternary compound, which melts incongruently at 235°C and crystallizes in the hexagonal structure with the lattice parameters a = 1223.9 ± 0.5, c = 439.0 ± 0.2 pm, and the calculated and experimental densities of 5.01 ± 0.01 and 4.88 g·cm^{-3}, respectively, is formed in the system (Thévet et al. 1972, 1973). This compound was prepared by direct interaction of SnS and SnBr$_2$ at 220°C–230C for 3 days. Its single crystals were obtained by annealing the powders for 3 days at 226°C followed by rapid quenching.

Two more compounds, **Sn$_7$S$_2$Br$_{10}$** and **Sn$_9$S$_2$Br$_{14}$**, are formed in the SnS–SnBr$_2$ system. First of them crystallizes in the hexagonal structure with the lattice parameters a = 1218.5 ± 0.5, c = 441.8 ± 0.4 pm, and a calculated density of 4.95 g·cm^{-3} (Valle et al. 1984). The yellow hexagonal prismatic needles of this compound was prepared as the product from a cooled 2SnS+5SnBr$_2$ melt under an atmosphere of oxygen-free nitrogen.

According to the data of Blachnik and Kasper (1974), Sn$_9$S$_2$Br$_{14}$ melts incongruently at 250°C and exists in the system instead of Sn$_2$SBr$_2$.

Interaction of SnS with liquid Br$_2$ at 0°C and 20°C occurs with the release of a large amount of heat and the formation of SnBr$_4$ and S$_2$Br$_2$ (Batsanov and Shestakova 1968). An alcoholic solution of bromine also causes a violent reaction with SnS.

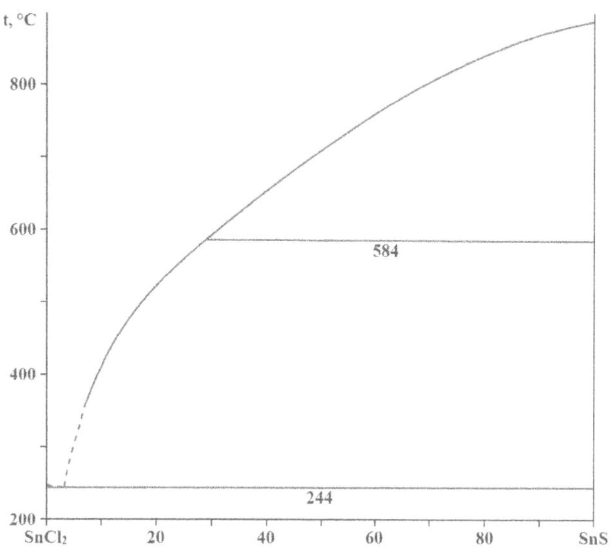

FIGURE 7.42 Phase diagram of the SnS–SnCl$_2$ system. (From Blachnik, R., and Kasper, F.-W., *Z. Naturforsch.*, **29B**(3–4), 159, 1974.)

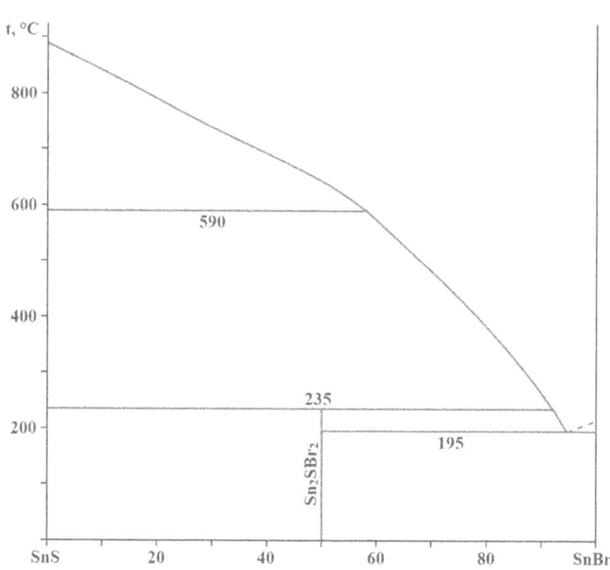

FIGURE 7.43 Phase diagram of the SnS–SnBr$_2$ system. (From Thevet, F., et al., *C. r. Acad. sci. Sér. C*, **276**(26), 1787, 1973.)

7.50 Tin–Iodine–Sulfur

$SnS–SnI_2$: The phase diagram of this system, constructed through DTA and XRD, is a eutectic type with peritectic transformations (Figure 7.44) (Novoselova et al. 1971b; Thévet et al. 1973, 1976). The eutectic contains 15 mol% SnS and crystallizes at 298°C (Thévet et al. 1973, 1976) [5 mol% SnS and 310°C (Novoselova et al. 1971b)]. Thermal effects at 590°C are due to the polymorphic transformation of SnS. The $\mathbf{Sn_2SI_2}$ ternary compound, which melts incongruently at 414°C and has polymorphic transformation at 342°C, is formed in the system (Thévet et al. 1972, 1973, 1976). α-Sn_2SI_2 crystallizes in the monoclinic structure with the lattice parameters a = 1430.5 ± 0.6, b = 1728.1 ± 0.4, c = 443.5 ± 0.5 pm, β = 110°28′ ± 3′, and the calculated and experimental densities of 5.08 and 5.21 g·cm^{-3}, respectively, and β-Sn_2SI_2 crystallizes in the orthorhombic structure with the lattice parameters a = 1744.7 ± 0.1, b = 2533.4 ± 0.2, c = 439.1 ± 0.1 pm, and the calculated and experimental densities of 5.32 and 5.32 ± 0.02 g·cm^{-3}, respectively (Thévet et al. 1976) [a = 1751 ± 1, b = 2543 ± 2, c = 440 ± 1 pm, and the calculated and experimental densities of 5.32 ± 0.02 and 5.325 g·cm^{-3}, respectively (Thévet et al. 1972, 1973)]. The title compound was synthesized by the interaction of SnS and SnI_2 at 380°C–390°C (Thévet et al. 1973, 1976). Single crystals of α-Sn_2SI_2 and β-Sn_2SI_2 were obtained by the annealing the powder at 315°C for 13 days or at 385°C for 7 days, respectively, with the next rapid cooling. The transformation from α- to β-Sn_2SI_2 was only partial in 7 days at 400°C; but the reverse was not possible around 200°C, even in the presence of an excess of iodide (Thévet et al. 1976). The α-Sn_2SI_2 ⇔ β-Sn_2SI_2 transformation was however observed under the following conditions: a mixture of α-Sn_2SI_2 and SnI_2 is introduced into a long Pyrex tube sealed under vacuum and brought to 296°C–297°C. After 24 h SnI_2 has distilled off in the cold part of the tube and α-Sn_2SI_2 is transformed into β-Sn_2SI_2.

According to the data of Novoselova et al. (1971b), instead of Sn_2SI_2 two another compounds, $\mathbf{Sn_3SI_4}$ and $\mathbf{Sn_7S_3I_8}$, are formed in the $SnS–SnI_2$ system. First of them melts incongruently at 330°C and has the polymorphic transformation at 266°C, and the second one also melts incongruently at 410°C.

Fenner (1978) noted that another ternary compound, $\mathbf{Sn_4SI_6}$, exists in this quasibinary system. It crystallizes in the monoclinic structure with the lattice parameters a = 1412.9 ± 0.5, b = 442.5 ± 0.2, c = 2514.8 ± 1.4 pm, β = 93.42°, and the calculated and experimental densities of 5.366 and 5.31 g·cm^{-3}, respectively. This compound was obtained by annealing a mixture of SnS and SnI_2 (molar ratio 1:3) at ca. 280°C for several days. The crystals grow as red needles along the crystallographic b-axis and are often twinned.

$SnS_2–SnI_2$: The phase diagram of this system, constructed through DTA and XRD, is a eutectic type (Figure 7.45) (Thévet et al. 1976). The eutectic contains 5 mol% SnS_2 and crystallizes at 305°C.

$SnS_2–SnI_4$: The phase diagram of this system, constructed through DTA and XRD, is a eutectic type with immiscibility region (Figure 7.46) (Tananaeva et al. 1977b). The eutectic contains 25 mol% SnS_2 and crystallizes at 110°C. In the range of 40–60 mol% SnS_2, immiscibility region exists in the system at a monotectic temperature of 160°C. The thermograms of samples containing 85–95 mol% SnS_2 showed a reversible endothermic effect at 380°C–400°C, which is due to the structural transformations. The ingots for investigations were annealed at 100°C for 3 months. To determine the region of the solid solutions based on SnS_2 the alloys containing 1–5 mol% SnI_2 were annealed at 350°C, 500°C, and 700°C for 1000 h at each temperature.

Another ternary compound, $\mathbf{SnI_4{\cdot}(S_8)_2}$, is formed in the Sn–I–S ternary system. It melts at 101°C (Hawes 1962; Laitinen et al. 1980) and crystallizes in the orthorhombic structure with the lattice parameters a = 2089.17 ± 0.12, b = 2183.37 ± 0.13, c = 1139.71 ± 0.06 pm, a calculated density of 2.911 g·cm^{-3}, and an energy gap of 2.17 eV (Guo et al. 2018a, b) [a = 2083,

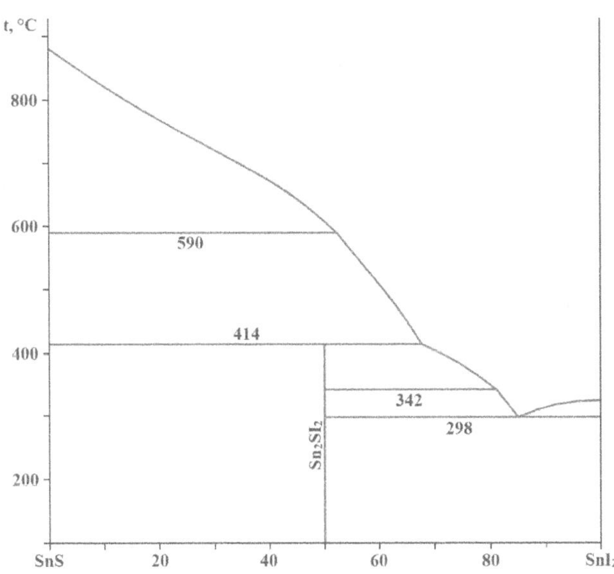

FIGURE 7.44 Phase diagram of the $SnS–SnI_2$ system. (From Thévet, F., et al., *J. Solid State Chem.*, **18**(2), 175, 1976.)

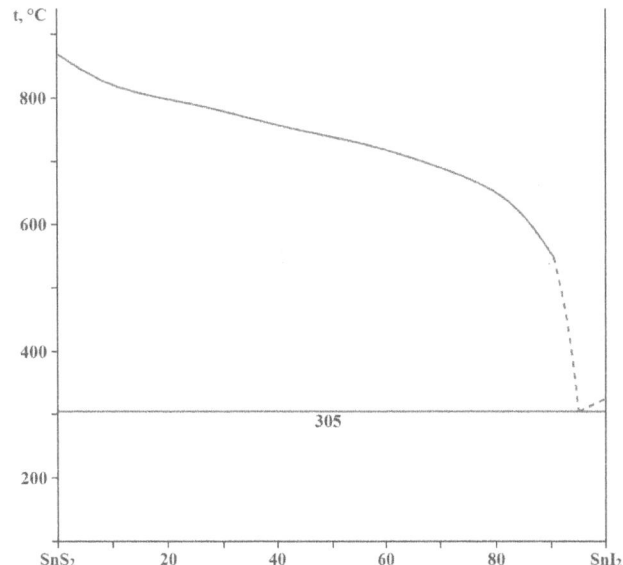

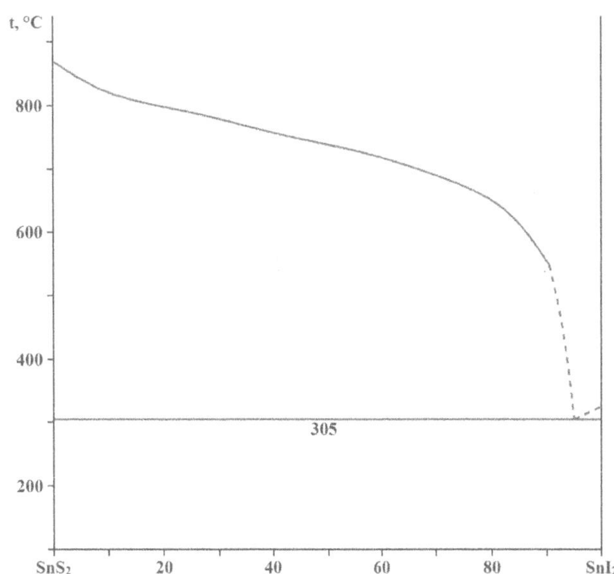

FIGURE 7.45 Phase diagram of the $SnS_2–SnI_2$ system. (From Thévet, F., et al., *J. Solid State Chem.*, **18**(2), 175, 1976.)

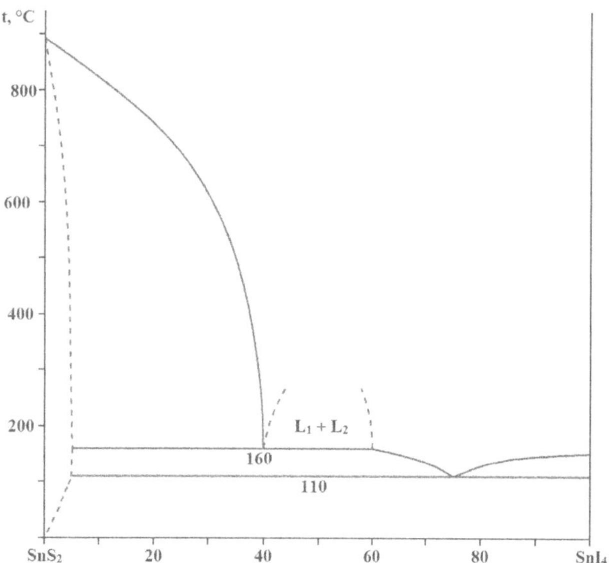

FIGURE 7.46 Phase diagram of the SnS_2–SnI_4 system. (From Tananaeva, O.I., et al., *Izv. AN SSSR. Neorgan. mater.*, **13**(3), 528, 1977.)

b = 2176, c = 1141 pm, and the calculated and experimental densities of 2.93 and 2.80 g·cm^{-3}, respectively (Hawes 1962); a = 2088.5 ± 0.6, b = 2180.5 ± 0.5, c = 1139.8 ± 0.3 pm, and a calculated density of 2.916 g·cm^{-3} (Laitinen et al. 1980)].

To prepare this compound, the mixture of SnI_2, I_2 and S (Sn/I/S molar ratios 1:5:16) was ground into fine powder in an agate mortar and pressed into a pellet, followed by being loaded into a quartz tube, which was evacuated to 0.13 Pa and flame-sealed (Guo et al. 2018a, b). The sample was placed in a furnace, heated from room temperature to 950°C (30°C·h^{-1}), and several intermediate equilibrated temperatures were set to homogenize the reactive system and prevent quartz tube's broken. The heat preservation at 950°C was maintained for 5 days, finally cooled down to 300°C (5°C·h^{-1}) and powered off. The block orange crystals of the title compound were obtained.

$SnI_4.(S_8)_2$ could be also obtained by dissolving sulfur and SnI_4 (molar ratio 2:1) in toluene at room temperature and crystallizing the product at −20°C (Laitinen et al. 1980). Sulfur and SnI_4 also combine readily in CS_2 or benzene solution forming a stable compound $SnI_4{\cdot}(S_8)_2$. A metastable monoclinic modification of this compound with an experimental density of 2.80 g·cm^{-3} was also observed during rapid cooling of benzene solutions. The title compound hydrolysis slowly in water and slowly decomposes on exposure to air.

When a mixture of SnS and I_2 is heated in a 1:1 molar ratio to 50°C in a vacuum, SnI_4 and partly SnS_2 are formed (Batsanov and Shestakova 1968). The formation of SnS_2 occurs, apparently, through intermediate compound **$SnSI_2$**.

7.51 Tin–Manganese–Sulfur

SnS_2–MnS: The **Mn_2SnS_4** ternary compound, which crystallizes in the orthorhombic structure with the lattice parameters a = 740.7 ± 0.1, b = 1047.5 ± 0.1, c = 366.7 ± 0.2 pm (Partik et al. 1995) [a = 739.7 ± 0.4, b = 1047.7 ± 0.7, c = 366.4 ± 0.3 pm (Jumas et al. 1978; Wintenberger and Jumas 1980); in the tetragonal structure with the lattice parameters a = 740.8 ± 0.9, c = 1041 ± 1 pm, and a calculated density of 4.15 g·cm^{-3} (Jumas et al. 1977b)], is formed in this system. Polycrystalline samples of this compound were prepared by heating MnS and SnS_2 (molar ratio 1:1) in fused silica ampoule at 700°C–800°C (two anneals for 4 weeks) and single crystals were grown by recrystallization of a polycrystalline products at 800°C for 4 weeks or through the chemical transport reactions (Jumas et al. 1977b, 1978; Partik et al. 1995). Mn_2SnS_4 is a semiconductor (Jumas et al. 1978) and undergoes antiferromagnetic ordering below 160 K (Partik et al. 1995).

Density-functional theory calculations were reported by Pielnhofer et al. (2014b) on a hypothetical **$Mn_3Sn_2S_2$** compound from scalar relativistic spin-polarized density-functional theory calculations including predictions on crystal and electronic structures. It was shown that this compound could crystallizes in the rhombohedral structure with the lattice parameters a = 550 pm and α = 58.8°.

7.52 Tin–Iron–Sulfur

SnS–FeS: The phase diagram of this system, constructed through DTA, XRA, metallography, and measuring of the microhardness, is a eutectic type (Figure 7.47) (Murach 1947; Sokolova 1949; Eric and Ergeneci 1992; Menshchikova et al. 1992; Raghavan 1998b). The eutectic contains 76.7 mol% SnS and crystallizes at 815°C (Eric and Ergeneci 1992) [77.8 mol% SnS and 785°C (Sokolova 1949); 75 mol% SnS and 790°C (Menshchikova et al. 1992)]. The ternary compound **Fe_4SnS_5** or **SnS·4FeS**, which according to the data of Pushkareva (1966) melts incongruently at 795°C, was not found in the system (Menshchikova et al. 1992). The ingots for the investigations were annealed at 650°C for 40 days and 750°C for 35 days.

The SnS activities in the SnS–FeS liquid alloys were obtained at 1000°C, 1100°C, and 1200°C by the dew-point method (Eric and Ergeneci 1992). Positive deviations were observed and the liquid solutions were modeled by the sub-regular treatment.

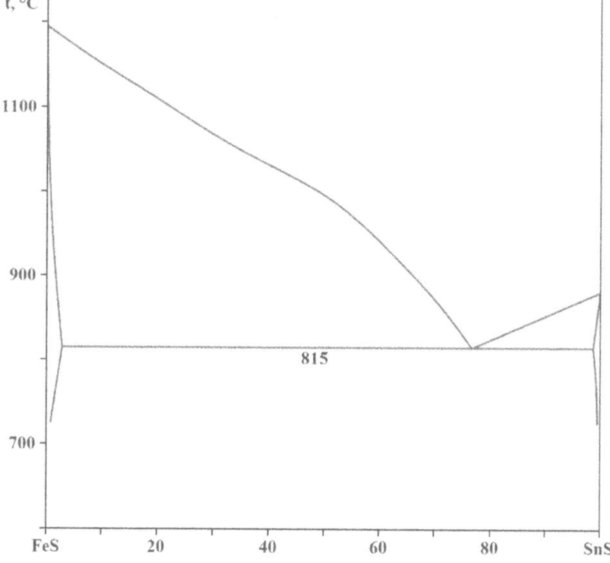

FIGURE 7.47 Phase diagram of the SnS–FeS system. (From Eric, R.H., and Ergeneci A., *Miner. Eng.*, **5**(8), 917, 1992.)

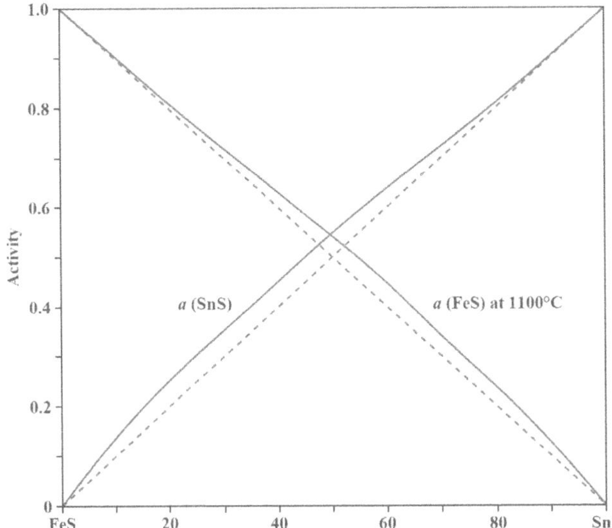

FIGURE 7.48 Activities of SnS and FeS in the SnS–FeS at 1100°C. (From Koike, K., *Shigen-to-Sozai*, **109**(1), 23, 1993.)

The vapor pressure of SnS of the SnS–FeS system was measured by means of the transportation method within a temperature range from 900°C to 1100°C, and activities of SnS and FeS was calculated (Koike 1993). It can be seen that both activities deviated slightly positively from Raoult's law (Figure 7.48). This system could be regarded as a regular solution, and the following equation relating to Raoultian activity coefficient of SnS with composition and temperature has been obtained: $\ln\gamma_{SnS} = 438(1-x_{SnS})^2 \cdot T^{-1}$. The activities coefficients at the infinite dilution of the components at 1100°C were estimated as $\gamma^0(SnS) = \gamma^0(FeS) = 1.38$.

SnS_2–FeS: The phase diagram of this system, constructed through DTA, XRA, metallography, and measuring of the microhardness, is a eutectic type (Figure 7.49) (Menshchikova et al. 1993; Raghavan 1998b). The eutectic contains 52 mol%

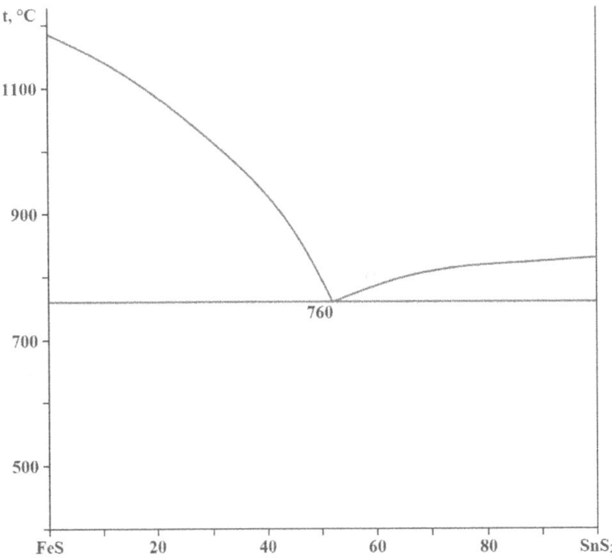

FIGURE 7.49 Phase diagram of the SnS_2–FeS system. (From Menshchikova, T.K., et al., *Neorgan. mater.*, **29**(1), 33, 1993.)

SnS_2 and crystallizes at 760°C. The ternary compounds were not found in this system, but according to the data of Jumas et al. (1977a, b), the **Fe_2SnS_4** ternary compound, which crystallizes in the tetragonal structure with the lattice parameters a = 730.8 ± 0.5 and c = 1033.8 ± 0.9 pm and a calculated density of 4.32 g·cm^{-3}, exists in the system. Polycrystalline samples of this compound were prepared by heating FeS and SnS_2 (molar ratio 1:1) in fused silica ampoule at 750°C–800°C for 1 week and single crystals were grown by the chemical transport reactions.

Density-functional theory calculations were reported by Pielnhofer et al. (2014b) on a hypothetical **$Fe_3Sn_2S_2$** compound from scalar relativistic spin-polarized density-functional theory calculations including predictions on crystal and electronic structures. It was shown that this compound could crystallizes in the rhombohedral structure with the lattice parameters a = 546 pm and α = 58.7°.

The isothermal section of the SnS–FeS–Fe–Sn subsystem of the Sn–Fe–S ternary system at 1200°C is presented in Figure 7.50 (Sokolova 1949; Eric and Ergeneci 1992; Raghavan 1998b). The immiscibility region together with the tie line distributions were established by quenching experiments. Activities of Fe, Sn, and S along the miscibility gap were calculated by utilizing the bounding binary thermodynamics, phase equilibria and tie lines.

The isothermal sections of the Sn–Fe–S ternary system at 450°C and 600°C were constructed by Moh (1974). At 600°C, SnS_2 and FeS_2 are coexisting phases and no measurable solid solution in either could be obtained. The $Fe_{1-x}S$ phase is stable with all other phases in the system, with the exception of S. SnS_2 is only stable with the most S-rich $Fe_{1-x}S$, and at the other extreme, almost stoichiometric FeS coexists with SnS. Sn_2S_3 is in equilibrium with any composition of $Fe_{1-x}S$. A little more than 0.5 mass% Sn_2S_3 can be dissolved in the $Fe_{1-x}S$ solid solution series near its median composition, whereas Sn_2S_3 contains 3 mass%, or a little more, of $Fe_{1-x}S$, but not FeS, in solid solution. SnS was found to take 2 mass% FeS in solid solution. The solubility of S, and of Fe, in liquid Sn is very low, but is greater for FeS. The composition of the maximum solubility in Sn melts in the SnS–Sn_{liq}–FeS monovariant region was found to lie at ~96 mass% Sn, 2 mass% Fe, and 2 mass% S, whereas in the monovariant field FeS–FeSn–Sn_{liq} it is about 96 mass% Sn, 3 mass% Fe, and 1 mass% S. The solubility of S in either FeSn or α-Fe is too small to be detected.

7.53 Tin–Cobalt–Sulfur

The **$Co_3Sn_2S_2$** ternary compound is formed in the Sn–Co–S system. It crystallizes in the rhombohedral structure with the lattice parameters a = 536.83 ± 0.02 and c = 1317.83 ± 0.05 pm in hexagonal settings (Kassem et al. 2016) [a = 537.8 ± 0.1 pm and α = 59.91 ± 0.03° or a = 537.1 ± 0.1 and c = 1318.3 ± 0.2 pm in hexagonal setting and the calculated and experimental densities of 7.00 and 6.91 g·cm^{-3}, respectively (Zabel et al. 1979b); a = 538, c = 1319 pm in hexagonal settings, and the calculated and experimental densities of 7.23 and 7.65 g·cm^{-3}, respectively (Natarajan et al. 1988); a = 538.0, c = 1318.2 pm in hexagonal settings (Weihrich et al. 2004b); a = 535.739 ± 0.006, 535.853 ± 0.005, 535.891 ± 0.005, 536.107 ± 0.004, and 536.8855 ± 0.003

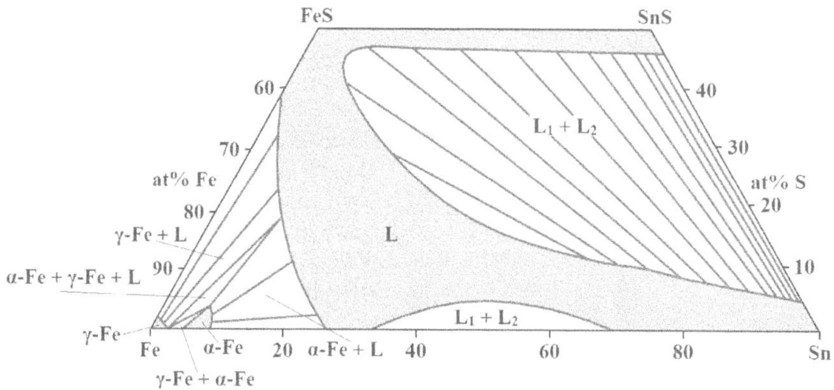

FIGURE 7.50 Isothermal section of the SnS–FeS–Fe–Sn subsystem of the Sn–Fe–S ternary system at 1200°C. (From Eric, R.H., and Ergeneci A., *Miner. Eng.*, **5**(8), 917, 1992.)

and c = 1312.74 ± 0.03, 1313.85 ± 0.03, 1314.10 ± 0.02, 1315.67 ± 0.02, and 1319.03 ± 0.01 at 50, 110, 120, 180, and 300 K, respectively (Vaqueiro and Sobany 2009); a = 536.72 ± 0.03 and c = 1317.65 ± 0.07 pm in hexagonal settings (Schnelle et al. 2013); a = 535.84 ± 0.02, 536.10 ± 0.08, and 536.94 ± 0.06 and c = 1313.55 ± 0.07, 1313.6 ± 0.2, and 1317.5 ± 0.01 at 20, 123, and 298 K, and a calculated density of 7.29 and 7.25 at 123 and 298K, respectively (Pielnhofer et al. 2014a); a = 536.994 ± 0.005 and c = 1319.34 ± 0.03 pm, and a = 538.1 and c = 1316.6 pm in hexagonal settings (the predicted values) (Corps et al. 2015); a = 537.066 and c = 1318.435 pm in hexagonal settings (Kassem et al. 2015); a = 537.4 ± 0.1 pm and α = 59.90 ± 0.03° or a = 536.6 ± 0.1 and c = 1317.3 ± 0.1 pm in hexagonal setting (the experimental values) and a = 537.62 pm and α = 59.97° or a = 537.38 and c = 1317.19 pm in hexagonal setting and a calculated density of 7.27 g·cm^{-3} and an energy gap of 0.37 eV (the predicted values) (Weihrich et al. 2015)]. This compound orders ferromagnetically at 172 K (Schnelle et al. 2013). The easy axis of the magnetic moment was found to be along the crystallographic c direction. A linear positive trend of the melting temperature as a function of pressure is observable in the pressure range of 7–11.5 GPa.

The liquid phase of $Co_3Sn_2S_2$ exists at temperatures higher than 1100°C and 7 GPa and at temperatures higher than 1350°C and 11.5 GPa (Sakai et al. 2015).

$Co_3Sn_2S_2$ was prepared by heating the reaction mixture of the elements in stoichiometric ratio at 500°C for 48 h, followed by a second firing at 700°C for another 48 h (Natarajan et al. 1988; Vaqueiro and Sobany 2009; Corps et al. 2015). The sample was cooled to room temperature at a rate of 0.5°C·min^{-1} after each firing. Single crystals of this compound were grown by a modified Bridgman technique (Schnelle et al. 2013).

According to the data of Bok and Boeyens (1957), the $\mathbf{Co_2SnS_4}$ and $\mathbf{CoSn_2S_4}$ ternary compounds were not found in the Sn–Co–S system but the first of them was prepared later by interaction of SnS_2 and CoS at 750°C–800°C for 1 week (Jumas et al. 1977b).

Density-functional theory calculations were reported by Pielnhofer et al. (2014b) on the $Co_3Sn_2S_2$ compound from scalar relativistic spin-polarized density-functional theory calculations including predictions on crystal and electronic structures. It was shown that this compound could crystallizes in the rhombohedral structure with the lattice parameters a = 536 pm and α = 60.0°.

7.54 Tin–Nickel–Sulfur

Several ternary compounds are formed in the Sn–Ni–S system. $\mathbf{Ni_3Sn_2S_2}$ crystallizes in the rhombohedral structure with the lattice parameters a = 546.06 ± 0.02 and c = 1318.8 ± 0.1 pm in hexagonal settings (Range et al. 1997b) [a = 540.91 ± 0.06 pm and α = 60.61 ± 0.02° or a = 545.89 ± 0.06 and c = 1318.8 ± 0.1 pm in hexagonal setting (Zabel et al. 1979b); a = 546.5 ± 0.1 and c = 1319.6 ± 0.1 pm in hexagonal setting and a calculated density of 6.97 g·cm^{-3} (Brower et al. 1974); in the monoclinic structure with the lattice parameters a = 933.1 ± 0.3, b = 540.3 ± 0.2, c = 545.8 ± 0.2 pm, and β = 124.55 ± 0.2° (Michelet and Collin 1976)].

This compound was prepared by heating the mixture of the elements in evacuated silica tube (Brower et al. 1974). Heat treatment varied from 500°C to 1100°C. In general, an attempt was made to obtain complete melting of the mixture, which was then heat treated in a rocking furnace in order to insure chemical homogeneity.

$\mathbf{Ni_6SnS_2}$ (mineral butianite) crystallizes in the tetragonal structure with the lattice parameters a = 365.00 ± 0.05, c = 1814.1 ± 0.2 pm, and a calculated density of 7.353 g·cm^{-3} (Baranov et al. 2003; Ma 2016). $\mathbf{Ni_9Sn_2S_2}$ also crystallizes in the tetragonal structure with the lattice parameters a = 367.10 ± 0.05, c = 2547.4 ± 0.2 pm, and a calculated density of 7.992 g·cm^{-3} at 173 K (Baranov et al. 2003). These two compounds could be prepared by the next way (Baranov et al. 2003); the stoichiometric mixtures (Ni/Sn/S = 6:1:2 or 9:2:2) of the elements were annealed in the evacuated silica ampoules at 540°C for 8 days. The products were then ground in an agate mortar, pressed into pellets, and further re-annealed under the same conditions several times (two for Ni_6SnS_2 and six for $Ni_9Sn_2S_2$). Single crystals were prepared from the vapor phase by means of the chemical transport reactions with I_2 as carrier reagent (~4 mM·L^{-1}) in the silica ampoules placed in a horizontal, two-temperature furnace with the temperature range 600°C (charge end) to 570°C (empty end). A 100 mg portion of annealed sample of the molar composition Ni/Sn/S = 35:3:12 containing Ni_6SnS_2, $Ni_9Sn_2S_2$, and a

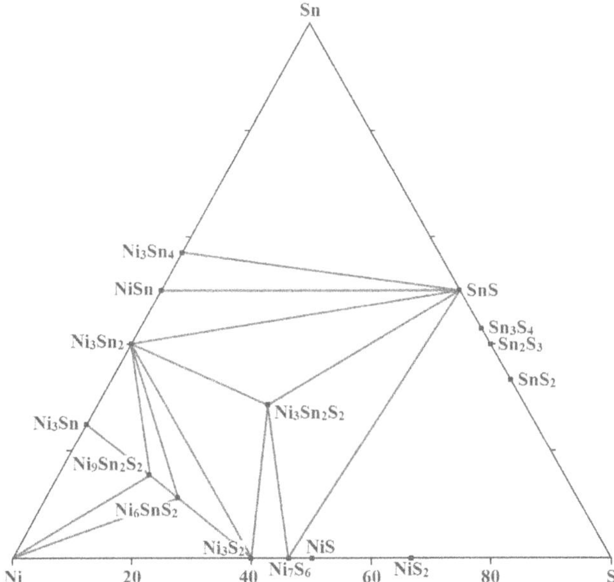

FIGURE 7.51 Isothermal section of the Sn–Ni–S ternary system at 540°C. (From Baranov, A.I., et al., *Inorg. Chem.*, **42**(21), 6667, 2003.)

small amount of Ni were used as a charge. After heating for 2 weeks, the formation of square and octagonal plate silvery was observed in the cold part of the ampoule.

According to the data of Bok and Boeyens (1957), the $\mathbf{Ni_2SnS_4}$ and $\mathbf{NiSn_2S_4}$ ternary compounds were not found in the Sn–Ni–S system but the first of them was prepared later by interaction of SnS_2 and NiS at 750°C–800°C for 1 week (Jumas et al. 1977b).

The isothermal section of the Sn–Ni–S ternary system at 540°C is presented in Figure 7.51 (Baranov et al. 2003).

7.55 Tin–Rhodium–Sulfur

The $\mathbf{Rh_3Sn_2S_2}$ ternary compound, which crystallizes in the trigonal structure with the lattice parameters $a = 562.68 \pm 0.03$, $c = 1330.67 \pm 0.07$, and a calculated density of 8.243 g·cm^{-3}, is formed in the Sn–Rh–S system (Zabel et al. 1979a; Anusca et al. 2009). It was prepared from the elements by high-temperature solid-state synthesis.

7.56 Tin–Palladium–Sulfur

The $\mathbf{Pd_3Sn_2S_2}$ ternary compound, which crystallizes in the trigonal structure, is formed in the Sn–Pd–S system (Zabel et al. 1979a, b). It was prepared from the elements as well as from SnS and Pd by high-temperature solid-state synthesis.

7.57 Tin–Iridium–Sulfur

The $\mathbf{IrSn_{1.5}S_{1.5}}$ ternary compound, which crystallizes in the cubic structure with the lattice parameter $a = 870.59$ pm, is formed in the Sn–Ir–S system (Lyons et al. 1978). It could be prepared by direct combination of the elements at 800°C under conditions of ambient pressure, or by heating the partially reacted elements at 600°C and 4GPa.

7.58 Tin–Platinum–Sulfur

The **PtSnS** ternary compound is formed in the Sn–Pt–S system (Barkov et al. 2001; Jambor and Roberts 2002). This compound crystallizes in the orthorhombic structure with the lattice parameters $a = 611.511 \pm 0.010$, $b = 612.383 \pm 0.010$, $c = 609.667 \pm 0.011$, and a calculated density of 10.06 g·cm^{-3} (Vymazalová et al. 2020b). According to the calculations, the title compound crystallizes in the trigonal structure with the lattice parameters $a = c = 607.95$ pm or in the orthorhombic structure with the lattice parameters $a = 609.16$, $b = 608.47$ ($b = 611.77 \pm 0.05$ from XRD), and $c = 606.29$ pm (Weihrich et al. 2004a). Mineral bowlesite, which have the composition PtSnS, crystallizes in the orthorhombic structure with the lattice parameters $a = b = 612$ and $c = 610$ pm (Vymazalová et al. 2019, 2020a).

The title compound was prepared in an evacuated silica glass tube using Pt, Sn, and S as starting materials (Vymazalová et al. 2020b). The evacuated tube with its charge was sealed, heated at 300°C for 1 month and afterwards annealed at 1000°C for 2 days. After cooling in a cold-water bath, the charge was ground to powder in acetone using an agate mortar, and thoroughly mixed to homogenize. The pulverized charge was sealed in an evacuated silica-glass tube again, and reheated at 600°C for 7 months. The experimental product was rapidly quenched in cold water.

It could be also synthesized from the elements by high-temperature solid state reactions (Weihrich et al. 2004a). Equimolar amounts of Pt, Sn, and S were heated in sealed and evacuated quartz tubes to 1000°C. The tubes were quenched after 1 week. PtSnS was received as dark polycrystallites. Its structure optimizations were performed with a first principles plane-wave code, which treats exchange and correlation in the local density approximation of the density functional theory.

From band structure calculations, PtSnS was predicted to be indirect band gap semiconductor with $E_g = 0.5$ eV (Weihrich et al. 2004a).

REFERENCES

All References are available as a downloadable eResource at www.routledge.com/9780367639235

8

Systems Based on Tin Selenides

8.1 Tin–Lithium–Selenium

Two ternary compounds, $^1_\infty\{Li_2[SnSe_3]\}$ and $\mathbf{Li_4SnSe_4}$, are formed in the Sn–Li–Se system (Kaib et al. 2013). First of them crystallizes in the monoclinic structure with the lattice parameters a = 1252.2 ± 0.3, b = 721.37 ± 0.14, c = 776.92 ± 0.16 pm, β = 120.96 ± 0.03°, and a calculated density of 4.078 g·cm^{-3} and the second crystallizes in the orthorhombic structure with the lattice parameters a = 1493.1 ± 0.3, b = 822.17 ± 0.16, c = 659.89 ± 0.13 pm, and a calculated density of 3.791 g·cm^{-3}. To prepare the yellow needles of these compounds, a stoichiometric mixture of Li_2Se (5.4 mM or 8.6 mM), Sn (5.4 mM or 4.3 mM), and Se (10.8 mM or 8.6 mM) was fused by using a gas burner in a silica ampoule and then heated for 30 min. For single-crystalline products, the reaction was performed in a furnace with a definite temperature program: (1) heating to 650°C with a heating rate of 70°C·h^{-1}, (2) keeping for 12 h at this temperature, (3) cooling to room temperature with a rate of 4°C·h^{-1}. All reaction steps were performed with strong exclusion of air and external moisture.

8.2 Tin–Sodium–Selenium

Three ternary compounds are formed in the Sn–Na–Se system. $\mathbf{Na_2SnSe_3}$ has two polymorphic modifications. First of them crystallizes in the orthorhombic structure with the lattice parameters a = 1397.5 ± 0.4, b = 2368.0 ± 0.6, c = 1251.9 ± 0.3 pm, and a calculated density of 3.86 g·cm^{-3} (Klepp 1995) [a = 1397.6 ± 0.4, b = 2372.5 ± 0.5, and c = 1251.0 ± 0.5 pm (Eisenmann and Hansa 1993b)]. The second modifications crystallizes in the monoclinic structure with the lattice parameters a = 724.4 ± 0.3, b = 1621.8 ± 0.4, c = 617.4 ± 0.3 pm, β = 112.6 ± 0.1°, and an experimental density of 3.920 g·cm^{-3} (Eisenmann and Hansa 1993a). For the synthesis of the orthorhombic modification, stoichiometric amounts of Na_2Se, Sn, and Se or Na, Sn, and Se were intimately mixed in an Ar-filled dry box, pelleted and sealed into an evacuated silica ampoule (Eisenmann and Hansa 1993b; Klepp 1995). The sample was allowed to react at 850°C–900°C for 4 days, and the thermal treatment was completed by a controlled cooling (2°C·h^{-1}) to ambient temperature. Dark red or orange-yellow crystals of prismatic shape were isolated from the crushed melt. They are sensitive to moisture and air; therefore all manipulations were performed under an Ar atmosphere. The monoclinic modification was obtained from stoichiometric amounts of the elements in evacuated graphitized silica ampoule at 800°C (Eisenmann and Hansa 1993a).

$\mathbf{Na_4SnSe_4}$ crystallizes in the tetragonal structure with the lattice parameters a = 816.9 and c = 726.5 pm (Palchik et al. 2002) [a = 816.5 ± 0.2, c = 726.8 ± 0.2 pm, and a calculated density of 3.61 g·cm^{-3} (Klepp 1992b)]. This Zintl phase was synthesized using the high-temperature method. Stoichiometric amounts of thoroughly mixed Na, Sn, and Se were heated in an evacuated quartz tube at 750°C for 24 h (Palchik et al. 2002). The heating rate was 30°C·h^{-1} and cooling was natural (total reaction time: 50 h). It could be also prepared through microwave-assisted solid state reaction method. All manipulations were conducted in an Ar-filled glove box with less than 1 ppm of O_2 and H_2O. Transparent yellow crystals of this compound were also prepared by reacting stoichiometric powdered mixture of Na_2Se, Sn, and Se at 680°C for 3 days with the next slow cooling (2°C·h^{-1}) to room temperature (Klepp 1992b).

$\mathbf{Na_6Sn_2Se_7}$ crystallizes in the monoclinic structure with the lattice parameters a = 971.0 ± 0.5, b = 1119.3 ± 0.5, c = 1606.4 ± 0.6 pm, β = 105.1 ± 0.1°, and an experimental density of 3.710 g·cm^{-3} (Eisenmann and Hansa 1993e). Yellow prisms of this compound were synthesized from stoichiometric amounts of the elements in evacuated graphitized silica ampoule at 700°C.

8.3 Tin–Potassium–Selenium

Several ternary compounds are formed in the Sn–K–Se system. $\mathbf{K_2Sn_2Se_4}$ crystallizes in the tetragonal structure with the lattice parameters a = 815.24 ± 0.12, c = 671.52 ± 0.13 pm, and an energy gap of 1.7 eV (Zhang et al. 2005a). This compound was synthesized by melting the mixture of K_2Se, Sn, and Se (molar ratio 1:1:3) in an evacuated quartz ampoule at 550° for 7 days. Then the sample was cooled to 200°C at a rate of 4°C·h^{-1} with the next cooling to room temperature by switching of the furnace. The obtain product was washed with Dimethylformamide (DMF).

$\mathbf{K_2Sn_2Se_5}$ crystallizes in the in the monoclinic structure with the lattice parameters a = 1161.3 ± 0.5, b = 818.9 ± 0.3, c = 1189.7 ± 0.6 pm, β = 108.28 ± 0.02°, and a calculated density of 6.55 g·cm^{-3} (Klepp 1992a). It was prepared by reacting stoichiometric mixture of K_2Se with Sn and Se at 800°C, followed by slow cooling of the melt.

$\mathbf{K_2Sn_4Se_8}$ also crystallizes in the monoclinic structure with the lattice parameters a = 643.6 ± 0.3, b = 1893.4 ± 0.9, c = 741.7 ± 0.5 pm, β = 96.26 ± 0.02°, and a calculated density of 4.38 g·cm^{-3} (Klepp and Fabian 1992). To prepare transparent, ruby red crystals of the title compound, the mixture of K_2Se, Sn, and Se taken in stoichiometric amounts was heated in evacuated quartz ampoule up to 800°C and then cooled to room temperature at a rate of 2°C·h^{-1}.

DOI: 10.1201/9781003123507-8

K_4SnSe_4 crystallizes in the orthorhombic structure with the lattice parameters $a = 1471.4 \pm 0.7$, $b = 1032.3 \pm 0.3$, $c = 824.8 \pm 0.6$ pm, and a calculated density of 3.13 g·cm^{-3} (Klepp 1992b). Transparent yellow crystals of this compound were prepared by reacting stoichiometric powdered mixture of K_2Se, Sn, and Se at 680°C for 3 days with the next slow cooling (2°C·h^{-1}) to room temperature.

$K_4Sn_2Se_6$ crystallizes in the triclinic structure with the lattice parameters $a = 862.4 \pm 0.5$, $b = 760.6 \pm 0.5$, $c = 695.2 \pm 0.5$ pm, $\alpha = 67.6 \pm 0.1°$, $\beta = 74.3 \pm 0.1°$, and $\gamma = 75.7 \pm 0.1°$ and a calculated density of 3.530 g·cm^{-3} (Eisenmann and Hansa 1993h) [$a = 862.5$, $b = 760.9$, $c = 695.1$ pm, $\alpha = 67.6°$, $\beta = 74.5°$, and $\gamma = 75.4°$ (Palchik et al. 2002)]. Yellow prisms of this compound were synthesized from stoichiometric amounts of the elements in evacuated graphitized quartz ampoule at 700°C (Eisenmann and Hansa 1993h). It could be also prepared through microwave-assisted solid state reaction method (Palchik et al. 2002). All manipulations were conducted in an Ar-filled glove box with less than 1 ppm of O_2 and H_2O.

$K_4Sn_3Se_8$ crystallizes in the orthorhombic structure with the lattice parameters $a = 2776.6 \pm 0.3$, $b = 819.1 \pm 0.1$, $c = 818.7 \pm 0.1$ pm, and the calculated and experimental densities of 4.08 and 4.06 g·cm^{-3}, respectively (Sheldrick 1988) [$a = 2777.1$, $b = 819.2$, $c = 818.5$ pm (Palchik et al. 2002)]. The title compound has been prepared by methanolothermal reaction of K_2CO_3 with elemental Sn and Se at 190°C (Sheldrick 1988). K_2CO_3 (1.82 mM), Sn (5.47 mM), and Se (10.94 mM) were suspended in MeOH (5 mL), washed and melted under Ar in a glass ampoule. The ampoule was first heated with an inclination of 5° to the horizontal plane at 100°C for 5 days and then at 190°C for 2 days. Degree of the ampoule filling was 10%. The ampoule was then cooled to room temperature at a rate of 5°C·h^{-1}. Red-brown prisms of $K_4Sn_3Se_8$ were formed on the upper boundary surface. Since this compound is immediately decomposed by oxygen with the separation of Se, the crystals were treated under Ar. It could be also prepared through microwave-assisted solid state reaction method (Palchik et al. 2002). All manipulations were conducted in an Ar-filled glove box with less than 1 ppm of O_2 and H_2O.

$K_6Sn_2Se_6$ crystallizes in the monoclinic structure with the lattice parameters $a = 905.6 \pm 0.2$, $b = 1277.4 \pm 0.3$, $c = 888.7 \pm 0.2$ pm, $\beta = 115.86 \pm 0.03°$, and a calculated density of 3.395 g·cm^{-3} at 203 K (Zimmermann and Dehnen 1999). To prepare this compound, K (2.54 mM), Sn (0.77 mM), and Se (2.31 mM) were weighed in an Ar protective atmosphere (glove box) in a quartz ampoule. The ampoule was then evacuated. For the solid-state reaction, the sample was heated to 900°C over a period of 25 h, held at this temperature for 20 h, and cooled to room temperature within 125 h. In addition to a small amount of Sn, a red-orange regulus of $K_6Sn_2Se_6$ was obtained, which formed and ocher-colored powder during grinding.

$K_6Sn_2Se_7$ also crystallizes in the monoclinic structure with the lattice parameters $a = 1705.2 \pm 0.6$, $b = 1232.8 \pm 0.6$, $c = 1970.9 \pm 0.6$ pm, and $\beta = 102.6 \pm 0.1°$ (Eisenmann and Hansa 1993c). Yellow, irregular broken crystals of this compound were synthesis from stoichiometric amounts of the elements in evacuated graphitized quartz ampoule at 700°C.

8.4 Tin–Rubidium–Selenium

Two ternary compounds, **$Rb_2Sn_2Se_5$** and **$Rb_4Sn_2Se_6$**, are formed in the Sn–Rb–Se system. First of them crystallizes in the monoclinic structure with the lattice parameters $a = 1155.5 \pm 0.4$, $b = 847.8 \pm 0.3$, $c = 1203.0 \pm 0.4$ pm, $\beta = 108.60 \pm 0.03°$, and the calculated and experimental densities of 4.78 and 4.75 g·cm^{-3}, respectively (Sheldrick and Braunbeck 1992). To prepare the title compound, Rb_2CO_3, $SnSe_2$, and BaSe (molar ratio 1:2:1) were slurried in MeOH in Duranglass ampoule (degree of filling 12.5%) under Ar atmosphere. The ampoule was heated to 160°C at 1.3 MPa and kept at this temperature for 3 days. It was then cooled to 90°C and annealed for 12 h. At the upper phase boundary, black cuboidal crystals of $Rb_2Sn_2Se_5$ were formed. This compound decomposes by atmospheric oxygen with deposition of Se; the crystals were therefore handled under Ar atmosphere.

$Rb_4Sn_2Se_6$ crystallizes in the triclinic structure with the lattice parameters $a = 722.1 \pm 0.2$, $b = 790.0 \pm 0.2$, $c = 898.3 \pm 0.2$ pm, $\alpha = 74.03 \pm 0.03°$, $\beta = 70.87 \pm 0.03°$, and $\gamma = 66.10 \pm 0.02°$ and a calculated density of 4.01 g·cm^{-3} (Sheldrick and Schaaf 1994c). This compound was obtained as follows. $SnCl_2$ (1.88 mM), Se (1.87 mM), and Rb_2CO_3 (7.49 mM) were slurried in MeOH (0.7 mL) and placed under Ar atmosphere in a Duranglass ampoule (degree of filling 8%). The ampoule was heated to 145°C for 110 h. Subsequently, it was cooled (2°C·h^{-1}) to 100°C and tempered for 24 h. Yellow crystals of $Rb_4Sn_2Se_6$ were obtained on the upper phase boundary.

8.5 Tin–Cesium–Selenium

Four ternary compounds are formed in the Sn–Cs–Se system. **$Cs_2Sn_2Se_5$** crystallizes in the monoclinic structure with the lattice parameters $a = 771.7 \pm 0.2$, $b = 1202.1 \pm 0.2$, $c = 2736.6 \pm 0.6$ pm, and $\beta = 96.23 \pm 0.03°$ (Loose and Sheldrick 1999). To synthesize this compound, Cs_2CO_3 (0.92 mM), Sn (1.32 mM), and Se (2.40 mM) were heated in a flame sealed glass tube to 130°C in a (0.5 mL H_2O + 1.0 mL MeOH) mixture. After 4 days, the contents were cooled to room temperature (1°C·h^{-1}) to afford light-red needles of **$Cs_2Sn_2Se_5·H_2O$** and darker crystals of $Cs_2Sn_2Se_5$. Crystalline product was separated manually and washed.

$Cs_2Sn_2Se_6$ or **$Cs_2Sn_2Se_4(Se_2)$** crystallizes in the triclinic structure with the lattice parameters $a = 712.9 \pm 0.1$, $b = 728.5 \pm 0.1$, $c = 796.5 \pm 0.2$ pm, $\alpha = 95.80 \pm 0.03°$, $\beta = 115.38 \pm 0.03°$, and $\gamma = 108.46 \pm 0.03°$ and a calculated density of 4.762 g·cm^{-3} (Loose and Sheldrick 1998) [$a = 728.4 \pm 0.6$, $b = 795 \pm 1$, $c = 713.5 \pm 0.9$ pm, $\alpha = 115.41 \pm 0.07°$, $\beta = 108.58 \pm 0.06°$, and $\gamma = 95.78 \pm 0.07°$ (Klepp and Fabian 1998)]. To prepare the title compound, Cs_2CO_3 (0.31 mM), SnSe (0.81 mM), and Se (0.94 mM) were slurried in 0.8 mL MeOH and heated in a sealed glass tube to 130°C (Loose and Sheldrick 1998). After tempered for 5 days, the content was allowed to cool to room temperature at 1°C·h^{-1}. Red platelets of $Cs_2Sn_2Se_6$ were obtained. Its single crystals were grown from a polyselenide melt prepared by reacting Cs_2Se, Sn, and Se (molar ratio

1:1:13) during 3 weeks at 350°C (Klepp and Fabian 1998). The melt was finally allowed to cool to ambient temperature at a controlled cooling rate of 2°C·h^{-1}.

$Cs_2Sn_3Se_7$ and **$Cs_4Sn_2Se_6$** crystallize in the monoclinic structure with the lattice parameters a = 2321.4 ± 0.4, b = 1367.9 ± 0.3, c = 1465.7 ± 0.3 pm, β = 111.92 ± 0.02°, and the calculated and experimental densities of 3.60 and 3.57 g·cm^{-3}, respectively, for the first compound (Sheldrick and Braunbeck 1990) and a = 1463.4 ± 0.4, b = 747.2 ± 0.3, c = 1282.1 ± 0.3 pm, β = 135.15 ± 0.02°, and a calculated density of 4.36 g·cm^{-3}, for the second one (Sheldrick and Braunbeck 1989). Methanothermal reaction of Cs_2CO_3 with $SnSe_2$ in the presence of SnSe at 160°C and 1.3 MPa was used to synthesize these compounds. Cs_2CO_3 (0.51 mM), $SnSe_2$ (0.75 mM), and SnSe (0.52 mM) were slurried in MeOH (2 mL) and melted under Ar into a glass ampoule (degree of filling 13.5%). The ampoule was heated to 160°C and held at this temperature for 3 days. It was then cooled to 110°C at a rate of 0.6°C·min^{-1} and heat-treated for 12 h. At the upper phase boundary, red platelet-shaped single crystals of $Cs_2Sn_3Se_7$ and yellow needles of $Cs_4Sn_2Se_6$ were formed, which could be separated from one another under the microscope. The crystals of $Cs_2Sn_3Se_7$ were handled under Ar atmosphere.

8.6 Tin–Copper–Selenium

SnSe–Cu_2Se: The phase diagram of this system, constructed through DTA, XRD, metallography, and measuring of the microhardness, is a eutectic type (Figure 8.1) (Rivet et al. 1970; Berger and Kotina 1973). The eutectic contains 46 mol% [53 mol% (Rivet et al. 1970)] SnSe and crystallizes at 540°C. Thermal effects at 140°C correspond to the polymorphic transformation of Cu_2Se. The mutual solubility of the initial components does not exceed 1 mol%. No ternary compounds were found in this system (Shemet et al. 2005, 2006d).

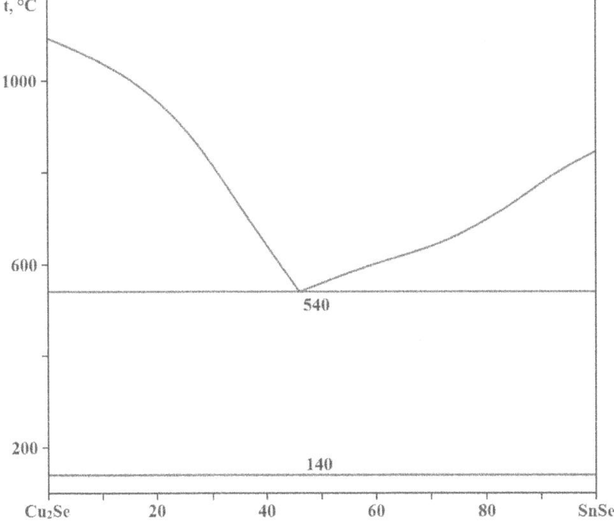

FIGURE 8.1 Phase diagram of the SnSe–Cu_2Se system. (From Berger, L.I., and Kotina E.G., *Izv. AN SSSR. Neorgan. mater.*, **9**(3), 368, 1973.)

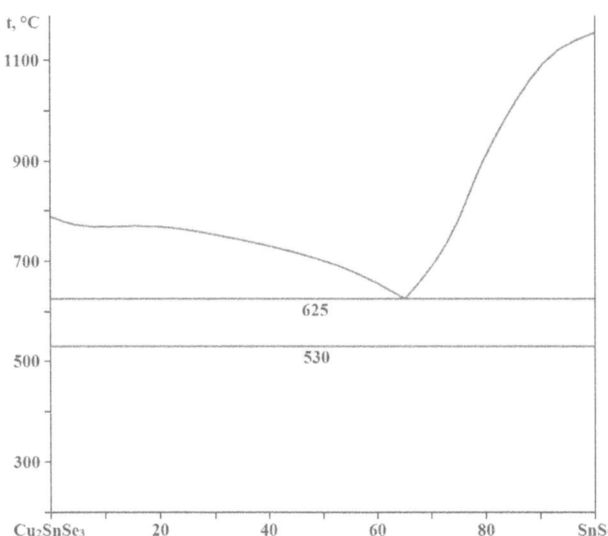

FIGURE 8.2 Phase diagram of the SnSe–Cu_2SnSe_3 system. (From Berger, L.I., and Kotina E.G., *Izv. AN SSSR. Neorgan. mater.*, **9**(3), 368, 1973.)

According to the data of Dovletov et al. (1977), the phase diagram of the SnSe–Cu_2Se system is a eutectic type with the formation of the **$Cu_4Sn_3Se_5$** ternary compound, which melts congruently at 600°C. The eutectics contain 45 and 75 mol% SnSe and crystallize at 550°C and 525°C, respectively. The solubility of Cu_2Se in SnSe at the eutectic temperature reaches 15 mol% and the solubility of SnSe in Cu_2Se is 8 mol% (Dovletov et al. 1974c, 1977).

SnSe–Cu_2SnSe_3: The phase diagram of this system, constructed through DTA, XRD, metallography, and measuring of the microhardness, is a eutectic type (Figure 8.2) (Berger and Kotina 1973). The eutectic crystallizes at 625°C. Thermal effects at 530°C correspond to the polymorphic transformation of SnSe. The mutual solubility of the initial components does not exceed 1 mol%.

$SnSe_2$–Cu_2Se: The phase diagram of this system, constructed through DTA, XRD, metallography, and measuring of the microhardness, is given in Figure 8.3 (Rivet et al. 1970; Berger and Kotina 1973; Zotova and Karagodin 1975b). The eutectics contain 16 and 78 mol% [20 and 76 mol% (Berger and Kotina 1973); 17 and 73 mol% (Rivet et al. 1970)] Cu_2Se and crystallize at 580°C and 665°C, respectively (Berger and Kotina 1973) [604°C and 668°C (Rivet et al. 1970)]. Thermal effects at 140°C correspond to the polymorphic transformation of Cu_2Se. The solid solution region based on $SnSe_2$ is not higher than 3 mol% Cu_2Se, and the solubility of $SnSe_2$ in Cu_2Se is 10 mol% (Zotova and Karagodin 1975b).

The **Cu_2SnSe_3** ternary compound is formed in the $SnSe_2$–Cu_2Se system. According to the earlier investigations, this compound has a homogeneity region not higher than 2 mol% (Berger and Kotina 1973) and melts congruently at 695°C (Berger and Kotina 1971; Berger and Kotina 1973; Zotova and Karagodin 1975b) [709°C (Goryunova et al. 1960a); 696°C (Palatnik et al. 1961a); 685°C (Rivet et al. 1963; Rivet 1965); 697°C (Averkieva et al. 1964; Sharma et al. 1977a); 698°C (Balanevskaya et al. 1966); 694°C (Sharma and

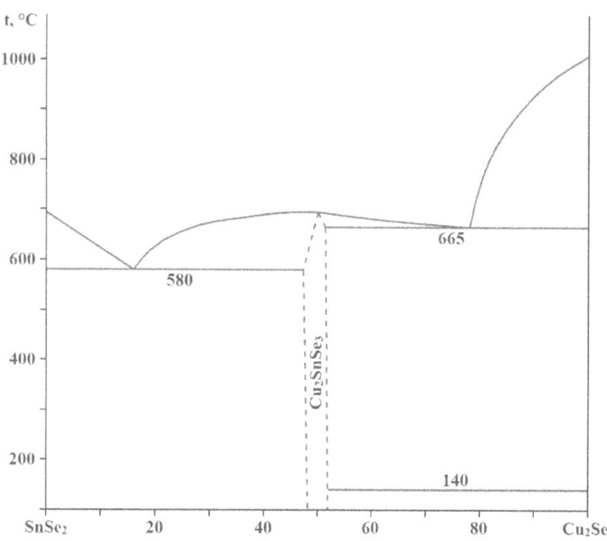

FIGURE 8.3 Phase diagram of the $SnSe_2$–Cu_2Se system. (From Berger, L.I., and Kotina E.G., *Izv. AN SSSR. Neorgan. mater.*, 9(3), 368, 1973.)

Chavada 1972); 691°C (Marcano et al. 2002a; Fan et al. 2013)]. This compound has polymorphic transformations at 545°C and 580°C (Zotova and Karagodin 1975b) [at 670°C–680°C (Berger et al. 1973a); at 657°C (Fan et al. 2013)] and is characterized by the presence of an ordering in the cation sublattice at 620°C (Berger et al. 1973a).

According to the data of Fan et al. (2014), this compound has a homogeneity region and its composition could be expressed as $\mathbf{Cu_{3-x}Sn_xSe_3}$ (x = 0.87–1.05). A compositionally induced evolvement from tetragonal via cubic to monoclinic crystal structures was observed, when the composition changes from a Cu-rich side of the compound to a Sn-rich one. The $Cu_{3-x}Sn_xSe_3$ materials show metal-to-semiconductor transition with increasing x. At the values of x up to 0.95, $Cu_{3-x}Sn_xSe_3$ crystallizes in the cubic structure with the lattice parameter $a = 567.35 \pm 0.03$, 567.58 ± 0.02, and 568.40 ± 0.01 for $x = 0.87$, 0.90, and 0.95, respectively (Fan et al. 2014).

Sharma et al. (1977a) noted, that the homogeneity region of $\mathbf{Cu_2Sn_{1-x}Se_3}$ lies within the interval from $x = 0$ to $x = 0.4$. The lattice parameter of the cubic structure and the melting point are the next: 568.8 pm and 697°C for $x = 0$; 566.5 pm and 676°C for $x = 0.1$; 565.4 pm and 672°C for $x = 0.2$; and 565.4 pm and 647°C for $x = 0.4$.

The lattice parameter of the cubic structure of Cu_2SnSe_3 is 568.77 ± 0.02 pm (Sharma and Chavada 1972) [a = 568 pm (Palatnik et al. 1961b); a = 567.6 pm (Averkieva et al. 1964); a = 570 pm (Goryunova et al. 1960a); a = 569 pm (Palatnik et al. 1961a; Bok and Wit 1963); a = 569.6 pm (Rivet et al. 1963, 1970; Rivet 1965)]. The calculated and experimental densities of cubic Cu_2SnSe_3 are, respectively, 5.79 and 5.94 ± 0.09 g·cm^{-3} (Rivet et al. 1963; Rivet 1965) [5.79 and 5.74 g·cm^{-3} (Palatnik et al. 1961a); experimental density is 5.22 g·cm^{-3} (Bok and Wit 1963)] and an energy gap is 0.6 eV (Balanevskaya et al. 1966).

There three monoclinic modifications of this compound, which are superstructures to the cubic sphalerite-type structure (Fan et al. 2013). First of them has the lattice parameters $a = 659.36 \pm 0.01$, $b = 1215.93 \pm 0.04$, $c = 660.84 \pm 0.03$ pm, $\beta = 108.56 \pm 0.02°$, and an energy gap of 0.843 eV (Marcano et al. 2001, 2002a) [$a = 696.70 \pm 0.03$, $b = 1204.93 \pm 0.07$, $c = 694.53 \pm 0.03$ pm, $\beta = 109.19 \pm 0.01°$, and a calculated density of 5.82 g·cm^{-3} (Delgado et al. 2003); $a = 696.4 \pm 0.6$, $b = 1205.6 \pm 0.6$, $c = 697.2 \pm 0.4$ pm, and $\beta = 109.49 \pm 0.01°$ (Nomura et al. 2013a); $a = 687.3$, $b = 1200.8$, $c = 692.1$ pm, and $\beta = 109.19$ (Morihama et al. 2014). The lattice parameters for the another monoclinic modifications are the next: $a = 697.14 \pm 0.02$, $b = 1207.87 \pm 0.05$, $c = 1339.35 \pm 0.05$ pm, $\beta = 99.865 \pm 0.005°$, and a calculated density of 5.77 g·cm^{-3} for the second modification of Cu_2SnSe_3 (Fan et al. 2013) [$a = 696.89 \pm 0.03$, $b = 1207.9 \pm 0.1$, $c = 1339.1 \pm 0.1$ pm, $\beta = 99.824 \pm 0.004°$ for $Cu_{3-x}Sn_xSe_3$ at $x = 0.975$; $a = 696.89 \pm 0.02$, $b = 1207.8 \pm 0.1$, $c = 1339.22 \pm 0.04$ pm, $\beta = 99.833 \pm 0.003°$ for $Cu_{3-x}Sn_xSe_3$ at $x = 1.00$; $a = 697.21 \pm 0.03$, $b = 1207.8 \pm 0.1$, $c = 1339.3 \pm 0.1$ pm, $\beta = 99.86 \pm 0.01°$ for $Cu_{3-x}Sn_xSe_3$ at $x = 1.025$ (Fan et al. 2014)] and $a = 696.12 \pm 0.14$, $b = 1204.3 \pm 0.2$, $c = 2648.1 \pm 0.5$ pm, $\beta = 99.97 \pm 0.03°$ (Gulay et al. 2010) [$a = 697.7 \pm 0.1$, $b = 1207.1 \pm 0.1$, $c = 2649.3 \pm 0.2$ pm, $\beta = 94.96 \pm 0.01°$ for $Cu_{3-x}Sn_xSe_3$ at $x = 1.05$ (Fan et al. 2014)] for the third one.

Cu_2SnSe_3 could also crystallize in the tetragonal structure with the lattice parameters $a = 569.68 \pm 0.01$ and $c = 1142.57 \pm 0.02$ pm (Fan et al. 2013) [$a = 568.9$ and $c = 1137$ pm and the calculated and experimental densities 5.79 and 5.72 g·cm^{-3}, respectively (Hahn et al. 1966).

A polycrystalline Cu_2SnSe_3 sample was synthesized from direct reaction of stoichiometric amounts of Cu, Sn, and Se powders (Siyar et al. 2018). The mixture was melted in an evacuated quartz ampoule at 1010°C for 3 h followed by periodic agitation to improve mixing of the reacting species. The molten sample was cooled at a rate of 1°C·min^{-1} down to 450°C and 550°C, annealed for 12 h at these temperatures and quenched in water. XRD shows that the Cu_2SnSe_3 sample annealed at 690°C crystallizes in the cubic structure. The samples, annealed at 550°C and 450°C, were found to have mostly the monoclinic structures. It is very difficult to synthesize pure phase Cu_2SnSe_3 at both low and high temperatures. The amount of cubic phase increases with increasing annealing temperature, and at high enough temperature, mostly cubic phase exists.

The ingots of Cu_2SnSe_3 with monoclinic modification were also prepared from the melt by the vertical Bridgman–Stockbarger technique (Marcano et al. 2001, 2002a; Delgado et al. 2003). They were obtained by heating the mixture of Cu, Sn, and Se (molar ratio 2:1:4) sealed in evacuated quartz ampoules. The ampoules were placed in a multiple zone vertical furnace. Initially they were heated from room temperature to 1150°C at a rate of 40°C/h. The molten mixture was kept at this temperature for 24 h. In order to assure a homogeneous mixing, the ampoule was agitated periodically. It was later cooled at a rate of 10°C/h up to 800°C, and then at 1°C/h to 640°C. The cooling rate from 640°C to 500°C was 5°C/h. The ingots were annealed at this temperature for 120 h. The furnace was then turned off and the ingot cooled down to room temperature.

Polycrystalline sample of new monoclinic modification of Cu_2SnSe_3 were prepared by direct reaction of stoichiometric mixture of the elements Cu, Sn, and Se in an evacuated fused quartz ampoule at 900°C for 12 h, cooled down to 580°C in 24 h, annealed at this temperature for 7 days, and finally quenched in cold water (Fan et al. 2013). This monoclinic modification crystallizes forming twinned domains.

Gulay et al. (2010) obtained one of the monoclinic modifications by melting the stoichiometric mixture of the elements in an evacuated silica tube. The ampoule was heated at a rate of 100°C·h^{-1} in a furnace to 500°C, then heated at a rate of 20°C·h^{-1} to 1110°C and kept at this temperature for 2 h. The ampoule was then cooled slowly (10°C·h^{-1}) to 350°C and annealed at this temperature for 500 h. After annealing, the sample was quenched in cold water.

$Cu_2Sn_{1-x}Se_3$ ($0 \le x \le 0.4$) was prepared from pure elements taken in stoichiometric ratio and melting them in an evacuated and sealed silica ampoules (Sharma et al. 1977a). The temperature was kept at 1000°C for 6 h and the furnace switched off afterwards.

Polycrystalline samples with nominal composition $Cu_{3-x}Sn_xSe_3$ (x = 0.87, 0.90, 0.95, 0.975, 1.00, 1.025, 1.05) were synthesized by direct reaction of elemental Cu (powder), Sn (foil), and Se (shot) in evacuated and sealed quartz ampoules (Fan et al. 2014). The starting mixtures were heated slowly up to 900°C and held for 12 h, then cooled down to 580°C in 24 h, followed by annealing at this temperature for 7 days. The obtained ingots were ground into fine powder, cold pressed into pellets, and then annealed at 580°C for additional 7 days to ensure homogeneity of the element distribution.

There two another ternary compounds in the Sn–Cu–Se system. **Cu_2SnSe_4** crystallizes in the cubic structure with the lattice parameter a = 568.46 ± 0.03 pm and an energy gap of 0.35 eV (Marcano et al. 2002b). The ingots of this compound were prepared in the same way as the first monoclinic modification of Cu_2SnSe_3 was obtained using the stoichiometric ratio of Cu, Sn, and Se as 2:1:4.

$Cu_6Sn_2Se_6$ decomposes at 546°C and has an experimental density of 6.23 g·cm^{-3} (Voroshilov et al. 1986).

The liquidus surface of the Sn–Cu–Se ternary system was constructed by Berger et al. (1973b,c) and Moroz et al. (2002) and is presented in Figure 8.4. The next invariant equilibria exist in the system (Moroz et al. 2002): U_1 – L + (Cu) ⇔ Cu_3Sn + Cu_2Se; U_2 – L + Cu_3Sn ⇔ γ + Cu_2Se; U_3 – L + Cu_2SnSe_3 ⇔ Cu_2Se + SnSe; U_4 – L + γ ⇔ ε + SnSe; U_5 – L + ε ⇔ η + SnSe; P_1 – L + γ + Cu_2Se ⇔ ε; P_2 – L + γ + SnSe ⇔ ε; E_1 – L ⇔ Cu_2Se + SnSe + ε; and E_2 – L ⇔ η + Sn + SnSe (γ, ε, and η are the binary phases of the Cu–Sn system).

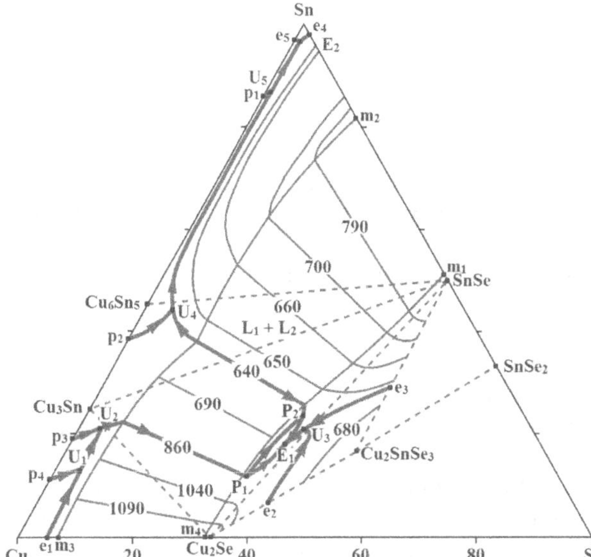

FIGURE 8.4 Liquidus surface of the Sn–Cu–Se ternary system. (From Moroz, V.M., et al., *Fiz. khim. tv. tila*, **3**(4), 654, 2002.)

8.7 Tin–Silver–Selenium

SnSe–Ag: The phase diagram of this system, constructed through DTA, XRD, and metallography, is a eutectic type (Ollitrault-Fichet et al. 1988). The eutectic crystallizes at 660°C. There is an immiscibility region in the system that leads to the appearance of two series of thermal effects at 220°C and 450°C, which are the temperatures of ternary eutectics in the SnSe–Ag–Sn and SnSe–Ag–Ag_2Se subsystems, respectively. The thermal effects of the Ag_2Se phase transformation exist on the phase diagram at 125°C.

According to the data of Yusibov et al. (2018), this system is a non-quasibinary section of the Sn–Ag–Se ternary system.

SnSe–"AgSe": This system is non-quasibinary section of the Sn–Ag–Se ternary system (Ollitrault-Fichet et al. 1988; Yusibov et al. 2018). The **$Ag_xSn_{1-x}Se$** solid solutions are formed in the system. They decompose peritectically at 590°C (Ollitrault-Fichet et al. 1988) [at 630°C (Gorochov 1968a); at 587°C (Yusibov et al. 2018)]. Thermal effects at 217°C and 220°C from the "AgSe"-side correspond to the crystallization of ternary eutectics.

The homogeneity region of $Ag_xSn_{1-x}Se$ is $0.45 < x < 0.49$ (Wold and Brec 1976). The composition of these solid solutions with x = 0.5 correspond to the **$AgSnSe_2$** ternary compound. This compound crystallizes in the cubic structure with the lattice parameter a = 562.83 ± 0.05 pm (Yusibov et al. 2018) [a = 567.7 ± 0.3 pm and the calculated and experimental densities 6.98 and 6.97 g·cm^{-3}, respectively (Wold and Brec 1976)]. Under 6.5 GPa and 1000°C, the region of the solid solutions enlarged, reaching the upper limit x = 0.66, corresponding to the **Ag_2SnSe_3** ternary compound, which crystallizes in the cubic structure with the lattice parameter a = 564.0 ± 0.4 pm and the calculated and experimental densities 7.05 and 7.02 g·cm^{-3}, respectively. At ambient conditions, the formation of Ag_2SnSe_3 was not confirmed (Yusibov et al. 2018), but Balanevskaya et al. (1966) noted that this compound is a semiconductor with E_g = 0.81 eV, melts at 490°C and has an experimental density of 4.86 g·cm^{-3}. At the preparation of Ag_2SnSe_3, the **$Ag_{1.68}Sn_{1.25}Se_{3.8}$** ternary compound, which melts at 506°C and crystallizes in the monoclinic structure with the lattice parameters a = 718, b = 1055, c = 670 pm, and β = 111.98°, was obtained (Pirela et al. 2007). Its sample was synthesized by direct melting of the mixture of Ag, Sn, and Se (molar ratio 2:1:3) encapsulated in pyrolized quartz ampoule.

The values of the thermodynamic functions in the temperature range of 210°C–310°C for saturated solid solution based on $AgSnSe_2$ in the standard state using the temperature dependence of galvanic cell were determined by Moroz and Prokhorenko (2015b). The next values were obtained: ΔG^0_{298} = –133.9 ± 1.6 kJ·M^{-1}, ΔH^0_{298} = –124.9 ± 1.3 kJ·M^{-1}, and ΔS^0_{298} = 30.1 ± 2.5 (J·M^{-1}K^{-1}).

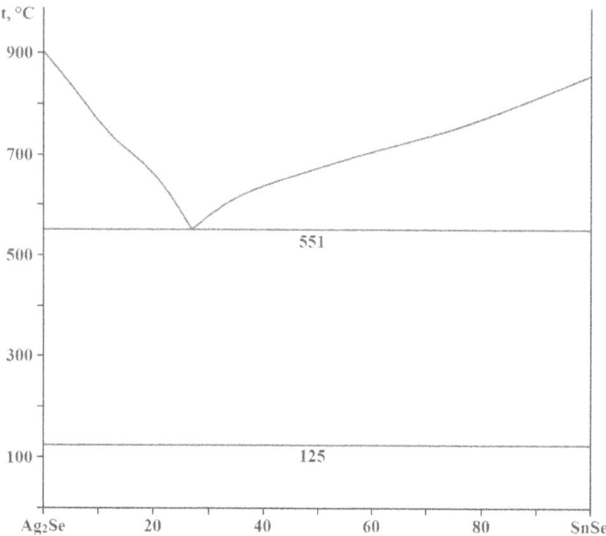

FIGURE 8.5 Phase diagram of the SnSe–Ag_2Se system. (From Ollitrault-Fichet, R., et al., *J. Less-Common Metals*, **138**(2), 241, 1988.)

SnSe–Ag_2Se: The phase diagram of this system, constructed through DTA, XRD, and metallography, is a eutectic type (Figure 8.5) (Gorochov 1968a; Dovletov et al. 1974c; Ollitrault-Fichet et al. 1988; Yusibov et al. 2018). The eutectic crystallizes at 551°C (Ollitrault-Fichet et al. 1988) [at 545°C (Gorochov 1968a); at 610°C (Dovletov et al. 1974c)]. Thermal effects at 125°C correspond to the polymorphic transformation of Ag_2Se. According to the data of Dovletov et al. (1974c), the solubility of Ag_2Se in SnSe reaches 10 mol% and the solubility of SnSe in Ag_2Se is 7–8 mol%.

$SnSe_2$–Ag_2Se: The phase diagram of this system, constructed through DTA and XRD, is presented in Figure 8.6 (Yusibov et al. 2018). The eutectics contain 15 and 56 mol% $SnSe_2$ and crystallize, respectively, at 722°C and 495°C.

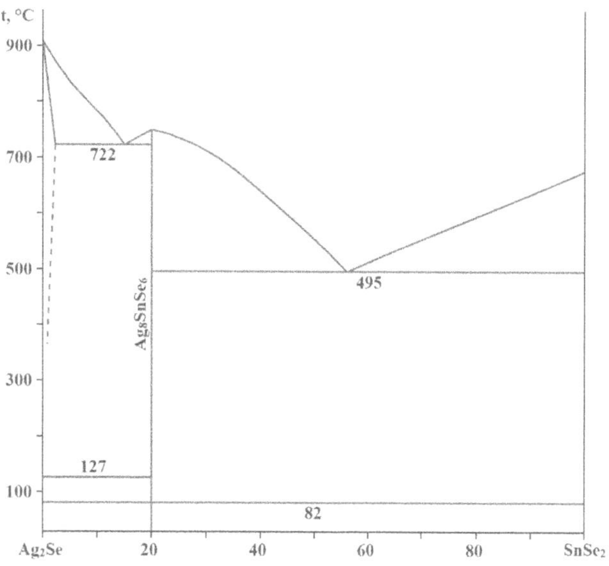

FIGURE 8.6 Phase diagram of the $SnSe_2$–Ag_2Se system. (From Yusibov, Yu. A., et al., *Russ. J. Inorg. Chem.*, **63**(12), 1622, 2018.)

Thermal effects at 82°C and 127°C correspond, respectively, to the polymorphic transformations of **Ag_8SnSe_6** and Ag_2Se. The Ag_8SnSe_6 ternary compound, which melts congruently at 742°C [according to the data of Gorochov et al. (1966), Gorochov and Flahaut (1967) and Gorochov (1968a,b), this compound melts incongruently at 735°C] and has a polymorphic transformation at 82°C (Zhdanova et al. 1977; Kolosov 1978), is formed in this system.

According to the data of Gorochov et al. (1966) and Gorochov (1968b), the phase diagram of the $SnSe_2$–Ag_2Se system is a eutectic type with peritectic transformation. The eutectic contains 43.5 mol% Ag_2Se and crystallizes at 505°C. The ingots for investigations were annealed at 480°C for 3 days.

The first modification of Ag_8SnSe_6 crystallize in the orthorhombic structure with the lattice parameters a = 791.35 ± 0.06, b = 782.44 ± 0.06, and c = 1104.67 ± 0.09 pm (Yusibov et al. 2018) [a = 791.68 ± 0.06, b = 782.19 ± 0.06, c = 1104.53 ± 0.08 pm, and a calculated densities of 7.072 ± 0.002 g·cm^{-3} (Gulay et al. 2002a); a = 791.25, b = 782.54, and c = 1104.65 pm (Alieva et al. 2014); a = 791.3 ± 0.1, b = 782.5 ± 0.1, c = 1104.7 ± 0.2 pm (Bagheri et al. 2014)].

The second modification of this compound crystallizes in the cubic structure with the lattice parameter a = 1105 pm (Li et al. 2000) [a = 1096 pm and the calculated and experimental densities of 7.34 and 7.06 g·cm^{-3}, respectively (Hahn et al. 1965); a = 1107 pm at room temperature and a = 1112 pm and the calculated and experimental densities of 7.12 and 7.01 g·cm^{-3} at 200°C, respectively (Gorochov et al. 1966; Gorochov and Flahaut 1967; Gorochov 1968b)].

The values of the thermodynamic functions in the temperature range of 210°C–310°C for saturated solid solution based on Ag_8SnSe_6 in the standard state using the temperature dependence of galvanic cell were determined by Moroz and Prokhorenko (2015b). The next values were obtained: ΔG^0_{298} = –352.5 ± 1.9 kJ·M^{-1}, ΔH^0_{298} = –323.1 ± 1.6 kJ·M^{-1}, and ΔS^0_{298} = 98.6 ± 3.1 (J·M^{-1}K^{-1}).

The orthorhombic modification of Ag_8SnSe_6 was prepared using high-purity elements in an evacuated quartz ampoule (Gulay et al. 2002a). The maximum temperature of synthesis was 800°C and its duration was 4 h. Uninterrupted vibration mixing was used. After cooling at a rate of 10°C·h^{-1} to 400°C, further annealing was carried out for 250 h. After annealing, the alloy was cooled to 200°C at a rate of 10°C·h^{-1}. At this temperature the furnace was turned off and the sample was cooled to room temperature. It was also synthesized from the elements by Bagheri et al. (2014). The mixture was placed into quartz ampoule evacuated to ~10^{-2} Pa and sealed. Synthesis was carried out in a two-zone furnace. The hot zone was heated up to temperature ~30–50°C above the melting point of the compound and the cold zone was heated to ca. 630°C.

The cubic modification can be prepared from stoichiometric mixtures of $AgNO_3$ (11.29 mM), $SnCl_2{\cdot}2H_2O$ (1.41 mM), and Se (8.48 mM) (Li et al. 2000). The mixture was loaded into a 100-mL Teflon liner autoclave, which was then filled with anhydrous ethylenediamine up to 90% of the total volume. The autoclave was sealed and maintained at 100°C for 10 h and then cooled to room temperature naturally. The product was filtered and washed with distilled water and absolute ethanol

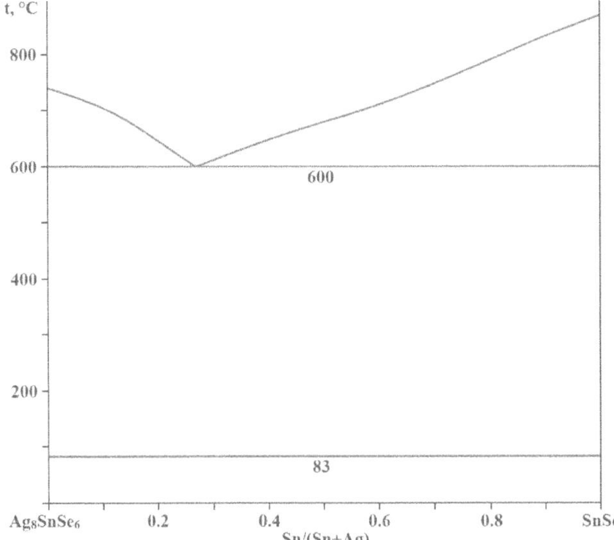

FIGURE 8.7 Phase diagram of the SnSe–Ag_8SnSe_6 system. (From Ollitrault-Fichet, R., et al., *J. Less-Common Metals*, **138**(2), 241, 1988.)

several times to remove the by-products. The precipitates were dried in vacuum at 60°C for 4 h.

SnSe–Ag_8SnSe_6: The phase diagram of this system, constructed through DTA, XRD, and metallography, is a eutectic type (Figure 8.7) (Ollitrault-Fichet et al. 1988; Yusibov et al. 2018). The eutectic crystallizes at 600°C. Thermal effects at 83°C correspond to the polymorphic transformation of Ag_8SnSe_6.

Ag_8SnSe_6–Se: The phase diagram of this system, constructed through DTA and XRD, is a eutectic type with immiscibility region (Figure 8.8) (Ollitrault-Fichet et al. 1988; Yusibov et al. 2018). The eutectic crystallizes at 220°C and is degenerated from the Se-side. Thermal effects at 82°C correspond to the polymorphic transformation of Ag_8SnSe_6.

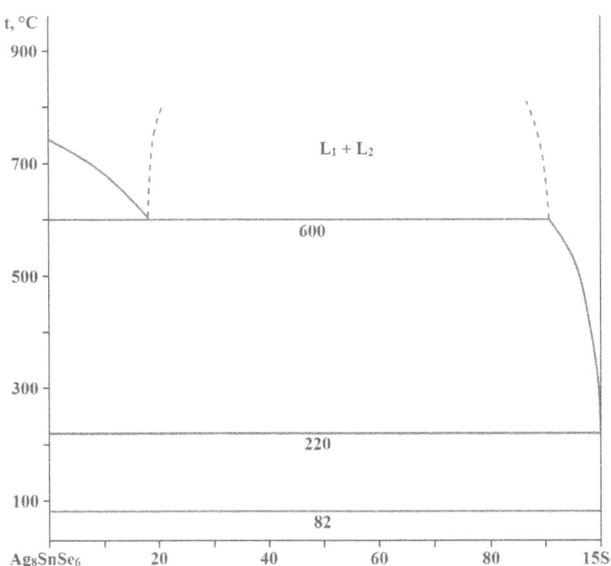

FIGURE 8.8 Phase diagram of the Ag_8SnSe_6–15Se (From Yusibov, Yu.A., et al., *Russ. J. Inorg. Chem.*, **63**(12), 1622, 2018.)

The isothermal sections of the Sn–Ag–Se ternary system at 400°C and 550°C are given in Figure 8.9 (Ramakrishnan et al. 2020). It is seen, that all the binary compounds have very limited ternary solubilities. The compositional regime of the liquid phase expands with higher temperature. The isothermal sections at 180°C–300°C were also constructed by Moroz and Prokhorenko (2015b) and at room temperature by Yusibov et al. (2018).

The liquidus surface of the Sn–Ag–Se ternary system (Figure 8.10) includes two immiscibility regions (Ollitrault-Fichet et al. 1988; Moroz et al. 2003; Yusibov et al. 2018). 3D modeling and visualization of the liquidus and phase-separation surfaces were performed by Yusibov et al. (2018). The next invariant equilibria exist in the

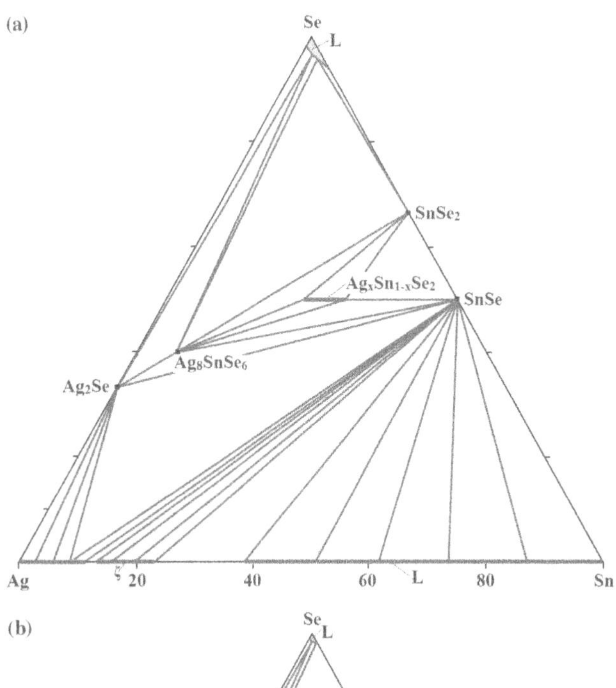

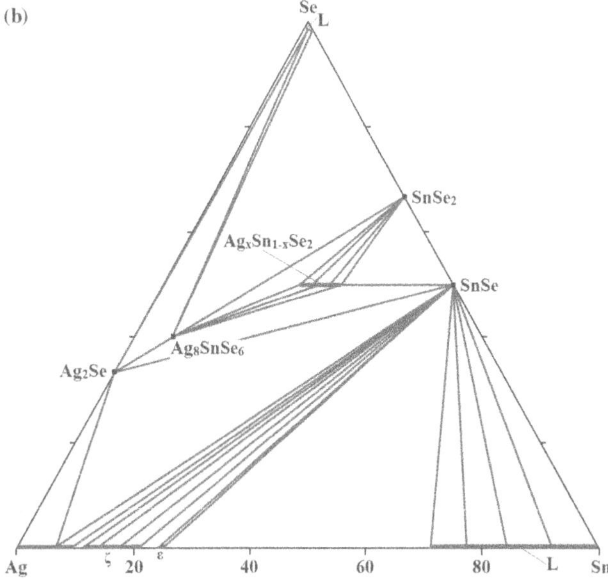

FIGURE 8.9 Isothermal sections of the Sn–Ag–Se ternary system at (a) 550°C and (b) 400°C. (From Ramakrishnan, A., et al., *J. Alloys Compd.*, **816**, 152670, 2020.)

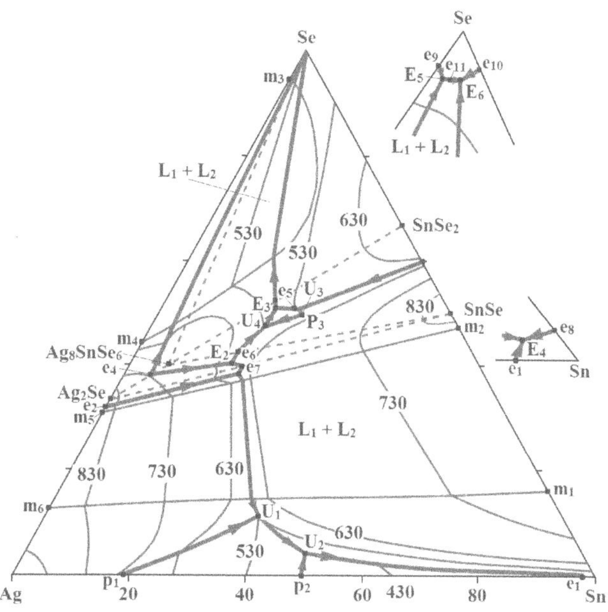

FIGURE 8.10 Liquidus surface of the Sn–Ag–Se ternary system. (From Yusibov, Yu. A., et al., *Russ. J. Inorg. Chem.*, **63**(12), 1622, 2018.)

system: U_1 (535°C) – L + α ⇔ β-SnSe + ξ; U_2 (460°C) – L + ξ ⇔ α-SnSe + ε; U_3 (560°C) – L + β-SnSe ⇔ $Ag_xSn_{1-x}Se$ + $SnSe_2$; U_4 (542°C) – L + β-SnSe ⇔ $Ag_xSn_{1-x}Se$ + β-Ag_8SnSe_6; E_1 (530°C) – L ⇔ α + α′ + α-SnSe; E_2 (547°C) – L ⇔ α′ + α-SnSe + β-Ag_8SnSe_6; E_3 (490°C) – L ⇔ α-SnSe + $Ag_xSn_{1-x}Se$ + β-Ag_8SnSe_6; E_4 (220°C) – L ⇔ α-SnSe + ε + Sn; E_5 (215°C) – L ⇔ α′ + β-Ag_8SnSe_6 + Se; E_6 (217°C) – L ⇔ β-Ag_8SnSe_6 + $SnSe_2$ + Se (α, α′, ξ and ε are the binary phases of the Ag–Sn system).

8.8 Tin–Gold–Selenium

SnSe–Au: The phase diagram of this system, constructed through DTA, XRD, and metallography, is a eutectic type (Figure 8.11) (Rouland et al. 1975, 1976). The most part of the areas of primary crystallization fields of SnSe and Au, including the eutectic composition, are located under the immiscibility region. The eutectic crystallizes at 730°C.

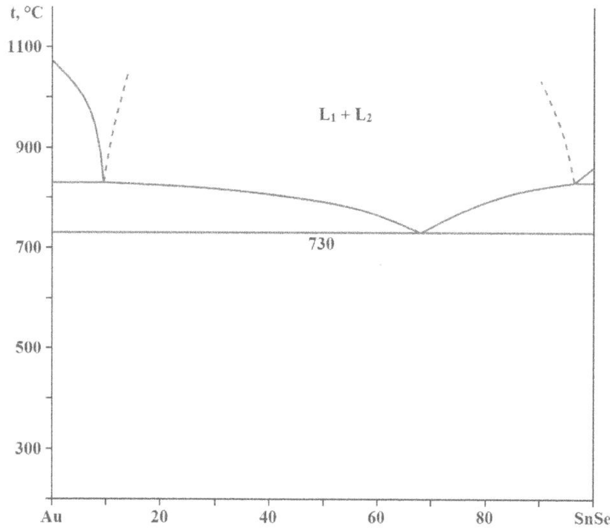

FIGURE 8.11 Phase diagram of the SnSe–Au system. (From Rouland, J.-C., et al., *Bull. Soc. chim. France*, (11–12), part 1, 1614, 1976.)

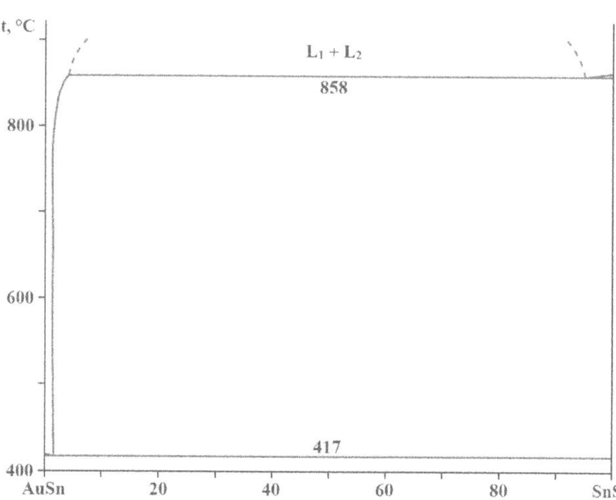

FIGURE 8.12 Phase diagram of the SnSe–AuSn system. (From Rouland J.-C., et al., *C. r. Acad. sci. Sér. C*, **281**(18), 719, 1975.)

SnSe–AuSn: The phase diagram of this system, constructed through DTA, XRD and metallography, is a eutectic type (Figure 8.12) (Rouland et al. 1975, 1976). The eutectic contains 1.6 mol% SnSe and crystallizes at 417°C. At 858°C, immiscibility is observed in the system, and the monotectic point is at 95 mol% SnSe.

$SnSe_2$–Au: The phase diagram of this system, constructed through DTA, XRD, and metallography, is a eutectic type (Figure 8.13) (Rouland et al. 1975, 1976). The eutectic contains 82 mol% $SnSe_2$ and crystallizes at 580°C. At 898°C, immiscibility is observed in the system, and the monotectic point is at 8.3 mol% $SnSe_2$.

The liquidus surface of the Sn–Au–Se ternary system (Figure 8.14) is characterized by the presence of two immiscibility regions, one of which is adjacent to the Au–Se binary

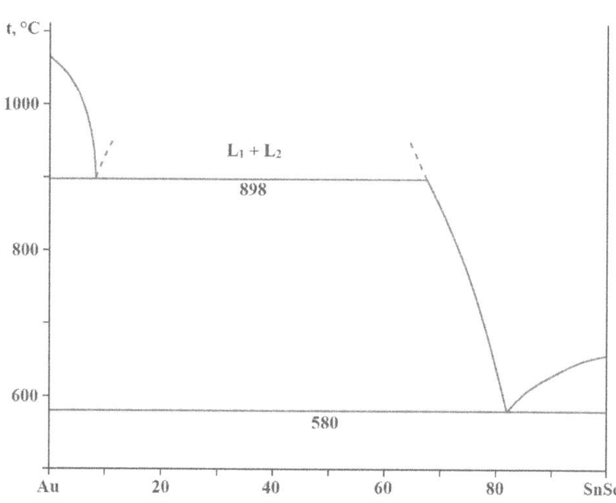

FIGURE 8.13 Phase diagram of the $SnSe_2$–Au system. (From Rouland J.-C., et al., *C. r. Acad. sci. Sér. C*, **281**(18), 719, 1975.)

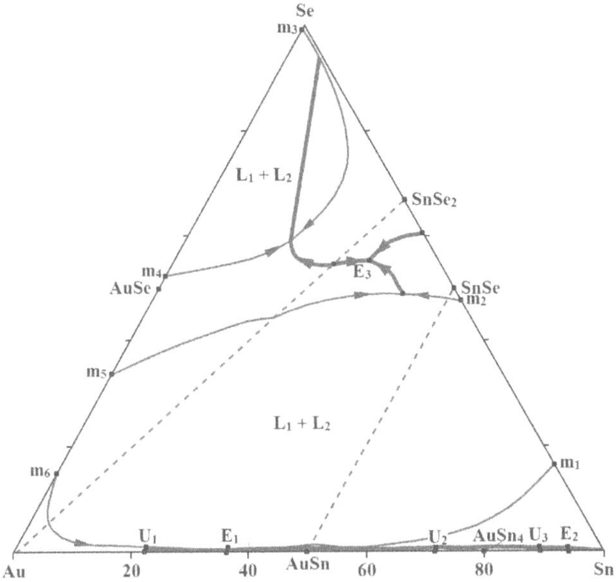

FIGURE 8.14 Liquidus surface of the Sn–Au–Se ternary system. (From Rouland, J.-C., et al., *Bull. Soc. chim. France*, (11–12), part 1, 1614, 1976.)

system and the other passes through the ternary system from the Au–Se system to the Sn–Se system (Rouland et al. 1976). The presence of immiscibility regions leads to poor homogenization of the initial samples and to the manifestation of "excess" thermal effects in the phase diagrams of the quasi-binary sections. The next invariant equilibria exist in the system: E_1 (276°C) – L ⇔ ξ + SnSe + AuSn; E_2 (217°C) – L ⇔ $AuSn_4$ + Sn + SnSe; E_3 (564°C) – L ⇔ Au + $SnSe_2$ + SnSe; degenerated near Se (214°C) – L ⇔ Au + $SnSe_2$ + Se; U_1 (530°C) – L + Au ⇔ ξ + SnSe; U_2 (309°C) – L + AuSn ⇔ $AuSn_2$ + SnSe; and U_3 (252°C) – L + $AuSn_2$ ⇔ $AuSn_4$ + SnSe (ξ is the binary phase of the Au–Sn system).

8.9 Tin–Magnesium–Selenium

The $\mathbf{Mg_2SnSe_4}$ ternary compound, which crystallizes in the orthorhombic structure with the lattice parameters a = 1341, b = 787, c = 643 pm, is formed in the Sn–Mg–Se system (Rocktäschel et al. 1964). It could be prepared by the interaction of Mg_2Sn and Se at 450°C–650°C.

8.10 Tin–Strontium–Selenium

Four ternary compounds are formed in the Sn–Sr–Se system. All these compounds crystallize in the orthorhombic structure with the lattice parameters a = 2189.2 ± 0.1, b = 2189.3 ± 0.1, c = 1354.70 ± 0.07 pm, a calculated density of 5.245 g·cm^{-3} and an energy gap 0.2 eV for $\mathbf{SrSn_2Se_4}$ and a = 1205.7 ± 0.2, b = 1658.3 ± 0.3, c = 863.9 ± 0.2 pm, a calculated density of 5.297 g·cm^{-3} and an energy gap 0.9 eV for $\mathbf{Sr_2SnSe_5}$ (Assoud et al. 2004b); a = 1197.3 ± 0.2, b = 1612.0 ± 0.3, c = 859.4 ± 0.2 pm, and a calculated density of 5.200 g·cm^{-3} at 200 ± 2 K for $\mathbf{Sr_4Sn_2Se_9}$ and a = 1202.8 ± 0.2, b = 1654.1 ± 0.3, c = 861.1 ± 0.2 pm, and a calculated density of 5.340 g·cm^{-3} at 200 ± 2 K for $\mathbf{Sr_4Sn_2Se_{10}}$ (Pocha and Johrendt 2004).

$SrSn_2Se_4$ and Sr_2SnSe_5 were obtained as follows (Assoud et al. 2004b). The two silica tubes containing Sr, Sn, and Se in the 1:2:4 and 2:1:5 molar ratios, respectively, were heated to 850°C within 24 h, kept at 850°C for 12 h, and then cooled to the annealing temperature of 650°C within 2 h. After sintering at 650°C for 300 h, the furnace with both tubes was allowed to cool to room temperature within 24 h. The samples looked homogeneous, comprising mostly microcrystalline black powder.

$Sr_4Sn_2Se_9$ and $Sr_4Sn_2Se_{10}$ were synthesized by heating stoichiometric amounts of the elements in alumina crucibles, sealed in quartz ampoules under atmospheres of purified Ar (Pocha and Johrendt 2004). At first, the temperature was raised slowly (50°C·h^{-1}) to 800°C and kept there for 15 h. After cooling, the samples were homogenized in an Ar-filled glove box and heated to 900°C for 24 h using the same crucibles. This procedure was repeated once more and resulted in homogeneous, dark red powders (almost black), which are not sensitive to air. Small single crystals have been selected from the powder.

8.11 Tin–Barium–Selenium

Several ternary compounds are formed in the Sn–Ba–Se system. $\mathbf{BaSnSe_3}$ crystallizes in the orthorhombic structure with the lattice parameters a = 1718, b = 1818, c = 1236 pm, and the calculated and experimental densities of 5.10 and 5.03 g·cm^{-3}, respectively (Hahn and Schulze 1965). It was prepared by the interaction of BaSe with $SnSe_2$ at the temperatures between 600°C and 1100°C.

$\mathbf{Ba_2SnSe_4}$ crystallizes in the monoclinic structure with the lattice parameters a = 858.19 ± 0.14, b = 925.54 ± 0.13, c = 1246.0 ± 0.2 pm, β = 114.276 ± 0.011°, and an energy gap of 1.90 eV (Wu et al. 2015c). This compound could be obtained by the melting the mixture of the elements in stoichiometric ratios (Bok and Wit 1963). It was also prepared by a molten flux method with Zn or Cd as a flux (Tampier and Johrendt 2001). It is stable in air for several weeks. Ba_2SnSe_4 has ferroelectric phase transition at 120 K (Tampier and Johrendt 2001).

$\mathbf{Ba_2SnSe_5}$ has three polymorphic modifications: low- and high-temperature α- and β-Ba_2SnSe_5 and metastable m-Ba_2SnSe_5. α-Ba_2SnSe_5 could be converted into β-Ba_2SnSe_5 at 750°C and β-Ba_2SnSe_5 transforms into α-Ba_2SnSe_5 at 650°C (Zelinska et al. 2010). The transition from m-Ba_2SnSe_5 into α-Ba_2SnSe_5 is an irreversible displacive phase transition between 250°C and 450°C (Graf et al. 2010).

α-Ba_2SnSe_5 crystallizes in the orthorhombic structure with the lattice parameters a = 1235.72 ± 0.08, b = 1723.5 ± 0.1, c = 1813.4 ± 0.1 pm, a calculated density of 5.422 g·cm^{-3}, and an energy gap of 1.2 eV (Assoud et al. 2005). This modification was prepared starting from the elements in the stoichiometric ratio. This mixture was placed into a fused silica tube, which was then sealed under vacuum. The tube was put into a furnace, heated to 750°C within 3 days, kept at 750°C for 4 days, cooled down to 700°C within 10 min, and then annealed at 650°C for 4 days. In the final step, the furnace was cooled down to room temperature within 4 days. The brown-black reaction mixture appeared to be homogeneous.

β-Ba_2SnSe_5 and *m*-Ba_2SnSe_5 crystallize in the monoclinic structure with the lattice parameters $a = 939.49 \pm 0.06$, $b = 886.56 \pm 0.06$, $c = 1257.45 \pm 0.07$ pm, $\beta = 113.299 \pm 0.004°$, a calculated density of 5.442 g·cm^{-3} and an energy gap of 2.0 eV for the first modification (Zelinska et al. 2010) and $a = 2117.19 \pm 0.08$, $b = 909.15 \pm 0.03$, $c = 2471.41 \pm 0.07$ pm, $\beta = 125.621 \pm 0.001°$, and a calculated density of 5.415 g·cm^{-3} (Graf et al. 2010) for the second one.

Shiny red prismatic crystals of β-Ba_2SnSe_5 can be obtained starting from the elements in the stoichiometric ratio (Zelinska et al. 2010). The reaction mixture was loaded into a silica tube within the glove box, and then sealed under vacuum (<0.1 Pa). The tube was placed into a furnace, heated to 250°C within 6 h, kept at 250°C for 12 h, followed by heating to 850°C within 24 h. Finally, the tube was slowly cooled to 750°C, then annealed at this temperature for 2 weeks, and then quenched in cold water.

Due to the air and moisture sensitivity of some starting materials, all preparative steps at the preparation of *m*-Ba_2SnSe_5 were carried out under argon in a glove box (Graf et al. 2010). The crystals were synthesized using Ba pieces, Sn shots, and Se pellets with the addition of LiCl/KCl flux. The reaction mixture (1 mM), weighed from the elements in the stoichiometric ratio, was loaded with the LiCl/KCl (45/55 mass%) flux mixture (1 g) into a glassy carbon crucible. The crucible was placed into a silica tube, which was evacuated and afterwards sealed under vacuum (<0.1 Pa). The tube was placed into a furnace, heated up to 500°C within 48 h and kept at this temperature for another 48 h. To achieve crystal growth, the tube was cooled down at a rate of 1°C·h^{-1} to room temperature. To remove the flux from the reactive sample, the tube was opened under air and the sample rinsed first with ethanol and then with acetone to yield prismatic reddish dark-brown crystals.

Pure powder samples of *m*-Ba_2SnSe_5 were obtained by loading a Ba/Sn/Se mixture (1 mM; molar ratio 2:1:5) with the LiCl/KCl (45/55 mass%) flux (1 g) into a glassy carbon crucible that was placed into a silica tube (Graf et al. 2010). The tube was evacuated and then sealed under vacuum (<0.1 Pa). The reaction mixture was heated up to 750°C within 72 h and kept at this temperature for 96 h. The ampoule was cooled down to 650°C within 12 h and held at 650°C for 96 h. Subsequently the tube was cooled down to room temperature within 96 h and then opened under air. Again, the sample was washed with ethanol and then dried with acetone.

$Ba_6Sn_6Se_{13}$ melts incongruently and crystallizes in the orthorhombic structure with the lattice parameters $a = 930.0 \pm 0.2$, $b = 1238.4 \pm 0.3$, $c = 2689.6 \pm 0.5$ pm, a calculated density of 5.495 g·cm^{-3} at 153 ± 2 K, and an energy gap of 1.52 ± 0.02 eV (Feng et al. 2013). A polycrystalline sample of this compound was synthesized by a solid-state reaction technique. A mixture of BaSe, Sn, and Se (molar ratio 6:6:7) was ground and loaded into fused silica tube under an Ar atmosphere in a glove box, which was sealed under 0.001 Pa and then placed in a furnace. The sample was heated to 750°C in 20 h, kept at that temperature for 48 h, and then the furnace was turned off.

To obtain the single crystals, a mixture of BaSe, Sn, and Se (molar ratio of 2:2:3) was ground and loaded into fused-silica tubes under an Ar atmosphere in a glove box, which was sealed under 0.001 Pa atmosphere and then placed in a furnace (Feng et al. 2013). The sample was heated to 1000°C in 20 h and kept at that temperature for 48 h, then cooled at a slow rate of 4°C·h^{-1} to 400°C, and finally cooled to room temperature. The product consisted of black crystals of $Ba_6Sn_6Se_{13}$.

$Ba_7Sn_3Se_{13}$ also crystallizes in the orthorhombic structure with the lattice parameters $a = 1274.8 \pm 0.1$, $b = 2489.1 \pm 0.3$, $c = 921.2 \pm 0.1$ pm, and a calculated density of 5.326 g·cm^{-3} (Assoud and Kleinke 2005). To prepare this compound, the reaction was started from the elements with a total sample mass of around 500 mg. $Ba_7Sn_3Se_{13}$ was first encountered by heating Ba, Sn, and Se (molar ratio 2:1:4) in an evacuated silica tube to 900°C, followed by slow cooling to 200°C within 5 days. Then, it was obtained using stoichiometric ratio of the elements.

8.12 Tin–Zinc–Selenium

SnSe–ZnSe: The phase diagram of this system, constructed through DTA, XRD, and metallography, is a eutectic type (Figure 8.15) (Dohnke et al. 1999). The eutectic composition and temperature are 14 mol% ZnSe and 830°C [850°C (Galiulin et al. 1982)]. At 519°C [517°C (Galiulin et al. 1982)], solid solutions based on SnSe undergo polymorphous transformation. Mutual solubility of SnSe and ZnSe is not higher than 0.5 mol% [according to the data of Dohnke et al. (1999), there is practically no mutual solubility of SnSe and ZnSe]. The ingots for investigations were annealed at 550°C for 250 h (Galiulin et al. 1982; Dohnke et al. 1999).

$SnSe_2$–ZnSe: The phase diagram of this system, constructed through DTA, XRD, and metallography, is a eutectic type (Figure 8.16) (Parasyuk et al. 2004). The eutectic composition and temperature are 13 mol% ZnSe and 622°C [16 mol% ZnSe and 642°C (Galiulin et al. 1982)], respectively. The solubility of $SnSe_2$ in ZnSe reaches 1 mol% and the solubility

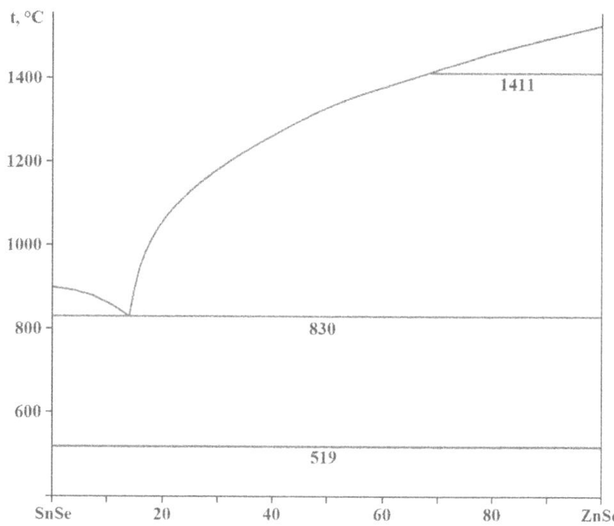

FIGURE 8.15 Phase diagram of the SnSe–ZnSe system. (From Dohnke, I., et al., *J. Cryst. Growth*, **198–199**, 287, 1999.)

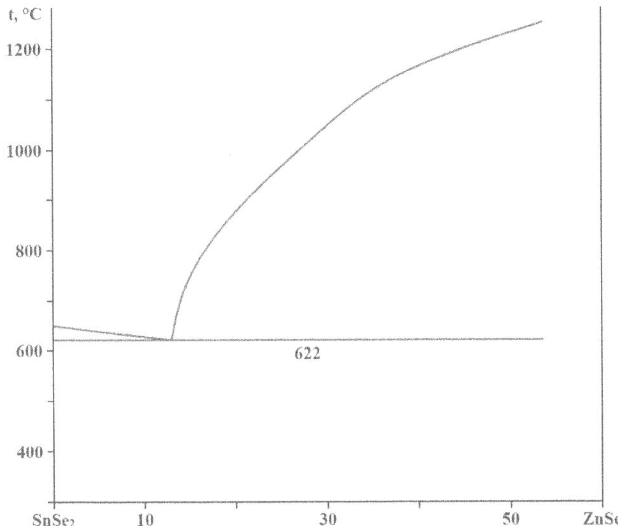

FIGURE 8.16 Phase diagram of the $SnSe_2$–ZnSe system. (From Parasyuk, O.V., et al., *J. Alloys Compd.*, **379**(1–2), 143, 2004.)

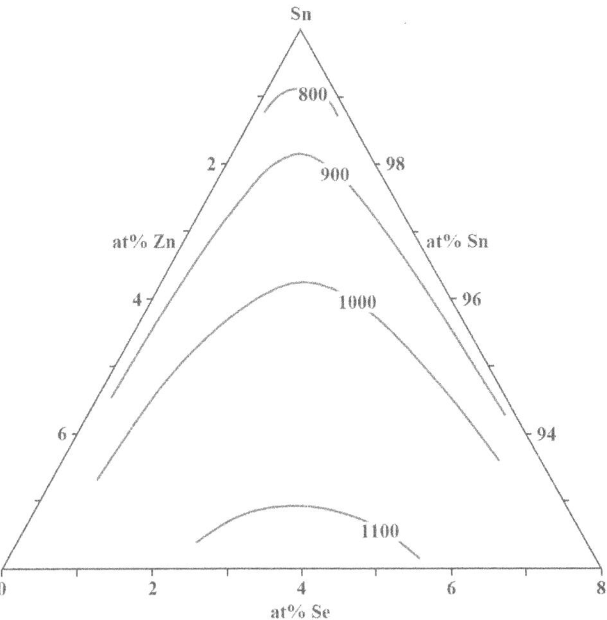

FIGURE 8.18 Part of the Sn–Zn–Se liquidus surface on the Sn-rich side. (From Kimura, S., *J. Chem. Thermodyn.*, **3**(1), 7, 1971.)

of ZnSe in $SnSe_2$ is not higher than 0.5 mol% (Galiulin et al. 1982; Parasyuk et al. 2004). The ingots for investigations were annealed at 550°C for 250 h (Galiulin et al. 1982).

Sn–ZnSe: The ZnSe solubility in liquid tin (Figure 8.17) was studied by Rubenstein (1968) and Kimura (1971) and a part of Sn–Zn–Se liquidus surface on the Sn-rich side (Figure 8.18) was constructed by Kimura (1971). The eutectic is degenerated from the Sn-rich side. Liquidus is described using the equation of regular solutions with constant interaction parameter (Vasil'ev and Novikova 1977). Liquidus temperatures were determined using high-temperature filtration (Rubenstein 1968) and visual observation of solid-phase appearance and disappearance for heating and cooling of ingots (accuracy of measurement is ±5°C) (Kimura 1971).

8.13 Tin–Cadmium–Selenium

SnSe–Cd: The solubility of Cd in SnSe at 550°C is 0.5 at% (Odin et al. 1983a).

SnSe–CdSe: The phase diagram of this system, constructed through DTA, XRD, and metallography, is a eutectic type (Figure 8.19) (Leute and Menge 1992). The eutectic composition and temperature are 21 mol% CdSe and 827°C [19 mol% CdSe and 822°C (Galiulin et al. 1981)]. The solubility of SnSe in CdSe could not be detected. According to the data of Galiulin et al. (1981) and Odin et al. (1983a), the solubility of SnSe in CdSe at the eutectic temperature reaches

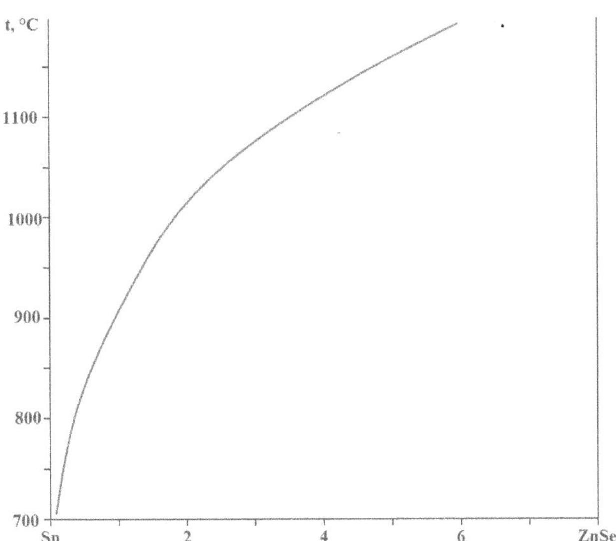

FIGURE 8.17 Temperature dependence of ZnSe solubility in liquid Sn. (From Kimura, S., *J. Chem. Thermodyn.*, **3**(1), 7, 1971.)

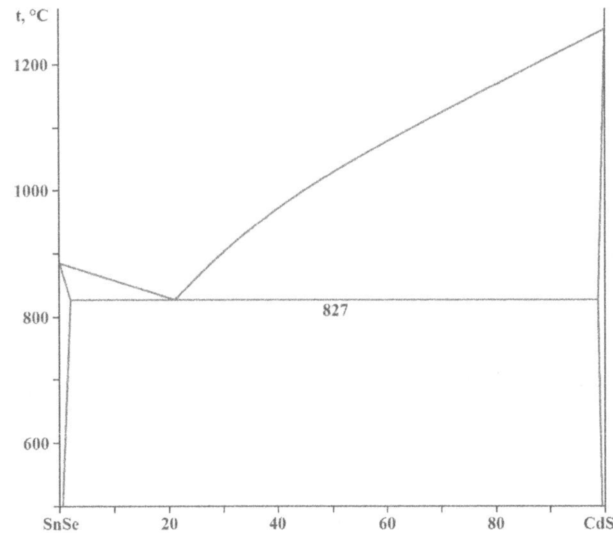

FIGURE 8.19 Phase diagram of the SnSe–CdSe system. (From Leute, V., and Menge, D., *Z. Phys. Chem. (Munchen)*, **176**(1), 47, 1992.)

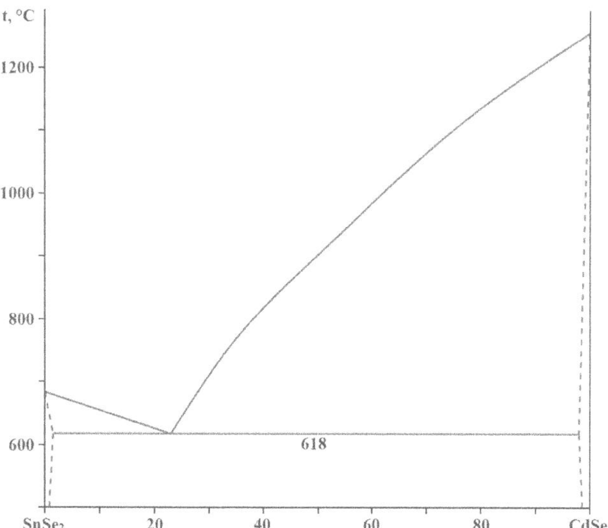

FIGURE 8.20 Phase diagram of the $SnSe_2$–CdSe system. (From Parasyuk, O.V., et al., *Zhurn. neorgan. khimii*, 44(8), 1363, 1999.)

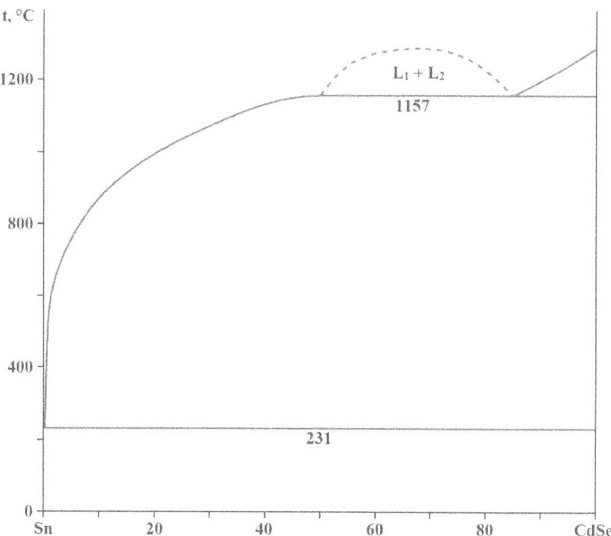

FIGURE 8.21 Phase diagram of the Sn–CdSe system. (From Odin, I.N., et al., *Zhurn. neorgan. khimii*, **28**(3), 764, 1983.)

15 mol% and decreases to 7 mol% at 650°C, and the solubility of CdSe in SnSe at the eutectic temperature is equal to 2 mol% and 1 mol% at 550°C. The ingots for investigations were annealed at 650°C [550°C (Odin et al. 1983a)] for 250 h (Galiulin et al. 1981).

$SnSe_2$–CdSe: The phase diagram of this system, constructed through DTA, XRD, and metallography, is a eutectic type (Figure 8.20) (Galiulin et al. 1982; Parasyuk et al. 1999). The eutectic composition and temperature are 23 mol% CdSe and 618°C [22 mol% CdSe and 620°C (Galiulin et al. 1982)]. The solubility of $SnSe_2$ in CdSe reaches 1 mol% and the solubility of CdSe in $SnSe_2$ is not higher than 0.5 mol% (Galiulin et al. 1982). **$CdSnSe_3$** ternary compound (Radautsan and Ivanova 1961) is not found in the CdSe–$SnSe_2$ system (Galiulin et al. 1982; Parasyuk et al. 1999). The ingots for investigations were annealed at 400°C (Parasyuk et al. 1999) [at 550°C (Galiulin et al. 1982)] for 250 h.

Sn–CdSe: The phase diagram of this system, constructed through DTA, XRD, and metallography, is a eutectic type (Figure 8.21) (Rubenstein 1968; Odin et al. 1983a). The eutectic is degenerated from the Sn-rich side and crystallizes at 231°C (Odin et al. 1983a). The immiscibility region exists in the CdS–Sn system within the interval of 50–85 mol% CdSe with monotectic temperature 1157°C.

The Sn content in the CdSe-doped single crystals reaches 0.76 at% (0.45 mass%) (Odin et al. 1999) (SnSe was used as a dopant). The ingots were annealed at 550°C for 250 h (Odin et al. 1983a). The solubility of CdSe in the liquid Sn was determined using high-temperature filtration (Rubenstein 1968).

At 550°C, the region of solid state is limited by the triangle CdSe–SnSe–$SnSe_2$, and the liquid regions are situated near Se-rich corner and Cd–Sn binary system (Figure 8.22) (Odin et al. 1983a).

There are seven fields of primary crystallization on the liquidus surface of the Sn–Cd–Se ternary system (Figure 8.23) (Odin et al. 1983a). The field of CdSe primary crystallization occupies almost all liquidus surface. Three immiscibility regions exist in this system. One of them begins in the Cd–Se binary system and

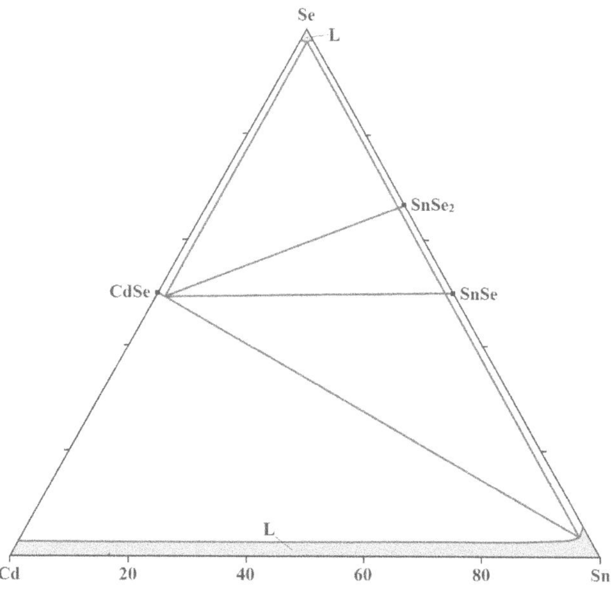

FIGURE 8.22 Isothermal sections of the Sn–Cd–Se ternary system at 550°C. (From Odin, I.N., et al., *Zhurn. neorgan. khimii*, **28**(3), 764, 1983.)

occupies a part of CdSe crystallization field, the second occupies a part of SnSe crystallization field and is situated near the Sn–Se binary system and the third is situated inside the ternary system in the field of CdSe crystallization. The next invariant equilibria exist in the system: E_1 (212°C) – L ⇔ CdSe + $SnSe_2$ + Se; E_2 (605°C) – L ⇔ CdSe + $SnSe_2$ + SnSe; E_3 (230°C) – L ⇔ CdSe + SnSe + Sn; E_4 (190°C) – L ⇔ CdSe + Cd + β; and U (220°C) – L + Sn ⇔ CdSe + β (β is the binary phase in the Cd–Sn system).

8.14 Tin–Mercury–Selenium

SnSe–HgSe: The phase diagram of this system, constructed through DTA, XRD, metallography, and the microhardness and density measurements, is a eutectic type (Figure 8.24)

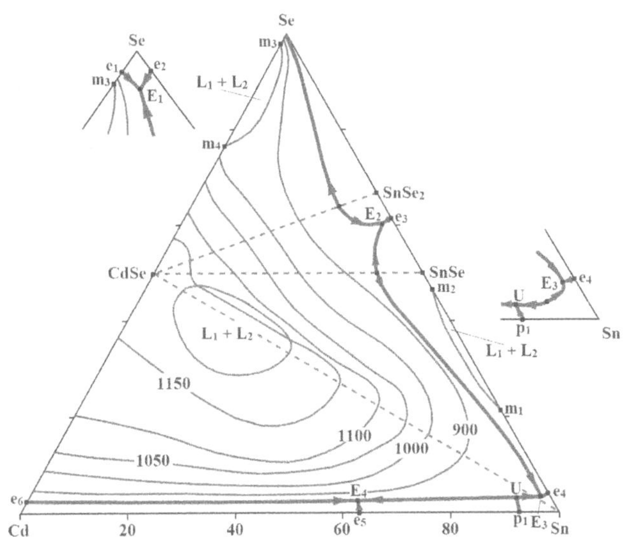

FIGURE 8.23 Liquidus surface of the Sn–Cd–Se ternary system. (From Odin, I.N., et al., *Zhurn. neorgan. khimii*, **28**(3), 764, 1983.)

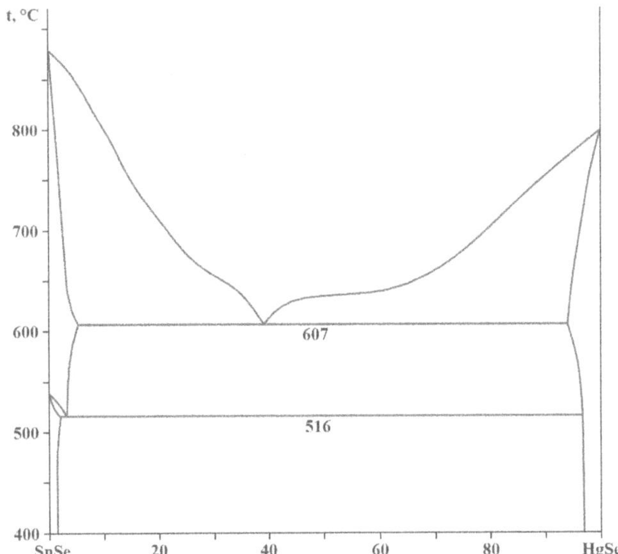

FIGURE 8.24 Phase diagram of the SnSe–HgSe system. (From Voroshilov, Yu.V., et al., *Zhurn. neorgan. khimii*, **38**(6), 1061, 1993.)

(Motrya 1991; Voroshilov et al. 1993). The eutectic composition and temperature are 39 mol% HgSe and 607°C, respectively. The solubility of SnSe in HgSe at eutectic temperature reaches 7 mol% and decreases to 5 mol% at 497°C. The solubility of HgSe in SnSe at the same temperatures is equal to 3 and not higher than 2 mol%, respectively. In the solid-solution region, the temperature of SnSe polymorphic transformation decreases to 516°C. The ingots for investigations were annealed at 400°C and 500°C for 1200 h.

$SnSe_2$–HgSe: The phase diagram of this system, constructed through DTA, XRD, metallography, and the microhardness and density measurements, is shown in Figure 8.25 (Motrya et al. 1991; Voroshilov et al. 1993). The eutectic composition and temperature are 49 mol% HgSe and 564°C, respectively. **Hg_2SnSe_4** ternary compound, which melts incongruently at 580°C (the composition of peritectic point is 64 mol% HgSe) and has

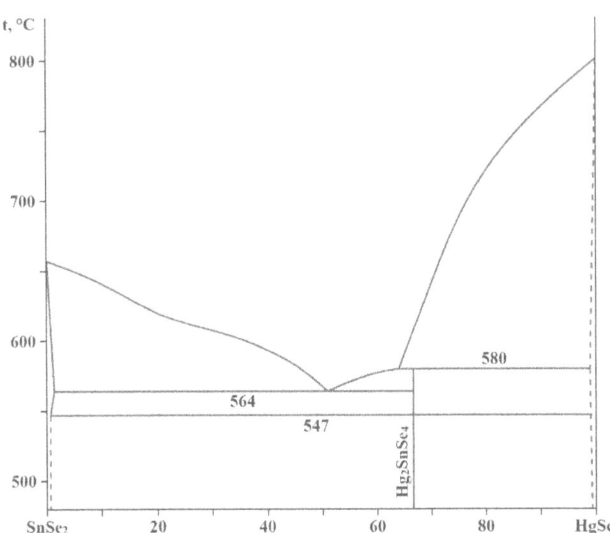

FIGURE 8.25 Phase diagram of the $SnSe_2$–HgSe system. (From Voroshilov, Yu.V., et al., *Zhurn. neorgan. khimii*, **38**(6), 1061, 1993.)

polymorphous transformation at 547°C, is formed in this system. This compound crystallizes in a thiogallate structure (defect chalcopyrite) with lattice parameters a = 577.70 ± 0.02 and c = 1155.70 ± 0.07 pm (Gulay and Parasyuk 2002) [a = 577.9 ± 0.2, c = 1155.8 ± 0.3 pm, and the calculated and experimental densities 7.13 and 7.03·cm^{-3}, respectively (Voroshilov et al. 1993), a = 578 ± 2 and c = 1155 ± 2 pm (Motrya et al. 1986b)].

The solubility of Hg_2SnSe_4 in HgSe and $SnSe_2$ is not higher than 2 mol%. The ingots for investigations were annealed at 400°C and 500°C for 1200 h (Motrya 1991; Voroshilov et al. 1993). Single crystals of Hg_2SnSe_4 were grown by crystallization from the solution in the melt (Voroshilov et al. 1993).

The isothermal section of the Sn–Hg–Se ternary system at 400°C is shown in Figure 8.26 (Motrya 1991; Voroshilov

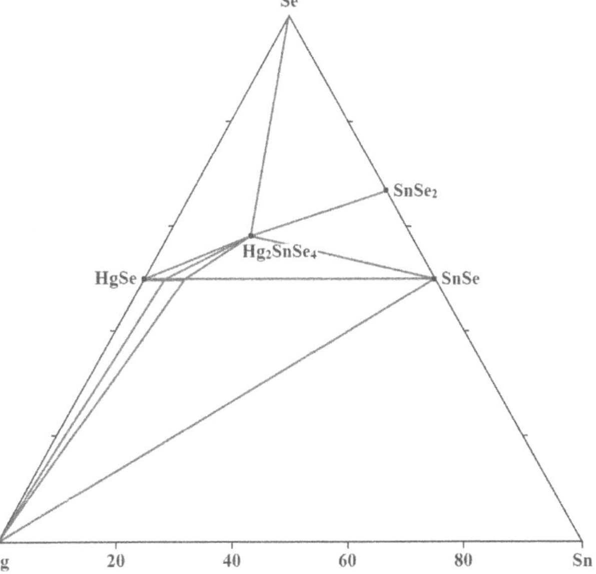

FIGURE 8.26 Isothermal sections of the Sn–Hg–Se ternary system at 400°C. (From Voroshilov, Yu.V., et al., *Zhurn. neorgan. khimii*, **38**(6), 1061, 1993.)

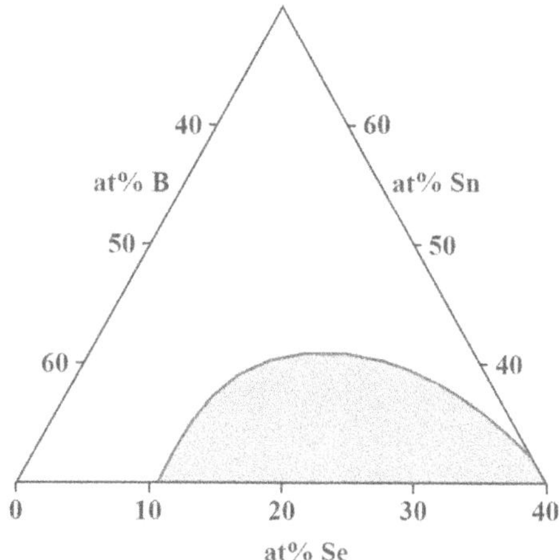

FIGURE 8.27 Glass forming region in the Sn–B–Se ternary system. (From Kirilenko, V.V. and Dembovskiy, S.A., *Izv. AN SSSR. Neorgan. mater.*, **10**(3), 542, 1974.)

et al. 1993). Hg_2SnSe_4 ternary compound exists at this temperature and it is in equilibria with HgSe, Se, $SnSe_2$, and SnSe. Homogeneity region of this compound is insignificant.

8.15 Tin–Boron–Selenium

The glass forming region in the Sn–B–Se ternary system is given in Figure 8.27 (Kirilenko and Dembovskiy 1974). The obtained black glass-like alloys are unstable in air.

8.16 Tin–Gallium–Selenium

SnSe–Ga: The solubility of Ga in SnSe reaches 3 at% (Rustamov et al. 1976b).

SnSe–GaSe: The phase diagram of this system, constructed through DTA, XRD, metallography, and the microhardness and density measurements, is a eutectic type (Figure 8.28) (Dovletov et al. 1978). The solubility of SnSe in GaSe reaches 20 mol% and the solubility of GaSe in SnSe is 50 mol%. The change in the unit cell parameters in both regions of solid solutions is insignificant. The ingots for investigations were annealed at 450°C–500°C for 300 h.

According to the data of Alapini et al. (1981), thermal effects were found in the SnSe–GaSe system at 670°C and 696°C, the first of which is due to the presence of a ternary eutectic in the SnSe–GaSe–Ga_4SnSe_7 subsystem, and the second corresponds to the crystallization of a binary eutectic at 65 mol% SnSe.

SnSe–Ga$_2$Se$_3$: The phase diagram of this system, constructed through DTA and XRD, is a eutectic type with peritectic transformation (Figure 8.29) (Alapini et al. 1980, 1981). The eutectic crystallizes at 680°C. The **Ga$_4$SnSe$_7$** ternary compound, which melts congruently at 715°C and decomposes at the cooling at 565°C, is formed in this system. It is thermally stable up to 600°C under a N_2 atmosphere and crystallizes in the monoclinic structure with the lattice parameters $a = 757.7 \pm 0.4$, $b = 666.6 \pm 0.3$, $c = 1302.3 \pm 0.8$ pm, $\beta = 106.680 \pm 0.007°$, a calculated density of 5.009 g·cm^{-3}, and an energy gap of 2.55 eV (Luo et al. 2014a) [in the orthorhombic structure with the lattice parameters $a = 569$, $b = 1237$, $c = 760$, and the calculated and experimental densities of 5.10 and 5.08 g·cm^{-3}, respectively (Alapini et al. 1980, 1981)]. To prepare the title compound, stoichiometric amounts of Sn (1.29 mM), Ga (1.29 mM), and Se (3.24 mM) were mixed roughly and loaded into a silica tube, which was sealed under vacuum (Luo et al. 2014a). The mixture was heated to 850°C within 48 h, kept at this temperature for 96 h, and cooled to 400°C in 100 h, followed by cooling to room temperature in 24 h. Yellow crystals were obtained, which were stable in air for months.

According to the data of Rustamov et al. (1973c), the phase diagram of the SnSe–Ga$_2$Se$_3$ system is a eutectic type. The eutectic contains 66 mol% SnSe and crystallizes at 690°C. The mutual solubility of the initial binary components at the eutectic temperature reaches 12–13 mol%.

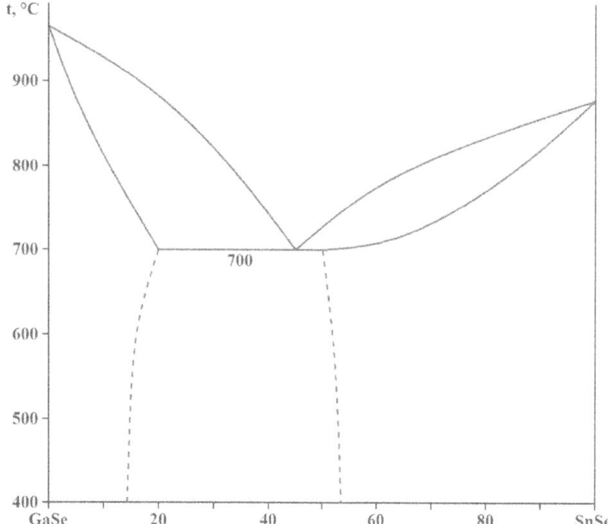

FIGURE 8.28 Phase diagram of the SnSe–GaSe system. (From Dovletov, K., et al, *Izv. AN SSSR. Neorgan. mater.*, **14**(1), 33, 1978.)

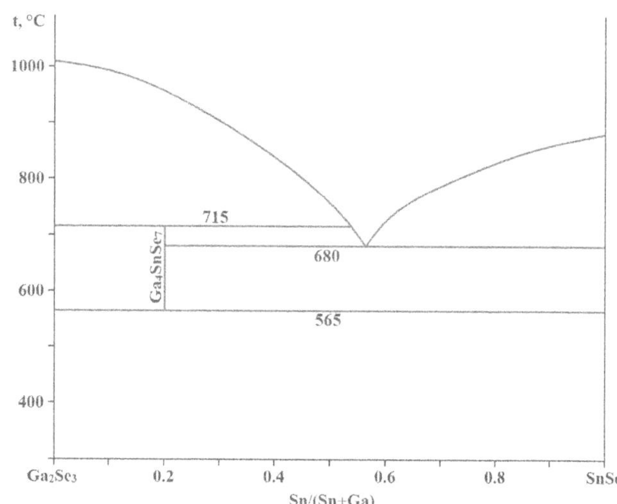

FIGURE 8.29 Phase diagram of the SnSe–Ga$_2$Se$_3$ system. (From Alapini, F., et al., *C. r. Acad. sci. Sér. C*, **290**(22), 433–435, 1980.)

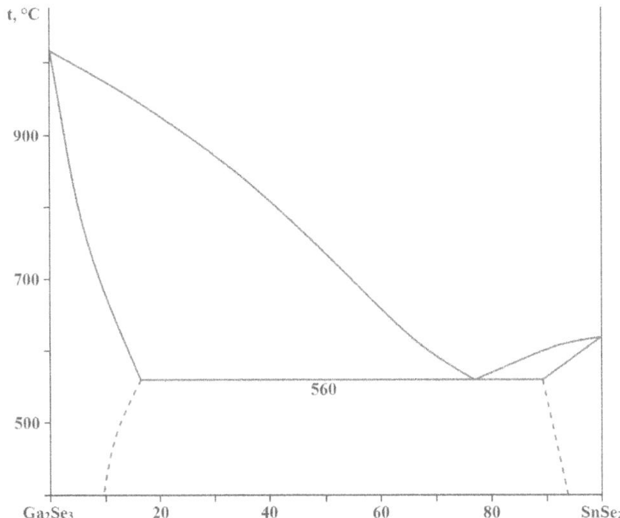

FIGURE 8.30 Phase diagram of the $SnSe_2$–Ga_2Se_3 system. (From Gadzhieva, A.Z., et al., *Uch. zap. Azerb. Univ. Ser. khim. nauk*, (1), 15, 1976.)

$SnSe_2$–Ga_2Se_3: The phase diagram of this system, constructed through DTA, XRD, metallography, and the microhardness and density measurements, is presented in Figure 8.30 (Gadzhieva et al. 1976). The eutectic contains 23 mol% Ga_2Se_3 and crystallizes at 560°C. The solubility of $SnSe_2$ in Ga_2Se_3 reaches 10 mol% and the solubility of Ga_2Se_3 in $SnSe_2$ is 5 mol%.

On the liquidus surface of the Sn–Ga–Se ternary system (Figure 8.31), the ternary eutectic E_1 (217°C) in the $SnSe_2$–Ga_2Se_3–Se subsystem is degenerate from the Se-side, and in the SnSe–$SnSe_2$–Ga_2Se_3 subsystem the eutectic crystallizes at 570°C and has a composition $Ga_{13}Sn_{28}Se_{58}$ (Alapini et al. 1981). In the SnSe–GaSe–Ga_2Se_3 subsystem the ternary eutectic crystallizes at 670°C. No ternary peritectics due to the formation of the Ga_4SnSe_7 compound was found in the system.

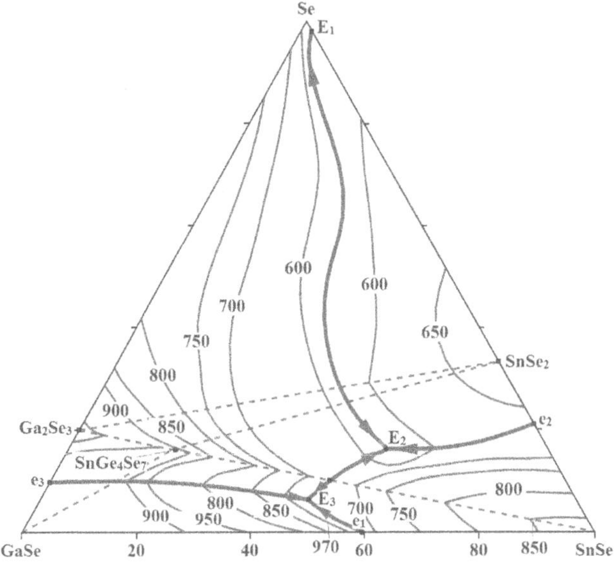

FIGURE 8.31 Liquidus surface of the Sn–Ga–Se ternary system. (From Alapini, F., et al., *Ann. chim. (France)*, **6**(6), 501, 1981.)

8.17 Tin–Indium–Selenium

SnSe–In: This system is non-quasibinary section of the Sn–In–Se ternary system (Rustamov et al. 1976c). In the range of 35–88 at% In, an immiscibility region is observed in the system. At 125°C, a peritectic reaction occurs with the formation of In_2Se, and the isothermal lines at 100°C and 180°C correspond to the melting temperatures of ternary eutectics.

SnSe–InSe: The phase diagram of this system, constructed through DTA, XRD, metallography, and the microhardness and density measurements, is a eutectic type (Figure 8.32) (Dovletov et al. 1974a, 1978). The eutectic contains 40 mol% SnSe and crystallizes at 540°C. The solubility of SnSe in InSe reaches 25 mol% [12 mol% (Rustamov et al. 1976b)] and the solubility of InSe in SnSe is 30 mol%. The unit cell parameters of solid solutions based on InSe change according to Vegard's law, and those of solid solutions based on SnSe change with a deviation from this law (Dovletov et al. 1974a; Rustamov et al. 1976c). The ingots for investigations were annealed at 450°C–500°C for 300 h (Dovletov et al. 1978).

SnSe–In_2Se_3: The phase diagram of this system, constructed through DTA, XRD, metallography, and the microhardness and density measurements, is given in Figure 8.33 (Rustamov et al. 1974c). The eutectic contains 63 mol% SnSe and crystallizes at 695°C. The temperature of phase transitions of In_2Se_3 increases with the addition of SnSe. All transitions are carried out by peritectoid reactions at 300°C, 670°C, and 820°C. The solubility of SnSe in In_2Se_3 reaches 5 mol% and the solubility of In_2Se_3 in SnSe is 17.5 mol%. The ingots for investigations were annealed at 500°C for 125 h.

$SnSe_2$–In_2Se_3: The phase diagram of this system, constructed through DTA, XRD, metallography, and the microhardness and density measurements, is given in Figure 8.34 (Gadzhieva et al. 1973). The eutectic contains 34 mol% In_2Se_3 and crystallizes at 590°C. The solubility of $SnSe_2$ in high-temperature δ-In_2Se_3 reaches 16 mol%. As the content of $SnSe_2$ increases, the δ-In_2Se_3 → γ-In_2Se_3 phase transition temperature

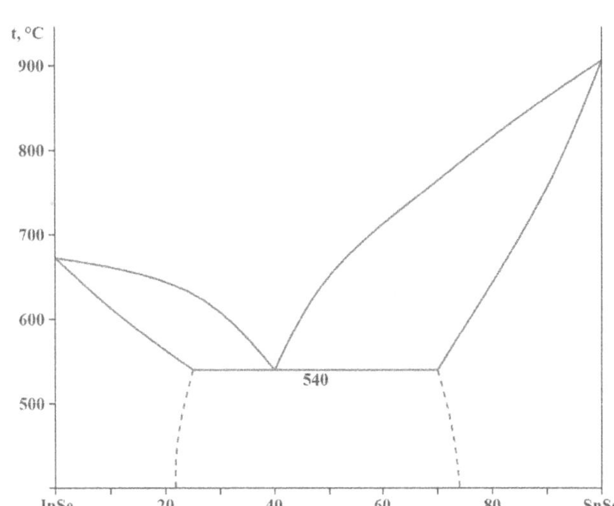

FIGURE 8.32 Phase diagram of the SnSe–InSe system. (From Dovletov, K., et al, *Izv. AN SSSR. Neorgan. mater.*, **14**(1), 33, 1978.)

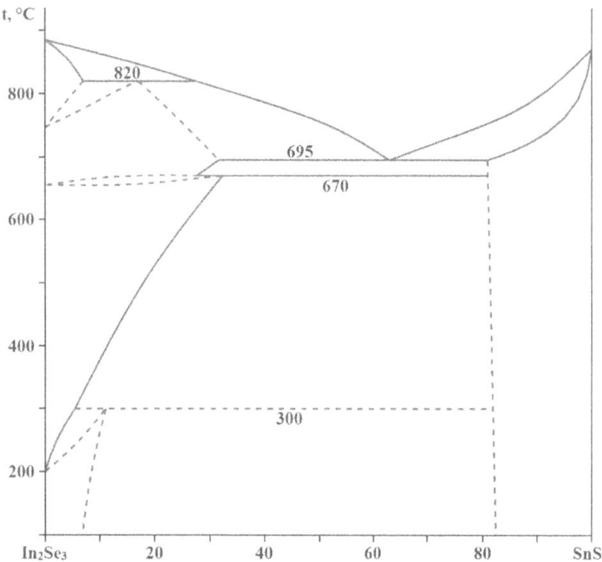

FIGURE 8.33 Phase diagram of the SnSe–In_2Se_3 system. (From Rustamov, P.G., et al., *Izv. AN SSSR. Neorgan. mater.*, **10**(10), 1796, 1974.)

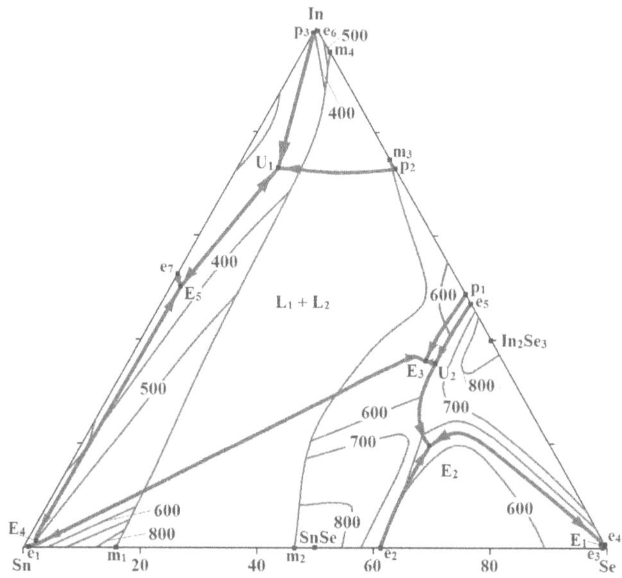

FIGURE 8.35 Liquidus surface of the Sn–In–Se ternary system. (From Rustamov, P.G., et al., *Zhurn. neorgan. khimii*, **21**(2), 518, 1976.)

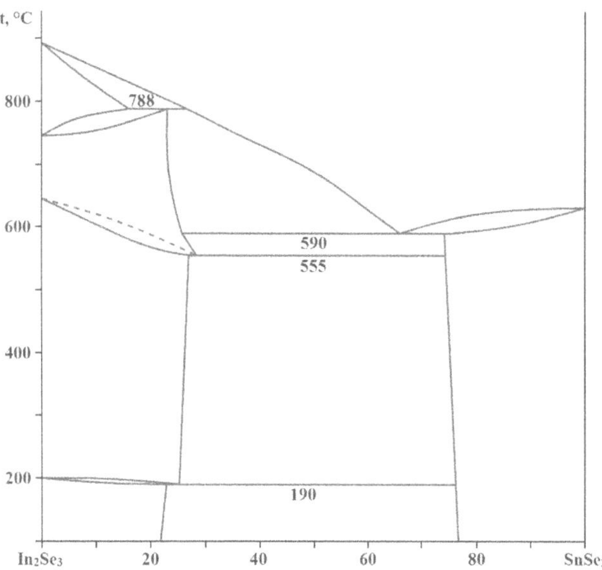

FIGURE 8.34 Phase diagram of the $SnSe_2$–In_2Se_3 system. (From Gadzhieva, A.Z., et al., *Azerb. khim. zhurn.*, (4), 138, 1973.)

also increases. At 555°C, β-In_2Se_3 dissolves up to 27 mol% $SnSe_2$. The low-temperature α-In_2Se_3 dissolves up to 21 mol% $SnSe_2$, while the β-In_2Se_3 → α-In_2Se_3 transformation temperature decreases. In alloys containing 22–80 mol% $SnSe_2$, this transformation takes place at 190°C. The ingots for investigations were annealed at 500°C for 200 h.

According to the data of Deiseroth and Pfeifer (1993b), **In_4SnSe_4** ternary compound is formed in the Sn–In–Se system. It crystallizes in the cubic structure with the lattice parameter $a = 1269.4 \pm 0.2$ pm and may be prepared from the pure elements in stoichiometric amounts or from SnSe, InSe, and In (molar ratio 1:3:1) in an evacuated quartz ampoule at 560°C.

The liquidus surface of the Sn–In–Se ternary system (Figure 8.35) includes nine fields of primary crystallization of Sn, In, Se, In_2Se_3, InSe, In_2Se, In_5Se_6, SnSe, and $SnSe_2$ and immiscibility region (Rustamov et al. 1976c). There are five ternary eutectics and two transition points in this system. The isothermal sections of the Sn–In–Se ternary system at 500°C and at room temperature were constructed by Mamedova et al. (1984).

8.18 Tin–Thallium–Selenium

SnSe–TlSe: The phase diagram of this system, constructed through DTA, XRD, metallography, and the microhardness and density measurements, is shown in Figure 8.36a (Houenou et al. 1979). The eutectics contain 64 and 90 mol% TlSe and crystallize, respectively, at 444°C and 300°C. Three compounds, **Tl_2SnSe_3**, **Tl_3SnSe_4**, and **$Tl_3Sn_2Se_5$**, are formed in this system. Tl_2SnSe_3 melts congruently at 472°C [at 462°C (Peresh et al. 1986)] and Tl_3SnSe_4, and $Tl_3Sn_2Se_5$ melts incongruently, respectively, at 360°C and 490°C.

According to the data of Kuliev and Kagramanyan (1978), Murguzov et al. (1986a), and Malakhovska-Rosokha et al. (2010), only two compounds, **$TlSnSe_2$** and Tl_2SnSe_3, are formed in the SnSe–TlSe system (Figure 8.36b). $TlSnSe_2$ melts incongruently at 397°C (Malakhovska-Rosokha et al. 2010) [at 490°C (Kuliev and Kagramanyan 1978); at 447°C (Murguzov et al. 1986a)]. Tl_2SnSe_3 melts congruently at 459°C (Houenou et al. 1979; Malakhovska-Rosokha et al. 2010) [melts incongruently at 380°C (Kuliev and Kagramanyan 1978); at 418°C (Murguzov et al. 1986a)].

$Tl_3Sn_2Se_5$ melts incongruently at 490°C (Houenou et al. 1979). $TlSnSe_2$ has a polymorphic transformation at 418°C (Kuliev and Kagramanyan 1978). The eutectics contain 17 and 36 mol% SnSe and crystallizes, respectively, at 280°C and 377°C, the peritectic point exists at 38 mol% SnSe and 395°C and metatectic point is at 40 mol% SnSe and 443°C (Malakhovska-Rosokha et al. 2010) [only one eutectic exists in the SnSe–TlSe system, which contains 83.5 mol% TlSe

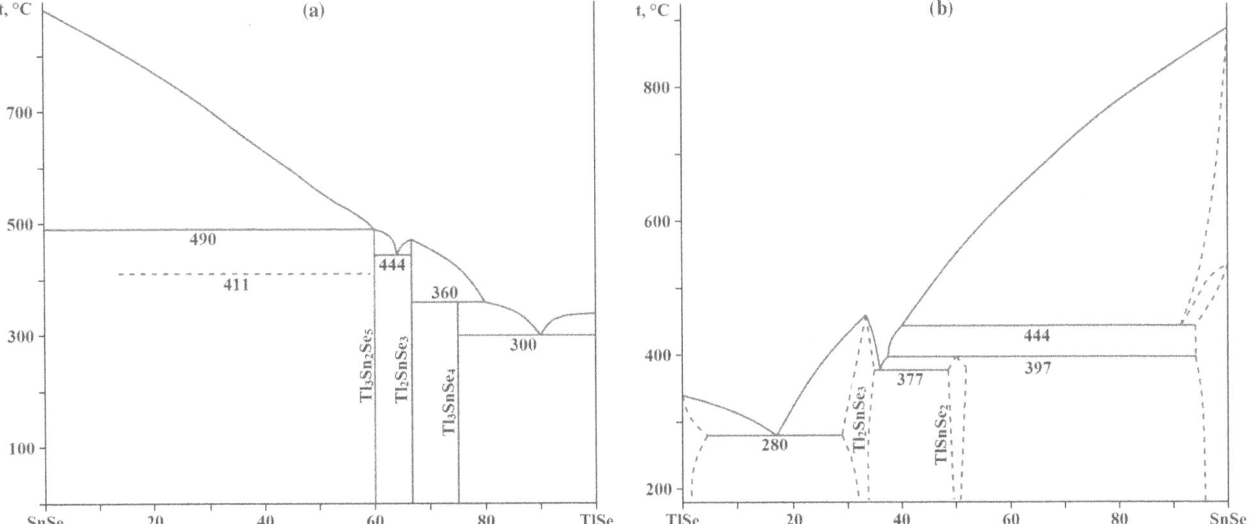

FIGURE 8.36 Phase diagram of the SnSe–TlSe system. (a, from Houenou, P., et al., *C. r. Acad. sci.*, **C288**(6), 193, 1979; b, from Malakhovska-Rosokha, T.O., *Nauk. visn. Uzhgorod. Univ., Ser. Khim.*, (23), 26, 2010.)

and crystallizes at 307°C (Kuliev and Kagramanyan 1978); at 247°C (Murguzov et al. 1986a)]. An increase in the SnSe content leads to a decrease in the temperature of the polymorphic transformation of solid solutions based on SnSe from 535°C to 443°C (Malakhovska-Rosokha et al. 2010).

$TlSnSe_2$ crystallizes in the monoclinic structure with the lattice parameters $a = 1086$, $b = 1064$, $c = 1553$ pm, and $\beta = 101.60°$ (Murguzov et al. 1986a). Single crystals of this compound were grown by the Bridgman technique (Lazarev et al. 1985).

Tl_2SnSe_3 crystallizes in the orthorhombic structure with the lattice parameters $a = 805.1 \pm 0.3$, $b = 816.9 \pm 0.3$, $c = 2124 \pm 1$ pm, and the calculated and experimental density of 7.2 and 7.1 g·cm^{-3}, respectively (Jaulmes and Houenou 1980) [$a = 805$, $b = 817$, $c = 2124$ pm (Houenou et al. 1979); in the tetragonal structure with the lattice parameters $a = 574$, $c = 2131$ pm, and a calculated density of 7.23 g·cm^{-3} (Houenou and Eholié 1976)]. The enthalpy and entropy of melting of this compound is equal to 23.4 kJ·M^{-1} and 31.8 J·(M·K)$^{-1}$, respectively, and an energy gap is 1.3 eV (Peresh et al. 1986).

$Tl_3Sn_2Se_5$ is nonstoichiometric compound and exhibits superstructure phenomenon (Akinocho et al. 1988). It crystallizes in the tetragonal structure with the lattice parameters $8.20 \le a \le 8.43$ and $10.66 \le c \le 10.68$ pm, and $a = 836$ and $c = 1060$ pm for $Tl_3Sn_2Se_5$ composition.

SnSe–Tl_2Se: The phase diagram of this system, constructed through DTA and XRD, is given in Figure 8.37 (Malakhovska et al. 2009a). The eutectic contains 47 mol% SnSe and crystallizes at 390°C, the peritectic point is at 50 mol% SnSe and 410°C and the phase transition of the solid solutions based on SnSe takes place at 491°C. At 300°C, the solubility of SnSe in Tl_2Se and Tl_2Se in SnSe is not higher than 37 and 10 mol%, respectively. The ingots for investigations were annealed at 300°C for 336 h. Two compounds, **$Tl_2Sn_2Se_3$** and **Tl_4SnSe_3**, are formed in this system. $Tl_2Sn_2Se_3$ melts incongruently at 410°C and decomposes at the cooling at 367°C. Tl_4SnSe_3 melts congruently at 433°C and forms continuous solid solutions with Tl_2Se and $Tl_2Sn_2Se_3$. The enthalpy and

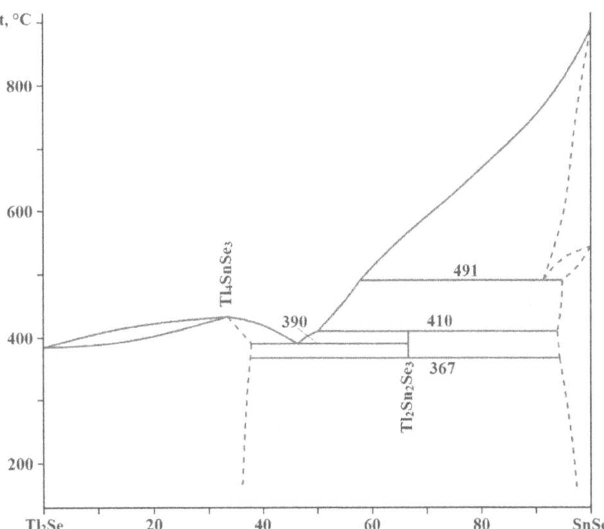

FIGURE 8.37 Phase diagram of the SnSe–Tl_2Se system. (From Malakhovska, T., et al., *Ukr. khim. zhurn.*, **75**(2), 89, 2009.)

entropy of melting of this compound is equal to 95 kJ·M^{-1} and 135 J·(M·K)$^{-1}$, respectively. Single crystals of Tl_4SnSe_3 were grown by Bridgman method.

Tl_4SnSe_3 is a semiconductor and crystallizes in the tetragonal structure with the lattice parameters $a = 852.2 \pm 0.2$, $c = 1272.2 \pm 0.6$ pm, and a calculated density of 8.593 g·cm^{-3} (Bradtmöller et al. 1994) [the calculated and experimental densities of this compound are of 8.43 and 8.25 g·cm^{-3}, respectively (Malakhovska et al. 2009a). It was prepared from a stoichiometric mixture of the elements in evacuated and graphitized quartz ampoule (Bradtmöller et al. 1994). The mixture was careful heated to 500°C followed by 2 months annealing at 380°C and then slowly cooled to room temperature. The needle-shaped crystals of the title compound were obtained. From case to case, traces of Se were deposited in the coldest part of the ampoule.

According to the data of Gotuk et al. (1978a), the phase diagram of the SnSe–Tl_2Se system, constructed through DTA, XRD, and measuring of the microhardness, is a peritectic type. The peritectic interaction takes place at 400°C. The solubility of SnSe in Tl_2Se at 300°C and 400°C is 10 and 12 mol%, respectively, and the solubility of Tl_2Se in SnSe at 300°C and 400°C reaches 2 and 3 mol%. The eutectoid due to the polymorphic transformation of SnSe exists at 92 mol% SnSe and 480°C. The ingots for investigations were annealed at 350°C for 350 h.

$SnSe_2$–TlSe: This system is non-quasibinary section of the Sn–Tl–Se ternary system and intersects three three-phase regions: $SnSe_2$–Tl_2SnSe_3–Se, Tl_2SnSe_3–Tl_3SnSe_4–Se and Tl_3SnSe_4–TlSe–Se (Gotuk et al. 1979a; Houenou et al. 1981). On the $SnSe_2$-side there are three invariant lines at 205°C, 320°C, and 375°C. At 375°C, eutectoid decomposition of Tl_2SnSe_3 occurs, at 320°C, immiscibility takes place, and at 205°C, the ternary eutectic crystallizes. On the TlSe–side there are 5 invariant lines at 290°C, 270°C, 217°C, 200°C, and 195°C.

$SnSe_2$–Tl_2Se: The phase diagram of this system, constructed through DTA and XRD, is presented in Figure 8.38 (Mucha et al. 2009). In spite of stirring the $SnSe_2$+Tl_2Se liquid alloys, supercolling easily occurred, which resulted in considerable scattering of the experimental points. One eutectic is degenerated from the Tl_2Se-side and crystallizes at 371°C. Three other eutectics contain 17.5, 45.5, and 66.0 mol% $SnSe_2$ and crystallize, respectively, at 354°C, 376°C, and 459°C. Five compounds, Tl_2SnSe_3, **$Tl_2Sn_2Se_5$**, **Tl_4SnSe_4**, **$Tl_4Sn_3Se_8$**, and **$Tl_{30}SnSe_{17}$**, are formed in this system [according to the data of Houenou and Eholié (1976) and Lazarev et al. (1985), only three first compounds were determined at the investigation of the $SnSe_2$–Tl_2Se system); the existence of the Tl_4SnSe_4 compound was also confirmed by Bahtiyarov and Vasil'ev (1974), Vasil'ev et al. (1976a), and Gotuk et al. (1977)].

$Tl_2Sn_2Se_5$ melts incongruently and exists within the temperature range 363°C–472°C (Mucha et al. 2009) [within the temperature range 375°C–460°C (Houenou and Eholié 1976)].

Tl_4SnSe_4 melts congruently at 425.9°C [at 445°C (Peresh et al. 1986); at 435°C (Akinocho et al. 1991); melts incongruently at 425°C (Houenou and Eholié 1976)] and is not a phase of variable composition (Mucha et al. 2009). This compound crystallizes in the monoclinic structure with the lattice parameters $a = 848.1 \pm 0.2$, $b = 841.1 \pm 0.2$, $c = 1580.0 \pm 0.5$ pm, $\beta = 102.39 \pm 0.02°$, and a calculated density of 7.56 g·cm^{-3} (Akinocho et al. 1991). The enthalpy and entropy of melting of this compound is equal to 27.1 kJ·M^{-1} and 37.6 J·(M·K)$^{-1}$, respectively, and an energy gap is 1.5 eV (Peresh et al. 1986). Tl_4SnSe_4 was prepared from the stoichiometric mixture of elements in tube sealed under vacuum (Akinocho et al. 1991). The mixture was brought to 435°C and is kept there for 70 days. Slow cooling down to room temperature enables single crystals to be obtained. These are gray in color, with a beautiful metallic luster and parallelepiped shape with truncations. Single crystals of this compound were grown by the Bridgman technique (Lazarev et al. 1985).

$Tl_4Sn_3Se_8$ melts congruently at 475.9°C and undergoes a polymorphic transition at 356°C (Mucha et al. 2009) [at the annealing of the mixture of $SnSe_2$ and Tl_2Se at 250°C for 250 h this compound was not obtained Olekseyuk et al. 2011b].

$Tl_{30}SnSe_{17}$ melts congruently at 389.5°C (Mucha et al. 2009) [at the annealing of the mixture of $SnSe_2$ and Tl_2Se at 250°C for 250 h this compound was not obtained Olekseyuk et al. 2011b].

SnSe–Tl_4SnSe_4: The phase diagram of this system, constructed through DTA and XRD, is a eutectic type (Figure 8.39) (Malakhovskaya-Rosokha et al. 2012). The eutectic contains 49 mol% SnSe and crystallizes at 391°C. The solubility of SnSe in Tl_4SnSe_4 at 300°C is 20 mol% and the solubility of Tl_4SnSe_4 in SnSe at this temperature does not exceed 10 mol%. The ingots for investigations were annealed at 300°C for 336 h.

The phase diagrams of the **Tl_2SnSe_3–Se**, **Tl_4SnSe_3–Tl**, **Tl_4SnSe_3–Sn**, **Tl_4SnSe_4–TlSe**, and **Tl_4SnSe_3–Tl_4SnSe_4**, constructed through DTA and XRD, are also a eutectic

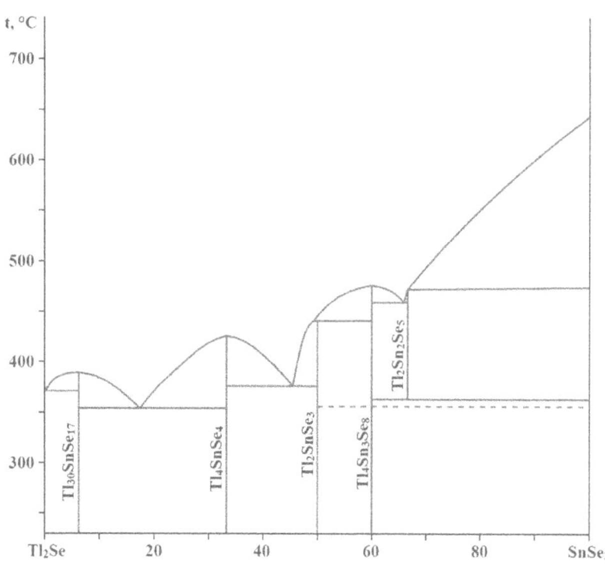

FIGURE 8.38 Phase diagram of the $SnSe_2$–Tl_2Se system. (From Mucha, I., et, *Calphad*, **33**(3), 545, 2009.)

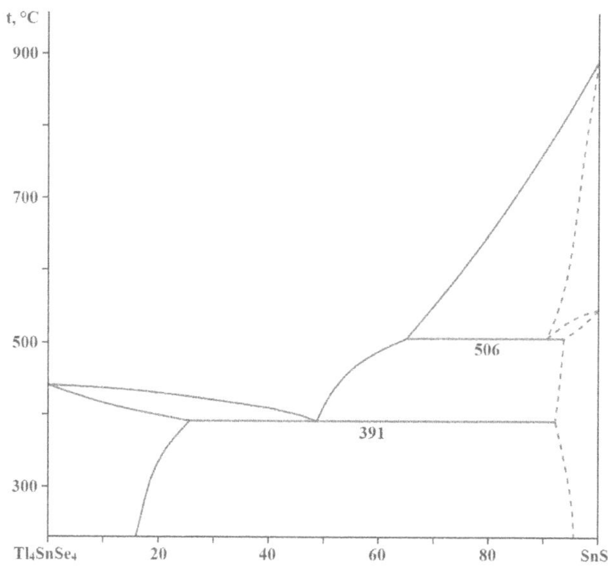

FIGURE 8.39 Phase diagram of the SnSe–Tl_4SnSe_4 system. (From Malakhovskaya-Rosokha, T.A., et al., *Russ. J. Inorg. Chem.*, **57**(3), 439, 2012.)

type (Houenou and Eholié 1976; Malakhovs'ka et al. 2008a; Malakhov'ska-Rosokha et al. 2010; Malakhovskaya-Rosokha et al. 2012).

On the liquidus surface of the $SnSe_2$–Tl_2SnSe_3–TlSe–Se subsystem, which is a part of the Sn–Tl–Se ternary system, there is an immiscibility region along the Tl_2SnSe_3–Se section (Houenou et al. 1981). One ternary eutectic and ternary peritectic are degenerated from the Se-side, and another ternary eutectic crystallizes at 200°C and has a composition $Tl_{25}Sn_{0.67}Se_{74.33}$. The liquidus surfaces of the $SnSe_2$–Tl_2SnSe_3–Se and Tl_2SnSe_3–TlSe–Se subsystem were also constructed by Barchii et al. (2013) and Malakhovs'ka et al. (2018).

8.19 Tin–Scandium–Selenium

SnSe–Sc_2Se_3: No ternary compounds were observed in this system (Gulay et al. 2008c).

8.20 Tin–Yttrium–Selenium

SnSe–Y_2Se_3: No ternary compounds were found in this system (Shemet et al. 2005, 2006d). The solubility of Y_2Se_3 in SnSe and SnSe in Y_2Se_3 was not observed.

8.21 Tin–Lanthanum–Selenium

SnSe–LaSe: The phase diagram of this system, constructed through DTA, XRD, metallography, and measuring of the microhardness, is a eutectic type (Gurshumov et al. 1984). The eutectic contains 5.8 mol% LaSe and crystallizes at 650°C. The solubility of LaSe in SnSe is 3.2 mol% at the eutectic temperature and 1.6 mol% at room temperature. The ingots for investigations were annealed at 200°C and 600°C for 250 h.

$SnSe_2$–La_2Se_3: The phase diagram of this system, constructed through DTA, XRD, metallography, and measuring of the microhardness and density, is given in Figure 8.40 (Rustamov et al. 1971; Samsonov et al. 1973; Nasibov et al. 1975b). The eutectics crystallize at 610°C and 1000°C. At the eutectic temperature, the region of the solid solution based on $SnSe_2$ reaches 3 mol%. Two compounds, **La_2SnSe_5** and **$La_2Sn_3Se_9$**, are formed in the system. Both compounds are semiconductors. La_2SnSe_5 melts congruently at 1120°C and crystallizes in the tetragonal structure with the lattice parameters $a = 615$, $c = 873$ pm, and the calculated and experimental densities of 6.22 and 6.23 $g{\cdot}cm^{-3}$, respectively. $La_2Sn_3Se_9$ melts incongruently at 650°C and has an experimental density of 6.10 $g{\cdot}cm^{-3}$. The ingots for investigations were annealed at 400°C–520°C for 600–650 h.

According to the data of Guittard et al. (1976), the La_2SnSe_5 and $La_2Sn_3Se_9$ compounds do not exist in the SnSe–La_2Se_3 system, and their detection could be explained by an incorrect interpretation of the DTA results. This system is nonquasibinary section of the Sn–La–Se ternary system, and the quasibinary system is the **SnSe–$LaSe_2$** section.

8.22 Tin–Cerium–Selenium

SnSe–CeSe: The phase diagram of this system, constructed through DTA, XRD, metallography, and measuring of the microhardness, is a eutectic type (Gurshumov et al. 1984). The eutectic contains 6.6 mol% CeSe and crystallizes at 670°C. The solubility of CeSe in SnSe is 2.0 mol% at room temperature. The ingots for investigations were annealed at 200°C and 600°C for 250 h.

SnSe–Ce_2Se_3: The phase diagram of this system, constructed through DTA, XRD, metallography, and measuring of the microhardness, is shown in Figure 8.41 (Murguzov 1986b). The eutectic crystallizes at 750°C and the peritectic is at 70 mol% SnSe and 830°C. The **Ce_2SnSe_4** ternary compound, which melts incongruently at 830°C and crystallizes in the orthorhombic structure with the lattice parameters $a = 1180$, $b = 1405$, and $c = 412$ pm, is formed in this system. The ingots for investigations were annealed at 700°C for

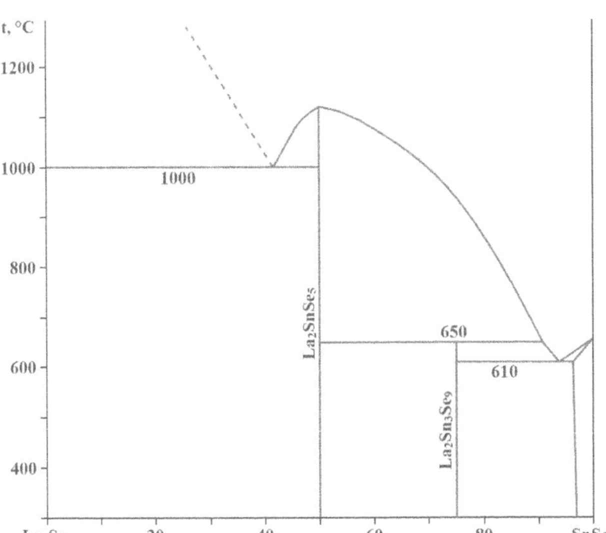

FIGURE 8.40 Phase diagram of the $SnSe_2$–La_2Se_3 system. (From Rustamov, P.G., et al., *Azerb. khim. zhurn.*, (3), 97, 1971.)

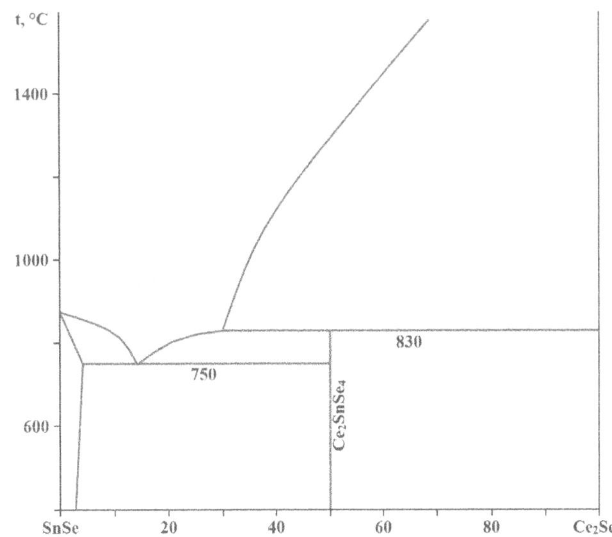

FIGURE 8.41 Phase diagram of the SnSe–Ce_2Se_3 system. (From Murguzov, M.I., et al., *Zhurn. neorgan. khimii*, **31**(7), 1906, 1986.)

200 h. Some of them were additionally annealed at 800°C for 240 h.

$SnSe_2$–Ce_2Se_3: The phase diagram of this system, constructed through DTA, XRD, metallography, and measuring of the microhardness and density, is a eutectic type (Samsonov et al. 1973; Nasibov et al. 1975b). The eutectics crystallize at 500°C, 800°C, and 910°C. At the eutectic temperature, the region of the solid solution based on $SnSe_2$ reaches 3 mol%. The energy gap increases from 0.92 eV for $SnSe_2$ to 1.56 eV for the solid solution containing 3 mol% Ce_2Se_3 (Nasibov et al. 1977). Two compounds, **Ce_2SnSe_5** and **$Ce_2Sn_3Se_9$**, are formed in the system. Both compounds are semiconductors. Ce_2SnSe_5 melts congruently at 1170°C and crystallizes in the tetragonal structure with the lattice parameters a = 612.6, c = 859.5 pm, and the calculated and experimental densities of 6.18 and 6.16 g·cm^{-3}, respectively. $Ce_2Sn_3Se_9$ melts congruently at 900°C, has an experimental density of 4.79 g·cm^{-3} and an energy gap of 0.30 eV. The ingots for investigations were annealed at 400°C–520°C for 600–650 h.

According to the data of Guittard et al. (1976), the Ce_2SnSe_5 and $Ce_2Sn_3Se_9$ compounds do not exist in the SnSe–Ce_2Se_3 system, and their detection could be explained by an incorrect interpretation of the DTA results.

8.23 Tin–Praseodymium–Selenium

SnSe–PrSe: The phase diagram of this system, constructed through DTA, XRD, metallography, and measuring of the microhardness, is a eutectic type (Gurshumov et al. 1986). The eutectic contains 6.0 mol% PrSe and crystallizes at 690°C. The solubility of PrSe in SnSe is 4.0 mol% at the eutectic temperature and 2.5 mol% at room temperature. The ingots for investigations were annealed at 200°C and 600°C for 250 h.

SnSe–Pr_2Se_3: The phase diagram of this system, constructed through DTA, XRD, metallography, and measuring of the microhardness, is shown in Figure 8.42 (Gurshumov et al. 1986a). The eutectic contains 8 mol% Pr_2Se_2 and crystallizes at 750°C. The **Pr_2SnSe_4** ternary compound, which melts incongruently at 860°C, is formed in this system. The ingots for investigations were annealed at 650°C–670°C for 200 h.

$SnSe_2$–Pr_2Se_3: The phase diagram of this system, constructed through DTA, XRD, metallography, and measuring of the microhardness and density, is a eutectic type with peritectic transformation (Sultanov et al. 1971b; Samsonov et al. 1973; Nasibov et al. 1975b). The eutectics crystallize at 600°C and 875°C. At the eutectic temperature, the region of the solid solution based on $SnSe_2$ reaches 3 mol%. Two compounds, **Pr_2SnSe_5** and **$Pr_2Sn_3Se_9$**, are formed in the system. Both compounds are semiconductors. Pr_2SnSe_5 melts congruently at 1025°C and crystallizes in the tetragonal structure with the lattice parameters a = 598.5, c = 847.0 pm, and the calculated and experimental densities of 6.02 and 6.00 g·cm^{-3}, respectively. $Pr_2Sn_3Se_9$ melts incongruently at 750°C, has an experimental density of 6.47 g·cm^{-3} and an energy gap of 0.30 eV. The ingots for investigations were annealed at 400°C–520°C for 600–650 h.

According to the data of Guittard et al. (1976), the Pr_2SnSe_5 and $Pr_2Sn_3Se_9$ compounds do not exist in the SnSe–Pr_2Se_3 system, and their detection could be explained by an incorrect interpretation of the DTA results.

8.24 Tin–Neodymium–Selenium

SnSe–NdSe: The phase diagram of this system, constructed through DTA, XRD, metallography, and measuring of the microhardness, is a eutectic type (Gurshumov et al. 1986). The eutectic contains 7.0 mol% NdSe and crystallizes at 680°C. The solubility of NdSe in SnSe is 5.0 mol% at the eutectic temperature. The ingots for investigations were annealed at 200°C and 600°C for 250 h.

SnSe–Nd_2Se_3: The phase diagram of this system, constructed through DTA, XRD, metallography, and measuring of the microhardness, is a eutectic type with peritectic transformation (Figure 8.43) (Gurshumov et al. 1986a; Orudzhev et al. 1988). The eutectic crystallizes at 730°C. The solubility of

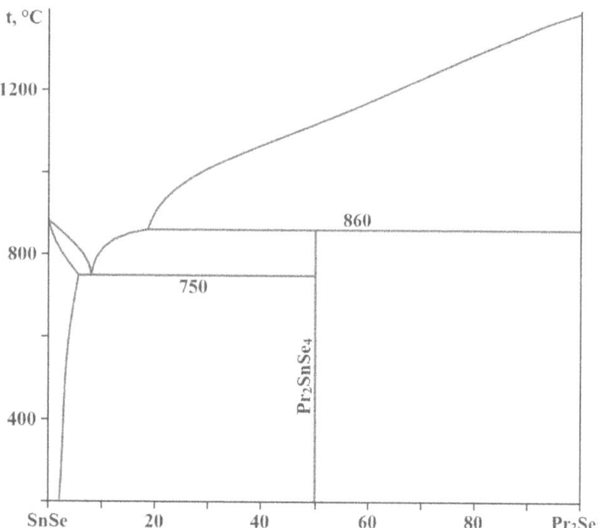

FIGURE 8.42 Phase diagram of the SnSe–Pr_2Se_3 system. (From Gurshumov, A.P., et al., *Izv. AN SSSR. Neorgan. mater.*, **22**(4), 587, 1986.)

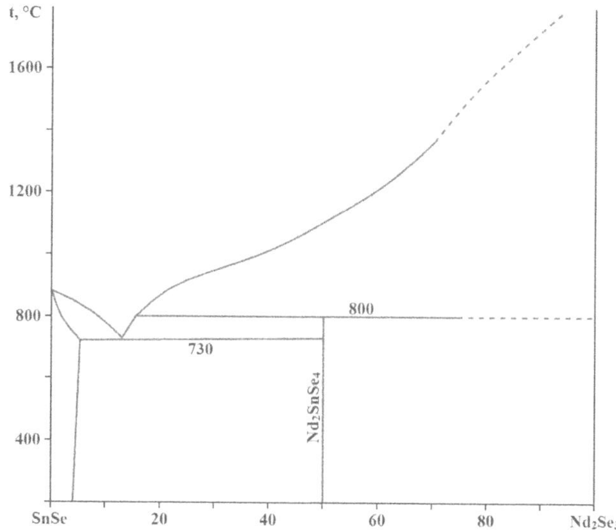

FIGURE 8.43 Phase diagram of the SnSe–Nd_2Se_3 system. (From Orudzhev, N.M., et al., *Izv. AN SSSR. Neorgan. mater.*, **24**(10), 1606, 1988.)

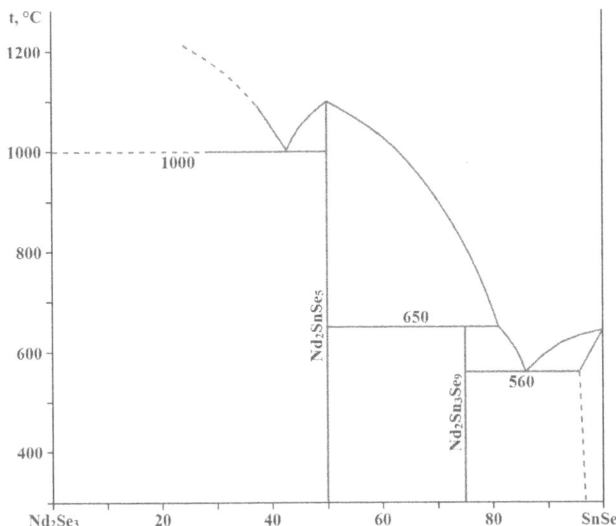

FIGURE 8.44 Phase diagram of the $SnSe_2$–Nd_2Se_3 system. (From Nasibov, I.O., et al., *Azerb. khim. zhurn.*, (1), 99, 1972.)

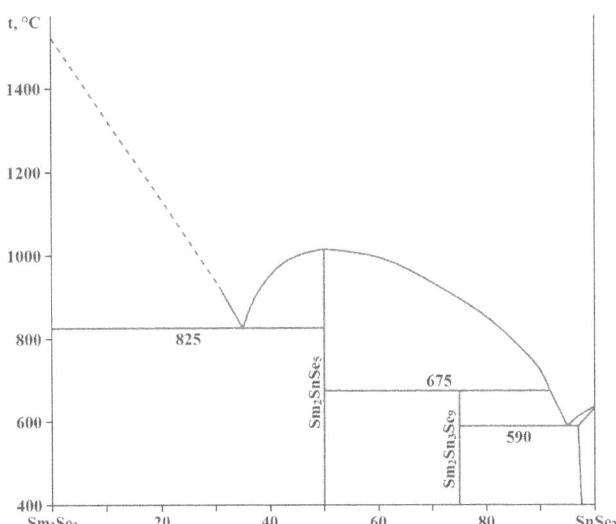

FIGURE 8.45 Phase diagram of the $SnSe_2$–Sm_2Se_3 system. (From Sultanov, T.I., et al., *Uch. zap. Azerb. Univ. Ser. khim. nauk*, (3), 3, 1971.)

Nd_2Se_3 in SnSe is 2 mol% at room temperature. The $\mathbf{Nd_2SnSe_4}$ ternary compound, which melts incongruently at 800°C and crystallizes in the orthorhombic structure with the lattice parameters $a = 1182$, $b = 1169$, $c = 659$ pm, and the calculated and experimental densities of 4.21 and 4.16 g·cm^{-3}, respectively, is formed in this system. The thermodynamic properties of Nd_2SnSe_4 are the next: at the formation from SnSe and Nd_2Se_3 $\Delta G^0_{298} = -19.22$ kJ·M^{-1}, $\Delta H^0_{298} = -21.96$ kJ·M^{-1}, $\Delta S^0_{298} = -9.15$ J·K^{-1}M^{-1} and at the formation from the elements $\Delta G^0_{298} = -34.99$ kJ·M^{-1}, $\Delta H^0_{298} = -31.76$ kJ·M^{-1}, $\Delta S^0_{298} = -14.31$ J·K^{-1}M^{-1} and $S^0_{298} = 39.89$ J·K^{-1}M^{-1}. The ingots for investigations were annealed at 650°C–700°C for 140 h.

$SnSe_2$–Nd_2Se_3: The phase diagram of this system, constructed through DTA, XRD, metallography, and measuring of the microhardness and density, is a eutectic type with peritectic transformation (Figure 8.44) (Nasibov et al. 1972, 1975b; Samsonov et al. 1973). The eutectics crystallize at 560°C and 1000°C. At the eutectic temperature, the region of the solid solution based on $SnSe_2$ reaches 3 mol%. Two compounds, $\mathbf{Nd_2SnSe_5}$ and $\mathbf{Nd_2Sn_3Se_9}$, are formed in the system. Both compounds are semiconductors. Nd_2SnSe_5 melts congruently at 1100°C and crystallizes in the tetragonal structure with the lattice parameters $a = 587.6$, $c = 837.0$ pm, and the calculated and experimental densities of 6.41 and 6.36 g·cm^{-3}, respectively. $Nd_2Sn_3Se_9$ melts incongruently at 650°C, has an experimental density of 5.68 g·cm^{-3} and an energy gap of 0.45–0.50 eV. The ingots for investigations were annealed at 400°C–520°C for 600–650 h.

According to the data of Guittard et al. (1976), the Nd_2SnSe_5 and $Nd_2Sn_3Se_9$ compounds do not exist in the SnSe–Nd_2Se_3 system, and their detection could be explained by an incorrect interpretation of the DTA results.

8.25 Tin–Samarium–Selenium

SnSe–SmSe: The phase diagram of this system, constructed through DTA, XRD, metallography, and measuring of the microhardness, is a eutectic type (Gurshumov et al. 1984). The eutectic contains 9.0 mol% SmSe and crystallizes at 700°C. The solubility of SmSe in SnSe is 7.0 mol% at the eutectic temperature and 5.0 mol% at room temperature. The ingots for investigations were annealed at 200°C and 600°C for 250 h.

SnSe–Sm_2Se_3: The phase diagram of this system was constructed through DTA, XRD, and metallography (Garayev 1968). The eutectics crystallize at 697°C and 957°C. Two compounds, $\mathbf{Sm_2SnSe_4}$ and $\mathbf{Sm_2Sn_3Se_6}$, are formed in the system. Sm_2SnSe_4 melts congruently at 1125°C and has an experimental density of 7.99 g·cm^{-3} and $Sm_2Sn_3Se_6$ melts congruently at 955°C and has an experimental density of 8.26 g·cm^{-3}. The ingots for investigations were annealed at 450°C and 700°C for 300 h.

$SnSe_2$–Sm_2Se_3: The phase diagram of this system, constructed through DTA, XRD, metallography, and measuring of the microhardness and density, is a eutectic type with peritectic transformation (Figure 8.45) (Sultanov et al. 1971a; Samsonov et al. 1973; Nasibov et al. 1975b). The eutectics crystallize at 590°C and 825°C. At the eutectic temperature, the region of the solid solution based on $SnSe_2$ reaches 3 mol%. Two compounds, $\mathbf{Sm_2SnSe_5}$ and $\mathbf{Sm_2Sn_3Se_9}$, are formed in the system. Both compounds are semiconductors. Sm_2SnSe_5 melts congruently at 1015°C and crystallizes in the tetragonal structure with the lattice parameters $a = 587.6$, $c = 827.7$ pm, and the calculated and experimental densities of 5.72 and 5.69 g·cm^{-3}, respectively. $Sm_2Sn_3Se_9$ melts incongruently at 675°C, has an experimental density of 5.76 g·cm^{-3} and an energy gap of 0.70 eV. The ingots for investigations were annealed at 400°C–520°C for 600–650 h.

According to the data of Guittard et al. (1976), the Sm_2SnSe_5 and $Sm_2Sn_3Se_9$ compounds do not exist in the SnSe–Sm_2Se_3 system, and their detection could be explained by an incorrect interpretation of the DTA results.

8.26 Tin–Europium–Selenium

$SnSe_2$–Eu_2Se_3: The phase diagram of this system, constructed through DTA, XRD, metallography, and measuring of the microhardness, is presented in Figure 8.46 (Nasibov and

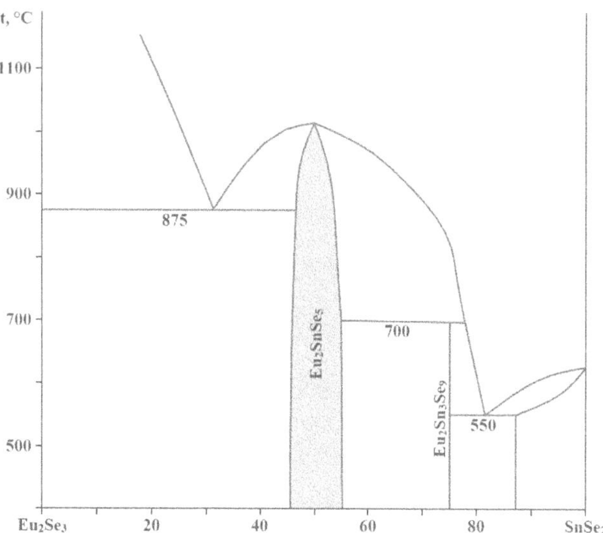

FIGURE 8.46 Phase diagram of the $SnSe_2$–Eu_2Se_3 system. (From Nasibov, I.O., et al., *Izv. AN SSSR. Neorgan. mater.*, **16**(3), 422, 1980.)

Sultanov 1980). The eutectics crystallize at 550°C and 875°C. Two compounds, $\mathbf{Eu_2SnSe_5}$ and $\mathbf{Eu_2Sn_3Se_9}$, are formed in the system. Both compounds are semiconductors. The ingots for investigations were annealed at 500°C for 650 h.

Eu_2SnSe_5 melts congruently and has a narrow homogeneity region. It crystallizes in the orthorhombic structure with the lattice parameters $a = 1199.0 \pm 0.2$, $b = 1642.5 \pm 0.4$, $c = 854.3 \pm 0.1$ pm, a calculated density of 6.454 g·cm^{-3} at 169 ± 2 K [the calculated and experimental densities of 6.45 and 6.43 g·cm^{-3}, respectively (Nasibov and Sultanov 1980)], and an energy gap of 1.07 eV (Evenson IV and Dorhout 2001) [E_g = 1.86 eV (Nasibov and Sultanov 1980); E_g = 1.30 eV (Yang et al. 2020)].

The title compound was prepared by reacting Se (0.5205 mM), K_2Se_4 (0.3470 mM), Sn (0.1735 mM), and Eu (0.1735 mM) (Evenson IV and Dorhout 2001). Reactants were loaded into a fused silica ampoule inside an inert atmosphere glove box. The ampoule was flame sealed and placed in a furnace. The furnace was heated to 725°C where it remained for 150 h. The furnace was then allowed to cool back to room temperature at 4°C·h^{-1}. Crystalline products were separated from excess flux with DMF yielding a yellow polycrystalline material and large, well-formed black rods of Eu_2SnSe_5.

Polycrystalline sample of this compound could be also synthesized by a high-temperature solid-state reaction with KI (Yang et al. 2020). Eu_2O_3, Sn, and Se were stoichiometrically weighted with a total mass of 500 mg with additional 600 mg KI. The mixture was firstly ground into fine powder by using an agate mortar and pressed into a pellet. Then the pellet was loaded into a quartz ampoule, which was evacuated to 0.01 Pa and sealed. The ampoule was put into a furnace and heated as follows. The reaction was heated from room temperature to 950°C, and kept at this temperature for 5 days, and then the temperature decreased slowly to 300°C (5°C·h^{-1}). Black block crystals of Eu_2SnSe_5 stable in air and water was obtained.

$Eu_2Sn_3Se_9$ melts incongruently at 700°C, has an experimental density of 6.30 g·cm^{-3} and an energy gap of 1.33 eV, and hydrolyzed in moisture air (Nasibov and Sultanov 1980).

8.27 Tin–Gadolinium–Selenium

SnSe–GdSe: The phase diagram of this system, constructed through DTA, XRD, metallography, and measuring of the microhardness, is a eutectic type (Gurshumov et al. 1984). The eutectic contains 7.5 mol% GdSe and crystallizes at 705°C. The solubility of GdSe in SnSe is 4.0 mol% at the eutectic temperature and 3.0 mol% at room temperature. The ingots for investigations were annealed at 200°C and 600°C for 250 h.

8.28 Tin–Terbium–Selenium

SnSe–TbSe: The phase diagram of this system, constructed through DTA, XRD, metallography, and measuring of the microhardness, is a eutectic type (Gurshumov et al. 1984). The eutectic contains 6.8 mol% TbSe and crystallizes at 720°C. The solubility of TbSe in SnSe is 3.2 mol% at room temperature. The ingots for investigations were annealed at 200°C and 600°C for 250 h.

$SnSe_2$–Tb_2Se_3: The phase diagram of this system, constructed through DTA, XRD, metallography, and measuring of the microhardness, is presented in Figure 8.47 (Nasibov et al. 1978b). The eutectics crystallize at 550°C and 850°C. Three compounds, $\mathbf{Tb_2SnSe_5}$, $\mathbf{Tb_2Sn_2Se_7}$, and $\mathbf{Tb_2Sn_3Se_9}$, are formed in the system. All compounds are semiconductors. Tb_2SnSe_5 melts congruently, has a narrow homogeneous region and an energy gap of 1.72 eV. $Tb_2Sn_2Se_7$ and $Tb_2Sn_3Se_9$ melts incongruently and have an energy gap 1.30 and 1.24 eV, respectively. All compounds are partly hydrolyzed in humid air. The ingots for investigations were annealed at 500°C–550°C for 600 h.

8.29 Tin–Dysprosium–Selenium

SnSe–DySe: The phase diagram of this system, constructed through DTA, XRD, metallography, and measuring of the microhardness and density, is a eutectic type with peritectic transformation (Figure 8.48) (Gurshumov et al. 1984; Aliev

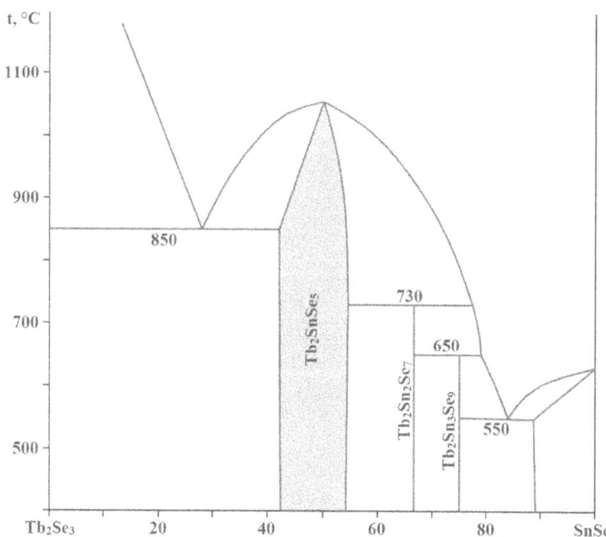

FIGURE 8.47 Phase diagram of the $SnSe_2$–Tb_2Se_3 system. (From Nasibov, I.O., et al., *Izv. AN SSSR. Neorgan. mater.*, **14**(7), 1265, 1978.)

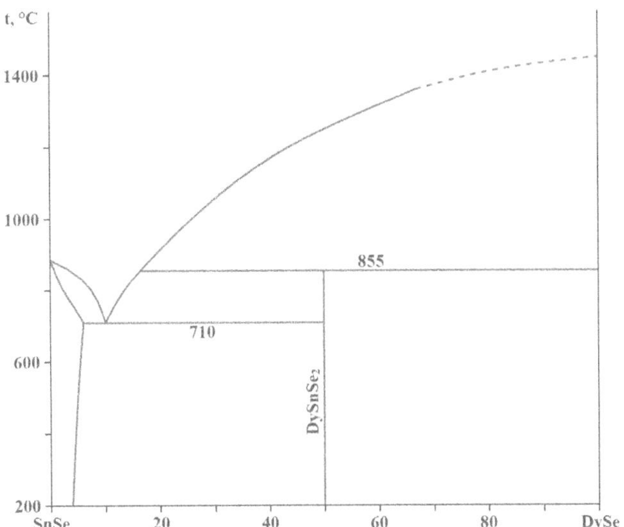

FIGURE 8.48 Phase diagram of the SnSe–DySe system. (From Aliev, I.I., et al., *Inorg. mater.*, **50**(3), 237, 2014.)

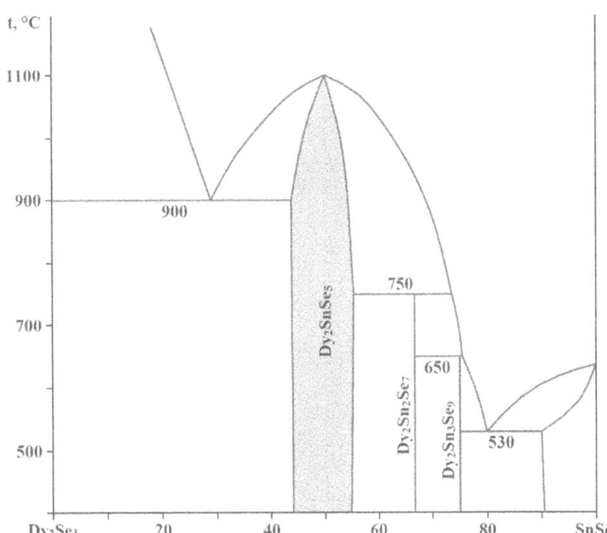

FIGURE 8.50 Phase diagram of the $SnSe_2$–Dy_2Se_3 system. (From Nasibov, I.O., et al., *Uch. zap. Azerb. Univ. Ser. khim. nauk*, (4), 11, 1978.)

et al. 2014). The eutectic contains 10 mol% DySe and crystallizes at 710°C (Aliev et al. 2014) [contains 8.0 mol% DySe and crystallizes at 720°C (Gurshumov et al. 1984)]. The solubility of DySe in SnSe is 6.0 mol% at the eutectic temperature (Gurshumov et al. 1984). The **$DySnSe_2$** ternary compound, which melt incongruently at 855°C and crystallizes in the orthorhombic structure with the lattice parameters $a = 574 \pm 2$, $b = 1049 \pm 3$, $c = 1166 \pm 5$ pm, and the calculated and experimental density of 7.29 and 7.02 $g{\cdot}cm^{-3}$, respectively, is formed in this system (Aliev et al. 2014). The ingots for investigations were annealed at 700°C for 240 h (Aliev et al. 2014) [at 200°C and 600°C for 250 h (Gurshumov et al. 1984)].

SnSe–Dy_2Se_3: The phase diagram of this system, constructed through DTA, XRD, metallography, and measuring of the microhardness and density, is a eutectic type with peritectic transformation (Figure 8.49) (Gurshumov et al. 1990). The eutectic contains 10 mol% Dy_2Se_3 and crystallizes at 780°C. SnSe dissolves up to 3.1 mol% Dy_2Se_3. The **Dy_2SnSe_4** ternary compound, which melt incongruently at 920°C, has polymorphic transformation at 850°C and crystallizes in the hexagonal structure with the lattice parameters $a = 1280$ and $c = 1462$ pm, is formed in this system. The ingots for investigations were annealed at 650°C for 380 h.

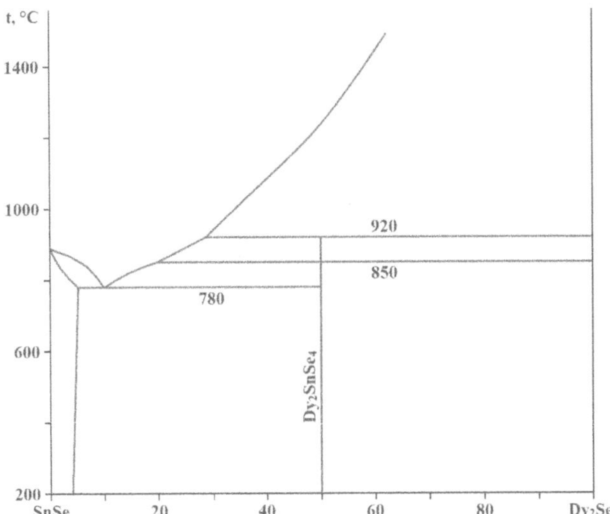

FIGURE 8.49 Phase diagram of the SnSe–Dy_2Se_3 system. (From Gurshumov, A.P., et al., *Izv. AN SSSR. Neorgan. mater.*, **26**(9), 1827, 1990.)

$SnSe_2$–Dy_2Se_3: The phase diagram of this system, constructed through DTA, XRD, metallography, and measuring of the microhardness and density, is shown in Figure 8.50 (Nasibov et al. 1978c, 1979a). The eutectics crystallize at 530°C and 900°C. Three compounds, **Dy_2SnSe_5**, **$Dy_2Sn_2Se_7$**, and **$Dy_2Sn_3Se_9$**, are formed in the system. All compounds are semiconductors. Dy_2SnSe_5 melts congruently at 1100°C, has a narrow homogeneous region and an energy gap of 1.70 eV. $Dy_2Sn_2Se_7$ and $Dy_2Sn_3Se_9$ melts incongruently at 750°C and 650°C and have an energy gap 1.48 and 1.12 eV, respectively. All compounds are partly hydrolyzed in humid air. The ingots for investigations were annealed at 500°C–550°C for 600 h.

8.30 Tin–Holmium–Selenium

SnSe–HoSe: The phase diagram of this system, constructed through DTA, XRD, metallography, and measuring of the microhardness, is a eutectic type (Gurshumov et al. 1984). The eutectic contains 7.8 mol% HoSe and crystallizes at 700°C. The solubility of HoSe in SnSe is 6.0 mol% at the eutectic temperature and 4.1 mol% at room temperature. The ingots for investigations were annealed at 200°C and 600°C for 250 h.

$SnSe_2$–Ho_2Se_3: The phase diagram of this system, constructed through DTA, XRD, metallography, and measuring of the microhardness, is presented in Figure 8.51 (Nasibov and Sultanov 1979). The eutectics crystallize at 950°C. Solid solution based on $SnSe_2$ decomposes peritectically at 700°C. Three compounds, **Ho_2SnSe_5**, **$Ho_2Sn_2Se_7$**, and **$Ho_2Sn_3Se_9$**, are formed in the system. All compounds are semiconductors. Ho_2SnSe_5 melts congruently at 1140°C, has a narrow homogeneous region and an energy gap of 1.71 eV. $Ho_2Sn_2Se_7$ and

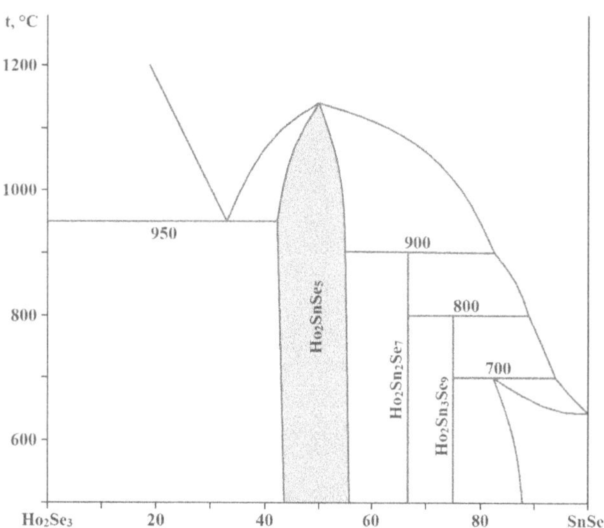

FIGURE 8.51 Phase diagram of the $SnSe_2$–Ho_2Se_3 system. (From Nasibov, I.O., and Sultanov, T.I., *Uch. zap. Azerb. Univ. Ser. khim. nauk*, (1), 34, 1979.)

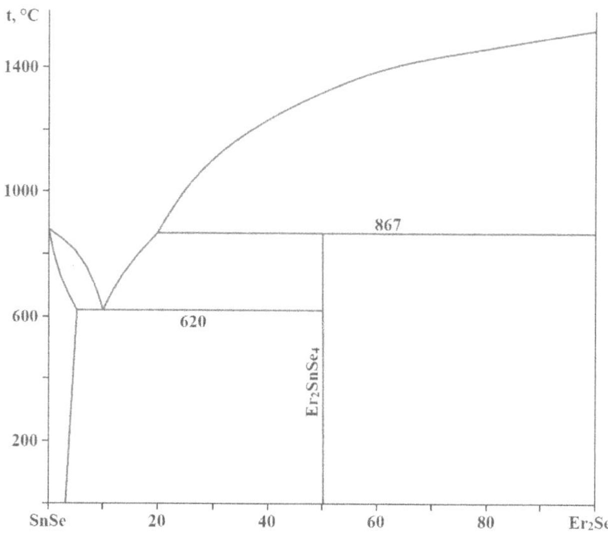

FIGURE 8.53 Phase diagram of the SnSe–Er_2Se_3 system. (From Bagieva, M.R., et al., *Zhurn. neorgan. khimii*, **47**(9), 1545, 2002.)

$Ho_2Sn_3Se_9$ melts incongruently at 900°C and 800°C and have an energy gap 1.45 and 1.10 eV, respectively. All compounds are partly hydrolyzed in humid air. The ingots for investigations were annealed at 500°C–550°C for 600 h.

8.31 Tin–Erbium–Selenium

SnSe–ErSe: The phase diagram of this system, constructed through DTA, XRD, metallography, and measuring of the microhardness, is a eutectic type (Figure 8.52) (Gurshumov et al. 1984; Aliyev et al. 2012). The eutectic contains 10 mol% ErSe (Aliyev et al. 2012) [8.0 mol% GdSe (Gurshumov et al. 1984; Aliev et al. 2011)] and crystallizes at 710°C (Gurshumov et al. 1984; Aliev et al. 2011; Aliyev et al. 2012). At room temperature, SnSe dissolves up to 5 mol% ErSe and the solubility based on ErSe is negligible (Aliev et al. 2011; Aliyev et al. 2012) [the solubility of ErSe in SnSe is 7.0 mol% at the eutectic temperature and 4.4 mol% at room temperature (Gurshumov et al. 1984)]. The **$ErSnSe_2$** ternary compound, which melts incongruently at 885°C (Aliev et al. 2011) [at 835°C (Aliev et al. 2005)] and crystallizes in the orthorhombic structure with the lattice parameters a = 427, b = 943, c = 1030 pm, the calculated and experimental density of 7.10 and 7.03 g·cm^{-3}, respectively, and an energy gap of 0.82 eV (Aliev et al. 2005), is formed in the SnSe–ErSe system. This compound was synthesizes by melting SnSe and ErSe (molar ratio 1:1) in an evacuated up to 0.133 Pa quartz ampoule (Aliev et al. 2005). The ingots for investigations were annealed at 500°C for 240 h (Aliyev et al. 2012) [at 200°C and 600°C for 250 h (Gurshumov et al. 1984)].

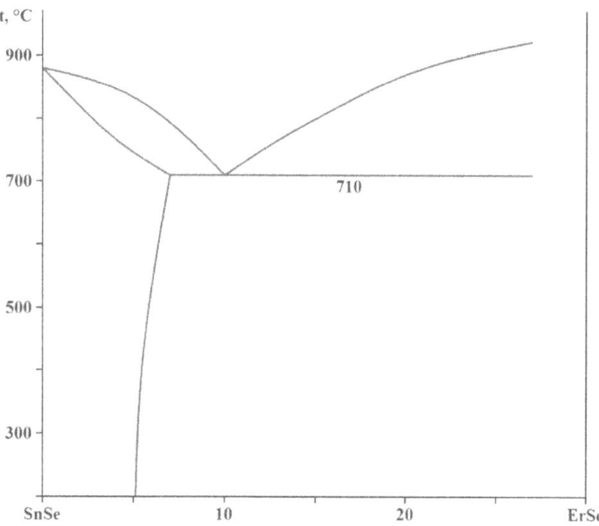

FIGURE 8.52 Part of the phase diagram of the SnSe–ErSe system. (From Aliyev, A.A., et al., *Azerb. khim. zhurn.*, (1), 104, 2012.)

SnSe–Er_2Se_3: The phase diagram of this system, constructed through DTA, XRD, and metallography, is given in Figure 8.53 (Bagieva et al. 2002). The eutectic contains 10 mol% Er_2Se_3 and crystallizes at 620°C. The solubility of Er_2Se_3 in SnSe reaches 3.5 mol%. The **Er_2SnSe_4** ternary compound, which melts incongruently at 867°C and crystallizes in the hexagonal structure with the lattice parameters a = 1276, c = 1454 pm, and an experimental density of 6.54 g·cm^{-3}, is formed in this system. Single crystals of this compound were grown by the chemical transport reactions using iodine as transport agent.

$SnSe_2$–ErSe: This system is a non-quasibinary section of the Sn–Er–Se ternary system (Aliev et al. 2007). The $SnSe_2$-based solid solution extends to 1.5 mol% ErSe.

$SnSe_2$–Er_2Se_3: The phase diagram of this system, constructed through DTA, XRD, metallography, and microhardness and density measurements, is given in Figure 8.54 (Bagieva et al. 2003). The eutectics contain 18 and 30 mol% Er_2Se_3 and crystallize at 600°C and 630°C, respectively. Two compounds are formed in this system: **Er_2SnSe_3** melts incongruently at 830°C and **$Er_2Sn_3Se_9$** melts congruently at 655°C. The ingot for investigations were annealed for 720 h at 600°C

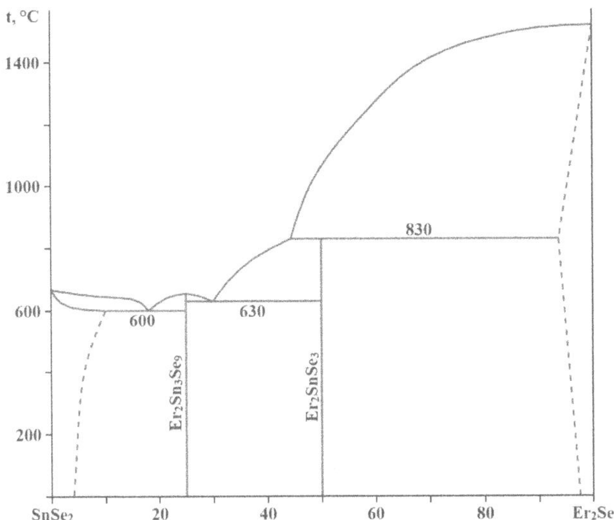

FIGURE 8.54 Phase diagram of the $SnSe_2$–Er_2Se_3 system. (From Bagieva, M.R., et al., *Inorg. mater.*, **39**(9), 927, 2003.)

in the composition range 0–50 mol% Er_2Se_3 and at 800°C between 50 and 100 mol% Er_2Se_3.

SnSe–$Er_2Sn_3Se_9$: The phase diagram of this system, constructed through DTA, XRD, metallography, and microhardness and density measurements, is a eutectic type (Figure 8.55) (Aliev et al. 2004). The eutectic contains 30 mol% SnSe and crystallizes at 620°C. The solid solution based on $Er_2Sn_3Se_9$ contains up to 5 mol% SnSe and solubility of $Er_2Sn_3Se_9$ in SnSe is 1.5 mol%. The ingots for investigations were annealed at 600°C for 500 h.

The liquidus surface of the Sn–Er–Se ternary system (Figure 8.56) consists of 14 fields of primary crystallization (Aliev et al. 2011). The crystallization region of ErSe occupies the biggest part of the ternary system. There are 16 invariant four-phase equilibria in this system, of which nine ternary eutectics and seven transition points.

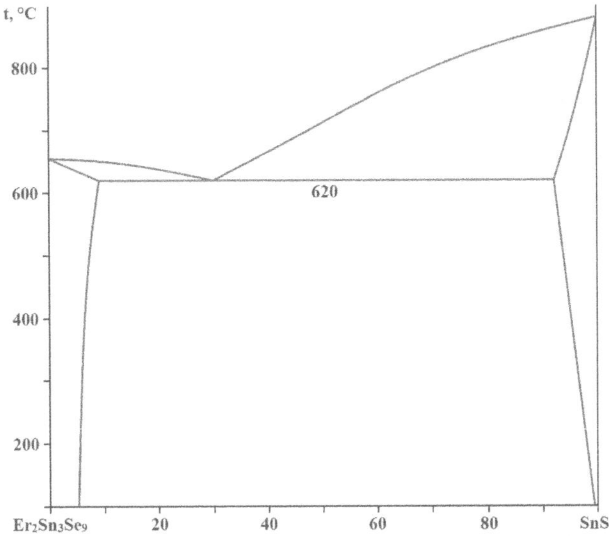

FIGURE 8.55 Phase diagram of the $SnSe_2$–$Er_2Sn_3Se_9$ system. (From Aliev, I.I., et al., *Zhurn. neorgan. khimii*, **49**(11), 1900, 2004.)

8.32 Tin–Ytterbium–Selenium

SnSe–YbSe: The phase diagram of this system, constructed through DTA, XRD, metallography, and measuring of the microhardness, is a eutectic type (Figure 8.57) (Gurshumov et al. 1984; Murguzov et al. 1984a). The eutectic contains 25 mol% YbSe and crystallizes at 685°C (Murguzov et al. 1984a) [at 670°C (Gurshumov et al. 1984)]. The solubility of YbSe in SnSe is 4.5 mol% at room temperature. The ingots for investigations were annealed at 200°C and 600°C for 250 h.

SnSe–Yb_2Se_3: The phase diagram of this system, constructed through DTA, XRD, metallography, and microhardness measurements, is a eutectic type with peritectic transformation (Figure 8.58) (Murguzov et al. 1984a). The eutectic crystallizes at 715°C. The **Yb_2SnSe_4** ternary compound, which melts incongruently at 880°C and crystallizes in the hexagonal structure with the lattice parameters a = 1268 and c = 1442 pm, is formed in the system. Single crystals of Yb_2SnSe_4 were grown through the chemical transport reactions.

$SnSe_2$–Yb_2Se_3: The phase diagram of this system, constructed through DTA, XRD, metallography, and microhardness measurements, is shown in Figure 8.59 (Murguzov et al. 1984a). The eutectics contain 30 and 88 mol% $SnSe_2$ and crystallize at 960°C and 550°C, respectively. Two compounds are formed in this system: **Yb_2SnSe_5** melts congruently at 1240°C and crystallizes in the tetragonal structure with the lattice parameters a = 526, c = 804 pm, and a calculated density of 6.85 g·cm^{-3}, and **$Yb_2Sn_2Se_7$** melts incongruently at 850°C. Single crystals of Yb_2SnSe_5 were grown through the chemical transport reactions.

8.33 Tin–Lutetium–Selenium

SnSe–LuSe: The phase diagram of this system, constructed through DTA, XRD, metallography, and measuring of the microhardness, is a eutectic type (Gurshumov et al. 1984). The eutectic contains 7.85 mol% LuSe and crystallizes at 750°C. The solubility of LuSe in SnSe is 5.0 mol% at the eutectic temperature. The ingots for investigations were annealed at 200°C and 600°C for 250 h.

8.34 Tin–Lead–Selenium

SnSe–PbSe: The phase diagram of this system, constructed through DTA, XRD, metallography, and measuring of the microhardness (Strauss 1968; Zlomanov et al. 1971; Shtanov et al. 1974; Latypov et al. 1976c; Shtanov et al. 1980; Corso et al. 1995) and optimized using a four parameter for the excess Gibbs energy by Volykhov et al. (2008), is a eutectic type (Figure 8.60). The eutectic contains 77 mol% SnSe and crystallizes at 861°C (Volykhov et al. 2008) [75 mol% SnSe and crystallizes at 870°C (Corso et al. 1995); 75 mol% SnSe and 855°C (Zlomanov et al. 1971; Shtanov et al. 1980); 70 mol% SnSe and 858°C (Shtanov et al. 1974); 72 mol% SnSe and 865°C (Latypov et al. 1976c); 870°C (Strauss 1968)]. The solubility limits were estimated

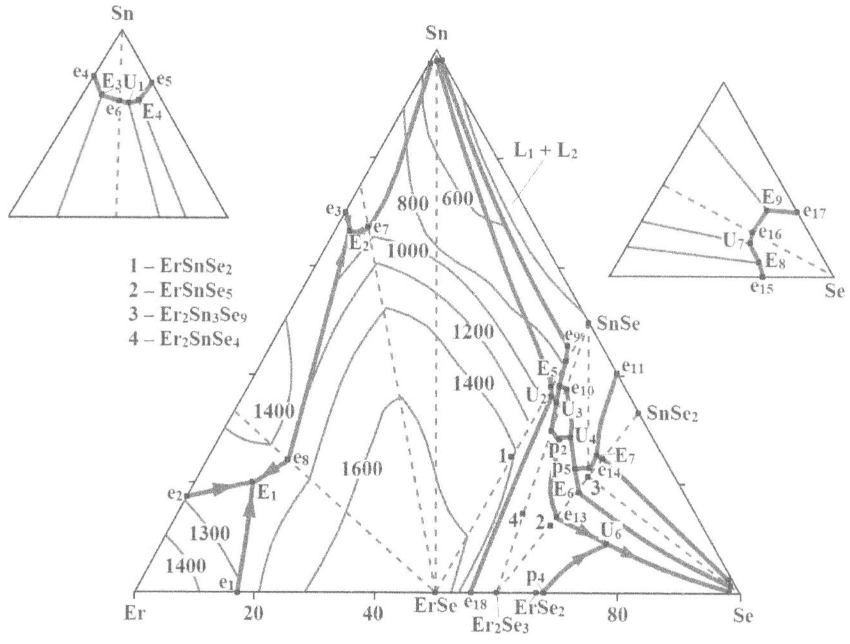

FIGURE 8.56 Liquidus surface of the Sn–Er–Se ternary system. (From Aliev, I.I., et al., *Russ. J. Inorg. Chem.*, **56**(8), 1324, 2011.)

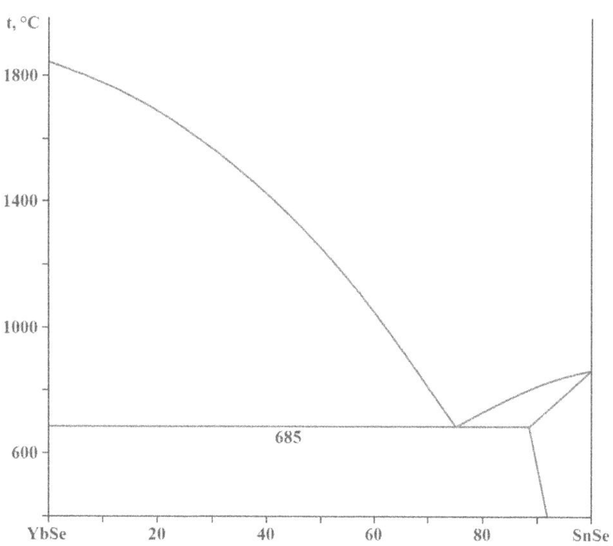

FIGURE 8.57 Phase diagram of the SnSe–YbSe system. (From Murguzov, M.I., et al., *Zhurn. neorgan. khimii*, **29**(10), 2696, 1984.)

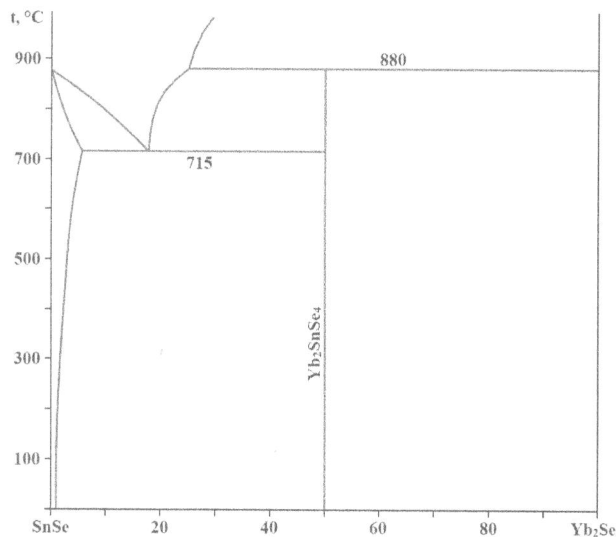

FIGURE 8.58 Phase diagram of the SnSe–Yb_2Se_3 system. (From Murguzov, M.I., et al., *Zhurn. neorgan. khimii*, **29**(10), 2696, 1984.)

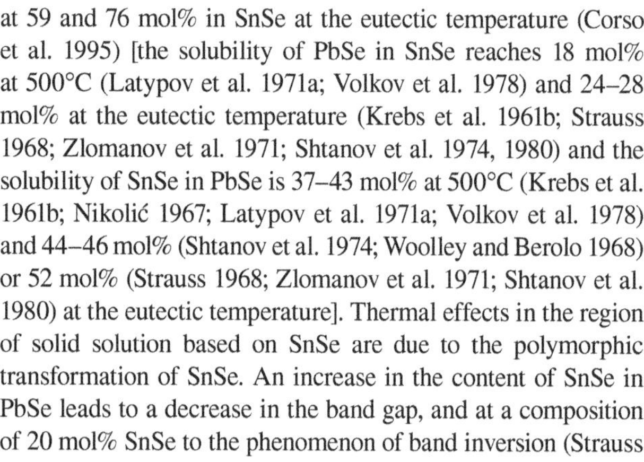

at 59 and 76 mol% in SnSe at the eutectic temperature (Corso et al. 1995) [the solubility of PbSe in SnSe reaches 18 mol% at 500°C (Latypov et al. 1971a; Volkov et al. 1978) and 24–28 mol% at the eutectic temperature (Krebs et al. 1961b; Strauss 1968; Zlomanov et al. 1971; Shtanov et al. 1974, 1980) and the solubility of SnSe in PbSe is 37–43 mol% at 500°C (Krebs et al. 1961b; Nikolić 1967; Latypov et al. 1971a; Volkov et al. 1978) and 44–46 mol% (Shtanov et al. 1974; Woolley and Berolo 1968) or 52 mol% (Strauss 1968; Zlomanov et al. 1971; Shtanov et al. 1980) at the eutectic temperature]. Thermal effects in the region of solid solution based on SnSe are due to the polymorphic transformation of SnSe. An increase in the content of SnSe in PbSe leads to a decrease in the band gap, and at a composition of 20 mol% SnSe to the phenomenon of band inversion (Strauss 1967). The ingots for investigations were annealed at 500°C for 3–6 weeks (Corso et al. 1995) [at 600°C–620°C for 520 h (Shtanov et al. 1974); at 790°C–810°C for 360 h (Zlomanov et al. 1971; Shtanov et al. 1980); at 500°C for 200 h (Latypov et al. 1971a); at 450°C–580°C for 80 h (Krebs et al. 1961b).

$SnSe_2$–PbSe: This system is a non-quasibinary section of the Sn–Pb–Se ternary system (Latypov et al. 1971b, 1976c; Shtanov et al. 1975; Corso et al. 1995). In the range of 0–65 mol% $SnSe_2$, PbSe primarily crystallizes, and in the range of 65–100 mol% $SnSe_2$, tin diselenide first crystallizes. The solidification of all alloys ends at 216°C.

Sn–PbSe: This system is non-quasibinary section of the Sn–Pb–Se ternary system (Latypov et al. 1973a). The main primary crystallizing phase is PbSe, since the liquidus line is

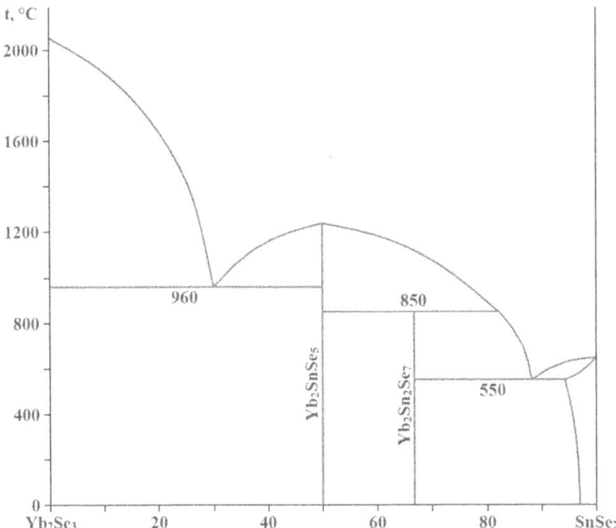

FIGURE 8.59 Phase diagram of the $SnSe_2$–Yb_2Se_3 system. (From Murguzov, M.I., et al., *Zhurn. neorgan. khimii*, **29**(10), 2696, 1984.)

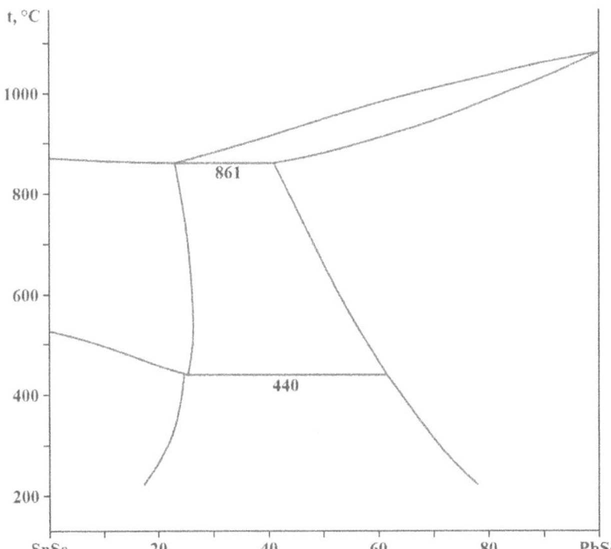

FIGURE 8.60 Phase diagram of the SnSe–PbSe system. (From Volykhov, A.A., et al., *Inorg. mater.*, **44**(4), 345, 2008.)

degenerate on the tin side. The solidification of all alloys ends at 180°C. Lead selenide, crystallizing from the melt, contains such an amount of selenium that its composition is closer to SnSe–PbSe section than to Sn–PbSe section. The ingots for investigations were annealed at 180°C for 600 h. The system was investigated through DTA, XRD, metallography, and measuring of the microhardness.

The vertical section $Sn_{0.35}Se_{0.65}$–$Pb_{.35}Se_{0.65}$ of the Sn–Pb–Se ternary system was constructed by Shtanov et al. (1978), and the total vapor pressure was measured by the static method for some alloys of this section.

Uniform crystals of $(Pb_{1-x}Sn_x)_{1-\delta}Se_{1+\delta}$ with $x = (0.03–0.15) \pm 0.01$ can be obtained from Se-rich melt (Se enrichment is 12–5 at%Se, 1010°C–930°C) (Shtanov and Yashina 2009). The use of metal-rich melt allows obtaining crystals with moderate

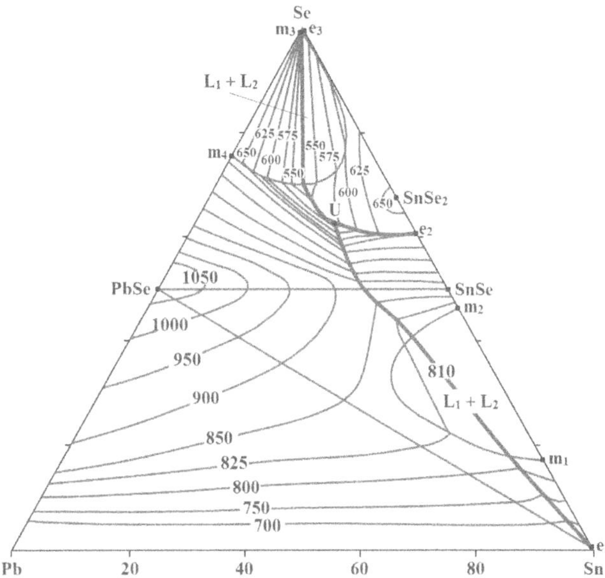

FIGURE 8.61 Liquidus surface of the Sn–Pb–Se ternary system. (From Savel'yev, V.P., et al., *Zhurn. neorgan. khimii*, **20**(7), 2006, 1975.)

segregation of 3–5 mol% SnSe for $x = 0.13–0.23$ at 3–9 at% of metal enrichment and 1000°C–930°C, and for $x = 0.35–0.45$ at 25–35 at% of metal enrichment and 850°C–780°C.

On the liquidus surface of the Sn–Pb–Se ternary system (Figure 8.61), there are two immiscibility regions, one of which occupies a significant part of the PbSe and $SnSe_2$ primary crystallization fields, and the other exists in the primary crystallization regions of PbSe and SnSe (Zlomanov et al. 1971, 1974b; Latypov et al. 1973b, 1976c; Laugier et al. 1974; Savel'yev et al. 1975; Petuhov et al. 1980a; Shtanov et al. 1980; Corso et al. 1995). The main area of the concentration triangle is occupied by the fields of primary crystallization of PbSe, which has the biggest field of primary crystallization, SnSe and $SnSe_2$. The eutectic line from the side Sn–Pb is degenerate. The ternary eutectic is degenerated from the Se-side.

8.35 Tin–Titanium–Selenium

The $\mathbf{Sn_{0.5}Ti_5Se_8}$ ternary compound, which crystallizes in the monoclinic structure with the lattice parameters $a = 1879.0 \pm 0.4$, $b = 357.5 \pm 0.1$, $c = 910.1 \pm 0.2$ pm, and $\beta = 104.03 \pm 0.02°$, is formed in the Sn–Ti–Se system (Novet et al. 1995). To prepare this compound, the ion-exchange reaction was used. $TlTi_5Se_8$ was mixed with SnI_2 at various molar ratios (1:1, 1:2, and 1:4) in a N_2-filled dry box and sealed in evacuated quartz tube approximately 20 cm in length. The tube was placed in a temperature gradient with the sample at the hot end (at temperatures 150°C–300°C) and the cold end of the tube extended out of the furnace assembly. The temperature of the hot end of the ampoule was raised until red crystals of TlI were observed on the cold end of the tube. The samples were annealed for approximately 12 h before quenching to room temperature.

8.36 Tin–Phosphorus–Selenium

$SnSe_2$–PSe: The part of the phase diagram of this system, constructed using DTA, XRD, and measuring of the microhardness, is given in Figure 8.62 (Potoriy et al. 1990). The eutectic contains 86 mol% $SnSe_2$ and crystallizes at 592°C. The **$Sn_2P_2Se_6$** ternary compound, which melts congruently and has a wide homogeneity region (at 500°C, it is within the interval from 50 to ~60 mol% $SnSe_2$), is formed in this system [according to the data of Prits et al. (1988), this compound has a homogeneity region of 55–60 mol% $SnSe_2$; the homogeneity region of $Sn_{2-x}P_2Se_{6+x}$ is at $x \le 0.4$ (Prits et al. 2011a)]. $Sn_2P_2Se_6$ crystallizes in the monoclinic structure with the lattice parameters a = 682.0, b = 767.9, c = 1170.6 pm, and β = 124.77° (Voroshilov et al. 1988b, 2016) [a = 697, b = 765, c = 1168, β = 124.5°, and the calculated and experimental densities of 5.14 and 5.11 g·cm^{-3}, respectively (Klingen et al. 1973); $a = 682.7 \pm 0.2$, $b = 770.0 \pm 0.2$, $c = 1171.8 \pm 0.3$ pm, β = 124.53 ± 0.03°, and the calculated and experimental densities of 5.06 and 4.99 g·cm^{-3}, respectively (Carpentier and Nitsche 1974); an experimental density of 5.08 g·cm^{-3} (Potoriy et al. 1990; Potoriy and Milyan 2016)]. This compound was synthesized by the sublimation method using SnSe, P, and Se as starting materials (Klingen et al. 1973; Brusilovets and Teplyakova 1974; Carpentier and Nitsche 1974) or through the chemical transport reactions using iodine as a transport agent (Voroshilov et al. 1988b). $Sn_2P_2Se_6$ decomposes at the heating in vacuum up to 500°C forming SnSe and volatile phosphorus selenides (Brusilovets and Teplyakova 1974).

The isothermal section of the Sn–P–S ternary system at room temperature is presented in Figure 8.63 (Potoriy and Milyan 2016).

In the Sn–P–Se system, glasses are formed in a narrow range of tin concentrations (Seregin et al. 1972b; Ignatyuk 1976). The maximum content of Sn, at which glass formation is still observed, is 9 at%, and the limiting contents of Se and P, at which the glassy state is preserved, depend on the concentration of Sn.

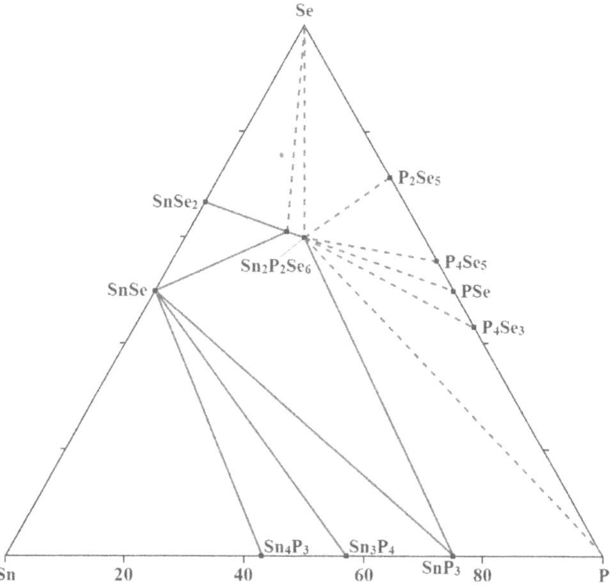

FIGURE 8.63 Isothermal sections of the Sn–P–Se ternary system at 400°C. (From Potoriy, M.V., and Milyan, P.M., *Ukr. khim. zhurn.*, **82**(2), 71, 2016.)

8.37 Tin–Arsenic–Selenium

SnSe–As_2Se_3: The phase diagram of this system, constructed using DTA, XRD, metallography, and measuring of the microhardness, is shown in Figure 8.64 (Dembovskiy et al. 1970, 1971b). The eutectic is degenerated from the As_2Se_3-side and crystallizes at 370°C. The **$SnAs_2Se_4$** ternary compound, which melts incongruently and is stable within the temperature interval from 375°C to 425°C, is formed in the system. According

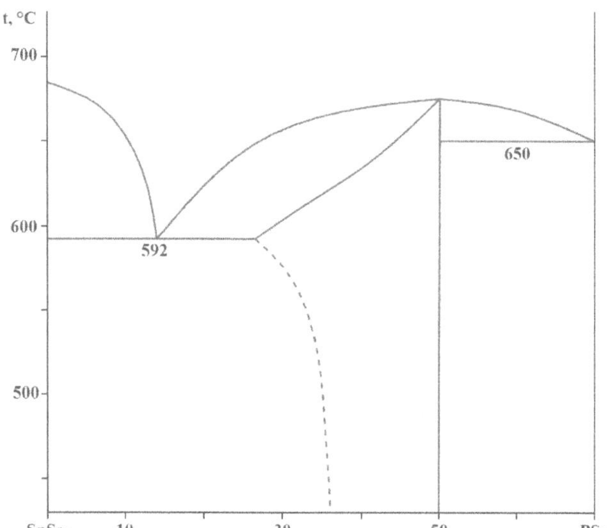

FIGURE 8.62 Part of the phase diagram of the $SnSe_2$–PSe system. (From Potoriy, M.V., et al., *Izv. AN SSSR. Neorgan. mater.*, **26**(11), 2363, 1990.)

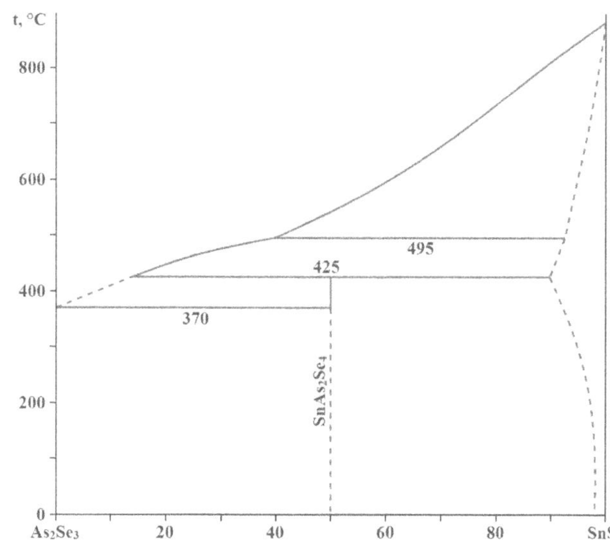

FIGURE 8.64 Phase diagram of the SnSe–As_2Se_3 system. (From Dembovskiy, S.A., et al., *Izv. AN SSSR. Neorgan. mater.*, **7**(10), 1859, 1971.)

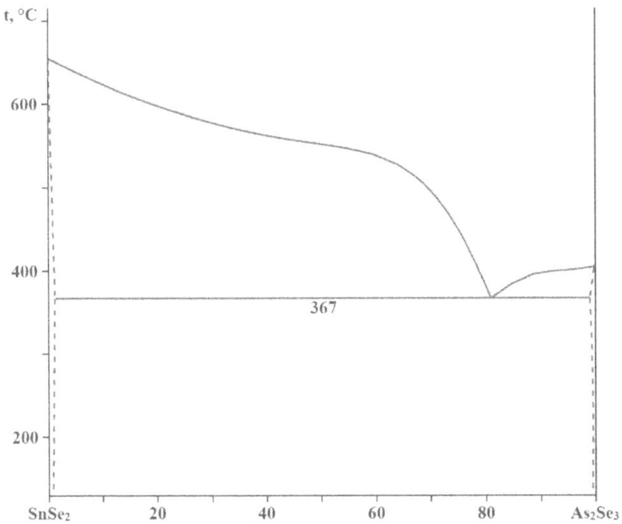

FIGURE 8.65 Phase diagram of the $SnSe_2$–As_2Se_3 system. (From Zmiy, O.F., et al., *Chem. Met. Alloys*, **1**(2), 115, 2008.)

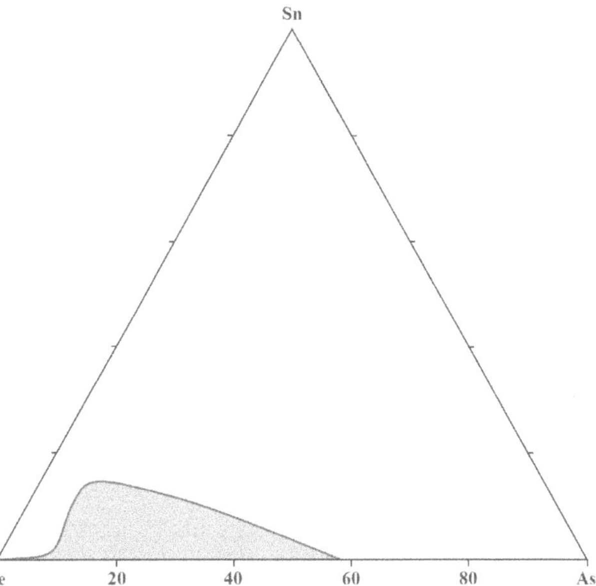

FIGURE 8.66 Glass forming region in the Sn–As–Se ternary system. (From Vasil'ev, L.N., et al., *Izv. AN SSSR. Neorgan. mater.*, **7**(11), 2069, 1971.)

to the data of Vasil'ev and Bahtiyarov (1975), the solubility of SnSe in As_2Se_3 reaches 33 mol%.

$SnSe_2$–As_2Se_3: The phase diagram of this system, constructed using DTA, XRD, metallography, and measuring of the microhardness, is a eutectic type (Figure 8.65) (Zmiy et al. 2008). The eutectic contains 19 mol% $SnSe_2$ and crystallizes at 367°C. No significant solid solubility ranges were detected for the initial components (<1–2 mol%).

SnAs–Se: The **$SnAs_{1-x}Se_x$** samples were synthesized up to $x = 0.13$ through a solid state reaction (He et al. 2017). Sn and Se powders and As particles were mixed according to necessary molar ratio. The mixtures were finely grounded and sealed in silica tubes under vacuum. Then the tubes were heated to 700°C in 10 h and kept there for next 10 h. Finally the tubes were cooled by turning off the furnace. Silver ingots were obtained by carefully breaking the silica tubes. The cell parameter a increases almost linearly with x until a sudden drop in 13 at% Se doped samples. The maximum a is 5.734 in $SnAs_{0.9}Se_{0.1}$.

The glass forming region in the Sn–As–Se ternary system is presented in Figure 8.66 (Borisova et al. 1970; Vasil'ev et al. 1971). An interesting feature of the system is that for glassy samples, the boundary of the $SnSe_2$ and SnSe + $SnSe_2$ regions at a low tin content differs from the boundary of crystallized samples, i.e. the process of "freezing" of structural units, including $SnSe_2$, is observed. Phase of SnSe precipitates during crystallization of such glasses. According to the data of Hilton et al. (1966a), only two compositions, **$SnAsSe_8$** and **$SnAsSe_{18}$**, produced glasses, glass-forming temperature of which is 150°C and 110°C, respectively.

8.38 Tin–Antimony–Selenium

SnSe–Sb: The phase diagram of this system, constructed through DTA, XRD, metallography, and measuring of the microhardness and density, is a eutectic type (Figure 8.67) (Alidzhanov et al. 1985). The eutectic contains 5 mol% $SnSe_2$ and crystallizes at 600°C. An immiscibility region exists at 800°C. The solubility of Sb in SnSe is 4 at% and SnSe practically not dissolved in Sb. The energy gap of the solid solution based on SnSe decreases from 0.96 eV for SnSe to 0.76 eV for the solid solution containing 4 at% Sb. The ingots for investigations were annealed at 550°C–570°C for 250 h.

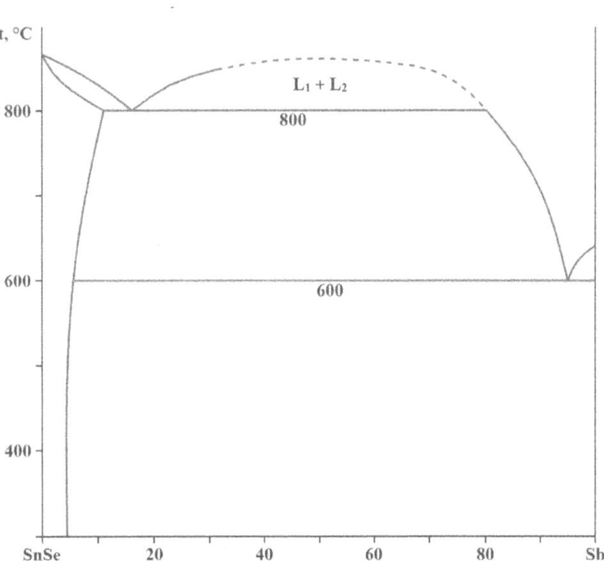

FIGURE 8.67 Phase diagram of the SnSe–Sb system. (From Alidzhanov, M.A., et al., *Izv. AN SSSR. Neorgan. mater.*, **21**(9), 1471, 1985.)

SnSe–Sb_2Se_3: The phase diagram of this system, constructed through DTA, XRD, metallography, and measuring of the microhardness, is shown in Figure 8.68 (Wobst 1968; Gospodinov et al. 1975). The eutectics contain 38.5 and 45.5 mol% SnSe and crystallize at 550°C and 553°C, respectively (Wobst 1968). Two compounds, **$Sn_2Sb_2Se_5$** and **$Sn_2Sb_6Se_{11}$**,

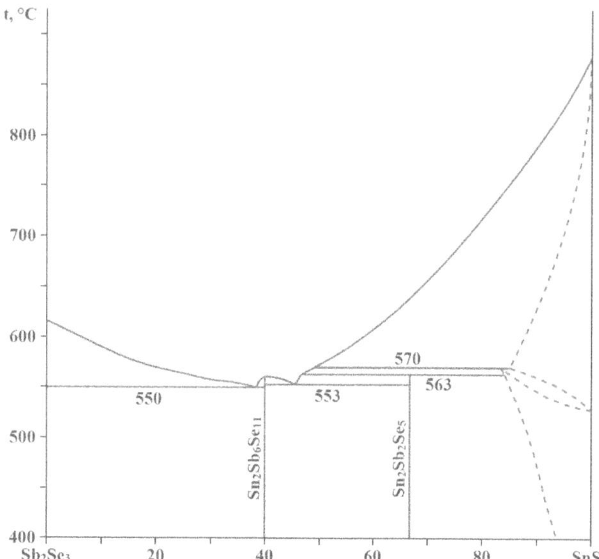

FIGURE 8.68 Phase diagram of the SnSe–Sb_2Se_3 system. (From Gospodinov, G.G., et al., *Izv. AN SSSR. Neorgan. mater.*, **11**(7), 1211, 1975.)

are formed in this system. First of them melts incongruently at 563°C (Gospodinov et al. 1974b, 1975), and the second compound melts congruently at 561°C (Wobst 1968; Gospodinov et al. 1974b, 1975). The temperature of SnSe phase transformation with an increase in the Sb_2Se_3 content rises to 570°C at 15 mol% Sb_2Se_3 (Gospodinov et al. 1975) [at 9 mol% Sb_2Se_3 (Wobst 1968)]. In the range of 50–70 mol% Sb_2Se_3, no equilibrium thermal effects were found upon cooling (Wobst 1968). The solubility of Sb_2Se_3 in SnSe at 400°C reaches 6 mol%, and the solubility of SnSe in Sb_2Se_3 is negligible. The ingots for the investigations were annealed at 400°C for 700 h (Gospodinov et al. 1974b, 1975) [for 240 h (Wobst 1968)]. No glass formation was observed in this system (Vassilev et al. 2009).

According to the data of Smith and Parise (1985), another ternary compound, **$SnSb_2Se_4$**, which crystallizes in the orthorhombic structure with the lattice parameters $a = 2661.0 \pm 0.3$, $a = 2106.6 \pm 0.3$, $c = 404.23 \pm 0.05$, and a calculated density of 5.96 g·cm^{-3}, is formed in the Sn–Sb–Se ternary system. This compound was synthesized from the elements combined in a stoichiometric proportion in an evacuated silica ampoule. The charge was held at 850°C for 4 h, and then at 540°C for 120 h.

$SnSe_2$–Sb_2Se_3: The phase diagram of this system, constructed through DTA, XRD, and metallography, is a eutectic type (Figure 8.69) (Ostapyuk et al. 2009). The eutectic contains 50 mol% $SnSe_2$ and crystallizes at 500°C. The ingots for the investigations were annealed at 350°C for 600 h with the next quenching in water.

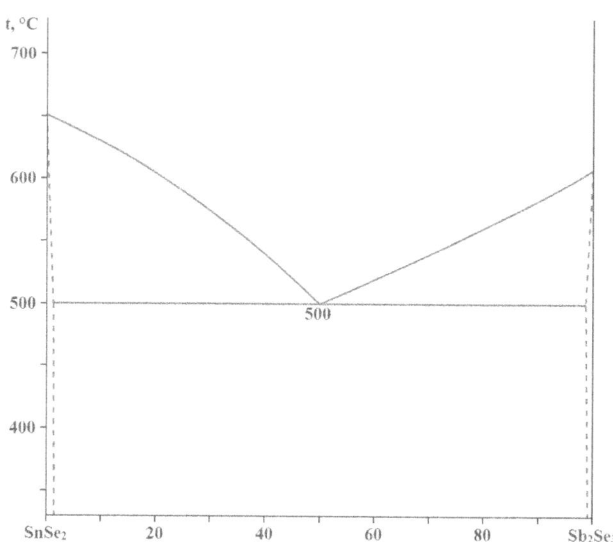

FIGURE 8.69 Phase diagram of the $SnSe_2$–Sb_2Se_3 system. (From Ostapyuk, T.A., et al., *Chem. Met. Alloys*, **2**(3–4), 164, 2009.)

8.39 Tin–Bismuth–Selenium

SnSe–Bi: The phase diagram of this system, constructed through DTA, XRD, metallography, and measuring of the microhardness, is a eutectic type with an immiscibility region (Figure 8.70) (Sher et al. 1978; Alidzhanov et al. 1986). The eutectic is degenerated from the Bi-side and crystallizes at 265°C (Sher et al. 1978) [contains 5 at% Bi and crystallizes

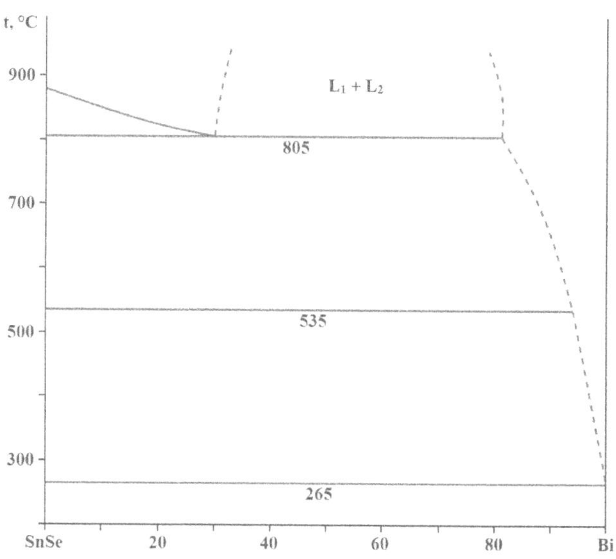

FIGURE 8.70 Phase diagram of the SnSe–Bi system. (From Sher, A.A., et al., *Izv. AN SSSR. Neorgan. mater.*, **14**(7), 1270, 1978.)

at 260°C (Alidzhanov et al. 1986)]. The monotectic reaction takes place at 805°C (Sher et al. 1978) [at 800°C (Alidzhanov et al. 1986)]. Thermal effects at 535°C correspond to the SnSe polymorphic transformation (Sher et al. 1978). The mutual solubility of SnSe and Bi is not higher than 1 mol%. The ingots for the investigations were annealed at 230°C for 1000 h.

SnSe–BiSe: This system is non-quasibinary section of the Sn–Bi–Se ternary system as BiSe melts incongruently (Sher et al. 1978, 1986). The solubility of SnSe in BiSe reaches 10 mol% (5 mol% at 450°C) and the solubility of BiSe in SnSe is 4 mol% (2 mol% at 450°C). The **$SnBiSe_2$** and **$SnBi_4Se_5$** ternary compounds and δ-phase, whose composition is close to **$Sn_3Bi_9Se_{13}$**, are formed in this system. $SnBiSe_2$ exists at 50–55 mol% BiSe and at temperatures 545°C–695°C, and δ-phase is stable at 50–55 mol% BiSe and at temperatures 560°C–610°C. $SnBiSe_2$ and δ-phase are formed upon cooling

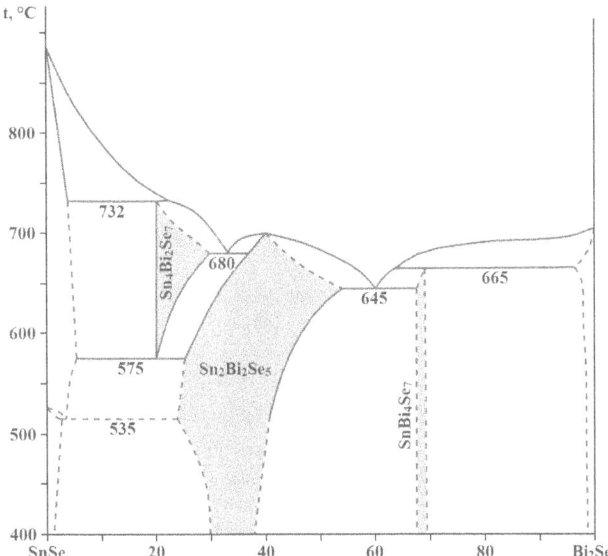

FIGURE 8.71 Phase diagram of the SnSe–Bi_2Se_3 system. (From Adouby, K., et al., *J. Alloys Compd.*, **453**(1–2), 161, 2008.)

by peritectic reactions and decompose by eutectoid reactions. Single-phase samples of δ-phase were not obtained. $SnBi_4Se_5$ crystallizes in the hexagonal structure with the lattice parameters $a = 422.3 \pm 0.3$ and $c = 5670 \pm 6$ pm. The ingots for the investigations were annealed at 450°C for 1000 h.

SnSe–Bi_2Se_3: The phase diagram of this system, constructed through DTA, XRD, and Mössbauer spectroscopy, is presented in Figure 8.71 (Adouby et al. 2008). It is characterized by the occurrence of two narrow solid solutions, six invariants and three new intermediate phases, **$SnBi_4Se_7$**, **$Sn_2Bi_2Se_5$**, and **$Sn_4Bi_2Se_7$**. $SnBi_4Se_7$ melts incongruently at 665°C and crystallizes in the trigonal structure with the lattice parameters $a = 416.02 \pm 0.05$ and $c = 3893.4 \pm 0.3$ pm [$a = 418.8 \pm 0.4$ and $c = 3946 \pm 4$ pm (Sher et al. 1978)] and the calculated and experimental densities of 6.366 and 6.34 g·cm^{-3}, respectively (Vicente et al. 1999; Adouby et al. 2008). $Sn_2Bi_2Se_5$ with a wide region of homogeneity (25–55 mol% Bi_2Se_3) melts congruently at 700°C. The structure of this compound has not been determined completely. $Sn_4Bi_2Se_7$ has a wide homogeneity region (20–33 mol% Bi_2Se_3) and is stable from 575°C to 732°C. It crystallizes in the cubic structure with the lattice parameter $a = 594.78 \pm 0.1$ pm and an experimental density of 6.254 g·cm^{-3} at room temperature and $a = 593.22 \pm 0.01$ at 130 K (Adouby et al. 1998, 2008). The lattice parameters of **$Sn_{1+x}Bi_{2x}Se_{1+2x}$** vary linearly with the composition in accordance with the Vegard's law. This leads to determine the boundary of the SnSe-based solid solution near $x = 0.04$ (4 mol% Bi_2Se_3). In the region between $Sn_4Bi_2Se_7$ and $SnBi_4Se_7$ there are two eutectics corresponding to 33 and 60 mol% Bi_2Se_3, and 680°C and 645°C, respectively (Adouby et al. 2008). The solubility of $SnSe_2$ in Bi_2Se_3 is 1.7 at% Sn at 450°C (Sher et al. 1986).

$SnBi_4Se_7$ was synthesized from SnSe and Bi_2Se_3 (Vicente et al. 1999). The stoichiometric mixture was placed into a silica tube and sealed under vacuum (~10^{-3} Pa). The mixture was heated at 50°C·h^{-1} up to 900°C for 3 days, then slowly cooled to 640°C and maintained at this temperature for 2 weeks. Finally, the sample was quenched in an ice-water bath.

$Sn_4Bi_2Se_7$ was prepared by a solid state reaction from SnSe and Bi_2Se_3 in the stoichiometric ratio (Adouby et al. 1998). The mixture was placed into a silica tube and sealed under vacuum (~10^{-3} Pa). It was heated at 30°C·h^{-1} up to 800°C, maintaining constant this temperature for four days. Two different ways have been used for cooling in order to obtain the final samples. The first one consists of a slow cooling to reach the room temperature in one day for obtaining the thermodynamically stable phases at room temperature. The second one consists of a quenching in an icy-water bath in order to trap the high-temperature form.

Two different procedures have been used to synthesize the samples of the SnSe–Bi_2Se_3 system (Adouby et al. 2008). The first series were heated at 50°C·h^{-1} to 900°C, kept at this temperature for 48 h, cooled to 400°C and annealed at this temperature during 4 months. Finally, the samples were quenched in an icy-water bath. For the second series, the samples were heated at 900°C, maintained at this temperature during 3 days, cooled to 640°C, annealed for 15 days and quenched again in an icy-water bath.

The earlier investigation on the SnSe–Bi_2Se_3 system by Hirai et al. (1967) has shown a complete miscibility of both Bi_2Se_3 and SnSe compounds. The equilibrium diagram of this system was refined by Odin et al. (1974b) by means of XRD, DTA, and metallography of alloys annealed at 450°C for 920 h, followed by quenching in icy-water bath. The phase diagram they have established shows a new non-stoichiometric ternary phase with a wide homogeneity region from 45 to 70 mol% Bi_2Se_3. The compound melts in a peritectic reaction at 670°C and at 660°C for the saturation of the compound with SnSe and Bi_2Se_3, respectively.

$SnSe_2$–Bi_2Se_3: The phase diagram of this system, constructed through DTA, XRD, and metallography, is a eutectic type (Figure 8.72) (Sher et al. 1978). The eutectic contains 30 mol% Bi_2Se_3 and crystallizes at 545°C. The solubility of Bi_2Se_3 in $SnSe_2$ at 450°C reaches 2 mol% and the solubility of $SnSe_2$ in Bi_2Se_3 at this temperature is 8 mol%. The ingots for the investigations were annealed at 450°C for 1000 h.

According to the data of Heinke et al. (2017), once more compound, **$Sn_{4.11}Bi_{22.60}Se_{38}$**, which crystallizes in the monoclinic

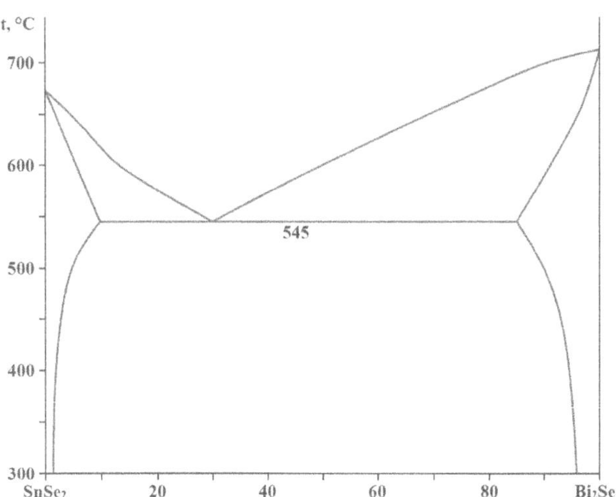

FIGURE 8.72 Phase diagram of the $SnSe_2$–Bi_2Se_3 system. (From Sher, A.A., et al., *Izv. AN SSSR. Neorgan. mater.*, **14**(7), 1270, 1978.)

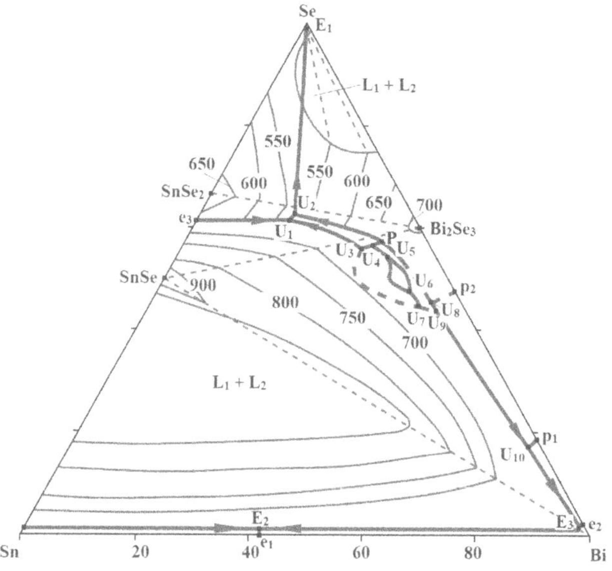

FIGURE 8.73 Liquidus surface of the Sn–Bi–Se ternary system. (From Sher, A.A., et al., *Zhurn. neorgan. khimii*, **31**(4), 1010, 1986.)

structure with the lattice parameters $a = 2814.1 \pm 0.2$, $b = 418.20 \pm 0.01$, $c = 2118.1 \pm 0.2$, $\beta = 131.082 \pm 0.009°$, and a calculated density of 7.2558 g·cm^{-3}, is formed in the Sn–Bi–Se ternary system. Samples of this compound were obtained by fusing stoichiometric amount of Sn, Bi, and Se in sealed silica glass ampoules under Ar atmosphere. The quenched melts were subsequently annealed, initially at 520°C for 4 days and then quenched in water. Single crystals were obtained from such samples. The content of the desired phase could be increased up to 90% by thermal treatment at higher temperature (650°C for 6 days).

The liquidus surface and the isothermal section at 450°C of the Sn–Bi–Se ternary system were constructed by Sher et al. (1986). There are 12 fields of primary crystallization and two immiscibility regions on the liquidus surface (Figure 8.73). Phase equilibria in the system are due to the next invariant equilibria: U_1 – L + (SnSe) ⇔ $SnBi_2Se_4$ + ($SnSe_2$); U_2 – L + (Bi_2Se_3) ⇔ $SnBi_2Se_4$ + ($SnSe_2$); U_3 – L + (SnSe) ⇔ $SnBi_2Se_4$ + $SnBiSe_2$; U_4 – L + $SnBiSe_2$ ⇔ $SnBi_2Se_4$ + $SnBi_4Se_5$; U_5 – L + δ-phase ⇔ $SnBiSe_2$ + $SnBi_4Se_5$; U_6 – L + δ-phase ⇔ $SnBiSe_2$ + $SnBi_4Se_5$; U_7 – L + $SnBiSe_2$ ⇔ $SnBi_4Se_5$ + (SnSe); U_8 – L + (Bi_2Se_3) ⇔ $SnBi_4Se_5$ + (BiSe); U_9 – L + $SnBi_4Se_5$ ⇔ (SnSe) + (BiSe); and P – L + $SnBi_2Se_4$ + (Bi_2Se_3) ⇔ $SnBi_4Se_5$. There are also three ternary degenerated eutectics (E_1, E_2 and E_3) on the liquidus surface. The lines of monovariant equilibria limiting the areas of primary crystallization of $SnBi_2Se_4$, $SnBiSe_2$, $SnBi_4Se_5$, and δ-phase are roughly plotted. At 450°C, the solid state region occupies the predominant part of the isothermal section. Small areas of liquid state are adjacent to the selenium corner and the Sn–Pb binary system.

8.40 Tin–Vanadium–Selenium

SnSe–V$_{1+x}$Se$_2$: Six misfit layer compounds are present in this system at the compositions $[(SnSe)_{1+\delta}]_m(VSe_2)_n$ (m = 1, 2, 3, 4, 5, and 6; n = 1) (Reisinger and Richter 2021a,b). The smallest integer ratio for m and n (e.g. $m = 2$ and $n = 1$) is reported to be more stable than a multiple of these values (e.g. $m = 4$ and $n = 2$). The phase equilibria involving these misfit layer compounds allow two different descriptions. On the one hand, each misfit layer compound can be treated as an individual phase with different c parameters. On the other hand, the misfit layer compounds can be treated like a homogeneous single-phase field with rather similar lattice parameters for a and b, but with modulation in c.

The **$([SnSe]_{1.15})_m(V_{1+x}Se_2)_n$** ternary intergrowth misfit layer compounds were synthesized using physical vapor deposition in a custom-built vacuum deposition chamber at base pressure as low as 10^{-6} Pa (Atkins et al. 2013, 2014; Falmbigl et al. 2015a,b). During the deposition process the low pressure was kept by a cryogenic absorber pump. Selenium was evaporated via an effusion cell, and an electron beam gun was used to evaporate Sn and V. The physical vapor deposition was carried out on (100) oriented Si-wafers. A short annealing duration at moderate temperatures is required to self-assemble the desired metastable compounds from the precursors. These compounds have a structure with alternating SnSe bilayers and VSe_2 trilayers, but there is extensive rotational disorder between the individual SnSe and VSe_2 constituents (Atkins et al. 2013).

These misfit layer compounds can be also synthesized using mineralizing agents (Reisinger and Richter 2021b). Best results were obtained when NH_4Cl was utilized. An off-stoichiometric sample composition (in the region between the three compositions of VSe_2, V_5Se_8, and misfit layer compound) larger qualities of misfit layer compound crystals could be obtained. A composition with larger vanadium content and close to V_5Se_8 was most favorable. An annealing duration of 1–3 months (excluding pre-alloying time) at approximately 550°C was found to be most successful for synthesizing large crystals.

The formation of the ferecrystalline compound $(SnSe)_{1.15}VSe_2$ from a modulated thin film precursor has been investigated by Falmbigl et al. (2017). Annealing initiates the simultaneous growth of three crystalline constituents, SnSe, $SnSe_2$, and VSe_2. The SnSe/$SnSe_2$ ratio increases with increasing temperature. At 400°C selenium evaporation causes a complete transformation of $SnSe_2$ into SnSe and Se vapor, resulting in the formation of ferecrystalline $(SnSe)_{1.15}VSe_2$.

A set of precursors $([SnSe]_{1.15})_1(V_{1+x}Se_2)_n$ was synthesized and subsequently annealed at 400°C for 20 min. For $([SnSe]_{1.15})_1(V_{1+x}Se_2)_n$ the next values of a-lattice parameters of SnSe and $V_{1+x}Se_2$ sublattices, c-lattice parameters and x in $V_{1+x}Se_2$ were obtained: a(SnSe) = 592.73 ± 0.04, 594.19 ± 0.05, 595.14 ± 0.04, 591.7 ± 0.1, 590.8 ± 0.4, and 594.12 ± 0.05; a($V_{1+x}Se_2$) = 340.82 ± 0.08, 342.40 ± 0.03, 343.35 ± 0.03, 340.37 ± 0.09, 339.85 ± 0.09, and 342.73 ± 0.04; c = 1203.0 ± 0.1, 1801.4 ± 0.1, 2400 ± 1, 3012 ± 1, 3618 ± 1, and 4228 ± 0.1; and x = 0.23, 0.14, 0.22, 0.42, 0.38, and 0.28 for n = 1, 2, 3, 4, 5, and 6, respectively (Falmbigl et al. 2015b).

For $([SnSe]_{1.15})_m(VSe_2)_1$ the experimental/theoretical c-lattice parameters are (1203.4 ± 0.01)/1222.58 pm for m = 1; (1782.8 ± 0.01)/1807.4 pm for m = 2; (2364.0 ± 0.01)/2410.6 pm for m = 3; and (2930 ± 10)/3002.3 pm for m = 4 (Atkins et al. 2014). The experimental a-lattice parameter for the SnSe constituent varied from 593.5 ± 0.4 to 599.76 ± 0.02 and 599.23 ± 0.07 pm for the m = 1, 2, and 3 compounds, respectively. The theoretical a-lattice parameter for the SnSe constituent in the commensurate approximates is 591 pm for m = 1 and 592 pm m = 2, 3, and 4.

The experimental *a*-lattice parameter of the VSe_2 constituent increases slightly from 341.4 ± 0.3 and 345.64 ± 0.04 to 346.30 ± 0.03 pm as *m* increases from 1 to 3, respectively. The theoretical *a*-lattice parameter for the VSe_2 constituent in the commensurate approximates is 336 pm for *m* = 1 and 338 pm *m* = 2, 3, and 4.

The **$SnVSe_3$** ternary misfit layer compound with the *c*-lattice parameter 2420.4 pm was synthesized by direct combination of the elements, mixed in pellet form and sealed in silica tube under vacuum with the next heating in a furnace at 500°C for 1 day and at 800°C for 7 days with heating rate of 2°C·min^{-1} (Hernán et al. 1991b, 1992). Polycrystalline black powder of the title compound with luster was obtained.

The phase relations of the ternary system Sn–V–Se was investigated over the whole composition range utilizing powder XRD and Scanning Electron Microscopy (SEM) coupled with Electron Microprobe Analysis (EPMA) by Reisinger and Richter (2021a). The results of the evaluations in the ternary system are presented by two isothermal sections at 550°C and 700°C. The binary phases with the highest solubilities into the ternary are SnSe with V-contents of 1.2 at%, VSe with Sn-contents of 1.3 at%, and SnV_3 with Se-contents of 0.6 at%. Below a Se content of 50 at% no ternary phase was observed in neither of the two isothermal sections. Regarding the solubilities of the binary compounds into the ternary, only slight deviations can be observed at the two temperatures and also the two-phase equilibria are alike. The phase equilibria VSe + V_3Se_4 + L_2 and SnSe + V_3Se_4 + L_2 were observed in both sections.

8.41 Tin–Niobium–Selenium

Several ternary compounds are formed in the Sn–Nb–Se system. **$Sn_{0.61}NbSe_2$** crystallizes in the hexagonal structure with the lattice parameters *a* = 348 and c = 1860 pm (Karnezos et al. 1975). This intercalation compound was synthesized by reaction of the powdered $NbSe_2$ with Sn vapor at 800°C–950°C in sealed silica tube. $NbSe_2$ powder and the condensed Sn phase were separated during the intercalation reaction so that it proceeded via the vapor rather than the liquid or solid metal phase. Only by repeated firing and subsequent grinding could reproducibly homogeneous material be produced by this method.

$SnNbSe_3$ crystallizes in the tetragonal structure with the lattice parameters *a* = 421 and *c* = 2454 pm (Maaren van 1972; Wiegers et al. 1989a)]. This compound was prepared by heating the stoichiometric mixture of the elements in an evacuated quartz ampoule at temperature between 600°C and 1000°C. Subsequently the material was ground and heated several times. Relatively long firing times, e.g. 300 h, were required to complete reaction. In order to obtain single crystals the vapor transport was used. It turned out that those crystals could be grown using Cl_2 as a carrier reagent in a temperature gradient of 840°C–770°C; the crystals grow at the low-temperature side of the tube over 10 days. To the sample to be transported (ca. 300 mg) about 5–10 mg of $(NH_4)_2PbCl_6$ was added. The crystals have a metallic luster.

Sample with nominal composition **$SnNb_2Se_5$** was prepared from the elements in the stoichiometric proportion by heating the pelleting mixture in evacuated silica ampoule (Hernán et al. 1991b, 1992). The heating rate was 1°C·min^{-1} and the temperature was maintained at 300°C for 1 day and at 900°C for 7 days. Polycrystalline black powder of the title compound with luster was obtained. The *c*-lattice parameter of this compound was 3705.1 pm.

Intergrowth compounds of **$([SnSe]_{1+\delta})_m(NbSe_2)_n$**, where $1 \leq m = n \leq 20$, with the same atomic composition but different *c*-axis lattice parameters and number of interfaces per volume were synthesized using the modulated elemental reactant technique (Alemayehu et al. 2015a,b,c). A *c*-axis lattice parameter change of 1217 ± 6 pm as a function of one unit of *m* = *n* was observed (Alemayehu et al. 2015c). In-plane diffraction shows independent crystal structures for both constituents, an increase in distortion of the SnSe layer as *m* increases becoming more similar to bulk-SnSe, and a broadening of the $NbSe_2$ peaks as a function of *n* indicating a smaller crystallite size and the presence of different coordination environments for the Nb atom: trigonal prismatic and octahedral. It was shown that the metastable higher order compounds convert into the thermodynamically stable (*m* = *n* = 1) compound upon annealing at 500°C.

For the compounds $([SnSe]_{1+\delta})_m(NbSe_2)_1$, where $1 \leq m \leq 10$, the *c*-axis lattice parameter systematically increases by 577 ± 5 pm as the value of *m* is increased, which indicates that an additional bilayer of rock salt structured SnSe is inserted for each unit of *m* (Alemayehu et al. 2015b). The in-plane structure of both constituents systematically changes as the thickness of SnSe increases. Both XRD and electron microscopy studies show the presence of turbostratic disorder between the different constituent layers.

A new polytype of the misfit layer compound **$(SnSe)_{1.16}(NbSe_2)_1$** with extensive rotational disorder was prepared via physical vapor deposition from designed modulated elemental reactants (Alemayehu et al. 2015a). This polytype, previously referred to as a ferecrystal due to the extensive rotational disorder, formed over a range of compositions and precursor thicknesses and the resulting *c*-axis lattice parameters ranged from 1221.0 ± 0.4 to 1236.0 ± 0.4 pm. These values bracket the value published for the crystalline misfit layer compound prepared at high temperature. The *a*- and *b*-axis in-plane lattice parameters of both the SnSe and $NbSe_2$ constituents are incommensurate, which differs from the misfit layer compound formed via high temperature reaction that has a common *b*-axis lattice parameter for the two constituents. The in-plane area per unit cell of the ferecrystal is 1–2% larger than the compound formed at high temperature.

The samples of the misfit layer compound $(SnSe)_{1.16}(NbSe_2)_1$ were also prepared from the mixture of the elements in the stoichiometric ratio (Wiegers and Zhou 1991). The mixture was enclosed in an evacuated quartz tube, heated at 450°C for 2 days and after that at 750°C for 5 days. Both sublattices of this compound crystallize in the orthorhombic structure with the lattice parameters *a* = 529.8 ± 0.1, *b* = 597.0 ± 0.2, *c* = 1228.2 ± 0.2 pm for SnSe-sublattice, and *a* = 344.1 ± 0.1, *b* = 597.1 ± 0.1, *c* = 2457.2 ± 0.5 pm for $NbSe_2$-sublattice.

8.42 Tin–Tantalum–Selenium

A modification of the modulated elemental reactants synthetic technique was developed and used to synthesize eleven members of the **$[(SnSe)_{1.15}]_m(TaSe_2)_n$** family of the ternary compounds, with m and n equal to integer values between 1 and 6 (Atkins et al. 2012). Each of the intergrowth compounds contained highly oriented intergrowths of SnSe bilayers and $TaSe_2$ monolayers with abrupt interfaces perpendicular to the c-axis. The c-lattice parameter increased 579 ± 1 pm per SnSe bilayer and 649 ± 1 pm per Se–Ta–Se trilayer ($TaSe_2$) as m and n were varied.

The atomic structure of the family of ferecrystals $\mathbf{[(SnSe)_{1.15}]_m(TaSe_2)}$ (m = 1, 3, and 6) was investigated by means of transmission electron microscopy (Grosse et al. 2013). The tantalum in the $TaSe_2$ layers was observed to have trigonal prismatic coordination similar to that found in the 2H polytype of bulk $TaSe_2$. The structure of the SnSe constituent was found to be similar to that of orthorhombic α-SnSe. In the compounds with m = 1 and m = 3, regions with a local ordering of the layers along a commensurate axis, similar to the ordering in conventional misfit layer compounds, were observed. However, on a longer range the ferecrystals were found to exhibit a turbostratically disordered structure. Stacking defects were occasionally found in the samples in which a layer is interrupted and the surrounding layers are bent around these defects, while maintaining abrupt interfaces instead of interdiffusing. Volume defects were found in one sample at m = 1 in which a SnSe layer locally substitutes a part of a $TaSe_2$ layer without interrupting the surrounding layers.

8.43 Tin–Oxygen–Selenium

Two ternary compounds, $\mathbf{Sn(SeO_3)_2}$ and $\mathbf{Sn_5O_6Se_4}$, are formed in the Sn–O–Se system. $Sn(SeO_3)_2$ has two polymorphic modifications (Steinhauser et al. 2006), is thermally stable up to 280°C and decomposes completely at 600°C (Ojkowa and Georgiew 1980; Gospodinov and Bogdanov 1989). α-$Sn(SeO_3)_2$ crystallizes in the cubic structure with the lattice parameter a = 817.25 ± 0.05 pm and a calculated density of 4.534 g·cm^{-3} and β-$Sn(SeO_3)_2$ crystallizes in the monoclinic structure with the lattice parameters a = 478.4 ± 0.1, b = 852.9 ± 0.1, c = 694.3 ± 0.1 pm, β = 111.01 ± 0.01°, and a calculated density of 4.680 g·cm^{-3} (Steinhauser et al. 2006). Single crystals of $Sn(SeO_3)_2$ were prepared by a low-hydrothermal technique using $SnCl_2$, SeO, and H_2O. Mixture of $SnCl_2$ (0.0340 g), SeO_2 (0.0395 g), and H_2O (1 mL) was filled into Teflon-lined steel vessels with a capacity of ca. 10 cm^3. The steel autoclave was heated to 222°C and kept at this temperature for 1 week. Finally, the autoclaves were cooled to room temperature within 12 h. Transparent colorless crystals of α- and β-$Sn(SeO_3)_2$ with octahedral or prismatic shapes, respectively, were obtained in low yield as coexisting phases.

α-$Sn(SeO_3)_2$ also forms at the interaction of H_2SeO_3 and SnO_2 as a finely crystalline precipitate (Ojkowa and Georgiew 1980) or by its continued crystallization at 100°C in aqueous solution of H_2SeO_3 (Gospodinov and Bogdanov 1989).

$Sn_5O_6Se_4$ has an experimental density of 6.62 g·cm^{-3} and could be prepared at the interaction of SnO and Se at 400°C for 4 h (Batsanov and Shestakova 1966; Vasil'ev et al. 1977). It also forms at the oxidation of SnSe as an intermediate compound.

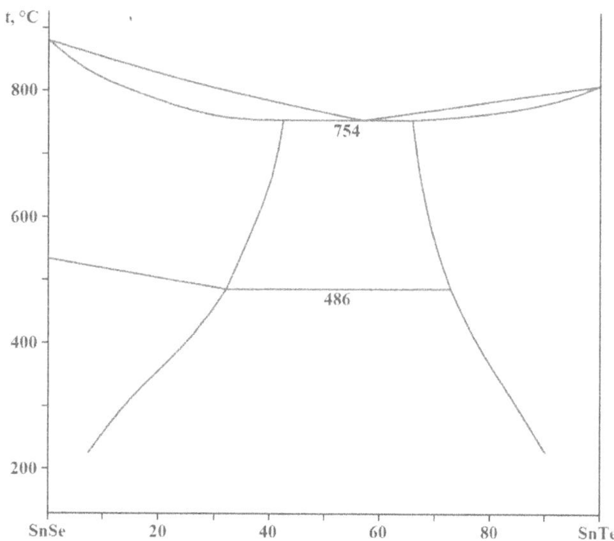

FIGURE 8.74 Phase diagram of the SnSe–SnTe system. (From Volykhov, A.A., et al., *Inorg. Mater.*, **44**(4), 345, 2008.)

8.44 Tin–Tellurium–Selenium

SnSe–SnTe: The phase diagram of this system, constructed through DTA and XRD (Totami et al. 1968; Leute and Menge 1992; Liu and Chang 1992; Majid and Legendre 1998), and optimized using a four parameter for the excess Gibbs energy by Volykhov et al. (2008), is a eutectic type (Figure 8.74). The eutectic contains 43 mol% SnSe and crystallizes at 754°C (Leute and Menge 1992) [at 45 mol% SnSe and 755°C (Totami et al. 1968); at 45 mol% SnSe and 762°C (Majid and Legendre 1998)]. With an increase in the content of SnTe in the solid solution based on SnSe, the temperature of the polymorphic transformation decreases. The solubility of SnSe in SnTe and SnTe in SnSe at the eutectic temperature is 36 and 41 mol%, respectively (Leute and Menge 1992) [the solubility of SnSe in SnTe reaches 22 mol% at 440°C (Krebs et al. 1961b), 25 mol% at 450°C (Vasil'ev et al. 1976b) and 30 mol% at 775°C (Nasirov and Feiziev 1969); and the solubility of SnTe in SnSe is 28 mol% at 440°C (Krebs et al. 1961b), 25 mol% at 450°C (Vasil'ev et al. 1976b) and 40 mol% at 775°C (Nasirov and Feiziev 1969)]. In the region of solid solutions based on SnSe, the band gap varies linearly from 1.2 eV (SnSe) to 0.2 eV ($SnTe_{0.4}Se_{0.6}$) (Nasirov and Feiziev 1969). The ingots for investigations were annealed at 300°C–700°C for 100–150 h (Totami et al. 1968) [at 450° for 40 h (Vasil'ev et al. 1976b)].

SnSe–Te: The phase diagram of this system, constructed through DTA and XRD, is a eutectic type (Figure 8.75) (Majid and Legendre 1998). The eutectic contains 10 mol% SnSe and crystallizes at 431°C.

$\mathbf{SnSe_2}$–Te: The phase diagram of this system, constructed through DTA and XRD, is a eutectic type (Figure 8.76) (Majid and Legendre 1998). The eutectic contains 6 mol% SnSe and crystallizes at 428°C.

Phase relations in the Sn–Te–Se ternary system at 500°C are shown in Figure 8.77 (Liu and Chang 1992). Tellurium

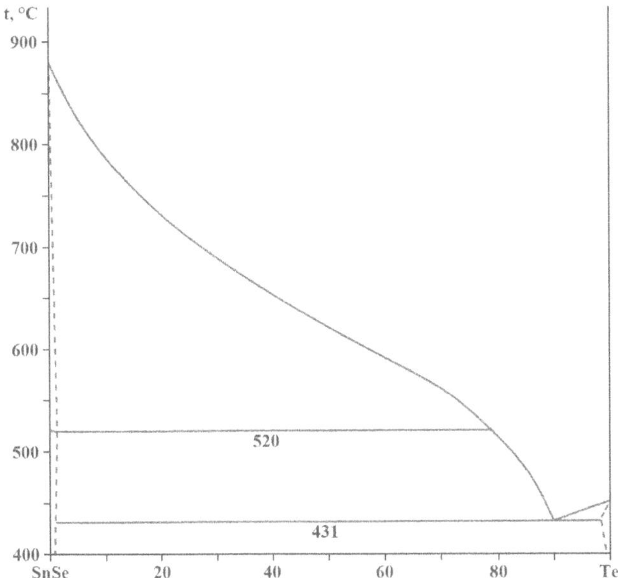

FIGURE 8.75 Phase diagram of the SnSe–Te system. (From Majid, M., and Legendre, B.J., *J. Therm. Anal.*, **54**(3), 963, 1998.)

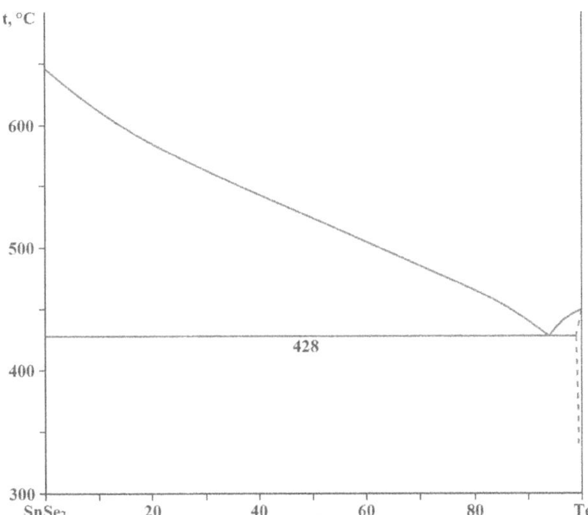

FIGURE 8.76 Phase diagram of the $SnSe_2$–Te system. (From Majid, M., and Legendre, B.J., *J. Therm. Anal.*, **54**(3), 963, 1998.)

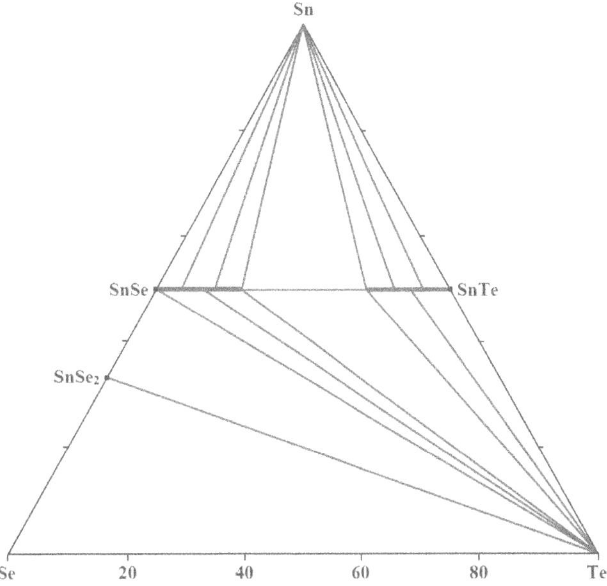

FIGURE 8.77 Isothermal sections of the Sn–Te–Se ternary system at 500°C. (From Liu, H., and Chang, L.L.Y., *J. Alloys Compd.*, **185**(1), 183, 1992.)

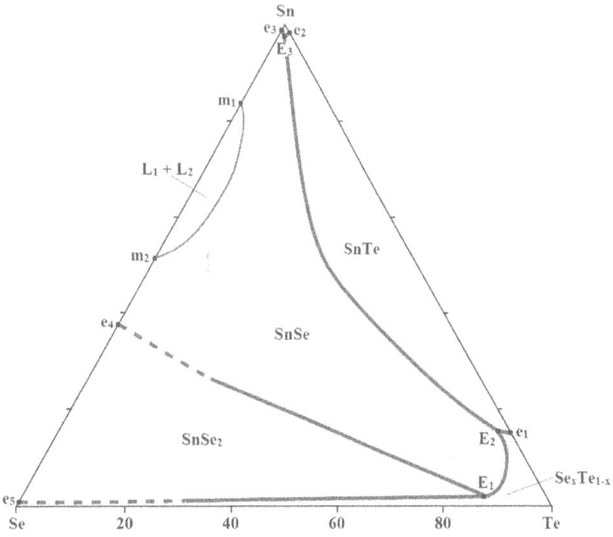

FIGURE 8.78 Projection of the liquidus surface of the Sn–Te–Se ternary system. (From Chen, C.-Y., et al., *J. Alloys Compd.*, **547**, 100, 2013.)

liquid forms equilibrium assemblages with all tin chalcogenides stable in the system. At 300°C, the phase relations differ from those shown in Figure 8.77 in two aspects: (1) Te is a solid phase at this temperature and there is no liquid phase in the region of $SnSe_2$–SnSe–SnTe–Te; (2) the regions occupied by the two-phase assemblages of (SnSe) + Te and (SnTe) + Te become smaller because of reduction in mutual solid solutions between SnSe and SnTe at 300°C.

The isothermal sections of this system at 231°C, 400°C, 427°C, 488°C, and 520°C were also constructed by Majid and Legendre (1998) and at 250°C by Chen et al. (2013).

The liquidus surface of the Sn–Te–Se ternary system is presented in Figure 8.78 (Chen et al. 2013). Three ternary eutectics exist in the system: E_1 (420°C) – L ⇔ SnSe + $SnSe_2$ + (Se,Te); E_2 (400°C) – L ⇔ SnSe + SnTe + (Se,Te); and E_3 (230°C) – L ⇔ SnSe + SnTe + Sn. Some vertical sections, the liquidus surface and the reaction scheme were also constructed by Majid and Legendre (1998) and the Sn–Te–Se ternary system was modeled using Calculation of Phase Diagrams (CALPHAD) technique by Cui et al. (2015).

Bulk glasses $Sn_xTe_{15}Se_{85-x}$ (Patial et al. 2011), $Sn_xTe_{20}Se_{80-x}$ ($0 \le x \le 9$) (Kaur et al. 2000), $Sn_xTe_{20}Se_{80-x}$ (x = 0, 2.5 and 5) (Abdel-Wahab 2011) and $Sn_xTe_{10}Se_{90-x}$ (x = 0, 2.5, 5, and 7.5) (Abdel-Rahim et al. 2013) were prepared by the conventional melt quenching technique using elemental Sn, Se, and Te. The glass transition temperature of the obtained glasses increases with the increasing in the heating rate and Sn content.

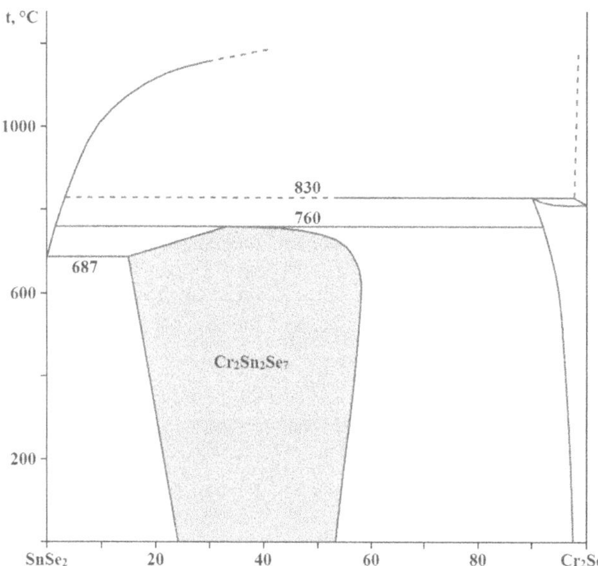

FIGURE 8.79 Phase diagram of the $SnSe_2$–Cr_2Se_3 system. (From Babitsyna, A.A., and Novotortsev, V.M., *Zhurn. neorgan. khimii*, **31**(7), 1825, 1986.)

8.45 Tin–Chromium–Selenium

SnSe–Cr_2Se_3: The phase diagram of this system, constructed through DTA, XRD, metallography, and measuring of the microhardness, is a eutectic type (Babitsyna and Novotortsev 1986). No ternary compounds were found in this system. The ingots for the investigations were annealed at 630°C for 300 h.

$SnSe_2$–Cr_2Se_3: The phase diagram of this system, constructed through DTA, XRD, metallography, and measuring of the microhardness, is presented in Figure 8.79 (Babitsyna and Novotortsev 1986). The eutectic is degenerated from the $SnSe_2$-side. On the basis of Cr_2Se_3, there is a small region of solid solutions. The **$Cr_2Sn_2Se_7$** ternary compound, which melts congruently at 760°C, has a wide homogeneity region and crystallizes in the cubic structure with the lattice parameter a = 1068 pm, is formed in this system.

Once more ternary compound, **$Cr_2Sn_3Se_7$**, which has two polymorphic modifications, is formed in the Sn–Cr–Se system. Fist modification crystallizes in the monoclinic structure with the lattice parameters a = 1276.5 ± 0.5, b = 383.5 ± 0.2, c = 1178.5 ± 0.4 pm, β = 105.21 ± 0.01°, and a calculated density of 6.05 g·cm^{-3} (Jobic et al. 1994) and the second one crystallizes in the orthorhombic structure with the lattice parameters a = 2309.7 ± 0.1, b = 383.07 ± 0.02, c = 1254.14 ± 0.06 pm, and a calculated density of 6.08 g·cm^{-3} (Jobic et al. 1995b). The monoclinic modification was obtained by the reaction between Cr, Sn, and Se (molar ratio 2:4:9) at 800°C for 9 days in an evacuated quartz tube. It leads to yield black needlelike air stable crystals after a few hours' cooling (Jobic et al. 1994). The orthorhombic modification was prepared by heating a mixture of Cr, Sn, and Se (molar ratio 2:4:9) at 800°C for 7 days in an evacuated quartz tube, followed by a 9-day cooling to room temperature. Black shiny needlelike air stable crystals were grown (Jobic et al. 1995b).

8.46 Tin–Molybdenum–Selenium

Several ternary compounds are formed in the Sn–Mo–Se system. During low-temperature diffusion of tin (480°C) into $Mo_{15}Se_9$, the compounds **$Sn_{2.1}Mo_{15}Se_9$**, **$Sn_{2.6}Mo_{15}Se_9$** (if the initial $Mo_{15}Se_9$ was obtained by the interaction of HCl and $In_3Mo_{15}Se_9$), **$Sn_2Mo_{15}Se_9$**, and **$Sn_{2.5}Mo_{15}Se_9$** (if the initial $Mo_{15}Se_9$ was obtained by the interaction of HCl and $In_2Mo_{15}Se_9$) are formed in the Sn–Mo–Se system, which crystallize in a hexagonal structure with lattice parameters a = 977.6 ± 0.3, 989.0 ± 0.1, 967.6 ± 0.2, 977.6 ± 0, and c = 1979 ± 0, 1957 ± 1, 5798 ± 2, 5779 ± 2, respectively (Tarascon et al. 1985). These compounds decompose at the temperatures higher than 800°C.

The **$[(SnSe)_{1.05}]_1(MoSe_2)_1$** ternary intergrowth misfit layer compound is also formed in the Sn–Mo–Se system (Smeller et al. 2012). The samples of this compound were prepared using a custom high vacuum deposition chamber. Layers of Sn and Mo were deposited using electron beam evaporation; Se was deposited with a Knudsen evaporation cell. Precursors were prepared by sequentially depositing each elemental reactant to form a repeating unit with a composition profile similar to the desired final product, which was then repeated to obtain a film of the desired total thickness. The relative thickness of the pair of deposited elemental layers to form each binary structural unit was calibrated to obtain the composition that corresponds to the desired stoichiometry. The thickness of each elemental pair was adjusted to yield individual structural units of SnSe or $MoSe_2$ upon annealing. The deposition produces a predominantly amorphous precursor, which is then gently annealed to form the desired product. Samples were annealed at 400°C in a nitrogen atmosphere for 0.5 h. The experimental a-lattice parameter for the SnSe constituent is 608 ± 2 pm, for the $MoSe_2$ constituent is 330.8 ± 0.4 and c-lattice parameter for this misfit compound is 1242.5 ± 0.6 pm.

Several **$[(SnSe)_{1+\delta}]_m[MoSe_2]_n$** metastable layered intergrowths compounds were prepared by the method of modulated elemental reactants (Beekman et al. 2012). Precise layering and digital control over layer sequence were confirmed by XRD, EPMA, and SEM. Based on in-plane synchrotron XRD data it was shown, that the structure of the SnSe component in these compounds at $m = n = 1$ adopts a tetragonal as opposed to orthorhombic structure, whereas $MoSe_2$ remains hexagonal. Analysis of in-plane diffraction originating from the individual components yields a structural misfit of δ = 0.06 and suggests turbostratic misorientation of the individual layers.

8.47 Tin–Chlorine–Selenium

SnSe–$SnCl_2$: The phase diagram of this system, constructed through DTA, is a eutectic type (Figure 8.80) (Blachnik and Kasper 1974). The eutectic contains 3.2 mol% SnSe and crystallizes at 244°C. At 600°C, 22.7 mol% SnSe dissolves in the melt $SnCl_2$ (Rodionov et al. 1972). The ingots for the investigation were annealed 300°C–400°C for 750 h (Blachnik and Kasper 1974).

The **$Sn_3Se_5Cl_2$** ternary compound with an experimental density of 5.32 g·cm^{-3} is formed in the Sn–Cl–Se system

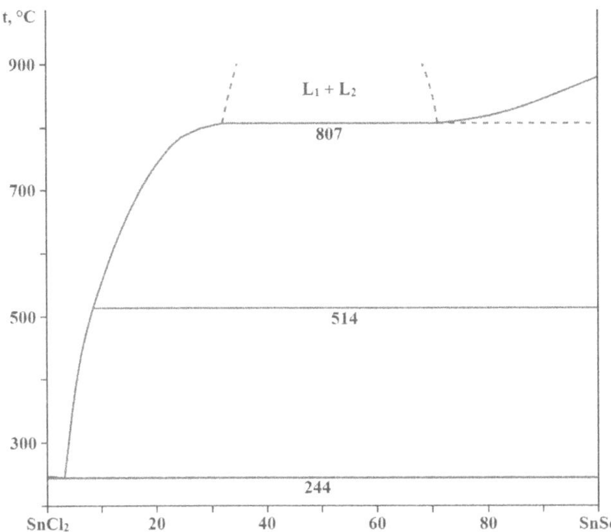

FIGURE 8.80 Phase diagram of the SnSe–$SnCl_2$ system. (From Blachnik, R., and Kasper, F.-W., *Z. Naturforsch.*, **29B**(3–4), 159, 1974.)

(Batsanov et al. 1963). It was obtained by the interaction of $SnCl_2$ and Se at 240°C for 20 h.

8.48 Tin–Bromine–Selenium

SnSe–$SnBr_2$: The phase diagram of this system, constructed through DTA, is a eutectic type (Figure 8.81) (Blachnik and Kasper 1974). The eutectic contains 4.5 mol% SnSe and crystallizes at 224°C. The ingots for the investigation were annealed at 300°C–400°C for 750 h. The **Sn_2SeBr_2** ternary compound was not found in this system (Thévet et al. 1972).

At room temperature, an alcoholic solution of bromine converts SnSe to $SnBr_4$ with the release of elemental selenium (Batsanov and Shestakova 1968).

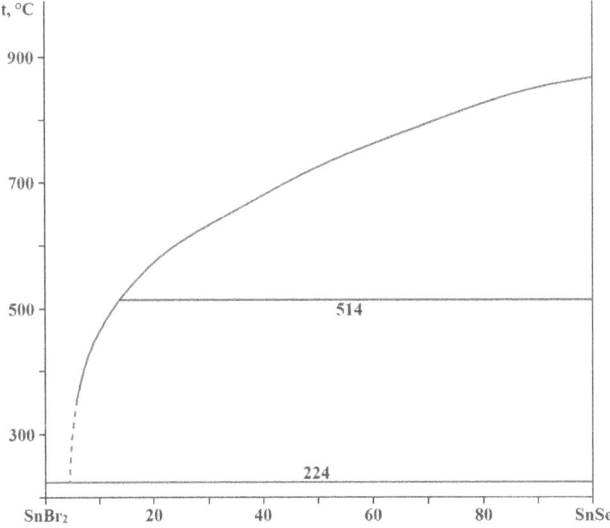

FIGURE 8.81 Phase diagram of the SnSe–$SnBr_2$ system. (From Blachnik, R., and Kasper, F.-W., *Z. Naturforsch.*, **29B**(3–4), 159, 1974.)

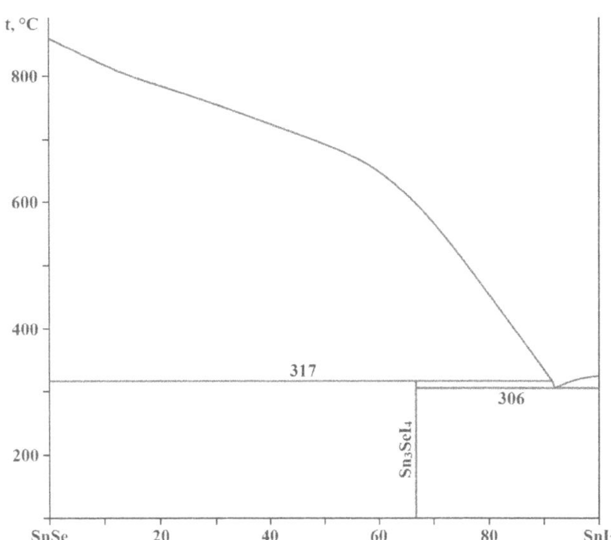

FIGURE 8.82 Phase diagram of the SnSe–SnI_2 system. (From Thévet, F., et al., *C. r. Acad. sci. Sér. C*, **281**(21), 865, 1975.)

8.49 Tin–Iodine–Selenium

SnSe–SnI_2: The phase diagram of this system, constructed through DTA and XRD, is a eutectic type (Figure 8.82) (Blachnik and Kasper 1974; Thévet et al. 1975). The eutectic contains 8 mol% SnSe and crystallizes at 306°C (Thévet et al. 1975) [5 mol% SnSe and 312°C (Blachnik and Kasper 1974)]. The **Sn_3SeI_4** ternary compound, which melts at 317°C [at 330°C (Blachnik and Kasper 1974)] and crystallizes in the tetragonal structure with the lattice parameters $a = 845.5 \pm 0.3$, $c = 1587 \pm 2$ pm, and the calculated and experimental densities of 5.52 ± 0.05 and 5.60 g·cm^{-3}, respectively (Thévet et al. 1975), is formed in this system. Single crystals of this compound were obtained by annealing stoichiometric mixtures of the starting components at 311°C for 7 days. The existence of compound **Sn_2SeI_2** discovered in Thévet et al. (1972) was later not confirmed (Thévet et al. 1975). The ingots for the investigation were annealed 300°C–400°C for 750 h.

At room temperature, an alcoholic solution of iodine converts SnSe to SnI_4 with the release of elemental selenium (Batsanov and Shestakova 1968).

8.50 Tin–Iron–Selenium

SnSe–FeSe: The phase diagram of this system, constructed through DTA, XRD, metallography, and measuring of the microhardness, is given in Figure 8.83 (Murguzov et al. 1984b). The eutectic contains 25 mol% FeSe and crystallizes at 740°C. The **$SnFeSe_2$** ternary compound, which melts incongruently at 780°C, is formed in this system. Thermal effects at 450°C are due to the peritectoid transformation of β-FeSe into γ-FeSe. The ingots for the investigations were annealed at 600°C–700°C for 350 h.

Another ternary compound, **$SnFe_2Se_4$**, which crystallizes in the orthorhombic structure with the lattice parameters

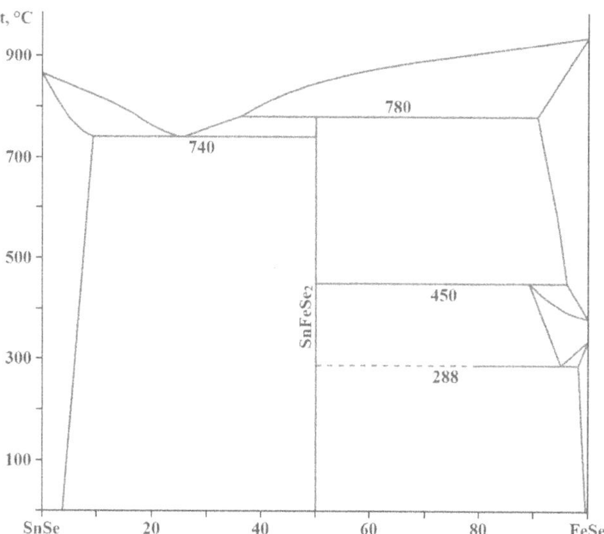

FIGURE 8.83 Phase diagram of the SnSe–FeSe system. (From Murguzov, M.I., et al., *Zhurn. neorgan. khimii*, **29**(10), 2692, 1984.)

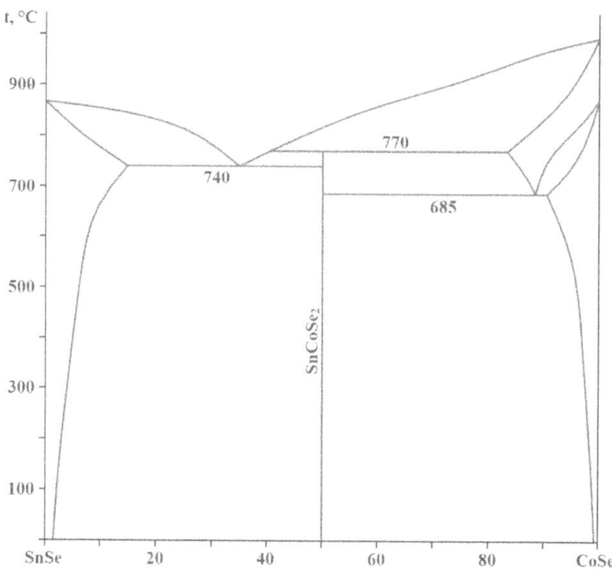

FIGURE 8.84 Phase diagram of the SnSe–CoSe system. (From Murguzov, M.I., et al., *Zhurn. neorgan. khimii*, **29**(10), 2692, 1984.)

$a = 1477.8 \pm 0.2$, $b = 1076.4 \pm 0.1$, $c = 606.1 \pm 0.1$ pm (Quintero et al. 2009) [a = 1480.3, b = 1076.8, c = 605.9 pm, and an energy gap of 0.005 eV (Wei et al. 2018)], is formed in the Sn Fe–Se system. The polycrystalline samples of this compound were prepared by the usual melt and anneal technique. The appropriate amounts of the component elements were melted together, at 1150°C for 3 h, then cooled to 550°C and after annealing at this temperature for 1 month the ingot was very slowly cooled to room temperature (Quintero et al. 2009) or heating the mixture of the elements to 600°C and subsequently held at this temperature for 10 days before it was air quenched to room temperature (Wei et al. 2018).

8.51 Tin–Cobalt–Selenium

SnSe–CoSe: The phase diagram of this system, constructed through DTA, XRD, metallography, and measuring of the microhardness, is shown in Figure 8.84 (Murguzov et al. 1984b). The eutectic contains 35 mol% CoSe and crystallizes at 740°C. The **$SnCoSe_2$** ternary compound, which melts incongruently at 770°C, is formed in this system. The ingots for the investigations were annealed at 600°C–700°C for 350 h.

Another ternary compounds, **$Sn_2Co_3Se_2$** and **$Sn_3Co_2Se_3$** or **$Sn_{1.5}Co_2Se_{1.5}$**, are formed in the Sn–Co–Se system. $Sn_2Co_3Se_2$ crystallizes in the trigonal structure with the lattice parameters $a = 556.35$ pm and $\alpha = 58.33°$ or $a = 543.22$ and $c = 1379.75$ pm in hexagonal setting, a calculated density of 8.17 g·cm^{-3}, and an energy gap of 0.24 eV (the predicted values) (Weihrich et al. 2015) and could be prepared by the interaction of Co and SnSe or Co_3Se_2 and Sn (Zabel et al. 1979a).

$Sn_3Co_2Se_3$ also crystallizes in the trigonal structure with the lattice parameters $a = 1232.78 \pm 0.03$, $c = 1512.67 \pm 0.06$, and an energy gap of 0.6–0.7 eV (Laufek et al. 2009a). It was synthesized from individual elements by high-temperature solid-state reactions. Stoichiometric amounts of Co, Sn, and Se were sealed into an evacuated silica glass tube and heated at 800°C for 48 h in a furnace. The material was then ground under acetone using agate mortar and pestle and heated at 600°C for 168 h. The resultant material was once again ground under acetone and heated at 600°C for 432 h. After annealing, the furnace was turned off and allowed to cool slowly to room temperature.

8.52 Tin–Nickel–Selenium

SnSe–NiSe: The phase diagram of this system, constructed through DTA, XRD, metallography, and measuring of the microhardness, is shown in Figure 8.85 (Murguzov et al. 1984b). The eutectic contains 20 mol% NiSe and crystallizes at 700°C. The **$SnNiSe_2$** ternary compound, which melts incongruently at

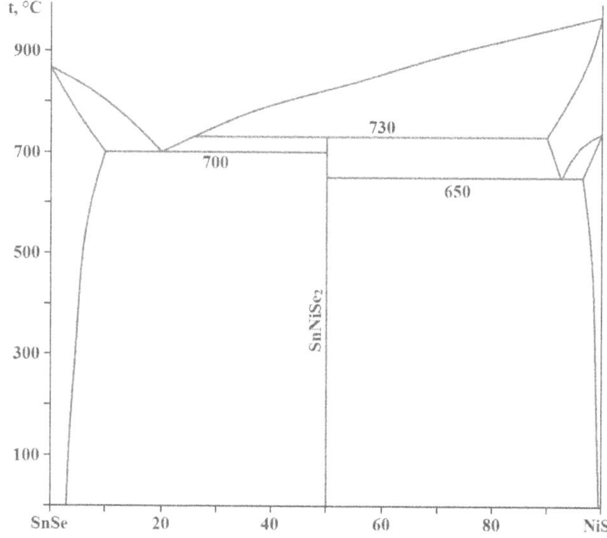

FIGURE 8.85 Phase diagram of the SnSe–NiSe system. (From Murguzov, M.I., et al., *Zhurn. neorgan. khimii*, **29**(10), 2692, 1984.)

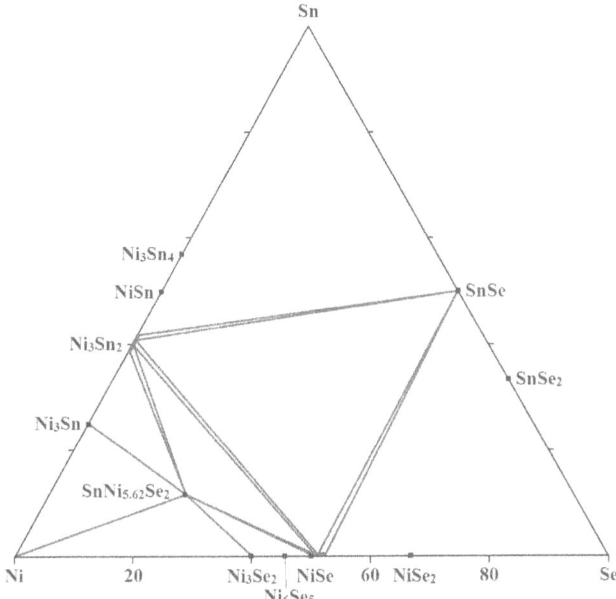

FIGURE 8.86 Isothermal sections of the Sn–Ni–Se ternary system at 540°C. (From Baranov, A.I., et al., *J. Solid State Chem.*, **177**(10), 3616, 2004.)

730°C, is formed in this system. The ingots for the investigations were annealed at 600°C–700°C for 350 h.

Another ternary compounds, $\mathbf{SnNi_{5.62}Se_2}$ and $\mathbf{Sn_2Ni_3Se_2}$ are formed in the Sn–Ni–Se system. $SnNi_{5.62}Se_2$ crystallizes in the tetragonal structure with the lattice parameters a = 368.90 ± 0.08, c = 1864.8 ± 0.3, and a calculated density of 7.942 g·cm^{-3} (Baranov et al. 2004). To prepare this compound, the stoichiometric mixture of the elements was annealed in evacuated silica ampoule at 540°C for 7 days. The product was then ground in an agate mortar, pressed into pellets and further re-annealed under the same conditions two times. The grayish powder of the title compound was obtained. The single crystals of $SnNi_{5.62}Se_2$ were prepared from the vapor phase by means of chemical transport reaction with I_2 as a transport agent in silica ampoule placed in a horizontal, two-temperature furnace.

$Sn_2Ni_3Se_2$ crystallizes in the trigonal structure and could be prepared by the interaction of Ni and SnSe or Ni_3Se_2 and Sn (Zabel et al. 1979a).

The isothermal section of the Sn–Ni–Se ternary system at 540°C is presented in Figure 8.86 (Baranov et al. 2004). The total annealing time (540°C, with several intermediate grindings in an agate mortar and re-annealing of products pressed into pellets) was about 10–20 days depending on whether the equilibrium had been reached.

8.53 Tin–Rhodium–Selenium

The $\mathbf{Sn_2Rh_3Se_2}$ ternary compound, which crystallizes in the trigonal structure and could be prepared by the interaction of Rh and SnSe or Rh_3Se_2 and Sn, is formed in the Sn–Rh–Se system (Zabel et al. 1979a).

8.54 Tin–Palladium–Selenium

The $\mathbf{Sn_2Pd_3Se_2}$ ternary compound, which crystallizes in the trigonal structure and could be prepared by the interaction of Pd and SnSe or Pd_3Se_2 and Sn, is formed in the Sn–Pd–Se system (Zabel et al. 1979a).

8.55 Tin–Iridium–Selenium

Three ternary compounds, $\mathbf{IrSn_{0.45}Se_{1.55}}$, $\mathbf{Ir_2SnSe_5}$, and $\mathbf{Ir_2Sn_3Se_3}$, are formed in the Sn–Ir–Se system. $IrSn_{0.45}Se_{1.55}$ crystallizes in the cubic structure with the lattice parameter a = 607.4419 ± 0.0004 pm (Trump et al. 2015).

Ir_2SnSe_5 crystallizes in the monoclinic structure with the lattice parameters a = 765.768 ± 0.005, b = 751.027 ± 0.005, c = 1249.737 ± 0.008 pm, and β = 102.0831 ± 0.0006° (Trump et al. 2015).

$Ir_2Sn_3Se_3$ crystallizes in the rhombohedral structure with the lattice parameters a = 895.5798 ± 0.0006 pm, α = 89.92625 ± 0.00005°, and an energy gap of ~0.4 eV (Trump et al. 2015) [a = 895.0 pm, α = 89.98°, a calculated density of 9.054 g·cm^{-3}, and an energy gap of 0.4 eV or a = 1265.6 ± 0.1 and c = 1550.7 ± 0.1 in hexagonal setting (Yan et al. 2015)].

The powders of all these compounds were grown by placing Ir, Sn, and Se in stoichiometric ratios in a fused silica tube (Trump et al. 2015). The tubes were filled with 0.03 MPa of Ar to minimize vaporization of Sn and Se. Each tube was heated quickly to 500°C, followed by a 50°C·h^{-1} ramp to an annealing temperature at which the samples were held for 4 days, before being furnace cooled. The resulting products were pulverized, pressed into pellets, heated at the same annealing temperature for 4 days, and furnace cooled again. $IrSn_{0.45}Se_{1.55}$ was annealed at 900°C and quenched in water after each treatment, Ir_2SnSe_5 was annealed at 780°C and $Ir_2Sn_3Se_3$ and was annealed at 750°C.

$Ir_2Sn_3Se_3$ could be also synthesized from the elements at 800°C with the next keeping at 600°C and 4 GPa for 1 h inside a belt apparatus (Yan et al. 2015). Microcrystalline powder of this compound was obtained from high temperature synthesis in sealed quartz ampoule. The mixture of Ir and SnSe was heated up to 1000°C and slowly cooled down after 5 days. $Ir_2Sn_3Se_3$ can also be prepared from the elements by multiple annealing to ensure phase purity. For generating suitable single crystals flux reactions of Ir in access of SnSe were performed. The excess of SnSe can be removed by dissolving the resulting product in aqua regia.

8.56 Tin–Platinum–Selenium

The **PtSnSe** ternary compound is formed in the Sn–Pt–Se. This compound melts at 1109.13°C and was synthesized from the elements by high-temperature solid state reactions (Weihrich et al. 2004a). Equimolar amounts of Pt, Sn, and Se were heated in sealed and evacuated quartz tubes to 1070°C. The tubes were quenched after 1 week. PtSnSe was received as dark polycrystallites. Its structure optimizations were performed with a first principles plane-wave code, which treats exchange and correlation in the local density approximation of the density functional theory. According to the calculations, the title compound crystallizes in the orthorhombic structure with the lattice parameters a = 624.88, b = 623.87, and c = 621.56 pm (a = 627.44 ± 0.13, b = 625.97 ± 0.15, and c = 623.51 ± 0.14 pm from XRD). From band structure calculations PtSnS was predicted to be indirect band gap semiconductor with E_g = 0.3 eV.

REFERENCES

All References are available as a downloadable eResource at www.routledge.com/9780367639235

9

Systems Based on Tin Telluride

9.1 Tin–Sodium–Tellurium

The $\mathbf{Na_4SnTe_4}$ ternary compound, which crystallizes in the orthorhombic structure with the lattice parameters $a = 865.8 \pm 0.5$, $b = 884.0 \pm 0.5$, $c = 1526.9 \pm 0.8$ pm, and the calculated and experimental densities of 4.097 and 4.15 $g{\cdot}cm^{-3}$, respectively, is formed in the Sn–Na–Te system (Eisenmann et al. 1983a). This compound was obtained by the interaction of the Na, Sn, and Te at 150°C in an Ar atmosphere.

9.2 Tin–Potassium–Tellurium

Three ternary compounds are formed in the Sn–K–Te system. $\mathbf{K_2SnTe_5}$ melts at 329°C and crystallizes in the tetragonal structure with the lattice parameters $a = 848.1 \pm 0.3$, $c = 1536.9 \pm 0.7$ pm, and the calculated and experimental densities of 5.02 and 5.08 $g{\cdot}cm^{-3}$, respectively (Eisenmann et al. 1983e). To prepare this compound, Sn and Te were used in stoichiometric proportion and placed into a quartz ampoule with a corundum inner crucible, pre-dried at 120°C, then rinsed several times with Ar and the potassium was added. Since the side reaction of the K with the corundum and quartz walls is unavoidable at the necessary reaction temperatures, empirical findings show that it is necessary to add 10 mass% of this element in excess. The evacuated and then melted ampoule was transferred in a corundum tube. The sample was heated to 750°C under argon with a rate of $3°C{\cdot}min^{-1}$, annealed at this temperature for 1 h and then the furnace was switched off. The plate-like crystals of the title compound were obtained.

$\mathbf{K_4Sn_2Te_6}$ crystallizes in the orthorhombic structure with the lattice parameters $a = 2315.11 \pm 0.04$, $b = 862.77 \pm 0.01$, and $c = 865.36 \pm 0.02$ pm at 170 K (Evenson IV and Dorhout 2000). This compound was synthesized by the next way. Te (0.666 mM), Sn (0.167 mM), and La (0.222 mM) were reacted in an evacuated fused silica ampoule at 825°C for 150 h. The product was ground up and combined with K_2Te_3 (0.221 mM) and reacted in a fused silica ampoule at 525°C for 150 h. This resulted in bronze colored plates of $K_4Sn_2Te_6$.

$\mathbf{K_6Sn_2Te_6}$ crystallizes in the monoclinic structure with the lattice parameters $a = 959.0 \pm 0.5$, $b = 1365.4 \pm 0.8$, $c = 960.6 \pm 0.5$ pm, $\beta = 116.84 \pm 0.05°$, and the calculated and experimental densities of 3.66 and 3.66 ± 0.01 $g{\cdot}cm^{-3}$, respectively (Dittmar 1979). To obtain this compound, stoichiometric amounts of the elements were heated to 630°C in the evacuated quartz tube for 1 h and then cooled to room temperature overnight. The crystal pieces of $K_6Sn_2Te_6$ with a metallic luster could be isolated from the gray-black regulus obtained.

9.3 Tin–Rubidium–Tellurium

The $\mathbf{Rb_2SnTe_5}$ ternary compound, which crystallizes in the tetragonal structure with the lattice parameters $a = 866.6 \pm 0.3$, $c = 1545.1 \pm 0.5$ pm, and a calculated density of 5.31 $g{\cdot}cm^{-3}$, is formed in the Sn–Rb–Te system (Brinkmann et al. 1985d). To prepare this compound, the mixture of Sn, Te, and $Rb(CH_3COO)$ (molar ratio 2:1:5) was heated slowly in corundum crucible under Ar atmosphere. The mixture was heated at a rate of $100°C{\cdot}h^{-1}$ to 850°C, annealed at 700°C for 24 h and then cooled to room temperature within 4 h. The resulting from the pyrolysis of the acetate water was absorbed in a P_2O_5-reservoir. A compact silver-gray regulus emerged, from which small black, strip-shaped crystals could be broken out. In humid air this compound decomposes very rapidly, must be therefore immediately coated with dry, thick-flowing paraffin oil after the sample had been opened.

9.4 Tin–Cesium–Tellurium

Four ternary compounds are formed in the Sn–Cs–Te system. $\mathbf{Cs_2SnTe_4}$ crystallizes in the orthorhombic structure with the lattice parameters $a = 1268.8 \pm 0.3$, $b = 1184.2 \pm 0.2$, $c = 784.5 \pm 0.2$ pm, and a calculated density of 5.04 $g{\cdot}cm^{-3}$ (Sheldrick and Schaaf 1994b). To obtain this compound, Cs_2CO_3 (0.933 mM) and Te (3.730 mM) were slurried in 0.6 mL of MeOH and melted in a glass ampoule (degree of filling 10%) under Ar atmosphere. The ampoule was located in a horizontal furnace at a 5° angle to the horizontal plane, heated to 190°C for 2 days, and then cooled to room temperature at a rate of $2°C{\cdot}h^{-1}$. The gray-black metallic shiny crystals were obtained.

Cs_2SnTe_4 was obtained by melting together Cs_2Te (10.2 mM), Te (10.2 mM), and Sn (5.1 mM) in a quartz ampoule in the N_2 current at 700°C for 15 min (Zimmermann and Dehnen 2003).

$\mathbf{Cs_4Sn_2Te_7}$ crystallizes in the monoclinic structure with the lattice parameters $a = 1649.1 \pm 0.7$, $b = 786.0 \pm 0.3$, $c = 1903.6 \pm 0.7$ pm, $\beta = 105.5 \pm 0.1°$, and the calculated and experimental densities of 4.64 and 4.62 $g{\cdot}cm^{-3}$, respectively (Brinkmann et al. 1985d). This compound was synthesized by the following way. The mixture of Sn, Te, and $Cs(CH_3COO)$ (molar ratio 4:2:7) was heated slowly in corundum crucible under Ar atmosphere. The mixture was heated at a rate of $100°C{\cdot}h^{-1}$ to 700°, kept at this temperature for 24 h and then cooled slowly to room temperature. The resulting from the pyrolysis of the acetate water was absorbed in a P_2O_5-reservoir. The compound looked as a compact regulus with a luster black surface out of which platelet-shaped diamonds had grown out. In humid air this compound decomposes very rapidly, must be therefore

DOI: 10.1201/9781003123507-9

immediately coated with dry, thick-flowing paraffin oil after the sample had been opened.

$Cs_6Sn_2Te_6$ also crystallizes in the monoclinic structure with the lattice parameters a = 1746.0 ± 0.4, b = 1433.0 ± 0.3, c = 1025.0 ± 0.2 pm, β = 96.73 ± 0.03°, and a calculated density of 4.695 g·cm^{-3} at 223 K (Friede and Jansen 1999a). It has been prepared at the reacting a stoichiometric ratio of the elements in an evacuated (0.1 Pa) silica glass tube. The mixture was heated to 650°C over a period of 5 h. It was held at this temperature for 1 h and then allowed to cool to room temperature overnight by switching off the furnace. The title compound crystallizes as black columns.

9.5 Tin–Copper–Tellurium

SnTe–Cu: According to the data of DTA, XRD, and metallography, this system is a non-quasibinary section of the Sn–Cu–Te ternary system (Carcaly et al. 1975, 1977). There two immiscibility regions at 1050°C and 625°C in the system (Carcaly et al. 1975, 1977; Bates et al. 1968). Thermal effects at 786°C, 724°C, 662°C, 648°C, 640°C, 606°C, and 560°C are due to the polymorphic transformations of $Cu_{2-x}Te$ and the formation of binary phases in the Cu–Sn system.

SnTe–Cu_2Te: According to the data of Dovletov et al. (1974b) and Zotova and Karagodin (1975a), the phase diagram of this system is a eutectic type. The eutectic contains 52 mol% Cu_2Te and crystallizes at 585°C (Zotova and Karagodin 1975a) [at 575°C (Dovletov et al. 1974b)]. Thermal effects at 570°C and 380°C are due to the polymorphic transformations of Cu_2Te. There are limited regions of the solid solutions based on the initial components.

According to the data of Carcaly et al. (1975), this system is a non-quasibinary section of the Sn–Cu–Te ternary system since with decreasing temperature, stoichiometry of Cu_2Te shifts towards an increase in the Te content. This section contains polymorphic transformations of $Cu_{2-x}Te$ at 150°C, 260°C, 305°C, 360°C, and 545°C, a ternary eutectic at 560°C, and invariant equilibria at 642°C, 662°C, 724°C, and 786°C with the participation of $Cu_{2-x}Te$. From the SnTe side, an insignificant area of solid solutions was found.

The molten samples of the SnTe–Cu_2Te system are liquid semiconductors (Dovletov et al. 1978).

The initial samples were annealed at temperatures below the temperature of the eutectic transformation for more than 500 h (Dovletov et al. 1974b). The system was investigated through DTA, XRD, metallography, and measuring of the microhardness (Dovletov et al. 1974b; Carcaly et al. 1975; Zotova and Karagodin 1975a; Dovletov et al. 1978).

The liquidus surface of the Cu–Sn–Te ternary system (Figure 9.1) is characterized by the formation of the **Cu_2SnTe_3** compound, which melts incongruently at 412°C [at 407°C (Palatnik et al. 1961a); at 411°C (Averkieva et al. 1964); at 410°C (Rivet 1965)]. Three ternary eutectics E_1 (337°C), E_2 (560°C), and E_3 (227°C), last of which is degenerated, and two immiscibility regions exist in this system (Carcaly et al. 1975, 1977).

Apparently, Cu_2SnTe_3 has three polymorphic modifications. One of them crystallizes in the cubic structure with the

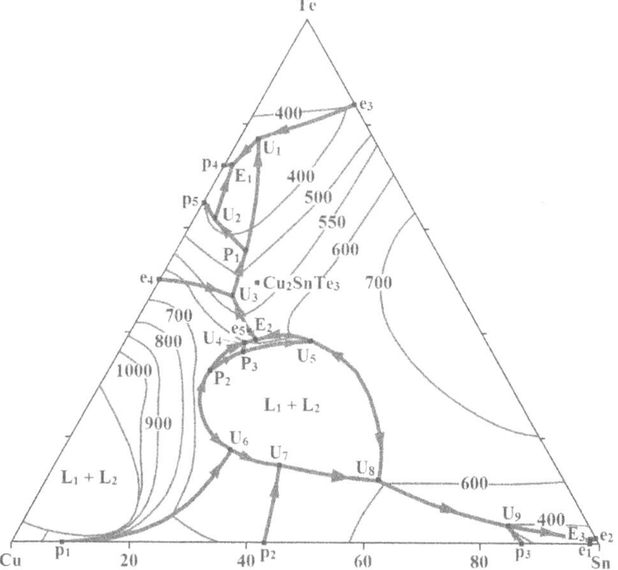

FIGURE 9.1 Liquidus surface of the Cu–Sn–Te ternary system. (From Carcaly, C., et al., *J. Less-Common Metals*, **41**(1), 1, 1975.)

lattice parameter a = 604,84 ± 0.06 pm and a calculated density of 6.28 ± 0.04 g·cm^{-3} (Alzahrani et al. 2020) [a = 604 pm (Palatnik et al. 1961a,b); a = 604.7 pm and the calculated and experimental densities 6.29 and 6.51 ± 0.08 g·cm^{-3}, respectively (Rivet et al. 1963; Rivet 1965); a = 603.6 pm and the calculated and experimental densities 6.31 and 6.23 g·cm^{-3}, respectively (Averkieva et al. 1964); a = 604.90 pm (Sharma et al. 1977a)]. Cubic Cu_2SnTe_3 was prepared by direct reaction of the elements (Alzahrani et al. 2020). The powders of Cu, Sn, and Te were loaded into a silica ampoule in the stoichiometric ratio and sealed in a quartz tube under vacuum. The specimen was reacted in a furnace at 700°C and held at this temperature for 7 days before it was air quenched to room temperature. It was then ground into fine powder, cold pressed into a pellet and annealed for 10 days at 400°C. The resulting pellet was ground into fine powder, sieved (~325 mesh) and loaded into a custom-designed tungsten carbide die and punch assembly that was lined with graphite foil, in order to prevent reaction with the surrounding material, for densification by spark plasma sintering at 300° and 400 MPa for 30 min under vacuum. A high-density polycrystalline specimen was obtained from densification.

Second modification crystallizes in the tetragonal structure with the lattice parameters a = 604.8 and c = 1211 pm and the calculated and experimental densities 6.27 and 6.21 g·cm^{-3}, respectively (Hahn et al. 1966).

Third modification crystallizes in the orthorhombic structure with the lattice parameters a = 1283.3 ± 0.4, b = 427.4 ± 0.1, and c = 604.3 ± 0.1 pm (Delgado et al. 2008). The samples of this modification were prepared by the vertical Bridgman-Stockbarger technique. Stoichiometric quantities of Cu, Sn, and Te were sealed in evacuated quartz ampoule. The ampoule was placed in a multiple zone vertical furnace. Initially they were heated from room temperature to 1150°C at a rate of 40°C·h^{-1}. The molten mixture was kept at this temperature for 24 h. In order to assure the homogeneous mixing, the ampoule was agitated periodically. It was later cooled at a rate of

10°C·h^{-1} up to 800°C, then at 1°C·h^{-1} to 640°C. The cooling rate from 640 to 500°C was 5°C·h^{-1}. The ingots were annealed at this temperature for 120 h. The furnace was then turned off and the ingot cooled down to room temperature.

The study of the "$SnTe_2$"–Cu_2Te vertical section, on which the Cu_2SnTe_3 compound should be located, does not confirm its existence: all samples of this section turned out to be three-phase (Glazov et al. 1976).

9.6 Tin–Silver–Tellurium

SnTe–AgTe: The phase diagram of this system is a eutectic type (Gorochov 1968a). The solubility of AgTe in SnTe reaches 36.5 mol%. Silver in these solid solutions is univalent, while tin exhibits valances +2 and +4.

Two compounds are formed in this system. **$AgSnTe_2$** melts incongruently at 600°C (Blachnik et al. 1978) and crystallizes in the cubic structure with the lattice parameter a = 604–610 pm (Hahn and Schulze 1964).

Ag_2SnTe_3 melts at 315°C, has an experimental density of 5.76 g·cm^{-3} and an energy gap of 0.08 eV (Balanevskaya et al. 1966).

The initial samples were annealed at 470°C for 2 weeks (Gorochov 1968a) [at 210°C for 60 days (Blachnik et al. 1978). The system was investigated through DTA, XRD, and metallography (Gorochov 1968a; Blachnik et al. 1978).

SnTe–Ag_2Te: The phase diagram of this system, constructed through DTA, XRD, and metallography, is a eutectic type (Figure 9.2) (Gorochov 1968a; Dovletov et al. 1974b; Blachnik et al. 1978; Blachnik and Gather 1978). The eutectic contains 55 mol% Ag_2Te and crystallizes at 596°C (Blachnik et al. 1978; Blachnik and Gather 1978) [at 620°C (Dovletov et al. 1974b)]. With an increase in the content of SnTe, the temperature of the α-β-transformation of Ag_2Te decreases from 147°C to 141°C, and of the β-γ-transformation increases from 803°C to 854°C. The solubility of SnTe in β-Ag_2Te reaches 15 mol% and in γ-Ag_2Te is 0.5–1 mol% (Dovletov et al. 1974b). The solubility of

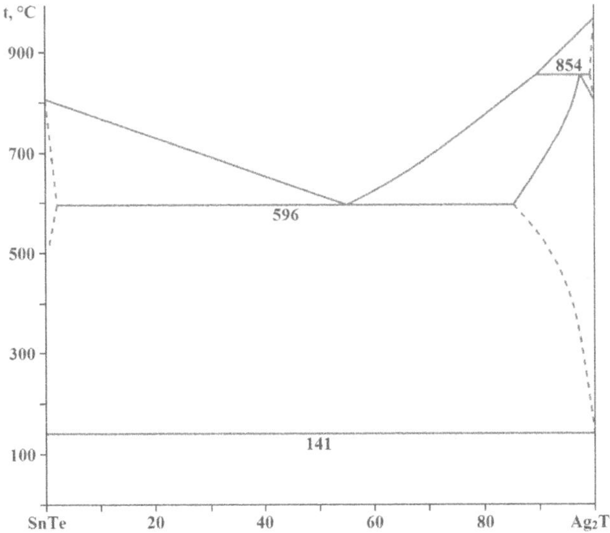

FIGURE 9.2 Phase diagram of the SnTe–Ag_2Te system. (From Blachnik, R., and Gather, B., *J. Less-Common Metals*, **60**(1), 25, 1978.)

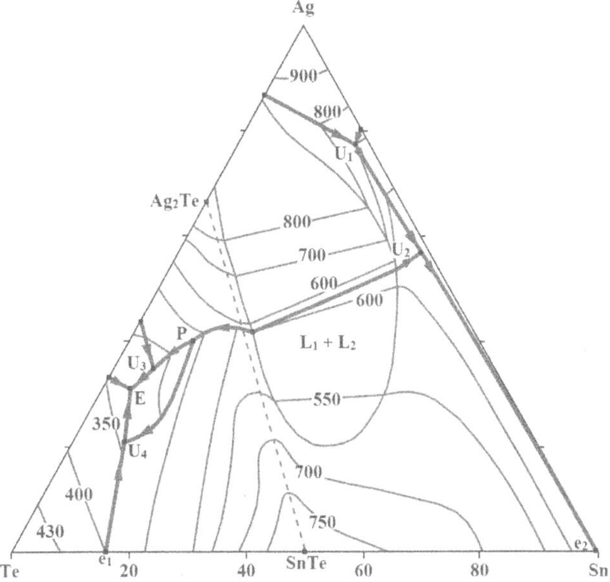

FIGURE 9.3 Liquidus surface of the Sn–Ag–Te ternary system. (From Blachnik, R., et al., *Z. Metallkdd.*, **69**(8), 530, 1978.)

α-Ag_2Te in SnTe at room temperature is 5 mol% (Yusibov et al. 2017b) and the solubility of β-Ag_2Te in SnTe is not higher than 3 mol% (Dovletov et al. 1974b). The ingots for the investigations were annealed at temperatures below the temperature of the eutectic transformation for 250 h (Dovletov et al. 1974b) [at 210°C for 60 days (Blachnik and Gather 1978; Blachnik et al. 1978); at 600°C for 110 h (Yusibov et al. 2017b)].

According to the data of Matye and Schön (1980), **Ag_8SnTe_6** ternary compound, which crystallizes in the cubic structure with the lattice parameter a = 1120 ± 5 pm and the calculated and experimental densities 8.26 and 7.98 ± 0.15 g·cm^{-3}, respectively, is formed in the Sn–Ag–Te system. However, other authors failed to obtain this compound: in all cases, either a mixture of $Ag_{5-x}Te$ and $Ag_xSn_{1-x}Te$ ($0 \le x \le 0.33$) (Gorochov 1968b), or a mixture of $Ag_{5-x}Te$, Ag_2Te and $AgSnTe_2$ (Gather and Blachnik 1983) was formed.

There is a ternary eutectic at 320°C in the SnTe–Ag_2Te–Te subsystem (Gorochov 1968b). The liquidus surface of the Sn–Ag–Te ternary system is presented in Figure 9.3 (Blachnik et al. 1978).

9.7 Tin–Gold–Tellurium

SnTe–Au: The phase diagram of this system, constructed through DTA, XRD, and metallography, is a eutectic type with a large region of immiscibility (Figure 9.4) (Legendre et al. 1975).

SnTe–AuSn: The phase diagram of this system, constructed through DTA, XRD, and metallography, is a eutectic type with a large region of immiscibility (Figure 9.5) (Legendre et al. 1972, 1975). The eutectic contains 98.7 mol% AuSn and crystallizes at 413°C. The immiscibility region exists in this system at 750°C.

SnTe–$AuTe_2$: The phase diagram of this system, constructed through DTA, XRD, and metallography, is a eutectic type (Figure 9.6) (Legendre et al. 1972, 1975). The eutectic contains 32.5 mol% AuSn and crystallizes at 402°C.

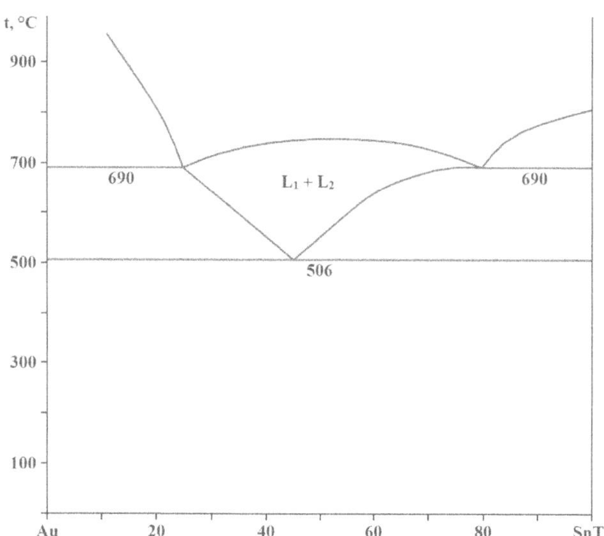

FIGURE 9.4 Phase diagram of the SnTe-Au system. (From Legendre, B., et al., *Bull. Soc. chim. France*, Pt 1, VI, (11–12), 2475, 1975.)

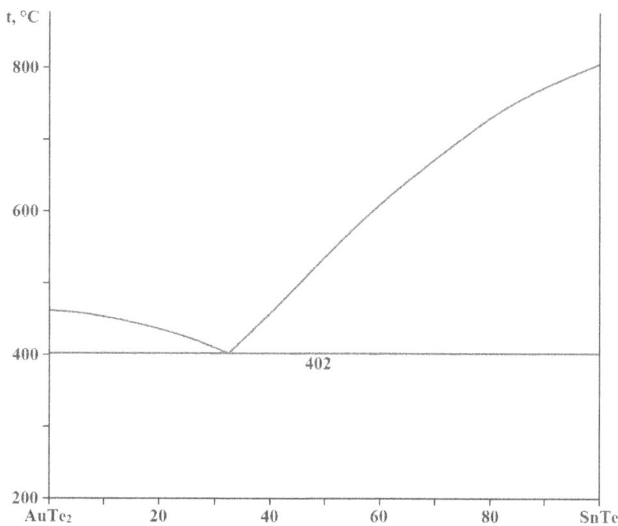

FIGURE 9.6 Phase diagram of the SnTe–$AuTe_2$ system. (From Legendre, B., et al., *C. r. Acad. sci. Sér. C*, **275**(15), 805, 1972.)

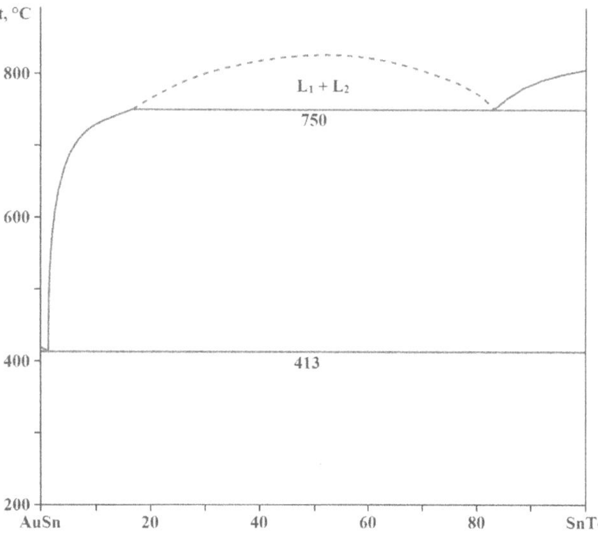

FIGURE 9.5 Phase diagram of the SnTe–AuSn system. (From Legendre, B., et al., *C. r. Acad. sci. Sér. C*, **275**(15), 805, 1972.)

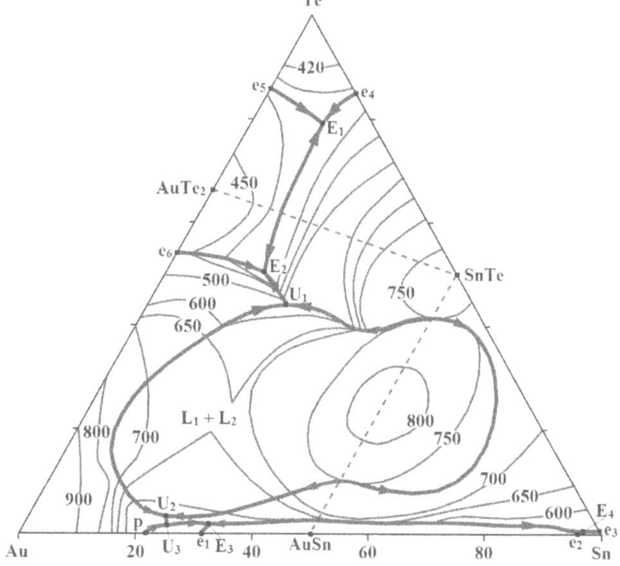

FIGURE 9.7 Liquidus surface of the Sn–Au–Te ternary system. (From Legendre, B., and Souleau, C., *C. r. Acad. sci. Sér. C*, **284**(18), 739, 1977.)

The liquidus surface of the Sn–Au–Te ternary system is shown in Figure 9.7 (Legendre et al. 1975; Legendre and Souleau 1977). The ternary eutectic in the SnTe–$AuTe_2$–Te subsystem crystallizes at 373°C and contains 9.1 at% Au, 78.6 at% Te, and 12.3 at% Sn; the ternary eutectics in the SnTe–$AuTe_2$–AuSn–Au subsystem crystallize at 398°C and 278°C; the transition point exists in the SnTe–Au–AuSn subsystem at 426°C; two transition points exist in the SnTe–AuSn–Sn subsystem at 309°C and 252°C and degenerated eutectic crystallizes at 217°C. A large immiscibility region exists on the liquidus surface.

9.8 Tin–Barium–Tellurium

The $\mathbf{Ba_2SnTe_5}$ ternary compound, which has two polymorphic modifications, is formed in the Sn–Ba–Te system. Fist modification of this compound crystallizes in the orthorhombic structure with the lattice parameters $a = 682.5 \pm 0.1$, $b = 2447.6 \pm 0.5$, $c = 688.9 \pm 0.1$ pm, and a calculated density of 5.953 $g{\cdot}cm^{-3}$ (Li et al. 1996). The bronze-colored, plates like crystals of this modification were grown from a mixture, containing K_2Te (0.5 mM), BaTe (1.5 mM), Sn (0.5 mM), and Te (2.5 mM). The starting materials were weighed and mixed in a glovebox and subsequently sealed in a Pyrex tube under vacuum. The temperature was raised slowly to 450°C (over 10 h) and the sample was kept at this temperature for 4 days. The tube was then cooled to 150°C at a rate of 4°C·h^{-1}. The final product was isolated from the excess flux by washing first with DMF followed by absolute ethanol and anhydrous diethyl ether. The crystals of this modification were also found in other reactions where a different temperature (525°C), different flux, and different ratio of the starting materials (Na_2Te/BaTe/Sn/6Te or Na_2Te/2BaTe/2Sn/6Te) were used.

Second modification of this compound crystallizes in the monoclinic structure with the lattice parameters a = 967.91 ± 0.05, b = 1316.96 ± 0.07, c = 2759.57 ± 0.15 pm, β = 91.585 ± 0.001°, a calculated density of 5.845 g·cm^{-3}, and an energy gap of 0.18 eV (Assoud et al. 2004a). Monoclinic Ba_2SnTe_5 was prepared by heating the mixture of the elements under exclusion of air to 900°C, followed by slow cooling to 200°C over a period of 1 week, and switching off the furnace thereafter. The monoclinic modification was also found by heating the mixture of the elements under a flow of argon to 1000°C, and subsequent cooling to ambient temperatures at a rate of 50°C·h^{-1}. Phase-pure samples of monoclinic Ba_2SnTe_5 were also obtained by prolonged heating for 3 weeks between 650°C and 750°C, finished by switching off the furnace.

All reactions were carried out in evacuated fused silica tubes placed into temperature-controlled resistance furnaces, starting directly from the elements.

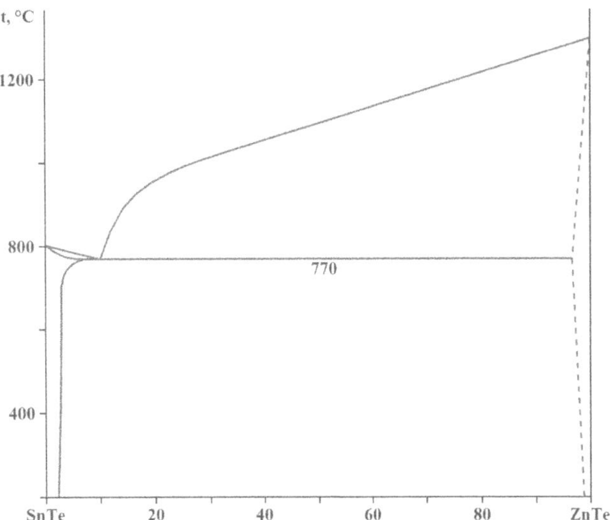

FIGURE 9.9 Phase diagram of the SnTe–ZnTe system. (From Sultanova, N.R., et al., *Izv. AN SSSR. Neorgan. mater.*, **10**(8), 1418, 1974.)

9.9 Tin–Zinc–Tellurium

Sn–ZnTe: The ZnTe solubility in liquid tin is given in Figure 9.8 (Rubenstein 1968). Liquidus temperatures were determined by high-temperature filtration.

SnTe–ZnTe: The phase diagram of this system, constructed through DTA, XRD, and metallography, is a eutectic type (Figure 9.9) (Sultanova et al. 1974). The eutectic composition and temperature are 10 mol. % ZnTe and 770°C, respectively. The solubility of ZnTe in SnTe is not higher than 2 mol%, and the solubility of SnTe in ZnTe corresponds to 1 mol%.

SnTe–Zn: When SnTe and Zn interact, an exchange reaction occurs with the formation of ZnTe and Sn (Bates et al. 1968). The ingots, containing 50 mol% SnTe, were annealed at 450°C and investigated through metallography.

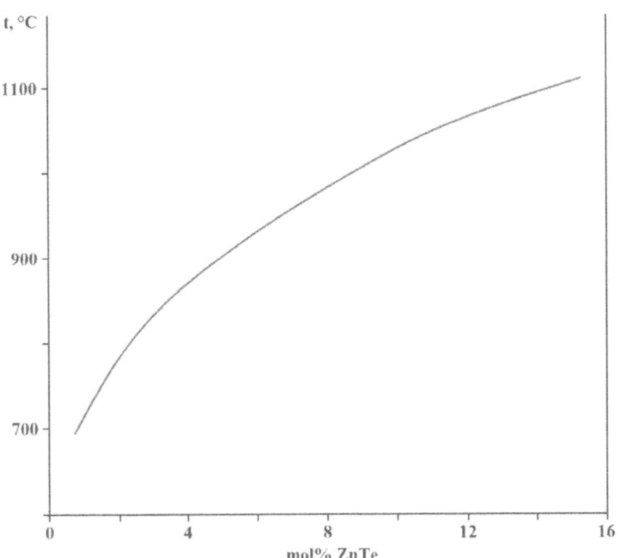

FIGURE 9.8 Temperature dependence of ZnTe solubility in liquid Sn. (From Rubenstein, M., *J. Cryst. Growth*, **3–4**, 309, 1968.)

9.10 Tin–Cadmium–Tellurium

Sn–CdTe: The phase diagram of this system, constructed through DTA, XRD, metallography, and measuring of the microhardness, is a eutectic type (Figure 9.10) (Panchuk et al. 1973; Tai and Hori 1974; Morgant et al. 1981). The eutectic is degenerated from the Sn-rich side, contains 99.995 at% Sn, and crystallizes at 232°C. The mutual solubility of CdTe and Sn is insignificant. At 700°C–750°C, tin occupies mainly cadmium vacancies and at 850°C–925°C, tellurium vacancies. Amphoteric behavior of Sn in CdTe leads to the complicated temperature dependences of Sn dissolution in CdTe at maximum and minimum cadmium vapor pressure (Figure 9.11) (Panchuk et al. 1978). These dependences are retrograde. Maximum solubility of Sn in the tellurium (p_{Cd}^{max}) and cadmium (p_{Cd}^{min}) sublattices is at 925°C and 775°C, respectively.

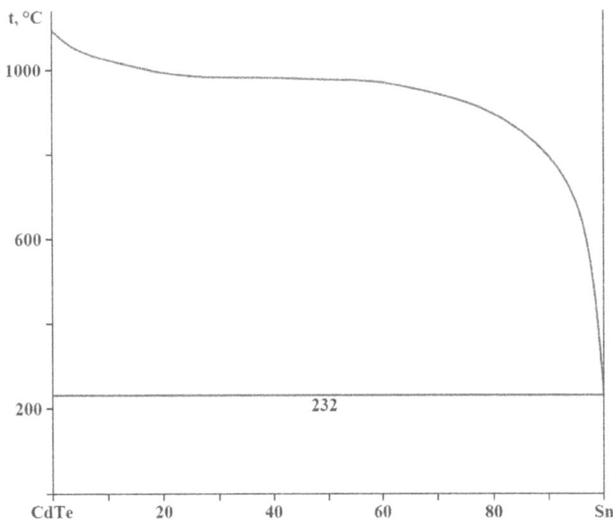

FIGURE 9.10 Phase diagram of the CdTe–Sn system. (From Panchuk, O.E., *Izv. AN SSSR. Neorgan. mater.*, **9**(4), 572, 1973.)

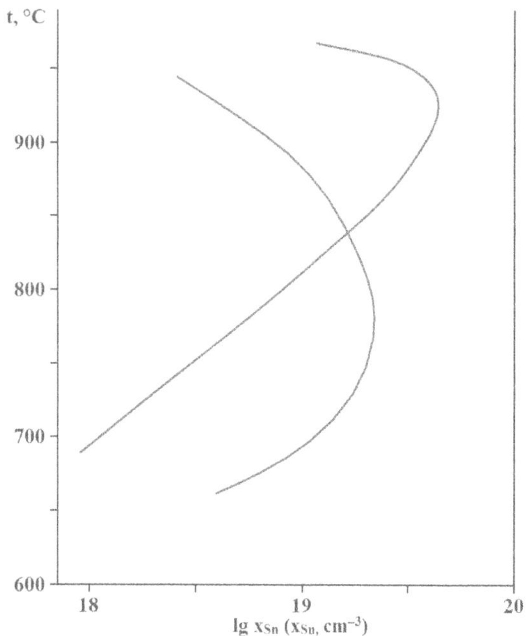

FIGURE 9.11 Temperature dependences of Sn solubility in CdTe: 1, p_{Cd}^{max}; 2, p_{Cd}^{min}. (From Panchuk, O.E., et al., *Izv. AN SSSR. Neorgan. mater.*, **14**(1), 50, 1978.)

The solubility of Sn in CdTe was determined using radioactive isotopes.

Equilibrium distribution coefficient of Sn in CdTe is equal to 0.066 ± 0.003 (Burachek et al. 1985).

SnTe–CdTe: The phase diagram of this system, constructed through DTA, XRD, metallography, and EPMA, is a eutectic type (Figure 9.12) (Rosenberg et al. 1964; Nasirov et al. 1970b; Morgant et al. 1981; Leute and Menge 1992). The eutectic composition and temperature are 18 mol. % CdTe and 792°C, respectively (Leute and Menge 1992) [20 mol. % CdTe and 792°C (Morgant et al. 1981), 12 mol. % CdTe and 780°C (Nasirov et al. 1970b), 784°C (Rosenberg et al. 1964)]. The solubility of CdTe in SnTe at 530°C, 630°C, and 730°C is equal to 4, 6.5, and 11 mol%, respectively (Leute and Menge 1992) [according to the data of Tairov et al. (1969) and Morgant et al. (1981), the solubility of CdTe in SnTe at the eutectic temperature is equal to 10 mol. % and decreases to 8 mol. % at 760°C]. The solubility of SnTe in CdTe at the eutectic temperature reaches 4 mol. % (Morgant et al. 1981).

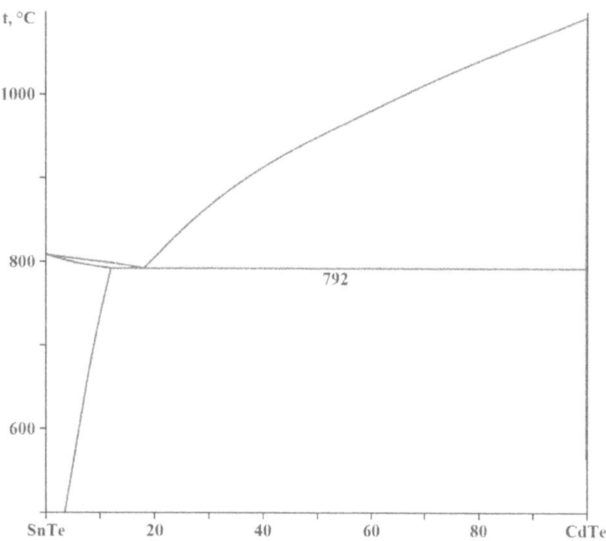

FIGURE 9.12 Phase diagram of the SnTe–CdTe system. (From Leute, V. and Menge, D., *Z. Phys. Chem. (Munchen)*, **176**(1), 47, 1992.)

"CdSnTe$_3$"–CdTe: The mixture of two or more phases with different crystal lattices is formed at the alloying of CdTe and "CdSnTe$_3$" (Radautsan and Ivanova 1961).

SnTe–Cd: This system is a non-quasibinary section of the Sn–Cd–Te ternary system (Bates et al. 1968; Morgant et al. 1981). At 175°C and 232°C invariant equilibria of the eutectic type appear, and at 140°C a eutectoid transformation occurs.

In the CdTe–SnTe–Te subsystem of the Sn–Cd–Te ternary system, there is a ternary eutectic at 405°C that is degenerated and practically coincides with the eutectic in the SnTe–Te system (Morgant et al. 1981).

9.11 Tin–Mercury–Tellurium

Sn–HgTe: This section is a non-quasibinary section of the Sn–Hg–Te ternary system. Tin telluride and Hg are formed at the interaction of HgTe with Sn (Vengel' et al. 1986).

SnTe–HgTe: The phase diagram of this system, constructed through DTA, XRD, metallography, and measuring of the microhardness, is a eutectic type (Figure 9.13) (Vengel' et al. 1983). The eutectic composition and temperature are 45 mol% SnTe and 584°C, respectively. The solubility of HgTe and SnTe at the eutectic temperature is equal to 7 mol%, and the solubility of SnTe in HgTe is not higher than 1 mol%.

9.12 Tin–Boron–Tellurium

The glass forming region in the Sn–B–Te ternary system was not found (Kirilenko and Dembovskiy 1974).

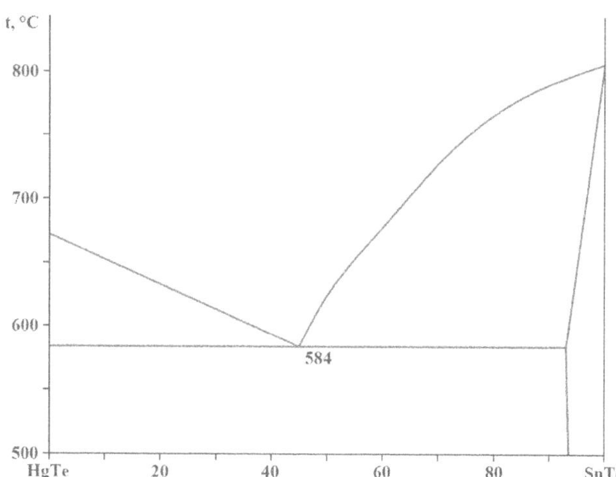

FIGURE 9.13 Phase diagram of the SnTe–HgTe system. (From Vengel', P.F., et al., *Ukr. khim. zhurn.*, **49**(12), 1247, 1983.)

9.13 Tin–Aluminum–Tellurium

The $\mathbf{Al_6SnTe_{10}}$ ternary compound, which crystallizes in the trigonal structure with the lattice parameters $a = 1450.3 \pm 0.2$, $c = 1775.0 \pm 0.4$ pm, and a calculated density of 4.820 g·cm^{-3}, is formed in the Sn–Al–Te ternary system (Kienle and Deiseroth 1998). Single crystals and well crystallized, homogeneous powder samples of the title compound were prepared by annealing for 1 day stoichiometric quantities of the pure elements in evacuated quartz ampoules at 1000°C up to a homogeneous melt. In order to avoid a possible reduction of SiO_2 by Al the inner surface of the ampoule was protected by a thin layer of glassy carbon, which was created by pyrolysis of acetone vapor. Subsequent cooling to room temperature resulted in grey metallic samples which were then grinded to a fine powder in an agate mortar, again filled into evacuated quartz ampoule and annealed for 6 weeks at 730°C.

9.14 Tin–Gallium–Tellurium

SnTe–GaTe: The phase diagram of this system, constructed through DTA, XRD, metallography, and measuring of the microhardness, is a eutectic type (Figure 9.14) (Alidzhanov et al. 1967; Rustamov et al. 1970; Glazov et al. 1973; Dovletov et al. 1976; Gather et al. 1987) [according to the data of Dzyubenko and Rogacheva (1988), this system is a non-quasibinary section of the Sn–Ga–Te ternary system]. The eutectic contains 55 mol% SnTe and crystallize at 687°C (Gather et al. 1987) [46.5 mol% SnTe and 655°C (Alidzhanov et al. 1967)]. According to the data of Dovletov et al. (1976), two eutectics in this system contain 30 and 60 mol% SnTe and crystallize at 660°C and 630°C [638°C (Glazov et al. 1973)], respectively.

Three compounds are formed in the SnTe–GaTe system. $\mathbf{GaSnTe_2}$ melts congruently at 730°C (Dovletov et al. 1976) [at 710°C (Glazov et al. 1973); at 728°C (Rustamov et al. 1970)]. $\mathbf{Ga_2SnTe_3}$ and $\mathbf{Ga_3Sn_7Te_{10}}$ melts incongruently at 700°C [at 685°C Glazov et al. 1973)] and 670°C, respectively (Dovletov et al. 1976). $GaSnTe_2$ and Ga_2SnTe_3 dissociate directly at the melting point (Berger et al. 1978). According to the data of Gather et al. (1987), the compounds $GaSnTe_2$ and Ga_2SnTe_3 do not exist.

The solubility of SnTe in GaTe reaches 15 mol% (Dovletov et al. 1976) [23 mol% at the eutectic temperature (Alidzhanov et al. 1967); 13 mol% at 635°C (Glazov et al. 1973)], and the solubility of GaTe in SnTe is 30–32 mol% (Dovletov et al. 1976) [47 mol% at the eutectic temperature (Alidzhanov et al. 1967); 20 mol% (Glazov et al. 1973)].

The ingots for the investigations were annealed for more than 400 h (Glazov et al. 1973) [at 550°C for 300 h (Dzyubenko and Rogacheva 1988)].

SnTe–Ga$_2$Te$_3$: The phase diagram of this system, constructed through DTA, XRD, metallography, and measuring of the microhardness, is presented in Figure 9.15 (Alidzhanov et al. 1971; Guittard et al. 1978a; Babayev et al. 1985; Gather et al. 1987; Dzyubenko and Rogacheva 1988). The eutectic contains 34 mol% Ga_2Te_3 and crystallizes at 663°C (Gather et al. 1987) [55 mol% Ga_2Te_3 and 650°C (Babayev et al. 1985; Babayev 2002); 33 mol% Ga_2Te_3 and 671°C (Dzyubenko and Rogacheva 1988); 46 mol% Ga_2Te_3 and 640°C (Guittard et al. 1978a); 33 mol% Ga_2Te_3 and 650°C (Alidzhanov et al. 1971)]. At 671°C, there is a polymorphic transformation of solid solutions based on Ga_2Te_3 (Babayev et al. 1985; Dzyubenko and Rogacheva 1988).

According to the data of Guittard et al. (1978a), Gather et al. (1987) and Deiseroth and Müller (1996), $\mathbf{Ga_6SnTe_{10}}$ ternary compound is formed in the SnTe–Ga_2Te_3 system. It melts incongruently at 710° C (Gather et al. 1987) [at 695°C (Guittard et al. 1978a)] and crystallizes in the rhombohedral structure with the lattice parameters a = 1020.3 pm and α = 89.94° or $a = 1441.2 \pm 0.6$ and $c = 1768.9 \pm 0.8$ pm in hexagonal setting and a calculated density of 5.68 g·cm^{-3} (Deiseroth and Müller 1996) [a = 1020.7 pm and α = 89.74° or a = 1440 and c = 1775 pm in hexagonal setting and a calculated density of 5.67 g·cm^{-3} (Guittard et al. 1978a); $a = 1440.8 \pm 0.2$, $c = 1767.8 \pm 0.4$ pm, and a calculated density of 5.684 g·cm^{-3} (Kienle and Deiseroth

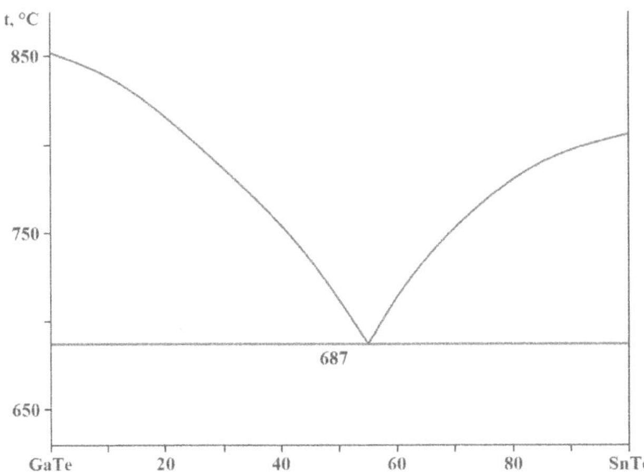

FIGURE 9.14 Phase diagram of the SnTe–GaTe system. (From Gather, B., et al., *J. Less-Common Metals*, **136**(1), 183, 1987.)

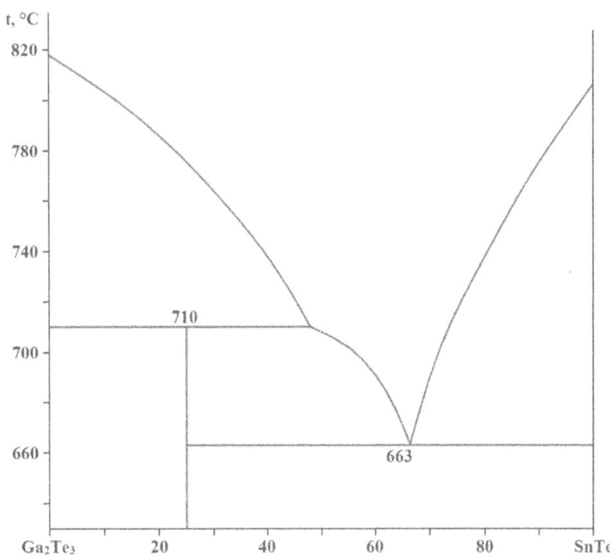

FIGURE 9.15 Phase diagram of the SnTe–Ga_2Te_3 system. (From Gather, B., et al., *J. Less-Common Metals*, **136**(1), 183, 1987.)

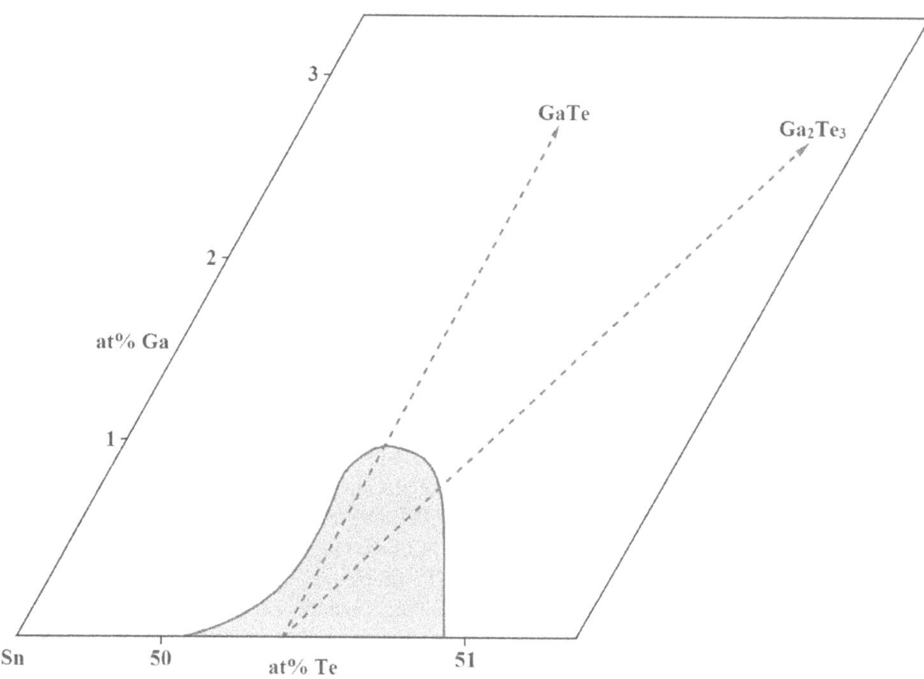

FIGURE 9.16 Region of the solid solutions based on SnTe at 550°C in the Sn–Ga–Te ternary system. (From Dzyubenko, N.I., and Rogacheva, E.I., *Izv. AN SSSR. Neorgan. mater.*, **24**(10), 1736, 1988.)

1998)]. This compound could be prepared directly from the elements by repeated grinding and annealing at 600°C for 2 weeks in the evacuated quartz ampoule. It appears as homogeneous, black and brittle polycrystalline sample, which is not significantly air sensitive.

According to the data of Alidzhanov et al. (1971), another compound, $\mathbf{Ga_4SnTe_7}$, which melts incongruently at 720°C, is formed in the SnTe–Ga_2Te_3 system. Dzyubenko and Rogacheva (1988) did not find this ternary compound.

The solubility of Ga_2Te_3 in SnTe reaches 2 mol% and SnTe dissolves 1 mol% Ga_2Te_3 (Alidzhanov et al. 1971; Babayev et al. 1985).

The ingots for investigations were annealed at 550°C for 500 h (Babayev et al. 1985) [for 300 h (Dzyubenko and Rogacheva 1988)] or at 600°C for 15 days (Guittard et al. 1978a).

The region of solid solutions based on SnTe at 550°C in the Sn–Ga–Te ternary system is shown in Figure 9.16 (Dzyubenko and Rogacheva 1988).

The excess enthalpies of the liquid Sn–Ge–Te alloys were measured in a heat-flow calorimeter at 930°C (Gather et al. 1987). The enthalpy surface (Figure 9.17) is characterized by a valley stretching from the exothermic minimum in the Sn–Te system to the minimum of the Ge–Te systems. The minimum in the ternary system was found in this valley, i.e. on the sections SnTe–Ga_2Te_3. A comparison of the experimental data with those calculated from the excess enthalpies of the constituent binaries with the aid of the Bonnier model, reveals only small deviations.

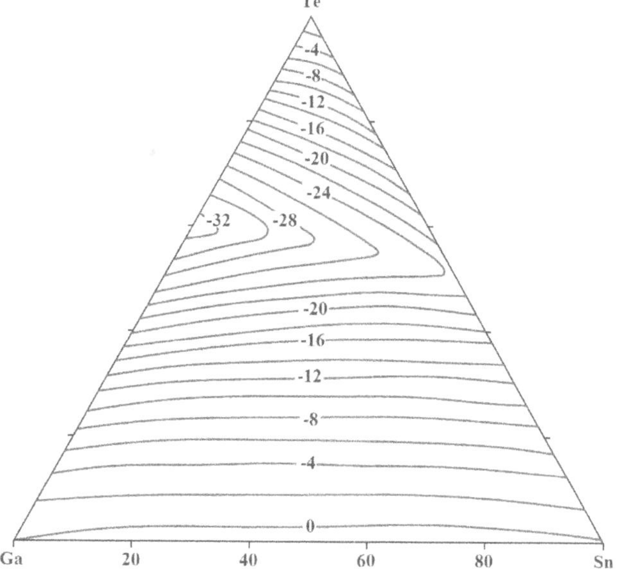

FIGURE 9.17 Isoenthalpic lines (in kJ·mol^{-1}) in the Sn–Ga–Te ternary system at 930°C. (From Gather, B., et al, *J. Less-Common Metals*, **136**(1), 183, 1987.)

9.15 Tin–Indium–Tellurium

SnTe–In: According to the data of DTA, XRD, metallography, and measuring of the microhardness, this system is a non-quasibinary section of the Sn–In–Te ternary system and intersects two ternary subsystems (Rustamov and Babayev 1985). In the first part of the section (50–100 mol% SnTe), passing through the triangle SnTe–InTe–Sn, the liquidus curve consists of two branches of (SnTe) and InTe primary crystallization. Thermal effects at 457°C correspond to the onset of (SnTe) and InTe co-crystallization. Crystallization of alloys in this part of the section ends at 182°C.

The liquidus of the second part of the section (0–50 mol% SnTe), passing through the InTe–In–Sn subsystem, consists of three branches, along which InTe, In_9Te_7 and In are primarily crystallized. The isothermal line at 152°C corresponds to the ternary peritectic equilibrium with the formation of In_9Te_7. The rest of the alloys finally crystallize according to the eutectic type at 117°C, where In_9Te_7, In, and Sn solidify.

The region of solid solutions based on SnTe is 3 mol% at room temperature and 13 mol% at 460°C (Rustamov and Babayev 1985).

The ingots for the investigations were annealed at 80°C for 400 h (1–55 mol% SnTe) or at 120°C for 300 h (55–99 mol% SnTe) (Rustamov and Babayev 1985).

When SnTe is doped with indium, the deviation from stoichiometry leads to a qualitative change in the structure of the solid solution due to the chemical interaction between In and matrix atoms with the formation of In_2Te_3 complexes (Gorne et al. 1987).

SnTe–InTe: The results of the study of this system are contradictory. According to the data of Rosenberg et al. (1964), Dovletov et al. (1976), and Dzyubenko et al. (1988), the phase diagram of this system, constructed through DTA, XRD, metallography, and measuring of the microhardness and density, is a eutectic type (Figure 9.18). The eutectic contains 30 mol% SnTe and crystallizes at 647°C [at 617°C (Beleites and Nieke 1966)] (Dzyubenko et al. 1988) [38 mol% SnTe and 500°C (Rustamov et al. 1969, 1972b); 630°C (Rosenberg et al. 1964)]. No ternary compounds were found in the system (Dzyubenko et al. 1988). The absence of ternary compounds in the system is also evidenced by the absence of singular points on the isotherms of electrical conductivity and thermo-EMF of molten samples of the system.

According to the data of Rustamov et al. (1969), Alidzhanov and Rustamov (1970), Kuliev et al. (1971), Rustamov et al. (1972b), and Glazov et al. (1973), **$InSnTe_2$**, **In_2SnTe_3**, and **In_3SnTe_4** ternary compounds, which melts incongruently, are formed in the SnTe–InTe system.

The solubility of InTe in SnTe reaches 50 mol% (Rosenberg et al. 1964; Rustamov et al. 1969, 1972b; Glazov et al. 1973; Dovletov et al. 1976) [30 mol% at 570°C (Abrikosov and Tskhadaya 1975); 35 mol% at 550°C (Baltrunas et al. 1988; Gorne et al. 1989)] and the solubility of SnTe in InTe is 6 mol% (Rustamov et al. 1969, 1972b) [15 mol% (Dovletov et al. 1976); 11–12 mol% (Glazov et al. 1973)]. The unit cell parameter of solid solutions based on SnTe decreases linearly to 625.2 pm at 40 mol% InTe (Rustamov et al. 1969). At 0.4 and 3 mol% InTe, singular points were found on the concentration dependences of the physical properties (Baltrunas et al. 1988).

The samples for investigations were annealed at 570°C for 350 h (Abrikosov and Tskhadaya 1975) [at temperatures exceeding the corresponding melting points by 100°C for 100 h (Beleites and Nieke 1966); at 400° for more than 2 months (Rustamov et al. 1969); at 550°C for 200–300 h (Baltrunas et al. 1988; Gorne et al. 1989)].

$SnTe–In_2Te_3$: The phase diagram of this system, constructed through DTA, XRD, metallography, and measuring of the microhardness, is a eutectic type (Figure 9.19) (Merino et al. 1968; Babayev et al. 1985; Dzyubenko et al. 1988; Babayev 2002). The eutectic contains 60 mol% In_2Te_3 and crystallizes at 607°C (Babayev et al. 1985; Babayev 2002) [62 mol% In_2Te_3 and 611°C (Merino et al. 1968); 33 mol% In_2Te_3 and 580°C (Il'yin et al. 1988)]. No ternary compounds were found in the system (Dzyubenko et al. 1988; Il'yin et al. 1988).

According to the data of Rustamov et al. (1973a), the **In_2SnTe_4** ternary compound, which melts incongruently at 610°C, is formed in this system.

The solubility of In_2Te_3 in SnTe is 12 mol% at 530°C (Il'yin et al. 1988) [7 mol% at 550°C and 9 mol% at 600°C (Merino et al. 1968); 5 mol% at room temperature and 7 mol% at the eutectic temperature (Rustamov et al. 1973a); 3 mol% at 570°C (Rustamov et al. 1977b); 3.5 mol% at 550°C (Gorne et al. 1989)] and the solubility of SnTe in In_2Te_3 is not higher than 1 mol% (Merino et al. 1968; Rustamov et al. 1977b). The solid solutions based on In_2Te_3 exist in three different polymorphic modifications.

Excess enthalpies of liquid alloys were determined in a heat-flow calorimeter at 800°C and 900°C for $In_{0.5}Sn_{0.5}$–Te section

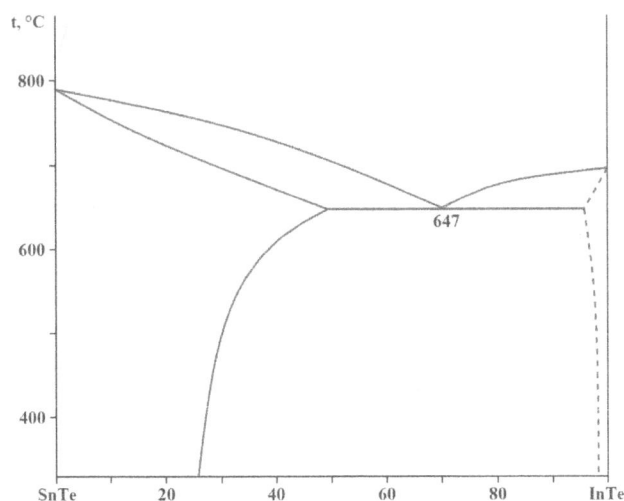

FIGURE 9.18 Phase diagram of the SnTe–InTe system. (From Dzyubenko, N.I., et al., *Izv. AN SSSR. Neorgan. mater.*, **24**(10), 1738, 1988.)

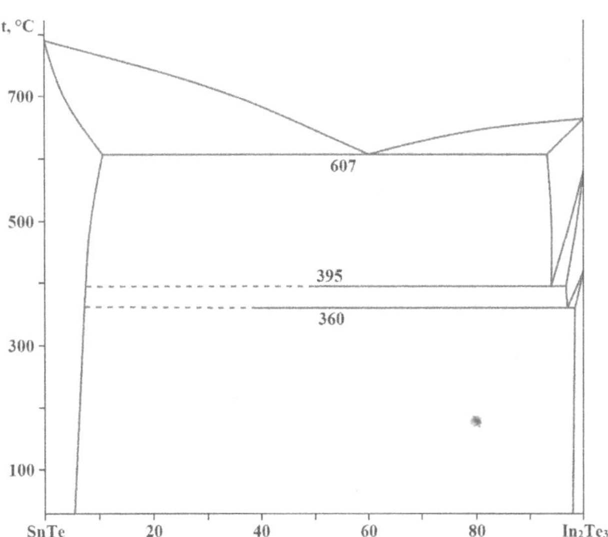

FIGURE 9.19 Phase diagram of the $SnTe–In_2Te_3$ system. (From Babayev, Ya.N., et al., *Zhurn. neorgan. khimii*, **30**(12), 3171, 1985.)

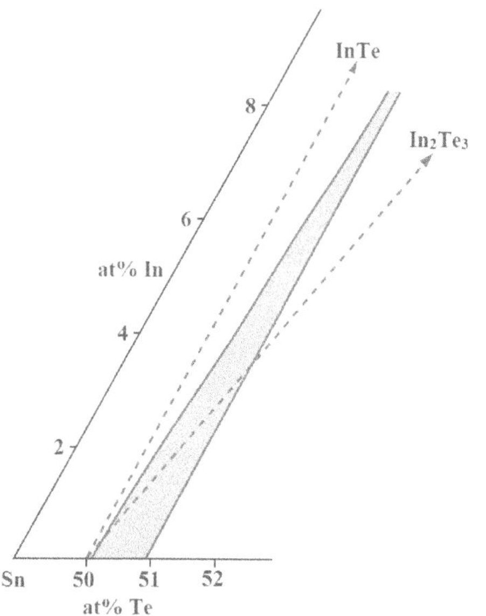

FIGURE 9.20 Region of the solid solutions based on SnTe at 550°C in the Sn–In–Te ternary system. (From Rogacheva, E.I., et al., *Izv. AN SSSR. Neorgan. mater.*, **19**(4), 573, 1983.)

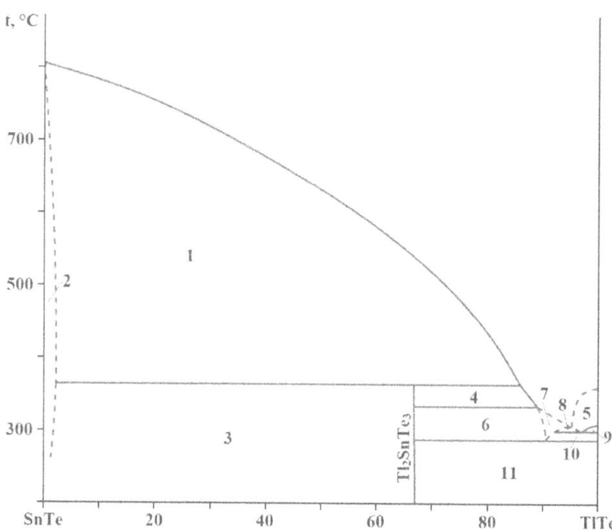

FIGURE 9.21 Phase relations in the SnTe–TlTe system: 1, L + SnTe; 2, (SnTe); 3, (SnTe) + Tl_2SnTe_3; 4, L + Tl_2SnTe_3; 5, L + (Tl_5Te_3); 6, L + Tl_2SnTe_3 + (Tl_4SnTe_3); 7, L + (Tl_4SnTe_3); 8, L + (Tl_4SnTe_3) + (Tl_5Te_3), 9, L + TlTe + (Tl_5Te_3); 10, L + TlTe + (Tl_4Te_3); 11, Tl_2SnTe_3 + TlTe. (From Dichi, E., et al., *J. Alloys Compd.*, **194**(1), 147, 1993.)

by Schlieper and Blachnik (2000). The association model with additional interaction parameters had to be used for the calculation of ternary excess enthalpies. They were obtained by fitting the calculated curves to the experimental data with the least square method. The minima of the exothermic enthalpies were found on the mention above section at 54 mol% Te. The values of excess enthalpies are –27.2 and –25.3 kJ·M^{-1} at 800°C and 900°C, respectively.

The samples for investigations were annealed at 530°C for 400 h (0–10 mol% In_2Te_3) or at 200°C for 250 h (10–100 mol% In_2Te_3) (Babayev et al. 1985) [at 605°C and 550°C for 500 h (Merino et al. 1968); at 600°C and 550°C for 400 and 500 h, respectively (Rustamov et al. 1973a); at 200°C for 300 h (Rustamov et al. 1977b); at 550°C for 200 h (Gorne et al. 1989)].

In the Sn–In–Te ternary system, the orientation of the SnTe homogeneity region is close to the direction towards InTe with some deviation towards Te (Figure 9.20) (Rogacheva et al. 1975a; Palatnik et al. 1977a,b; Rogacheva et al. 1976, 1983a).

9.16 Tin–Thallium–Tellurium

SnTe–Tl: This system is a non-quasibinary section of the Sn–Tl–Te ternary system which intersects SnTe–Tl_2Te–Sn and Sn–Tl_2Te–Tl subsystem (Rustamov and Babayev 1987). The isothermal line at 480°C corresponds to the onset of joint crystallization of SnTe and Tl_2Te. In this part of the section, crystallization ends at the point of the ternary eutectic at 157°C. On the Tl side, crystallization ends at the point of another ternary eutectic at 97°C. The maximum in the section corresponds to 627°C and 34 mol% SnTe. The solubility of Tl in SnTe is 1.5 mol%. The ingots for the investigations were annealed at 140°C for 150 h (40–100 mol% SnTe) or at 80°C for 200 h (1.5–35 mol% SnTe).

SnTe–TlTe: This system is a non-quasibinary section of the Sn–Tl–Te ternary system as TlTe melts incongruently (Figure 9.21) (Gotuk et al. 1979b; Dichi et al. 1993d). The **Tl_2SnTe_3** ternary compound, which melts incongruently at 364°C (Dichi et al. 1993d) [at 370°C (Gotuk et al. 1979b)] and crystallizes in the orthorhombic structure with the lattice parameters a = 841.0 ± 0.1, b = 2225.1 ± 0.6, c = 849.9 ± 0.4 pm, and the calculated and experimental densities of 7.60 and 7.53 g·cm^{-3}, respectively (Agafonov et al. 1991a; Dichi et al. 1993d), is formed in this system. Single crystals of Tl_2SnTe_3 were prepared by prolonged annealing at 270°C from a stoichiometric mixture of previously melted Sn, Tl, and Te.

This system was investigated through DTA, DSC, XRD, and measuring of the microhardness (Gotuk et al. 1979b; Dichi et al. 1993d). The ingots for the investigations were annealed at 200°C for 15 days (Dichi et al. 1993d).

SnTe–Tl_2Te: The phase diagram of this system, constructed through DTA, XRD, and measuring of the microhardness, is presented in Figure 9.22 (Gotuk et al. 1978b; Gawel et al. 1990; Malakhovs'ka-Rosokha 2011). The eutectic contains 35 mol% SnTe and crystallizes at 500°C (Malakhovs'ka-Rosokha 2011) [49 mol% SnTe and 505°C (Gotuk et al. 1978b, 1979b); 51.4 mol% SnTe and 515.3°C (Gawel et al. 1990)].

Thermodynamic data for liquid phase of the SnTe–Tl_2Te system have been determined by concentration cell EMF measurements (Zaleska and Małachowicz 1991). The dependence of partial thermodynamic functions of Tl on the composition of the liquid and solid phases of the system has been discussed. The effect of the presence of ternary compound on the character of partial molar change of thermodynamics functions enables to suggest that association phenomenon occurs also in the liquid phase of the ternary system.

According to the data of Gotuk et al. (1978b, 1979b), samples containing up to 7 mol% SnTe are characterized by invariant equilibrium at 300°C and contain three phases: Tl_2Te, (Tl), and (Tl_5Te_3), and in the range of 7–12 mol% SnTe, the samples consist of (Tl_5Te_3) and (Tl). In the opinion of

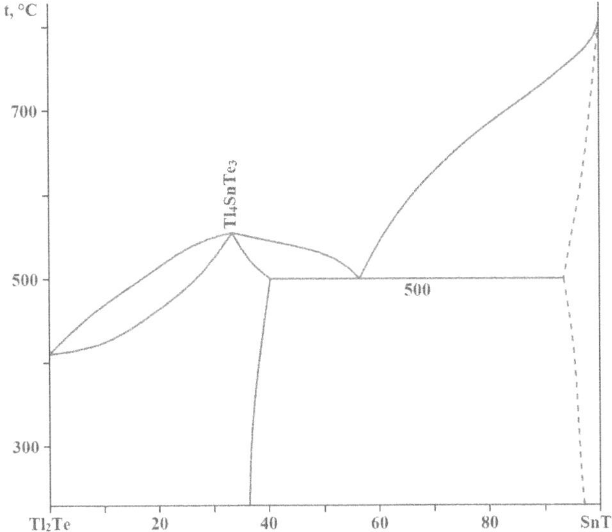

FIGURE 9.22 Phase diagram of the SnTe–Tl_2Te_3 system. (From Malakhovs'ka-Rosokha, T.O., *Nauk. visn. Uzhgorod. Univ., Ser. Khim.*, [2(26)], 16, 2011.)

these authors, the system is a non-quasibinary section of the Sn–Tl–Te ternary system.

The $\mathbf{Tl_4SnTe_3}$ ternary compound is formed in this system. It melts congruently at 544°C (Malakhovs'ka et al. 2007; Malakhovs'ka-Rosokha 2011; Peresh et al. 2015) [at 555°C (Gotuk et al. 1979b); at 554°C (Bradtmöller and Böttcher 1993); at 548°C (Dichi et al. 1993d)] and forms continuous solid solutions with Tl_2Te (Gotuk et al. 1978b; Gawel et al. 1990; Malakhovs'ka-Rosokha 2011). At 300°C, the solid solution based on Tl_2Te does not exceed 35 mol% SnTe, and based on SnTe is not higher than 10 mol% Tl_2Te (Malakhovs'ka-Rosokha 2011).

Tl_4SnTe_3 crystallizes in the tetragonal structure with the lattice parameters a = 881.9 ± 0.2, c = 1301.3 ± 0.3 pm, and a calculated density of 8.773 g·cm^{-3} (Bradtmöller and Böttcher 1993) [a = 898 and c = 1241 pm (Babanly et al. 1979b; Gotuk et al. 1979b); a = 883.6, c = 1305.2 pm, and an experimental density of 8.60 g·cm^{-3} (Voroshilov et al. 1988a); a = 881.3, c = 1304.4 pm, and experimental density of 8.48 g·cm^{-3} (Dichi et al. 1993a); a = 883.03 and c = 1302.35 pm (Chami et al. 1983b); a = 882.2 and c = 1301.9 pm (the predicted values) (Piasecki et al. 2017) and the calculated and experimental densities of 8.65 and 8.53 g·cm^{-3}, respectively (Malakhovs'ka et al. 2007)].

According to the data of Kuropatwa et al. (2011), the Tl_4SnTe_3 ternary compound has a homogeneity region and its true composition could be presented as $\mathbf{Tl_{10-x}Sn_xTe_6}$. The alloys of this compound were obtained by heating the mixtures of Tl, Sn, and Te in the respective stoichiometric molar ratios inside evacuated (ca. 10^{-1} Pa) silica tubes. These tubes were sealed under a H_2/O_2-flame, and samples were heated in a furnace to 650°C, followed by slow cooling to 470°C within 5 days. Those typically mixed-phase samples were crushed into fine powder, resealed in silica tubes and annealed at 450°C for 15 days. In order to explore the full phase range of the alloys, x was increased in increments of 0.2 such that $0 < x < 3.4$. There were obtained the next lattice parameters: a = 884.16 ± 0.06, c = 1301.14 ± 0.09 pm for $Tl_{9.05}Sn_{0.95\pm0.08}Te_6$; a = 885.0 ± 0.1, c = 1305.9 ± 0.2 pm for $Tl_{8.37}Sn_{1.63\pm0.09}Te_6$; a = 883.84 ± 0.04, c = 1305.25 ± 0.06 pm for $Tl_{8.0}Sn_{2.0\pm0.3}Te_6$; and a = 883.75 ± 0.01, c = 1305.85 ± 0.03 pm for $Tl_{7.5}Sn_{2.5\pm0.3}Te_6$.

The melting entropy of Tl_4SnS_3 is 99 kJ·M^{-1} (Malakhovs'ka et al. 2007). This compound is a narrow band gap semiconductor (Piasecki et al. 2017).

The single crystals of the title compound were obtained using directional crystallization of stoichiometric composition performed by Bridgman–Stockbarger method (Bradtmöller and Böttcher 1993; Malakhovs'ka et al. 2007; Piasecki et al. 2017). For conversion to the liquid state, the sample was heated up to 580°C, maintained at this temperature for 24 h, and then was slowly cooled to the annealing temperature in the growth zone decreased to 250°C and maintained for 72 h. The crystallization front displacement rate was equal to 0.1–0.3 mm·h^{-1}, a temperature gradient in the zone of crystallization was 2–4°C·mm^{-1}, and the rate of cooling to ambient temperature after annealing was 30°C·h^{-1}.

The ingots for the investigations were annealed at 300°C–450°C for 400–450 h (Gotuk et al. (1978b, 1979b) or at 300°C for 336 h (1.5–35 mol% SnTe) (Gawel et al. 1990).

SnTe–$\mathbf{Tl_2Te_3}$: This system is a non-quasibinary section of the Sn–Tl–Te ternary system as Tl_2Te_3 melts incongruently (Rustamov et al. 1976a). Thermal effects at 250°C correspond to the onset of joint crystallization of SnTe and TlTe. When TlTe interacts with a liquid, Tl_2Te_3 is formed. Solidification of the alloys of this section occurs at the temperature of the transition point (200°C). The range of solid solutions based on SnTe does not exceed 1 mol%. The system was investigated through DTA, metallography, and measuring of the microhardness. The ingots for the investigations were annealed at 160°C for 400 h.

SnTe–$\mathbf{Tl_5Te_3}$: According to the data of Gotuk et al. (1979b), this system is a non-quasibinary section of the Sn–Tl–Te ternary system. Thermal effects at 500°C correspond to the crystallization of the (Tl_5Te_3)+SnTe eutectic, and at 370°C and 295°C to the four-phase reactions with formation of Tl_2SnTe_3 and TlTe, respectively. The system was investigated through DTA, metallography, and measuring of the microhardness.

SnTe–$\mathbf{Tl_4SnTe_3}$: The phase diagram of this system, constructed through DTA, XRD, and DSC, is a eutectic type (Figure 9.23) (Dichi et al. 1993a). The eutectic contains 27.1 at% Sn + 42.4 at% Te + 30.5 at% Tl (N_{Tl} = 0.53; N_{Tl} = at% Tl/ (at% Tl + at% Sn) and crystallizes at 517°C. The limit of solid solution based on Tl_4SnTe_3 corresponds to the composition N_{Tl} = 0.68 for the eutectic temperature. The domain of homogeneity based on SnTe is very narrow (N_{Tl} = 0.02).

$\mathbf{Tl_5Te_3}$–$\mathbf{Tl_4SnTe_3}$: The phase equilibrium diagram in this section shows a eutectic type (Dichi et al. 1993a). The eutectic temperature and composition are N_{Tl} = 0.99 and 435°C, respectively.

$\mathbf{Tl_2SnTe_3}$–Te: The phase relation in this system was determined by (Dichi et al. 1993d). The $\mathbf{Tl_2SnTe_5}$ ternary compound, which melts incongruently at 286°C [285°C–290°C (Sharp et al. 1999)] and crystallizes in the tetragonal structure with the lattice parameters a = 830.6 ± 0.2, c = 1516.1 ± 0.5 pm, and the calculated and experimental densities of 7.40 and 7.54 g·cm^{-3}, respectively, is formed in this system (Agafonov et al. 1991b; Dichi et al. 1993d). Single crystals of this compound were obtained by heating at 270°C a molten

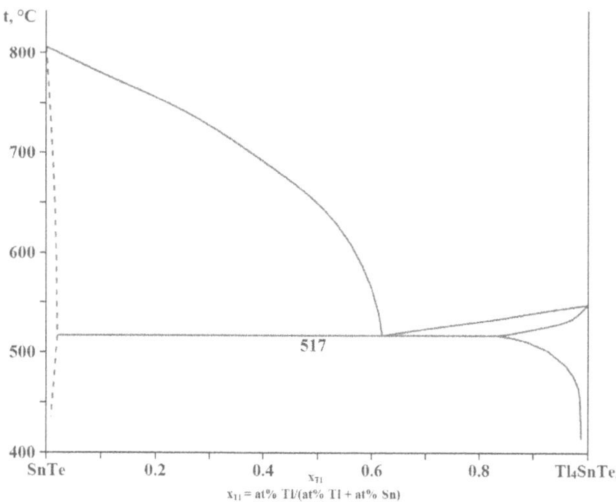

FIGURE 9.23 Phase diagram of the SnTe–Tl_4SnTe_3 system. (From Dichi, E., et al., *J. Alloys Compd.*, **194**(1), 155, 1993.)

stoichiometric mixture of Tl, Sn, and Te in an evacuated silica tube for several weeks.

Voroshilov et al. (1986) noted, that another ternary compound, $\mathbf{Tl_6Sn_2Te_6}$, which melts at 370°C and has an experimental density of 7.81 g·cm^{-3}, is formed in the Sn–Tl–Te system. According to the data of (Dichi et al. 1993d), $Tl_6Sn_2Te_6$ is a mixture of Tl_2SnTe_3 and Tl_4SnTe_3.

The liquidus surface of the Sn–Tl–Te ternary system was constructed by Gotuk et al. (1979b) and is presented in Figure 9.24. The SnTe–Tl_4SnTe_3–Tl_5Te_3–Te and SnTe–Sn–Tl–Tl_5Te_3–Tl_4SnTe_3 subsystems of mentioned above ternary system were investigated also by (Dichi et al. 1993b,c). The invariant equilibria in these subsystems as well as some vertical sections and the reaction schemes are given. The isothermal section of the Sn–Tl–Te system at room temperature was constructed by Gotuk et al. (1979b).

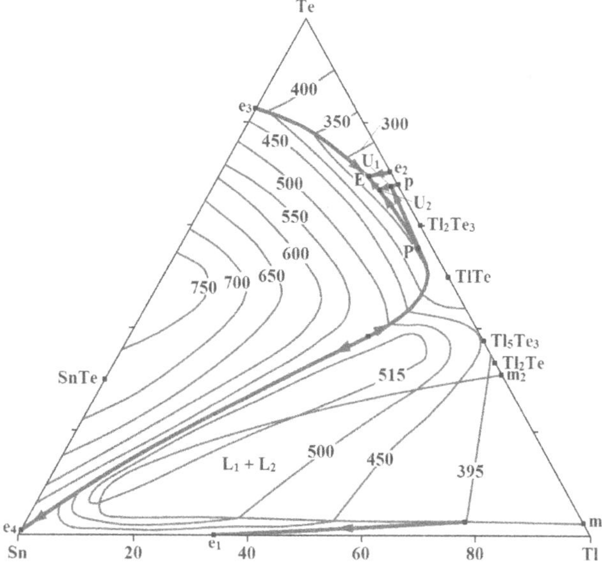

FIGURE 9.24 Liquidus surface of the Sn–Tl–Te ternary system. (From Gotuk, A.A., et al., *Izv. AN SSSR. Neorgan. mater.*, **15**(8), 1356, 1979.)

9.17 Tin–Yttrium–Tellurium

SnTe–Y_2Te_3: The solubility of Y_2Te_3 in SnTe is 4 mol% at 600°C (Shemet et al. 2006a). No ternary compounds were found in the system. The ingots for the investigations were annealed at 600°C for 240 h with the next quenching in the cold water.

9.18 Tin–Lanthanum–Tellurium

SnTe–LaTe: The solubility of LaTe in SnTe is 4–10 mol% (Erofeev et al. 1974; Alizade et al. 1985). The ingots for the investigations were annealed at 500°C–600°C for 500–700 h (Erofeev et al. 1974) or at the temperatures which are equal to 2/3 melting temperature of the corresponding alloy for 50–100 h (Alizade et al. 1985).

9.19 Tin–Cerium–Tellurium

The isothermal section of the Sn–Ce–Te ternary system at Te-corner (≥50 at% Te) at room temperature consists of 4 three-phase regions and one two-phase region: SnTe + $CeTe_3$ + Te, SnTe + $CeTe_2$ + $CeTe_3$, SnTe + $CeTe_2$ + Ce_2Te_3, $Sn_{1-x}Ce_xTe$ + CeTe + Ce_2Te_3, and $Sn_{1-x}Ce_xTe$ + Ce_2Te_3 (Zhang et al. 2010). The maximum solubility of Ce in SnTe was determined to be about 3 at%. None of the other phases in this system has no a remarkable homogeneity range at room temperature. No ternary compound was observed.

To prepare the samples for the investigation, the melted samples were annealed at 820°C for 24 h and then slowly cooled down at 1°C·h^{-1} to 800°C for 10–14 days. They were then cooled down to 200°C and kept for 5–7 days. Finally they were slowly cooled to room temperature without breaking the containers spontaneously.

9.20 Tin–Praseodymium–Tellurium

SnTe–PrTe: The phase diagram of this system, constructed through DTA, XRD, metallography, and measuring of the microhardness and density, is a eutectic type (Figure 9.25) (Murguzov et al. 2006; Gojaev et al. 2010). The eutectic contains 13 mol% PrTe and crystallizes at 725°C. The solubility of PrTe in SnTe is 4 mol% at room temperature and 7 mol% at 725°C. The ingots for the investigations were equilibrated by annealing at 500°C and 700°C for 250°C.

The isothermal section of the Sn–Pr–Te ternary system at Te-corner (≥ 50 at% Te) at room temperature consists of 4 three-phase regions and one two-phase region: SnTe + $PrTe_3$ + Te, SnTe + $PrTe_3$ + Pr_4Te_7, SnTe + Pr_3Te_4 + Pr_4Te_7, $Sn_{1-x}Pr_xTe$ + PrTe + Pr_3Te_4, and $Sn_{1-x}Pr_xTe$ + Pr_3Te_4 (Zhang et al. 2010). The maximum solubility of Pr in SnTe was determined to be about 2.5 at%. None of the other phases in this system has no a remarkable homogeneity range at room temperature. No ternary compound was observed.

To prepare the samples for the investigation, the melted samples were annealed at 820°C for 24 h and then slowly cooled

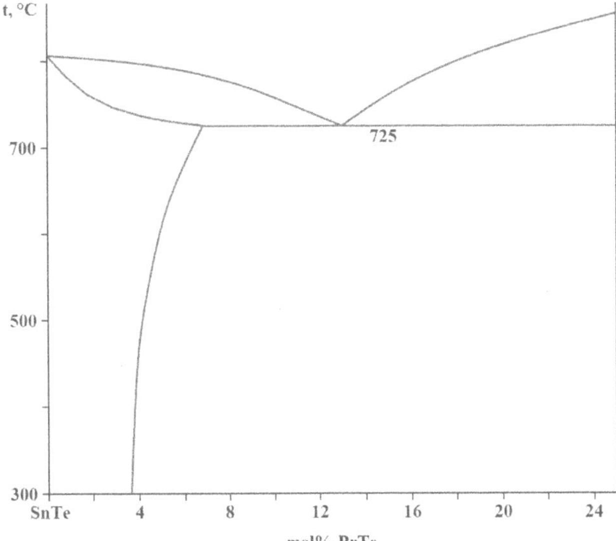

FIGURE 9.25 Part of the phase diagram of the SnTe–PrTe system. (From Gojaev, E.M., et al., *Inorg. Mater.*, **46**(10), 1061, 2010.)

down at 1°C·h^{-1} to 800°C for 10–14 days. These samples were then cooled down to 200°C and kept for 5–7 days. Finally they were slowly cooled to room temperature without breaking the containers spontaneously.

9.21 Tin–Neodymium–Tellurium

SnTe–NdTe: The phase diagram of this system, constructed through DTA, XRD, metallography, and measuring of the microhardness and density, is a eutectic type with peritectic transformation (Figure 9.26) (Murguzov et al. 2011). The eutectic contains 15 mol% NdTe and crystallizes at 725°C. At room temperature, the solubility of NdTe in SnTe reaches 5 mol% [4–10 mol% (Erofeev et al. 1974; Alizade et al. 1985)] and

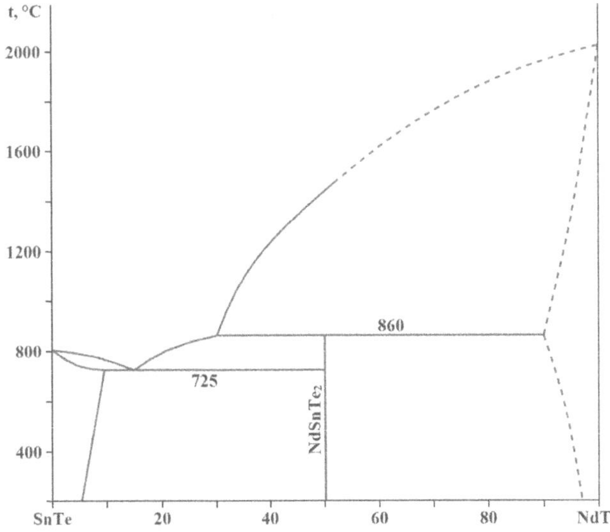

FIGURE 9.26 Phase diagram of the SnTe–NdTe system. (From Murguzov, M.I., et al., *Azerb. chim. zhurn.*, (2), 146, 2011.)

the solubility SnTe in NdTe is 3 mol%. The **$NdSnTe_2$** ternary compound, which melts incongruently at 860°C and has an experimental density of 7.10 g·cm^{-3}, is formed in this system. The ingots for the investigations were annealed at 700°C for 110 h (Murguzov et al. 2011) [at 500°C–600°C for 500–700 h (Erofeev et al. 1974) or at the temperatures which are equal to 2/3 melting temperature of the corresponding alloy for 50–100 h (Alizade et al. 1985)].

The isothermal section of the Sn–Nd–Te ternary system at Te-corner (≥50 at% Te) at room temperature consists of 5 three-phase regions and one two-phase region: SnTe + $NdTe_3$ + Te, SnTe + $NdTe_3$ + Nd_2Te_5, SnTe + $NdTe_2$ + Nd_2Te_5, SnTe + $NdTe_2$ + Nd_3Te_4, $Sn_{1-x}Nd_xTe$ + NdTe + Nd_3Te_4, and $Sn_{1-x}Nd_xTe$ + Nd_3Te_4 (Zhang et al. 2010). The maximum solubility of Nd in SnTe was determined to be about 3.5 at%. None of the other phases in this system has no a remarkable homogeneity range at room temperature. No ternary compound was observed.

To prepare the samples for the investigation, the melted samples were annealed at 820°C for 24 h and then slowly cooled down at 1°C·h^{-1} to 800°C for 10–14 days. They were then cooled down to 200°C and kept for 5–7 days. Finally they were slowly cooled to room temperature without breaking the containers spontaneously.

9.22 Tin–Samarium–Tellurium

SnTe–SmTe: The solubility of SmTe in SnTe is 4–10 mol% (Erofeev et al. 1974; Alizade et al. 1985). The ingots for the investigations were annealed at 500°C–600°C for 500–700 h (Erofeev et al. 1974) or at the temperatures which are equal to 2/3 melting temperature of the corresponding alloy for 50–100 h (Alizade et al. 1985).

9.23 Tin–Europium–Tellurium

SnTe–EuTe: The phase diagram of this system, constructed through DTA, XRD, metallography, and measuring of the microhardness, is a eutectic type (Figure 9.27) (Bakhshaliyeva and Aliyev 2004). The eutectic contains 42 mol% EuTe and crystallizes at 650°C. At 200°C, EuTe dissolves 10 mol% SnTe and SnTe dissolves 30 mol% EuTe. According to the data of Erofeev et al. (1974) and Alizade et al. (1985), the solubility of EuTe in SnTe is 4–10 mol%. The ingots for the investigations were annealed at 450°C for 2 weeks [at 500°C–600°C for 500–700 h (Erofeev et al. 1974) or at the temperatures which are equal to 2/3 melting temperature of the corresponding alloy for 50–100 h (Alizade et al. 1985)].

9.24 Tin–Gadolinium–Tellurium

SnTe–GdTe: The solubility of GdTe in SnTe is 4–10 mol% (Erofeev et al. 1974; Alizade et al. 1985). The ingots for the investigations were annealed at 500°C–600°C for 500–700 h (Erofeev et al. 1974) or at the temperatures which are equal to 2/3 melting temperature of the corresponding alloy for 50–100 h (Alizade et al. 1985).

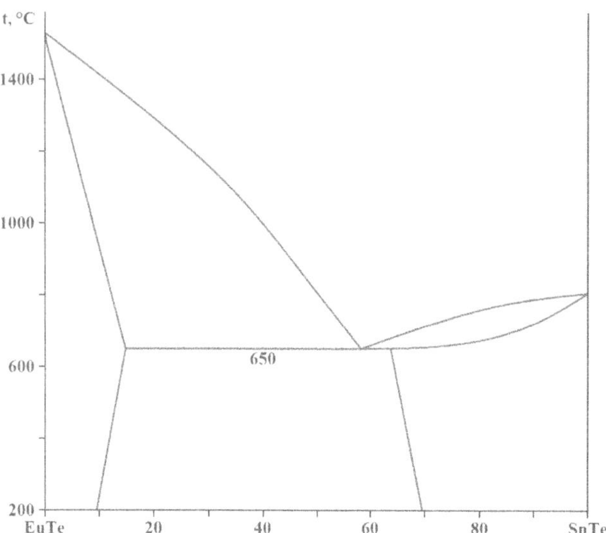

FIGURE 9.27 Phase diagram of the SnTe–EuTe system. (From Bakhshaliyeva, E.A., et al., *Zhurn. khim. probl.*, (1), 82, 2004.)

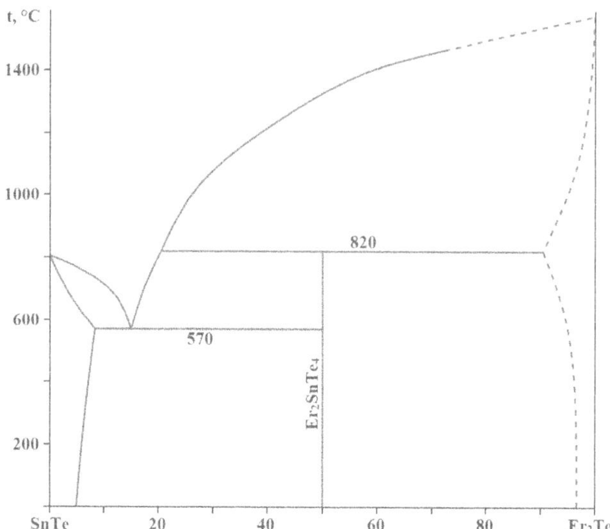

FIGURE 9.28 Phase diagram of the SnTe–Er_2Te_3 system. (From Babanly, M.B., et al., *Zhurn. neorgan. khimii*, **48**(2), 326, 2003.)

The isothermal section of the Sn–Gd–Te ternary system at Te-corner (≥50 at% Te) at room temperature was constructed by Zhan et al. (2009). There are four three-phase regions (SnTe + $GdTe_3$ + Te, SnTe + $GdTe_2$ + $GdTe_3$, SnTe + $GdTe_2$ + Gd_2Te_3, and $Sn_{1-x}Gd_xTe$ + GdTe + Gd_2Te_3) and one two-phase region ($Sn_{1-x}Gd_xTe$ + Gd_2Te_3) in this part of the system. The lattice parameters of the $Sn_{1-x}Gd_xTe$ solid solution decrease with the increase of the Gd content. The melted samples for the investigations were annealed at 820°C for 2 h, then slowly cooled down at 1°C·h^{-1} to 790°C for 48 h, and finally cooled down to 200°C for 5–15 days. Finally they were slowly cooled to room temperature.

9.25 Tin–Dysprosium–Tellurium

The isothermal section of the Sn–Dy–Te ternary system at Te-corner (≥50 at% Te) at room temperature was determined by Zhang et al. (2009). There are 5 three-phase regions (SnTe + $DyTe_3$ + Te, SnTe + $DyTe_3$ + Dy_4Te_9, SnTe + $DyTe_{1.75}$ + Dy_4Te_9, SnTe + $DyTe_{1.75}$ + Dy_2Te_3, and $Sn_{1-x}Dy_xTe$ + DyTe + Dy_2Te_3) and one two-phase region ($Sn_{1-x}Gd_xTe$ + Dy_2Te_3) in this part of the system. The solid solubility of DyTe in SnTe was determined to be about 9 mol%. The lattice parameters of the $Sn_{1-x}Dy_xTe$ solid solution decrease with the increase of the Dy content. The melted samples for the investigations were annealed at 820°C for 6 h, and then slowly cooled down at 2°C·h^{-1} to 800°C for 48–72 h. Then, they were cooled down to 200°C for 5–15 days. Finally they were cooled down to 200° for 10–15 days and then slowly cooled to room temperature.

9.26 Tin–Erbium–Tellurium

SnTe–ErTe: The **$ErSnTe_2$** ternary compound, which melts incongruently at 845°C and crystallizes in the orthorhombic structure with the lattice parameters a = 467, b = 940, c = 1113 pm, the calculated and experimental density of 7.35 and 7.32 g·cm^{-3}, respectively, and an energy gap of 0.94 eV, is formed in this system (Aliev et al. 2005). This compound was synthesized by melting SnTe and ErTe (molar ratio 1:1) in an evacuated up to 0.133 Pa quartz ampoule.

SnTe–Er_2Te_3: The phase diagram of this system, constructed through DTA, XRD, metallography, and measuring of the microhardness and density, is a eutectic type with peritectic transformation (Figure 9.28) (Babanly et al. 2003). The eutectic contains 15 mol% Er_2Te_3 and crystallizes at 570°C. SnTe dissolves 6 mol% Er_2Te_3 at room temperature and 8 mol% at 300°C. The **Er_2SnTe_4** ternary compound, which melts incongruently at 820°C and has an experimental density of 6.82 g·cm^{-3}, is formed in this system.

9.27 Tin–Ytterbium–Tellurium

SnTe–YbTe: The phase diagram of this system, constructed through DTA, XRD, and SEM data, is presented in Figure 9.29 (Aliev et al. 2014). Apparently, it is quasibinary section of the

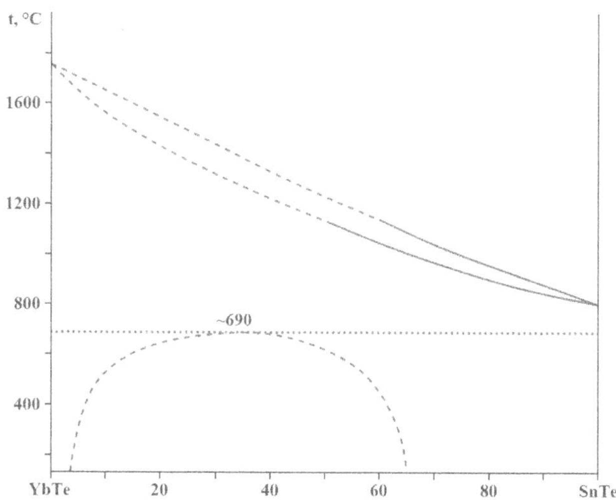

FIGURE 9.29 Phase diagram of the SnTe–YbTe system. (From Aliev, Z.S., et al., *J. Alloys Compd.*, **602**, 248, 2014.)

Yb–Sn–Te ternary system and belongs to type I within the Roseboom's classification above 730°C. Because of both YbTe and SnTe have very close lattice parameters and same space group, they are completely miscible in the solid state and form continuous solid solution series below solidus. The **$YbSnTe_2$** ternary compound was not detected. Below ca. 690°C, a spinodal-type decomposition occurs which extends from ~3 to 65 mol% at room temperature. All alloys were heated up to 1030°C and held at this temperature for about 10–12 h and were then annealed at 430°C and 730°C for ~1200 h. In most cases alloys were then quenched in cold water.

The phase diagram of this system completely different from those given above was constructed by Aliev et al. (2006b, 2009) using DTA, XRD, and measuring of the microhardness and density. This phase diagram is a eutectic type and has two eutectics that contain 38 and 75 mol% SnTe and crystallize at 877°C and 667°C, respectively. According their data, the $YbSnTe_2$ ternary compound, which melts congruently at 1062°C and crystallizes in the orthorhombic structure with the lattice parameters a = 472, b = 950, c = 1116 pm, and an experimental density of 6.95 g·cm^{-3}, is formed in this system. The solubility of YbTe in SnTe reaches 8–10 mol%. The ingots for the investigations were annealed at 230°C–530°C for 1 month.

This system has been studied using the EMF measurements of the concentration cells in the 30°C–160°C temperature interval by Ibadova et al. (2013a). It was established that the solubility of YbTe in SnTe in this temperature interval is about 35 mol%. The existence the $YbSnTe_2$ ternary compound was not confirmed. The partial molar functions of YbTe and ytterbium were calculated in the alloys, from which the standard thermodynamic functions of formation and standard entropy of the solid solutions were calculated. The ingots for the investigations were annealed at 730°C for 600 h and then at 180°C for 300 h.

The analytical expressions for the free energy of mixing of $Sn_xYb_{1-x}Te$ solid solutions in the temperature range 330°C–825°C have been derived and the location of their phase boundaries has been determined by Mamedov et al. (2016).

SnTe–YbSn: According to the data of Aliev et al. (2006b, 2009), the phase diagram of this system is a eutectic type. The eutectic contains 50 mol% SnTe and crystallizes at 477°C. The solubility of SnTe in YbSn reaches 5 mol% and YbSn dissolves 15 mol% SnTe. The ingots for the investigations were annealed at 230°C–530°C for 1 month.

The liquidus surface of the Sn–Yb–Te ternary system was constructed by Aliev et al. (2009). There are 16 monovariant curves and eight invariant points: seven ternary eutectic and one transition point. But since there are large differences in the phase diagram of the SnTe–YbTe system, this liquidus surface must be reconstructed.

9.28 Tin–Uranium–Tellurium

SnTe–UTe: The phase diagram of this system, constructed through DTA, XRD, and metallography, is a eutectic type (Terehov et al. 1972). The eutectic crystallizes at 775°C. The solidus temperature from the SnTe-side first increases to 805°C at 4.3 mol% UTe, and then decreases. The solubility of UTe in SnTe is 6.3 mol% at 650°C and 6.1 mol% at 450°C. The ingots for the investigations were annealed at 650°C and 450°C for 3 and 72 h, respectively.

The **USnTe** ternary compound, which crystallizes in the tetragonal structure with the lattice parameters a = 425.96 ± 0.03 and c = 913.13 ± 0.04 pm and a calculated density of 9.70 g·cm^{-3}, is formed in the Sn–U–Te system (Haneveld and Jellinek 1969). This compound was prepared in the following way. A small rod of uranium was placed in a quartz tube and heated in a current of hydrogen at 230°C. The voluminous uranium hydride powder thus produced was dehydrated by heating at 400°C under vacuum. An equivalent proportion of SnTe was added immediately and the quartz tube was evacuated and sealed. The mixture was heated at 800°C–900°C for 2 or 3 days. Then, the product was slowly cooled to room temperature. This compound is stable in air.

9.29 Tin–Lead–Tellurium

SnTe–PbTe: The phase diagram of this system, constructed through DTA, XRD, metallography, and measuring of the microhardness (Abrikosov et al. 1958a; Grimes 1965; Wagner and Woolley 1967; Hiscocks and West 1968; Laugier 1973) and optimized using a four parameter for the excess Gibbs energy by Volykhov et al. (2006) and Yashina et al. (2006), belongs to the type I according the Rosebom's classification (Figure 9.30). The agreement of the calculation with the experimental data for both the liquidus and solidus is very good (Steininger 1970; Harris et al. 1975; Yashina et al. 2006). A continuous series of solid solutions with the NaCl-type structure forms in the system (Abrikosov et al. 1958a; Mazelsky et al. 1962; Nikolić 1967; Hiscocks and West 1968; Short 1968; Bis 1969; Laugier 1973). The cell parameter of the solid solutions, depending on the composition, changes according to the Vegard's law (Mazelsky et al. 1962; Bis 1969) [there is a slight positive deviation from the Vegard's law (Abrikosov et al. 1958a; Wagner and Woolley 1967; Short 1968). According to the data of Volykhov et al. (2006), $Sn_xPb_{1-x}Te$ solid solutions are nearly ideal, with a slight negative deviation from ideality.

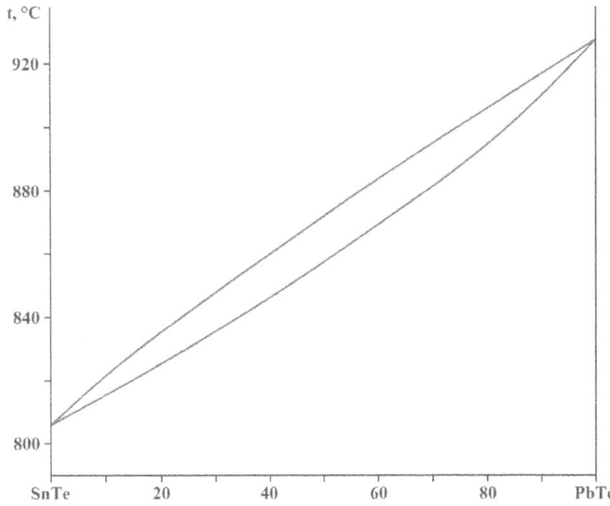

FIGURE 9.30 Phase diagram of the SnTe–PbTe system. (From Yashina, L.V., et al., *J. Alloys Compd.*, **413**(1–2), 133, 2006.)

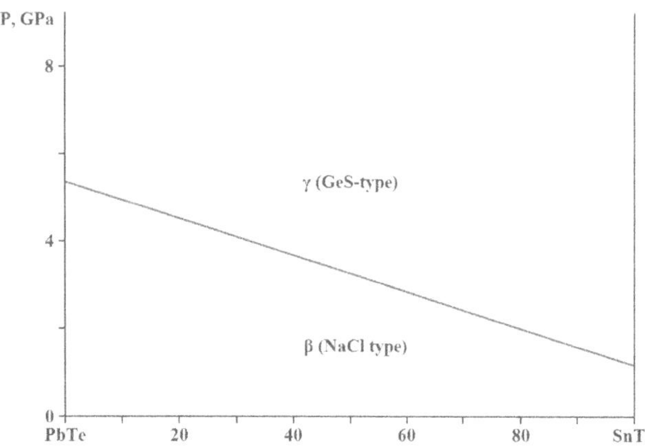

FIGURE 9.31 *p-x*-Phase diagrams of SnTe–PbTe system. (From Serebryanaya, A.R., *Izv. AN SSSR. Neorgan. mater.*, **27**(8), 1611, 1991.)

The *p-x*-phase diagram of the SnTe–PbTe system is given in Figure 9.31 (Serebryanaya 1991). It was established that within the interval of the pressure from 1.2 to 11 GPa only one $\beta \Leftrightarrow \gamma$ phase transition takes place, where γ-phase is the phase with the orthorhombic structure of the GeS-type.

The vaporization of the alloys in the PbTe–SnTe system has been investigated by Northop (1971). It was shown, that the vaporization is incongruent as SnTe, the more volatile component, is lost preferentially. Due to the similarity of the PbTe and SnTe vapor pressures, the observed differences in total effusion rates between the alloys and changes in rate expected due to incongruency are of the magnitude of the experimental error and the vaporization appears to be congruent in gravimetric experiments. The partial pressures of Te_2, PbTe, and SnTe have been measured by Huang and Brebrick (1988) over $(Sn_xPb_{1-x})_{1-y}Te_y$ for $x = 0.13$, 0.20, and 1.0. It was shown that the Gibbs energy of formation of these solid solutions from the binary compounds is slightly different than that expected for an ideal solution.

The vapor composition was studied by high-temperature mass spectroscopy and the partial pressures of the vapor components of the Sn–Pb–Te system were determined by Demidova et al. (1989). It was found that the $(Sn_xPb_{1-x})_{1-y}Te_y$ solid solutions sublimate incongruently with respect to the composition parameter *y* and quasi-congruently with respect to the composition parameter *x*. The following ions were found in the vapor mass spectrum of the Sn–Pb–Te ternary system: $PbTe^+$, $SnTe^+$, Pb^+, Sn^+, Te^+, Te_2^+, $PbTe_2^+$, $SnTe_2^+$, $Sn_2Te_2^+$, and $PbSnTe_2^+$.

The solid–liquid tie lines of the $Sn_xPb_{1-x}Te$ ($0.12 \le x \le 0.22$) crystals grown by liquid phase epitaxy technique were studied by Tamari and Shtrikman (1978). The concentration of SnTe in the layer (x_S) was found to rise monotonically versus that of the melt (x_L). No measurable influence of cooling rate upon composition was found up to a rate of about 9°C·mm^{-1}.

The liquidus surface of the Sn–Pb–Te ternary system (Figure 9.32) was constructed as a result of the investigation of numerous vertical sections (Linden and Kennedy 1969; Thompson and Wagner 1971; Cadoz et al. 1973; Shelimova and Abrikosov 1973; Laugier et al. 1974; Zlomanov et al. 1974a; Tamm et al. 1975; Shotov and Davarashvili 1977; Mikhaylov et al. 1978; Astles et al. 1979; Muszynski et al. 1979; Novoselova et al. 1980; Petuhov et al. 1980b). On the Figure 9.32, the solidus isoconcentrates (Mikhaylov et al. 1978) [isomolarity

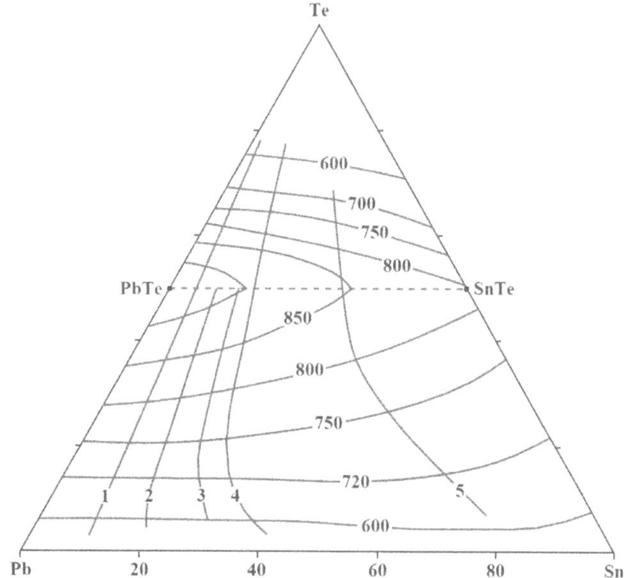

FIGURE 9.32 Liquidus surface of the Sn–Pb–Te ternary system with the solidus isoconcentrates for the $Sn_xPb_{1-x}Te$ solid solution: 1, $x = 0.06$, 2, $x = 0.10$, 3, $x = 0.14$, 4, $x = 0.18$, 5, $x = 0.50$. (From Novoselova, A.V., et al., *Vestn. Mosk. Univ. Ser. Khimia*, **21**(2), 107, 1980.)

(Novoselova et al. 1980), isosolidus (Laugier et al. 1974; Tamm et al. 1975; Shotov and Davarashvili 1977)] are plotted, connecting points of the liquidus surface that are in equilibrium with a solid phase with the same Sn content. At low concentrations of Te, these lines bend sharply in the direction of tin content increasing (Tamm et al. 1975).

The results of determining the coordinates of the surfaces of vapor, solidus, and liquidus of the phase diagram of the Sn–Pb–Te ternary system are generalized and analyzed by Kuznetsov and Zlomanov (1999). The character of sublimation and thermodynamic properties of solid phases were also considered, and diagrams of partial pressures were constructed.

Solidus lines of the $Sn_xPb_{1-x}Te$ solid solutions from the side of excess metals up to SnTe–PbTe section are shown in Figure 9.33 (Gas'kov et al. 1980). The homogeneity region of the $Sn_xPb_{1-x}Te$ solid solutions with an increase in the Sn content is shifted towards the Te excess, and at a certain concentration of tin the section goes beyond the homogeneity region (Tamm et al. 1975). At 600°C, the homogeneity region in the $Sn_{0.5}Pb_{0.5}Te$ samples is in the range 50.007–50.11 at% Te, and in the $Sn_{0.75}Pb_{0.25}Te$ in the range of 50.035–50.3 at% Te (Tairov et al. 1970). At 500°C, the homogeneity region along the $Sn_{0.4}Pb_{0.6}$–Te section is in the range of 50–50.3 at% Te (Shelimova and Abrikosov 1973).

Anomalies were found on the concentration dependence of microhardness in the range of 1.5–5 at% SnTe, which are explained by the manifestation of inter-impurity interaction in solid solutions with increasing impurity content (Rogacheva et al. 1989).

The coordinates of the solidus surface of the $Sn_xPb_{1-x}Te$ solid solutions are given by Hatto et al. (1982) and Novoselova et al. (1986).

In the SnTe–PbTe system at 653°C, significant deviations from Raoult's law are observed only in the range of 70–90 mol% PbTe (Sokolov et al. 1969b).

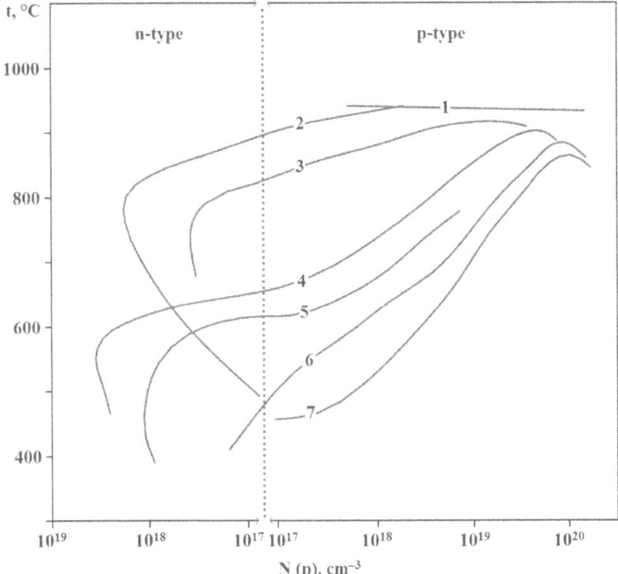

FIGURE 9.33 Solidus lines of the $Sn_xPb_{1-x}Te$ solid solutions: 1, liquidus, 2–7, solidus; 2, $x = 0$, 3, $x = 0.1$, 4, $x = 0.13$, 5, $x = 0.17$, 6, $x = 0.2$, 7, $x = 0.26$. (From Gas'kov, A.M., et al. *Izv. AN SSSR. Neorgan. mater.*, **16**(2), 268, 1980.)

The $Sn_xPb_{1-x}Te$ solid solutions are known narrow-gap materials with an inverted band structure (Dimmock 1966). An increase in the SnTe content leads to a decrease in E_g and zone inversions at $x \approx 0.35$ and ≈ 0.62 at −261°C and 27°C, respectively.

The samples for the investigations were annealed at 350°C and 500°C in an Ar atmosphere for 4.5 months (Shelimova and Abrikosov 1973) [at 600°C for 500 h (Taïrov et al. 1970); at 700°C for 21 days (Nikolić 1967); at 600°C for 100 h (Short 1968)]. Single crystals of the $Sn_xPb_{1-x}Te$ solid solutions were grown by the Bridgman method (Shelimova and Abrikosov 1973; Gas'kov et al. 1980), the Czochralski method under excess Ar pressure of 686 kPa (Kurbanov et al. 1976) and recrystallization through the gas phase (Ickert et al. 1976). The crystallization temperature, in addition to the DTA method, was also determined by visual observation of the appearance of the first crystals on the liquid surface (Cadoz et al. 1973; Astles et al. 1979). The coordinates of some points of the liquidus surface of the Sn–Pb–Te ternary system were determined by liquid-phase epitaxy (Tamm et al. 1975; Shotov and Davarashvili 1977). The liquidus surface was also constructed using simplex simulation (Muszynski et al. 1979).

9.30 Tin–Zirconium–Tellurium

The **ZrSnTe** ternary compound, which crystallizes in the tetragonal structure with the lattice parameters $a = 405.49 \pm 0.06$, $c = 871.1 \pm 0.2$ pm, and a calculated density of 7.826 g·cm^{-3}, is formed in the Sn–Zr–Te system. (Wang and Hughbanks 1995). This compound was synthesized when a Sn flux was used in a reaction mixture that had been loaded with elemental starting materials at the "Zr_3Te_2B" composition. Sn and the "Zr_3Te_2B" mixture were combined in a 1: 1 molar ratio and loaded into a Nb tube which was in turn sealed in an evacuated (10^{-2} Pa) silica ampoule. The reaction temperature was uniformly raised to 550°C over 36 h and maintained at 550°C for 2 days. The temperature was then uniformly raised to 950°C over a 3-day interval and maintained at 950°C for 21 days. The reaction mixture was then cooled to 400°C at a rate of 4°C·h^{-1} and the furnace was subsequently turned off to cool to room temperature. Many large square-faceted crystals were found in the product. The product was always found to be contaminated with a minor amount of $ZrTe_2$.

9.31 Tin–Phosphorus–Tellurium

The $\mathbf{Sn_2P_2Te_6}$ ternary compound, which melts at 380°C and has an experimental density of 6.18 g·cm^{-3}, is formed in the Sn–P–Te system (Voroshilov et al. 1986).

9.32 Tin–Arsenic–Tellurium

SnTe–As: The phase diagram of this system, constructed through DTA, XRD, and metallography, is a eutectic type (Arkoosh and Peretti 1969b). The eutectic contains 74 at% As and crystallizes at 806°C. No mutual solubility of the initial components was found (Krebs et al. 1961a; Arkoosh and Peretti 1969b).

$\mathbf{SnTe–As_2Te_3}$: The phase diagram of this system, constructed through DTA, XRD, and metallography, is a eutectic type (Figure 9.34) (Paulsen and Peretti 1968). The eutectic contains 15.5 mol% (7.8 mass%) SnTe and crystallizes at 378°C. At the eutectic temperature, the solubility of SnTe in As_2Te_3 reaches 12 mol% (6 mass%), and the solubility of As_2Te_3 in SnTe does not exceed 0.25 mol% (0.5 mass%).

SnTe–SnAs: The phase diagram of this system, constructed through DTA, XRD, and metallography, is a eutectic type (Figure 9.35) (Arkoosh and Peretti 1969a; Vdovina

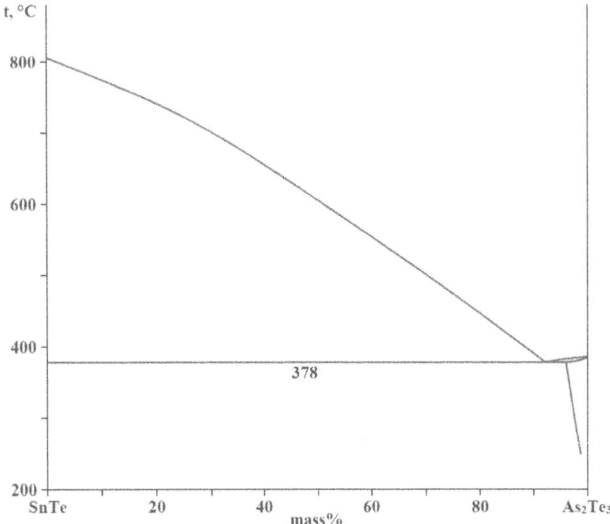

FIGURE 9.34 Phase diagram of the SnTe–As_2Te_3 system. (From Paulsen, J.K., and Peretti, E.A., *J. Mater. Sci.*, **3**(5), 565, 1968.)

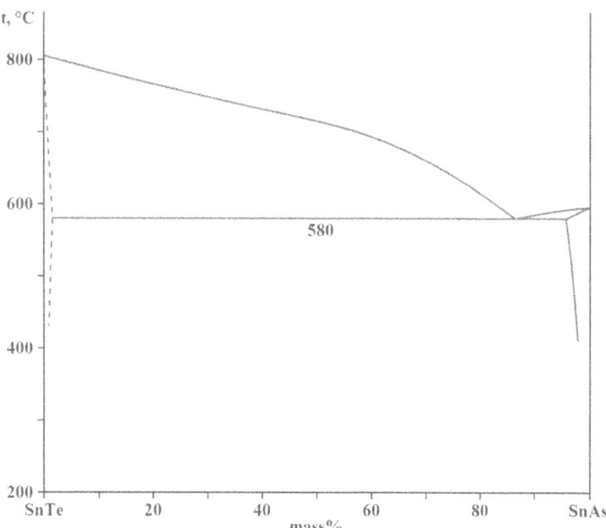

FIGURE 9.35 Phase diagram of the SnTe–SnAs system. (From Arkoosh, M.A., and Peretti, E.A., *J. Inorg. Nucl. Chem.*, **31**(12), 3759, 1969.)

and Medvedeva 1974). The eutectic contains 89.1 mol% (86.5 mass%) SnTe and crystallizes at 580°C. The solubility of SnTe in SnAs at 550°C is not higher than 3.6 mol% (4.5 mass%) (Arkoosh and Peretti 1969a; Kao and Peretti 1970a,b) [the maximum solubility reaches 1 mol% (Vdovina and Medvedeva 1974)], and the solubility of SnAs in SnTe is 1.3 mol% (1 mass%) (Arkoosh and Peretti 1969a; Kao and Peretti 1970a,b) [less than 1 mol% (Vdovina and Medvedeva 1974)]. The ingots were annealed at 520°C for 720 h (Arkoosh and Peretti 1969a; Vdovina and Medvedeva 1974).

SnTe–Sn_4As_3: The phase diagram of this system, constructed through DTA, XRD, and metallography, is a eutectic type (Figure 9.36) (Kao and Peretti 1970a; Vdovina and Medvedeva 1974). The eutectic contains 28.9 mol% (12.5 mass%) SnTe and crystallizes at 575°C–576.5°C. The solubility of SnTe in Sn_4As_3 at the eutectic temperature reaches 8 mol% (Vdovina and Medvedeva 1974) [2.8 mol% (Arkoosh and Peretti 1969a; Kao and Peretti 1970a)], and the solubility of Sn_4As_3 in SnTe is 0.4 mol% (Arkoosh and Peretti 1969a; Kao and Peretti 1970a,b; Vdovina and Medvedeva 1974)]. The ingots were annealed at 520°C for 720 h (Kao and Peretti 1970a; Vdovina and Medvedeva 1974).

The liquidus surfaces of the SnTe–Sn–Sn_4As_3, SnTe–SnAs–Sn_4As_3 and SnTe–As_2Te_3–Te subsystems (Figure 9.37), which are the parts of the Sn–As–Te ternary system, were constructed as a result of studying six vertical sections in each of the subsystems (Kao and Peretti 1970a,b, 1971). In the SnTe–Sn–Sn_4As_3 subsystem (Figure 9.37a), the ternary eutectic is practically degenerate from the Sn side and crystallizes at 231.3°C (Kao and Peretti 1970b). In the SnTe–SnAs–Sn_4As_3 subsystem (Figure 9.37b) the crystallization of all alloys will end in a ternary eutectic, which crystallizes at 576.5°C and practically coincides with the binary eutectic in the SnTe–Sn_4As_3 system (Kao and Peretti 1970a,b). In the SnTe–As_2Te_3–Te subsystem (Figure 9.37c) the ternary eutectic contains 70.2 mol% Te, 12.2 mol% SnTe, and 17.6 mol% As_2Te_3 (42 mass% Te, 14 mass% SnTe, and 44 mass% As_2Te_3) and crystallizes at 358°C (Kao and Peretti 1971). On the liquidus surfaces of the mentioned above subsystems, the field of SnTe primary crystallization occupies the biggest part of these surfaces. No ternary compounds were found in the system (Kao and Peretti 1970a,b, 1971; Vdovina and Medvedeva 1974).

The glass forming region in the Sn–As–Te ternary system is presented in Figure 9.38 (Minaev 1983). Synthesis of the glassy alloys was carried out from the elements in quartz ampoules under vacuum at 1000°C for 12 h with vigorous stirring and the next quenching in water. The vitreous state was identified by a characteristic glass wedge, a shell fracture, as well as through XRD, DTA, and metallography. The boundary of the glass formation region was fixed with an accuracy of 2 at%.

9.33 Tin–Antimony–Tellurium

SnTe–Sb: The phase diagram of this system, constructed through DTA, XRD, metallography, and measuring of the microhardness, is a eutectic type (Figure 9.39) (O'Kane and Stemple 1966; Abrikosov et al. 1968). The eutectic contains 15 mol% SnTe and crystallizes at 577°C. The solubility of SnTe in Sb at 550°C is 1.5 mol% [3 mol% at 525°C (O'Kane and Stemple 1966); 13 mol% (Krebs et al. 1961a)], and the solubility of Sb in SnTe at 550°C is within the interval of 1–3 mol% (Abrikosov et al. 1968) [10 mol% at 525°C and 1.8 mol% at 777°C (O'Kane and Stemple 1966)]. At a content of 5, 10, and 20 mol% SnTe, the cooling curves showed small thermal effects at 551°C, which are due to the presence of a eutectic in

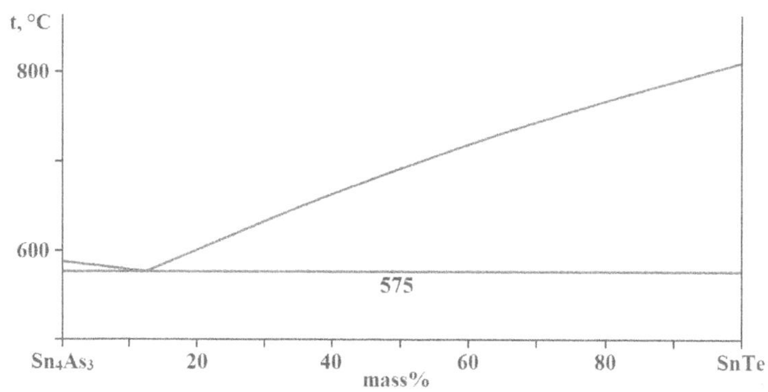

FIGURE 9.36 Phase diagram of the SnTe–Sn_4As_3 system. (From Kao, W.-H., and Peretti E.A., *J. Mater. Sci.*, **5**(12), 1047, 1970.)

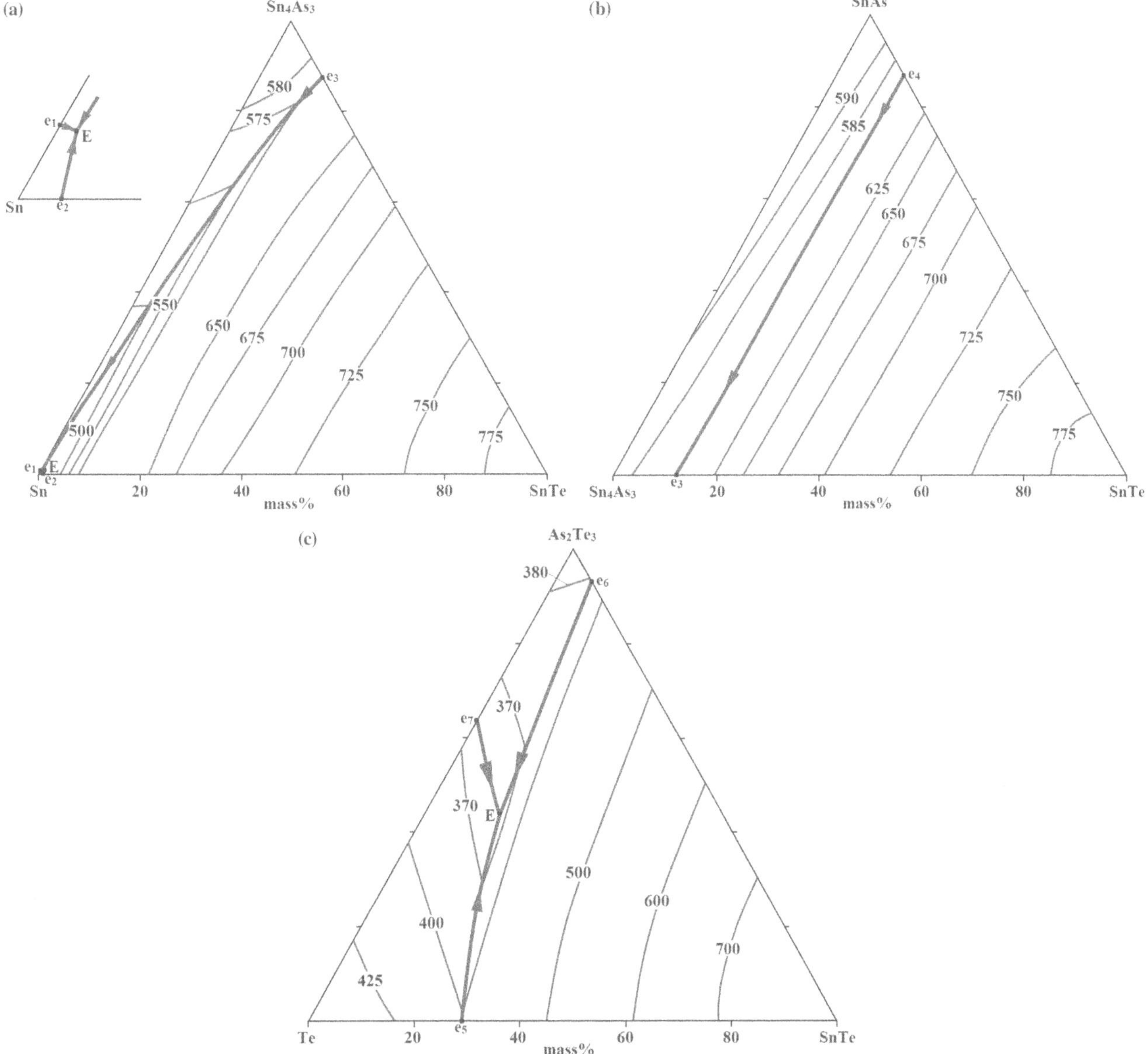

FIGURE 9.37 Liquidus surface of the (a) SnTe–Sn–Sn_4As_3, (b) SnTe–SnAs–Sn_4As_3 and (c) SnTe–As_2Te_3–Te subsystems (a, From Kao, W.-H., and Peretti E.A., *J. Mater. Sci.*, **5**(12), 1047, 1970; b,Kao, W.-H., and Peretti E.A., *J. Less-Common Metals*, **22**(1), 39, 1970; c, Kao, W.-H., and Peretti E.A., *J. Less-Common Metals*, **24**(2), 211, 1971.)

the Sb_2Te_3–Sb system and are explained by the nonequilibrium of the samples (O'Kane and Stemple 1966).

The ingots for the investigations were annealed at 550°C for ca. 3 months (Abrikosov et al. 1968) [at 530°C for 21 days (O'Kane and Stemple 1966)].

SnTe–Sb_2Te_3: The phase diagram of this system, constructed through DTA, XRD, metallography, and measuring of the microhardness, is shown in Figure 9.40) (Elagina and Abrikosov 1959a,b; Stegherr 1969). The eutectic contains 42 mol% (78 mass%) Sb_2Te_3 and crystallizes at 592°C. The solubility of Sb_2Te_3 in SnTe at 550°C is not higher than 10 mol% (Rogacheva et al. 1985a) [9 mol% or 20 mass% at 500°C (Elagina and Abrikosov 1959a,b)]. The ingots for the investigations were annealed at 400°C and 500°C for 500 h (Elagina and Abrikosov 1959a,b) [at 550°C for 400 h (Rogacheva et al. 1985a)].

The earlier investigation on the SnTe–Bi_2Te_3 system by Hirai et al. (1967) noted that the phase diagram of this system is a eutectic type without the formation of ternary compound. The eutectic occurs at about 50 mol% SnTe and a temperature of 600°C. The solid-solubility of Sb_2Te_3 in SnTe is expected to be more than 10 mol % at the eutectic temperature.

The **$SnSb_2Te_4$** ternary compound, which melts incongruently at 603°C, is formed in this system. This compound crystallizes in the trigonal structure with the lattice parameters $a = 429.8 \pm 0.1$, $c = 4157 \pm 1$ pm, and a calculated density of 6.493 g·cm^{-3}(Oeckler et al. 2011) [in the hexagonal structure with the lattice parameters $a = 431$ and $c = 4170$ pm (Talybov 1961); in the tetragonal structure with the lattice parameters $a = 429.4$, $c = 4154.8$ pm and an experimental density of 6.34 g·cm^{-3} (Zhukova and Zaslavskiy 1970a, 1971)].

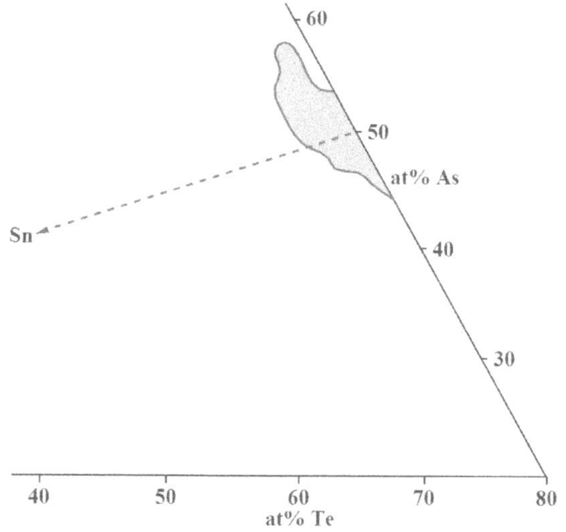

FIGURE 9.38 Glass forming region in the Sn–As–Te ternary system. (From Minaev, V.S., *Fiz. i khim. stekla*, **9**(4), 432, 1983.)

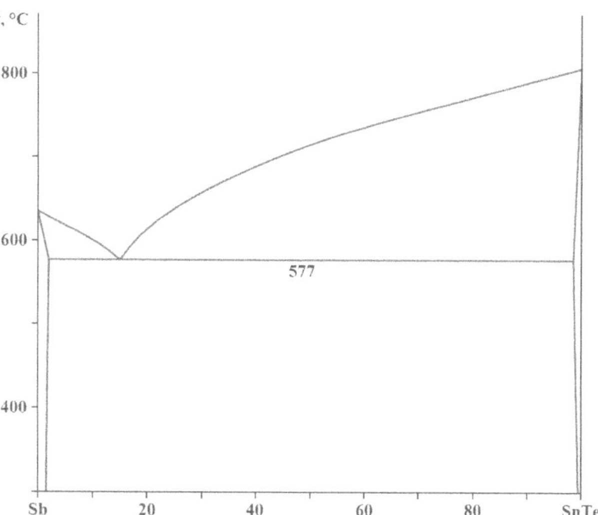

FIGURE 9.39 Phase diagram of the SnTe–Sb system. (From Abrikosov, N.H., et al., *Izv. AN SSSR. Neorgan. mater.*, **4**(10), 1670, 1968.)

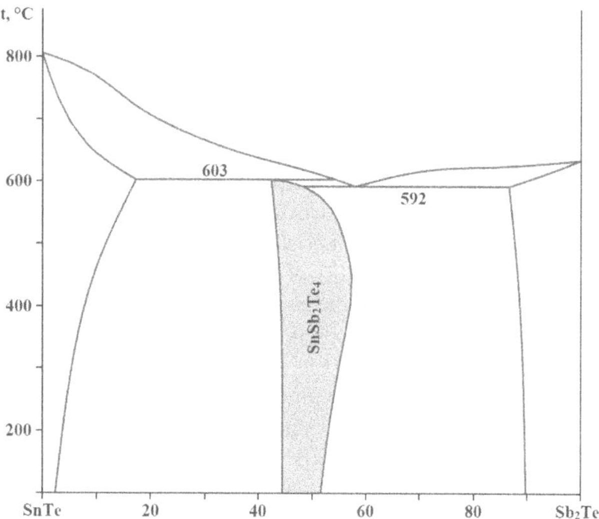

FIGURE 9.40 Phase diagram of the SnTe–Sb_2Te_3 system. (From Elagina, E.I., and Abrikosov, N.H., *Zhurn. neorgan. khimii*, **4**(7), 1638, 1959.)

The Gibbs free energy, enthalpy, and entropy of formation for the $SnSb_2Te_4$ ternary compound at 370°C from SnTe and Sb_2Te_3 were determined by measuring the EMF of galvanic cells (Goncharuk and Sidorko 1998). The thermodynamic properties of formation for this compound from the elements were calculated: $\Delta G_f^0 = -129.30 \pm 27.85$ kJ·M^{-1}, $\Delta H_f^0 = -108.40 \pm 4.25$ kJ·M^{-1}, $\Delta S_f^0 = 32.66 \pm 27.52$ J·(M·K)$^{-1}$. Using EMF measurements, the self-consistent data on the standard thermodynamic functions of $SnSb_2Te_4$ were calculated by Guseinov et al. (2017). The next values were obtained: standard Gibbs energy of formation $\Delta G_{298}{}^{\circ} = -145.4 \pm 1.9$ kJ·M^{-1}, standard enthalpy of formation $\Delta H_{298}{}^{\circ} = -140.4 \pm 1.6$ kJ·M^{-1}, and standard entropy $S_{298}{}^{\circ} = 357.2 \pm 7.7$ J·(M·K)$^{-1}$.

Bulk samples of the $SnSb_2Te_4$ compound were prepared by melting stoichiometric amounts of the Sn, Sb, and Te at 950°C in sealed silica glass ampoules under Ar atmosphere and subsequent annealing at 450°C–500°C for 2–5 days (Oeckler et al. 2011). Single crystals have been grown within 3 days from powdered $SnSb_2Te_4$ by chemical transport in evacuated silica glass ampoules using 10 mass% SbI_3 as transport agent; a temperature gradient from 600°C to 400°C was employed. Hexagonal plate-like single crystals were obtained and residual SbI_3 was washed off with acetone.

SnTe–SnSb: This system is a non-quasibinary section of the Sn–Sb–Te ternary system that intersects the fields of SnTe and Sb primary crystallization (Abrikosov et al. 1968). On the SnSb side, at temperatures below 425°C, a peritectic reaction of SnSb formation occurs. In the area of SnTe primary crystallization, the eutectic SnTe + Sb is crystallized, and in the area adjacent to the solid solution based on SnTe, the eutectic SnTe + SnSb is crystallized. At 420°C, an L + Sb ⇔ SnSb + SnTe invariant reaction takes place in the system. The crystallization of the alloys ends with the formation of the SnTe + SnSb two-phase region. At 312°C in a small temperature range, a polymorphic transformation of solid solutions based on SnSb is observed. The solubility of SnSb in SnTe is slightly more than 1 mol%.

The ingots for the investigations were annealed at 320°C, 420°C, and 500°C for ca. 3 months. The system was investigated through DTA and XRD.

The biggest part of the liquidus surface of the Sn–Sb–Te ternary system is occupied by the primary crystallization field of SnTe (Stegherr 1969).

The region of the solid solutions based on SnTe at 550°C in the Sn–Sb–Te ternary system (Figure 9.41) is oriented along the SnTe–Sb_2Te_3 section (Rogacheva et al. 1987). It was found that the solubility of Sb in SnTe increases linearly with an increase in the concentration of cation vacancies (Rogacheva et al. 1988a). Partial substitution (up to 1 mol%) of Sn with antimony leads to an increase in the equilibrium concentration of vacancies in SnTe.

9.34 Tin–Bismuth–Tellurium

SnTe–Bi: The phase diagram of this system, constructed through DTA, XRD, and Differential Scanning Calorimetry (DSC), is a eutectic type (Figure 9.42) (Adouby et al. 2000). The eutectic reaction occurs at 267°C and 20 mol% SnTe. The solubility of SnTe in Bi reaches 10 mol% at the eutectic

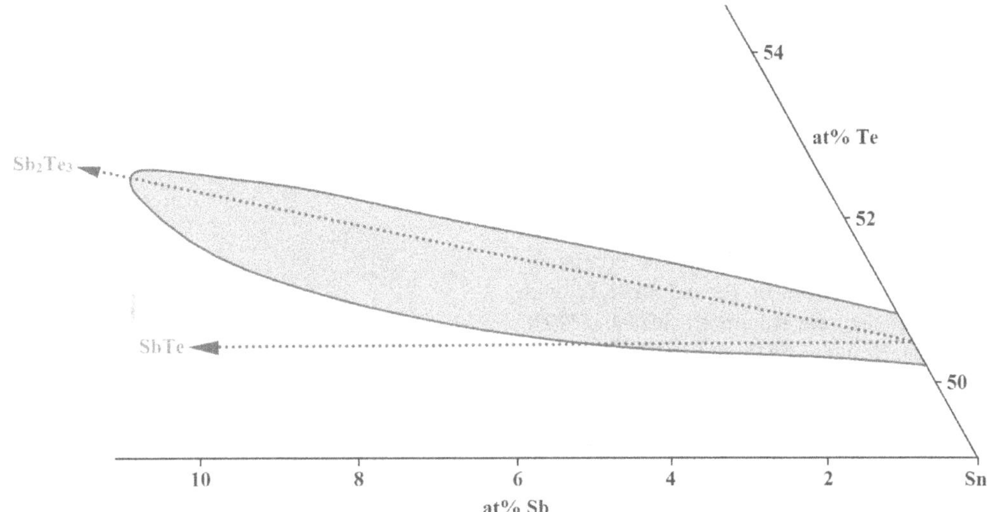

FIGURE 9.41 Region of the solid solutions based on SnTe at 550°C in the Sn–Sb–Te ternary system. (From Rogacheva, E.I., et al., *Izv. AN SSSR. Neorgan. mater.*, **23**(11), 1830, 1987.)

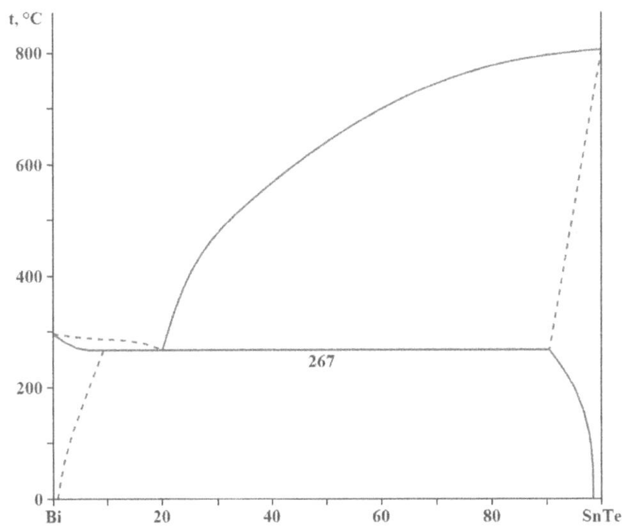

FIGURE 9.42 Phase diagram of the SnTe–Bi system. (From Adouby, K., et al., *C. r. Acad. sci. Sér. 2. Fasc. C*, **3**(1), 51, 2000.)

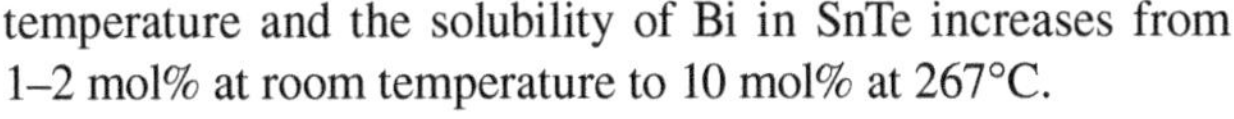

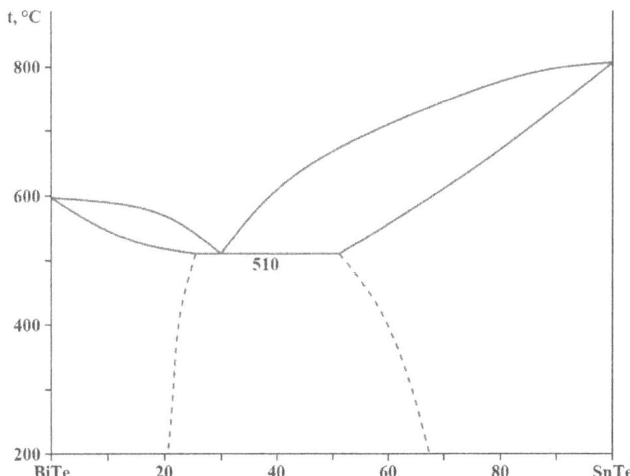

FIGURE 9.43 Phase diagram of the SnTe–BiTe system. (From Dovletov, K., et al., *Izv. AN SSSR. Neorgan. mater.*, **11**(7), 1215, 1975.)

temperature and the solubility of Bi in SnTe increases from 1–2 mol% at room temperature to 10 mol% at 267°C.

According to the thermodynamic calculations, the phase diagram of this system is also a eutectic type but the eutectic is degenerated from the Bi side (Schlieper et al. 1999).

SnTe–BiTe: The phase diagram of this system, constructed through DTA, XRD, and metallography, is a eutectic type (Figure 9.43) (Dovletov et al. 1974d, 1975). The eutectic contains 30 mol% SnTe and crystallizes at 510°C. The solubility of SnTe in BiTe is 25 mol% and the solubility of BiTe in SnTe reaches 28 mol%. The lattice parameters of the solid solutions, based on the binary components, change according to Vegard's law. The ingots for the investigations were annealed at 275°C for 300 h.

SnTe–Bi_2Te_3: The phase diagram of this system, constructed through DTA, XRD, and DSC, is given in Figure 9.44 (Adouby et al. 2000). The eutectic contains ca. 40 mol% SnTe

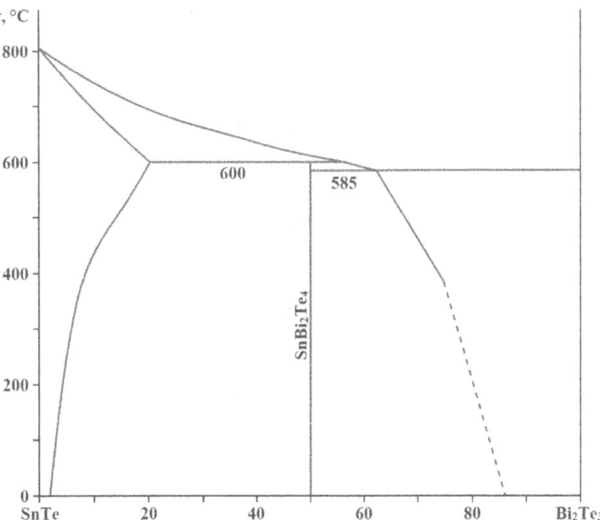

FIGURE 9.44 Phase diagram of the SnTe–Bi_2Te_3 system. (From Adouby, K., et al., *C. r. Acad. sci. Sér. 2. Fasc. C*, **3**(1), 51, 2000.)

and crystallizes at 585°C [43 mol% SnTe and 570°C (Abrikosov et al. 1975)]. The maximum of Bi_2Te_3 solubility in SnTe of ca. 20 mol% is reached at 600°C (Adouby et al. 2000) [the solubility of Bi_2Te_3 in SnTe reaches 14 mol% (Abrikosov et al. 1975); 15 mol% at 605°C and 12 mol% at 450°C (Odin et al. 1974b)]. For the Bi_2Te_3-rich alloys, the range of solid solution existence as a function of the temperature has not been well established (Adouby et al. 2000). According to the data of Reynolds (1967) and Odin et al. (1974b), the continuous solid solutions are formed between Bi_2Te_3 and $\mathbf{SnBi_4Te_7}$ with a minimum on the phase diagram at 18 mol% SnTe and 579°C (Odin et al. 1974b) [at 15 mol% SnTe and 581°C (Reynolds 1967)]. No evidence of $SnBi_4Te_7$ was found by Adouby et al. (2000). The ingots for the investigations were annealed at 450°C for 60 days (Adouby et al. 2000) [at 500°C and 550°C for 2000 h (Abrikosov et al. 1975); at 450°C for 920 h (Odin et al. 1974b)].

The SnTe–Bi_2Te_3 system contains a homologous series of nSnTe·$m$$Bi_2Te_3$ layered compounds ($n = 1, 2$; $m = 1–3$) (Karpinskii et al. 2003; Shelimova et al. 2004a, 2006). Judging from the liquidus relations in the composition range 45–100 mol% SnTe, this system contains no SnTe-rich ($n/m > 1$) compounds. Such compounds could only be obtained in thin films. $SnBi_2Te_4$ seems to be the most stable compound in this system.

$SnBi_2Te_4$ melts incongruently at 600°C (Adouby et al. 2000) [at 596°C (Abrikosov et al. 1975); at 605°C (Odin et al. 1974b)]. This compound crystallizes in the trigonal structure with the lattice parameters $a = 439.54 \pm 0.04$, $c = 4160.6 \pm 0.1$ pm, and a calculated density of 7.49 ± 0.01 g·cm^{-3} (Adouby et al. 2000) [$a = 441.1$, $c = 4151.1$ pm, and an experimental density of 7.36 g·cm^{-3} (Zhukova and Zaslavskiy 1970a, 1971); $a = 440.4 \pm 0.1$, $c = 4161.2 \pm 0.5$ pm, and an energy gap of 0.21 eV (Kuznetsova et al. 2000); $a = 440.5 \pm 0.1$ and $c = 4160 \pm 1$ pm (Karpinskii et al. 2003; Shelimova et al. 2004a, 2006); $a = 440.387 \pm 0.003$ and $c = 4160.03 \pm 0.04$ pm (Kuropatwa and Kleinke 2012); $a = 440.283 \pm 0.010$ and $c = 4171.39 \pm 0.22$ pm or $a = 439.15$ and $c = 4119.48$ pm (calculated values) (Vilaplana et al. 2016)].

$SnBi_2Te_4$ exhibits a pressure-induced electronic topological transition or Lifshitz transition between 3.5 and 5.0 GPa (Vilaplana et al. 2016). A reversible phase transition was also observed above 7.0 GPa in this compound. This compound at low temperatures has p-type conductivity, and at high temperatures it becomes n-type semiconductor (Zhukova et al. 1974).

$SnBi_4Te_7$ melts incongruently at 590°C (Karpinskii et al. 2003) [at 615°C(Abrikosov et al. 1975)] and crystallizes in the trigonal structure with the lattice parameters $a = 440.37 \pm 0.01$ and $c = 2406.71 \pm 0.4$ pm (Kuropatwa and Kleinke 2012) [$a = 439.8$ and $c = 4149.3$ pm (Zhukova and Zaslavskiy 1970a, 1971); $a = 439.0 \pm 0.1$ and $c = 2398 \pm 1$ pm or $a = 439.3 \pm 0.2$ and $c = 4189 \pm 3$ pm (Kuznetsova et al. 2000); $a = 439.2 \pm 0.1$ and $c = 2399 \pm 1$ pm (Karpinskii et al. 2003; Shelimova et al. 2004a, 2006)].

$\mathbf{SnBi_6Te_{10}}$ also crystallizes in the trigonal structure with the lattice parameters $a = 438.35 \pm 0.09$ and $c = 10242 \pm 2$ pm (Kuropatwa and Kleinke 2012) [$a = 439.1 \pm 0.1$ and $c = 10264 \pm 5$ pm (Karpinskii et al. 2003; Shelimova et al. 2006); $a = 439.0 \pm 0.1$ and $c = 10247 \pm 1$ pm (Shelimova et al. 2004a)].

These three compounds were synthesized using classic solid state techniques from the elements or the binary compounds in stoichiometric ratios in an Ar-filled glove box with annealing at 500°C for 160–1000 h and the next quenching in ice water (Kuznetsova et al. 2000; Karpinskii et al. 2003; Shelimova et al. 2004a; Kuropatwa and Kleinke 2012; Vilaplana et al. 2016).

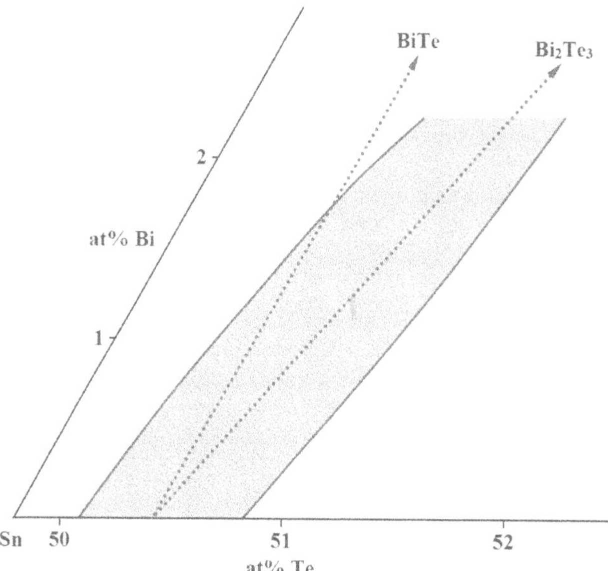

FIGURE 9.45 Region of the solid solutions based on SnTe at 550°C in the Sn–Bi–Te ternary system. (From Gorne, G.V., et al., *Izv. AN SSSR. Neorgan. mater.*, **24**(7), 1214, 1988.)

According to the data of Hirai et al. (1967), one more compound, which has an approximately composition $\mathbf{Sn_3Bi_4Te_9}$, melts incongruently at 595°C and forms an eutectic with Bi_2Te_3 at 570°C, is formed in the SnTe–Bi_2Te_3 system.

The region of solid solutions based on SnTe in the Sn–Bi–Te ternary system (Figure 9.45) is situated along the SnTe–Bi_2Te_3 section (Gorne et al. 1988).

The Sn–Bi–Te system has been studied in the composition region SnTe–Bi_2Te_3–Te at temperatures from 27°C to 157°C using EMF measurements on reversible concentration cells (Babanly et al. 2011a). The subsolidus phase diagram includes the ternary compounds $SnBi_2Te_4$, $SnBi_4Te_7$, and $SnBi_6Te_{10}$ and a broad (0–25 mol% SnTe) range of Bi_2Te_3-based solid solutions. The three ternary compounds and solid solution based on Bi_2Te_3 are in equilibrium with elemental tellurium. The tellurium solubility in these phases is insignificant. The standard Gibbs energy and enthalpy of formation and standard entropies of the ternary phases are the next: $\Delta G^0_{298} = -173.1 \pm 2.7$ kJ·M^{-1}, $\Delta H^0_{298} = -170.5 \pm 2.1$ kJ·M^{-1}, $S^0_{298} = 374 \pm 14$ kJ·(K·M)$^{-1}$ for $SnBi_2Te_4$; $\Delta G^0_{298} = -258.8 \pm 4.3$ kJ·M^{-1}, $\Delta H^0_{298} = -255.8 \pm 2.6$ kJ·M^{-1}, $S^0_{298} = 640 \pm 25$ kJ·(K·M)$^{-1}$ for $SnBi_4Te_7$; $\Delta G^0_{298} = -341.8 \pm 5.7$ kJ·M^{-1}, $\Delta H^0_{298} = -338.6 \pm 3.0$ kJ·M^{-1}, $S^0_{298} = 904 \pm 30$ kJ·(K·M)$^{-1}$ for $SnBi_6Te_{10}$; $\Delta G^0_{298} = -423.0 \pm 7.2$ kJ·M^{-1}, $\Delta H^0_{298} = -421.1 \pm 3.4$ kJ·M^{-1}, $S^0_{298} = 1164 \pm 39$ kJ·(K·M)$^{-1}$ for $\mathbf{SnBi_8Te_{13}}$; and $\Delta G^0_{298} = -471.1 \pm 8.0$ kJ·M^{-1}, $\Delta H^0_{298} = -468.8 \pm 3.5$ kJ·M^{-1}, $S^0_{298} = 1297 \pm 42$ kJ·(K·M)$^{-1}$ for $\mathbf{SnBi_9Te_{14.5}}$. The alloys for investigations were equilibrated by long-term thermal annealing. To this, as-cast, non-homogenized alloys were ground into powder, which was thoroughly mixed and pressed into pellets, which were annealed at 530°C for 1000 h.

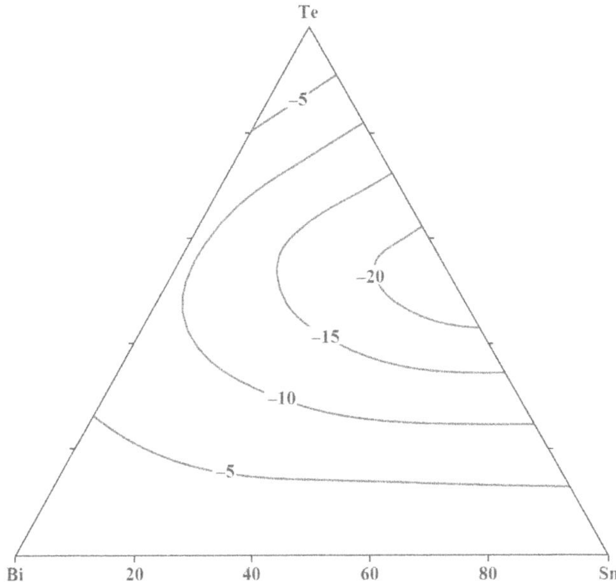

FIGURE 9.46 Isoenthalpic lines (in kJ·mol^{-1}) in the Sn–Bi–Te ternary system at 900°C. (From Schlieper, A., et al., *Z. Metallkdd.*, **90**(4), 250, 1999.)

The excess enthalpies of liquid alloys were determined in a heat-flow calorimeter (Schlieper et al. 1999). The measurements were carried out along the sections Bi_xSn_{1-x}–Te with $x = 0.2$, 0.5 and 0.8 at 900°C and with $x = 0.5$ at 800°C. The maxima of the exothermic enthalpies were found on a line stretching from SnTe to Bi_2Te_3 (Figure 9.46). The association model with additional ternary interaction parameters was applied for the analytical description of the thermodynamic data of the melt.

9.35 Tin–Niobium–Tellurium

SnTe-Nb: Annealing of the samples, containing equimolar amounts of SnTe and Nb, at 1100°C leads to exchange interaction; however, the phase composition of the resulting products has not been established (Bates et al. 1968).

9.36 Tin–Tantalum–Tellurium

SnTe–Ta: Annealing of the samples, containing 25, 33.3, and 50 at% Ta, at 950°C leads to the exchange interaction with the formation of tin and various tantalum tellurides (Bates et al. 1968).

9.37 Tin–Oxygen–Tellurium

SnO_2–Te: The phase diagram of this system is characterized by the degenerate eutectic near Te at 451°C ± 8°C and the immiscibility region within the interval 40–98 at% Te (Mikhaylov et al. 1991a). The liquidus temperatures are higher than 1500°C.

SnO_2–TeO_2: The phase diagram of this system was constructed in the region of 20–40 mol% SnO_2 (Mikhaylov et al. 1991a). The eutectic contains 4 mol% SnO_2 and crystallizes at 709°C. The **$SnTe_3O_8$** ternary compound, which melts incongruently at 878°C (Mikhaylov et al. 1991a) [at 905°C (Gospodinov 1985); at 917°C (Galy and Meunier 1969)] and crystallizes in the cubic structure with the lattice parameter $a = 1112.2$ pm and a calculated density of 6.020 ± 0.001 g·cm^{-3} (Gospodinov 1985) [a = 1114.4 pm and the calculated and experimental densities of 6.04 and 6.02 g·cm^{-3}, respectively (Galy and Meunier 1969)], is formed in this system (Vasil'ev et al. 1979). The composition of the peritectic point is 8 mol% SnO_2 (Mikhaylov et al. 1991a).

Vaporization studies along the SnO_2–TeO_2 ($SnTe_3O_8$ + SnO_2) quasibinary system were conducted using Knudsen effusion mass spectrometry in the temperature range 570°C–695°C by Trinadh et al. (2019). It was shown that $SnTe_3O_8$ vaporizes incongruently and TeO_2(g), TeO(g), Te_2(g), and O_2(g) were the neutral species observed over the equilibrium vapor. Partial pressure–temperature relations of TeO_2(g), TeO(g), Te_2(g) over $SnTe_3O_8$ + SnO_2 two-phase region were determined in the mention above temperature range. Partial pressures of O_2(g) were calculated by considering gas phase equilibrium reactions involving these species and TeO_2(g), TeO(g), Te_2(g). Using the p-T relations, enthalpy changes of the following reactions were evaluated: $SnTe_3O_8$(s) ⇔ SnO_2(s) +3TeO_2(g); $SnTe_3O_8$(s) + 2TeO(g) ⇔ 4TeO_2(g) +0.5 Te_2(g) +SnO_2(s) and 0.5Te_2(g) + TeO_2(g) ⇔ 2TeO(g). Subsequently, the enthalpy and Gibbs energy of formation of $SnTe_3O_8$ phase at 23°C were deduced as ΔH^0_f=−1613.7 ± 67.8 kJ M^{-1} and ΔG^0_f=−1386.7 ± 67.8 kJ M^{-1}.

The standard Gibbs energy of formation of $SnTe_3O_8$ was also determined from its vapor pressure measurements over the temperature range 700°C–885°C by employing thermogravimetry-based transpiration method (Jain et al. 2013). The temperature dependence of vapor pressure of TeO_2 over the mixture $SnTe_3O_8$(s) + SnO_2(s) generated by the incongruent vaporization of $SnTe_3O_8$(s) could be represented as: log (p (TeO_2, g)/Pa ± 0.03) = 13.943–14181 (K/T) (700°C–885°C). The standard Gibbs energy of formation of $SnTe_3O_8$(s) was also determined by measuring the oxygen potential of SnO_2(s)–Te(s)–$SnTe_3O_8$(s) phase mixture by the EMF method. Enthalpy increments of $SnTe_3O_8$(s) were determined by inverse drop calorimetric method in the temperature range 250°C–700°C. The thermodynamic functions (heat capacity, entropy and free energy functions) were derived from the measured values of enthalpy increments. A mean value of −1642 ± 2.0 kJ·M^{-1} was obtained for ΔH^0_{298} ($SnTe_3O_8$, s) by combining the value of ΔG^0_{298} ($SnTe_3O_8$, s) derived from vapor pressure data and the free energy functions obtained from the drop calorimetric data.

$SnTe_3O_8$ starts to dissociate slightly in the solid state at 880°C (Gospodinov 1985). Dissociation in the melted state is still negligible and the mass loss of the sample at 950°C is as little as 5%.

The title compound was prepared by a solid-state reaction between SnO_2 and TeO_2 (molar ratio 1:3) (Galy and Meunier 1969; Jain et al. 2013). This mixture was prepared by homogenizing the two oxides intimately followed by cold compaction into cylindrical pellets. These green compacts were heated in air at 400°C for 24 h and then at 700°C for 60 h. The pellets were withdrawn intermittently from the furnace, powdered, re-compacted and heat treated again. This cycle was repeated twice in order to ensure complete formation of $SnTe_3O_8$.

Rubbing SnTe in air in an agate mortar leads to the formation of SnO_2 (Seregin et al. 1972a).

One more ternary compound, **$Sn_5Te_3O_7$**, which has an experimental density of 7.34 g·cm^{-3}, is formed in the Sn–O–Te system (Batsanov and Shestakova 1966). It can be synthesized by melting the mixture of SnO and Te at 600°C.

On the liquidus surface of the SnO_2–TeO_2–Te subsystem of the Sn–O–Te ternary system, an extensive immiscibility region is observed, extending from TeO_2–Te system to SnO_2–Te system (Mikhaylov et al. 1991b). The ternary eutectic is degenerated from Te corner and crystallizes at 451°C ± 5°C.

9.38 Tin–Chromium–Tellurium

SnTe–CrTe: The solubility of CrTe in SnTe at 230°C is 2 mol% and increases up to 3.5 mol% at 730°C (Stavrianidis et al. 1984). The ingots for investigations were annealed at 230°C, 530°C, and 730°C for 278, 192, and 192 h, respectively, and studied through DTA, XRD, and metallography.

SnTe–Cr_3Te_4: The solubility of Cr_3Te_4 in SnTe at 230°C and 730°C is 0.5 and 1 mol%, respectively (Keyyan et al. 1987). The ingots for investigations were annealed at 230°C, 530°C, and 730°C for 278, 192, and 192 h, respectively, and studied through DTA, XRD, and metallography.

The **$SnCrTe_3$** ternary compound, which crystallizes in the hexagonal structure with the lattice parameters a = 701 pm according to the first principles calculations within the density-functional theory (Zhuang et al. 2015), is formed in the Sn–Cr–Te system.

9.39 Tin–Molybdenum–Tellurium

SnTe–Mo: Annealing of the samples, containing 20 and 33.3 at% of Mo, at 870°C for 21 h leads to the formation of new phase, the composition of which has not been established and which does not formed at the annealing of the samples at 1100°C for 15 h (Bates et al. 1968).

9.40 Tin–Tungsten–Tellurium

SnTe–W: Annealing of the samples, containing 20 at% of W, at 1000°C for 16.5 h does not lead to the formation of new phases, which indicated on the quasibinarity of this system (Bates et al. 1968).

9.41 Tin–Chlorine–Tellurium

SnTe–$SnCl_2$: The phase diagram of this system, constructed through DTA, is a eutectic type with immiscibility region (Figure 9.47) (Blachnik and Kasper 1974). The eutectic contains 1 mol% SnTe and crystallizes at 246°C. According to the data of Rodionov et al. (1972), 2.3 mol% SnTe dissolves in the $SnCl_2$ melt at 600°C. The ingots for the investigations were annealed at 200°C–400°C for 750 h (Blachnik and Kasper 1974).

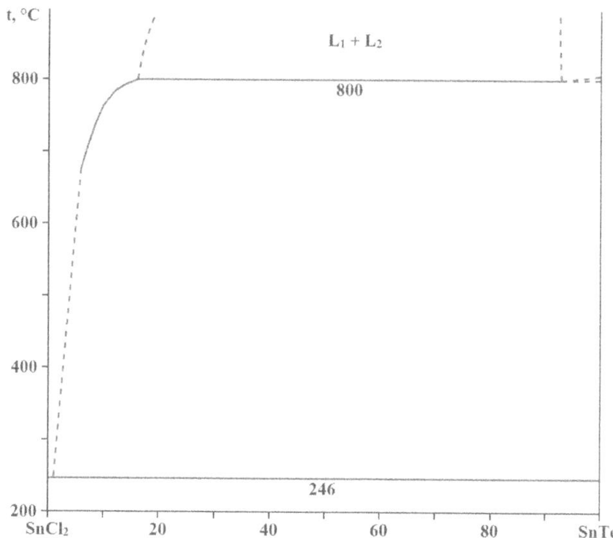

FIGURE 9.47 Phase diagram of the SnTe–$SnCl_2$ system. (From Blachnik, R., and Kasper, F.-W., *Z. Naturforsch.*, **29B**(3–4), 159, 1974.)

The **$Sn_4Cl_2Te_7$** ternary compound, which has an experimental density of 6.15 g·cm^{-3}, is formed in the Sn–Cl–Te system (Batsanov et al. 1963). It was prepared by the interaction of $SnCl_2$ and Te at 480°C for 20 h.

9.42 Tin–Bromine–Tellurium

SnTe–$SnBr_2$: The phase diagram of this system, constructed through DTA, is a eutectic type with immiscibility region (Figure 9.48) (Blachnik and Kasper 1974). The eutectic contains 3 mol% SnTe and crystallizes at 227°C. The ingots for the investigations were annealed at 200°C–400°C for 750 h.

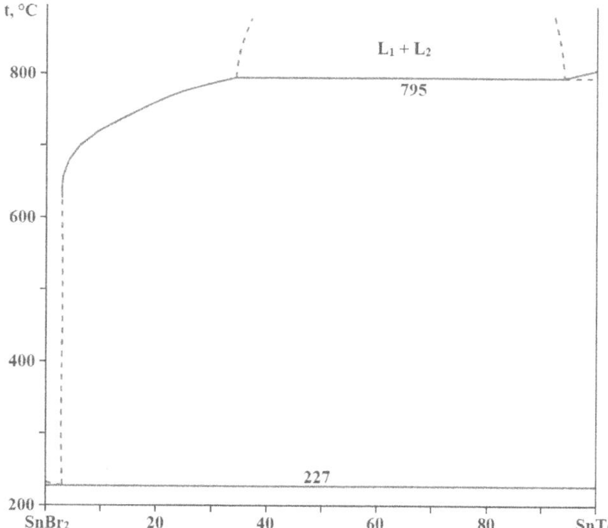

FIGURE 9.48 Phase diagram of the SnTe–$SnBr_2$ system. (From Blachnik, R., and Kasper, F.-W., *Z. Naturforsch.*, **29B**(3–4), 159, 1974.)

The interaction of SnTe with liquid Br_2 at 0°C and 20°C occurs with a large heat releasing (Batsanov and Shestakova 1968). The interaction is slow and accelerates with increasing temperature. As a result of the interaction, $SnBr_4$ and $TeBr_4$ are formed. When SnTe interacts with an alcohol solution of Br_2, a $SnBr_4{\cdot}TeBr_4$ salt is formed. Bromination with liquid bromine was carried out at 0°C and 20°C, and with bromine vapor at 40°C.

9.43 Tin–Iodine–Tellurium

SnTe–SnI_2: The phase diagram of this system, constructed through DTA, is a eutectic type with immiscibility region (Figure 9.49) (Blachnik and Kasper 1974). The eutectic contains 5 mol% SnTe and crystallizes at 309°C. The ingots for the investigations were annealed at 200–400°C for 750 h. According to the data of Rodionov and Klokman (1972), SnI_2 melt dissolves 26.2 mol% SnTe at 600°C.

SnTe–TeI_4: The phase diagram of this system, constructed through DTA and XRD, is presented in (Figure 9.50) (Adenis and Lindqvist 1991). One peritectic point is located at 25 mol% TeI_4, while three eutectic points appeared at 10, 40, and 70 mol% TeI_4. The lines at 140°C and 320°C correspond to the formation of SnI_4 and SnI_2, respectively. DTA indicates the existence of two ternary compounds, **A** and **B**, at 30 and 50 mol% TeI_4, respectively. However, no new phases could be detected by XRD at room temperature. Compound **A** decomposes before melting and is affected by the action of several redox species between 300°C and 425°C. This phase is metastable and therefore its existence depends on the decomposition kinetics. Compound **B**, corresponding to the special composition **$SnTe_2I_4$**, probably exists at temperatures higher than 175°C and melts congruently at 556°C.

SnI_2–Te: The phase diagram of this system is a eutectic type without any ternary intermediate phases (Adenis and Lindqvist 1991). The eutectic point is located at 12.5 at% Te, corresponding to a temperature of 317°C.

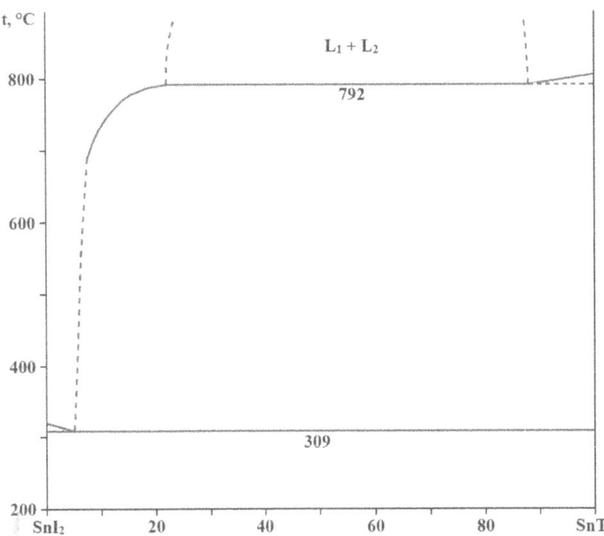

FIGURE 9.49 Phase diagram of the SnTe–SnI_2 system. (From Blachnik, R., and Kasper, F.-W., *Z. Naturforsch.*, **29B**(3–4), 159, 1974.)

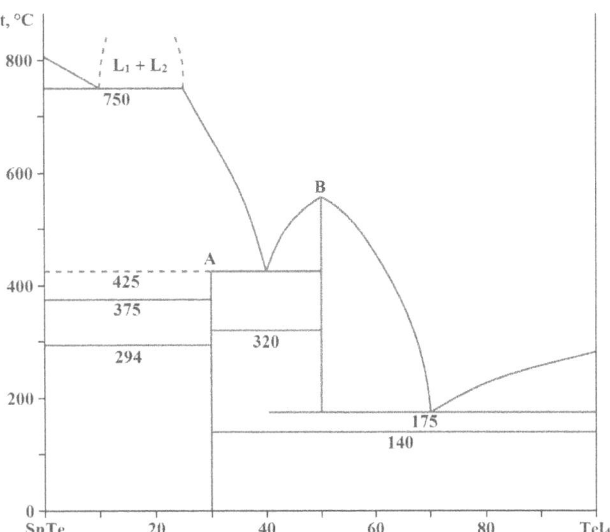

FIGURE 9.50 Phase diagram of the SnTe–TeI_4 system. (From Adenis, C., and Lindqvist, O., *Thermochim. Acta*, **179**, 247, 1991.)

SnI_2–TeI_4: According to the data of Adenis and Lindqvist (1991), **$5SnI_2{\cdot}3TeI_2$** with small homogeneity region is formed in this system. It melts congruently at 438°C and decomposes at 153°C. Two eutectics contain 50 and 91 mol% SnI_2. First of them crystallizes at 187°C.

SnI_4–TeI_4: The phase diagram of this system, constructed through DTA and XRD, is a eutectic type (Katryniok and Kniep 1979). The eutectic contains ca. 5 mol% TeI_4 and crystallizes at 137°C.

The liquidus surface of the Sn–I–Te ternary system was constructed by Katryniok and Kniep (1979). The lowest melting temperature of the ternary system is near the eutectic at 88 at% I and 77°C in the Sn–I binary system. No ternary compounds were found. The possibly existing ternary compound **$SnTeI_6$** (Batsanov and Shestakova 1968) should be metastable with regard to the phase diagram (Katryniok and Kniep 1979).

The interaction of SnTe and I_2 in vacuum at 60°C leads to the formation of SnI_4, and if the initial substances are taken in a 1:1 molar ratio, then SnTe and Te remain in the reaction products (Batsanov and Shestakova 1968). When SnTe interacts with iodine alcohol solution, the **$SnTeI_6{\cdot}H_2O$** compound is formed.

9.44 Tin–Manganese–Tellurium

SnTe–Mn: Annealing of the samples, containing 50 at% Mn, at 1100°C for 15 h leads to exchange interaction with the formation of tin and MnTe (Bates et al. 1968).

The **$SnMn_2Te_4$** ternary compound, which crystallizes in the orthorhombic structure with the lattice parameters $a = 1402.03 \pm 0.04$, $b = 814.68 \pm 0.03$, and $c = 660.75 \pm 0.04$ pm (Delgado et al. 2009b) [$a = 1402.0 \pm 0.2$, $b = 814.7 \pm 0.1$, $c = 660.7 \pm 0.1$ pm, and a calculated density of 6.58 g·cm^{-3} (Delgado et al. 2010)], is formed in the Sn–Mn–Te system. The samples of the title compound were synthesized by direct fusion of stoichiometric quantities of Mn, Ge, and Te in sealed, evacuated quartz ampoules (Delgado et al. 2009b, 2010). The fusion process

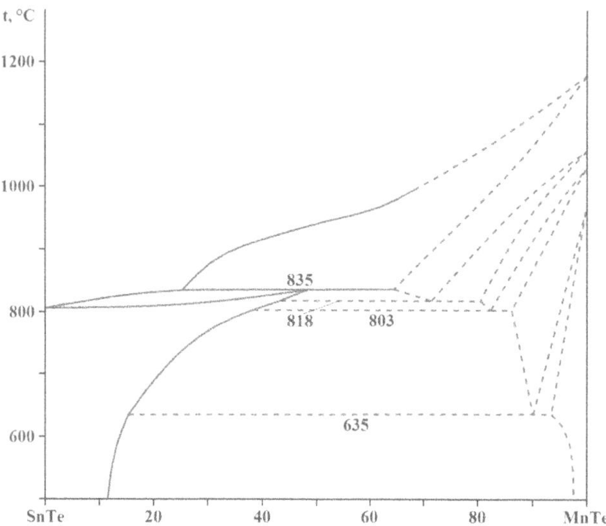

FIGURE 9.51 Phase diagram of the SnTe–MnTe system. (From Dudkin, L.D., et al., *Izv. AN SSSR. Neorgan. mater.*, 7(9), 1503, 1971.)

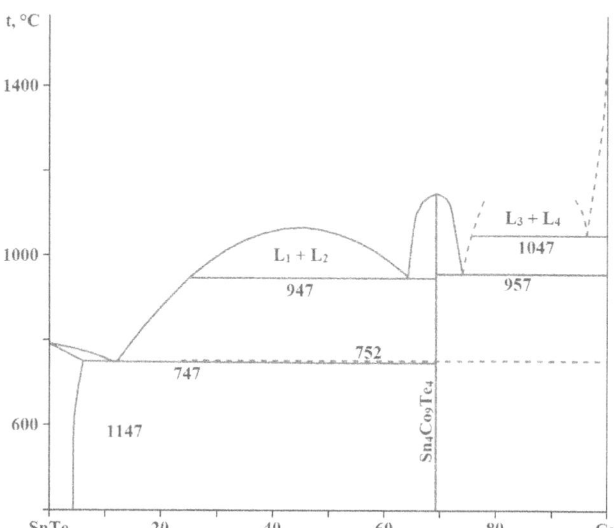

FIGURE 9.52 Phase diagram of the SnTe–Co system. (From Asadova, S.Yu., and Allazov, M.R., *Zhurn. neorgan. khimii*, **33**(2), 541, 1988.)

(14 days) was carried out into a furnace in vertical position heated up to 1050°C. Then, the temperature was gradually lowered to 500°C and the furnace was turned off and the ingots were cooled to room temperature.

SnTe–MnTe: The phase diagram of this system, constructed through DTA, XRD, and metallography, is a peritectic type (Figure 9.51) (Dudkin et al. 1971). The solid solutions based on SnTe are formed according to the peritectic reaction at 835°C as a result of the interaction of high-temperature modification of MnTe and liquid. Below the solidus there is a raw of thermal effects due to polymorphic transformations of MnTe. The solubility of SnTe in MnTe at 600°C is not higher than 5 mol%. The ingots for the investigations were annealed at 600°C for 100 h.

9.45 Tin–Iron–Tellurium

SnTe–Fe: The phase diagram of this system, constructed through DTA, metallography, and measuring of the microhardness, is a eutectic type (Bates et al. 1968). The eutectic crystallizes at 734°C. When SnTe and Fe interacted, no new phases were found in the system (Kuliev et al. 1974; Markholiya et al. 1975). The ingots for the investigations were annealed at 860°C–1100°C for 15–45 h (Markholiya et al. 1975) [at 450°C–700°C for 6–192 h (Bates et al. 1968)].

SnTe–Fe_2Te_3: The solubility of Fe_2Te_3 in SnTe reaches 5 mol % (Asadov et al. 1998). The ingots for the investigations were annealed at 500°C for 300 h.

9.46 Tin–Cobalt–Tellurium

SnTe–Co: The phase diagram of this system, constructed through DTA, metallography, and measuring of the microhardness, is presented in Figure 9.52 (Asadova and Allazov 1988). The **$Sn_4Co_9Te_4$** ternary compound, which melts congruently at 1147°C and has a polymorphic transformation at 752°C, is formed in this system. The eutectics contain 12 and 74 at% Co and crystallize at 747°C and 957°C, respectively. In the concentration ranges of 30–68 and 76–96 at% Co, immiscibility regions are observed in the liquid state. The temperature of monotectic reactions is 947°C and 1047°C, respectively. The region of solid solutions based on SnTe reaches 5 mol%.

One more ternary compound, **$Sn_{1.5}CoTe_{1.5}$**, which exists in two polymorphic modifications, is formed in the SnTe–Co system. First modification of this compound crystallizes in the trigonal structure with the lattice parameters a = 1290.916 ± 0.002 and c = 1578.585 ± 0.004 (Vaqueiro et al. 2010) [a = 1290.63 ± 0.02, c = 1578.37 ± 0.03, and a calculated density of 7.50 g·cm^{-3} (Laufek et al. 2007a, 2008a). Second modification crystallizes in the cubic structure with the lattice parameter a = 911.65 ± 0.01 pm (Vaqueiro and Sobany 2007) [a = 921.8 pm and an energy gap of 0.51 eV (Bertini and Cenedese 2007)].

To synthesize this compound, mixtures of the elements corresponding to its stoichiometry were ground in an agate mortar prior to sealing into an evacuated (< 0.01 Pa) silica tube (Vaqueiro and Sobany 2007; Vaqueiro et al. 2010). The inner wall of the silica tube was coated with a thin layer of carbon by pyrolysis of acetone. The reaction mixtures were heated at 430°C for 2 days, followed by another 2 days at 580°C. Samples were cooled at 0.1°C·min^{-1} to room temperature, prior to removal from the furnace. Following re-grinding, samples were sealed into a second carbon-coated silica tube and re-fired at 580°C for 2 days with the next cooling to room temperature at 0.1°C·min^{-1}. One or two additional firings at 580°C were required to ensure complete reaction.

SnTe–CoSn: This system is non-quasibinary section of the Sn–Co–Te ternary system (Asadova et al. 1987). The liquidus lines characterizing the primary crystallization of SnTe and Co_3Sn_2 intersect at 6 mol% CoSn. The complete crystallization of the samples ends at 662°C. There are two phases in the solid state: SnTe and CoSn. The mutual solubility of the initial components does not exceed 1 mol%. The system was investigated through DTA, XRD, metallography, and measuring of the microhardness.

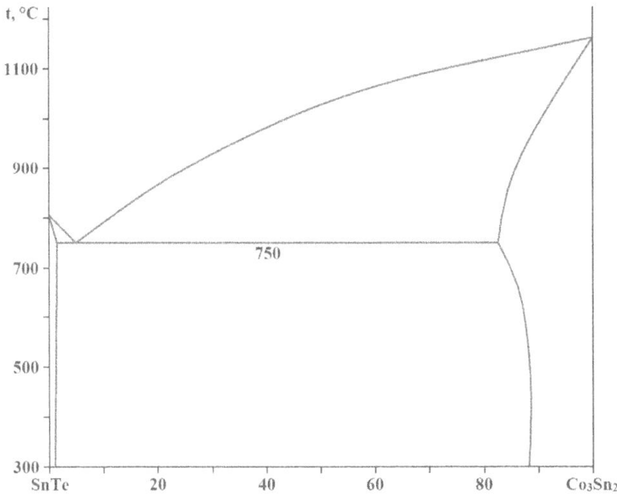

FIGURE 9.53 Phase diagram of the SnTe–Co_3Sn_2 system. (From Asadova, S.Yu., et al., *Zhurn. neorgan. khimii*, **31**(9), 2362, 1986.)

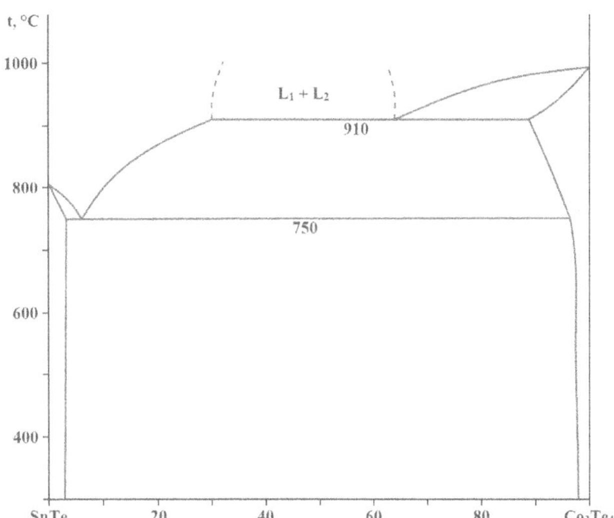

FIGURE 9.54 Phase diagram of the SnTe–Co_3Te_4 system. (From Asadova, S.Yu., et al., *Zhurn. neorgan. khimii*, **31**(9), 2362, 1986.)

SnTe–$CoSn_2$: This system is non-quasibinary section of the Sn–Co–Te ternary system (Asadova et al. 1987). In the system, SnTe and Co_3Sn_2 primary crystallize. The section has an isothermal line at 557°C, corresponding to the L + Co_3Sn_2 ⇔ CoSn incongruent process, and at 352°C, due to the incongruent formation of $CoSn_2$. The mutual solubility of the initial components does not exceed 1 mol%. The system was investigated through DTA, XRD, metallography, and measuring of the microhardness.

SnTe–Co_3Sn_2: The phase diagram of this system, constructed through DTA, metallography, and measuring of the microhardness, is a eutectic type (Figure 9.53) (Asadova et al. 1986). The eutectic contains 5 mol% Co_3Sn_2 and crystallizes at 750°C. The solubility of Co_3Sn_2 in SnTe at room temperature is 1 mol% and the solubility of SnTe in Co_3Sn_2 at the same temperature reaches 11 mol%. The ingots for the investigations were annealed at 700°C for 100 h.

SnTe–CoTe: The solubility of CoTe in SnTe at room temperature is 10 mol% and increases up to 12 mol% at 550°C (Alidzhanov et al. 1982c). The lattice parameter of the solid solution decreases up to 622.5 pm at 8 mol% CoTe.

SnTe–$CoTe_2$: This system is a non-quasibinary section of the Sn–Co–Te ternary system (Asadova et al. 1985). On the $CoTe_2$ side, the solid solution based on Co_3Te_4 primary crystallizes. There is an immiscibility region in the system at 900°C–828°C within the interval of 9–31.5 at% Sn. The lines of the primary crystallization of the solid solution based on Co_3Te_4 and SnTe intersect at 13 at% Sn and 730°C. Below this isothermal line there is a region of secondary precipitation, and incongruent crystallization of the solid solution based on $CoTe_2$ occurs. Crystallization ends at 3 at% and 635°C. The ingots for the investigations were annealed at 580°C for 200 h. The system was investigated through DTA, metallography, and measuring of the microhardness.

SnTe–Co_3Te_4: The phase diagram of this system, constructed through DTA, metallography, and measuring of the microhardness, is a eutectic type with monotectic transformation (Figure 9.54) (Asadova et al. 1986). The eutectic contains 6 mol% Co_3Te_4 and crystallizes at 750°C. In the range of 30–64 mol% Co_3Te_4, an immiscibility region is observed in the system with a monotectic equilibrium temperature of 910°C. The solubility of SnTe in Co_3Te_4 at 910°C is 19 mol%, and at 500°C and 700°C it is 3 mol%. The solubility of Co_3Te_4 in SnTe at 500°C and 700°C reaches 3 mol%. The ingots for the investigations were annealed at 700°C for 100 h.

The liquidus surface of the Sn–Co–Te ternary system has ten fields of the primary crystallization (Figure 9.55) (Allazov and Asadova 1988). Three fields correspond to the crystallization of the initial components, and the crystallization field of Sn is degenerate. The fields of Co_3Sn_2 and $Co_9Sn_4Te_4$ are the biggest in the system. There are nine invariant four-phase equilibria in this system, of which six ternary eutectics and three

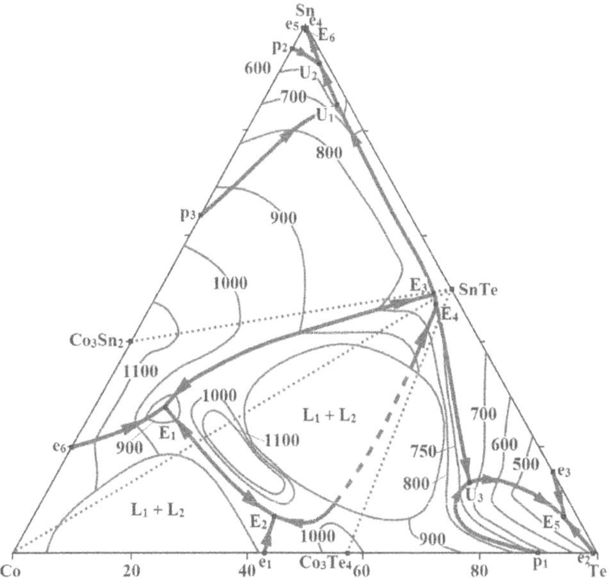

FIGURE 9.55 Liquidus surface of the Sn–Co–Te ternary system. (From Allazov, M.R., and Asadova. S.Yu., *Zhurn. neorgan. khimii*, **33**(9), 2407, 1988.)

transition points in this system: E_1 (800°C) – L ⇔ Co + Co_3Sn_2 + $Sn_4Co_9Te_4$; E_2 (850°C) – L ⇔ Co + Co_3Te_4 + $Sn_4Co_9Te_4$; E_3 (750°C) – L ⇔ Co_3Sn_2 + SnTe + $Sn_4Co_9Te_4$; E_4 (740°C) – L ⇔ Co_3Te_4 + SnTe + $Sn_4Co_9Te_4$; E_5 (380°C) – L ⇔ $CoTe_2$ + SnTe + Te; E_6 (230°C) – L ⇔ $CoSn_2$ + SnTe + Sn; U_1 (570°C) – L + Co_3Sn_2 ⇔ CoSn + SnTe; U_2 (370°C) – L + CoSn ⇔ $CoSn_2$ + SnTe; U_3 (680°C) – L + Co_3Te_4 ⇔ $CoTe_2$ + SnTe.

Using the data on binary systems, melting points of the starting components, coordinates of eutectics delimiting quasibinary systems, the liquidus surface of the SnTe–Sn–Co_3Sn_2 subsystem of the Sn–Co–Te ternary system was calculated by Mamedov et al. (1990).

9.47 Tin–Nickel–Tellurium

SnTe–Ni: When SnTe and Ni interacted, NiTe and Ni_3Sn_2 are formed (Bates et al. 1968; Markholiya et al. 1975). The interaction begins at 470°C (Kuliev et al. 1974). The ingots for the investigations were annealed at 450°C–700°C for 6–192 h (Markholiya et al. 1975) [at 1100°C for 15 h (Bates et al. 1968)].

SnTe–Ni_3Te_2: The phase diagram of this system, constructed through DTA, metallography, and measuring of the microhardness, is presented in Figure 9.56 (Movsum-zade et al. 1986a). Two ternary compounds, **$SnNi_6Te_5$** and **$Sn_3Ni_3Te_5$**, which melts incongruently at 867°Cand 727°C, respectively, are formed in this system. The eutectic contains 14 mol% Ni_3Te_2 and crystallizes at 597°C. The solubility of SnTe in Ni_3Te_2 and Ni_3Te_2 in SnTe at room temperature is 2 and 4 mol%, respectively. The ingots for the investigations were annealed at 550°C for 150 h.

SnTe–Ni_3Te_4: The phase diagram of this system, constructed through DTA, metallography, and measuring of the microhardness, is given in Figure 9.57 (Movsum-zade et al. 1984a). Two ternary compounds, **$SnNi_3Te_5$** and **$SnNi_9Te_{13}$**, which melts incongruently at 700°C and 850°C, respectively, are formed in this system. The eutectic contains 17 mol%

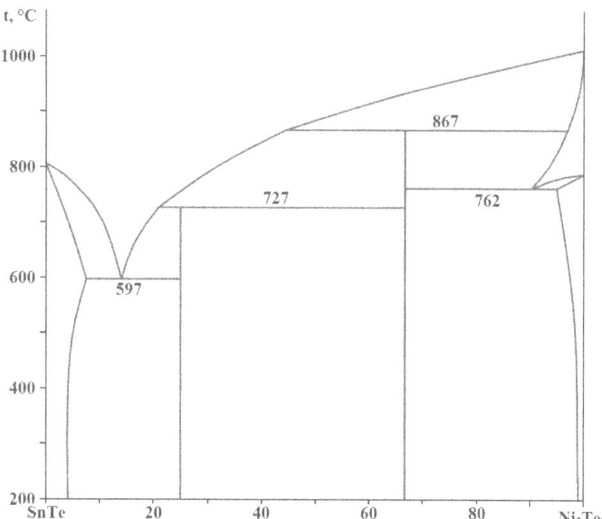

FIGURE 9.56 Phase diagram of the SnTe–Ni_3Te_2 system. (From Movsum-zade, S.A., et al., *Izv. AN SSSR. Neorgan. mater.*, **22**(3), 415, 1986.)

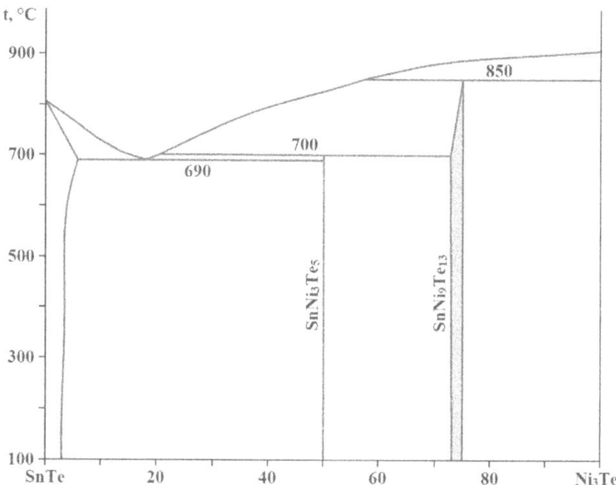

FIGURE 9.57 Phase diagram of the SnTe–Ni_3Te_4 system. (From Movsum-zade, S.A., et al., *Zhurn. neorgan. khimii*, **29**(8), 2090, 1984.)

Ni_3Te_4 and crystallizes at 690°C. The solubility of Ni_3Te_4 in SnTe reaches 3 mol%, and the solubility of SnTe in Ni_3Te_4 is negligible. The ingots for the investigations were annealed at 550°C for 150 h.

Two more ternary compounds, **$SnNi_{3-x}Te_2$** and **$SnNi_{6-x}Te_2$**, are formed in the Sn–Ni–Te system. First of them crystallizes in the hexagonal structure with the lattice parameters a = 399.65 ± 0.04 and c = 1582.0 ± 0.2 pm for $SnNi_{2.87}Te_2$ (Deiseroth et al. 2007b) and a = 393.10 ± 0.06 and c = 1579.5 ± 0.3 pm for $SnNi_{2.58}Te_2$ (Litvinenko et al. 2007). According to the data of Litvinenko et al. (2007), the phases of this type can be regarded as members of the homologous series where x ranges from 0 to 1. $SnNi_{2.87}Te_2$ was prepared in a solid-state reaction starting from intimate mixture of Ni, Sn, and Te (molar ratio 3:1:2) (Deiseroth et al. 2007b). Together with a small quantity of iodine as transport agent the educts was filled into a dry quartz ampoule, evacuated and sealed. The ampoule was placed in the furnace at 800°C for 10 days. The product consists of black hexagonally shaped crystals and forms preferably at the cold end of the ampoule at an estimated temperature of 600°C. The title compound can be handled in air.

The samples of $SnNi_{2.58}Te_2$ were prepared by high-temperature ampoule synthesis (Litvinenko et al. 2007). The mixture of Ni, Sn, and Te (molar ratio 2:1:2) were placed into quartz ampoules, evacuated to ~1 Pa, and sealed off. The sample heated to 750°C over a period of 5 h, annealed at this temperature for 10 days, and cooled in the switched off the furnace. The samples after annealing were air stable gray-colored polycrystalline cakes. The single crystals were grown under the same heating conditions but with subsequent slow (2°C·h^{-1}) cooling.

$SnNi_{6-x}Te_2$ crystallizes in the tetragonal structure with the lattice parameters a = 375.9 ± 0.1, c = 1941.0 ± 0.2 pm, and a calculated density of 8.68 g·cm^{-3} at 153 K for $SnNi_{5.81}Te_2$ (Isaeva et al. 2007b) and a = 376.80 ± 0.05, c = 1941.9 ± 0.4 pm, and a calculated density of 8.581 g·cm^{-3} at 153 K for $SnNi_{5.78\pm0.02}Te_2$ at room temperature (Baranov et al. 2004). To prepare this compound, the stoichiometric mixture of the elements was annealed in evacuated silica ampoule at 540°C for

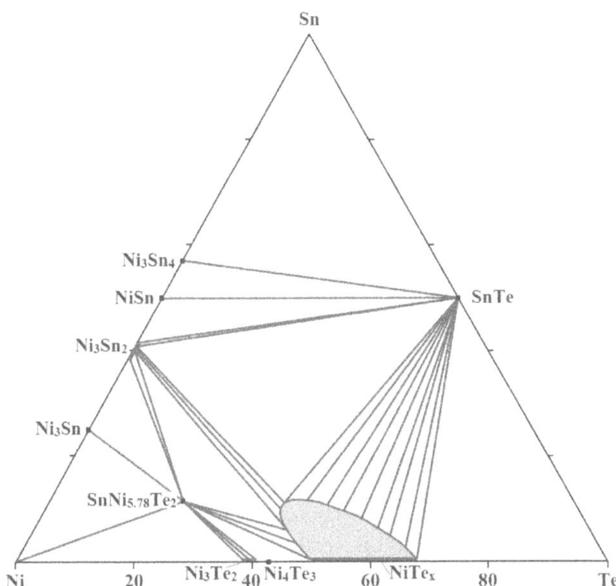

FIGURE 9.58 Isothermal sections of the Sn–Ni–Te ternary system at 540°C. (From Baranov, A.I., et al., *J. SolidState Chem.*, **177**(10), 3616, 2004.)

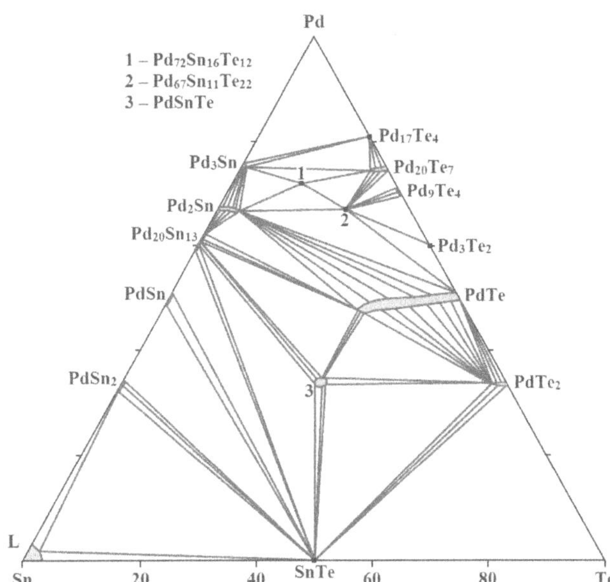

FIGURE 9.59 Isothermal sections of the Sn–Pd–Te ternary system at 400°C. (From Vymazalová, A., and Drábek, M., *Canad. Mineralog.*, **48**(5), 1051, 2010.)

7 days (Baranov et al. 2004). The product was then ground in an agate mortar, pressed into pellets and further re-annealed under the same conditions four times. The grayish powder of the title compound was obtained. The single crystals of $SnNi_{5.78\pm0.02}Te_2$ were prepared from the vapor phase by means of chemical transport reaction with I_2 as a transport agent in silica ampoule placed in a horizontal, two-temperature furnace (Baranov et al. 2004; Isaeva et al. 2007b).

The isothermal section of the Sn–Ni–Se ternary system at 540°C is given in Figure 9.58 (Baranov et al. 2004). The total annealing time (540°C, with several intermediate grindings in an agate mortar and re-annealing of products pressed into pellets) was about 10–20 days depending on whether the equilibrium had been reached.

9.48 Tin–Rhodium–Tellurium

The $\mathbf{RhSn_{1.5}Te_{1.5}}$ ternary compound, which exists in two polymorphic modifications, is formed in the Sn–Rh–Te system. First modification of this compound crystallizes in the trigonal structure with the lattice parameters $a = 1316.485 \pm 0.006$ and $c = 1612.41 \pm 0.01$ (Vaqueiro et al. 2010). Second modification crystallizes in the cubic structure with the lattice parameter $a = 930.64 \pm 0.01$ pm (Vaqueiro and Sobany 2007).

This compound was synthesized in the same way as $CoSn_{1.5}Te_{1.5}$ was prepared (Vaqueiro and Sobany 2007; Vaqueiro et al. 2010).

9.49 Tin–Palladium–Tellurium

The isothermal section of the Sn–Pd–Te ternary system at 400°C was constructed by Vymazalová and Drábek (2010) and is shown in Figure 9.59. Three ternary phases, **PdSnTe**, $\mathbf{Pd_{67}Sn_{11}Te_{22}}$, and $\mathbf{Pd_{72}Sn_{16}Te_{12}}$, exist in the system. $\mathbf{Pd_{73}Sn_{10}Te_{17}}$ and $\mathbf{Pd_{71}Sn_7Te_{22}}$ were not found. $PdTe_2$ dissolves up to 3.5 at% Sn, PdTe dissolves up to 19 at% Sn, $Pd_{20}Te_7$ dissolves up to 4 at% Sn, Pd_2Sn dissolves up to 4 at% Te, and Pd_3Sn dissolves up to 0.5 at% Te.

The composition of PdSnTe actually varies in the ranges 33.3–34.3 at% Pd, 30.8–32.6 at% Sn, and 33.6–35.6 at% Te (Vymazalová and Drábek 2010). This phase breaks down at 588°C to SnTe, $Pd_{20}Sn_{13}$ and Pd(Te,Sn) solid solution. This compound melts at 1005.59°C and crystallizes in the orthorhombic structure with the calculated lattice parameters a = 647.61, b = 646.11, and c = 644.57 pm ($a = 649.74 \pm 0.12$, $b = 647.72 \pm 0.12$, and $c = 646.96 \pm 0.14$ pm from XRD) and an energy gap of 0.1 eV (Weihrich et al. 2004a) [$a = 656.87 \pm 0.02$, $b = 660.28 \pm 0.02$, and $c = 1288.49 \pm 0.04$ pm (Laufek et al. 2009b)]. Its structure optimizations were performed with a first principles plane-wave code, which treats exchange and correlation in the local density approximation of the density functional theory (Weihrich et al. 2004a).

PdSnTe was synthesized from the elements by high-temperature solid state reactions. Equimolar amounts of Pd, Sn, and Te were heated in sealed and evacuated quartz tubes to 980°C for 1 week (Weihrich et al. 2004a) or at 400°C for 48 h with the next regrinding under acetone and new heating at 400°C for 720 h (Laufek et al. 2009b). After that, the product was rapidly quenched in a cold-water bath. The title compound was received as black powder.

$Pd_{67}Sn_{11}Te_{22}$ forms a narrow region of the solid solution (Vymazalová and Drábek 2010). The Pd content varies in the range 66.5–68.1 at%, the Sn content varies between 10.5 and 11.4 at%, and the Te content varies in the range 21.0–23.1 at%. The upper stability limit of this phase lies at ~596°C, where it decomposes peritectically forming $Pd_{72}Sn_{16}Te_{12}$ and PdTe. It crystallizes in the tetragonal structure with the lattice parameters $a = 400.6 \pm 0.1$ and $c = 2081.21 \pm 0.04$ pm (Vymazalová and Drábek 2010) [$a = 400.1 \pm 0.1$ and 400.5 ± 0.1,

$c = 2092.9 \pm 0.3$ and 2093.0 ± 0.7 pm, and a calculated density of 10.24 and 11.07 g·cm^{-3} for **$Pd_{7-\delta}SnTe_2$** at $\delta = 0.8$ and 0.5, respectively (Savilov et al. 2005); $a = 400.1 \pm 0.1$ and $c = 2092.9 \pm 0.3$ pm for mineral kojonenite, which has the same composition (Stanley and Vymazalová 2014, 2015)]. Bulk samples of this compound ($0.3 \leq \delta \leq 0.8$) were synthesized from the elements in evacuated (ca. 4 Pa) silica ampoules (Savilov et al. 2005). Two different preparation routes were used: (1) mixtures of Pd, Sn, and Te (molar ratio 6:1:2) were annealed for 1 week at 700°C; after cooling to room temperature at 4°C·h^{-1}, single crystals of the series could be isolated from the ingots; (2) single crystals of this phase were also prepared from a tin flux by heating mixtures of Pd, Sn, and Te (molar ratio 7:2:1) at 850°C for 24 h and cooling to 100°C at 2°C·h^{-1}. Square prismatic gray-colored crystals were found under both conditions.

$Pd_{72}Sn_{16}Te_{12}$ forms a limited solid solution at 400°C (Vymazalová and Drábek 2010). The Pd content varies between 71.7 and 72.2 at%, the Sn content is between 15.0 and 16.8 at%, and the Te content varies in the range 11.5–13.3 at%. This phase undergoes a polymorphic transformation at ~753°C and does not melt up to ~900°C. High-temperature modification crystallizes in the tetragonal structure with the lattice parameters $a = 764.56 \pm 0.03$, $c = 1395.75 \pm 0.09$ pm, and a calculated density of 10.71 g·cm^{-3} for **$Pd_{73}Sn_{14}Te_{13}$** composition (Laufek et al. 2007b). The evacuated silica glass tube method was used for synthesis high-temperature modification of this phase (Laufek et al. 2007b). Pd, Sn, and Te were used as starting materials for synthesis. Carefully weighed sample was loaded into a high purity silica tube, and a tightly fitting silica glass rod was placed on the top of the reagents. The evacuated tube with its charge was annealed at 1300°C for 72 h. After regrinding under acetone, the sample was heated at 800°C for 552 h. The experimental product was rapidly quenched in a cold-water bath to obtain the high-temperature phase of $Pd_{73}Sn_{14}Te_{13}$.

According to the data of Vymazalová and Drábek (2004), two more compounds, **$Pd_{45}Sn_{16}Te_{39}$** and **$Pd_{68}Sn_{10}Te_{21}$**, exist in the Sn–Pd–Te ternary system. First of them crystallizes in the monoclinic structure with the lattice parameters $a = 866.1$, $b = 972.0$, $c = 768.7$ pm, and $\beta = 72.42°$. The upper stability limit of this compound lies at 650°C. Second compound crystallizes in the orthorhombic structure with the lattice parameters $a = 1937.3$, $b = 1039.1$, and $c = 628.9$ pm. Its upper stability limit lies at 620°C.

9.50 Tin–Iridium–Tellurium

The **$IrSn_{1.5}Te_{1.5}$** ternary compound, which exists in two polymorphic modifications, is formed in the Sn–Ir–Te system. First modification of this compound crystallizes in the trigonal structure with the lattice parameters $a = 1319.386 \pm 0.004$ and $c = 1616.72 \pm 0.01$ (Vaqueiro et al. 2010) [$a = 1318.32 \pm 0.05$ and $c = 1616.00 \pm 0.05$ (Bos and Cava 2007)]. Second modification crystallizes in the cubic structure with the lattice parameter $a = 934.98 \pm 0.01$ pm (Vaqueiro and Sobany 2007).

This compound was synthesized in the same way as $CoSn_{1.5}Te_{1.5}$ was prepared (Bos and Cava 2007; Vaqueiro and Sobany 2007; Vaqueiro et al. 2010).

REFERENCES

All References are available as a downloadable eResource at www.routledge.com/9780367639235

10

Systems Based on Lead Sulfide

10.1 Lead–Sodium–Sulfur

PbS–Na_2S: The phase diagram of this system, constructed through DTA, XRD, and metallography, is given in Figure 10.1 (Smirnov and Kudryashov 1956; Kopylov 1967). The eutectic contains 54.4 mol% (28 mass%) Na_2S (Smirnov and Kudryashov 1956) [58.5 mol% (31.5 mass%) Na_2S (Kopylov 1967) and crystallizes at 525°C (Smirnov and Kudryashov 1956). The **$Na_2Pb_3S_4$** ternary compound, which melts incongruently at 575°C and crystallizes in the cubic structure with the lattice parameter a = 586–587 pm, is formed in the system. The solubility of Na_2S in PbS at the eutectic temperature reaches 9 mol% (7 mass%).

10.2 Lead–Copper–Sulfur

PbS–Cu: This system is non-quasibinary section of the Pb–Cu–S ternary system (Wagner and Wagner 1957; Craig and Kullerud 1968; Barton and Skipper 1970). At temperatures above 250°C, copper enters into an exchange chemical reaction with PbS. The system was investigated through DTA, XRD, and metallography.

PbS–Cu_2S: The phase diagram of this system, constructed through DTA, XRD, and metallography, is a eutectic type with formation of the **$7Cu_2S{\cdot}2PbS$** [or **$Cu_{14}Pb_2S_{9-x}$** ($0 < x < 0.15$) (Craig and Kullerud 1965–1966, 1968)] ternary compound (Figure 10.2) (Godovikov and Fedorova 1969; Kopylov et al. 1976). This compound melts incongruently at 528°C, decomposes at 486°C, and has an experimental density of 8.42 g·cm^{-3} (Craig and Kullerud 1968). According to the data of Johto and Taskinen (2013), the composition of this compound is **$Cu_{16}Pb_2S_{10}$** or **$8Cu_2S{\cdot}2PbS$** and it exists within the temperature interval 490°C–515°C. The eutectic contains 37.5 mol% PbS and crystallizes at 525°C. The solubility of PbS in Cu_2S at 500°C is not higher than 1 mol%.

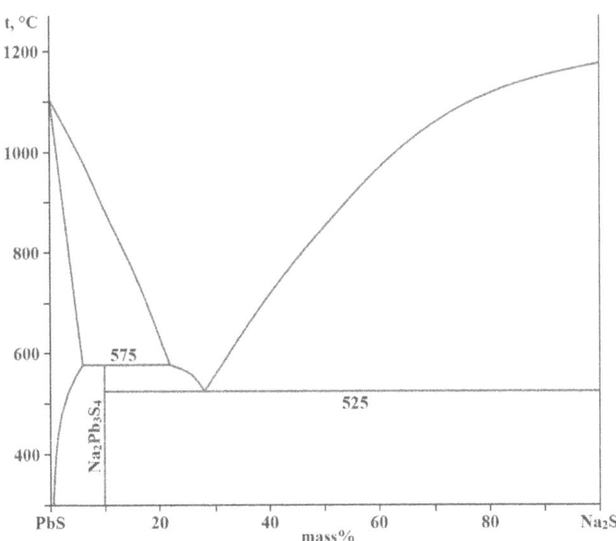

FIGURE 10.1 Phase diagram of the PbS–Na_2S system. (From Smirnov, M.P., and Kudryashov, L.N., *Tsvet. Met.*, (12), 36, 1956.)

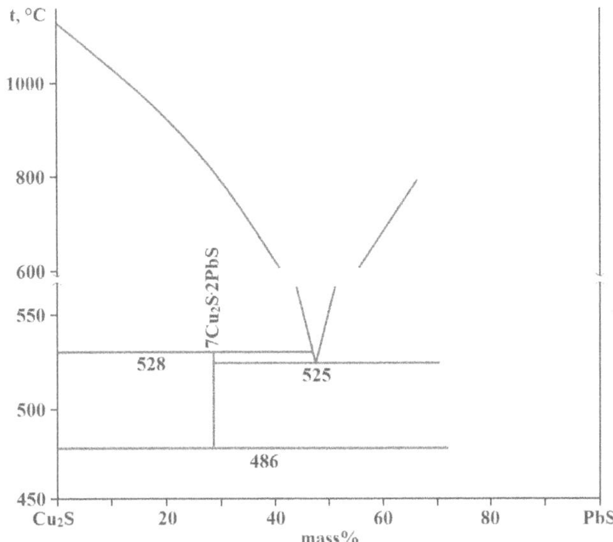

FIGURE 10.2 Phase diagram of the PbS–Cu_2S system. (From Kopylov, N.I., et al., *Izv. AN SSSR. Ser. Metally*, (3), 226, 1976.)

Measurements of vapor pressure above the molten PbS–Cu_2S have been made by the effusive Knudsen method (Botor et al. 1990). The measurements were carried out under isothermal conditions, at 769°C, 808°C, and 830°C in the PbS concentration range 80–20 mass%. High-temperature mass spectrometric measurements proved that the predominant gas phase constituent comprised molecules of PbS(g). The temperature and concentration dependence of the PbS activity in molten PbS–Cu_2S has been described by the Redlich-Kister equation. The results obtained indicate negative deviation from an ideal solution (Eriç and Timuçin 1981; Botor et al. 1990). The value of the activity coefficient of PbS evaporation increases with an increase in its concentration in the melt and decreases with increasing temperature (Isakova et al. 1964). The enthalpy of its evaporation is 215.1 kJ·M^{-1}, and the entropy of this process changes for the alloys from 95.23 to 133.09 J·M^{-1}K^{-1}.

Thermodynamics of copper mattes in the PbS–Cu_2S system and copper saturated-mattes in PbS–Cu–Cu_2S system were studied in the temperature range from 1150°C to 1250°C

DOI: 10.1201/9781003123507-10

(Azuma et al. 1970). The vapor pressures of PbS or Pb in these mattes of various compositions were measured by the dew point method and the thermodynamic functions were calculated from this experimental data. In this system, the activity curve of PbS deviates positively from Raoult's law in the mole fraction range of PbS from 0 to 0.3, negatively from Raoult's law in the mole fraction range of PbS from 0.3 to 1 and the activity curve of Cu_2S deviates positively from Raoult's law. The values of activity coefficient, γ are 0.3–3.0 for PbS and 1.0–1.4 for Cu_2S. By using these activity data, free energy change, enthalpy and entropy change of mixing, and chemical potentials of PbS and Cu_2S in the PbS–Cu_2S system were obtained. In the PbS–Cu–Cu_2S subsystem, the activity curve of Pb deviates from Raoult's law and activity coefficient was found to be 1.4–3.7 at 1200°C. The distribution of Pb into two liquid phases, matte phase and metal phase, in this subsystem was measured by chemical analysis and it was shown that at 1200°C the distribution coefficients are 3.8 and 9.6, respectively.

The parts of the phase diagrams of the **PbS–$Cu_{1.96}S$** and **PbS–$Cu_{1.8}S$** systems from the side of the copper chalcogenides were constructed by Kopylov et al. (1976). The eutectics in these systems crystallize at 550°C and 565°C, respectively.

The isothermal section of the Pb–Cu–S ternary system at 300°C was constructed by Gulay (2008). The formation of ternary compounds has been not observed. It was shown that at this temperature the system is divided for four subsystems: PbS–CuS–S, PbS–CuS–$Cu_{2-x}S$, PbS–$Cu_{2-x}S$–Pb, and $Cu_{2-x}S$–Cu–Pb.

At 1130°C, three homogeneous liquid fields, which are divided by two immiscibility regions, exist in the Pb–Cu–S ternary system (Craig and Kullerud 1968). The part of the liquidus surface in the area of the PbS–Cu_2S–$Cu_{1.8}S$ subsystem (Figure 10.3) is characterized by the presence of a distinct maximum (saddle point **M**) at 570°C (Kopylov 1976).

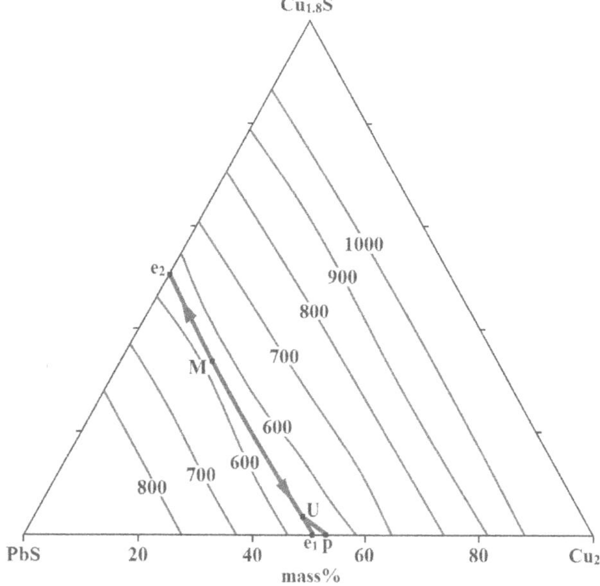

FIGURE 10.3 Liquidus surface of the PbS–Cu_2S–$Cu_{1.8}S$ subsystem of the Pb–Cu–S ternary system. (From Kopylov, N.I., et al., *Izv. AN SSSR. Ser. Metally*, (3), 226, 1976.)

Phase relations in the ternary Pb–Cu–S system have been experimentally investigated by Goto et al. (1983) at temperatures from 600°C to 1200°C through 100°C within the area between Pb–Cu and PbS–Cu_2S joins. At 1100°C and 1200°C, a miscibility gap between a sulfide liquid and a metallic liquid dominates the area (Goto et al. 1983; Eric and Timucin 1989). Tie line distributions over the miscibility gap have been obtained at 1200°C. The slope of the immiscibility surface is very steep in the temperature range investigated, especially the surface standing almost perpendicular to the isothermal planes on the liquid sulfide side of the miscibility gap in the close vicinity of the PbS–Cu_2S join. A solid solution region with Cu_2S–Cu_9S_5 as the main constituent has been confirmed to exist in the area.

The ternary miscibility gap and tie line distributions in the Cu-poor region of the Pb–Cu–S ternary system were investigated at 1150°C and 1200°C (Goto et al. 1982). The result obtained at 1150°C was almost the same as obtained at 1200°C within the experimental errors and it was concluded that the miscibility gap boundary rises sharply at the temperature range from 1150°C to 1200°C. Thermodynamic activities of Pb, Cu, and S were calculated from the tie line distributions at 1200°C (Goto et al. 1982; Goto and Ogawa 1983).

A consistent set of thermodynamic activities of Pb, Cu, and S in the Pb–Cu–S system was derived at 1200°C from a knowledge of the high-temperature phase relations, the boundary binary thermodynamics and the estimated limiting activity coefficients of S in Cu–Pb melts (Choudary et al. 1977). The derived and the experimental values for the activity of Pb are in good agreement. The tie lines distribution in the Pb-rich portion of the ternary miscibility gap has been slightly modified to be consistent with the thermodynamic properties of the Pb–S system.

The thermodynamic properties of the Pb–Cu–S ternary system between 490°C and 1300°C were critically assessed and optimized using the experimental results and critically evaluated literature data by Johto and Taskinen (2013). The isothermal sections of this system at 600°C, 1000°C, and 1200°C were calculated.

The spatial diagram was constructed for the PbS–Cu_2S–Pb subsystem, and for it, as well as for the Cu–Cu_2S–Pb subsystem of the Pb–Cu–S ternary system, the equilibrium ratios of all possible alloys from the beginning to the end of crystallization were determined by Guertler and Landau (1934).

The new thermodynamic modeling of this ternary system is presented by Shishin et al. (2020). All available experimental data were collected, assessed and used to optimize the model parameters. The isothermal sections at 900°C, 1000°C, 1100°C, and 1200°C were calculated as well as a liquidus projection in the Pb corner of the system.

The effect of Cu upon the thermodynamic activity of S in Pb solution was determined in the temperature range from 600°C to 900°C by a two-furnace closed gas circulation system using a H_2+H_2S gas mixture (Grant and Russell 1970). At 600°C, no significant departure from a linear dependence of $lg\gamma_S$ upon the atom fraction in Pb was evident up to 3 at% Cu and within the sensitivity of the experiments.

Equilibrium experiments were also performed between molten Pb and PbS+Cu_2S mixed sulfides in the range from

550°C to 800°C by Michimoto et al. (2007). After cooling, a lead sample was analyzed for Cu and S. A good linear relationship between logarithms of Cu concentration and reciprocal values of absolute temperatures was obtained. In the case of sulfur, the similar tendency was obtained.

10.3 Lead–Silver–Sulfur

PbS–Ag: This system is non-quasibinary section of Pb–Ag–S ternary system (Vogel 1953; Craig 1967). At 732°C, four phases are in equilibrium: $L_1 + L_2 \Leftrightarrow PbS + Ag$. The system was investigated through DTA, XRD, and metallography.

The temperature dependences of the sulfur solubility determined by the radioisotope method for PbS samples containing an excess of sulfur (c_1) and an excess of lead (c_2) have the following form (Slinkina et al. 1985): c_1 (cm^{-3}) = 1.8×10^{20} exp[–(0.22 ± 0.16) kJ·M^{-1}/RT] and c_2 (cm^{-3}) = 4.1×10^{25} exp[–(109 ± 2) kJ·M^{-1}/RT].

PbS–Ag$_2$S: The phase diagram of this system, constructed through DTA, XRD, and metallography, is a eutectic type (Figure 10.4) (Urazov and Sokolova 1941; Murach 1947; Van Hook 1960). The eutectic contains 25 mol% PbS and crystallizes at 605°C. The temperature of the Ag_2S high-temperature transformation decreases from 590°C to 455°C, and the temperature of low-temperature transformation (175°C) does not change. The solubility of PbS in Ag_2S at the eutectic temperature is 3 mol% and decreases to 2 mol% at 455°C, and the solubility of Ag_2S in PbS at 700°C is not higher than 0.4 mol%.

The isothermal section of the Pb–Ag–S ternary system at 300°C was constructed by Gulay (2008). The formation of ternary compounds has been not observed. It was shown that at this temperature the system is divided for three subsystems: PbS–Ag_2S–S, PbS–Ag_2S–Ag, and PbS–Ag––Pb.

The scheme of the liquidus surface of the PbS–Ag_2S–Ag–Pb subsystem, which is the part of the Pb–Ag–S ternary system, is characterized by the intersection of the immiscibility region by a eutectic line (Vogel 1953).

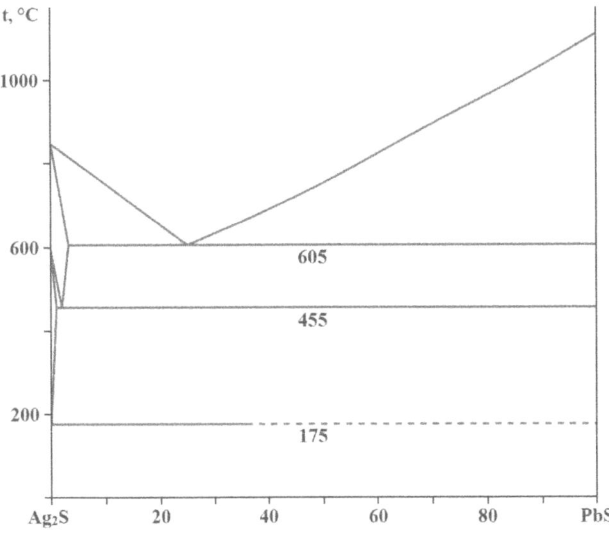

FIGURE 10.4 Phase diagram of the PbS–Ag_2S system. (From Van Hook, H.J., *Econ. Geol.*, **55**(4), 759, 1960.)

10.4 Lead–Magnesium–Sulfur

PbS–MgS: The structural, electronic, and optical properties of **$Mg_xPb_{1-x}S$** solid solutions for $0 \le x \le 1$ in their cubic structure have been calculated using the full-potential linearized augmented plane wave method under the framework of density functional theory (Chattopadhyaya and Bhattacharjee 2017b). The calculated lattice parameters of these solid solutions exhibit significant upward deviations from Vegard's law with bowing parameters –26.8 pm. The non-liner behavior of the energy gap versus concentration for the $Mg_xPb_{1-x}S$ solid solutions can be fitted by the next expression: $E_g = 1.094 - 3.903x + 6.398x^2$.

10.5 Lead–Calcium–Sulfur

PbS–CaS: The CaS phase can accommodate very little solubility of PbS (only about 1 mol% at 630°C); on the contrary, the solid-state solubility of CaS in the PbS phase can be as high as 6 mol% at 330°C (Hao et al. 2014). The bulk thermodynamic equilibrium in the PbS–CaS phase diagram is an incoherent immiscibility gap. The **$Ca_xPb_{1-x}S$** (x = 0, 0.10, 0.25, 0.50, 0.75, 0.90, and 1.0) samples were synthesized by a melting reaction using elemental Pb, Ca, and S inside carbon-coated fused quartz tube in a N_2-filled glove box. The tube was then evacuated to a pressure of ca. 10^{-2} Pa, flame-sealed, slowly heated to 450°C in 12 h, then to 1150°C in 7 h, soaked at this temperature for 6 h, and subsequently furnace cooled to room temperature.

10.6 Lead–Strontium–Sulfur

PbS–SrS: There are three predicted ordered phases in the phase diagram, corresponding to three T = 0 K ground states with stoichiometry: **$SrPb_3S_4$**, **$SrPbS_2$**, and **Sr_3PbS_4** (Hao et al. 2014). At finite temperatures, each of these phases is stable for a range of off-stoichiometric compositions. Also, each phase undergoes and order–disorder transformation into a stable, disordered rock-salt solid solution at high temperature. There are also several regions of stable two-phase coexistence. The existence of the $SrPbS_2$ and $SrPb_3S_4$ ordered phases were confirmed by successful experimental synthesis and validation using transmission electron microscopy and band gap measurements. The energy gap is 1.02, 0.68 and 1.47 eV for $SrPb_3S_4$, $SrPbS_2$, and Sr_3PbS_4, respectively. The **$Sr_xPb_{1-x}S$** (x = 0, 0.10, 0.25, 0.50, 0.75, 0.90, and 1.0) samples were synthesized in the same way as $Ca_xPb_{1-x}S$ samples were prepared.

The structural, electronic and optical properties of $Sr_xPb_{1-x}S$ solid solutions for $0 \le x \le 1$ in their cubic structure have been calculated using the full-potential linearized augmented plane wave method under the framework of density functional theory (Chattopadhyaya and Bhattacharjee 2017a). The calculated lattice parameter of these solid solutions exhibits marginal downward deviation from Vegard's law with a bowing parameter 0.29 pm. The non-liner behavior of the energy gap versus concentration was determined, and it was shown that the energy gap increases with increasing x.

10.7 Lead–Barium–Sulfur

PbS–BaS: The structural, electronic, and optical properties of $\mathbf{Ba_xPb_{1-x}S}$ solid solutions for $0 \le x \le 1$ in their cubic structure have been calculated using the full-potential linearized augmented plane wave method under the framework of density functional theory (Chattopadhyaya and Bhattacharjee 2017c). The calculated lattice parameter of these solid solutions exhibits its tendency to Vegard's law with a marginal downward bowing parameter 2.97 pm. The non-liner behavior of the energy gap vs. concentration was determined, and it was shown that the energy gap increases with increasing x.

10.8 Lead–Zinc–Sulfur

PbS–ZnS: The phase diagram of this system, constructed through DTA, XRD, and metallography, is a eutectic type (Figure 10.5) (Murach 1947; Dutrizac 1980). The eutectic composition and temperature are 13 mol% and 1050°C, respectively. The solubility of ZnS in PbS, determined by diffusion saturation, is equal to 0.6 and 1.5 mol% at 800°C and 900°C, respectively. The estimation of solubility according to the Vegard's law gives the value of 2.1 mol% ZnS at 1000°C (Bundel' et al. 1969).

Equilibrium between ZnS crystals and PbS vapor at 800°C–1000°C was investigated by Zubkovskaya and Vishniakov (1981). It was shown that the absorption of PbS is a complex process that includes both the dissolution of PbS in the bulk of the crystal and its binding by their surface. The adsorption isotherms of PbS vapor on ZnS were measured, and the value of the solubility of PbS in ZnS was determined: 1.5×10^{-5} g Pb/g ZnS at 975°C and (6–8) $\times 10^{-6}$ g Pb/g ZnS at 800°C.

10.9 Lead–Cadmium–Sulfur

PbS–CdS: The phase diagram of this system, constructed through DTA, XRD, and metallography, is a eutectic type (Figure 10.6) (Calawa et al. 1972; Stetiu 1973; Oleinik et al. 1983). The eutectic composition and temperature are 38 mol% CdS and 1050°C, respectively (Oleinik et al. 1983) [40 ± 2 mol% CdS and 1054°C (Odin 2001), 42 mol% CdS, and 1052°C (Calawa et al. 1972)]. The solubility of CdS in PbS at 600°C and 1000°C is equal correspondingly to 4 and 31.5 mol. % (Oleinik et al. 1983). According to the data of Bethke and Barton (1971), the solubility of CdS in PbS within the interval of 400°C–930°C can be described as follows: log $x_{\mathrm{CdS}} = 5.216$ log $T - 14.677$. Lattice parameters of solid solutions change according to Vegard's law (Sood et al. 1978). The solubility of PbS in CdS is insignificant (Calawa et al. 1972).

The solid solutions over the entire range of concentrations form in this system when they are synthesized under 2–3 GPa and 600°C–800°C for 1–4 h (Kobayashi et al. 1979; Susa et al. 1980). Supersaturated and metastable $Cd_xPb_{1-x}S$ solid solutions at $0.03 < x < 0.18$ could be synthesized at low temperature (Maskayeva et al. 2003). The higher-temperature limit of the stability of the $Cd_xPb_{1-x}S$ solid-solution-deposited films is 132°C–137°C.

Deshmukh et al. (1994) employed a modified chemical deposition for the preparation of $(CdS)_x(PbS)_{1-x}$ thin-film composites with $0.2 \le x \le 0.8$. $CdSO_4$, $Pb(CH_3COO)_2$, and thiourea were used as the basic source materials. It was shown that these composites include hexagonal and cubic CdS, cubic CdO, and PbS, tetragonal PbO, and PbO_2, and free elemental Cd and S. The content of hexagonal CdS was increased, while the content of cubic PbS was decreased as x increased from 0.2 to 0.8.

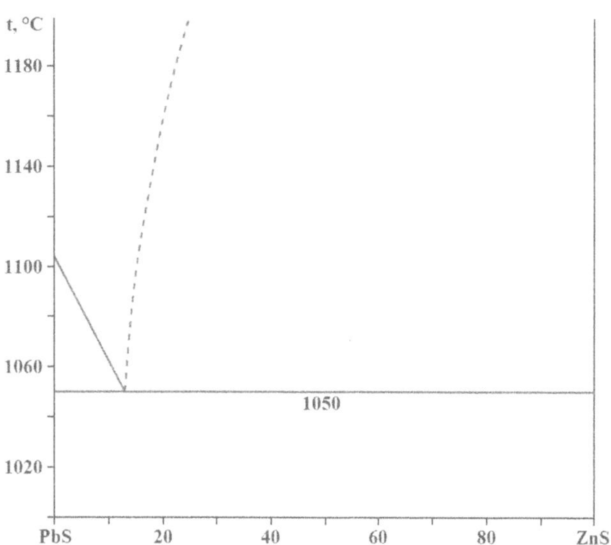

FIGURE 10.5 Phase diagram of the PbS–ZnS system. (From Dutrizac, J.E., *Can. J. Chem.*, **58**(7), 739, 1980.)

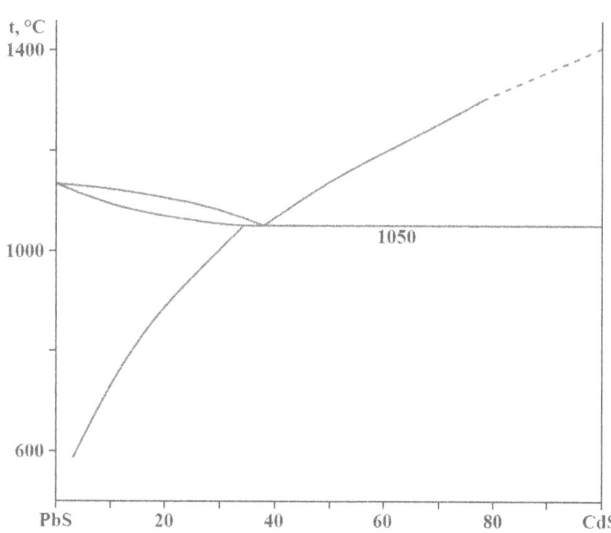

FIGURE 10.6 Phase diagram of the PbS–CdS system. (From Oleinik, G.S., et al., *Izv. AN SSSR. Neorgan. mater.*, **19**(11), 1799, 1983.)

The solubility of CdS in PbS was determined by saturation of PbS from mixtures CdS + PbS or Cd + Pb + S (Bethke and Barton 1971). The ingots were annealed at 1000°C for 10 days (Calawa et al. 1972).

10.10 Lead–Mercury–Sulfur

PbS–HgS: The phase diagram of this system, constructed through DTA, XRD, and metallography, is a eutectic type (Figure 10.7) (Kulakov and Sokolovskaya 1975). The eutectic composition and temperature are 32 mol% PbS and 676°C, respectively. The solubility of HgS in PbS increases from 5 mol% at 340°C to 11 mol% at the eutectic temperature. The solubility of PbS in β-HgS is less than 0.5 mol%, but such small solubility increases the temperature of polymorphous transformation for HgS from 344°C (Kulakov 1975) to 351 ± 2°C (Kulakov and Sokolovskaya 1975).

Sharma et al. (1977b) obtained a metastable solid-solution $Hg_xPb_{1-x}S$ in the form of thin films within the interval of $0 \le x \le 0.33$.

10.11 Lead–Boron–Sulfur

$PbS–B_2S_3$: The **$Pb_2B_2S_5$** ternary compound, which decomposes at temperatures higher than 600°C and crystallizes in the tetragonal structure with the lattice parameters a = 960, c = 1547 pm, and the calculated and experimental densities of 5.55 and 5.40 g·cm^{-3}, respectively (Hagenmuller et al. 1966) [a = 961.5 ± 0.5 and c = 1551.2 ± 0.5 pm and the calculated and experimental densities of 5.55 and 5.40 g·cm^{-3}, respectively (Hardy 1968)], is formed in the Pb–B–S system. This compound in finely dispersed form is a red powder that slowly hydrolyzes in air.

If this compound is formed at temperatures above 600°C, it crystallizes in the monoclinic structure with the lattice parameters $a = 650 \pm 2$, $b = 665 \pm 2$, $c = 1768 \pm 5$ pm, $\beta = 115.8 \pm 0.3°$, and the calculated and experimental densities of 5.75 and 5.6 g·cm^{-3}, respectively (Thomas and Tridot 1964).

10.12 Lead–Aluminum–Sulfur

$PbS–Al_2S_3$: The **$PbAl_2S_4$** ternary compound, which has two polymorphic modifications, is formed in the Pb–Al–S system (Eholié et al. 1971). One of them crystallizes in the orthorhombic structure with the lattice parameters a = 1018, b = 1030, c = 604.8 pm (Eholié et al. 1971). Second modification of this compound crystallizes in the tetragonal structure with the lattice parameters a = 1032, c = 595.7 pm, and a calculated density of 4.08 g·cm^{-3} (Eholié et al. 1971) [$a = 516.0 \pm 0.5$, $c = 595.8 \pm 0.3$ pm, and the calculated and experimental densities of 4.05 and 4.17 g·cm^{-3}, respectively, (Flahaut 1952)]. This compound was obtained by the heating of the mixture of PbS and Al in a stream of H_2S at 1200°C (Eholié et al. 1971). The annealing of $PbAl_2S_4$ at 1000°C or 600°C with the next quenching gives the possibility to obtain its high- and low-temperature modification.

10.13 Lead–Gallium–Sulfur

PbS–GaS: The phase diagram of this system, constructed through DTA, metallography, and measuring of the microhardness, is a eutectic type (Figure 10.8) (Melikova and Rustamov 1982). The eutectic contains 44 mol% PbS and crystallizes at 750°C. The solubility of PbS in GaS reaches 4 mol% and the solubility of GaS in PbS is 2 mol%. The samples for the investigation were annealed at temperatures 330°C–350°C below the solidus temperature for 500 h.

$PbS–Ga_2S_3$: The phase diagram of this system, constructed through DTA, XRD, metallography, and measuring of the microhardness, is presented in Figure 10.9 (Golovey et al.

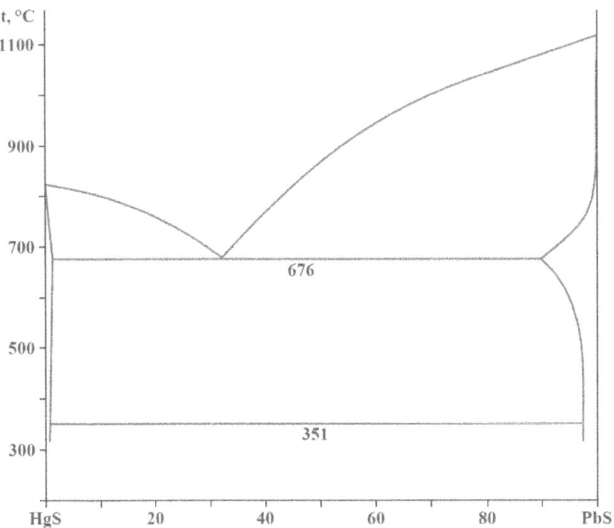

FIGURE 10.7 Phase diagram of the PbS–HgS system. (From Kulakov, M.P., and Sokolovskaya, Zh.D., *Zhurn. neorgan. khimii*, **20**(8), 2290, 1975.)

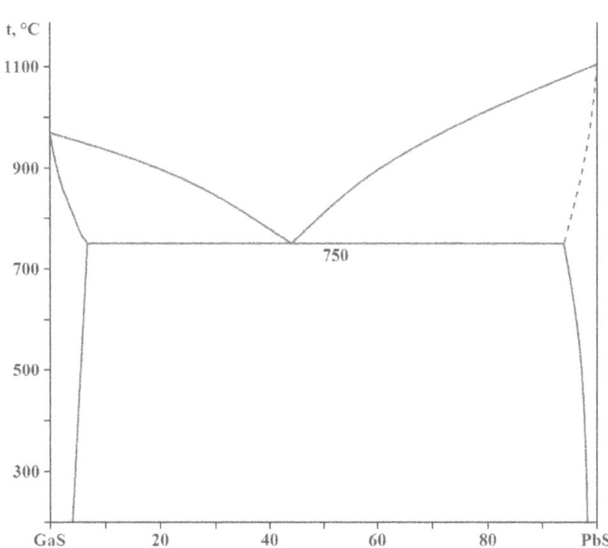

FIGURE 10.8 Phase diagram of the PbS–GaS system. (From Melikova, Z.D., and Rustamov, P.G., *Zhurn. neorgan. khimii*, **27**(5), 1330, 1982.)

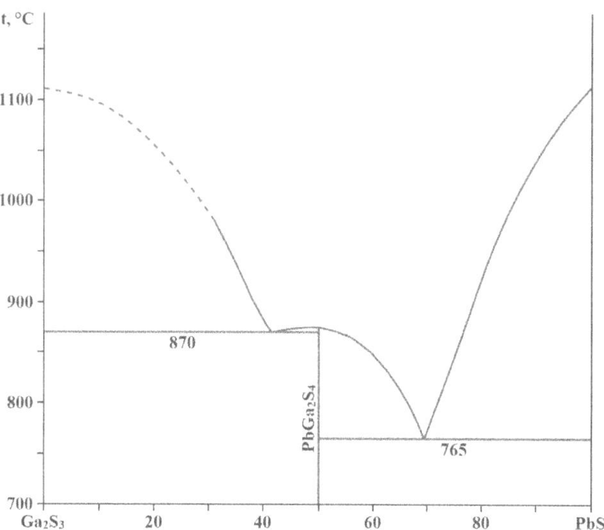

FIGURE 10.9 Phase diagram of the PbS–Ga_2S_3 system. (From Golovey, V.M., et al., *Zhurn. neorgan. khimii*, **26**(7), 1976, 1981.)

1981a; Melikova and Rustamov 1983). The eutectics contain 41.5 and 69.5 mol% PbS and crystallize at 870°C and 765°C, respectively (Golovey et al. 1981a) [42 and 66 mol% PbS and crystallize at 750°C and 660°C, respectively (Melikova and Rustamov 1983). The solubility of PbS in Ga_2S_3 and Ga_2S_3 in PbS is 2 and 1 mol%, respectively. The samples for the investigation were annealed at temperatures 50°C below the eutectic temperatures (Golovey et al. 1981a) [at temperatures 330°C–350°C below the solidus temperature for 500 h (Melikova and Rustamov 1983)].

The $\mathbf{PbGa_2S_4}$ ternary compound, which melts congruently at 875°C, is formed in this system (Golovey et al. 1981a,b; Melikova and Rustamov 1983). The homogeneity region of this compound is within the interval from 49.4 to 50.9 mol% PbS (Figure 10.10), and the maximum melting temperature corresponds to its stoichiometric composition (Golovey et al. 1981b). $PbGa_2S_4$ crystallizes in the orthorhombic structure with

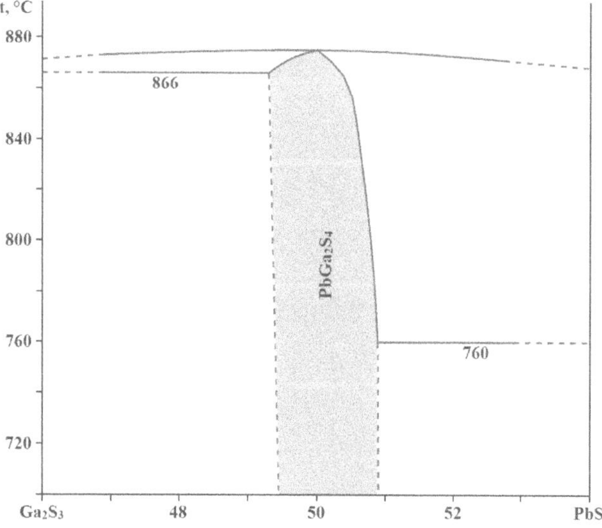

FIGURE 10.10 Homogeneity region of $PbGa_2S_4$. (From Golovey, V.M., et al., *Izv. AN SSSR. Neorgan. mater.*, **17**(3), 540, 1981.)

the lattice parameters a = 2071.2 ± 0.6, b = 2043.1 ± 0.4, and c = 1216.3 ± 0.3 pm (Filyuk et al. 2007) [a = 2070.6 ± 0.8, b = 2038.0 ± 0.9, c = 1315.6 ± 0.9 pm, and the calculated and experimental densities of 4.92 and 4.6 g·cm^{-3}, respectively (Peters and Baglio 1972); a = 2044, b = 2064, c = 1209 pm, and a calculated density of 4.94 g·cm^{-3} (Chilouer et al. 1979); a = 2069, b = 2051, c = 1223 pm, an experimental density of 4.80 g·cm^{-3} and an energy gap of 2.67 eV (Golovey et al. 1981b); a = 1216 ± 3, b = 2040 ± 5, c = 2070 ± 5 pm, a calculated density of 4.915 g·cm^{-3} and an energy gap of 2.38 eV (Wu et al. 2015a)].

To prepare the crystals of $PbGa_2S_4$, the mixtures of PbS (1 mM) and Ga_2S_3 (1 mM) were heated to 1000°C in 30 h, left for 72 h, cooled to 500°C at a rate of 2°C·h^{-1}, and finally cooled to room temperature by switching off the furnace (Wu et al. 2015a). Many block-shaped crystals with yellow color were found in the ampoule and the color from yellow to yellow-brown as it increases in thickness. These crystals are stable in air. Single crystals of this compound were also obtained by vertical crystallization (Golovey et al. 1981b).

Polycrystalline samples were synthesized by solid-state reaction techniques (Wu et al. 2015a). The mixture of PbS and Ga_2S_3 (molar ratio 1:1) was heated to 800°C in 20 h and kept at that temperature for 72 h, and then the furnace was turned off.

This compound could be prepared if stoichiometric quantities of PbS and Ga_2S_3 was thoroughly mixed by grinding in a glass mortar and the resulting blend was subjected to a solid-state reaction at 900°C–1000°C in H_2S atmosphere (Peters and Baglio 1972). Under these conditions, the reaction usually resulted in the formation of $PbGa_2S_4$ in about 2 h. The compound was obtained in the form of a sintered cake.

The vapor pressure over the PbS–Ga_2S_3 system was measured by the Knudsen-effusion and torsion-effusion methods in the temperature range of 625°C–994°C (Williamson and Edwards 1986.) The four ternary compounds, $PbGa_2S_4$, $\mathbf{Pb_2Ga_2S_5}$, $\mathbf{Pb_2Ga_6S_{11}}$, and $\mathbf{Pb_4Ga_6S_{13}}$, were found to vaporize incongruently. $Pb_2Ga_6S_{11}$ and $Pb_4Ga_6S_{13}$ exist only at the high temperatures of the experiments. They were not found at room temperature because, as the samples were quenched, the compounds disproportionate into $PbGa_2S_4$ and Ga_2S_3 in the case of $Pb_2Ga_6S_{11}$, and $PbGa_2S_4$ and $Pb_2Ga_2S_5$ in the case of $Pb_4Ga_6S_{13}$. Enthalpies of ternary compounds with respect to those of the binary constituents PbS(s) and Ga_2S_3(s) were (per mol of PbS): ΔH^0_{298} = –35,43 ± 3.0 kJ·M^{-1} for $PbGa_2S_4$, ΔH^0_{298} = –20,18 ± 2.0 kJ·M^{-1} for $Pb_2Ga_2S_5$, ΔH^0_{298} = –39,90 ± 4.0 kJ·M^{-1} for $Pb_2Ga_6S_{11}$, and ΔH^0_{298} = –28,58 ± 3.0 kJ·M^{-1} for $Pb_4Ga_6S_{13}$.

The existence of $PbGa_2S_4$ and $Pb_2Ga_2S_5$ were also confirmed by Chilouer et al. (1979), Mazurier et al. (1980), Filyuk et al. (2007), and Li et al. (2015). First of them melts incongruently and crystallizes in the monoclinic structure with the lattice parameters a = 728.1 ± 0.1, b = 633.9 ± 0.1, c = 1238.5 ± 0.3 pm, β = 105.97 ± 0.03°, and a calculated density of 4.293 g·cm^{-3} at 153 K and an energy gap of 3.08 eV (Li et al. 2015). This compound was synthesized by the crystallization from the Bi_2S_3 flux. For this, powders of PbS, Ga_2S_3, and Bi_2S_3 were mixed (molar ratio 1:2:1) and then sealed in a fused-silica tube under an Ar atmosphere in a glove box. The sample was heated to 1000°C in 24 h and kept at this temperature for 48 h, then cooled at a slow rate of 3°C·h^{-1} to 350°C, and finally cooled to room temperature naturally. Yellow crystals of the title compound were obtained.

$Pb_2Ga_2S_5$ crystallizes in the orthorhombic structure with the lattice parameters a = 1240.7 ± 0.2, b = 1196.6 ± 0.9, and c = 1101.8 ± 0.1 pm (Filyuk et al. 2007) [a = 1239, b = 1190, c = 1103 pm, and the calculated and experimental densities of 5.85 and 5.99 g·cm^{-3}, respectively (Chilouer et al. 1979; Mazurier et al. 1980)].

10.14 Lead–Indium–Sulfur

PbS–In$_2$S$_3$: The phase diagram of this system, constructed through DTA, XRD, metallography, and measuring of the microhardness, is shown in Figure 10.11 (Rustamov et al. 1976d; Krämer and Berroth 1980). The solubility of PbS in In_2S_3 reaches 2–3 mol% and the solubility of In_2S_3 in PbS is 1 mol%. With an increase in the PbS content, the temperature of the phase transformations of solid solutions based on In_2S_3 increases from 420°C to 810°C and from 755°C to 850°C (Krämer and Berroth 1980). The eutectics contain 40 and 70 mol% PbS and crystallize 890°C and 810°C, respectively (Krämer and Berroth 1980) [30 and 70 mol% PbS and crystallize 880°C and 815°C, respectively (Rustamov et al. 1976d). The ingots for investigations were annealed at 700°C or 1000 h (Rustamov et al. 1976d).

Two compounds, **PbIn$_2$S$_4$** and **Pb$_6$In$_{10}$S$_{21}$**, are formed in this system (Hahn and Klingler 1950; Rustamov et al. 1976d; Krämer and Berroth 1980). $PbIn_2S_4$ melts congruently at 895°C (Krämer and Berroth 1980) [at 930°C (Rustamov et al. 1976d)] and crystallizes in the orthorhombic structure with the lattice parameters a = 1168.8 ± 0.1, b = 385.28 ± 0.01, and c = 1376.3 ± 0.1 pm (Krämer and Berroth 1980; Likforman et al. 1989) [in the monoclinic structure with the lattice parameters a = 1267, b = 398, c = 1435 pm, β = 95°, and the calculated and experimental densities of 5.23 and 5.5 g·cm^{-3}, respectively (Rustamov et al. 1976d)].

$Pb_6In_{10}S_{21}$ melts congruently at 905°C and crystallizes in the monoclinic structure with the lattice parameters a = 2762.9 ± 0.3, b = 386.30 ± 0.05, c = 1570.5 ± 0.2 pm, and β = 95.9° (Krämer and Berroth 1980; Likforman et al. 1989).

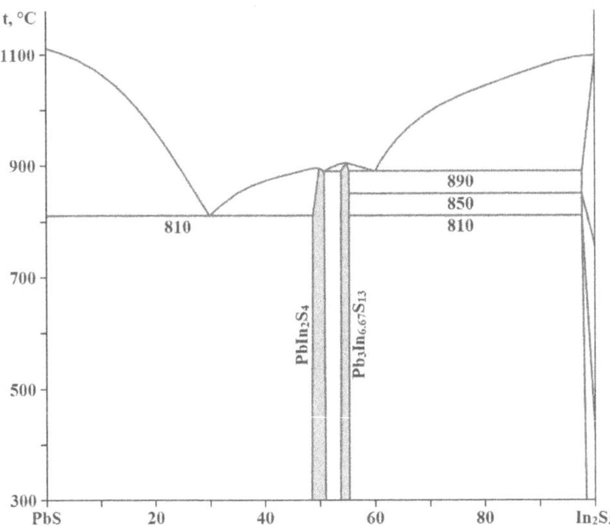

FIGURE 10.11 Phase diagram of the PbS–In$_2$S$_3$ system. (From Krämer, V., and Berroth, K., *Mater. Res. Bull.*, **15**(3), 299, 1980.)

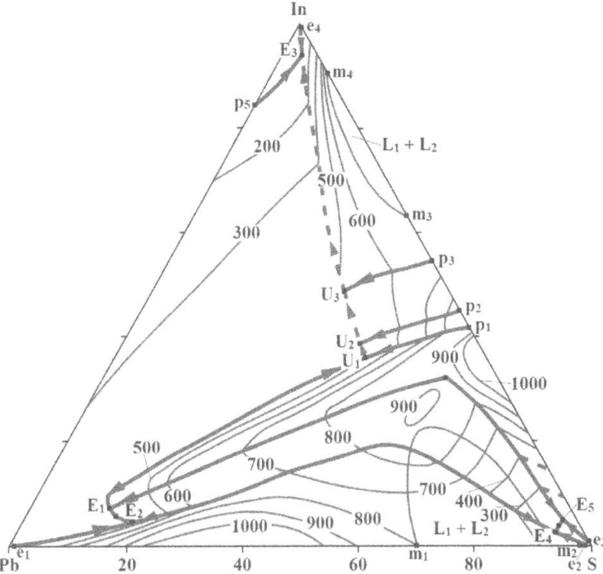

FIGURE 10.12 Liquidus surface of the Pb–In–Te ternary system. (From Rustamov, P.G., and Melikova, Z.D., *Zhurn. neorgan. khimii*, **26**(5), 1432, 1981.)

The width of the homogeneity region of both phases does not exceed 1.5 mol% (Krämer and Berroth 1980) [the homogeneity region of $PbIn_2S_4$ is within the interval of 49.5–50 mol% In_2S_3 (Rustamov et al. 1976d).

Two more ternary compounds, **Pb$_3$In$_{6.67}$S$_{13}$** and **Pb$_4$In$_9$S$_{17}$**, are formed in the Pb–In–S system. $Pb_3In_{6.67}S_{13}$ crystallizes in the monoclinic structure with the lattice parameters a = 3812.9 ± 0.2, b = 1380.9 ± 0.5, c = 386.9 ± 0.02 pm, β = 91.25 ± 0.2°, and an experimental density of 5.7 ± 0.1 g·cm^{-3} (Ginderow 1978; Likforman et al. 1989) [a = 3813.2 ± 0.7, b = 1376.3 ± 0.2, c = 386.43 ± 0.08 pm, β = 90.856 ± 0.002°, and a calculated density of 5.9 g·cm^{-3} (Krämer and Berroth 1980)].

$Pb_4In_9S_{17}$ crystallizes in the orthorhombic structure with the lattice parameters a = 2278.0 ± 0.2, b = 383.39 ± 0.01, and c = 1521.3 ± 0.2 pm an experimental density of 5.9 ± 0.1 g·cm^{-3} (Krämer and Berroth 1980) [a = 2276.1 ± 0.9, b = 1520.1 ± 0.6, c = 385.8 ± 0.02 pm, and an experimental density of 6.0 g·cm^{-3} (Ginderow 1978; Likforman et al. 1989).

The projection of the liquidus surface of the Pb–In–S ternary system (Figure 10.12) consists of nine fields of primary crystallization of phases, four of which are in equilibrium with $PbIn_2S_4$ (Rustamov and Melikova 1981). Invariant curves intersect at eight invariant points, five of which are ternary eutectics, and three are ternary transition points. According to the data of Massalski (1990), an immiscibility region is absent in the In–Pb binary system. Therefore this region was not indicated on the liquidus surface.

10.15 Lead–Thallium–Sulfur

PbS–Tl: This system is a non-quasibinary section of Pb–Tl–S ternary system (Latypov et al. 1979b). Along the liquidus line, in addition to PbS, the **Tl$_6$PbS$_4$** ternary compound is crystallized in the system. Thermal effects at 460°C are due to the peritectic decomposition of this compound. At a content of more than 70 at% Tl, an immiscibility region is observed. The solidification of all alloys ends with the crystallization of solid solutions containing Pb. The solubility of Tl in PbS at room temperature is 0.23 at% and increases up to 0.46 at% at 460°C (Fayzullina et al. 1983b). The system was investigated using DTA, XRD, metallography, and measuring of the microhardness (Latypov et al. 1979b). The samples for the investigations were annealed at temperatures below the solidus temperature for 300–500 h.

PbS–TlS: This a system is non-quasibinary section of Pb–Tl–S ternary system as TlS melts incongruently (Latypov et al. 1979b; Gotuk et al. 1979d). On the TlS side, the primary crystallizing phase is Tl_4S_3. Solidification of all alloys ends at 235°C (Gotuk et al. 1979d) [at 226°C (Latypov et al. 1979b)]. In the solid state, the samples of the system consist of a mixture of PbS and TlS. The solubility of TlS in PbS at 230°C is 7.5 mol% and decreases to 4.8 mol% at room temperature (Latypov et al. 1979b; Fayzullina et al. 1983b). The system was investigated using DTA, XRD, metallography, and measuring of the microhardness (Latypov et al. 1979b; Gotuk et al. 1979d). The samples for the investigations were annealed at temperatures below the solidus temperature for 300–500 h.

PbS–Tl$_2$S: The phase diagram of this system, constructed through DTA, XRD, metallography, and measuring of the microhardness, is shown in Figure 10.13 (Gotuk et al. 1978c; Latypov et al. 1979a; Filep et al. 2011). The eutectic crystallizes at 412°C (Filep et al. 2011) [13 mol% PbS and crystallizes at 430°C (Latypov et al. 1979a); 10 mol% PbS and crystallizes at 416°C (Gotuk et al. 1978c)]. The solubility of PbS in Tl_2S is 2.8 mol% at 430°C and 0.5 mol% at room temperature (Latypov et al. 1979a) [8 mol% at 350°C and ca. 9 mol% at 416°C (Gotuk et al. 1978c)] and the solubility of Tl_2S in PbS is 0.72 mol% at 470°C and 0.54 mol% at room temperature (Fayzullina et al. 1983b). According to Latypov et al. (1979a), the thermal effects at 420°C are due to the melting of the eutectic in the Tl_2S–Tl_6PbS_4 system, and they appears on thermograms due to non-equilibrium crystallization.

The **Tl$_4$PbS$_3$** ternary compound, which melts incongruently at 452°C (Filep et al. 2011; Peresh et al. 2017) [at 460°C (Gotuk et al. 1978c); at 437°C (Peresh et al. 2015)] and has a polymorphic transformation at ca. 416°C (Filep et al. 2011) [at 420°C (Gotuk et al. 1978c)], is formed in this system. α-Tl_4PbS_3 crystallizes in the tetragonal structure with the lattice parameters $a = 834.6$ and $c = 1252.6$ pm, and β-Tl_4PbS_3 crystallizes in the orthorhombic structure with the lattice parameters $a = 891.6$ $b = 879.5$ and $c = 821.1$ pm (Filep et al. 2011). Single crystals of this compound were grown using Bridgman method (Malakhovs'ka et al. 2007).

According to the data of Latypov et al. (1979a), Tl_6PbS_4 (instead of Tl_4PbS_3) ternary compound, which melts incongruently at 465°C, is formed in the system.

The samples for the investigations were annealed at 370°C for 168 h (Filep et al. 2011) [at temperatures below the solidus temperature for 300–500 h (Gotuk et al. 1978c)].

PbS–Tl$_4$S$_3$: This system is non-quasibinary section of Pb–Tl–S ternary system as Tl_4S_3 melts incongruently (Gotuk et al. 1979d). In the solid state (below 295°C), the samples of the system consist of a mixture of PbS and Tl_4S_3. The system was investigated using DTA, XRD, metallography, and measuring of the microhardness. The samples for the investigations were annealed at temperatures below the solidus temperature for 300–500 h.

The liquidus surface of the Pb–Tl–S ternary system consists from the fields of primary crystallization of PbS, Tl_4PbS_3, Tl_4S_3, and solid solution based on Tl_2S (Figure 10.14) (Gotuk et al. 1979d). Primary crystallization fields of Pb, Tl, S, TlS, and solid solution of the Pb–Tl system are practically degenerate. There are two immiscibility regions in this system. At room temperature, a noticeable region of homogeneity in the ternary system exists only on the basis of Tl_2S. The region of solid solutions based on PbS is elongated in a narrow strip along the PbS–TlS section (Fayzullina et al. 1983b).

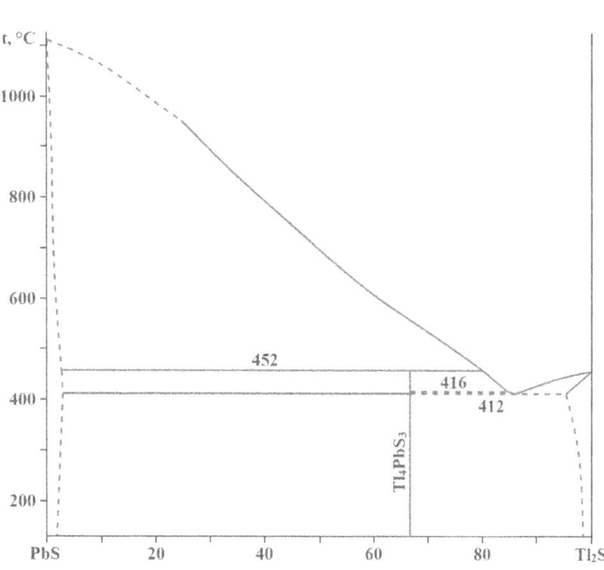

FIGURE 10.13 Phase diagram of the PbS–Tl$_2$S system. (From Filep, M.Y., et al., *Nauk. visn. Uzhgorod. Univ., Ser. Khim.*, **2**(26), 9, 2011.)

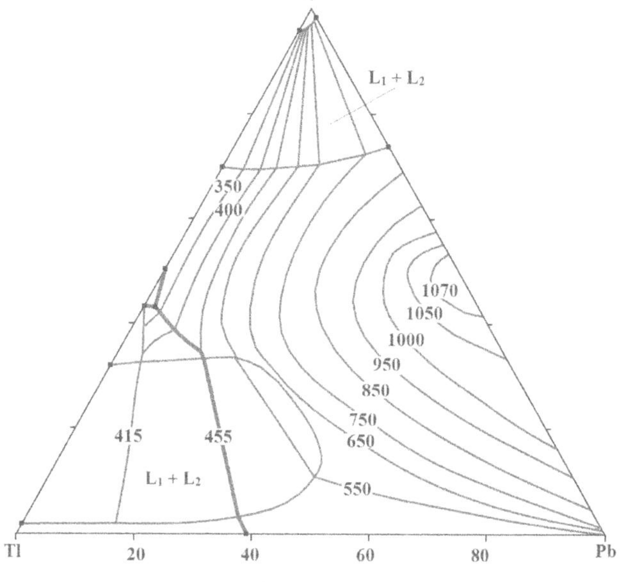

FIGURE 10.14 Liquidus surface of the Pb–Tl–S ternary system. (From Gotuk, A.A., et al., *Zhurn. neorgan. khimii*, **24**(9), 2497, 1979.)

10.16 Lead–Scandium–Sulfur

PbS–Sc_2S_3: The **Sc_2PbS_4** ternary compound, which crystallizes in the orthorhombic structure with the lattice parameters $a = 1164.2 \pm 0.4$, $b = 375.7 \pm 0.1$, $c = 1371.1 \pm 0.4$ pm, and a calculated density of 4.711 g·cm^{-3} (Shemet et al. 2006b) [$a = 1165.95 \pm 0.05$, $b = 1369.33 \pm 0.05$, $c = 375.31 \pm 0.01$ pm (IJdo 1982)], is formed in this system. To prepare this compound, calculated amounts of the elements were sealed in an evacuated silica ampoule, which was heated to 1150°C at a rate of 30°C·h^{-1} (Shemet et al. 2006b). The sample was kept at this temperature for 3 h. Subsequently, it was cooled slowly (10°C·h^{-1}) to 600°C and annealed at this temperature for 240 h. After annealing the ampoule was quenched in cold water. It can be also synthesized by heating a mixture of PbS and Sc_2S_3 in an evacuated quartz tube at 830°C for a week.

10.17 Lead–Yttrium–Sulfur

PbS–Y_2S_3: The solubility of Y_2S_3 in PbS at 500°C is ca. 3 mol% (Marchuk et al. 2007; Gulay et al. 2007b; Filyuk et al. 2008). The **Y_2PbS_4** ternary compound is formed in this system.

10.18 Lead–Lanthanum–Sulfur

PbS–La_2S_3: The solid solution is formed in this system within the interval 0–50 mol% PbS: from La_2S_3 to **La_2PbS_4** (Patrie et al. 1969). La_2PbS_4 crystallizes in the cubic structure with the lattice parameter $a = 876.7$ pm.

10.19 Lead–Cerium–Sulfur

PbS–Ce_2S_3: The solid solution is formed in this system within the interval 0–50 mol% PbS: from Ce_2S_3 to **Ce_2PbS_4** (Patrie et al. 1969). Ce_2PbS_4 crystallizes in the cubic structure with the lattice parameter $a = 870.5$ pm.

10.20 Lead–Praseodymium–Sulfur

PbS–Pr_2S_3: The solid solution is formed in this system within the interval 0–50 mol% PbS: from Pr_2S_3 to **Pr_2PbS_4** (Patrie et al. 1969). Pr_2PbS_4 crystallizes in the cubic structure with the lattice parameter $a = 865.13 \pm 0.01$ pm and a calculated density of 6.33 g·cm^{-3} (Marchuk et al. 2006a) [$a = 867.5$ pm (Patrie et al. 1969)].

10.21 Lead–Neodymium–Sulfur

PbS–Nd_2S_3: The solid solution is formed in this system within the interval 0–50 mol% PbS: from Nd_2S_3 to **Nd_2PbS_4** (Patrie et al. 1969). Nd_2PbS_4 crystallizes in the cubic structure with the lattice parameter $a = 863.2$ pm.

10.22 Lead–Samarium–Sulfur

PbS–Sm_2S_3: The solid solution is formed in this system within the interval 0–50 mol% PbS: from Sm_2S_3 to **Sm_2PbS_4** (Patrie et al. 1969). Sm_2PbS_4 crystallizes in the cubic structure with the lattice parameter $a = 857.2$ pm.

10.23 Lead–Gadolinium–Sulfur

PbS–Gd_2S_3: The solid solution is formed in this system within the interval 0–50 mol% PbS: from Gd_2S_3 to **Gd_2PbS_4** (Patrie et al. 1969). There is also the region of solid solution based on PbS with the NaCl-type cubic structure. Gd_2PbS_4 crystallizes in the cubic structure with the lattice parameter $a = 852.2$ pm.

10.24 Lead–Terbium–Sulfur

PbS–Tb_2S_3: At 600°C, the solid solution between the **Tb_2PbS_4** ternary compound and **$Tb_{2.4}Pb_{0.4}S_4$** composition, which crystallize in the cubic structure with the lattice parameter $a = 850.21 \pm 0.01$ pm and a calculated density of 7.059 g·cm^{-3} for Tb_2PbS_4 and $a = 839.54 \pm 0.01$ pm for $Tb_{2.4}Pb_{0.4}S_4$ (Khvaleba et al. 2006), is formed in this system. According to the data of Patrie et al. (1969), the solid solution based on Tb_2S_3 reaches the composition of PbS.1.70Tb_2S_3. There is also the region of solid solution based on PbS with the NaCl-type cubic structure.

10.25 Lead–Dysprosium–Sulfur

PbS–Dy_2S_3: The solubility of Dy_2S_3 in PbS at 500°C is ca. 8 mol% (Gulay et al. 2007b). Two ternary compounds, **Dy_2PbS_4** and **$Dy_{2+2x/3}Pb_{1-x}S_4$** ($x = 0.44 - 0.50$), are formed in this system. $Dy_{2+2x/3}Pb_{1-x}S_4$ crystallizes in the cubic structure with the lattice parameter $a = 842.50 \pm 0.03$ pm for $x = 0.44$ and $a = 840.62 \pm 0.03$ pm and a calculated density of 6.831 g·cm^{-3} for $x = 0.50$.

According to the data of Patrie et al. (1969), the solid solution based on Dy_2S_3 reaches the composition of PbS·1.25Dy_2S_3. There is also the region of solid solution based on PbS with the NaCl-type cubic structure.

10.26 Lead–Holmium–Sulfur

PbS–Ho_2S_3: The solubility of Ho_2S_3 in PbS at 500°C is ca. 3 mol% (Gulay et al. 2007b). According to the data of Patrie et al. (1969), the solid solution based on Dy_2S_3 reaches the composition of PbS·2Ho_2S_3. There is also the region of solid solution based on PbS with the NaCl-type cubic structure.

The **Ho_2PbS_4** ternary compound, which crystallizes in the orthorhombic structure with the lattice parameters $a = 1189$, $b = 1425$, and $c = 401$ pm (Patrie et al. 1969; Gasymov et al. 1980), is formed in this system.

10.27 Lead–Erbium–Sulfur

PbS–Er_2S_3: The solubility of Er_2S_3 in PbS at 500°C is ca. 4 mol% (Gulay et al. 2007b). There is also the region of solid solution based on PbS with the NaCl-type cubic structure in this system (Patrie et al. 1969).

The **Er_2PbS_4** ternary compound, which crystallizes in the orthorhombic structure with the lattice parameters a = 1185, b = 1417, and c = 400 pm (Patrie et al. 1969; Gasymov et al. 1980), is formed in this system.

10.28 Lead–Thulium–Sulfur

PbS–Tm_2S_3: There is the region of solid solution based on PbS with the NaCl-type cubic structure (Patrie et al. 1969).

The **Tm_2PbS_4** ternary compound, which crystallizes in the orthorhombic structure with the lattice parameters a = 1183, b = 1410, and c = 398 pm (Patrie et al. 1969; Gasymov et al. 1980), is formed in this system.

10.29 Lead–Ytterbium–Sulfur

PbS–Yb_2S_3: There is the region of solid solution based on PbS with the NaCl-type cubic structure (Patrie et al. 1969).

The **Yb_2PbS_4** ternary compound, which crystallizes in the orthorhombic structure with the lattice parameters a = 1189.9 ± 0.2, b = 390.15 ± 0.08, c = 1412.7 ± 0.2 pm, and a calculated density of 6.902 g·cm^{-3} (Gulay et al. 2008b) [a = 1178, b = 1408, and c = 396 pm (Patrie et al. 1969; Gasymov et al. 1980)], is formed in this system. To prepare this compound, the calculated amount of the elements was sealed in an evacuated silica ampoule (Gulay et al. 2008b). The synthesis was realized in a tube furnace. The ampoule was gradually heated with 30°C·h^{-1} to 1150°C and the sample was kept at this temperature for 4 h. After that, it was cooled slowly (10°C·h^{-1}) to 600°C and annealed at this temperature for 240 h. After annealing the ampoule with the sample was quenched in cold water.

10.30 Lead–Lutetium–Sulfur

PbS–Lu_2S_3: There is the region of solid solution based on PbS with the NaCl-type cubic structure (Patrie et al. 1969).

The **Lu_2PbS_4** ternary compound, which crystallizes in the orthorhombic structure with the lattice parameters a = 1191.9 ± 0.2, b = 388.90 ± 0.08, c = 1410.3 ± 0.3 pm, and a calculated density of 6.964 g·cm^{-3} (Gulay et al. 2008b) [a = 1178, b = 1407, and c = 396 pm (Patrie et al. 1969; Gasymov et al. 1980)], is formed in this system.

10.31 Lead–Uranium–Sulfur

The **PbU_2S_5** ternary compound, which crystallizes in the monoclinic structure with the lattice parameters a = 823.76 ± 0.04, b = 744.34 ± 0.03, c = 1172.36 ± 0.05 pm, β = 90.132 ± 0.002°, and a calculated density of 7.795 g·cm^{-3} (Prakash et al. 2014) [in the orthorhombic structure with the lattice parameters a = 744 ± 1, b = 827 ± 1, c = 1175 ± 2 pm, and the calculated and experimental densities of 7.75 and 7.65 ± 0.05 g·cm^{-3}, respectively (Brochu et al. 1970, 1972)] is formed in the Pb–U–S system.

Crystals of PbU_2S_5 were obtained from the reaction of Pb (0.482 mM), U (0.084 mM), and S (0.343 mM) (Prakash et al. 2014). The reaction mixture was heated to 500°C in 12 h, held there for 12 h, and then heated to 1000°C in 24 h, where it remained for 72 h. The reaction mixture was cooled to 750°C in 48 h and to room temperature in 12 h. The reaction product contained black blocks of PbU_2S_5, PbS and black plates of UOS.

This compound was also obtained by the reaction in the solid state between PbS and β-US_2 (PbS + 2β-US_2) at 1100°C for 48 h (Brochu et al. 1970, 1972)]. The stoichiometric mixture was ground and pelletized under a pressure of 19.6 MPa, in a dry argon atmosphere. The reaction was carried out in a silica tube sealed under vacuum.

10.32 Lead–Titanium–Sulfur

PbS–TiS_2: Several ternary compounds are formed in this system. **$PbTiS_3$** has apparently two polymorphic modifications. First of them crystallizes in the monoclinic structure with the lattice parameters a = 582 ± 3, b = 575 ± 3, c = 2374 ± 9 pm, and β = 98.4 ± 0.5° (Hernán et al. 1991a) [a = 589, b = 582, c = 1144 pm, and β = 95.24° (Wiegers et al. 1989a)]. It was obtained by direct synthesis from the elements and grown by the chemical transport method using Cl_2 as carrier reagent.

Second modification crystallizes in the tetragonal structure a = 411 and c = 1175 pm (Schmidt 1970) [a = 416 and c = 1175.2 pm and the calculated and experimental densities of 5.74 and 5.6 g·cm^{-3}, respectively (Sterzel 1966; Novoselova and Aslanov 1967; Sterzel and Horn 1970)]. The samples of this polymorph were synthesized by firing the elements in evacuated quartz tubes in the temperature range from 600°C to 1100°C with repeated grinding and pressing into pellets. Long firing times of at least 200 h were required in order to achieve complete reaction.

$PbTi_2S_5$ crystallizes in the orthorhombic structure with the lattice parameters a = 2282 ± 5, b = 587 ± 1, c = 3490 ± 1 pm (Hernán et al. 1991a). It was prepared from its elements in a ratio of 1:2:5 in evacuated quartz ampoules heated at 900°C for 14 days. The application of crystal growth techniques with Cl_2 derived from $(NH_4)_2PbCl_6$ resulted in non-uniform distributions of the elements.

$(PbS)_{1.18}TiS_2$ is a misfit layer compound built of alternate double layers of PbS and sandwiches of TiS_2 and has two polymorphic modification. First of them crystallizes in the monoclinic structure with the lattice parameters a = 580.0 ± 0.2, b = 588.1 ± 0.2, c = 1175.9 ± 0.9 pm, and β = 95.27 ± 0.02° for PbS-sublattice and a = 340.9 ± 0.1, b = 588.0 ± 0.2, c = 1176.0 ± 0.2 pm, and β = 95.29 ± 0.02° for TiS_2-sublattice (Wiegers et al. 1990b; Smaalen van et al. 1991; Wiegers and Meerschaut 1992). Second modification crystallizes in the orthorhombic structure with the lattice parameters a = 580.20 ± 0.05,

$b = 588.22 \pm 0.08$, and $c = 2341.3 \pm 0.3$ pm for PbS-sublattice and $a = 340.86 \pm 0.03$, $b = 588.14 \pm 0.07$, and $c = 2340.6 \pm 0.2$ pm for TiS_2-sublattice (Smaalen van and Boer de 1992).

A powder sample of this compound was synthesized from the elements in the stoichiometric ratio (Wiegers et al. 1989, 1990b; Wiegers and Meerschaut 1992). The mixture was heated at 800°C in an evacuated quartz tube for 7 days. Single crystals were obtained by vapor transport using chlorine as a carrier reagent, for which about 1 mass% $(NH_4)_2PbCl_6$ was used. Crystals grow as thin platelets at the cold part of the gradient of 720°C–650°C. It could be also prepared at the reaction of the ternary oxides with CS_2 or H_2S vapor at 1000°C–1300°C.

$(PbS)_{1.18}(TiS_2)_2$ is also a misfit layer compound and takes the next preferred orientation: a two-atom-thick layer of PbS and two adjacent three-atom-thick sandwiches of TiS_2 are stacked alternately (Gotoh et al. 1990a). It crystallizes in the monoclinic structure with the lattice parameters $a = 575.5 \pm 0.9$, $b = 586.9 \pm 0.6$, $c = 1735 \pm 5$ pm, and $\beta = 93.74 \pm 0.05°$ for PbS-sublattice and $a = 339.6 \pm 0.6$, $b = 586.6 \pm 0.7$, $c = 1741 \pm 3$ pm, and $\beta = 93.60 \pm 0.06°$ for TiS_2-sublattice (Meerschaut et al. 1992).

To prepare this compound, Pb, Ti, and S (molar ratio 1:2:5) were mixed and heated in a quartz tube, sealed under vacuum, at 850°C–900°C (gradient ca. 50°C) for 10 days. The product of the reaction was ground and reheated at the same temperature, for 1 week, in the presence of a small amount of iodine. Platelet-shaped single crystals were obtained (Meerschaut et al. 1992).

10.33 Lead–Zirconium–Sulfur

PbS–ZrS_2: The **$PbZrS_3$** ternary compound, which crystallizes in the orthorhombic structure with the lattice parameters $a = 901.34 \pm 0.07$, $b = 376.60 \pm 0.02$, and $c = 1392.37 \pm 0.10$ pm (Lelieveld and Ijdo 1978) [$a = 903.7 \pm 0.5$, $b = 1392.6 \pm 0.5$, and $c = 377.1 \pm 0.2$ pm (Yamaoka 1972); $a = 1517$, $b = 1177$, and $c = 1816$ pm and an experimental density of 6.1 g·cm^{-3} (Sterzel and Horn 1970)], is formed in this system.

To prepare this compound, a mixture of PbS, Zr, and S (molar ratio 1:1:2) was sealed in a Pt capsule; a piston-cylinder type high-pressure apparatus was used with the furnace assembly in a pressure cell (Yamaoka 1972). Pressure was applied to the sample, and the temperature was then raised to the desired value at a rate of 2–3°C·min^{-1} and held for a predetermined time. The specimen was quenched to room temperature before the pressure was released. Brown material with a slight odor similar to that of H_2S was obtained from runs of at least 40 h at 2 MPa and 800°C–900°C. It can be also obtained by firing stoichiometric amounts of the binary sulfides in an evacuated sealed quartz tube at 800°C for a week (Lelieveld and Ijdo 1978).

10.34 Lead–Hafnium–Sulfur

PbS–HfS_2: The **$PbHfS_3$** ternary compound, which crystallizes in the orthorhombic structure with the lattice parameters $a = 898.8 \pm 0.2$, $b = 373.9 \pm 0.1$, $c = 1392.4 \pm 0.2$ pm, and a calculated density of 6.840 g·cm^{-3} (Wiegers et al. 1989b) [$a = 899.5$, $b = 374.61$, $c = 1392.8$ pm (Wiegers et al. 1989a); $a = 1514$, $b = 1174$, and $c = 1816$ pm and an experimental density of 7.8 g·cm^{-3} (Sterzel and Horn 1970)], is formed in this system. Powder sample of this compound was obtained by heating the elements in an evacuated quartz ampoule (Wiegers et al. 1989a,b). An intimate mixture of the elements was first slowly heated to about 400°C for 2 days and then heated at 750°C for 10 days. Single crystals were grown by vapor transport using chlorine as a transport agent in a temperature gradient of 770°C–710°C over 14 days. The obtained crystals have a metallic luster.

10.35 Lead–Nitrogen–Sulfur

The **$Pb(NS)_2$** ternary compound is formed in the Pb–N–S system (Goehring et al. 1955). To prepare this compound, PbI_2 or $Pb(NO_3)_2$ was interacted with S_4N_4 in liquid ammonia to give the green, poorly soluble **$Pb(NS)_2 \cdot NH_3$**. This substance was pulverized well and heated for about 8 h at 80°C–90°C in high vacuum (10^{-2} Pa). During this treatment, the ammonia could be removed from the formed compound, and dark red-brown $Pb(NS)_2$ with an experimental density of 4.2 g·cm^{-3} was formed. This compound is not thermally stable.

10.36 Lead–Phosphorus–Sulfur

PbS–"P_2S_4": The phase diagram of this system, constructed through DTA, XRD, and measuring of the microhardness and the density, is shown in Figure 10.15 (Prits et al. 1989; Potoriy et al. 1990). The eutectic contains 75 mol% PbS and crystallizes at 782°C. No mutual solubility was found in the system. The **$Pb_2P_2S_6$** ternary compound, which melts congruently at 920°C and has a polymorphic transformation at 422°C, is formed in this system (Prits et al. 1989; Potoriy et al. 1990; Potoriy and Milyan 2016). It crystallizes in the monoclinic structure with the lattice parameters $a = 660.0$, $b = 744.5$, $c = 1163.2$ pm, and

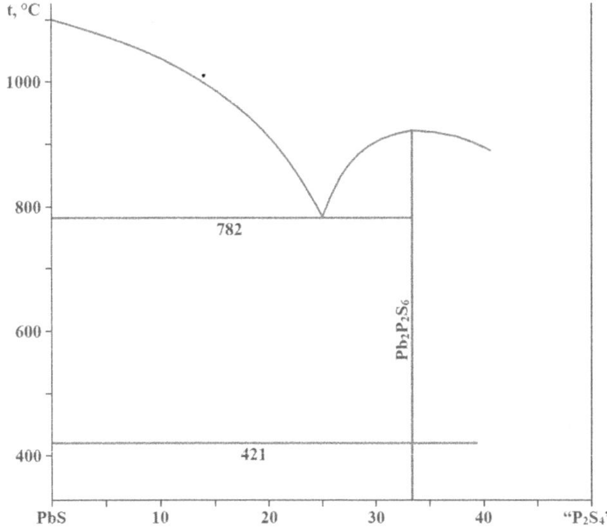

FIGURE 10.15 Part of the phase diagram of the PbS–"P_2S_4" system. (From Potoriy, M.V., et al., *Izv. AN SSSR. Neorgan. mater.*, **26**(11), 2363, 1990.)

β = 126.19° (Voroshilov et al. 2016) and an experimental density of 4.80–4.81 g·cm^{-3} (Potoriy et al. 1990; Potoriy and Milyan 2016) [a = 662, b = 746, c = 1137 pm, β = 124.5°, and the calculated and experimental densities of 4.83 and 4.78 g·cm^{-3}, respectively (Klingen et al. 1973); a = 660.6 ± 0.2, b = 746.4 ± 0.2, c = 1134.6 ± 0.3 pm, β = 124.10 ± 0.03°, and the calculated and experimental densities of 4.79 and 4.66 g·cm^{-3}, respectively (Carpentier and Nitsche 1974); a = 657.2, b = 744.1, c = 1156.2 pm, and β = 126.21° (Prits et al. 1989)]. Single crystals of $Pb_2P_2S_6$ were grown through the chemical transport reactions (Klingen et al. 1973; Carpentier and Nitsche 1974). The ingots for the investigations were annealed at 500°C for 2 weeks (Prits et al. 1989; Potoriy et al. 1990).

PbS–P_4S_{10}: The phase diagram of this system, constructed through DTA, XRD, and measuring of the density, is presented in Figure 10.16 (Prits et al. 1989). The eutectic contains 79 mol% PbS and crystallizes at 635°C. The solubility of P_4S_{10} in PbS reaches 2 mol%. The **$Pb_3P_2S_8$** ternary compound, which melts congruently at 635°C (Potoriy and Milyan 2016) [at 720°C (Post and Kramer 1984); at 655°C (Prits et al. 1989)] and has a polymorphic transformation at 435°C, is formed in this system. It crystallizes in the cubic structure with the lattice parameter a = 1093.9 pm (Voroshilov et al. 2016) and an experimental density of 4.79 g·cm^{-3} (Potoriy and Milyan 2016) [a = 1093.93 ± 0.07 pm and the calculated and experimental densities of 4.769 and 4.794 g·cm^{-3}, respectively (Post and Kramer 1984); a = 1092.72 pm and an experimental density of 4.78 g·cm^{-3} (Prits et al. 1989)].

It should be noted that the phase diagram of the PbS–P_4S_{10} system is reconstructed, since the $Pb_3P_2S_8$ compound contains not 75 mol%, but 83.3 mol% PbS.

The synthesis of $Pb_3P_2S_8$ was performed in sealed quartz tube with stoichiometric amounts of the elements; the ampoule was slowly heated up to 600°C and annealed for 1 week (Post and Kramer 1984). Single crystals of this compound were grown by chemical vapor transport using iodine as a transport agent.

The ingots for the investigations were annealed at 500°C for 2 weeks (Prits et al. 1989).

Two more compounds, **$Pb_2P_2S_7$** and **$Pb(PS_3)_2$**, were prepared by the melting the mixture of PbS and P_4S_{10} in evacuated and sealed ampoules (Soklakov and Nechaeva 1969).

10.37 Lead–Arsenic–Sulfur

PbS–AsS: Three compounds, **PbS·AsS**, **PbS·2AsS**, and **3PbS·2AsS**, are formed in this system (Isabaev et al. 1980; Zhumashev 1983). They decompose at 460°C–490°C forming initial binary compounds. PbS·2AsS and 3PbS·2AsS can be prepared by sulfidizing lead arsenates.

PbS–As_2S_3: Phase relations along the PbS-rich portion of the phase diagram of this system are shown in Figure 10.17 (Roland 1968). The phase diagram of the PbS–As_2S_3 system was also constructed through DTA, XRD, and measuring of the microhardness by Kutoglu (1969). It was shown that two eutectics, one of which contains 72 mol% PbS and crystallizes at 525°C (Kutoglu 1969) and another is degenerated from the As_2S_3-side, exist in the system (Kolomiets and Polyakov 1965; Kutoglu 1969). According to the data of Kutoglu (1969), five compounds, **$PbAs_2S_4$**, **$Pb_2As_2S_5$**, **$Pb_3As_4S_9$**, **$Pb_9As_4S_{15}$**, and **$Pb_9As_{13}S_{28}$**, are formed in the PbS–As_2S_3 system.

$PbAs_2S_4$ melts incongruently at 305°C (Kutoglu 1969) and crystallizes in the monoclinic structure with the lattice parameters given in Table 10.1. According to the data of Bannister et al. (1939), this compound (mineral sartorite) crystallizes in the monoclinic structure with an orthorhombic subcell (a = 1946, b = 779, and c = 417 pm). As can be seen from Table 10.1, there is great disagreement in the definition of the lattice parameters of this compound.

$Pb_2As_2S_5$ melts congruently at 600°C–650°C (Kutoglu 1969) and also crystallizes in the monoclinic structure with the lattice parameters given in Table 10.2. This compound

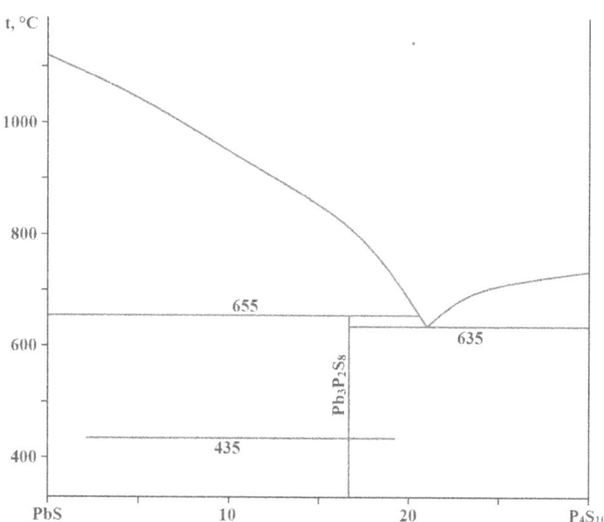

FIGURE 10.16 Part of the phase diagram of the PbS–P_4S_{10} system. (From Prits, I.P., et al., *Ukr. khim. zhurn.*, **55**(2), 135, 1989.)

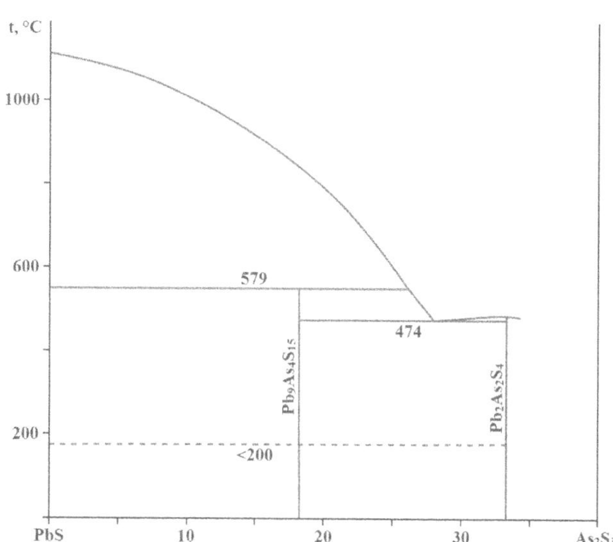

FIGURE 10.17 Part of the phase diagram of the PbS–As_2S_3 system. (From Roland, G.W., *Mineral. Deposita*, **3**(3), 249, 1968.)

TABLE 10.1

Crystallographic Data for $PbAs_2S_4$

a, pm	*b*, pm	*c*, pm	β, °	$d_{calc.}$	$d_{meas.}$	References
				g·cm^{-3}		
5838	779	8330	90	–	–	[a]Vaux and Bannister (1938); Bannister et al. (1939)
1940	830	782	90	–	–	Rösch and Hellner (1959)
1962 ± 2	789 ± 1	419 ± 5	90	5.05	–	[a]Nowacki et al. (1960, 1961); Iitaka and Nowacki (1961); Pring et al. (1993)
1956	791.1	418.1	90	–	–	Kutoglu (1969)
7480	789	4610	128.1	–	–	[a]Ozawa and Takéuchi (1993)
3771 ± 2	789.8 ± 0.3	2010.6 ± 0.8	101.993 ± 0.007	–	–	[a,b]Berlepsch et al. (2003)

[a] Mineral sartorite.
[b] Sartorite, containing up to 6.5 mass% Tl (**$Pb_8Tl_{1.5}As_{17.5}S_{35}$**).

TABLE 10.2

Crystallographic Data for $Pb_2As_2S_5$

a, pm	*b*, pm	*c*, pm	β, °	$d_{calc.}$	$d_{meas.}$	References
				g·cm^{-3}		
841	2585	788	90.5	5.61	5.53	[a]Berry (1953); Nowacki et al. (1960)
843 ± 1	2580 + 5	791 ± 1	90.00 ± 0.25	5.34 ± 0.03	5.41 ± 0.08	[a]Le Bihan (1959, 1962); Fleischer (1969b)
838	2561	789	90.42	–	–	[a]Nowacki et al. (1961)
790	2574	837	90.35	–	–	[a]Marumo and Nowacki (1967); Ribár et al. (1969)
840.3	2586	788	90.58	–	–	[a]Kutoglu (1969)
788.5 ± 0.1	2572 ± 1	836.5 ± 0.1	90.4 ± 0.1	–	–	[a]Shimizu et al. (2005)

[a] Mineral dufrénoysite.

is characterized by the presence of a polymorphic transformation at 487°C, and both polymorphic phases differ only slightly from each other.

$Pb_3As_4S_9$ melts incongruently at 458°C (Kutoglu 1969) and also crystallizes in the monoclinic structure with the lattice parameters given in Table 10.3.

$Pb_9As_4S_{15}$ melts incongruently at 549°C and is characterized by the presence of a polymorphic transformation at 509°C, and both polymorphic phases differ only slightly from each other (Kutoglu 1969). First modification crystallizes in the monoclinic structure with the lattice parameters a = 895.1, b = 3190, c =848.9 pm, and β = 117.82° (Kutoglu 1969) and the second modification (mineral gratonite) crystallizes in the rhombohedral structure with the lattice parameters a = 1058 ± 1 and α = 114°10′ ± 10′ or a = 1775.8 ± 1.4, and c = 780.7 ± 0.6 in hexagonal setting and the calculated and experimental densities of 6.18 and 6.22 g·cm^{-3}, respectively (Ribár and Nowacki 1969).

$Pb_9As_{13}S_{28}$ melts incongruently at 549°C (Kutoglu 1969) and crystallizes in the monoclinic structure with the lattice parameters presented Table 10.4.

$Pb_{32}As_{40}S_{92}$ (mineral marumoite) crystallizes in the orthorhombic structure with the lattice parameters a = 11575, b = 790, c = 837 pm (Pring 2001; Shimizu et al. 2005).

TABLE 10.3

Crystallographic Data for $Pb_3As_4S_9$

a, pm	*b*, pm	*c*, pm	β, °	$d_{calc.}$	$d_{meas.}$	References
				g·cm^{-3}		
2274	833	789	97.42	–	–	[a,b]Berry (1953)
4551	830.5	790.2	97.5	–	–	[a]Rösch and Hellner (1959)
2278 ± 4	833 ± 1	790 ± 1	97.4 ± 0.2	5.13 ± 0.03	5.24 ± 0.08	[a,b]Le Bihan (1961a, 1962)
2280 ± 1	835.7 ± 0.5	789.4 ± 0.5	97.27 ± 0.07	–	–	[a,b,c]Engel and Nowacki (1969); Pring and Graeser (1994)
2280	833.9	790.1	97.49	–	–	[a]Kutoglu (1969)
2286 ± 2	836.2 ± 0.8	789.2 ± 0.6	97.83	5.41	5.44 ± 0.02	[a,b]Graeser et al. (1986)
2277 ± 4	835.3 ± 1.0	787.7 ± 1.0	97.27			[a,b]Graeser et al. (1986)
4474 ± 4	847.7 ± 0.4	791.0 ± 0.6	93.37 ± 0.05	–	–	[a,d]Pring et al. (1990)

[a] Mineral baumhauerite.
[b] Triclinic structure with α ≈ γ ≈ 90°.
[c] For the $Pb_{11.6}As_{15.7}Ag_{0.6}S_{36}$ composition.
[d] For the $Pb_{11}As_{17.2}Ag_{0.7}Sb_{0.4}S_{36}$ composition.

TABLE 10.4

Crystallographic Data for $Pb_9As_{13}S_{28}$

a, pm	*b*, pm	*c*, pm	β, °	$d_{calc.}$	$d_{meas.}$	References
				g·cm⁻³		
832	7090	791	90	5.52	5.42	[a]Berry (1953)
843 ± 1	7090 ± 10	791 ± 1	90,00 ± 0.25	5.27 ± 0.03	5.33 ± 0.08	[a]Le Bihan (1961b, 1962); Fleischer (1969b)
842.9	7091	791.6	90	–	–	[a]Kutoglu (1969)
837.1 ± 0.5	7049 ± 5	791.4 ± 0.5	90.13 ± 0.03	–	–	[a]Engel and Nowacki (1970)

[a] Mineral liveingite.

$PbAs_2S_4$ and $Pb_3As_4S_9$ have been prepared from the mixtures of PbS and As_2S_3 (molar ratio 1:1 and 1:8) using the hydrothermal reactions at a temperature of 480°C and 280°C and a pressure of 0.15 GPa and 20 MPa for 6 and 23 days, respectively (Rösch and Hellner 1959).

According to the data of Isabaev et al. (1980), one more compound, $\mathbf{Pb_3As_2S_6}$, is formed in the PbS–As_2S_3 system. This compound, like $PbAs_2S_4$ and $Pb_2As_2S_5$, decomposes at 435°C with the formation of sulfur and lead thioarsenates (PbS·AsS, PbS·2AsS and 3PbS·2AsS).

The formation of ternary chemical compounds in the PbS–As_2S_3 system is also confirmed by the study of co-precipitation of lead with As_2S_3 (Toptygina et al. 1979).

In the Pb–As–S ternary system, solid solutions based on As were not found (Krebs et al. 1961a).

10.38 Lead–Antimony–Sulfur

PbS–Sb: The phase diagram of this system, constructed through DTA, XRD, metallography, and measuring of the microhardness, is a eutectic type (Figure 10.18) (Zargarova et al. 1975). The eutectic contains 4 mol% (7.5 mass%) PbS and crystallizes at 590°C. The solubility of Sb in PbS reaches 4 mol% (2 mass%). The samples for the investigation were annealed at temperatures 500°C for 100 h.

PbS–$\mathbf{Sb_2S_3}$: The phase diagram of this system, constructed through DTA, XRD, metallography, and measuring of the microhardness, is presented in Figure 10.19 (Jaeger and Klooster van 1912; Salanci and Moh 1970; Craig et al. 1973; Garvin 1973; Dmytriv 1983; Kitakaze et al. 1995). The eutectic contains 20 mol% PbS and crystallizes at 516°C (Kitakaze et al. 1995) [at 495°C (Jaeger and Klooster van 1912); at 523°C (Craig et al. 1973)]. The range of solid solutions based on PbS does not exceed 2 mol% at 400°C, 3 mol% at 600°C–640°C, and reaches 5 mol% at the eutectic temperature and the solubility of PbS in Sb_2S_3 at 520°C is not higher than 2 mol% (Salanci and Moh 1970; Craig et al. 1973; Garvin 1973) [1.5 mol% at 636°C (Kitakaze et al. 1995)]. A number of incongruently melting ternary compounds is formed in this system.

$\mathbf{Pb_2Sb_2S_5}$ is stable at temperatures from 490°C to 584°C (Kitakaze et al. 1995) [melts incongruently at 570°C and has a phase transformation at 523°C (Jaeger and Klooster van 1912)]. Below 490°C, it has more S-rich compositions than PbS–Sb_2S_3 join, and disappears on the quasibinary section. This compound crystallizes in the monoclinic structure with the lattice parameters *a* = 5030, *b* = 401, *c* = 2078 pm, and β = 114.50° (Kitakaze et al. 1995) [*a* = 1616, *b* = 1375, *c* = 430 pm, β = 91°24′, and the calculated and experimental densities of 5.692 and 5.78 g·cm⁻³, respectively (Dmytriv 1975);

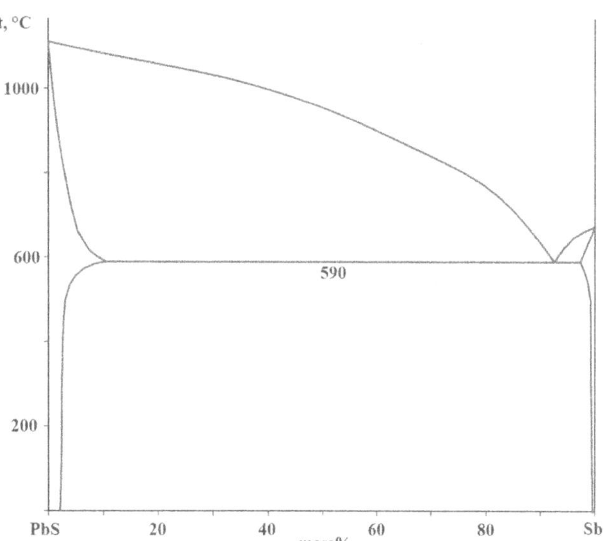

FIGURE 10.18 Phase diagram of the PbS–Sb system. (From Zargarova, M.I., et al., *Izv. AN SSSR. Neorgan. mater.*, **11**(6), 1138, 1975.)

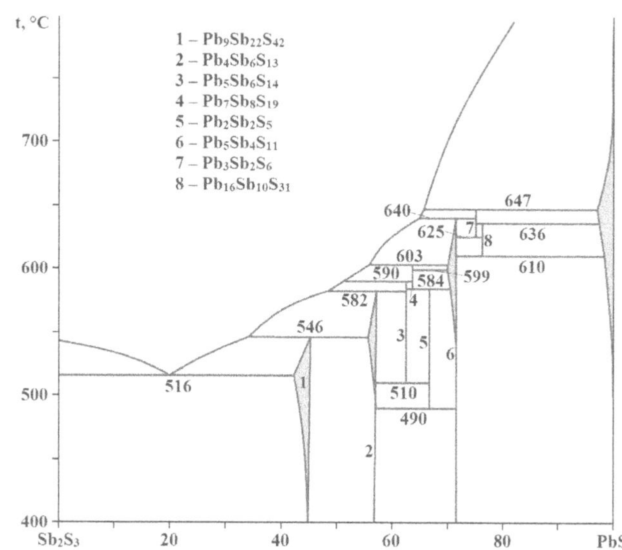

FIGURE 10.19 Phase diagram of the PbS–Sb_2S_3 system. (From Kitakaze, A., et al., *Mineralog. J.*, **17**(6), 282, 1995.)

in the orthorhombic structure with the lattice parameters a = 1980.8 ± 0.5, b = 404.2 ± 0.2, c = 1135.3 ± 0.4 pm (Smith and Hyde 1983); a = 1135.5 ± 0.4, b = 1978.3 ± 0.8, c = 404.2 ± 0.1 pm, and a calculated density of 5.95 g·cm^{-3} (Skowron and Brown 1990d); a = 1980, b = 1140, c = 404 pm (Bente and Anton 1995)]. The experimental density of this compound is 5.62 g·cm^{-3} (Jaeger and Klooster van 1912). The crystals of $Pb_2Sb_2S_5$ were prepared by the annealing at 590°C in the presence of iodine in vacuum-sealed ampoules (Skowron and Brown 1990d).

$Pb_3Sb_2S_6$ is stable at temperatures from 625°C to 647°C [621°C–637°C (Salanci and Moh 1970); 605°C–642°C (Craig et al. 1973)], at which it melts incongruently (Kitakaze et al. 1995). It crystallizes in the monoclinic structure with the lattice parameters a = 2162.6, b = 1140.9, c = 787.6 pm, and β = 96.19° (Pruseth et al. 2001) [a = 2493, b = 810, c = 1451 pm, β = 79°10′, and the calculated and experimental densities of 6.06 and 6.2 g·cm^{-3}, respectively, for mineral falkmanite (Hiller 1940); a = 2163, b = 2352, c = 812 pm, and β = 100.83° for mineral falkmanite (Dunn et al. 1984); a = 2154.7 ± 0.6, b = 2347.5 ± 0.9, c = 809.0 ± 0.1 pm, and β = 100.75 ± 0.03° for mineral falkmanite (McQueen 1987); in the orthorhombic structure with the lattice parameters a = 1140, b = 2368, c = 409 pm (Kitakaze et al. 1995); a = 1136, b = 2398, c = 410 pm (Wang 1977)]. This compound was synthesized from a mixture of PbS and Sb_2S_3 (molar ratio 3:1) at 500°C–560°C for 36–50 days (Pruseth et al. 2001).

$Pb_{3+x}Sb_2S_6$ (orthorhombic structure; a = 2409.9, b = 1138.0, and c = 827.9 pm) was prepared from the samples containing a mixture of PbS and Sb_2S_3 (molar ratio 3.2:1), which was heated in evacuated silica tube at 630°C and ground three times within the first week (Pruseth et al. 2001). The sample was then brought to melt at 640°C, ground once more, and held in the temperature range 628°C–635°C for three additional weeks. The quenched product consisted mainly of well-formed, prismatic crystals of this compound.

$Pb_4Sb_6S_{13}$ [(**$Pb_7Sb_{12}S_{25}$** according to the data of Garvin (1973)] melts incongruently at 582°C and has a very narrow homogeneity region from 56.25 to 57.14 mol% PbS at 540°C (Kitakaze et al. 1995). Early investigations indicated that this compound (mineral robinsonite) crystallizes in the triclinic structure with the lattice parameters a = 1651.9 ± 0.2, b = 1764.1 ± 0.2, c = 397.1 ± 0.1 pm, α = 96.12 ± 0.02°, β = 96.32 ± 0.02°, γ = 91.15 ± 0.01°, and the calculated and experimental densities of 5.74 and 5.63 g·cm^{-3}, respectively (Ayora and Gali 1981) [a = 1651, b = 1762, c = 397 pm, α = 96°04′, β = 96°22′, γ = 91°12′, and the calculated and experimental densities of 5.40 and 5.20–5.34 g·cm^{-3}, respectively (Berry et al. 1952); a = 1649, b = 1762, c = 397.6 pm, α = 95°42′, β = 96°11′, γ = 91°14′ (Jambor and Lachance 1968)]. This compound obtained from the mixture of PbS and Sb_2S_3 in the aqueous solution of 10 mass% NH_4Cl at 400°C in titanium lined autoclaves also crystallizes in the triclinic structure with the lattice parameters a = 1656 ± 1, b = 1769 ± 1, c = 398.2 ± 0.1 pm, α = 91.09 ± 0.05°, β = 96.46 ± 0.04°, γ = 96.77 ± 0.04°, and the calculated and experimental densities of 5.7 and 5.41 g·cm^{-3}, respectively (Petrova et al. 1978a).

Later it was shown that the mineral robinsonite also crystallizes in the monoclinic structure with the lattice parameters a = 2364.2 ± 0.4, b = 397.61 ± 0.06, c = 2442.0 ± 0.4 pm, β = 93.808 ± 0.003°, and a calculated densities of 5.743 g·cm^{-3} (Makovicky et al. 2004) [a = 2369.8 ± 0.8, b = 398.0 ± 0.8, c = 2446.6 ± 0.8 pm, β = 93.9 ± 0.3°, and a calculated densities of 5.7 g·cm^{-3} (Jambor and Owens 1982; Jambor and Burke 1990; Skowron and Brown 1990b)]. Monoclinic $Pb_4Sb_6S_{13}$ was synthesized from Pb, Sb, and S, taken in the stoichiometric ratio (Skowron and Brown 1990b). The elements were combined, sealed in an evacuated silica tube, melted at 800°C and quenched. Single crystals were grown by the chemical transport reaction using iodine as a transport agent.

$Pb_4Sb_6S_{13}$ could be also synthesized from a mixture of the elements, which was sealed in evacuated silica tube, heated at 560°C for 4 days, quenched, ground and pelletized, and reheated at the same temperature for 3 weeks (Jambor and Plant 1975).

$Pb_5Sb_4S_{11}$ melts incongruently at 640°C (Kitakaze et al. 1995) [at 642°C (Frumar 1969); at 631°C (Salanci and Moh 1970); at 638°C (Craig et al. 1973); at 634°C (Garvin 1973); melts congruently at 620°C (Rustamov et al. 1977a)], has a polymorphic transformation at 460°C [at 350°C–400°C (Schultz et al. 2017)] and an energy gap of 1.32 eV (Frumar 1969) [1.47 eV for mineral boulangerite and 1.43 eV for synthetic material (Lomelino and Mozurkewich 1989)]. The homogeneity region of this compound is richer in Sb_2S_3 by 2.5 mol% at 598°C [at 505°C (Salanci and Moh 1970)], and 1.5 mol% at 580°C and 1.0 mol% PbS at 570°C, than that of ideal composition (Kitakaze et al. 1995).

α-$Pb_5Sb_4S_{11}$ crystallizes in the monoclinic structure with the lattice parameters given in Table 10.5 and β-$Pb_5Sb_4S_{11}$ crystallizes in the orthorhombic structure with the lattice parameters given in Table 10.6. The standard entropy of formation (S^0_{298} = 824.7 J·M^{-1}K^{-1}), entropy (ΔS^0_{298} = 33.36 J·M^{-1}K^{-1}), enthalpy

TABLE 10.5

Crystallographic Data for α-$Pb_5Sb_4S_{11}$

a, pm	*b*, pm	*c*, pm	β, °	$d_{calc.}$ g·cm^{-3}	$d_{meas.}$ g·cm^{-3}	References
2114	2346	807	100.8	–	5.98	[a]Palache and Berman (1942)
2150	2354	809	100.83	–	–	[a]Robinson (1948)
2150	2347	805	99.26	–	–	[a]Hoda and Chang (1975)
2160	2350	810	100.8	–	–	Bente and Anton (1995)
2155.4 ± 0.4	2345.4 ± 0.4	807.9 ± 0.2	100.76 ± 0.01	6.21	–	[a]Ventruti et al. (2012)

[a] Mineral boulangerite.

TABLE 10.6

Crystallographic Data for β-$Pb_5Sb_4S_{11}$

			$d_{calc.}$	$d_{meas.}$	
a, pm	*b*, pm	*c*, pm	g·cm^{-3}		References
4228	2346	807	–	–	[a]Born and Hellner (1960)
2124 ± 1	2351 ± 1	403.6 ± 0.1	–	–	[a]Petrova et al. (1978b)
2130 ± 2	2347 ± 2	404 ± 7	–	–	Petrova et al. (1978b)
2349.0 ± 0.5	2124.5 ± 0.5	402.0 ± 0.1	6.21	–	Skowron and Brown (1990a)
2129 ± 1	2353 ± 2	401.9 ± 0.2	–	–	Bente and Anton (1995)
2344.3	2118.3	450.5	–	–	Pruseth et al. (2001)
2353.0 ± 0.2	403.85 ± 0.08	2127.3 ± 0.2	6.163	–	Schultz et al. (2017)

[a] Mineral boulangerite.

(ΔH^0_{298} = 1058.1 kJ·M^{-1}) and Gibbs energy of formation (ΔG^0_{298} = 1048.1 kJ·M^{-1}) for $Pb_5Sb_4S_{11}$ were calculated by Mammadov et al. (2014).

Single-crystal needles of $Pb_5Sb_4S_{11}$ were grown via chemical vapor reactions (Kaden et al. 2012; Schultz et al. 2017). The source material was synthesized from Pb, Sb, and S. In a first step, the mixture was slowly heated to 250°C and annealed for 5 days. In a second step, further slow heating (50°C per day) up to 850°C and annealing for 2 days was followed by quenching to room temperature. This material was powdered and pressed to pellets that were annealed at 750°C for 7 days, followed by quenching in air. For chemical vapor reactions, ca. 100 mg of this source material and ca. 0.03 mg·mL^{-1} NH_4Cl as transport agent were used. Chemical transport took place for 7 d from 637°C to 600°C–570°C. The crystals of this compound could be prepared by annealing of the powder at 590°C in the presence of iodine in vacuum-sealed ampoule (Skowron and Brown 1990a).

Polycrystalline bulk material was synthesized by slowly heating a stoichiometric mixture of the elements to 900°C, keeping it there for several hours (up to 2 days) and subsequently quenching the melt to room temperature (Schultz et al. 2017). In order to obtain the high-temperature modification, the material was annealed at 550°C for 6–7 days and then quenched in air again. In some cases, the product was ground, pressed to a pellet and annealed at 550°C again. The low-temperature modification can be obtained by subsequent annealing at 300°C–330°C for 3–7 days or longer times.

$Pb_5Sb_6S_{14}$ melts incongruently at 590°C and decomposes at 510°C (Kitakaze et al. 1995). No evidence of homogeneity region has been observed. It crystallizes in the triclinic structure with the lattice parameters a = 2836.7 ± 0.8, b = 402.4 ± 0.1, c = 2204.4 ± 0.7 pm, α = 89.59 ± 0.02°, β = 92.28 ± 0.03°, γ = 89.93 ± 0.02°, and a calculated density of 5.86 g·cm^{-3} (Skowron et al. 1992) or in the monoclinic structure with the lattice parameters a = 2198, b = 3512, c = 402 pm, and β = 126.25° (Kitakaze et al. 1995). This compound was obtained from a mixture of the elements in the stoichiometric ratio, which was melted in an evacuated silica tube at 880°C for 1 day, and subsequently annealing at 450°C for 2 weeks (Skowron et al. 1992). Single crystals were grown through chemical transport reactions using iodine as a transport agent.

$Pb_7Sb_8S_{19}$ (mineral heteromorphite) is stable at temperatures from 586°C to 603°C, at which it melts incongruently (Kitakaze et al. 1995). It crystallizes in the monoclinic structure with the lattice parameters a = 1362.8 ± 0.5, b = 1194.3 ± 0.4, c = 2128.5 ± 0.8 pm, β = 90°55′ ± 7′, and the calculated and experimental densities of 5.81 and 5.73 g·cm^{-3}, respectively (Edenharter 1980) [a = 1360, b = 1193, c = 2122 pm, β = 90.83°, and the calculated and experimental densities of 5.86 and 5.73 g·cm^{-3}, respectively (Jambor 1969; Kohatsu and Wuensch 1974a)] or in the orthorhombic structure with the lattice parameters a = 1131, b = 1978, and c = 404 pm (Kitakaze et al. 1995) and an energy gap of 1.4 eV (Lomelino and Mozurkewich 1989).

$Pb_9Sb_{22}S_{42}$ melts incongruently at 590°C and is characterized by a narrow homogeneity region (1.0 mol% at 530°C, 2.7 mol% at 512°C, and 2.0 mol% at 500°C in more Sb_2S_3-rich region) (Kitakaze et al. 1995). It crystallizes in the hexagonal structure with the lattice parameters given in Table 10.7.

$Pb_{16}Sb_{10}S_{31}$ melts incongruently at 636°C, decomposes at 610°C and crystallizes the orthorhombic structure with

TABLE 10.7

Crystallographic Data for $Pb_9Sb_{22}S_{42}$

		$d_{calc.}$	$d_{meas.}$	
a, pm	*c*, pm	g·cm^{-3}		References
4406	860	–	–	[a]Vaux and Bannister (1938)
4418	864.1	–	–	[a]Takeda and Horiuchi (1971)
4424	864	5.282	5.28 ± 0.05	[a]Lebas and Bihan Le (1976)
2210.8 ± 0.2	432.65 ± 0.04	5.34	–	[a]Smith (1986)
4411 ± 3	862.5 ± 0.2	–	–	[a]Johan and Mantienne (2000)

[a] Mineral zinkenite.

the lattice parameters a = 1424, b = 2550, and c = 404 pm (Kitakaze et al. 1995).

Some other compounds were also found in the PbS–Sb_2S_3 system.

$PbSb_2S_4$ (Salanci and Moh 1970; Craig et al. 1973; Tilley and Wright 1987) [**$Pb_6Sb_{14}S_{27}$** (Harris 1965; Garvin 1973; Portheine and Nowacki 1975)] melts incongruently at 545°C (Salanci and Moh 1970; Craig et al. 1973) [at 540°C (Garvin 1973); melts congruently at 800°C (Rustamov et al. 1977a)] and crystallizes in the hexagonal structure with the lattice parameters a = 2212.6 and c = 434.2 pm (Tilley and Wright 1987) [a = 4413 and c = 864 pm (Hoda and Chang 1975); a = 2214.8 ± 0.6, c = 433.3 ± 0.6 pm, and the calculated and experimental densities of 5.16 and 5.36 g·cm^{-3}, respectively (Portheine and Nowacki 1975)]. The homogeneity region of this compound is not higher than 1.5 mol% (Salanci and Moh 1970).

The standard entropy of formation (S^0_{298} = 274.2 J·M^{-1}K^{-1}), entropy (ΔS^0_{298} = 9.84 J·M^{-1}K^{-1}), enthalpy (ΔH^0_{298} = 342.3 kJ·M^{-1}) and Gibbs energy of formation (ΔG^0_{298} = 339.4 kJ·M^{-1}) for $PbSb_2S_4$ were calculated by Mammadov et al. (2014).

According to the lattice parameters of this compound, it is the $Pb_9Sb_{22}S_{42}$ compound (mineral zinkenite) described above.

$Pb_3Sb_4S_9$ melts incongruently at 603°C (Craig et al. 1973) [at 600°C (Salanci and Moh 1970)] and decomposes at 425°C (Craig et al. 1973; Garvin 1973) [at 505°C (Salanci and Moh 1970)].

$Pb_3Sb_8S_{15}$ (mineral fülöppite) crystallizes in the monoclinic structure with the lattice parameters given in Table 10.8.

$Pb_5Sb_8S_{17}$ (mineral plagionite) melts incongruently at 690°C (Jaeger and Klooster van 1912) and crystallizes in the monoclinic structure with the lattice parameters given in Table 10.9.

$Pb_6Sb_{10}S_{21}$ melts incongruently at 582°C, decomposes at 318°C and (Salanci and Moh 1970; Craig et al. 1973) and crystallizes in the triclinic structure with the lattice parameters a = 1641, b = 1789, c = 402 pm, α = 99.24°, β = 95.76°, γ = 89.46° (Hoda and Chang 1975). Apparently, this compound is a robinsonite mineral ($Pb_4Sb_6S_{13}$) since it is characterized by the same melting point, close component ratio, and close lattice parameters.

$Pb_7Sb_4S_{13}$ crystallizes in the orthorhombic structure with the lattice parameters a = 2348.8 ± 0.3, b = 2522.0 ± 0.3, and c = 408.21 ± 0.04 pm and a calculated density of 4.466 g·cm^{-3} (Skowron and Boswell 1994b). Single crystals of this compound were synthesized from a mixture of Pb, Sb, and S. The sample was sealed in evacuated silica tube, melted at 880°C for 2 days, and then annealed at 500°C for 7 days. Separation of the ingot occurred in the ampoule. Separate part of the ingot was ground to a fine powder from which the pellets were made. Each pellet was placed in silica tube with a small quantity of iodine. The tube was evacuated, sealed and annealed in a two-zone horizontal furnace. The pellets were kept at the cooler end (625°C); the other end was at 638°C. Needle-like crystals grew on the walls of the ampoules.

$Pb_9Sb_8S_{21}$ (mineral semseyite) also crystallizes in the monoclinic structure with the lattice parameters given in Table 10.10.

$Pb_{16}Sb_{18}S_{43}$ (mineral playfairite) also crystallizes in the monoclinic structure with the lattice parameters a = 4540 ± 5, b = 829 ± 6, c = 2130 ± 5 pm, β = 92°30′ ± 30′, and a calculated density of 5.72 g·cm^{-3} (Jambor 1967; Fleischer et al. 1968; Pierrot 1968).

$Pb_{22}Sb_{26}S_{61}$ (mineral launayite) also crystallizes in the monoclinic structure with the lattice parameters a = 4261 ± 8, b = 804 ± 5, c = 3230 ± 6 pm, β = 102°05′ ± 45′, and a calculated density of 5.83 g·cm^{-3} (Jambor 1967; Fleischer et al. 1968; Pierrot 1968).

Under non-equilibrium conditions, various defect structures are formed on the Sb_2S_3 side (Tilley and Wright 1986a).

One more compound, **$Pb_{11}Sb_{13}S_{30}(S_2)$** (mineral disulfodadsonite), which not belong to the PbS–Sb_2S_3 system, was found in the Pb–Sb–S ternary system. It crystallizes in the triclinic structure with the lattice parameters a = 411.92 ± 0.03, b = 1741.67 ± 0.14, c = 1916.64 ± 0.16 pm, α = 96.127 ± 0.006°, β = 90.015 ± 0.007°, γ = 91.229 ± 0.007°, and a calculated densities

TABLE 10.8

Crystallographic Data for $Pb_3Sb_8S_{15}$

a, pm	*b*, pm	*c*, pm	β, °	$d_{calc.}$ g·cm^{-3}	$d_{meas.}$ g·cm^{-3}	References
1341	1171	1690	94.72	5.21	5.22	Jambor 1969
1339	1169	1691	94.68	–	–	Cho and Wuensch 1970; Kohatsu and Wuensch 1974a
1343.5 ± 0.5	1172.7 ± 0.4	1693.4 ± 0.5	94.70 ± 0.08	5.18	–	Edenharter and Nowacki 1975
1344.1 ± 1.5	1172.6 ± 1.5	1693.0 ± 1.5	94.71 ± 0.08	5.19	5.22	Nuffield 1975
1344.4 ± 0.4	1172.7 ± 0.2	1693.4 ± 0.3	94.7 ± 0.4	–	–	Sejkora et al. 2017

TABLE 10.9

Crystallographic Data for $Pb_5Sb_8S_{17}$

a, pm	*b*, pm	*c*, pm	β, °	$d_{calc.}$ g·cm^{-3}	$d_{meas.}$ g·cm^{-3}	References
–	–	–	–	–	5.47	Jaeger and Klooster van (1912)
1347	1182	1999	107.33	5.58	5.54	Jambor (1969)
1348.57 ± 0.08	1186.56 ± 0.04	1998.34 ± 0.07	107.168 ± 0.004	5.55	5.54	Cho and Wuensch (1970, 1974); Kohatsu and Wuensch (1974a)
1346.2 ± 0.7	1186.8 ± 0.6	2003 ± 1	107.16 ± 0.03	–	–	Sejkora et al. (2017)

TABLE 10.10

Crystallographic Data for $Pb_9Sb_8S_{21}$

a, pm	*b*, pm	*c*, pm	β, °	$d_{calc.}$	$d_{meas.}$	References
				g·cm^{-3}		
1364	1196	2446	105.87	6.08	6.03	Jambor (1969)
1364	1201	2457	105.82	–	–	Cho and Wuensch (1970)
1360.3 ± 0.3	1193.6 ± 0.8	2443.5 ± 0.7	106.047 ± 0.010	6.12	6.03	Kohatsu and Wuensch (1974a,b)
1362.67 ± 0.10	1197.42 ± 0.09	2458.91 ± 0.18	105.997 ± 0.003	6.048	–	Matsushita et al. (1997); Matsushita (2018)

of 5.898 g·cm^{-3} (Orlandi et al. 2013; Belakovskiy et al. 2014) [$a = 412.27 \pm 0.02$, $b = 1742.74 \pm 0.12$, $c = 1917.04 \pm 0.13$ pm, $\alpha = 96.196 \pm 0.006°$, $\beta = 89.960 \pm 0.004°$, and $\gamma = 91.405 \pm 0.005°$ (Orlandi et al. 2012)].

The isothermal sections of the Pb–Sb–S ternary system were constructed by Craig et al. (1973) and Garvin (1973). Some vertical sections and the liquidus surface of this system (Figure 10.20) were constructed by Bakhtiyarly et al. (2013) [the $Pb_{0.9}Sb_{0.1}$–$Sb_{0.1}S_{0.9}$ vertical section was also constructed by Azhdarova et al. (2007)]. There are 7 fields of primary crystallization of phases on the liquidus surface. Three ternary eutectics (E_1, E_2, and E_3) in the S-corner are degenerated. The $Pb_5Sb_4S_{11}$–Sb, $PbSb_2S_4$–Sb, $Pb_5Sb_4S_{11}$–S, and $PbSb_2S_4$–S sections are quasibinary and a eutectic type (Azhdarova et al. 2009, 2011; Bakhtiyarly et al. 2013).

10.39 Lead–Bismuth–Sulfur

PbS–Bi_2S_3: Phase equilibria in this system have been studied by many authors. The results of constructing of the phase diagram are contradictory and it seems impossible to single out the most reliable ones. According to the data of some authors, the phase diagram is a eutectic type with peritectic transformations.

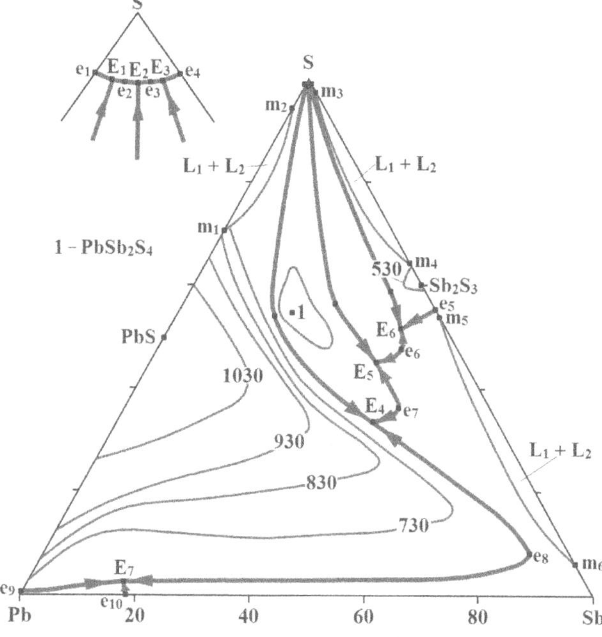

FIGURE 10.20 Liquidus surface of the Pb–Sb–S ternary system. (From Bakhtiyarly, I.B., et al., *Russ. J. Inorg. Chem.*, **58**(6), 728, 2013.)

The eutectic contains 24 mol% PbS and crystallizes at 722°C (Gospodinov et al. 1974) [23 mol% PbS and 730°C (Salanci 1965; Salanci and Moh 1969); 21 mol% PbS and 713°C (Godovikov et al. 1967a; Godovikov and Fedorova 1969)]. These data were confirmed by the experiments of Malevskiy et al. (1963), Craig (1967) and Dolgikh (1985a,b).

According to the data of Van Hook (1960) and Sadyhova et al. (1977), the phase diagram of this system is a eutectic type with peritectic transformations and formation of the **$PbBi_4S_7$** ternary compound, which melts congruently at 790°C. The eutectics contain 40 and 84 mol% and crystallize at 590°C and 680°C, respectively.

The solubility of Bi_2S_3 in PbS reaches 10 mol% at 835°C and 2 mol% at 400°C (Godovikov et al. 1967a; Godovikov and Fedorova 1969; Gospodinov et al. 1974) [11 mol% at 825°C (Salanci and Moh 1969); 9 mol% at 800°C and 1–3 mol% at 400°C (Van Hook 1960); 7.5 mol% at 780°C and 5 mol% at 500°C (Sadyhova et al. 1977)] and the solubility of PbS in Bi_2S_3 at the eutectic temperature is 6 mol% (Sadyhova et al. 1977) [0.8 mol% (Godovikov et al. 1967a)].

The ingots for the investigations were annealed at 400°C for 1000 h (Gospodinov et al. 1974) [at 500°C for 350 h (Sadyhova et al. 1977)] and the system was investigated through DTA, XRD, metallography, and measuring of the microhardness and density.

Several ternary compounds are formed in the PbS–Bi_2S_3 system. **$PbBi_2S_4$** melts incongruently at 750°C (Craig 1967; Godovikov et al. 1967a; Godovikov and Fedorova 1969; Gospodinov et al. 1974) [at 755°C (Frumar 1969; Dmytriv 1983); at 725°C (Sadyhova et al. 1977)] and has a defined homogeneity region (Salanci 1965; Craig 1967; Godovikov et al. 1967a; Gospodinov et al. 1974). Its composition is shifted towards a higher content of Bi_2S_3 (Gospodinov et al. 1974) and at 736°C $PbBi_2S_4$ contains about 2 mol% Bi_2S_3 (Salanci 1965). This compound crystallizes in the orthorhombic structure with the lattice parameters given in Table 10.11. It has a lower stability limit between 375°C and 390°C (Chang and Hoda 1977). The structure of $PbBi_2S_4$ remains stable until at least 9 GPa (Olsen et al. 2007). Bulk modulus and its first derivative are characterized by the next values: $B_0 = 43.9 \pm 0.7$ GPa and $B' = 6.3 \pm 0.3$.

$PbBi_4S_7$ [mineral mozgovaite (Miyawaki et al. 2019a,b)] melts incongruently at 732°C [at 736°C (Salanci 1965; Godovikov et al. 1967a; Salanci and Moh 1969; Godovikov and Fedorova 1969; Gospodinov et al. 1974); at 730°C (Craig 1967)], is only stable at high temperatures within the ranges from 675°C to 735°C and formed a solid solution between $Pb_{1.07}Bi_{3.96}S_7$ and $Pb_{0.91}Bi_{4.06}S_7$ at 700°C (Sugaki et al. 1974). According to the data of Craig (1967) and Gospodinov et al. (1974), $PbBi_4S_7$ contains 33–35 mol% PbS, and at 730°C

TABLE 10.11

Crystallographic Data for $PbBi_2S_4$

a, pm	*b*, pm	*c*, pm	$d_{calc.}$ g·cm^{-3}	$d_{meas.}$ g·cm^{-3}	References
1172 ± 3	1452 ± 3	407 ± 2	7.18	7.04	[a]Berry (1940)
1179.0 ± 0.7	1459.0 ± 0.8	410.0 ± 0.5	–	–	[a]Iitaka and Nowacki (1962)
1173	1447	408	–	–	Craig (1967)
1161	1357	408	–	–	Hoda and Chang (1975)
1159	1436	414	7.18	–7.12	Sadyhova et al. (1977)
1159	1459	410	–	–	Gasymov et al. (1980)
1175.3	1461.2	408.7	–	–	Tilley and Wright (1986b)
1184	1455	408	–	–	[a]Bente and Anton (1995)
1181.5 ± 0.2	1459.3 ± 0.2	408.14 ± 0.06	–	–	[a]Pinto et al. (2006)
1180.21 ± 0.17	1456.90 ± 0.12	407.58 ± 0.05	7,140	–	[a]Olsen et al. (2007)
1128.63 ± 0.13	1389.07 ± 0.14	392.33 ± 0.02	8,136	–	[a]Olsen et al. (2007) (at 9 GPa)
1174.2 ± 0.2	408.14 ± 0.08	1452.2 ± 0.3	7.19	–	[a]Topa et al. (2010b)

[a] Mineral galenobismutite.

dissolves up to 8 mol% Bi_2S_3 (Salanci 1965). At 675°C, this compound decomposes with the formation of Bi_2S_3 and $PbBi_2S_4$ (Salanci 1965; Craig 1967; Godovikov et al. 1967a; Godovikov and Fedorova 1969; Salanci and Moh 1969; Gospodinov et al. 1974).

This compound crystallizes in the monoclinic structure with the lattice parameters $a = 1346.1$, $b = 1237.1$, $c = 402.8$ pm, and $\beta = 92.19°$ (Tilley and Wright 1986b) [in the orthorhombic structure with the lattice parameters $a = 1327$, $b = 3684$, $c = 404$ pm, and an experimental density of 6.06 g·cm^{-3} for $Pb_{1.07}Bi_{3.96}S_7$ and $a = 1327$, $b = 5970$, and $c = 404$ pm for $Pb_{0.87}Bi_{4.08}S_7$ (Sugaki et al. 1974); $a = 774$, $b = 692$, $c = 600$ pm, and the calculated and experimental densities of 6.78 and 6.97 g·cm^{-3}, respectively (Sadyhova et al. 1977)].

$PbBi_4S_7$ was synthesized by a solid state reaction of PbS and Bi_2S_3, which were mixed in an agate mortar under acetone (Sugaki et al. 1974). The mixture was sealed in an evacuated silica tube and heated in a furnace at 700°C for 7 days. Generally, homogeneous products were obtained by the first heating. After the heating, the products were taken out and ground under acetone. It was again sealed in the evacuated silica tube and heated at 730°C for 7 days. After the second heating, it was quenched in cold water.

Starting from a material having the composition of PbS + $2Bi_2S_3$, crystals of $PbBi_4S_7$, which are stable above 675°C and below 736°C, have been synthesized by the evaporation method (Takéuchi et al. 1974). The syntheses which were carried out by applying various temperature gradients yielded three different types of crystals. One of these type crystallizes in the monoclinic structure ($a = 1324.7 \pm 0.3$, $b = 1204.2 \pm 0.2$, $c = 403.0 \pm 0.1$ pm, and $\beta = 105.02 \pm 0.02°$ and a calculated density of 6.74 g·cm^{-3}), and two another crystallize in the orthorhombic structure ($a = 1330 \pm 10$, $b = 3680 \pm 20$, $c = 403 \pm 1$ pm, and $a = 1334.3 \pm 0.2$, $b = 6002.2 \pm 0.9$, $c = 403.3 \pm 0.1$ pm). The composition of this compound lies between PbS·1·96Bi_2S_5 and PbS·2·86Bi_2S_5.

$\mathbf{Pb_2Bi_2S_5}$ is thermally stable up to 500°C (Sugaki et al. 1974) [according to the data of Craig (1967), the upper limit of stability of this compound in the presence of sulfur vapors is 424°C]. Salanci (1965) noted that when the temperature drops below 755°C, this compound changes its composition to $\mathbf{Pb_8Bi_6S_{17}}$, and then to $\mathbf{Pb_5Bi_4S_{11}}$. $Pb_2Bi_2S_5$ crystallizes in the orthorhombic structure with the lattice parameters given in Table 10.12.

Craig (1967) and Gospodinov et al. (1974) were unable to obtain this compound by annealing the starting components at 200°C–600°C, although, according to the data of Sugaki et al. (1974), it can be obtained by solid-state reactions. For this, PbS and Bi_2S_3 (molar ratio 2:1) were mixed uniformly under acetone in agate mortar. The mixture was sealed in the Hario glass tube under vacuum (0.13 Pa), put into a furnace and heated at 450°C

TABLE 10.12

Crystallographic Data for $Pb_2Bi_2S_5$

a, pm	*b*, pm	*c*, pm	$d_{calc.}$ g·cm^{-3}	$d_{meas.}$ g·cm^{-3}	References
1910.1	2391.3	406.1	–	–	[a]Weitz and Hellner (1960)
1909	2389	405.8	6.86	6.86–6.99	[a]Hayashi (1961); Bente and Anton (1995)
1909.8 ± 0.4	2389.0 ± 0.5	405.7 ± 0.1	7.17	–	[a]Srikrishnan and Nowacki (1974)
2384.05 ± 0.11	404.79 ± 0.02	1904.36	–	–	[a]Izumino et al. (2014)
2378.78 ± 0.09	405.66 ± 0.03	1910.26 ± 0.08	–	–	[a]Kovač et al. (2019)

[a] Mineral cosalite.

TABLE 10.13

Crystallographic Data for $Pb_3Bi_2S_6$

a, pm	*b*, pm	*c*, pm	$d_{calc.}$	$d_{meas.}$	References
			g·cm^{-3}		
1587	1905	411	6.58	7.10	[a]Fleischer (1962)
3100 ± 10	1280 ± 10	415 ± 2	7.34	7.16 ± 0.06	[a]Fleischer (1965); Syritso and Senderova (1964)
1347	2075	413	–	–	[a]Hoda and Chang (1975)
1280	3112	478	5.92	5.95	Sadyhova et al. (1977)
1352	2061	411	–	–	[a]Bente and Anton (1995)
1354 ± 10	2074 ± 10	415.8 ± 0.5	7.09	7.06–7.14	Klyakhin and Dmitrieva (1968); Fleischer (1969a)
1353.5 ± 0.3	2045.1 ± 0.5	410.4 ± 0.1	–	–	[a]Takagi and Takéuchi (1972)
1357.6 ± 0.9	2060.6 ± 0.8	411.9 ± 0.2	–	–	[a]Borodayev et al. (2001) (for the crystal with Pb/Bi ≈ 1.5)
1356 ± 1	2057 ± 1	411.5 ± 0.2	–	–	[a]Borodayev et al. (2001) (for the crystal with Pb/Bi ≈ 1.41)
1356.7 ± 0.9	2065.5 ± 0.2	412.16 ± 0.04	–	–	[a]Pinto et al. (2011)

[a] Mineral lillianite.

for 5 days. The product was taken out from the glass tube after cooling in air and again mixed under acetone. The mixture was sealed in the evacuated glass tube again and reheated at 450°C for 15 days with the next cooling in air.

Mineral cosalite ($Pb_2Bi_2S_5$) could contain some quantities of Cu, Ag, and Fe (Topa and Makovicky 2010; Izumino et al. 2014; Kovač et al. 2019).

$Pb_3Bi_2S_6$ melts incongruently at 816°C (Craig 1967; Godovikov et al. 1967a; Godovikov and Fedorova 1969) [at 805°C (Salanci and Moh 1969); at 804°C (Gospodinov et al. 1974); at 780°C (Sadyhova et al. 1977)]. The homogeneity region of this compound at 750°C is within the interval 25–29 mol% Bi_2S_3 (Craig 1967) [26–31 mol% Bi_2S_3 (Godovikov et al. 1967a)]. Apparently, this compound has two polymorphic modifications. First of them (mineral lillianite) crystallizes in the orthorhombic structure with the lattice parameters given in Table 10.13.

The second modification of this compound (mineral xilingolite) crystallizes in the monoclinic structure with the lattice parameters a = 1351.1 ± 0.3, b = 408.50 ± 0.11, c = 2064.9 ± 0. 2 pm, and β = 92.15 ± 0.02° (Berlepsch et al. 2001; Jambor et al. 2002) [a = 1365, b = 407.8, c = 2068 pm, β = 93.0°, and the calculated and experimental densities of 7.07 and 7.08 g·cm^{-3}, respectively (Dunn et al. 1984a)].

The synthetic compound $Pb_3Bi_2S_6$ was investigated by XRD on single crystals in a diamond-anvil cell between 0.1 MPa and 10.5 GPa (Olsen et al. 2008). It was shown, that this compound undergoes a first-order phase transition at hydrostatic pressure between 3.7 and 4.9 GPa. The transition is strongly anisotropic, with a contraction along one of the crystal axes by 16% and expansion along another one by 14%. This is a piezoplastic phase transition, a displacive pressure-induced phase transition with systematic shearing of atomic planes and a migration of chemical bonds in the structure. The lattice parameters a and c of the orthorhombic cell decrease from 1354.0 ± 0.3 and 411.03 ± 0.07 pm at 0.1 MPa to 1089.1 ± 0.4 and 399.26 ± 0.09 pm at 7.92 ± 0.05 GPa, while the lattice parameter b and a calculated density increase from 2063.7 ± 0.4 pm and 7.125 g·cm^{-3} to 2268.3 ± 0.9 pm and 8.295 g·cm^{-3} at the same pressures.

Samples of $Pb_3Bi_2S_6$ were prepared in two different ways (Prodan et al. 1982). (1) Appropriate stoichiometric amounts of PbS and Bi_2S_3 were mixed, evacuated in a quartz tube, and heated for a week at 800°C. After opening the tube and repeating the procedure, the sample was slowly cooled to room temperature in the switched-off furnace. The resulting product consisted of a uniform mass of crystallites with well-developed faces. (2) The sample was also synthesized in an open H_2S stream. Pressed tablets were heated by gradually increasing the temperature between 400°C and 800°C (24 h at 400°C, 24 h at 500°C, and 6 h at 800°C) to prevent Bi evaporation.

The title compound could be also obtained by the hydrothermal method (Klyakhin and Dmitrieva 1968; Fleischer 1969a).

$Pb_4Bi_2S_7$ melts incongruently at 831°C (Gospodinov et al. 1974) [at 834°C (Godovikov et al. 1967a; Godovikov and Fedorova 1969); at 829°C (Craig 1967)] and decomposes at the temperatures below 475°C (Craig 1967; Gospodinov et al. 1974). Sometimes this compound is attributed to the formula **$Pb_9Bi_4S_{15}$** (Craig 1967; Gospodinov et al. 1974; Dmytriv 1975) or **$Pb_6Bi_2S_9$** (Van Hook 1960; Godovikov and Fedorova 1969; Prodan et al. 1982), the first of which better describes the composition of the compound at high temperatures, and the second at low temperatures. $Pb_4Bi_2S_7$ crystallizes in the orthorhombic structure with the lattice parameters a = 1371, b = 3146, and c = 413 pm (Hoda and Chang 1975) [a = 1351, b = 2341, and c = 411 pm (Gasymov and Mamedov 1978)].

$Pb_5Bi_4S_{11}$ (or **$Pb_8Bi_6S_{17}$**, or **$Pb_{13}Bi_{10}S_{28}$**) is the new mineral sulfosalt, which was described in Nowacki and Stalder (1969) and Fleischer (1970). It crystallizes in the orthorhombic structure with the lattice parameters a = 3960 ± 10, b = 410 ± 5, and c = 1428 ± 5 pm. Later, this mineral was called bursaite (Jambor and Burke 1989) and it was shown that its XRD pattern can be indexed as a mixture of two orthorhombic phases with a = 1342, b = 2037, c = 412 pm, and a = 1355, b = 1979, c = 404 pm. Then, this mineral was discredited as it is a mixture of two phases (Burke 2006).

$Pb_6Bi_2S_9$ has apparently two polymorphic modifications. First of them (mineral heyrovskyite) crystallizes in the orthorhombic structure with the lattice parameters given in

TABLE 10.14
Crystallographic Data for $Pb_6Bi_2S_9$

a, pm	*b*, pm	*c*, pm	$d_{calc.}$ g·cm^{-3}	$d_{meas.}$ g·cm^{-3}	References
1370.5 ± 1.3	3119.4 ± 3.3	412.1 ± 0.3	7.18	7.17	Klomínský et al. (1971); Fleischer (1972)
1371.2 ± 0.2	3121.0 ± 0.5	413.1 ± 0.1	–	–	Takéuchi and Takagi (1974)
1358.3 ± 0.5	3053.3 ± 1.0	412.9 ± 1.0	–	–	Karup-Møller and Makovicky (1981) (enriched with Ag and Bi)
411.0 ± 0.1	1360.0 ± 0.3	3048.5 ± 1.2	–	–	Makovicky et al. (1991) (enriched with Ag and Bi)
1369	3156	413	–	–	Bente and Anton (1995)
1371.9 ± 0.4	3139.3 ± 0.9	413.19 ± 0.10	7.274	–	Olsen et al. (2010) (at 0.1 MPa)
2530.2 ± 0.7	3081.9 ± 0.9	406.40 ± 0.13	8.1551	–	Olsen et al. (2010) (at 5.06 GPa)
1374.98 ± 0.04	3150.53 ± 0.08	414.75 ± 0.01	7.294	–	Pinto et al. (2011)
414.57 ± 0.01	1374.21 ± 0.05	3150.05 ± 0.12	7.216	–	Pervukhina et al. (2012)

Table 10.14. The crystal structure of this modification was investigated at pressures between 0 and 5.6 GPa with XRD on single-crystals (Olsen et al. 2010). Heyrovskyite is stable until at least 3.9 GPa and a first-order phase transition occurs between 3.92 and 4.84 GPa. A single-crystal is retained after the reversible phase transition despite an anisotropic contraction of the unit cell and a volume decrease of 4.2%.

Second modification of $Pb_6Bi_2S_9$ (mineral aschamalmite) crystallizes in the monoclinic structure with the lattice parameters $a = 1371.9 \pm 0.1$, $b = 413.2 \pm 0.1$, $c = 3141.9 \pm 0.3$ pm, and $\beta = 90.94 \pm 0.01°$ (Callegari and Boiocchi 2009) [$a = 1371$, $b = 409$, $c = 3143$ pm, and $\beta = 91.0°$ and a calculated density of 7.33 g·cm^{-3} (Dunn et al. 1984b)].

$Pb_8Bi_{10}S_{23}$ or **$Pb_{46}Bi_{54}S_{127}$** (mineral cannizzarite) crystallizes in the monoclinic structure with the lattice parameters $a = 18980 \pm 20$, $b = 409 \pm 1$, $c = 7406 \pm 7$ pm, $\beta = 101.93 \pm 0.08°$, and the calculated and experimental densities of 6.95 and 6.7 g·cm^{-3}, respectively (Matzat 1979) [$a = 706$, $b = 408$, $c = 1530$ pm, and $\beta = 99°00'$ (Nowacki and Stalder 1969; Fleischer 1970)].

In addition to the above compounds, the existence of **$PbBi_6S_{10}$**, **$Pb_3Bi_8S_{15}$**, **$Pb_5Bi_6S_{14}$** (Van Hook 1960), **$Pb_2Bi_6S_{11}$** (orthorhombic structure with the lattice parameters $a = 1327.8$, $b = 3676.8$, and $c = 403.4$ pm), and **$Pb_3Bi_{10}S_{18}$** (orthorhombic structure with the lattice parameters $a = 1336.6$, $b = 5954.2$, and $c = 406.0$ pm) (Tilley and Wright 1986b), **$Pb_{4.65}Bi_{20.90}S_{36}$** (orthorhombic structure with the lattice parameters $a = 1334.3 \pm 0.2$, $b = 6002.2 \pm 0.9$, and $c = 403.3 + 0.1$ pm) (Takéuchi et al. 1979) and **$Pb_5Bi_2S_8$** (orthorhombic structure with the lattice parameters $a = 1356$, $b = 2805$, and $c = 414$ pm) (Gasymov and Mamedov 1978) are also noted. To obtain $Pb_{4.65}Bi_{20.90}S_{36}$, denoted as phase V-3, a mixture having a composition $PbBi_4S_7$, with 7% excess sulfur, was put in a sealed, evacuated silica tube, 60 mm in length (Takéuchi et al. 1979). The tube was then heated with a temperature gradient, 725°C at one end and 690°C at the other end. In the product, thin tabular crystals of V-3 were found at that portion of the tube which had been in the temperature range from 700°C to 715°C; at other portions of the tube crystals of $PbBi_2S_3$ and Bi_2S_3 were identified.

It is assumed that some of these compounds exist in the solid phase at sufficiently low temperatures, and therefore, when studying them, an equilibrium state is not achieved (Van Hook 1960).

Some isothermal sections of the Pb–Bi–S ternary system were constructed by Craig (1967) and Salanci and Moh (1969).

10.40 Lead–Vanadium–Sulfur

One compound is formed in the Pb–V–S system. **$PbVS_3$** crystallizes in the monoclinic structure with the lattice parameters $a = 571.9 \pm 0.3$, $b = 576 \pm 2$, $c = 2392.4 \pm 0.8$ pm, and $\beta = 98.83 \pm 0.07°$ (Hernán et al. 1991a). To prepare the title compound, Pb, V, and S were mixed in the 1:1:3 proportions in relation with the ratios PbS/VS_2 of 1:1 (Maaren van 1972; Guemas et al. 1988; Wiegers et al. 1989a). The mixtures were put into silica tubes, sealed under vacuum and subsequently heated up to 800°C–900°C for about 1–2 weeks. In order to obtain single crystals the vapor transport method was used. It turned out that crystals could be grown using chlorine as transport agent in a temperature gradient from 920°C to 800°C for 7 days. The crystals grow at the low-temperature side of the tube. To the sample to be transported (ca. 300 mg) about 5–10 mg $(NH_4)_2PbCl_6$ was added. The crystals have a metallic luster.

According to the data of later investigations, this compound is a misfit layer compound and could be described by the **$(PbS)_{1.12}VS_2$** formula. It crystallizes in the monoclinic structure with the lattice parameters $a = 580.7 \pm 0.2$, $b = 573.6 \pm 0.2$, $c = 1180.9 \pm 0.5$ pm, and $\beta = 94.76 \pm 0.03°$ (Wiegers et al. 1989c, 1990b; Wiegers and Meerschaut 1992) [$a = 572.0$, $b = 576.6$, $c = 2389.6$ pm, and $\beta = 98.91°$ (Gotoh et al. 1989); $a = 572.8$, $b = 578.9$, $c = 2393.9$ pm, and $\beta = 98.95°$ (Gotoh et al. 1990b); $a = 581.0$, $b = 573.3$, $c = 1179$ pm, and $\beta = 94.7°$ (Wiegers et al. 1989c, 1990b] for PbS-sublattice, and $a = 327.9 \pm 0.4$, $b = 571.8 \pm 0.6$, $c = 2357.9 \pm 0.2$ pm, and $\beta = 94.61 \pm 0.08°$ (Wiegers et al. 1989c, 1990b; Wiegers and Meerschaut 1992); $a = 573.0$, $b = 325.6$, $c = 2394.8$, and $\beta = 98.97°$ (Gotoh et al. 1990b); $a = 572.79 \pm 0.03$, $b = 578.86 \pm 0.07$, $c = 2393.9 \pm 0.2$ pm, and $\beta = 98.947 \pm 0.003°$ (Onoda et al. 1990)] for VS_2-sublattice. The calculated density of this compound is 5.788 g·cm^{-3} (Onoda et al. 1990).

10.41 Lead–Niobium–Sulfur

Several ternary compounds are formed in the Pb–Nb–S system. $\mathbf{Pb_{0.15}Nb_3S_4}$ crystallizes in the hexagonal structure with the lattice parameters a = 962.6 ± 0.4, c = 339.0 ± 0.2 pm, and a calculated density of 5.37 ± 0.03 g·cm^{-3} (Amberger et al. 1985). It was prepared from a mixture of Nb, S, and Pb powders (molar ratio 3:4:1), which was heated to 750°C for 24 h in a sealed silica tube and later transported for 1 week between 950°C and 1050°C with Cl_2 as transporting agent.

The $\mathbf{Pb_{1/3}NbS_2}$ intercalate compound crystallizes in the hexagonal structure with the lattice parameters a = 575.9 ± 0.1, c = 1481.3 ± 0.2 pm, and a calculated density of 7.356 g·cm^{-3} (Fang et al. 1996) [a = 579.2 and c = 1480 pm (Eppinga and Wiegers 1980)]. The preparation of this compound from the elements and from 2*H*-NbS_2 and Pb lead to the same product. The mixtures were heated in evacuated quarts ampoules at 700°C–900°C for 1–4 weeks (Eppinga and Wiegers 1980).

Another intercalate compound, $\mathbf{PbNbS_2}$, also crystallizes in the hexagonal structure with the lattice parameters a = 334.3 and c = 1773 pm (Eppinga and Wiegers 1980). It was obtained in the same way as $Pb_{1/3}NbS_2$ was prepared.

$\mathbf{PbNbS_3}$ crystallizes in the orthorhombic structure with the lattice parameters a = 2316.2 ± 0.3, b = 579.0 ± 0.1, c = 2390.1 ± 0.5 pm [c = 2385 pm (Hernán et al. 1992)] (Guemas et al. 1988) [in the tetragonal structure a = 407 and c = 1192 pm (Schmidt 1970)]. To prepare this compound, Pb, Nb, and S were mixed (molar ratio 1:1:3). The mixture was put into silica tubes, sealed under vacuum and subsequently heated up to 900°C for about 15 days after 2 days at 300°C. Cooling down took place within a few hours. Samples were then ground and re-heated up to 900°C with a small amount of iodine (ca. 5 mg·cm^{-3}).

$\mathbf{PbNb_2S_5}$ also crystallizes in the orthorhombic structure with the lattice parameters a = 2328.0 ± 0.6, b = 577.7 ± 0.1, c = 3584 ± 1 pm [c = 3567 pm (Hernán et al. 1992)] (Guemas et al. 1988). This compound was obtained in the same way as $PbNbS_3$ was synthesized using the molar ratio Pb, Nb, and S of 1:2:5.

$\mathbf{(PbS)_{1.14}NbS_2}$ is a misfit layer compounds and crystallizes in the orthorhombic structure with the lattice parameters a = 583.4 ± 0.2, b = 580.1 ± 0.1, c = 1190.2 ± 0.3 pm for PbS-sublattice and a = 331.3 ± 0.1, b = 580.1 ± 0.1, c = 2380.7 ± 0.1 pm for NbS_2-sublattice (Wiegers et al. 1989a,c, 1990a; Kuypers et al. 1990; Wiegers and Meerschaut 1992). Two methods were used for syntheses of this compound, namely direct combination of the elements or the binary compounds at temperatures of 800°C–1200°C and reaction of the ternary oxides with CS_2 or H_2S vapor at 1000°C–1300°C. Single crystals were grown by vapor transport using $(NH_4)_2PbCl_2$ (2 mass%) as a source for chlorine. The crystals grow at the low-temperature side (gradient 800°C–700°C) of a quartz ampoule. They have a metallic luster.

$\mathbf{(PbS)_{1.14}(NbS_2)_2}$ is also a misfit layer compounds and has two polytypes. First polytype crystallizes in the orthorhombic structure with the lattice parameters a = 582.9, b = 577.5, c = 3586 pm for PbS-sublattice a = 332.6, b = 577.5, c = 3587 pm for NbS_2-sublattice (Wiegers and Meerschaut 1992).

Second polytype of this compound crystallizes in the monoclinic structure with the lattice parameters a = 583.7 ± 0.4, b = 578.8 ± 0.2, c = 3604 ± 1 pm, and β = 96.13 ± 0.02° for PbS-sublattice and a = 332.0 ± 0.1, b = 578.1 ± 0.2, c = 3604.2 ± 0.8 pm, and β = 96.12 ± 0.01° for NbS_2-sublattice (Auriel et al. 1993). This compound was obtained in the same way as $(PbS)_{1.14}NbS_2$ was synthesized (Gotoh et al. 1990a; Wiegers and Meerschaut 1992; Auriel et al. 1993).

10.42 Lead–Tantalum–Sulfur

Several ternary compounds are formed in the Pb–Ta–S system. The $\mathbf{Pb_{1/3}TaS_2}$ intercalate compound crystallizes in the hexagonal structure with the lattice parameters a = 575.0 pm and c = 1484 pm (Di Salvo et al. 1973; Eppinga and Wiegers 1980). The $\mathbf{PbTaS_2}$ intercalate compound also crystallizes in the hexagonal structure with the lattice parameters a = 332 pm and c = 170 pm (Eppinga and Wiegers 1980) [a = 330 pm, c = 1750 pm (Di Salvo et al. 1973)]. The superlattice of these compounds is quite stable. They were prepared by direct exposure of TaS_2 to the Pb vapor (this method yields the stoichiometry) or by heating the mixture of elements (Di Salvo et al. 1973; Eppinga and Wiegers 1980). Temperatures from 600°C–900°C for periods from 1–4 weeks were applied.

$\mathbf{PbTaS_3}$ crystallizes in the tetragonal structure a = 408 and c = 1197 pm (Schmidt 1970; Wiegers et al. 1989a) [in the orthorhombic structure with the lattice parameters a = 1144, b = 629, c = 1197 pm, and the calculated and experimental densities of 7.51 and 7.32 g·cm^{-3}, respectively (Novoselova and Aslanov 1967)]. The samples of this compound were synthesized by firing the elements in evacuated quartz tubes in the temperature range from 600°C to 1100°C with repeated grinding and pressing into pellets (Novoselova and Aslanov 1967; Schmidt 1970; Wiegers et al. 1989a). Long firing times of at least 200 h were required in order to achieve complete reaction. In order to obtain single crystals the vapor transport was used. It turned out that those crystals could be grown using Cl_2 as a carrier reagent in a temperature gradient of 850°C–780°C; the crystals grow at the low-temperature side of the tube over 10 days. To the sample to be transported (about 300 mg) about 5–10 mg of $(NH_4)_2PbCl_6$ was added. The crystals have a metallic luster.

$\mathbf{(PbS)_{1.13}TaS_2}$ is a misfit layer compounds and crystallizes in the orthorhombic structure with the lattice parameters a = 582.5 ± 0.1, b = 577.8 ± 0.1, c = 2396.2 ± 0.2 pm for PbS-sublattice, and a = 330.44 ± 0.04, b = 577.9 ± 0.4, c = 2395.4 ± 0.4 pm for TaS_2-sublattice (Wulff et al. 1990) [a = 580.3, b = 577.2, c = 2400 pm for PbS-sublattice, and a = 330.6, b = 577.2, c = 2400 pm for TaS_2-sublattice (Wiegers et al. 1989c, 1990b; Kuypers et al. 1990; Wiegers and Meerschaut 1992)]. The compound was prepared by heating the elements in evacuated quartz ampoule at 850°C. Crystals were grown by vapor transport in a gradient of 930°C–850°C. To about 300 mg of the compound to be transported 5–10 mg $(NH_4)_2PbCl_6$, which acts as a convenient source for chlorine, was added. The crystals grow at the low-temperature side of the tube as thin platelets.

The $\mathbf{(PbS)(TaS_2)_2}$ ternary compound has some layered structure and take preferred orientation: a two-atom-thick layer of PbS and three-atom-thick sandwiches of TaS_2 are stacked alternately (Gotoh et al. 1990a). It was prepare as follows. Powders of Pb, Ta, and S were mixed together in the molar ratio 1:2:5 in a silica tube. The tube was heated at first at 500°C and then at 800°C.

10.43 Lead–Oxygen–Sulfur

Several ternary compounds are formed in the Pb–O–S system. **$PbSO_3$** has apparently two polymorphic modifications and decomposes rapidly above 400°C forming PbS, **$PbSO_4$**, and **PbO·$PbSO_4$** (Lutz and El Suradi 1976). First modification crystallizes in the orthorhombic structure with the lattice parameters $a = 792.5 \pm 0.1$, $b = 548.5 \pm 0.1$, $c = 681.6 \pm 0.1$ pm, and the calculated and experimental densities of 6.42 and 6.28 g·cm^{-3}, respectively (Christensen and Hewat 1990) [$a = 790.3 \pm 0.1$, $b = 548.8 \pm 0.1$, and $c = 680.2 \pm 0.1$ pm (Lutz et al. 1983)]. It was prepared by mixing a solution of $Pb(CH_3COO)_2$ (26 g) in water (250 mL) with a solution of Na_2SO_3 (10 g) in water (250 mL) (Lutz and El Suradi 1976; Christensen and Hewat 1990). The precipitate was washed with water and dried at 90°C. Single crystals of the title compound have been obtained by gel crystallization technique (Lutz et al. 1983). The crystallization medium was a silica gel adjusted to pH 6.8–7.0 with HCl or CH_3COOH. Diamond-shaped single crystals could be obtained using 0.1 M solutions of $Pb(CH_3COO)_2$ and Na_2SO_3 at room temperature.

Second modification of $PbSO_3$ (mineral scotlandite) crystallizes in the monoclinic structure with the lattice parameters $a = 454.2 \pm 0.2$, $b = 533.3 \pm 0.2$, $c = 641.3 \pm 0.2$ pm, $\beta = 106.22 \pm 0.04°$, and the calculated and experimental densities of 6.40 and 6.37 ± 0,02 g·cm^{-3}, respectively (Dunn et al. 1985; Paar et al. 1984) [$a = 672$, $b = 534$, $c = 451$ pm, $\beta = 114°$, and a calculated density of 6.44 g·cm^{-3} (Sarp and Burri 1984); $a = 450.5 \pm 0.2$, $b = 533.3 \pm 0.2$, $c = 640.5 \pm 0.6$ pm, $\beta = 106.24 \pm 0.03°$, and a calculated density of 6.396 g·cm^{-3} (Pertlik and Zemann 1985)].

$PbSO_4$ has a polymorphic transformation at 864°C [at 840°C (Margulis and Kopylov 1964; Tridot et al. 1969)] and melts congruently at 1170°C (Jaeger and Germs 1921). The quenching tests do not allow preserving the high temperature modification (Tridot et al. 1969). This compound exists in a nature as a mineral anglesite, which crystallizes in the orthorhombic structure with the lattice parameters $a = 695.802 \pm 0.001$, $b = 848.024 \pm 0.003$, $c = 539.754 \pm 0.001$ pm (Antao 2012) [$a = 845$, $b = 538$, $c = 693$ pm (James and Wood 1925); $a = 695.9 \pm 0.2$, $b = 848.2 \pm 0.2$, $c = 539.8 \pm 0.2$ pm (Miyake et al. 1978); $a = 695.49 \pm 0.09$, $b = 847.23 \pm 0.11$, $c = 539.73 \pm 0.8$ pm (Jacobsen et al. 1998)].

PbS_2O_3 also crystallizes in the orthorhombic structure with the lattice parameters $a = 718.02 \pm 0.03$, $b = 691.99 \pm 0.03$, $c = 1611.56 \pm 0.08$ pm, and a calculated density of 5.30 g·cm^{-3} (Christensen et al. 1991). A white crystalline precipitate of this compound was formed when a solution of $Pb(NO_3)_2$ (22.7 g in 500 mL H_2O) was added drop wise to a solution of $Na_2S_2O_3 \cdot 5H_2O$ (19.0 g in 500 mL H_2O). The solution was stirred, and the addition of $Pb(NO_3)_2$ resulted in a precipitate that was redissolved immediately until approximately 30% of the $Pb(NO_3)_2$ solution has been added. When all the $Pb(NO_3)_2$ solution had been added, the precipitate formed was washed by decanting, filtered, washed on the filter and dried in air at room temperature. Instead of lead nitrate $Pb(CH_3COO)_2 \cdot 3H_2O$ (26 g) can be used.

In early works it was shown that the PbO·$PbSO_4$, **2PbO·$PbSO_4$**, and **4PbO·$PbSO_4$** ternary compounds are formed in the Pb–O–S system (Schenck and Rassbach 1908; Reinders 1915; Schenck and Albers 1919; Jaeger and Germs 1921; Schenck and Borkenstein 1925).

PbO·$PbSO_4$ melts congruently at 965°C (Margulis and Kopylov 1964) [at 977°C (Jaeger and Germs 1921); at 960°C (Tridot et al. 1969)]. No phase transformations between room and melting temperature were determined for this compound (Tridot et al. 1969). This compound crystallizes in the monoclinic structure with the lattice parameters given in Table 10.15.

2PbO·$PbSO_4$ melts congruently at 961°C [at 932°C (Margulis and Kopylov 1964)], is unstable below 450°C and decomposes at higher temperatures into a mixture of PbO·$PbSO_4$ and 4PbO·$PbSO_4$ (Jaeger and Germs 1921; Lander 1949; Bode and Voss 1959; Boivin et al. 1968); the decomposition temperature in air is shown to be 631°C ± 8°C (Esdaile 1966)]. It has a polymorphic transformation at 450°C (Esdaile 1966; Tridot et al. 1969). According to the data of Billhardt (1970) and Tridot et al. (1969), this compound is unstable below about 640°C and decomposes at this temperature to PbO·$PbSO_4$ and 4PbO·$PbSO_4$. This decomposition takes place very slowly. Even at cooling rates of 10°C·h^{-1}, β-2PbO·$PbSO_4$ is stable down to 450°C and then it forms metastable α-2PbO·$PbSO_4$.

TABLE 10.15

Crystallographic Data for PbO·$PbSO_4$

a, pm	*b*, pm	*c*, pm	β, °	$d_{calc.}$ g·cm^{-3}	$d_{meas.}$ g·cm^{-3}	References
1373	568	707	116.22	–	6.92	[a]Richmond and Wolfe (1938)
1375 ± 4	568 ± 2	705 ± 2	116.2 ± 0.2	–	6.92	[a]Binnie (1951)
1376.9 ± 0.5	569.8 ± 0.3	707.9 ± 0.2	115.93 ± 0.17	–	–	[a]Sahl (1969, 1970)
1375.3 ± 0.2	569.7 ± 0.2	706.4 ± 0.2	115.9 ± 0.2	7.02	7.08 ± 0.10	Billhardt (1970)
1374.6 ± 0.3	569.64 ± 0.09	706.6 ± 0.1	115.79 ± 0.01	–	–	[a]Mentzen and Latrach (1983)
1375.76 ± 0.01	565.78 ± 0.01	703.22 ± 0.01	115.885 ± 0.001	–	–	Mentzen et al. (1984b) (at 5 K)

[a] Mineral lanarkite.

The calorimetric study shows that the normal process of evolution of 2PbO·$PbSO_4$, that is to say the one which comes closest to an evolution at equilibrium, is as follows (Boher et al. 1987): 3α-2PbO·$PbSO_4$ → 2(PbO·$PbSO_4$) + 4PbO·$PbSO_4$ at 440°C, then 2(PbO·$PbSO_4$) + 4PbO·$PbSO_4$ → 3β-2PbO·$PbSO_4$ at 620°C. For higher heating rates (>40°C·h^{-1}), a kinetic process appears as follows: α-2PbO·$PbSO_4$ → β-2PbO·$PbSO_4$ at 440°C. β-2PbO·$PbSO_4$ is unstable and break down into PbO·$PbSO_4$ and 4PbO·$PbSO_4$.

A new low-temperature phase of 2PbO·$PbSO_4$ has been observed by Boher et al. (1985). The transition is of the ferroelastic type; monoclinic (>55 K) ⇔ triclinic (<55 K). The evolution of the cell parameters has been determined by neutron diffraction. Below 55 K paraelastic domains are locked between the ferroelastic ones respectively (Boher et al. 1985; Latrach et al. 1985a).

α-2PbO·$PbSO_4$ (low-temperature phase) crystallizes in the triclinic structure with the lattice parameters a = 714.45 ± 0.04 and 714.65 ± 0.04, b = 575.47 ± 0.03 and 575.56 ± 0.03, c = 804.62 ± 0.05 and 804.53 ± 0.04 pm, α = 89.393 ± 0.005° and 89.361 ± 0.005°, β = 102.260 ± 0.005° and 102.258 ± 0.005°, γ = 88.614 ± 0.004° and 88.544 ± 0.004° at 45 and 2 K, respectively (Boher et al. 1985; Latrach et al. 1985a).

β-2PbO·$PbSO_4$ (second low-temperature phase) crystallizes in the monoclinic structure with the lattice parameters a = 714.71 ± 0.01, 714.54 ± 0.04 and 719.40 ± 0.01, b = 575.47 ± 0.03, and 579.424 ± 0.008, c = 804.84 ± 0.01, 804.62 ± 0.05 and 806.42 ± 0.01 pm, β = 102.283 ± 0.001°, 102.278 ± 0.004° and 102.4016 ± 0.0009° at 5, 45, and 300 K, respectively (Boher et al. 1985; Latrach et al. 1985a,b) [a = 781.4, b = 580.3, c = 803.5 pm, β = 102.64°, and the calculated and experimental densities of 6.98 and 7.08 g·cm^{-3}, respectively (Boivin et al. 1968; Tridot et al. 1969); a = 806 ± 2, b = 579 ± 2, c = 717.5 ± 2.0 pm, β = 103.00 ± 0.02°, and the calculated and experimental densities of 7.63 and 7.60 ± 0.02 g·cm^{-3}, respectively (Billhardt 1970); a = 716.8 ± 0.2, b = 577.1 ± 0.1, c = 803.6 ± 0.2 pm, β = 102.40 ± 0.02° (Sahl 1981); a = 718.16 ± 0.03, b = 578.48 ± 0.02, c = 805.28 ± 0.03 pm, and β = 102.395 ± 0.002° (Mentzen et al. 1984a)]. According to the data of Bode and Voss (1959), an experimental density of this compound is 7.59 g·cm^{-3}.

γ-2PbO·$PbSO_4$ (high-temperature phase) crystallizes in the orthorhombic structure with the lattice parameters a = 969.03 ± 0.09, b = 1196.88 ± 0.08, and c = 609.89 ± 0.07 pm at 700°C (Boher et al. 1984) [a = 967.82 ± 0.04, b = 1195.65 ± 0.05, and c = 609.40 ± 0.03 pm at 700°C (Mentzen et al. 1984a)]. The evolution of cell parameters has been determined for the low-temperature β-phase below 440°C and for the γ-phase in the metastable range 440°C–630°C and up to 700°C by Boher et al. (1984).

The 4PbO·$PbSO_4$ ternary compound is also stable in the PbO–$PbSO_4$ system above 210°C (Jones and Rothschild 1958; Esdaile 1966). It decomposes incongruently at 800°C with the formation of 2PbO·$PbSO_4$ and PbO (Margulis and Kopylov 1964) [melts at 890°C–895°C (Bode and Voss 1959; Dapo 1974); at 920°C (Tridot et al. 1969); according to the data of Jaeger and Germs (1921), this compound has a composition of 3PbO·$PbSO_4$ and melts incongruently at 897°C]. At 840°C there may be an allotropic transformation for this compound (Tridot et al. 1969). 4PbO·$PbSO_4$ crystallizes in the monoclinic structure with the lattice parameters a = 729.7 ± 0.1, b = 1169.8 ± 0.2, c = 1149.8 ± 0.2 pm, β = 90.93 ± 0.01°, and a calculated density of 8.095 g·cm^{-3} (Steele and Pluth 1998) [a = 1450, b = 1168, c = 1153 pm, β = 90.4°, and the calculated and experimental densities of 8.15 and 8.0 g·cm^{-3}, respectively (Bode and Voss 1959); a = 1144.3 ± 2.0, b = 1166.6 ± 2.0, c = 731.6 ± 2.0 pm, β = 90.82 ± 0.20°, and the calculated and experimental densities of 8.135 and 8.1± 0.1 g·cm^{-3}, respectively (Billhardt 1970); a = 1461.0 ± 0.4, b = 1170.3 ± 0.3, c = 1152.6 ± 0.3 pm, β = 91.00 ± 0.20° (Sahl 1975); a = 728.2 ± 0.3, b = 1166.2 ± 0.3, c = 1148.2 ± 0.3 pm, β = 90.90 ± 0.03°, and the calculated and experimental densities of 8.05 and 8.10 [8.04 ± 0.6 (Dapo 1974)] g·cm^{-3}, respectively (Mentzen et al. 1981)]. Zimkina and Shchemelev (1960) gave for 4PbO·$PbSO_4$ the following orthorhombic, pseudo-tetragonal crystal data: a = 1170, b = 1150 and c = 730 pm, obtained from electron diffraction patterns, and Pinsker and Farmakovskaya (1961) determined the hexagonal cell parameters a = 898 and c = 2480 pm for this compound. However, Billhardt (1970) has never observed the hexagonal modification of this compound, nor any other basic lead sulfate with hexagonal cell parameters.

Various wet methods of formation of all basic lead sulfates are given by Clark et al. (1936). All three lead sulfates can be prepared from PbO and $PbSO_4$ taken in stoichiometric ratios (Lander 1949; Esdaile 1966). The mixtures were heated in Pt crucibles for 0.5 h at temperatures 30°C above the liquidus temperatures, quenched, finely ground, reheated at 800°C and finally quenched. PbO·$PbSO_4$ and 4PbO·$PbSO_4$ could be also obtained by boiling water suspensions of PbO and $PbSO_4$ in the proper molar ratio (Lander 1949) or by interaction of the appropriate amounts of PbO and $PbSO_4$ at 500°C for PbO·$PbSO_4$ and at 600°C–700°C for 4PbO·$PbSO_4$ (Lamb and Niebylski 1951). PbO·$PbSO_4$ and 4PbO·$PbSO_4$ were also synthesized from PbO and the stoichiometric quantity of H_2SO_4 (Burbank 1966). One mole of powdered PbO was mixed in a liter of distilled water, and the required amount of H_2SO_4 solution (d = 1.250 g·cm^{-3}) was added drop wise with stirring which was continued 4 h after addition of the acid. The mixture for 4PbO·$PbSO_4$ was heated during the initial reaction period to approximately 80°C and then allowed to cool to room temperature. The mixtures were digested for 4 days at room temperature, the solids collected on a filter, and air dried.

4PbO·$PbSO_4$ could be also synthesized as follows (Dapo 1974): 250 g of yellow PbO was suspended in 1.0 L of deionized water by means of rapid stirring. To this was added 31.0 mL of H_2SO_4 (d = 1.400 g·cm^{-3}). The suspension was stirred and heated at 80°C for 4 h. The solid was separated by filtration, washed and dried at 110°C. This intermediate solid was ground and transferred to a Ni crucible. The solid was heated at 550°C for 4 h to obtain the title compound (Dapo 1974; Mentzen et al. 1981).

4PbO·$PbSO_4$ can also form at room temperature by the interaction of PbO and H_2SO_4, although the formation rate is low (Chang and Wright 1981). Given the necessary conditions and sufficient time, this compound can be the only sulfate in the product. The necessary conditions are evidently that PbO must be in the orthorhombic form and of large particle size and low crystallinity and that the reaction mixture is quiescent. In the presence of excess PbO, 4PbO·$PbSO_4$ is the stable sulfate at room temperature.

Acicular single crystals of this compound were prepared by an original technique of flow growth in the binary system $PbO–PbSO_4$: a mixture of these two constituents (15 mol% $PbSO_4$) was melted at 900°C in an alumina crucible and was cooled slowly between 900°C and 830°C in the presence of a low thermal gradient, and then was quenched in air (Mentzen et al. 1981).

The end product of the PbS interaction with a gaseous medium containing O_2 and some percent of SO_2 is $PbSO_4$ (Yazawa and Gubčová 1967). At 900°C, the interaction of PbS with $PbSO_4$ or PbO formed in the system proceeds with the formation of Pb and the release of SO_2. At 1100°C, all compounds, with the exception of $PbSO_4$, are in a molten state. At this temperature, the following reaction becomes possible: $PbS + O_2 = Pb + SO_2$.

Phase equilibria in the Pb–O–S system were investigated in the temperature range 600°C–1050°C at one atmospheric pressure (Fukatsu and Kozuka 1982). Nine monovariant equilibria were determined by the measurement of the equilibrium oxygen partial pressure using the galvanic cell. A consistent potential phase diagram for the system was constructed based on experimental results and available thermo-chemical reactions. The Gibbs energy of formation of $PbSO_4$, $PbO{\cdot}PbSO_4$, and $2PbO{\cdot}PbSO_4$ were evaluated. Phase boundary between liquid Pb and liquid $PbO{\cdot}PbSO_4$ was estimated. Calculated sulfur partial pressures at the liquid Pb/solid PbS boundary is in good agreement with the experimentally determined values.

Detailed calculations on the Pb–O–S vapor system containing a total of 20 species were presented by Chakraborti and Jha (2004).

10.44 Lead–Selenium–Sulfur

PbS–Se: The phase diagram of this system, constructed through DTA, XRD, and metallography, is a eutectic type (Figure 10.21) (Filep and Sabov 2017a). The eutectic contains 5 mol% PbS and crystallizes at 129°C. In this system an immiscibility region exists within the interval from 5 to 75 mol% PbS, a monotectic temperature is 684°C. The ingots for the investigation were annealed at 80°C for 168 h.

PbS–PbSe: The phase diagram of this system, constructed through DTA, XRD, metallography, and measuring of the microhardness (Simpson 1964; Strauss and Harman 1973) and optimized using a four parameter for the excess Gibbs energy by Volykhov et al. (2006), is given in Figure 10.22. The system is characterized by unlimited mutual solubility of the constituent chalcogenides in both the solid and liquid states. A continuous series of solid solutions with a cubic structure of NaCl-type is formed in the system (Sindeeva and Godovikov 1959; Gromakov et al. 1964; Wright 1964; Darrow et al. 1969b; Liu and Chang 1994b). The minimum is at 59 mol% PbSe and 1061°C (Volykhov et al. 2006) [at 73 mol% PbSe and 1076°C (Strauss and Harman 1973)]. A critical temperature along the join PbS–PbSe at ~100°C was obtained by Liu and Chang (1994b) from interaction parameter calculations, which suggest immiscibility region in this system.

The PbS_xSe_{1-x} solid solution forms easily and is stable (Sushkova et al. 2004). According to the *ab initio* calculations, a linear behavior of the lattice parameter versus the composition was determined (Kacimi et al. 2008). The Vegard's law is valid for this solid solution. The band gap increases with increasing x and its variation is nonlinear. According to the data of Volykhov et al. (2006), PbS_xSe_{1-x} solid solutions exhibit positive deviation from ideality. The ingots for the investigations were annealed at 400°C–650°C for 1–6 months (Simpson 1964).

Figure 10.23 shows the liquidus surface of the PbS–Pb–PbSe subsystem of the Pb–Se–S ternary system, with the dotted lines showing the "isomolarity" lines connecting points of the liquidus surface in equilibrium with the $PbSe_xS_{1-x}$ solid phase with the same sulfur content (Novoselova et al. 1986; Kuznetsova et al. 1987).

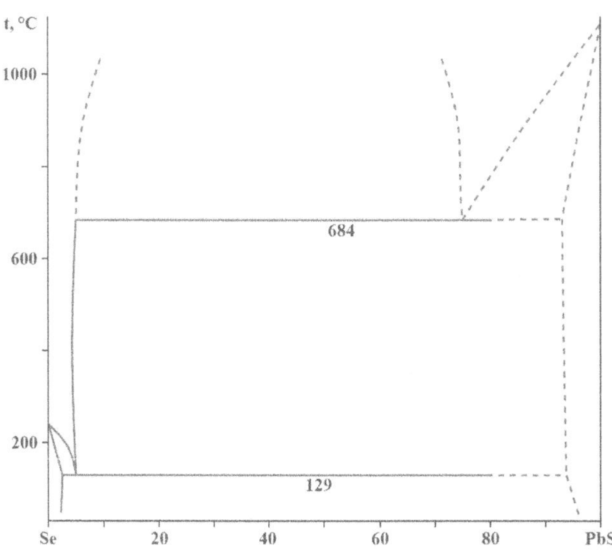

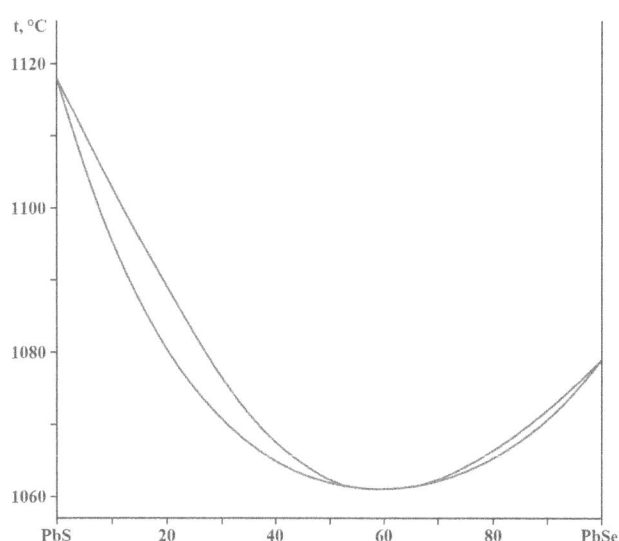

FIGURE 10.21 Phase diagram of the PbS–Se system. (From Filep, and Sabov, M., *Chem. Met. Alloys*, **10**(3–4), 120, 2017.)

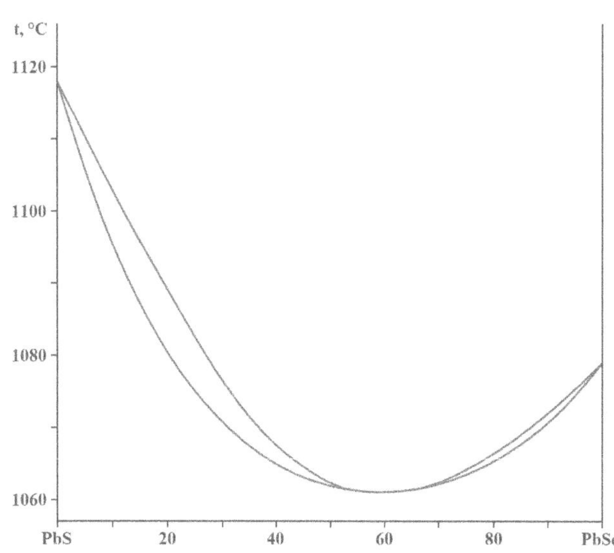

FIGURE 10.22 Phase diagram of the PbS–PbSe system. (From Volykhov, A.A., et al., *Inorg. Mater.*, **42**(6), 596, 2006.)

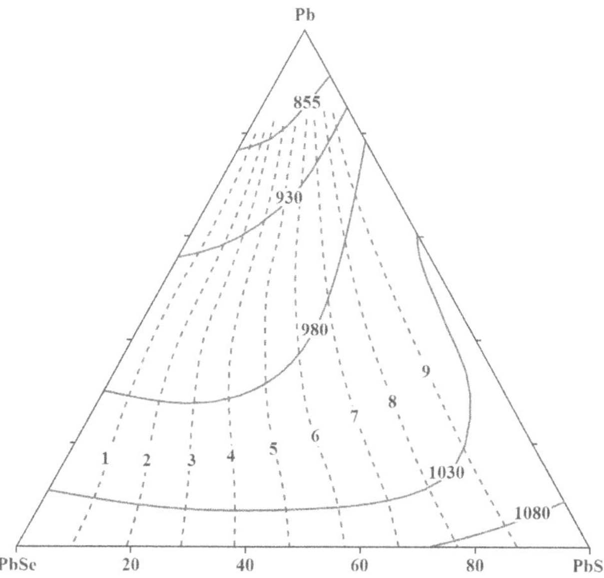

FIGURE 10.23 Liquidus surface of the PbS–Pb–PbSe subsystem of the Pb–Se–S ternary system with the solidus isoconcentrates for the PbS_xSe_{1-x} solid solution: 1, $x = 0.1$, 2, $x = 0.2$, …, 9, $x = 0.9$. (From Kuznetsova, L.A., et al., *Izv. AN SSSR. Neorgan. mater.*, **23**(3), 390, 1987.)

10.45 Lead–Tellurium–Sulfur

PbS–Te: The phase diagram of this system, constructed through DTA, XRD, and metallography, is a eutectic type (Figure 10.24) (Filep and Sabov 2017a). The eutectic contains 3 mol% PbS and crystallizes at 424°C. The ingots for the investigation were annealed at 80°C for 168 h.

PbS–PbTe: The phase diagram of this system, constructed through DTA, XRD, metallography, and measuring of the microhardness (Darrow et al. 1969a; Leute and Volkmer 1985; Liu and Chang 1994b) and optimized using a four parameter for the excess Gibbs energy by Volykhov et al. (2006), is presented in Figure 10.25. Above 801°C, this system contains a continuous series of solid solutions, which undergoes spinodal decomposition on cooling. The minimum is at 59 mol% PbTe and 895°C (Volykhov et al. 2006) [at 65 mol% PbTe and 871°C (Darrow et al. 1969a; Leute and Volkmer 1985)]. In the solid state, the decomposition of solid solutions is observed (Sindeeva and Godovikov 1959; Malevskiy 1963; Gromakov et al. 1964; Darrow et al. 1966, 1969b). The critical point of spinodal decomposition is at 37 mol% PbTe and 801°C (Volykhov et al. 2006) [at 30 mol% PbTe and 805°C (Darrow et al. 1966; Leute and Volkmer 1985); at 35 mol% PbTe and 845°C (Romanenko and Sidorov 1975)]. According to the data of Darrow et al. (1966) and Leute and Volkmer (1985), the two-phase region at 400°C is within the interval 1–96.5 mol% PbTe. The PbS_xTe_{1-x} solid solution is unstable even at high temperatures (Sushkova et al. 2004). The ingots for the investigations were annealed at 500°C–800°C for 360 h (Leute and Volkmer 1985) [at 850°C for 50–100 h (Malevskiy 1963); at 500°C–800°C for 150 and 720 h (Romanenko and Sidorov 1975).

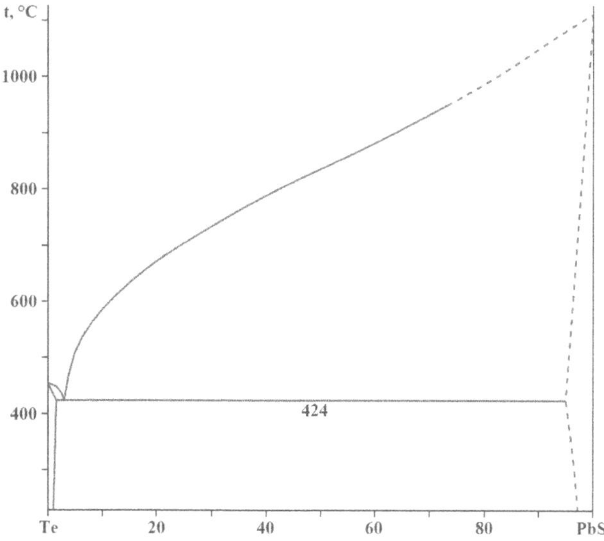

FIGURE 10.24 Phase diagram of the PbS–Te system (From Filep, and Sabov, M., *Chem. Met. Alloys*, **10**(3–4), 120, 2017.)

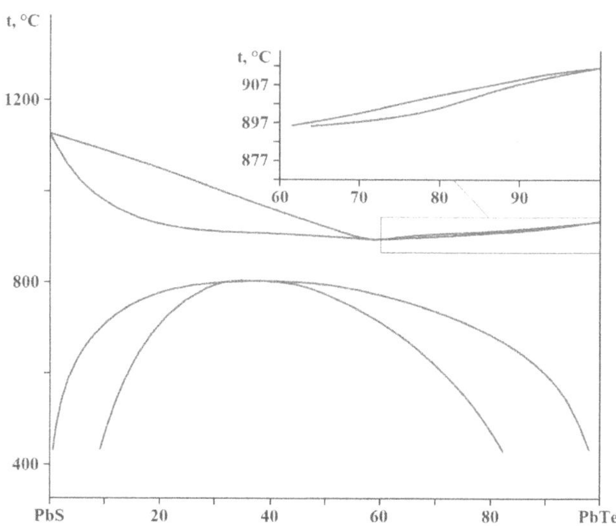

FIGURE 10.25 Phase diagram of the PbS–PbTe system. (From Volykhov, A.A., et al., *Inorg. Mater.*, **42**(6), 596, 2006.)

The liquidus surface of the PbS–Pb–PbTe subsystem of the Pb–Te–S ternary system consists of the regions of primary crystallization of solid solutions based on PbTe and solid solutions based on PbS (Figure 10.26) (Talakin et al. 1987, 1988). The homogeneity region of the $Pb_{1-x}(S_{0.05}Te_{0.95})_x$ was investigated within the interval 590°C–770°C by Demin et al. (1986) and Novoselova et al. (1986). In the single-phase region, the unit cell parameter (Darrow et al. 1966, 1969b) and the band gap (Kasimov et al. 1967), depending on the composition, change almost linearly.

10.46 Lead–Chromium–Sulfur

The $\mathbf{PbCr_2S_4}$ ternary compound, which crystallizes in the hexagonal structure with the lattice parameters $a = 2141.1 \pm 0.2$, $c = 347.8 \pm 0.1$ pm, the calculated and experimental densities of 4.756 and 4.6 g·cm^{-3}, respectively, and an energy gap

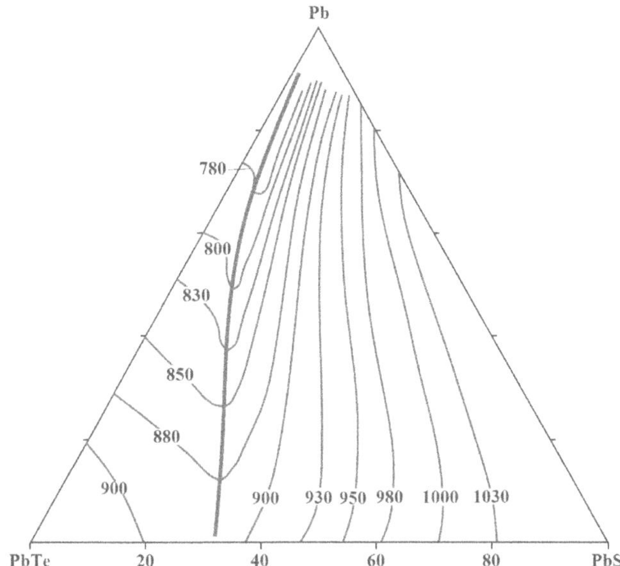

FIGURE 10.26 Liquidus surface of the PbS–Pb–PbTe subsystem of the Pb–Te–S ternary system. (From Talakin, K.N., et al., *Rasplavy*, **1**(3), 123, 1987.)

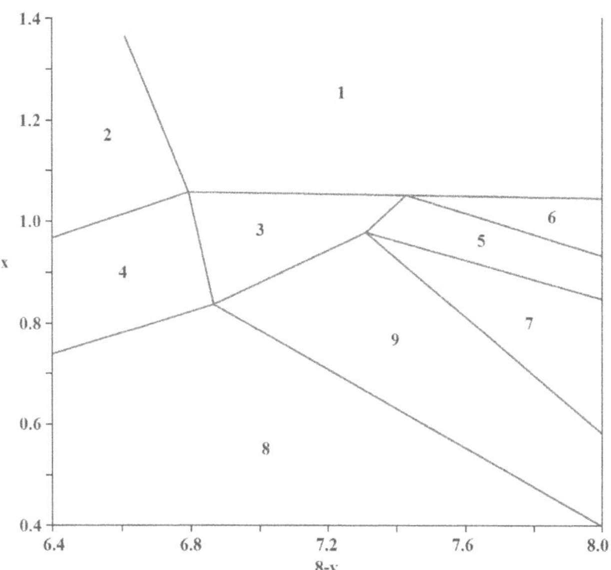

FIGURE 10.27 Composition and phase relations of $Pb_xMo_6S_{8-y}$ at 1100°C for variable Pb and S content: 1, L + $Pb_xMo_6S_{8-y}$; 2, L + Mo + $Pb_xMo_6S_{8-y}$; 3, $Pb_xMo_6S_{8-y}$; 4, $Pb_xMo_6S_{8-y}$ + Mo; 5, $Pb_xMo_6S_{8-y}$ + MoS_2; 6, L + $Pb_xMo_6S_{8-y}$ + MoS_2; 7, $Pb_xMo_6S_{8-y}$ + Mo_2S_3 + MoS_2; 8, $Pb_xMo_6S_{8-y}$ + Mo + Mo_2S_3; 9, $Pb_xMo_6S_{8-y}$ + Mo_2S_3. (From Hauck, J., *Mater. Res. Bull.*, **12**(10), 1015, 1977.)

of ≈1 eV (Omloo and Jellinek 1968; Omloo et al. 1971) [a = 2141.0 and c = 347.55 pm (Sleight and Frederick 1973)], is formed in the Pb–Cr–S. No phase transition for this compound exists in the range of 170–850 K (Omloo and Jellinek 1968). This compound was prepared by heating a stoichiometric mixture of the three elements or mixture of PbS and Cr_2S_3 in evacuated quartz tubes at 1100°C during 1 week (Omloo and Jellinek 1968; Omloo et al. 1971). The product was slowly cooled, annealed at 1000°C for 2 weeks and slowly cooled or quenched to room temperature. Small, black, needle-shaped crystals of $PbCr_2S_4$ were obtained. They were also obtained by transport reactions with iodine as a transport agent.

10.47 Lead–Molybdenum–Sulfur

The $\mathbf{Pb_xMo_6S_{8-y}}$ ternary compound, which composition can vary between $0.85 < x < 1.05$ and $0.6 < y < 1.2$ at 1100°C and is shifted to lower Pb content at higher temperatures, is formed in the Pb–Mo–S system (Figure 10.27) (Hauck 1977; Krabbes and Oppermann 1981a,b). The phase melts incongruently at about 1530°C (Hauck 1977) and crystallizes in the rhombohedral structure with the lattice parameters a = 655.5 pm and α = 89.55° (Viswanath and Ramasamy 1994) [a = 654 pm and α = 89°28′ or a = 920 and c = 1143 pm in hexagonal setting (Chevrel et al. 1971); $a = 655.1 \pm 0.1$ pm and $\alpha = 89.33 \pm 0.02°$ (Marezio et al. 1973); $a = 654.4 \pm 0.01$ pm and $\alpha = 89.48 \pm 0.02°$ and the calculated and experimental densities of 6.06 and 6.10 g·cm^{-3}, respectively (Guillevic et al. 1976); $a = 653.8 \pm 0.2$ pm and a calculated density of 6.15 g·cm^{-3} (Fedorov et al. 1982); $a = 920.7 \pm 0.3$ and $c = 1147.0 \pm 0.2$ pm in hexagonal setting (Nanjundaswamy et al. 1987)]. At the approximate composition line $Pb_xMo_6S_7$, the rhombohedral lattice constant a shows maximum while the rhombohedral angle α has a minimum (Hauck 1977). This compound is characterized by a transition to a superconducting state at 14.1–14.5 K (Marezio et al. 1973).

To prepare this phase, stoichiometric mixture of finely powdered Pb, Mo, and S were ground under acetone (Hauck 1977). The dry powder sample was heated in fused-silica tube slightly above the melting point of sulfur and then compressed with the aid of a tight fitting silica rod. Then, the tube was evacuated and sealed, and the samples were pre-reacted by slowly increasing the temperature from 400°C to 1100°C within about 1 week. This slow increase is necessary to avoid separation of the phases and destruction of the fused-silica tubes. After final equilibration for about 30 h at 1100°C–1250°C, the sample was quenched in water.

The powder of $Pb_xMo_6S_{8-y}$ could be also obtained from the PbS, Mo, and S, which was ground and mixed in an agate mortar (Chevrel et al. 1971; Marezio et al. 1973). This powder was pressed at about 29.5 MPa·m^{-2}. The pellet was placed in an evacuated quartz tube, heated to 575°C, held 16 h, and then the temperature was increased to 850°C and held 22 h. Finally, the temperature was increased to 1100°C, held 3 h, and the sample was removed from the furnace and air-quenched.

This phase can be also synthesized in powder form by the high-temperature reaction technique from a mixture of Pb, MoS_2, and Mo (Viswanath and Ramasamy 1994). First, a mixture of the starting materials in a vacuum (1.3×10^{-6} Pa) sealed quartz ampoule was heated at 500°C for 48 h. Then the temperature was raised slowly to 1020°C and held at that temperature for 48 h, followed by furnace-cooling. The product was mixed well and treated again at 1020°C for 72 h, followed by furnace-cooling.

Single crystals of $Pb_xMo_6S_{8-y}$ were grown through the chemical transport reaction with $PbBr_2$ as a transport reagent

(Krabbes and Oppermann 1981a,b) or using the crystallization from the solution in the melt (Hauck 1977).

The isothermal sections of the Pb–Mo–S ternary system at 900°C, 1000°C, and 1050°C were constructed by Rikel' and Alekseeva (1981).

10.48 Lead–Chlorine–Sulfur

PbS–PbCl$_2$: The phase diagram of this system, constructed through DTA, XRD, and metallography, is a eutectic type (Figure 10.28) (Urazov and Sokolova 1941; Bell and Flengas 1966b; Rabenau and Rau 1969; Novoselova et al. 1970b). The eutectic contains 24.7 mol% (22 mass%) PbS and crystallizes at 442°C (Urazov and Sokolova 1941; Bell and Flengas 1966b) [18 mol% and 450°C (Rabenau and Rau 1969); at 453°C (Morozov and Li 1963)]. The **$Pb_7S_2Cl_{10}$** ternary compound, which melts incongruently at 550°C (Bell and Flengas 1966b) [decomposes at 180°C (Novoselova et al. 1970b)], is formed in this system. According to the data of Rabenau and Rau (1969), one more compound, **Pb_4SCl_6**, which crystallizes in the orthorhombic structure with the lattice parameters $a = 423$, $b = 922$, $c = 1496$ pm, and the calculated and experimental densities of 6.10 and 6.08 g·cm^{-3}, respectively, exists in the PbS–PbCl$_2$ system. The mutual solubility of the initial components of the system is insignificant (Bell and Flengas 1966b). The solubility of Pb in the PbS–PbCl$_2$ mixtures increases with increasing temperature and PbS concentration to a value of 0.65 at% Pb at 50 mol% PbS and 886°C (Bell and Flengas 1966a).

10.49 Lead–Bromine–Sulfur

PbS–PbBr$_2$: The phase diagram of this system, constructed through DTA, XRD, and metallography, is a eutectic type (Figure 10.29) (Bell and Flengas 1966b; Rabenau and Rau 1969; Novoselova et al. 1970b,c). The eutectic contains 9 mol% PbS and crystallizes at 350°C (Rabenau and Rau 1969). The ingots for the investigations were annealed at 340°C for 1000 h (Novoselova et al. 1970c).

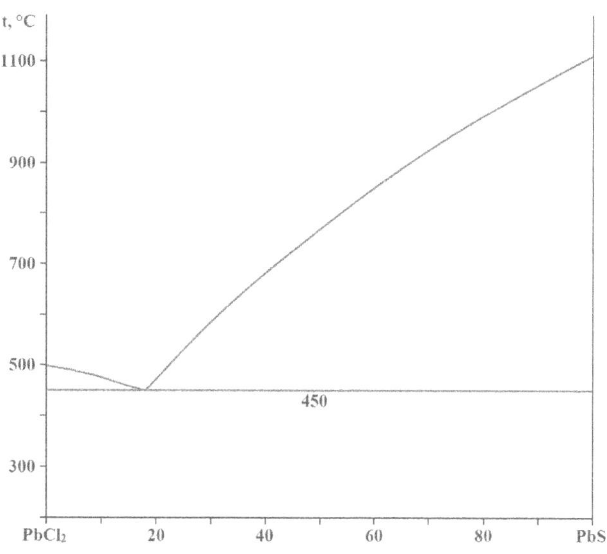

FIGURE 10.28 Phase diagram of the PbS–PbCl$_2$ system. (From Rabenau, A., and Rau H., *Z. anorg. und allg. Chem.*, **369**(3–6), 295, 1969.)

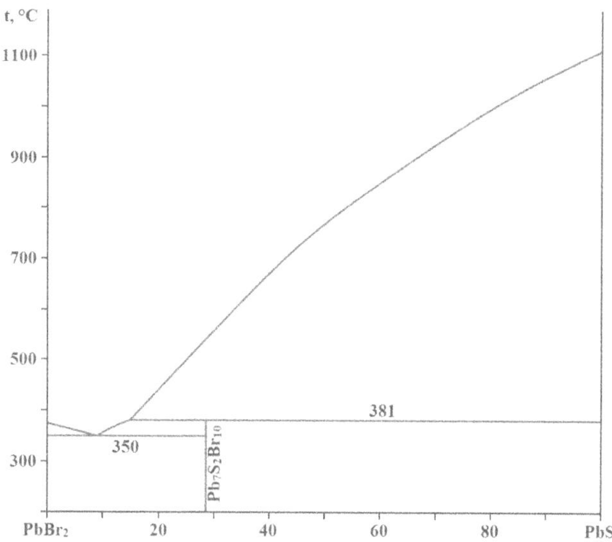

FIGURE 10.29 Phase diagram of the PbS–PbBr$_2$ system. (From Rabenau, A., and Rau H., *Z. anorg. und allg. Chem.*, **369**(3–6), 295, 1969.)

The **$Pb_7S_2Br_{10}$** ternary compound, which melts incongruently at 381°C [at 394°C (Novoselova et al. 1970c)] and crystallizes in the hexagonal structure with the lattice parameters $a = 1227.3 \pm 0.006$, $c = 431.9 \pm 0.002$ pm, and the calculated and experimental densities of 6.817 and 6.77 g·cm^{-3}, respectively (Krebs 1973) [$a = 1230$, $c = 433$ pm, and the calculated and experimental densities of 6.76 and 6.60 g·cm^{-3}, respectively (Rabenau et al. 1968); $a = 1227$, $c = 431.8$ pm, and the calculated and experimental densities of 6.82 and 6.77 g·cm^{-3}, respectively (Rabenau and Rau 1969); the calculated and experimental densities of 6.79 and 6.70 g·cm^{-3}, respectively (Novoselova et al. 1970c)], is formed in this system. According to the data of Novoselova et al. (1970c) and Popolitov and Mininzon (1982), the homogeneity region of this compound is ca. 7 mol%. $Pb_7S_2Br_{10}$ is air stable but decomposes by the cold water (Novoselova et al. 1970c).

$Pb_7S_2Br_{10}$ was obtained at the annealing of the mixture of PbS and PbBr$_2$ at 320°C (Rabenau et al. 1968). Its single crystals were grown by hydrothermal synthesis, as well as by crystallization from the melt under the pressure of an inert gas (Popolitov and Mininzon 1982). Single crystals can also be obtained by separating the starting components of the charge in independent solvents (Popolitov et al. 1982).

One more ternary compound, **Pb_2SBr_2**, can be prepared by treating the ether suspension of PbBr$_2$ in benzene with hydrogen sulfide (Hardt and Scheepker 1970). The reflection spectra of this compound correspond to that of $Pb_7S_2Br_{10}$.

10.50 Lead–Iodine–Sulfur

PbS–PbI$_2$: The phase diagram of this system, constructed through DTA, XRD, and metallography, is presented in Figure 10.30 (Rabenau and Rau 1969; Novoselova et al. 1970a; Odin 2000). The eutectic contains 17 mol% PbS and crystallizes at 380°C

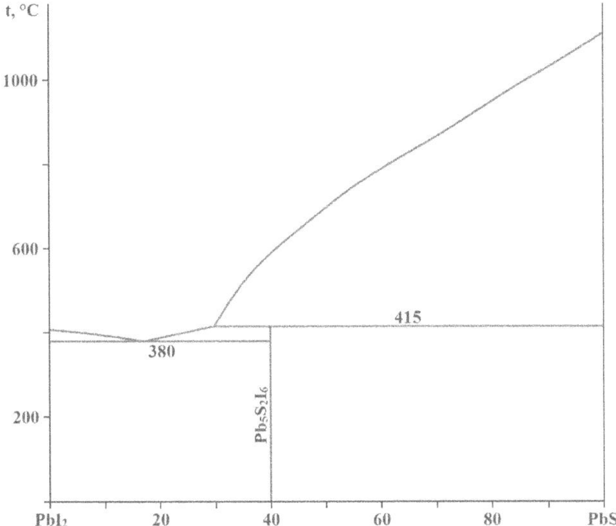

FIGURE 10.30 Phase diagram of the PbS–PbI_2 system. (From Novoselova, A.V., et al., *Izv. AN SSSR. Neorgan. mater.*, **6**(1), 135, 1970.)

[at 374°C (Rabenau and Rau 1969) (Novoselova et al. 1970a)]. From the PbI_2-side in the system, 2*H*-PbI_2 polytype is primarily crystallized (Odin 2000). The solubility of PbI_2 in PbS at 420°C is 0.8 mol% (Novoselova et al. 1970a). The ingots for the investigations were annealed at 370°C for 1000 h.

The $\mathbf{Pb_5S_2I_6}$ ternary compound, which melts incongruently at 415°C–418°C, is formed in this system. This compound is stable and prolonged annealing at 350°C–400°C does not lead to its decomposition (Odin 2000). It crystallizes in the monoclinic structure with the lattice parameters a = 1433.8 ± 0.7, b = 443.7 ± 0.2, c = 1455.2 ± 0.7 pm, β = 98.04 ± 0.05°, and the calculated and experimental densities of 6.743 and 6.61 g·cm^{-3}, respectively (Krebs 1970, 1973) [a = 1428, b = 445, c = 1448 pm, β = 98.7°, and the calculated and experimental densities of 6.79 and 6.61 g·cm^{-3}, respectively (Rabenau et al. 1968); a = 1433, b = 443.4, c = 1453 pm, β = 98.0°, and the calculated and experimental densities of 6.76 and 6.61 g·cm^{-3}, respectively (Rabenau and Rau 1969). This compound has a polymorphic transformation at 290°C and is a semiconductor with an energy gap of 1.8 eV (Popolitov and Mininzon et al. 1982; Popolitov and Plakhov 1989).

$Pb_5S_2I_6$ was obtained at the annealing of the mixture of PbS and PbI_2 at 350°C (Rabenau et al. 1968). Its single crystals could be grown by a reaction of PbS with concentration hydroiodine acid under hydrothermal conditions (Krebs 1970, 1973) or by crystallization from the melt under the pressure of inert gas (Popolitov and Mininzon et al. 1982).

Two more ternary compounds, $\mathbf{Pb_2SI_2}$ and $\mathbf{Pb_4S_3I_2}$, exist in the PbS–PbI_2 system. First of them decomposes at 67°C with formation of $Pb_5S_2I_6$ and PbS and crystallizes in the tetragonal structure with the lattice parameters a = 1110 and c = 1320 pm (Hardt and Scheepker 1970). This compound was prepared by introducing hydrogen sulfide into acetone solutions of $KPbI_3$ as bright red precipitate. The reflection spectra of this compound are practically identical to those of $Pb_5S_2I_6$.

$Pb_4S_3I_2$ is a high-pressure phase and crystallizes in the orthorhombic structure with the lattice parameters a = 812.93 ± 0.06, b = 1556.13 ± 0.11, and c = 818.20 ± 0.06 pm, a calculated density of 7.564 g·cm^{-3} and an energy gap of 1.84 eV for a direct transition and 1.65 eV for an indirect one (Ni et al. 2019b). It was prepared using a high-pressure solid-state method. The PbI_2 crystals were mixed stoichiometrically with Pb powder and S pieces. The mixture was ground into a fine powder and loaded in a BN crucible. The crucible was inserted into a pyrophyllite cube assembly, pressed to 4 GPa using a cubic multi-anvil system, and heated to 600°C at 50°C·min^{-1}. The sample was kept at 600°C for 3 h and then quenched to room temperature before decompression. The product obtained displayed a dark red color and was relatively stable in air. The obtained samples were largely polycrystalline.

10.51 Lead–Manganese–Sulfur

PbS–MnS: This system is a non-quasibinary section of the Pb–Mn–S ternary system (Vanyarkho et al. 1973). The solubility of MnS in PbS at 930°C is 2 mol%, at 870°C – 6 mol%, at 720°C – 3 mol%, and at 580°C – 1 mol%. The XRD patterns of the samples containing 65 mol% MnS showed reflections of a phase not described in the literature. With an increase in the MnS content, the intensity of these reflections increases, and the intensity of the lines belonging to the $Pb_{1-x}Mn_xS$ solid solution decreases. Elemental lead was found on the walls of the ampoules. The ingots for the investigations were annealed at 580°C–930°C for 1100–150 h.

10.52 Lead–Iron–Sulfur

PbS–Fe: This system is a non-quasibinary section of the Pb–Fe–S ternary system (Urazov et al. 1936; Vanyarkho et al. 1973). It was studied through DTA, XRD, and metallography, and the ingots for the investigation were annealed at 700°C for 30 days.

PbS–FeS: The phase diagram of this system, constructed through DTA, XRD, metallography, and measuring of the microhardness, is a eutectic type (Figure 10.31) (Urazov

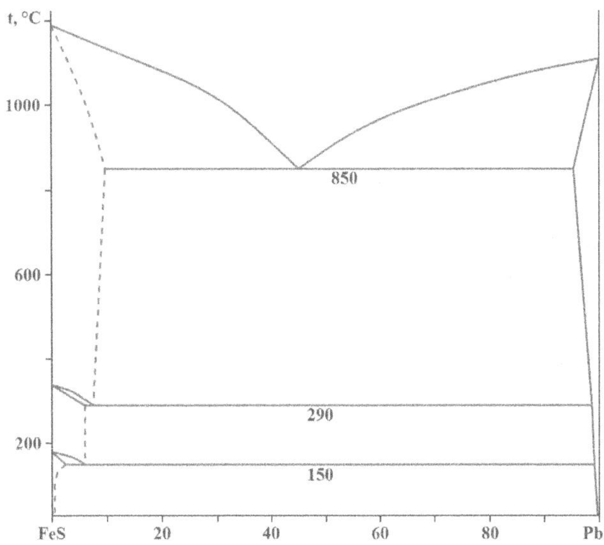

FIGURE 10.31 Phase diagram of the PbS–FeS system. (From Gasanova, U.A., et al., *Russ. J. Inorg. Chem.*, **64**(2), 242, 2019.)

et al. 1936; Murach 1947; Dutrizac 1980; Eric and Ozok 1994: Raghavan 1998a; Gasanova et al. 2019). The eutectic contains 55 mol% FeS and crystallizes at 850°C (Gasanova et al. 2019) [53.8 mol% (30 mass%) FeS and 865°C (Urazov et al. 1936); at 863°C (Murach 1947); at 848°C (Brett and Kullerud 1967); at 860°C (Kopylov et al. 1977); at 850°C (Dutrizac 1980); at 54 mol% FeS and 842°C (Eric and Ozok 1994; Raghavan 1998a)]. The mutual solubility of PbS and FeS extends to 1–2 mol% at room temperature (Gasanova et al. 2019). According to the data of Eric and Ozok (1994) and Brett and Kullerud (1967), no terminal solid solution ranges could be found.

The vapor pressure of PbS in the system in the range 1050°C–1250°C deviates only slightly from Raoult's law (Isakova et al. 1964). The activities of PbS and FeS at 1100°C and 1200°C were determined by Eric and Ozok (1994). It was shown that these activities show moderate deviations from ideality. According to the data of Eriç and Timuçin (1981), the PbS–FeS melts are ideal solutions at 1200°C.

The ingots for the investigation were annealed at 780°C for 200 h (Gasanova et al. 2019) [at 700°C for 30 days (Brett and Kullerud 1967); at 840°C for 2–3 days (Dutrizac 1980)].

PbS–FeS$_2$: A strong thermal effect was recorded at 719°C for the sample containing 54 mol% PbS, which is due to the peritectic interaction in the system (Brett and Kullerud 1967). A second thermal effect was recorded for this sample at 727°C. The solubility of PbS in FeS$_2$ is negligible. The system was studied through DTA, XRD, and metallography, and the ingots for the investigation were annealed at 700°C for 30 days.

The liquidus surface of the Pb–PbS–FeS–Fe subsystem of the Pb–Fe–S ternary system contains the fields of PbS, FeS and Fe primary crystallization (Vanyarkho et al. 1973). The exchange reaction PbS + Fe = FeS + Pb is mutually reversible. The ternary eutectic in the PbS–FeS–Fe subsystem contains 21.1 at% Pb, 17.4 at% Fe and 61.5 at% S (60 mass% Pb, 13 mass% Fe and 27 mass% S) (Urazov et al. 1936).

Within the Pb–Fe–S ternary system, the boundaries of the immiscibility region together with the tie line distributions were established at 1200°C by Eric and Ozok (1994). Activities of Pb were also measured by the dew-point technique along the metal-rich boundary of the miscibility gap. Activities of Fe, Pb, and S along the miscibility gap were calculated by utilizing the bounding binary thermodynamics, phase equilibria, and tie lines.

10.53 Lead–Cobalt–Sulfur

The **Co$_3$Pb$_2$S$_2$** ternary compound, which crystallizes in the rhombohedral structure, is formed in the Pb–Co–S system (Zabel et al. 1979a,b). It was prepared from the elements as well as from PbS and Co by high-temperature solid-state synthesis. According the data of Kassem et al. (2015), this compound is not a stable phase and cannot be synthesized even if Pb flux is used. Co$_3$Pb$_2$S$_2$ is also absent on the isothermal section of the Pb–Co–S ternary system at 400°C (Figure 10.32), constructed by Gulay (2008). No ternary compounds were also found on the isothermal sections of this system at 340°C, 476°C, and 700°C (Mariolacos 1986).

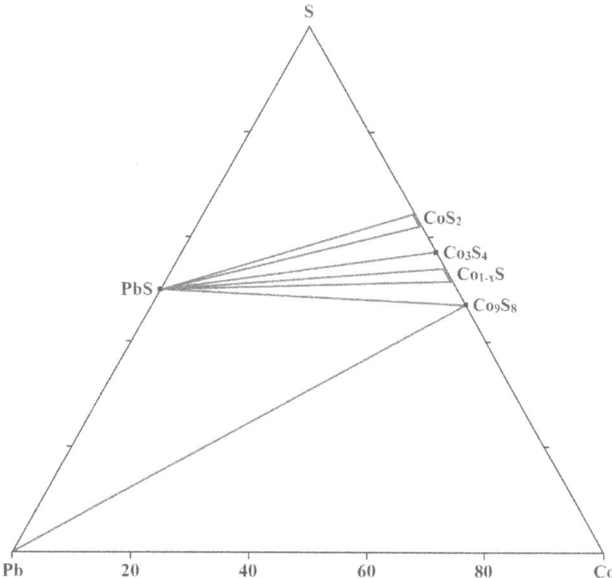

FIGURE 10.32 Isothermal section of the Pb–Co–S ternary system at 400°C. (From Gulay, L.D., *Nauk. Visnyk Volyns'k. Nats. Univ. im. Lesi Ukrainky, Ser. Khim. nauky*, (16), 6, 2008.)

10.54 Lead–Nickel–Sulfur

PbS–Ni: The phase diagram of this system was constructed through DTA, XRD, metallography, and measuring of the microhardness by Zargarova et al. (1976a). The eutectic contains 46.4 mol% (77.9 mass%) PbS and crystallizes at 730°C. The solubility of Ni in PbS at the eutectic temperature is 2.8 mol% (0.7 mass%). The **Ni$_3$Pb$_2$S$_2$** ternary compound, which melts incongruently at 760°C, is formed in this system. Complete melting of the alloy corresponding to the composition of the compound occurs at 800°C. On the thermogram of Ni$_3$Pb$_2$S$_2$, in addition to the effects of melting and decomposition, a thermal effect is recorded at 560°C, which apparently reflects a polymorphic transformation that occurs in all alloys with higher Ni content and is presented in the phase diagram by eutectoid equilibrium at 550°C. The compound has a narrow homogeneity region from the Ni side.

One of the Ni$_3$Pb$_2$S$_2$ modifications crystallizes in the rhombohedral structure with the lattice parameters 557.6 pm and α = 60.20° [558 pm and α = 60° (Mariolacos 1986)] or a = 559.50 ± 0.06° and c = 1362.60 ± 0.24 pm in hexagonal settings and a calculated density of 8.828 g·cm^{-3} (Weihrich et al. 2007) [556.5 pm and α = 60° and the calculated and experimental densities of 8.86 and 8.72 g·cm^{-3}, respectively, for mineral shandite of the same composition (Peacock and McAndrew 1950); a = 559.1 ± 0.1 and c = 1357.9 ± 0.1 pm in hexagonal settings and the calculated and experimental densities of 8.87 and 8.65 g·cm^{-3}, respectively (Brower et al. 1974); a = 556.5 pm and α = 60° (Michelet and Collin 1976); a = 557.02 ± 0.05 pm and α = 60.28 ± 0.02° or a = 559.40 ± 0.05 and c = 1316.5 ± 0.1 pm in hexagonal settings (Zabel et al. 1979a,b)].

Second modification of Ni$_3$Pb$_2$S$_2$ crystallizes in the monoclinic structure with the lattice parameters a = 964.5 ± 0.4, b = 557.7 ± 0.2, c = 560.2 ± 0.3 pm, β = 125.00 ± 0.03° (Michelet and Collin 1976).

This compound was prepared by heating the mixture of the elements as well as from PbS and Ni or Ni_3S_2 and Pb in evacuated silica tube (Brower et al. 1974; Michelet and Collin 1976; Zabel et al. 1979a,b). Heat treatment of the mixture of Pb, Ni, and S was varied from 500°C–600°C to 1100°C–1100° (Brower et al. 1974). In general, an attempt was made to obtain complete melting of the mixture, which was then heat treated in a rocking furnace in order to insure chemical homogeneity. To avoid the formation of PbS, synthesis of $Ni_3Pb_2S_2$ was also done in two steps (Weihrich et al. 2007). Ni_3S_2 was prepared first followed by reaction with Pb. The mixture was heated in an evacuated quartz tube to 850°C and annealed for 5 days.

One more ternary compound, $\mathbf{Ni_{60}Pb_9S_{31}}$, which is stable in the temperature range from 490°C to 578°C [554°C–610°C (Mariolacos 1986)] and crystallizes in the cubic structure with the lattice parameter $a = 1551.349 \pm 0.005$ pm, is formed in the Pb–Ni–S system (Baranov et al. 2002). Single crystals of this compound were grown through the chemical transport reactions with the use of I_2, $PbCl_2$, NH_4Cl, Br_2, $(NH_4)_2PbCl_6$, and $AlCl_3$ as the carriers. The growth time was 300–400 h. Then the tube was quenched into water.

The isothermal section of the Pb–Ni–S ternary system at 400°C is presented in Figure 10.33 (Gulay 2008). It is seen that seven three-phase regions exist in this section. There are one compound, $Ni_3Pb_2S_2$, on the isothermal sections of the Pb–Ni–S ternary system at 340°C and 538°C and two compounds, $Ni_3Pb_2S_2$ and $Ni_{60}Pb_9S_{31}$, on the isothermal sections at 554°C and 600°C (Mariolacos 1986).

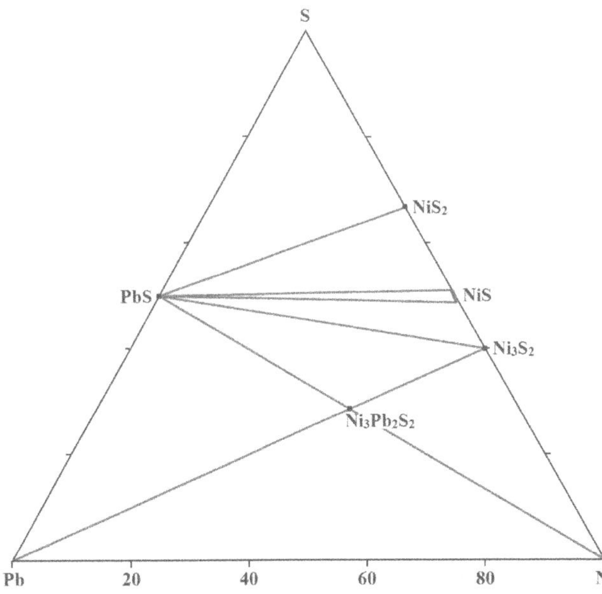

FIGURE 10.33 Isothermal section of the Pb–Ni–S ternary system at 400°C. (From Gulay, L.D., *Nauk. Visnyk Volyns'k. Nats. Univ. im. Lesi Ukrainky, Ser. Khim. nauky*, (16), 6, 2008.)

10.55 Lead–Rhodium–Sulfur

PbS–Rh: The $\mathbf{Rh_3Pb_2S_2}$ ternary compound is formed in this system. It crystallizes in the rhombohedral structure with the lattice parameters $a = 570.15 \pm 0.03$, $c = 1379.45 \pm 0.06$ in hexagonal settings, and a calculated density of 9.963 g·cm^{-3} (Anusca et al. 2009) [$a = 565.73 \pm 0.04$ pm and $\alpha = 60.55 \pm 0.02°$ or $a = 570.41 \pm 0.04$, and $c = 1380.0 \pm 0.1$ pm in hexagonal settings and the calculated and experimental densities of 10.08 and 10.01 g·cm^{-3}, respectively (Zabel et al. 1979a,b; Natarajan et al. 1988); $a = 573 \pm 1$, $c = 1400 \pm 1$ in hexagonal settings, and a calculated density of 9.74 g·cm^{-3} for mineral rhodoplumsite with the same composition (Genkin et al. 1983)].

This compound has been synthesized by the direct combination of stoichiometric quantities of high-purity elements in evacuated (10^{-3} Pa) and sealed quartz tube (Natarajan et al. 1988). The quartz tube containing the reactants was heated initially at 400°C–500°C for 2–3 days until all the sulfur was reacted. The temperature was then raised to 800°C–1000°C and kept for 24 h. In most cases, a molten mass was noted inside the quartz tube after this treatment. The compound was recovered, ground thoroughly in an agate mortar and pressed in the form of discs with a WC-lined stainless steel die and sintered at 600°C for 1–2 days in an evacuated quartz tube. Single-phase material has thus been obtained. $Rh_3Pb_2S_2$ is black in color and is stable to exposure to air and moisture. It oxidized by heating in air but is stable when heated to temperatures higher than 900°C in evacuated and sealed quartz tube and it sinter well.

10.56 Lead–Palladium–Sulfur

PbS–Pd: The $\mathbf{Pd_3Pb_2S_2}$ ternary compound, which crystallizes in the rhombohedral structure, is formed in this system (Zabel et al. 1979a,b). It was prepared from the elements as well as from PbS and Pd by high-temperature solid-state synthesis. Apparently, this compound has two polymorphic modifications, since the mineral laflammeite of the same composition crystallizes in the monoclinic structure with the lattice parameters $a = 1152.1 \pm 1.1$, $b = 829.4 \pm 1.0$, $c = 832.1 \pm 0.6$ pm, $\beta = 134.38 \pm 0.05°$, and a calculated density of 9.41 g·cm^{-3} (Barkov et al. 2002; Jambor and Roberts 2003).

REFERENCES

All References are available as a downloadable eResource at www.routledge.com/9780367639235

11

Systems Based on Lead Selenide

11.1 Lead–Copper–Selenium

PbSe–Cu_2Se: The phase diagram of this system, constructed through DTA, XRD, and metallography, is a eutectic type (Figure 11.1) (Marchuk et al. 2006b). The eutectic reaction takes place at 613°C and corresponds to 46 mol% PbSe. The polymorphic transformation of the solid solution based on Cu_2Se takes place at 147°C. No ternary compounds exist in this system (Marchuk et al. 2006b; Gulay et al. 2008c).

The isothermal section of the Pb–Cu–Se ternary system at 300°C was constructed by Gulay (2008).

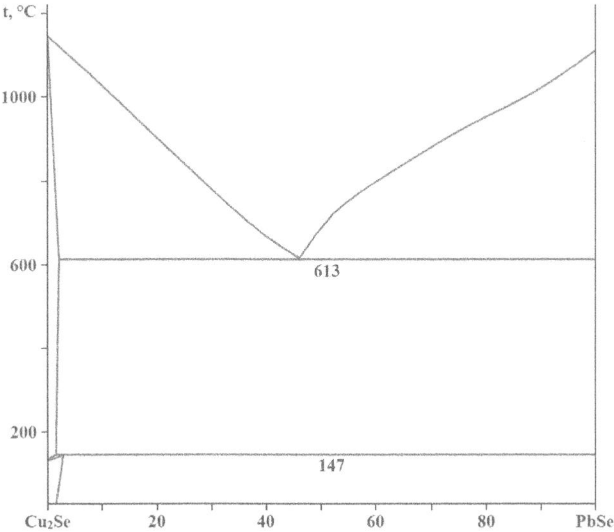

FIGURE 11.1 Phase diagram of the PbSe–Cu_2Se system. (From Marchuk, O.V., et al., *J. Alloys Compd.*, **416**(1–2), 106, 2006.)

T-x-y and *p*(Se_2)-*T-x* phase diagrams of this ternary system were determined and 3D modeled along the PbSe filed of the primary crystallization (Mamedov et al. 2018). The excess free energies of Pb and Se mixing and the Se saturated vapor pressure were calculated for Pb–Cu–Se liquid alloys saturated with PbSe. It is shown that the vapor phase consists mainly of the Se_2 molecules. The analytical dependencies were obtained for the PbSe–Se and PbSe–Pb liquidus regions.

11.2 Lead–Silver–Selenium

PbSe–Ag: The phase diagram of this system, constructed through DTA, XRD, metallography and measuring of the microhardness, is given in Figure 11.2 (Legendre and Souleau 1972b). The eutectic crystallizes at 742°C, and the immiscibility region exists within the temperature interval from 740°C to 816°C. No ternary compounds were found in this system (Marchuk et al. 2006b).

PbSe–AgSe: This system is non-quasibinary section of the Pb–Ag–Se ternary system (Legendre and Souleau 1972b). Thermal effects at 128°C correspond to the phase transformation of Ag_2Se, at 216°C to the ternary eutectic, and at 466°C to the monotectic reaction. The solubility of AgSe in PbSe at 600°C is 0.25 mol%. The composition of AgSe is in the immiscibility region of the Ag–Se binary system. This system was investigated through DTA, XRD, metallography, and measuring of the microhardness. The ingots for the investigations were annealed at 600°C for 48 h.

PbSe–Ag_2Se: The phase diagram of this system, constructed through DTA, XRD, metallography and measuring of the microhardness, is a eutectic type (Figure 11.3) (Novoselova et al. 1967b; Legendre and Souleau 1972b). The eutectic

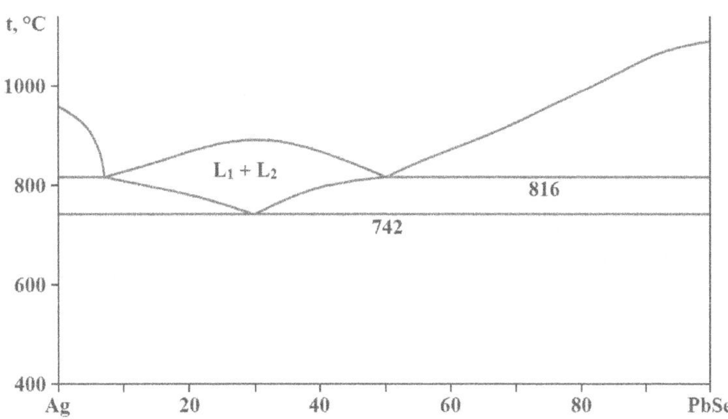

FIGURE 11.2 Phase diagram of the PbSe–Ag system. (From Legendre, B., and Souleau, C., *Bull. Soc. chim. France*, IV, (2), 463, 1972.)

DOI: 10.1201/9781003123507-11

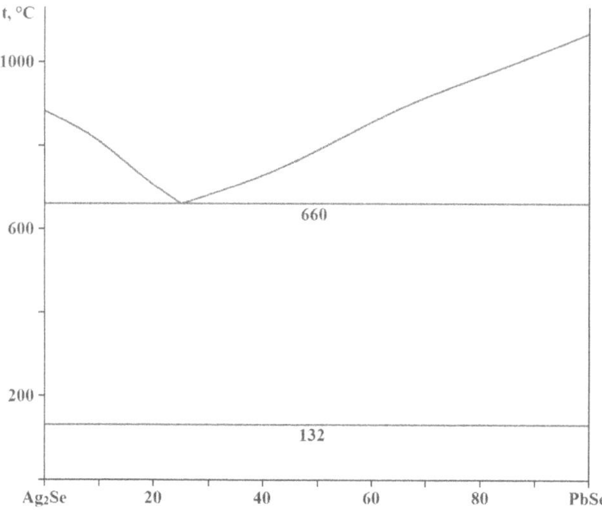

FIGURE 11.3 Phase diagram of the PbSe–Ag_2Se system. (From Legendre, B., and Souleau, C., *Bull. Soc. chim. France*, IV, (2), 463, 1972.)

contains 25 mol% PbSe and crystallizes at 660°C. The solubility of Ag_2Se in PbSe at 600°C–650°C is 0.20–0.25 mol%. The ingots for the investigations were annealed at 650°C for 220–250 h (Novoselova et al. 1967b) or at 600°C for 48 h (Legendre and Souleau 1972b).

In the Pb–Ag–Se ternary system, there are two immiscibility regions separated by the PbSe–Ag_2Se quasibinary section (Legendre and Souleau 1972b). The ternary eutectic in the PbSe–Ag_2Se–Se subsystem is degenerated and practically coincide with the melting temperature of Se. The ternary eutectics in the PbSe–Ag_2Se–Ag and PbSe–Ag–Pb subsystems crystallize at 656°C and 290°C, respectively. Based on the existing data of the boundary binary systems, using DTA and computer programs, the problem of 3D–visualization of Ag_2Se and PbSe primary crystallization surfaces was solved by Ibragimova et al. (2016).

The isothermal section of the Pb–Ag–Se ternary system at 300°C was constructed by Gulay (2008).

11.3 Lead–Gold–Selenium

PbSe–Au: The phase diagram of this system, constructed through DTA, XRD, and metallography, is a eutectic type with monotectic transformation (Rouland et al. 1977). The eutectic crystallizes at 955°C, and the immiscibility region exists within the temperature interval from 955°C to 1018°C. In the range from 350°C to 600°C, the solubility of gold in PbS continuously increases with temperature (Kharakhorin et al. 1965). Beginning at 650°C, the Au-coated surfaces of PbSe melt, which indicates on the formation of a new phase on the surface. A scheme of the liquidus surface projection of the Pb–Au–Se ternary system is given in Rouland et al. (1977).

11.4 Lead–Magnesium–Selenium

PbSe–MgSe: The solubility of MgSe in PbSe at 800°C reaches 6 mol% and at 400°C it is 0.5 mol% (Sealy and Crocker 1973b). In the region of the solid solutions, the unit cell parameter, depending on the composition, changes according to Vegard's law. At the content of 3 and 6 mol% MgSe, the equilibrium distribution coefficient is equal to 0.99 and 0.70, respectively, which indicates a decrease in the liquidus temperature with an increase in the content of MgSe. Single crystals containing 6 mol% MgSe were grown using the Bridgman method, and the samples for the investigations were annealed at 400°C–750°C

The structural, electronic and optical properties of $\mathbf{Mg_xPb_{1-x}Se}$ solid solutions for $0 \le x \le 1$ in their cubic structure have been calculated using the full-potential linearized augmented plane wave method under the framework of density functional theory (Chattopadhyaya and Bhattacharjee 2017b). The calculated lattice parameters of these solid solutions exhibit significant upward deviations from Vegard's law with bowing parameters −29.5 pm. The non-liner behavior of the energy gap versus concentration for the $Mg_xPb_{1-x}Se$ solid solutions can be fitted by the next expression: $E_g = 0.834 - 2.346x + 4.081x^2$.

11.5 Lead–Strontium–Selenium

PbSe–SrSe: The structural, electronic and optical properties of $\mathbf{Sr_xPb_{1-x}Se}$ solid solutions for $0 \le x \le 1$ in their cubic structure have been calculated using the full-potential linearized augmented plane wave method under the framework of density functional theory (Chattopadhyaya and Bhattacharjee 2017a). The calculated lattice parameter of these solid solution vary almost linearly with x having negligible upward bowing parameter −0.057 pm. The non-liner behavior of the energy gap versus concentration was determined, and it was shown that the energy gap increases with increasing x.

11.6 Lead–Barium–Selenium

PbSe–BaSe: The structural, electronic and optical properties of $\mathbf{Ba_xPb_{1-x}Se}$ solid solutions for $0 \le x \le 1$ in their cubic structure have been calculated using the full-potential linearized augmented plane wave method under the framework of density functional theory (Chattopadhyaya and Bhattacharjee 2017c). The calculated lattice parameter of these solid solutions exhibits tendency to Vegard's law with a marginal downward bowing parameter 0.12 pm. The non-liner behavior of the energy gap versus concentration was determined, and it was shown that the energy gap increases with increasing x.

11.7 Lead–Zinc–Selenium

PbSe–Zn: The solubility of zinc in PbSe, determined using radioactive isotopes, at the temperatures from 700°C to 900°C is retrograde and lies in the range 8×10^{18}–5×10^{20} cm^{-3} (Kharakhorin et al. 1966). The ingots for the investigations were annealed at 700°C–875°C for 20–123 h.

PbSe–ZnSe: The phase diagram of this system, constructed through DTA, XRD, and metallography, is a eutectic type (Figure 11.4) (Oleinik et al. 1982). The eutectic contains 22 mol% ZnSe and crystallizes at 1010°C. Mutual solubility of ZnSe and PbSe is not higher than 1 mol%.

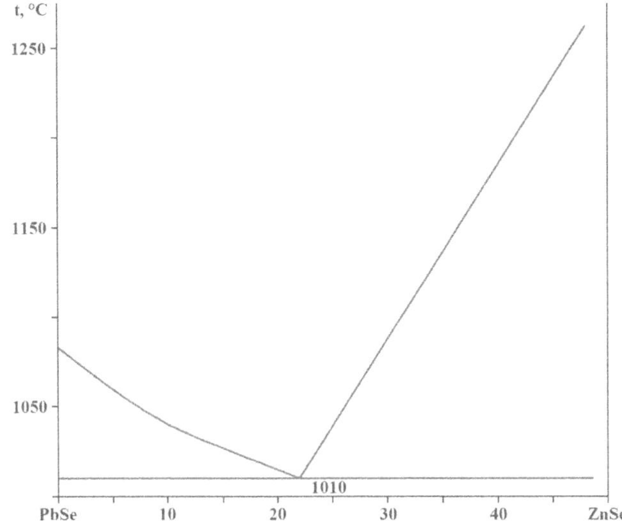

FIGURE 11.4 Phase diagram of the PbSe–ZnSe system. (From Oleinik, G.S., et al., *Izv. AN SSSR. Neorgan. mater.*, **18**(5), 873, 1982.)

11.8 Lead–Cadmium–Selenium

PbSe–Cd: The solubility of cadmium in PbSe, determined using radioactive isotopes, at the temperatures 700°C–900°C is retrograde and lies in the range 8 × 10^{18}–5 × 10^{20} cm^{-3} (Kharakhorin et al. 1966). The ingots for the investigations were annealed at 700°C–875°C for 12.5–93 h.

PbSe–CdSe: The phase diagram of this system, constructed through DTA, XRD, metallography and measuring of the microhardness, is a eutectic type (Figure 11.5) (Tomashik et al. 1980). The eutectic contains 46 mol% CdSe and crystallizes at 995°C. The data about nonquasibinarity of this system (Wald and Rosenberg 1965) were not confirmed (Tomashik et al. 1980). The solubility of CdSe in PbSe at the eutectic temperature is 30 mol% (Tomashik et al. 1980) and at 940°C, 800°C, and 600°C, respectively, 26, 12, and 6 mol% (Wald and Rosenberg 1965; Sealy and Crocker 1973b). The solubility of PbSe in CdSe at the eutectic temperature is not higher than 1 mol% (Tomashik et al. 1980).

The $Cd_xPb_{1-x}Se$ solid solutions with substitution content from 20 to 80 mol% have been obtained by a high-pressure (2–9 GPa) and high-temperature (900°C) method (Chufarov et al. 2016). The obtained solid solutions have the NaCl-type structure. It was found that the presence of CdSe in the cubic modification in the course of the synthesis initiates stabilization of the cubic modification of the solid solutions, regardless of the CdSe content in the solid solution, although this modification is not equilibrium for CdSe under normal conditions.

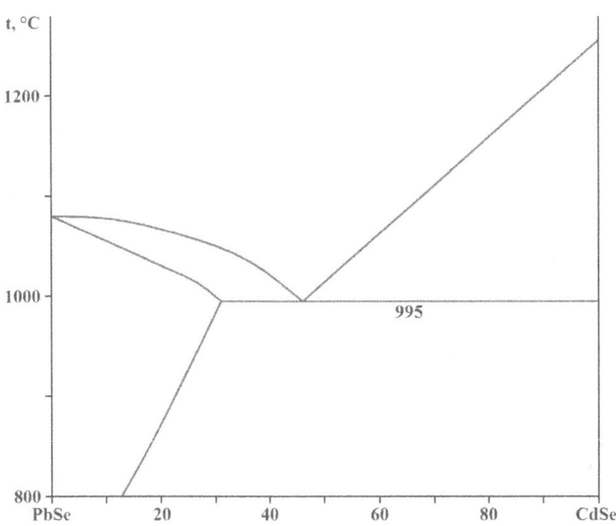

FIGURE 11.5 Phase diagram of the PbSe–CdSe system. (From Tomashik, V.N., et al., *Izv. AN SSSR. Neorgan. mater.*, **16**(2), 261, 1980.)

11.9 Lead–Mercury–Selenium

PbSe–HgSe: The phase diagram of this system, constructed through DTA, XRD, metallography, and measuring of the microhardness, is a eutectic type (Figure 11.6) (Vanyarho et al. 1967, 1968, 1969b; Leute and Köller 1986). The eutectic contains 70 mol% HgSe and crystallizes at 690°C. At the eutectic temperature, the solubility of HgSe in PbSe reaches 6.7 mol% and decreases to 5, 3, 2.5, and 1.5 mol% at 630°C, 530°C, 480°C, and 430°C, respectively, and the solubility of PbSe in HgSe at the same temperatures is not higher than 0.5 mol. % (Leute and Köller 1986). According to the data of Vanyarho et al. (1967, 1968, 1969b), solid solutions based on PbSe contain 9 mol% at 820°C, 12 mol% at 690°C, and 7 mol at 650°C, and the solubility of PbSe in HgSe at the eutectic temperature is less than 5 mol%. The ingots for investigations were annealed at 820°C (Vanyarho et al. 1967, 1968, 1969b).

11.10 Lead–Aluminum–Selenium

PbSe–Al_2Se_3: The **$PbAl_2Se_4$** ternary compound, which crystallizes in the orthorhombic structure with the lattice parameters a = 636 ± 1, b = 1063 ± 1, c = 1078 ± 1 pm, and the calculated and experimental densities of 5.26 and 5.23 g·cm^{-3},

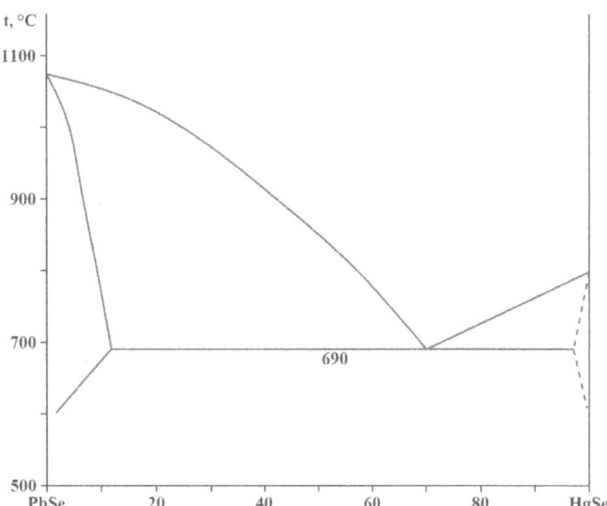

FIGURE 11.6 Phase diagram of the PbSe–HgSe system. (From Vanyarkho, V.G., et al., *Izv. AN SSSR. Neorgan. mater.*, **5**(11), 2025, 1969.)

respectively (Klee and Schäfer 1980) [a = 1070, b = 1080, c = 628.8 pm and a calculated density of 5.06 g·cm^{-3} (Eholié et al. 1971)], is formed in this system. To prepare this compound, stoichiometric amounts of the elements were weighed into quartz ampoules with a corundum inner crucible and rinsed several times with dry Ar (Klee and Schäfer 1980). Then, the mixture was slowly heated to 1100°C over a period of 8 h under a pressure of 13.3 Pa, left at this temperature for one hour and then slowly cooled overnight. The compound is extremely sensitive to humid air and instantly becomes covered with a dark layer of decomposition products of previously unknown stoichiometry, with a strong smell of H_2Se. $PbAl_2Se_4$ must be stored under a layer of paraffin oil. $PbAl_2Se_4$ could be also obtained by the heating of the mixture of PbSe and Al in a stream of H_2S at 1200°C (Eholié et al. 1971).

11.11 Lead–Gallium–Selenium

PbSe–GaSe: The phase diagram of this system, constructed through DTA, XRD, metallography, and measuring of the microhardness, is a eutectic type (Figure 11.7) (Eholié et al. 1969; Balitskiy et al. 1981). The eutectic contains 50 mol% GaSe and crystallizes at 700°C (Eholié et al. 1969) [40 mol% GaSe and 672°C (Balitskiy et al. 1981)]. The solubility of GaSe in PbSe reaches 4 mol% and the solubility of PbSe in GaSe is 1 mol% (Balitskiy et al. 1981). The detection of **$PbGa_2Se_4$** (Eholié et al. 1969; Eholié and Flahaut 1972) in the system is apparently explained by the nonequilibrium state of the samples under study. The ingots for investigations were annealed at 350°C at 150 h.

PbSe–Ga_2Se_3: The phase diagram of this system, constructed through DTA, XRD, and metallography, is a eutectic type with peritectic transformation (Figure 11.8) (Eholié et al. 1969; Eholié and Flahaut 1972; Sosovs'ka et al. 2006). The eutectic contains 68 mol% PbSe and crystallizes at 723°C (Sosovs'ka et al. 2006) [at 735°C (Eholié and Flahaut 1972)].

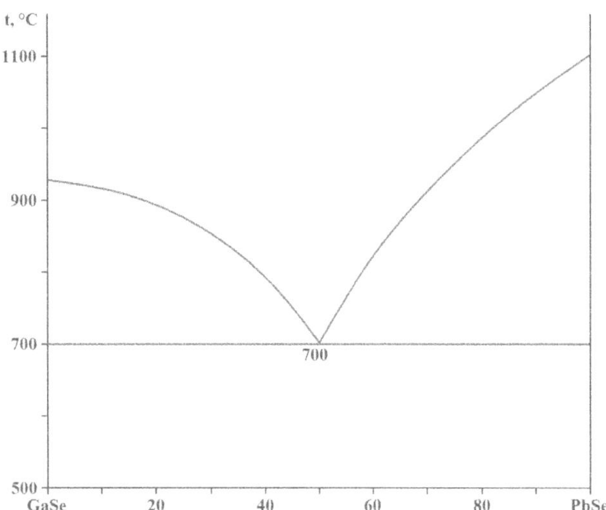

FIGURE 11.7 Phase diagram of the PbSe–GaSe system. (From Eholié, R., et al., *C. r. Acad. sci. Sér. C*, **268**(8), 700, 1969.)

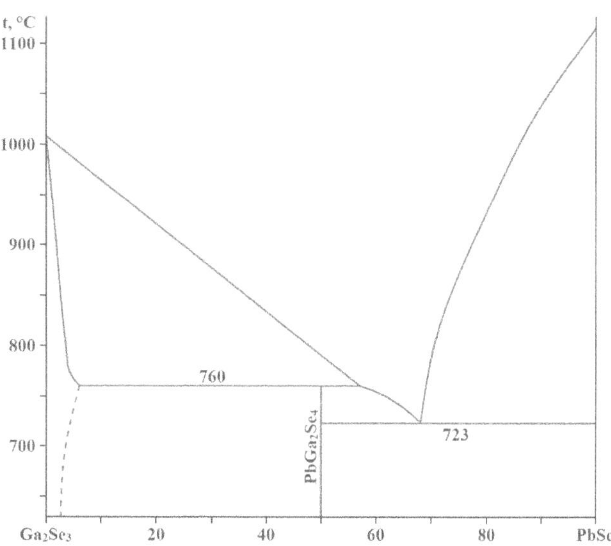

FIGURE 11.8 Phase diagram of the PbSe–Ga_2Se_3 system. (From Sosovs'ka, S.M., et al., *Nauk. Visnyk Volyns'k. Derzh. Univ. im. Lesi Ukrainky, Ser. Khim. nauky*, (4), 71, 2006.)

The coordinates of peritectic point are 57 mol% PbSe and 760°C. The solubility of PbSe in Ga_2Se_3 is not higher than 5 mol% [7 mol% at 500°C (Filyuk et al. 2008)]. The ingots for investigations were annealed at 600°C for 250 h (Sosovs'ka et al. 2006) [at 700°C for 4 days (Eholié and Flahaut 1972)].

The $PbGa_2Se_4$ ternary compound, which melts incongruently at 760°C (Sosovs'ka et al. 2006) [at 780°C (Eholié and Flahaut 1972); at 786°C ± 5°C (Rigan 2005); according to the data of Wu et al. (2015a), this compound melts congruently at ca. 779°C], is formed in the PbSe–Ga_2Se_3 system. It crystallizes in the orthorhombic structure with the lattice parameters a = 1273 ± 2, b = 2126 ± 4, c = 2155 ± 5 pm, a calculated density of 6.036 g·cm^{-3}, and an energy gap of 1.83 eV (Wu et al. 2015a) [a = 1064, b = 1077, c = 635.9 pm, the calculated and experimental density of 6.04 and 6.36 g·cm^{-3}, respectively, and an energy gap of 2.38 eV (Eholié et al. 1969, 1971; Eholié and Flahaut 1972; Georgobiani et al. 1997); a = 2137 ± 2, b = 2147 ± 2, c = 1273 ± 1 pm, and the calculated and experimental density of 6.03 and 5.97 g·cm^{-3}, respectively (Klee and Schäfer 1980); a = 1063.8, b = 1076.6 and, c = 635.7 pm (Rigan 2005); a = 2141 ± 2, b = 2129 ± 2, and c = 1266.9 ± 0.9 pm (Sosovs'ka et al. 2006)].

$PbGa_2Se_4$ vaporizes quantitatively starting at 680°C (Rigan 2005). Below 800°C, the dominant vapor species is $PbGa_2$. At higher temperatures, the vapor also contains Ga_2Se, Se_2, and Pb. In addition, the mass spectrum shows signals from Ga, Ga_2, Se, and $GaSe_2$. Between 930°C and 1030°C, the vapor composition remains unchanged.

To prepare the crystals of this compound, the mixtures of PbSe (1 mM) and Ga_2Se_3 (1 mM) were heated to 900°C in 30 h, left for 72 h, cooled to 400°C at a rate of 3°C·h^{-1}, and finally cooled to room temperature by switching off the furnace (Wu et al. 2015a). Many block-shaped crystals with yellowish color were found in the ampoule. Polycrystalline samples were synthesized by solid-state reaction techniques. The mixture of PbSe and Ga_2Se_3 (molar ratio 1:1) was heated to 800°C in 20 h and kept at that temperature for 72 h, and then the furnace was turned off.

Single crystals of $PbGa_2Se_4$ were grown by the Bridgman-Stockbarger method in silica ampoule evacuated to 0.1–0.01 Pa (Rigan 2005) or through the chemical transport reactions using iodine as carrier agent (Georgobiani et al. 1997).

In the Pb–Ga–S system, there is a wide immiscibility region adjacent to the Ga–Se binary system (Eholié and Flahaut 1972).

11.12 Lead–Indium–Selenium

PbSe–In: This system is a non-quasibinary section of the Pb–In–Se ternary system which intersects the fields of the PbSe, InSe, In_2Se, and In primary crystallization (Melikova and Rustamov 1977). Thermal effects at 600° C are due to the beginning of joint crystallization of PbSe and InSe, at 275°C to the crystallization of the ternary eutectic, and at 250°C to the existence of a ternary transition point. The solubility of In in PbSe at 600°C and at room temperature is 1.5 and 0.1 at%, respectively.

According to the data of Kharakhorin et al. (1966), the solubility of In in PbSe within the temperature interval from 700°C to 900°C, determined using radioactive isotopes, is retrograde and is in the range 8×10^{18}–5×10^{20} cm^{-3}. The ingots for the investigations were annealed at 700°C–880°C for 100–136 h.

PbSe–InSe: The phase diagram of this system, constructed through DTA, XRD, metallography, and measuring of the microhardness, is a eutectic type (Figure 11.9) (Wald and Rosenberg 1965; Rustamov et al. 1974d; Eddike et al. 1997; Tedenac et al. 1997). The eutectic crystallizes at 572°C and contains 23 mol% PbSe (Eddike et al. 1997; Tedenac et al. 1997) [at 550°C and contains 9 mol% PbSe (Rustamov et al. 1974d); at 545°C (Wald and Rosenberg 1965)]. The solid solution based on InSe is very limited (1.5 mol%) at 572°C while the PbSe one accepts 20.5 mol% InSe and is reduced to 5 mol% at room temperature (Eddike et al. 1997; Tedenac et al. 1997). According to the data of Wald and Rosenberg (1965) and Rustamov et al. (1974d), the solubility of InSe in PbSe at the eutectic temperature is 4.5–5 mol% and 3.3 mol% at 600°C.

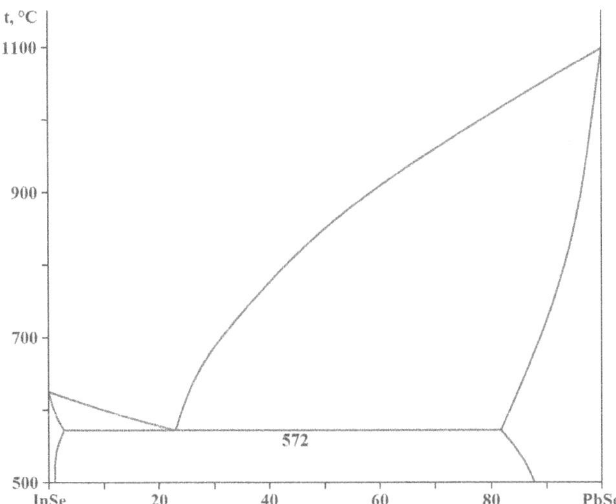

FIGURE 11.9 Phase diagram of the PbSe–InSe system. (From Tedenac, J.C., et al., *Powder Metall. Met. Ceram.*, **36**(1–2), 3, 1997.)

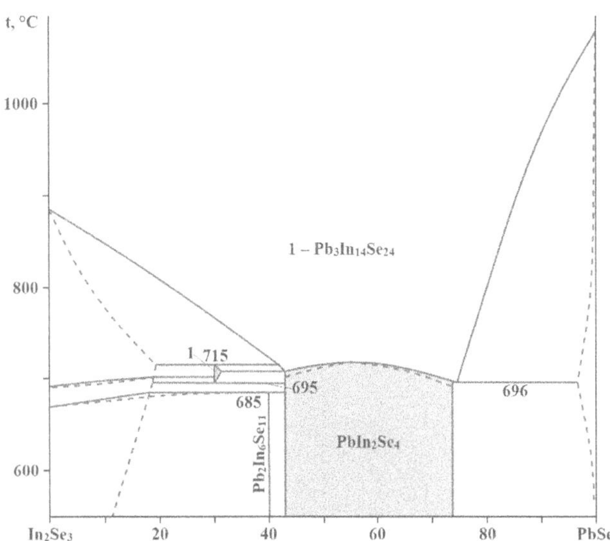

FIGURE 11.10 Phase diagram of the PbSe–In_2Se_3 system. (From Tedenac, J.C., et al., *Powder Metall. Met. Ceram.*, **36**(1–2), 3, 1997.)

PbSe–In_2Se_3: The phase diagram of this system, constructed through DTA, XRD, metallography and measuring of the microhardness, is presented in Figure 11.10 (Eddike et al. 1997; Tedenac et al. 1997). Three ternary compounds, **$PbIn_2Se_4$**, **$Pb_2In_6Se_{11}$**, and **$Pb_3In_{14}Se_{24}$**, are formed in the PbSe–In_2Se_3 system. $PbIn_2Se_4$ has a large homogeneity region, melts at 718°C for 55 mol% PbSe and crystallizes in the orthorhombic structure with the lattice parameters $a = 2368$, $b = 1576$, and $c = 2386$ pm (Eddike et al. 1997; Tedenac et al. 1997) [in the monoclinic structure with the lattice parameters $a = 1333$, $b = 404.8$, $c = 1448$ pm, $\beta = 94°$, and an experimental density of 5.9 g·cm^{-3} (Melikova et al. 1976)]. Single crystals of this compound were grown using the chemical transport reactions (Melikova et al. 1976). $PbIn_2Se_4$ is a *p*-type semiconductor (Alidzhanov et al. 1987).

$Pb_2In_6Se_{11}$ decomposes peritectically at 685°C and $Pb_3In_{14}Se_{24}$ is stable between 695°C and 715°C (Eddike et al. 1997; Tedenac et al. 1997). First of these two compounds crystallizes in the monoclinic structure with the lattice parameters $a = 1368 \pm 4$, $b = 406 \pm 1$, $c = 2908 \pm 10$ pm, and $\beta = 95.4 \pm 0.3°$. The existence of this compound was not confirmed by Pashyns'kyi et al. (2008).

According to the data of (Melikova et al. 1976), only $PbIn_2Se_4$ ternary compound, which melts congruently at 725°C and has a homogeneity region within the interval 49.5–50.5 mol% In_2Se_3, is formed in this system. The eutectics contain 43.5 and 58.5 mol% In_2Se_3 and crystallize at 715°C and 700°C, respectively. The solubility of PbSe in In_2Se_3 is 1 mol% and the solubility of In_2Se_3 in PbSe is 1.5 mol% at room temperature. In the presence of PbSe, all polymorphic transformations of solid solutions based on In_2Se_3 occur eutectoidally. The ingots for the investigations were annealed at 500°C for 1000 h.

The triangulation of the Pb–In–Se ternary system at room temperature (Figure 11.11) has been performed using XRD and EDX measurements (Daouchi et al. 2000). In addition to six binary phases, two ternary phases, $PbIn_2Se_4$ and

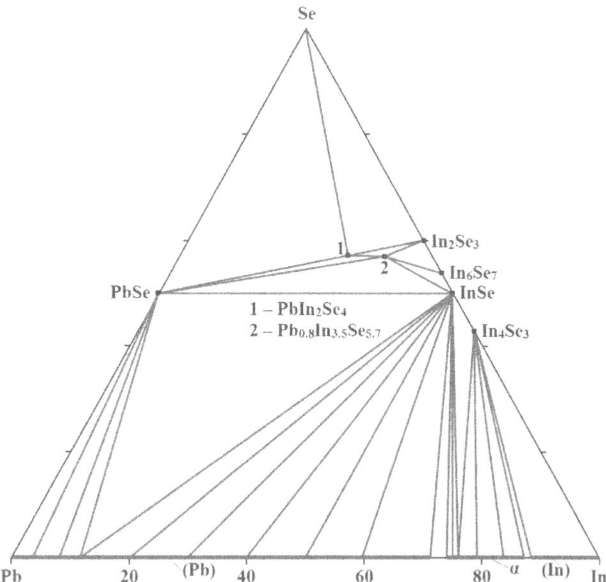

FIGURE 11.11 Isothermal section of the Pb–In–Se ternary system at room temperature (From Daouchi, B., et al., *J. Alloys Compd.*, **296**(1–2), 229, 2000.)

$\mathbf{Pb_{0.8}In_{3.5}Se_{5.7}}$, were observed; the ternary diagram is thus divided at room temperature into 11 secondary triangles. Two forms have been obtained at room temperature for In_2Se_3, but the corresponding secondary triangle, because of its narrowness, could not be observed. No ternary solubility in the binary intermediate and terminal phases has been observed.

Eight ternary invariants have been observed in the PbSe–InSe–In–Pb subsystem of the Pb–In–Se ternary system using DSC, XRD, and energy dispersive X-ray spectroscopy measurements (Record et al. 2003). These results allow suggesting the existence of one ternary eutectic reaction E_1 (312°C), and three transition reactions U_1 (157°C), U_2 (160°C), U_3 (170°C), generating, respectively, the three-phase fields PbSe + InSe + (Pb), In_4Se_3 + α + (In), InSe + In_4Se_3 + α and InSe + α + (Pb), where α is the phase in the In–Pb binary system. The approximate locations for E_1, U_1, U_2, U_3 are, respectively, close to Pb_8InSe, $Pb_{0.5}In_9Se_{0.5}$, $Pb_{1.25}In_{7.5}Se_{1.25}$, $Pb_{1.5}In_7Se_{1.5}$ compositions.

In the PbSe–InSe–Se subsystem, two ternary phases, $Pb_{0.8}In_{3.5}Se_{5.7}$ (melts congruently) and $\mathbf{Pb_{1.46}In_{2.86}Se_{5.71}}$ (melts incongruently), were observed (Record et al. 2001). The existence of the $Pb_3In_{14}Se_{24}$ ternary compound was not confirmed. Miscibility gaps which emerge from the Pb–Se and In–Se binary systems extend into the ternary system to form one gap across this subsystem. Reaction scheme for the PbSe–InSe–Se subsystem was proposed. The next invariant equilibria were determined: P_1 (707°C) – L + $Pb_{1.46}In_{2.86}Se_{5.71}$ + In_2Se_3 ⇔ $Pb_{0.8}In_{3.5}Se_{5.7}$; U_4 (650°C) – L + $Pb_{1.46}In_{2.86}Se_{5.71}$ ⇔ PbSe + $Pb_{0.8}In_{3.5}Se_{5.7}$; U_5 (640°C) – L + In_2Se_3 ⇔ $Pb_{0.8}In_{3.5}Se_{5.7}$ + In_6Se_7; E_2 (585°C) – L_1 ⇔ L_2 + $Pb_{1.46}In_{2.86}Se_{5.71}$ + In_2Se_3; U_6 (574°C) – L + In_6Se_7 ⇔ $Pb_{0.8}In_{3.5}Se_{5.7}$ + InSe; E_3 (572°C) – L ⇔ PbSe + $Pb_{0.8}In_{3.5}Se_{5.7}$ + InSe; E_4 (571°C) – L_1 ⇔ L_2 + $Pb_{1.46}In_{2.86}Se_{5.71}$ + PbSe; E_5 (215°C) – L ⇔ PbSe + Se + S; E_6 (215°C) – L ⇔ Se + S + In_2Se_3.

According to the data of Melikova and Rustamov (1977, 1979), the liquidus surface of the Pb–In–Se ternary system consists from nine fields of primary crystallization of phases and an immiscibility region. Phase equilibria in the system are defined by six ternary eutectics and two transition points.

11.13 Lead–Thallium–Selenium

PbSe–Tl: This system is a non-quasibinary section of the Pb–Tl–Se ternary system (Latypov and Fayzullina 1978). The PbSe primary crystallization takes place within the temperature interval from 1085°C to 540°C and ends at 60 at% Tl. Further, up to pure thallium, a bertholide-type δ-phase is formed. Thermal effects at 540°C correspond to the beginning of PbSe and δ-phase joint crystallization. Solidification of all alloys ends with crystallization of Pb-based solid solutions. Thermal effects at 232°C are due to the polymorphic transformation of thallium. This system was investigated through DTA, XRD, and metallography, and the ingots for the investigations were annealed at 300°C for 300 h.

PbSe–TlSe: The phase diagram of this system, constructed through DTA, XRD, metallography, and measuring of the microhardness, is a eutectic type (Figure 11.12) (Latypov et al. 1976a; Gotuk et al. 1980). The eutectic is degenerated from the TlSe-side and crystallizes at 336°C. No ternary compounds were found in the system. The ingots for the investigations were annealed at 300°C for 100 h.

PbSe–Tl₂Se: The phase diagram of this system, constructed through DTA, XRD, metallography, and measuring of the microhardness, is shown in Figure 11.13 (Fayzullina et al. 1983a; Malakhovs'ka-Rosokha et al. 2011). The eutectic contains 40 mol% PbSe and crystallizes at 527°C. The $\mathbf{Tl_4PbSe_3}$ ternary compound, which melts congruently at 530°C (Malakhovs'ka-Rosokha et al. 2011) [at 550°C (Fayzullina et al. 1983a); at 519°C (Malakhovs'ka-Rosokha et al. 2009b)] and forms continuous solid solution with Tl_2Se, is formed in this system. The solubility of Tl_2Se in PbSe reaches 20 mol% at 527°C; at 300°C solid solution based on Tl_2Se contains 35 mol% PbSe,

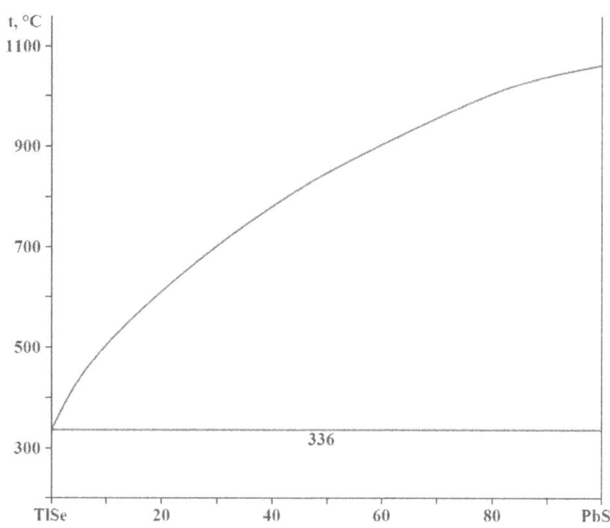

FIGURE 11.12 Phase diagram of the PbSe–TlSe system. (From Latypov, Z.M., et al., *Izv. AN SSSR. Neorgan. mater.*, **12**(9), 1671, 1976.)

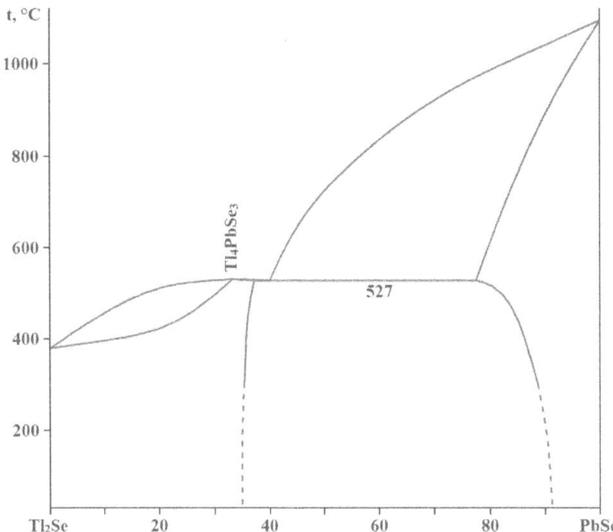

FIGURE 11.13 Phase diagram of the PbSe–Tl_2Se system. (From Malakhovs'ka-Rosokha, T.A., et al., *Inorg. Mater.*, **47**(7), 700, 2011.)

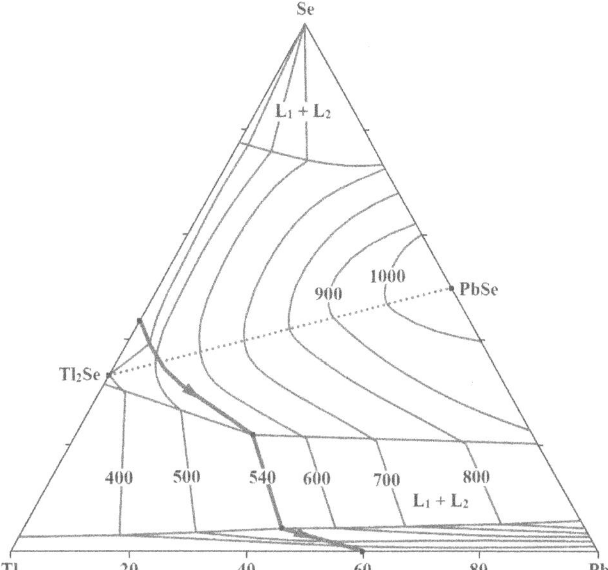

FIGURE 11.14 Liquidus surface of the Pb–Tl–Se ternary system. (From Gotuk, A.A., et al, *Izv. AN SSSR. Neorgan. mater.*, **16**(9), 1519, 1980.)

and PbSe dissolves 10 mol% Tl_2Se (Malakhovs'ka-Rosokha et al. 2011) [7 mol% Tl_2Se (Filep and Sabov 2018)] [the solubility of PbSe in Tl_2Se at 300°C is not higher than 1 mol% (Gotuk et al. 1978a)]. The ingots for the investigations were annealed at 300°C for 336 h (Malakhovs'ka-Rosokha et al. 2011).

According to the data of Latypov et al. (1976a) and Gotuk et al. (1978a), the phase diagram of the PbSe–Tl_2Se is a peritectic type with a peritectic horizontal at 545°C–550°C. The ingots for the investigations were annealed at 450°C for 350 h (Gotuk et al. 1978a) [at 300°C for 100 h (Latypov et al. 1976a)].

Tl_4PbSe_3 crystallizes in the tetragonal structure with the lattice parameters $a = 851.1$ and $c = 1264.0$ pm (Filep and Sabov 2017b, 2018) [$a = 853.46 \pm 0.02$, $c = 1268.71 \pm 0.07$ pm and the calculated and experimental densities of 9.067 and 9.05 g·cm^{-3}, respectively (Malakhovs'ka-Rosokha et al. 2009b, 2011)]. This compound was prepared from Tl, Pb, and Se by encapsulating them under vacuum in quartz ampoule (Malakhovs'ka-Rosokha et al. 2009b). The sample was annealed at 580°C for 24 h and then at 300°C for 1 week. Single crystals were grown by the Bridgman technique. The enthalpy and entropy of melting of Tl_4PbSe_3 are 64 kJ·M^{-1} and 81 J·(M·K)$^{-1}$, respectively.

The liquidus surface of the Pb–Tl–Se ternary system (Figure 11.14) consists from the fields of PbSe and (Tl_2Se) primary crystallization and two immiscibility regions (Gotuk et al. 1980). It is necessary to note that the immiscibility region near the Tl–Pb binary system must be ended within the ternary system as in the Pb–Se binary system the immiscibility region from the Pb-side is absent (Massalski 1990). The fields of Pb, Tl, Se and TlSe primary crystallization are degenerated. At room temperature, a wide homogeneity region based on Tl_2Se exists in the system.

11.14 Lead–Scandium–Selenium

PbSe–Sc_2Se_3: The $\mathbf{Sc_2PbSe_4}$ ternary compound, which crystallizes in the orthorhombic structure with the lattice parameters $a = 1220.29 \pm 0.05$, $b = 390.61 \pm 0.02$, $c = 1428.01 \pm 0.06$ pm, and a calculated density of 5.9808 g·cm^{-3} (Shemet et al. 2006b), is formed in this system (Gulay et al. 2008c). To prepare this compound, calculated amounts of the elements were sealed in an evacuated silica ampoule, which was heated to 1150°C at a rate of 30°C·h^{-1} (Shemet et al. 2006b). The sample was kept at this temperature for 3 h. Subsequently, it was cooled slowly (10°C·h^{-1}) to 600°C and annealed at this temperature for 240 h. After annealing the ampoule was quenched in cold water.

11.15 Lead–Yttrium–Selenium

PbSe–Y_2Se_3: The solubility of Y_2Se_3 in PbSe at 500°C reaches 4 mol% (Filyuk et al. 2008). Two compounds, $\mathbf{Y_{4.2}Pb_{0.7}Se_7}$ and $\mathbf{Y_6Pb_2Se_{11}}$, are formed in this system (Pashyns'kyi et al. 2008). $Y_6Pb_2Se_{11}$ crystallizes in the orthorhombic structure with the lattice parameters $a = 406.20 \pm 0.08$, $b = 1346.7 \pm 0.2$, $c = 376.24 \pm 0.07$ pm, and a calculated density of 5.862 g·cm^{-3} (Gulay et al. 2005). To prepare this compound, calculated amounts of the components were sealed in evacuated silica ampoule (Gulay et al. 2005, 2006a). The synthesis was realized in a tube furnace. The ampoule was heated with the heating rate of 30°C·h^{-1} to 1350°C and the sample was kept at this temperature for 4 h. After that, it was cooled slowly (10°C·h^{-1}) to 600°C and annealed at this temperature during 240 h. After annealing, the ampoule with the sample was quenched in cold water.

The existence of $\mathbf{Y_2PbSe_4}$ ternary compound (Patrie et al. 1969; Filyuk et al. 2008) was not confirmed by Pashyns'kyi et al. (2008).

11.16 Lead–Lanthanum–Selenium

PbSe–LaSe: According to the data of Shafagatova et al. (2005), $\mathbf{LaPbSe_2}$ ternary compound, which crystallizes in the cubic structure with the lattice parameter $a = 609.2$ pm

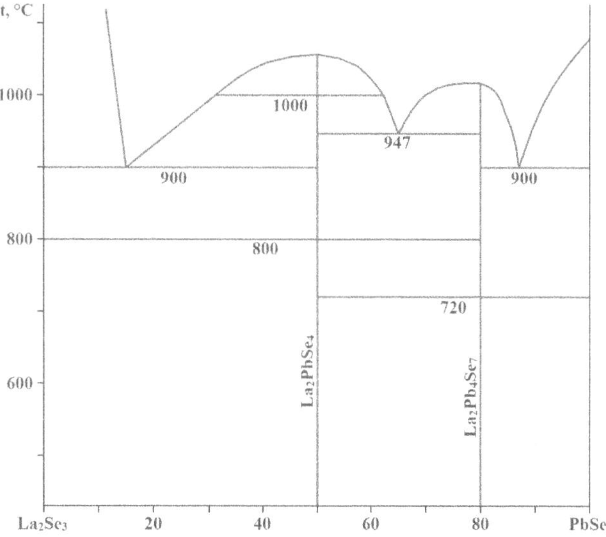

FIGURE 11.15 Phase diagram of the PbSe–La_2Se_3 system. (From Nasibov, I.O., et al., *Nauch. tr. Azerb. Univ. Ser. khim. nauk*, (3), 11, 1979.)

and forms continuous solid solution both with PbSe and LaSe, is formed in this system. This system was investigated using DTA, XRD, metallography, and measuring of the microhardness, and the alloys were homogenized at 700°C for 240 h.

PbSe–La_2Se_3: The phase diagram of this system, constructed through DTA, XRD, metallography, and measuring of the microhardness, is given in Figure 11.15 (Nasibov et al. 1979c). Two from three eutectics crystallize at 900°C and one of them contains 13 mol% La_2Se_3. Third eutectic crystallizes at 947°C. The solubility of La_2Se_3 in PbSe reaches 5 mol% (Shafagatova and Imanov 2015). Two ternary compounds, **La_2PbSe_4** and **$La_2Pb_4Se_7$**, which melt congruently, are formed in this system. First of them has polymorphic transformations at 800°C and 1000°C (Nasibov et al. 1979c) and crystallizes in the cubic structure with the lattice parameter a = 910.6 pm (Patrie et al. 1969). $La_2Pb_4Se_7$ has a polymorphic transformation at 720°C (Nasibov et al. 1979c). Both compounds are p-type semiconductors with an energy gap of 1.26 (La_2PbSe_4) and 0.86 eV ($La_2Pb_4Se_7$) (Nasibov et al. 1987a). The existence of $La_2Pb_4Se_7$ ternary compound was not confirmed by Pashyns'kyi et al. (2008).

According to the data of Patrie et al. (1969), Filyuk et al. (2008) and Pashyns'kyi et al. (2008), the solid solution is formed in this system within the interval 0–50 mol% PbSe: from La_2Se_3 to La_2PbSe_4.

11.17 Lead–Cerium–Selenium

PbSe–Ce_2Se_3: The phase diagram of this system, constructed through DTA, XRD, metallography, and measuring of the microhardness and density, is similar to the phase diagram of the PbSe–La_2Se_3 system (Nasibov et al. 1979e). The eutectics crystallize at 900°C, 950°C, and 1000°C. Two ternary compounds, **Ce_2PbSe_4** and **$Ce_2Pb_4Se_7$**, are formed in this system. First of them melts congruently at 1100°C, has polymorphic transformations at 850°C and 1050°C (Nasibov et al. 1979e) and crystallizes in the cubic structure with the lattice parameter a = 904.5 pm (Patrie et al. 1969). $Ce_2Pb_4Se_7$ melts congruently at 1075°C and has polymorphic transformations at 760°C (Nasibov et al. 1979e). Ce_2PbSe_4 and $Ce_2Pb_4Se_7$ are p-type semiconductors with an energy gap of 1.28 and 1.01 eV, respectively (Shafagatova et al. 1990). The ingots for the investigations were annealed at 500°C–550°C for 600 h (Nasibov et al. 1979e).

According to the data of Patrie et al. (1969), the solid solution, which crystallizes in the cubic structure, is formed in this system within the interval 0–50 mol% PbSe.

11.18 Lead–Praseodymium–Selenium

PbSe–Pr_2Se_3: The phase diagram of this system, constructed through DTA, XRD, metallography, and measuring of the microhardness and density, is similar to the phase diagram of the PbSe–La_2Se_3 system (Nasibov et al. 1979b). The eutectics crystallize at 900°C and 950°C. Two ternary compounds, **Pr_2PbSe_4** and **$Pr_2Pb_4Se_7$**, are formed in this system. First of them melts congruently, has polymorphic transformation at 800°C (Nasibov et al. 1979b) and crystallizes in the cubic structure with the lattice parameter $a = 899.16 \pm 0.02$ pm, and a calculated density of 7.3531 g·cm^{-3} (Marchuk et al. 2006b) [a = 899.6 pm (Patrie et al. 1969)]. $Pr_2Pb_4Se_7$ melts incongruently (Nasibov et al. 1979b). The existence of this compound was not confirmed by Marchuk et al. (2006b). No solubility of Pr_2Se_3 in PbSe was observed (Marchuk et al. 2006b) [the solubility of Pr_2Se_3 in PbSe reaches 5 mol% (Nasibov et al. 1989)]. The ingots for the investigations were annealed at 500°C–550°C for 600 h (Nasibov et al. 1979b).

According to the data of Patrie et al. (1969) and Marchuk et al. (2006b), the solid solution, which crystallizes in the cubic structure, is formed in this system within the interval 0–50 mol% PbSe (between Pr_2Se_3 and Pr_2PbSe_4).

11.19 Lead–Neodymium–Selenium

PbSe–Nd_2Se_3: The phase diagram of this system, constructed through DTA, XRD, metallography, and measuring of the microhardness and density, is similar to the phase diagram of the PbSe–La_2Se_3 system (Nasibov et al. 1978a, 1980). The eutectics crystallize at 900°C and 950°C and the latter eutectic contains 25 mol% PbSe. Two ternary compounds, **Nd_2PbSe_4** and **$Nd_2Pb_4Se_7$**, are formed in this system. First of them melts congruently and has polymorphic transformation at 830°C (Nasibov et al. 1978a, 1980). α-Nd_2PbSe_4 crystallizes in the cubic structure with the lattice parameter a = 888 pm, the calculated and experimental densities of 7.69 and 7.62 g·cm^{-3}, respectively, and an energy gap of 1.35 eV (Nasibov et al. 1992) [a = 896.8 pm (Patrie et al. 1969)]. β-Nd_2PbSe_4 crystallizes in the tetragonal structure with the lattice parameter a = 821 and c = 1120 pm (Nasibov et al. 1992).

$Nd_2Pb_4Se_7$ melts incongruently (Nasibov et al. 1980) and crystallizes in the cubic structure with the lattice parameter a = 801 pm, the calculated and experimental densities of 7.98

and 7.92 g·cm^{-3}, respectively, and an energy gap of 1.11 eV (Nasibov et al. 1992). The ingots for the investigations were annealed at 500°C–550°C for 600 h (Nasibov et al. 1978a, 1980).

Both compounds were prepared from the mixtures of the elements in stoichiometric ratios by melting in the evacuated quartz ampoules at 1030°C–1130°C for 4–5 h with the next annealing at 500°C–580°C for 20 days (Nasibov et al. 1992).

According to the data of Patrie et al. (1969), the solid solution, which crystallizes in the cubic structure, is formed in this system within the interval 0–50 mol% PbSe (between Nd_2Se_3 and Nd_2PbSe_4).

11.20 Lead–Samarium–Selenium

PbSe–SmSe: According to the data of XRD and SEM with EDX, the solubility of SmSe in PbSe is 9 mol% (Ibrahim et al. 2007). The solid solution was prepared using conventional solid-state reaction. Stoichiometric amounts of the constituent materials were reacted at high temperatures (1100°C) in a closed silica tube.

PbSe–Sm_2Se_3: The phase diagram of this system, constructed through DTA, XRD, metallography, and measuring of the microhardness and density, is similar to the phase diagram of the PbSe–La_2Se_3 system (Shafagatova et al. 1979). The eutectics crystallize at 930°C and 900°C and the latter eutectic contains 35 mol% PbSe. Two ternary compounds, **Sm_2PbSe_4** and **$Sm_2Pb_4Se_7$**, are formed in this system. First of them melts congruently at 1195°C and has polymorphic transformation at 800°C (Shafagatova et al. 1979). α-Sm_2PbSe_4 crystallizes in the cubic structure with the lattice parameter a = 884 pm, the calculated and experimental densities of 7.91 and 7.75 g·cm^{-3}, respectively, and an energy gap of 1.37 eV (Nasibov et al. 1992) [a = 890.9 pm (Patrie et al. 1969)]. β-Sm_2PbSe_4 crystallizes in the tetragonal structure with the lattice parameter a = 817 and c = 1117 pm (Shafagatova et al. 1979).

$Sm_2Pb_4Se_7$ melts incongruently (Shafagatova et al. 1979) and crystallizes in the cubic structure with the lattice parameter a = 797 pm, the calculated and experimental densities of 8.01 and 7.96 g·cm^{-3}, respectively, and an energy gap of 1.16 eV (Nasibov et al. 1992). The ingots for the investigations were annealed at 500°C–550°C for 600 h (Shafagatova et al. 1979).

Both compounds were prepared from the mixtures of the elements in stoichiometric ratios by melting in the evacuated quartz ampoules at 1030°C–1130°C for 4–5 h with the next annealing at 500°C–580°C for 20 days (Nasibov et al. 1992).

According to the data of Patrie et al. (1969), the solid solution, which crystallizes in the cubic structure, is formed in this system within the interval 0–50 mol% PbSe (between Sm_2Se_3 and Sm_2PbSe_4).

11.21 Lead–Gadolinium–Selenium

PbSe–GdSe: The solubility of GdSe in PbSe is ~4.5 mol% (Shvangiradze and Kutsiya 1999).

PbSe–Gd_2Se_3: The phase diagram of this system, constructed through DTA, XRD, metallography, and measuring of the microhardness and density, is presented in Figure 11.16

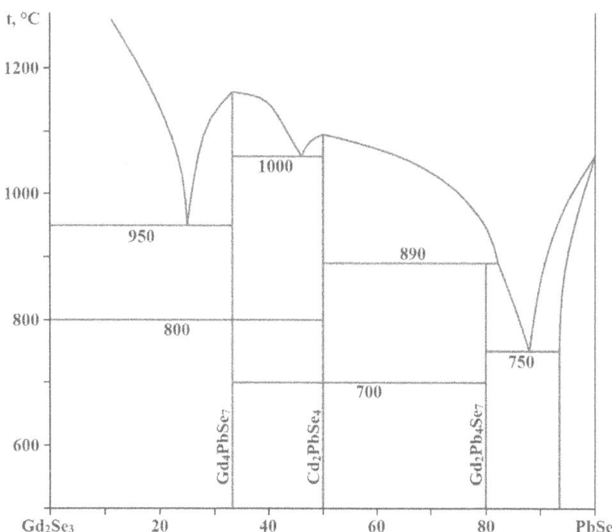

FIGURE 11.16 Phase diagram of the PbSe–Gd_2Se_3 system. (From Nasibov, I.O., et al., *Nauch. tr. Azerb. Univ. Ser. khim. nauk*, (4), 3, 1979.)

(Nasibov et al. 1979d). The eutectics crystallize at 750°C, 950°C, and 1060°C. Three ternary compounds, **Gd_2PbSe_4**, **$Gd_2Pb_4Se_7$**, and **Gd_4PbSe_7**, are formed in this system. First of them melts congruently and has a polymorphic transformation at 700°C. $Gd_2Pb_4Se_7$ melts incongruently and Gd_4PbSe_7 melts congruently. All these compounds are stable up in vacuum to their melting temperature.

According to the data of Patrie et al. (1969), the solid solution based on Gd_2Se_3 is formed up to the PbSe·1.6Gd_2Se_3 composition. At 500°C, $Gd_{2+2/3x}Pb_{1-x}Se_4$ (x = 0.5–0.9) solid solution is formed in the PbSe–Gd_2Se_3 system (Marchuk et al. 2014).

11.22 Lead–Terbium–Selenium

PbSe–TbSe: The solubility of TbSe in PbSe is ~4.5 mol% (Shvangiradze and Kutsiya 1999).

PbSe–Tb_2Se_3: The phase diagram of this system, constructed through DTA, XRD, metallography, and measuring of the microhardness and density, is similar to the phase diagram of the PbSe–Gd_2Se_3 system (Nasibov et al. 1981f). The eutectics crystallize at 800°C, 1000°C, and 1100°C. Three ternary compounds, **Tb_2PbSe_4**, **$Tb_2Pb_4Se_7$**, and **Tb_4PbSe_7**, are formed in this system. First of them melts congruently and has a polymorphic transformation at 750°C. $Tb_2Pb_4Se_7$ melts incongruently and Tb_4PbSe_7 melts congruently and has a polymorphic transformation at 870°C. All these compounds are stable up in vacuum to their melting temperature. The solubility of PbSe in Tb_2Se_3 reaches 20 mol% (Patrie et al. 1969) and the solubility of Tb_2Se_3 in PbSe at 600°C is 4 mol% (Gulay et al. 2006d).

According to the data of Gulay et al. (2006d), the **$Tb_{2+2x}Pb_{1-3x}Se_4$** ($0.07 \le x \le 0.19$) solid solution is formed in this system at 600°C. It crystallizes in the cubic structure with the lattice parameter a = 876.72 ± 0.02 and a calculated density of 7.7872 g·cm^{-3} at x = 0.165.

11.23 Lead–Dysprosium–Selenium

PbSe–DySe: The solubility of DySe in PbSe is ~4.5 mol% (Shvangiradze and Kutsiya 1999).

PbSe–Dy_2Se_3: The solubility of Dy_2Se_3 in PbSe at 600°C is 7 mol% (Gulay et al. 2006d). The **$Dy_6Pb_2Se_{11}$** ternary compound is formed in this system. It crystallizes in the orthorhombic structure with the lattice parameters a = 407.72 ± 0.08, b = 1345.8 ± 0.3, and c = 375.89 ± 0.09 pm (Gulay et al. 2005). To prepare this compound, calculated amounts of the components were sealed in evacuated silica ampoule (Gulay et al. 2005, 2006a). The synthesis was realized in a tube furnace. The ampoule was heated with the heating rate of 30°C·h^{-1} to 1350°C and the sample was kept at this temperature for 4 h. After that, it was cooled slowly (10°C·h^{-1}) to 600°C and annealed at this temperature during 240 h. After annealing, the ampoule with the sample was quenched in cold water.

The existence of **Dy_2PbSe_4** ternary compound (Patrie et al. 1969) was not confirmed by Gulay et al. (2006d).

11.24 Lead–Holmium–Selenium

PbSe–HoSe: The solubility of HoSe in PbSe is ~4.5 mol% (Shvangiradze and Kutsiya 1999).

PbSe–Ho_2Se_3: The solubility of Ho_2Se_3 in PbSe at 600°C is 6 mol% (Gulay et al. 2006c). The **$Ho_6Pb_2Se_{11}$** ternary compound, which crystallizes in the orthorhombic structure with the lattice parameters a = 405.61 ± 0.02, b = 1340.18 ± 0.06, c = 375.25 ± 0.01 pm, and a calculated density of 7.3992 g·cm^{-3} (Gulay et al. 2005, 2006c) [a = 406.03 ± 0.07, b = 1339.7 ± 0.3, c = 375.42 ± 0.05 pm, and a calculated density of 7.392 g·cm^{-3} (Gulay et al. 2006a)] is formed in this system. To prepare this compound, calculated amounts of the components were sealed in evacuated silica ampoule (Gulay et al. 2005, 2006a). The synthesis was realized in a tube furnace. The ampoule was heated with the heating rate of 30°C·h^{-1} to 1350°C and the sample was kept at this temperature for 4 h. After that, it was cooled slowly (10°C·h^{-1}) to 600°C and annealed at this temperature during 240 h. After annealing, the ampoule with the sample was quenched in cold water.

The existence of **Ho_2PbSe_4** ternary compound (Patrie et al. 1969) was not confirmed by Gulay et al. (2006c).

11.25 Lead–Erbium–Selenium

PbSe–Er_2Se_3: The solubility of Er_2Se_3 in PbSe at 600°C is 7 mol% (Gulay et al. 2006c). The **Er_2PbSe_4** ternary compound, which crystallizes in the orthorhombic structure with the lattice parameters a = 1254.1 ± 0.3, b = 408.10 ± 0.08, c = 1486.5 ± 0.3 pm, and a calculated density of 7.487 g·cm^{-3} (Gulay et al. 2007a) [a = 1245, b = 1485, c = 5129 pm (Patrie et al. 1969); a = 1245, b = 1485, c = 412 pm (Gasymov et al. 1980); a = 1255.4 ± 0.2, b = 407.78 ± 0.05, c = 1488.5 ± 0.5 pm, and a calculated density of 7.474 g·cm^{-3} (Gulay et al. 2006c)], is formed in this system.

11.26 Lead–Thulium–Selenium

PbSe–TmSe: The solubility of TmSe in PbSe is ~4.5 mol% (Shvangiradze and Kutsiya 1999).

PbSe–Tm_2Se_3: The solubility of Tm_2Se_3 in PbSe at 600°C is ca. 3 mol% (Gulay et al. 2006e). The **Tm_2PbSe_4** ternary compound, which crystallizes in the orthorhombic structure with the lattice parameters a = 1250.5 ± 0.3, b = 406.30 ± 0.08, c = 1482.0 ± 0.3 pm, and a calculated density of 7.594 g·cm^{-3} (Gulay et al. 2006e) [a = 1259, b = 1475, c = 5262 pm (Patrie et al. 1969); a = 1259, b = 1475, c = 410 pm (Gasymov et al. 1980)], is formed in this system. This compound was prepared as follows. The calculated amount of the elements was sealed in an evacuated silica ampoule (Gulay et al. 2007a). The synthesis was realized in a tube furnace. The ampoule was gradually heated with 30°C·h^{-1} to 1150°C and the sample was kept at this temperature for 4 h. After that, it was cooled slowly (10°C·h^{-1}) to 600°C and annealed at this temperature for 240 h. After annealing the ampoule with the sample was quenched in cold water.

11.27 Lead–Ytterbium–Selenium

PbSe–Yb_2Se_3: The **Yb_2PbSe_4** ternary compound, which crystallizes in the orthorhombic structure with the lattice parameters a = 1250.1 ± 0.2, b = 403.80 ± 0.08, c = 1470.7 ± 0.3 pm, and a calculated density of 7.776 g·cm^{-3} (Gulay et al. 2007a) [a = 1283, b = 1464, c = 5235 pm (Patrie et al. 1969); a = 1283, b = 1464, c = 408 pm (Gasymov et al. 1980)], is formed in this system. To prepare this compound, the calculated amount of the elements was sealed in an evacuated silica ampoule (Gulay et al. 2007a). The synthesis was realized in a tube furnace. The ampoule was gradually heated with 30°C·h^{-1} to 1150°C and the sample was kept at this temperature for 4 h. After that, it was cooled slowly (10°C·h^{-1}) to 600°C and annealed at this temperature for 240 h. After annealing the ampoule with the sample was quenched in cold water.

11.28 Lead–Lutetium–Selenium

PbSe–LuSe: The solubility of LuSe in PbSe is ~4.5 mol% (Shvangiradze and Kutsiya 1999).

PbSe–Lu_2Se_3: The phase diagram of this system, constructed through DTA, XRD, metallography, and measuring of the microhardness, is similar to the phase diagram of the PbSe–La_2Se_3 system (Nasibov et al. 1981c). The eutectics crystallize at 850°C and 900°C and the solubility of Lu_2Se_3 in PbSe at 600°C is 5–6 mol% (Nasibov et al. 1981c; Gulay et al. 2006e). Two ternary compounds, **Lu_2PbSe_4** and **$Lu_2Pb_4Se_7$**, are formed in this system. First of them melts congruently at 1200°C and has polymorphic transformation at 1000°C (Nasibov et al. 1981c). This compound crystallizes in the orthorhombic structure with the lattice parameters a = 1247.18 ± 0.05, b = 403.45 ± 0.01, c = 1473.38 ± 0.06 pm, and a calculated density of 7.8206 g·cm^{-3} (Gulay et al. 2006e) [a = 1287, b = 1459, c = 5238 pm (Patrie et al. 1969); a = 1287, b = 1459,

$c = 407$ pm (Gasymov et al. 1980)]. $Lu_2Pb_4Se_7$ melts incongruently at 950°C and has polymorphic transformation at 750°C (Nasibov et al. 1981c). Both compounds are *p*-type semiconductors. The ingots for the investigations were annealed at 500°C–550°C for 600 h.

11.29 Lead–Uranium–Selenium

The $\mathbf{PbU_2Se_5}$ ternary compound, which crystallizes in the orthorhombic structure with the lattice parameters $a = 860.5 \pm 0.5$, $b = 778.8 \pm 0.5$, $c = 1227 \pm 1$ pm (Potel et al. 1975) [$a = 779 \pm 1$, $b = 861.0 \pm 0.5$, $c = 1227 \pm 2$ pm, and the calculated and experimental densities of 8.750 and 8.65 g·cm^{-3}, respectively (Brochu et al. 1970, 1972)], is formed in the Pb–U–Se system. This compound was prepared from a PbSe + 2U + 4Se mixture, pelletized, then placed in a vacuum sealed silica tube; after a treatment of 12 h at 600°C, the temperature was maintained above 800°C for 24 h. The manipulation is carried out in an Ar-filled glove box. The single crystals of PbU_2Se_5 can be isolated, if the preparation is carried out at 1150°C., after a rapid increasing the temperature.

The solid solution $\mathbf{Pb_xU_{3-x}Se_5}$ ($0 \le x \le 1$) is formed in the $\mathbf{PbU_2Se_5}$–$\mathbf{U_3Se_5}$ system (Brochu et al. 1970, 1972). The lattice parameter *b* varies linearly with composition, while the lattice parameter *a* practically does not change.

11.30 Lead–Titanium–Selenium

Two ternary compounds are formed in the Pb–Ti–Se system. $\mathbf{Pb_{0.5}Ti_5Se_8}$ crystallizes in the monoclinic structure with the lattice parameters $a = 1878.0 \pm 0.4$, $b = 357.9 \pm 0.1$, $c = 910.6 \pm 0.2$ pm, and $\beta = 104.09 \pm 0.02°$ (Novet et al. 1995). To prepare this compound, the ion-exchange reaction was used. $TlTi_5Se_8$ was mixed with PbI_2 at various molar ratios (1:1, 1:2, and 1:4) in a N_2-filled dry box and sealed in evacuated quartz tube approximately 20 cm in length. The tube was placed in a temperature gradient with the sample at the hot end (at temperatures 150°C–325°C) and the cold end of the tube extended out of the furnace assembly. The temperature of the hot end of the ampoule was raised until red crystals of TlI were observed on the cold end of the tube. The samples were annealed for approximately 12 h before quenching to room temperature.

$\mathbf{(PbSe)_{1.18}(TiSe_2)_2}$ was synthesized using modulated elemental reactants and structurally characterized as a disordered variant of a misfit layered compound. (Moore et al. 2013). Their structure consists of an intergrowth between one distorted rock-salt structured PbSe bilayer and two Se–Ti–Se trilayers. In addition to the lattice mismatch, there is extensive rotational disorder between these constituents.

11.31 Lead–Phosphorus–Selenium

PbSe–"$\mathbf{P_2Se_4}$": The phase diagram of this system, constructed through DTA, XRD, and measuring of the microhardness and the density, is shown in Figure 11.17 (Prits et al. 1989; Potoriy et al. 1990). The eutectic contains 74 mol% PbSe and

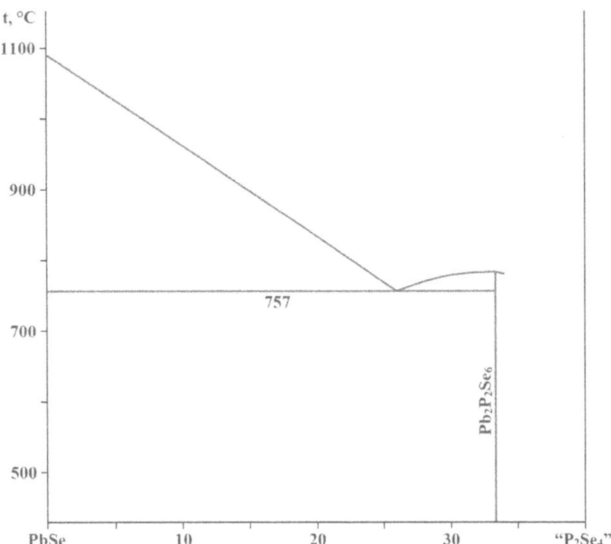

FIGURE 11.17 Part of the phase diagram of the PbSe–"P_2Se_4" system. (From Potoriy, M.V., et al., *Izv. AN SSSR. Neorgan. mater.*, **26**(11), 2363, 1990.)

crystallizes at 757°C. No mutual solubility was found in the system. The $\mathbf{Pb_2P_2Se_6}$ ternary compound, which melts congruently at 784°C, is formed in this system (Prits et al. 1989; Potoriy et al. 1990; Potoriy and Milyan 2016). It crystallizes in the monoclinic structure with the lattice parameters given in Table 11.1. Single crystals of $Pb_2P_2Se_6$ were grown through the chemical transport reactions (Klingen et al. 1973; Carpentier and Nitsche 1974). When heated in vacuum up to 500°C, this compound decomposes into PbSe and volatile phosphorus selenides (Brusilovets and Teplyakova 1974). The ingots for the investigations were annealed at 500°C for 2 weeks (Prits et al. 1989; Potoriy et al. 1990).

11.32 Lead–Arsenic–Selenium

PbSe–As: No solid solutions based on As were detected by X-ray phase analysis (Krebs et al. 1961a).

PbSe–$\mathbf{As_2Se_3}$: The phase diagram of this system, constructed through DTA, XRD, and measuring of the microhardness, is presented in Figure 11.18 (Dembovskiy et al. 1970, 1971b). The eutectic is degenerated and its crystallization temperature practically coincides with the melting temperature of As_2Se_3. The $\mathbf{PbAs_2Se_4}$ ternary compound, which melts incongruently and exists within the interval of 375°C–475°C, is formed in this system. An insignificant region of the solid solution was found on the basis of PbSe. The ingots for the investigations were annealed at 350°C–500°C until an equilibrium state was reached.

The glass formation region in the Pb–As–Se ternary system was defined by Borisova et al. (1973). The glass transition temperature for these glasses is within the interval of 80°C–180°C. The criteria for the glassy state were the presence of a conchoidal fracture and the homogeneity of the samples during metallographic studies. No more than 5–6 at% Pb can be added to the composition of glassy arsenic selenides. Arsenic selenides

TABLE 11.1

Crystallographic Data for $Pb_2P_2Se_6$

a, pm	*b*, pm	*c*, pm	β, °	$d_{calc.}$ g·cm^{-3}	$d_{meas.}$ g·cm^{-3}	References
690	765	1184	124.5	6.13	6.08	Klingen et al. (1973)
691.0 ± 0.2	767.0 ± 0.2	1181.6 ± 0.3	124.35 ± 0.3	6.10	5.98	Carpentier and Nitsche (1974)
974.2 ± 0.4	766.2 ± 0.3	689.8 ± 0.3	91.44 ± 0.5	6.13	6.08	Becker et al. (1984)
689.7 ± 0.3	764.2 ± 0.3	969.6 ± 0.4	91.51 ± 0.1	6.174	–	Yun and Ibers (1987)
691.1	766.1	1210.1	126.247	–	6.10	Prits et al. (1989)
669.4	764.9	1208	126.27	–	–	Voroshilov et al. (2016)
–	–	–	–	6.14	5.98	Potoriy et al. (1990); Potoriy and Milyan (2016)

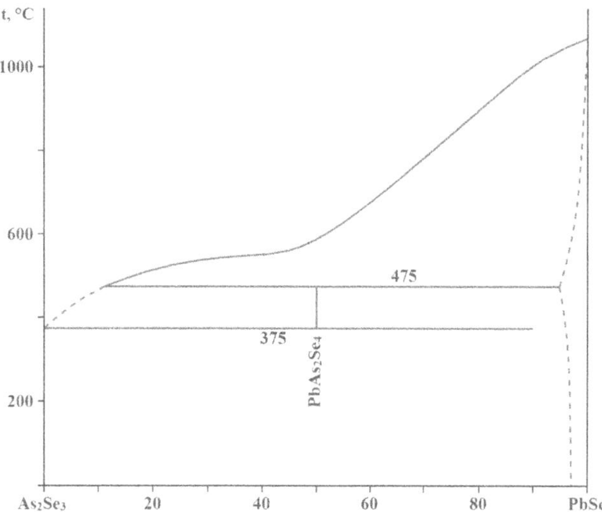

FIGURE 11.18 Phase diagram of the PbSe–As_2Se_3 system. (From Dembovskiy, S.A., et al., *Izv. AN SSSR. Neorgan. mater.*, 7(10), 1859, 1971.)

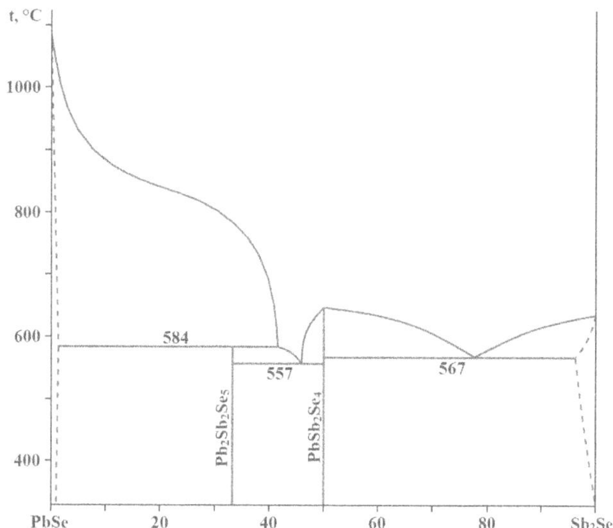

FIGURE 11.19 Phase diagram of the PbSe–Sb_2Se_3 system. (From Olekseyuk, et al., *Nauk. Visnyk Volyns'k. Nats. Univ. im. Lesi Ukrainky, Ser. Khim. nauky*, (16), 38, 2010.)

of compositions As_2Se_3 and As_2Se_5 have the maximum capacity for glass formation with lead. The glasses of the system have a very high crystalline capacity. All glassy alloys crystallize already during thermal analysis. The maximum synthesis temperature of the samples was 800°C. The alloys were cooled by quenching in air and in ice water.

11.33 Lead–Antimony–Selenium

PbSe–Sb$_2$Se$_3$: The phase diagram of this system, constructed through DTA, XRD and measuring of the microhardness, is presented in Figure 11.19 (Elagina 1961b; Olekseyuk et al. 2010b). One eutectic contains 20 mol% PbSe and crystallizes at 567°C and the second eutectic crystallizes at 557°C (Olekseyuk et al. 2010b) [the eutectic crystallize at 590°C and 547°C, the later containing 68 mol% Sb_2Se_3 (Elagina 1961b)]. The ingots for the investigations were annealed at 350°C for 600 h with the next quenching in water (Olekseyuk et al. 2010b) [at 500°C for 1–4 months (Elagina 1961b)].

Two ternary compounds, **PbSb$_2$Se$_4$** and **Pb$_2$Sb$_2$Se$_5$**, are formed in this system (Olekseyuk et al. 2010b) [only $PbSb_2Se_4$ exist in the system according to the data of Elagina (1961b)]. $PbSb_2Se_4$ melts congruently at 647°C (Olekseyuk et al. 2010b) [at 618°C (Elagina 1961b); melts incongruently at 590°C (Frumar et al. 1972a)] and crystallizes in the orthorhombic structure with the lattice parameters a = 2120.6 ± 0.9, b = 2666.0 ± 0.9, c = 406.8 ± 0.1 pm, a calculated density of 6.641 g·cm^{-3} (Skowron et al. 1994a), and an energy gap of 1.8 eV (Elagina 1961b). At 500°C and 550°C, the homogeneity region of this compound is in the range of 50–53 and 49–56 mol% Sb_2Se_3, respectively (Elagina 1961b).

Single crystals of $PbSb_2Se_4$ were synthesized from Pb, Sb, and Se in a stoichiometric ratio (Skowron et al. 1994a). The sample was sealed in an evacuated silica ampoule, melted at 854°C for 2 days, slowly cooled to 500°C over 24 h, annealed for 14 days and quenched. The gray needle-shaped crystals grew in various region of the ampoule.

$Pb_2Sb_2Se_5$ melts incongruently at 584°C (Olekseyuk et al. 2010b) and also crystallizes in the orthorhombic structure with the lattice parameters a = 2459.1 ± 0.8, b = 1975.7 ± 0.8, c = 416.6 ± 0.8 pm, and a calculated density of 6.91 g·cm^{-3} (Skowron and Brown 1990c). Single crystals of this compound were synthesized from Pb, Sb, and Se in a stoichiometric ratio by slow cooling the mixture in an evacuated silica tube from 800°C over a period of 2 weeks to 550°C.

Two another ternary compounds were found in the Pb–Sb–Se system. **$Pb_4Sb_6Se_{13}$** (the composition of this compound lies on the $PbSe–Sb_2Se_3$ section) crystallizes in the monoclinic structure with the lattice parameters a = 2459.1 ± 0.1, b = 409.10 ± 0.02, c = 2521.2 ± 0.1 pm, β = 93.943 ± 0.001°, a calculated density of 6.783 g·cm^{-3}, and an energy gap of 0.9 eV (Derakhshan et al. 2006). **$Pb_6Sb_6Se_{17}$** melts incongruently and crystallizes in the orthorhombic structure with the lattice parameters a = 1587.2 ± 0.4, b = 2406.1 ± 0.7, c = 413.82 ± 0.09 pm, and a calculated density of 6.991 g·cm^{-3} (Derakhshan et al. 2006) [a = 1583.5 ± 0.3, b = 2404.3 ± 0.5, c = 413.4 ± 0.2 pm, and a calculated density of 6.997 g·cm^{-3} (Emirdag-Eanes and Kolis 2002)]. By using the EMF method, the self-consistent data on the standard integral function of this compound were obtained as follows: $\Delta G^0_{f,298}$ = −1064 ± 28 kJ·M^{-1}, $\Delta H^0_{f,298}$ = −1080 ± 20 kJ·M^{-1} and S^0_{298} = 1387 ± 35 J·(K·M)$^{-1}$ (Mansimova et al. 2017).

Single crystals of both compounds were obtained by heating stoichiometric mixtures of the elements in evacuated quartz tubes to 600°C ($Pb_4Sb_6Se_{13}$) and 800°C ($Pb_6Sb_6Se_{17}$) (Derakhshan et al. 2006). At these temperatures the products were molten, and subsequent slow cooling (2°C·h^{-1}) yielded needlelike crystals. Black needlelike crystals of $Pb_6Sb_6Se_{17}$ were also prepared from a reaction mixture of PbSe (0.1 mM), Sb_2Se_3 (0.1 mM), K_2Se_2 (0.29 mM), NH_4Cl (0.1 g) as mineralizer, and H_2O (0.7 mL) in a quartz tube (Emirdag-Eanes and Kolis 2002). The tube was then put into an autoclave with a counter pressure of 17.2 MPa of Ar. The reaction mixture was heated for 3 days at 375°C and cooled in air to room temperature. The tube was opened and the crude product was washed with water and acetone.

A series of **$Pb_mSb_{2n}Se_{m+3n}$** nanocrystals (m = 2, 4, 6 and 8; n = 1) are demonstrated that exist only as a distinct phase on the nanoscale (Soriano et al. 2015). The new materials tend to phase separate exothermically at a high temperature of 385°C into the binary PbSe and Sb_2Se_3 end members. They are new phases stabilized only on the nanoscale and crystallize in the cubic structure with the lattice parameter a = 613.7 ± 0.3 pm for $Pb_2Sb_2Se_5$, a = 614.5 ± 0.2 pm for **$Pb_4Sb_2Se_7$**, a = 615.1 ± 0.1 pm for **$Pb_6Sb_2Se_9$**, and a = 616.3 ± 0.2 pm for **$Pb_8Sb_2Se_{11}$**. The band gap onset values are around 0.27 eV for all the prepared compounds. The syntheses of these compounds were carried out under air free condition using Schlenk line. In a typical synthesis, stoichiometric amounts of $Pb(CH_3COO)_2·3H_2O$, $Sb(CH_3COO)_3$, 20.0 mL of 1-octadecene and 5.00 mL of oleic acid were placed in a 100 mL Schlenk flask and dried under vacuum at 100°C for 3–4 h to obtain a pale yellow solution. The reaction flask was flushed with N_2, and the temperature was subsequently raised to 180°C. At 180°C, an appropriate amount of 1.0 M solution of Te in tri-n-octylphosphine was swiftly injected. Upon injection a dark brown colored mixture results. The temperature of the resulting mixture was maintained at 170°C–180°C for 2 min and the reaction was then rapidly cooled to room temperature using an ice-water bath. The resulting nanocrystals were precipitated by the addition of a solution of ethanol/hexanes (volume ratio 1:1) followed by centrifugation. A black pellet of $Pb_mSb_{2n}Se_{m+3n}$ nanocrystal aggregates was isolated by discarding the supernatant liquid. The resulting nanocrystal aggregates were re-suspended in hexanes and re-precipitated with excess ethanol. The purified nanocrystals can be dispersed in non-polar organic solvents to form stable colloidal suspension. $Pb_mSb_{2n}Se_{m+3n}$ nanocrystals of different atomic compositions were prepared by varying the molar ratio of Pb, Sb and Se precursors.

The liquidus projection and the isothermal section 400°C of the Pb–Sb–Se ternary system were determined by Chang and Chen (2016). A new ternary compound **$PbSb_5Se_6$** was found. There are eight primary solidification phase regions in the liquidus projection which are (Pb), (Sb), (Se), PbSe, Sb_2Se_3, $PbSb_2Se_4$, $Pb_6Sb_6Se_{17}$, and $PbSb_5Se_6$. There are nine ternary invariant reactions involving liquid. The reactions with the lowest and highest reaction temperatures are L ⇔ PbSe + (Sb) + (Pb) at 250°C and L + PbSe + (Sb) ⇔ $PbSb_5Se_6$ at 578.3°C respectively. In the isothermal section at 400°C, there are nine tie triangles: PbSe–(Sb)–L, $PbSe–PbSb_2Se_4$–(Sb), $PbSb_2Se_4–PbSb_5Se_6$–(Sb), $PbSb_2Se_4–PbSb_5Se_6–Sb_2Se_3$, $PbSb_5Se_6–Sb_2Se_3$–(Sb), $PbSe–Pb_6Sb_6Se_{17}–PbSb_2Se_4$, $PbSe–Pb_6Sb_6Se_{17}$–L, $Pb_6Sb_6Se_{17}–PbSb_2Se_4–Sb_2Se_3$, and $Pb_6Sb_6Se_{17}–Sb_2Se_3$–L.

11.34 Lead–Bismuth–Selenium

PbSe–BiSe: The **$PbBiSe_2$** ternary compound, which crystallizes in the tetragonal structure with the lattice parameters a = 526, c = 384 pm, and the calculated and experimental densities of 8.9 and 8.5 g·cm^{-3}, respectively, is formed in this system (Palatnik et al. 1961c). The solubility of PbSe in BiSe at 450°C reaches 5 mol% (Abrikosov et al. 1972) and the solubility of BiSe in PbSe is not higher than 1 mol% (Efimova et al. 1968). This system was investigated through XRD, metallography, and measuring of the microhardness (Palatnik et al. 1961c; Efimova et al. 1968; Abrikosov et al. 1972). The ingots for the investigations were annealed at 600°C–800°C for 200 h (Efimova et al. 1968) [at 250°C and 450°C for 30 and 10, respectively (Abrikosov et al. 1972).

$PbSe–Bi_2Se_3$: The phase diagram of this system is a eutectic type with peritectic transformations (Elagina 1961a; Malevskiy et al. 1963; Godovikov 1967b). The eutectic contains 21 mol% PbSe and crystallizes at 685°C (Godovikov 1967b) [at 657°C (Elagina 1961a)]. The solubility of Bi_2Se_3 in PbSe reaches 11 mol% (Godovikov 1967b) [20 mol% (Elagina 1961a)]. Three ternary compounds were found in this system: **$PbBi_2Se_4$** and **$PbBi_4Se_7$**, which melt incongruently at 725°C and 705°C (Godovikov 1967b) [at 700°C and 675°C (Elagina 1961a)], respectively, and a phase which has a homogeneous region in the range 69–72 mol% Bi_2Se_3 at 725°C, and melts incongruently at 750°C (Godovikov 1967b) [the **$Pb_3Bi_4Se_9$** compound with incongruent melting at 720°C (Elagina 1961a)]. When studying thin films, compound **$Pb_2Bi_2Se_5$** was found in the system (Agaev et al. 1966). $PbBi_2Se_4$, $PbBi_4Se_7$, and $Pb_3Bi_4Se_9$ have an energy gap of 0.32, 0.12, and 0.24 eV, respectively (Elagina 1961a).

According to the data of Liu and Chang (1994a), three phases were found in the $PbSe–Bi_2Se_3$ system: **$Pb_9Bi_4Se_{15}$**, which melts congruently at 560°C, **$Pb_8Bi_6Se_{17}$** which melts congruently at 575°C, and a $PbBi_2Se_4$-based solid solution, which melts congruently at 655°C and has a homogeneity region of 47–57 mol% Bi_2Se_3 at 500°C and 47–53 mol% Bi_2Se_3 at 400°C.

The formation of other three compounds, $\mathbf{Pb_5Bi_6Se_{14}}$, $\mathbf{Pb_5Bi_{12}Se_{23}}$, and $\mathbf{Pb_5Bi_{18}Se_{32}}$, which melt incongruently at 720°C, 700°C, and 675°C, respectively, was determined by Shelimova et al. (2008, 2010). They crystallizes in the monoclinic structure with the lattice parameters $a = 1600.96 \pm 0.02$, $b = 420.148 \pm 0.004$, $c = 2156.89 \pm 0.03$ pm, and $\beta = 97.537 \pm 0.001°$ (Zhang et al. 2005b) [$a = 1599.9 \pm 0.2$, $b = 420.0 \pm 0.3$, $c = 2157.0 \pm 0.3$ pm, and $\beta = 97.54°$ (Shelimova et al. 2008); $a = 1599.8 \pm 0.4$, $b = 420.0 \pm 0.6$, $c = 2156.4 \pm 0.6$ pm, and $\beta = 97.61 \pm 0.03°$ (Shelimova et al. 2010)] for $Pb_5Bi_6Se_{14}$; $a = 2639.2 \pm 0.6$, $b = 420.0 \pm 0.7$, $c = 2149.2 \pm 0.7$ pm, and $\beta = 106.40 \pm 0.04°$ (Shelimova et al. 2008); $a = 2641.5 \pm 0.9$, $b = 419.9 \pm 0.5$, $c = 2154.2 \pm 0.5$ pm, and $\beta = 106.35 \pm 0.04°$ (Shelimova et al. 2010)] for $Pb_5Bi_{12}Se_{23}$. These compounds constitute a $\mathbf{[(PbSe)_5]_m [(Bi_2Se_3)_3]_n}$ homologous series (Shelimova et al. 2008, 2010). Their layered structures are made up of two types of blocks, [5(PbSe)] and [3(Bi_2Se_3)], stacked along the *a* axis. Since the blocks differ in symmetry and periodicity, the structures of these compounds are incommensurate.

As a result of the PbSe–Bi_2Se_3–Se subsystem studying by measuring EMF of the concentration chains and XRD, the formation of the $Pb_5Bi_6Se_{14}$, $Pb_5Bi_{12}Se_{23}$, and $Pb_5Bi_{18}Se_{32}$ compounds was confirmed (Guseinov et al. 2012). The alloys were annealed at 530°C for 1000 h and at 200°C for 300 h. The self-consistent data on the standard integral function of these compounds were obtained as follows: $\Delta G^0_{f,298} = -1027 \pm 39$ kJ·M^{-1}, $\Delta H^0_{f,298} = -1028 \pm 38$ kJ·M^{-1} and $S^0_{298} = 1268 \pm 46$ J·(K·M)$^{-1}$ for $Pb_5Bi_6Se_{14}$; $\Delta G^0_{f,298} = -1500 \pm 45$ kJ·M^{-1}, $\Delta H^0_{f,298} = -1500 \pm 40$ kJ·M^{-1} and $S^0_{298} = 2005 \pm 60$ J·(K·M)$^{-1}$ for $Pb_5Bi_{12}Se_{23}$; and $\Delta G^0_{f,298} = -1939 \pm 54$ kJ·M^{-1}, $\Delta H^0_{f,298} = -1940 \pm 50$ kJ·M^{-1} and $S^0_{298} = 2736 \pm 62$ J·(K·M)$^{-1}$ for $Pb_5Bi_{18}Se_{32}$.

This system was investigated through DTA, XRD, metallography, and measuring of the microhardness (Elagina 1961a; Godovikov 1967b). The ingots for the investigations were annealed at 650°C–750°C for 1824–720 h (Godovikov 1967b) [at 550°C–720°C for 1–5 months [(Elagina 1961a); at 500°C for 200–400 h with next quenching in ice water (Shelimova et al. 2008, 2010)]. $PbBi_2Se_4$ and $Pb_2Bi_2Se_5$ were obtained by recrystallization through the gas phase (Agaev et al. 1966; Agaev and Semiletov 1968).

Several other compounds in the PbSe–Bi_2Se_3 system were found. Single crystalline samples of $\mathbf{(PbSe)_5(Bi_2Se_3)_{3m}}$ with a series of *m* values were grown by a combination of modified Bridgman and self-flux methods (Segawa et al. 2015). The growth process is complicated, because these compounds do not exhibit congruent melting. To grow high-quality single crystals, the following steps were taken: high-purity Pb, Bi, and Se (molar ratio of PbSe/Bi_2Se_3 = 45:55) were sealed in a quartz tube that was heat-treated beforehand. The raw materials were reacted and homogenized in a melt treated at 900°C for 6 h, and the crystal growth was fostered by slowly decreasing the temperature from 750°C to 650°C at a rate of 2°C·h^{-1} in a temperature gradient of roughly 1°C·cm^{-1}. At the beginning of the growth, $m = 1$ crystals were formed at the low-temperature end of the melt, which causes the composition of the melt to become PbSe-poor. As the temperature is lowered and the crystals growth proceeds, the melt becomes more and more PbSe-poor and the composition of the grown crystals changes from $m = 1$ to higher *m* values. $\mathbf{Pb_5Bi_{24}Se_{41}}$ ($m = 1$) crystallizes in the monoclinic structure with the lattice parameters $a = 2167 \pm 3$, $b = 412.3 \pm 0.5$, $c = 4431 \pm 8$ pm, and $\beta = 93.35 \pm 0.05°$.

$\mathbf{Pb_{46}Bi_{54}Se_{137}}$ (mineral cannizzarite) also crystallizes in the monoclinic structure with the lattice parameters $a = 3886.9 \pm 2.9$, $b = 409.0 \pm 0.3$, $c = 3983.5 \pm 3.3$ pm, $\beta = 102.30 \pm 0.03°$, and a calculated density of 6.97 g·cm^{-3} (Topa et al. 2010a).

11.35 Lead–Vanadium–Selenium

PbSe–$\mathbf{V_{1+x}Se_2}$: The $\mathbf{(PbSe)_{1+\delta}(V_{1+x}Se_2)}$ misfit layer compound is formed in this system (Reisinger and Richter 2021b). This compound can be synthesized using mineralizing agents. Best results were obtained when NH_4Cl was utilized. An off-stoichiometric sample composition (in the region between the three compositions of VSe_2, V_5Se_8, and misfit layer compound) larger qualities of crystals could be obtained. A composition with larger vanadium content and close to V_5Se_8 was most favorable. An annealing duration of one to three months (excluding pre-alloying time) at approximately 550°C was found to be most successful for synthesizing large crystals.

One partial at 375°C and five full isothermal sections at 550°C, 700°C, 800°C, 1000°C, and 1100°C were constructed in the Pb–V–Se ternary system by Reisinger and Richter (2020). No ternary compounds were observed. Two impurity stabilized phases might be present. In contrast to PbV_3, V_5Se_4 was observed but neglected in the construction of the isothermal section. Apart from the Se-rich liquid, the binary compounds in these sections exhibit only small solid solubilities into the ternary. In order to investigate the solubility of the ternary Pb-rich liquid, further experiments in the binary Pb–V system were conducted. It can be conclude on low V-contents in the liquid till 1100°C (e.g. max. 1.3 at% V) and similar low Pb-contents in vanadium that is in equilibrium with the respective liquid (e.g. max. 2 at% Pb). A new $(PbSe)_{1+\delta}(V_{1+x}Se_2)$ misfit layer compound, containing 22.6 ± 0.2 at% Pb, 56.6 ± 0.2 at% Se, and 20.8 ± 0.3 at% V, was observed in this ternary system. It was obtainable only by the addition of NH_4Cl to the powder pills. Due to several difficulties this phase was not included in the equilibrium phase diagram.

11.36 Lead–Niobium–Selenium

Several ternary compounds are formed in the Pb–Nb–Se system. The $\mathbf{Pb_{0.40}Nb_6Se_8}$ intercalate compound crystallizes in the hexagonal structure with the lattice parameters $a = 1000.4 \pm 0.1$ and $c = 935$ pm (Huan and Greenblatt 1987). It was prepared by means of molten salt ion exchange using $Tl_xNb_6Se_8$ and PbI_2. A high $PbI_2/Tl_xNb_6Se_8$ molar ratio (4:1) was necessary for completion of the reaction; the reactant mixture was initially kept at 450°C (~10°C above the melting point of TlI) for 24–36 h and then annealed at this temperature for 14 days.

The $\mathbf{Pb_{0.63}NbSe_2}$ intercalate compound also crystallizes in the hexagonal structure with the lattice parameters $a = 346$ and c = 1860 pm (Karnezos et al. 1975). This compound was synthesized by reaction of the powdered $NbSe_2$ with Pb vapor at 800°C–950°C in sealed silica tube. $NbSe_2$ powder and the condensed Pb phase were separated during the intercalation reaction so that it proceeded via the vapor rather than via the liquid or solid metal phase. Only by repeated firing and subsequent grinding could reproducibly homogeneous material be produced by this method.

$PbNb_2Se_5$ crystallizes in the orthorhombic structure with the lattice parameter $c = 3737$ pm (Hernán et al. 1992)]. This compound was obtained by direct combination of elements mixed in pellet form and sealed in silica tube under vacuum. The tube was heated first at 300°C for 1 day and then at 900°C for 7 days. The sample was reground and reheated at 900°C for another 7 days.

The powder samples of the **$(PbSe)_{1.14}(NbSe_2)_n$** ($n = 1, 2, 3$) layered compounds were prepared as follows (Oosawa et al. 1992, 1994). The mixtures of the starting materials (Pb, Nb, and Se) in the stoichiometric ratios were sealed in silica tubes under vacuum. Each tube was heated at 500°C for 12 h and then at 800°C for 24 h. The product were crushed, mixed and sealed again in silica tubes. The tubes were isothermally heated at series of temperatures from 500 to 1000°C for 96 to 6 h. The compound with $n = 1$ was formed at 1000°C and above, with $n = 2$ was formed from 900°C to 500°C and with $n = 3$ was formed at 900°C. All the samples contained some amounts of PbSe. These compounds were black-grayish microcrystalline powder with luster.

$(PbSe)_{1.14}(NbSe_2)_1$ crystallizes in the orthorhombic structure with the lattice parameters $a = 601.3 \pm 0.7$, $b = 605 \pm 1$, and $c = 2491 \pm 2$ pm for PbSe sublattice and $a = 601.5 \pm 0.6$, $b = 345.6 \pm 0.3$, and $c = 2490 \pm 2$ pm for $NbSe_2$ sublattice (Oosawa et al. 1992, 1994). $(PbSe)_{1.14}(NbSe_2)_2$ crystallizes in the monoclinic structure with the lattice parameters $a = 599.1 \pm 0.3$, $b = 602.5 \pm 0.6$, $c = 1875.2 \pm 0.6$ pm, and $\beta = 93.10 \pm 0.04°$ for PbSe sublattice and $a = 599.0 \pm 0.4$, $b = 344.9 \pm 0.2$, $c = 1874.6 \pm 0.8$ pm, and $\beta = 93.09 \pm 0.05°$ for $NbSe_2$ sublattice. $(PbSe)_{1.14}(NbSe_2)_3$ also crystallizes in the monoclinic structure with the lattice parameters $a = 598.6 \pm 0.2$, $b = 601.1 \pm 0.4$, $c = 2508.7 \pm 0.5$ pm, and $\beta = 92.22 \pm 0.02°$ for PbSe sublattice and $a = 598.6 \pm 0.2$, $b = 344.1 \pm 0.1$, $c = 2508.6 \pm 0.5$ pm, and $\beta = 92.22 \pm 0.02°$ for $NbSe_2$ sublattice.

$[(PbSe)_{1.10}]_m(NbSe_2)_n$ were synthesized using the self-assembly of designed reactants to target particular *m*, *n* compounds (Heideman et al. 2008). *m* and *n* values as large as 10 and 15, respectively, have been prepared, but the limit of this approach was not be probed. Very regular changes in lattice parameters were observed as *m* and *n* were varied, as expected from the insertion of regular structural units. The compounds were found to be very crystalline perpendicular to the layering (i.e. along the *c*-axis) and in the *a*–*b* plane. The structural coherence between the layers was found to be smaller. All compounds were metallic. An estimated misfit between the constituents was obtained from the lattice parameters of the binary components. For the **$[(PbSe)_{1+\delta}]_m(NbSe_2)_n$** family of compounds, this calculation yielded a value of 0.07 for δ, compared to a range between 0.10 and 0.14 reported in the literature for the compound with $m = n = 1$.

$[(PbSe)_{1.14}]_m(NbSe_2)_n$ compounds with $1 \le m \le 6$ and $n = 1$ were synthesized using the modulated elemental reactants method (Alemayehu et al. 2014). XRD showed that the desired compounds self-assembled during annealing of the precursors with their *c*-axis crystallographically aligned normal to the substrate. The *c*-axis lattice parameter increased by 612 pm as *m*, the number of PbSe bilayers, increased by one. Analysis of the in-plane diffraction patterns indicated that the *a*-lattice parameters remained constant as *m* was varied.

11.37 Lead–Tantalum–Selenium

The **Pb_xTaSe_2** intercalate compound, which crystallizes in the hexagonal structure with the lattice parameters $a = 344$ and $c = 935$ pm is formed in the Pb–Ta–Se system (Eppinga and Wiegers 1980). It was prepared by direct exposure of $NbSe_2$ to the lead vapor (this method yields the stoichiometry) or by heating the mixture of elements. Temperatures from 700°C to 900°C for periods from 1 to 4 weeks were applied.

$[(PbSe)_{1.12}]_m(TaSe_2)_n$ were synthesized using the self-assembly of designed reactants to target particular *m*, *n* compounds (Heideman et al. 2008). *m* and *n* values as large as 10 and 15, respectively, have been prepared, but the limit of this approach was not be probed. Very regular changes in lattice parameters were observed as *m* and *n* were varied, as expected from the insertion of regular structural units. The compounds were found to be very crystalline perpendicular to the layering (i.e. along the *c*-axis) and in the *a*–*b* plane. The structural coherence between the layers was found to be smaller. All compounds were metallic.

11.38 Lead–Oxygen–Selenium

PbSe–PbO: The phase diagram of this system, constructed through DTA and XRD, is presented in Figure 11.20 (Popovkin et al. 1963b). The eutectic contains 20 mol% PbSe and crystallizes at 760°C.

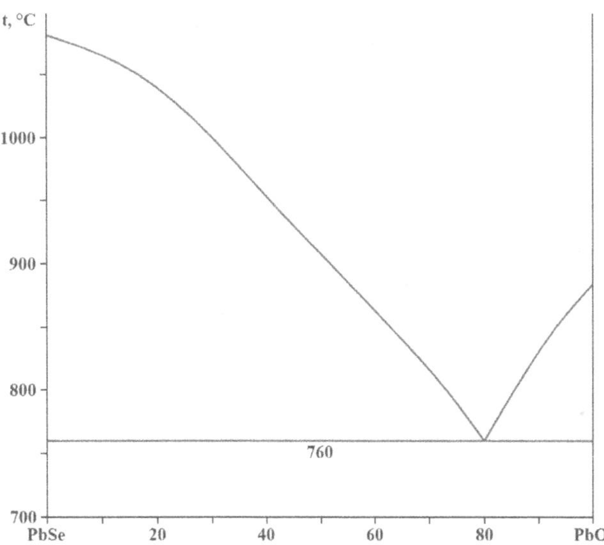

FIGURE 11.20 Phase diagram of the PbSe–PbO system. (From Popovkin, B.A., et al., *Zhurn. neorgan. khimii*, **8**(5), 1224, 1963.)

Several ternary compounds are formed in the Pb–O–Se system. **$PbSeO_3$** was first obtained by Marino (1908), and then its existence was confirmed by Oykova and Gospodinov (1981). This compound melts incongruently at 675°C ± 10°C (Markovskiy and Sapozhnikov 1960; Popovkin et al. 1960; Popovkin and Novoselova 1961; Bakeeva et al. 1971; Gospodinov and Bogdanov 1989) with a melting enthalpy of 37.7 kJ·M^{-1} (Bakeeva et al. 1971) [36.28 ± 0.13 kJ·M^{-1} (Gospodinov and Bogdanov 1989)] and has a polymorphic transformation: α-$PbSeO_3$ transforms slowly to the β-$PbSeO_3$ between 315°C and 355°C and the transformation process was finished at 375°C (Lahtinen and Valkonen 2002). One of the modifications crystallizes in the monoclinic structure with the lattice parameters given in Table 11.2. Another modification crystallizes in the orthorhombic structure with the lattice parameters a = 789.3 ± 0.7, b = 679.6 ± 0.7, c = 548.7 ± 0.5 pm, and the calculated and experimental densities of 6.56 and 6.54 g·cm^{-3}, respectively (Odin and Popovkin 1967). The dissociation pressure of the $PbSeO_3$ melt at 750°C is 4.32 kPa of SeO_2 (Bakeeva et al. 1971).

$PbSeO_3$ is formed during the oxidation of PbSe with oxygen at 500°C–600°C (Zlomanov et al. 1961). It was also synthesized by melting together PbO and SeO_2, largely in the absence of air (Fischer 1972). To obtain this compound, the interaction of lead nitrate or acetate with the solution K_2SeO_3, Na_2SeO_3 or H_2SeO_3 in a stoichiometric ratio can be used as well as the interaction of lead carbonate with the solution of H_2SeO_3 (Selivanova et al. 1958; Markovskiy and Sapozhnikov 1960; Odin and Popovkin 1967; Lahtinen and Valkonen 2002). It can be also prepared by its continued crystallization at 100°C in aqueous solution of H_2SeO_3 (Gospodinov and Bogdanov 1989).

$PbSeO_4$ was first obtained by Marino (1908), and then its existence was confirmed by Oykova and Gospodinov. (1981). It melts incongruently at 715°C [at 930°C (Selivanova and Shneider 1958)], has a polymorphic transformation at 645°C (Popovkin et al. 1960) and begins to decompose at 500°C forming $PbSeO_3$ (the end product of decomposition is PbO) (Selivanova and Shneider 1958). This compound crystallizes in the monoclinic structure with the lattice parameters a = 715.4 ± 0.4, b = 740.7 ± 0.4, c = 695.4 ± 0.4 pm, β = 103.14 ± 0.04°, and a calculated density of 6.48 g·cm^{-3} (Effenberger and Pertlik 1986). $PbSeO_4$ can be prepared by neutralization of selenious acid with $PbCO_3$ (Lenher and Kao 1925) or by slow sedimentation using solutions of $PbCl_2$ or $Pb(NO_3)_2$ and K_2SeO_4 (Selivanova and Shneider 1958; Sauka and Apinit 1959). It was also prepared by a hydrothermal reaction in a steel vessel lined with Teflon (Effenberger and Pertlik 1986). Colorless to light brown crystals were obtained from an equimolar mixture (2 g) of SeO_2 and Pb_3O_4, to which H_2O_2 (2 mL) was added. The vessel was filled up with H_2O to about 80 mol%. The heating period was 48 h at 220°C ± 10°C.

$PbSe_2O_5$ is thermally stable up to 380°C ± 10°C and decomposes at the heating to 500°C with formation of $PbSeO_3$ (Markovskiy and Sapozhnikov 1960). It also crystallizes in the monoclinic structure with the lattice parameters a = 451.5 ± 0.1, b = 950.3 ± 0.3, c = 1161.8 ± 0.2 pm, β = 90.33 ± 0.01°, and a calculated density of 5.93 g·cm^{-3} (Koskenlinna and Valkonen 1995) [(an experimental density was determined to be 5.35 ± 0.02 g·cm^{-3} (Markovskiy and Sapozhnikov 1960)]. The title compound was prepared by the precipitation of amorphous $PbSeO_3$ from an aqueous solution of $Pb(NO_3)_2$ with selenous acid (Koskenlinna and Valkonen 1995). The precipitate was added to a concentrated (>1.0 M) aqueous solution of SeO_2 with a stoichiometric ratio of Pb to SeO_2 of 1:2. Well-formed crystals with yellowish tint were grown by allowing the suspension to stand at temperatures between 57°C and 100°C for a few days. $PbSe_2O_5$ could be also obtained as a result of the heating of **$Pb(HSeO_3)$** at a temperature higher than 130°C (Markovskiy and Sapozhnikov 1960).

$Pb(SeO_3)_2$ also crystallizes in the monoclinic structure with the lattice parameters a = 481.1 ± 0.2, b = 876.8 ± 0.4, c = 707.9 ± 0.3 pm, β = 110.10 ± 0.02°, and a calculated density of 5.461 g·cm^{-3} (Steinhauser et al. 2006). The synthesis of this compound was performed using SeO_2 and $Pb(CH_3COO)_4$ (Frydrych 1976; Steinhauser et al. 2006). SeO_2 (6.76 g) were dissolved in water (20 mL). This solution was heated to a temperature of 70°C. $Pb(CH_3COO)_4$ (4.71 g) was dissolved in pure acetic acid (60 mL) at 90°C. The solution of selenic acid was slowly added to the lead acetate solution, precipitating amorphous $Pb(SeO_3)_2$, which crystallizes within 5 min, when the suspension is stirred. The triple molar excess of selenic acid is important to get a crystalline product. The light-yellow precipitate is fine-grained. The crystals of the title compound are unstable at ambient conditions, starting decomposition within

TABLE 11.2

Crystallographic Data for Monoclinic $PbSeO_3$

a, pm	b, pm	c, pm	β, °	$d_{calc.}$ g·cm^{-3}	$d_{meas.}$ g·cm^{-3}	References
–	–	–	–	–	6.375	Selivanova et al. (1958)
–	–	–	–	–	6.86 ± 0.04	Markovskiy and Sapozhnikov (1960)
715.3 ± 0.3	740.3 ± 0.3	695.7 ± 0.3	103.27 ± 0.05	6.482	6.37	Pistorius and Pistorius (1962)
453 ± 1	551 ± 1	686 ± 2	112.98 ± 0.03	7.08	6.87	Popovkin et al. (1963a)
686	548	450	112.75	7.12	7.07 ± 0.05	[a]Fleischer (1965); Mandarino (1965)
691	548	451	112.83	–	–	[a]Fischer (1972)
915.87 ± 0.01	809.02 ± 0.01	879.32 ± 0.01	103.032 ± 0.001	–	–	[a]Lahtinen and Valkonen (2002)
454.37 ± 0.07	551.37 ± 0.08	663.4 ± 0.1	106.547 ± 0.007	–	–	[a]Pasero and Rotiroti (2003)

[a] Mineral molybdomenite.

a few days [it decomposes at 150°C with formation of $PbSeO_3$ (Frydrych 1976)].

PbO·PbSeO$_3$ (Pb$_2$SeO$_4$) also crystallizes in the monoclinic structure with the lattice parameters a = 1401 ± 1, b = 574.8 ± 0.5, c = 720.4 ± 0.5 pm, β = 114°52′ ± 10′, and the calculated and experimental densities of 7.23 and 6.99 g·cm^{-3}, respectively (Popovkin and Simanov 1965). This compound was obtained by annealing a stoichiometric mixture of PbO and $PbSeO_3$ in an evacuated and sealed quartz ampoule at 650°C for 20 h.

PbO·PbSeO$_4$ (Pb$_2$SeO$_5$) also crystallizes in the monoclinic structure with the lattice parameters a = 744, b = 1182, c = 1168 pm, β = 90°52′ ± 7′, and the calculated and experimental densities of 8.03 and 7.93 g·cm^{-3}, respectively (Popovkin and Simanov 1965) [a = 1394, b = 578, c = 725 pm, and β = 115.9° (Jones and Rothschild 1958)]. It was prepared by annealing a stoichiometric mixture of PbO and $PbSeO_4$ in an evacuated and sealed quartz ampoule at 650°C for 20 h (Popovkin and Simanov 1965).

Pb$_2$(SeO$_3$)(SeO$_4$) (Pb$_2$Se$_2$O$_7$) was first obtained by Marino (1908) and is stable up to 232°C (Shang and Halasyamani 2020). It crystallizes in the orthorhombic structure with the lattice parameters a = 1350.80 ± 0.04, b = 562.380 ± 0.010, c = 927.06 ± 0.02 pm, a calculated density of 6.454 g·cm^{-3}, and an energy gap of 3.49 eV (Shang and Halasyamani 2020). The crystals of this compound were grown by hydrothermal method. They were grown by combining PbO (2 mM), SeO_2 (2 mM) with H_2SeO_4 (2 mL), and H_2O (8 mL) (Shang and Halasyamani 2020). The mixture was placed in 23 mL Teflon-lined autoclave that was subsequently sealed. The autoclave were gradually heated to 200°C, held for 120 h, and then slowly cooled to room temperature at a rate of 6°C·h^{-1}. Colorless crystals were isolated from the mother liquor by vacuum filtration and washed with deionized water.

2PbO·PbSeO$_3$ (Pb$_3$SeO$_5$) is stable up to ca. 600°C, after which the compound rapidly loses SeO_2 and PbO, and undergoes an irreversible phase transition at ca. 440°C (Kim et al. 2009). According to the data of Popovkin and Novoselova (1961), this compound melts incongruently at 755°C [at 740°C ± 10°C (Zlomanov et al. 1961); at 760°C (Gospodinov and Bogdanov 1989)] and forms a eutectic with $PbSeO_3$ at 605°C and 33.33 mol% PbO in the PbO–$PbSeO_3$ system]. It also crystallizes in the orthorhombic structure with the lattice parameters a = 1052.11 ± 0.13, b = 1071.51 ± 0.13, c = 574.52 ± 0.07 pm, and a calculated density of 8.007 g·cm^{-3} (Kim et al. 2009) [a = 2216, b = 1076, c = 1670 pm, and the calculated and experimental densities of 7.81 and 7.71 g·cm^{-3}, respectively (Popovkin and Simanov 1965); a = 1052.9 ± 0.2, b = 1072.2 ± 0.2, c = 575.27 ± 0.12 pm, and a calculated density of 7.98 g·cm^{-3} (Krivovichev et al. 2004); a = 1053.84 ± 0.11, b = 1074.52 ± 0.13, c = 575.77 ± 0.07 pm, and a calculated density of 7.952 g·cm^{-3} for mineral plumboselite (Kampf et al. 2011; Welch et al. 2013); in the tetragonal structure with the lattice parameters a = 392 ± 1, and c = 537 ± 1 pm and the calculated and experimental densities of 7.91 and 8.08 g·cm^{-3}, respectively (Popovkin et al. 1960; Popovkin and Novoselova 1961; Zlomanov et al. 1961)]. The dissociation pressure of this compound at 750°C is ca. 130 Pa (Bakeeva et al. 1971).

This compound was hydrothermally synthesized by combining PbO (4.50 mM) and of SeO_2 (4.50 mM) with 1 M NaOH(aq) (8 mL) (Krivovichev et al. 2004; Kim et al. 2009). The mixture was placed in a 23 mL Teflon-lined autoclave that was subsequently sealed. The autoclave was gradually heated to 230°C, held for 2 days, and cooled slowly to room temperature at a rate 6°C·h^{-1}. The mother liquor was decanted, and the product, transparent colorless rod-shaped crystals, was recovered by filtration and washed with excess amounts of distilled water and acetone. 2PbO·$PbSeO_3$ was also obtained by annealing a stoichiometric mixture of PbO and $PbSeO_3$ in an evacuated and sealed quartz ampoule at 560°C–650°C for 20–100 h (Popovkin and Simanov 1965; Gospodinov and Bogdanov 1989). It is also formed during the oxidation of PbSe with oxygen at 600°C–600°C (Zlomanov et al. 1961).

4PbO·PbSeO$_3$ (Pb$_5$SeO$_7$) melts congruently at 805°C [at 780°C (Zlomanov et al. 1961); at 812°C (Gospodinov and Bogdanov 1989)] and forms a eutectic with PbO at 800°C and 83.3 mol% PbO in the PbO–$PbSeO_3$ system (Popovkin and Novoselova 1961). The dissociation pressure of this compound at 750°C is ca. 130 Pa (Bakeeva et al. 1971). It also crystallizes in the orthorhombic structure with the lattice parameters a = 742.8, b = 1171, c = 1137 pm, and the calculated and experimental densities of 8.23 and 8.03 g·cm^{-3}, respectively (Popovkin and Simanov 1965) [a = 392.1 ± 0.1, b = 373.1 ± 0.1, and c = 572.1 ± 0.1 pm (Popovkin et al. 1960; Popovkin and Novoselova 1961); Zlomanov et al. 1961]. It was synthesized by annealing a stoichiometric mixture of PbO and $PbSeO_3$ in an evacuated and sealed quartz ampoule at 800°C–900°C for 20 h (Popovkin and Simanov 1965) [at 630°C–640°C for 100 h (Gospodinov and Bogdanov 1989)]. This compound is also formed during the oxidation of PbSe with oxygen at 600°C–650°C or at the heating of 2PbO·$PbSeO_3$ at 600°C for 25 h in vacuum (Zlomanov et al. 1961).

According to the data of Jones and Rothschild (1958), the **4PbO·PbSeO$_4$ (Pb$_5$SeO$_8$)** ternary compound, which crystallizes in the monoclinic structure with the lattice parameters a = 1470, b = 1177, c = 1163 pm, β = 90.2°, and a calculated density of 8.22 g·cm^{-3} (Bode and Voss 1959), is formed in the Pb–O–Se system.

X-ray diffraction studies have shown that after annealing at 500°C–600°C for 10 h, the equilibrium of reactions PbSe + 2$PbSeO_3$ ⇔ 3Pb = 3SeO_2 and 3PbSe + 2(2PbO·$PbSeO_3$) ⇔ 9Pb = 5SeO_2 is shifted towards the formation of PbSe and $PbSeO_3$ or 2PbO·$PbSeO_3$ (Zlomanov et al. 1962)

11.39 Lead–Tellurium–Selenium

PbSe–Te: The phase diagram of this system, constructed through DTA, XRD, and metallography, is a eutectic type (Figure 11.21) (Filep and Sabov 2017a). The eutectic contains 3 mol% PbSe and crystallizes at 415°C. The ingots for the investigation were annealed at 80°C for 168 h.

PbSe–PbTe: The phase diagram of this system, constructed through DTA, XRD, metallography, and measuring of the microhardness (Grimes 1965; Steininger 1970a,b; Leute and Köller 1986a; Kuznetsov et al. 1987) and optimized using a four parameter for the excess Gibbs energy by Volykhov et al. (2006), is presented in Figure 11.22. A continuous series of solid solutions forms in the system (Elagina and Abrikosov

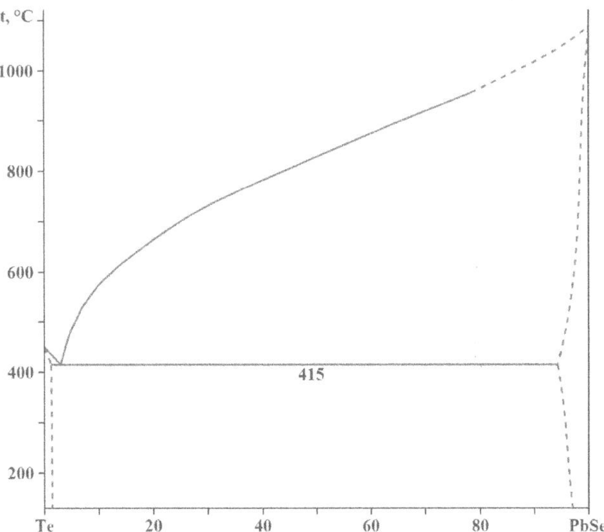

FIGURE 11.21 Phase diagram of the PbSe–Te system. (From Filep, M., and Sabov, M., *Chem. Met. Alloys*, **10**(3–4), 120, 2017.)

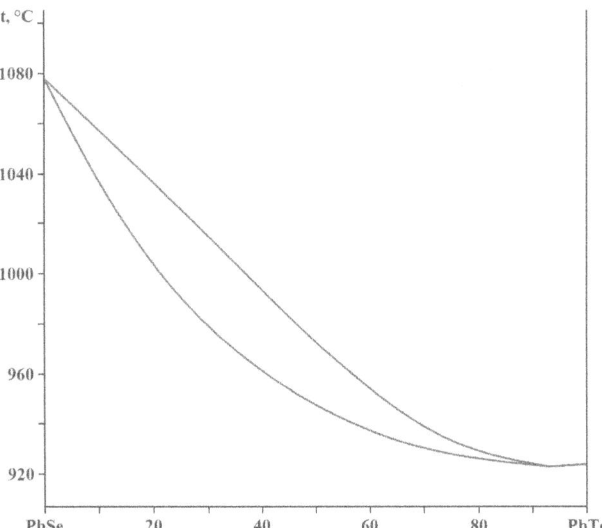

FIGURE 11.22 Phase diagram of the PbSe–PbTe system. (From Volykhov, A.A., et al., *Inorg. Mater.*, **42**(6), 596, 2006.)

1956; Grimes 1965; Darrow et al. 1969b; Gangulee 1969; Steininger 1970a; Kuznetsov et al. 1987). According to the data of Elagina and Abrikosov (1956), the system contains a minimum at 18 mol% PbSe and 900°C, which was no detected later (Grimes 1965; Steininger 1970a). The calculated liquidus and solidus curves (Volykhov et al. 2006) also have a minimum at 7 mol% PbSe and 923°C. The unit cell parameter of solid solutions, depending on the composition, changes linearly (Gangulee 1969; Leute and Köller 1986a) [with a slight negative deviation from linearity (Elagina and Abrikosov 1956; Sokolov et al. 1969a)]. The PbSe–PbTe system tends to decompose the $PbSe_xTe_{1-x}$ solid solution at a sufficiently low temperature (Sushkova et al. 2004). According to the data of Sokolov et al. (1969a) and Volykhov et al. (2006), $PbSe_xTe_{1-x}$ solid solutions exhibit positive deviation from ideality. The composition dependences of the thermodynamic functions evaluated by EMF measurements for the saturated solid solutions of this system indicate on the formation of the intermediate bertholide phase near the $PbSe_{0.7}Te_{0.3}$ composition (Moroz and Prokhorenko 2015c).

The temperature dependences of the Pb, PbSe, and PbTe partial pressure for the $Pb_{1-y}(Se_{1-x}Te_x)_y$ solid solutions were calculated in (Kuznetsov et al. 1984b), and the vapor pressure at temperatures below the solidus line was measured by Sokolov et al. (1969a), where the activities, partial and integral Gibbs free energies were also calculated.

The vapor composition was studied by high-temperature mass spectroscopy and the partial pressures of the vapor components of the Pb–Se–Te system were determined by Demidova et al. (1989). It was found that the $Pb_{1-y}(Se_{1-x}Te_x)_y$ solid solutions sublimate congruently with respect to the composition parameter *y* and quasi-congruently with respect to the composition parameter *x*. The following ions were found in the vapor mass spectrum of the Pb–Se–Te ternary system: $PbTe^+$, $PbSe^+$, Pb^+, Se^+, Te^+, Se_2^+, Te_2^+, and $PbSeTe_2^+$.

The ingots for the investigations were annealed at 800°C and 900°C in the Ar atmosphere for 720 h (Elagina and Abrikosov 1956) [at 800°C for 14 days (Gangulee 1969); at 750°C for 800 h (Sokolov et al. 1969a)].

The liquidus surface of the Pb–Se–Te ternary system (Figure 11.23) has a sharp maximum in the area of the PbSe–PbTe quasibinary section (Krapukhin et al. 1984; Novoselova et al. 1984, 1986; Kuznetsov et al. 1987). At the crystallization from the melt, the composition of the $PbSe_xTe_{1-x}$ solid solution enriches with lead selenide (Novoselova et al. 1984, 1986; Kuznetsov et al. 1987). The coordinates of the solidus surface of the $Pb_{1-y}(Se_{1-x}Te_x)_y$ solid solution were determined by Kuznetsov et al. (1984a) and Novoselova et al. (1986). On the side of Pb excess (*n*-type), the liquidus surface smoothly changes from PbTe to PbSe throughout the studied temperature

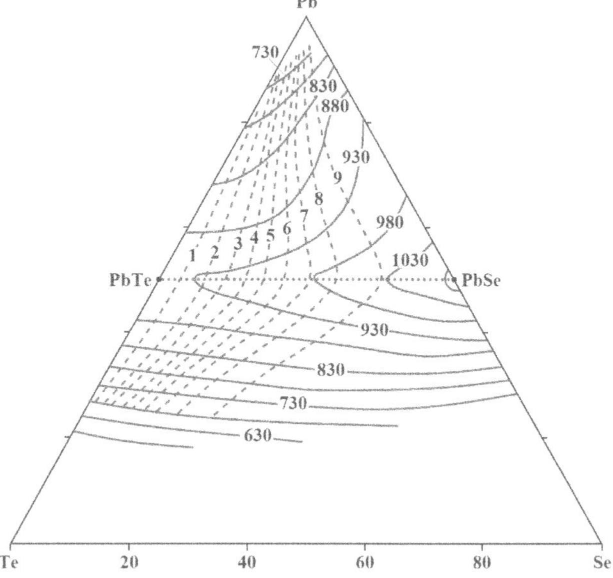

FIGURE 11.23 Liquidus surface of the Pb–Te–Se ternary system with the solidus isoconcentrates for the $PbSe_xTe_{1-x}$ solid solution: 1, $x = 0.2$, 2, $x = 0.2$, …, 9, $x = 0.9$. (From Novoselova, A.V., et al., *Zhurn. neorgan. khimii*, **31**(11), 2957, 1986.)

range, and in the *p*-type region, a discrepancy between the obtained results and the literature data for PbTe and PbSe was found, which may be associated with the high decomposition rate of the supersaturated $Pb_{1-y}(Se_{1-x}Te_x)_y$ solid solution.

The activities of PbTe and PbSe in liquid Te-saturated PbSe–PbTe solid solutions in the temperature range 460°C–560°C exhibit positive deviation from Raoult's law and decrease with increasing temperature (Shamsuddin and Misra 1979; Shamsuddin et al. 1994). The continuous variation of the activity and activity coefficient of PbSe and PbTe with composition indicates that this system is completely miscible and consists of a single-phase field throughout the entire range of composition in the studied temperature range. The PbSe–PbTe system does not follow a regular solution model. The thermodynamic stability of the system was seen to increase with increasing of temperature.

11.40 Lead–Chromium–Selenium

Two ternary compounds, $\mathbf{Pb_{0.5}Cr_5Se_8}$ and $\mathbf{PbCr_2Se_4}$, are formed in the Pb–Cr–Se system. First of them crystallizes in the monoclinic structure with the lattice parameters $a = 1872 \pm 1$, $b = 718.8 \pm 0.2$, $c = 895.1 \pm 0.2$ pm, and $\beta = 104.65 \pm 0.04°$ (Novet et al. 1995). To prepare this compound, the ion-exchange reaction was used. $TlCr_5Se_8$ was mixed with PbI_2 at various molar ratios (1:1, 1:2, and 1:4) in a N_2-filled dry box and sealed in evacuated quartz tube approximately 20 cm in length. The tube was placed in a temperature gradient with the sample at the hot end (at temperatures 150°C–450°C) and the cold end of the tube extended out of the furnace assembly. The temperature of the hot end of the ampoule was raised until red crystals of TlI were observed on the cold end of the tube. The samples were annealed for approximately 12 h before quenching to room temperature.

$PbCr_2Se_4$ crystallizes in the hexagonal structure with the lattice parameters $a = 2232.7 \pm 1.5$ and $c = 364.8 \pm 0.1$ pm (Omloo et al. 1971) [$a = 2234.2$ and $c = 359.46$ pm (Sleight and Frederick 1973)]. This compound was prepared by heating a stoichiometric mixture of the three elements or mixture of PbSe and Cr_2Se_3 in evacuated quartz tubes at temperature between 1000°C and 1200°C during 1 week (Omloo et al. 1971). After cooling to room temperature, the sample was ground to powder. This powder was annealed at temperature between 1000°C and 1200°C during one to several weeks, slowly cooled to 750°C during 3 weeks and quenched or cooled during 1 day to room temperature.

11.41 Lead–Molybdenum–Selenium

Several ternary compounds are formed in the Pb–Mo–Se system. $\mathbf{Pb_{0.5}Mo_3Se_4}$ crystallizes in the rhombohedral structure with the lattice parameters $a = 681.0 \pm 0.2$ pm and $\alpha = 89.23 \pm 0.04°$, the calculated and experimental densities of 7.44 and 7.42 $g{\cdot}cm^{-3}$ (Guillevic et al. 1976).

The synthesis through low temperature of Pb diffusion in $Mo_{15}Se_{19}$ at 480°C leads to the formation of $\mathbf{Pb_{2.3}Mo_{15}Se_{19}}$ (if the original $Mo_{15}Se_{19}$ was obtained by the interaction of HCl and $In_3Mo_{15}Se_{19}$), $\mathbf{Pb_2Mo_{15}Se_{19}}$ and $\mathbf{Pb_{2.15}Mo_{15}Se_{19}}$ (if the original $Mo_{15}Se_{19}$ was obtained by the interaction of HCl and $In_2Mo_{15}Se_{19}$) (Tarascon et al. 1985). These phases decompose at the temperatures higher than 800°C and crystallize in the hexagonal structure with the lattice parameters $a = 961.1 \pm 0.1$ and $c = 1959 \pm 1$ pm for $Pb_{2.3}Mo_{15}Se_{19}$, $a = 968.3 \pm 0.2$ and $c = 5778 \pm 2$ pm for $Pb_2Mo_{15}Se_{19}$, and $a = 968.0 \pm 0.1$ and $c = 5777 \pm 2$ pm $Pb_{2.15}Mo_{15}Se_{19}$.

The $\mathbf{([PbSe]_{1.00})_1(MoSe_2)_1}$ ternary intergrowth misfit layer compound is also formed in the Pb–Mo–Se system (Smeller et al. 2012). The samples of this compound were prepared using a custom high vacuum deposition chamber. Layers of Pb and Mo were deposited using electron beam evaporation; Se was deposited with a Knudsen evaporation cell. Precursors were prepared by sequentially depositing each elemental reactant to form a repeating unit with a composition profile similar to the desired final product, which was then repeated to obtain a film of the desired total thickness. The relative thickness of the pair of deposited elemental layers to form each binary structural unit was calibrated to obtain the composition that corresponds to the desired stoichiometry. The thickness of each elemental pair was adjusted to yield individual structural units of PbSe or $MoSe_2$ upon annealing. The deposition produces a predominantly amorphous precursor, which is then gently annealed to form the desired product. Samples were annealed at 400°C in a nitrogen atmosphere for 0.5 h. The experimental *a*-lattice parameter for the PbSe constituent is 618.2 ± 0.2 pm, for the $MoSe_2$ constituent is 330.8 ± 0.4 and *c*-lattice parameter for this misfit compound is 1265.8 ± 0.2 pm. According to the data of Heideman et al. (2010), $([PbSe]_{1.00})_1(MoSe_2)_1$ forms over a range of composition leading to a spread of lattice parameters (the *c*-lattice parameter is varied from 1246 to 1275 pm), which suggest the presence of volume defects in the obtained films of this compound.

18 metastable compounds $\mathbf{[(PbSe)_{1.00}]_m(MoSe_2)_n}$ were synthesized using precursors each designed to self-assemble into a specific compound (Heideman et al. 2013). To form a compound with specific values for *m* and *n*, the number of atoms within each deposited elemental layer was carefully controlled to provide the correct absolute number of atoms to form complete layers of each component structural unit. On low-temperature annealing, these structures self-assemble with a specific crystallographic orientation between the component structural units with atomically abrupt interfaces. There is rotational disorder between the component structural units and between $MoSe_2$ basal plane units within the $MoSe_2$ layers themselves. The PbSe constituent has a distorted rock-salt structure exactly *m* bilayers thick leading to peaks in the off-axis diffraction pattern as a result of the finite size of and rotational disorder between the crystallites. The in-plane lattice parameters of the PbSe and $MoSe_2$ components are independent of the value of *m* and *n*, suggesting little or no strain caused by the interface between them. These compounds are small bandgap semiconductors.

11.42 Lead–Tungsten–Selenium

The $\mathbf{([PbSe]_{0.99})_1(WSe_2)_1}$ ternary intergrowth misfit layer compound is formed in the Pb–W–Se system (Smeller et al. 2012). It was prepared in the same way as $([PbSe]_{1.00})_1(MoSe_2)_1$ using

W instead of Mo and the samples were annealed at 400°C in a nitrogen atmosphere for 1 h. The experimental *a*-lattice parameter for the PbSe constituent is 618 ± 2 pm and for the WSe_2 constituent it is 331 ± 1 and *c*-lattice parameter for this misfit compound is 1269.7 ± 0.3 pm.

64 members of the $\mathbf{[(PbSe)_{1-\delta}]_m(WSe_2)_n}$ family of compounds, where *m* and *n* are integers that were systematically varied from 1 to 8 were synthesized by annealing reactant precursors containing *m* layers of alternating elemental Pb and Se followed by *n* layers of alternating elemental W and Se, in which the thickness of each pair of elemental layers was calibrated to yield a structural bilayer of rock-salt structured PbSe and a trilayer of hexagonal WSe_2 (Lin et al. 2010). The compounds are kinetically trapped by the similarity of the composition profiles and modulation lengths in the precursor and the targeted compounds. The structural evolution from initial reactant of layer elements to crystalline misfit layer compounds was tracked using XRD. The crystal structures of these compounds were probed using both analytical electron microscopy and XRD. The *c*-axis of the misfit layer compound is perpendicular to the substrate, with a *c*-axis lattice parameter that changes linearly with a slope of 612–615 pm as *m* is changed and *n* is held constant and with a slope of 654–656 pm as *n* is varied and *m* is held constant. The in-plane lattice parameters did not change as the individual layer thicknesses were increased and a misfit parameter of $\delta = -0.01$ was calculated.

11.43 Lead–Chlorine–Selenium

PbSe–$PbCl_2$: The phase diagram of this system, constructed through DTA, XRD, and metallography, is a eutectic type (Figure 11.24) (Novoselova et al. 1970b). The eutectic contains 7 mol% PbSe and crystallizes at 478°C. The ternary compounds were not found in this system. According to the data of Rodionov et al. (1972), $PbCl_2$ melt dissolves 12.8 mol% PbSe at 600°C. The ingots for the investigations were annealed at 450°C for 200 h (Novoselova et al. 1970b). The solubility of PbSe in $PbCl_2$ melt was determined by the isothermal saturation method followed by analysis of the salt phase (Rodionov et al. (1972).

11.44 Lead–Bromine–Selenium

PbSe–$PbBr_2$: The phase diagram of this system, constructed through DTA, XRD, and metallography, is shown in Figure 11.25 (Novoselova et al. 1970c). The eutectic contains 6 mol% PbSe and crystallizes at 353°C. The $\mathbf{Pb_{7+x}Se_{2-2x}Br_6}$ at $0 < x < 0.25$ [$\mathbf{Pb_4SeBr_6}$ (Rabenau and Rau 1969; Krebs 1970, 1973)] ternary compound, which melts incongruently at 383°C and has a homogeneity region of 6 mol% PbSe, is formed in this system (Novoselova et al. 1970c). Pb_4SeBr_6 crystallizes in the orthorhombic structure with the lattice parameters $a = 436.1 \pm 0.2$, $b = 1578.0 \pm 0.8$, $c = 972.0 \pm 0.5$ pm, and the calculated and experimental densities of 6.886 and 6.84 g·cm^{-3}, respectively (Krebs 1970, 1973) [$a = 436$, $b = 972$, $c = 1578$ pm, and the calculated and experimental densities of 6.88 and 6.84 g·cm^{-3}, respectively (Rabenau and Rau 1969)]. Single crystals of this compound could be obtained by a reaction of PbSe with HBr under hydrothermal conditions (Krebs 1970, 1973).

The ingots for the investigations were annealed at 340°C for 1000 h (Novoselova et al. 1970c).

Another ternary compound, $\mathbf{Pb_3Se_2Br_2}$, crystallizes in the cubic structure with the lattice parameter $a = 932.00 \pm 0.04$ pm, a calculated density of 7.707 g·cm^{-3}, and an energy gap of 1.48 eV (Ni et al. 2019a). It was synthesized through a high-pressure solid-state method. $PbBr_2$ powder was mixed stoichiometrically with elemental Se shot and Pb powder, and ground into a finely powdered mixture. It was then loaded into a BN crucible and inserted into a pyrophillite cube assembly, which was pressed to 4 GPa using a cubic multi-anvil system, and heated to 700°C at 50°C·min^{-1}. The sample was annealed

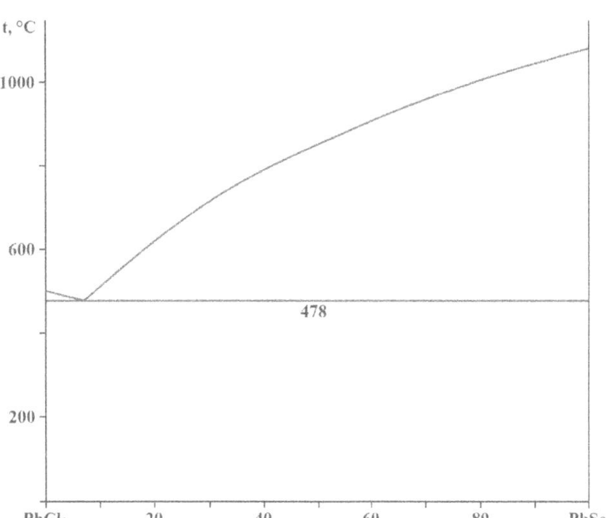

FIGURE 11.24 Phase diagram of the PbSe–$PbCl_2$ system. (From Novoselova, A.V., et al., *Izv. AN SSSR. Neorgan. mater.*, **6**(2), 381, 1970.)

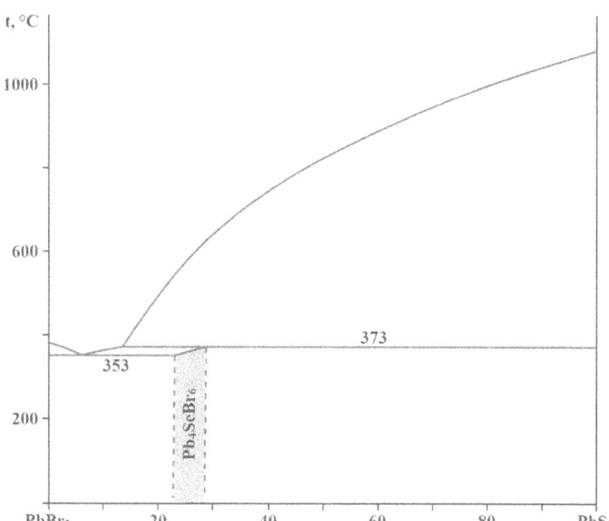

FIGURE 11.25 Phase diagram of the PbSe–$PbBr_2$ system. (From Novoselova, A.V., et al., *Izv. AN SSSR. Neorgan. mater.*, **6**(2), 257, 1970.)

at 700°C and 4 GPa for 3 h and then quenching to room temperature before decompression. The product obtained was a dark gray color and relatively stable in air.

11.45 Lead–Iodine–Selenium

PbSe–PbI$_2$: The phase diagram of this system, constructed through DTA, XRD, metallography, and measuring of the microhardness, is a eutectic type (Figure 11.26) (Novoselova et al. 1966). The eutectic contains 14 mol% PbSe and crystallizes at 384°C. The solubility of PbI$_2$ in PbSe at the eutectic temperature reaches 0.8 mol% (Novoselova et al. 1966; Odin et al. 1970a). According to the data of Rodionov et al. (1972), PbI$_2$ melt dissolves 15.1 and 22.5 mol% PbSe at 510°C and 600°C, respectively. The ingots for the investigations were annealed at 384°C, 500°C, 600°C, and 750°C for respectively 800, 600, 500, and 350 h (Odin et al. 1970a). The solubility of PbSe in PbI$_2$ melt was determined by the isothermal saturation method followed by analysis of the salt phase (Rodionov et al. (1972).

On the cooling curves in the range of 13–100 mol% PbSe, in addition to the effects corresponding to liquidus, two satisfactorily resolving effects were observed at 388°C and 373°C, which is due to the no equilibrium nature of crystallization (Novoselova et al. 1966). According to the cooling curves, a metastable compound of approximate composition **PbSe·4PbI$_2$** is formed in the system (Figure 11.27) (Odin 2000). From the PbI$_2$-side in the system, 2*H*-PbI$_2$ polytype is primarily crystallized.

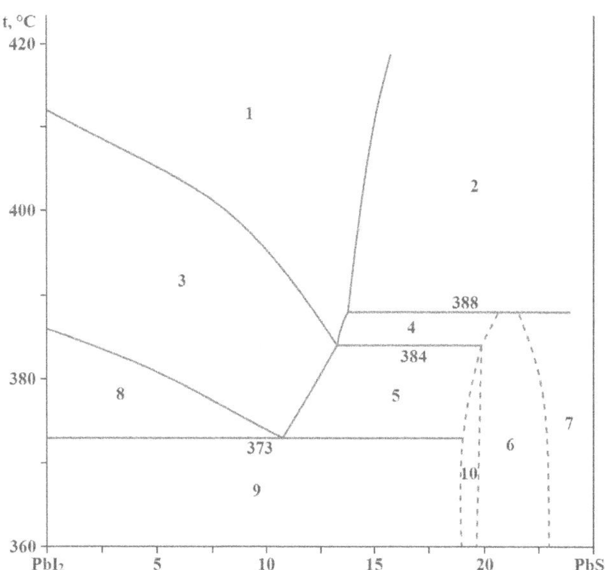

FIGURE 11.27 Part of the phase diagram of the PbSe–PbI$_2$ system with metastable phases according to the cooling curves: 1, L; 2, L + PbSe; 3, L + 2*H*-PbI$_2$; 4, L + PbSe·4PbI$_2$; 5, L + 2*H*-PbI$_2$ + PbSe·4PbI$_2$; 6, PbSe. 4PbI$_2$; 7, PbSe·4PbI$_2$ + PbSe; 8, L + 2*H*-PbI$_2$ + 6*R*-PbI$_2$; 9, 2*H*-PbI$_2$ + 6*R*-PbI$_2$ + PbSe·4PbI$_2$; 10, 2*H*-PbI$_2$ + PbSe·4PbI$_2$. (From Odin, I.N., *Zhurn. neorgan. khimii*, **45**(4), 698, 2000.)

11.46 Lead–Manganese–Selenium

PbSe–MnSe: The phase diagram of this system, constructed through DTA, XRD, metallography, and measuring of the microhardness, is a eutectic type (Figure 11.28) (Vanyarkho et al. 1969a). The eutectic contains 30 mol% MnSe and crystallizes at 1025°C. The solubility of α-MnSe in PbSe increases from 10 mol% at 720°C to 17 mol% at 930°C (Vanyarkho et al. 1967, 1969a). The solubility of PbSe in α-MnSe at the eutectic temperature is not higher than 5 mol%. The ingots for the investigations were annealed at 720°C, 870°C, and 930°C for respectively 820, 520, and 450 h.

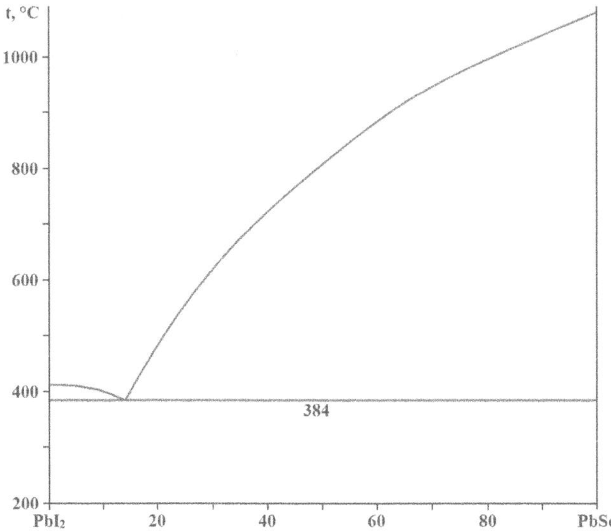

FIGURE 11.26 Phase diagram of the PbSe–PbI$_2$ system. (From Novoselova, A.V., et al., *Izv. AN SSSR. Neorgan. mater.*, **2**(8), 1397, 1966.)

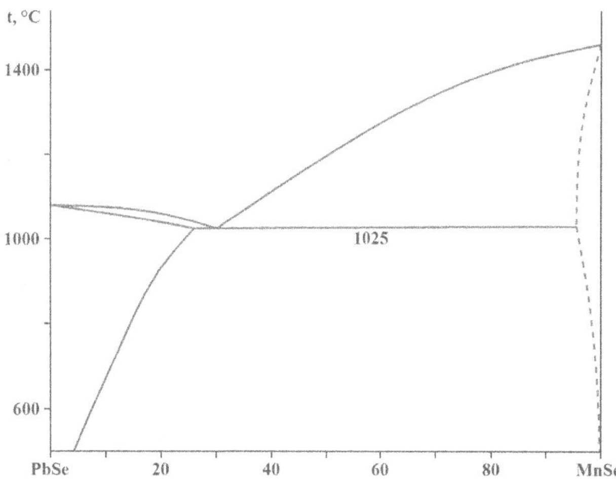

FIGURE 11.28 Phase diagram of the PbSe–MnSe system. (From Vanyarkho, V.G., et al., *Izv. AN SSSR. Neorgan. mater.*, **5**(10), 1699, 1969.)

11.47 Lead–Iron–Selenium

PbSe–Fe: The solubility of iron in PbSe, determined using radioactive isotopes, in the range 700°C–900°C is slightly retrograde and lies in the range 8×10^{18}–5×10^{20} cm^{-3}

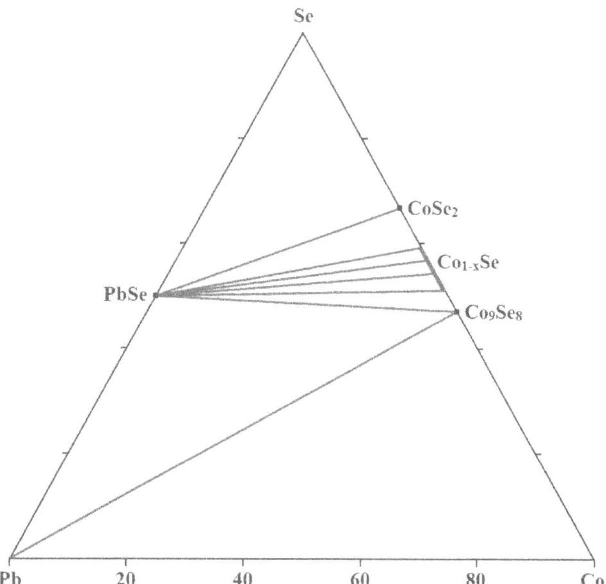

FIGURE 11.29 Isothermal section of the Pb–Co–Se ternary system at 400°C. (From Gulay, L.D., *Nauk. Visnyk Volyns'k. Nats. Univ. im. Lesi Ukrainky. Ser. Khim. nauky*, (16), 6, 2008.)

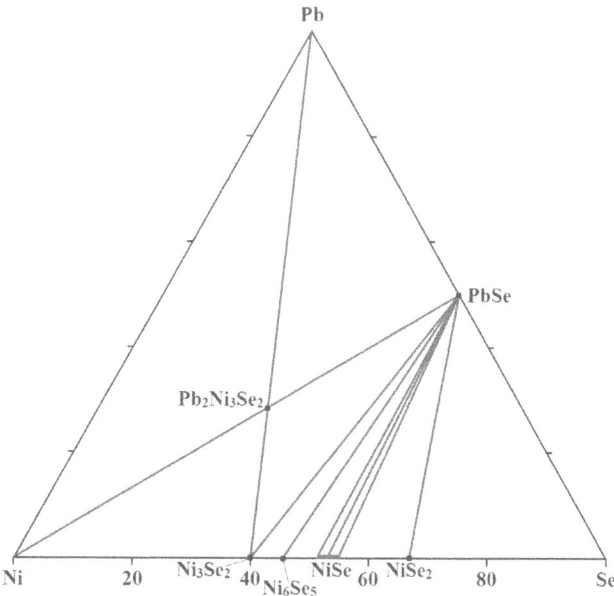

FIGURE 11.30 Isothermal section of the Pb–Ni–Se ternary system at 540°C. (From Baranov, A.I., et al., *Russ. Chem. Bull., Int. Ed.*, **51**(12), 2139, 2002.)

(Kharakhorin et al. 1966). The ingots for the investigations were annealed at 700°C–875°C for 100–136 h.

11.48 Lead–Cobalt–Selenium

PbSe–Co: The $\mathbf{Pb_2Co_3Se_2}$ ternary compound, which crystallizes in the hexagonal structure, is formed in this system (Zabel et al. 1979a,b). It can be synthesized both by the interaction of elementary components and by the interaction of Co_3Se_2 with Pb or PbSe with Co.

The isothermal section of the Pb–Co–Se ternary system at 400°C is given in Figure 11.29 (Gulay 2008). The formation of the ternary compounds at this temperature has been not observed.

11.49 Lead–Nickel–Selenium

PbSe–Ni: The phase diagram of this system, constructed through DTA, XRD, metallography, and measuring of the microhardness, is a eutectic type with peritectic transformation (Muller and Sotnikova 1969). The eutectic contains 55 at% (20 mass%) Ni and crystallizes at 670°C. The mutual solubility of the starting components is negligible. The ingots for the investigations were annealed at 630°C for 500 h.

The $\mathbf{Pb_2Ni_3Se_2}$ ternary compound, which melts incongruently at 700°C (Muller and Sotnikova 1969) and crystallizes in the trigonal structure with the lattice parameters $a = 561.67 \pm 0.05$ and $c = 1428.6 \pm 0.2$ pm (Range et al. 1997a) [$a = 561.61 \pm 0.08$ and $c = 1429.2 \pm 0.1$ pm (Zabel et al. 1979b); in the monoclinic structure with the lattice parameters $a = 971.9 \pm 0.2$, $b = 575.5 \pm 0.1$, $c = 560.8 \pm 0.1$ pm, and $\beta = 124.24 \pm 0.02°$ (Michelet and Collin 1976)] is formed in this system.

The isothermal sections of the Pb–Ni–Se ternary system at 400°C and 540°C were constructed by Gulay (2008) and Baranov et al. (2002), respectively, the latter being shown in Figure 11.30.

11.50 Lead–Rhodium–Selenium

PbSe–Rh: The $\mathbf{Rh_3Pb_2Se_2}$ ternary compound, which crystallizes in the tetragonal structure with the lattice parameters $a = 580$ and $c = 1392$ pm in hexagonal settings and the calculated and experimental densities of 10.76 and 10.63 g·cm^{-3}, respectively (Natarajan et al. 1988), is formed in this system (Zabel et al. 1979a,b).

This compound has been synthesized by the direct combination of stoichiometric quantities of high-purity elements in evacuated (10^{-3} Pa) and sealed quartz tube (Natarajan et al. 1988). The quartz tube containing the reactants was heated initially at 400°C–500°C for 2–3 days until all Se was reacted. The temperature was then raised to 800°C–1000°C and kept for 24 h. In most cases, a molten mass was noted inside the quartz tube after this treatment. The compound was recovered, ground thoroughly in an agate mortar and pressed in the form of discs with a WC-lined stainless steel die and sintered at 600°C for 1–2 days in an evacuated quartz tube. Single-phase material has thus been obtained. $Rh_3Pb_2Se_2$ is black in color and is stable to exposure to air and moisture. It is oxidized by heating in the air but is stable when heated to temperatures higher than 900°C in evacuated and sealed quartz tube and it sinters well.

11.51 Lead–Palladium–Selenium

PbSe–Pd: The $\mathbf{Pd_3Pb_2Se_2}$ ternary compound, which crystallizes in the tetragonal structure with the lattice parameters $a = 588.8 \pm 0.1$, $c = 1460.1 \pm 0.1$ pm, and a calculated densities of

10.131 g·cm^{-3} (Seidlmayer et al. 2010) [in the tetragonal structure with the lattice parameters a = 589.3 ± 0.1 and c = 1462.1 ± 0.2 pm (Zabel et al. 1979a,b) or a = 588 and c = 1460 pm in hexagonal settings and the calculated and experimental densities of 10.10 and 9.89 g·cm^{-3}, respectively (Natarajan et al. 1988)], is formed in this system. This compound was prepared in the same way as $Rh_3Pb_2Se_2$ (Natarajan et al. 1988).

11.52 Lead–Iridium–Selenium

The $\mathbf{Ir_4PbSe_8}$ ternary compound, which crystallizes in the monoclinic structure with the lattice parameters a = 1590.1 ± 0.3, b = 373.00 ± 0.01, c = 1103.5 ± 0.4 pm, β = 125.190 ± 0.008°, and an energy gap of 0.76 ± 0.11 eV, is formed in the Pb–Ir–Se system (Trump and McQueen 2016). Polycrystalline Ir_4PbSe_8 was grown by placing Ir, Pb, and Se in the stoichiometric ratio of PbSe/$IrSe_2$ = 1.1:2 in a fused silica tube, which was backfilled with 34 Pa of Ar to minimize vaporization of Se. The tube was heated quickly to 500°C, followed by a 50°C·h^{-1} ramp to an annealing temperature of 950°C, and held for 4 days before quenching in water. The resulting boule was pulverized, and the heat sequence and quench were repeated a second time. This resulted in shiny, silver PbSe micro-rings as well as loose, gray powder. After PbSe was mechanically removed, the resulting phase pure powder was the title compound.

11.53 Lead–Platinum–Selenium

PbSe–Pt: The $\mathbf{Pt_3Pb_2Se_2}$ ternary compound, which crystallizes in the orthorhombic structure with the lattice parameters a = 898.0 ± 0.2, b = 1230.8 ± 0.2, c = 552.7 ± 0.1 pm, is formed in the Pb–Pt–Se system (Wang et al. 2021). Single crystals of this compound were synthesized via solid-state reaction of a mixture of Pb, Pt, and Se powders (molar ratio 0.67:1:0.67) at 850°C. The mixture was loaded into a carbon-coated fused silica tube, which was then flame-sealed under vacuum (0.1 Pa) and heated slowly to 850°C in a furnace. The tube was kept at this temperature for 2 days and then slowly cooled to 400°C at the rate of 1°C·h^{-1}. Then, the furnace was turned off to cool the tube to room temperature. The products were taken out and transferred into deionized water, followed by ultrasonic washing several times and drying by acetone. Finally, the black crystals of $Pt_3Pb_2Se_2$ were obtained. The powder of this compound was synthesized at 700°C for 2 days by a same proportion of the Pb, Pt, and Se powders. The samples need to be compressed before sintering. In the synthesis process, excess Pb powders were necessary. Lead acts as not only the reactant in the reaction to form compounds, but also serving as a flux.

REFERENCES

All References are available as a downloadable eResource at www.routledge.com/9780367639235

12

Systems Based on Lead Telluride

12.1 Lead–Sodium–Tellurium

PbTe–NaTe: This system is a non-quasibinary section of the Pb–Na–Te ternary system as NaTe melts incongruently (Finogenova et al. 1971). PbTe and Na_2Te primarily crystallize in the system. Thermal effects at 360°C correspond to the joint crystallization PbTe and Na_2Te and at 320°C to the peritectic formation of NaTe. The coefficient of NaTe equilibrium distribution in PbTe is 0.20 (Nensberg et al. 1969). This system was investigated through DTA and XRD and the ingots for the investigations were annealed at 280°C 170 h.

PbTe–Na_2Te: The phase diagram of this system, constructed through DTA and XRD, is presented in Figure 12.1 (Finogenova et al. 1971). The eutectic contains 45 mol% PbTe and crystallizes at 650°C. The **Na_2PbTe_2** ternary compound, which melts incongruently at 720°C, is formed in this system.

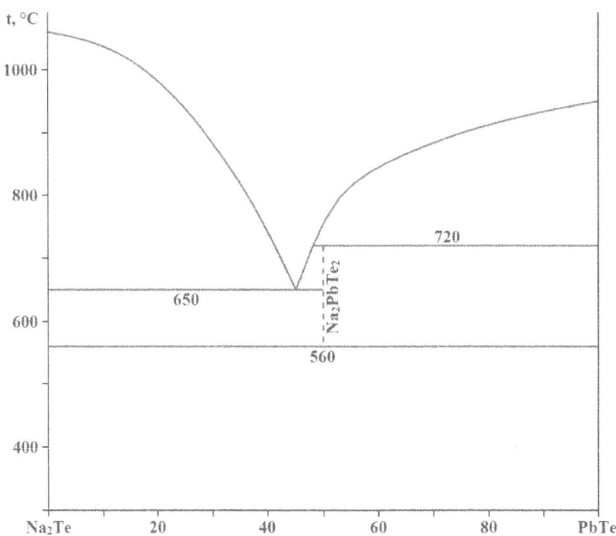

FIGURE 12.1 Phase diagram of the PbTe–Na_2Te system. (From Finogenova, V.K., et al., *Izv. AN SSSR. Neorgan. mater.*, 7(10), 1862, 1971.)

12.2 Lead–Copper–Tellurium

PbTe–Cu: This system is a non-quasibinary section of the Pb–Cu–Te ternary system (Gravermann and Wallbaum 1956; Bates et al. 1968; Grytsiv and Vengel' 1984; Shemet et al. 2006c). The system was investigated through DTA, XRD, metallography, and measuring of the microhardness.

PbTe–Cu_2Te: The phase diagram of this system, constructed through DTA, XRD, and metallography, is a eutectic type (Gravermann and Wallbaum 1956; Grytsiv and Vengel' 1984). The eutectic contains 38.8 mol% (45.5 mass%) PbTe and crystallizes at 649°C. Thermal effects at 260°C, 316°C, 362°C, and 560°C correspond to the polymorphic transformations of Cu_2Te. The mutual solubility of PbTe and Cu_2Te is negligible. No ternary compounds were found in the PbTe–Cu_2Te system (Shemet et al. 2006c).

There are three immiscibility regions on the liquidus surface of the PbTe–Cu_2Te–Cu–Pb subsystem of the Pb–Cu–Te ternary system (Gravermann and Wallbaum 1956). Two of them are adjacent to the Cu–Cu_2Te and Cu–Pb binaries and the third is situated inside of the concentration region. The ternary eutectic L ⇔ PbTe + Cu + Pb crystallizes at 325.5°C and the transition point L + Cu_2Te ⇔ Cu + PbTe exists at 595°C.

The fields of the primary crystallization of the phases, the types and coordinates of non- and monovariant equilibria in the Pb–Cu–Te system were determined by Akhmedova et al. (2019). The ternary compounds were not found in the system, which is divided in the solid state into the next subsystems: PbTe–Cu–Pb, PbTe–$Cu_{2-x}Te$–Cu, PbTe–$Cu_{2-x}Te$–Cu_5Te_3, PbTe–Cu_5Te_3–Cu_4Te_3, PbTe–CuTe–$Cu_{3-x}Te_2$, and PbTe–CuTe–Te. The liquidus surface was also constructed. The equations were obtained for calculation and 3D modeling of the PbTe, CuTe, $Cu_{2-x}Te$, $Cu_{3-x}Te_2$, and Cu_5Te_3 fields of primary crystallization and the surface of the liquid immiscibility in the ternary system. It was shown, that the critical temperature of the immiscibility decreases with an increase quantity of the third component.

12.3 Lead–Silver–Tellurium

PbTe–Ag: This system is a non-quasibinary section of the Pb–Ag–Te ternary system (Bates et al. 1968; Blachnik and Bolte 1978). PbTe, Ag_2Te and Ag primarily crystallize in the system. The solubility of Ag in PbTe corresponds to 0.6–0.7 at% (Sharov 2008b, 2009). The crystals of the $Pb_{1-x}Ag_xTe$ solid solution were grown by the direct crystallization method with homogenization at 1100°C for 12 h.

This system was investigated through DTA, XRD, and metallography. The ingots for the investigations were annealed at 270°C for 4 weeks (Blachnik and Bolte 1978) [at 650°C and 950°C for respectively 46 and 20 h (Bates et al. 1968)].

PbTe–Ag_2Te: The phase diagram of this system, constructed through DTA, XRD, and metallography, is a eutectic type (Figure 12.2) (Wald 1967; Blachnik and Gather 1978). The eutectic contains 62 mol% Ag_2Te and crystallizes at 694°C. Two polymorphic transformations of the solid solution based on Ag_2Te exist in the system. With the addition of PbTe to Ag_2Te, the β-Ag_2Te phase stabilizes the temperature of the α-β-transformation which decreases from 147°C

DOI: 10.1201/9781003123507-12

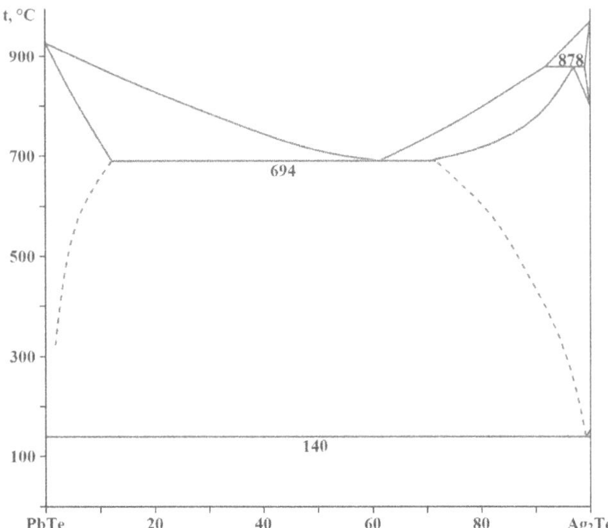

FIGURE 12.2 Phase diagram of the PbTe–Ag_2Te system. (From Blachnik, R., and Gather, B., *J. Less-Common Metals*, **60**(1), 25, 1978.)

to 140°C, and the β-γ-transformation temperature rises from 803°C to 878°C. Thermal effects at 475°C (Wald 1967) were not detected (Blachnik and Gather 1978). The solubility of PbTe in α-Ag_2Te was not detected, while the solubility of PbTe in β-Ag_2Te and γ-Ag_2Te is 29 and 1.5 mol%, respectively (Blachnik and Gather 1978). The solubility of β-Ag_2Te in PbTe is 12 mol%.

The solvus lines of the PbTe and Ag_2Te phases in the PbTe–Ag_2Te system have been determined by diffusion couple experiments and a unidirectional solidification experiment (Bergum et al. 2011). The solubilities of both Ag_2Te in PbTe and PbTe in Ag_2Te decrease with decrease in temperature from 14.9 (694°C) to 0.5 at% Ag (375°C) for the Ag_2Te dissolution in PbTe and from 12.4 (650°C) to 3.1 at% Pb (375°C) for the PbTe dissolution in Ag_2Te.

A number of polythermal sections and an isothermal section at room temperature of the Pb–Ag–Te ternary system, as well as a projection of the liquidus surface, have been constructed by Tagiev et al. (2016). The fields of primary crystallization of phases, types and coordinates of non- and monovariant equilibria have been determined. No ternary compounds were found in the system. The ternary system is divided into three stable quasiternary subsystems: Ag–Ag_2Te–PbTe, Ag_2Te–Te–PbTe, and Ag–PbTe–Pb. Equations were obtained for the calculation and 3D visualization of the crystallization surfaces of Ag_2Te, PbTe, and the immiscibility surface in the ternary system.

The thermodynamic modeling of the Pb–Ag–Te ternary system was performed based on the literature information and a set of thermodynamic parameters was obtained which were then used to reproduce known data (Gierlotka et al. 2010). Good agreement between calculations and experimental results was found. With the obtained thermodynamic description of the liquid phase, lead influence on Te pressure over Pb–Ag–Te alloys was shows that an increase in lead content of an alloy increases the Te vapor pressure.

In the Pb–Ag–Te ternary system, PbTe dissolves about 3.9 at% Ag into solid solution along the PbTe–Ag_5Te_3 joint and about 1.5 at% Ag along the PbTe–"AgTe" joint (Wald 1967).

PbTe occupies the biggest field of primary crystallization on the liquidus surface of the Pb–Ag–Te ternary system (Blachnik and Bolte 1978).

The excess enthalpies of liquid alloys in the ternary system were experimentally determined at 700°C, 800°C, and 900°C in a heat flow calorimeter (Römermann and Blachnik 1998). Five sections Ag_xPb_{1-x}–Te with constant concentration ratios Ag/Pb were measured. The enthalpy surface in the ternary system is characterized by a valley stretching from the exothermic minimum in the system Pb–Te (226.9 kJ·M^{-1} at 33.7 at% Te) to the exothermic minimum in the system Ag–Te (221.0 kJ·M^{-1} at 52.0 at% Te) at 900°C. The association model was applied for the analytical description of the ternary liquid. Additional ternary interaction parameters had to be used for the calculation of the excess enthalpies which were obtained by adjusting the curves to the experimental data using the least squares method.

12.4 Lead–Gold–Tellurium

PbTe–Au: The phase diagram of this system, constructed through DTA, XRD, and metallography, is a eutectic type (Figure 12.3) (Mottern and Wald 1968; Legendre and Souleau 1972a). The eutectic contains 57.15 at% Au and crystallizes at 748°C. The mutual solubility of the initial components is negligible.

PbTe–AuTe$_2$: The phase diagram of this system, constructed through DTA, XRD, and metallography, is a eutectic type (Figure 12.4) (Legendre and Souleau 1972a). The eutectic contains 22.5 mol% PbTe and crystallizes at 426°C. The mutual solubility of the initial components is negligible.

The liquidus surface of the Pb–Au–Te ternary system is presented in Figure 12.5 (Legendre and Souleau 1972a). In the PbTe–$AuTe_2$–Te subsystem, there is a ternary eutectic E_1, which contains 5.6 at% Pb, 81.7 at% Te, 12.7 at% Au and crystallizes at 388°C. The ternary eutectic E_2 in the PbTe–$AuTe_2$–Au

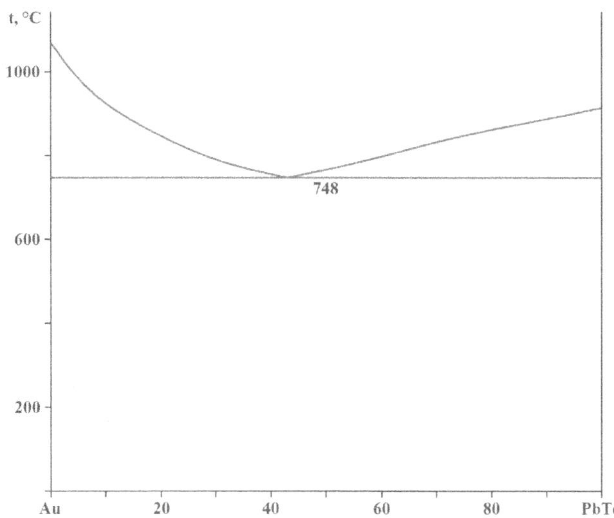

FIGURE 12.3 Phase diagram of the PbTe–Au system. (From Legendre, B., et al., *Bull. Soc. chim. France*, IV, (2), 473, 1972.)

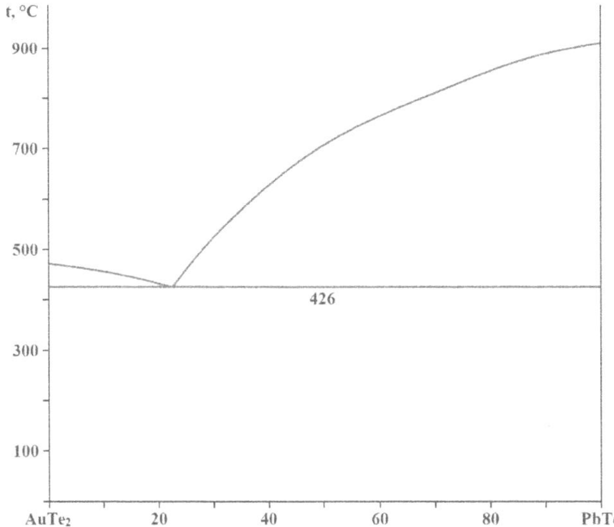

FIGURE 12.4 Phase diagram of the PbTe–AuTe$_2$ system. (From Legendre, B., et al., *Bull. Soc. chim. France*, IV, (2), 473, 1972.)

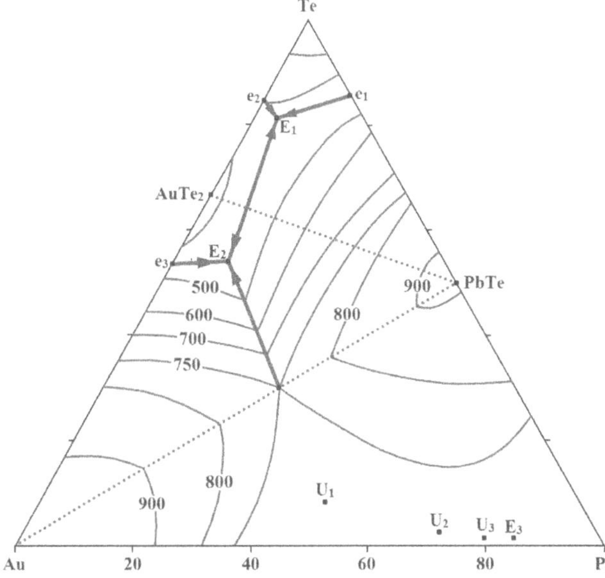

FIGURE 12.5 Liquidus surface of the Pb–Au–Te ternary system. (From Legendre, B., et al., *Bull. Soc. chim. France*, IV, (2), 473, 1972.)

subsystem contains 10.7 at% Pb, 53.6 at% Te, 35.7 at% Au, and crystallizes at 402°C. There are also three transition points (U_1 at 420°C, U_2 at 249°C, and U_3 at 215°C) and one ternary eutectic (E_3 at 210°C), which practically coincides with a eutectic in the Au–Pb binary system.

12.5 Lead–Magnesium–Tellurium

PbTe–MgTe: A part of the phase diagram of this system has an anomalous maximum at 7–8 mol% MgTe (Figure 12.6) (Crocker and Sealy 1972). The limit solubility of MgTe in PbTe reaches 6.5 mol%. The segregation coefficient of MgTe in PbTe is $k \geq 1$ (Sealy and Crocker 1973a).

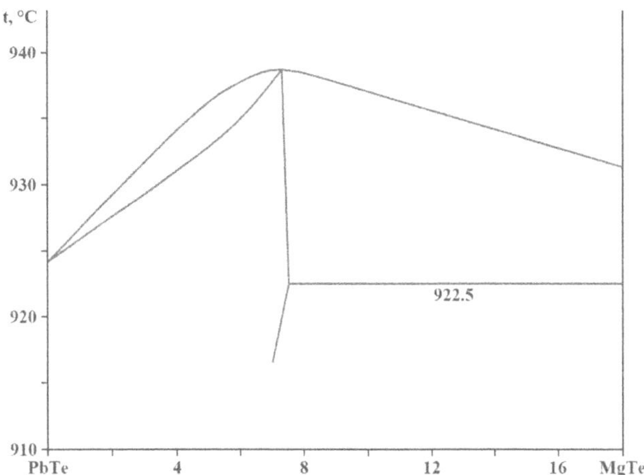

FIGURE 12.6 Part of the PbTe–MgTe phase diagram. (From Crocker, A.J., and Sealy B.J., *J. Phys. Chem. Solids*, **33**(12), 2183, 1972.)

The structural, electronic and optical properties of **$Mg_xPb_{1-x}Te$** solid solutions for $0 \leq x \leq 1$ in their cubic structure have been calculated using the full-potential linearized augmented plane wave method under the framework of density functional theory (Chattopadhyaya and Bhattacharjee 2017b). The calculated lattice parameters of these solid solutions exhibit significant upward deviations from Vegard's law with bowing parameters –34.4 pm. The non-liner behavior of the energy gap versus concentration for the $Mg_xPb_{1-x}Te$ solid solutions can be fitted by the next expression: $E_g = 0.971 - 1.247x + 1.418x^2$.

12.6 Lead–Strontium–Tellurium

PbTe–SrTe: In a typical process of the **$Sr_xPb_{1-x}Te$** solid solutions preparation, KOH (1.0 g) and distilled water (5 mL) were put into a beaker (Wang et al. 2018). Then, $Pb(CH_3COO)_2$ (1.36 mM) and triethanolamine (5 mL) was introduced into the above solution. After constantly stirring for 10 min and adding $SrCl_2$ (1.12 mM) and ethylenediaminetetraacetic acid disodium salt (1 g), Na_2TeO_3 (1.44 mM) and concentrated $N_2H_4{\cdot}H_2O$ (80 mass%, 0.5 mL) were put into above solution. After being stirred for 15 min, the chemicals were good mixed. The resulting suspension was transferred into a 15 mL stainless Teflon-lined, kept heating at 180°C for 4 h and cooled to room temperature. Finally, the sample was washed by distilled water and ethanol for several times respectively, dried in a vacuum oven at 60°C for 6 h. Using this procedure, much higher level amount of Sr alloyed into the PbTe matrix (5.2–6.5 mol%) can be achieved, which is beyond its thermodynamic solubility limit of 1 mol% in bulk PbTe.

The structural, electronic and optical properties of $Sr_xPb_{1-x}Te$ solid solutions for $0 \leq x \leq 1$ in their cubic structure have been calculated using the full-potential linearized augmented plane wave method under the framework of density functional theory (Chattopadhyaya and Bhattacharjee 2017a). The calculated lattice parameter of these solid solutions exhibits marginal upward deviation from Vegard's law with a bowing parameter 0.23 pm. The non-liner behavior of the energy gap versus concentration was determined, and it was shown that the energy gap increases with increasing x.

12.7 Lead–Barium–Tellurium

PbTe–BaTe: The structural, electronic and optical properties of **Ba$_x$Pb$_{1-x}$Te** solid solutions for $0 \le x \le 1$ in their cubic structure have been calculated using the full-potential linearized augmented plane wave method under the framework of density functional theory (Chattopadhyaya and Bhattacharjee 2017c). A large deviation of the lattice parameter from Vegard's law with an upward bowing parameter of –26.9 pm was observed. The non-liner behavior of the energy gap versus concentration was determined, and it was shown that the energy gap increases with increasing x.

12.8 Lead–Zinc–Tellurium

PbTe–Zn: This system is a non-quasibinary section of the Pb–Zn–Te ternary system (Bates et al. 1968; Movsum-zade et al. 1986b). Crystallization of ZnTe occurs in the region of immiscibility, which exists in the range of 13–100 at % Zn. Complete crystallization of all phases in the PbTe–ZnTe–Pb subsystem ends at 300°C and 49 at% Zn, and in the Pb–Zn–ZnTe subsystem, at 310°C and 50.5 at% Zn (Movsum-zade et al. 1986b). The system was investigated through DTA, XRD, metallography, and measuring of the microhardness.

Pb–ZnTe: The phase diagram of this system, constructed through DTA, XRD, metallography, and measuring of the microhardness, is a eutectic type (Figure 12.7) (Movsum-zade et al. 1986b). The eutectic composition and temperature are 3 mol% ZnTe and 315°C, respectively. An immiscibility region within the interval from 25 to 95 at% Pb exists in this system at 1100°C. The ingots for the investigations were annealed at a temperature 50°C lower than solidus temperature for 200 h.

PbTe–ZnTe: The phase diagram of this system, constructed through DTA, XRD, metallography, and measuring of the microhardness, is a eutectic type (Figure 12.8) (Grytsiv et al. 1980; Movsum-zade et al. 1986b). The eutectic composition and temperature are 17 mol% ZnTe and 887°C, respectively (Grytsiv et al. 1980) [10 mol% ZnTe and 875°C (Movsum-zade et al. 1986b)]. The solubility of ZnTe in PbTe at 800°C, 720°C, and 250°C is equal correspondingly to 1.5, 1, and 1 mol% (Rosenberg et al. 1964; Sealy and Crocker 1973a).

Liquidus surface of the Zn–Pb–Te ternary system includes five fields of primary crystallization (Figure 12.9) (Movsum-zade et al. 1986b). The field of ZnTe primary crystallization occupies almost all liquidus surface. The immiscibility region from the Zn–Pb binary system penetrates deeply in the ternary system and occupies the great part of the ZnTe field of primary crystallization. The ternary eutectics E_1, E_2, and E_3 crystallize at 400°C, 300°C, and 310°C, respectively.

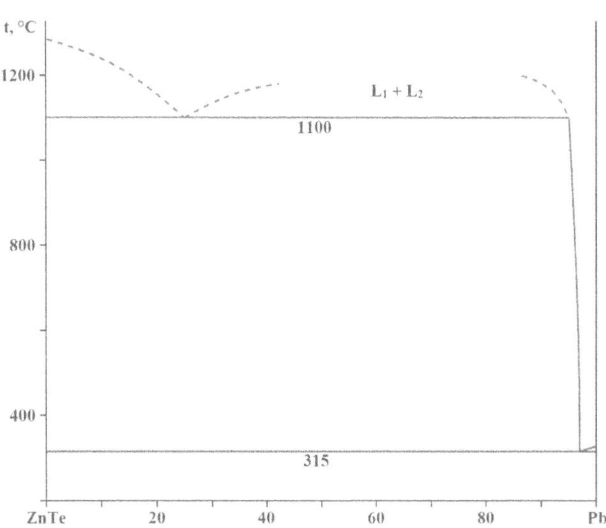

FIGURE 12.7 Phase diagram of the Pb–ZnTe system. (From Movsum-zade, A.A., et al., *Zhurn. neorgan. khimii*, **31**(1), 198, 1986.)

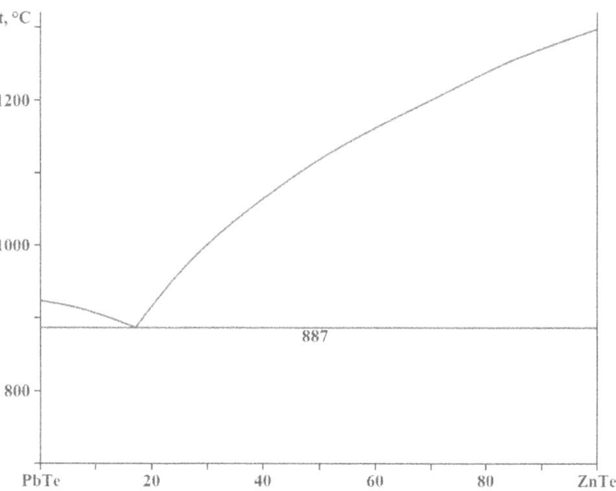

FIGURE 12.8 Phase diagram of the PbTe–ZnTe system. (From Grytsiv, V.I., et al., *Izv. AN SSSR. Neorgan. mater.*, **16**(3), 543, 1980.)

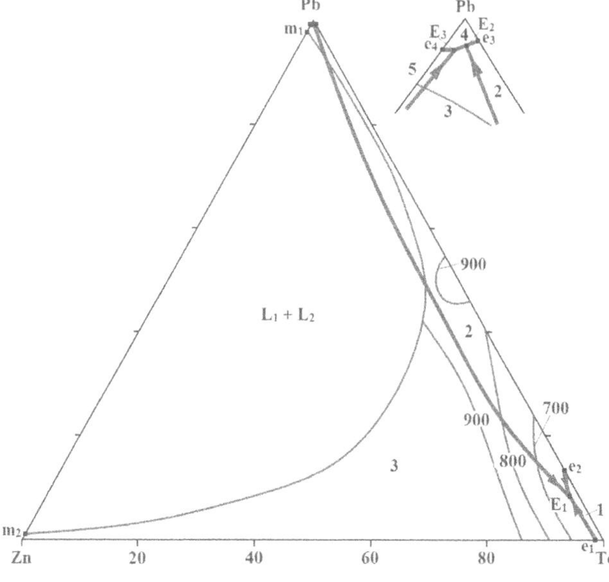

FIGURE 12.9 Liquidus surface of the Pb–Zn–Te ternary system: Te (1), PbTe (2), ZnTe (3), Pb (4), and Zn (5) primary crystallization fields. (From Movsum-zade, A.A. et al., *Zhurn. neorgan. khimii*, **31**(1), 198, 1986.)

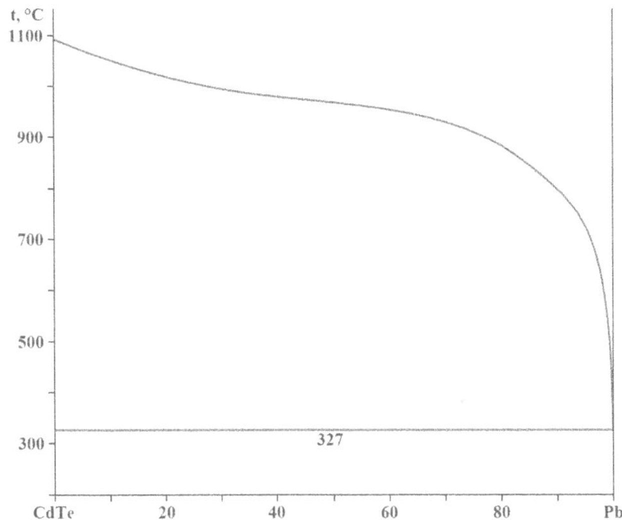

FIGURE 12.10 Phase diagram of the Pb–CdTe system. (From Tai, H. and Hori, S., *J. Jpn. Inst. Met.*, **38**(5), 451, 1974.)

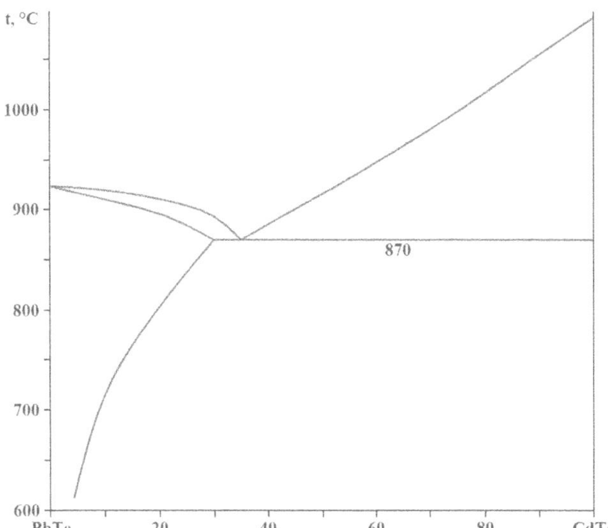

FIGURE 12.11 Phase diagram of the PbTe–CdTe system. (From Tomashik, Z.F. and Tomashik, V.N., *Izv. AN SSSR. Neorgan. mater.*, **18**(12), 1994, 1982.)

12.9 Lead–Cadmium–Tellurium

PbTe–Cd: This system is a non-quasibinary section of the Pb–Cd–Te ternary system (Bates et al. 1968; Silberg and Zemel 1977; Grytsiv 1980; Morgant et al. 1980). At the interaction of PbTe and Cd, lead and CdTe were found. Thermal effects at 248°C and 327'C correspond to the crystallization of the ternary eutectics in the CdTe–Cd–Pb and PbTe–CdTe–Pb subsystems. The solubility of Cd in PbTe increases monotonically at the increasing of the Cd vapor pressure. The system was investigated through DTA, XRD, metallography, and measuring of the microhardness. The ingots for the investigations were annealed at 950°C for 15 h and then at 600°C for 48 h (Bates et al. 1968).

Pb–CdTe: The phase diagram of this system, constructed through DTA, XRD, and metallography, is a eutectic type (Figure 12.10) (Tai and Hori 1974). The eutectic is degenerated from Pb rich side, contains 99.94 at% Pb, and crystallizes at 327°C. The mutual solubility of CdTe and Pb is insignificant.

PbTe–CdTe: The phase diagram of this system, constructed through DTA, XRD, metallography, and measuring of the microhardness, is a eutectic type (Figure 12.11) (Hirai and Kurata 1968; Morgant et al. 1980; Tomashik and Tomashik 1982). The eutectic composition and temperature are 35 mol% CdTe and 870°C, respectively (Tomashik and Tomashik 1982) [36 mol% CdTe and 884°C (Morgant et al. 1980); 40 mol% CdTe and 840°C (Hirai and Kurata 1968); 866°C (Rosenberg et al. 1964)]. The solubility of CdTe in PbTe at 250°C, 630°C, 720°C, and 800°C is correspondingly equal to 3, 4.6, 10, and 17 mol% (Rosenberg et al. 1964; Nikolić 1966; Crocker 1968; Hirai and Kurata 1968) [at 450°C, 550°C, and 650°C correspondingly 2, 3, and 4 mol% (Rogacheva et al. 1988b)].

Non-monotonic nature of concentration dependences of electrophysical properties was observed in the region of diluted solid solutions based on PbTe (Rogacheva et al. 1988b). It can be explained by the transition of diluted solid solutions, where inter-impurity interaction can be ignored, to concentrated solid solutions, where the interaction between particles transfers the crystal in the qualitatively new state. In the CdTe–PbTe system, this transition takes place at 0.4 mol% CdTe. The ingots were annealed at 450°C, 550°C, and 650°C for 400 and 100 h.

The regions of phase secondary crystallization in the subsystem CdTe–PbTe–Te of the Cd–Pb–Te ternary system have been determined by Lesyna et al. (1998). It was noted that the ternary eutectic in this subsystem crystallizes at 355°C and contains 2 mol% CdTe, 94 mol% Te, and 4 mol% PbTe.

12.10 Lead–Mercury–Tellurium

Pb–HgTe: This section is a non-quasibinary section of the Pb–Hg–Te ternary system. Lead telluride and Hg are formed at the interaction of HgTe with Pb (Tomashik et al. 1986).

PbTe–HgTe: The phase diagram of this system, constructed through DTA, XRD, metallography, and measuring of the microhardness, is a eutectic type (Figure 12.12) (Vaniarkho

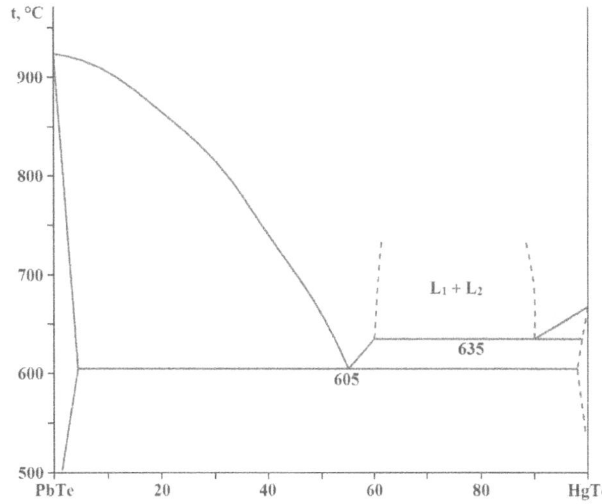

FIGURE 12.12 Phase diagram of the PbTe–HgTe system. (From Vaniarkho, V.G., et al., *Izv. AN SSSR. Neorgan. mater.*, **6**(1), 133, 1970.)

et al. 1968, 1970; Leute and Köller 1986a,b). The eutectic composition and temperature are 55 mol% HgTe and 605°C, respectively (Vaniarkho et al. 1968, 1970) [76 mol% HgTe and 610°C (Leute and Köller 1986a,b)]. The immiscibility region within the interval of 60–90 mol% HgTe exists in the HgTe–PbTe system at 635°C (Vaniarkho et al. 1968, 1970) [according to the data of Leute and Köller (1986a,b), there is not any immiscibility region in this system]. The solubility of HgTe in PbTe at 430°C, 530°C, 610°C, and 630°C is equal correspondingly to 1.5, 3, 4.6, and 4.0 mol%, and the solubility of PbTe in HgTe is not higher than 0.5 mol% (Leute and Köller 1986a,b).

12.11 Lead–Gallium–Tellurium

PbTe–Ga: This system is a non-quasibinary section of the Pb–Ga–Te ternary system (Bates et al. 1968; Babayev 2005). At the interaction of PbTe and Ga, Ga_2Te_3, Pb, and nonidentified phase (apparently, solid solution of GaTe in PbTe) were detected. The system was investigated through XRD, metallography, and measuring of the microhardness. The ingots for the investigations were annealed at 950°C for 15 h and then at 600°C for 48 h.

PbTe–GaTe: The phase diagram of this system, constructed through DTA, XRD, metallography, and measuring of the microhardness, is a eutectic type (Figure 12.13) (Abilov et al. 1974). The eutectic composition and temperature are 40 mol% PbTe and 630°C, respectively. The solubility of GaTe in PbTe reaches 5 mol% (Abilov et al. 1974) [15 mol% (Rustamov et al. 1974a); 1.25 mol% at 600°C (Bushmarina et al. 1980)]. The region of solid solutions based on GaTe is insignificant (Abilov et al. 1974). The ingots for the investigations were annealed at 600°C for 500 h (Abilov et al. 1974). Single crystals of the solid solutions based on PbTe were grown using Bridgman method (Sealy and Crocker 1973a).

PbTe–Ga_2Te_3: The phase diagram of this system, constructed through DTA, XRD, metallography, and measuring of the microhardness, is given in Figure 12.14 (Baranets et al. 1987). The eutectic contains 43 mol% Ga_2Te_3 and crystallizes

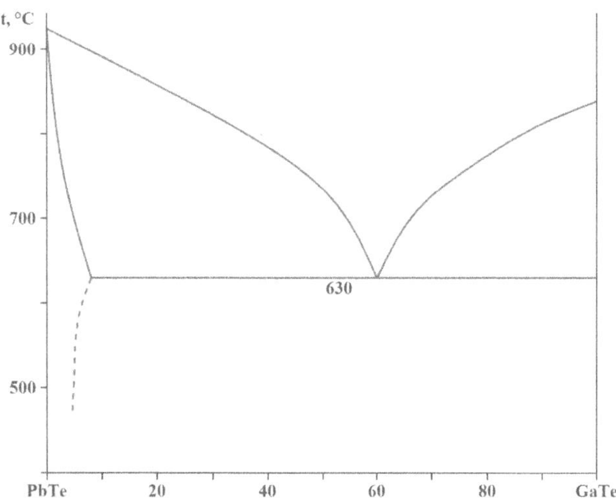

FIGURE 12.13 Phase diagram of the PbTe–GaTe system. (From Abilov, Ch.I., et al., *Izv. AN SSSR. Neorgan. mater.*, **10**(1), 142, 1974.)

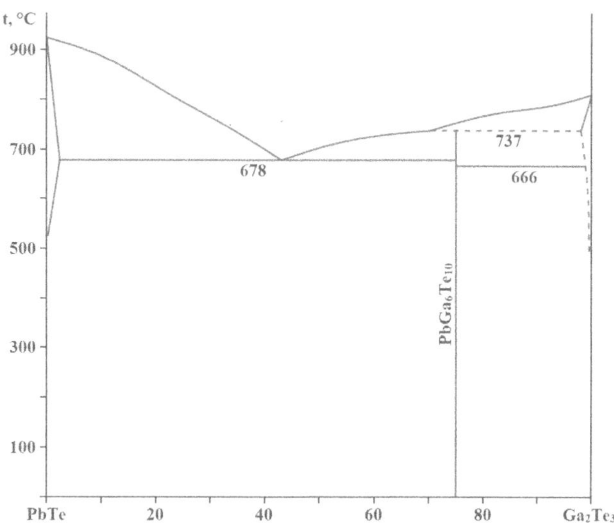

FIGURE 12.14 Phase diagram of the PbTe–Ga_2Te_3 system. (From Baranets, S.M., et al., *Izv. AN SSSR. Neorgan. mater.*, **23**(7), 1221, 1987.)

at 678°C. The solubility of PbTe in Ga_2Te_3 reaches 4 mol% and the solubility of Ga_2Te_3 in PbTe is less than 1 mol% (Rustamov et al. 1979; Baranets et al. 1987). In the range of compositions 80–95 mol% Ga_2Te_3, there are thermal effects at 666°C, caused by the polymorphic transformation of Ga_2Te_3, which can be associate with the thermal effects at 400°C with a shift in the composition of the studied samples from the PbTe–Ga_2Te_3 section (Baranets et al. 1987).

The phase diagram of this system was also constructed through DTA, XRD, and metallography by Avanesov et al. (2014). The eutectics contain 43 and 87 mol% Ga_2Te_3 and crystallize at 678°C and 670°C, respectively. The **$PbGa_6Te_{10}$** ternary compound, which melts congruently and has the homogeneity region from 19 to 27 mol% PbTe, is formed in this system. The maximum melting temperature (730°C) is shifted towards 1.5 mol% PbTe. It should be noted that the authors did not indicated the temperature scale; therefore, it is not possible to reproduce the phase diagram.

According to the data of Baranets et al. (1987), $PbGa_6Te_{10}$ melts incongruently at 737°C [at 725°C (Guittard et al. 1978a)] and apparently has a polymorphic transformation. First modification of this compound crystallizes in the cubic structure with the lattice parameter a = 1023 pm (Dedegkaev et al. 1986). Second modification crystallizes in the rhombohedral structure with the lattice parameters a = 1445.2 and c = 1773.2 pm in hexagonal setting (Avanesov et al. 2014) [a = 1446.5 ± 0.2 and c = 1771.8 ± 0.4 pm in hexagonal setting and a calculated density of 5.898 g·cm^{-3} (Kienle and Deiseroth 1998)]. This compound could be prepared directly from the elements by repeated grinding and annealing at 600°C for 2 weeks in the evacuated quartz ampoule (Kienle and Deiseroth 1998). It appears as homogeneous, black and brittle polycrystalline sample, which is not significantly air sensitive.

According to the data of Davletshin et al. (1979), Rustamov et al. (1979), and Babayev (2002), the phase diagram of the PbTe–Ga_2Te_3 system is a eutectic type. The eutectic contains 40 mol% Ga_2Te_3 and crystallizes at 550°C (Rustamov et al. 1979) [40 mol% Ga_2Te_3 and 696°C (Davletshin et al. 1979);

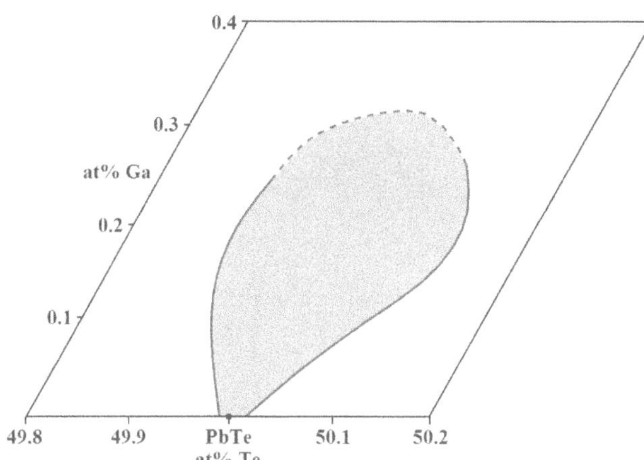

FIGURE 12.15 Region of the solid solution based on PbTe in the Pb–Ga–Te ternary system at 550°C. (From Rogacheva, E.I., et al., *Izv. AN SSSR. Neorgan. mater.*, **19**(2), 204, 1983.)

40 mol% Ga_2Te_3 and 670°C (Babayev 2002)]. The addition of PbTe to Ga_2Te_3 increases the temperature of the polymorphic transformation of solid solutions based on Ga_2Te_3 from 680°C to 700°C (Rustamov et al. 1979). According to Davletshin et al. (1979), polymorphic transformation of Ga_2Te_3 is observed at 670°C.

The ingots for the investigations were annealed at 350°C (for the compositions with less than 60 mol% Ga_2Te_3) and 660°C (for the compositions with bigger than 60 mol% Ga_2Te_3) for 150 h (Baranets et al. 1987) [at 400°C for 500 h (Rustamov et al. 1979); at 650°C for 350 h (Davletshin et al. 1979); at 500°C, 600°C, 700°C, and 800°C (Avanesov et al. 2014)].

The region of the solid solutions based on PbTe in the Pb–Ga–Te ternary system at 550°C is elongated in the direction of the PbTe–GaTe section (Figure 12.15), and the introduction of Ga enlarges the PbTe homogeneity region both towards Pb and Te (Rogacheva et al. 1983b; Bushmarina et al. 1987).

The liquidus surface of this ternary system was constructed using the data on the phase diagram of the binary systems and some vertical sections of this ternary system (Babayev 2004b). The mono- and invariant reactions in the system were determined. It was shown, that six ternary eutectics and one transition point exist on the liquidus surface.

The glass forming region in the Pb–Ga–Te ternary system is presented in Figure 12.16 (Minaev 1983). Synthesis of the glassy alloys was carried out from the elements in quartz ampoules under vacuum at 1000°C for 12 h with vigorous stirring and the next quenching in water. The vitreous state was identified by a characteristic glass wedge, a shell fracture, as well as through XRD, DTA, and metallography. The boundary of the glass formation region was fixed with an accuracy of 2 at%.

12.12 Lead–Indium–Tellurium

PbTe–In: This system is a non-quasibinary section of the Pb–In–Te ternary system (Bates et al. 1968). At the interaction of PbTe and In, In_2Te_3 and Pb were detected. The system

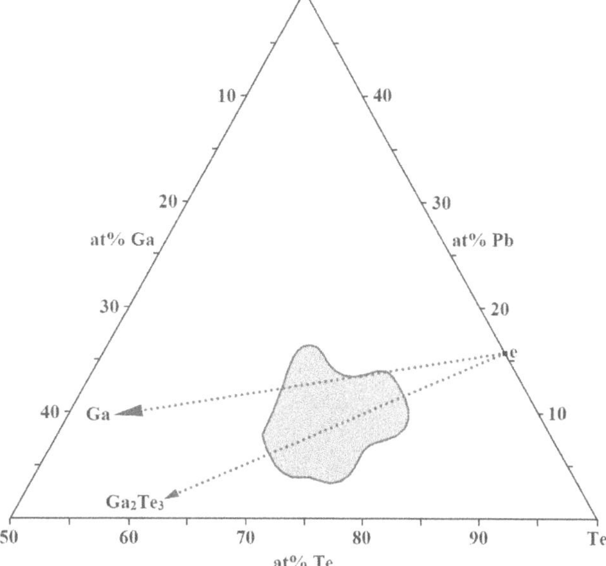

FIGURE 12.16 Glass forming region in the Pb–Ga–Te ternary system. (From Minaev, V.S., *Fiz. i khim. stekla*, **9**(4), 432, 1983.)

was investigated through XRD, metallography, and measuring of the microhardness. The ingots for the investigations were annealed at 950°C for 15 h and then at 600°C for 48 h.

PbTe–InTe: The phase diagram of this system, constructed through DTA, XRD, metallography, and measuring of the microhardness, is a eutectic type (Figure 12.17) (Rosenberg et al. 1964; Niehuus and Nieke 1966; Rogacheva et al. 1979). The eutectic contains 25–26 mol% PbTe and crystallizes 640°C (Niehuus and Nieke 1966; Rogacheva et al. 1979) [at 646°C (Rosenberg et al. 1964)]. The solubility of InTe in PbTe reaches 15 mol% (Rustamov et al. 1974a) [35 mol% (Rosenberg et al. 1964); 7 mol% (Alidzhanov et al. 1973; Rustamov et al. 1973b)], and the solubility of PbTe in InTe is 3 mol% (Rustamov et al. 1973b, 1976b). The unit cell parameter

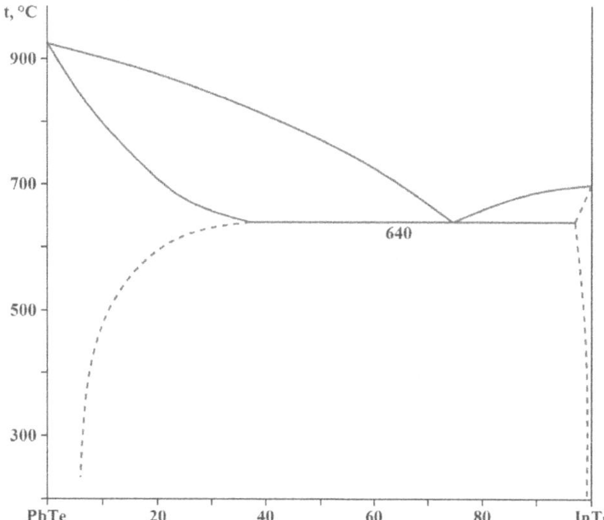

FIGURE 12.17 Phase diagram of the PbTe–InTe system. (From Rogacheva, E.I., *Izv. AN SSSR. Neorgan. mater.*, **15**(8), 1366, 1979.)

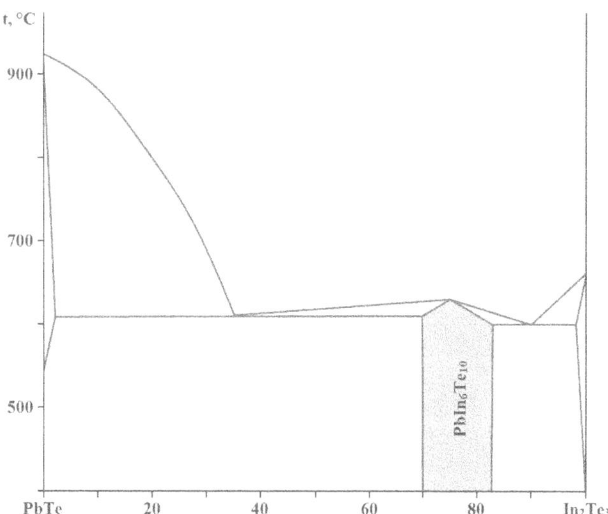

FIGURE 12.18 Phase diagram of the PbTe–In_2Te_3 system. (From Avanesov, S.A., et al., *J. Alloys Compd.*, **612**, 386, 2014.)

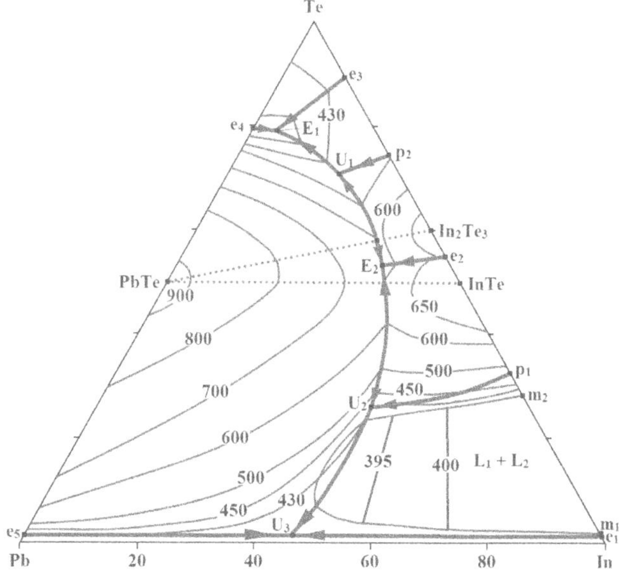

FIGURE 12.19 Liquidus surface of the Pb–In–Te ternary system. (From Latypov, Z.M., et al., *Izv. AN SSSR. Neorgan. mater.*, **13**(5), 824,1977.)

of solid solution based on PbTe changes linearly depending on the composition (Rosenberg et al. 1964).

According to the data of Rustamov et al. (1973b), the phase diagram of this system is a eutectic type with peritectic transformation. The eutectic contains 15 mol% PbTe and crystallizes at 400°C. The **$InPbTe_2$** ternary compound, which melts incongruently at 650°C and contains 52 mol% PbTe, is formed in the system.

$Pb_{1-x}In_xTe$ single crystals were grown using vapor–liquid–solid technique (Sergeev et al. 1998).

The ingots for the investigations were annealed at 300°C–700°C for 150–500 h (Rogacheva et al. 1979) [at 600°C for 400 h (Rustamov et al. 1973b)].

PbTe–In_2Te_3: The phase diagram of this system, constructed through DTA, XRD, and metallography, is presented in Figure 12.18 (Avanesov et al. 2014). The eutectics contain 35 and 90 mol% In_2Te_3 and crystallize at 610°C and 600°C, respectively. The **$PbIn_6Te_{10}$** ternary compound, which melts congruently at 630°C and has the homogeneity region from 17 to 30 mol% PbTe, is formed in this system.

The phase diagram of this system was also constructed through DTA, XRD, metallography, and measuring of the microhardness by Latypov et al. (1977) and Il'yin et al. (1988). According to this data, the eutectic contains 38 mol% PbTe and crystallizes at 600°C [19 mol% PbTe and 607°C (Babayev 2002)]. The temperature of In_2Te_3 polymorphic transformation with the introduction of PbTe decreases from 548°C to 544°C. The solubility of PbTe in In_2Te_3 at 500°C reaches 1.5 mol% and decreases to 0.05 mol% at room temperature.

The $PbIn_6Te_{10}$ ternary compound, which apparently has a polymorphic transformation, is formed in this system. First modification of this compound crystallizes in the cubic structure with the lattice parameter a = 1939 pm (Rogacheva et al. 1980). Second modification crystallizes in the rhombohedral structure with the lattice parameters a = 1496.1 and c = 1861.0 pm in hexagonal setting (Avanesov et al. 2014) [a = 1061.9 pm, α = 89.65°, and a calculated density of 6.03 g·cm^{-3} or a = 1497.1 ± 0.3 and c = 1850.5 ± 0.6 pm in hexagonal setting (Deiseroth and Müller 1996)]. This compound could be prepared directly from the elements by repeated grinding and annealing at 500°C for 2 weeks in dry, evacuated quartz ampoules. $PbIn_6Te_{10}$ appear as homogeneous, black and brittle polycrystalline samples, which are not significantly air sensitive.

The ingots for the investigations were annealed at 450°C for 300 h (Latypov et al. 1977) [at 300°C–780°C for 200–500 h (Rogacheva et al. 1980), at 500°C, 550°C, 610°C, 700°C, and 800°C (Avanesov et al. 2014)].

The liquidus surface of the Pb–In–Te ternary system (Figure 12.19) consists from six fields of primary crystallization of the phases (Latypov et al. 1989). The largest area is occupied by the field of PbTe primary crystallization. There are five invariant points in the system: two ternary eutectics and three transition points.

Excess enthalpies of liquid alloys of the Pb–In–Te ternary system were determined in a heat-flow calorimeter at 800°C and 900°C for In_xPb_{1-x}–Te (x = 0.2, 0.5 and 0.8) sections by Schlieper and Blachnik (2000). The association model with additional interaction parameters had to be used for the calculation of ternary excess enthalpies. They were obtained by fitting the calculated curves to the experimental data with the least square method. The minima of the exothermic enthalpies were found of mention above sections at 54 mol% Te. The values of excess enthalpies are –30.7 and –28.5 kJ·M^{-1} at 800°C and 900°C, respectively.

12.13 Lead–Thallium–Tellurium

PbTe–Tl: This system is a non-quasibinary section of the Pb–Tl–Te ternary system (Berg and Latypov 1970a). The liquidus line, corresponding to the PbTe primary crystallization, continues up to 55 at% Tl. Thermal effects at 592°C correspond to the beginning of the joint crystallization of PbTe and **Tl_4PbTe_3**. Crystallization of alloys ends with solidification of solid solutions Tl in Pb. The **Tl_2PbTe** ternary compound was not found in the system. The system was investigated through DTA, XRD, metallography, and measuring of the microhardness and the ingots for the investigations were annealed at 300°C for 800 h.

PbTe–TlTe: This system is also a non-quasibinary section of the Pb–Tl–Te ternary system (Berg and Latypov 1972; Rustamov et al. 1974b). On the heating curves of samples containing 5–90 mol% TlTe, there are two endothermic effects at 180°C and 450°C and one high-temperature endothermic effect, the temperature of which decreases with an increase in the TlTe content (Berg and Latypov 1972). In the range of 90–100 mol% TlTe, a thermal effect appears at 300°C. The thermograms of unannealed alloys show effects at 224°C and 240°C. The annealed samples consist of the PbTe and TlTe phases, while the unannealed ones contain the third phase, which, apparently, has the composition Tl_4PbTe_3. According to the data of Rustamov et al. (1974b), at 250°C the nonvariant reaction L + Tl_5Te_3 ⇔ TlTe + (PbTe) takes place. Thermal effects at 450°C (Berg and Latypov 1972) were not detected in the system (Rustamov et al. 1974b). The solubility of TlTe in PbTe at 250°C reaches 17 mol% (Rustamov et al. 1974a,b) [1.2 mol% at 200°C (Berg and Latypov 1970b)].

The system was investigated through DTA, XRD, metallography, and measuring of the microhardness and the ingots for the investigations were annealed at 200°C for 500–600 h (Berg and Latypov 1970b, 1972; Rustamov et al. 1974a,b).

PbTe–Tl_2Te: The phase diagram of this system, constructed through DTA, XRD, metallography, and measuring of the microhardness, is presented in Figure 12.20 (Malakhovs'ka et al. 2008b). The eutectic contains 47 mol% PbTe and crystallizes at 590°C [43 mol% PbTe and 600°C (Gotuk et al. 1978b); the eutectic is degenerated from the Tl_2Te side and crystallizes at 414°C (Chami et al. 1983b)]. The solubility of Tl_2Te in PbTe at the eutectic temperature is not higher than 10 mol%

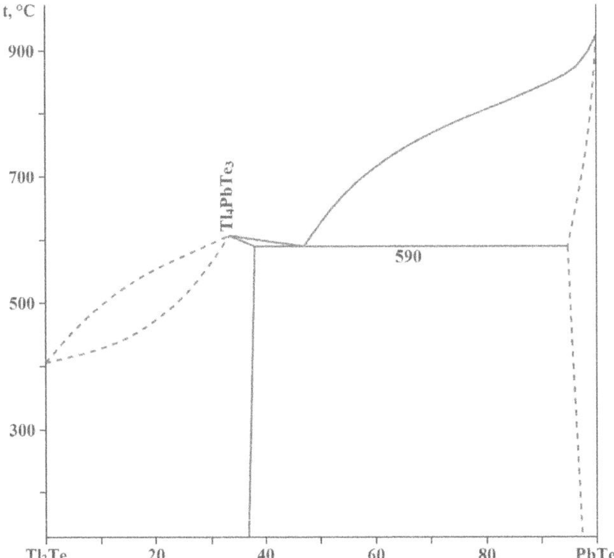

FIGURE 12.20 Phase diagram of the PbTe–Tl_2Te system. (From Malakhovs'ka, T.O., et al., *Nauk. visn. Uzhgorod. Univ., Ser. Khim.*, (19), 8, 2008.)

(Malakhovs'ka et al. 2008b) [1 mol% at 200°C (Berg and Latypov 1970b); 7 mol% at 624°C (Chami et al. 1983b)]. It is possible that continuous solid solutions are formed between Tl_2Te_3 and Tl_4PbTe_3 (Malakhovs'ka et al. 2008b).

The Tl_4PbTe_3 ternary compound, which formed in the system, melts congruently at 607°C (Barchiy et al. 2010) [at 594°C (Chami et al. 1983b); at 575°C (Bradtmöller and Böttcher 1993)] and crystallizes in the tetragonal structure with the lattice parameters a = 884.1 ± 0.2, c = 1305.6 ± 0.3 pm, and a calculated density of 9.161 g·cm^{-3} (Bradtmöller and Böttcher 1993) [a = 885.6, c = 1308.5 pm (Malakhovs'ka et al. 2008b)]. The experimental density of this compound and enthalpy and entropy of its melting are 9.07 g·cm^{-3}, 81 kJ·M^{-1} and 93 J·(K·M)$^{-1}$, respectively (Malakhovs'ka et al. 2007; Barchiy et al. 2010).

Thermodynamic data for the liquid phase of the PbTe–Tl_2Te system were determined by EMF measurements of concentration cells (Zaleska and Małachowicz 1990). An influence of the Tl_2Te associates presence on changes of the values of partial thermodynamic functions of Tl in the wide temperature range has been discussed. The obtain results indicate on the incongruent melting of Tl_4PbTe_3.

This compound was obtained from stoichiometric mixture of the elements (Bradtmöller and Böttcher 1993). The finely powdered mixture was filled under protective gas in quartz ampoule, which was then melted under vacuum at about 700°C for a homogenization and repeatedly shaken. After this, it was slowly cooled to 545°C and held there for about 3–4 weeks. Single crystals of Tl_4PbTe_3 were grown using Bridgman method (Voroshilov et al. 1988a; Malakhovs'ka et al. 2007).

According to the data of Berg et al. (1969) and Gotuk et al. (1978b), this system is non-quasibinary section of the Pb–Tl–Te ternary system.

The ingots for the investigations were annealed at 200°C for 480 h (Malakhovs'ka et al. 2008b) [at 200°C–300C for 500 h (Berg et al. 1969; Berg and Latypov 1970b); at 300°C–550°C for 400–450 h (Gotuk et al. 1978b)].

PbTe–Tl_2Te_3: This system is a non-quasibinary section of the Pb–Tl–Te ternary system as Tl_2Te_3 melts incongruently (Berg and Latypov 1977). In the range of 0–1 mol% PbTe, the primary crystallization of TlTe occurs, and with a higher content of PbTe, the primary crystallization of PbTe exists. The joint crystallization of TlTe and PbTe begins at 280°C, and at 240°C, the final crystallization of alloys occurs. The solubility of Tl_2Te_3 in PbTe reaches 0.5 mol%. The system was investigated through DTA, XRD, metallography, and measuring of the microhardness and the ingots for the investigations were annealed at 200°C for 500 h (Berg and Latypov 1970b, 1977).

PbTe–Tl_5Te_3: This system is also a non-quasibinary section of the Pb–Tl–Te ternary (Berg and Latypov 1977; Voroshilov et al. 1988a). The section contains a phase of variable composition based on Tl_4PbTe_3, formed by a peritectic reaction at 577°C. A polymorphic transformation of the ternary phase was found, the temperature of which decreases with an increase in the PbTe content from 317°C to 292°C. The region of homogeneity based on Tl_5Te_3 does not exceed 3 mol%, and based on PbTe – 5 mol%. The system was investigated through DTA, XRD, metallography, and measuring of the microhardness and the ingots for the investigations were annealed at 250°C for 500 h.

The liquidus surface of the Pb–Tl–Te ternary system includes five fields of primary crystallization of PbTe, Te, Tl_2Te, Tl_4PbTe_3, Pb_xTl_{1-x}, and an immiscibility region (Berg and Latypov 1977). The boundaries of the Tl_5Te_3 crystallization region could not be established. It is assumed that this area is in the form of a narrow strip along the Tl–Te binary system.

12.14 Lead–Yttrium–Tellurium

PbTe–Y_2Te_3: No ternary compounds were found in this system (Shemet et al. 2006c). PbTe dissolves 6 mol% Y_2Te_3 at 600°C.

12.15 Lead–Lanthanum–Tellurium

PbTe–LaTe: The phase diagram of this system, constructed through DTA, XRD, metallography, and measuring of the microhardness, is a eutectic type (Figure 12.21) (Iskender-zade at al. 1990). The eutectic contains 13 mol% LaTe and crystallizes at 827°C. The solubility of LaTe in PbTe at room temperature reaches 3 mol% and 7 mol% at the eutectic temperature.

12.16 Lead–Neodymium–Tellurium

PbTe–Nd_2Te_3: The phase diagram of this system, constructed through DTA, metallography, and measuring of the microhardness, is given in Figure 12.22 (Gabib-zade and Melikova 1978). The eutectic contains 7 mol% Nd_2Te_3 and crystallizes at 875°C. The solubility of Nd_2Te_3 in PbTe is 3 mol%. The **$PbNd_2Te_4$** ternary compound, which melts incongruently at 950°C, is formed in the system. Thermal effects at 990°C correspond to the polymorphic transformation of the solid solution based on Nd_2Te_3.

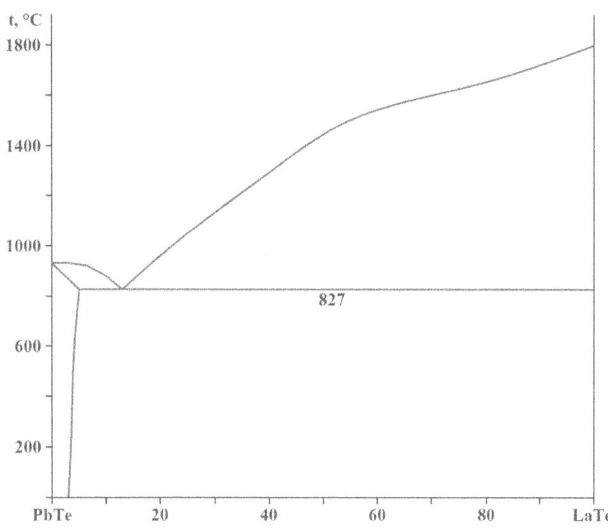

FIGURE 12.21 Phase diagram of the PbTe–LaTe system. (From Iskender-zade, Z.A., et al., *Izv. AN SSSR. Neorgan. mater.*, **26**(2), 435, 1990.)

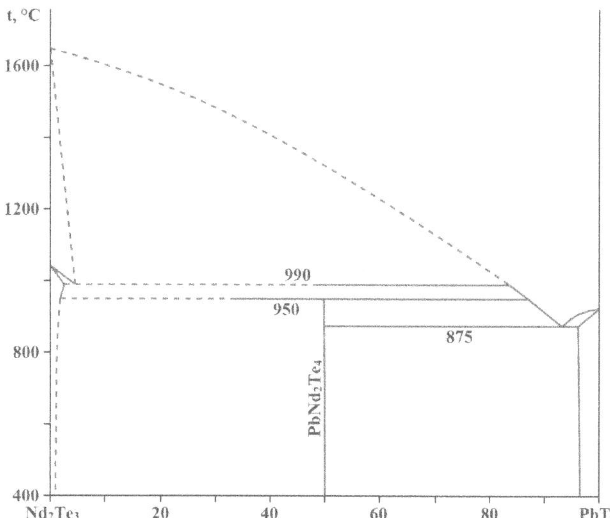

FIGURE 12.22 Phase diagram of the PbTe–Nd_2Te_3 system. (From Gabib-zade, S.A., and Melikova, Z.D., *Uch. zap. Azerb. Univ. Ser. khim. nauk*, (1), 14, 1978.)

12.17 Lead–Samarium–Tellurium

PbTe–SmTe: The part of the phase diagram of this system, constructed through DTA, XRD, and metallography, is presented in Figure 12.23 (Klanichka et al. 2002). The peritectic transformation at 1000°C takes place in the system (Erofeev et al. 1974; Erofeev and Solomatnikova 1974). The solubility of SmTe in PbTe at 700°C is less than 10 mol% and increases at the temperature increasing up to ca. 20 mol%. The dependence of the unit cell of the lattice parameter on the SmTe content has a minimum at 3 mol% SmTe. An unusual donor behavior of Sm in PbTe has been established, characterized by the presence of maximum electrical activity in the concentration range up to 5 mol% SmTe. The ingots for the investigations were annealed at 700°C for 500 h (Klanichka et al. 2002) [at 500°C–700°C for 500–700 h (Erofeev et al. 1974; Erofeev and Solomatnikova 1974).

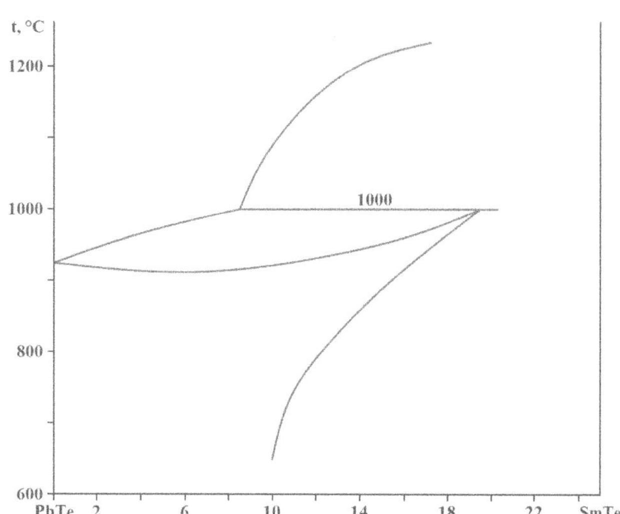

FIGURE 12.23 Part of the phase diagram of the PbTe–EuTe system. (From Klanichka, V.M., et al., *Fiz. khim. tv. tila*, **3**(1), 93, 2002.)

12.18 Lead–Europium–Tellurium

PbTe–EuTe: The $Pb_{1-x}Eu_xTe$ solid solution exists over all the composition $0 \le x \le 1$ (Ravot 1990). According to the data of Erofeev and Solomatnikova (1974), the solubility of EuTe in PbTe reaches 25 mol%, and only in the thin-film state a continuous solid solution is formed (Suryanarayanan and Paparoditis 1968). The variation of the lattice parameter is linear in the concentration range $0 \le x \le 0.75$ with a slope higher than the Vergard's law prediction (Ravot 1990). At higher Eu concentrations the x dependence of the lattice parameter becomes weaker.

All samples of $Pb_{1-x}Eu_xTe$ were synthesized by direct reaction of the required amounts of Pb, Eu, and Te in arc-welded Mo or Ta crucibles under an Ar atmosphere (Ravot 1990). Owing to the high reactivity of Eu and its compounds all manipulations took place in a glove box under a continuously purified Ar atmosphere. Thermal treatment consisted of rapid heating to 950°C, a step at this temperature for 7 days followed by a rapid cooling (usually furnace inertia controlled).

PbTe–Eu$_2$Te$_3$: The phase diagram of this system, constructed through DTA, XRD, and metallography, is a eutectic type (Nasibov et al. 1987b). The eutectic contains 9.8 mol% Nd_2Te_3 and crystallizes at 750°C. The ingots for the investigations were annealed at 450°C–500°C for 250–300 h.

12.19 Lead–Gadolinium–Tellurium

PbTe–GdTe: The solubility of GdTe in PbTe is ~1.2 mol% (Shvangiradze and Kutsiya 1999).

PbTe–Gd$_2$Te$_3$: The phase diagram of this system, constructed through DTA, XRD, metallography, and measuring of the microhardness, is shown in Figure 12.24 (Nasibov et al. 1985a; Valiev et al. 1989). The eutectics crystallize at 950°C and 750°C. The solubility of Gd_2Te_3 in PbTe is 6 mol% at 830°C and decreases to 4 mol% at room temperature. Two compounds, **PbGd$_2$Te$_4$** and **Pb$_4$Gd$_2$Te$_7$**, are formed in the system. First of them melts incongruently at 1100°C, has an energy gap of 0.46 eV and a small homogeneity region. Second compound also melts incongruently at 850°C and has an energy gap of 0.58 eV. Both compounds are stable when stored for a long time in dry air, but they hydrolyze in water and in humid air.

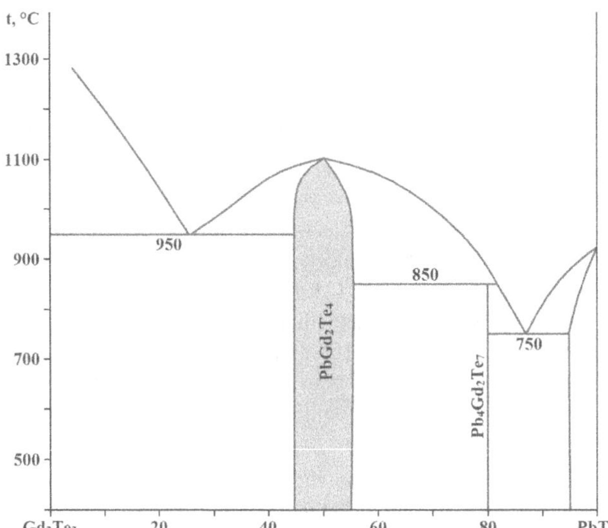

FIGURE 12.24 Phase diagram of the PbTe–Gd_2Te_3 system. (From Valiev, V.K., et al., *Zhurn. neorgan. khimii*, **34**(4), 979, 1989.)

The ingots for the investigations were annealed at 600°C–650°C for 500 h (Valiev et al. 1989) [at 500°C for 600 h (Nasibov et al. 1985a)].

12.20 Lead–Terbium–Tellurium

PbTe–TbTe: The solubility of TbTe in PbTe is ~1.2 mol% (Shvangiradze and Kutsiya 1999).

PbTe–Tb$_2$Te$_3$: The phase diagram of this system, constructed through DTA, XRD, metallography, and measuring of the microhardness and density, is similar to the phase diagram of the PbTe–Gd_2Te_3 system (Nasibov et al. 1987c). The eutectics crystallize at 950°C and 820°C. The solubility of Tb_2Te_3 in PbTe is 5 mol% at room temperature and 10 mol% at 940°C (Nasibov et al. 1985b). Two compounds, **PbTb$_2$Te$_4$** and **Pb$_4$Tb$_2$Te$_7$**, are formed in the system. First of them melts incongruently at 1125°C, has an energy gap of 0.49 eV and a small homogeneity region. Second compound also melts incongruently at 873°C and has an energy gap of 0.58 eV. Both compounds are stable when stored for a long time in dry air, but they hydrolyze in water and in humid air.

The ingots for the investigations were annealed at 650°C–700°C for 500 h (Nasibov et al. 1987c) [at 550°C for 550 h (Nasibov et al. 1985b)].

12.21 Lead–Dysprosium–Tellurium

PbTe–DyTe: The solubility of DyTe in PbTe is ~1.2 mol% (Shvangiradze and Kutsiya 1999).

PbTe–Dy$_2$Te$_3$: The phase diagram of this system was constructed through DTA, XRD, metallography, and microhardness and density measurements by Nasibov et al. (1990). Eutectics crystallize at 760°C and 967°C. The solubility of Dy_2Te_3 in PbTe reaches 5 mol%. Two ternary compounds are formed in the system: **Dy$_2$PbTe$_4$** and **Dy$_2$Pb$_4$Te$_7$**. The former melts congruently at 1175°C and the latter decomposes peritectically at 820°C. The alloys for research were annealed at 650°C for 450 h.

12.22 Lead–Holmium–Tellurium

PbTe–HoTe: The solubility of HoTe in PbTe is ~1.2 mol% (Shvangiradze and Kutsiya 1999).

PbTe–Ho$_2$Te$_3$: No ternary compounds were found in this system (Gulay and Olekseyuk 2006). The solubility of Ho_2Te_3 in PbTe at 600°C is 7 mol%. The ingots for the investigations were annealed at 600°C for 240 h with the next quenching in cold water.

12.23 Lead–Erbium–Tellurium

PbTe–Er$_2$Te$_3$: No ternary compounds were found in this system (Gulay and Olekseyuk 2006). The solubility of Er_2Te_3 in PbTe at 600°C is 4 mol%. The ingots for the investigations were annealed at 600°C for 240 h with the next quenching in cold water.

12.24 Lead–Thulium–Tellurium

PbTe–TmTe: The phase diagram of this system was constructed through DTA, XRD, metallography, and microhardness and density measurements by Rustamov et al. (1999). The solubility of TmTe in PbTe is approximately 17 mol% at 860°C and decreases to ≈10 mol% at room temperature [is ~1.2 mol% (Shvangiradze and Kutsiya 1999)] and the solubility of PbTe in TmTe is less than 1 mol%. Peritectic transformation takes place in the system at 957°C. The ingots for the investigations were annealed at 780°C for 400 h.

PbTe–Tm$_2$Te$_3$: The part of the phase diagram of this system, constructed through DTA and XRD, is given in Figure 12.25 (Dorofeev et al. 2005). The solubility of Tm_2Te_3 in PbTe at 600°C is 6 mol% and 8 mol% at 800°C (Dorofeev et al. 1999, 2005; Gulay and Olekseyuk 2006). The lattice parameter of the solid solution decreases with an increase in the content of Tm_2Te_3 (Dorofeev et al. 1999). The distribution coefficient of thulium in PbTe is >1. PbTe dissolves 1.5 at% Tm at room temperature (Dorofeev et al. 2005). No ternary compounds were found in this system.

The quasibinary sections of the Pb–TmTe–Te subsystem at 850°C are PbTe–Tm_2Te_3 and Tm_2Te_3–Pb, while PbTe–Tm_4Te_5, PbTe–TmTe, Pb–Tm_4Te_5, and Pb–TmTe sections are non-quasibinary (Dorofeev et al. 2005).

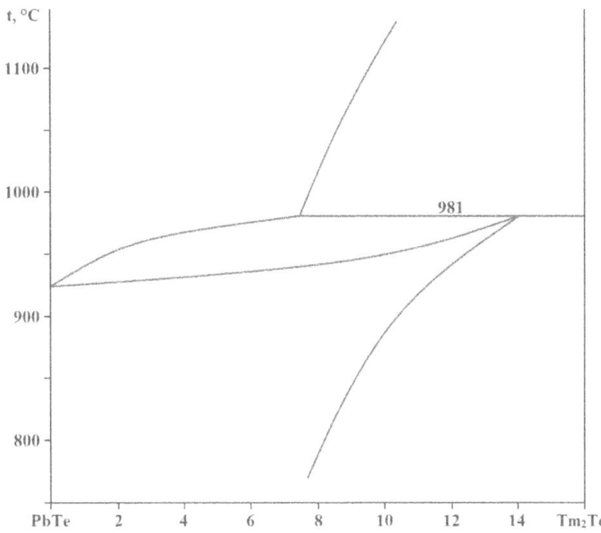

FIGURE 12.25 Part of the phase diagram of the PbTe–Tm_2Te_3 system. (From Dorofeev, S.G., et al., *Inorg. Mater.*, **41**(6), 570, 2005.)

12.25 Lead–Ytterbium–Tellurium

PbTe–YbTe: The results of the investigation of this system are contradictory. According to the data of Aliev et al. (2006a) and Aliyev et al. (2013a), the phase diagram of this system, constructed through DTA, XRD, metallography, and measuring of the microhardness and density, is a eutectic type. The eutectics contain 35 and 60 mol% PbTe and crystallize at 927°C and 777°C, respectively. The solubility of YbTe in PbTe is 35 mol% [33 mol% (Erofeev et al. 1974) and practically does not depend from temperature (Aliyev et al. 2013a)]. The **YbPbTe$_2$** ternary compounds, which melts congruently at 1077°C and crystallizes in the orthorhombic structure with the lattice parameters a = 476, b = 964, and c = 1128 pm, is formed in this system (Aliev et al. 2006a; Aliyev et al. 2013a). The ingots for the investigations were annealed at 730°C for 1 week.

According to the data of Ibadova et al. (2013b) and Aliev et al. (2017), the phase diagram of the PbTe–YbTe system, constructed through DTA, XRD, SEM, and EMF measurements using concentration cells, is given in Figure 12.26. $YbPbTe_2$ was not found in the system. Above 730°C, the phase diagram belongs to the type I within the Roseboom's classification. The lattice parameter of the solid solution varies continuously with a slight positive deviation from Vegard's law. High-temperature solid solution undergoes spinodal decomposition below ca. 780°C. At room temperature PbTe dissolves ca. 35 mol% YbTe and YbTe dissolves ca. 5 mol% PbTe.

Thoroughly ground mixtures of PbTe and YbTe of various compositions were heated to ~1030°C and held at this temperature for 5–6 h (Aliev et al. 2017). Next, they were ground into powder, which was then thoroughly mixed and pressed into pellets. The pellets were annealed for 600 h at 830°C (the samples containing more than 70 mol% SnTe were annealed at 780°C).

The analytical expressions for the free energy of mixing of the $Pb_xYb_{1-x}Te$ solid solutions in the temperature range 330°C–825°C have been derived and the location of their phase boundaries has been determined by Mamedov et al. (2016).

PbTe–YbPb: The phase diagram of this system, constructed through DTA, XRD, metallography, and measuring of the microhardness and density, is a eutectic type (Figure 12.27) (Aliev et al. 2006a; Bakhshalieva et al. 2008). The eutectic contains 50 mol% PbTe and crystallizes at 687°C. The solubility of PbTe in YbTe and YbTe in PbTe at room temperature is correspondingly 5 and 15 mol%. The ingots for the investigations were annealed at 830°C for 125 h.

PbTe–YbPb$_3$: The phase diagram of this system, constructed through DTA, XRD, metallography, and measuring of the microhardness and density, is a eutectic type (Aliev et al. 2006a). The eutectic contains 40 mol% PbTe and crystallizes at 527°C.

According to the data of Bakhshalieva et al. (2008), the phase diagram of the **Yb$_5$Pb$_3$–YbTe** system, constructed through DTA, XRD, and metallography, is a eutectic type. The eutectic contains 30 mol% YbTe and crystallizes at 922°C. It should be noted, that Yb_5Pb_3 melts incongruently (Massalski

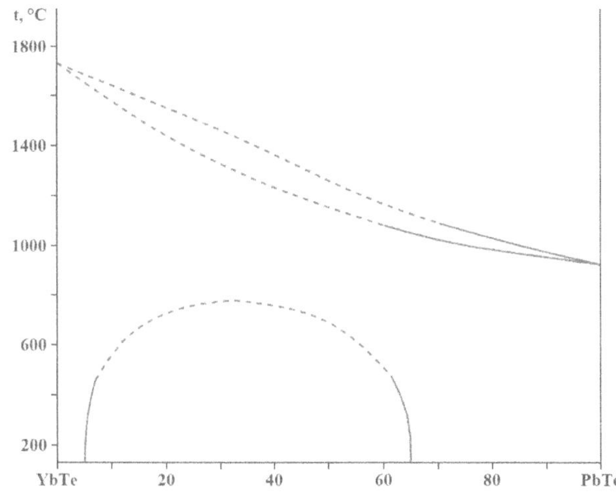

FIGURE 12.26 Phase diagram of the PbTe–YbTe system. (From Aliev, Z.S., et al., *Inorg. Mater.*, **53**(8), 797, 2017.)

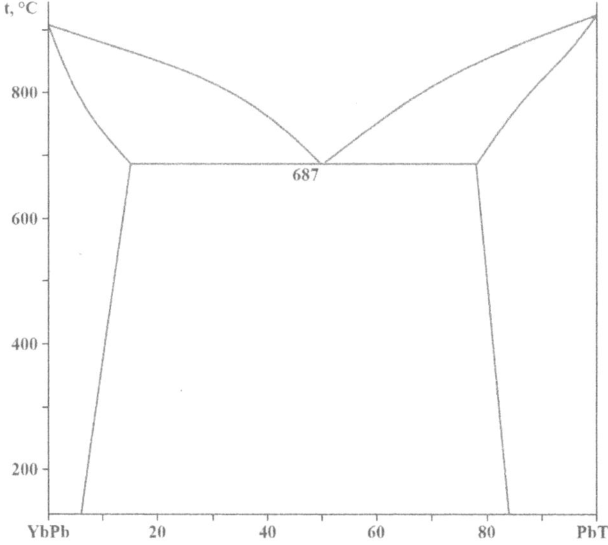

FIGURE 12.27 Phase diagram of the PbTe–YbPb system. (From Bakhshalieva, E.A., et al., *Azerb. khim. zhurn.*, (1), 80, 2008.)

1990). Therefore, this system cannot be a quasibinary section of the Pb–Yb–Te ternary system.

The Yb_5Pb_3–$YbPbTe_2$, YbPb–$YbPbTe_2$ and $YbPbTe_2$–Te sections of the Pb–Yb–Te ternary system are quasibinary and PbTe–Yb and $YbPbTe_2$–Pb are non-quasibinary sections (Aliev et al. 2006a; Aliyev et al. 2013b). On the liquidus surface of the Pb–Yb–Te ternary system, there are the fields of the Pb, Yb, Te, Yb_2Pb, YbPb, Yb_5Pb_3, $YbPb_3$, YbTe, PbTe, and $YbPbTe_2$ primary crystallization (Aliev et al. 2006a). Eight ternary eutectics and one transition point exist in this ternary system: E_1 (627°C) – L ⇔ Yb + Yb_2Pb + YbTe; E_2 (577°C) – L ⇔ Yb_5Pb_3 + YbTe + $YbPbTe_2$; E_3 (627°C) – L ⇔ Yb_5Pb_3 + YbPb + $YbPbTe_2$; E_4 (677°C) – L ⇔ YbPb + $YbPbTe_2$ + PbTe; E_5 (477°C) – L ⇔ YbPb + PbTe + $YbPb_3$; E_6 (202°C) – L ⇔ Pb + $YbPb_3$ + PbTe; E_7 (312°C) – L ⇔ PbTe + Te + $YbPbTe_2$; E_8 (297°C) – L ⇔ $YbPbTe_2$ + Te + YbTe; and U (560°C) L + Yb_5Pb_3 ⇔ Yb_2Pb + YbTe.

12.26 Lead–Lutetium–Tellurium

PbTe–LuTe: The solubility of LuTe in PbTe is ~1.2 mol% (Shvangiradze and Kutsiya 1999).

12.27 Lead–Uranium–Tellurium

PbTe–UTe: The phase diagram of this system, constructed through DTA, XRD, and metallography, is a eutectic type (Terehov et al. 1972). The eutectic crystallizes at 900°C. The solubility of UTe in PbTe changes insignificantly with temperature and at 500°C and 800°C is 5.2 and 5.8 mol%, respectively. In the region of solid solutions, the unit cell parameter does not change depending on the composition. The ingots for the investigations were annealed at 800°C, 650°C, and 500°C for 3, 7, and 96 h, respectively.

12.28 Lead–Titanium–Tellurium

PbTe–Ti: This system is a non-quasibinary section of the Pb–Ti–Te ternary system (Bates et al. 1968). At the interaction of PbTe and Ti, TiTe and Pb were detected. The system was investigated through XRD, metallography, and measuring of the microhardness. The ingots for the investigations were annealed at 1100°C for 14 h.

12.29 Lead–Zirconium–Tellurium

PbTe–Zr: This system is a non-quasibinary section of the Pb–Zr–Te ternary system (Bates et al. 1968). The phase composition of the interaction products has not been established. The system was investigated through XRD, metallography, and measuring of the microhardness. The ingots for the investigations were annealed at 1100°C for 21 h.

12.30 Lead–Phosphorus–Tellurium

The $\mathbf{Pb_2P_2Te_6}$ ternary compound, which melts at 406°C and has an experimental density of 6.73 g·cm^{-3}, is formed in the Pb–P–Te system (Voroshilov et al. 1986). This compound was synthesized from the elemental components.

12.31 Lead–Arsenic–Tellurium

PbTe–As_2Te_3: The phase diagram of this system, constructed through DTA and XRD, is a eutectic type (Figure 12.28) (Koudelka and Frumar 1972; Gaümann 1974). The eutectic contains 10 mol% PbTe and crystallizes at 350°C (Koudelka and Frumar 1972) [5.3 mass% PbTe and 359°C (Gaümann 1974)]. The solubility of PbTe in As_2Te_3 at 290°C is not higher than 2 mol% (Koudelka and Frumar 1972) [2–3 mass% at the eutectic temperature (Gaümann 1974)], and the solubility of As_2Te_3 in PbTe reaches 0.5 mol% (Koudelka and Frumar 1972)

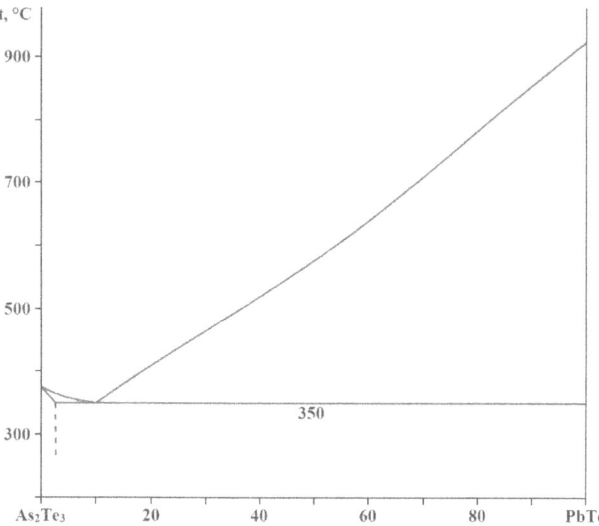

FIGURE 12.28 Phase diagram of the PbTe–As_2Te_3 system. (From Koudelka, L., and Frumar, M., *J. Therm. Anal.*, **4**(4), 471, 1972.)

[8 mass% at the eutectic temperature (Gaümann 1974)]. The ingots for the investigations were annealed at 290°C for 500 h (Koudelka and Frumar 1972).

The glass forming region in the Pb–As–Te ternary system was determined by Minaev (1983). Synthesis of the glassy alloys was carried out from the elements in quartz ampoules under vacuum at 1000°C for 12 h with vigorous stirring and the next quenching in water. The vitreous state was identified by a characteristic glass wedge, a shell fracture, as well as through XRD, DTA, and metallography. The boundary of the glass formation region was fixed with an accuracy of 2 at%.

12.32 Lead–Antimony–Tellurium

PbTe–Sb: The phase diagram of this system, constructed through DTA, XRD, metallography, and measuring of the microhardness, is a eutectic type (Figure 12.29) (Henger and Peretti 1965a; Abrikosov et al. 1969b). The eutectic contains 94.5 mol% Sb and crystallizes at 605°C (Abrikosov et al. 1969b) [85.9 mass% Sb and 601°C (Henger and Peretti 1965a; crystallizes at 243°C–250°C (Raevski 1979)]. The solubility of Sb in PbTe at 550°C is 1.5 mol% (Abrikosov et al. 1969b) [less than 0.1 mass% at 590°C (Henger and Peretti 1965a)], and the solubility of PbTe in Sb at 450°C is not higher than 0.5 mol% (Abrikosov et al. 1969b) [less than 0.5 mass% at 590°C (Henger and Peretti 1965a)]. According to the data of Raevski (1979), the solubility of Sb in PbTe, saturated with Pb and Te, reaches its maximum value (2×10^{19} cm^{-3}) in the region of 650°C, and the solubility of Sb in the presence of Te significantly increases. This indicates that the concentration of soluble Sb is directly related to the electron-hole equilibrium existing in the semiconductor.

The thermodynamic functions of the quasi-binary sections PbTe–Sb were optimized and the phase equilibria calculated by Schlieper et al. (1998).

The ingots for the investigations were annealed at 450°C for 1 month and then at 550°C for 2 months (Abrikosov et al. 1969b) [at 590°C (Henger and Peretti 1965a)].

PbTe–SbTe: This system is a non-quasibinary section of the Pb–Sb–Te ternary system as SbTe melts incongruently (Abrikosov et al. 1969b). The section intersects the fields of the primary crystallization of the solid solutions based on PbTe and Sb_2Te_3. In the middle part of the section, a binary eutectic crystallizes, and at 570°C there is a transition point. The solubility of SbTe in PbTe is not higher than 2 mol% and the solubility of PbTe in SbTe reaches 20 mol%. This system was investigated using DTA and metallography and the ingots for the investigations were annealed first at 450°C for one month. Then, the ingots enriched by SbTe were annealed at 500°C for 1.5 months and the ingots enriched by PbTe at 580°C for 2 months.

PbTe–Sb_2Te_3: The phase diagram of this system, constructed through DTA, XRD, metallography, and measuring of the microhardness, is a eutectic type (Figure 12.30) (Hirai et al. 1967; Reynolds 1967). The eutectic contains 40 mol% PbTe and crystallizes at 590°C (Hirai et al. 1967) [39 mol% PbTe and 582°C (Abrikosov et al. 1965c); at 588°C (Reynolds 1967)]. The solubility of Sb_2Te_3 in PbTe at 550°C reaches

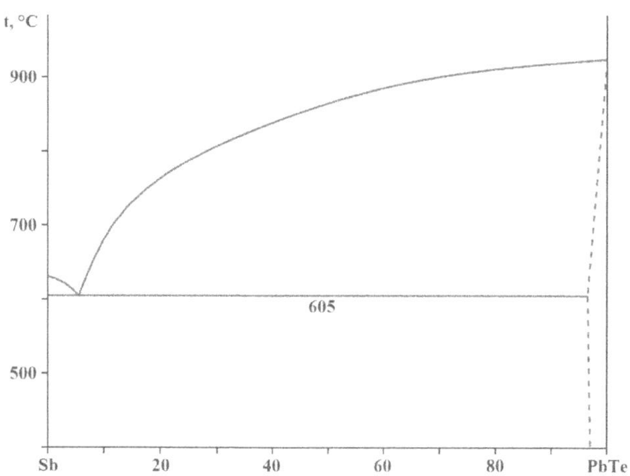

FIGURE 12.29 Phase diagram of the PbTe–Sb system. (From Abrikosov, N.H., et al., *Izv. AN SSSR. Neorgan. mater.*, **5**(4), 741, 1969.)

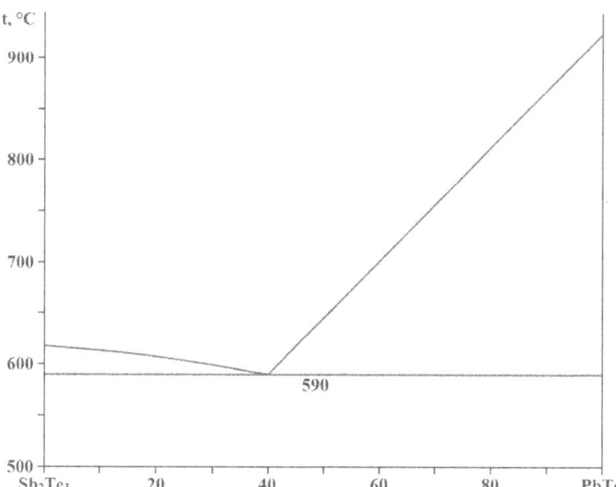

FIGURE 12.30 Phase diagram of the PbTe–Sb_2Te_3 system. (From Hirai, T., et al., *J. Less-Common Metals*, **13**(3), 352, 1967.)

3 mol% [4.5 mol% (Rogacheva and Laptev 1984)] and the solubility of PbTe is within the interval 2–5 mol% and at the eutectic temperature reaches 7 mol% (Abrikosov et al. 1965c). The ingots for the investigations were annealed at 550°C for 600 h (Abrikosov et al. 1965c).

According to the data of Abrikosov et al. (1965c), the $\mathbf{Pb_2Sb_6Te_{11}}$ ternary compound, which melts incongruently at 587°C, is formed in the $PbTe–Sb_2Te_3$ system. The existence of this compound was not confirmed by the next investigations (Hirai et al. 1967; Reynolds 1967; Schlieper et al. 1998; Buyanov et al. 1999).

The activity of PbTe in the $PbTe–Sb_2Te_3$ system was calculated using the results of the electrochemical measurements: a_{PbTe} = 0.735, 0.700, 0.669, and 0.642 at 300°C, 330°C, 360°C, and 390°C, respectively (Buyanov et al. 1999).

The thermodynamic functions of the quasi-binary sections $PbTe–Sb_2Te_3$ were optimized and the phase equilibria calculated by Schlieper et al. (1998).

The $PbTe–Sb_2Te_3$ system contains a homologous series of $\boldsymbol{n}\mathbf{PbTe{\cdot}}\boldsymbol{m}\mathbf{Sb_2Te_3}$ layered compounds (Shelimova et al. 2004b, 2006). Two of them, $\mathbf{PbSb_2Te_4}$ and $\mathbf{PbSb_4Te_7}$, crystallize in the trigonal structure with the lattice parameters a = 435.1 ± 0.1, c = 4171.2 ± 0.2, and a = 430.6, c = 2401.7 ± 0.4 pm, respectively. $PbSb_2Te_4$ melts incongruently at 587°C.

A series of novel rock-salt-type $n\text{PbTe}{\cdot}m\text{Sb}_2\text{Te}_3$ nanocrystals (n = 2, 3, 4, 6, 8, and 10; m = 1 and 2) were successfully prepared using a colloidal synthesis route (Soriano et al. 2012a). These materials are stable only on the nanoscale and have no bulk analogues. They behave as metastable homogeneous solid solutions at room temperature and tend to phase separate into PbTe and Sb_2Te_3 at 300°C. Such nanocrystals are distorted with respect to the NaCl-type structure. The syntheses of these nanocrystals were carried out under air-free conditions using a Schlenk line. In a typical synthesis, stoichiometric amounts of $Pb(CH_3COO)_2{\cdot}3H_2O$ and $Sb(CH_3COO)_3$, octadecene (20.0 mL), oleic acid (5.00 mL), and oleylamine (5.00 mL) were placed in a 100 mL Schlenk flask and dried under vacuum at 100°C for 3–4 h to obtain a homogeneous pale yellow solution. The reaction flask was flushed with N_2, and the temperature was subsequently raised to 160°C. At this temperature, an appropriate amount of Te/tri-n-octylphosphine solution was swiftly injected. Upon injection, a dark brown colored mixture resulted. The temperature of the resulting mixture was maintained at 150°C–160°C for 1 min, and the reaction mixture was then rapidly cooled to room temperature using an ice-water bath. The resulting nanocrystals were precipitated by the addition of an ethanol/hexanes solution (20.0 mL; volume ratio 1:1) and an additional 20.0 mL of ethanol followed by centrifugation. A black pellet of $Pb_mSb_{2n}Te_{m+3n}$ nanocrystals was isolated by discarding the supernatant liquid. The resulting nanocrystals were re-suspended in hexanes and re-precipitated with excess ethanol. The purified nanocrystals can be dispersed in nonpolar organic solvents to form stable colloidal suspension. The $Pb_mSb_{2n}Te_{m+3n}$ nanocrystals of different atomic compositions were prepared by varying the molar ratios of the Pb, Sb, and Te precursors.

The region of the solid solutions based on PbTe at 550°C in the Pb–Sb–Te ternary system is elongated in the direction of the $PbTe–Sb_2Te_3$ section (Figure 12.31) (Rogacheva and Laptev 1984).

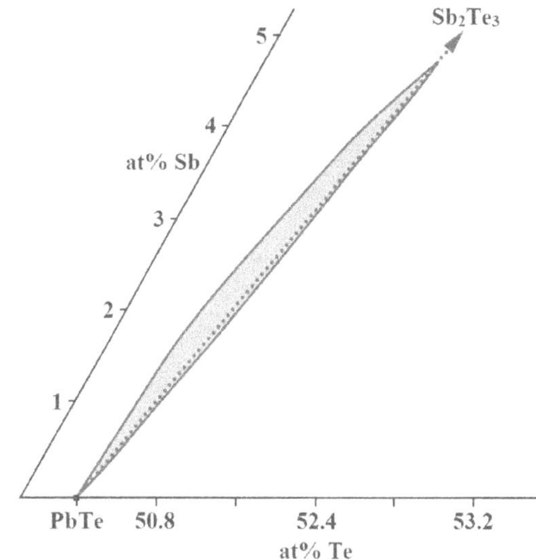

FIGURE 12.31 Region of the solid solutions based on PbTe at 550°C in the Pb–Sb–Te ternary system. (From Rogacheva, E.I., and Laptev S.A., *Izv. AN SSSR. Neorgan. mater.*, **20**(8), 1347, 1984.)

The biggest part of the liquidus surface of the Pb–Sb–Te ternary system is occupied by the field of PbTe primary crystallization (Henger and Peretti 1965b; Abrikosov et al. 1969b). As this liquidus surface contains the field of primary crystallization of the no existing $Pb_2Sb_6Te_{11}$ ternary compound, it needs to be reconstructed.

The excess enthalpies of the Pb–Sb–Te liquid ternary system were determined in a heat-flow calorimeter along the Pb_xSb_{1-x}–Te sections with x = 0.2, 0.5, and 0.8 at 900°C and with x = 0.5 at 800°C by Schlieper et al. (1998). The largest values of the exothermic enthalpies were found on a line stretching from PbTe to Sb_2Te_3. The association model with additional ternary interaction parameters was applied for the analytical description of the ternary melt. In addition, the phase equilibria were determined by DTA and XRD measurements.

The liquidus surface of the Pb–Sb–Te ternary system is characterized by a broad primary crystallization range of PbTe and divided into three subternaries by the $PbTe–Sb_2Te_3$ and PbTe–Sb quasibinary systems (Schlieper et al. 1998). The alloys on the PbTe–Sb quasibinary system have the highest melting points which decrease smoothly from PbTe into the direction of the Pb–Sb and Sb–Te binaries, respectively. The ternary eutectics are found for the $PbTe–Sb_2Te_3$–Te subsystem at the composition 87 at% Pb, 5 at% Sb, and 8 at% Te with a eutectic temperature of 399°C and a degenerated for PbTe–Pb–Sb with a temperature of 254°C, very close to that in the Pb–Sb binary system.

12.33 Lead–Bismuth–Tellurium

PbTe–Bi: The phase diagram of this system, constructed through DTA, XRD, metallography, and measuring of the microhardness, is a eutectic type (Figure 12.32) (Abrikosov et al. 1969c, 1970b; O'Shea et al. 1961). The eutectic contains less than 1 mol% PbTe and crystallizes at 266°C–267°C. The solubility of Bi in PbTe reaches 1 mol% and the solid solution based on Bi was not found (Krebs et al. 1961a).

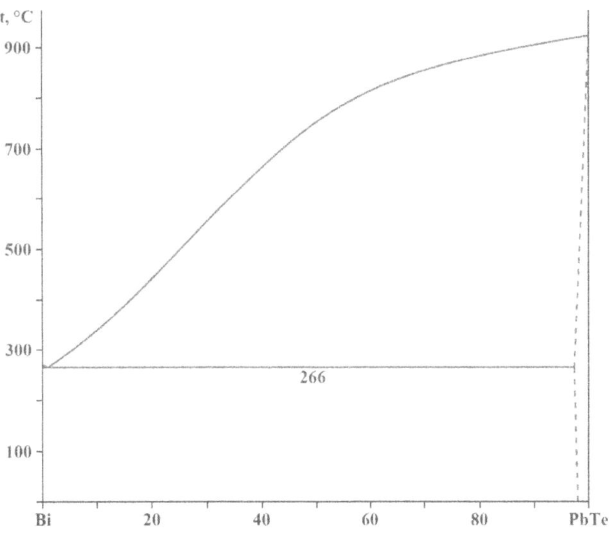

FIGURE 12.32 Phase diagram of the PbTe–Bi system. (From Abrikosov, N.H., et al., *Izv. AN SSSR. Neorgan. mater.*, **5**(10), 1682, 1969.)

PbTe–BiTe: This system is a non-quasibinary section of the Pb–Bi–Te ternary system as BiTe melts incongruently (Abrikosov et al. 1969c, 1970b; Efimova et al. 1968). The section intersects the fields of the primary crystallization of the solid solutions based on PbTe and Bi_2Te_3. The solid solution based on BiTe crystallizes through a transition reaction. The solid solution based on PbTe contains not higher than 1.5 mol% BiTe and the solubility of PbTe in BiTe reaches 30 mol% (Abrikosov et al. 1969c, 1970b). This system was investigated through DTA and metallography and the ingots for the investigations were annealed at 600°C–800°C for 200 h (Abrikosov et al. 1969c, 1970b; Efimova et al. 1968).

PbTe–Bi_2Te_3: The phase diagram of this system, constructed through DTA, XRD, metallography, and measuring of the microhardness, is given in Figure 12.33 (Golovanova et al. 1983; Liu and Chang 1994a; Karpinskii et al. 2002; Shelimova et al. 2004a). The eutectics contain 65 and 91 mol% Bi_2Te_3 and

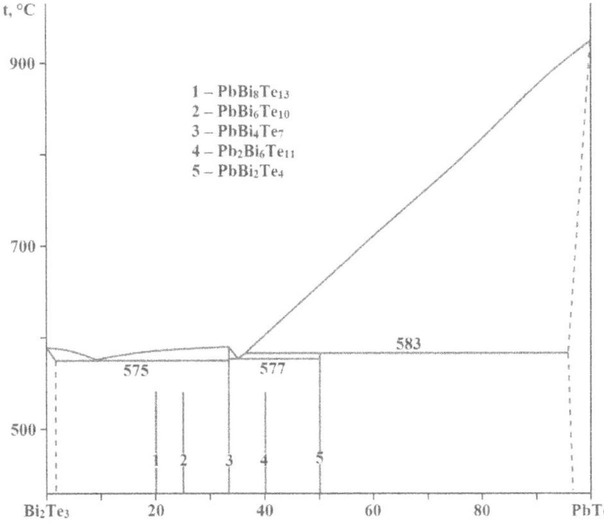

FIGURE 12.33 Phase diagram of the PbTe–Bi_2Te_3 system. (From Shelimova, L.E., et al., *Inorg. Mater.*, **40**(5), 451, 2004.)

crystallize respectively at 577°C and 575°C. The solubility of PbTe in Bi_2Te_3 reaches 5 mol% at 564°C (Hirai et al. 1967) [solid solution based on Bi_2Te_3 was not found (Chami et al. 1983a; Dumas et al. 1985)], and the solubility of Bi_2Te_3 in PbTe is 5 mol% at 550°C (Rogacheva et al. 1984) and 12 mol% at 583°C (Chami et al. 1983a; Dumas et al. 1985) [is not higher than 5 mol% (Efimova et al. 1968)].

The ingots for the investigations were annealed at 500°C for 30–60 days, and in the region 0.5–33 mol% Bi_2Te_3, additionally at 700°C for 14 days (Golovanova et al. 1983) [at 500°C for 1 month (Chami et al. 1983a; Dumas et al. 1985)].

According to the some literature data, the **$PbBi_4Te_7$** ternary compound melts incongruently. Therefore, only one eutectic, which contains 19 mol% (9 mass%) PbTe and crystallizes at 566°C (Elagina and Abrikosov 1959a,b) [10 mol% PbTe and 564°C (Hirai et al. 1967); 18 mol% PbTe and 572°C (Chami et al. 1983a; Dumas et al. 1985)], can exist in the PbTe–Bi_2Te_3 system.

In this system, several Bi_2Te_3-rich ternary compounds with the ***n*PbTe·*m*Bi_2Te_3** ($n/m < 1$) general formula are formed (Karpinskii et al. 2002; Shelimova et al. 2004a, 2006). The compounds with $n/m > 1$ can only be obtained in thin films.

$PbBi_2Te_4$ melts incongruently at 585°C (Liu and Chang 1994a; Shelimova et al. 2004a) and is characterized by the enthalpies of the formation and melting of –150 and 154 kJ·M^{-1}, respectively (Skoropanov et al. 1985a,b). It crystallizes in the trigonal structure with the lattice parameters given in Table 12.1. The standard thermodynamic function and standard entropy of this compound are the next: $\Delta G^0_f = -170.4 \pm 3.9$ kJ·M^{-1}, $\Delta H^0_f = -168.5 \pm 3.2$ kJ·M^{-1}, $S^0_{298} = 385 \pm 14$ J·(K·M)$^{-1}$ (Babanly et al. 2011b) [$\Delta G^{578}_f = -9.37 \pm 0.35$ kJ·M^{-1}, $\Delta H^{578}_f = 8.47 \pm 2.05$ kJ·M^{-1}, $S^{578}_f = 30.87 \pm 3.50$ J·(K·M)$^{-1}$ (Sidorko et al. 1994)].

$PbBi_4Te_7$ melts congruently at 585°C (Golovanova et al. 1983; Shelimova et al. 2004a,b) [at ca. 580°C (Spiridonov et al. 1989); Jambor and Vanko 1991); melts incongruently at 577°C (Elagina and Abrikosov 1959a,b); at 610°C (Datsenko

TABLE 12.1

Crystallographic Data for $PbBi_2Te_4$

			$d_{calc.}$	$d_{meas.}$	
***a*, pm**	***c*, pm**	**α, °**	**g·cm^{-3}**		**References**
445.2	4153.1	–	–	7.92	Zhukova and Zaslavskiy (1970a, 1971, 1976)
442.2	4149	–	8.06	7.739 ± 0.001	[a]Fleischer and Cabri (1978); Zav'yalov and Begizov (1977)
444.1	4180	–	7.93	–	Chami et al. (1983a); Dumas et al. (1985)
1416	–	18.04			
441.80 ± 0.13	4151.3 ± 1.5	–	–	–	[a]Bayliss (1991)
439.7	4143	–	–	–	[a]Kase et al. (1993)
443.9 ± 0.1	4167.7 ± 0.6	–	–	–	Kuznetsova et al. (2000)
443.56 ± 0.06	4177 ± 1	–	–	–	Karpinskii et al. 2002; Shelimova et al. (2004a, 2006)
–	4149 ± 6	–	–	–	Shelimova et al. (2004a)

[a] Mineral rucklidgeite.

TABLE 12.2

Crystallographic Data for $PbBi_4Te_7$

a, pm	*c*, pm	α, °	$d_{calc.}$ g·cm^{-3}	$d_{meas.}$ g·cm^{-3}	References
450	1760	–	–	–	Talybov and Vaynshtein (1961)
442	2360	–	–	–	Petrov and Imamov (1969)
445.1	4153.2	–	–	7.73	Zhukova and Zaslavskiy (1970a,b, 1971)
441.7	24.09	–	7.91	–	Chami et al. (1983a); Dumas et al. (1985)
842.5	–	30.39			
428.5	7708.6	–	–	–	Golovanova et al. (1983)
441.6 ± 0.8	7209 ± 10	–	7.89	7.94 ± 0.3	[a]Spiridonov et al. (1989); Jambor and Vanko (1991)
442.1 ± 0.1	2403.5 ± 0.8	–	–	–	Kuznetsova et al. (2000)
444.5 ± 0.2	4173 ± 5	–	–	–	
441.08 ± 0.07	2407 ± 1	–	–	–	Karpinskii et al. 2002; Shelimova et al. (2004a,b)
442.6	2389.2 ± 0.1				Shelimova et al. (2006)

[a] Mineral kochkarite.

et al. 1981); at 583°C (Chami et al. 1983a; Dumas et al. 1985); at 570°C (Liu and Chang 1994a)] and is characterized by the enthalpies of the formation and melting of –223 and 292 kJ·M^{-1}, respectively (Skoropanov et al. 1985a,b). It crystallizes in the trigonal structure with the lattice parameters given in Table 12.2. According to the data of Talybov (1964), the superstructure-III of this compound crystallizes in the hexagonal structure with the lattice parameters a = 444 ± 2 and c = 10740 ± 30 pm. The standard thermodynamic function and standard entropy of this compound are the next: $\Delta G^0_f = -256.6 \pm 5.6$ kJ·M^{-1}, $\Delta H^0_f = -253.0 \pm 4.0$ kJ·M^{-1}, S^0_{298} = 655 ± 22 J·(K·M)$^{-1}$ (Babanly et al. 2011b) [$\Delta G^{578}_f = -14.350 \pm 0.283$ kJ·M^{-1}, $\Delta H^{578}_f = 12.54 \pm 0.75$ kJ·M^{-1}, $S^{578}_f = 46.53 \pm 1.14$ J·(K·M)$^{-1}$ (Sidorko et al. 1994)].

$PbBi_6Te_{10}$ also crystallizes in the trigonal structure with the lattice parameters a = 440.6 and c = 10254 ± 1 pm (Karpinskii et al. 2002; Shelimova et al. 2004a, 2006). The standard thermodynamic function and standard entropy of this compound are the next: $\Delta G^0_f = -337.5 \pm 7.3$ kJ·M^{-1}, $\Delta H^0_f = -333.0 \pm 4.7$ kJ·M^{-1}, S^0_{298}= 922 ± 30 J·(K·M)$^{-1}$ (Babanly et al. 2011b).

$PbBi_2Te_4$, $PbBi_4Te_7$ and $PbBi_6Te_{10}$ were synthesized using classic solid state techniques from the elements or the binary compounds in stoichiometric ratios at 800°C for 5 h in an evacuated quartz ampoule with annealing at 510°C for 750–1000 h and the next rapid quenching (Kuznetsova et al. 2000; Shelimova et al. 2004a).

$PbBi_8Te_{13}$ and **$Pb_2Bi_6Te_{11}$** also crystallize in the trigonal structure with the lattice parameters a = 439.8, c = 13289 ± 3 pm, and a = 441.9, c = 11337 ± 2 pm [c = 11270 ± 3 pm (Shelimova et al. 2004b)], respectively (Karpinskii et al. 2002; Shelimova et al. 2004a, 2006).

$Pb_2Bi_2Te_5$ is characterized by the enthalpies of the formation and melting of –225 and 185 kJ·M^{-1}, respectively (Skoropanov et al. 1985a,b) and crystallizes in the hexagonal structure with the lattice parameters a = 446 and c = 1750 pm (Petrov and Imamov 1969; Vaynshteyn et al. 1969).

The standard thermodynamic function and standard entropy of $Pb_2Bi_6Te_{11}$ are the next: $\Delta G^0_f = -429.3 \pm 9.5$ kJ·M^{-1}, $\Delta H^0_f = -423.6 \pm 7.2$ kJ·M^{-1}, S^0_{298} = 1041 ± 35 J·(K·M)$^{-1}$ (Babanly et al. 2011b).

$Pb_3Bi_4Te_9$ melts incongruently at 578°C and decomposes at 395°C (Hirai et al. 1967).

The region of the solid solutions based on PbTe at 550°C in the Pb–Bi–Te ternary system is elongated in the direction of the PbTe–Bi_2Te_3 section (Figure 12.34) (Rogacheva et al. 1984). On the liquidus surface of the Pb–Bi–Te ternary system (Figure 12.35), the biggest part is occupied by the field of PbTe primary crystallization (Abrikosov et al. 1969c, 1970b; Donovan et al. 1963).

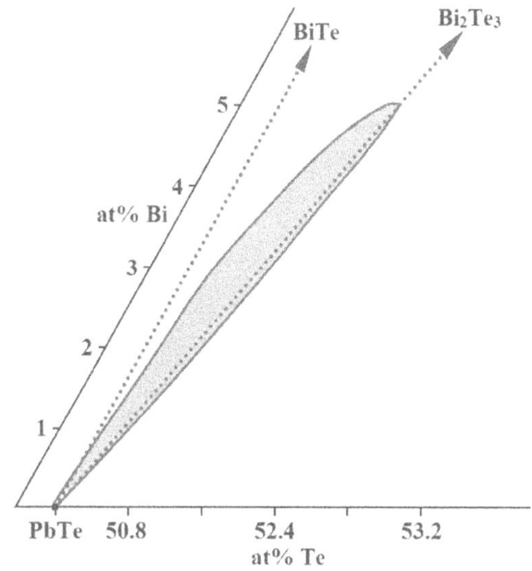

FIGURE 12.34 Region of the solid solutions based on PbTe at 550°C in the Pb–Bi–Te ternary system. (From Rogacheva, E.I., et al., *Izv. AN SSSR. Neorgan. mater.*, **20**(8), 1350, 1984.)

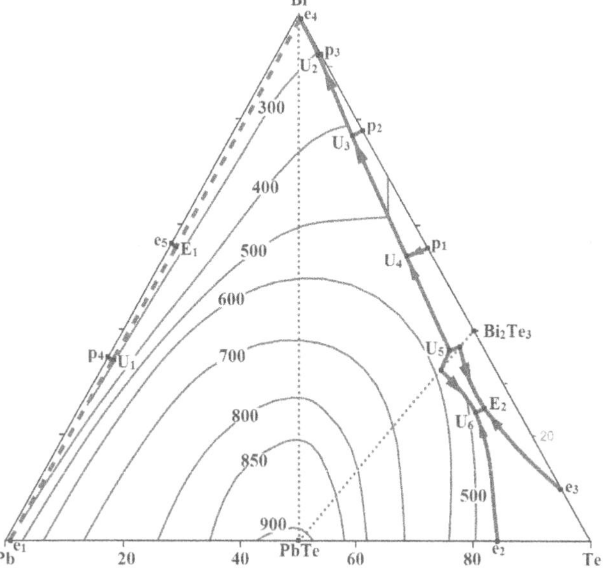

FIGURE 12.35 Liquidus surface of the Pb–Bi–Te ternary system. (From Abrikosov, N.H., et al., *Izv. AN SSSR. Neorgan. mater.*, **5**(10), 1682, 1969.)

The excess enthalpies of the Pb–Bi–Te liquid alloys were measured at 900°C and 800°C in a heat-flow calorimeter (Blachnik et al. 1997b). Five sections Bi_xPb_{1-x}–Te with constant [Bi]/[Pb] concentration ratios were measured. The enthalpy surface is characterized by a valley stretching from exothermic minimum in the system Pb–Te to the exothermic minimum in the system Bi–Te. For the description of the values in the ternary system the equation of Bonnier was taken using additional ternary coefficients. These were obtained by adjusting the curves to the experimental data using the least square method.

12.34 Lead–Vanadium–Tellurium

PbTe-V: This system is a non-quasibinary section of the Pb–V–Te ternary system (Bates et al. 1968). In the samples containing 20 at% V, after annealing at 1100°C for 15 h interaction products were found through XRD, metallography, and measuring the microhardness, the phase composition of which has not been established. The solubility limit of vanadium in PbTe is less than 1 at% (Vinokurov et al. 2006).

PbTe–VTe_2: This system is also a non-quasibinary section of the Pb–V–Te ternary system (Vinokurov et al. 2008). In the composition range 0–15 mol% VTe_2, the primary phase is the VTe_2-based solid solution. At 14–15 mol% VTe_2 and ca. 890°C, this solid solution and high-temperature VTe_2 begin to co-crystallize. At higher VTe_2 contents, the liquidus temperature exceeds 1000°C. The value of the high-temperature heat effect decreases with decreasing vanadium content. The thermal event at ca. 310°C seems to be the phase transformation of VTe_2. Solid solution based on PbTe exists in a narrow composition range (≤1 and 0.5 mol % VTe_2 at 600°C and 530°C, respectively), presumably, with a retrograde solubility.

PbTe–V_3Te_4: This system is also a non-quasibinary section of the Pb–V–Te ternary system (Vinokurov et al. 2006). The solubility limit of V_3Te_4 in PbTe is less than 2 mol%.

12.35 Lead–Niobium–Tellurium

PbTe-Nb: This system is a non-quasibinary section of the Pb-Nb-Te ternary system (Bates et al. 1968). In the samples containing 20 at% Nb, after annealing at 1100°C for 15 h interaction products were found through XRD, metallography, and measuring the microhardness, the phase composition of which has not been established.

12.36 Lead–Tantalum–Tellurium

PbTe–Ta: This system is a non-quasibinary section of the Pb–Ta–Te ternary system (Bates et al. 1968). At the interaction of PbTe and Ta, Pb and tantalum tellurides were detected. The system was investigated through XRD, metallography, and measuring of the microhardness. The ingots for the investigations were annealed at 965°C for 20 h.

TABLE 12.3

Crystallographic Data for Tetragonal $PbTeO_3$

a, pm	*c*, pm	$d_{calc.}$ g·cm^{-3}	$d_{meas.}$ g·cm^{-3}	References
531.5 ± 0.8	596.5 ± 0.8	7.54	7.45	Tananaeva et al. (1977a)
532.3 ± 0.2	1195.0 ± 0.8	–	–	Kosse et al. (1983a,b)
530.4 ± 0.3	1190.0 ± 0.6	7.59	–	Sciau et al. (1986)
531.5	1139.0	7.54	7.49	Gaitán et al. (1987)
532.5 ± 0.6	1195.3 ± 0.8	–	–	Mylian et al. (2011)
531.93 ± 0.06	1193.26 ± 0.15	7.531	–	Weil et al. (2018)

12.37 Lead–Oxygen–Tellurium

PbO–TeO_2: The phase diagram of this system was constructed through DTA, XRD, thermogravimetric analysis, and metallography (Marinov et al. 1974; Robertson et al. 1976; Young 1979; Stavrakieva et al. 1988). Several ternary compounds were found at the investigation of PbO and TeO_2 interaction. **$PbTeO_3$** melts congruently (Marinov et al. 1974) at 566°C (Weil et al. 2018) [at 565°C (Mylian et al. 2011); at 560°C (Tananaeva and Novoselova 1967; Young 1979)], has a homogeneity region (Young 1979) and several polymorphic modifications. It recrystallizes with a great hysteresis at much lower temperatures (Weil et al. 2018). The oxidation of $PbTeO_3$ to **$PbTeO_4$** takes place at 780°C (Spiridonov and Tananayeva 1982). At the heating, the transformation of triclinic modification in tetragonal one is completed at about 450°C. The tetragonal polymorph is stable up to 495°C and then transforms into another modification. This phase remains stable in oven chamber on direct cooling from 525°C back to room temperature (Weil et al. 2018).

First modification of this compound crystallizes in the tetragonal structure with the lattice parameters given in Table 12.3.

Second modification of $PbTeO_3$ crystallizes in the monoclinic structure with the lattice parameters a = 2655.5 ± 0.5, b = 459.3 ± 0.1, c = 1795.8 ± 0.4 pm, β = 106.97 ± 0.03°, and a calculated density of 7.282 g·cm^{-3} (Zavodnik et al. 2008) [a = 2759, b = 461, c = 1797 pm, and β = 112.90° at 52 mol% PbO (Young 1979; Kosse et al. 1983a)].

Third modification crystallizes in the triclinic structure with the lattice parameters a = 701.85 ± 0.03, b = 1061.66 ± 0.04, c = 1196.16 ± 0.05 pm, α = 78.548 ± 0.003°, β = 82.992 ± 0.002°, γ = 84.048 ± 0.002°, and a calculated density of 7.356 g·cm^{-3} (Weil et al. 2018).

Fourth modification crystallizes in the orthorhombic structure with the lattice parameters a = 842.3, b = 1373.9, c = 919.9 pm, and a calculated density of 7.16 g·cm^{-3} (mineral plumbotellurite) (Fleischer et al. 1982; Spiridonov and Tananayeva 1982)

Fifth modification crystallizes in the cubic structure with the lattice parameter a = 879.4 pm and the calculated and experimental densities of 7.51 and 7.39 g·cm^{-3}, respectively (Gaitán et al. 1987).

Polycrystalline samples of $PbTeO_3$ were prepared by precipitation (Weil et al. 2018). A diluted aqueous solution of $Pb(NO_3)_2$ (ca. 25°C, 5 mass%) was added drop wise to a warm aqueous Na_2TeO_3 solution (ca. 60°C, 10 mass%) which resulted in an immediate precipitation of a colorless flocculent material. The solid became crystalline after keeping the solution between 60°C and 70°C under constant stirring for several hours. The as-obtained product was filtered off, washed with mother liquor, water, and ethanol, and dried in a desiccator over $CaCl_2$ for several days. Single crystals of triclinic polymorph were grown by hydrothermal treatment of the precipitated material obtained as described above; 300 mg of the precipitate were loaded together with approximately 5 mL of the mother liquor in a Teflon container (10 mL). The container was sealed with a Teflon lid, placed in a steel autoclave, and heated at 210°C for 1 week at autogenous pressure. The autoclave was then cooled to room temperature over the course of several hours. Colorless to light yellow translucent crystals were isolated from the reaction product. Crystals of tetragonal modification of $PbTeO_3$ appeared as a minority phase after hydrothermal reaction of $Pb(NO_3)_2$ (267 mg), TeO_2 (84 mg), H_2SeO_4 (0.17 mL), and NaOH (64 mg) in a stoichiometric ratio of 1.5:1:2:3.

Tetragonal $PbTeO_3$ was also obtained by the solid state reactions from the mixture of $Pb(CH_3COO)_2.3H_2O$ and $Te(OH)_6$ (molar ratio 1:1) (Gaitán et al. 1987). The mixture was heated in air at 710°C for 24 h. If this mixture was heated in air at 800°C for 72 h, cubic $PbTeO_3$ is formed. Cubic modification can be also obtained at the same conditions if TeO_3 instead of $Te(OH)_6$ was used.

$PbTeO_3$ was also obtained by the heating of $\mathbf{3PbTeO_3.2H_2O}$ in the inert atmosphere at 550°C or by heating a stoichiometric mixture of PbO and TeO_2 in evacuated quartz ampoule with platinum insert or in air in corundum crucible at 500°C–550°C (Tananaeva and Novoselova 1967; Mylian et al. 2011).

Single crystals of tetragonal $PbTeO_3$ were obtained by the method of spontaneous crystallization from a solution in a melt, and monoclinic crystals were obtained by the Czochralski method (Kosse et al. 1983a,b). At 257°C, a reversible phase transition is recorded in the crystals, which is apparently associated with the appearance of a ferroelectric state.

$\mathbf{PbTe_2O_5}$ and $\mathbf{PbTe_4O_9}$ melt incongruently at 480°C and 500°C [at 518°C (Robertson et al. 1976)] (Marinov et al. 1974), respectively. According to the data of Stavrakieva et al. (1988), $PbTe_4O_9$ has a homogeneity region from 20 to 22.2 mol% PbO.

$\mathbf{PbTe_5O_{11}}$ melts incongruently at 520°C and crystallizes in the monoclinic structure with the lattice parameters $a = 1895.9 \pm 0.4$, $b = 441.4 \pm 0.1$, $c = 2579.8 \pm 0.6$ pm, $\beta = 98.12 \pm 0.01°$, and the calculated and experimental densities of 6.34 and 6.30 ± 0.05 g·cm^{-3}, respectively (Oufkir et al. 2001). A mixture of glass and of $PbTe_5O_{11}$ crystals was obtained by heating an intimate mixture of PbO and TeO_2 (molar ratio 1:4) in a sealed gold tube at 730°C for 12 h, and then slowly cooling it (0.1°C·min^{-1}) to 400°C (7 days of annealing at this temperature). The obtained crystals were prismatic.

The formation of $\mathbf{Pb_2TeO_4}$ and $\mathbf{Pb_5TeO_7}$ was established by Stavrakieva et al. (1988). It was assumed that these compounds have polymorphic transformations.

$\mathbf{Pb_2Te_3O_8}$ melts congruently (Marinov et al. 1974; Stavrakieva et al. 1988) at 597°C (Kosse et al. 1983a) [at 595°C (Robertson et al. 1976); at 500°C (Young 1979)]. It crystallizes in the orthorhombic structure with the lattice parameters $a = 1952.2 \pm 0.4$, $b = 712.1 \pm 0.1$, $c = 1881.3 \pm 0.4$ pm, and an experimental density of 7.03 ± 0.03 g·cm^{-3} (Champarnaud-Mesjard et al. 2001) [$a = 1883 \pm 2$, $b = 714 \pm 1$, $c = 1937 \pm 2$ pm, and the calculated and experimental densities of 7.08 and 7.16 g·cm^{-3}, respectively (Dewan et al. 1978); $a = 714$, $b = 1877$ and $c = 1949$ pm (Robertson et al. 1976; Kosse et al. 1983a; in the tetragonal structure with the lattice parameters $a = 513.2$, $c = 1286.2$ pm, and the calculated and experimental densities of 7.82 and 7.6 g·cm^{-3}, respectively (Gaitán et al. 1987)]. This compound has no phase transformation in the temperature region from –150°C to 530°C (Kosse et al. 1983a). It was prepared by the solid state reactions from the mixture of $Pb(CH_3COO)_2{\cdot}3H_2O$ and TeO_3 (molar ratio 1:1) (Gaitán et al. 1987). The mixture was heated in air at 800°C for 24 h. $Pb_2Te_3O_8$ and cubic $PbTeO_3$ were obtained. The attempts to synthesized $Pb_2Te_3O_8$ alone were unsuccessful. Single crystals of this compound were grown by the Czochralski method (Kosse et al. 1983a) or by heating crystalline $Pb_2Te_3O_8$ in a gold crucible at 700°C for 12 h under pure N_2 flowing and then cooling (1°C·h^{-1}) it down to room temperature (Champarnaud-Mesjard et al. 2001).

According to the data of Stavrakieva et al. (1988), the phase $\mathbf{Pb_2Te_5O_{12}}$ corresponds to the phase $PbTe_2O_5$.

$\mathbf{Pb_3TeO_5}$ melts congruently at 690°C and has a polymorphic transition at 665°C (Young 1979). Stavrakieva et al. (1988) also assume that this compound has a polymorphic transformation, but according to the data of Kosse et al. (1983a), it has no phase transformation in the temperature region from –150°C to 530°C. Pb_3TeO_5 crystallizes in the cubic structure with the lattice parameters $a = 1183.0$ pm and the calculated and experimental densities of 6.65 and 6.48 g·cm^{-3}, respectively (Gaitán et al. 1987) or in the orthorhombic structure with the lattice parameters $a = 550$, $b = 660$, and $c = 1150$ pm (Kosse et al. 1983a). Therefore, this compound obviously has a phase transformation. A white crystalline powder of Pb_3TeO_5 was obtained by the solid state reactions from the mixture of $Pb(CH_3COO)_2.3H_2O$ and $Te(OH)_6$ (molar ratio 1:3) (Gaitán et al. 1987). The mixture was heated in air at 800°C for 24 h.

$\mathbf{Pb_5Te_2O_9}$ has a homogeneity region from 61.5 to 75 mol% PbO (Stavrakieva et al. 1988).

$\mathbf{Pb_6TeO_8}$ melts incongruently at 864°C (Robertson et al. 1976).

According to the data of Robertson et al. (1976), the $\mathbf{Pb_8Te_3O_{14}}$ ternary compound also exists in this system.

Transparent glass-ceramic and crystalline samples of $PbO–TeO_2$, containing 15 and 20 mol% PbO were prepared by quenching their melts from different temperatures in the range of 650°C–950°C (Khanna et al. 2021). It is found that the melts solidify into transparent glass-ceramics on cooling from lower temperatures of 650°C–750°C, however when the melts are cooled from higher temperatures in the range of 750°C–950°C, the glass forming ability decreases drastically and anti-glass and/or crystalline phases are produced on solidification. The high-temperature melts freeze faster into crystalline phases as compared to the melts at a lower temperature; the latter supercool and solidify into glassy phases.

$\mathbf{PbTe–PbTeO_3}$**:** The phase diagram of this system, constructed through DTA and XRD, is a eutectic type (Figure 12.36)

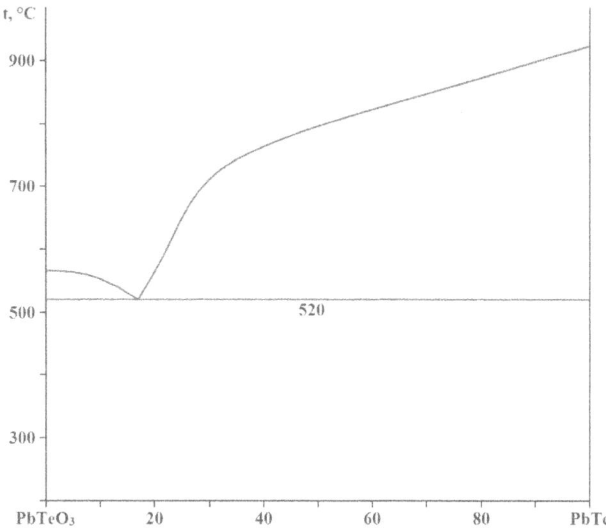

FIGURE 12.36 Phase diagram of the PbTe–$PbTeO_3$ system. (From Tananaeva, O.I., *Izv. AN SSSR. Neorgan. mater.*, **13**(2), 386, 1977.)

(Tananaeva et al. 1977a). The eutectic contains 17 mol% PbTe and crystallizes at 520°C.

PbTe–$PbTeO_4$: Three ternary compounds, **PbO·$PbTeO_4$** or **Pb_2TeO_5**, **2PbO·$PbTeO_4$** or **Pb_3TeO_6** and **4PbO·$PbTeO_4$** or **Pb_5TeO_8**, were found in this system. Pb_2TeO_5 crystallizes in the monoclinic structure with the lattice parameters a = 1308.53 ± 0.06, b = 570.71 ± 0.03, c = 752.25 ± 0.03 pm, and β = 123.772 ± 0.002° (Artner and Weil 2013); [a = 1309.9 ± 0.3, b = 571.4 ± 0.1, c = 752.0 ± 0.2 pm, and β = 123.80 ± 0.03° (Wedel et al. 1998); a = 753.53 ± 0.06, b = 571.42 ± 0.05, c = 1089.81 ± 0.12 pm, β = 91.330 ± 0.006°, and a calculated density of 8.806 g·cm^{-3} for mineral ottoite (Kampf et al. 2010)]. To prepare this compound, a well-ground and pressed mixture of PbO (4.0 mM) and TeO_2 (2.0 mM) was heated in a corundum crucible according to the following temperature program: room temperature → 800°C (6 h) → 600°C (12 h) → 400°C (12 h) → 400°C (2 h) → room temperature (24 h) (Artner and Weil 2013). The sample had a grey-yellowish color, containing few small yellowish single crystals of Pb_2TeO_5. Single crystals of this compound were produced by heating an intimate mixture of PbO, TeO_2, and PbF_2 (molar ratio 2:1:5) (Wedel et al. 1998). PbF_2 largely evaporated within 2 days when heated to 750°C in air. The batch was removed from the furnace without cooling. It contains largely the yellow Pb_2TeO_5 in addition to small portions of white, previously unknown crystals.

Pb_3TeO_6 melts at 870°C (Tananaeva et al. 1977a), has a phase transformation at 217°C and crystallizes in the monoclinic structure with the lattice parameters a = 744, b = 1202, c = 1315 pm, and β = 125.8° (Kosse et al. 1983a). According to the data of Knyazeva et al. (1977), this compound is thermally unstable and is reduced at temperatures above 200°C, releasing oxygen: $2Pb_3TeO_6 \Leftrightarrow 4PbO + 2PbTeO_3 + O_2$. For preparing this compound, Te (1 g) was mixed with PbO (13 g) and Na_2CO_3 (8 g) (Knyazeva et al. 1977). This mixture was thoroughly ground, poured into a porcelain crucible, which was placed in an oven preheated to 800°C, and calcined at this temperature for 1.5 h. The cake was transferred into a beaker and treated with 200 mL of NH_4Cl hot saturated solution. The contents of the glass were boiled for 5–10 min. The insoluble precipitate was separated from the solution, washed with water and dried at 100°C–110°C. Single crystals of this compound were grown by the method of spontaneous crystallization from a solution in a melt (Kosse et al. 1983a).

During systematic phase formation studies of lead oxotellurates (VI) under ambient pressure and at temperatures above 700°C, Artner and Weil (2013) failed to reproduce the preparation of Pb_3TeO_6. Inconsistencies with respect to given crystallographic data and physical properties of this compound were clarified on basis of structure determination from single crystal XRD, revealing the true composition as Pb_5TeO_8.

Pb_5TeO_8 melts incongruently at 828°C [at 875°C (Tananaeva et al. 1977a)] and crystallizes in the monoclinic structure with the lattice parameters a = 744.26 ± 0.02, b = 1201.07 ± 0.03, c = 1065.67 ± 0.02 pm, and β = 91.040 ± 0.001° (Artner and Weil 2013). No first order phase transition was revealed for this compound in the temperature range 30°C–1000°C. Its crystals were obtained from a sample containing a mixture of PbO (6.05 mM) and TeO_3 (1.25 mM), which underwent a heat treated treatment in a corundum crucible under the same condition as given above to obtain Pb_2TeO_5 (Artner and Weil 2013). A phase-pure sample of Pb_2TeO_5 was obtained by heating a well-ground and pressed mixture of PbO (9.4 mM) and TeO_2 (1.8 mM), heated at 750°C for 100 h.

One more ternary compound, $PbTeO_4$, which melts incongruently at 665°C (Tananaeva and Novoselova 1967), is formed in the Pb–O–Te system. It is thermally stable up to 700°C–750°C, when it loses 1/2O_2 giving cubic $PbTeO_3$ (Gaitán et al. 1987). It was prepared by the solid state reactions from the mixture of PbO·$x$$H_2O$ and $Te(OH)_6$ (molar ratio 1:1), which was heated in vacuum at 150°C for 24 h.

Oxidation of PbTe is a complex process that proceeds in different ways depending on temperature (Tananaeva et al. 1969). In the range 400°C–600°C, the oxidation products are two polymorph of $PbTeO_3$, at 600°C–650°C, $PbTeO_4$ is also formed, and above 800°C, an oxide phase of variable composition, containing Te^{4+} and Te^{6+}, were found.

The isothermal section of the Pb–O–Te ternary system at room temperature (Figure 12.37) was calculated taking into account the change of the Gibbs energies of formation with temperature (Berchenko et al. 2014). It predicts, in good agreement with XRD studies, the formation of $PbTeO_3$ at relative low temperatures (<400°C). According to SIMS, the ternary oxides appear at PbTe even at room temperature. At higher temperatures (>400°C), the diagram predicts the formation of other ternary oxides in PbTe sintered material after high-temperature oxidation (Figure 12.38).

12.38 Lead–Chromium–Tellurium

PbTe–Cr: This system is a non-quasibinary section of the Pb–Cr–Te ternary system (Bates et al. 1968; Abilov and Velieva 1990). It intersects two subsystems: PbTe–Pb–Cr_3Te_4 and Pb–Cr_3Te_4–Cr. At room temperature, PbTe dissolves up to 1 at% Cr and the solubility of PbTe in Cr was not found. The system was investigated through DTA, XRD, metallography, and measuring of the microhardness and density. The ingots for the investigations

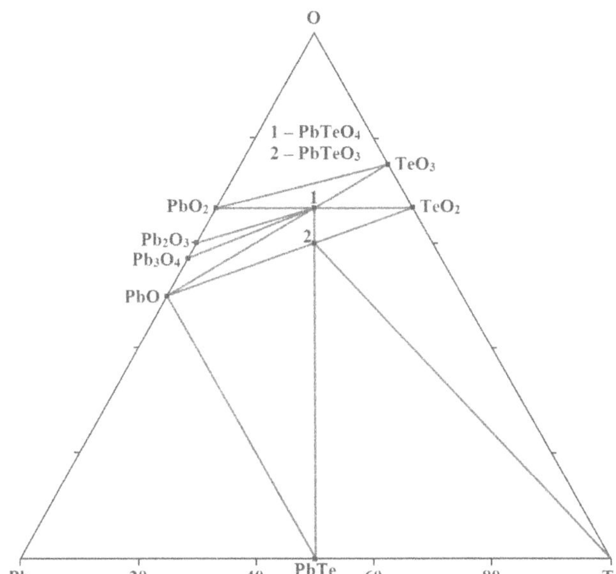

FIGURE 12.37 Isothermal section of the Pb–O–Te ternary system at room temperature. (From Berchenko, N., et al., *Thermochim. Acta*, **579**, 64, 2014.)

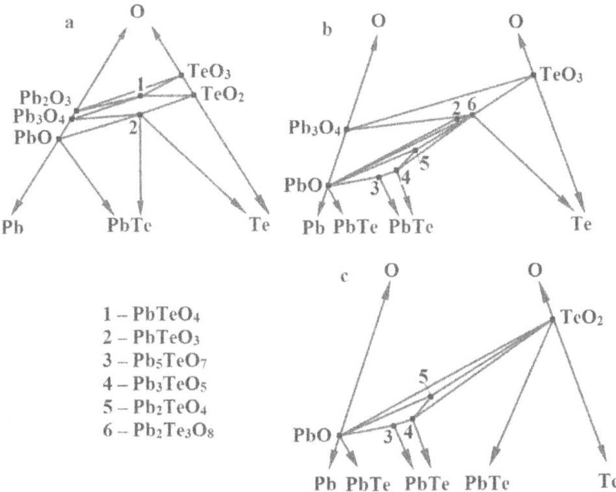

FIGURE 12.38 Central parts of the isothermal sections of the Pb–O–Te ternary system at (a) 400°C, (b) 500°C, and (c) 600°C. (From Berchenko, N., et al., *Thermochim. Acta*, **579**, 64, 2014.)

were annealed at 330°C for 450 h (Abilov and Velieva 1990) [at 900°C for 15 h and then at 600°C for 48 h, at 1100°C for 15 and 72 h and at 1000°C for 88 h (Bates et al. 1968)].

PbTe–CrTe: The solubility of CrTe in PbTe is 1 mol% at 230°C, 1.3 mol% at 500°C, and 2 mol% at 830°C (Chomentowski et al. 1965; Stavrianidis et al. 1986) [0.7 mol % at 800°C (Vulchev et al. 1986)]. This system was investigated through XRD and metallography and the ingots for the investigations were annealed at 830°C, 555°C, and 230°C for correspondingly 200, 250, and 280 h (Stavrianidis et al. 1986) [at 800°C for 250 h (Vulchev et al. 1986); at 500°C for 5 days (Chomentowski et al. 1965).

PbTe–$CrTe_3$: This system is also a non-quasibinary section of the Pb–Cr–Te ternary system as $CrTe_3$ melts incongruently (Abilov and Iskender-zade 1991). PbTe and Cr_5Te_8 crystallize primarily in the system. The solubility of Cr_5Te_8 in PbTe

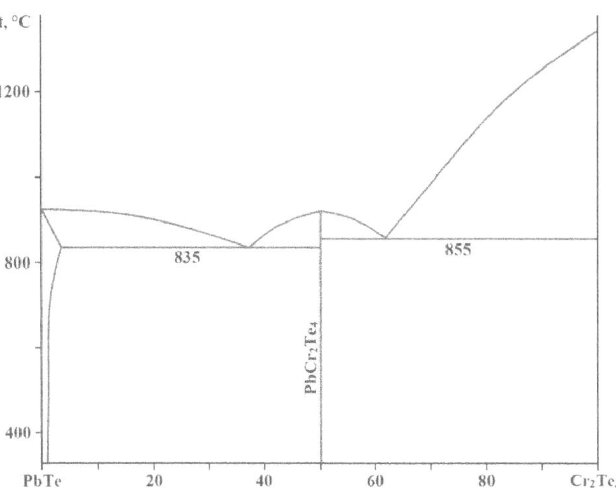

FIGURE 12.39 Phase diagram of the PbTe–Cr_2Te_3 system. (From Rustamov, P.G., et al., *Azerb. khim. zhurn.*, (4), 99, 1985.)

reaches 1.5 mol% at room temperature. The system was investigated through DTA, XRD, metallography, and measuring of the microhardness and density. The ingots for the investigations were annealed at 280°C for 400 h.

PbTe–Cr_2Te_3: The phase diagram of this system, constructed through DTA, XRD, metallography, and measuring of the microhardness and density, is shown in Figure 12.39 (Rustamov et al. 1985a). The **$PbCr_2Te_4$** ternary compound, which melts congruently at 920°C, is formed in this system. The solubility of Cr_2Te_3 in PbTe at room temperature reaches 2 mol% and the solubility of PbTe in Cr_2Te_3 is negligible. The ingots for the investigations were annealed at 750°C for 500 h.

PbTe–Cr_3Te_4: The phase diagram of this system, constructed through DTA, XRD, metallography, and measuring of the microhardness and density, is presented in Figure 12.40 (Abilov and Iskender-zade 1991). The immiscibility region at 877°C exists in the system. The solubility of Cr_3Te_4 in PbTe is ca. 1 mol% at room temperature (Abilov and Iskender-zade 1991) [0.5 mol% at 830°C (Keyyan et al. 1988)]. The ingots

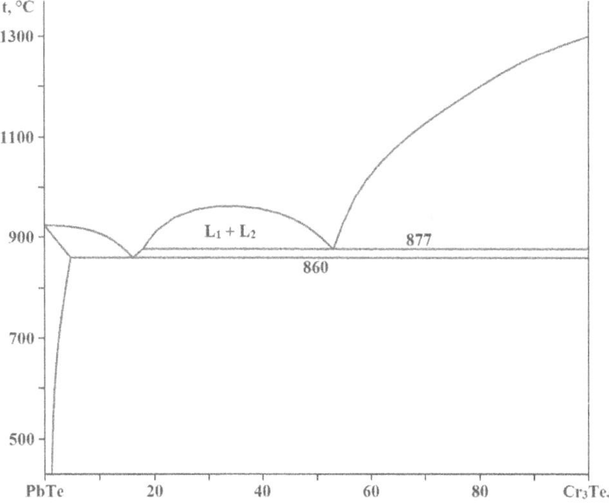

FIGURE 12.40 Phase diagram of the PbTe–Cr_3Te_4 system. (From Abilov, Ch.I., and Iskender-zade, Z.A., *Neorgan. mater.*, **27**(11), 2295, 1991.)

for the investigations were annealed at ca. 730°C for 400 h (Abilov and Iskender-zade 1991) [at 830°C, 555°C, and 230°C for correspondingly 200, 250, and 280 h (Keyyan et al. 1988)].

Another ternary compound, **$Pb_{0.5}Cr_5Te_8$**, which crystallizes in the monoclinic structure with the lattice parameters a = 2027 ± 2, b = 784.5 ± 0.5, c = 945.8 ± 0.6 pm, and β = 104.68 ± 0.04°, is formed in the Pb–Cr–Te system (Novet et al. 1995). To prepare this compound, the ion-exchange reaction was used. $TlCr_5Te_8$ was mixed with PbI_2 at various molar ratios (1:1, 1:2 and 1:4) in a N_2-filled dry box and sealed in evacuated quartz tube approximately 20 cm in length. The tube was placed in a temperature gradient with the sample at the hot end (at temperatures 150°C–450°C) and the cold end of the tube extended out of the furnace assembly. The temperature of the hot end of the ampoule was raised until red crystals of TlI were observed on the cold end of the tube. The samples were annealed for approximately 12 h before quenching to room temperature.

12.39 Lead–Molybdenum–Tellurium

PbTe–Mo: In samples containing 20 and 33.3 at% Mo and annealed at 1100°C for 15 h, only PbTe and Mo were found (Bates et al. 1968). In the case of annealing such mixtures at 970°C for 21 h, the formation of a new phase was detected, which disappears after annealing at 1100°C. It is possible that the new phase is peritectic. This system was investigated through XRD and metallography.

12.40 Lead–Tungsten–Tellurium

PbTe–W: This system is a quasibinary section of the Pb–W–Te ternary system (Bates et al. 1968). Annealing of the samples containing 33.3 at% W at 950°C–1200°C for 15–100 h did not reveal the formation of new phases by XRD.

12.41 Lead–Fluorine–Tellurium

The **$PbTeF_6$** ternary compound, which crystallizes in the monoclinic structure with the lattice parameters a = 462.5 ± 0.1, b = 1281.2 ± 0.2, c = 851.9 ± 0.1 pm, and β = 102.36 ± 0.01° and a calculated density of 6.04 g·cm^{-3}, is formed in the Pb–F–Te system (Ider et al. 1996). This compound was prepared by heating an intimate equimolar mixture of PbF_2 and TeF_4 in a sealed Pt tube for 12 h at 150°C and for 24 h at 200°C with the next water quenching. All the starting products and the compounds were handled and stored in a glove box under a strictly dried and deoxygenated argon atmosphere. $PbTeF_6$ so prepared was obtained as a very hygroscopic white powder. It is stable only below 290°C.

12.42 Lead–Chlorine–Tellurium

PbTe–$PbCl_2$: The phase diagram of this system, constructed through DTA, XRD, and metallography, is a eutectic type (Figure 12.41) (Novoselova et al. 1968, 1969). The eutectic is degenerated from the $PbCl_2$ side and crystallized at 496°C.

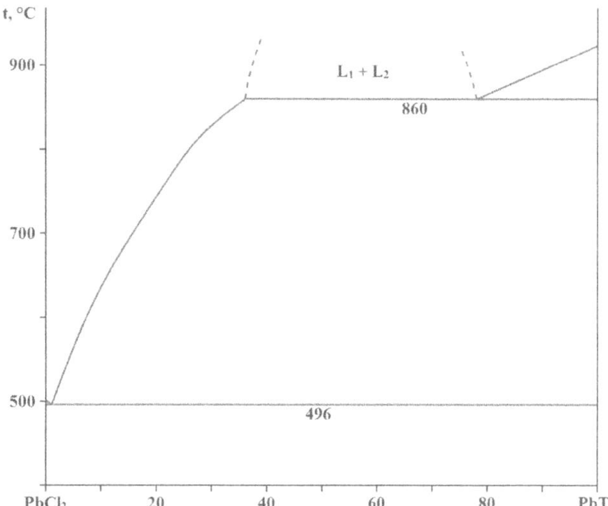

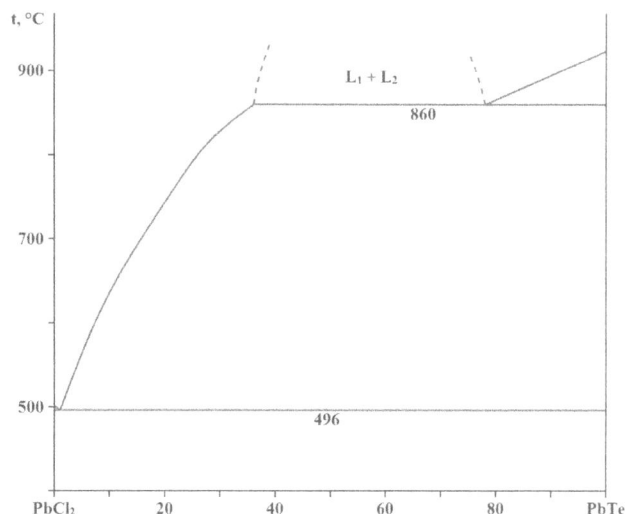

FIGURE 12.41 Phase diagram of the PbTe–$PbCl_2$ system. (From Novoselova, A.V., et al., *Izv. AN SSSR. Neorgan. mater.*, **4**(5), 777 (1968.)

Within the interval 36–78 mol% PbTe there is an immiscibility region at 860°C. The solubility of chlorine in PbTe is 2–3 at% (Ugay et al. 2004b; Sharov 2006). It was found that the calculated density decreases linearly with increasing the chlorine content, and the measured density is non-monotonic (Ugay et al. 2004b). Increasing the chlorine content in the **$PbTe_{1-x}Cl_x$** solid solution increases the lattice parameter fluctuations and atomic displacement from equilibrium positions (Sharov 2008a). For x > 0.02–0.03, the fluctuations are composition-independent. At 600°C, 3.1 mol% PbTe dissolves in the $PbCl_2$ melt (Rodionov et al. 1972).

The ingots for the investigations were annealed at 500°C, 600°C, and 750°C correspondingly for 600, 500, and 350 h (Novoselova et al. 1968). The solubility of PbTe in $PbCl_2$ was determined by isothermal saturation followed by analysis of the salt phase (Rodionov et al. 1972).

The **$PbTeCl_2$** ternary compound, (mineral kolarite), which crystallizes in the orthorhombic structure with the lattice parameters a = 593 ± 5, b = 325 ± 5, c = 389 ± 5 pm, and a calculated density of 9.14 g·cm^{-3}, is formed in the Pb–Cl–Te ternary system (Genkin et al. 1985; Hawthorne et al. 1986).

12.43 Lead–Bromine–Tellurium

PbTe–$PbBr_2$: The phase diagram of this system, constructed through DTA, XRD and metallography, is a eutectic type (Figure 12.42) (Novoselova et al. 1969, 1970b). The eutectic is degenerated from the $PbBr_2$ side and crystallized at 368°C. The solubility of bromine in PbTe is ca. 7 at% (Sharov 2006). The ingots for the investigations were annealed at 340°C for 1000 h (Novoselova et al. 1970b).

12.44 Lead–Iodine–Tellurium

PbTe–I_2: This system is a non-quasibinary section of the Pb–I–Te ternary system which intersects the fields of PbTe, PbI_2 and Te primary crystallization (Malkova and Latypov

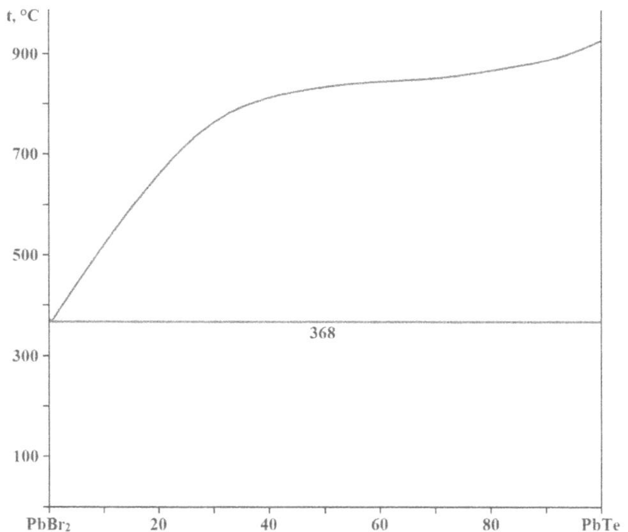

FIGURE 12.42 Phase diagram of the PbTe–$PbBr_2$ system. (From Novoselova, A.V., et al., *Izv. AN SSSR. Neorgan. mater.*, **6**(2), 257, 1970.)

1974). The wide immiscibility region exists in this system. At temperatures below 410°C, joint crystallization of PbTe and Te begins, below 405°C – PbTe, PbI_2 and Te, below 165°C in the range of 15–25 mol% PbTe – PbI_2, TeI_4, and TeI, and in the range of 25–50 mol% PbTe – Te, TeI, and PbI_2, and below 100°C – PbI_2, TeI, and I_2. The system was investigated through DTA, XRD, and metallography.

PbTe–PbI$_2$: The phase diagram of this system, constructed through DTA, XRD, and metallography, is a eutectic type (Figure 12.43) (Berg et al. 1967; Novoselova et al. 1967a, 1969; Odin 2000). The eutectic contains 6 mol% PbTe and crystallizes at 398°C (Novoselova et al. 1967a) [the eutectic is degenerated (Berg et al. 1967)]. The solubility of PbI_2 in PbTe at the eutectic temperature is 0.3 mol% (Novoselova et al. 1967a; Odin et al. 1970b) [the solubility of iodine in PbTe is ca. 9 at% (Sharov 2006); 3 at% (Ugay et al. 2004b)]. The lattice

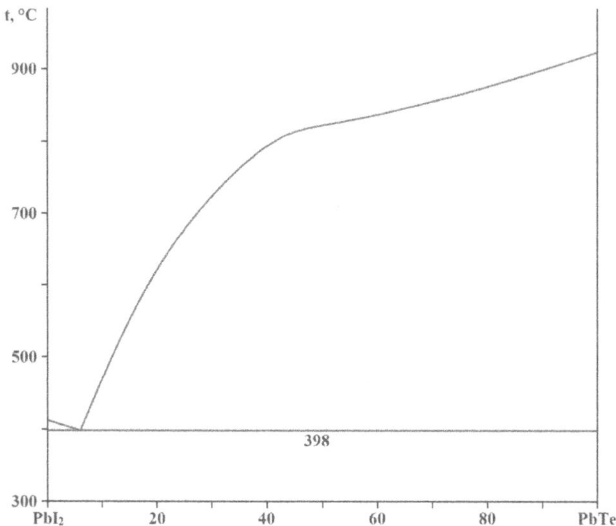

FIGURE 12.43 Phase diagram of the PbTe–PbI_2 system. (From Novoselova, A.V., et al., *Izv. AN SSSR. Neorgan. mater.*, **3**(11), 2101, 1967.)

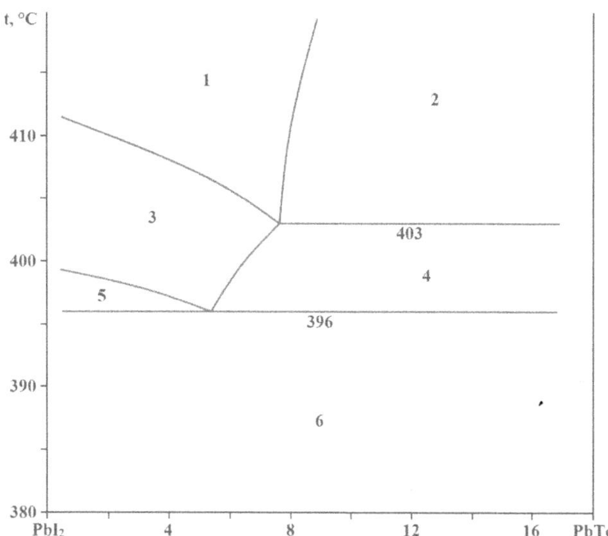

FIGURE 12.44 Part of the phase diagram of the PbTe–PbI_2 system with metastable phases according to the cooling curves: 1, L; 2, L + PbTe; 3, L + 2*H*-PbI_2; 4, L + 2*H*-PbI_2 + PbTe; 5, L + 2*H*-PbI_2 + 6*R*-PbI_2; 6, -PbI_2 + 6*R*-PbI_2 + PbTe. (From Odin, I.N., *Zhurn. neorgan. khimii*, **45**(4), 698, 2000.)

parameter of the **PbTe$_{1-x}$I$_x$** solid solution is shown to be a non-monotonic function, with a sharp minimum, of iodine content (Ugay et al. 2004a). It was found that the calculated density decreases linearly with increasing iodine content, and the measured density is non-monotonic (Ugay et al. 2004b).

According to the heating curves for long-term annealed samples, PbTe and PbI_2 participate in equilibria, the latter being a mixture of two polytypes (2*H* and 6*R*) (Odin 2000). The data on the cooling curves (Figure 12.44) differ sharply. From the PbI_2 side, 2*H*-PbI_2 crystallizes primarily, and at 396°C the secondary crystallization of 6*R*-PbI_2 takes place.

At 600°C, 10.5 mol% PbTe dissolves in the PbI_2 melt (Rodionov et al. 1972).

The ingots for the investigations were annealed at 500°C, 600°C, and 750°C for correspondingly 600, 500, and 350 h (Novoselova et al. 1967a) [at 398°C for 1100 h (Odin et al. 1970b]. The solubility of PbTe in PbI_2 was determined by isothermal saturation followed by analysis of the salt phase (Rodionov et al. 1972).

PbTe–TeI$_4$: The phase diagram of this system, constructed through DTA and XRD, is given in Figure 12.45 (Berg and Malkova 1969). The eutectic contains 3 mol% PbTe and crystallized at 175°C. The **Pb$_3$Te$_5$I$_8$** ternary compound, which melts incongruently at 385°C (the peritectic point contains 40 mol% PbTe), is formed in this system. The ingots for the investigations were annealed at 165°C for 1400 h.

12.45 Lead–Manganese–Tellurium

PbTe–Mn: This system is a non-quasibinary section of the Pb–Mn–Te ternary system (Bates et al. 1968). At the interaction of PbTe and Mn, MnTe, Pb, and **Pb$_x$Mn$_{1-x}$Te** were detected. PbTe–Mn vertical section was constructed by Abilov et al. (1991) and it was shown that it intersects the PbTe–MnTe–Pb

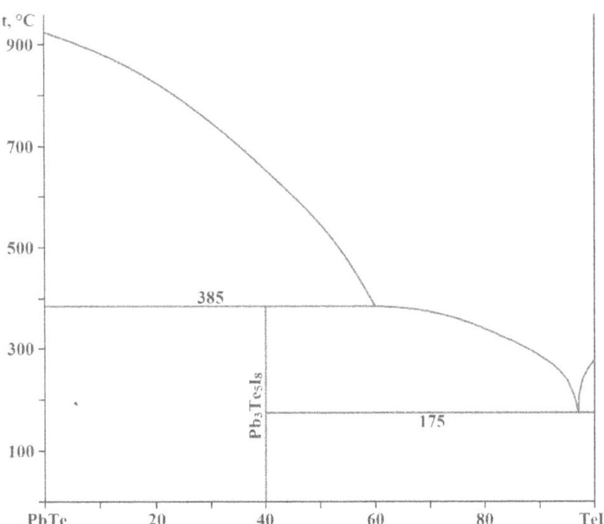

FIGURE 12.45 Phase diagram of the PbTe–TeI_4 system. (From Berg, L.G., et al., *ChemChemTech [Izv. Vuzov. Khim. Khim. Tekhnol.]*, **12**(6), 691, 1969.)

and Pb–MnTe–Mn subsystems. The solubility of Mn in PbTe reaches 1.5 at% and the solubility of PbTe in Mn is negligible.

The system was investigated through DTA, XRD, metallography, and measuring of the microhardness and density (Bates et al. 1968; Abilov et al. 1991). The ingots for the investigations were annealed at 950°C for 20 h and then at 600°C for 46 h and at 1100°C for 15 h (Bates et al. 1968) [at 310°C for 400 h and the solid solution based on PbTe at 550°C for 400 h (Abilov et al. 1991).

PbTe–MnTe: This system is also a non-quasibinary section of the Pb–Mn–Te ternary system as MnTe melts incongruently (Figure 12.46) (Vanyarkho et al. 1970b). Mn crystallizes primarily from the MnTe-side. The eutectic contains 30 mol% MnTe and crystallizes at 900°C. The boundary between the solid solution based on PbTe and MnTe was able to be determined only up to a temperature of 650°C. The solubility of PbTe in MnTe is not higher than 5 mol% and the solubility of MnTe in PbTe at 500°C, 620°C, 720°C, and 900°C is correspondingly 3, 10, 12, and 20 mol% (Chomentowski et al. 1965; Vanyarkho et al. 1970b). Single crystals of the $Pb_xMn_{1-x}Te$ solid solution with $x \le 0.2$ were grown by the Bridgman method and its lattice constants agreed with Vegard's law (Golubović et al. 2004). The concentration dependence of the microhardness of the $Pb_{1-x}Mn_xTe$ (x = 0–0.04) solid solutions at room temperature was determined by Rogacheva et al. (1998b). An abnormal decrease in microhardness in the region x = 0.0075–0.0125 was found. The introduction of manganese into PbTe leads to a linear decrease in the lattice parameter to x = 0.0225, after which a practically does not change.

The system was investigated through DTA, XRD, and metallography and the ingots for the investigations were annealed at 620°C, 720°C, and 875°C for 720, 450, and 300 h, respectively (Vanyarkho et al. 1970b) [at 500°C for 5 days (Chomentowski et al. 1965); at 550°C for 200 h (Rogacheva et al. 1998b)].

PbTe–$MnTe_2$: This system is also a non-quasibinary section of the Pb–Mn–Te ternary system (Rustamov et al. 1986). Peritectic transformation takes place in the system within the temperature interval 735°C–360°C which ends by the next reaction: L + MnTe ⇔ $MnTe_2$ + (PbTe). The solubility of $MnTe_2$ in PbTe at room temperature reaches 3 mol% and the solubility of PbTe in $MnTe_2$ is negligible. The system was investigated through DTA, XRD, and metallography and the ingots for the investigations were annealed at 300°C for 500 h.

Pb–MnTe: The phase diagram of this system is a eutectic type (Figure 12.47) (Abilov and Akhmedova 2002). The eutectic contains 3 mol% MnTe and crystallizes at 317°C. The solubility of Pb in MnTe at room temperature reaches 1 at%.

The liquidus surface of the Pb–Mn–Te ternary system consists of six fields of primary crystallization of phases and two immiscibility regions (Figure 12.48) (Abilov and Akhmedova 2002). The next invariant equilibria exist in the system: E_1 (297°C) – L ⇔ MnTe + Mn + Pb; E_2 (307°C) – L ⇔ MnTe + Pb + PbTe; E_3 (377°C) – L ⇔ $MnTe_2$ + Te + PbTe; U (560°C) L + MnTe ⇔ $MnTe_2$ + PbTe. It should be mention that according to Massalski (1990), there is no invariant point in the Mn–Te system between MnTe and m_4 and peritectic point p_1 contains 71 at% Te. Beside, an immiscibility region in the Pb–Mn system from Pb-rich side contains more lead.

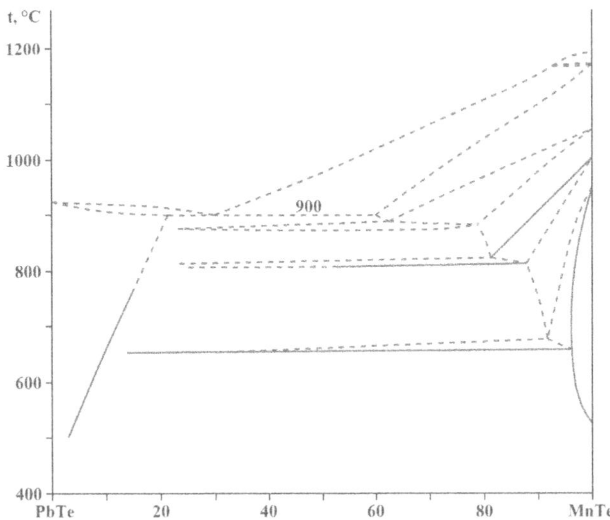

FIGURE 12.46 Phase relations in the PbTe–MnTe system. (From Vanyarkho, V.G., et al., *Izv. AN SSSR. Neorgan. mater.*, **6**(8), 1534, 1970.)

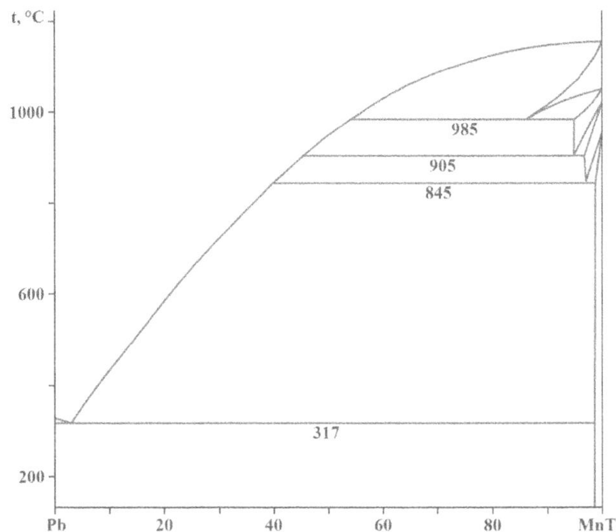

FIGURE 12.47 Phase diagram of the MnTe–Pb system. (From Abilov, Ch.I., and Akhmedova, Dzh.A., *Zhurn. neorgan. khimii*, **47**(10), 1716, 2002.)

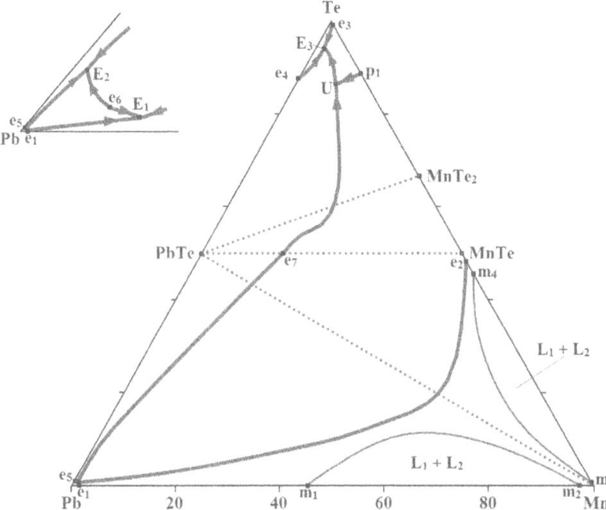

FIGURE 12.48 Projection of the liquidus surface of the Pb–Mn–Te ternary system. (From Abilov, Ch.I., and Akhmedova, Dzh.A., *Zhurn. neorgan. khimii*, **47**(10), 1716, 2002.)

12.46 Lead–Iron–Tellurium

PbTe–Fe: The phase diagram of this system, constructed through DTA, XRD, metallography, and measuring of the microhardness, is a eutectic type (Figure 12.49) (Bates et al. 1968; Wald and Stormont 1968a; Rustamov and Abilov 1987). The eutectic contains 27 at% Fe and crystallizes at 857°C (Rustamov and Abilov 1987) [14 at% Fe and 875°C (Bates et al. 1968)]. Iron-based solid solutions undergo polymorphic transformations at 747°C, 857°C, and 1367°C. The solubility of Fe in PbTe reaches 3 at% at room temperature [0.3 at% (Wald and Stormont 1968a)]. Some doubt is caused by the coincidence of the temperature of one of the phase transitions of the iron-based solid solution with the eutectic temperature.

According to the data of Bates et al. (1968) and Wald and Stormont (1968a), an immiscibility region exists in this system.

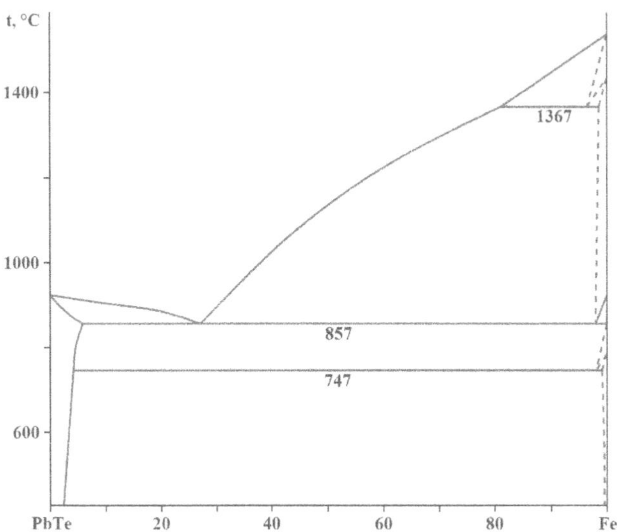

FIGURE 12.49 Phase diagram of the PbTe–Fe system. (From Rustamov, P.G., and Abilov, Ch.I., *Zhurn. neorgan. khimii*, **32**(7), 1710, 1987.)

Small deviations from quasi-binary behavior due to the presence of small amounts of iron telluride were found in the range of 5–10 at% Fe (Wald and Stormont 1968a).

When deposited a layer of PbTe on a bar made of iron at 1000°C in an argon atmosphere, the formation of a solid solution of tellurium in iron is observed at the interface (Ambartsumyan et al. 1968). Penetration of lead into the iron layer is practically absent, and the diffusion of iron into the deposited PbTe layer is insignificant.

The ingots for the investigations were annealed at 650°C for 400 h (Rustamov and Abilov 1987) [at 950°C–1100°C for 15–41 h (Bates et al. 1968)].

PbTe–FeTe: This system is a non-quasibinary section of the Pb–Fe–Te ternary system as FeTe melts incongruently (Rustamov et al. 1982). The solubility of FeTe in PbTe at room temperature reaches 4 mol% (Rustamov et al. 1982) [at 500°C is not higher than 3 mol% (Chomentowski et al. 1965)]. The homogeneity region based on FeTe was not found (Rustamov et al. 1982). The system was investigated through DTA, XRD, metallography, and measuring of the microhardness (Rustamov et al. 1982) and the ingots for the investigations were annealed at 450°C for 500 h (Rustamov et al. 1982) [at 500°C for 5 days (Chomentowski et al. 1965)].

PbTe–FeTe$_2$: This system is also a non-quasibinary section of the Pb–Fe–Te ternary system as $FeTe_2$ melts incongruently (Abilov et al. 1986). In this section the δ'-phase of the Fe–Te binary system, containing 59.2–65.1 mol% Te, is primarily crystallized. The system was investigated through DTA, XRD, metallography, and measuring of the microhardness.

PbTe–Fe$_2$Te: According to the data of Abilov and Agdamskaya (1987), the phase diagram of this system, constructed through DTA, XRD, metallography, and measuring of the microhardness, is a eutectic type. The eutectic contains 57 at% Fe_2Te and crystallizes at 747°C. The solubility of Fe_2Te in PbTe at room temperature reaches 5 mol% and the solubility of PbTe in Fe_2Te is 2 mol%. The ingots for the investigations were annealed at 500°C for 400 h. It should be noted that, according to Massalski (1990), Fe_2Te does not form in the Fe–Te binary system.

PbTe–Fe$_2$Te$_3$: Agdamskaya et al. (1989) shown that this system is a quasibinary section of the Pb–Fe–Te ternary system with two eutectics and one ternary compound (**PbFe$_2$Te$_4$**), which melts congruently at 872°C and has an energy gap of 0.55 eV. At room temperature, the solubility of Fe_2Te_3 in PbTe reaches 3 mol% and the solubility of PbTe in Fe_2Te_3 is 0.5 mol%. The system was investigated through DTA, XRD, metallography, and measuring of the microhardness and density and the ingots were annealed at 630°C for 400 h. It should be noted, that composition of the Fe_2Te_3 compound corresponds to the δ′-phase of the Fe–Te system (Massalski 1990), which melts incongruently. Therefore, this system cannot be a quasibinary section of the Pb–Fe–Te ternary system.

12.47 Lead–Cobalt–Tellurium

PbTe–Co: The phase diagram of this system, constructed through DTA, XRD, and metallography, is a eutectic type with an immiscibility region (Figure 12.50) (Wald and Stormont

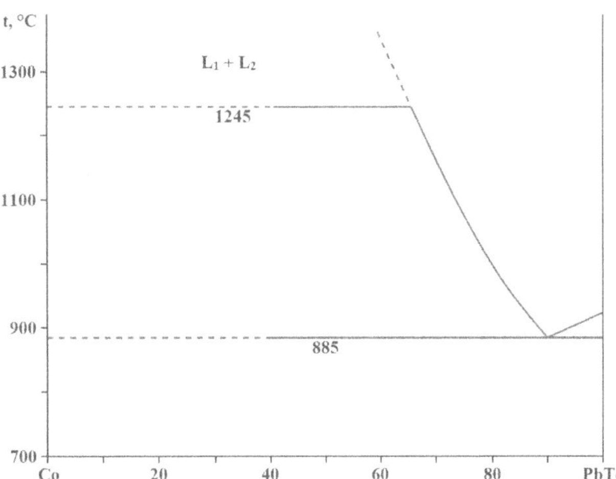

FIGURE 12.50 Phase diagram of the PbTe–Co system. (From Wald, F., and Stormont, R.W, *Trans. Metallurg. Soc. AIME*, **242**(1), 72, 1968; used with permission of the Minerals, Metals & Materials Society)

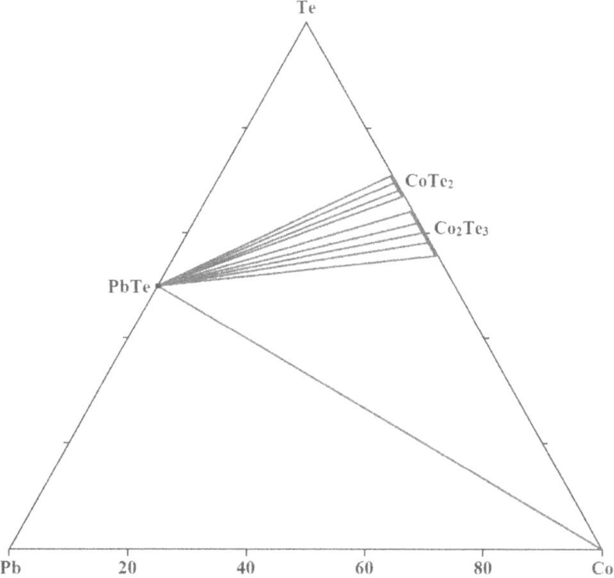

FIGURE 12.51 Isothermal section of the Pb–Co–Te ternary system at 400°C. (From Gulay, L.D., *Nauk. Visnyk Volyns'k. Nats. Univ. im. Lesi Ukrainky. Ser. Khim. nauky*, (16), 6, 2008.)

1968b). The solubility of Co in PbTe is less than 1 at%. According to the data of Abilov (1991), this system is non-quasibinary section of the Pb–Co–Te ternary system and intersects the PbTe–Co_3Te_4–Pb and PbTe–Co_3Te_4–Co subsystem.

PbTe–CoTe$_2$: This system is also a non-quasibinary section of the Pb–Co–Te ternary system as $CoTe_2$ melts incongruently (Rustamov et al. 1985c). The solubility of $CoTe_2$ in PbTe reaches 8 mol% and solid solution based on $CoTe_2$ was not found. The system was investigated through DTA, XRD, metallography, and measuring of the microhardness and the ingots for the investigations were annealed at 630°C for 500 h.

PbTe–Co$_3$Te$_4$: The phase diagram of this system, constructed through DTA, XRD, metallography, and measuring of the microhardness and density, is a eutectic type (Abilov 1991). The eutectic contains 34 mol% Co_3Te_4 and crystallizes at 782°C. The solubility of Co_3Te_4 in PbTe reaches ca. 5 mol% at the eutectic temperature and decreases to 2 mol% at room temperature. The samples were annealed at temperatures 50°C–100°C lower than the solidus temperatures for 400–600 h.

Pb–Co$_3$Te$_4$: The phase diagram of this system, constructed through DTA, XRD, metallography, and measuring of the microhardness and density, is a eutectic type (Abilov 1991). The eutectic contains 14 mol% Co_3Te_4 and crystallizes at 312°C. The solubility of Pb in Co_3Te_4 is 3 at% at room temperature. The samples were annealed at temperatures 50°C–100°C lower than the solidus temperatures for 400–600 h.

The isothermal section of the Pb–Co–Te ternary system at 400°C (Figure 12.51) was constructed using XRD (Gulay 2008). The formations of new ternary compounds were not observed.

The liquidus surface consists of six fields of primary crystallization of phases (Abilov 1991).

12.48 Lead–Nickel–Tellurium

PbTe–Ni: This system is a non-quasibinary section of the Pb–Ni–Te ternary system (Ambartsumyan et al. 1968; Bates et al. 1968; Abilov 1989). The solubility of Ni in PbTe at room temperature reaches 2 at% and solid solution based on Ni was not found. The system was investigated through DTA, XRD, metallography, and measuring of the microhardness and density and the ingots for the investigations were annealed at 630°C for 400 h (Abilov 1989) [at 950°C and 600°C for 15 and 45 h, respectively (Bates et al. 1968)].

PbTe–NiTe$_2$: Part of the phase diagram of this system in the region 0–10 mol% $NiTe_2$ (Figure 12.52) was constructed by Alidzhanov et al. (1991). The solubility of $NiTe_2$ in PbTe at

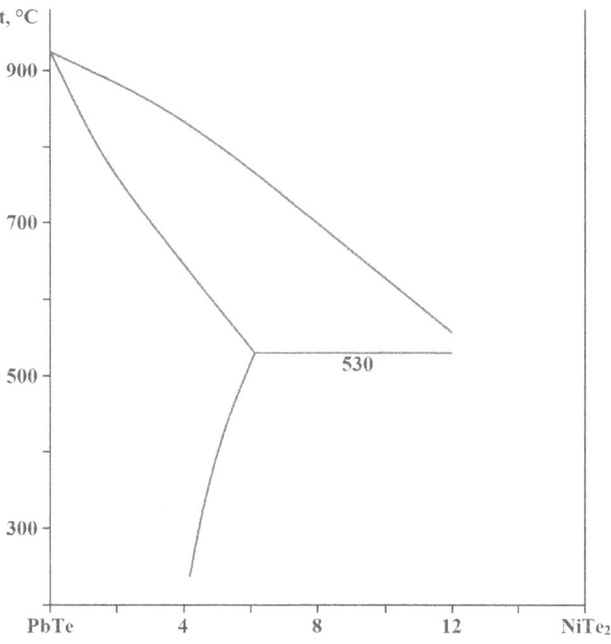

FIGURE 12.52 Phase diagram of the PbTe–NiTe$_2$ system. (From Alidzhanov, M.A., et al., *Neorgan. mater.*, **27**(11), 2437, 1991.)

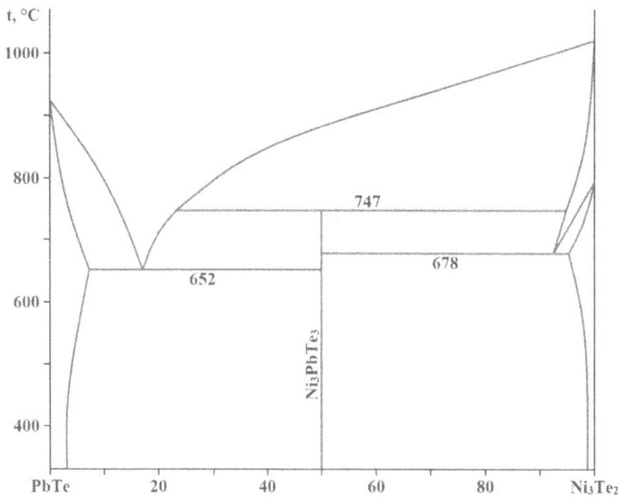

FIGURE 12.53 Phase diagram of the PbTe-Ni_3Te_2 system. (From Abilov, Ch.I., and Iskender-zade, Z.A., *Izv. AN SSSR. Neorgan. mater.*, **25**(2), 250, 1989.)

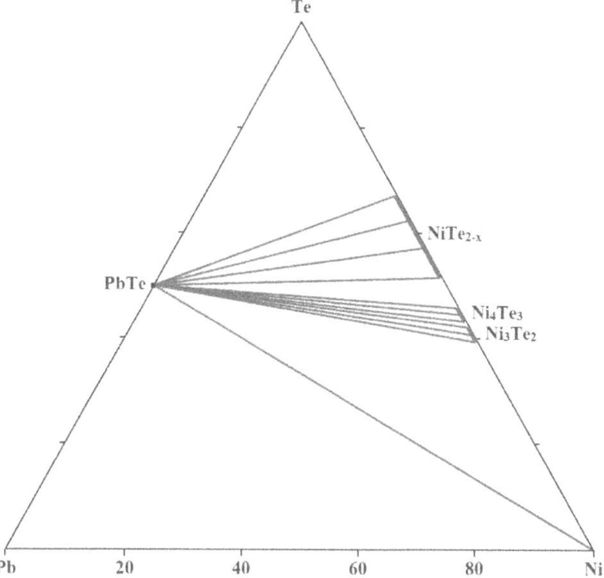

FIGURE 12.54 Isothermal section of the Pb–Ni–Te ternary system at 400°C. (From Gulay, L.D., *Nauk. Visnyk Volyns'k. Nats. Univ. im. Lesi Ukrainky*, (16), 6, 2008.)

room temperature is 4 mol%. The ingots for the investigations were annealed at 430°C for 250–280 h.

PbTe–Ni_3Te_2: The phase diagram of this system, constructed through DTA, XRD, metallography, and measuring of the microhardness, is presented in Figure 12.53 (Abilov and Iskender-zade 1989). The eutectic contains 17 mol% Ni_3Te_2 and crystallizes at 652°C. At room temperature the solubility of Ni_3Te_2 in PbTe and PbTe in Ni_3Te_2 is correspondingly 3 and 1 mol%. The **Ni_3PbTe_3** ternary compound, which melts incongruently at 747°C, is formed in this system. The ingots for the investigations were annealed at 630°C for 350 h.

PbTe–Ni_3Te_4: According to the data of Abilov and Rustamov (1990), the phase diagram of this system, constructed through DTA, XRD, metallography, and measuring of the microhardness, is a eutectic type. The eutectic contains 42 at% Ni_3Te_4 and crystallizes at 782°C. The solubility of Ni_3Te_4 in PbTe at room temperature is 5 mol%. The ingots for the investigations were annealed at 630°C for 500 h. It should be noted that, according to Massalski (1990), Ni_3Te_4 does not form in the Ni–Te binary system: the composition of this compound is in the homogeneity region of the $NiTe_{2-x}$ phase. Therefore, this system cannot be a quasibinary section of the Pb–Ni–Te ternary system.

The isothermal sections of the Pb–Ni–Te ternary system at 400°C and 540°C were constructed by Gulay (2008) and Baranov et al. (2002), respectively, the first of which is shown in Figure 12.54. The formations of new ternary compounds, including Ni_3PbTe_3, were not observed.

The liquidus surface of the Pb–Ni–Te ternary system includes eight fields of the phase primary crystallization and an immiscibility region (Abilov 1989). Four ternary eutectics (E_1 at 402°C, E_2 at 627°C, E_3 at 307°C, and E_4 at 302°C) and three transition points (U_1 at 677°C, U_2 at 642°C, and U_3 at 577°C) exist in this system.

12.49 Lead–Palladium–Tellurium

The **$Pb_2Pd_3Te_2$** ternary compound (mineral pašavaite), which crystallizes in the orthorhombic structure with the lattice parameters $a = 859.9 \pm 0.1$, $b = 593.81 \pm 0.06$, $c = 631.73 \pm 0.08$ pm, and the calculated and experimental densities of 10.18 and 9.9 g·cm^{-3}, respectively, is formed in the Pb–Pd–Te ternary system (Piilonen et al. 2009; Vymazalová et al. 2009). The synthetic $Pd_3Pb_2Te_2$ phase was prepared using evacuated silica glass tube method. A mixture of the elements in the stoichiometric ratio was loaded into the silica tube and a tightly fitting silica glass rod was placed on top of the reagents in order to keep the charge in place and also to reduce the volume of vapor on heating. The evacuated tube was sealed with its charge and then annealed at 1200°C for 3 days. After cooling in a cold-water bath, the charge was carefully taken out of the tube so as not to lose any of the products, which were ground into powder in acetone using an agate mortar, and thoroughly mixed so as to become homogeneous. The pulverized charge was sealed in an evacuated silica-glass tube again, and reheated at 400°C for five months.

REFERENCES

All References are available as a downloadable eResource at www.routledge.com/9780367639235

Index

A

B

C

D

E

K

L

M

N

P

R

S

T

U

V

Y

Z

For Product Safety Concerns and Information please contact our EU
representative GPSR@taylorandfrancis.com
Taylor & Francis Verlag GmbH, Kaufingerstraße 24, 80331 München, Germany

www.ingramcontent.com/pod-product-compliance
Lightning Source LLC
Chambersburg PA
CBHW080710220326
41598CB00033B/5368
9780367643072